Bhushan (Ed.) **Nanotribology and Nanomechanics**

Bharat Bhushan (Ed.)

Nanotribology and Nanomechanics

An Introduction

With 624 Figures

Springer

Professor Bharat Bhushan
Nanotribology Laboratory
for Information storage and MEMS/NEMS
The Ohio State University
206 W. 18th Ave.
Coumbus, Ohio 43210-1107
USA

Library of Congress Control Number: 2004117507

ISBN-10 3-540-24267-8 Springer Berlin Heidelberg New York
ISBN-13 978-3-540-24267-3 Springer Berlin Heidelberg New York

This work is subject to copyright. All rights are reserved, whether the whole or part of the material is concerned, specifically the rights of translation, reprinting, reuse of illustrations, recitation, broadcasting, reproduction on microfilm or in any other way, and storage in data banks. Duplication of this publication or parts thereof is permitted only under the provisions of the German Copyright Law of September 9, 1965, in its current version, and permission for use must always be obtained from Springer. Violations are liable for prosecution under the German Copyright Law.

Springer is a part of Springer Science+Business Media
springeronline.com

© Springer-Verlag Berlin Heidelberg 2005
Printed in Germany

The use of general descriptive names, registered names, trademarks, etc. in this publication does not imply, even in the absence of a specific statement, that such names are exempt from the relevant protective laws and regulations and therefore free for general use.

Typesetting: Da-T_EX Gerd Blumenstein, Leipzig
Production: LE-T_EX Jelonek, Schmidt & Vöckler GbR, Leipzig
Cover Design: *design & production* GmbH, Heidelberg

Printed on acid-free paper   41/3141/YL - 5 4 3 2 1 0

Foreword

The invention of the scanning tunneling microscope in 1981 has led to an explosion of a family of instruments called scanning probe microscopes (SPMs). One of the most popular instruments in this family is the atomic force microscope (AFM), which was introduced to the scientific community in 1986. The application of SPMs has penetrated numerous science and engineering fields. The proliferation of SPMs in science and technology labs is similar to optical microscopes fifty years ago. SPMs have even made it into some high school science labs. Evolution of nanotechnology has accelerated the use of SPMs and vice versa. The scientific and industrial applications include quality control in the semiconductor industry and related research, molecular biology and chemistry, medical studies, materials science, and the field of information storage systems.

AFMs were developed initially for imaging with atomic or near atomic resolution. After their invention, they were modified for tribological studies. AFMs are now intensively used in this field and have lead to the development of the field of nanotribology. Researchers can image single lubricant molecules and their agglomeration and measure surface topography, adhesion, friction, wear, lubricant film thickness, and mechanical properties all on a micrometer to nanometer scale. SPMs are also used for nanofabrication and nanomachining. Beyond use as an analytical instrument, SPMs are being developed as industrial tools such as for data storage.

With the advent of more powerful computers, atomic-scale simulations have been conducted of tribological phenomena. Simulations have been able to predict the observed phenomena. Development of the field of nanotribology and nanomechanics has attracted numerous physicists and chemists. I am very excited that SPMs have had such an immense impact on the field of tribology.

I congratulate Professor Bharat Bhushan in helping to develop this field of nanotribology and nanomechanics. Professor Bhushan has harnessed his own knowledge and experience, gained in several industries and universities, and has assembled a large number of internationally recognized authors. The authors come from both academia and industry.

Professor Bhushan's comprehensive book is intended to serve both as a textbook for university courses as well as a reference for researchers. It is a timely addition to the literature on nanotribology and nanomechanics, which I anticipate will stimulate further interest in this important new field. I expect that it will be well received by the international scientific community.

IBM Research Division Rueschlikon,   Prof. Dr. Gerd Binnig
Switzerland Nobel Laureate Physics, 1986

Contents

**1 Introduction –
Measurement Techniques and Applications**
Bharat Bhushan 1
1.1 Definition and History of Tribology 1
1.2 Industrial Significance of Tribology 3
1.3 Origins and Significance of Micro/Nanotribology 4
1.4 Measurement Techniques 6
    1.4.1 Scanning Probe Microscopy 6
    1.4.2 Surface Force Apparatus (SFA) 13
    1.4.3 Vibration Isolation 23
1.5 Magnetic Storage Devices and MEMS/NEMS 25
    1.5.1 Magnetic Storage Devices 25
    1.5.2 MEMS/NEMS 27
1.6 Role of Micro/Nanotribology and Micro/Nanomechanics
     in Magnetic Storage Devices and MEMS/NEMS 33
1.7 Organization of the Book 34
References 34

Part I Scanning Probe Microscopy

**2 Scanning Probe Microscopy –
Principle of Operation, Instrumentation, and Probes**
Bharat Bhushan, Othmar Marti 41
2.1 Introduction 41
2.2 Scanning Tunneling Microscope 43
    2.2.1 Binnig et al.'s Design 45
    2.2.2 Commercial STMs 45
    2.2.3 STM Probe Construction 49
2.3 Atomic Force Microscope 50
    2.3.1 Binnig et al.'s Design 54

	2.3.2	Commercial AFM	54
	2.3.3	AFM Probe Construction	64
	2.3.4	Friction Measurement Methods	70
	2.3.5	Normal Force and Friction Force Calibrations of Cantilever Beams	76
2.4	AFM Instrumentation and Analyses		78
	2.4.1	The Mechanics of Cantilevers	78
	2.4.2	Instrumentation and Analyses of Detection Systems for Cantilever Deflections	84
	2.4.3	Combinations for 3-D-Force Measurements	99
	2.4.4	Scanning and Control Systems	99
References			108

3 Probes in Scanning Microscopies

Jason H. Hafner 117

3.1	Introduction		118
3.2	Atomic Force Microscopy		119
	3.2.1	Principles of Operation	119
	3.2.2	Standard Probe Tips	120
	3.2.3	Probe Tip Performance	121
	3.2.4	Oxide-Sharpened Tips	124
	3.2.5	FIB tips	125
	3.2.6	EBD tips	125
	3.2.7	Carbon Nanotube Tips	125
3.3	Scanning Tunneling Microscopy		136
	3.3.1	Mechanically Cut STM Tips	136
	3.3.2	Electrochemically Etched STM Tips	137
References			137

4 Noncontact Atomic Force Microscopy and Its Related Topics

Seizo Morita, Franz J. Giessibl, Yasuhiro Sugawara, Hirotaka Hosoi, Koichi Mukasa, Akira Sasahara, Hiroshi Onishi 141

4.1	Introduction		141
4.2	Principles of Noncontact Atomic Force Microscope (NC-AFM)		142
	4.2.1	Imaging Signal in AFM	142
	4.2.2	Experimental Measurement and Noise	143
	4.2.3	Static AFM Operating Mode	144
	4.2.4	Dynamic AFM Operating Mode	145
	4.2.5	The Four Additional Challenges Faced by AFM	146
	4.2.6	Frequency-Modulation AFM (FM-AFM)	147
	4.2.7	Relation Between Frequency Shift and Forces	148
	4.2.8	Noise in Frequency-Modulation AFM – Generic Calculation	151
	4.2.9	Conclusion	152

4.3	Applications to Semiconductors		152
	4.3.1	Si(111)7×7 Surface	152
	4.3.2	Si(100)2×1 and Si(100)2×1:H Monohydride Surfaces	155
	4.3.3	Metal-Deposited Si Surface	157
4.4	Applications to Insulators		161
	4.4.1	Alkali Halides, Fluorides, and Metal Oxides	162
	4.4.2	Atomically Resolved Imaging of a NiO(001) Surface	168
	4.4.3	Atomically Resolved Imaging Using Noncoated and Fe-Coated Si Tips	170
4.5	Applications to Molecules		172
	4.5.1	Why Molecules and What Molecules?	172
	4.5.2	Mechanism of Molecular Imaging	173
	4.5.3	Perspectives	177
References			178

5 Low Temperature Scanning Probe Microscopy
Markus Morgenstern, Alexander Schwarz, Udo D. Schwarz 185

5.1	Introduction		186
5.2	Microscope Operation at Low Temperatures		187
	5.2.1	Drift	187
	5.2.2	Noise	187
	5.2.3	Stability	188
	5.2.4	Piezo Relaxation and Hysteresis	188
5.3	Instrumentation		188
	5.3.1	A Simple Design for a Variable Temperature STM	189
	5.3.2	A Low Temperature SFM Based on a Bath Cryostat	191
5.4	Scanning Tunneling Microscopy and Spectroscopy		194
	5.4.1	Atomic Manipulation	194
	5.4.2	Imaging Atomic Motion	195
	5.4.3	Detecting Light from Single Atoms and Molecules	198
	5.4.4	High Resolution Spectroscopy	198
	5.4.5	Imaging Electronic Wave Functions	206
	5.4.6	Imaging Spin Polarization: Nanomagnetism	215
5.5	Scanning Force Microscopy and Spectroscopy		218
	5.5.1	Atomic-Scale Imaging	219
	5.5.2	Force Spectroscopy	224
	5.5.3	Electrostatic Force Microscopy	226
	5.5.4	Magnetic Force Microscopy	227
References			233

6 Dynamic Force Microscopy
André Schirmeisen, Boris Anczykowski, Harald Fuchs 243

6.1	Motivation: Measurement of a Single Atomic Bond	244
6.2	Harmonic Oscillator: A Model System for Dynamic AFM	250
6.3	Dynamic AFM Operational Modes	252

	6.3.1	Amplitude-Modulation/ Tapping-Mode AFMs	254
	6.3.2	Self-Excitation Modes	262
6.4	Q-Control		268
6.5	Dissipation Processes Measured with Dynamic AFM		272
6.6	Conclusion		278
References			279

7 Molecular Recognition Force Microscopy

Peter Hinterdorfer 283

7.1	Introduction	283
7.2	Ligand Tip Chemistry	284
7.3	Fixation of Receptors to Probe Surfaces	287
7.4	Single-Molecule Recognition Force Detection	289
7.5	Principles of Molecular Recognition Force Spectroscopy	293
7.6	Recognition Force Spectroscopy: From Isolated Molecules to Biological Membranes	296
	7.6.1 Forces, Energies, and Kinetic Rates	296
	7.6.2 Complex Bonds and Energy Landscapes	299
	7.6.3 Live Cells and Membranes	303
7.7	Recognition Imaging	304
7.8	Concluding Remarks	307
References		308

Part II Nanotribology and Nanomechanics

8 Micro/Nanotribology and Materials Characterization Studies Using Scanning Probe Microscopy

Bharat Bhushan 315

8.1	Introduction	315
8.2	Description of AFM/FFM and Various Measurement Techniques	318
	8.2.1 Surface Roughness and Friction Force Measurements	318
	8.2.2 Adhesion Measurements	323
	8.2.3 Scratching, Wear and Fabrication/Machining	323
	8.2.4 Surface Potential Measurements	324
	8.2.5 In Situ Characterization of Local Deformation Studies	325
	8.2.6 Nanoindentation Measurements	326
	8.2.7 Localized Surface Elasticity and Viscoelasticity Mapping	327
	8.2.8 Boundary Lubrication Measurements	329
8.3	Friction and Adhesion	329
	8.3.1 Atomic-Scale Friction	329
	8.3.2 Microscale Friction	331
	8.3.3 Directionality Effect on Microfriction	339
	8.3.4 Velocity Dependence on Microfriction	341

		8.3.5	Effect of Tip Radii and Humidity on Adhesion and Friction	341

	8.3.5	Effect of Tip Radii and Humidity on Adhesion and Friction . .	341
	8.3.6	Scale Dependence on Friction	348
8.4	Scratching, Wear, Local Deformation, and Fabrication/Machining		349
	8.4.1	Nanoscale Wear	349
	8.4.2	Microscale Scratching	350
	8.4.3	Microscale Wear	354
	8.4.4	In Situ Characterization of Local Deformation	359
	8.4.5	Nanofabrication/Nanomachining	362
8.5	Indentation		364
	8.5.1	Picoindentation	364
	8.5.2	Nanoscale Indentation	365
	8.5.3	Localized Surface Elasticity and Viscoelasticity Mapping	368
8.6	Boundary Lubrication		371
	8.6.1	Perfluoropolyether Lubricants	371
	8.6.2	Self-Assembled Monolayers	378
	8.6.3	Liquid Film Thickness Measurements	378
8.7	Closure		383
References			384

9 Surface Forces and Nanorheology of Molecularly Thin Films
Marina Ruths, Alan D. Berman, Jacob N. Israelachvili 389

9.1	Introduction: Types of Surface Forces		389
9.2	Methods Used to Study Surface Forces		392
	9.2.1	Force Laws	392
	9.2.2	Adhesion Forces	393
	9.2.3	The SFA and AFM	394
	9.2.4	Some Other Force-Measuring Techniques	397
9.3	Normal Forces Between Dry (Unlubricated) Surfaces		398
	9.3.1	Van der Waals Forces in Vacuum and Inert Vapors	398
	9.3.2	Charge Exchange Interactions	400
	9.3.3	Sintering and Cold Welding	402
9.4	Normal Forces Between Surfaces in Liquids		403
	9.4.1	Van der Waals Forces in Liquids	403
	9.4.2	Electrostatic and Ion Correlation Forces	405
	9.4.3	Solvation and Structural Forces	408
	9.4.4	Hydration and Hydrophobic Forces	411
	9.4.5	Polymer-Mediated Forces	415
	9.4.6	Thermal Fluctuation Forces	417
9.5	Adhesion and Capillary Forces		419
	9.5.1	Capillary Forces	419
	9.5.2	Adhesion Mechanics	422
	9.5.3	Effects of Surface Structure, Roughness, and Lattice Mismatch	423

	9.5.4	Nonequilibrium and Rate-Dependent Interactions: Adhesion Hysteresis	425
9.6		Introduction: Different Modes of Friction and the Limits of Continuum Models	427
9.7		Relationship Between Adhesion and Friction Between Dry (Unlubricated and Solid Boundary Lubricated) Surfaces	429
	9.7.1	Amontons' Law and Deviations from It Due to Adhesion: The Cobblestone Model	429
	9.7.2	Adhesion Force and Load Contribution to Interfacial Friction	431
	9.7.3	Examples of Experimentally Observed Friction of Dry Surfaces	441
	9.7.4	Transition from Interfacial to Normal Friction with Wear	444
9.8		Liquid Lubricated Surfaces	445
	9.8.1	Viscous Forces and Friction of Thick Films: Continuum Regime	445
	9.8.2	Friction of Intermediate Thickness Films	448
	9.8.3	Boundary Lubrication of Molecularly Thin Films: Nanorheology	449
9.9		Role of Molecular Shape and Surface Structure in Friction	463
References			467

10 Friction and Wear on the Atomic Scale
Enrico Gnecco, Roland Bennewitz, Oliver Pfeiffer, Anisoara Socoliuc, Ernst Meyer 483

10.1	Friction Force Microscopy in Ultra-High Vacuum		483
	10.1.1	Friction Force Microscopy	484
	10.1.2	Force Calibration	485
	10.1.3	The Ultra-high Vacuum Environment	489
	10.1.4	A Typical Microscope in UHV	489
10.2	The Tomlinson Model		491
	10.2.1	One-dimensional Tomlinson Model	491
	10.2.2	Two-dimensional Tomlinson Model	492
	10.2.3	Friction Between Atomically Flat Surfaces	493
10.3	Friction Experiments on Atomic Scale		495
	10.3.1	Anisotropy of Friction	499
10.4	Thermal Effects on Atomic Friction		501
	10.4.1	The Tomlinson Model at Finite Temperature	501
	10.4.2	Velocity Dependence of Friction	506
	10.4.3	Temperature Dependence of Friction	508
10.5	Geometry Effects in Nanocontacts		508
	10.5.1	Continuum Mechanics of Single Asperities	508
	10.5.2	Load Dependence of Friction	511
	10.5.3	Estimation of the Contact Area	512
10.6	Wear on the Atomic Scale		515
	10.6.1	Abrasive Wear on the Atomic Scale	515

		Contents	XIII

	10.6.2	Wear Contribution to Friction	516
10.7		Molecular Dynamics Simulations of Atomic Friction and Wear	518
	10.7.1	Molecular Dynamics Simulation of Friction Processes	519
	10.7.2	Molecular Dynamics Simulations of Abrasive Wear	521
10.8		Energy Dissipation in Noncontact Atomic Force Microscopy	524
10.9		Conclusion	527
References			528

11 Nanoscale Mechanical Properties – Measuring Techniques and Applications

Andrzej J. Kulik, András Kis, Gérard Gremaud, Stefan Hengsberger, Philippe K. Zysset, Lásló Forró 535

11.1		Introduction	535
11.2		Local Mechanical Spectroscopy by Contact AFM	537
	11.2.1	The Variable-Temperature SLAM (T-SLAM)	538
	11.2.2	Example One: Local Mechanical Spectroscopy of Polymers	540
	11.2.3	Example Two: Local Mechanical Spectroscopy of NiTi	541
11.3		Static Methods – Mesoscopic Samples	544
	11.3.1	Carbon Nanotubes – Introduction to Basic Morphologies and Production Methods	544
	11.3.2	Measurements of the Mechanical Properties of Carbon Nanotubes by SPM	546
	11.3.3	Microtubules and Their Elastic Properties	555
11.4		Scanning Nanoindentation: An Application to Bone Tissue	556
	11.4.1	Scanning Nanoindentation	557
	11.4.2	Application of Scanning Nanoindentation	557
	11.4.3	Example: Study of Mechanical Properties of Bone Lamellae Using SN	558
	11.4.4	Conclusion	567
11.5		Conclusions and Perspectives	568
References			569

12 Nanomechanical Properties of Solid Surfaces and Thin Films

Adrian B. Mann 575

12.1		Introduction	575
12.2		Instrumentation	576
	12.2.1	AFM and Scanning Probe Microscopy	576
	12.2.2	Nanoindentation	577
	12.2.3	Adaptations of Nanoindentation	580
	12.2.4	Complimentary Techniques	582
	12.2.5	Bulge Tests	582
	12.2.6	Acoustic Methods	583
	12.2.7	Imaging Methods	585
12.3		Data Analysis	586
	12.3.1	Elastic Contacts	586

	12.3.2	Indentation of Ideal Plastic Materials	588
	12.3.3	Adhesive Contacts	588
	12.3.4	Indenter Geometry	589
	12.3.5	Analyzing Load/Displacement Curves	590
	12.3.6	Modifications to the Analysis	595
	12.3.7	Alternative Methods of Analysis	598
	12.3.8	Measuring Contact Stiffness	599
	12.3.9	Measuring Viscoelasticity	600
12.4	Modes of Deformation		601
	12.4.1	Defect Nucleation	601
	12.4.2	Variations with Depth	603
	12.4.3	Anisotropic Materials	604
	12.4.4	Fracture and Delamination	604
	12.4.5	Phase Transformations	605
12.5	Thin Films and Multilayers		608
	12.5.1	Thin Films	609
	12.5.2	Multilayers	613
12.6	Developing Areas		615
References			616

13 Computational Modeling of Nanometer-Scale Tribology
Seong-Jun Heo, Susan B. Sinnott, Donald W. Brenner, Judith A. Harrison ... 623

13.1	Introduction		623
13.2	Computational Molecular Dynamics Simulations		625
	13.2.1	Interatomic Potentials	626
	13.2.2	Temperature Regulation	630
13.3	Indentation		634
	13.3.1	Metals	636
	13.3.2	Ceramics	641
	13.3.3	Molecular Systems	646
13.4	Friction		651
	13.4.1	Friction Between Bare Sliding Surfaces	651
	13.4.2	Friction in the Presence of a Third Body	659
13.5	Lubrication		663
	13.5.1	Liquid Films	664
	13.5.2	Nanoparticles	668
	13.5.3	Self-Assembled Thin Films	674
	13.5.4	Solid Thin Films	676
13.6	Conclusions		680
References			680

14 Mechanics of Biological Nanotechnology
Rob Phillips, Prashant K. Purohit, Jané Kondev 693
- 14.1 Introduction 693
- 14.2 Science at the Biology–Nanotechnology Interface 694
  - 14.2.1 Biological Nanotechnology 694
  - 14.2.2 Self-Assembly as Biological Nanotechnology 694
  - 14.2.3 Molecular Motors as Biological Nanotechnology 695
  - 14.2.4 Molecular Channels and Pumps as Biological Nanotechnology 696
  - 14.2.5 Biologically Inspired Nanotechnology 697
  - 14.2.6 Nanotechnology and Single Molecule Assays in Biology 698
  - 14.2.7 The Challenge of Modeling the Bio-Nano Interface 702
- 14.3 Scales at the Bio-Nano Interface 704
  - 14.3.1 Spatial Scales and Structures 705
  - 14.3.2 Temporal Scales and Processes 708
  - 14.3.3 Force and Energy Scales: The Interplay of Deterministic and Thermal Forces 710
- 14.4 Modeling at the Nano-Bio Interface 713
  - 14.4.1 Tension Between Universality and Specificity.............. 713
  - 14.4.2 Atomic-Level Analysis of Biological Systems 714
  - 14.4.3 Continuum Analysis of Biological Systems................ 715
- 14.5 Nature's Nanotechnology Revealed: Viruses as a Case Study 717
- 14.6 Concluding Remarks... 725
- References 727

15 Mechanical Properties of Nanostructures
Bharat Bhushan 731
- 15.1 Introduction 732
- 15.2 Experimental Techniques for Measurement of Mechanical Properties of Nanostructures...................... 733
  - 15.2.1 Indentation and Scratch Tests Using Micro/Nanoindenters ... 733
  - 15.2.2 Bending Tests of Nanostructures Using an AFM............ 735
  - 15.2.3 Bending Tests Using a Nanoindenter 740
- 15.3 Experimental Results and Discussion 742
  - 15.3.1 Indentation and Scratch Tests of Various Materials Using Micro/Nanoindenters 742
  - 15.3.2 Bending Tests of Nanobeams Using an AFM 748
  - 15.3.3 Bending Tests of Microbeams Using a Nanoindenter 753
- 15.4 Finite Element Analysis of Nanostructures with Roughness and Scratches 755
  - 15.4.1 Stress Distribution in a Smooth Nanobeam 758
  - 15.4.2 Effect of Roughness in the Longitudinal Direction 758
  - 15.4.3 Effect of Roughness in the Transverse Direction and Scratches....................................... 758

| | 15.4.4 | Effect on Stresses and Displacements for Materials That Are Elastic, Elastic-Plastic, or Elastic-Perfectly Plastic .... | 766 |

15.5 Closure 767
References 768

16 Scale Effect in Mechanical Properties and Tribology
Bharat Bhushan and Michael Nosonovsky 773
16.1 Nomenclature 773
16.2 Introduction 775
16.3 Scale Effect in Mechanical Properties 779
    16.3.1 Yield Strength and Hardness 779
    16.3.2 Shear Strength at the Interface 782
16.4 Scale Effect in Surface Roughness and Contact Parameters 786
    16.4.1 Scale Dependence of Roughness and Contact Parameters 787
    16.4.2 Dependence of Contact Parameters on Load 789
16.5 Scale Effect in Friction 792
    16.5.1 Adhesional Friction 792
    16.5.2 Two-Body Deformation 795
    16.5.3 Three-Body Deformation Friction 796
    16.5.4 Ratchet Mechanism 800
    16.5.5 Meniscus Analysis 801
    16.5.6 Total Value of Coefficient of Friction and Transition from Elastic to Plastic Regime 803
    16.5.7 Comparison with the Experimental Data 805
16.6 Scale Effect in Wear 811
16.7 Scale Effect in Interface Temperature 811
16.8 Closure 813
References 815
A Statistics of Particle Size Distribution 818
    A.1 Statistical Models of Particle Size Distribution 818
    A.2 Typical Particle Size Distribution Data 824

Part III Molecularly-Think Films for Lubrication

17 Nanotribology of Ultrathin and Hard Amorphous Carbon Films
Bharat Bhushan 827
17.1 Introduction 828
17.2 Description of Commonly Used Deposition Techniques 832
    17.2.1 Filtered Cathodic Arc Deposition Technique 833
    17.2.2 Ion Beam Deposition Technique 835
    17.2.3 Electron Cyclotron Resonance Chemical Vapor Deposition Technique 836
    17.2.4 Sputtering Deposition Technique 837

		17.2.5 Plasma-Enhanced Chemical Vapor Deposition Technique	837
17.3	Chemical Characterization and Effect of Deposition Conditions on Chemical Characteristics and Physical Properties		838
	17.3.1	EELS and Raman Spectroscopy	839
	17.3.2	Hydrogen Concentrations	843
	17.3.3	Physical Properties	844
	17.3.4	Summary	845
17.4	Micromechanical and Tribological Characterizations of Coatings Deposited by Various Techniques		846
	17.4.1	Micromechanical Characterization	846
	17.4.2	Microscratch and Microwear Studies	858
	17.4.3	Macroscale Tribological Characterization	868
	17.4.4	Coating Continuity Analysis	878
References			879

18 Self-Assembled Monolayers for Controlling Adhesion, Friction and Wear
Bharat Bhushan, Huiwen Liu 885

18.1	Introduction		885
18.2	A Primer to Organic Chemistry		889
	18.2.1	Electronegativity/Polarity	890
	18.2.2	Classification and Structure of Organic Compounds	891
	18.2.3	Polar and Nonpolar Groups	895
18.3	Self-Assembled Monolayers: Substrates, Head Groups, Spacer Chains, and End Groups		896
18.4	Tribological Properties of SAMs		900
	18.4.1	Surface Roughness and Friction Images of SAMs Films	905
	18.4.2	Adhesion, Friction, and Work of Adhesion	906
	18.4.3	Stiffness, Molecular Spring Model, and Micropatterned SAMs	910
	18.4.4	Influence of Humidity, Temperature, and Velocity on Adhesion and Friction	914
	18.4.5	Wear and Scratch Resistance of SAMs	917
18.5	Closure		922
References			924

19 Nanoscale Boundary Lubrication Studies
Bharat Bhushan, Huiwen Liu 929

19.1	Introduction		929
19.2	Lubricants Details		930
19.3	Nanodeformation, Molecular Conformation, and Lubricant Spreading		933
19.4	Boundary Lubrication Studies		936
	19.4.1	Friction and Adhesion	937
	19.4.2	Rest Time Effect	941

XVIII Contents

      19.4.3 Velocity Effect 945
      19.4.4 Relative Humidity and Temperature Effect 949
      19.4.5 Tip Radius Effect... 952
      19.4.6 Wear Study.. 956
19.5 Closure 959
References 960

Part IV Applications

20 Micro/Nanotribology and Micro/Nanomechanics of Magnetic Storage Devices
Bharat Bhushan 965
20.1 Introduction 965
      20.1.1 Magnetic Storage Devices 965
      20.1.2 Micro/Nanotribology and Micro/Nanomechanics
             and Their Applications 969
20.2 Experimental 970
      20.2.1 Description of AFM/FFM 970
      20.2.2 Test Specimens 974
20.3 Surface Roughness 974
20.4 Friction and Adhesion... 979
      20.4.1 Magnetic Head Materials 980
      20.4.2 Magnetic Media 981
20.5 Scratching and Wear 985
      20.5.1 Nanoscale Wear... 985
      20.5.2 Microscale Scratching 986
      20.5.3 Microscale Wear 996
20.6 Indentation 1007
      20.6.1 Picoscale Indentation 1007
      20.6.2 Nanoscale Indentation 1009
      20.6.3 Localized Surface Elasticity........................... 1013
20.7 Lubrication.. 1013
      20.7.1 Boundary Lubrication Studies 1021
20.8 Closure 1025
References 1026

21 Micro/Nanotribology of MEMS/NEMS Materials and Devices
Bharat Bhushan 1031
21.1 Introduction 1032
21.2 Introduction to MEMS 1034
21.3 Introduction to NEMS.. 1038
21.4 Tribological Issues in MEMS/NEMS 1039
      21.4.1 MEMS 1040
      21.4.2 NEMS.. 1046

	21.4.3	Tribological Needs	1049
21.5		Tribological Studies of Silicon and Related Materials	1049
	21.5.1	Tribological Properties of Silicon and the Effect of Ion Implantation	1050
	21.5.2	Effect of Oxide Films on Tribological Properties of Silicon	1053
	21.5.3	Tribological Properties of Polysilicon Films and SiC Film	1057
21.6		Lubrication Studies for MEMS/NEMS	1058
	21.6.1	Perfluoropolyether Lubricants	1058
	21.6.2	Self-Assembled Monolayers (SAMs)	1065
	21.6.3	Hard Diamond-like Carbon (DLC) Coatings	1069
21.7		Component-Level Studies	1070
	21.7.1	Surface Roughness Studies of Micromotor Components	1070
	21.7.2	Adhesion Measurements	1071
	21.7.3	Static Friction Force (Stiction) Measurements in MEMS	1079
	21.7.4	Mechanisms Associated with Observed Stiction Phenomena in Micromotors	1082
References			1083

22 Mechanical Properties of Micromachined Structures
Harold Kahn 1091

22.1		Measuring Mechanical Properties of Films on Substrates	1091
	22.1.1	Residual Stress Measurements	1091
	22.1.2	Mechanical Measurements Using Nanoindentation	1092
22.2		Micromachined Structures for Measuring Mechanical Properties	1093
	22.2.1	Passive Structures	1093
	22.2.2	Active Structures	1098
22.3		Measurements of Mechanical Properties	1108
	22.3.1	Mechanical Properties of Polysilicon	1109
	22.3.2	Mechanical Properties of Other Materials	1113
References			1114

The Editor 1117

Index 1119

List of Contributors

Boris Anczykowski
nanoAnalytics GmbH,
Gievenbecker Weg 11,
48149, Münster,
Germany
anczykowski@nanoanalytics.com

Roland Bennewitz
Physics Department,
McGill University,
3600 rue University,
H3A 2T8, Montreal, QC,
Canada
roland@physics.mcgill.ca

Alan D. Berman
Monitor Venture Enterprises,
241 S. Figueroa St. Suite 300,
Los Angeles, CA, 90012,
USA
alan.berman.2001@anderson.ucla.edu

Bharat Bhushan
Nanotribology Laboratory for Information Storage and MEMS/NEMS,
The Ohio State University,
206 W. 18th Avenue,
Columbus, OH, 43210-1107,
USA
bhushan.2@osu.edu

Donald W. Brenner
Department of Materials Science and Engineering,
North Carolina State University,
Raleigh, North Carolina, 27695-7909,
USA
brenner@ncsu.edu

Lásló Forró
Institute of Physics of Complex Matter,
Swiss Federal Institute of Technology
(EPFL), Ecublens, 1015, Lausanne,
Switzerland
laszlo.forro@epfl.ch

Harald Fuchs
Physikalisches Institut,
Universität Münster,
Wilhelm-Klemm-Straße 10,
48149, Münster,
Germany
fuchsh@uni-muenster.de

Franz J. Giessibl
Lehrstuhl für Experimentalphysik VI,
Universität Augsburg,
Universitätsstraße 1,
86135, Augsburg,
Germany
franz.giessibl@physik.uni-augsburg.de

Enrico Gnecco
Department of Physics,
University of Basel,
Klingelbergstraße 82,
4056, Basel,
Switzerland
Enrico.Gnecco@unibas.ch

Gérard Gremaud
Institute of Physics of Complex Matter,
Swiss Federal Institute of Technology
(EPFL), Ecublens, 1015, Lausanne,
Switzerland
gremaud@epfl.ch

Jason H. Hafner
Department of Physics & Astronomy,
Rice University,
Houston, TX, 77251-1892,
USA
hafner@rice.edu

Judith A. Harrison
Department of Chemistry,
U.S. Naval Academy,
Annapolis, Maryland, 21402,
USA
jah@usna.edu

Stefan Hengsberger
University of Applied Science of
Fribourg,
Bd de Pérolles,
1705, Fribourg,
Switzerland
stefan.hengsberger@eif.ch

Seong-Jun Heo
Department of Materials Science and
Engineering,
University of Florida,
Gainesville, FL, 32611-6400,
USA
heogyver@ufl.edu

Peter Hinterdorfer
Institute for Biophysics,
Johannes Kepler University of Linz,
Altenbergerstraße 69,
4040, Linz,
Austria
peter.hinterdorfer@jku.at

Hirotaka Hosoi
Japan Science and Technology
Corporation,
Innovation Plaza, Hokkaido,
Sapporo, 060-0819,
Japan
hosoi@sapporo.jst-plaza.jp

Jacob N. Israelachvili
Department of Chemical Engineering
and Materials Department,
University of California,
Santa Barbara, CA, 93106,
USA
Jacob@engineering.ucsb.edu

Harold Kahn
Department of Materials Science and
Engineering, Case Western Reserve
University, 10900 Euclid Avenue,
Cleveland, OH, 44106-7204,
USA
kahn@cwru.edu

András Kis
Institute of Physics of Complex Matter,
Swiss Federal Institute of Technology
(EPFL), 1015, Ecublens, Lausanne,
Switzerland
andras@igahpse.epfl.ch

Jané Kondev
Physics Department, Brandeis
University, Waltham, MA, 02454,
USA
kondev@brandeis.edu

Andrzej J. Kulik
Institute of Physics of Complex Matter,
Swiss Federal Institute of Technology
(EPFL), 1015, Lausanne,
Switzerland
andrzej.kulik@epfl.ch

Huiwen Liu
Nanotribology Laboratory for Information Storage and MEMS/NEMS,
Ohio State University,
3070 St John Ct 7,
Columbus, OH, 43210-1107,
USA
liu.403@osu.edu

Adrian B. Mann
Department of Ceramics and Materials Engineering, Rutgers University,
607 Taylor Road,
Piscataway, NJ, 08854,
USA
abmann@rci.rutgers.edu

Othmar Marti
Department of Experimental Physics,
University of Ulm,
Albert-Einstein-Allee 11, 89069, Ulm,
Germany
Othmar.Marti@physik.uni-ulm.de

Ernst Meyer
Institute of Physics, University of Basel,
Klingelbergstraße 82, 4056, Basel,
Switzerland
Ernst.Meyer@unibas.ch

Markus Morgenstern
Institute of Applied Physics,
University of Hamburg,
Jungiusstraße 11, 20355, Hamburg,
Germany
mmorgens@physnet.uni-hamburg.de

Seizo Morita
Department of Electronic Engineering,
Osaka University,
Yamada-Oka 2-1,
Suita-Citiy, Osaka, 565-0871,
Japan
smorita@ele.eng.osaka-u.ac.jp

Koichi Mukasa
Nanoelectronics Laboratory,
Hokkaido University,
Nishi-8, Kita-13, Kita-ku,
Sapporo, 060-8628,
Japan
mukasa@nano.eng.hokudai.ac.jp

Michael Nosonovsky
Nanotribology Labartory for Information Storage and NEMS/NEMS,
The Ohio State Unversity,
650 Ackerman Road, suite 255,
Columbus, Ohio, 43202,
USA
nosonovsky.1@osa.edu

Hiroshi Onishi
Surface Chemistry Laboratory,
Kanagawa Academy of Science and Technology,
KSP East 404, 3-2-1 Sakado, Takatsu-ku, Kawasaki-shi,
Kanagawa, 213-0012,
Japan
oni@net.ksp.or.jp

Oliver Pfeiffer
Institute of Physics,
University of Basel,
Klingelbergstraß 82,
4056, Basel,
Switzerland
Oliver.Pfeiffer@stud.unibas.ch

Rob Phillips
Mechanical Engineering and Applied Physics,
California Institute of Technology,
1200 California Boulevard,
Pasadena, CA, 91125,
USA
phillips@aero.caltech.edu

Prashant K. Purohit
Mechanical Engineering,
California Institute of Technology,
1200 California Boulevard,
Pasadena, CA, 91125,
USA
prashant@caltech.edu

Marina Ruths
Department of Physical Chemistry,
Åbo Akademi University,
Porthansgatan 3–5,
20500, Åbo,
Finland
mruths@abo.fi

Akira Sasahara
Surface Chemistry Laboratory,
Kanagawa Academy of Science and Technology,
KSP East 404, 3-2-1 Sakado, Takatsu-ku, Kawasaki-shi,
Kanagawa, 213-0012,
Japan
ryo@net.ksp.or.jp

André Schirmeisen
University of Münster,
Institute of Physics,
Whilhem-Klemm-Straße 10,
48149, Münster,
Germany
schira@uni-muenster.de

Alexander Schwarz
Institute of Applied Physics,
University of Hamburg,
Jungiusstraße 11,
20355, Hamburg,
Germany
aschwarz@physnet.uni-hamburg.de

Udo D. Schwarz
Department of Mechanical Engineering,
Yale University,
15 Prospect Street,
New Haven, CT, 06510,
USA
udo.schwarz@yale.edu

Susan B. Sinnott
Department of Materials Science and Engineering,
University of Florida,
Gainesville, FL, 32611-6400,
USA
ssinn@mse.ufl.edu

Anisoara Socoliuc
Institute of Physics,
University of Basel,
Klingelbergstraße 82,
4056, Basel,
Switzerland
A.Socoliuc@unibas.ch

Yasuhiro Sugawara
Department of Applied Physics,
Osaka University,
Yamada-Oka 2-1,
Suita, 565-0871,
Japan
sugawara@ap.eng.osaka-u.ac.jp

Philippe K. Zysset
Institut für Leichtbau und Flugzeugbau (ILFB), Technische Universität Wien,
Gußhausstraße 27–29, 1040, Wien,
Austria
philippe.zysset@epfl.ch

List of Abbrevations

μCP	microcontact printing
μTAS	micro-total analysis systems
2-DEG	two-dimensional electron gas
2-DES	two-dimensional electron gas
AFAM	atomic force acoustic microscopy
AFM	atomic force microscope/microscopy
AM	amplitude modulation
APCVD	atmospheric pressure chemical vapor deposition
ATP	adenosine triphosphat
BIOMEMS	Biological or Biomedical Microelectromechanical Systems
BSA	bovine serum albumin
CA	constant amplitude
CBA	cantilever beam array
CDW	charge density wave
CFM	chemical force microscopy
CG	controlled geometry
CNT	carbon nanotube
CSM	continuous stiffness measurement
CVD	chemical vapor deposition
DAS	dimer adatom stacking
DFM	dynamic force microscopy
DLC	diamond-like carbon
DLP	digital light processing
DLVO	Derjaguin–Landau–Verwey–Overbeek
DMD	digital micromirror device
DMT	Derjaguin–Muller–Toporov
DOS	density of states
DSC	differential scanning calorimetry
DSP	digital signal processor
EBD	electron beam deposition

ECR-CVD	electron cyclotron resonance chemical vapor deposition
EELS	electron energy loss spectrometer/spectroscopy
EFC	electrostatic force constant
EFM	electric field gradient microscopy
EHD	electrohydrodynamic
FCA	filtered cathodic arc
FCP	force calibration plot
FEM	finite element method/modeling
FET	field-effect transistor
FFM	friction force microscope/microscopy
FIB	focused ion beam
FIM	field-ion microscope/microscopy
FKT	Frenkel-Kontorova-Tomlinson
FM	frequency modulation
FM-AFM	frequency modulation AFM
FM-SFM	frequency-modulation SFM
FMM	force modulation mode
FS	force spectroscopy
GMR	giant magnetoresistance
HARMEMS	high-aspect-ratio MEMS
HF	hydrofluoric acid
HOP	highly oriented pyrolytic
HOPG	highly oriented pyrolytic graphite
HtBDC	hexa-tert-butyl-decacyclene
HTCS	high temperature superconductivity
IBD	ion beam deposition
IC	integrated circuit
ICAM-1	intercellular adhesion molecule-1
ISE	indentation size effect
JKR	Johnson–Kendall–Roberts
KPFM	Kelvin probe force microscopy
LB	Langmuir–Blodgett
LDOS	local density of states
LFA-1	leukocyte function-associated antigen-1
LFM	lateral force microscope
LN	liquid nitrogen
LPCVD	low pressure chemical vapor deposition
LTSPM	low-temperature SPM
LVDT	linear variable differential transformers
MD	molecular dynamics
MDS	molecular dynamics simulation
ME	metal evaporated
MEMS	microelectromechanical systems
MFM	magnetic field microscope/microscopy
MHA	16-mercaptohexadecanoic acid thiol

MP	metal particle
mrecognitionfm	molecular recognition force microscopy
mresonancefm	magnetic resonance force microscopy
NC-AFM	noncontact atomic force microscopy
NEMS	nanoelectromechanical systems
NSOM	near-field scanning optical microscope/microscopy
NTA	nitrilotriacetate
OMVPE	organometallic vapor phase epitaxy
OT	optical tweezers
OTS	octadecyltrichlorosilane
P/W	power to weight
PDMS	polydimethylsiloxane
PDP	2-pyridyldithiopropionyl
PECVD	plasma enhanced CVD
PEG	poly(ethylene glycol)
PES	photoemission spectroscopy
pesspectroscopy	photoemission spectroscopy
PET	poly(ethylene terephthalate)
PFDA	perfluorodecanoic acid
PFPE	perfluoropolyether
PMMA	poly(methylmethacrylate)
PS	polystyrene
PSGL-1	P-selection glycoprotein ligand-1
PTFE	polytetrafluoroethylene
PVC	
PZT	lead zirconate titanate
RF	radiofrequency
RH	relative humidity
RICM	reflection interface contrast microscopy
SACA	static advancing contact angle
SAM	self-assembling monolayer
sammicroscopy	scanning acoustic microscopy
SCM	scanning capitance microscopy
SCPM	scanning chemical potential
SEcM	scanning electrochemical microscopy
SEFM	scanning electrostatic force microscopy
SEM	scanning electron microscope/microscopy
SFA	surface forces apparatus
SFAM	scanning force acoustic microscopy
SFD	shear flow detachment
SFM	scanning force microscopy
SFS	scanning force spectroscopy
SICM	scanning ion conductance microscopy
SKPM	scanning Kelvin probe microscopy
SLAM	scanning local-acceleration microscopy

SMM	scanning magnetic microscopy
SN	scanning nanoindentation
SNOM	scanning near-field optical microscopy
SPM	scanning probe microscopy
ssDNA	single stranded DNA
SThM	scanning thermal microscopy
STM	scanning tunneling microscope/microscopy
SWNT	single-wall nanotubes
T-SLAM	variable-temperature SLAM
TEM	transmission electron microscopy
TESP	tapping-mode etched silicon probe
TIRM	total internal reflection microscopy
TTF	tetrathiofulvane
UHV	ultrahigh vacuum

1

Introduction – Measurement Techniques and Applications

Bharat Bhushan

Summary. In this introductory chapter, the definition and history of tribology and their industrial significance and origins and significance of an emerging field of micro/nanotribology are described. Next, various measurement techniques used in micro/nanotribological and micro/nanomechanical studies are described. The interest in micro/nanotribology field grew from magnetic storage devices and latter the applicability to emerging field micro/nanoelectromechanical systems (MEMS/NEMS) became clear. A few examples of magnetic storage devices and MEMS/NEMS are presented where micro/nanotribological and micro/nanomechanical tools and techniques are essential for interfacial studies. Finally, reasons why micro/nanotribological and micro/nanomechanical studies are important in magnetic storage devices and MEMS/NEMS are presented. In the last section, organization of the book is presented.

1.1 Definition and History of Tribology

The word tribology was first reported in a landmark report by *Jost* [1]. The word is derived from the Greek word tribos meaning rubbing, so the literal translation would be "the science of rubbing". Its popular English language equivalent is friction and wear or lubrication science, alternatively used. The latter term is hardly all-inclusive. Dictionaries define tribology as the science and technology of interacting surfaces in relative motion and of related subjects and practices. Tribology is the art of applying operational analysis to problems of great economic significance, namely, reliability, maintenance, and wear of technical equipment, ranging from spacecraft to household appliances. Surface interactions in a tribological interface are highly complex, and their understanding requires knowledge of various disciplines including physics, chemistry, applied mathematics, solid mechanics, fluid mechanics, thermodynamics, heat transfer, materials science, rheology, lubrication, machine design, performance and reliability.

It is only the name tribology that is relatively new, because interest in the constituent parts of tribology is older than recorded history [2]. It is known that drills made during the Paleolithic period for drilling holes or producing fire were fitted with bearings made from antlers or bones, and potters' wheels or stones for grinding

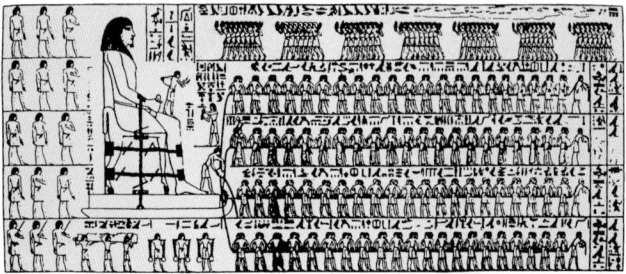

Fig. 1.1. Egyptians using lubricant to aid movement of colossus, El-Bersheh, circa 1800 BC

cereals, etc., clearly had a requirement for some form of bearings [3]. A ball thrust bearing dated about AD 40 was found in Lake Nimi near Rome.

Records show the use of wheels from 3500 BC, which illustrates our ancestors' concern with reducing friction in translationary motion. The transportation of large stone building blocks and monuments required the know-how of frictional devices and lubricants, such as water-lubricated sleds. Figure 1.1 illustrates the use of a sledge to transport a heavy statue by egyptians circa 1880 BC [4]. In this transportation, 172 slaves are being used to drag a large statue weighing about 600 kN along a wooden track. One man, standing on the sledge supporting the statue, is seen pouring a liquid (most likely water) into the path of motion; perhaps he was one of the earliest lubrication engineers. (*Dowson* [2] has estimated that each man exerted a pull of about 800 N. On this basis, the total effort, which must at least equal the friction force, becomes 172×800 N. Thus, the coefficient of friction is about 0.23.) A tomb in Egypt that was dated several thousand years BC provides the evidence of use of lubricants. A chariot in this tomb still contained some of the original animal-fat lubricant in its wheel bearings.

During and after the glory of the Roman empire, military engineers rose to prominence by devising both war machinery and methods of fortification, using tribological principles. It was the renaissance engineer-artist Leonardo da Vinci (1452–1519), celebrated in his days for his genius in military construction as well as for his painting and sculpture, who first postulated a scientific approach to friction. Da Vinci deduced the rules governing the motion of a rectangular block sliding over a flat surface. He introduced for the first time, the concept of coefficient of friction as the ratio of the friction force to normal load. His work had no historical influence, however, because his notebooks remained unpublished for hundreds of years. In 1699, the French physicist Guillaume Amontons rediscovered the rules of friction after he studied dry sliding between two flat surfaces [5]. First, the friction force that resists sliding at an interface is directly proportional to the normal load. Second, the amount of friction force does not depend on the apparent area of contact. These observations were verified by French physicist Charles-Augustin Coulomb (better known for his work on electrostatics [6]). He added a third rule that the friction force is independent of velocity once motion starts. He also made a clear distinction between static friction and kinetic friction.

Many other developments occurred during the 1500s, particularly in the use of improved bearing materials. In 1684, Robert Hooke suggested the combination of steel shafts and bell-metal bushes as preferable to wood shod with iron for wheel bearings. Further developments were associated with the growth of industrialization in the latter part of the eighteenth century. Early developments in the petroleum industry started in Scotland, Canada, and the United States in the 1850s [2–7].

Though essential laws of viscous flow were postulated by Sir Isaac Newton in 1668; scientific understanding of lubricated bearing operations did not occur until the end of the nineteenth century. Indeed, the beginning of our understanding of the principle of hydrodynamic lubrication was made possible by the experimental studies of *Tower* [8] and the theoretical interpretations of *Reynolds* [9] and related work by *Petroff* [10]. Since then developments in hydrodynamic bearing theory and practice were extremely rapid in meeting the demand for reliable bearings in new machinery.

Wear is a much younger subject than friction and bearing development, and it was initiated on a largely empirical basis. Scientific studies of wear developed little until the mid-twentieth century. *Holm* made one of the earliest substantial contributions to the study of wear [11].

The industrial revolution (1750–1850 A.D.) is recognized as a period of rapid and impressive development of the machinery of production. The use of steam power and the subsequent development of the railways in the 1830s led to promotion of manufacturing skills. Since the beginning of the twentieth century, from enormous industrial growth leading to demand for better tribology, knowledge in all areas of tribology has expanded tremendously [11–17].

1.2 Industrial Significance of Tribology

Tribology is crucial to modern machinery which uses sliding and rolling surfaces. Examples of productive friction are brakes, clutches, driving wheels on trains and automobiles, bolts, and nuts. Examples of productive wear are writing with a pencil, machining, polishing, and shaving. Examples of unproductive friction and wear are internal combustion and aircraft engines, gears, cams, bearings, and seals.

According to some estimates, losses resulting from ignorance of tribology amount in the United States to about 4% of its gross national product (or about $ 200 billion dollars per year in 1966), and approximately one-third of the world's energy resources in present use appear as friction in one form or another. Thus, the importance of friction reduction and wear control cannot be overemphasized for economic reasons and long-term reliability. According to *Jost* [1, 18], savings of about 1% of gross national product of an industrialized nation can be realized by research and better tribological practices. According to recent studies, expected savings are expected to be on the order of 50 times the research costs. The savings are both substantial and significant, and these savings can be obtained without the deployment of large capital investment.

The purpose of research in tribology is understandably the minimization and elimination of losses resulting from friction and wear at all levels of technology where the rubbing of surfaces is involved. Research in tribology leads to greater plant efficiency, better performance, fewer breakdowns, and significant savings.

Tribology is not only important to industry, it also affects day-to-day life. For example, writing is a tribological process. Writing is accomplished by a controlled transfer of lead (pencil) or ink (pen) to the paper. During writing with a pencil there should be good adhesion between the lead and paper so that a small quantity of lead transfers to the paper and the lead should have adequate toughness/hardness so that it does not fracture/break. Objective during shaving is to remove hair from the body as efficiently as possible with minimum discomfort to the skin. Shaving cream is used as a lubricant to minimize friction between a razor and the skin. Friction is helpful during walking and driving. Without adequate friction, we would slip and a car would skid! Tribology is also important in sports. For example, a low friction between the skis and the ice is desirable during skiing.

1.3 Origins and Significance of Micro/Nanotribology

At most interfaces of technological relevance, contact occurs at numerous asperities. Consequently, the importance of investigating a single asperity contact in studies of the fundamental tribological and mechanical properties of surfaces has been long recognized. The recent emergence and proliferation of proximal probes, in particular tip-based microscopies (e.g., the scanning tunneling microscope and the atomic force microscope) and of computational techniques for simulating tip-surface interactions and interfacial properties, has allowed systematic investigations of interfacial problems with high resolution as well as ways and means for modifying and manipulating nanoscale structures. These advances have led to the development of the new field of microtribology, nanotribology, molecular tribology, or atomic-scale tribology [15, 16, 19–22]. This field is concerned with experimental and theoretical investigations of processes ranging from atomic and molecular scales to microscales, occurring during adhesion, friction, wear, and thin-film lubrication at sliding surfaces.

The differences between the conventional or macrotribology and micro/nanotribology are contrasted in Fig. 1.2. In macrotribology, tests are conducted on components with relatively large mass under heavily loaded conditions. In these tests, wear is inevitable and the bulk properties of mating components dominate the tribological performance. In micro/nanotribology, measurements are made on components, at least one of the mating components, with relatively small mass under lightly loaded conditions. In this situation, negligible wear occurs and the surface properties dominate the tribological performance.

The micro/nanotribological studies are needed to develop fundamental understanding of interfacial phenomena on a small scale and to study interfacial phenomena involving ultrathin films (as low as 1–2 nm) and in micro/nanostructures, both used in magnetic storage systems, micro/nanoelectromechanical systems (MEMS/NEMS) and other industrial applications. The components used in micro-

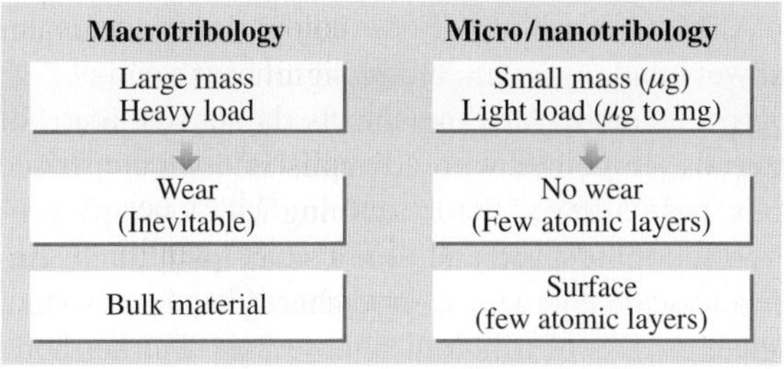

Macrotribology	Micro/nanotribology
Large mass Heavy load	Small mass (μg) Light load (μg to mg)
Wear (Inevitable)	No wear (Few atomic layers)
Bulk material	Surface (few atomic layers)

Fig. 1.2. Comparisons between macrotribology and micro/nanotribology

and nanostructures are very light (on the order of few micrograms) and operate under very light loads (smaller than one microgram to a few milligrams). As a result, friction and wear (on a nanoscale) of lightly loaded micro/nanocomponents are highly dependent on the surface interactions (few atomic layers). These structures are generally lubricated with molecularly thin films. Micro/nanotribological techniques are ideal to study the friction and wear processes of ultrathin films and micro/nanostructures. Although micro/nanotribological studies are critical to study ultrathin films and micro/nanostructures, these studies are also valuable in fundamental understanding of interfacial phenomena in macrostructures to provide a bridge between science and engineering.

The probe-based microscopes (scanning tunneling microscope, the atomic force and friction force microscopes) and the surface force apparatus are widely used for micro/nanotribological studies [16, 19–23]. To give a historical perspective of the field, the scanning tunneling microscope (STM) developed by *Binnig* and *Rohrer* and their colleagues in 1981 at the IBM Zurich Research Laboratory, Forschungslabor, is the first instrument capable of directly obtaining three-dimensional (3D) images of solid surfaces with atomic resolution [24]. STMs can only be used to study surfaces which are electrically conductive to some degree. Based on their design of STM, in 1985, *Binnig* et al. [25, 26] developed an atomic force microscope (AFM) to measure ultrasmall forces (less than 1 µN) present between the AFM tip surface and the sample surface. AFMs can be used for measurement of all engineering surfaces which may be either electrically conducting or insulating. AFM has become a popular surface profiler for topographic measurements on micro- to nanoscale. AFMs modified to measure both normal and friction forces, generally called friction force microscopes (FFMs) or lateral force microscopes (LFMs), are used to measure friction on micro- and nanoscales. AFMs are also used for studies of adhesion, scratching, wear, lubrication, surface temperatures, and for measurements of elastic/plastic mechanical properties (such as indentation hardness and modulus of elasticity [14, 16, 19, 21].

Surface force apparatuses (SFAs), first developed in 1969, are used to study both static and dynamic properties of the molecularly thin liquid films sandwiched between two molecularly smooth surfaces [16, 20–22, 27]. However, the liquid under study has to be confined between molecularly-smooth optically-transparent or sometimes opaque surfaces with radii of curvature on the order of 1 mm (leading to poorer lateral resolution as compared to AFMs). Only AFMs/FFMs can be used to study engineering surfaces in the dry and wet conditions with atomic resolution.

Meanwhile, significant progress in understanding the fundamental nature of bonding and interactions in materials, combined with advances in computer-based modeling and simulation methods, have allowed theoretical studies of complex interfacial phenomena with high resolution in space and time [16, 20–22]. Such simulations provide insights into atomic-scale energetics, structure, dynamics, thermodynamics, transport and rheological aspects of tribological processes. Furthermore, these theoretical approaches guide the interpretation of experimental data and the design of new experiments, and enable the prediction of new phenomena based on atomistic principles.

1.4 Measurement Techniques

1.4.1 Scanning Probe Microscopy

Family of instruments based on STMs and AFMs, called Scanning Probe Microscopes (SPMs), have been developed for various applications of scientific and industrial interest. These include – STM, AFM, FFM (or LFM), scanning electrostatic force microscopy (SEFM), scanning force acoustic microscopy (SFAM) (or atomic force acoustic microscopy, AFAM), scanning magnetic microscopy (SMM) (or magnetic force microscopy, MFM), scanning near field optical microscopy (SNOM), scanning thermal microscopy (SThM) scanning electrochemical microscopy (SEcM), scanning Kelvin Probe microscopy (SKPM), scanning chemical potential microscopy (SCPM), scanning ion conductance microscopy (SICM), and scanning capacitance microscopy (SCM). Family of instruments which measure forces (e.g. AFM, FFM, SEFM, SFAM, and SSM) are also referred to as scanning force microscopies (SFM). Although these instruments offer atomic resolution and are ideal for basic research, yet these are used for cutting edge industrial applications which do not require atomic resolution.

STMs, AFMs and their modifications can be used at extreme magnifications ranging from $10^3\times$ to $10^9\times$ in x-, y-, and z-directions for imaging macro to atomic dimensions with high-resolution information and for spectroscopy. These instruments can be used in any environment such as ambient air, various gases, liquid, vacuum, low temperatures, and high temperatures. Imaging in liquid allows the study of live biological samples and it also eliminates water capillary forces present in ambient air present at the tip–sample interface. Low temperature imaging is useful for the study of biological and organic materials and the study of low-temperature phenomena such as superconductivity or charge-density waves. Low-temperature operation is also advantageous for high-sensitivity force mapping due to the reduction in thermal vibration. These instruments also have been used to image liquids such as liquid crystals and lubricant molecules on graphite surfaces. While the pure imaging capabilities of SPM techniques dominated the application of these methods at their early development stages, the physics and chemistry of probe–sample interactions and the quantitative analyses of tribological, electronic, magnetic, biological, and chemical surfaces are commonly carried out. Nanoscale science and technology are strongly driven by SPMs which allow investigation and manipulation of surfaces down to the atomic scale. With growing understanding of the underlying interaction mechanisms, SPMs have found applications in many fields outside basic research fields. In addition, various derivatives of all these methods have been developed for special applications, some of them targeting far beyond microscopy.

A detailed overview of scanning probe microscopy – principle of operation, instrumentation, and probes is presented in a later chapter (also see [16, 20–23]). Here, a brief description of commercial STMs and AFMs follows.

Commercial STMs

There are a number of commercial STMs available on the market. Digital Instruments, Inc. located in Santa Barbara, CA introduced the first commercial STM, the Nanoscope I, in 1987. In a recent Nanoscope IV STM for operation in ambient air, the sample is held in position while a piezoelectric crystal in the form of a cylindrical tube (referred to as PZT tube scanner) scans the sharp metallic probe over the surface in a raster pattern while sensing and outputting the tunneling current to the control station, Fig. 1.3. The digital signal processor (DSP) calculates the desired separation of the tip from the sample by sensing the tunneling current flowing between the sample and the tip. The bias voltage applied between the sample and the tip encourages the tunneling current to flow. The DSP completes the digital feedback loop by outputting the desired voltage to the piezoelectric tube. The STM operates in both the "constant height" and "constant current" modes depending on a parameter selection in the control panel. In the constant current mode, the feedback gains are set high, the tunneling tip closely tracks the sample surface, and the variation in the tip height required to maintain constant tunneling current is measured by the change in the voltage applied to the piezo tube. In the constant height mode, the feedback gains are set low, the tip remains at a nearly constant height as it sweeps over the sample surface, and the tunneling current is imaged.

Physically, the Nanoscope STM consists of three main parts: the head which houses the piezoelectric tube scanner for three dimensional motion of the tip and the preamplifier circuit (FET input amplifier) mounted on top of the head for the tunneling current, the base on which the sample is mounted, and the base support, which supports the base and head [16, 21]. The base accommodates samples up to 10 mm by 20 mm and 10 mm in thickness. Scan sizes available for the STM are $0.7\,\mu m \times 0.7\,\mu m$ (for atomic resolution), $12\,\mu m \times 12\,\mu m$, $75\,\mu m \times 75\,\mu m$ and $125\,\mu m \times 125\,\mu m$.

The scanning head controls the three dimensional motion of tip. The removable head consists of a piezo tube scanner, about 12.7 mm in diameter, mounted into an invar shell used to minimize vertical thermal drifts because of good thermal match between the piezo tube and the Invar. The piezo tube has separate electrodes for X, Y and Z which are driven by separate drive circuits. The electrode configuration (Fig. 1.3) provides x and y motions which are perpendicular to each other, minimizes horizontal and vertical coupling, and provides good sensitivity. The vertical

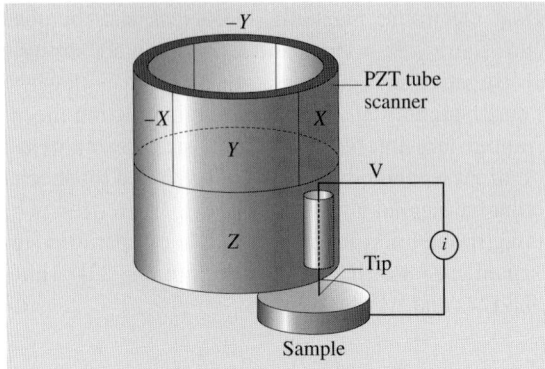

Fig. 1.3. Principle of operation of a commercial STM, a sharp tip attached to a piezoelectric tube scanner is scanned on a sample (from [16])

motion of the tube is controlled by the Z electrode which is driven by the feedback loop. The x and y scanning motions are each controlled by two electrodes which are driven by voltages of same magnitudes, but opposite signs. These electrodes are called $-Y$, $-X$, $+Y$, and $+X$. Applying complimentary voltages allows a short, stiff tube to provide a good scan range without large voltages. The motion of the tip due to external vibrations is proportional to the square of the ratio of vibration frequency to the resonant frequency of the tube. Therefore, to minimize the tip vibrations, the resonant frequencies of the tube are high about 60 kHz in the vertical direction and about 40 kHz in the horizontal direction. The tip holder is a stainless steel tube with a 300 μm inner diameter for 250 μm diameter tips, mounted in ceramic in order to keep the mass on the end of the tube low. The tip is mounted either on the front edge of the tube (to keep mounting mass low and resonant frequency high) (Fig. 1.3) or the center of the tube for large range scanners, namely 75 and 125 μm (to preserve the symmetry of the scanning.) This commercial STM accepts any tip with a 250 μm diameter shaft. The piezotube requires X–Y calibration which is carried out by imaging an appropriate calibration standard. Cleaved graphite is used for the small-scan length head while two dimensional grids (a gold plated ruling) can be used for longer range heads.

The Invar base holds the sample in position, supports the head, and provides coarse x–y motion for the sample. A spring-steel sample clip with two thumb screws holds the sample in place. An x–y translation stage built into the base allows the sample to be repositioned under the tip. Three precision screws arranged in a triangular pattern support the head and provide coarse and fine adjustment of the tip height. The base support consists of the base support ring and the motor housing. The stepper motor enclosed in the motor housing allows the tip to be engaged and withdrawn from the surface automatically.

Samples to be imaged with STM must be conductive enough to allow a few nanoamperes of current to flow from the bias voltage source to the area to be scanned. In many cases, nonconductive samples can be coated with a thin layer of a conductive material to facilitate imaging. The bias voltage and the tunneling current depend on the sample. Usually they are set at a standard value for engagement and fine tuned to enhance the quality of the image. The scan size depends on the sample and the features of interest. Maximum scan rate of 122 Hz can be used. The maximum scan rate is usually related to the scan size. Scan rate above 10 Hz is used for small scans (typically 60 Hz for atomic-scale imaging with a 0.7 μm scanner). The scan rate should be lowered for large scans, especially if the sample surfaces are rough or contain large steps. Moving the tip quickly along the sample surface at high scan rates with large scan sizes will usually lead to a tip crash. Essentially, the scan rate should be inversely proportional to the scan size (typically 2–4 Hz for 1 μm, 0.5–1 Hz for 12 μm, and 0.2 Hz for 125 μm scan sizes). Scan rate in length/time, is equal to scan length divided by the scan rate in Hz. For example, for 10 μm × 10 μm scan size scanned at 0.5 Hz, the scan rate is 10 μm/s. The 256 × 256 data formats are most commonly used. The lateral resolution at larger scans is approximately equal to scan length divided by 256.

Commercial AFM

A review of early designs of AFMs is presented by *Bhushan* [21]. There are a number of commercial AFMs available on the market. Major manufacturers of AFMs for use in ambient environment are: Digital Instruments Inc., a subsidiary of Veeco Instruments, Inc., Santa Barbara, California; Topometrix Corp., a subsidiary of Veeco Instruments, Inc., Santa Clara, California; and other subsidiaries of Veeco Instruments Inc., Woodbury, New York; Molecular Imaging Corp., Phoenix, Arizona; Quesant Instrument Corp., Agoura Hills, California; Nanoscience Instruments Inc., Phoenix, Arizona; Seiko Instruments, Japan; and Olympus, Japan. AFM/STMs for use in UHV environment are primarily manufactured by Omicron Vakuumphysik GMBH, Taunusstein, Germany.

We describe here two commercial AFMs – small sample and large sample AFMs – for operation in the contact mode, produced by Digital Instruments, Inc., Santa Barbara, CA, with scanning lengths ranging from about 0.7 μm (for atomic resolution) to about 125 μm [28–31]. The original design of these AFMs comes from *Meyer* and *Amer* [32]. Basically the AFM scans the sample in a raster pattern while outputting the cantilever deflection error signal to the control station. The cantilever deflection (or the force) is measured using laser deflection technique, Fig. 1.4. The DSP in the workstation controls the z-position of the piezo based on the cantilever deflection error signal. The AFM operates in both the "constant height" and "constant force" modes. The DSP always adjusts the height of the sample under the tip based on the cantilever deflection error signal, but if the feedback gains are low the piezo remains at a nearly "constant height" and the cantilever deflection data is collected. With the high gains, the piezo height changes to keep the cantilever deflection nearly constant (therefore the force is constant) and the change in piezo height is collected by the system.

To further describe the principle of operation of the commercial small sample AFM shown in Fig. 1.4a, the sample, generally no larger than 10 mm × 10 mm, is

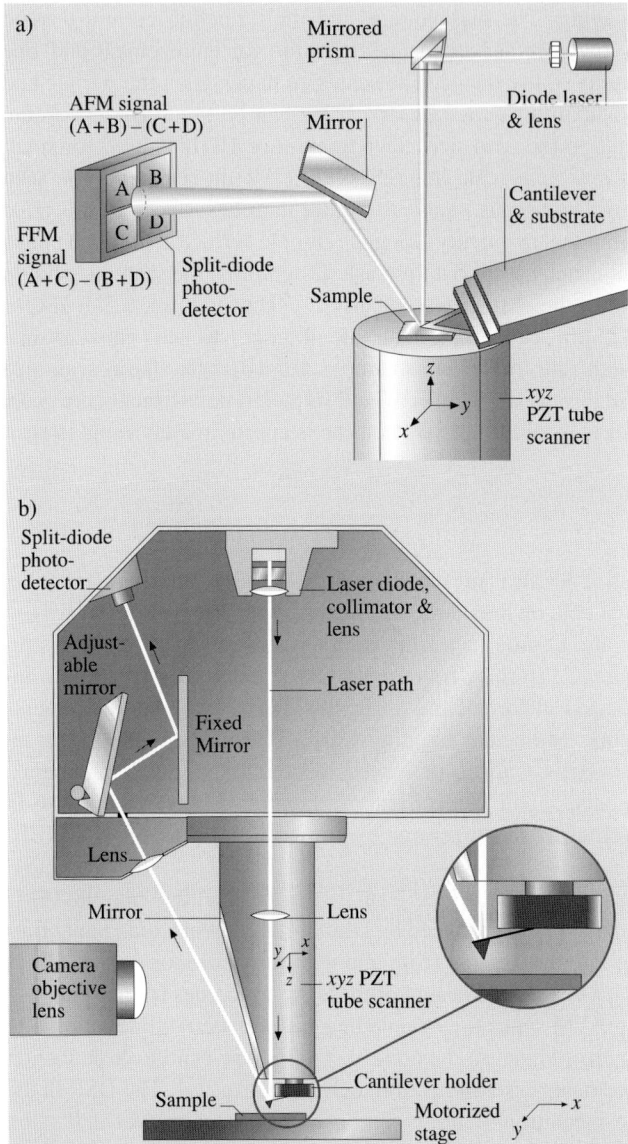

Fig. 1.4. Principles of operation of (**a**) a commercial small sample AFM/FFM, and (**b**) a large sample AFM/FFM (from [16])

mounted on a PZT tube scanner which consists of separate electrodes to scan precisely the sample in the x-y plane in a raster pattern and to move the sample in the vertical (z) direction. A sharp tip at the free end of a flexible cantilever is brought in contact with the sample. Features on the sample surface cause the cantilever to deflect in the vertical and lateral directions as the sample moves under the tip. A laser beam from a diode laser (5 mW max peak output at 670 nm) is directed by a prism onto the back of a cantilever near its free end, tilted downward at about 10° with respect to the horizontal plane. The reflected beam from the vertex of the cantilever is directed through a mirror onto a quad photodetector (split photodetector with four quadrants, commonly called position-sensitive detector or PSD, produced by Silicon Detector Corp., Camarillo, California). The differential signal from the top and bottom photodiodes provides the AFM signal which is a sensitive measure of the cantilever vertical deflection. Topographic features of the sample cause the tip to deflect in the vertical direction as the sample is scanned under the tip. This tip deflection will change the direction of the reflected laser beam, changing the intensity difference between the top and bottom sets of photodetectors (AFM signal). In the AFM operating mode called the height mode, for topographic imaging or for any other operation in which the applied normal force is to be kept a constant, a feedback circuit is used to modulate the voltage applied to the PZT scanner to adjust the height of the PZT, so that the cantilever vertical deflection (given by the intensity difference between the top and bottom detector) will remain constant during scanning. The PZT height variation is thus a direct measure of the surface roughness of the sample.

In a large sample AFM, both force sensors using optical deflection method and scanning unit are mounted on the microscope head, Fig. 1.4b. Because of vibrations added by cantilever movement, lateral resolution of this design is somewhat poorer than the design in Fig. 1.4a in which the sample is scanned instead of cantilever beam. The advantage of the large sample AFM is that large samples can be measured readily.

Most AFMs can be used for topography measurements in the so-called tapping mode (intermittent contact mode), also referred to as dynamic force microscopy. In the tapping mode, during scanning over the surface, the cantilever/tip assembly is sinusoidally vibrated by a piezo mounted above it, and the oscillating tip slightly taps the surface at the resonant frequency of the cantilever (70–400 Hz) with a constant (20–100 nm) oscillating amplitude introduced in the vertical direction with a feedback loop keeping the average normal force constant, Fig. 1.5. The oscillating amplitude is kept large enough so that the tip does not get stuck to the sample because of adhesive attractions. The tapping mode is used in topography measurements to minimize effects of friction and other lateral forces and/or to measure topography of soft surfaces.

Topographic measurements are made at any scanning angle. At a first instance, scanning angle may not appear to be an important parameter. However, the friction force between the tip and the sample will affect the topographic measurements in a parallel scan (scanning along the long axis of the cantilever). Therefore a perpendicular scan may be more desirable. Generally, one picks a scanning angle which

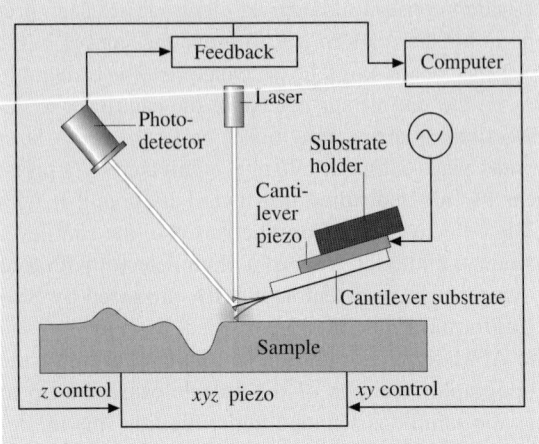

Fig. 1.5. Schematic of tapping mode used for surface roughness measurement (from [16])

gives the same topographic data in both directions; this angle may be slightly different than that for the perpendicular scan.

For measurement of friction force being applied at the tip surface during sliding, left hand and right hand sets of quadrants of the photodetector are used. In the so-called friction mode, the sample is scanned back and forth in a direction orthogonal to the long axis of the cantilever beam. A friction force between the sample and the tip will produce a twisting of the cantilever. As a result, the laser beam will be reflected out of the plane defined by the incident beam and the beam reflected vertically from an untwisted cantilever. This produces an intensity difference of the laser beam received in the left hand and right hand sets of quadrants of the photodetector. The intensity difference between the two sets of detectors (FFM signal) is directly related to the degree of twisting and hence to the magnitude of the friction force. This method provides three-dimensional maps of friction force. One problem associated with this method is that any misalignment between the laser beam and the photodetector axis would introduce error in the measurement. However, by following the procedures developed by *Ruan* and *Bhushan* [30], in which the average FFM signal for the sample scanned in two opposite directions is subtracted from the friction profiles of each of the two scans, the misalignment effect is eliminated. By following the friction force calibration procedures developed by *Ruan* and *Bhushan* [30], voltages corresponding to friction forces can be converted to force unites. The coefficient of friction is obtained from the slope of friction force data measured as a function of normal loads typically ranging from 10 to 150 nN. This approach eliminates any contributions due to the adhesive forces [33]. For calculation of the coefficient of friction based on a single point measurement, friction force should be divided by the sum of applied normal load and intrinsic adhesive force. Furthermore, it should be pointed out that for a single asperity contact, the coefficient of friction is not independent of load.

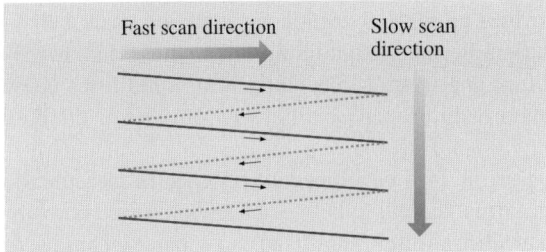

Fig. 1.6. Schematic of triangular pattern trajectory of the AFM tip as the sample is scanned in two dimensions. During imaging, data are recorded only during scans along the solid scan lines (from [16])

The tip is scanned in such a way that its trajectory on the sample forms a triangular pattern, Fig. 1.6. Scanning speeds in the fast and slow scan directions depend on the scan area and scan frequency. Scan sizes ranging from less than 1 nm × 1 nm to 125 µm × 125 µm and scan rates from less than 0.5 to 122 Hz typically can be used. Higher scan rates are used for smaller scan lengths. For example, scan rates in the fast and slow scan directions for an area of 10 µm × 10 µm scanned at 0.5 Hz are 10 µm/s and 20 nm/s, respectively.

1.4.2 Surface Force Apparatus (SFA)

Surface Force Apparatuses (SFAs) are used to study both static and dynamic properties of the molecularly-thin liquid films sandwiched between two molecularly smooth surfaces. The SFAs were originally developed by *Tabor* and *Winterton* [27] and later by *Israelachvili* and *Tabor* [34] to measure van der Waals forces between two mica surfaces as a function of separation in air or vacuum. *Israelachvili* and *Adams* [35] developed a more advanced apparatus to measure normal forces between two surfaces immersed in a liquid so thin that their thickness approaches the dimensions of the liquid molecules themselves. A similar apparatus was also developed by *Klein* [36]. The SFAs, originally used in studies of adhesive and static interfacial forces were first modified by *Chan* and *Horn* [37] and later by *Israelachvili* et al. [38] and *Klein* et al. [39] to measure the dynamic shear (sliding) response of liquids confined between molecularly smooth optically-transparent mica surfaces. Optically transparent surfaces are required because the surface separation is measured using an optical interference technique. *Van Alsten* and *Granick* [40] and *Peachey* et al. [41] developed a new friction attachment which allow for the two surfaces to be sheared past each other at varying sliding speeds or oscillating frequencies while simultaneously measuring both the friction force and normal force between them. *Israelachvili* [42] and *Luengo* et al. [43] also presented modified SFA designs for dynamic measurements including friction at oscillating frequencies. Because the mica surfaces are molecularly smooth, the actual area of contact is well defined and measurable, and asperity deformation do not complicate the analysis. During sliding experiments, the area of parallel surfaces is very large compared to the thickness of the sheared film and this provides an ideal condition for studying shear behavior because it permits one to study molecularly-thin liquid films whose thickness is well defined to the resolution of an angstrom. Molecularly thin liquid films cease to behave as a structural continuum with properties different from that of the bulk material [40, 44–47].

Tonck et al. [48] and *Georges* et al. [49] developed a SFA used to measure the static and dynamic forces (in the normal direction) between a smooth fused borosilicate glass against a smooth and flat silicon wafer. They used capacitance technique to measure surface separation; therefore, use of optically-transparent surfaces was not required. Among others, metallic surfaces can be used at the interface. *Georges* et al. [50] modified the original SFA so that a sphere can be moved towards and away from a plane and can be sheared at constant separation from the plane, for interfacial friction studies.

For a detailed review of various types of SFAs, see *Israelachvili* [42, 51], *Horn* [52], and *Homola* [53]. SFAs based on their design are commercially available from SurForce Corporation, Santa Barbara, California.

Israelachvili's and Granick's Design

Following review is primarily based on the papers by *Israelachvili* [42] and *Homola* [53]. Israelachvili et al.'s design later followed by Granick et al. for oscillating shear studies, is most commonly used by researchers around the world.

Classical SFA The classical apparatus developed for measuring equilibrium or static intersurface forces in liquids and vapors by *Israelachvili* and *Adams* [35], consists of a small, air-tight stainless steel chamber in which two molecularly smooth curved mica surfaces can be translated towards or away from each other, see Fig. 1.7. The distance between the two surfaces can also be independently controlled to within ± 0.1 nm and the force sensitivity is about 10 nN. The technique utilizes two molecularly smooth mica sheets, each about 2 μm thick, coated with a semi reflecting 50–60 nm layer of pure silver, glued to rigid cylindrical silica disks of radius about 10 mm (silvered side down) mounted facing each other with their axes mutually at right angles (crossed cylinder position), which is geometrically equivalent to a sphere contacting a flat surface. The adhesive glue which is used to affix the mica to the support is sufficiently compliant, so the mica will flatten under the action of adhesive forces or applied load to produce a contact zone in which the surfaces are locally parallel and planar. Outside of this contact zone the separation between surfaces increases and the liquid, which is effectively in a bulk state, makes a negligible contribution to the overall response. The lower surface is supported on a cantilever spring which is used to push the two surfaces together with a known load. When the surfaces are forced into contact, they flatten elastically so that the contact zone is circular for duration of the static or sliding interactions. The surface separation is measured using optical interference fringes of equal chromatic order (FECO) which enables the area of molecular contact and the surface separation to be measured to within 0.1 nm. For measurements, white light is passed vertically up through the

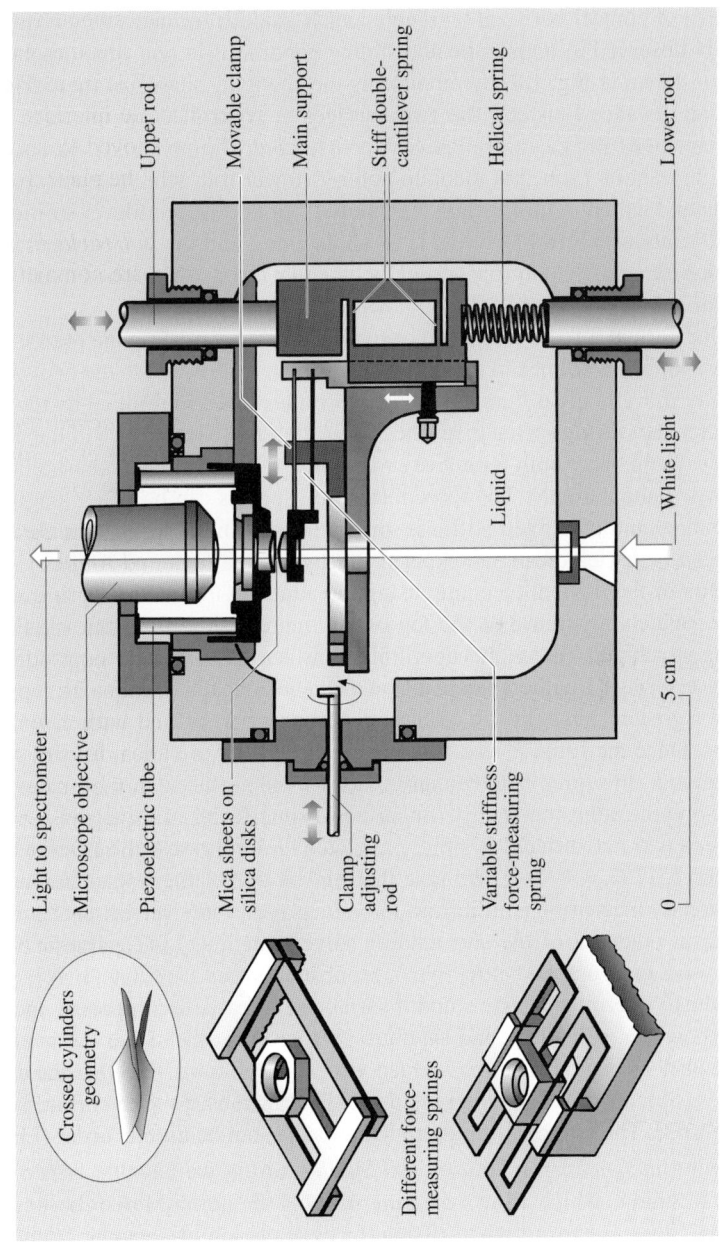

Fig. 1.7. Schematic of the surface force apparatus that employs the cross cylinder geometry [35, 42]

two mica surfaces and the emerging beam is then focused onto the slit of a grating spectrometer. From the positions and shapes of the colored FECO fringes in the spectrogram, the distance between the two surfaces and the exact shape of the two surfaces can be measured (as illustrated in Fig. 1.8), as can the refractive index of the liquid (or material) between them. In particular, this allows for reasonably accurate determinations of the quantity of material deposited or adsorbed on the surfaces and the area of contact between two molecularly smooth surfaces. Any changes may be readily observed in both static and sliding conditions in real time (applicable to the design shown in Fig. 1.8) by monitoring the changing shapes of these fringes.

The distance between the two surfaces is controlled by use of a three-stage mechanism of increasing sensitivity: coarse control (upper rod) allows positioning of within about 1 µm, the medium control (lower rod, which depresses the helical spring and which in turn, bends the much stiffer double-cantilever spring by 1/1000 of this amount) allows positioning to about 1 nm, and the piezoelectric crystal tube – which expands or controls vertically by about 0.6 nm/V applied axially across the cylindrical wall – is used for final positioning to 0.1 nm.

The normal force is measured by expanding or contracting the piezoelectric crystal by a known amount and then measuring optically how much the two surfaces have actually moved; any difference in the two values when multiplied by the stiffness of the force measuring spring gives the force difference between the initial and final positions. In this way both repulsive and attractive forces can be measured with a sensitivity of about 10 nN. The force measuring springs can be either single-cantilever or double-cantilever fixed-stiffness springs (as shown in Fig. 1.7), or the spring stiffness can be varied during an experiment (by up to a factor of 1000) by shifting the position of the dovetailed clamp using the adjusting rod. Other spring attachments, two of which are shown at the top of the figure, can replace the variable stiffness spring attachment (top right: nontilting nonshearing spring of fixed stiffness). Each of these springs are interchangeable and can be attached to the main support, allowing for greater versatility in measuring strong or weak and attractive or repulsive forces. Once the force F as a function of distance D is known for the two surfaces of radius R, the force between any other curved surfaces simply scales by R. Furthermore, the adhesion energy (or surface or interfacial free energy) E per unit area between two flat surfaces is simply related to F by the so-called Derjaguin approximation [51] $E = F/2\pi R$. We note that SFA is one of the few techniques available for directly measuring equilibrium force-laws (i.e., force versus distance at constant chemical potential of the surrounding solvent medium) [42]. The SFA allows for both weak or strong and attractive or repulsive forces.

Mostly the molecularly smooth surface of mica is used in these measurements [55], however, silica [56] and sapphire [57] have also been used. It is also possible to deposit or coat each mica surface with metal films [58, 59], carbon and metal oxides [60], adsorbed polymer layers [61], surfactant monolayers and bilayers [51, 58, 62, 63]. The range of liquids and vapors that can be used is almost endless.

Sliding Attachments for Tribological Studies So far we have described a measurement technique which allows measurements of the normal forces between surfaces, that is, those occurring when two surfaces approach or separate from each other.

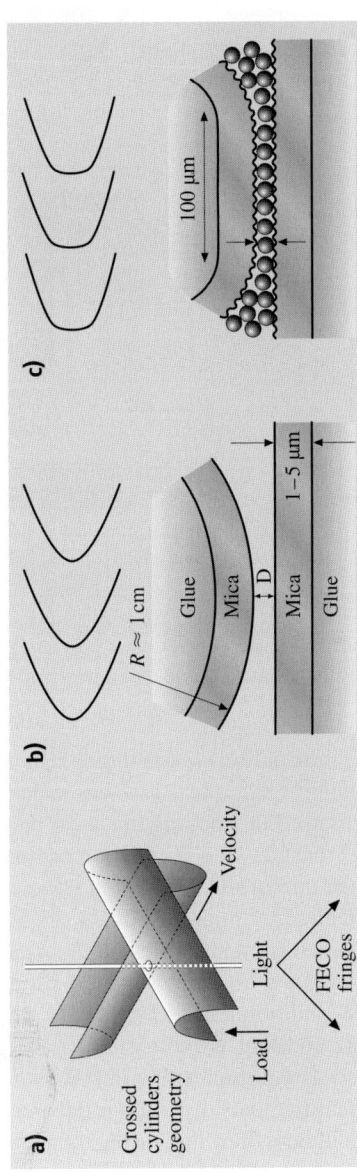

Fig. 1.8. (a) Cross cylinder configuration of mica sheet, showing formation of contact area. Schematic of the fringes of equal chromatic order (FECO) observed when two mica surfaces are (b) separated by distance D and (c) are flattened with a monolayer of liquid between them [54]

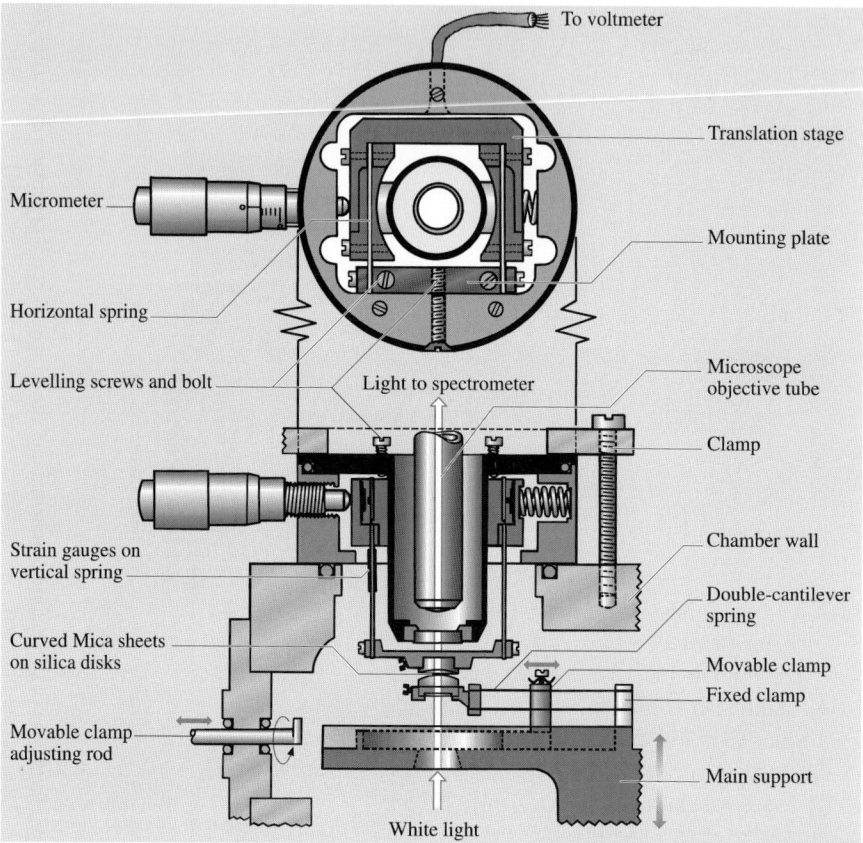

Fig. 1.9. Schematic of shear force apparatus. Lateral motion is initiated by a variable speed motor-driven micrometer screw that presses against the translation stage which is connected through two horizontal double-cantilever strip springs to the rigid mounting plate [38, 42]

However, in tribological situations, it is the transverse or shear forces that are of primary interest when two surfaces slide past each other. There are essentially two approaches used in studying the shear response of confined liquid films. In the first approach (constant velocity friction or steady-shear attachment), the friction is measured when one of the surfaces is traversed at a constant speed over a distance of several hundreds of microns [38, 39, 45, 54, 60, 64, 65]. The second approach (oscillatory shear attachment) relies on the measurement of viscous dissipation and elasticity of confined liquids by using periodic sinusoidal oscillations over a range of amplitudes and frequencies [40, 41, 44, 66, 67].

For the constant velocity friction (steady-shear) experiments, the surface force apparatus was outfitted with a lateral sliding mechanism [38, 42, 46, 54, 64, 65] allowing measurements of both normal and shearing forces (Fig. 1.9). The piezoelectric crystal tube mount supporting the upper silica disk of the basic apparatus shown

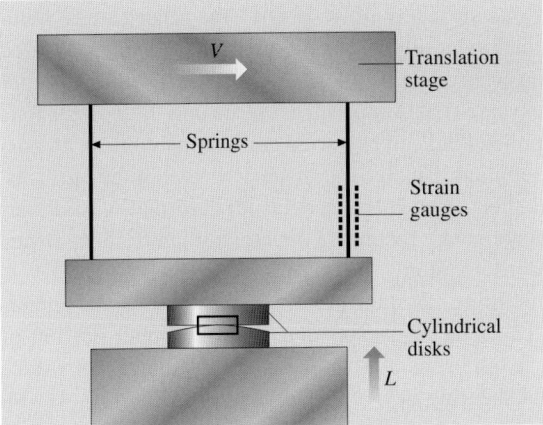

Fig. 1.10. Schematic of the sliding attachment. The translation stage also supports two vertical double-cantilever springs, which at their lower end are connected to a steel plate supporting the upper silica disk [46]

in Fig. 1.7, is replaced. Lateral motion is initiated by a variable speed motor-driven micrometer screw that presses against the translation stage, which is connected via two horizontal double-cantilever strip springs to the rigid mounting plate. The translation stage also supports two vertical double-cantilever springs (Fig. 1.10) that at their lower end are connected to a steel plate supporting the upper silica disk. One of the vertical springs acts as a frictional force detector by having four resistance strain gages attached to it, forming the four arms of a Wheatstone bridge and electrically connected to a chart recorder. Thus, by rotating the micrometer, the translation stage deflects, causing the upper surface to move horizontally and linearly at a steady rate. If the upper mica surface experiences a transverse frictional or viscous shearing force, this will cause the vertical springs to deflect, and this deflection can be measured by the strain gages. The main support, force-measuring double-cantilever spring, movable clamp, white light, etc., are all parts of the original basic apparatus (Fig. 1.7), whose functions are to control the surface separation, vary the externally applied normal load, and measure the separation and normal force between the two surfaces, as already described. Note that during sliding, the distance between the surfaces, their true molecular contact area, their elastic deformation, and their lateral motion can all be simultaneously monitored by recording the moving FECO fringe pattern using a video camera and recording it on a tape [46].

The two surfaces can be sheared past each other at sliding speeds which can be varied continuously from 0.1 to 20 µm/s while simultaneously measuring both the transverse (frictional) force and the normal (compressive or tensile) force between them. The lateral distances traversed are on the order of a several hundreds of micrometers which correspond to several diameters of the contact zone.

With an oscillatory shear attachment, developed by *Granick* et al., viscous dissipation and elasticity and dynamic viscosity of confined liquids by applying periodic sinusoidal oscillations of one surface with respect to the other can be studied [40, 41, 44, 66, 67]. This attachment allows for the two surfaces to be sheared past each other at varying sliding speeds or oscillating frequencies while simultaneously measuring both the transverse (friction or shear) force and the normal load between them. The externally applied load can be varied continuously, and both positive and negative loads can be applied. Finally the distance between the surfaces, their true molecular contact area, their elastic (or viscoelastic) deformation and their lateral motion can all be simultaneously by recording the moving interference fringe pattern using a video camera-recorder system.

To produce shear while maintaining constant film thickness or constant separation of the surfaces, the top mica surface is suspended from the upper portion of the apparatus by two piezoelectric bimorphs. A schematic description of the surface force apparatus with the installed shearing device is shown in Fig. 1.11 [40, 41, 44, 66, 67]. *Israelachvili* [42] and *Luengo* et al. [43] have also presented similar designs. The lower mica surface, as in the steady-shear sliding attachment, is stationary and sits at the tip of a double cantilever spring attached at the other end to a stiff support. The externally applied load can be varied continuously by displacing the lower surface vertically. An AC voltage difference applied by a signal generator (driver) across one of the bimorphs tends to bend it in oscillatory fashion while the frictional force resists that motion. Any resistance to sliding induces an output voltage across the other bimorphs (receiver) which can be easily measured by a digital oscilloscope. The sensitivity in measuring force is on the order of a few μN and the amplitudes of measured lateral displacement can range from a few nm to 10 μm. The design is flexible and allows to induce time-varying stresses with different characteristic wave shapes simply by changing the wave form of the input electrical signal. For example, when measuring the apparent viscosity, a sine wave input is convenient to apply.

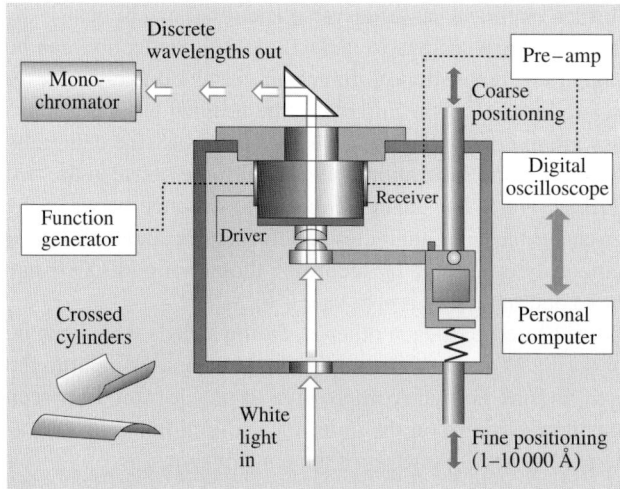

Fig. 1.11. Schematic of the oscillatory shearing apparatus [40]

Figure 1.12a shows an example of the raw data, obtained with a hexadecane film at a moderate pressure, when a sine wave was applied to one of the bimorphs [66]. By comparing the calibration curve with the response curve, which was attenuated in amplitude and lagged in phase, an apparent dynamic viscosity can be inferred. On the other hand, a triangular waveform is more suitable when studying the yield stress behavior of solid-like films as of Fig. 1.12b. The triangular waveform, showing a linear increase and decrease of the applied force with time, is proportional to the driving force acting on the upper surface. The response waveform, which represents a resistance of the interface to shear, remains very small indicating that the surfaces are in a stationary contact with respect to each other until the applied stress reaches a yield point. At the yield point the slope of the response curve increases dramatically, indicating the onset of sliding.

Homola [53] compared the two approaches – steady shear attachment and oscillating shear attachment. In experiments conducted by Israelachvili and his co-workers, the steady-shear attachment was employed to focus on the dynamic fric-

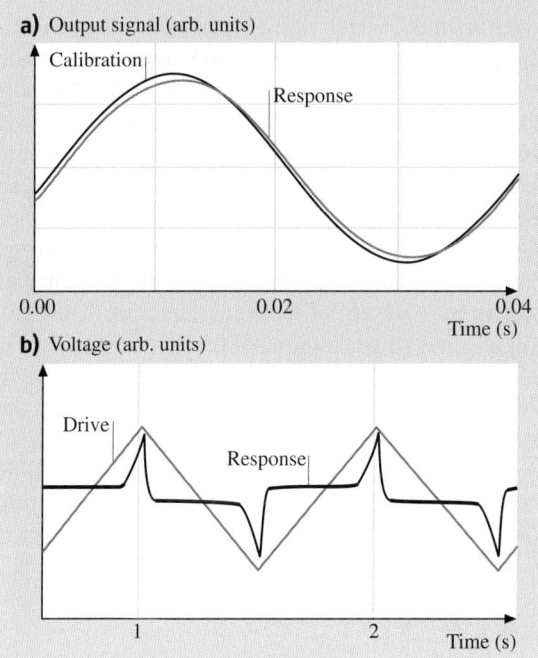

Fig. 1.12. (**a**) Two output signals induced by an applied sine wave (not shown) are displaced. The "calibration" waveform is obtained with the mica sheets completely separated. The response waveform is obtained with a thin liquid film between the sheets, which causes it to lag the calibration waveform, (**b**) the oscilloscope trace of the drive and response voltages used to determine critical shear stress. The drive waveform shows voltage proportional to induced stress on the sheared film and the response waveform shows voltage proportional to resulting velocity. Spikes in the response curve correspond to the stick–slip event [66]

tional behavior of the film after a sufficiently high shear stress was applied to exceed the yield stress and to produce sliding at a constant velocity. In these measurements, the film was subjected to a constant shearing force for a time sufficiently long to allow them to reach a dynamic equilibrium, i.e., the molecules, within the film, had enough time to order and align with respect to the surface, both normally and tangentially. Under these conditions, dynamic friction was observed to be "quantized" according to the number of liquid layers between the solid surfaces and independent of the shear rate [38]. Clearly, in this approach, the molecular ordering is optimized by a steady shear which imposes a preferred orientation on the molecules in the direction of shear.

The above mode of sliding is particularly important when the sheared film is made of a long chain lubricant molecules requiring a significantly long sliding time to order and align and even a longer time to relax (disorder) when sliding stops. This suggests that a steady-state friction is realized only when the duration of sliding exceeds the time required for an ensemble of the molecules to fully order in a specific shear field. It also suggests, that static friction should depend critically on the sliding time and the extend of the shear induced ordering [53].

In contrast, the oscillatory shear method, which utilizes periodic sinusoidal oscillations over a range of amplitudes and frequencies, addresses a response of the system to rapidly varying strain rates and directions of sliding. Under these conditions, the molecules, especially those exhibiting a solid-like behavior, cannot respond sufficiently fast to stress and are unable to order fully during duration of a single pass, i.e., their dynamic and static behavior reflects and oscillatory shear induced ordering which might or might not represent an equilibrium dynamic state. Thus, the response of the sheared film will depend critically on the conditions of shearing, i.e., the strain, the pressure, and the sliding conditions (amplitude and frequency of oscillations) which in turn will determine a degree of molecular ordering. This may explain the fact that the layer structure and "quantization" of the dynamic and static friction was not observed in these experiments in contrast to results obtained when velocity was kept constant. Intuitively, this behavior is expected considering that the shear-ordering tendency of the system is frequently disturbed by a shearing force of varying magnitude and direction. Nonetheless, the technique is capable of providing an invaluable insight into the shear behavior of molecularly thin films subjected to non-linear stresses as it is frequently encountered in practical applications. This is especially true under conditions of boundary lubrication where interacting surface asperities will be subjected to periodic stresses of varying magnitudes and frequencies [53].

Georges et al.'s Design

The SFA developed by *Tonck* et al. [48] and *Georges* et al. [49] to measure static and dynamic forces in the normal direction, between surfaces in close proximity, is shown in Fig. 1.13. In their apparatus, a drop of liquid is introduced between a macroscopic spherical body and a plane. The sphere is moved towards and away from a plane using the expansion and the vibration of a piezoelectric crystal. Piezoelectric crystal is vibrated at low amplitude around an average separation for dynamic measurements to provide dynamic function of the interface. The plane specimen is supported by a double-cantilever spring. Capacitance sensor C_1 measures the elastic deformation of the cantilever and thus the force transmitted through the liquid to the plane. Second capacitance sensor C_2 is designed to measure the relative displacement between the supports of the two solids. The reference displacement signal is the sum of two signals: first, a ramp provides a constant normal speed from 50 to 0.01 nm/s, and, second, the piezoelectric crystal is designed to provide a small sinusoidal motion, in order to determine the dynamic behavior of sphere-plane interactions. A third capacitance sensor C measures the electrical capacitance between the sphere and the plane. In all cases, the capacitance is determined by incorporating the signal of an oscillator in the inductive–capacitance (L–C) resonant input stage of an oscillator to give a signal-dependent frequency in the range of 5–12 MHz. The resulting fluctuations in oscillation frequency are detected using a low noise frequency discriminator. Simultaneous measurements of sphere-plane displacement, surface force, and the damping of the interface allows an analysis of all regimes of the interface [49]. *Loubet* et al. [68] used SFA in the crossed-cylinder geometry using two freshly-cleaved mica sheets similar to the manner used by Israelachvili and coworkers.

Georges et al. [50] modified their original SFA to measure friction forces. In this apparatus, in addition to having sphere move normal to the plane, sphere can be sheared at constant separation from the plane. The shear force apparatus is shown in Fig. 1.14. Three piezoelectric elements controlled by three capacitance sensors permit accurate motion control and force measurement along three orthogonal axes with displacement sensitivity of 10^{-3} nm and force sensitivity of 10^{-8} N. Adhesion and normal deformation experiments are conducted in the normal approach (z-axis). Friction experiments are conducted by introducing displacement in the X-direction at a constant normal force. In one of the experiment, *Georges* et al. [50] used 2.95 mm diameter sphere made of cobalt-coated fused borosilicate glass and a silicon wafer for the plane.

1.4.3 Vibration Isolation

STM, AFM and SFA should be isolated from sources of vibration in the acoustic and sub-acoustic frequencies especially for atomic-scale measurements. Vibration isolation is generally provided by placing the instrument on a vibration isolation air table. For further isolation, the instrument should be placed on a pad of soft silicone rubber. A cheaper alternative consists of a large mass of 100 N or more, suspended from elastic "bungee" cords. The mass should stretch the cords at least 0.3 m, but not so much that the cords reach their elastic limit. The instrument should be placed on the large mass. The system, including the microscope, should have a natural frequency of about 1 Hz or less both vertically and horizontally. Test this by gently pushing on the mass and measure the rate at which its swings or bounces.

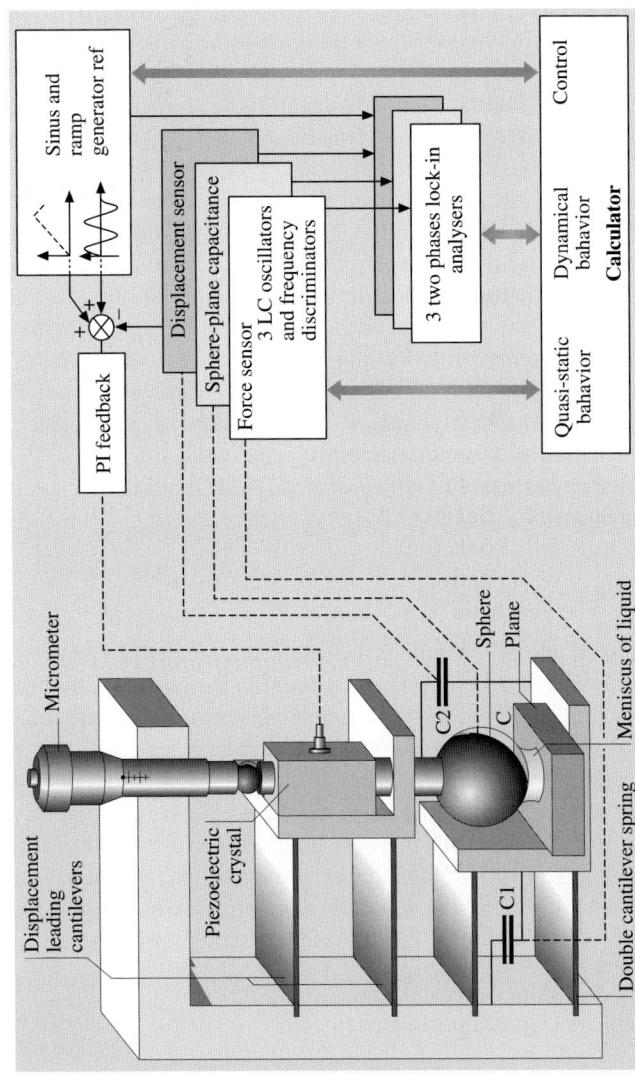

Fig. 1.13. Schematic of the surface force apparatus that employs a sphere–plane arrangement [49]

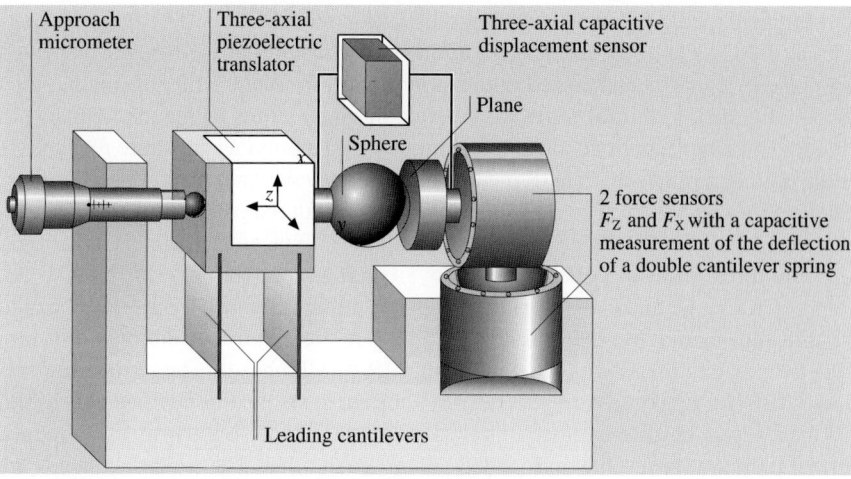

Fig. 1.14. Schematic of shear force apparatus [50]

1.5 Magnetic Storage Devices and MEMS/NEMS

1.5.1 Magnetic Storage Devices

Magnetic storage devices used for storage and retrieval are tape, flexible (floppy) disk and rigid disk drives. These devices are used for audio, video and data storage applications. Magnetic storage industry is some $ 60 billion a year industry with $ 20 billion for audio and video recording (almost all tape drives/media) and $ 40 billion for data storage. In the data storage industry, magnetic rigid disk drives/media, tape drives/media, flexible disk drives/media, and optical disk drive/media account for about $ 25 B, $ 6 B, $ 3 B, and $ 6 B, respectively. Magnetic recording and playback involves the relative motion between a magnetic medium (tape or disk) against a read-write magnetic head. Heads are designed so that they develop a (load-carrying) hydrodynamic air film under steady operating conditions to minimize head–medium contact. However, physical contact between the medium and head occurs during starts and stops, referred to as contact-start-stops (CSS) technology [13, 14, 69]. In the modern magnetic storage devices, the flying heights (head-to-medium separation) are on the order of 5 to 20 nm and roughnesses of head and medium surfaces are on the order of 1–2 nm RMS. The need for ever-increasing recording densities requires that surfaces be as smooth as possible and the flying heights be as low as possible. High stiction (static friction) and wear are the limiting technology to future of this industry. Head load/unload (L/UL) technology has recently been used as an alternative to CSS technology in rigid disk drives that eliminates stiction and wear failure mode associated with CSS. Several contact or near contact recording devices are at various stages of development. High stiction and wear are the major impediments to the commercialization of the contact recording.

Magnetic media fall into two categories: particulate media, where magnetic particles (γ-Fe_2O_3, Co-γFe_2O_3, CrO_2, Fe or metal (MP), or barium ferrite) are dispersed in a polymeric matrix and coated onto a polymeric substrate for flexible media (tape and flexible disks); thin-film media, where continuous films of magnetic materials are deposited by vacuum deposition techniques onto a polymer substrate for flexible media or onto a rigid substrate (typically aluminium and more recently glass or glass ceramic) for rigid disks. The most commonly used thin magnetic films for tapes are evaporated Co−Ni (82−18 at.%) or Co−O dual layer. Typical magnetic films for rigid disks are metal films of cobalt-based alloys (such as sputtered Co−Pt−Ni, Co−Ni, Co−Pt−Cr, Co−Cr and Co−NiCr). For high recording densities, trends have been to use thin-film media. Magnetic heads used to date are either conventional thin-film inductive, magnetoresistive (MR) and giant MR (GMR) heads. The air-bearing surfaces (ABS) of tape heads are generally cylindrical in shape. For dual-sided flexible-disk heads, two heads are either spherically contoured and slightly offset (to reduce normal pressure) or are flat and loaded against each other. The rigid-disk heads are supported by a leaf spring (flexure) suspension. The ABS of heads are almost made of Mn−Zn ferrite, Ni−Zn ferrite, Al_2O_3−TiC and calcium titanate. The ABS of some conventional heads are made of plasma sprayed coatings of hard materials such as Al_2O_3−TiO_2 and ZrO_2 [13, 14, 69].

Figure 1.15 shows the schematic illustrating the tape path with details of tape guides in a data-processing linear tape drive (IBM LTO Gen1) which uses a rectangular tape cartridge. Figure 1.16a shows the sectional views of particulate and thin-film magnetic tapes. Almost exclusively, the base film is made of semicrystalline biaxially-oriented poly (ethylene terephthalate) (or PET) or poly (ethylene 2,6 naphthalate) (or PEN) or Aramid. The particulate coating formulation consists of binder (typically polyester polyurethane), submicron accicular shaped magnetic particles (about 50 nm long with an aspect ratio of about 5), submicron head cleaning agents (typically alumina) and lubricant (typically fatty acid ester). For protection against wear and corrosion and low friction/stiction, the thin-film tape is first coated with a diamondlike carbon (DLC) overcoat deposited by plasma enhanced chemical vapor deposition, topically lubricated with primarily a perfluoropolyether lubricant. Figure 1.16b shows the schematic of an 8-track (along with 2 servo tracks) thin-film read-write head with MR read and inductive write. The head steps up and down to provide 384 total data tracks across the width of the tape. The ABS is made of Al_2O_3−TiC. A tape tension of about 1 N over a 12.7 mm wide tape (normal pressure $\approx$ 14 kPa) is used during use. The RMS roughnesses of ABS of the heads and tape surfaces typically are 1−1.5 nm and 5−8 nm, respectively.

Figure 1.17 shows the schematic of a data processing rigid disk drive with 21.6, 27.4, 48, 63.5, 75, and 95 mm form factor. Nonremovable stack of multiple disks mounted on a ball bearing or hydrodynamic spindle, are rotated by an electric motor at constant angular speed ranging from about 5000 to in excess of 15 000 RPM, dependent upon the disk size. Head slider-suspension assembly (allowing one slider for each disk surface) is actuated by a stepper motor or a voice coil motor using a rotary actuator. Figure 1.18a shows the sectional views of a thin-film rigid disk. The substrate for rigid disks is generally a non heat-treatable aluminium–magnesium alloy

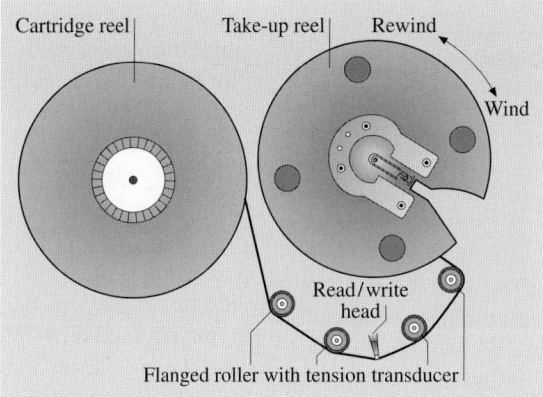

Fig. 1.15. Schematic of tape path in an IBM Linear Tape Open (LTO) tape drive

5086, glass or glass ceramic. The protective overcoat commonly used for thin-film disks is sputtered DLC, topically lubricated with perfluoropolyether type of lubricants. Lubricants with polar-end groups are generally used for thin-film disks in order to provide partial chemical bonding to the overcoat surface. The disks used for CSS technology are laser textured in the landing zone. Figure 1.18b shows the schematic of two thin-film head picosliders with a step at the leading edge, and GMR read and inductive write. "Pico" refers to the small sizes of 1.25 mm × 1 mm. These sliders use Al_2O_3–TiC (70–30 wt%) as the substrate material with multilayered thin-film head structure coated and with about 3.5 nm thick DLC coating to prevent the thin film structure from electrostatic discharge. The seven pads on the padded slider are made of DLC and are about 40 µm in diameter and 50 nm in height. A normal load of about 3 g is applied during use.

1.5.2 MEMS/NEMS

The advances in silicon photolithographic process technology led to the development of MEMS in the mid-1980s [16]. More recently, lithographic and nonlithographic processes have been developed to process nonsilicon (plastics or ceramics) materials. MEMS for mechanical applications include acceleration, pressure, flow, and gas sensors, linear and rotary actuators, and other microstructures of microcomponents such as electric motors, gear trains, gas turbine engines, nozzles, fluid pumps, fluid valves, switches, grippers, and tweezers. MEMS for chemical applications include chemical sensors and various analytical instruments. Microoptoelectromechanical systems (or MOEMS) include micromirror arrays and fiber optic connectors. Radio frequency MEMS or RF-MEMS include inductors, capacitors, and antennas. High-aspect ratio MEMS (HARMEMS) have also been introduced. BioMEMS include biofluidic chips (microfluidic chips or bioflips or simple biochips) for chemical and biochemical analyses (biosensors in medical diagnostics, e.g., DNA, RNA, proteins, cells, blood pressure and assays, and toxin identification), and implantable drug de-

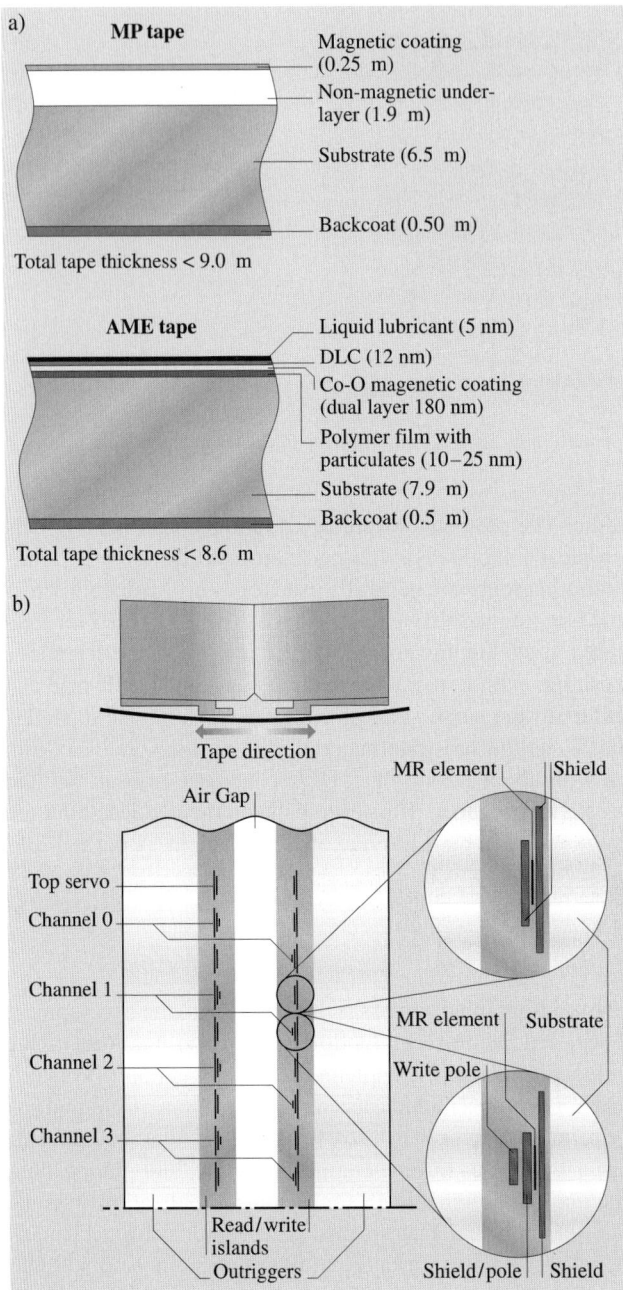

Fig. 1.16. (a) Sectional views of particulate and thin-film magnetic tapes, and (b) schematic of a magnetic thin-film read/write head for an IBM LTO Gen 1 tape drive

Fig. 1.17. Schematic of a data-processing magnetic rigid disk drive

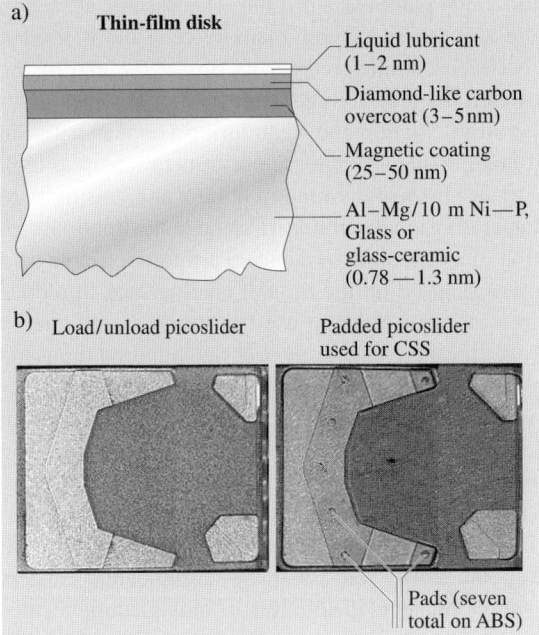

Fig. 1.18. (**a**) Sectional views of a thin-film magnetic rigid disk, and (**b**) schematic of two picosliders – load/unload picoslider and padded picoslider used for CSS

livery. Killer applications include capacitive-type silicon accelerometers for automotive sensory applications and digital micromirror devices for projection displays. Any component requiring relative motions needs to be optimized for stiction and wear [16, 20, 22, 69, 70].

Figure 1.19 also shows two digital micromirror device (DMD) pixels used in digital light processing (DLP) technology for digital projection displays in portable and home theater projectors as well as table top and projection TVs [16, 71, 72]. The entire array (chip set) consists of a large number of rotatable aluminium micromirrors (digital light switches) which are fabricated on top of a CMOS static random access memory integrated circuit. The surface micromachined array consists of half of a million to more than two million of these independently controlled reflective, micromirrors (mirror size on the order of $14\,\mu m \times 14\,\mu m$ and $15\,\mu m$ pitch) which flip backward and forward at a frequency of on the order of 5000 times a second. For the binary operation, micromirror/yoke structure mounted on torsional hinges is rotated $\pm 10°$ (with respect to the horizontal plane) as a result of electrostatic attraction between the micromirror structure and the underlying memory cell, and is limited by a mechanical stop. Contact between cantilevered spring tips at the end of the yoke (four present on each yoke) with the underlying stationary landing sites is required for true digital (binary) operation. Stiction and wear during a contact between aluminium alloy spring tips and landing sites, hinge memory (metal creep at high operating temperatures), hinge fatigue, shock and vibration failure, and sensitivity to particles in the chip package and operating environment are some of the important issues affecting the reliable operation of a micromirror device. Perfluorodecanoic acid (PFDA) self-assembled monolayers are used on the tip and landing sites to reduce stiction and wear. The spring tip is used in order to use the spring stored energy to pop up the tip during pull-off. A lifetime estimate of over one hundred thousand operating hours with no degradation in image quality is the norm.

NEMS are produced by nanomachining in a typical top-down approach (from large to small) and bottom-up approach (from small to large) largely relying on nanochemistry [16]. The top-down approach relies on fabrication methods including advanced integrated-circuit (IC) lithographic methods – electron-beam lithography, and STM writing by removing material atom by atom. The bottom-up approach includes chemical synthesis, the spontaneous "self-assembly" of molecular clusters (molecular self-assembly) from simple reagents in solution, or biological molecules (e.g., DNA) as building blocks to produce three dimensional nanostructures, quantum dots (nanocrystals) of arbitrary diameter (about $10-10^5$ atoms), molecular beam epitaxy (MBE) and organometallic vapor phase epitaxy (OMVPE) to create specialized crystals one atomic or molecular layer at a time, and manipulation of individual atoms by an atomic force microscope or atom optics. The self-assembly must be encoded, that is, one must be able to precisely assemble one object next to another to form a designed pattern. A variety of nonequilibrium plasma chemistry techniques are also used to produce layered nanocomposites, nanotubes, and nanoparticles. NEMS field, in addition to fabrication of nanosystems, has provided impetus to development of experimental and computation tools.

1 Introduction – Measurement Techniques and Applications

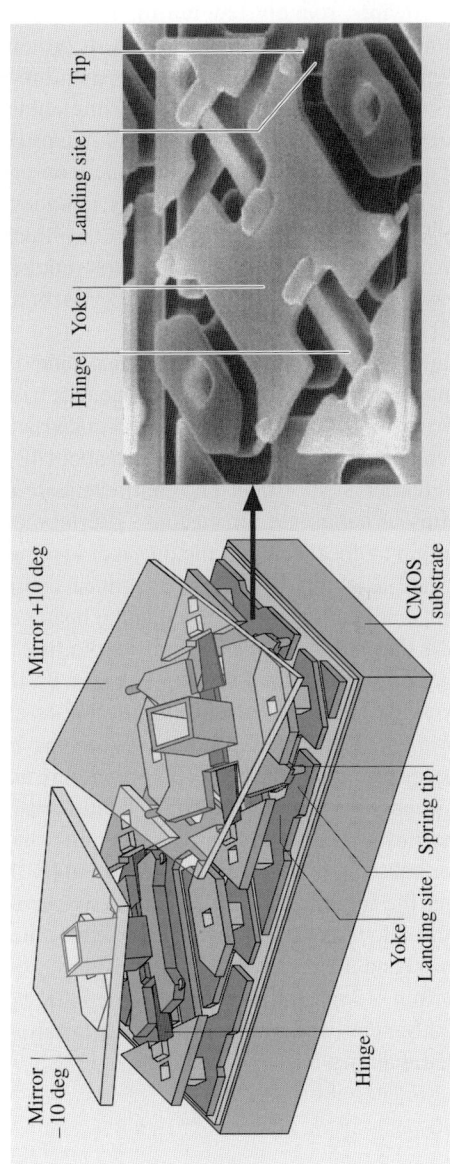

Fig. 1.19. Digital micromirror devices for projection displays (from [16])

Examples of NEMS include nanocomponents, nanodevices, nanosystems, and nanomaterials such as microcantilever with integrated sharp nanotips for STM and AFM, AFM array (Millipede) for data storage, AFM tips for nanolithography, dip-pen nanolithography for printing molecules, biological (DNA) motors, molecular gears, molecularly-thick films (e.g., in giant magnetioresistive or GMR heads and magnetic media), nanoparticles (e.g., nanomagnetic particles in magnetic media), nanowires, carbon nanotubes, quantum wires (QWRs), quantum boxes (QBs), and quantum transistors [16]. BIONEMS include nanobiosensors – microarray of silicon nanowires, roughly few nm in size, to selectively bind and detect even a single biological molecule such as DNA or protein by using nanoelectronics to detect the slight electrical charge caused by such binding, or a microarray of carbon nanotubes to electrically detect glucose, implantable drug-delivery devices – e.g., micro/nanoparticles with drug molecules encapsulated in functionized shells for a site-specific targeting applications, and a silicon capsule with a nanoporous membrane filled with drugs for long term delivery, nanodevices for sequencing single molecules of DNA in the Human Genome Project, cellular growth using carbon nanotubes for spinal cord repair, nanotubes for nanostructured materials for various applications such as spinal fusion devices, organ growth, and growth of artificial tissues using nanofibers.

Figure 1.20 shows AFM based nanoscale data storage system for ultrahigh density magnetic recording which experiences tribological problems [73]. The system uses arrays of several thousand silicon microcantilevers ("Millipede") for thermomechanical recording and playback on an about 40-nm thick polymer (PMMA) medium with a harder Si substrate. The cantilevers are integrated with integrated tip heaters with tips of nanoscale dimensions. Thermomechanical recording is a combination of applying a local force to the polymer layer and softening it by local heating. The tip heated to about 400 °C is brought in contact with the polymer for recording. Imaging and reading are done using the heater cantilever, originally used for recording, as a thermal readback sensor by exploiting its temperature-dependent resistance. The principle of thermal sensing is based on the fact that the thermal conductivity between the heater and the storage substrate changes according to the spacing between them. When the spacing between the heater and sample is reduced as the tip moves into a bit, the heater's temperature and hence its resistance will decrease. Thus, changes in temperature of the continuously heated resistor are monitored while the cantilever is scanned over data bits, providing a means of detecting the bits. Erasing for subsequent rewriting is carried out by thermal reflow of the storage field by heating the medium to 150 °C for a few seconds. The smoothness of the reflown medium allows multiple rewriting of the same storage field. Bit sizes ranging between 10 and 50 nm have been achieved by using a 32×32 (1024) array write/read chip (3 mm × 3 mm). It has been reported that tip wear occurs by the contact between tip and Si substrate during writing. Tip wear is considered a major concern for the device reliability.

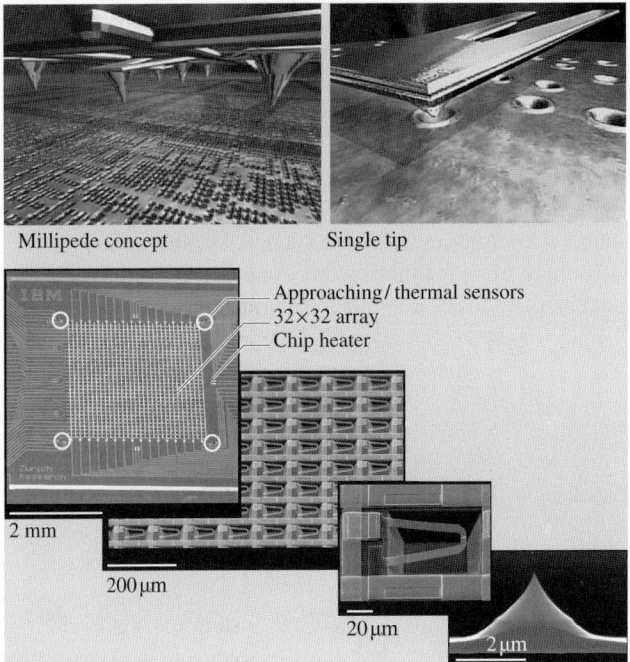

Fig. 1.20. AFM based nanoscale data storage system with 32 × 32 tip array – that experiences a tribological problem (from [16])

1.6 Role of Micro/Nanotribology and Micro/Nanomechanics in Magnetic Storage Devices and MEMS/NEMS

The magnetic storage devices and MEMS/NEMS are the two examples where micro/nanotribological and micro/nanomechanical tools and techniques are essential for studies of micro/nano scale phenomena. Magnetic storage components continue to shrink in physical dimensions. Thicknesses of hard solid coating and liquid lubricant coatings on the magnetic disk surface continue to decrease. Number of contact recording devices are at various stages of development. Surface roughnesses of the storage components continue to decrease and are expected to approach to about 0.5 nm RMS or lower. Interface studies of components with ultra-thin coatings can be ideally performed using micro/nanotribological and micro/nanomechanical tools and techniques.

In the case of MEMS/NEMS, the friction and wear problems of ultrasmall moving components generally made of single-crystal silicon, polysilicon films or polymers need to be addressed for high performance, long life, and reliability. Molecularly-thin films of solid and/or liquids are used for low friction and wear in many applications. Again, interfacial phenomena in MEMS/NEMS can be ideally studied using micro/nanotribological and micro/nanomechanical tools and techniques.

1.7 Organization of the Book

The introductory book integrates knowledge of nanotribology and nanomechanics. The book starts with the definition of tribology, history of tribology and micro/nanotribology, its industrial significance, various measurement techniques employed, followed by various industrial applications. The remaining book is divided into four parts. The first part introduces scanning probe microscopy. The second part provides an overview of nanotechnology and nanomechanics. The third part provides an overview of molecularly-thick films for lubrication. And the last part focuses on nanotribology and nanomechanics studies conducted for various industrial applications.

References

1. P. Jost. *Lubrication (Tribology) – A Report on the Present Position and Industry's Needs*. Dept. of Education and Science (H.M. Stationary Office), 1966.
2. D. Dowson. *History of Tribology*. Inst. Mech. Engineers, 2nd edition, 1998.
3. C. S. C. Davidson. Bearings since the stone age. *Eng.*, 183:2–5, 1957.
4. A. G. Layard. *Discoveries in the Ruins of Nineveh and Babylon*, volume I, II. Murray, 1853.
5. G. Amontons. De la resistance causee dans les machines. *Mem. Acad. R. A*, volume]Is there a volume?:257–282, 1699.
6. C. A. Coulomb. Theorie des machines simples, en ayant regard an frottement de leurs parties et a la roideur des cordages. *Mem. Math. Phys. X, Paris*, :161–342, 1785.
7. W. F. Parish. Three thousand years of progress in the development of machinery and lubricants for the hand crafts. *Mill Factory*, 16, 17:pages]Please give the pages., 1935.
8. B. Tower. Report on friction experiments. *Proc. Inst. Mech. Eng.*, :632, 1884.
9. O. O. Reynolds. On the theory of lubrication and its applications to mr. beauchamp tower's experiments. *Philos. Trans. R. Soc. (London)*, 117:157–234, 1886.
10. N. P. Petroff. Friction in machines and the effects of the lubricant. *Eng. J.*, , 1883.
11. R. Holm. *Electrical Contacts*. Springer, 1946.
12. F. P. Bowden and D. Tabor. *The Friction and Lubrication of Solids*, volume 1, 2. Clarendon, 1950, 1964.
13. B. Bhushan. *Tribology and Mechanics of Magnetic Storage Devices*. Springer, 2nd edition, 1996.
14. B. Bhushan. *Mechanics and Reliability of Flexible Magnetic Media*. Springer, 2nd edition, 2000.
15. B. Bhushan. *Introduction to Tribology*. Wiley, 2002.
16. B. Bhushan (Ed.). *Springer Handbook of Nanotechnology*. Springer, 2004.
17. B. Bhushan and B. K. Gupta. *Handbook of Tribology: Materials, Coatings, and Surface Treatments*. McGraw-Hill, 1991.
18. P. Jost. *Economic Impact of Tribology*, volume 423 of *NBS Spec. Pub*. Proc. Mechanical Failures Prevention Group, 1976.
19. B. Bhushan, J. N. Israelachvili, and U. Landman. Nanotribology: Friction, wear and lubrication at the atomic scale. *Nature*, 374:607–616, 1995.
20. B. Bhushan. *Micro/Nanotribology and its Applications*. NATO ASI Series E: Applied Sciences **330**. Kluwer Academic, 1997.

21. B. Bhushan. *Handbook of Micro/Nanotribology*. CRC, 2nd edition, 1999.
22. B. Bhushan. *Fundamentals of Tribology and Bridging the Gap Between the Macro- and Micro/Nanoscales*, volume 10 of *NATO Science Series II: Mathematics, Physics, and Chemistry*. Kluwer Academic, 2001.
23. B. Bhushan, H. Fuchs, and S. Hosaka. *Applied Scanning Probe Methods*. Springer, 2004.
24. G. Binnig, H. Rohrer, Ch. Gerber, and E. Weibel. Surface studies by scanning tunnelling microscopy. *Phys. Rev. Lett.*, 49:57–61, 1982.
25. G. Binnig, C. F. Quate, and Ch. Gerber. Atomic force microscope. *Phys. Rev. Lett.*, 56:930–933, 1986.
26. G. Binnig, Ch. Gerber, E. Stoll, T. R. Albrecht, and C. F. Quate. Atomic resolution with atomic force microscope. *Europhys. Lett.*, 3:1281–1286, 1987.
27. D. Tabor and R. H. S. Winterton. The direct measurement of normal and retarded van der waals forces. *Proc. R. Soc. Lond. A*, 312:435–450, 1969.
28. S. Alexander, L. Hellemans, O. Marti, J. Schneir, V. Elings, and P. K. Hansma. An atomic-resolution atomic-force microscope implemented using an optical lever. *J. Appl. Phys.*, 65:164–167, 1989.
29. B. Bhushan and J. Ruan. Atomic-scale friction measurements using friction force microscopy: Part ii – application to magnetic media. *ASME J. Tribol.*, 116:389–396, 1994.
30. J. Ruan and B. Bhushan. Atomic-scale friction measurements using friction force microscopy: Part i – general principles and new measurement techniques. *ASME J. Tribol.*, 116:378–388, 1994.
31. J. Ruan and B. Bhushan. Atomic-scale and microscale friction of graphite and diamond using friction force microscopy. *J. Appl. Phys.*, 76:5022–5035, 1994.
32. G. Meyer and N. M. Amer. Novel optical approach to atomic force microscopy. *Appl. Phys. Lett.*, 53:1045–1047, 1988.
33. B. Bhushan, V. N. Koinkar, and J. Ruan. Microtribology of magnetic media. *Proc. Inst. Mech. Engineers. Part J: J. Eng. Tribol.*, 208:17–29, 1994.
34. J. N. Israelachvili and D. Tabor. The measurement of van der waals dispersion forces in the range of 1.5 to 130 nm. *Proc. R. Soc. Lond. A*, 331:19–38, 1972.
35. J. N. Israelachvili and G. E. Adams. Measurement of friction between two mica surfaces in aqueous electrolyte solutions in the range 0–100 nm. *Chem. Soc. J., Faraday Trans. I*, 74:975–1001, 1978.
36. J. Klein. Forces between mica surfaces bearing layers of adsorbed polystyrene in cyclohexane. *Nature*, 288:248–250, 1980.
37. D. Y. C. Chan and R. G. Horn. The drainage of thin liquid films between solid surfaces. *J. Chem. Phys.*, 83:5311–5324, 1985.
38. J. N. Israelachvili, P. M. McGuiggan, and A. M. Homola. Dynamic properties of molecularly thin liquid films. *Science*, 240:189–190, 1988.
39. J. Klein, D. Perahia, and S. Warburg. Forces between polymer-bearing surfaces undergoing shear. *Nature*, 352:143–145, 1991.
40. J. Van Alsten and S. Granick. Molecular tribology of ultrathin liquid films. *Phys. Rev. Lett.*, 61:2570–2573, 1988.
41. J. Peachey, J. Van Alsten, and S. Granick. Design of an apparatus to measure the shear response of ultrathin liquid films. *Rev. Sci. Instrum.*, 62:463–473, 1991.
42. J. N. Israelachvili. Techniques for direct measurements of forces between surfaces in liquids at the atomic scale. *Chemtracts – Anal. Phys. Chem.*, 1:1–12, 1989.
43. G. Luengo, F. J. Schmitt, R. Hill, and J. N. Israelachvili. Thin film bulk rheology and tribology of confined polymer melts: Contrasts with build properties. *Macromol.*, 30:2482–2494, 1997.

44. J. Van Alsten and S. Granick. Shear rheology in a confined geometry – polysiloxane melts. *Macromol.*, 23:4856–4862, 1990.
45. A. M. Homola, J. N. Israelachvili, M. L. Gee, and P. M. McGuiggan. Measurement of and relation between the adhesion and friction of two surfaces separated by thin liquid and polymer films. *ASME J. Tribol.*, 111:675–682, 1989.
46. M. L. Gee, P. M. McGuiggan, J. N. Israelachvili, and A. M. Homola. Liquid to solid-like transitions of molecularly thin films under shear. *J. Chem. Phys.*, 93:1895–1906, 1990.
47. S. Granick. Motions and relaxations of confined liquids. *Science*, 253:1374–1379, 1991.
48. A. Tonck, J. M. Georges, and J. L. Loubet. Measurements of intermolecular forces and the rheology of dodecane between alumina surfaces. *J. Colloid Interface Sci.*, 126:1540–1563, 1988.
49. J. M. Georges, S. Millot, J. L. Loubet, and A. Tonck. Drainage of thin liquid films between relatively smooth surfaces. *J. Chem. Phys.*, 98:7345–7360, 1993.
50. J. M. Georges, A. Tonck, and D. Mazuyer. Interfacial friction of wetted monolayers. *Wear*, 175:59–62, 1994.
51. J. N. Israelachvili. *Intermolecular and Surface Forces*. Academic, 2nd edition, 1992.
52. R. G. Horn. Surface forces and their action in ceramic materials. *Am. Ceram. Soc.*, 73:1117–1135, 1990.
53. A. M. Homola. *Interfacial friction of molecularly thin liquid films*, pages 271–298. World Scientific, 1993.
54. A. M. Homola, J. N. Israelachvili, P. M. McGuiggan, and M. L. Gee. Fundamental experimental studies in tribology: The transition from interfacial friction of undamaged molecularly smooth surfaces. *Wear*, 136:65–83, 1990.
55. R. M. Pashley. Hydration forces between solid surfaces in aqueous electrolyte solutions. *J. Colloid Interface Sci.*, 80:153–162, 1981.
56. R. G. Horn, D. T. Smith, and W. Haller. Surface forces and viscosity of water measured between silica sheets. *Chem. Phys. Lett.*, 162:404–408, 1989.
57. R. G. Horn and J. N. Israelachvili. Molecular organization and viscosity of a thin film of molten polymer between two surfaces as probed by force measurements. *Macromol.*, 21:2836–2841, 1988.
58. H. K. Christenson. Adhesion between surfaces in unsaturated vapors – a reexamination of the influence of meniscus curvature and surface forces. *J. Colloid Interface Sci.*, 121:170–178, 1988.
59. C. P. Smith, M. Maeda, L. Atanasoska, and H. S. White. Ultrathin platinum films on mica and measurement of forces at the platinum/water interface. *J. Phys. Chem.*, 95:199–205, 1988.
60. S. J. Hirz, A. M. Homola, G. Hadzioannou, and S. W. Frank. Effect of substrate on shearing properties of ultrathin polymer films. *Langmuir*, 8:328–333, 1992.
61. S. S. Patel and M. Tirrell. Measurement of forces between surfaces in polymer fluids. *Annu. Rev. Phys. Chem.*, 40:597–635, 1989.
62. J. N. Israelachvili. Solvation forces and liquid structure – as probed by direct force measurements. *Accounts Chem. Res.*, 20:415–421, 1987.
63. J. N. Israelachvili and P. M. McGuiggan. Forces between surface in liquids. *Science*, 241:795–800, 1988.
64. A. M. Homola. Measurement of and relation between the adhesion and friction of two surfaces separated by thin liquid and polymer films. *ASME J. Tribol.*, 111:675–682, 1989.
65. A. M. Homola, H. V. Nguyen, and G. Hadzioannou. Influence of monomer architecture on the shear properties of molecularly thin polymer melts. *J. Chem. Phys.*, 94:2346–2351, 1991.

66. J. Van Alsten and S. Granick. Tribology studied using atomically smooth surfaces. *Tribol. Trans.*, 33:436–446, 1990.
67. W. W. Hu, G. A. Carson, and S. Granick. Relaxation time of confined liquids under shear. *Phys. Rev. Lett.*, 66:2758–2761, 1991.
68. J. L. Loubet, M. Bauer, A. Tonck, S. Bec, and B. Gauthier-Manuel. *Nanoindentation with a surface force apparatus*, pages 429–447. Kluwer Academic, 1993.
69. B. Bhushan. *Macro- and microtribology of magnetic storage devices*, volume 2, pages 1413–1513. CRC, 2001editor]Please give the editor(s).
70. B. Bhushan. *Tribology Issues and Opportunities in MEMS*. Kluwer Academic, 1998.
71. L. J. Hornbeck and W. E. Nelson. *Bistable deformable mirror devices*, volume 8 of *OSA Technical Digest Series*, pages 107–110. OSA, 1988.
72. L. J. Hornbeck. A digital light processing^{TM} update – status and future applications. *Proc. SPIE*, 3634:158–170, 1999.
73. P. Vettinger, J. Brugger, M. Despont, U. Dreschier, U. Duerig, and W. Haeberie. Ultrahigh density, high data-rate nems based afm data storage systems. *Microelectron. Eng.*, 46:11–27, 1999.

Part I

Scanning Probe Microscopy

2

Scanning Probe Microscopy – Principle of Operation, Instrumentation, and Probes

Bharat Bhushan and Othmar Marti

Summary. Since the introduction of the STM in 1981 and AFMin 1985, many variations of probe based microscopies, referred to as SPMs, have been developed. While the pure imaging capabilities of SPM techniques is dominated by the application of these methods at their early development stages, the physics of probe–sample interactions and the quantitative analyses of tribological, electronic, magnetic, biological, and chemical surfaces have now become of increasing interest. Nanoscale science and technology are strongly driven by SPMs which allow investigation and manipulation of surfaces down to the atomic scale. With growing understanding of the underlying interaction mechanisms, SPMs have found applications in many fields outside basic research fields. In addition, various derivatives of all these methods have been developed for special applications, some of them targeted far beyond microscopy.

This chapter presents an overview of STM and AFM and various probes (tips) used in these instruments, followed by details on AFM instrumentation and analyses.

2.1 Introduction

The Scanning Tunneling Microscope (STM) developed by *Dr. Gerd Binnig* and his colleagues in 1981 at the IBM Zurich Research Laboratory, Rueschlikon, Switzerland, is the first instrument capable of directly obtaining three-dimensional (3-D) images of solid surfaces with atomic resolution [1]. *Binnig* and *Rohrer* received a Nobel Prize in Physics in 1986 for their discovery. STMs can only be used to study surfaces which are electrically conductive to some degree. Based on their design of the STM, in 1985, *Binnig* et al. developed an Atomic Force Microscope (AFM) to measure ultrasmall forces (less than 1 µN) present between the AFM tip surface and the sample surface [2] (also see [3]). AFMs can be used for measurement of all engineering surfaces which may be either electrically conductive or insulating. The AFM has become a popular surface profiler for topographic and normal force measurements on the micro to nanoscale [4]. AFMs modified in order to measure both normal and lateral forces, are called Lateral Force Microscopes (LFMs) or Friction Force Microscopes (FFMs) [5–11]. FFMs have been further modified to measure lateral forces in two orthogonal directions [12–16]. A number of researchers have continued to improve the AFM/FFM designs and used them to measure adhesion and friction of solid and liquid surfaces on micro- and nanoscales [4, 17–30]. AFMs have been used for scratching, wear, and measurement of elastic/plastic mechanical properties (such as indentation hardness and the modulus of elasticity) [4, 10, 11, 21, 23, 26–29, 31–36]. AFMs have been used for manipulation of individual atoms of Xenon [37], molecules [38], silicon surfaces [39], and polymer surfaces [40]. STMs have been used for formation of nanofeatures by localized heating or by inducing chemical reactions under the STM tip [41–43] and nanomachining [44]. AFMs have been used for nanofabrication [4, 10, 45–47] and nanomachining [48].

STMs and AFMs are used at extreme magnifications ranging from 10^3 to 10^9 in x, y and z directions for imaging macro to atomic dimensions with high resolution information and for spectroscopy. These instruments can be used in any environment such as ambient air [2, 49], various gases [17], liquid [50–52], vacuum [1, 53], at low temperatures (lower than about 100 K) [54–58] and high temperatures [59, 60]. Imaging in liquid allows the study of live biological samples and it also eliminates water capillary forces present in ambient air present at the tip–sample interface. Low temperature (liquid helium temperatures) imaging is useful for the study of biological and organic materials and the study of low-temperature phenomena such as superconductivity or charge-density waves. Low-temperature operation is also advantageous for high-sensitivity force mapping due to the reduction in thermal vibration. They also have been used to image liquids such as liquid crystals and lubricant molecules on graphite surfaces [61–64]. While the pure imaging capabilities of SPM techniques dominated the application of these methods at their early development stages, the physics and chemistry of probe–sample interactions and the quantitative analyses of tribological, electronic, magnetic, biological, and chemical surfaces have now become of increasing interest. Nanoscale science and technology are strongly driven by SPMs which allow investigation and manipulation of surfaces down to the atomic scale. With growing understanding of the underlying interaction mechanisms, SPMs have found applications in many fields outside basic research fields. In addition, various derivatives of all these methodshave been developed for special applications, some of them targeting far beyond microscopy.

Families of instruments based on STMs and AFMs, called Scanning Probe Microscopes (SPMs), have been developed for various applications of scientific and industrial interest. These include – STM, AFM, FFM (or LFM), scanning electrostatic force microscopy (SEFM) [65, 66], scanning force acoustic microscopy (SFAM) (or atomic force acoustic microscopy (AFAM)) [21, 22, 36, 67–69], scanning magnetic microscopy (SMM) (or magnetic force microscopy (MFM)) [70–73], scanning near field optical microscopy (SNOM) [74–77], scanning thermal microscopy (SThM) [78–80], scanning electrochemical microscopy (SEcM) [81], scanning Kelvin Probe microscopy (SKPM) [82–86], scanning chemical potential microscopy (SCPM) [79], scanning ion conductance microscopy (SICM) [87, 88], and scanning capacitance microscopy (SCM) [82, 89–91]. Families of instruments which measure forces (e.g., AFM, FFM, SEFM, SFAM, and SMM) are also referred to as scanning force microscopy (SFM). Although these instruments offer atomic resolution and are ideal for basic research, they are used for cutting edge industrial applications which do not require atomic resolution. Commercial production of SPMs started with the STM in 1987 and the AFM in 1989 by Digital Instruments Inc. For comparisons of SPMs with other microscopes, see Table 2.1 (Veeco Instruments, Inc.). Numbers of these instruments are equally divided between the U.S., Japan and Europe, with the following industry/university and Government labs. splits: 50/50, 70/30, and 30/70, respectively. It is clear that research and industrial applications of SPMs are rapidly expanding.

Table 2.1. Comparison of Various Conventional Microscopes with SPMs

	Optical	SEM/TEM	Confocal	SPM
Magnification	10^3	10^7	10^4	10^9
Instrument Price (U.S. $)	$10 k	$250 k	$30 k	$100 k
Technology Age	200 yrs	40 yrs	20 yrs	20 yrs
Applications	Ubiquitous	Science and technology	New and unfolding	Cutting edge
Market 1993	$800 M	$400 M	$80 M	$100 M
Growth Rate	10%	10%	30%	70%

2.2 Scanning Tunneling Microscope

The principle of electron tunneling was proposed by *Giaever* [92]. He envisioned that if a potential difference is applied to two metals separated by a thin insulating film, a current will flow because of the ability of electrons to penetrate a potential barrier. To be able to measure a tunneling current, the two metals must be spaced no more than 10 nm apart. *Binnig* et al. [1] introduced vacuum tunneling combined with lateral scanning. The vacuum provides the ideal barrier for tunneling. The lateral scanning allows one to image surfaces with exquisite resolution, lateral-less than 1 nm and vertical-less than 0.1 nm, sufficient to define the position of single atoms. The very high vertical resolution of the STM is obtained because the tunnel current varies exponentially with the distance between the two electrodes, that is, the metal tip and the scanned surface. Typically, tunneling current decreases by a factor of 2 as the separation is increased by 0.2 nm. Very high lateral resolution depends upon sharp tips. *Binnig* et al. overcame two key obstacles for damping external vibrations and for moving the tunneling probe in close proximity to the sample. Their instrument is called the scanning tunneling microscope (STM). Today's STMs can be used in the ambient environment for atomic-scale image of surfaces. Excellent reviews on this subject are presented by *Hansma* and *Tersoff* [93], *Sarid* and *Elings* [94], *Durig* et al. [95]; *Frommer* [96], *Güntherodt* and *Wiesendanger* [97], *Wiesendanger* and *Güntherodt* [98], *Bonnell* [99], *Marti* and *Amrein* [100], *Stroscio* and *Kaiser* [101], and *Güntherodt* et al. [102].

The principle of the STM is straightforward. A sharp metal tip (one electrode of the tunnel junction) is brought close enough (0.3–1 nm) to the surface to be investi-

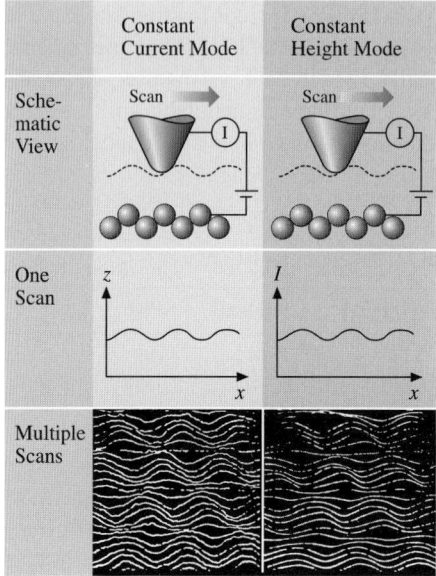

Fig. 2.1. STM can be operated in either the constant-current or the constant-height mode. The images are of graphite in air [93]

gated (the second electrode) that, at a convenient operating voltage (10 mV–1 V), the tunneling current varies from 0.2 to 10 nA which is measurable. The tip is scanned over a surface at a distance of 0.3–1 nm, while the tunneling current between it and the surface is measured. The STM can be operated in either the constant current mode or the constant height mode, Fig. 2.1. The left-hand column of Fig. 2.1 shows the basic constant current mode of operation. A feedback network changes the height of the tip z to keep the current constant. The displacement of the tip given by the voltage applied to the piezoelectric drives then yields a topographic map of the surface. Alternatively, in the constant height mode, a metal tip can be scanned across a surface at nearly constant height and constant voltage while the current is monitored, as shown in the right-hand column of Fig. 2.1. In this case, the feedback network responds only rapidly enough to keep the average current constant. A current mode is generally used for atomic-scale images. This mode is not practical for rough surfaces. A three-dimensional picture $[z(x,y)]$ of a surface consists of multiple scans $[z(x)]$ displayed laterally from each other in the y direction. It should be noted that if different atomic species are present in a sample, the different atomic species within a sample may produce different tunneling currents for a given bias voltage. Thus the height data may not be a direct representation of the topography of the surface of the sample.

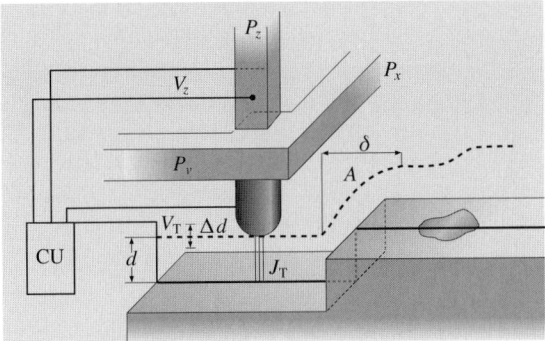

Fig. 2.2. Principle of operation of the STM made by *Binnig* and *Rohrer* [103]

2.2.1 Binnig et al.'s Design

Figure 2.2 shows a schematic of one of Binnig and Rohrer's designs for operation in ultrahigh vacuum [1, 103]. The metal tip was fixed to rectangular piezodrives P_x, P_y, and P_z made out of commercial piezoceramic material for scanning. The sample is mounted via either superconducting magnetic levitation or a two-stage spring system to achieve a stability of the gap width of about 0.02 nm. The tunnel current J_T is a sensitive function of the gap width d that is $J_T \propto V_T \exp(-A\phi^{1/2}d)$, where V_T is the bias voltage, ϕ is the average barrier height (work function) and the constant $A = 1.025 \, \text{eV}^{-1/2}\text{Å}^{-1}$. With a work function of a few eV, J_T changes by an order of magnitude for every angstrom change of d. If the current is kept constant to within, for example, 2%, then the gap d remains constant to within 1 pm. For operation in the constant current mode, the control unit CU applies a voltage V_z to the piezo P_z such that J_T remains constant when scanning the tip with P_y and P_x over the surface. At constant work function ϕ, $V_z(V_x, V_y)$ yields the roughness of the surface $z(x, y)$ directly, as illustrated at a surface step at A. Smearing the step, δ (lateral resolution) is on the order of $(R)^{1/2}$, where R is the radius of the curvature of the tip. Thus, a lateral resolution of about 2 nm requires tip radii on the order of 10 nm. A 1-mm-diameter solid rod ground at one end at roughly 90 °C yields overall tip radii of only a few hundred nm, but with closest protrusion of rather sharp microtips on the relatively dull end yields a lateral resolution of about 2 nm. In-situ sharpening of the tips by gently touching the surface brings the resolution down to the 1-nm range; by applying high fields (on the order of 10^8 V/cm) during, for example, half an hour, resolutions considerably below 1 nm could be reached. Most experiments were done with tungsten wires either ground or etched to a radius typically in the range of 0.1–10 µm. In some cases, in-situ processing of the tips was done for further reduction of tip radii.

2.2.2 Commercial STMs

There are a number of commercial STMs available on the market. Digital Instruments, Inc. located in Santa Barbara, CAconstant amplitude introduced the first com-

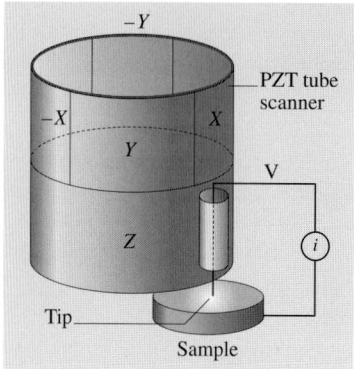

Fig. 2.3. Principle of operation of a commercial STM, a sharp tip attached to a piezoelectric tube scanner is scanned on a sample

mercial STM, the Nanoscope I, in 1987. In a recent Nanoscope IV STM for operation in ambient air, the sample is held in position while a piezoelectric crystal in the form of a cylindrical tube (referred to as PZT tube scanner) scans the sharp metallic probe over the surface in a raster pattern while sensing and outputting the tunneling current to the control station (Fig. 2.3). The digital signal processor (DSP) calculates the desired separation of the tip from the sample by sensing the tunneling current flowing between the sample and the tip. The bias voltage applied between the sample and the tip encourages the tunneling current to flow. The DSP completes the digital feedback loop by outputting the desired voltage to the piezoelectric tube. The STM operates in both the "constant height" and "constant current" modes depending on a parameter selection in the control panel. In the constant current mode, the feedback gains are set high, the tunneling tip closely tracks the sample surface, and the variation in the tip height required to maintain constant tunneling current is measured by the change in the voltage applied to the piezo tube. In the constant height mode, the feedback gains are set low, the tip remains at a nearly constant height as it sweeps over the sample surface, and the tunneling current is imaged.

Physically, the Nanoscope STM consists of three main parts: the head which houses the piezoelectric tube scanner for three dimensional motion of the tip and the preamplifier circuit (FET input amplifier) mounted on top of the head for the tunneling current, the base on which the sample is mounted, and the base support, which supports the base and head [4]. The base accommodates samples up to 10 mm by 20 mm and 10 mm in thickness. Scan sizes available for the STM are 0.7 µm (for atomic resolution), 12 µm, 75 µm and 125 µm square.

The scanning head controls the three dimensional motion of the tip. The removable head consists of a piezo tube scanner, about 12.7 mm in diameter, mounted into an invar shell used to minimize vertical thermal drifts because of a good thermal match between the piezo tube and the Invar. The piezo tube has separate electrodes for X, Y and Z which are driven by separate drive circuits. The electrode configuration (Fig. 2.3) provides x and y motions which are perpendicular to each other, minimizes horizontal and vertical coupling, and provides good sensitivity. The vertical motion of the tube is controlled by the Z electrode which is driven by the feedback loop. The x and y scanning motions are each controlled by two electrodes which are driven by voltages of the same magnitude, but opposite signs. These electrodes are called $-Y$, $-X$, $+Y$, and $+X$. Applying complimentary voltages allows a short, stiff tube to provide a good scan range without large voltages. The motion of the tip due to external vibrations is proportional to the square of the ratio of vibration frequency to the resonant frequency of the tube. Therefore, to minimize the tip vibrations, the resonant frequencies of the tube are high at about 60 kHz in the vertical direction and about 40 kHz in the horizontal direction. The tip holder is a stainless steel tube with a 300 μm inner diameter for 250 μm diameter tips, mounted in ceramic in order to keep the mass on the end of the tube low. The tip is mounted either on the front edge of the tube (to keep mounting mass low and resonant frequency high) (Fig. 2.3) or the center of the tube for large range scanners, namely 75 and 125 μm (to preserve the symmetry of the scanning). This commercial STM accepts any tip with a 250 μm diameter shaft. The piezotube requires X-Y calibration which is carried out by imaging an appropriate calibration standard. Cleaved graphite is used for the small-scan length head while two dimensional grids (a gold plated ruling) can be used for longer range heads.

The Invar base holds the sample in position, supports the head, and provides coarse X-Y motion for the sample. A spring-steel sample clip with two thumb screws holds the sample in place. An x-y translation stage built into the base allows the sample to be repositioned under the tip. Three precision screws arranged in a triangular pattern support the head and provide coarse and fine adjustment of the tip height. The base support consists of the base support ring and the motor housing. The stepper motor enclosed in the motor housing allows the tip to be engaged and withdrawn from the surface automatically.

Samples to be imaged with the STM must be conductive enough to allow a few nanoamperes of current to flow from the bias voltage source to the area to be scanned. In many cases, nonconductive samples can be coated with a thin layer of a conductive material to facilitate imaging. The bias voltage and the tunneling current depend on the sample. Usually they are set at a standard value for engagement and fine tuned to enhance the quality of the image. The scan size depends on the sample and the features of interest. A maximum scan rate of 122 Hz can be used. The maximum scan rate is usually related to the scan size. Scan rates above 10 Hz are used for small scans (typically 60 Hz for atomic-scale imaging with a 0.7 μm scanner). The scan rate should be lowered for large scans, especially if the sample surfaces are rough or contain large steps. Moving the tip quickly along the sample surface at high scan rates with large scan sizes will usually lead to a tip crash. Essentially, the scan rate should be inversely proportional to the scan size (typically 2–4 Hz for 1 μm, 0.5–1 Hz for 12 μm, and 0.2 Hz for 125 μm scan sizes). Scan rate in length/time, is equal to scan length divided by the scan rate in Hz. For example, for 10 μm × 10 μm scan size scanned at 0.5 Hz, the scan rate is 10 μm/s. Typically, 256 × 256 data formats are most commonly used. The lateral resolution at larger scans is approximately equal to scan length divided by 256.

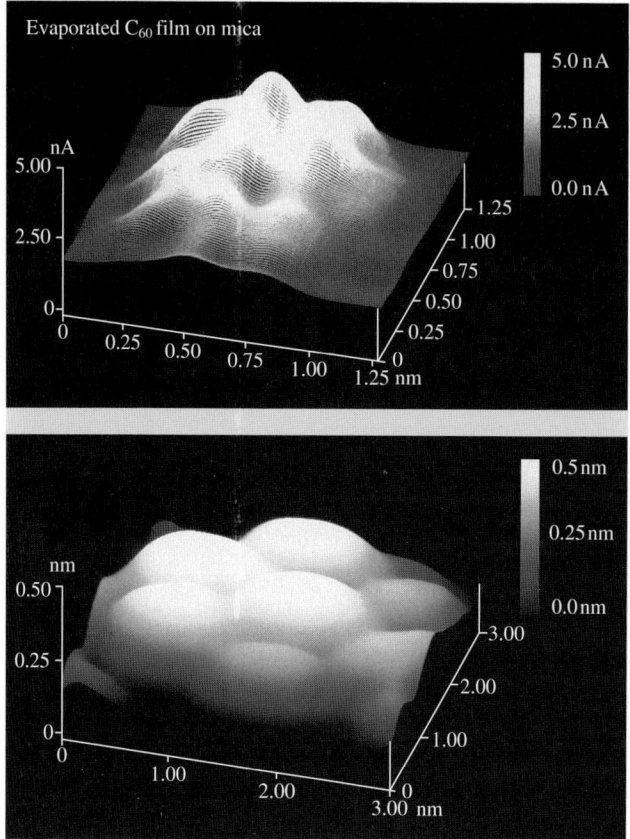

Fig. 2.4. STM images of evaporated C_{60} film on a gold-coated freshly-cleaved mica using a mechanically sheared Pt-Ir (80-20) tip in constant height mode [104]

Figure 2.4 shows an example of STM images of an evaporated C_{60} film on a gold-coated freshly-cleaved mica taken at room temperature and ambient pressure [104]. Images with atomic resolution at two scan sizes are obtained. Next we describe STM designs which are available for special applications.

Electrochemical STM

The electrochemical STM is used to perform and monitor the electrochemical reactions inside the STM. It includes a microscope base with an integral potentiostat, a short head with a 0.7 μm scan range and a differential preamp and the software required to operate the potentiostat and display the result of the electrochemical reaction.

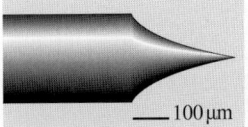

Fig. 2.5. Schematic of a typical tungsten cantilever with a sharp tip produced by electrochemical etching

Standalone STM

The stand alone STMs are available to scan large samples which rest directly on the sample. From Digital instruments, it is available in 12 and 75 μm scan ranges. It is similar to the standard STM except the sample base has been eliminated.

2.2.3 STM Probe Construction

The STM probe should have a cantilever integrated with a sharp metal tip with a low aspect ratio (tip length/tip shank) to minimize flexural vibrations. Ideally, the tip should be atomically sharp, but in practice most tip preparation methods produce a tip with a rather ragged profile that consists of several asperities with the one closest to the surface responsible for tunneling. STM cantilevers with sharp tips are typically fabricated from metal wires of tungsten (W), platinum-iridium (Pt-Ir), or gold (Au) and sharpened by grinding, cutting with a wire cutter or razor blade, field emission/evaporator, ion milling, fracture, or electrochemical polishing/etching [105, 106]. The two most commonly used tips are made from either a Pt-Ir (80/20) alloy or tungsten wire. Iridium is used to provide stiffness. The Pt-Ir tips are generally mechanically formed and are readily available. The tungsten tips are etched from tungsten wire with an electrochemical process, for example by using 1 molar KOH solution with a platinum electrode in a electrochemical cell at about 30 V. In general, Pt-Ir tips provide better atomic resolution than tungsten tips, probably due to the lower reactivity of Pt. But tungsten tips are more uniformly shaped and may perform better on samples with steeply sloped features. The tungsten wire diameter used for the cantilever is typically 250 μm with the radius of curvature ranging from 20 to 100 nm and a cone angle ranging from 10 to 60 °C (Fig. 2.5). The wire can be bent in an L shape, if so required, for use in the instrument. For calculations of normal spring constant and natural frequency of round cantilevers, see *Sarid* and *Elings* [94].

For imaging of deep trenches, high-aspect-ratio, controlled geometry (CG) Pt-Ir probes are commercially available (Fig. 2.6). These probes are electrochemically etched from Pt-Ir (80/20) wire and polished to a specific shape which is consistent from tip to tip. Probes have a full cone angle of approximately 15 °C, and a tip radius of less than 50 nm. For imaging of very deep trenches (> 0.25 μm) and nanofeatures, focused ion beam (FIB) milled CG probes with an extremely sharp tip (radius < 5 nm) are used. For electrochemistry, Pt/Ir probes are coated with a nonconducting

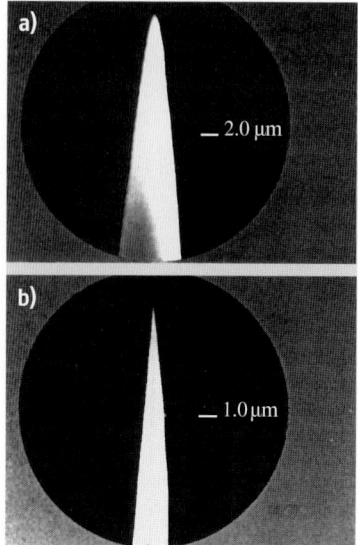

Fig. 2.6. Schematics of (**a**) CG Pt-Ir probe, and (**b**) CG Pt-Ir FIB milled probe

film (not shown in the figure). These probes are available from Materials Analytical Services, Raleigh, North Carolina.

Pt alloy and W tips are very sharp and have high resolution, but are fragile and sometimes break when contacting a surface. Diamond tips have been used by *Kaneko* and *Oguchi* [107]. The diamond tip made conductive by boron ion implantation is found to be chip resistant.

2.3 Atomic Force Microscope

Like the STM, the AFM relies on a scanning technique to produce very high resolution, 3-D images of sample surfaces. The AFM measures ultrasmall forces (less than 1 nN) present between the AFM tip surface and a sample surface. These small forces are measured by measuring the motion of a very flexible cantilever beam having an ultrasmall mass. While STMs require that the surface to be measured be electrically conductive, AFMs are capable of investigating surfaces of both conductors and insulators on an atomic scale if suitable techniques for measurement of cantilever motion are used. In the operation of a high resolution AFM, the sample is generally scanned instead of the tip as with the STM, because the AFM measures the relative displacement between the cantilever surface and reference surface and any cantilever movement would add vibrations. For measurements of large samples, AFMs are available where the tip is scanned and the sample is stationary. As long as the AFM is operated in the so-called contact mode, little if any vibration is introduced.

The AFM combines the principles of the STM and the stylus profiler (Fig. 2.7). In an AFM, the force between the sample and tip is detected, rather than the tunnel-

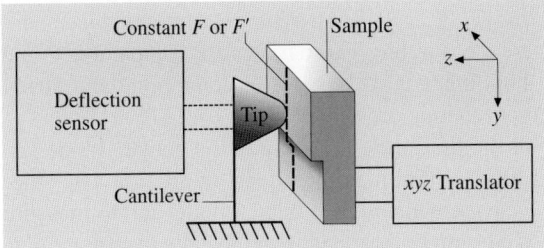

Fig. 2.7. Principle of operation of the AFM. Sample mounted on a piezoelectric tube scanner is scanned against a short tip and the cantilever deflection is measured, mostly, using a laser deflection technique. Force (contact mode) or force gradient (noncontact mode) is measured during scanning

ing current, to sense the proximity of the tip to the sample. The AFM can be used either in a static or dynamic mode. In the static mode, also referred to as repulsive mode or contact mode [2], a sharp tip at the end of a cantilever is brought in contact with a sample surface. During initial contact, the atoms at the end of the tip experience a very weak repulsive force due to electronic orbital overlap with the atoms in the sample surface. The force acting on the tip causes a cantilever deflection which is measured by tunneling, capacitive, or optical detectors. The deflection can be measured to within 0.02 nm, so for typical cantilever spring constant of 10 N/m a force as low as 0.2 nN (corresponding normal pressure $\sim$ 200 MPa for a Si_3N_4 tip with radius of about 50 nm against single-crystal silicon) can be detected. (To put these number in perspective, individual atoms and human hair are typically a fraction of a nanometer and about 75 μm in diameter, respectively, and a drop of water and an eyelash have a mass of about 10 μN and 100 nN, respectively.) In the dynamic mode of operation for the AFM, also referred to as attractive force imaging or noncontact imaging mode, the tip is brought in close proximity (within a few nm) to, and not in contact with the sample. The cantilever is deliberately vibrated either in amplitude modulation (AM) mode [65] or frequency modulation (FM) mode [65, 94, 108, 109]. Very weak van der Waals attractive forces are present at the tip–sample interface. Although in this technique, the normal pressure exerted at the interface is zero (desirable to avoid any surface deformation), it is slow, and is difficult to use, and is rarely used outside research environments. In the two modes, surface topography is measured by laterally scanning the sample under the tip while simultaneously measuring the separation-dependent force or force gradient (derivative) between the tip and the surface (Fig. 2.7). In the contact (static) mode, the interaction force between tip and sample is measured by measuring the cantilever deflection. In the noncontact (or dynamic) mode, the force gradient is obtained by vibrating the cantilever and measuring the shift of resonant frequency of the cantilever. To obtain topographic information, the interaction force is either recorded directly, or used as a control parameter for a feedback circuit that maintains the force or force derivative at a constant value. With an AFM operated in the contact mode, topographic images with a vertical resolution of less than 0.1 nm (as low as 0.01 nm) and a lateral resolution of about 0.2 nm have been obtained [3, 50, 110–114]. With a 0.01 nm displacement sensitivity, 10 nN to 1 pN forces are measurable. These forces are comparable to the forces associated with chemical bonding, e.g., 0.1 µN for an ionic bond and 10 pN for a hydrogen bond [2]. For further reading, see [94–96, 100, 102, 115–119].

Lateral forces being applied at the tip during scanning in the contact mode, affect roughness measurements [120]. To minimize the effects of friction and other lateral forces in the topography measurements in the contact-mode, and to measure the topography of soft surfaces, AFMs can be operated in the so called tapping mode or force modulation mode [32, 121].

The STM is ideal for atomic-scale imaging. To obtain atomic resolution with the AFM, the spring constant of the cantilever should be weaker than the equivalent spring between atoms. For example, the vibration frequencies ω of atoms bound in a molecule or in a crystalline solid are typically 10^{13} Hz or higher. Combining this with the mass of the atoms m, on the order of 10^{-25} kg, gives an interatomic spring constants k, given by $\omega^2 m$, on the order of 10 N/m [115]. (For comparison, the spring constant of a piece of household aluminium foil that is 4 mm long and 1 mm wide is about 1 N/m.) Therefore, a cantilever beam with a spring constant of about 1 N/m or lower is desirable. Tips have to be as sharp as possible. Tips with a radius ranging from 5 to 50 nm are commonly available.

Atomic resolution cannot be achieved with these tips at the normal load in the nN range. Atomic structures at these loads have been obtained from lattice imaging or by imaging of the crystal periodicity. Reported data show either perfectly ordered periodic atomic structures or defects on a larger lateral scale, but no well-defined, laterally resolved atomic-scale defects like those seen in images routinely obtained with a STM. Interatomic forces with one or several atoms in contact are 20–40 or 50–100 pN, respectively. Thus, atomic resolution with an AFM is only possible with a sharp tip on a flexible cantilever at a net repulsive force of 100 pN or lower [122]. Upon increasing the force from 10 pN, *Ohnesorge* and *Binnig* [122] observed that monoatomic steplines were slowly wiped away and a perfectly ordered structure was left. This observation explains why mostly defect-free atomic resolution has been observed with AFM. Note that for atomic-resolution measurements, the cantilever should not be too soft to avoid jumps. Further note that measurements in the noncontact imaging mode may be desirable for imaging with atomic resolution.

The key component in an AFM is the sensor for measuring the force on the tip due to its interaction with the sample. A cantilever (with a sharp tip) with extremely low spring constant is required for high vertical and lateral resolutions at small forces (0.1 nN or lower) but at the same time a high resonant frequency is desirable (about 10 to 100 kHz) in order to minimize the sensitivity to vibration noise from the building which is near 100 Hz. This requires a spring with extremely low vertical spring constant (typically 0.05 to 1 N/m) as well as low mass (on the order of 1 ng). Today, the most advanced AFM cantilevers are microfabricated from silicon or silicon nitride using photolithographic techniques. (For further details on cantilevers, see a later section). Typical lateral dimensions are on the order of 100 µm, with the thicknesses on the order of 1 µm. The force on the tip due to its interaction with the sample is sensed by detecting the deflection of the compliant lever with a known

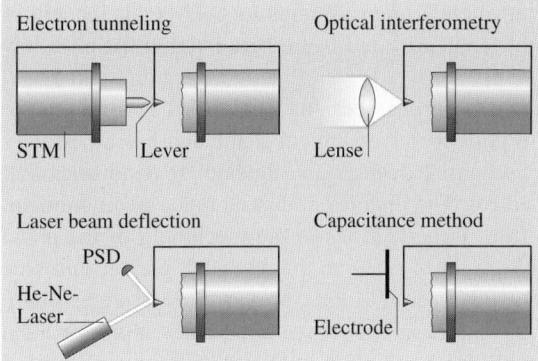

Fig. 2.8. Schematics of the four more commonly used detection systems for measurement of cantilever deflection. In each set up, the sample mounted on piezoelectric body is shown on the right, the cantilever in the middle, and the corresponding deflection sensor on the left [118]

spring constant. This cantilever deflection (displacement smaller than 0.1 nm) has been measured by detecting a tunneling current similar to that used in the STM in the pioneering work of *Binnig* et al. [2] and later used by *Giessibl* et al. [56], by capacitance detection [123, 124], piezoresistive detection [125, 126], and by four optical techniques namely (1) by optical interferometry [5, 6, 127, 128] with the use of optical fibers [57, 129] (2) by optical polarization detection [72, 130], (3) by laser diode feedback [131] and (4) by optical (laser) beam deflection [7, 8, 53, 111, 112]. Schematics of the four more commonly used detection systems are shown in Fig. 2.8. The tunneling method originally used by *Binnig* et al. [2] in the first version of the AFM, uses a second tip to monitor the deflection of the cantilever with its force sensing tip. Tunneling is rather sensitive to contaminants and the interaction between the tunneling tip and the rear side of the cantilever can become comparable to the interaction between the tip and sample. Tunneling is rarely used and is mentioned first for historical purposes. *Giessibl* et al. [56] have used it for a low temperature AFM/STM design. In contrast to tunneling, other deflection sensors are far away from the cantilever at distances of microns to tens of mm. The optical techniques are believed to be more sensitive, reliable and easily implemented detection methods than others [94, 118]. The optical beam deflection method has the largest working distance, is insensitive to distance changes and is capable of measuring angular changes (friction forces), therefore, it is most commonly used in the commercial SPMs.

Almost all SPMs use piezo translators to scan the sample, or alternatively, to scan the tip. An electric field applied across a piezoelectric material causes a change in the crystal structure, with expansion in some directions and contraction in others. A net change in volume also occurs [132]. The first STM used a piezo tripod for scanning [1]. The piezo tripod is one way to generate three-dimensional movement of a tip attached to its center. However, the tripod needs to be fairly large (~ 50 mm) to get a suitable range. Its size and asymmetric shape makes it susceptible to thermal drift. The tube scanners are widely used in AFMs [133]. These provide ample scanning range with a small size. Control electronics systems for AFMs can use either analog or digital feedback. Digital feedback circuits are better suited for ultra low noise operation.

Images from the AFMs need to be processed. An ideal AFM is a noise free device that images a sample with perfect tips of known shape and has a perfectly linear scanning piezo. In reality, scanning devices are affected by distortions and these distortions must be corrected for. The distortions can be linear and nonlinear. Linear distortions mainly result from imperfections in the machining of the piezo translators causing cross talk between the Z-piezo to the X- and Y-piezos, and vice versa. Nonlinear distortions mainly result because of the presence of a hysteresis loop in piezoelectric ceramics. These may also result if the scan frequency approaches the upper frequency limit of the X- and Y-drive amplifiers or the upper frequency limit of the feedback loop (z-component). In addition, electronic noise may be present in the system. The noise is removed by digital filtering in real space [134] or in the spatial frequency domain (Fourier space) [135].

Processed data consists of many tens of thousand of points per plane (or data set). The output of the first STM and AFM images were recorded on an X-Y chart recorder, with z-value plotted against the tip position in the fast scan direction. Chart recorders have slow response so computers are used to display the data. The data are displayed as a wire mesh display or gray scale display (with at least 64 shades of gray).

2.3.1 Binnig et al.'s Design

In the first AFM design developed by *Binnig* et al. [2], AFM images were obtained by measurement of the force on a sharp tip created by the proximity to the surface of the sample mounted on a 3-D piezoelectric scanner. Tunneling current between STM tip and the backside of the cantilever beam with attached tip was measured to obtain the normal force. This force was kept at a constant level with a feedback mechanism. The STM tip was also mounted on a piezoelectric element to maintain the tunneling current at a constant level.

2.3.2 Commercial AFM

A review of early designs of AFMs is presented by *Bhushan* [4]. There are a number of commercial AFMs available on the market. Major manufacturers of AFMs for use in an ambient environment are: Digital Instruments Inc., a subsidiary of Veeco Instruments, Inc., Santa Barbara, California; Topometrix Corp., a subsidiary of Veeco Instruments, Inc., Santa Clara, California; other subsidiaries of Veeco Instruments Inc., Woodbury, New York; Molecular Imaging Corp., Phoenix, Arizona; Quesant Instrument Corp., Agoura Hills, California; Nanoscience Instruments Inc., Phoenix, Arizona; Seiko Instruments, Japan; and Olympus, Japan. AFM/STMs for use in UHV environments are manufactured by Omicron Vakuumphysik GMBH, Taunusstein, Germany.

We describe here two commercial AFMs – small sample and large sample AFMs – for operation in the contact mode, produced by Digital Instruments, Inc., Santa Barbara, CAconstant amplitude, with scanning lengths ranging from about 0.7 μm (for atomic resolution) to about 125 μm [9, 111, 114, 136]. The original design of these AFMs comes from *Meyer* and *Amer* [53]. Basically the AFM scans the sample in a raster pattern while outputting the cantilever deflection error signal to the control station. The cantilever deflection (or the force) is measured using laser deflection technique (Fig. 2.9). The DSP in the workstation controls the z position of the piezo based on the cantilever deflection error signal. The AFM operates in both the "constant height" and "constant force" modes. The DSP always adjusts the height of the sample under the tip based on the cantilever deflection error signal, but if the feedback gains are low the piezo remains at a nearly "constant height" and the cantilever deflection data is collected. With the high gains the piezo height changes to keep the cantilever deflection nearly constant (therefore the force is constant) and the change in piezo height is collected by the system.

To further describe the principle of operation of the commercial small sample AFM shown in Fig. 2.9a, the sample, generally no larger than 10 mm × 10 mm, is mounted on a PZT tube scanner which consists of separate electrodes to scan precisely the sample in the x-y plane in a raster pattern and to move the sample in the vertical (z) direction. A sharp tip at the free end of a flexible cantilever is brought in contact with the sample. Features on the sample surface cause the cantilever to deflect in the vertical and lateral directions as the sample moves under the tip. A laser beam from a diode laser (5 mW max peak output at 670 nm) is directed by a prism onto the back of a cantilever near its free end, tilted downward at about 10 °C with respect to the horizontal plane. The reflected beam from the vertex of the cantilever is directed through a mirror onto a quad photodetector (split photodetector with four quadrants) (commonly called position-sensitive detector or PSD, produced by Silicon Detector Corp., Camarillo, California). The differential signal from the top and bottom photodiodes provides the AFM signal which is a sensitive measure of the cantilever vertical deflection. Topographic features of the sample cause the tip to deflect in the vertical direction as the sample is scanned under the tip. This tip deflection will change the direction of the reflected laser beam, changing the intensity difference between the top and bottom sets of photodetectors (AFM signal). In the AFM operating mode called the height mode, for topographic imaging or for any other operation in which the applied normal force is to be kept a constant, a feedback circuit is used to modulate the voltage applied to the PZT scanner to adjust the height of the PZT, so that the cantilever vertical deflection (given by the intensity difference between the top and bottom detector) will remain constant during scanning. The PZT height variation is thus a direct measure of the surface roughness of the sample.

In a large sample AFM, both force sensors using optical deflection methods and scanning unit are mounted on the microscope head (Fig. 2.9b). Because of vibrations added by cantilever movement, lateral resolution of this design is somewhat poorer than the design in Fig. 2.9a in which the sample is scanned instead of the cantilever beam. The advantage of the large sample AFM is that large samples can be measured readily.

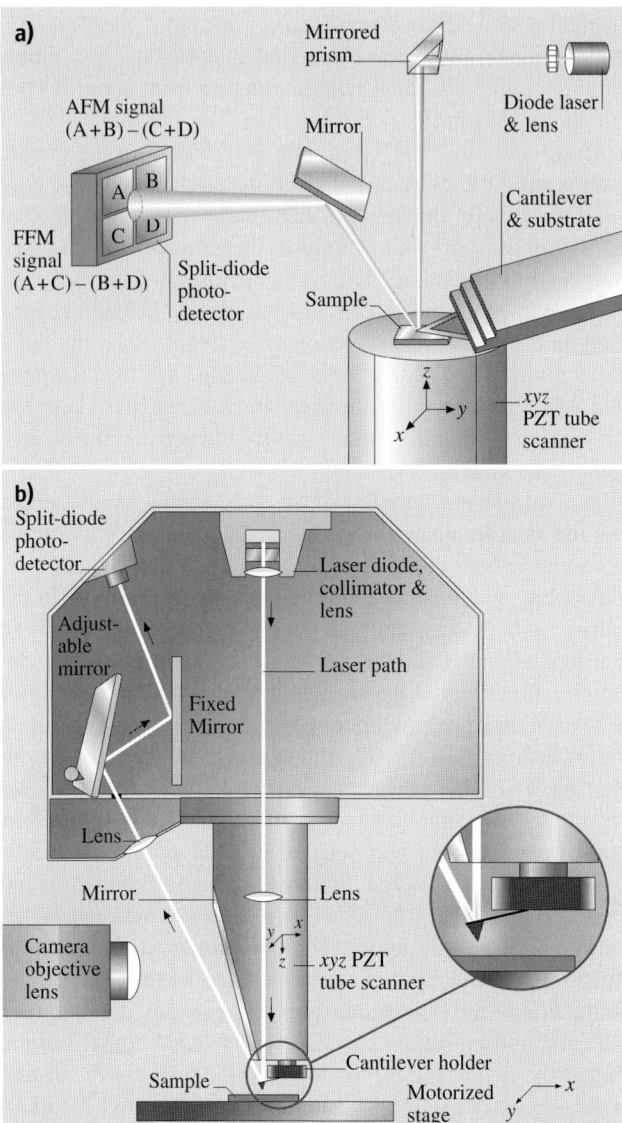

Fig. 2.9. Principles of operation of (**a**) a commercial small sample AFM/FFM, and (**b**) a large sample AFM/FFM

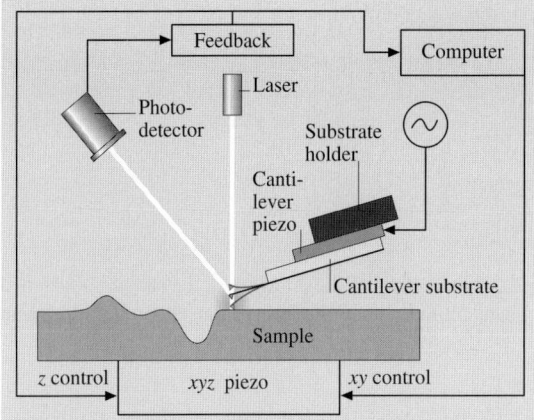

Fig. 2.10. Schematic of tapping mode used for surface roughness measurements

Most AFMs can be used for topography measurements in the so-called tapping mode (intermittent contact mode), also referred to as dynamic force microscopy. In the tapping mode, during scanning over the surface, the cantilever/tip assembly is sinusoidally vibrated by a piezo mounted above it, and the oscillating tip slightly taps the surface at the resonant frequency of the cantilever (70–400 Hz) with a constant (20–100 nm) oscillating amplitude introduced in the vertical direction with a feedback loop keeping the average normal force constant (Fig. 2.10). The oscillating amplitude is kept large enough so that the tip does not get stuck to the sample because of adhesive attractions. The tapping mode is used in topography measurements to minimize effects of friction and other lateral forces to measure the topography of soft surfaces.

Topographic measurements are made at any scanning angle. At a first instance, scanning angle may not appear to be an important parameter. However, the friction force between the tip and the sample will affect the topographic measurements in a parallel scan (scanning along the long axis of the cantilever). Therefore a perpendicular scan may be more desirable. Generally, one picks a scanning angle which gives the same topographic data in both directions; this angle may be slightly different than that for the perpendicular scan.

For measurement of the friction force being applied at the tip surface during sliding, the left hand and right hand sets of quadrants of the photodetector are used. In the so-called friction mode, the sample is scanned back and forth in a direction orthogonal to the long axis of the cantilever beam. A friction force between the sample and the tip will produce a twisting of the cantilever. As a result, the laser beam will be reflected out of the plane defined by the incident beam and the beam reflected vertically from an untwisted cantilever. This produces an intensity difference of the laser beam received in the left hand and right hand sets of quadrants of the photodetector. The intensity difference between the two sets of detectors (FFM signal) is directly related to the degree of twisting and hence to the magnitude of the

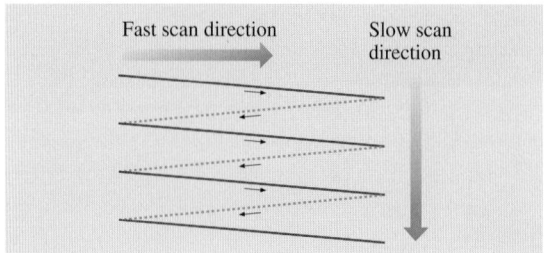

Fig. 2.11. Schematic of triangular pattern trajectory of the AFM tip as the sample is scanned in two dimensions. During imaging, data are recorded only during scans along the solid scan lines

friction force. This method provides three-dimensional maps of the friction force. One problem associated with this method is that any misalignment between the laser beam and the photodetector axis would introduce error in the measurement. However, by following the procedures developed by *Ruan* and *Bhushan* [136], in which the average FFM signal for the sample scanned in two opposite directions is subtracted from the friction profiles of each of the two scans, the misalignment effect is eliminated. By following the friction force calibration procedures developed by *Ruan* and *Bhushan* [136], voltages corresponding to friction forces can be converted to force units. The coefficient of friction is obtained from the slope of friction force data measured as a function of normal loads typically ranging from 10 to 150 nN. This approach eliminates any contributions due to the adhesive forces [10]. For calculation of the coefficient of friction based on a single point measurement, friction force should be divided by the sum of applied normal load and intrinsic adhesive force. Furthermore, it should be pointed out that for a single asperity contact, the coefficient of friction is not independent of load. For further details, refer to a later section.

The tip is scanned in such a way that its trajectory on the sample forms a triangular pattern (Fig. 2.11). Scanning speeds in the fast and slow scan directions depend on the scan area and scan frequency. Scan sizes ranging from less than 1 nm × 1 nm to 125 μm × 125 μm and scan rates from less than 0.5 to 122 Hz typically can be used. Higher scan rates are used for smaller scan lengths. For example, scan rates in the fast and slow scan directions for an area of 10 μm × 10 μm scanned at 0.5 Hz are 10 μm/s and 20 nm/s, respectively.

We now describe the construction of a small sample AFM in more detail. It consists of three main parts: the optical head which senses the cantilever deflection, a PZT tube scanner which controls the scanning motion of the sample mounted on its one end, and the base which supports the scanner and head and includes circuits for the deflection signal (Fig. 2.12a). The AFM connects directly to a control system. The optical head consists of laser diode stage, photodiode stage preamp board, cantilever mount and its holding arm, and deflection beam reflecting mirror (Fig. 2.12b). The laser diode stage is a tilt stage used to adjust the position of the laser beam relative to the cantilever. It consists of the laser diode, collimator, focusing lens, baseplate, and the X and Y laser diode positioners. The positioners are used to place the laser spot on the end of the cantilever. The photodiode stage is an adjustable stage used to position the photodiode elements relative to the reflected laser beam. It consists of the split photodiode, the base plate, and the photodiode positioners. The deflection beam reflecting mirror is mounted on the upper left in the interior of the head which reflects the deflected beam toward the photodiode. The cantilever mount is a metal (for operation in air) or glass (for operation in water) block which holds the cantilever firmly at the proper angle (Fig. 2.12d). Next, the tube scanner consists of an Invar cylinder holding a single tube made of piezoelectric crystal which provides the necessary three-dimensional motion to the sample. Mounted on top of the tube is a magnetic cap on which the steel sample puck is placed. The tube is rigidly held at one end with the sample mounted on the other end of the tube. The scanner also contains three fine-pitched screws which form the mount for the optical head. The optical head rests on the tips of the screws which are used to adjust the position of the head relative to the sample. The scanner fits into the scanner support ring mounted on the base of the microscope (Fig. 2.12c). The stepper motor is controlled manually with the switch on the upper surface of the base and automatically by the computer during the tip engage and tip-withdraw processes.

The scan sizes available for these instruments are 0.7 μm, 12 μm and 125 μm. The scan rate must be decreased as the scan size is increased. A maximum scan rate of 122 Hz can be used. Scan rates of about 60 Hz should be used for small scan lengths (0.7 μm). Scan rates of 0.5 to 2.5 Hz should be used for large scans on samples with tall features. High scan rates help reduce drift, but they can only be used on flat samples with small scan sizes. Scan rate, or scanning speed in length/time in the fast scan direction, is equal to twice the scan length times the scan rate in Hz, and in the slow direction, it is equal to scan length times the scan rate in Hz divided by number of data points in the transverse direction. For example, for 10 μm × 10 μm scan size scanned at 0.5 Hz, the scan rates in the fast and slow scan directions are 10 μm/s and 20 nm/s, respectively. Normally 256 × 256 data points are taken for each image. The lateral resolution at larger scans is approximately equal to scan length divided by 256. The piezo tube requires x-y calibration which is carried out by imaging an appropriate calibration standard. Cleaved graphite is used for small scan heads while two-dimensional grids (a gold plating ruling) can be used for longer range heads.

Examples of AFM images of freshly-cleaved highly-oriented pyrolytic (HOP) graphite and mica surfaces are shown in Fig. 2.13 [50, 110, 114]. Images with near atomic resolution are obtained.

The force calibration mode is used to study interactions between the cantilever and the sample surface. In the force calibration mode, the X and Y voltages applied to the piezo tube are held at zero and a sawtooth voltage is applied to the Z electrode of the piezo tube, Fig. 2.14a. The force measurement starts with the sample far away and the cantilever in its rest position. As a result of the applied voltage, the sample is moved up and down relative to the stationary cantilever tip. As the piezo moves the sample up and down, the cantilever deflection signal from the photodiode is monitored. The force–distance curve, a plot of the cantilever tip deflection signal as a function of the voltage applied to the piezo tube, is obtained. Figure 2.14b shows

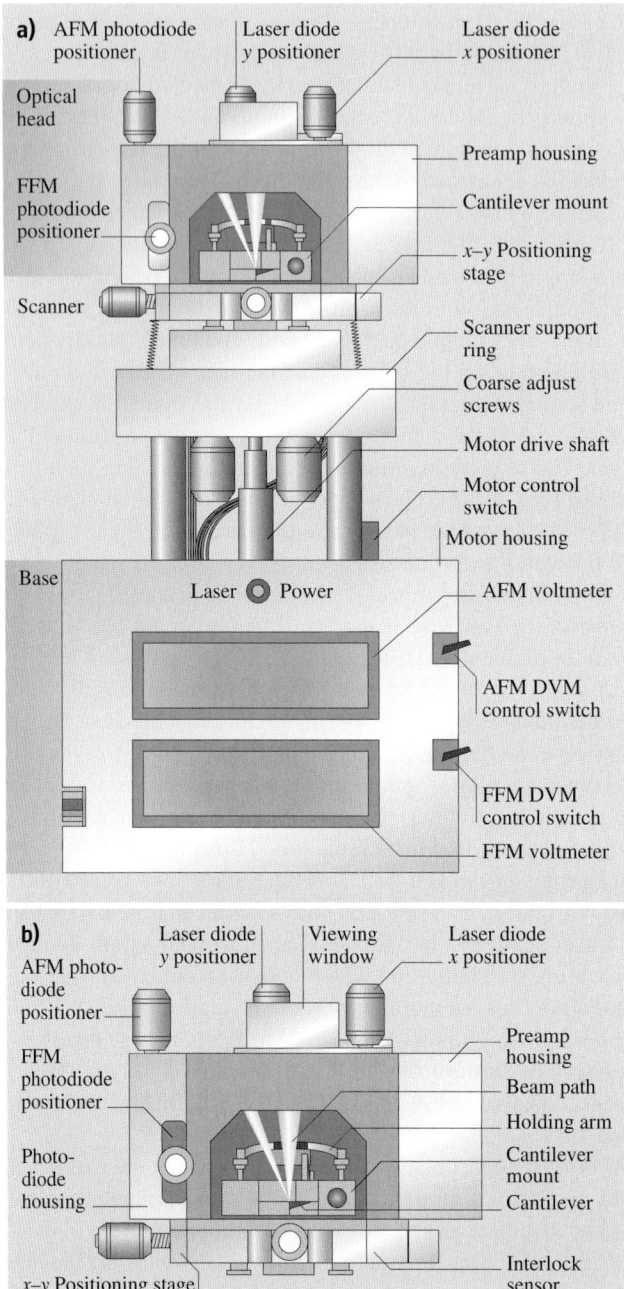

Fig. 2.12. Schematics of a commercial AFM/FFM made by Digital Instruments Inc. (**a**) front view, (**b**) optical head, (**c**) base, and (**d**) cantilever substrate mounted on cantilever mount (not to scale)

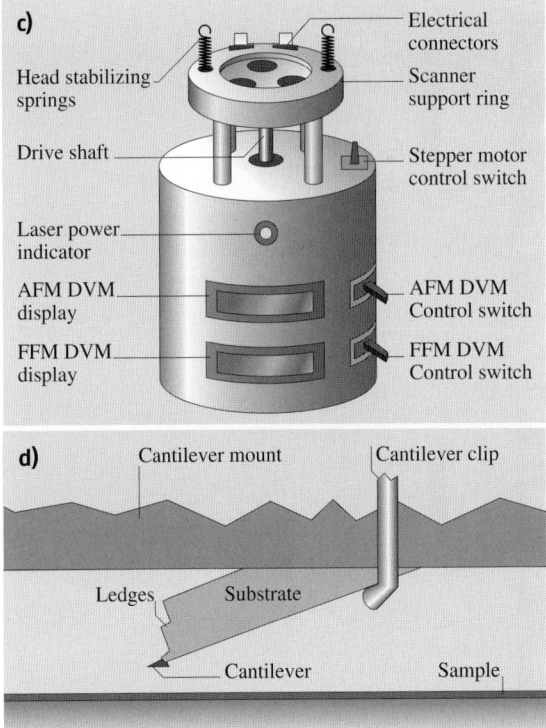

Fig. 2.12. continue

a typical force–distance curve showing the various features of the curve. The arrow heads reveal the direction of piezo travel. As the piezo extends, it approaches the tip, which is at this point in free air and hence shows no deflection. This is indicated by the flat portion of the curve. As the tip approaches the sample within a few nanometers (point A), an attractive force exists between the atoms of the tip surface and the atoms of the sample surface. The tip is pulled towards the sample and contact occurs at point B on the graph. From this point on, the tip is in contact with the surface and as the piezo further extends, the tip gets further deflected. This is represented by the sloped portion of the curve. As the piezo retracts, the tip goes beyond the zero deflection (flat) line because of attractive forces (van der Waals forces and long range meniscus forces), into the adhesive regime. At point C in the graph, the tip snaps free of the adhesive forces, and is again in free air. The horizontal distance between point B and C along the retrace line gives the distance moved by the tip in the adhesive regime. This distance multiplied by the stiffness of the cantilever gives the adhesive force. Incidentally, the horizontal shift between the loading and unloading curves results from the hysteresis in the PZT tube [4].

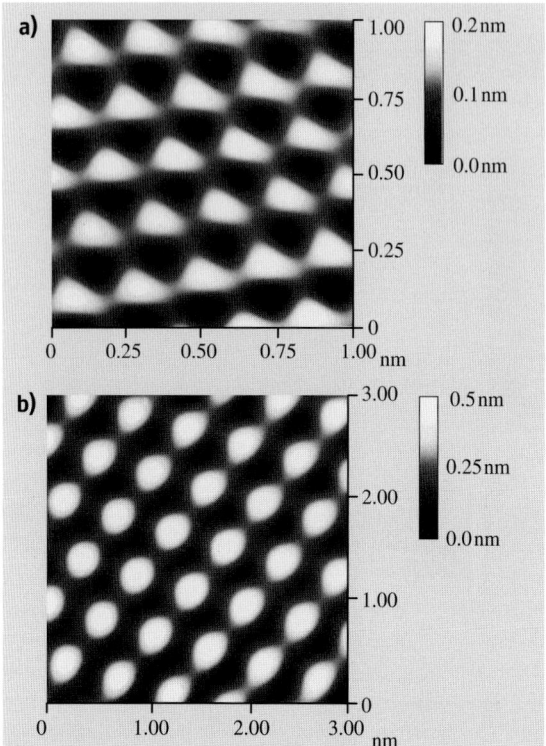

Fig. 2.13. Typical AFM images of freshly-cleaved (**a**) highly-oriented pyrolytic graphite and (**b**) mica surfaces taken using a square pyramidal Si_3N_4 tip

Multimode Capabilities

The multimode AFM can be used for topography measurements in the contact mode and tapping mode, described earlier, and for measurements of lateral (friction) force, electric force gradients and magnetic force gradients.

The multimode AFM, used with a grounded conducting tip, can measure electric field gradients by oscillating the tip near its resonant frequency. When the lever encounters a force gradient from the electric field, the effective spring constant of the cantilever is altered, changing its resonant frequency. Depending on which side of the resonance curve is chosen, the oscillation amplitude of the cantilever increases or decreases due to the shift in the resonant frequency. By recording the amplitude of the cantilever, an image revealing the strength of the electric field gradient is obtained.

In the magnetic force microscope (MFM) used with a magnetically-coated tip, static cantilever deflection is detected which occurs when a magnetic field exerts a force on the tip and the MFM images of magnetic materials can be produced. MFM sensitivity can be enhanced by oscillating the cantilever near its resonant frequency. When the tip encounters a magnetic force gradient, the effective spring constant, and

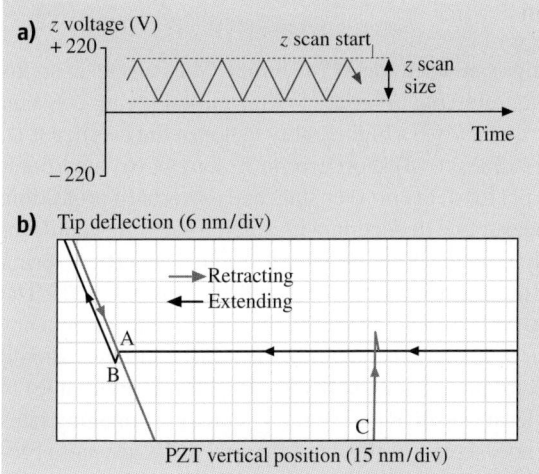

Fig. 2.14. (**a**) Force calibration Z waveform, and (**b**) a typical force-distance curve for a tip in contact with a sample. Contact occurs at point B; tip breaks free of adhesive forces at point C as the samples moves away from the tip

hence the resonant frequency, is shifted. By driving the cantilever above or below the resonant frequency, the oscillation amplitude varies as the resonance shifts. An image of magnetic field gradients is obtained by recording the oscillation amplitude as the tip is scanned over the sample.

Topographic information is separated from the electric field gradients and magnetic field images by using a so-called lift mode. Measurements in lift mode are taken in two passes over each scan line. On the first pass, topographical information is recorded in the standard tapping mode where the oscillating cantilever lightly taps the surface. On the second pass, the tip is lifted to a user-selected separation (typically 20–200 nm) between the tip and local surface topography. By using the stored topographical data instead of the standard feedback, the separation remains constant without sensing the surface. At this height, cantilever amplitudes are sensitive to electric field force gradients or relatively weak but long-range magnetic forces without being influenced by topographic features. Two-pass measurements are taken for every scan line, producing separate topographic and magnetic force images.

Electrochemical AFM

This option allows one to perform electrochemical reactions on the AFM. It includes a potentiostat, a fluid cell with a transparent cantilever holder and electrodes, and the software required to operate the potentiostat and display the results of the electrochemical reaction.

2.3.3 AFM Probe Construction

Various probes (cantilevers and tips) are used for AFM studies. The cantilever stylus used in the AFM should meet the following criteria: (1) low normal spring constant (stiffness), (2) a high resonant frequency, (3) a high quality factor of the cantilever Q, (4) high lateral spring constant (stiffness), (5) short cantilever length, (6) incorporation of components (such as mirror) for deflection sensing, and (7) a sharp protruding tip [137]. In order to register a measurable deflection with small forces, the cantilever must flex with a relativly low force (on the order of few nN) requiring vertical spring constants of 10^{-2} to 10^2 N/m for atomic resolution in the contact profiling mode. The data rate or imaging rate in the AFM is limited by the mechanical resonant frequency of the cantilever. To achieve a large imaging bandwidth, the AFM cantilever should have a resonant frequency greater than about 10 kHz (30–100 kHz is preferable) in order to make the cantilever the least sensitive part of the system. Fast imaging rates are not just a matter of convenience, since the effects of thermal drifts are more pronounced with slow-scanning speeds. The combined requirements of a low spring constant and a high resonant frequency is met by reducing the mass of the cantilever. The quality factor Q ($= \omega_R/(c/m)$, where ω_R is the resonant frequency of the damped oscillator and c is the damping constant and m is the mass of the oscillator) should have a high value for some applications. For example, resonance curve detection is a sensitive modulation technique for measuring small force gradients in noncontact imaging. Increasing the Q increases the sensitivity of the measurements. Mechanical Q values of 100–1000 are typical. In contact modes, the Q is of less importance. High lateral spring constant in the cantilever is desirable to reduce the effect of lateral forces in the AFM as frictional forces can cause appreciable lateral bending of the cantilever. Lateral bending results in error in the topography measurements. For friction measurements, cantilevers with less lateral rigidity is preferred. A sharp protruding tip must be formed at the end of the cantilever to provide a well defined interaction with sample over a small area. The tip radius should be much smaller than the radii of corrugations in the sample in order for these to be measured accurately. The lateral spring constant depends critically on the tip length. Additionally, the tip should be centered at the free end.

In the past, cantilevers have been cut by hand from thin metal foils or formed from fine wires. Tips for these cantilevers were prepared by attaching diamond fragments to the ends of the cantilevers by hand, or in the case of wire cantilevers, electrochemically etching the wire to a sharp point. Several cantilever geometries for wire cantilevers have been used. The simplest geometry is the L-shaped cantilever, usually made by bending a wire at a 90 °C angle. Other geometries include single-V and double-V geometries with a sharp tip attached at the apex of V, and double-X configuration with a sharp tip attached at the intersection [31, 138]. These cantilevers can be constructed with high vertical spring constants. For example, double-cross cantilever with an effective spring constant of 250 N/m was used by *Burnham* and *Colton* [31]. The small size and low mass needed in the AFM make hand fabrication of the cantilever a difficult process with poor reproducibility. Conventional microfabrication techniques are ideal for constructing planar thin-film structures which have submicron lateral dimensions. The triangular (V-shaped) cantilevers have improved (higher) lateral spring constant in comparison to rectangular cantilevers. In terms of spring constants, the triangular cantilevers are approximately equivalent to two rectangular cantilevers in parallel [137]. Although the macroscopic radius of a photolithographically patterned corner is seldom much less than about 50 nm, microscopic asperities on the etched surface provide tips with near atomic dimensions.

Cantilevers have been used from a whole range of materials. Most commonly are cantilevers made of Si_3N_4, Si, and diamond. Young's modulus and the density are the material parameters which determine the resonant frequency, besides the geometry. Table 2.2 shows the relevant properties and the speed of sound, indicative of the resonant frequency for a given shape. Hardness is important to judge the durability of the cantilevers, and is also listed in the table. Materials used for STM cantilevers are also included.

Table 2.2. Relevant properties of materials used for cantilevers

Property	Young's Modulus (E) (GPa)	Density (ρg) (kg/m^3)	Microhardness (GPa)	Speed of sound ($\sqrt{E/\rho}$) (m/s)
Diamond	900–1050	3515	78.4–102	17 000
Si_3N_4	310	3180	19.6	9900
Si	130–188	2330	9–10	8200
W	350	19 310	3.2	4250
Ir	530	–	~ 3	5300

Silicon nitride cantilevers are less expensive than those made of other materials. They are very rugged and well suited to imaging in almost all environments. They are especially compatible to organic and biological materials. Microfabricated silicon nitride triangular beams with integrated square pyramidal tips made of plasma-enhanced chemical vapor deposition (PECVD) are most commonly used [137]. Four cantilevers with different sizes and spring constants on each cantilever substrate made of boron silicate glass (Pyrex), marketed by Digital Instruments, are shown in Figs. 2.15a and 2.16. Two pairs of the cantilevers on each substrate measure about 115 and 193 μm from the substrate to the apex of the triangular cantilever with base widths of 122 and 205 μm, respectively. Both cantilever legs, with the same thickness (0.6 μm) of all the cantilevers, are available with wide and narrow legs. Only one cantilever is selected and used from each substrate. Calculated spring constant and measured natural frequencies for each of the configurations are listed in Table 2.3. The most commonly used cantilever beam is the 115-μm long, wide-legged cantilever (vertical spring constant = 0.58 N/m). Cantilevers with smaller spring constants should be used on softer samples. The pyramidal tips are highly symmetric with its end having a radius of about 20–50 nm. The tip side walls have a slope of 35 deg and the length of the edges of the tip at the cantilever base is about 4 μm.

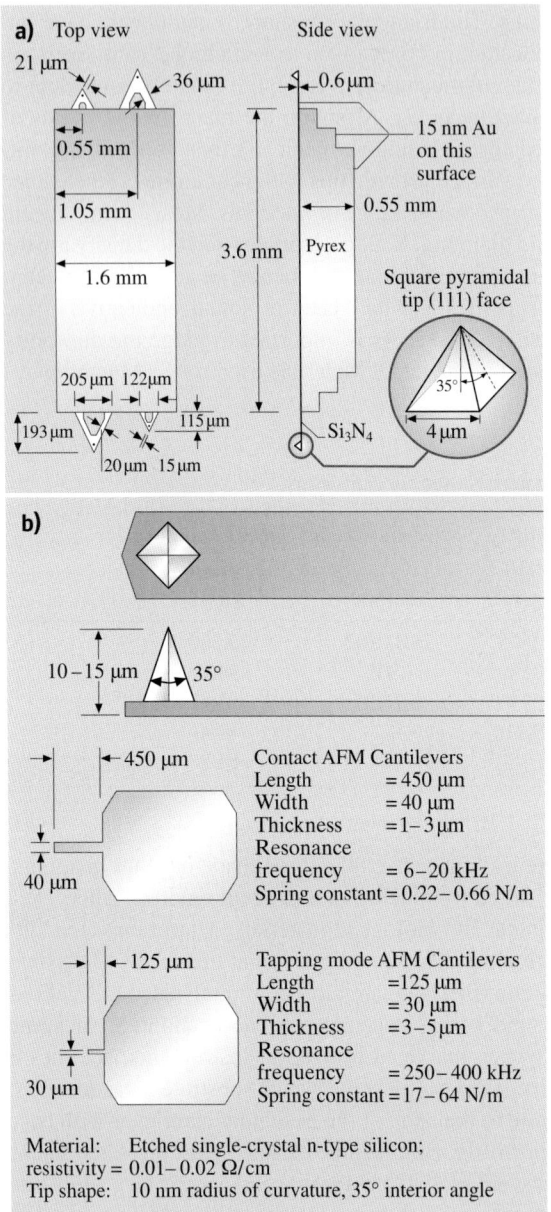

Fig. 2.15. Schematics of (**a**) triangular cantilever beam with square pyramidal tips made of PECVD Si_3N_4, (**b**) rectangular cantilever beams with square pyramidal tips made of etched single-crystal silicon, and (**c**) rectangular cantilever stainless steel beam with three-sided pyramidal natural diamond tip

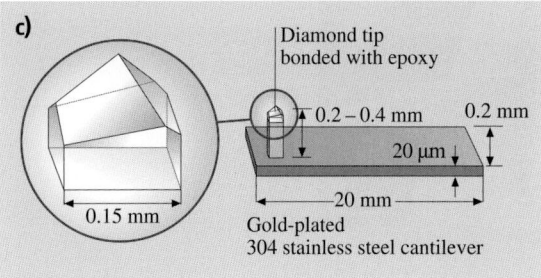

Fig. 2.15. continue

An alternative to silicon nitride cantilevers with integrated tips are microfabricated single-crystal silicon cantilevers with integrated tips. Si tips are sharper than Si_3N_4 tips because they are directly formed by the anisotropic etch in single-crystal Si rather than using an etch pit as a mask for deposited materials [139]. Etched single-crystal *n*-type silicon rectangular cantilevers with square pyramidal tips with a lower radius of less than 10 nm for contact and tapping mode (tapping-mode etched silicon probe or TESP) AFMs are commercially available from Digital Instruments and Nanosensors GmbH, Aidlingen, Germany, Figs. 2.15b and 2.16. Spring constants and resonant frequencies are also presented in the Fig. 2.15b.

Commercial triangular Si_3N_4 cantilevers have a typical width-thickness ratio of 10 to 30 which results in 100 to 1000 times stiffer spring constants in the lateral direction compared to the normal direction. Therefore these cantilevers are not well suited for torsion. For friction measurements, the torsional spring constant should be minimized in order to be sensitive to the lateral forces. Rather long cantilevers with small thickness and large tip length are most suitable. Rectangular beams have lower torsional spring constants in comparison to the triangular (V-shaped) cantilevers. Table 2.4 lists the spring constants (with full length of the beam used) in three directions of the typical rectangular beams. We note that lateral and torsional

Table 2.3. Measured vertical spring constants and natural frequencies of triangular (V-shaped) cantilevers made of PECVD Si_3N_4 (Data provided by Digital Instruments, Inc.)

Cantilever dimension	Spring constant (k_z)	Natural frequency (ω_0)
115-µm long, narrow leg	0.38	40
115-µm long, wide leg	0.58	40
193-µm long, narrow leg	0.06	13–22
193-µm long, wide leg	0.12	13–22

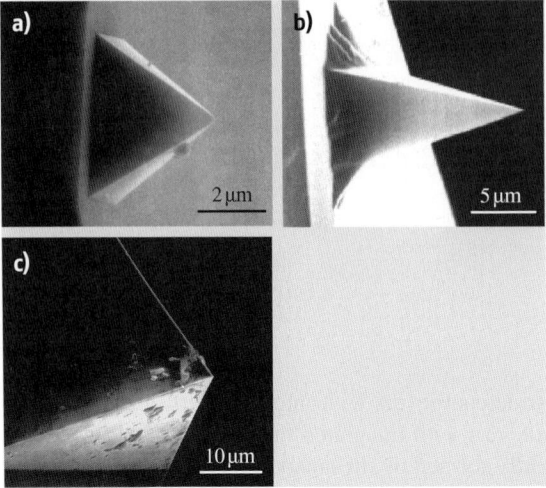

Fig. 2.16. SEM micrographs of a square-pyramidal PECVD Si_3N_4 tip (**a**), a square pyramidal etched single-crystal silicon tip (**b**), and a three-sided pyramidal natural diamond tip (**c**)

spring constants are about two orders of magnitude larger than the normal spring constants. A cantilever beam required for the tapping mode is quite stiff and may not be sensitive enough for friction measurements. *Meyer* et al. [140] used a specially designed rectangular silicon cantilever with length = 200 µm, width = 21 µm, thickness = 0.4 µm, tip length = 12.5 µm, and shear modulus = 50 GPa, giving a normal spring constant of 0.007 N/m and torsional spring constant of 0.72 N/m which gives a lateral force sensitivity of 10 pN and an angle of resolution of 10^{-7} rad. With this particular geometry, sensitivity to lateral forces could be improved by about a factor of 100 compared with commercial V-shaped Si_3N_4 or rectangular Si or Si_3N_4 cantilevers used by *Meyer* and *Amer* [8] with torsional spring constant of ~ 100 N/m. *Ruan* and *Bhushan* [136] and *Bhushan* and *Ruan* [9] used 115-µm long, wide-legged V-shaped cantilevers made of Si_3N_4 for friction measurements.

For scratching, wear, and indentation studies, single-crystal natural diamond tips ground to the shape of a three-sided pyramid with an apex angle of either 60 °C or 80 °C whose point is sharpened to a radius of about 100 nm are commonly used [4, 10] (Figs. 2.15c and 2.16). The tips are bonded with conductive epoxy to a gold-plated 304 stainless steel spring sheet (length = 20 mm, width = 0.2 mm, thickness = 20 to 60 µm) which acts as a cantilever. The free length of the spring is varied to change the beam stiffness. The normal spring constant of the beam ranges from about 5 to 600 N/m for a 20 µm thick beam. The tips are produced by R-DEC Co., Tsukuba, Japan.

For imaging within trenches by an AFM, high aspect ratio tips are used. Examples of the two probes are shown in Fig. 2.17. The high-aspect ratio tip (Hart) probes are produced by starting with a conventional Si_3N_4 pyramidal probe. Through a combination of focused ion beam (FIB) and high resolution scanning electron mi-

Table 2.4. Vertical (k_z), lateral (k_y), and torsional (k_{yT}) spring constants of rectangular cantilevers made of Si (IBM) and PECVD Si_3N_4 (Veeco Instruments, Inc.)

Dimensions/stiffness	Si cantilever	Si_3N_4 cantilever
Length (L) (μm)	100	100
Width (b) (μm)	10	20
Thickness (h) (μm)	1	0.6
Tip length (ℓ) (μm)	5	3
k_z (N/m)	0.4	0.15
k_y (N/m)	40	175
k_{yT} (N/m)	120	116
ω_0 (kHz)	~ 90	~ 65

Note: $k_z = Ebh^3/4L^3$, $k_y = Eb^3h/4\ell^3$, $k_{yT} = Gbh^3/3L\ell^2$, and $\omega_0 = [k_z/(m_c + 0.24bhL\rho)]^{1/2}$, where E is Young's modulus, G is the modulus of rigidity [= $E/2(1+\nu)$, ν is Poisson's ratio], ρ is the mass density of the cantilever, and m_c is the concentrated mass of the tip (~ 4 ng) [94]. For Si, $E = 130$ GPa, $\rho g = 2300$ kg/m^3, and $\nu = 0.3$. For Si_3N_4, $E = 150$ GPa, $\rho g = 3100$ kg/m^3, and $\nu = 0.3$

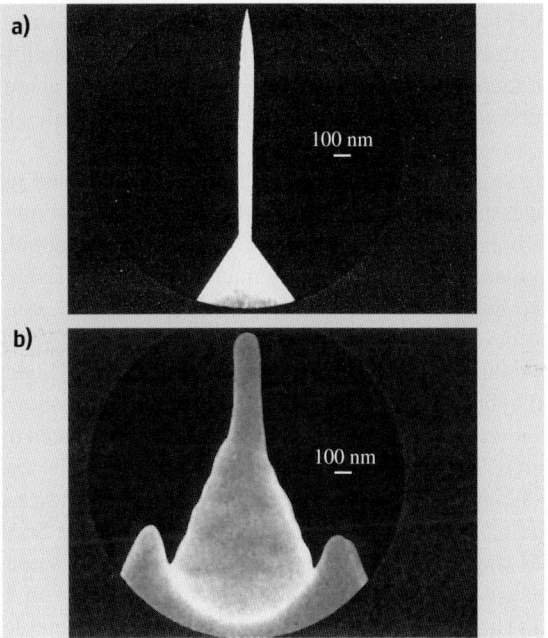

Fig. 2.17. Schematics of (**a**) HART Si_3N_4 probe, and (**b**) FIB milled Si_3N_4 probe

croscopy (SEM) techniques, a thin filament is grown at the apex of the pyramid. The probe filament is approximately 1 μm long and 0.1 μm in diameter. It tapers to an extremely sharp point (radius better than the resolution of most SEMs). The long thin

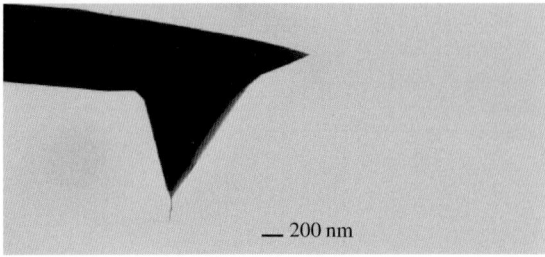

Fig. 2.18. SEM micrograph of a multi-walled carbon nanotube (MWNT) tip physically attached on the single-crystal silicon, square-pyramidal tip (Courtesy Piezomax Technologies, Inc.)

shape and sharp radius make it ideal for imaging within "vias" of microstructures and trenches (> 0.25 µm). Because of flexing of the probe, it is unsuitable for imaging structures at the atomic level since the flexing of the probe can create image artifacts. For atomic-scale imaging, a FIB-milled probe is used which is relatively stiff yet allows for closely spaced topography. These probes start out as conventional Si_3N_4 pyramidal probes but the pyramid is FIB milled until a small cone shape is formed which has a high aspect ratio with 0.2–0.3 µm in length. The milled probes allow nanostructure resolution without sacrificing rigidity. These types of probes are manufactured by various manufacturers including Materials Analytical Services, Raleigh, North Carolina.

Carbon nanotube tips having small diameter and high aspect ratio are used for high resolution imaging of surfaces and of deep trenches, in the tapping mode or non-contact mode. Single-walled carbon nanotubes (SWNT) are microscopic graphitic cylinders that are 0.7 to 3 nm in diameter and up to many microns in length. Larger structures called multi-walled carbon nanotubes (MWNT) consist of nested, concentrically arranged SWNT and have diameters ranging from 3 to 50 nm. MWNT carbon nanotube AFM tips are produced by manual assembly [141], chemical vapor deposition (CVD) synthesis, and hybrid fabrication process [142]. Figure 2.18 shows TEM micrograph of a carbon nanotube tip, ProbeMax^{TM}, commercially produced by mechanical assembly by Piezomax Technologies, Inc., Middleton, Wisconsin. For production of these tips, MWNT nanotubes are produced by carbon arc. They are physically attached on the single-crystal silicon, square-pyramidal tips in the SEM using a manipulator and the SEM stage to control the nanotubes and the tip independently. Once the nanotube is attached to the tip, it is usually too long to image with. It is shortened by using an AFM and applying voltage between the tip and the sample. Nanotube tips are also commercially produced by CVD synthesis by NanoDevices, Santa Barbara, California.

2.3.4 Friction Measurement Methods

Based on the work by *Ruan* and *Bhushan* [136], the two methods for friction measurements are now described in more detail. (Also see [8].) A scanning angle is

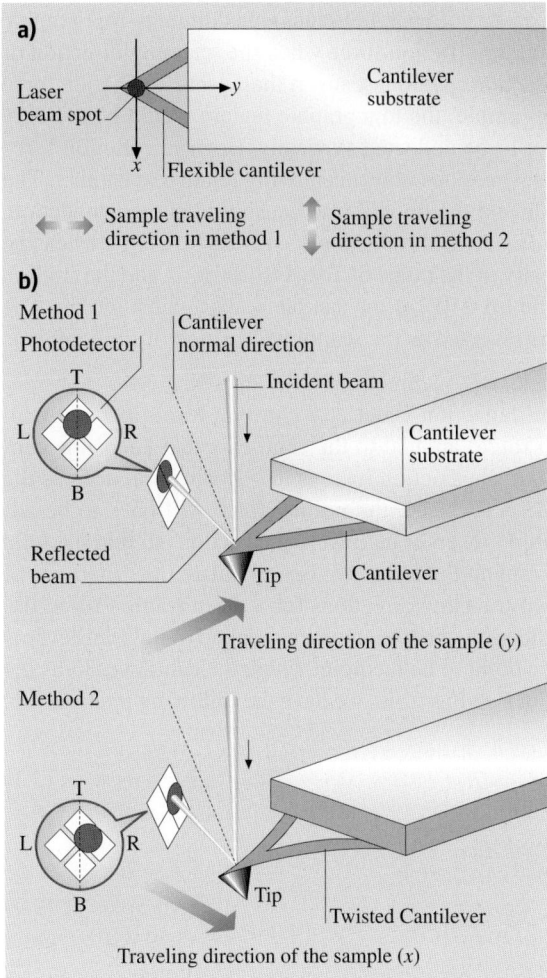

Fig. 2.19. (a) Schematic defining the x- and y-directions relative to the cantilever, and showing the sample traveling direction in two different measurement methods discussed in the text, (b) schematic of deformation of the tip and cantilever shown as a result of sliding in the x- and y-directions. A twist is introduced to the cantilever if the scanning is in the x-direction ((b), *lower part*) [136]

defined as the angle relative to the y-axis in Fig. 2.19a. This is also the long axis of the cantilever. A zero degree scanning angle corresponds to the sample scanning in the y direction, and a 90 degree scanning angle corresponds to the sample scanning perpendicular to this axis in the xy plane (along x axis). If the scanning direction is in both the y and $-y$ directions, we call this "parallel scan". Similarly, a "perpendicular scan" means the scanning direction is in the x and $-x$ directions. The sample traveling direction for each of these two methods is illustrated in Fig. 2.19b.

In method 1 (using "height" mode with parallel scans) in addition to topographic imaging, it is also possible to measure friction force when the scanning direction of the sample is parallel to the y direction (parallel scan). If there were no friction force between the tip and the moving sample, the topographic feature would be the only factor which cause the cantilever to be deflected vertically. However, friction force does exist on all contact surfaces where one object is moving relative to another. The friction force between the sample and the tip will also cause a cantilever deflection. We assume that the normal force between the sample and the tip is W_0 when the sample is stationary (W_0 is typically in the range of 10 nN to 200 nN), and the friction force between the sample and the tip is W_f as the sample scans against the tip. The direction of friction force (W_f) is reversed as the scanning direction of the sample is reversed from positive (y) to negative ($-y$) directions ($\vec{W}_{f(y)} = -\vec{W}_{f(-y)}$).

When the vertical cantilever deflection is set at a constant level, it is the total force (normal force and friction force) applied to the cantilever that keeps the cantilever deflection at this level. Since the friction force is in opposite directions as the traveling direction of the sample is reversed, the normal force will have to be adjusted accordingly when the sample reverses its traveling direction, so that the total deflection of the cantilever will remain the same. We can calculate the difference of the normal force between the two traveling directions for a given friction force W_f. First, by means of a constant deflection, the total moment applied to the cantilever is constant. If we take the reference point to be the point where the cantilever joins the cantilever holder (substrate), point P in Fig. 2.20, we have the following relationship:

$$(W_0 - \Delta W_1)L + W_f \ell = (W_0 + \Delta W_2)L - W_f \ell \tag{2.1}$$

or

$$(\Delta W_1 + \Delta W_2)L = 2W_f \ell. \tag{2.2}$$

Thus

$$W_f = (\Delta W_1 + \Delta W_2)L/(2\ell), \tag{2.3}$$

where ΔW_1 and ΔW_2 are the absolute value of the changes of normal force when the sample is traveling in $-y$ and y directions, respectively, as shown in Fig. 2.20; L is the length of the cantilever; ℓ is vertical distance between the end of the tip and point P. The coefficient of friction (μ) between the tip and the sample is then given as

$$\mu = \frac{W_f}{W_0} = \left(\frac{(\Delta W_1 + \Delta W_2)}{W_0}\right)\left(\frac{L}{2\ell}\right). \tag{2.4}$$

In all circumstances, there are adhesive and interatomic attractive forces between the cantilever tip and the sample. The adhesive force can be due to water from the capillary condensation and other contaminants present at the surface which form meniscus bridges [4, 143, 144] and the interatomic attractive force includes van der

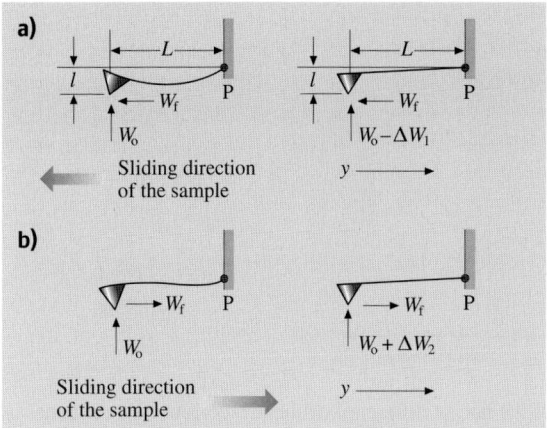

Fig. 2.20. (**a**) Schematic showing an additional bending of the cantilever – due to friction force when the sample is scanned in the *y*- or −*y*-direction (*left*). (**b**) This effect will be canceled by adjusting the piezo height by a feedback circuit (*right*) [136]

Waals attraction [18]. If these forces (and indentation effect as well, which is usually small for rigid samples) can be neglected, the normal force W_0 is then equal to the initial cantilever deflection H_0 multiplied by the spring constant of the cantilever. $(\Delta W_1 + \Delta W_2)$ can be measured by multiplying the same spring constant by the height difference of the piezo tube between the two traveling directions (*y* and −*y* directions) of the sample. This height difference is denoted as $(\Delta H_1 + \Delta H_2)$, shown schematically in Fig. 2.21. Thus, (2.4) can be rewritten as

$$\mu = \frac{W_f}{W_0} = \left(\frac{(\Delta H_1 + \Delta H_2)}{H_0}\right)\left(\frac{L}{2\ell}\right). \tag{2.5}$$

Since the piezo tube vertical position is affected by the surface topographic profile of the sample in addition to the friction force being applied at the tip, this difference has to be taken point by point at the same location on the sample surface as shown in Fig. 2.21. Subtraction of point by point measurements may introduce errors, particularly for rough samples. We will come back to this point later. In addition, precise measurement of L and ℓ (which should include the cantilever angle) are also required.

If the adhesive forces between the tip and the sample are large enough that it can not be neglected, one should include it in the calculation. However, there could be a large uncertainty in determining this force, thus an uncertainty in using (2.5). An alternative approach is to make the measurements at different normal loads and to use $\Delta(H_0)$ and $\Delta(\Delta H_1 + \Delta H_2)$ from the measurements in (2.5). Another comment on (2.5) is that, since only the ratio between $(\Delta H_1 + \Delta H_2)$ and H_0 comes into this equation, the piezo tube vertical position H_0 and its position difference $(\Delta H_1 + \Delta H_2)$ can be in the units of volts as long as the vertical traveling distance of the piezo tube and the voltage applied to it has a linear relationship. However, if there is a large

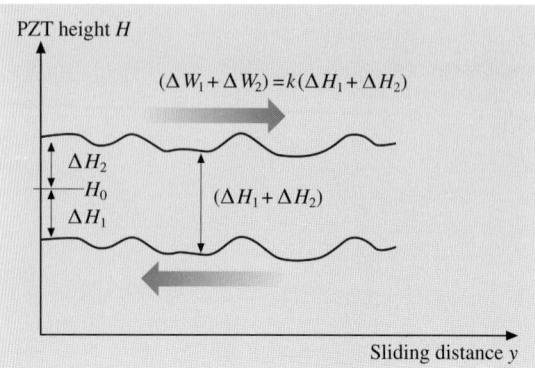

Fig. 2.21. Schematic illustration of the height difference of the piezoelectric tube scanner as the sample is scanned in y and $-y$ directions

nonlinearity between the piezo tube traveling distance and the applied voltage, this nonlinearity must be included in the calculation.

It should also be pointed out that (2.4) and (2.5) are derived under the assumption that the friction force W_f is the same for the two scanning directions of the sample. This is an approximation since the normal force is slightly different for the two scans and there may also be a directionality effect in friction. However, this difference is much smaller than W_0 itself. We can ignore the second order correction.

Method 2 ("aux" mode with perpendicular scan) to measure friction was suggested by *Meyer* and *Amer* [8]. The sample is scanned perpendicular to the long axis of the cantilever beam (i.e., to scan along the x or $-x$ direction in Fig. 2.19a) and the output of the horizontal two quadrants of the photodiode-detector is measured. In this arrangement, as the sample moves under the tip, the friction force will cause the cantilever to twist. Therefore the light intensity between the left and right (L and R in Fig. 2.19b, right) detectors will be different. The differential signal between the left and right detectors is denoted as FFM signal [(L − R)/(L + R)]. This signal can be related to the degree of twisting, hence to the magnitude of friction force. Again, because of a possible error in determining normal force due to the presence of an adhesive force at the tip–sample interface, the slope of the friction data (FFM signal vs. normal load) needs to be taken for an accurate value of the coefficient of friction.

While friction force contributes to the FFM signal, friction force may not be the only contributing factor in commercial FFM instruments (for example, NanoScope IV). One can notice this fact by simply engaging the cantilever tip with the sample. Before engaging, the left and right detectors can be balanced by adjusting the position of the detectors so that the intensity difference between these two detectors is zero (FFM signal is zero). Once the tip is engaged with the sample, this signal is no longer zero even if the sample is not moving in the xy plane with no friction force applied. This would be a detrimental effect. It has to be understood and eliminated from the data acquisition before any quantitative measurement of friction force becomes possible.

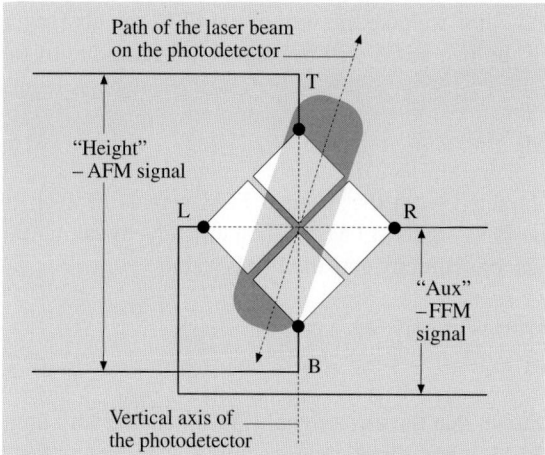

Fig. 2.22. The trajectory of the laser beam on the photodetectors in as the cantilever is vertically deflected (with no torsional motion) for a misaligned photodetector with respect to the laser beam. For a change of normal force (vertical deflection of the cantilever), the laser beam is projected at a different position on the detector. Due to a misalignment, the projected trajectory of the laser beam on the detector is not parallel with the detector vertical axis (the line joint T-B) [136]

One of the fundamental reasons for this observation is the following. The detectors may not have been properly aligned with respect to the laser beam. To be precise, the vertical axis of the detector assembly (the line joining T-B in Fig. 2.22) is not in the plane defined by the incident laser beam and the beam reflected from an untwisted cantilever (we call this plane the "beam plane"). When the cantilever vertical deflection changes due to a change of applied normal force (without having the sample scanned in the xy plane), the laser beam will be reflected up and down and form a projected trajectory on the detector. (Note that this trajectory is in the defined beam plane.) If this trajectory is not coincident with the vertical axis of the detector, the laser beam will not evenly bisect the left and right quadrants of the detectors, even under the condition of no torsional motion of the cantilever, see Fig. 2.22. Thus when the laser beam is reflected up and down due a change of the normal force, the intensity difference between the left and right detectors will also change. In other words, the FFM signal will change as the normal force applied to the tip is changed, even if the tip is not experiencing any friction force. This (FFM) signal is unrelated to friction force or to the actual twisting of the cantilever. We will call this part of FFM signal "FFM_F", and the part which is truly related to friction force "FFM_T".

The FFM_F signal can be eliminated. One way of doing this is as follows. First the sample is scanned in both x and $-x$ directions and the FFM signal for scans in each direction is recorded. Since friction force reverses its directions when the scanning direction is reversed from x to $-x$ direction, the FFM_T signal will have opposite signs as the scanning direction of the sample is reversed ($FFM_T(x) = -FFM_T(-x)$). Hence the FFM$_T$ signal will be canceled out if we take the sum of the FFM signals for the two scans. The average value of the two scans will be related to FFM$_F$ due to the misalignment,

$$\text{FFM}(x) + \text{FFM}(-x) = 2\text{FFM}_F . \tag{2.6}$$

This value can therefore be subtracted from the original FFM signals of each of these two scans to obtain the true FFM signal (FFM$_T$). Or, alternately, by taking the difference of the two FFM signals, one directly gets the FFM$_T$ value

$$\begin{aligned}\text{FFM}(x) - \text{FFM}(-x) &= \text{FFM}_T(x) - \text{FFM}_T(-x) \\ &= 2\text{FFM}_T(x) .\end{aligned} \tag{2.7}$$

Ruan and *Bhushan* [136] have shown that the error signal (FFM$_F$) can be very large compared to the friction signal FFM$_T$, thus correction is required.

Now we compare the two methods. The method of using "height" mode and parallel scan (method 1) is very simple to use. Technically, this method can provide 3-D friction profiles and the corresponding topographic profiles. However, there are some problems with this method. Under most circumstances, the piezo scanner displays a hysteresis when the traveling direction of the sample is reversed. Therefore the measured surface topographic profiles will be shifted relative to each other along the y-axis for the two opposite (y and $-y$) scans. This would make it difficult to measure the local height difference of the piezo tube for the two scans. However, the average height difference between the two scans and hence the average friction can still be measured. The measurement of average friction can serve as an internal means of friction force calibration. Method 2 is a more desirable approach. The subtraction of FFM$_F$ signal from FFM for the two scans does not introduce error to local friction force data. An ideal approach in using this method would be to add the average value of the two profiles in order to get the error component (FFM$_F$) and then subtract this component from either profiles to get true friction profiles in either directions. By making measurements at various loads, we can get the average value of the coefficient of friction which then can be used to convert the friction profile to the coefficient of friction profile. Thus any directionality and local variations in friction can be easily measured. In this method, since topography data are not affected by friction, accurate topography data can be measured simultaneously with friction data and a better localized relationship between the two can be established.

2.3.5 Normal Force and Friction Force Calibrations of Cantilever Beams

Based on *Ruan* and *Bhushan* [136], we now discuss normal force and friction force calibrations. In order to calculate the absolute value of normal and friction forces in Newtons using the measured AFM and FFM$_T$ voltage signals, it is necessary to first have an accurate value of the spring constant of the cantilever (k_c). The spring constant can be calculated using the geometry and the physical properties of the cantilever material [8, 94, 137]. However, the properties of the PECVD Si$_3$N$_4$ (used in fabricating cantilevers) could be different from those of the bulk material. For example, by using an ultrasonic measurement, we found the Young's modulus of the cantilever beam to be about 238 ± 18 GPa which is less than that of bulk Si_3N_4 (310 GPa). Furthermore the thickness of the beam is nonuniform and difficult to measure precisely. Since the stiffness of a beam goes as the cube of thickness, minor errors in precise measurements of thickness can introduce substantial stiffness errors. Thus one should experimentally measure the spring constant of the cantilever. *Cleveland* et al. [145] measured the normal spring constant by measuring resonant frequencies of the beams.

For normal spring constant measurement, *Ruan* and *Bhushan* [136] used a stainless steel spring sheet of known stiffness (width = 1.35 mm, thickness = 15 μm, free hanging length = 5.2 mm). One end of the spring was attached to the sample holder and the other end was made to contact with the cantilever tip during the measurement, see Fig. 2.23. They measured the piezo traveling distance for a given cantilever deflection. For a rigid sample (such as diamond), the piezo traveling distance Z_t (measured from the point where the tip touches the sample) should equal the cantilever deflection. To keep the cantilever deflection at the same level using a flexible spring sheet, the new piezo traveling distance $Z_{t'}$ would be different from Z_t. The difference between $Z_{t'}$ and Z_t corresponds to the deflection of the spring sheet. If the spring constant of the spring sheet is k_s, the spring constant of the cantilever k_c can be calculated by

$$(Z_{t'} - Z_t)k_s = Z_t k_c$$

or

$$k_c = k_s(Z_{t'} - Z_t)/Z_t. \tag{2.8}$$

The spring constant of the spring sheet (k_s) used in this study is calculated to be 1.54 N/m. For a wide-legged cantilever used in our study (length = 115 μm, base width = 122 μm, leg width = 21 μm, and thickness = 0.6 μm), k_c was measured to be 0.40 N/m instead of 0.58 N/m reported by its manufacturer – Digital Instruments Inc. To relate photodiode detector output to the cantilever deflection in nm, they used the same rigid sample to push against the AFM tip. Since for a rigid sample the cantilever vertical deflection equals the sample traveling distance measured from the point where the tip touches the sample, the photodiode output as the tip is pushed by the sample can be converted directly to cantilever deflection. For these measurements, they found the conversion factor to be 20 nm/V.

The normal force applied to the tip can be calculated by multiplying the cantilever vertical deflection by the cantilever spring constant for samples which have very small adhesive force with the tip. If the adhesive force between the sample and the tip is large, it should be included in the normal force calculation. This is particularly important in atomic-scale force measurement because in this region, the typical normal force that is measured is in the range of a few hundreds of nN to a few mN. The adhesive force could be comparable to the applied force.

The conversion of friction signal (from FFM_T to friction force) is not as straightforward. For example, one can calculate the degree of twisting for a given friction force using the geometry and the physical properties of the cantilever [53, 144]. One would need the information on the detectors such as the quantum efficiency of the detector, the laser power, the instrument's gain, etc. in order to be able convert the signal into the degree of twisting. Generally speaking, this procedure can not be accomplished without having some detailed information about the instrument. This information is not usually provided by the manufactures. Even if this information is readily available, error may still occur in using this approach because there will always be variations as a result of the instrumental set up. For example, it has been noticed that the measured FFM_T signal could be different for the same sample when different AFM microscopes of the same kind are used. The essence is that, one can not calibrate the instrument experimentally using this calculation. *O'Shea* et al. [144] did perform a calibration procedure in which the torsional signal was measured as the sample is displaced a known distance laterally while ensuring that the tip does not slide over the surface. However, it is difficult to verify that tip sliding does not occur.

Apparently, a new method of calibration is required. There is a more direct and simpler way of doing this. The first method described (method 1) to measure friction can directly provide an absolute value of the coefficient of friction. It can therefore be used just as an internal means of calibration for the data obtained using method 2. Or for a polished sample which introduces least error in friction measurement using method 1, method 1 can be used to obtain calibration for friction force for method 2. Then this calibration can be used for measurement on all samples using method 2. In method 1, the length of the cantilever required can be measured using an optical microscope; the length of the tip can be measured using a scanning electron microscope. The relative angle between the cantilever and the horizontal sample surface can be measured directly. Thus the coefficient of friction can be measured with few unknown parameters. The friction force can then be calculated by multiplying the coefficient of friction by the normal load. The FFM_T signal obtained using method 2 can then be converted into friction force. For their instrument, they found the conversion to be 8.6 nN/V.

2.4 AFM Instrumentation and Analyses

The performance of AFMs and the quality of AFM images greatly depend on the instrument available and the probes (cantilever and tips) in use. This section describes the mechanics of cantilevers, instrumentation and analysis of force detection systems for cantilever deflections, and scanning and control systems.

2.4.1 The Mechanics of Cantilevers

Stiffness and Resonances of Lumped Mass Systems

Any one of the building blocks of an AFM, be it the body of the microscope itself or the force measuring cantilevers, are mechanical resonators. These resonances can be excited either by the surroundings or by the rapid movement of the tip or the sample. To avoid problems due to building or air induced oscillations, it is of paramount importance to optimize the design of AFMs for high resonant frequencies. This usually means decreasing the size of the microscope [146]. By using cube-like or sphere-like structures for the microscope, one can considerably increase the lowest eigenfrequency. The fundamental natural frequency, ω_0, of any spring is given by

$$\omega_0 = \frac{1}{2\pi}\sqrt{\frac{k}{m_{\text{eff}}}}, \tag{2.9}$$

where k is the spring constant (stiffness) in the normal direction and m_{eff} is the effective mass. The spring constant k of a cantilever beam with uniform cross section (Fig. 2.24) is given by [147]

$$k = \frac{3EI}{L^3}, \tag{2.10}$$

where E is the Young's modulus of the material, L is the length of the beam and I is the moment of inertia of the cross section. For a rectangular cross section with a width b (perpendicular to the deflection) and a height h one obtains an expression for I

$$I = \frac{bh^3}{12}. \tag{2.11}$$

Combining (2.9), (2.10) and (2.11) we get an expression for ω_0

$$\omega_0 = \sqrt{\frac{Ebh^3}{4L^3 m_{\text{eff}}}}. \tag{2.12}$$

The effective mass can be calculated using Raleigh's method. The general formula using Raleigh's method for the kinetic energy T of a bar is

$$T = \frac{1}{2}\int_0^L \frac{m}{L}\left(\frac{\partial z(x)}{\partial t}\right)^2 dx. \tag{2.13}$$

For the case of a uniform beam with a constant cross section and length L one obtains for the deflection $z(x) = z_{\max}\left[1 - (3x/2L) + (x^3/2L^3)\right]$. Inserting $z_{\max}$ into (2.13) and solving the integral gives

$$T = \frac{1}{2}\int_0^L \frac{m}{L}\left[\frac{\partial z_{\max}(x)}{\partial t}\left(1 - \frac{3x}{2L}\right) + \left(\frac{x^3}{L^3}\right)\right]^2 dx$$

$$= \frac{1}{2}m_{\text{eff}}(z_{\max}t)^2,$$

which gives

$$m_{\text{eff}} = \frac{9}{20}m. \tag{2.14}$$

Substituting (2.14) into (2.12) and noting that $m = \rho Lbh$, where ρ is the mass density, one obtains the following expression

$$\omega_0 = \left(\frac{\sqrt{5}}{3}\right)\sqrt{\frac{E}{\rho}}\frac{h}{L^2}. \tag{2.15}$$

It is evident from (2.15), that one way to increase the natural frequency is to choose a material with a high ratio E/ρ; see Table 2.2 for typical values of $\sqrt{E/\rho}$ of various commonly used materials. Another way to increase the lowest eigenfrequency is also evident in (2.15). By optimizing the ratio h/L^2, one can increase the resonant frequency. However it does not help to make the length of the structure smaller than the width or height. Their roles will just be interchanged. Hence the optimum structure is a cube. This leads to the design rule, that long, thin structures like sheet metal should be avoided. For a given resonant frequency, the quality factor Q should be as low as possible. This means that an inelastic medium such as rubber should be in contact with the structure to convert kinetic energy into heat.

Stiffness and Resonances of Cantilevers

Cantilevers are mechanical devices specially shaped to measure tiny forces. The analysis given in the previous section is applicable. However, to understand better the intricacies of force detection systems we will discuss the example of a cantilever beam with uniform cross section, Fig. 2.24. The bending of a beam due to a normal load on the beam is governed by the Euler equation [147]

$$M = EI(x)\frac{d2z}{dx^2}, \tag{2.16}$$

where M is the bending moment acting on the beam cross section. $I(x)$ the moment of inertia of the cross section with respect to the neutral axis defined by

$$I(x) = \int_z \int_y z^2 dy dz. \tag{2.17}$$

For a normal force F_z acting at the tip,

$$M(x) = (L-x)F_z \tag{2.18}$$

since the moment must vanish at the endpoint of the cantilever. Integrating (2.16) for a normal force F_z acting at the tip and observing that EI is a constant for beams with a uniform cross section, one gets

$$z(x) = \frac{L^3}{6EI}\left(\frac{x}{L}\right)^2\left(3-\frac{x}{L}\right)F_z. \tag{2.19}$$

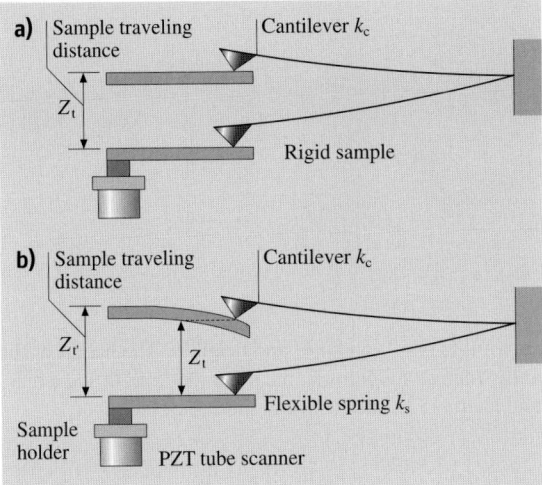

Fig. 2.23. Illustration showing the deflection of cantilever as it is pushed by (**a**) a rigid sample or by (**b**) a flexible spring sheet [136]

The slope of the of the beam is

$$z'(x) = \frac{Lx}{2EI}\left(2 - \frac{x}{L}\right) F_z. \tag{2.20}$$

From (2.19) and (2.20), at the end of the cantilever, i.e. for $x = L$, for a rectangular beam, and by using an expression for I in (2.11), one gets,

$$z(L) = \frac{4}{Eb}\left(\frac{L}{h}\right)^3 F_z, \tag{2.21}$$

$$z'(L) = \frac{3}{2}\left(\frac{z}{L}\right). \tag{2.22}$$

Now the stiffness in the normal (z) direction, k_z, is

$$k_z = \frac{F_z}{z(L)} = \frac{Eb}{4}\left(\frac{h}{L}\right)^3. \tag{2.23}$$

and a change in angular orientation of the end of cantilever beam is

$$\Delta\alpha = \frac{3}{2}\frac{z}{L} = \frac{6}{Ebh}\left(\frac{L}{h}\right)^2 F_z. \tag{2.24}$$

Now we ask what will, to first order, happen if we apply a lateral force F_y to the end of the tip (Fig. 2.24). The cantilever will bend sideways and it will twist. The stiffness in the lateral (y) direction, k_y, can be calculated with (2.23) by exchanging b and h

$$k_y = \frac{Eh}{4}\left(\frac{b}{L}\right)^3. \tag{2.25}$$

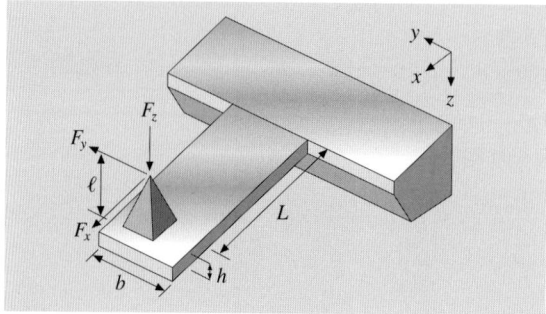

Fig. 2.24. A typical AFM cantilever with length L, width b, and height h. The height of the tip is ℓ. The material is characterized by Young's modulus E, the shear modulus G and a mass density ρ. Normal (F_z), axial (F_x), and lateral (F_y) forces exist at the end of the tip

Therefore the bending stiffness in lateral direction is larger than the stiffness for bending in the normal direction by $(b/h)^2$. The twisting or torsion on the other hand is more complicated to handle. For a wide, thin cantilever ($b \gg h$) we obtain torsional stiffness along y axis, k_{yT}

$$k_{yT} = \frac{Gbh^3}{3L\ell^2}, \tag{2.26}$$

where G is the modulus of rigidity ($= E/2(1+\nu)$, where ν is the Poisson's ratio). The ratio of the torsional stiffness to the lateral bending stiffness is

$$\frac{k_{yT}}{k_y} = \frac{1}{2}\left(\frac{\ell b}{hL}\right)^2, \tag{2.27}$$

where we assume $\nu = 0.333$. We see that thin, wide cantilevers with long tips favor torsion while cantilevers with square cross sections and short tips favor bending. Finally we calculate the ratio between the torsional stiffness and the normal bending stiffness,

$$\frac{k_{yT}}{k_z} = 2\left(\frac{L}{\ell}\right)^2. \tag{2.28}$$

Equations (2.26) to (2.28) hold in the case where the cantilever tip is exactly in the middle axis of the cantilever. Triangular cantilevers and cantilevers with tips not on the middle axis can be dealt with by finite element methods.

The third possible deflection mode is the one from the force on the end of the tip along the cantilever axis, F_x (Fig. 2.24). The bending moment at the free end of the cantilever is equal to the $F_x \ell$. This leads to the following modification of (2.18) for the case of forces F_z and F_x

$$M(x) = (L-x)F_z + F_x \ell. \tag{2.29}$$

Integration of (2.16) now leads to

$$z(x) = \frac{1}{2EI}\left[Lx^2\left(1 - \frac{x}{3L}\right)F_z + \ell x^2 F_x\right] \tag{2.30}$$

and

$$z'(x) = \frac{1}{EI}\left[\frac{Lx}{2}\left(2 - \frac{x}{L}\right)F_z + \ell x F_x\right]. \tag{2.31}$$

Evaluating (2.30) and (2.31) at the end of the cantilever, we get the deflection and the tilt

$$z(L) = \frac{L^2}{EI}\left(\frac{L}{3}F_z - \frac{\ell}{2}F_x\right),$$

$$z'(L) = \frac{L}{EI}\left(\frac{L}{2}F_z + \ell F_x\right). \tag{2.32}$$

From these equations, one gets

$$F_z = \frac{12EI}{L^3}\left[z(L) - \frac{Lz'(L)}{2}\right],$$

$$F_x = \frac{2EI}{\ell L^2}[2Lz'(L) - 3z(L)]. \tag{2.33}$$

A second class of interesting properties of cantilevers is their resonance behavior. For cantilever beams one can calculate the resonant frequencies [147, 148]

$$\omega_n^{\text{free}} = \frac{\lambda_n^2}{2\sqrt{3}}\frac{h}{L^2}\sqrt{\frac{E}{\rho}} \tag{2.34}$$

with $\lambda_0 = (0.596864\ldots)\pi$, $\lambda_1 = (1.494175\ldots)\pi$, $\lambda_n \to (n + 1/2)\pi$. The subscript n represents the order of the frequency, e.g., fundamental, second mode, and the nth mode.

A similar equation to (2.34) holds for cantilevers in rigid contact with the surface. Since there is an additional restriction on the movement of the cantilever, namely the location of its end point, the resonant frequency increases. Only the λ_n's terms change to [148]

$$\lambda_0' = (1.2498763\ldots)\pi, \quad \lambda_1' = (2.2499997\ldots)\pi,$$
$$\lambda_n' \to (n + 1/4)\pi. \tag{2.35}$$

The ratio of the fundamental resonant frequency in contact to the fundamental resonant frequency not in contact is 4.3851.

For the torsional mode we can calculate the resonant frequencies as

$$\omega_0^{\text{tors}} = 2\pi\frac{h}{Lb}\sqrt{\frac{G}{\rho}}. \tag{2.36}$$

For cantilevers in rigid contact with the surface, we obtain the expression for the fundamental resonant frequency [148]

$$\omega_0^{\text{tors, contact}} = \frac{\omega_0^{\text{tors}}}{\sqrt{1 + 3(2L/b)^2}}. \tag{2.37}$$

The amplitude of the thermally induced vibration can be calculated from the resonant frequency using

$$\Delta z_{\text{therm}} = \sqrt{\frac{k_B T}{k}}, \tag{2.38}$$

where k_B is Boltzmann's constant and T is the absolute temperature. Since AFM cantilevers are resonant structures, sometimes with rather high Q, the thermal noise is not evenly distributed as (2.38) suggests. The spectral noise density below the peak of the response curve is [148]

$$z_0 = \sqrt{\frac{4k_B T}{k\omega_0 Q}} \quad (\text{in m}/\sqrt{\text{Hz}}), \tag{2.39}$$

where Q is the quality factor of the cantilever, described earlier.

2.4.2 Instrumentation and Analyses of Detection Systems for Cantilever Deflections

A summary of selected detection systems was provided in Fig. 2.8. Here we discuss in detail pros and cons of various systems.

Optical Interferometer Detection Systems

Soon after the first papers on the AFM [2] appeared, which used a tunneling sensor, an instrument based on an interferometer was published [149]. The sensitivity of the interferometer depends on the wavelength of the light employed in the apparatus. Figure 2.25 shows the principle of such an interferometeric design. The light incident from the left is focused by a lens on the cantilever. The reflected light is collimated by the same lens and interferes with the light reflected at the flat. To separate the reflected light from the incident light a $\lambda/4$ plate converts the linear polarized incident light to circular polarization. The reflected light is made linear polarized again by the $\lambda/4$-plate, but with a polarization orthogonal to that of the incident light. The polarizing beam splitter then deflects the reflected light to the photo diode.

Homodyne Interferometer To improve the signal to noise ratio of the interferometer the cantilever is driven by a piezo near its resonant frequency. The amplitude Δz of the cantilever as a function of driving frequency Ω is

$$\Delta z(\Omega) = \Delta z_0 \frac{\Omega_0^2}{\sqrt{\left(\Omega^2 - \Omega_0^2\right)^2 + \frac{\Omega^2 \Omega_0^2}{Q^2}}}, \tag{2.40}$$

where Δz_0 is the constant drive amplitude and Ω_0 the resonant frequency of the cantilever. The resonant frequency of the cantilever is given by the effective potential

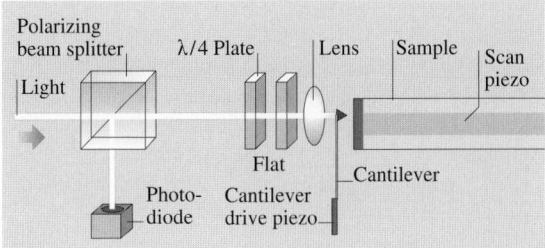

Fig. 2.25. Principle of an interferometric AFM. The light of the laser light source is polarized by the polarizing beam splitter and focused on the back of the cantilever. The light passes twice through a quarter wave plate and is hence orthogonally polarized to the incident light. The second arm of the interferometer is formed by the flat. The interference pattern is modulated by the oscillating cantilever

$$\Omega_0 = \sqrt{\left(k + \frac{\partial^2 U}{\partial z^2}\right)\frac{1}{m_{\text{eff}}}}, \tag{2.41}$$

where U is the interaction potential between the tip and the sample. Equation (2.41) shows that an attractive potential decreases Ω_0. The change in Ω_0 in turn results in a change of the Δz (see (2.40)). The movement of the cantilever changes the path difference in the interferometer. The light reflected from the cantilever with the amplitude $A_{\ell,0}$ and the reference light with the amplitude $A_{r,0}$ interfere on the detector. The detected intensity $I(t) = [A_\ell(t) + A_r(t)]^2$ consists of two constant terms and a fluctuating term

$$2A_\ell(t) A_r(t) = \\ A_{\ell,0} A_{r,0} \sin\left[\omega t + \frac{4\pi\delta}{\lambda} + \frac{4\pi\Delta z}{\lambda}\sin(\Omega t)\right]\sin(\omega t). \tag{2.42}$$

Here ω is the frequency of the light, λ is the wavelength of the light, δ is the path difference in the interferometer, and Δz is the instantaneous amplitude of the cantilever, given according to (2.40) and (2.41) as a function of Ω, k, and U. The time average of (2.42) then becomes

$$\langle 2A_\ell(t) A_r(t)\rangle_T \propto \cos\left[\frac{4\pi\delta}{\lambda} + \frac{4\pi\Delta z}{\lambda}\sin(\Omega t)\right]$$

$$\approx \cos\left(\frac{4\pi\delta}{\lambda}\right) - \sin\left[\frac{4\pi\Delta z}{\lambda}\sin(\Omega t)\right]$$

$$\approx \cos\left(\frac{4\pi\delta}{\lambda}\right) - \frac{4\pi\Delta z}{\lambda}\sin(\Omega t). \tag{2.43}$$

Here all small quantities have been omitted and functions with small arguments have been linearized. The amplitude of Δz can be recovered with a lock-in technique.

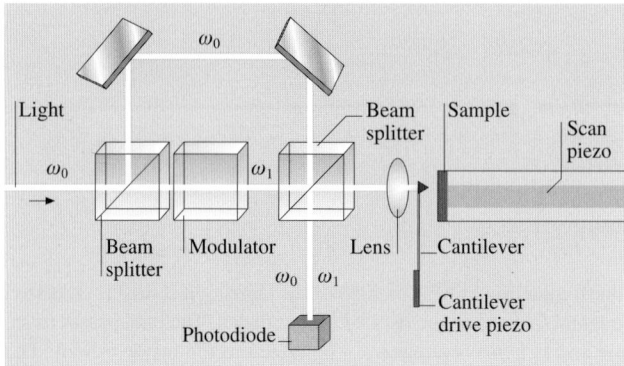

Fig. 2.26. Principle of a heterodyne interferometric AFM. Light with the frequency ω_0 is split into a reference path (upper path) and a measurement path. The light in the measurement path is frequency shifted to ω_1 by an acousto-optical modulator (or an electro-optical modulator) The light reflected from the oscillating cantilever interferes with the reference beam on the detector

However, (2.43) shows that the measured amplitude is also a function of the path difference δ in the interferometer. Hence this path difference δ must be very stable. The best sensitivity is obtained when $\sin(4\delta/\lambda) \approx 0$.

Heterodyne Interferometer This influence is not present in the heterodyne detection scheme shown in Fig. 2.26. Light incident from the left with a frequency ω is split in a reference path (upper path in Fig. 2.26) and a measurement path. Light in the measurement path is shifted in frequency to $\omega_1 = \omega + \Delta\omega$ and focused on the cantilever. The cantilever oscillates at the frequency Ω, as in the homodyne detection scheme. The reflected light $A_\ell(t)$ is collimated by the same lens and interferes on the photo diode with the reference light $A_r(t)$. The fluctuating term of the intensity is given by

$$2A_\ell(t) A_r(t) = A_{\ell,0} A_{r,0} \sin\left[(\omega + \Delta\omega)t + \frac{4\pi\delta}{\lambda} + \frac{4\pi\Delta z}{\lambda}\sin(\Omega t)\right]\sin(\omega t), \quad (2.44)$$

where the variables are defined as in (2.42). Setting the path difference $\sin(4\pi\delta/\lambda) \approx 0$ and taking the time average, omitting small quantities and linearizing functions with small arguments we get

$\langle 2A_\ell(t) A_r(t) \rangle_T$

$$\propto \cos\left[\Delta\omega t + \frac{4\pi\delta}{\lambda} + \frac{4\pi\Delta z}{\lambda}\sin(\Omega t)\right]$$

$$= \cos\left(\Delta\omega t + \frac{4\pi\delta}{\lambda}\right)\cos\left[\frac{4\pi\Delta z}{\lambda}\sin(\Omega t)\right]$$

$$- \sin\left(\Delta\omega t + \frac{4\pi\delta}{\lambda}\right)\sin\left[\frac{4\pi\Delta z}{\lambda}\sin(\Omega t)\right]$$

$$\approx \cos\left(\frac{4\pi\delta}{\lambda}\right) - \sin\left[\frac{4\pi\Delta z}{\lambda}\sin(\Omega t)\right]$$

$$\approx \cos\left(\Delta\omega t + \frac{4\pi\delta}{\lambda}\right)\left[1 - \frac{8\pi^2\Delta z^2}{\lambda^2}\sin(\Omega t)\right]$$

$$- \frac{4\pi\Delta z}{\lambda}\sin\left(\Delta\omega t + \frac{4\pi\delta}{\lambda}\right)\sin(\Omega t)$$

$$= \cos\left(\Delta\omega t + \frac{4\pi\delta}{\lambda}\right) - \frac{8\pi^2\Delta z^2}{\lambda^2}\cos\left(\Delta\omega t + \frac{4\pi\delta}{\lambda}\right)$$

$$\times \sin(\Omega t) - \frac{4\pi\Delta z}{\lambda}\sin\left(\Delta\omega t + \frac{4\pi\delta}{\lambda}\right)\sin(\Omega t)$$

$$= \cos\left(\Delta\omega t + \frac{4\pi\delta}{\lambda}\right) - \frac{4\pi^2\Delta z^2}{\lambda^2}\cos\left(\Delta\omega t + \frac{4\pi\delta}{\lambda}\right)$$

$$+ \frac{4\pi^2\Delta z^2}{\lambda^2}\cos\left(\Delta\omega t + \frac{4\pi\delta}{\lambda}\right)\cos(2\Omega t)$$

$$- \frac{4\pi\Delta z}{\lambda}\sin\left(\Delta\omega t + \frac{4\pi\delta}{\lambda}\right)\sin(\Omega t)$$

$$= \cos\left(\Delta\omega t + \frac{4\pi\delta}{\lambda}\right)\left(1 - \frac{4\pi^2\Delta z^2}{\lambda^2}\right)$$

$$+ \frac{2\pi^2\Delta z^2}{\lambda^2}\left\{\cos\left[(\Delta\omega + 2\Omega)t + \frac{4\pi\delta}{\lambda}\right]\right.$$

$$\left. + \cos\left[(\Delta\omega - 2\Omega)t + \frac{4\pi\delta}{\lambda}\right]\right\}$$

$$+ \frac{2\pi\Delta z}{\lambda}\left\{\cos\left[(\Delta\omega + \Omega)t + \frac{4\pi\delta}{\lambda}\right]\right.$$

$$\left. + \cos\left[(\Delta\omega - \Omega)t + \frac{4\pi\delta}{\lambda}\right]\right\}. \qquad (2.45)$$

Multiplying electronically the components oscillating at $\Delta\omega$ and $\Delta\omega+\Omega$ and rejecting any product except the one oscillating at Ω we obtain

$$\begin{aligned}
A &= \frac{2\Delta z}{\lambda}\left(1 - \frac{4\pi^2\Delta z^2}{\lambda^2}\right)\cos\left[(\Delta\omega + 2\Omega)t + \frac{4\pi\delta}{\lambda}\right] \\
&\quad \times \cos\left(\Delta\omega t + \frac{4\pi\delta}{\lambda}\right) \\
&= \frac{\Delta z}{\lambda}\left(1 - \frac{4\pi^2\Delta z^2}{\lambda^2}\right)\left\{\cos\left[(2\Delta\omega + \Omega)t + \frac{8\pi\delta}{\lambda}\right]\right. \\
&\quad \left. + \cos(\Omega t)\right\} \\
&\approx \frac{\pi\Delta z}{\lambda}\cos(\Omega t)\,.
\end{aligned} \qquad (2.46)$$

Unlike in the homodyne detection scheme the recovered signal is independent from the path difference δ of the interferometer. Furthermore a lock-in amplifier with the reference set $\sin(\Delta\omega t)$ can measure the path difference δ independent of the cantilever oscillation. If necessary, a feedback circuit can keep $\delta = 0$.

Fiber-optical Interferometer The fiber-optical interferometer [129] is one of the simplest interferometers to build and use. Its principle is sketched in Fig. 2.27. The light of a laser is fed into an optical fiber. Laser diodes with integrated fiber pigtails are convenient light sources. The light is split in a fiber-optic beam splitter into two fibers. One fiber is terminated by index matching oil to avoid any reflections back into the fiber. The end of the other fiber is brought close to the cantilever in the AFM. The emerging light is partially reflected back into the fiber by the cantilever. Most of the light, however, is lost. This is not a big problem since only 4% of the light is reflected at the end of the fiber, at the glass-air interface. The two reflected light waves interfere with each other. The product is guided back into the fiber coupler and again split into two parts. One half is analyzed by the photodiode. The other half is fed back into the laser. Communications grade laser diodes are sufficiently resistant against feedback to be operated in this environment. They have, however, a bad coherence length, which in this case does not matter, since the optical path difference is in any case no larger than 5 µm. Again the end of the fiber has to be positioned on a piezo drive to set the distance between the fiber and the cantilever to $\lambda(n + 1/4)$.

Nomarski-Interferometer Another solution to minimize the optical path difference is to use the Nomarski interferometer [130]. Figure 2.28 shows a schematic of the microscope. The light of a laser is focused on the cantilever by lens. A birefringent crystal (for instance calcite) between the cantilever and the lens with its optical axis 45 °C off the polarization direction of the light splits the light beam into two paths, offset by a distance given by the length of the crystal. Birefringent crystals have varying indexes of refraction. In calcite, one crystal axis has a lower index than the other two. This means, that certain light rays will propagate at a different speed

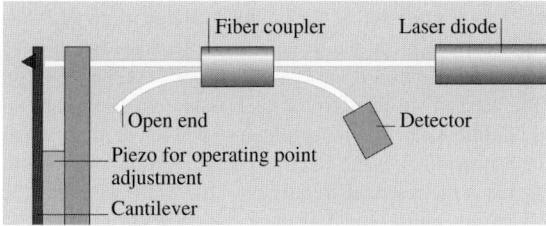

Fig. 2.27. A typical setup for a fiber optic interferometer readout

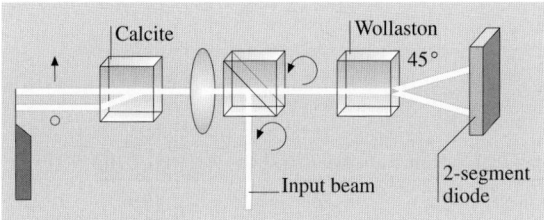

Fig. 2.28. Principle of Nomarski AFM. The circular polarized input beam is deflected to the left by a non-polarizing beam splitter. The light is focused onto a cantilever. The calcite crystal between the lens and the cantilever splits the circular polarized light into two spatially separated beams with orthogonal polarizations. The two light beams reflected from the lever are superimposed by the calcite crystal and collected by the lens. The resulting beam is again circular polarized. A Wollaston prism produces two interfering beams with $\pi/2$ phase shift between them. The minimal path difference accounts for the excellent stability of this microscope

through the crystal than the others. By choosing a correct polarization, one can select the ordinary ray, the extraordinary ray or one can get any distribution of the intensity amongst those two rays. A detailed description of birefringence can be found in textbooks (e.g., [150]). A calcite crystal deflects the extraordinary ray at an angle of 6 °C within the crystal. By choosing a suitable length of the calcite crystal, any separation can be set.

The focus of one light ray is positioned near the free end of the cantilever while the other is placed close to the clamped end. Both arms of the interferometer pass through the same space, except for the distance between the calcite crystal and the lever. The closer the calcite crystal is placed to the lever, the less influence disturbances like air currents have.

Sarid [116] has given values for the sensitivity of the different interferometeric detection systems. Table 2.5 presents a summary of his results.

Optical Lever

The most common cantilever deflection detection system is the optical lever [53, 111]. This method, depicted in Fig. 2.29, employs the same technique as light beam

Table 2.5. Noise in Interferometers. F is the finesse of the cavity in the homodyne interferometer, P_i the incident power, P_d is the power on the detector, η is the sensitivity of the photodetector and RIN is the relative intensity noise of the laser. P_R and P_S are the power in the reference and sample beam in the heterodyne interferometer. P is the power in the Nomarski interferometer, $\delta\theta$ is the phase difference between the reference and the probe beam in the Nomarski interferometer. B is the bandwidth, e is the electron charge, λ is the wavelength of the laser, k the cantilever stiffness, ω_0 is the resonant frequency of the cantilever, Q is the quality factor of the cantilever, T is the temperature, and δi is the variation of current i

	Homodyne interferometer, fiber optic interferometer	Heterodyne interferometer	Nomarski interferometer
Laser noise $\langle \delta i^2 \rangle_L$	$\dfrac{1}{4}\eta^2 F^2 P_i^2$ RIN	$\eta^2 \left(P_R^2 + P_S^2\right)$ RIN	$\dfrac{1}{16}\eta^2 P^2 \delta\theta$
Thermal noise $\langle \delta i^2 \rangle_T$	$\dfrac{16\pi^2}{\lambda^2}\eta^2 F^2 P_i^2 \dfrac{4 k_B T B Q}{\omega_0 k}$	$\dfrac{4\pi^2}{\lambda^2}\eta^2 P_d^2 \dfrac{4 k_B T B Q}{\omega_0 k}$	$\dfrac{\pi^2}{\lambda^2}\eta^2 P^2 \dfrac{4 k_B T B Q}{\omega_0 k}$
Shot Noise $\langle \delta i^2 \rangle_S$	$4 e \eta P_d B$	$2 e \eta \left(P_R + P_S\right) B$	$\dfrac{1}{2} e \eta P B$

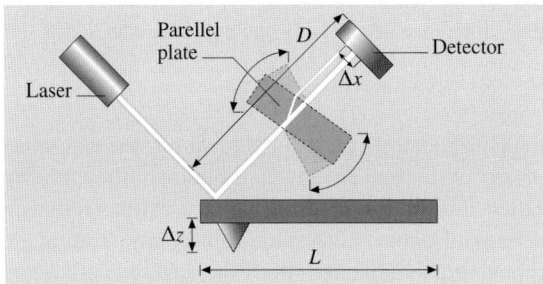

Fig. 2.29. The setup of optical lever detection microscope

deflection galvanometers. A fairly well collimated light beam is reflected off a mirror and projected to a receiving target. Any change in the angular position of the mirror will change the position, where the light ray hits the target. Galvanometers use optical path lengths of several meters and scales projected to the target wall as a read-out help.

For the AFM using the optical lever method a photodiode segmented into two (or four) closely spaced devices detects the orientation of the end of the cantilever. Initially, the light ray is set to hit the photodiodes in the middle of the two sub-diodes. Any deflection of the cantilever will cause an imbalance of the number of photons reaching the two halves. Hence the electrical currents in the photodiodes will be unbalanced too. The difference signal is further amplified and is the input signal to the feedback loop. Unlike the interferometric AFMs, where often a modulation technique is necessary to get a sufficient signal to noise ratio, most AFMs employing the optical lever method are operated in a static mode. AFMs based on the optical lever method are universally used. It is the simplest method for constructing an optical readout and it can be confined in volumes smaller than 5 cm on the side.

The optical lever detection system is a simple yet elegant way to detect normal and lateral force signals simultaneously [7, 8, 53, 111]. It has the additional advantage that it is the fact that it is a remote detection system.

Implementations Light from a laser diode or from a super luminescent diode is focused on the end of the cantilever. The reflected light is directed onto a quadrant diode that measures the direction of the light beam. A Gaussian light beam far from its waist is characterized by an opening angle β. The deflection of the light beam by the cantilever surface tilted by an angle α is 2α. The intensity on the detector then shifts to the side by the product of 2α and the separation between the detector and the cantilever. The readout electronics calculates the difference of the photocurrents. The photocurrents in turn are proportional to the intensity incident on the diode.

The output signal is hence proportional to the change in intensity on the segments

$$I_{sig} \propto 4\frac{\alpha}{\beta} I_{tot}. \tag{2.47}$$

For the sake of simplicity, we assume that the light beam is of uniform intensity with its cross section increasing proportional to the distance between the cantilever and the quadrant detector. The movement of the center of the light beam is then given by

$$\Delta x_{Det} = \Delta z \frac{D}{L}. \tag{2.48}$$

The photocurrent generated in a photodiode is proportional to the number of incoming photons hitting it. If the light beam contains a total number of N_0 photons then the change in difference current becomes

$$\Delta (I_R - I_L) = \Delta I = \text{const } \Delta z \, D \, N_0. \tag{2.49}$$

Combining (2.48) and (2.49), one obtains that the difference current ΔI is independent of the separation of the quadrant detector and the cantilever. This relation is true, if the light spot is smaller than the quadrant detector. If it is greater, the difference current ΔI becomes smaller with increasing distance. In reality the light beam has a Gaussian intensity profile. For small movements Δx (compared to the diameter of the light spot at the quadrant detector), (2.49) still holds. Larger movements Δx, however, will introduce a nonlinear response. If the AFM is operated in a constant force mode, only small movements Δx of the light spot will occur. The feedback loop will cancel out all other movements.

The scanning of a sample with an AFM can twist the microfabricated cantilevers because of lateral forces [5, 7, 8] and affect the images [120]. When the tip is subjected to lateral forces, it will twist the cantilever and the light beam reflected from the end of the cantilever will be deflected perpendicular to the ordinary deflection direction. For many investigations this influence of lateral forces is unwanted. The design of the triangular cantilevers stems from the desire, to minimize the torsion effects. However, lateral forces open up a new dimension in force measurements. They allow, for instance, a distinction of two materials because of the different friction coefficient, or the determination of adhesion energies. To measure lateral forces the original optical lever AFM has to be modified. The only modification compared with Fig. 2.29 is the use of a quadrant detector photodiode instead of a two-segment photodiode and the necessary readout electronics, see Fig. 2.9a. The electronics calculates the following signals:

$$U_{\text{Normal Force}} = \alpha \left[\left(I_{\text{Upper Left}} + I_{\text{Upper Right}} \right) - \left(I_{\text{Lower Left}} + I_{\text{Lower Right}} \right) \right],$$
$$U_{\text{Lateral Force}} = \beta \left[\left(I_{\text{Upper Left}} + I_{\text{Lower Left}} \right) - \left(I_{\text{Upper Right}} + I_{\text{Lower Right}} \right) \right]. \tag{2.50}$$

The calculation of the lateral force as a function of the deflection angle does not have a simple solution for cross-sections other than circles. An approximate formula for the angle of twist for rectangular beams is [151]

$$\theta = \frac{M_t L}{\beta G b^3 h}, \tag{2.51}$$

where $M_t = F_y \ell$ is the external twisting moment due to lateral force, F_y, and β a constant determined by the value of h/b. For the equation to hold, h has to be larger than b.

Inserting the values for a typical microfabricated cantilever with integrated tips

$$b = 6 \times 10^{-7} \, \text{m},$$
$$h = 10^{-5} \, \text{m},$$
$$L = 10^{-4} \, \text{m},$$
$$\ell = 3.3 \times 10^{-6} \, \text{m},$$
$$G = 5 \times 10^{10} \, \text{Pa},$$
$$\beta = 0.333 \tag{2.52}$$

into (2.51) we obtain the relation

$$F_y = 1.1 \times 10^{-4} \, \text{N} \times \theta. \tag{2.53}$$

Typical lateral forces are of order 10^{-10} N.

Sensitivity The sensitivity of this setup has been calculated in various papers [116, 148, 152]. Assuming a Gaussian beam the resulting output signal as a function of the deflection angle is dispersion like. Equation (2.47) shows that the sensitivity can be increased by increasing the intensity of the light beam I_{tot} or by decreasing the divergence of the laser beam. The upper bound of the intensity of the light I_{tot} is given by saturation effects on the photodiode. If we decrease the divergence of a laser beam we automatically increase the beam waist. If the beam waist becomes larger than the width of the cantilever we start to get diffraction. Diffraction sets a lower bound on the divergence angle. Hence one can calculate the optimal beam waist w_{opt} and the optimal divergence angle β [148, 152]

$$w_{opt} \approx 0.36 b,$$
$$\theta_{opt} \approx 0.89 \frac{\lambda}{b}. \tag{2.54}$$

The optimal sensitivity of the optical lever then becomes

$$\varepsilon\,[\mathrm{mW/rad}] = 1.8 \frac{b}{\lambda} I_{tot}\,[\mathrm{mW}]. \tag{2.55}$$

The angular sensitivity optical lever can be measured by introducing a parallel plate into the beam. A tilt of the parallel plate results in a displacement of the beam, mimicking an angular deflection.

Additional noise sources can be considered. Of little importance is the quantum mechanical uncertainty of the position [148, 152], which is, for typical cantilevers at room temperature

$$\Delta z = \sqrt{\frac{\hbar}{2 m \omega_0}} = 0.05\,\mathrm{fm}, \tag{2.56}$$

where $\hbar$ is the Planck constant ($= 6.626 \times 10^{-34}$ J s). At very low temperatures and for high frequency cantilevers this could become the dominant noise source. A second noise source is the shot noise of the light. The shot noise is related to the particle number. We can calculate the number of photons incident on the detector

$$n = \frac{I\tau}{\hbar \omega} = \frac{I\lambda}{2\pi B \hbar c} = 1.8 \times 10^9 \frac{I[\mathrm{W}]}{B[\mathrm{Hz}]}, \tag{2.57}$$

where I is the intensity of the light, τ the measurement time, $B = 1/\tau$ the bandwidth, and c the speed of light. The shot noise is proportional to the square root of the number of particles. Equating the shot noise signal with the signal resulting for the deflection of the cantilever one obtains

$$\Delta z_{shot} = 68 \frac{L}{w} \sqrt{\frac{B\,[\mathrm{kHz}]}{I\,[\mathrm{mW}]}}\,[\mathrm{fm}], \tag{2.58}$$

where w is the diameter of the focal spot. Typical AFM setups have a shot noise of 2 pm. The thermal noise can be calculated from the equipartition principle. The amplitude at the resonant frequency is

$$\Delta z_{therm} = 129 \sqrt{\frac{B}{k\,[\mathrm{N/m}]\,\omega_0 Q}}\,[\mathrm{pm}]. \tag{2.59}$$

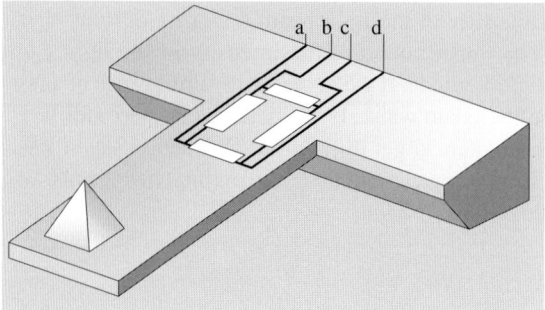

Fig. 2.30. A typical setup for a piezoresistive readout

A typical value is 16 pm. Upon touching the surface, the cantilever increases its resonant frequency by a factor of 4.39. This results in a new thermal noise amplitude of 3.2 pm for the cantilever in contact with the sample.

Piezoresistive Detection

Implementation An alternative detection system which is not as widely used as the optical detection schemes are piezoresistive cantilevers [125, 126, 132]. These cantilevers are based on the fact that the resistivity of certain materials, in particular of Si, changes with the applied stress. Figure 2.30 shows a typical implementation of a piezo-resistive cantilever. Four resistances are integrated on the chip, forming a Wheatstone bridge. Two of the resistors are in unstrained parts of the cantilever, the other two are measuring the bending at the point of the maximal deflection. For instance when an AC voltage is applied between terminals a and c one can measure the detuning of the bridge between terminals b and d. With such a connection the output signal varies only due to bending, but not due to changing of the ambient temperature and thus the coefficient of the piezoresistance.

Sensitivity The resistance change is [126]

$$\frac{\Delta R}{R_0} = \Pi \delta, \tag{2.60}$$

where Π is the tensor element of the piezo-resistive coefficients, δ the mechanical stress tensor element and R_0 the equilibrium resistance. For a single resistor they separate the mechanical stress and the tensor element in longitudinal and transverse components

$$\frac{\Delta R}{R_0} = \Pi_t \delta_t + \Pi_l \delta_l. \tag{2.61}$$

The maximum value of the stress components are $\Pi_t = -64.0 \times 10^{-11}$ m^2/N and $\Pi_l = -71.4 \times 10^{-11}$ m^2/N for a resistor oriented along the (110) direction in silicon [126]. In the resistor arrangement of Fig. 2.30 two of the resistors are subject to the longitudinal piezo-resistive effect and two of them are subject to the transversal piezo-resistive effect. The sensitivity of that setup is about four times that of a single resistor, with the advantage that temperature effects cancel to first order. The resistance change is then calculated as

$$\frac{\Delta R}{R_0} = \Pi \frac{3Eh}{2L^2}\Delta z = \Pi \frac{6L}{bh^2} F_z, \qquad (2.62)$$

where $\Pi = 67.7 \times 10^{-11}$ m^2/N is the averaged piezo-resistive coefficient. Plugging in typical values for the dimensions (Fig. 2.24) ($L = 100\,\mu$m, $b = 10\,\mu$m, $h = 1\,\mu$m) one obtains

$$\frac{\Delta R}{R_0} = \frac{4 \times 10^{-5}}{\text{nN}} F_z. \qquad (2.63)$$

The sensitivity can be tailored by optimizing the dimensions of the cantilever.

Capacitance Detection

The capacitance of an arrangement of conductors depends on the geometry. Generally speaking, the capacitance increases for decreasing separations. Two parallel plates form a simple capacitor (see Fig. 2.31, upper left), with the capacitance

$$C = \frac{\varepsilon\varepsilon_0 A}{x}, \qquad (2.64)$$

where A is the area of the plates, assumed equal, and x is the separation. Alternatively one can consider a sphere versus an infinite plane (see Fig. 2.31, lower left). Here the capacitance is [116]

$$C = 4\pi\varepsilon_0 R \sum_{n=2}^{\infty} \frac{\sinh(\alpha)}{\sinh(n\alpha)} \qquad (2.65)$$

where R is the radius of the sphere, and α is defined by

$$\alpha = \ln\left(1 + \frac{z}{R} + \sqrt{\frac{z^2}{R^2} + 2\frac{z}{R}}\right). \qquad (2.66)$$

One has to keep in mind that capacitance of a parallel plate capacitor is a nonlinear function of the separation. Using a voltage divider one can circumvent this problem. Figure 2.32a shows a low pass filter. The output voltage is given by

$$U_{\text{out}} = U_\approx \frac{\frac{1}{j\omega C}}{R + \frac{1}{j\omega C}} = U_\approx \frac{1}{j\omega C R + 1} \qquad (2.67)$$
$$\cong \frac{U_\approx}{j\omega C R}.$$

Here C is given by (2.64), ω is the excitation frequency and j is the imaginary unit. The approximate relation in the end is true when $\omega C R \gg 1$. This is equivalent to

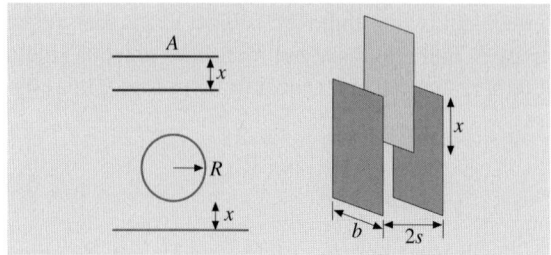

Fig. 2.31. Three possible arrangements of a capacitive readout. The *upper left* shows the cross section through a parallel plate capacitor. The *lower left* shows the geometry of sphere versus plane. The *right side* shows the more complicated, but linear capacitive readout

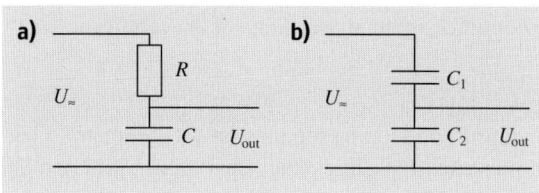

Fig. 2.32. Measuring the capacitance. The *left side*, (**a**), shows a low pass filter, the *right side*, (**b**), shows a capacitive divider. C (*left*) or C_2 (*right*) are the capacitances under test

the statement that C is fed by a current source, since R must be large in this setup. Plugging (2.64) into (2.67) and neglecting the phase information one obtains

$$U_{\text{out}} = \frac{U_\approx x}{\omega R \varepsilon \varepsilon_0 A}, \tag{2.68}$$

which is linear in the displacement x.

Figure 2.32b shows a capacitive divider. Again the output voltage U_{out} is given by

$$U_{\text{out}} = U_\approx \frac{C_1}{C_2 + C_1} = U_\approx \frac{C_1}{\frac{\varepsilon \varepsilon_0 A}{x} + C_1}. \tag{2.69}$$

If there is a stray capacitance C_s then (2.69) is modified as

$$U_{\text{out}} = U_\approx \frac{C_1}{\frac{\varepsilon \varepsilon_0 A}{x} + C_s + C_1}. \tag{2.70}$$

Provided $C_s + C_1 \ll C_2$ one has a system which is linear in x. The driving voltage $U_\approx$ has to be large (more than 100 V) to have the output voltage in the range of 1 V. The linearity of the readout depends on the capacitance C_1 (Fig. 2.33).

Another idea is to keep the distance constant and to change the relative overlap of the plates (see Fig. 2.31, right side). The capacitance of the moving center plate versus the stationary outer plates becomes

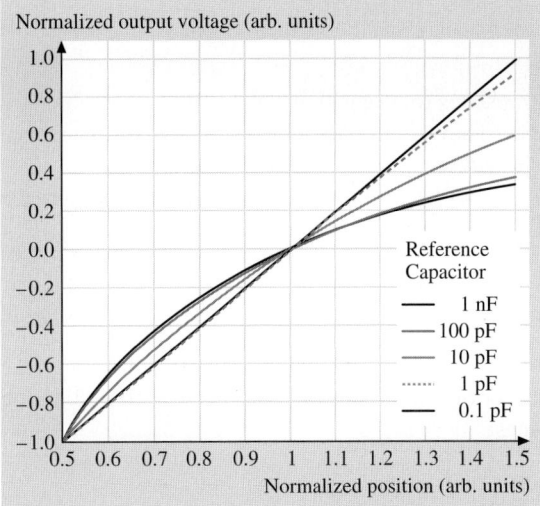

Fig. 2.33. Linearity of the capacitance readout as a function of the reference capacitor

$$C = C_s + 2\frac{\varepsilon\varepsilon_0 bx}{s}, \tag{2.71}$$

where the variables are defined in Fig. 2.31. The stray capacitance comprises all effects, including the capacitance of the fringe fields. When the length x is comparable to the width b of the plates one can safely assume that the stray capacitance is constant and independent of x. The main disadvantage of this setup is that it is not as easily incorporated in a microfabricated device as the others.

Sensitivity The capacitance itself is not a measure of the sensitivity, but its derivative is indicative of the signals one can expect. Using the situation described in Fig. 2.31, upper left, and in (2.64) one obtains for the parallel plate capacitor

$$\frac{dC}{dx} = -\frac{\varepsilon\varepsilon_0 A}{x^2}. \tag{2.72}$$

Assuming a plate area A of 20 μm by 40 μm and a separation of 1 μm one obtains a capacitance of 31 fF (neglecting stray capacitance and the capacitance of the connection leads) and a dC/dx of 3.1×10^{-8} F/m = 31 fF/μm. Hence it is of paramount importance to maximize the area between the two contacts and to minimize the distance x. The latter however is far from being trivial. One has to go to the limits of microfabrication to achieve a decent sensitivity.

If the capacitance is measured by the circuit shown in Fig. 2.32 one obtains for the sensitivity

$$\frac{dU_{\text{out}}}{U_\approx} = \frac{dx}{\omega R \varepsilon\varepsilon_0 A}. \tag{2.73}$$

Using the same value for A as above, setting the reference frequency to 100 kHz, and selecting $R = 1$ GΩ, we get the relative change of the output voltage U_{out} to

$$\frac{dU_\text{out}}{U_\approx} = \frac{22.5 \times 10^{-6}}{\text{Å}} \times dx. \tag{2.74}$$

A driving voltage of 45 V then translates to a sensitivity of 1 mV/Å. A problem in this setup is the stray capacitances. They are in parallel to the original capacitance and decrease the sensitivity considerably.

Alternatively one could build an oscillator with this capacitance and measure the frequency. RC-oscillators typically have an oscillation frequency of

$$f_\text{res} \propto \frac{1}{RC} = \frac{x}{R\varepsilon\varepsilon_0 A}. \tag{2.75}$$

Again the resistance R must be of the order of 1 GΩ, when stray capacitances C_s are neglected. However C_s is of the order of 1 pF. Therefore one gets $R = 10\,\text{M}\Omega$. Using these values the sensitivity becomes

$$df_\text{res} = \frac{C\,dx}{R(C+C_s)^2 x} \approx \frac{0.1\,\text{Hz}}{\text{Å}} dx. \tag{2.76}$$

The bad thing is that the stray capacitances have made the signal nonlinear again. The linearized setup in Fig. 2.31 has a sensitivity of

$$\frac{dC}{dx} = 2\frac{\varepsilon\varepsilon_0 b}{s}. \tag{2.77}$$

Substituting typical values, $b = 10\,\mu\text{m}$, $s = 1\,\mu\text{m}$ one gets $dC/dx = 1.8 \times 10^{-10}\,\text{F/m}$. It is noteworthy that the sensitivity remains constant for scaled devices.

Implementations The readout of the capacitance can be done in different ways [123, 124]. All include an alternating current or voltage with frequencies in the 100 kHz to the 100 MHz range. One possibility is to build a tuned circuit with the capacitance of the cantilever determining the frequency. The resonance frequency of a high quality Q tuned circuit is

$$\omega_0 = (LC)^{-1/2}. \tag{2.78}$$

where L is the inductance of the circuit. The capacitance C includes not only the sensor capacitance but also the capacitance of the leads. The precision of a frequency measurement is mainly determined by the ratio of L and C

$$Q = \left(\frac{L}{C}\right)^{1/2} \frac{1}{R}. \tag{2.79}$$

Here R symbolizes the losses in the circuit. The higher the quality the more precise the frequency measurement. For instance a frequency of 100 MHz and a capacitance of 1 pF gives an inductance of 250 μH. The quality becomes then 2.5×10^8. This value is an upper limit, since losses are usually too high.

Using a value of $dC/dx = 31\,\text{fF}/\mu\text{m}$ one gets $\Delta C/\text{Å} = 3.1\,\text{aF/Å}$. With a capacitance of 1 pF one gets

$$\frac{\Delta\omega}{\omega} = \frac{1}{2}\frac{\Delta C}{C},$$

$$\Delta\omega = 100\,\text{MHz} \times \frac{1}{2}\frac{3.1\,\text{aF}}{1\,\text{pF}} = 155\,\text{Hz}. \tag{2.80}$$

This is the frequency shift for 1 Å deflection. The calculation shows, that this is a measurable quantity. The quality also indicates that there is no physical reason why this scheme should not work.

2.4.3 Combinations for 3-D-Force Measurements

Three dimensional force measurements are essential if one wants to know all the details of the interaction between the tip and the cantilever. The straightforward attempt to measure three forces is complicated, since force sensors such as interferometers or capacitive sensors need a minimal detection volume, which often is too large. The second problem is that the force-sensing tip has to be held by some means. This implies that one of the three Cartesian axes is stiffer than the others.

However by the combination of different sensors one can achieve this goal. Straight cantilevers are employed for these measurements, because they can be handled analytically. The key observation is, that the optical lever method does not determine the position of the end of the cantilever. It measures the orientation. In the previous sections, one has always made use of the fact, that for a force along one of the orthogonal symmetry directions at the end of the cantilever (normal force, lateral force, force along the cantilever beam axis) there is a one to one correspondence of the tilt angle and the deflection. The problem is, that the force along the cantilever beam axis and the normal force create a deflection in the same direction. Hence what is called the normal force component is actually a mixture of two forces. The deflection of the cantilever is the third quantity, which is not considered in most of the AFMs. A fiber optic interferometer in parallel to the optical lever measures the deflection. Three measured quantities then allow the separation of the three orthonormal force directions, as is evident from (2.27) and (2.33) [12–16].

Alternatively one can put the fast scanning direction along the axis of the cantilever. Forward and backward scans then exert opposite forces F_x. If the piezo movement is linearized, both force components in AFM based on the optical lever detection can be determined. In this case, the normal force is simply the average of the forces in the forward and backward direction. The force, F_x, is the difference of the forces measured in forward and backward direction.

2.4.4 Scanning and Control Systems

Almost all SPMs use piezo translators to scan the tip or the sample. Even the first STM [1, 103] and some of the predecessor instruments [153, 154] used them. Other materials or setups for nano-positioning have been proposed, but were not successful [155, 156].

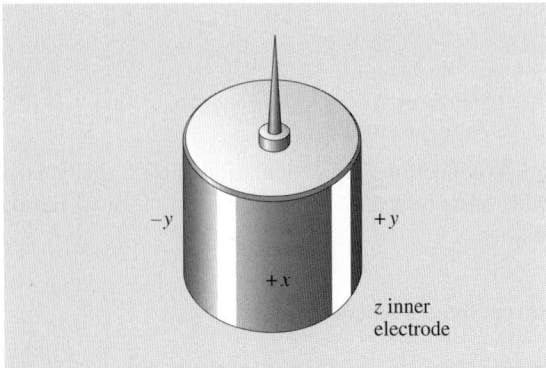

Fig. 2.34. Schematic drawing of a piezoelectric tube scanner. The piezo ceramic is molded into a tube form. The outer electrode is separated into four segments and connected to the scanning voltages. The z-voltage is applied to the inner electrode

Piezo Tubes

A popular solution is tube scanners (Fig. 2.34). They are now widely used in SPMs due to their simplicity and their small size [133, 157]. The outer electrode is segmented in four equal sectors of 90 degrees. Opposite sectors are driven by signals of the same magnitude, but opposite sign. This gives, through bending, a two-dimensional movement on, approximately, a sphere. The inner electrode is normally driven by the z signal. It is possible, however, to use only the outer electrodes for scanning and for the z-movement. The main drawback of applying the z-signal to the outer electrodes is, that the applied voltage is the sum of both the x- or y-movement and the z-movement. Hence a larger scan size effectively reduces the available range for the z-control.

Piezo Effect

An electric field applied across a piezoelectric material causes a change in the crystal structure, with expansion in some directions and contraction in others. Also, a net volume change occurs [132]. Many SPMs use the transverse piezo electric effect, where the applied electric field **E** is perpendicular to the expansion/contraction direction.

$$\Delta L = L(\mathbf{E} \cdot \mathbf{n})\, d_{31} = L\frac{V}{t} d_{31}, \tag{2.81}$$

where d_{31} is the transverse piezoelectric constant, V is the applied voltage, t is the thickness of the piezo slab or the distance between the electrodes where the voltage is applied, L is the free length of the piezo slab, and **n** is the direction of polarization. Piezo translators based on the transverse piezoelectric effect have a wide range of sensitivities, limited mainly by mechanical stability and breakdown voltage.

Scan Range

The calculation of the scanning range of a piezotube is difficult [157–159]. The bending of the tube depends on the electric fields and the nonuniform strain induced. A finite element calculation where the piezo tube was divided into 218 identical elements was used [158] to calculate the deflection. On each node the mechanical stress, stiffness, strain and piezoelectric stress was calculated when a voltage was applied on one electrode. The results were found to be linear on the first iteration and higher-order corrections were very small even for large electrode voltages. It was found that to first order the x- and z-movement of the tube could be reasonably well approximated by assuming that the piezo tube is a segment of a torus. Using this model one obtains

$$dx = (V_+ - V_-)|d_{31}|\frac{L^2}{2td}, \tag{2.82}$$

$$dz = (V_+ + V_- - 2V_z)|d_{31}|\frac{L}{2t}, \tag{2.83}$$

where $|d_{31}|$ is the coefficient of the transversal piezoelectric effect, L is tube's free length, t is tube's wall thickness, d is tube's diameter, V_+ is voltage on positive outer electrode while V_- is voltage of the opposite quadrant negative electrode, and V_z is voltage of the inner electrode.

The cantilever or sample mounted on the piezotube has an additional lateral movement because the point of measurement is not in the end plane of the piezotube. The additional lateral displacement of the end of the tip is $\ell \sin \varphi \approx \ell \varphi$, where ℓ is the tip length and φ is the deflection angle of the end surface. Assuming that the sample or cantilever are always perpendicular to the end of the walls of the tube and calculating with the torus model one gets for the angle

$$\varphi = \frac{L}{R} = \frac{2dx}{L}, \tag{2.84}$$

where R is the radius of curvature the piezo tube. Using the result of (2.84) one obtains for the additional x-movement

$$dx_{add} = \ell \varphi = \frac{2dx\ell}{L}$$
$$= (V_+ - V_-)|d_{31}|\frac{\ell L}{td} \tag{2.85}$$

and for the additional z-movement due to the x-movement

$$dz_{add} = \ell - \ell \cos \varphi = \frac{\ell \varphi^2}{2} = \frac{2\ell (dx)^2}{L^2}$$
$$= (V_+ - V_-)^2 |d_{31}|^2 \frac{\ell L^2}{2t^2 d^2}. \tag{2.86}$$

Carr [158] assumed for his finite element calculations that the top of the tube was completely free to move and, as a consequence, the top surface was distorted, leading to a deflection angle about half that of the geometrical model. Depending on the attachment of the sample or the cantilever this distortion may be smaller, leading to a deflection angle in-between that of the geometrical model and the one of the finite element calculation.

Nonlinearities and Creep

Piezo materials with a high conversion ratio, i.e. a large d_{31} or small electrode separations, with large scanning ranges are hampered by substantial hysteresis resulting in a deviation from linearity by more than 10%. The sensitivity of the piezo ceramic material (mechanical displacement divided by driving voltage) decreases with reduced scanning range, whereas the hysteresis is reduced. A careful selection of the material for the piezo scanners, the design of the scanners, and of the operating conditions is necessary to get optimum performance.

Passive Linearization: Calculation The analysis of images affected by piezo nonlinearities [160–163] shows that the dominant term is

$$x = AV + BV^2, \tag{2.87}$$

where x is the excursion of the piezo, V the applied voltage and A and B two coefficients describing the sensitivity of the material. Equation (2.87) holds for scanning from $V = 0$ to large V. For the reverse direction the equation becomes

$$x = \tilde{A}V - \tilde{B}(V - V_{\max})^2, \tag{2.88}$$

where $\tilde{A}$ and $\tilde{B}$ are the coefficients for the back scan and $V_{\max}$ is the applied voltage at the turning point. Both equations demonstrate that the true x-travel is small at the beginning of the scan and becomes larger towards the end. Therefore images are stretched at the beginning and compressed at the end.

Similar equations hold for the slow scan direction. The coefficients, however, are different. The combined action causes a greatly distorted image. This distortion can be calculated. The data acquisition systems record the signal as a function of V. However the data is measured as a function of x. Therefore we have to distribute the x-values evenly across the image this can be done by inverting an approximation of (2.87). First we write

$$x = AV\left(1 - \frac{B}{A}V\right). \tag{2.89}$$

For $B \ll A$ we can approximate

$$V = \frac{x}{A}. \tag{2.90}$$

We now substitute (2.90) into the nonlinear term of (2.89). This gives

$$x = AV\left(1 + \frac{Bx}{A^2}\right),$$
$$V = \frac{x}{A}\frac{1}{(1 + Bx/A^2)} \approx \frac{x}{A}\left(1 - \frac{Bx}{A^2}\right). \tag{2.91}$$

Hence an equation of the type

$$x_{\text{true}} = x(\alpha - \beta x/x_{\text{max}})$$
$$\text{with} \quad 1 = \alpha - \beta \tag{2.92}$$

takes out the distortion of an image. α and β are dependent on the scan range, the scan speed and on the scan history and have to be determined with exactly the same settings as for the measurement. x_{max} is the maximal scanning range. The condition for α and β guarantees that the image is transformed onto itself.

Similar equations as the empirical one shown above (2.92) can be derived by analyzing the movements of domain walls in piezo ceramics.

Passive Linearization: Measuring the Position An alternative strategy is to measure the position of the piezo translators. Several possibilities exist.

1. The interferometers described above can be used to measure the elongation of the piezo elongation. The fiber optic interferometer is especially easy to implement. The coherence length of the laser only limits the measurement range. However the signal is of a periodic nature. Hence a direct use of the signal in a feedback circuit for the position is not possible. However as a measurement tool and, especially, as a calibration tool the interferometer is without competition. The wavelength of the light, for instance in a HeNe laser, is so well defined that the precision of the other components determines the error of the calibration or measurement.
2. The movement of the light spot on the quadrant detector can be used to measure the position of a piezo [164]. The output current changes by 0.5 A/cm $\times$ $P(\text{W})/R(\text{cm})$. Typical values ($P = 1$ mW, $R = 0.001$ cm) give 0.5 A/cm. The noise limit is typically 0.15 nm $\times \sqrt{\Delta f(\text{Hz})/H(\text{W/cm}^2)}$. Again this means that the laser beam above would have a 0.1 nm noise limitation for a bandwidth of 21 Hz. The advantage of this method is that, in principle, one can linearize two axes with only one detector.
3. A knife-edge blocking part of a light beam incident on a photodiode can be used to measure the position of the piezo. This technique, commonly used in optical shear force detection [75, 165], has a sensitivity of better than 0.1 nm.
4. The capacitive detection [166, 167] of the cantilever deflection can be applied to the measurement of the piezo elongation. Equations (2.64) to (2.79) apply to the problem. This technique is used in some commercial instruments. The difficulties lie in the avoidance of fringe effects at the borders of the two plates. While conceptually simple, one needs the latest technology in surface preparation to get a decent linearity. The electronic circuits used for the readout are often proprietary.
5. Linear Variable Differential Transformers (LVDT) are a convenient means to measure positions down to 1 nm. They can be used together with a solid state joint setup, as often used for large scan range stages. Unlike the capacitive detection there are few difficulties in implementation. The sensors and the detection circuits LVDTs are available commercially.

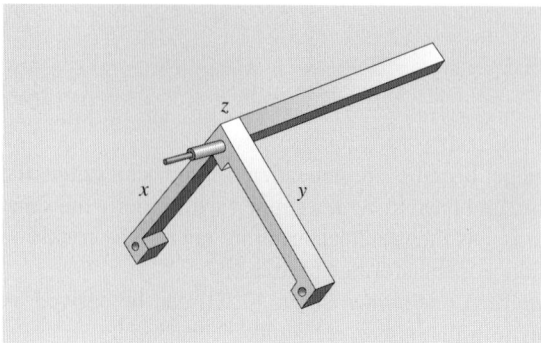

Fig. 2.35. An alternative type of piezo scanners: the tripod

6. A popular measurement technique is the use of strain gauges. They are especially sensitive when mounted on a solid state joint where the curvature is maximal. The resolution depends mainly on the induced curvature. A precision of 1 nm is attainable. The signals are low – a Wheatstone bridge is needed for the readout.

Active Linearization Active linearization is done with feedback systems. Sensors need to be monotonic. Hence all the systems described above, with the exception of the interferometers are suitable. The most common solutions include the strain gauge approach, the capacitance measurement or the LVDT, which are all electronic solutions. Optical detection systems have the disadvantage that the intensity enters into the calibration.

Alternative Scanning Systems

The first STMs were based on piezo tripods [1]. The piezo tripod (Fig. 2.35) is an intuitive way to generate the three dimensional movement of a tip attached to its center. However, to get a suitable stability and scanning range, the tripod needs to be fairly large (about 50 mm). Some instruments use piezo stacks instead of monolithic piezoactuators. They are arranged in the tripod arrangement. Piezo stacks are thin layers of piezoactive materials glued together to form a device with up to 200 μm of actuation range. Preloading with a suitable metal casing reduces the nonlinearity.

If one tries to construct a homebuilt scanning system, the use of linearized scanning tables is recommended. They are built around solid state joints and actuated by piezo stacks. The joints guarantee that the movement is parallel with little deviation from the predefined scanning plane. Due to the construction it is easy to add measurement devices such as capacitive sensors, LVDTs or strain gauges which are essential for a closed loop linearization. Two-dimensional tables can be bought from several manufacturers. They have a linearity of better than 0.1% and a noise level of 10^{-4} to 10^{-5} of the maximal scanning range.

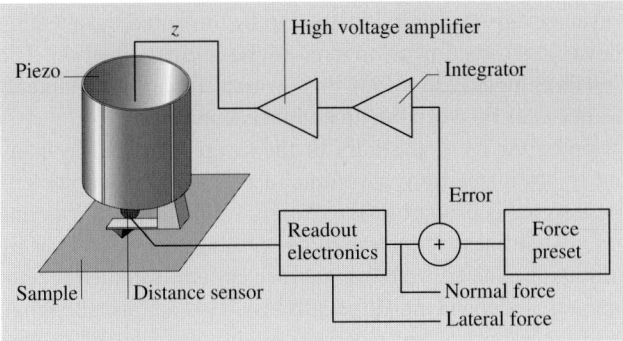

Fig. 2.36. Block schematic of the feedback control loop of an AFM

Control Systems

Basics The electronics and software play an important role in the optimal performance of an SPM. Control electronics and software are supplied with commercial SPMs. Control electronic systems can use either analog or digital feedback. While digital feedback offers greater flexibility and the ease of configuration, analog feedback circuits might be better suited for ultralow noise operation. We will describe here the basic setups for AFMs.

Figure 2.36 shows a block schematic of a typical AFM feedback loop. The signal from the force transducer is fed into the feedback loop consisting mainly of a subtraction stage to get an error signal and an integrator. The gain of the integrator (high gain corresponds to short integration times) is set as high as possible without generating more than 1% overshoot. High gain minimizes the error margin of the current and forces the tip to follow the contours of constant density of states as well as possible. This operating mode is known as Constant Force Mode. A high voltage amplifier amplifies the outputs of the integrator. As AFMs using piezotubes usually require ±150 V at the output, the output of the integrator needs to be amplified by a high voltage amplifier.

In order to scan the sample, additional voltages at high tension are required to drive the piezo. For example, with a tube scanner, four scanning voltages are required, namely $+V_x$, $-V_x$, $+V_y$ and $-V_y$. The x- and y-scanning voltages are generated in a scan generator (analog or computer controlled). Both voltages are input to the two respective power amplifiers. Two inverting amplifiers generate the input voltages for the other two power amplifiers. The topography of the sample surface is determined by recording the input-voltage to the high voltage amplifier for the z-channel as a function of x and y (Constant Force Mode).

Another operating mode is the Variable Force Mode. The gain in the feedback loop is lowered and the scanning speed increased such that the force on the cantilever is no longer constant. Here the force is recorded as a function of x and y.

Force Spectroscopy Four modes of spectroscopic imaging are in common use with force microscopes: measuring lateral forces, $\partial F/\partial z$, $\partial F/\partial x$ spatially resolved, and measuring force versus distance curves. Lateral forces can be measured by detecting the deflection of a cantilever in a direction orthogonal to the normal direction. The optical lever deflection method most easily does this. Lateral force measurements give indications of adhesion forces between the tip and the sample.

$\partial F/\partial z$ measurements probe the local elasticity of the sample surface. In many cases the measured quantity originates from a volume of a few cubic nanometers. The $\partial F/\partial z$ or local stiffness signal is proportional to Young's modulus, as far as one can define this quantity. Local stiffness is measured by vibrating the cantilever by a small amount in the z-direction. The expected signal for very stiff samples is zero: for very soft samples one gets, independent of the stiffness, also a constant signal. This signal is again zero for the optical lever deflection and equal to the driving amplitude for interferometeric measurements. The best sensitivity is obtained when the compliance of the cantilever matches the stiffness of the sample.

A third spectroscopic quantity is the lateral stiffness. It is measured by applying a small modulation in the x-direction on the cantilever. The signal is again optimal when the lateral compliance of the cantilever matches the lateral stiffness of the sample. The lateral stiffness is, in turn, related to the shear modulus of the sample.

Detailed information on the interaction of the tip and the sample can be gained by measuring force versus distance curves. It is necessary to have cantilevers with high enough compliance to avoid instabilities due to the attractive forces on the sample.

Using the Control Electronics as a Two-Dimensional Measurement Tool Usually the control electronics of an AFM is used to control the x- and y-piezo signals while several data acquisition channels record the position dependent signals. The control electronics can be used in another way: it can be viewed as a two-dimensional function generator. What is normally the x- and y-signal can be used to control two independent variables of an experiment. The control logic of the AFM then ensures that the available parameter space is systematically probed at equally spaced points. An example is friction force curves measured along a line across a step on graphite.

Figure 2.37 shows the connections. The z-piezo is connected as usual, like the x-piezo. However the y-output is used to command the desired input parameter. The offset of the y-channel determines the position of the tip on the sample surface, together with the x-channel.

Some Imaging Processing Methods

The visualization and interpretation of images from AFMs is intimately connected to the processing of these images. An ideal AFM is a noise-free device that images a sample with perfect tips of known shape and has perfect linear scanning piezos. In reality, AFMs are not that ideal. The scanning device in AFMs is affected by distortions. The distortions are both linear and nonlinear. Linear distortions mainly result from imperfections in the machining of the piezotranslators causing crosstalk from the Z-piezo to the X- and Y-piezos, and vice versa. Among the linear distortions, there are two kinds which are very important. First, scanning piezos invariably have different sensitivities along the different scan axes due to the variation of the piezo material and uneven sizes of the electrode areas. Second, the same reasons might

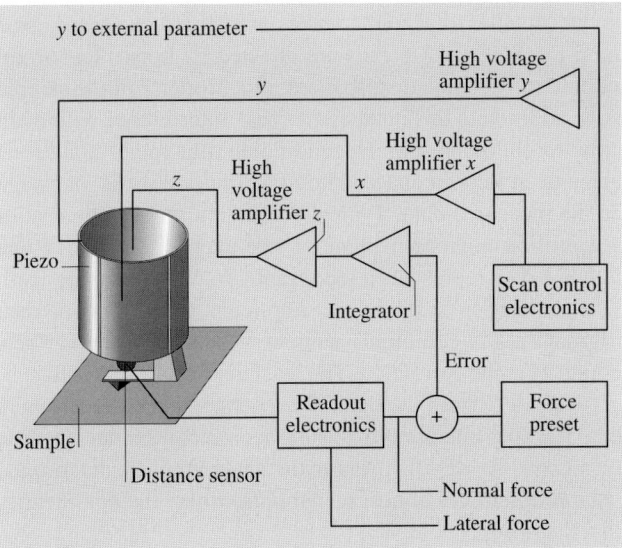

Fig. 2.37. Wiring of an AFM to measure friction force curves along a line

cause the scanning axes not to be orthogonal. Furthermore, the plane in which the piezoscanner moves for constant height z is hardly ever coincident with the sample plane. Hence, a linear ramp is added to the sample data. This ramp is especially bothersome when the height z is displayed as an intensity map.

The nonlinear distortions are harder to deal with. They can affect AFM data for a variety of reasons. First, piezoelectric ceramics do have a hysteresis loop, much like ferromagnetic materials. The deviations of piezoceramic materials from linearity increase with increasing amplitude of the driving voltage. The mechanical position for one voltage depends on the previously applied voltages to the piezo. Hence, to get the best positional accuracy, one should always approach a point on the sample from the same direction. Another type of nonlinear distortion of the images occurs when the scan frequency approaches the upper frequency limit of the X- and Y-drive amplifiers or the upper frequency limit of the feedback loop (z-component). This distortion, due to feedback loop, can only be minimized by reducing the scan frequency. On the other hand, there is a simple way to reduce distortions due to the X- and Y-piezo drive amplifiers. To keep the system as simple as possible, one normally uses a triangular waveform for driving the scanning piezos. However, triangular waves contain frequency components as multiples of the scan frequency. If the cutoff frequency of the X- and Y-drive electronics or of the feedback loop is too close to the scanning frequency (two or three times the scanning frequency), the triangular drive voltage is rounded off at the turning points. This rounding error causes, first, a distortion of the scan linearity and, second, through phase lags, the projection of part of the backward scan onto the forward scan. This type of distortion can be minimized by carefully selecting the scanning frequency and by using driving voltages for the

X- and Y-piezos with waveforms like trapezoidal waves, which are closer to a sine wave. The values measured for X, Y, or Z piezos are affected by noise. The origin of this noise can be either electronic, some disturbances, or a property of the sample surface due to adsorbates. In addition to this incoherent noise, interference with main and other equipment nearby might be present. Depending on the type of noise, one can filter it in the real space or in Fourier space. The most important part of image processing is to visualize the measured data. Typical AFM data sets can consist of many thousands to over a million points per plane. There may be more than one image plane present. The AFM data represents a topography in various data spaces.

Most commercial data acquisition systems use implicitly some kind of data processing. Since the original data is commonly subject to slopes on the surface, most programs use some kind of slope correction. The least disturbing way is to substrate a plane $z(x,y) = Ax + By + C$ from the data. The coefficients are determined by fitting $z(x,y)$ to the data. Another operation is to subtract a second order function such as $z(x,y) = Ax^2 + By^2 + Cxy + Dx + Ey + F$. Again, the parameters are determined with a fit. This function is appropriate for almost planar data, where the nonlinearity of the piezos caused such a distortion.

In the image processing software from Digital Instruments, up to three operations are performed on the raw data. First, a zero-order flatten is applied. The flatten operation is used to eliminate image bow in the slow scan direction (caused by a physical bow in the instrument itself), slope in the slow scan direction, bands in the image (caused by differences in the scan height from one scan line to the next). The flattening operation takes each scan line and subtracts the average value of the height along each scan line from each point in that scan line. This brings each scan line to the same height. Next, a first-order plane-fit is applied in the fast scan direction. The plane-fit operation is used to eliminate bow and slope in the fast scan direction. The plane-fit operation calculated a best-fit plane for the image and subtracts it from the image. This plane has a constant non-zero slope in the fast scan direction. In some cases, a higher-order polynomial "plane" may be required. Depending upon the quality of the raw data, the flattening operation and/or the planefit operation may not be required at all.

References

1. G. Binnig, H. Rohrer, Ch. Gerber, and E. Weibel. Surface studies by scanning tunneling microscopy. *Phys. Rev. Lett.*, 49:57–61, 1982.
2. G. Binnig, C. F. Quate, and Ch. Gerber. Atomic force microscope. *Phys. Rev. Lett.*, 56:930–933, 1986.
3. G. Binnig, Ch. Gerber, E. Stoll, T. R. Albrecht, and C. F. Quate. Atomic resolution with atomic force microscope. *Europhys. Lett.*, 3:1281–1286, 1987.
4. B. Bhushan. *Handbook of Micro/Nanotribology*. CRC, 2nd edition, 1999.
5. C. M. Mate, G. M. McClelland, R. Erlandsson, and S. Chiang. Atomic-scale friction of a tungsten tip on a graphite surface. *Phys. Rev. Lett.*, 59:1942–1945, 1987.
6. R. Erlandsson, G. M. McClelland, C. M. Mate, and S. Chiang. Atomic force microscopy using optical interferometry. *J. Vac. Sci. Technol. A*, 6:266–270, 1988.

7. O. Marti, J. Colchero, and J. Mlynek. Combined scanning force and friction microscopy of mica. *Nanotechnol.*, 1:141–144, 1990.
8. G. Meyer and N. M. Amer. Simultaneous measurement of lateral and normal forces with an optical-beam-deflection atomic force microscope. *Appl. Phys. Lett.*, 57:2089–2091, 1990.
9. B. Bhushan and J. Ruan. Atomic-scale friction measurements using friction force microscopy: Part ii – application to magnetic media. *ASME J. Tribol.*, 116:389–396, 1994.
10. B. Bhushan, V. N. Koinkar, and J. Ruan. Microtribology of magnetic media. *Proc. Inst. Mech. Eng., Part J: J. Eng. Tribol.*, 208:17–29, 1994.
11. B. Bhushan, J. N. Israelachvili, and U. Landman. Nanotribology: Friction, wear, and lubrication at the atomic scale. *Nature*, 374:607–616, 1995.
12. S. Fujisawa, M. Ohta, T. Konishi, Y. Sugawara, and S. Morita. Difference between the forces measured by an optical lever deflection and by an optical interferometer in an atomic force microscope. *Rev. Sci. Instrum.*, 65:644–647, 1994.
13. S. Fujisawa, E. Kishi, Y. Sugawara, and S. Morita. Fluctuation in 2-dimensional stick-slip phenomenon observed with 2-dimensional frictional force microscope. *Jpn. J. Appl. Phys.*, 33:3752–3755, 1994.
14. S. Grafstrom, J. Ackermann, T. Hagen, R. Neumann, and O. Probst. Analysis of lateral force effects on the topography in scanning force microscopy. *J. Vac. Sci. Technol. B*, 12:1559–1564, 1994.
15. R. M. Overney, H. Takano, M. Fujihira, W. Paulus, and H. Ringsdorf. Anisotropy in friction and molecular stick-slip motion. *Phys. Rev. Lett.*, 72:3546–3549, 1994.
16. R. J. Warmack, X. Y. Zheng, T. Thundat, and D. P. Allison. Friction effects in the deflection of atomic force microscope cantilevers. *Rev. Sci. Instrum.*, 65:394–399, 1994.
17. N. A. Burnham, D. D. Domiguez, R. L. Mowery, and R. J. Colton. Probing the surface forces of monolayer films with an atomic force microscope. *Phys. Rev. Lett.*, 64:1931–1934, 1990.
18. N. A. Burham, R. J. Colton, and H. M. Pollock. Interpretation issues in force microscopy. *J. Vac. Sci. Technol. A*, 9:2548–2556, 1991.
19. C. D. Frisbie, L. F. Rozsnyai, A. Noy, M. S. Wrighton, and C. M. Lieber. Functional group imaging by chemical force microscopy. *Science*, 265:2071–2074, 1994.
20. V. N. Koinkar and B. Bhushan. Microtribological studies of unlubricated and lubricated surfaces using atomic force/friction force microscopy. *J. Vac. Sci. Technol. A*, 14:2378–2391, 1996.
21. V. Scherer, B. Bhushan, U. Rabe, and W. Arnold. Local elasticity and lubrication measurements using atomic force and friction force microscopy at ultrasonic frequencies. *IEEE Trans. Mag.*, 33:4077–4079, 1997.
22. V. Scherer, W. Arnold, and B. Bhushan. Lateral force microscopy using acoustic friction force microscopy. *Surf. Interface Anal.*, 27:578–587, 1999.
23. B. Bhushan and S. Sundararajan. Micro/nanoscale friction and wear mechanisms of thin films using atomic force and friction force microscopy. *Acta Mater.*, 46:3793–3804, 1998.
24. U. Krotil, T. Stifter, H. Waschipky, K. Weishaupt, S. Hild, and O. Marti. Pulse force mode: A new method for the investigation of surface properties. *Surf. Interface Anal.*, 27:336–340, 1999.
25. B. Bhushan and C. Dandavate. Thin-film friction and adhesion studies using atomic force microscopy. *J. Appl. Phys.*, 87:1201–1210, 2000.
26. B. Bhushan. *Micro/Nanotribology and its Applications*. Kluwer, 1997.
27. B. Bhushan. *Principles and Applications of Tribology*. Wiley, 1999.

28. B. Bhushan. *Modern Tribology Handbook Vol. 1: Principles of Tribology*. CRC, 2001.
29. B. Bhushan. *Introduction to Tribology*. Wiley, 2002.
30. M. Reinstaedtler, U. Rabe, V. Scherer, U. Hartmann, A. Goldade, B. Bhushan, and W. Arnold. On the nanoscale measurement of friction using atomic force microscope cantilever torsional resonances. *Appl. Phys. Lett.*, 82:2604–2606, 2003.
31. N. A. Burnham and R. J. Colton. Measuring the nanomechanical properties and surface forces of materials using an atomic force microscope. *J. Vac. Sci. Technol. A*, 7:2906–2913, 1989.
32. P. Maivald, H. J. Butt, S. A. C. Gould, C. B. Prater, B. Drake, J. A. Gurley, V. B. Elings, and P. K. Hansma. Using force modulation to image surface elasticities with the atomic force microscope. *Nanotechnol.*, 2:103–106, 1991.
33. B. Bhushan, A. V. Kulkarni, W. Bonin, and J. T. Wyrobek. Nano/picoindentation measurements using capacitive transducer in atomic force microscopy. *Philos. Mag. A*, 74:1117–1128, 1996.
34. B. Bhushan and V. N. Koinkar. Nanoindentation hardness measurements using atomic force microscopy. *Appl. Phys. Lett.*, 75:5741–5746, 1994.
35. D. DeVecchio and B. Bhushan. Localized surface elasticity measurements using an atomic force microscope. *Rev. Sci. Instrum.*, 68:4498–4505, 1997.
36. S. Amelio, A. V. Goldade, U. Rabe, V. Scherer, B. Bhushan, and W. Arnold. Measurements of mechanical properties of ultra-thin diamond-like carbon coatings using atomic force acoustic microscopy. *Thin Solid Films*, 392:75–84, 2001.
37. D. M. Eigler and E. K. Schweizer. Positioning single atoms with a scanning tunnelling microscope. *Nature*, 344:524–528, 1990.
38. A. L. Weisenhorn, J. E. MacDougall, J. A. C. Gould, S. D. Cox, W. S. Wise, J. Massie, P. Maivald, V. B. Elings, G. D. Stucky, and P. K. Hansma. Imaging and manipulating of molecules on a zeolite surface with an atomic force microscope. *Science*, 247:1330–1333, 1990.
39. I. W. Lyo and Ph. Avouris. Field-induced nanometer-to-atomic-scale manipulation of silicon surfaces with the stm. *Science*, 253:173–176, 1991.
40. O. M. Leung and M. C. Goh. Orientation ordering of polymers by atomic force microscope tip-surface interactions. *Science*, 225:64–66, 1992.
41. D. W. Abraham, H. J. Mamin, E. Ganz, and J. Clark. Surface modification with the scanning tunneling microscope. *IBM J. Res. Dev.*, 30:492–499, 1986.
42. R. M. Silver, E. E. Ehrichs, and A. L. de Lozanne. Direct writing of submicron metallic features with a scanning tunnelling microscope. *Appl. Phys. Lett.*, 51:247–249, 1987.
43. A. Kobayashi, F. Grey, R. S. Williams, and M. Ano. Formation of nanometer-scale grooves in silicon with a scanning tunneling microscope. *Science*, 259:1724–1726, 1993.
44. B. Parkinson. Layer-by-layer nanometer scale etching of two-dimensional substrates using the scanning tunneling microscopy. *J. Am. Chem. Soc.*, 112:7498–7502, 1990.
45. A. Majumdar, P. I. Oden, J. P. Carrejo, L. A. Nagahara, J. J. Graham, and J. Alexander. Nanometer-scale lithography using the atomic force microscope. *Appl. Phys. Lett.*, 61:2293–2295, 1992.
46. B. Bhushan. Micro/nanotribology and its applications to magnetic storage devices and mems. *Tribol. Int.*, 28:85–96, 1995.
47. L. Tsau, D. Wang, and K. L. Wang. Nanometer scale patterning of silicon(100) surface by an atomic force microscope operating in air. *Appl. Phys. Lett.*, 64:2133–2135, 1994.
48. E. Delawski and B. A. Parkinson. Layer-by-layer etching of two-dimensional metal chalcogenides with the atomic force microscope. *J. Am. Chem. Soc.*, 114:1661–1667, 1992.

49. B. Bhushan and G. S. Blackman. Atomic force microscopy of magnetic rigid disks and sliders and its applications to tribology. *ASME J. Tribol.*, 113:452–458, 1991.
50. O. Marti, B. Drake, and P. K. Hansma. Atomic force microscopy of liquid-covered surfaces: atomic resolution images. *Appl. Phys. Lett.*, 51:484–486, 1987.
51. B. Drake, C. B. Prater, A. L. Weisenhorn, S. A. C. Gould, T. R. Albrecht, C. F. Quate, D. S. Cannell, H. G. Hansma, and P. K. Hansma. Imaging crystals, polymers and processes in water with the atomic force microscope. *Science*, 243:1586–1589, 1989.
52. M. Binggeli, R. Christoph, H. E. Hintermann, J. Colchero, and O. Marti. Friction force measurements on potential controlled graphite in an electrolytic environment. *Nanotechnol.*, 4:59–63, 1993.
53. G. Meyer and N. M. Amer. Novel optical approach to atomic force microscopy. *Appl. Phys. Lett.*, 53:1045–1047, 1988.
54. J. H. Coombs and J. B. Pethica. Properties of vacuum tunneling currents: Anomalous barrier heights. *IBM J. Res. Dev.*, 30:455–459, 1986.
55. M. D. Kirk, T. Albrecht, and C. F. Quate. Low-temperature atomic force microscopy. *Rev. Sci. Instrum.*, 59:833–835, 1988.
56. F. J. Giessibl, Ch. Gerber, and G. Binnig. A low-temperature atomic force/scanning tunneling microscope for ultrahigh vacuum. *J. Vac. Sci. Technol. B*, 9:984–988, 1991.
57. T. R. Albrecht, P. Grutter, D. Rugar, and D. P. E. Smith. Low temperature force microscope with all-fiber interferometer. *Ultramicroscopy*, 42–44:1638–1646, 1992.
58. H. J. Hug, A. Moser, Th. Jung, O. Fritz, A. Wadas, I. Parashikor, and H. J. Güntherodt. Low temperature magnetic force microscopy. *Rev. Sci. Instrum.*, 64:2920–2925, 1993.
59. C. Basire and D. A. Ivanov. Evolution of the lamellar structure during crystallization of a semicrystalline-amorphous polymer blend: Time-resolved hot-stage spm study. *Phys. Rev. Lett.*, 85:5587–5590, 2000.
60. H. Liu and B. Bhushan. Investigation of nanotribological properties of self-assembled monolayers with alkyl and biphenyl spacer chains. *Ultramicroscopy*, 91:185–202, 2002.
61. J. Foster and J. Frommer. Imaging of liquid crystal using a tunneling microscope. *Nature*, 333:542–547, 1988.
62. D. Smith, H. Horber, C. Gerber, and G. Binnig. Smectic liquid crystal monolayers on graphite observed by scanning tunneling microscopy. *Science*, 245:43–45, 1989.
63. D. Smith, J. Horber, G. Binnig, and H. Nejoh. Structure, registry and imaging mechanism of alkylcyanobiphenyl molecules by tunnelling microscopy. *Nature*, 344:641–644, 1990.
64. Y. Andoh, S. Oguchi, R. Kaneko, and T. Miyamoto. Evaluation of very thin lubricant films. *J. Phys. D*, 25:A71–A75, 1992.
65. Y. Martin, C. C. Williams, and H. K. Wickramasinghe. Atomic force microscope-force mapping and profiling on a sub 100-A scale. *J. Appl. Phys.*, 61:4723–4729, 1987.
66. J. E. Stern, B. D. Terris, H. J. Mamin, and D. Rugar. Deposition and imaging of localized charge on insulator surfaces using a force microscope. *Appl. Phys. Lett.*, 53:2717–2719, 1988.
67. K. Yamanaka, H. Ogisco, and O. Kolosov. Ultrasonic force microscopy for nanometer resolution subsurface imaging. *Appl. Phys. Lett.*, 64:178–180, 1994.
68. K. Yamanaka and E. Tomita. Lateral force modulation atomic force microscope for selective imaging of friction forces. *Jpn. J. Appl. Phys.*, 34:2879–2882, 1995.
69. U. Rabe, K. Janser, and W. Arnold. Vibrations of free and surface-coupled atomic force microscope: Theory and experiment. *Rev. Sci. Instrum.*, 67:3281–3293, 1996.
70. Y. Martin and H. K. Wickramasinghe. Magnetic imaging by force microscopy with 1000 Å resolution. *Appl. Phys. Lett.*, 50:1455–1457, 1987.

71. D. Rugar, H. J. Mamin, P. Guethner, S. E. Lambert, J. E. Stern, I. McFadyen, and T. Yogi. Magnetic force microscopy – general principles and application to longitudinal recording media. *J. Appl. Phys.*, 63:1169–1183, 1990.
72. C. Schoenenberger and S. F. Alvarado. Understanding magnetic force microscopy. *Z. Phys. B*, 80:373–383, 1990.
73. U. Hartmann. Magnetic force microscopy. *Annu. Rev. Mater. Sci.*, 29:53–87, 1999.
74. D. W. Pohl, W. Denk, and M. Lanz. Optical stethoscopy-image recording with resolution lambda/20. *Appl. Phys. Lett.*, 44:651–653, 1984.
75. E. Betzig, J. K. Troutman, T. D. Harris, J. S. Weiner, and R. L. Kostelak. Breaking the diffraction barrier – optical microscopy on a nanometric scale. *Science*, 251:1468–1470, 1991.
76. E. Betzig, P. L. Finn, and J. S. Weiner. Combined shear force and near-field scanning optical microscopy. *Appl. Phys. Lett.*, 60:2484, 1992.
77. P. F. Barbara, D. M. Adams, and D. B. O'Connor. Characterization of organic thin film materials with near-field scanning optical microscopy (nsom). *Annu. Rev. Mater. Sci.*, 29:433–469, 1999.
78. C. C. Williams and H. K. Wickramasinghe. Scanning thermal profiler. *Appl. Phys. Lett.*, 49:1587–1589, 1986.
79. C. C. Williams and H. K. Wickramasinghe. Microscopy of chemical-potential variations on an atomic scale. *Nature*, 344:317–319, 1990.
80. A. Majumdar. Scanning thermal microscopy. *Annu. Rev. Mater. Sci.*, 29:505–585, 1999.
81. O. E. Husser, D. H. Craston, and A. J. Bard. Scanning electrochemical microscopy – high resolution deposition and etching of materials. *J. Electrochem. Soc.*, 136:3222–3229, 1989.
82. Y. Martin, D. W. Abraham, and H. K. Wickramasinghe. High-resolution capacitance measurement and potentiometry by force microscopy. *Appl. Phys. Lett.*, 52:1103–1105, 1988.
83. M. Nonnenmacher, M. P. O'Boyle, and H. K. Wickramasinghe. Kelvin probe force microscopy. *Appl. Phys. Lett.*, 58:2921–2923, 1991.
84. J. M. R. Weaver and D. W. Abraham. High resolution atomic force microscopy potentiometry. *J. Vac. Sci. Technol. B*, 9:1559–1561, 1991.
85. D. DeVecchio and B. Bhushan. Use of a nanoscale kelvin probe for detecting wear precursors. *Rev. Sci. Instrum.*, 69:3618–3624, 1998.
86. B. Bhushan and A. V. Goldade. Measurements and analysis of surface potential change during wear of single-crystal silicon (100) at ultralow loads using kelvin probe microscopy. *Appl. Surf. Sci.*, 157:373–381, 2000.
87. P. K. Hansma, B. Drake, O. Marti, S. A. C. Gould, and C. B. Prater. The scanning ion-conductance microscope. *Science*, 243:641–643, 1989.
88. C. B. Prater, P. K. Hansma, M. Tortonese, and C. F. Quate. Improved scanning ion-conductance microscope using microfabricated probes. *Rev. Sci. Instrum.*, 62:2634–2638, 1991.
89. J. Matey and J. Blanc. Scanning capacitance microscopy. *J. Appl. Phys.*, 57:1437–1444, 1985.
90. C. C. Williams. Two-dimensional dopant profiling by scanning capacitance microscopy. *Annu. Rev. Mater. Sci.*, 29:471–504, 1999.
91. D. T. Lee, J. P. Pelz, and B. Bhushan. Instrumentation for direct, low frequency scanning capacitance microscopy, and analysis of position dependent stray capacitance. *Rev. Sci. Instrum.*, 73:3523–3533, 2002.
92. I. Giaever. Energy gap in superconductors measured by electron tunneling. *Phys. Rev. Lett.*, 5:147–148, 1960.

93. P. K. Hansma and J. Tersoff. Scanning tunneling microscopy. *J. Appl. Phys.*, 61:R1–R23, 1987.
94. D. Sarid and V. Elings. Review of scanning force microscopy. *J. Vac. Sci. Technol. B*, 9:431–437, 1991.
95. U. Durig, O. Zuger, and A. Stalder. Interaction force detection in scanning probe microscopy: Methods and applications. *J. Appl. Phys.*, 72:1778–1797, 1992.
96. J. Frommer. Scanning tunneling microscopy and atomic force microscopy in organic chemistry. *Angew. Chem. Int. Ed. Engl.*, 31:1298–1328, 1992.
97. H. J. Güntherodt and R. Wiesendanger, editors. *Scanning Tunneling Microscopy I: General Principles and Applications to Clean and Adsorbate-Covered Surfaces*. Springer, 1992.
98. R. Wiesendanger and H. J. Güntherodt, editors. *Scanning Tunneling Microscopy, II: Further Applications and Related Scanning Techniques*. Springer, 1992.
99. D. A. Bonnell, editor. *Scanning Tunneling Microscopy and Spectroscopy – Theory, Techniques, and Applications*. VCH, 1993.
100. O. Marti and M. Amrein, editors. *STM and SFM in Biology*. Academic, 1993.
101. J. A. Stroscio and W. J. Kaiser, editors. *Scanning Tunneling Microscopy*. Academic, 1993.
102. H. J. Güntherodt, D. Anselmetti, and E. Meyer, editors. *Forces in Scanning Probe Methods*. Kluwer, 1995.
103. G. Binnig and H. Rohrer. Scanning tunnelling microscopy. *Surf. Sci.*, 126:236–244, 1983.
104. B. Bhushan, J. Ruan, and B. K. Gupta. A scanning tunnelling microscopy study of fullerene films. *J. Phys. D*, 26:1319–1322, 1993.
105. R. L. Nicolaides, W. E. Yong, W. F. Packard, and H. A. Zhou. Scanning tunneling microscope tip structures. *J. Vac. Sci. Technol. A*, 6:445–447, 1988.
106. J. P. Ibe, P. P. Bey, S. L. Brandon, R. A. Brizzolara, N. A. Burnham, D. P. DiLella, K. P. Lee, C. R. K. Marrian, and R. J. Colton. On the electrochemical etching of tips for scanning tunneling microscopy. *J. Vac. Sci. Technol. A*, 8:3570–3575, 1990.
107. R. Kaneko and S. Oguchi. Ion-implanted diamond tip for a scanning tunneling microscope. *Jpn. J. Appl. Phys.*, 28:1854–1855, 1990.
108. F. J. Giessibl. Atomic resolution of the silicon(111)–(7×7) surface by atomic force microscopy. *Science*, 267:68–71, 1995.
109. B. Anczykowski, D. Krueger, K. L. Babcock, and H. Fuchs. Basic properties of dynamic force spectroscopy with the scanning force microscope in experiment and simulation. *Ultramicroscopy*, 66:251–259, 1996.
110. T. R. Albrecht and C. F. Quate. Atomic resolution imaging of a nonconductor by atomic force microscopy. *J. Appl. Phys.*, 62:2599–2602, 1987.
111. S. Alexander, L. Hellemans, O. Marti, J. Schneir, V. Elings, and P. K. Hansma. An atomic-resolution atomic-force microscope implemented using an optical lever. *J. Appl. Phys.*, 65:164–167, 1989.
112. G. Meyer and N. M. Amer. Optical-beam-deflection atomic force microscopy: The NaCl(001) surface. *Appl. Phys. Lett.*, 56:2100–2101, 1990.
113. A. L. Weisenhorn, M. Egger, F. Ohnesorge, S. A. C. Gould, S. P. Heyn, H. G. Hansma, R. L. Sinsheimer, H. E. Gaub, and P. K. Hansma. Molecular resolution images of langmuir–blodgett films and dna by atomic force microscopy. *Langmuir*, 7:8–12, 1991.
114. J. Ruan and B. Bhushan. Atomic-scale and microscale friction of graphite and diamond using friction force microscopy. *J. Appl. Phys.*, 76:5022–5035, 1994.
115. D. Rugar and P. K. Hansma. Atomic force microscopy. *Phys. Today*, 43:23–30, 1990.

116. D. Sarid. *Scanning Force Microscopy*. Oxford Univ. Press, 1991.
117. G. Binnig. Force microscopy. *Ultramicroscopy*, 42–44:7–15, 1992.
118. E. Meyer. Atomic force microscopy. *Surf. Sci.*, 41:3–49, 1992.
119. H. K. Wickramasinghe. Progress in scanning probe microscopy. *Acta Mater.*, 48:347–358, 2000.
120. A. J. den Boef. The influence of lateral forces in scanning force microscopy. *Rev. Sci. Instrum.*, 62:88–92, 1991.
121. M. Radmacher, R. W. Tillman, M. Fritz, and H. E. Gaub. From molecules to cells: Imaging soft samples with the atomic force microscope. *Science*, 257:1900–1905, 1992.
122. F. Ohnesorge and G. Binnig. True atomic resolution by atomic force microscopy through repulsive and attractive forces. *Science*, 260:1451–1456, 1993.
123. G. Neubauer, S. R. Coben, G. M. McClelland, D. Horne, and C. M. Mate. Force microscopy with a bidirectional capacitance sensor. *Rev. Sci. Instrum.*, 61:2296–2308, 1990.
124. T. Goddenhenrich, H. Lemke, U. Hartmann, and C. Heiden. Force microscope with capacitive displacement detection. *J. Vac. Sci. Technol. A*, 8:383–387, 1990.
125. U. Stahl, C. W. Yuan, A. L. Delozanne, and M. Tortonese. Atomic force microscope using piezoresistive cantilevers and combined with a scanning electron microscope. *Appl. Phys. Lett.*, 65:2878–2880, 1994.
126. R. Kassing and E. Oesterschulze. *Sensors for scanning probe microscopy*, pages 35–54. Kluwer, 1997.
127. C. M. Mate. Atomic-force-microscope study of polymer lubricants on silicon surfaces. *Phys. Rev. Lett.*, 68:3323–3326, 1992.
128. S. P. Jarvis, A. Oral, T. P. Weihs, and J. B. Pethica. A novel force microscope and point contact probe. *Rev. Sci. Instrum.*, 64:3515–3520, 1993.
129. D. Rugar, H. J. Mamin, and P. Guethner. Improved fiber-optical interferometer for atomic force microscopy. *Appl. Phys. Lett.*, 55:2588–2590, 1989.
130. C. Schoenenberger and S. F. Alvarado. A differential interferometer for force microscopy. *Rev. Sci. Instrum.*, 60:3131–3135, 1989.
131. D. Sarid, D. Iams, V. Weissenberger, and L. S. Bell. Compact scanning-force microscope using laser diode. *Opt. Lett.*, 13:1057–1059, 1988.
132. N. W. Ashcroft and N. D. Mermin. *Solid State Physics*. Holt Reinhart and Winston, 1976.
133. G. Binnig and D. P. E. Smith. Single-tube three-dimensional scanner for scanning tunneling microscopy. *Rev. Sci. Instrum.*, 57:1688, 1986.
134. S. I. Park and C. F. Quate. Digital filtering of stm images. *J. Appl. Phys.*, 62:312, 1987.
135. J. W. Cooley and J. W. Tukey. An algorithm for machine calculation of complex fourier series. *Math. Computation*, 19:297, 1965.
136. J. Ruan and B. Bhushan. Atomic-scale friction measurements using friction force microscopy: Part i – general principles and new measurement techniques. *ASME J. Tribol.*, 116:378–388, 1994.
137. T. R. Albrecht, S. Akamine, T. E. Carver, and C. F. Quate. Microfabrication of cantilever styli for the atomic force microscope. *J. Vac. Sci. Technol. A*, 8:3386–3396, 1990.
138. O. Marti, S. Gould, and P. K. Hansma. Control electronics for atomic force microscopy. *Rev. Sci. Instrum.*, 59:836–839, 1988.
139. O. Wolter, T. Bayer, and J. Greschner. Micromachined silicon sensors for scanning force microscopy. *J. Vac. Sci. Technol. B*, 9:1353–1357, 1991.
140. E. Meyer, R. Overney, R. Luthi, and D. Brodbeck. Friction force microscopy of mixed langmuir–blodgett films. *Thin Solid Films*, 220:132–137, 1992.
141. H. J. Dai, J. H. Hafner, A. G. Rinzler, D. T. Colbert, and R. E. Smalley. Nanotubes as nanoprobes in scanning probe microscopy. *Nature*, 384:147–150, 1996.

142. J. H. Hafner, C. L. Cheung, A. T. Woolley, and C. M. Lieber. Structural and functional imaging with carbon nanotube afm probes. *Prog. Biophys. Mol. Biol.*, 77:73–110, 2001.
143. G. S. Blackman, C. M. Mate, and M. R. Philpott. Interaction forces of a sharp tungsten tip with molecular films on silicon surface. *Phys. Rev. Lett.*, 65:2270–2273, 1990.
144. S. J. O'Shea, M. E. Welland, and T. Rayment. Atomic force microscope study of boundary layer lubrication. *Appl. Phys. Lett.*, 61:2240–2242, 1992.
145. J. P. Cleveland, S. Manne, D. Bocek, and P. K. Hansma. A nondestructive method for determining the spring constant of cantilevers for scanning force microscopy. *Rev. Sci. Instrum.*, 64:403–405, 1993.
146. D. W. Pohl. Some design criteria in stm. *IBM J. Res. Dev.*, 30:417, 1986.
147. W. T. Thomson and M. D. Dahleh. *Theory of Vibration with Applications*. Prentice Hall, 5th edition, 1998.
148. J. Colchero. Reibungskraftmikroskopie, 1993.
149. G. M. McClelland, R. Erlandsson, and S. Chiang. *Atomic force microscopy: General principles and a new implementation*, volume 6B, pages 1307–1314. Plenum, 1987.
150. Y. R. Shen. *The Principles of Nonlinear Optics*. Wiley, 1984.
151. T. Baumeister and S. L. Marks. *Standard Handbook for Mechanical Engineers*. McGraw-Hill, 7th edition, 1967.
152. J. Colchero, O. Marti, H. Bielefeldt, and J. Mlynek. Scanning force and friction microscopy. *Phys. Stat. Sol.*, 131:73–75, 1991.
153. R. Young, J. Ward, and F. Scire. Observation of metal-vacuum-metal tunneling, field emission, and the transition region. *Phys. Rev. Lett.*, 27, 1971.
154. R. Young, J. Ward, and F. Scire. The topographiner: An instrument for measuring surface microtopography. *Rev. Sci. Instrum.*, 43:999, 1972.
155. C. Gerber and O. Marti. Magnetostrictive positioner. *IBM Tech. Disclosure Bull.*, 27:6373, 1985.
156. R. Garcìa Cantù, M. A. Huerta Garnica. Long-scan imaging by stm. *J. Vac. Sci. Technol. A*, 8:354, 1990.
157. C. J. Chen. In situ testing and calibration of tube piezoelectric scanners. *Ultramicroscopy*, 42–44:1653–1658, 1992.
158. R. G. Carr. *J. Microscopy*, 152:379, 1988.
159. C. J. Chen. Electromechanical deflections of piezoelectric tubes with quartered electrodes. *Appl. Phys. Lett.*, 60:132, 1992.
160. N. Libioulle, A. Ronda, M. Taborelli, and J. M. Gilles. Deformations and nonlinearity in scanning tunneling microscope images. *J. Vac. Sci. Technol. B*, 9:655–658, 1991.
161. E. P. Stoll. Restoration of stm images distorted by time-dependent piezo driver aftereffects. *Ultramicroscopy*, 42–44:1585–1589, 1991.
162. R. Durselen, U. Grunewald, and W. Preuss. Calibration and applications of a high precision piezo scanner for nanometrology. *Scanning*, 17:91–96, 1995.
163. J. Fu. In situ testing and calibrating of z-piezo of an atomic force microscope. *Rev. Sci. Instrum.*, 66:3785–3788, 1995.
164. R. C. Barrett and C. F. Quate. Optical scan-correction system applied to atomic force microscopy. *Rev. Sci. Instrum.*, 62:1393, 1991.
165. R. Toledo-Crow, P. C. Yang, Y. Chen, and M. Vaez-Iravani. Near-field differential scanning optical microscope with atomic force regulation. *Appl. Phys. Lett.*, 60:2957–2959, 1992.
166. J. E. Griffith, G. L. Miller, and C. A. Green. A scanning tunneling microscope with a capacitance-based position monitor. *J. Vac. Sci. Technol. B*, 8:2023–2027, 1990.
167. A. E. Holman, C. D. Laman, P. M. L. O. Scholte, W. C. Heerens, and F. Tuinstra. A calibrated scanning tunneling microscope equipped with capacitive sensors. *Rev. Sci. Instrum.*, 67:2274–2280, 1996.

3

Probes in Scanning Microscopies

Jason H. Hafner

Summary. Scanning probe microscopy (SPM) provides nanometer-scale mapping of numerous sample properties in essentially any environment. This unique combination of high resolution and broad applicability has lead to the application of SPM to many areas of science and technology, especially those interested in the structure and properties of materials at the nanometer scale. SPM images are generated through measurements of a tip-sample interaction. A well-characterized tip is the key element to data interpretation and is typically the limiting factor.

Commercially available atomic force microscopy (AFM) tips, integrated with force sensing cantilevers, are microfabricated from silicon and silicon nitride by lithographic and anisotropic etching techniques. The performance of these tips can be characterized by imaging nanometer-scale standards of known dimension, and the resolution is found to roughly correspond to the tip radius of curvature, the tip aspect ratio, and the sample height. Although silicon and silicon nitride tips have a somewhat large radius of curvature, low aspect ratio, and limited lifetime due to wear, the widespread use of AFM today is due in large part to the broad availability of these tips. In some special cases, small asperities on the tip can provide resolution much higher than the tip radius of curvature for low-Z samples such as crystal surfaces and ordered protein arrays.

Several strategies have been developed to improve AFM tip performance. Oxide sharpening improves tip sharpness and enhances tip asperities. For high-aspect-ratio samples such as integrated circuits, silicon AFM tips can be modified by focused ion beam (FIB) milling. FIB tips reach three-degree cone angles over lengths of several microns and can be fabricated at arbitrary angles. Other high resolution and high-aspect-ratio tips are produced by electron beam deposition (EBD) in which a carbon spike is deposited onto the tip apex from the background gases in an electron microscope. Finally, carbon nanotubes have been employed as AFM tips. Their nanometer-scale diameter, long length, high stiffness, and elastic buckling properties make carbon nanotubes possibly the ultimate tip material for AFM. Nanotubes can be manually attached to silicon or silicon nitride AFM tips or "grown" onto tips by chemical vapor deposition (CVD), which should soon make them widely available. In scanning tunneling microscopy (STM), the electron tunneling signal decays exponentially with tip-sample separation, so that in principle only the last few atoms contribute to the signal. STM tips are, therefore, not as sensitive to the nanoscale tip geometry and can be made by simple mechanical cutting or electrochemical etching of metal wires. In choosing tip materials, one prefers hard, stiff metals that will not oxidize or corrode in the imaging environment.

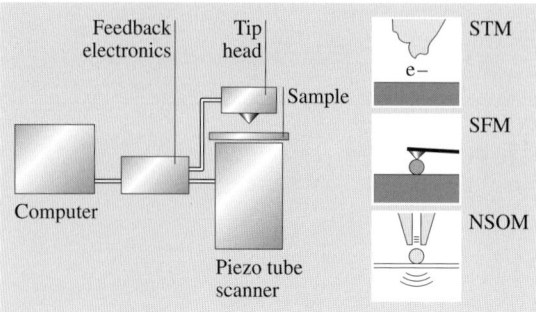

Fig. 3.1. A schematic of the components of a scanning probe microscope and the three types of signals observed: STM senses electron tunneling currents, AFM measures forces, and NSOM measures near-field optical properties via a sub-wavelength aperture

3.1 Introduction

In scanning probe microscopy (SPM), an image is created by raster scanning a sharp probe tip over a sample and measuring some highly localized tip-sample interaction as a function of position. SPMs are based on several interactions, the major types including scanning tunneling microscopy (STM), which measures an electronic tunneling current; atomic force microscopy (AFM), which measures force interactions; and near-field scanning optical microscopy (NSOM), which measures local optical properties by exploiting near-field effects (Fig. 3.1). These methods allow the characterization of many properties (structural, mechanical, electronic, optical) on essentially any material (metals, semiconductors, insulators, biomolecules) and in essentially any environment (vacuum, liquid, or ambient air conditions). The unique combination of nanoscale resolution, previously the domain of electron microscopy, *and broad applicability* has led to the proliferation of SPM into virtually all areas of nanometer-scale science and technology.

Several enabling technologies have been developed for SPM, or borrowed from other techniques. Piezoelectric tube scanners allow accurate, sub-angstrom positioning of the tip or sample in three dimensions. Optical deflection systems and microfabricated cantilevers can detect forces in AFM down to the picoNewton range. Sensitive electronics can measure STM currents less than 1 picoamp. High transmission fiber optics and sensitive photodetectors can manipulate and detect small optical signals of NSOM. Environmental control has been developed to allow SPM imaging in UHV, cryogenic temperatures, at elevated temperatures, and in fluids. Vibration and drift have been controlled such that a probe tip can be held over a single molecule for hours of observation. Microfabrication techniques have been developed for the mass production of probe tips, making SPMs commercially available and allowing the development of many new SPM modes and combinations with other characterization methods. However, of all this SPM development over the past 20 years, what has received the least attention is perhaps the most important aspect: the probe tip.

Interactions measured in SPMs occur at the tip-sample interface, which can range in size from a single atom to tens of nanometers. The size, shape, surface chemistry, electronic and mechanical properties of the tip apex will directly influence the data signal and the interpretation of the image. Clearly, the better characterized the tip the more useful the image information. In this chapter, the fabrication and performance of AFM and STM probes will be described.

3.2 Atomic Force Microscopy

AFM is the most widely used form of SPM, since it requires neither an electrically conductive sample, as in STM, nor an optically transparent sample or substrate, as in most NSOMs. Basic AFM modes measure the topography of a sample with the only requirement being that the sample is deposited on a flat surface and rigid enough to withstand imaging. Since AFM can measure a variety of forces, including van der Waals forces, electrostatic forces, magnetic forces, adhesion forces and friction forces, specialized modes of AFM can characterize the electrical, mechanical, and chemical properties of a sample in addition to its topography.

3.2.1 Principles of Operation

In AFM, a probe tip is integrated with a microfabricated force-sensing cantilever. A variety of silicon and silicon nitride cantilevers are commercially available with micron-scale dimensions, spring constants ranging from 0.01 to 100 N/m, and resonant frequencies ranging from 5 kHz to over 300 kHz. The cantilever deflection is detected by optical beam deflection, as illustrated in Fig. 3.2. A laser beam bounces off the back of the cantilever and is centered on a split photodiode. Cantilever deflections areproportional to the difference signal $V_A - V_B$. Sub-angstrom deflections can be deflected and, therefore, forces down to tens of picoNewtons can be measured. A more recently developed method of cantilever deflection measurement is through a piezoelectric layer on the cantilever that registers a voltage upon deflection [1].

A piezoelectric scanner rasters the sample under the tip while the forces are measured through deflections of the cantilever. To achieve more controlled imaging conditions, a feedback loop monitors the tip-sample force and adjusts the sample Z-position to hold the force constant. The topographic image of the sample is then taken from the sample Z-position data. The mode described is called contact mode, in which the tip is deflected by the sample due to repulsive forces, or "contact". It is generally only used for flat samples that can withstand lateral forces during scanning. To minimize lateral forces and sample damage, two AC modes have been developed. In these, the cantilever is driven into AC oscillation near its resonant frequency (tens to hundreds of kHz) with amplitudes of 5 to tens of s. When the tip approaches the sample, the oscillation is damped, and the reduced amplitude is the feedback signal, rather than the DC deflection. Again, topography is taken from the varying Z-position of the sample required to keep the tip oscillation amplitude constant. The two AC modes differ only in the nature of the interaction. In intermittent

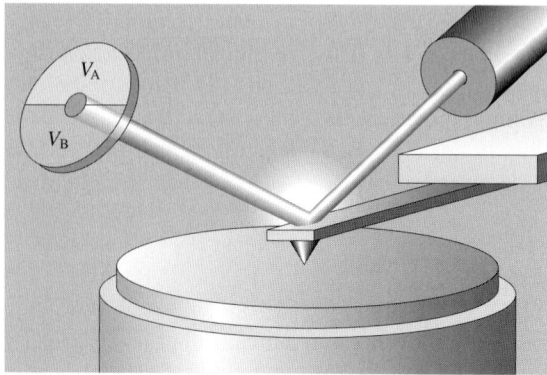

Fig. 3.2. An illustration of the optical beam deflection system that detects cantilever motion in the AFM. The voltage signal $V_A - V_B$ is proportional to the deflection

contact mode, also called tapping mode, the tip contacts the sample on each cycle, so the amplitude is reduced by ionic repulsion as in contact mode. In noncontact mode, long-range van der Waals forces reduce the amplitude by effectively shifting the spring constant experienced by the tip and changing its resonant frequency.

3.2.2 Standard Probe Tips

In early AFM work, cantilevers were made by hand from thin metal foils or small metal wires. Tips were created by gluing diamond fragments to the foil cantilevers or electrochemically etching the wires to a sharp point. Since these methods were labor intensive and not highly reproducible, they were not amenable to large-scale production. To address this problem, and the need for smaller cantilevers with higher resonant frequencies, batch fabrication techniques were developed (see Fig. 3.3). Building on existing methods to batch fabricate Si_3N_4 cantilevers, *Albrecht* et al. [2] etched an array of small square openings in an SiO_2 mask layer over a (100) silicon surface. The exposed square (100) regions were etched with KOH, an anisotropic etchant that terminates at the (111) planes, thus creating pyramidal etch pits in the silicon surface. The etch pit mask was then removed and another was applied to define the cantilever shapes with the pyramidal etch pits at the end. The Si wafer was then coated with a low stress Si_3N_4 layer by LPCVD. The Si_3N_4 fills the etch pit, using it as a mold to create a pyramidal tip. The silicon was later removed by etching to free the cantilevers and tips. Further steps resulting in the attachment of the cantilever to a macroscopic piece of glass are not described here. The resulting pyramidal tips were highly symmetric and had a tip radius of less than 30 nm, as determined by scanning electron microscopy (SEM). This procedure has likely not changed significantly, since commercially available Si_3N_4 tips are still specified to have a curvature radius of 30 nm.

Wolter et al. [3] developed methods to batch fabricate single-crystal Si cantilevers with integrated tips. Microfabricated Si cantilevers were first prepared using

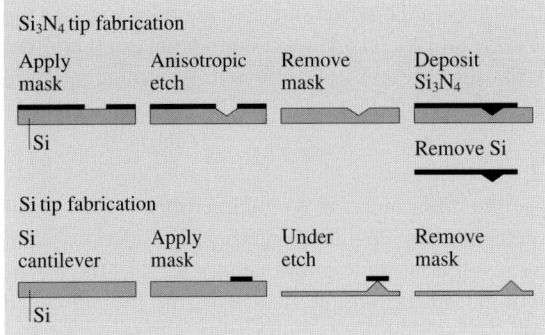

Fig. 3.3. A schematic overview of the fabrication of Si and Si$_3$N$_4$ tip fabrication as described in the text

previously described methods, and a small mask was formed at the end of the cantilever. The Si around the mask was etched by KOH, so that the mask was under cut. This resulted in a pyramidal silicon tip under the mask, which was then removed. Again, this partial description of the full procedure only describes tip fabrication. With some refinements the silicon tips were made in high yield with curvature radii of less than 10 nm. Si tips are sharper than Si$_3$N$_4$ tips, because they are directly formed by the anisotropic etch in single-crystal Si, rather than using an etch pit as a mask for deposited material. Commercially available silicon probes are made by similar refined techniques and provide a curvature typical radius of < 10 nm.

3.2.3 Probe Tip Performance

In atomic force microscopy the question of resolution can be a rather complicated issue. As an initial approximation, resolution is often considered strictly in geometrical terms that assume rigid tip-sample contact. The topographical image of a feature is broadened or narrowed by the size of the probe tip, so the resolution is approximately the width of the tip. Therefore, the resolution of AFM with standard commercially available tips is on the order of 5 to 10 nm. *Bustamante* and *Keller* [4] carried the geometrical model further by drawing an analogy to resolution in optical systems. Consider two sharp spikes separated by a distance d to be point objects imaged by AFM (see Fig. 3.4). Assume the tip has a parabolic shape with an end radius R. The tip-broadened image of these spikes will appear as inverted parabolas. There will be a small depression between the images of depth Δz. The two spikes are considered "resolved" if Δz is larger than the instrumental noise in the z direction. Defined in this manner, the resolution d, the minimum separation at which the spikes are resolved, is

$$d = 2\sqrt{2R(\Delta z)}, \tag{3.1}$$

where one must enter a minimal detectable depression for the instrument (Δz) to determine the resolution. So for a silicon tip with radius 5 nm and a minimum detectable

Δz of 0.5 nm, the resolution is about 4.5 nm. However, the above model assumes the spikes are of equal height. *Bustamante* and *Keller* [4] went on to point out that if the height of the spikes is not equal, the resolution will be affected. Assuming a height difference of Δh, the resolution becomes:

$$d = \sqrt{2R}\left(\sqrt{\Delta z} + \sqrt{\Delta z + \Delta h}\right). \tag{3.2}$$

For a pair of spikes with a 2 nm height difference, the resolution drops to 7.2 nm for a 5 nm tip and 0.5 nm minimum detectable Δz. While geometrical considerations are a good starting point for defining resolution, they ignore factors such as the possible compression and deformation of the tip and sample. *Vesenka* et al. [5] confirmed a similar geometrical resolution model by imaging monodisperse gold nanoparticles with tips characterized by transmission electron microscopy (TEM).

Noncontact AFM contrast is generated by long-range interactions such as van der Waals forces, so resolution will not simply be determined by geometry because the tip and sample are not in rigid contact. *Bustamante* and *Keller* [4] have derived an expression for the resolution in noncontact AFM for an idealized, infinitely thin "line" tip and a point particle as the sample (Fig. 3.4). Noncontact AFM is sensitive to the gradient of long-range forces, so the van der Waals force gradient was calculated as a function of position for the tip at height h above the surface. If the resolution d is defined as the full width at half maximum of this curve, the resolution is:

$$d = 0.8h. \tag{3.3}$$

This shows that even for an ideal geometry, the resolution is fundamentally limited in noncontact mode by the tip-sample separation. Under UHV conditions, the tip-sample separation can be made very small, so atomic resolution is possible on flat, crystalline surfaces. Under ambient conditions, however, the separation must be larger to keep the tip from being trapped in the ambient water layer on the surface. This larger separation can lead to a point where further improvements in tip sharpness do not improve resolution. It has been found that imaging 5 nm gold nanoparticles in noncontact mode with carbon nanotube tips of 2 nm diameter leads to particle widths of 12 nm, larger than the 7 nm width one would expect assuming rigid contact [6]. However, in tapping mode operation, the geometrical definition of resolution is relevant, since the tip and sample come into rigid contact. When imaging 5 nm gold particles with 2 nm carbon nanotube tips in tapping mode, the expected 7 nm particle width is obtained [7].

The above descriptions of AFM resolution cannot explain the sub-nanometer resolution achieved on crystal surfaces [8] and ordered arrays of biomolecules [9] in contact mode with commercially available probe tips. Such tips have nominal radii of curvature ranging from 5 nm to 30 nm, an order of magnitude larger than the resolution achieved. A detailed model to explain the high resolution on ordered membrane proteins has been put forth by [10]. In this model, the larger part of the silicon nitride tip apex balances the tip-sample interaction through electrostatic forces, while a very small tip asperity interacts with the sample to provide contrast (see Fig. 3.5). This model is supported by measurements at varying salt concentrations to vary the electrostatic interaction strength and the observation of defects in the ordered samples.

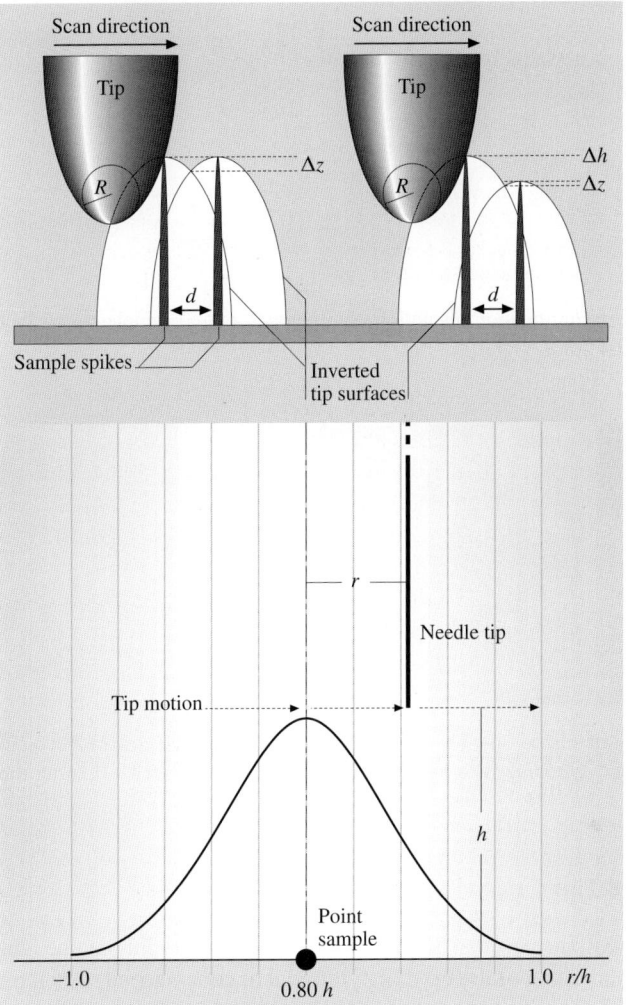

Fig. 3.4. The factors that determine AFM imaging resolution in contact mode (*top*) and non-contact mode (*bottom*), adapted from [4]

However, the existence of such asperities has never been confirmed by independent electron microscopy images of the tip. Another model, considered especially applicable to atomic resolution on crystal surfaces, assumes the tip is in contact with a region of the sample much larger than the resolution observed, and that force components matching the periodicity of the sample are transmitted to the tip, resulting in an "averaged" image of the periodic lattice. Regardless of the mechanism, the structures determined are accurate and make this a highly valuable method for membrane proteins. However, this level of resolution should not be expected for most biological systems.

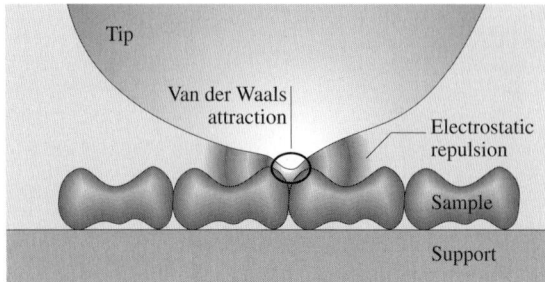

Fig. 3.5. A tip model to explain the high resolution obtained on ordered samples in contact mode, from [10]

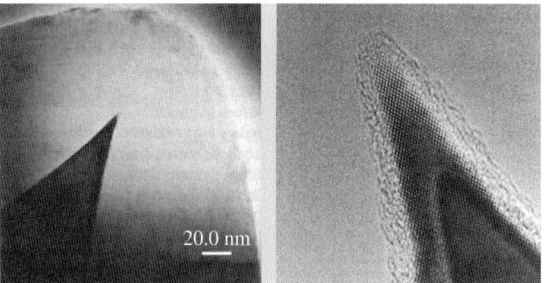

Fig. 3.6. Oxide sharpening of silicon tips. The *left image* shows a sharpened core of silicon in an outer layer of SiO_2. The *right image* is a higher magnification view of such a tip after the SiO_2 is removed. Adapted from [11]

3.2.4 Oxide-Sharpened Tips

Both Si and Si_3N_4 tips with increased aspect ratio and reduced tip radius can be fabricated through oxide sharpening of the tip. If a pyramidal or cone-shaped silicon tip is thermally oxidized to SiO_2 at low temperature ($< 1050\,°C$), Si-SiO_2 stress formation reduces the oxidation rate at regions of high curvature. The result is a sharper, higher-aspect-ratio cone of silicon at the high curvature tip apex inside the outer pyramidal layer of SiO_2 (see Fig. 3.6). Etching the SiO_2 layer with HF then leaves tips with aspect ratios up to 10:1 and radii down to 1 nm [11], although 5–10nm is the nominal specification for most commercially available tips. This oxide sharpening technique can also be applied to Si_3N_4 tips by oxidizing the silicon etch pits that are used as molds. As with tip fabrication, oxide sharpening is not quite as effective for Si_3N_4. Si_3N_4 tips were reported to have an 11 nm radius of curvature [12], while commercially available oxide-sharpened Si_3N_4 tips have a nominal radius of < 20 nm.

3.2.5 FIB tips

A common AFM application in integrated circuit manufacture and MEMs microelectromechanical systemsis to image structures with very steep sidewalls such as trenches. To accurately image these features, one must consider the micron-scale tip structure, rather than the nanometer-scale structure of the tip apex. Since tip fabrication processes rely on anisotropic etchants, the cone half-angles of pyramidal tips are approximately 20 degrees. Images of deep trenches taken with such tips display slanted sidewalls and may not reach the bottom of the trench due to the tip broadening effects. To image such samples more faithfully, high-aspect-ratio tips are fabricated by focused ion beam (FIB) machining a Si tip to produce a sharp spike at the tip apex. Commercially available FIB tips have half cone angles of < 3 degrees over lengths of several microns, yielding aspect ratios of approximately 10:1. The radius of curvature at the tip end is similar to that of the tip before the FIB machining. Another consideration for high-aspect-ratio tips is the tip tilt. To ensure that the pyramidal tip is the lowest part of the tip-cantilever assembly, most AFM designs tilt the cantilever about 15 degrees from parallel. Therefore, even an ideal "line tip" will not give an accurate image of high steep sidewalls, but will produce an image that depends on the scan angle. Due to the versatility of the FIB machining, tips are available with the spikes at an angle to compensate for this effect.

3.2.6 EBD tips

Another method of producing high-aspect-ratio tips for AFM is called electron beam deposition (EBD). First developed for STM tips [13, 14], EBD tips were introduced for AFM by focusing an SEM onto the apex of a pyramidal tip arranged so that it pointed along the electron beam axis (see Fig. 3.7). Carbon material was deposited by the dissociation of background gases in the SEM vacuum chamber. *Schiffmann* [15] systematically studied the following parameters and how they affected EBD tip geometry:

Deposition time : 0.5 to 8 min
Beam current : 3 − −300pA
Beam energy : 1 − −30keV
Working distance: 8 − −48mm .

EBD tips were cylindrical with end radii of 20–40nm, lengths of 1 to 5 µm, and diameters of 100 to 200 nm. Like FIB tips, EBD tips were found to achieve improved imaging of steep features. By controlling the position of the focused beam, the tip geometry can be further controlled. Tips were fabricated with lengths over 5 µm and aspect ratios greater than 100:1, yet these were too fragile to use as a tip in AFM [14].

3.2.7 Carbon Nanotube Tips

Carbon nanotubes are microscopic graphitic cylinders that are nanometers in diameter, yet many microns in length. Single-walled carbon nanotubes (SWNT) consist

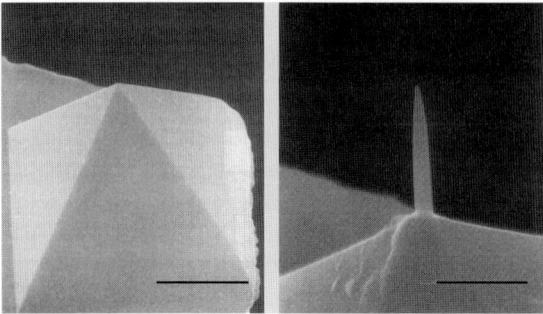

Fig. 3.7. A pyramidal tip before (*left*, 2-μm-scale bar) and after (*right*, 1-μm-scale bar) electron beam deposition, adapted from [14]

of single sp^2 hybridized carbon sheets rolled into seamless tubes and have diameters ranging from 0.7 to 3 nm.

Carbon Nanotube Structure

Larger structures called multiwalled carbon nanotubes (MWNT) consist of nested, concentrically arranged SWNT and have diameters ranging from 3 to 50 nm. Figure 3.8 shows a model of nanotube structure, as well as TEM images of a SWNT and a MWNT. The small diameter and high aspect ratio of carbon nanotubes suggests their application as high resolution, high-aspect-ratio AFM probes.

Carbon Nanotube Mechanical Properties

Carbon nanotubes possess exceptional mechanical properties that impact their use as probes. Their lateral stiffness can be approximated from that of a solid elastic rod:

$$k_{\text{lat}} = \frac{3\pi Y r^4}{4l^3}, \tag{3.4}$$

where the spring constant k_{lat} represents the restoring force per unit lateral displacement, r is the radius, l is the length, and Y is the Young's modulus (also called the elastic modulus) of the material. For the small diameters and extreme aspect ratios of carbon nanotube tips, the thermal vibrations of the probe tip at room temperature can become sufficient to degrade image resolution. These thermal vibrations can be approximated by equating $\frac{1}{2}k_B T$ of thermal energy to the energy of an oscillating nanotube:

$$\frac{1}{2}k_B T = \frac{1}{2}k_{\text{lat}} a^2, \tag{3.5}$$

where k_B is Boltzmann's constant, T is the temperature, and a is the vibration amplitude. Substituting for k_{lat} from (3.4) yields:

$$a = \sqrt{\frac{4k_B T l^3}{3\pi Y r^4}}. \tag{3.6}$$

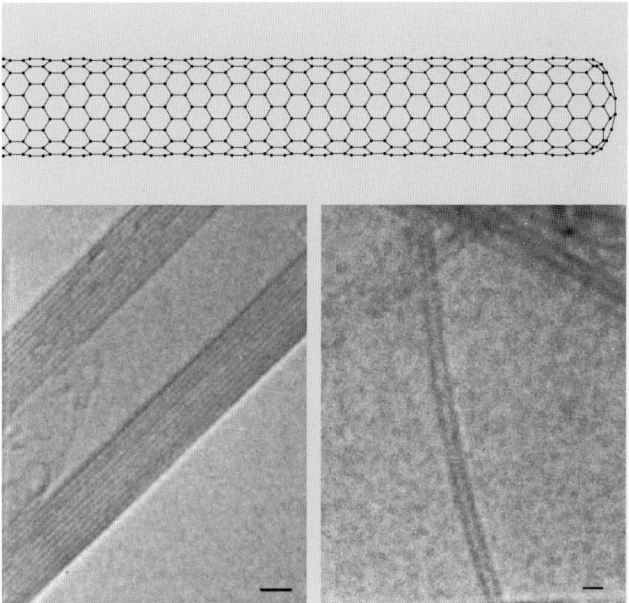

Fig. 3.8. The structure of carbon nanotubes, including TEM images of a MWNT (*left*) and a SWNT (*right*), from [16]

The strong dependence on radius and length reveals that one must carefully control the tip geometry at this size scale. (3.6) implies that the stiffer the material, i.e., the higher its Young's modulus, the smaller the thermal vibrations and the longer and thinner a tip can be. The Young's moduli of carbon nanotubes have been determined by measurements of the thermal vibration amplitude by TEM [17, 18] and by directly measuring the forces required to deflect a pinned carbon nanotube in an AFM [19]. These experiments revealed that the Young's modulus of carbon nanotubes is 1–2 TPa, in agreement with theoretical predictions [20]. This makes carbon nanotubes the stiffest known material and, therefore, the best for fabricating thin, high-aspect-ratio tips. A more detailed and accurate derivation of the thermal vibration amplitudes was derived for the Young's modulus measurements [17, 18].

Carbon nanotubes elastically buckle under large loads, rather than fracture or plastically deform like most materials. Nanotubes were first observed in the buckled state by transmission electron microscopy [21], as shown in Fig. 3.9. The first experimental evidence that nanotube buckling is elastic came from the application of nanotubes as probe tips [22], described in detail below. A more direct experimental observation of elastic buckling was obtained by deflecting nanotubes pinned to a low friction surface with an AFM tip [19]. Both reports found that the buckling force could be approximated with the macroscopic Euler buckling formula for an elastic column:

$$F_{\text{Euler}} = \frac{\pi^3 Y r^4}{4 l^2} . \tag{3.7}$$

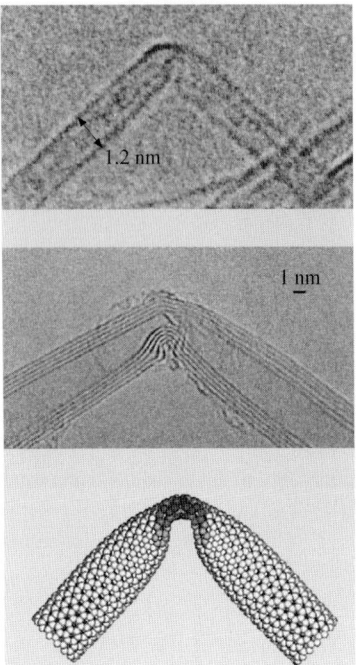

Fig. 3.9. TEM images and a model of a buckled nanotube, adapted from [21]

The buckling force puts another constraint on the tip length: If the nanotube is too long the buckling force will be too low for stable imaging. The elastic buckling property of carbon nanotubes has significant implications for their use as AFM probes. If a large force is applied to the tip inadvertently, or if the tip encounters a large step in sample height, the nanotube can buckle to the side, then snap back without degraded imaging resolution when the force is removed, making these tips highly robust. No other tip material displays this buckling characteristic.

Manually Assembled Nanotube Probes

The first carbon nanotube AFM probes [22] were fabricated by techniques developed for assembling single-nanotube field emission tips [23]. This process, illustrated in Fig. 3.10, used purified MWNT material synthesized by the carbon arc procedure. The raw material, which must contain at least a few percent of long nanotubes (> 10 μm) by weight, purified by oxidation to approximately 1% of its original mass. A torn edge of the purified material was attached to a micromanipulator by carbon tape and viewed under a high power optical microscope. Individual nanotubes and nanotube bundles were visible as filaments under dark field illumination. A commercially available AFM tip was attached to another micromanipulator opposing the nanotube material. Glue was applied to the tip apex from high vacuum carbon tape supporting the nanotube material. Nanotubes were then manually attached to the

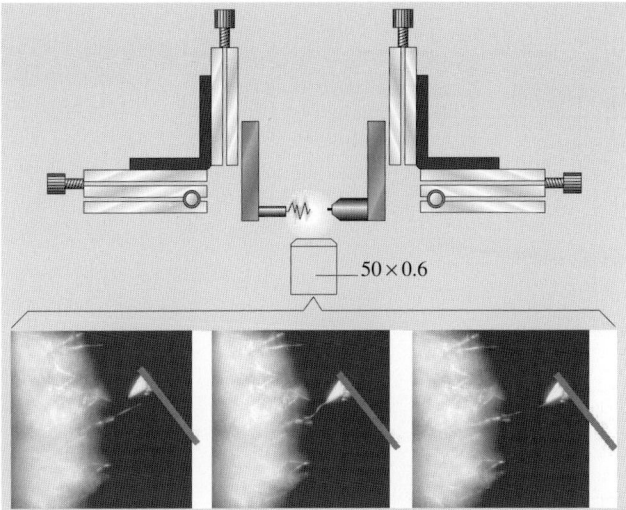

Fig. 3.10. A schematic drawing of the setup for manual assembly of carbon nanotube tips (*top*) and optical microscopy images of the assembly process (the cantilever was drawn in for clarity)

tip apex by micromanipulation. As assembled, MWNT tips were often too long for imaging due to thermal vibrations and low buckling forces described in Sect. 3.2.7. Nanotubes tips were shortened by applying 10 V pulses to the tip while it was near a sputtered niobium surface. This process etched ~ 100 nm lengths of nanotube per pulse.

The manually assembled MWNT tips demonstrated several important nanotube tip properties [22]. First, the high aspect ratio of the MWNT tips allowed the accurate imaging of trenches in silicon with steep sidewalls, similar to FIB and EBD tips. Second, elastic buckling was observed indirectly through force curves (see Fig. 3.11). Note that as the tip taps the sample, the amplitude drops to zero and a DC deflection is observed, because the nanotube is unbuckled and is essentially rigid. As the tip moves closer, the force on the nanotube eventually exceeds the buckling force. The nanotube buckles, allowing the vibration amplitude to partially recover, and the deflection remains constant. Numeric tip trajectory simulations could only reproduce these force curves if elastic buckling was included in the nanotube response. Finally, the nanotube tips were highly robust. Even after "tip crashes" or hundreds of controlled buckling cycles, the tip retained its resolution and high aspect ratio.

Manual assembly of carbon nanotube probe tips is straightforward, but has several limitations. It is labor intensive and not amenable to mass production. Although MWNT tips have been made commercially available by this method, they are about ten times more expensive than silicon probes. The manual assembly method has also been carried out in an SEM, rather than an optical microscope [24]. This eliminates the need for pulse-etching, since short nanotubes can be attached to the tip, and the

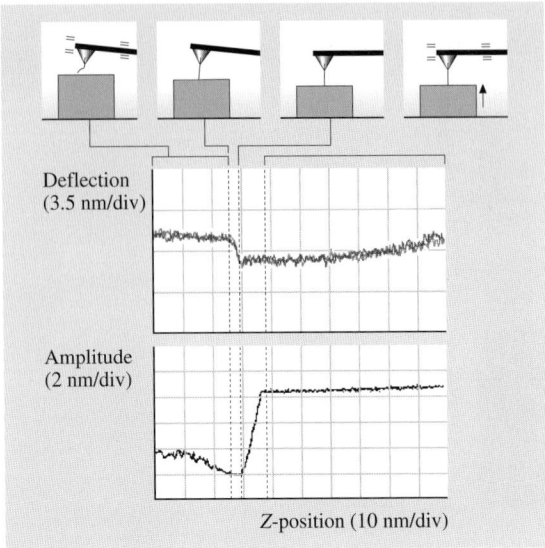

Fig. 3.11. Nanotube tip buckling. *Top* diagrams correspond to labeled regions of the force curves. As the nanotube tip buckles, the deflection remains constant and the amplitude increases, from [16]

"glue" can be applied by EBD. But this is still not the key to mass production, since nanotube tips are made individually. MWNT tips provided a modest improvement in resolution on biological samples, but typical MWNT radii are similar to that of silicon tips, so they cannot provide the ultimate resolution possible with a SWNT tip. SWNT bundles can be attached to silicon probes by manual assembly. Pulse etching at times produces very high resolution tips that likely result from exposing a small number of nanotubes from the bundle, but this is not reproducible [25]. Even if a sample could be prepared that consisted of individual SWNT for manual assembly, such nanotubes would not be easily visible by optical microscopy or SEM.

CVD Nanotube Probe Synthesis

The problems of manual assembly of nanotube probes discussed above can largely be solved by directly growing nanotubes onto AFM tips by metal-catalyzed chemical vapor deposition (CVD). The key features of the nanotube CVD process are illustrated in Fig. 3.12. Nanometer-scale metal catalyst particles are heated in a gas mixture containing hydrocarbon or CO. The gas molecules dissociate on the metal surface, and carbon is adsorbed into the catalyst particle. When this carbon precipitates, it nucleates a nanotube of similar diameter to the catalyst particle. Therefore, CVD allows control over nanotube size and structure, including the production of SWNTs [26] with radii as low as 3.5 Angstrom [27].

Several key issues must be addressed to grow nanotube AFM tips by CVD: (1) the alignment of the nanotubes at the tip, (2) the number of nanotubes that grow

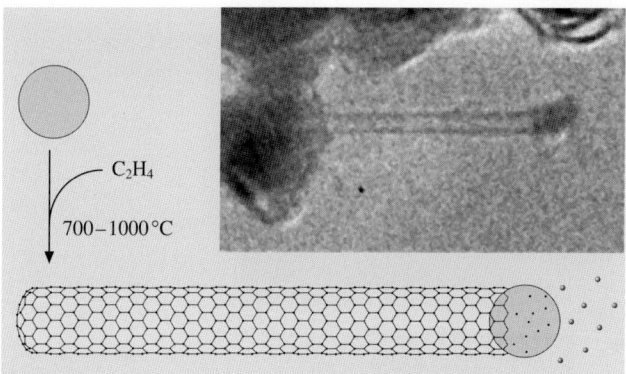

Fig. 3.12. CVD nanotube synthesis. Ethylene reacts with a nanometer-scale iron catalyst particle at high temperature to form a carbon nanotube. The *inset* in the upper right is a TEM image showing a catalyst particle at the end of a nanotube, from [16]

at the tip, and (3) the length of the nanotube tip. *Li* et al. [28] found that nanotubes grow perpendicular to a porous surface containing embedded catalyst. This approach was exploited to fabricate nanotube tips by CVD [29] with the proper alignment, as illustrated in Fig. 3.13. A flattened area of approximately 1–5µm² was created on Si tips by scanning in contact mode at high load (1 µN) on a hard, synthetic diamond surface. The tip was then anodized in HF to create 100 nm-diameter pores in this flat surface [30]. It is important to only anodize the last 20–40µm of the cantilever, which includes the tip, so that the rest of the cantilever is still reflective for use in the AFM. This was achieved by anodizing the tip in a small drop of HF under the view of an optical microscope. Next, iron was electrochemically deposited into the pores to form catalyst particles [31]. Tips prepared in this way were heated in low concentrations of ethylene at 800 °C, which is known to favor the growth of thin nanotubes [26]. When imaged by SEM, nanotubes were found to grow perpendicular to the surface from the pores as desired (Fig. 3.13). TEM revealed that the nanotubes were thin, individual, multiwalled nanotubes with typical radii ranging from 3–5nm. If nanotubes did not grow in an acceptable orientation, the carbon could be removed by oxidation, and then CVD repeated to grow new nanotube tips.

These "pore-growth" CVD nanotube tips were typically several microns in length – too long for imaging – and were pulse-etched to a usable length of < 500 nm. The tips exhibited elastic buckling behavior and were very robust in imaging. In addition, the thin, individual nanotube tips enabled improved resolution [29] on isolated proteins. The pore-growth method demonstrated the potential of CVD to simplify the fabrication of nanotube tips, although there were still limitations. In particular, the porous layer was difficult to prepare and rather fragile.

An alternative approach for CVD fabrication of nanotube tips involves direct growth of SWNTs on the surface of a pyramidal AFM tip [32, 33]. In this "surface-growth" approach, an alumina/iron/molybdenum-powdered catalyst known to produce SWNT [26] was dispersed in ethanol at 1 mg/mL. Silicon tips were dipped in

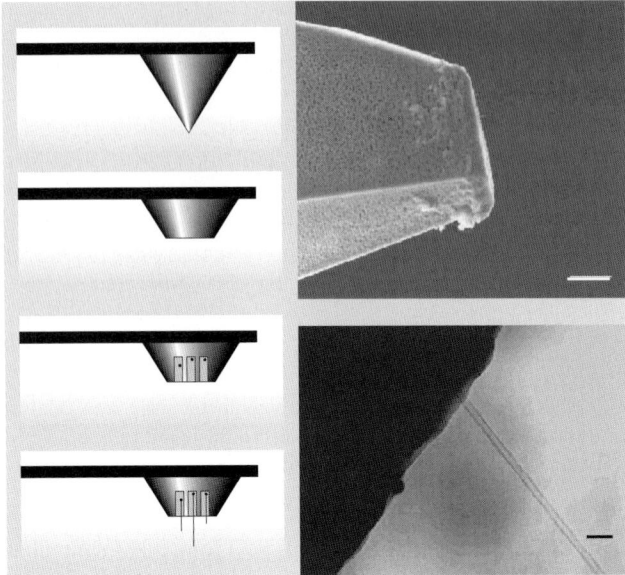

Fig. 3.13. Pore-growth CVD nanotube tip fabrication. The *left panel*, from top to bottom, shows the steps described in the text. The *upper right* is an SEM image of such a tip with a small nanotube protruding from the pores (scale bar is 1 μm). The *lower right* is a TEM of a nanotube protruding from the pores (scale bar is 20 nm), from [16]

this solution and allowed to dry, leaving a sparse layer of ~ 100 nm catalyst clusters on the tip. When CVD conditions were applied, single-walled nanotubes grew along the silicon tip surface. At a pyramid edge, nanotubes can either bend to align with the edge, or protrude from the surface. If the energy required to bend the tube and follow the edge is less than the attractive nanotube-surface energy, then the nanotube will follow the pyramid edge to the apex. Therefore, nanotubes were effectively steered toward the tip apex by the pyramid edges. At the apex, the nanotube protruded from the tip, since the energetic cost of bending around the sharp silicon tip was too high. The high aspect ratio at the oxide-sharpened silicon tip apex was critical for good nanotube alignment. A schematic of this approach is shown in Fig. 3.14. Evidence for this model came from SEM investigations that show that a very high yield of tips contains nanotubes only at the apex, with very few protruding elsewhere from the pyramid. TEM analysis demonstrated that the tips typically consist of small SWNT bundles that are formed by nanotubes coming together from different edges of the pyramid to join at the apex, supporting the surface growth model described above (Fig. 3.14). The "surface growth" nanotube tips exhibit a high aspect ratio and high resolution imaging, as well as elastic buckling.

The surface growth method has been expanded to include wafer-scale production of nanotube tips with yields of over 90% [34], yet one obstacle remains to the mass production of nanotube probe tips. Nanotubes protruding from the tip are several microns long, and since they are so thin, they must be etched to less than 100 nm.

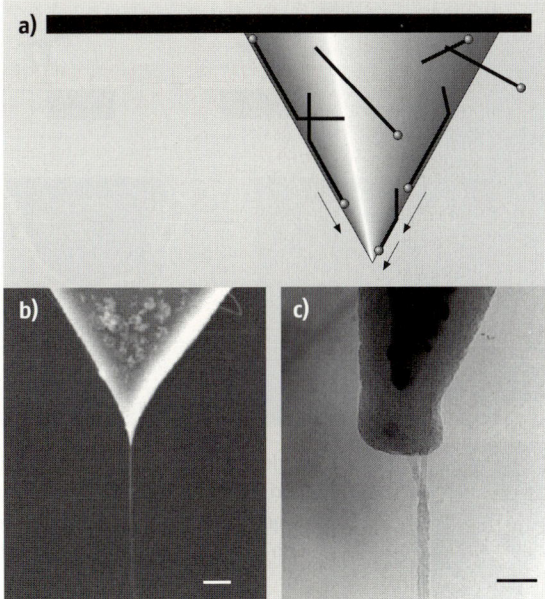

Fig. 3.14. Surface-growth nanotube tip fabrication. (**a**) Schematic represents the surface growth process in which nanotubes growing on the pyramidal tip are guided to the tip apex. (**b**),(**c**) Images show (**b**) SEM (200-nm-scale bar) and (**c**) TEM (20-nm-scale bar) images of a surface growth tip, from [16]

While the pulse-etching step is fairly reproducible, it must be carried out on nanotube tips in a serial fashion, so surface growth does not yet represent a true method of batch nanotube tip fabrication.

Hybrid Nanotube Tip Fabrication: Pick-up Tips

Another method of creating nanotube tips is something of a hybrid between assembly and CVD. The motivation was to create AFM probes that have an *individual* SWNT at the tip to achieve the ultimate imaging resolution. In order to synthesize isolated SWNT, they must be nucleated at sites separated farther than their typical length. The alumina-supported catalyst contains a high density of catalyst particles per 100 nm cluster, so nanotube bundles cannot be avoided. To fabricate completely isolated nanotubes, isolated catalyst particles were formed by dipping a silicon wafer in an isopropyl alcohol solution of $Fe(NO_3)_3$. This effectively left a submonolayer of iron on the wafer, so that when it was heated in a CVD furnace, the iron became mobile and aggregated to form isolated iron particles. During CVD conditions, these particles nucleated and grew SWNTs. By controlling the reaction time, the SWNT lengths were kept shorter than their typical separation, so that the nanotubes never had a chance to form bundles. AFM analysis of these samples revealed 1–3nm-diameter SWNT and un-nucleated particles on the surface (Fig. 3.15). However, there of 5 nm gold particles that have a *full width* of ~ 7 nm, the expected geometrical resolution for a 2 nm cylindrical probe [7]. Although the pick-up method is serial in nature, it may still be the key to the mass production of nanotube tips. Note that the original nanotube tip length can be measured electronically from the size of the Z-piezo step. The shortening can be electronically controlled through the hysteresis in the force curves. Therefore, the entire procedure (including tip exchange) can be automated by computer.

3.3 Scanning Tunneling Microscopy

Scanning tunneling microscopy (STM) was the original scanning probe microscopy and generally produces the highest resolution images, routinely achieving atomic resolution on flat, conductive surfaces. In STM, the probe tip consists of a sharpened metal wire that is held 0.3 to 1 nm from the sample. A potential difference of 0.1 V to 1 V between the tip and sample leads to tunneling currents on the order of 0.1 to 1 nA. As in AFM, a piezo-scanner rasters the sample under the tip, and the Z-position is adjusted to hold the tunneling current constant. The Z-position data represents the "topography", or in this case the surface, of constant electron density. As with other SPMs, the tip properties and performance greatly depend on the experiment being carried out. Although it is nearly impossible to prepare a tip with a known atomic structure, a number of factors are known to affect tip performance, and several preparation methods have been developed that produce good tips.

The nature of the sample being investigated and the scanning environment will affect the choice of the tip material and how the tip is fabricated. Factors to consider are mechanical properties – a hard material that will resist damage during tip-sample contact is desired. Chemical properties should also be considered – formation of oxides, or other insulating contaminants will affect tip performance. Tungsten is a common tip material because it is very hard and will resist damage, but its use is limited to ultrahigh vacuum (UHV) conditions, since it readily oxidizes. For imaging under ambient conditions an inert tip material such as platinum or gold is preferred. Platinum is typically alloyed with iridium to increase its stiffness.

3.3.1 Mechanically Cut STM Tips

STM tips can be fabricated by simple mechanical procedures such as grinding or cutting metal wires. Such tips are not formed with highly reproducible shapes and have a large opening angle and a large radius of curvature in the range of 0.1 to 1 µm (see Fig. 3.18a). They are not useful for imaging samples with surface roughness above a few nanometers. However, on atomically flat samples, mechanically cut tips can achieve atomic resolution due to the nature of the tunneling signal, which drops exponentially with tip-sample separation. Since mechanically cut tips contain many small asperities on the larger tip structure, atomic resolution is easily achieved as long as one atom of the tip is just a few angstroms lower than all of the others.

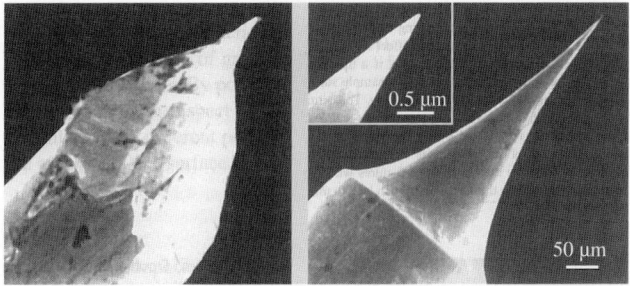

Fig. 3.18. A mechanically cut STM tip (*left*) and an electrochemically etched STM tip (*right*), from [35]

3.3.2 Electrochemically Etched STM Tips

For samples with more than a few nanometers of surface roughness, the tip structure in the nanometer-size range becomes an issue. Electrochemical etching can provide tips with reproducible and desirable shapes and sizes (Fig. 3.18), although the exact atomic structure of the tip apex is still not well controlled. The parameters of electrochemical etching depend greatly on the tip material and the desired tip shape. The following is an entirely general description. A fine metal wire (0.1–1mm diameter) of the tip material is immersed in an appropriate electrochemical etchant solution. A voltage bias of 1–10V is applied between the tip and a counterelectrode such that the tip is etched. Due to the enhanced etch rate at the electrolyte-air interface, a neck is formed in the wire. This neck is eventually etched thin enough so that it cannot support the weight of the part of the wire suspended in the solution, and it breaks to form a sharp tip. The widely varying parameters and methods will be not be covered in detail here, but many recipes are found in the literature for common tip materials [36–39].

References

1. R. Linnemann, T. Gotszalk, I. W. Rangelow, P. Dumania, and E. Oesterschulze. Atomic force microscopy and lateral force microscopy using piezoresistive cantilevers. *J. Vac. Sci. Technol. B*, 14(2):856–860, 1996.
2. T. R. Albrecht, S. Akamine, T. E. Carver, and C. F. Quate. Microfabrication of cantilever styli for the atomic force microscope. *J. Vac. Sci. Technol. A*, 8(4):3386–3396, 1990.
3. O. Wolter, T. Bayer, and J. Greschner. Micromachined silicon sensors for scanning force microscopy. *J. Vac. Sci. Technol. B*, 9(2):1353–1357, 1991.
4. C. Bustamante and D. Keller. Scanning force microscopy in biology. *Phys. Today*, 48(12):32–38, 1995.
5. J. Vesenka, S. Manne, R. Giberson, T. Marsh, and E. Henderson. Colloidal gold particles as an incompressible atomic force microscope imaging standard for assessing the compressibility of biomolecules. *Biophys. J.*, 65:992–997, 1993.
6. C. L. Cheung C. M. Lieber J. H. Hafner. Unpublished results, 2001.

7. J. H. Hafner, C. L. Cheung, T. H. Oosterkamp, and C. M. Lieber. High-yield assembly of individual single-walled carbon nanotube tips for scanning probe microscopies. *J. Phys. Chem. B*, 105(4):743–746, 2001.
8. F. Ohnesorge and G. Binnig. True atomic resolution by atomic force microscopy through repulsive and attractive forces. *Science*, 260:1451–1456, 1993.
9. D. J. Muller, D. Fotiadis, and A. Engel. Mapping flexible protein domains at subnanometer resolution with the atomic force microscope. *FEBS Lett.*, 430(1–2 Special Issue SI):105–111, 1998.
10. D. J. Muller, D. Fotiadis, S. Scheuring, S. A. Muller, and A. Engel. Electrostatically balanced subnanometer imaging of biological specimens by atomic force microscope. *Biophys. J.*, 76(2):1101–1111, 1999.
11. R. B. Marcus, T. S. Ravi, T. Gmitter, K. Chin, D. Liu, W. J. Orvis, D. R. Ciarlo, C. E. Hunt, and J. Trujillo. Formation of silicon tips with < 1 nm radius. *Appl. Phys. Lett.*, 56(3):236–238, 1990.
12. S. Akamine, R. C. Barrett, and C. F. Quate. Improved atomic force microscope images using microcantilevers with sharp tips. *Appl. Phys. Lett.*, 57(3):316–318, 1990.
13. T. Ichihashi and S. Matsui. In situ observation on electron beam induced chemical vapor deposition by transmission electron microscopy. *J. Vac. Sci. Technol. B*, 6(6):1869–1872, 1988.
14. D. J. Keller and C. Chih-Chung. Imaging steep, high structures by scanning force microscopy with electron beam deposited tips. *Surf. Sci.*, 268:333–339, 1992.
15. K. I. Schiffmann. Investigation of fabrication parameters for the electron-beam-induced deposition of contamination tips used in atomic force microscopy. *Nanotechnology*, 4:163–169, 1993.
16. J. H. Hafner, C. L. Cheung, A. T. Woolley, and C. M. Lieber. Structural and functional imaging with carbon nanotube afm probes. *Prog. Biophys. Mol. Biol.*, 77(1):73–110, 2001.
17. M. M. J. Treacy, T. W. Ebbesen, and J. M. Gibson. Exceptionally high young's modulus observed for individual carbon nanotubes. *Nature*, 381:678–680, 1996.
18. A. Krishnan, E. Dujardin, T. W. Ebbesen, P. N. Yianilos, and M. M. J. Treacy. Young's modulus of single-walled nanotubes. *Phys. Rev. B*, 58(20):14013–14019, 1998.
19. E. W. Wong, P. E. Sheehan, and C. M. Lieber. Nanobeam mechanics – elasticity, strength, and toughness of nanorods and nanotubes. *Science*, 277(5334):1971–1975, 1997.
20. J. P. Lu. Elastic properties of carbon nanotubes and nanoropes. *Phys. Rev. Lett.*, 79(7):1297–1300, 1997.
21. S. Iijima, C. Brabec, A. Maiti, and J. Bernholc. Structural flexibility of carbon nanotubes. *J. Chem. Phys.*, 104(5):2089–2092, 1996.
22. H. J. Dai, J. H. Hafner, A. G. Rinzler, D. T. Colbert, and R. E. Smalley. Nanotubes as nanoprobes in scanning probe microscopy. *Nature*, 384(6605):147–150, 1996.
23. A. G. Rinzler, Y. H. Hafner, P. Nikolaev, L. Lou, S. G. Kim, D. Tomanek, D. T. Colbert, and R. E. Smalley. Unraveling nanotubes: Field emission from atomic wire. *Science*, 269:1550, 1995.
24. H. Nishijima, S. Kamo, S. Akita, Y. Nakayama, K. I. Hohmura, S. H. Yoshimura, and K. Takeyasu. Carbon-nanotube tips for scanning probe microscopy: Preparation by a controlled process and observation of deoxyribonucleic acid. *Appl. Phys. Lett.*, 74(26):4061–4063, 1999.
25. S. S. Wong, A. T. Woolley, T. W. Odom, J. L. Huang, P. Kim, and D. V. Vezenov, C. M. Lieber. Single-walled carbon nanotube probes for high-resolution nanostructure imaging. *Appl. Phys. Lett.*, 73(23):3465–3467, 1998.

26. J. H. Hafner, M. J. Bronikowski, B. R. Azamian, P. Nikolaev, A. G. Rinzler, D. T. Colbert, K. A. Smith, and R. E. Smalley. Catalytic growth of single-wall carbon nanotubes from metal particles. *Chem. Phys. Lett.*, 296(1–2):195–202, 1998.
27. P. Nikolaev, M. J. Bronikowski, R. K. Bradley, F. Rohmund, D. T. Colbert, K. A. Smith, and R. E. Smalley. Gas-phase catalytic growth of single-walled carbon nanotubes from carbon monoxide. *Chem. Phys. Lett.*, 313(1–2):91–97, 1999.
28. W. Z. Li, S. S. Xie, L. X. Qian, B. H. Chang, B. S. Zou, W. Y. Zhou, R. A. Zhao, and G. Wang. Large-scale synthesis of aligned carbon nanotubes. *Science*, 274(5293):1701–1703, 1996.
29. J. H. Hafner, C. L. Cheung, and C. M. Lieber. Growth of nanotubes for probe microscopy tips. *Nature*, 398(6730):761–762, 1999.
30. V. Lehmann. The physics of macroporous silicon formation. *Thin Solid Films*, 255:1–4, 1995.
31. F. Ronkel, J. W. Schultze, and R. Arensfischer. Electrical contact to porous silicon by electrodeposition of iron. *Thin Solid Films*, 276(1–2):40–43, 1996.
32. J. H. Hafner, C. L. Cheung, and C. M. Lieber. Direct growth of single-walled carbon nanotube scanning probe microscopy tips. *J. Am. Chem. Soc.*, 121(41):9750–9751, 1999.
33. E. B. Cooper, S. R. Manalis, H. Fang, H. Dai, K. Matsumoto, S. C. Minne, T. Hunt, and C. F. Quate. Terabit-per-square-inch data storage with the atomic force microscope. *Appl. Phys. Lett.*, 75(22):3566–3568, 1999.
34. E. Yenilmez, Q. Wang, R. J. Chen, D. Wang, and H. Dai. Wafer scale production of carbon nanotube scanning probe tips for atomic force microscopy. *Appl. Phys. Lett.*, 80(12):2225–2227, 2002.
35. A. Stemmer, A. Hefti, U. Aebi, and A. Engel. Scanning tunneling and transmission electron microscopy on identical areas of biological specimens. *Ultramicroscopy*, 30(3):263, 1989.
36. R. Nicolaides, L. Yong, W. E. Packard, W. F. Zhou, H. A. Blackstead, K. K. Chin, J. D. Dow, J. K. Furdyna, M. H. Wei, R. C. Jaklevic, W. J. Kaiser, A. R. Pelton, M. V. Zeller, and J. J. Bellina. Scanning tunneling microscope tip structures. *J. Vac. Sci. Technol. A*, 6(2):445–447, 1988.
37. J. P. Ibe, P. P. Bey, S. L. Brandow, R. A. Brizzolara, N. A. Burnham, D. P. DiLella, K. P. Lee, C. R. K. Marrian, and R. J. Colton. On the electrochemical etching of tips for scanning tunneling microscopy. *J. Vac. Sci. Technol. A*, 8:3570–3575, 1990.
38. L. Libioulle, Y. Houbion, and J.-M. Gilles. Very sharp platinum tips for scanning tunneling microscopy. *Rev. Sci. Instrum.*, 66(1):97–100, 1995.
39. A. J. Nam, A. Teren, T. A. Lusby, and A. J. Melmed. Benign making of sharp tips for stm and fim: Pt, Ir, Au, Pd, and Rh. *J. Vac. Sci. Technol. B*, 13(4):1556–1559, 1995.

4

Noncontact Atomic Force Microscopy and Its Related Topics

Seizo Morita, Franz J. Giessibl, Yasuhiro Sugawara, Hirotaka Hosoi, Koichi Mukasa, Akira Sasahara, and Hiroshi Onishi

Summary. The scanning probe microscopy (SPM) such as the STM and the NC-AFM is the basic technology for the nanotechnology and also for the future bottom-up process. In Sect. 4.2, the principles of AFM such as operating modes and the frequency modulation method of the NC-AFM are fully explained. Then, in Sect. 4.3, applications of NC-AFM to semiconductors that make clear its potentials such as spatial resolution and functions are introduced. Next, in Sect. 4.4, applications of NC-AFM to insulators such as alkali halides, fluorides and transition metal oxides are introduced. At last, in Sect. 4.5, applications of NC-AFM to molecules such as carboxylate (RCOO$^-$) with R = H, CH$_3$, C(CH$_3$)$_3$ and CF$_3$ are introduced. Thus, the NC-AFM can observe atoms and molecules on various kinds of surfaces such as semiconductor, insulator and metal oxide with atomic/molecular resolutions. These sections are essential to understand the status of the art and the future possibility of NC-AFM that is the second generation of atom/molecule technology.

4.1 Introduction

The scanning tunneling microscope (STM) is an atomic tool based on an electric method that measures the tunneling current between a conductive tip and a conductive surface. It can electrically observe individual atoms/molecules. It can characterize or analyze the electronic nature around surface atoms/molecules. In addition, it can manipulate individual atoms/molecules. Hence, the STM is the first generation of atom/molecule technology. On the other hand, the atomic force microscope (AFM) is a unique atomic tool based on a mechanical method that can deal with even the insulator surface. Since the invention of noncontactAFM (NC-AFM) in 1995, the NC-AFM and the NC-AFM-based method have rapidly developed into a powerful surface tool on atomic/molecular scales, because the NC-AFM has the following characteristics: (1) it has true atomic resolution, (2) it can measure atomic force (so-called atomic force spectroscopy), (3) it can observe even insulators, and (4) it can measure mechanical responses such as elastic deformation. Thus, the NC-AFM is the second generation of atom/molecule technology. The scanning probe microscopy (SPM) such as the STM and the NC-AFM is the basic technology for nanotechnology and also for the future bottom-up process.

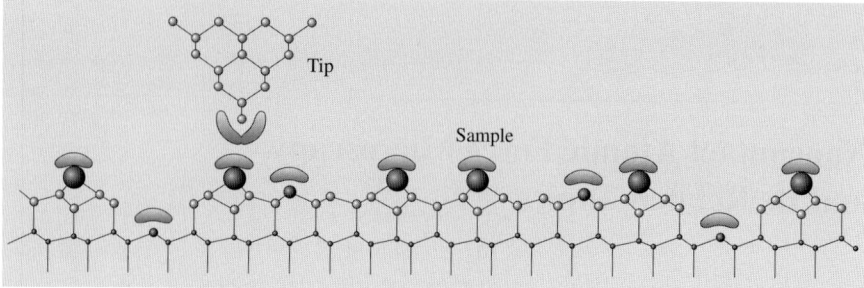

Fig. 4.1. Schematic view of an AFM tip close to a sample

In Sect. 4.2, principles of NC-AFM will be fully introduced. Then, in Sect. 4.3, applications to semiconductors will be presented. Next, in Sect. 4.4, applications to insulators will be described. And, in Sect. 4.5, applications to molecules will be introduced. These sections are essential to understanding the status of the art and the future possibility of the NC-AFM.

4.2 Principles of Noncontact Atomic Force Microscope (NC-AFM)

The atomic force microscope (AFM), invented by *Binnig* [1] and introduced in 1986 by *Binnig*, *Quate*, and *Gerber* [2], is an offspring of the scanning tunneling microscope (STM) [3]. The STM is covered in books and review articles, e.g., [4–9]. Early on in the development of STM it became evident that relatively strong forces act between a tip in close proximity to a sample. It was found that these forces could be put to good use in the atomic force microscope (AFM). Detailed information about the noncontact AFM can be found in [10–12].

4.2.1 Imaging Signal in AFM

Figure 4.1 shows a sharp tip close to a sample. The potential energy between the tip and sample V_{ts} causes a z component of the tip-sample force $F_{ts} = -\partial V_{ts}/\partial z$. Depending on the mode of operation, the AFM uses F_{ts} or some entity derived from F_{ts} as the imaging signal.

Unlike the tunneling current, which has a very strong distance dependence, F_{ts} has long- and short-range contributions. We can classify the contributions by their range and strength. In vacuum, there are van der Waals, electrostatic, and magnetic forces with long-range (up to 100 nm) and short-range chemical forces (fractions of a nm).

The van der Waals interaction is caused by fluctuations in the electric dipole moment of atoms and their mutual polarization. For a spherical tip with radius R next to a flat surface (z is the distance between the plane connecting the centers of the surface atoms and the center of the closest tip atom), the van der Waals potential is given by [13]:

$$V_{\text{vdW}} = -\frac{A_{\text{H}}}{6z}. \tag{4.1}$$

The "Hamaker constant", A_{H}, depends on the type of materials (atomic polarizability and density) of the tip and sample and is on the order of 1 eV for most solids [13].

When the tip and sample are both conductive and have an electrostatic potential difference, $U \neq 0$, electrostatic forces are important. For a spherical tip with radius R, the force is given by [14]:

$$F_{\text{electrostatic}} = -\frac{\pi \varepsilon_0 R U^2}{z}. \tag{4.2}$$

Chemical forces are more complicated. Empirical model potentials for chemical bonds are the Morse potential (see e.g., [13]).

$$V_{\text{Morse}} = -E_{\text{bond}} \left(2 e^{-\kappa(z-\sigma)} - e^{-2\kappa(z-\sigma)} \right) \tag{4.3}$$

and the Lennard-Jones potential [13]:

$$V_{\text{Lennard-Jones}} = -E_{\text{bond}} \left(2 \frac{\sigma^6}{z^6} - \frac{\sigma^{12}}{z^{12}} \right). \tag{4.4}$$

These potentials describe a chemical bond with bonding energy E_{bond} and equilibrium distance σ. The Morse potential has an additional parameter – a decay length κ.

4.2.2 Experimental Measurement and Noise

Forces between the tip and sample are typically measured by recording the deflection of a cantilever beam that has a tip mounted to its end (see Fig. 4.2). Today's microfabricated silicon cantilevers were first created by the *Calvin F. Quate* group [15–17] and *Wolter* et al. at IBM [18].

The cantilever is characterized by its spring constant k, eigenfrequency f_0, and quality factor Q.

For a rectangular cantilever with dimensions w, t, and L (see Fig. 4.2), the spring constant k is given by [6]:

$$k = \frac{E_Y w t^3}{4 L^3}, \tag{4.5}$$

where E_Y is Young's modulus. The eigenfrequency f_0 is given by [6]:

$$f_0 = 0.162 \frac{t}{L^2} \sqrt{E_Y/\rho}, \tag{4.6}$$

where ρ is the mass density of the cantilever material. The Q-factor depends on the damping mechanisms present in the cantilever. For micromachined cantilevers

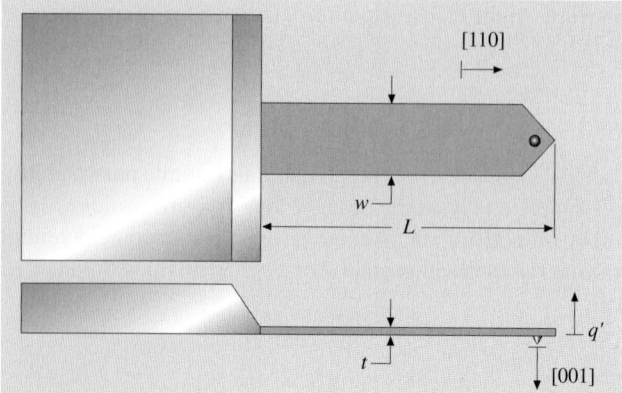

Fig. 4.2. Top view and side view of a microfabricated silicon cantilever (schematic)

operated in air, Q is typically a few hundred but can reach hundreds of thousands in vacuum.

In the first AFM, the deflection of the cantilever was measured with an STM – the backside of the cantilever was metalized, and a tunneling tip was brought close to it to measure the deflection [2]. Today's designs use optical (interferometer, beam-bounce) or electrical methods (piezoresistive, piezoelectric) for measuring the cantilever deflection. A discussion of the various techniques can be found in [19], descriptions of piezoresistive detection schemes are found in [17, 20], and piezoelectric methods are explained in [21–24].

The quality of the cantilever deflection measurement can be expressed in a schematic plot of the deflection noise density versus frequency, as in Fig. 4.3.

The noise density has a $1/f$ dependence for low frequency and merges into a constant noise density ("white noise") above the "$1/f$ corner frequency".

4.2.3 Static AFM Operating Mode

In the static mode of operation, the force translates into a deflection $q' = F_{ts}/k$ of the cantilever, yielding images as maps $z(x, y, F_{ts} = \text{const.})$. The noise level of the force measurement is then given by the cantilever's spring constant k times the noise level of the deflection measurement. In this respect, a small value for k increases force sensitivity. On the other hand, instabilities are more likely to occur with soft cantilevers (see Sect. 4.2.5). Because the deflection of the cantilever should be significantly larger than the deformation of the tip and sample, the cantilever should be much softer than the bonds between the bulk atoms in the tip and sample. Interatomic force constants in solids are in a range from $10\,\text{N/m}$ to about $100\,\text{N/m}$ – in biological samples, they can be as small as $0.1\,\text{N/m}$. Thus, typical values for k in the static mode are $0.01–5\,\text{N/m}$.

Even though it has been demonstrated that atomic resolution is possible with static AFM, the method can only be applied in certain cases. The detrimental effects of

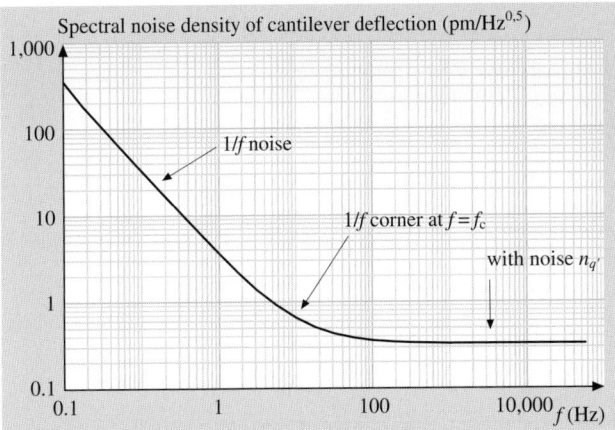

Fig. 4.3. Schematic view of $1/f$ noise apparent in force detectors. Static AFMs operate in a frequency range from 0.01 Hz to a few hundred Hz, while dynamic AFMs operate at frequencies around 10 kHz to a few hundred kHz. The noise of the cantilever deflection sensor is characterized by the $1/f$ corner frequency f_c and the constant deflection noise density $n_{q'}$ for the frequency range where white noise is dominating

$1/f$ noise can be limited by working at low temperatures [25], where the coefficients of thermal expansion are very small, or by building the AFM of a material with a low thermal expansion coefficient [26]. The long-range attractive forces have to be cancelled by immersing the tip and sample in a liquid [26], or by partly compensating the attractive force by pulling at the cantilever after jump-to-contact has occurred [27]. *Jarvis* et al. have cancelled the long-range attractive force with an electromagnetic force applied to the cantilever [28]. Even with these restrictions, static AFM does not produce atomic resolution on reactive surfaces like silicon, as the chemical bonding of the AFM tip and sample pose an unsurmountable problem [29, 30].

4.2.4 Dynamic AFM Operating Mode

In the dynamic operation modes, the cantilever is deliberately vibrated. There are two basic methods of dynamic operation: amplitude modulation (AM) and frequency modulation (FM) operation. In AM-AFM [31], the actuator is driven by a fixed amplitude A_{drive} at a fixed frequency f_{drive}, where f_{drive} is close to f_0. When the tip approaches the sample, elastic and inelastic interactions cause a change in both the amplitude and the phase (relative to the driving signal) of the cantilever. These changes are used as the feedback signal. While the AM mode was initially used in a noncontact mode, it was later implemented very successfully at a closer distance range in ambient conditions involving repulsive tip–sample interactions.

The change in amplitude in AM mode does not occur instantaneously with a change in the tip-sample interaction, but on a time scale of $\tau_{AM} \approx 2Q/f_0$, and the AM mode is slow with high-Q cantilevers. However, the use of high Q-factors

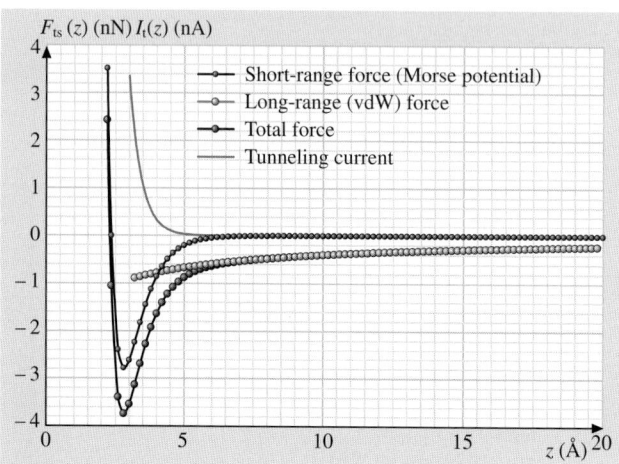

Fig. 4.4. Plot of tunneling current I_t and force F_{ts} (typical values) as a function of distance z between front atom and surface atom layer

reduces noise. *Albrecht* et al. found a way to combine the benefits of high Q and high speed by introducing the frequency modulation (FM) mode [32], where the change in the eigenfrequency settles on a time scale of $\tau_{FM} \approx 1/f_0$.

Using the FM mode, the resolution was improved dramatically, and finally atomic resolution [33, 34] was obtained by reducing the tip-sample distance and working in vacuum. For atomic studies in vacuum, the FM mode (see Sect. 4.2.6) is now the preferred AFM technique. However, atomic resolution in vacuum can also be obtained with the AM mode, as demonstrated by *Erlandsson* et al. [35].

4.2.5 The Four Additional Challenges Faced by AFM

Some of the inherent AFM challenges are apparent by comparing the tunneling current and tip-sample force as a function of distance (Fig. 4.4).

The tunneling current is a monotonic function of the tip-sample distance and has a very sharp distance dependence. In contrast, the tip-sample force has long- and short-range components and is not monotonic.

Jump-to-Contact and Other Instabilities

If the tip is mounted on a soft cantilever, the initially attractive tip-sample forces can cause a sudden "jump-to-contact" when the tip approaches the sample. This instability occurs in the quasi-static mode if [36, 37]

$$k < \max\left(-\frac{\partial^2 V_{ts}}{\partial z^2}\right) = k_{ts}^{max} . \tag{4.7}$$

Jump-to-contact can be avoided even for soft cantilevers by oscillating it at a large enough amplitude A [38]:

$$kA > \max(-F_{ts}) \ . \tag{4.8}$$

If hysteresis occurs in the $F_{ts}(z)$-relation, the energy ΔE_{ts} needs to be supplied to the cantilever for each oscillation cycle. If this energy loss is large compared to the intrinsic energy loss of the cantilever, amplitude control can become difficult. An additional approximate criterion for k and A is then

$$\frac{kA^2}{2} \geq \frac{\Delta E_{ts} Q}{2\pi} \ . \tag{4.9}$$

Contribution of Long-Range Forces

The force between tip and sample is composed of many contributions: electrostatic, magnetic, van der Waals, and chemical forces in vacuum. All of these force types except for the chemical forces have strong long-range components that conceal the atomic-force components. For imaging by AFM with atomic resolution, it is desirable to filter out the long-range force contributions and only measure the force components that vary at the atomic scale. While there is no way to discriminate between long- and short-range forces in static AFM, it is possible to enhance the short-range contributions in dynamic AFM through the proper choice of the oscillation amplitude A of the cantilever.

Noise in the Imaging Signal

Measuring the cantilever deflection is subject to noise, especially at low frequencies ($1/f$ noise). In static AFM, this noise is particularly problematic because of the approximate $1/f$ dependence. In dynamic AFM, the low-frequency noise is easily discriminated when using a bandpass filter with a center frequency around f_0.

Non-monotonic Imaging Signal

The tip-sample force is not monotonic. In general, the force is attractive for large distances, and upon decreasing the distance between tip and sample, the force turns repulsive (see Fig. 4.4). Stable feedback is only possible on a monotonic subbranch of the force curve.

Frequency-modulation AFM (FM-AFM) helps to overcome challenges. The non-monotonic imaging signal in AFM is a remaining complication for FM-AFM.

4.2.6 Frequency-Modulation AFM (FM-AFM)

In FM-AFM, a cantilever with eigenfrequency f_0 and spring constant k is subject to controlled positive feedback such that it oscillates with a constant amplitude A [32], as shown in Fig. 4.5.

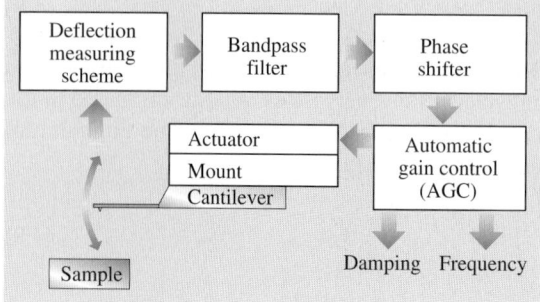

Fig. 4.5. Block diagram of a frequency-modulation force sensor

Experimental Setup

The deflection signal is phase shifted, routed through an automatic gain control circuit and fed back to the actuator. The frequency f is a function of f_0, its quality factor Q, and the phase shift ϕ between the mechanical excitation generated at the actuator and the deflection of the cantilever. If $\phi = \pi/2$, the loop oscillates at $f = f_0$.

Three physical observables can be recorded: 1) a change in resonance frequency, Δf, 2) the control signal of the automatic gain control unit as a measure of the tip-sample energy dissipation and 3) an average tunneling current (for conducting cantilevers and tips).

Applications

FM-AFM was introduced by *Albrecht* and coworkers in magnetic force microscopy [32]. The noise level and imaging speed were enhanced significantly compared to amplitude modulation techniques. Achieving atomic resolution on the Si(111)-(7 × 7) surface has been an important step in the development of the STM [39], and in 1994, this surface was imaged by AFM with true atomic resolution for the first time [33] (see Fig. 4.6).

The initial parameters that provided *true atomic resolution* (see caption of Fig. 4.6) were found empirically. Surprisingly, the amplitude necessary for obtaining good results was very large compared to atomic dimensions. It turned out later that the amplitudes had to be that large in order to fulfill the stability criteria listed in Sect. 4.2.5. Cantilevers with $k \approx 2000\,\text{N/m}$ can be operated with amplitudes in the Å-range [24].

4.2.7 Relation Between Frequency Shift and Forces

The cantilever (spring constant k, effective mass m^*) is a macroscopic object and its motion can be described by classical mechanics. Figure 4.7 shows the deflection $q'(t)$ of the tip of the cantilever: It oscillates with an amplitude A at a distance $q(t)$ to a sample.

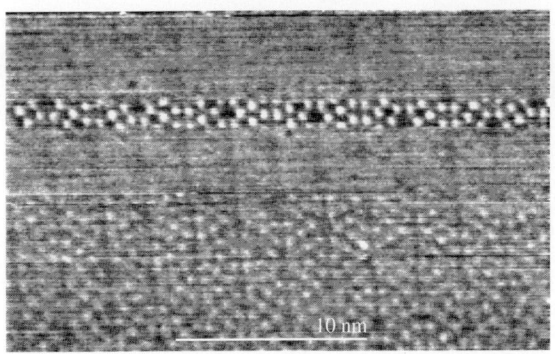

Fig. 4.6. First AFM image of the Si(111)-(7×7) surface. Parameters $k = 17\,\text{Nm}$, $f_0 = 114\,\text{kHz}$, $Q = 28\,000$, $A = 34\,\text{nm}$, $\Delta f = -70\,\text{Hz}$, $V_t = 0\,\text{V}$

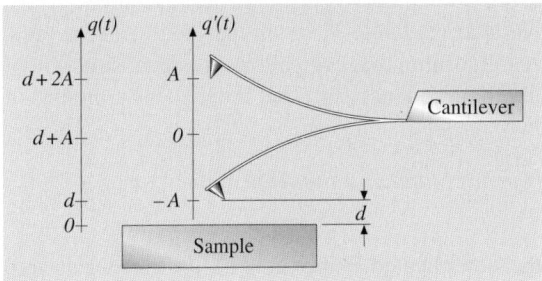

Fig. 4.7. Schematic view of an oscillating cantilever and definition of geometric terms

Generic Calculation

The Hamiltonian of the cantilever is:

$$H = \frac{p^2}{2m^*} + \frac{kq'^2}{2} + V_{ts}(q), \quad (4.10)$$

where $p = m^* dq'/dt$. The unperturbed motion is given by:

$$q'(t) = A\cos(2\pi f_0 t), \quad (4.11)$$

and the frequency is:

$$f_0 = \frac{1}{2\pi}\sqrt{\frac{k}{m^*}}. \quad (4.12)$$

If the force gradient $k_{ts} = -\partial F_{ts}/\partial z = \partial^2 V_{ts}/\partial z^2$ is constant during the oscillation cycle, the calculation of the frequency shift is trivial:

$$\Delta f = \frac{f_0}{2k} k_{ts}. \quad (4.13)$$

However, in classic FM-AFM, k_{ts} varies in orders of magnitude during one oscillation cycle, and a perturbation approach, as shown below, has to be employed for the calculation of the frequency shift.

Hamilton-Jacobi Method

The first derivation of the frequency shift in FM-AFM was achieved in 1997 [38] using canonical perturbation theory [40]. The result of this calculation is:

$$\Delta f = -\frac{f_0}{kA^2} \langle F_{ts} q' \rangle \qquad (4.14)$$
$$= -\frac{f_0}{kA^2} \int_0^{1/f_0} F_{ts}(d + A + q'(t)) q'(t) \, dt \; .$$

The applicability of first-order perturbation theory is justified, because in FM-AFM, E is typically in the range of several keVs, while V_{ts} is on the order of a few eVs. Dürig [41] has found a generalized algorithm that even allows the reconstruction of the tip-sample potential if not only the frequency shift, but the higher harmonics of the cantilever oscillation are known.

A Descriptive Expression for Frequency Shifts as a Function of the

Tip-Sample Forces

With integration by parts, the complicated expression (4.14) transforms into a very simple expression that resembles (4.13) [42].

$$\Delta f = \frac{f_0}{2k} \int_{-A}^{A} k_{ts}(z - q') \frac{\sqrt{A^2 - q'^2}}{\frac{\pi}{2} k A^2} \, dq' \; . \qquad (4.15)$$

This expression is closely related to (4.13): The constant k_{ts} is replaced by a weighted average, where the weight function $w(q', A)$ is a semicircle with radius A divided by the area of the semicircle $\pi A^2/2$ (see Fig. 4.8). For $A \to 0$, $w(q', A)$ is a representation of Dirac's delta function and the trivial zero-amplitude result of (4.13) is immediately recovered. The frequency shift results from a convolution between the tip-sample force gradient and the weight function. This convolution can easily be reversed with a linear transformation, and the tip-sample force can be recovered from the frequency shift versus distance curve [42].

The dependence of the frequency shift with amplitude confirms an empirical conjecture: Small amplitudes increase the sensitivity to short-range forces! Adjusting the amplitude in FM-AFM resembles tuning an optical spectrometer to a passing wavelength. When short-range interactions are to be probed, the amplitude should be in the range of the short-range forces. While using amplitudes in the Å-range has been elusive with conventional cantilevers, because of the instability problems described in Sect. 4.2.5, cantilevers with a stiffness on the order of 1000 N/m like the ones introduced in [23] are well suited for small-amplitude operation.

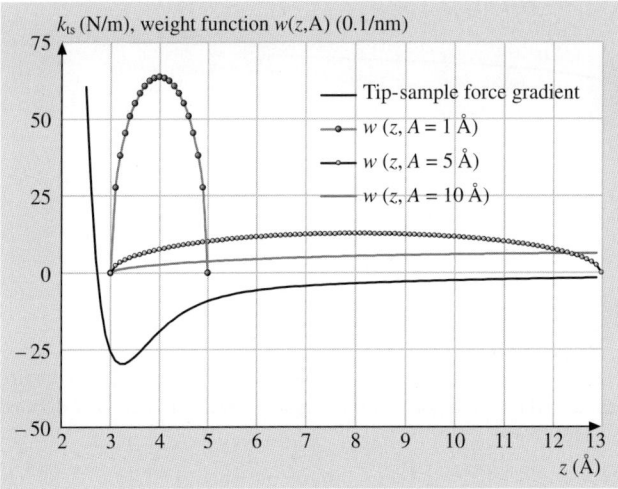

Fig. 4.8. Tip-sample force gradient k_{ts} and weight function for the calculation of the frequency shift

4.2.8 Noise in Frequency-Modulation AFM – Generic Calculation

The vertical noise in FM-AFM is given by the ratio between the noise in the imaging signal and the slope of the imaging signal with respect to z:

$$\delta z = \frac{\delta \Delta f}{\left| \frac{\partial \Delta f}{\partial z} \right|} . \qquad (4.16)$$

Figure 4.9 shows a typical frequency shift versus distance curve. Because the distance between the tip and sample is measured indirectly through the frequency shift, it is clearly evident from Fig. 4.9 that the noise in the frequency measurement $\delta \Delta f$ translates into vertical noise δz and is given by the ratio between $\delta \Delta f$ and the slope of the frequency shift curve $\Delta f(z)$ (4.16). Low vertical noise is obtained for a low-noise frequency measurement and a steep slope of the frequency shift curve.

The frequency noise $\delta \Delta f$ is typically inversely proportional to the cantilever amplitude A [32, 43]. The derivative of the frequency shift versus distance is constant for $A \ll \lambda$, where λ is the range of the tip-sample interaction and proportional to $A^{-1.5}$ for $A \gg \lambda$ [38]. Thus, minimal noise occurs if [44]

$$A_{\text{optimal}} \approx \lambda , \qquad (4.17)$$

for chemical forces, $\lambda \approx 1$ Å. However, for stability reasons (Sect. 4.2.5), extremely stiff cantilevers are needed for small amplitude operation. The excellent noise performance of the stiff cantilever and small amplitude technique has been verified experimentally [24].

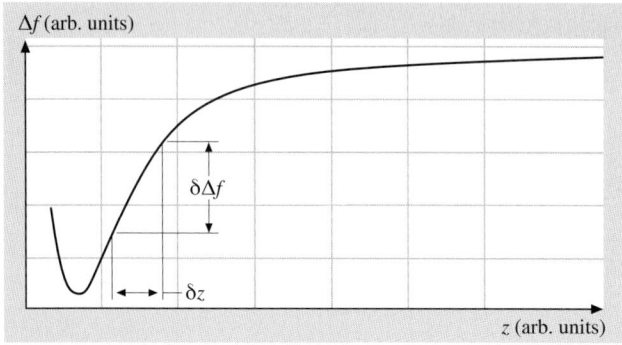

Fig. 4.9. Plot of frequency shift Δf as a function of tip-sample distance z. The noise in the tip-sample distance measurement is given by the noise of the frequency measurement $\delta \Delta f$ divided by the slope of the frequency shift curve

4.2.9 Conclusion

Dynamic force microscopy, in particular frequency-modulation atomic force microscopy, has matured into a viable technique that allows true atomic resolution of conducting and insulating surfaces and spectroscopic measurements on individual atoms [10, 45]. Even true atomic resolution in lateralforce microscopy is now possible [46]. Challenges remain in the chemical composition and structural arrangement of the AFM tip.

4.3 Applications to Semiconductors

For the first time, corner holes and adatoms on Si(111)7 × 7 surface have been observed at a very local area by *Giessibl* using the pure noncontact AFM in ultrahigh vacuum (UHV) [33]. It became the breakthrough for the true atomic resolution imaging on a well-defined clean surface using noncontact AFM. Then, Si(111)7 × 7 [34, 35, 45, 47], InP(110) [48], and Si(100)2 × 1 [34] surfaces were successively resolved with true atomic resolution. Furthermore, thermally induced motion of atoms or atomic-scale point defects on InP(110) surface have been observed at room temperature [48]. In this section, we will describe typical results of atomically resolved noncontact AFM imaging of semiconductor surfaces.

4.3.1 Si(111)7×7 Surface

Figure 4.10 shows the atomic resolution images of the Si(111)7 × 7 surface [49]. Here, Fig. 4.10a (TYPE I) was obtained using the Si tip without dangling, which is covered with an inert oxide layer. Fig. 4.10b (TYPE II) was obtained using the Si tip with dangling bond on which the Si atoms were deposited, due to the mechanical soft contact between the tip and the Si surface. The variable frequency shift mode

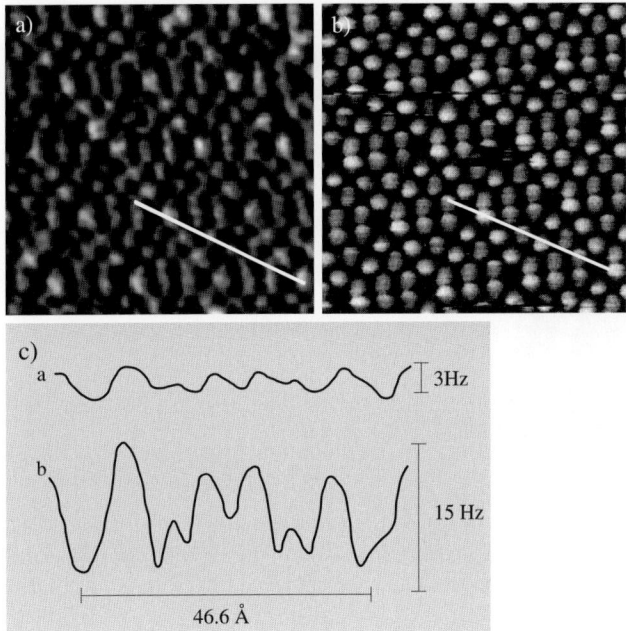

Fig. 4.10. Noncontact mode AFM images of Si(111) 7 × 7 reconstructed surface obtained using the Si tips (**a**) without and (**b**) with dangling bond. The scan area is 99 Å × 99 Å. (**c**) The cross sectional profiles along long diagonal of the 7 × 7 unit cell indicated by *white lines* in (**a**) and (**b**)

was used. We can see not only adatoms and corner holes, but also missing adatoms described by the dimer-adatom-stacking fault (DAS) model. We can see that the image contrast in Fig. 4.10b is clearly stronger than that in Fig. 4.10a.

Interestingly, by using the Si tip with dangling bond, we observed contrast between inequivalent halves and between inequivalent adatoms of the 7 × 7 unit cell. Namely, as shown in Fig. 4.11a, the faulted halves surrounded with a solid line show brighter than the unfaulted halves surrounded with a broken line. Here, the positions of the faulted and unfaulted halves were decided from the step direction. From the cross-sectional profile along the long diagonal of the 7 × 7 unit cell in Fig. 4.11b, the heights of the corner adatoms are slightly higher than those of the adjacent center adatoms in the faulted and unfaulted halves of the unit cell. The measured corrugations are in the following decreasing order:

Co-F > Ce-F > Co-U > Ce-U ,

where Co-F and Ce-F indicate the corner and center adatoms in faulted halves, and Co-U and Ce-U indicate the corner and center adatoms in unfaulted halves, respectively. As averaging over several units, the corrugation height differences are estimated to be 0.25 Å, 0.15 Å, and 0.05 Å for Co-F, Ce-F, and Co-U, respectively, referring to Ce-U. This tendency, that the heights of the corner adatoms are higher than

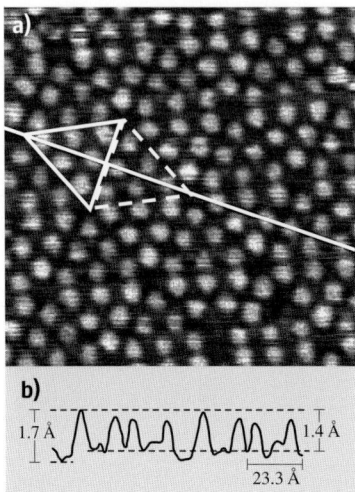

Fig. 4.11. (a) Noncontact mode AFM image with contrast of unequivalent adatoms and (b) a cross-sectional profile indicated by *white line*. The halves of the 7×7 unit cell surrounded with *solid line* and *broken line* correspond to faulted and unfaulted halves, respectively. The scan area is 89 Å × 89 Å

those of the center adatoms, is consistent with the experimental results using a silicon tip [47], although they could not determine faulted and unfaulted halves of the unit cell in the measured AFM images. However, this tendency is completely contrary to the experimental results using a tungsten tip [35]. This difference may originate from the difference in the materials of the tip, which seems to affect the interaction between the tip and the reactive sample surface. Another possibility is that the tip is in contact with the surface during the small fraction of the oscillating cycle in their experiments [35].

We consider that contrast between inequivalent adatoms is not caused by tip artifacts for the following reasons: 1) Each adatom, corner hole and defect were clearly observed; 2) the apparent heights of adatoms are the same whether they are located adjacent to defects or not; 3) the same contrast in several images for the different tips has been observed.

It should be noted that the corrugation amplitude of adatoms ∼ 1.4 Å in Fig. 4.11b is higher than that of 0.8–1.0 Å obtained with the STM, although the depth of corner holes obtained with noncontact AFM is almost the same as that observed with STM. Besides, in noncontact mode AFM images, the corrugation amplitude of adatoms was frequently larger than the depth of the corner holes. The origin of such a large corrugation of adatoms may be due to the effect of the chemical interaction, but that is yet unclear.

The atom positions, surface energies, dynamic properties, and chemical reactivities on the Si(111)7 × 7 reconstructed surface have been extensively investigated theoretically and experimentally. From these investigations, the possible origins of the contrast between inequivalent adatoms in AFM images are the following: the true

Table 4.1. Comparison between the adatom heights observed in an AFM image and the variety of properties for inequivalent adatoms

	Decreasing order	Agreement
AFM image	Co–F > Ce–F > Co–U > Ce–U	–
Calculated height	Co–F > Co–U > Ce–F > Ce–U	×
Stiffness of inter-atomic bonding	Ce–U > Co–U > Ce–F > Co–F	×
Amount of charge of adatom	Co–F > Ce–F > Co–U > Ce–U	○
Calculated chemical reactivity	Faulted > Unfaulted	○
Experimental chemical reactivity	Co–F > Ce–F > Co–U > Ce–U	○

atomic heights that correspond to the adatom core positions, the stiffness (spring constant) of inter-atomic bonding with the adatoms that corresponds to the frequencies of the surface mode, the amount of charge of adatom, and the chemical reactivity of adatoms. Table 4.1 summarizes the decreasing orders of the inequivalent adatoms for individual property. From Table 4.1, we can see that the calculated adatom heights and the stiffness of inter-atomic bonding cannot explain the AFM data, while the amount of charge of adatom and the chemical reactivity of adatoms can explain our data. The contrast due to the amount of charge of adatom means that the AFM image originated from the difference of the vdW, or the electrostatic physical interactions between the tip and the valence electrons at the adatoms. The contrast due to the chemical reactivity of adatoms means that the AFM image originated from the difference of covalent bonding chemical interaction between the atoms at the tip apex and the dangling bond of adatoms. Thus, we can see there are two possible interactions that explain the strong contrast between inequivalent adatoms of the 7×7 unit cell observed using the Si tip with dangling bond.

A weak contrast image in Fig. 4.10a is due to vdW and/or electrostatic force interactions. On the other hand, strong contrast images in Figs. 4.10b and 4.11a are due to a covalent bonding formation between the AFM tip with Si atoms and Si adatoms. These results signify the capability of the noncontact mode AFM to image the variation in chemical reactivity of Si adatoms. In the future, by controlling an atomic species at the tip apex, the study of chemical reactivity on an atomic scale will be possible using the noncontact AFM.

4.3.2 Si(100)2×1 and Si(100)2×1:H Monohydride Surfaces

In order to investigate the imaging mechanism of the noncontact AFM, a comparative study between a reactive surface and an insensitive surface using the same tip is very useful. Si(100)2×1:H monohydride surface is a Si(100)2×1 reconstructed surface that is terminated by a hydrogen atom. It does not reconstruct as metal is deposited on the semiconductor surface. The surface structure hardly changes. Thus, Si(100)2×1:H monohydride surface is one of most useful surfaces for a model system to investigate the imaging mechanism, experimentally and theoretically. Furthermore, whether the interaction between a very small atom such as hydrogen and a tip

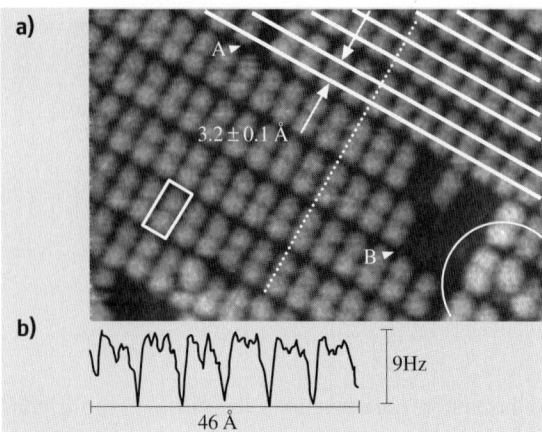

Fig. 4.12. (a) Noncontact AFM image of a Si(001)2 × 1 reconstructed surface. The scan area was 69×46 Å. One 2×1 unit cell is outlined with a *box*. *White rows* are superimposed to show the bright spots arrangement. The distance between bright spots on dimer row is 3.2 ± 0.1 Å. Into *white arc*, the alternative bright spots are shown. (b) Cross-sectional profile indicated by a *white dotted line*

apex is observable with noncontact AFM is interesting. Here, we show noncontact AFM images measured on a Si(100)2 × 1 reconstructed surface with a dangling bond and a Si(100)2 × 1:H monohydride surface with a dangling bond that is terminated by a hydrogen atom [50].

Figure 4.12a shows the atomic resolution image of a Si(100)2 × 1 reconstructed surface. The paired bright spots arranged like rows with a 2 × 1 symmetry was observed with clear contrast. Missing pairs of bright spots were observed, as indicated by arrows. Further, pairs of bright spots are shown by the white dashed arc and appear to be the stabilize-buckled asymmetric dimer structure. The distance between paired bright spots is 3.2 ± 0.1 Å.

Figure 4.13a shows the atomic resolution image of a Si(100)2×1:H monohydride surface. Paired bright spots arranged like rows were observed. Missing paired bright spots, as well as those paired in rows and single bright spots were observed, as indicated by arrows. The distance between paired bright spots is 3.5±0.1 Å. This distance is 0.2 Å larger than that of a Si(100)2 × 1 reconstructed surface. It was found that the distance between bright spots increases in size due to the hydrogen termination.

The bright spots in Fig. 4.12 do not merely image the silicon atom site, because the distance between bright spots forming the dimer structure of Fig. 4.12a (3.2 ± 0.1 Å) is larger than the distance between silicon atoms of every dimer structure model (2.9 Å is the maximum distance between upper silicons on asymmetric dimer structures). This seems to be due to the contribution to imaging of the chemical bonding interaction between the dangling bond out of the silicon tip apex and the dangling bond on the Si(100)2 × 1 reconstructed surface. The chemical bonding interaction works with strong direction dependence between the dangling bond

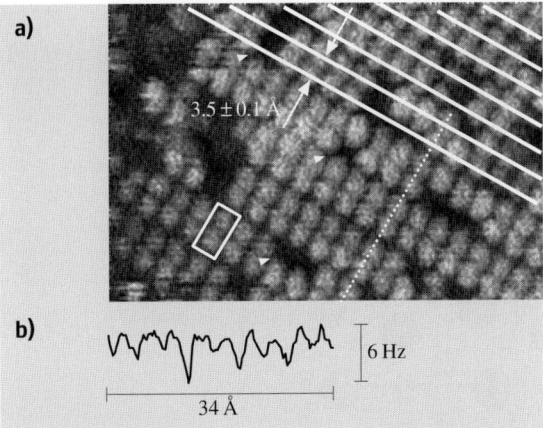

Fig. 4.13. (a) Noncontact AFM image of Si(001)2×1-H surface. The scan area was 69×46 Å. One 2 × 1 unit cell is outlined with a *box*. *White rows* are superimposed to show the bright spots arrangement. The distance between bright spots on dimer row is 3.5 ± 0.1 Å. (b) Cross-sectional profile indicated by a *white dotted line*

out of the silicon dimer structure on the Si(100)2 × 1 reconstructed surface and the dangling bond out of the silicon tip apex, and the dimer structure is obtained with a larger distance than between silicons on the surface.

The bright spots of Fig. 4.13 seem to be located at hydrogen atom sites on a Si(100)2 × 1:H monohydride surface, because the distance between bright spots forming the dimer structures (3.5 ± 0.1 Å) agrees approximately with the distance between the hydrogen, i.e., 3.52 Å. Thus, noncontact AFM atomically resolves the individual hydrogen atoms on the topmost layer. On this surface, the dangling bond is terminated by a hydrogen atom, and the hydrogen atom on the topmost layer does not have a chemical reactivity. Therefore, the interaction between the hydrogen atom on the top most layer and the silicon tip apex does not contribute to the chemical bonding interaction with strong direction dependence as in a silicon surface, and the bright spots of a noncontact AFM image correspond to the hydrogen atom sites on the topmost layer.

4.3.3 Metal-Deposited Si Surface

In this section, we will introduce the comparative study of force interaction between a Si tip and a metal deposited Si surface and between a metal adsorbed Si tip and a metal-deposited Si surface [51, 52]. As for the metal-deposited Si surface, a Si(111) $\sqrt{3} \times \sqrt{3}$-Ag (hereafter referred to as $\sqrt{3}$-Ag) surface was used.

For the $\sqrt{3}$-Ag surface, the honeycomb-chained trimer (HCT) model has been accepted as the appropriate model. As shown in Fig. 4.14, this structure contains Si trimer at the second layer, 0.75 Å below the Ag trimer at the topmost layer. The topmost Ag atoms and the lower Si atoms form covalent bonds. The inter-atomic

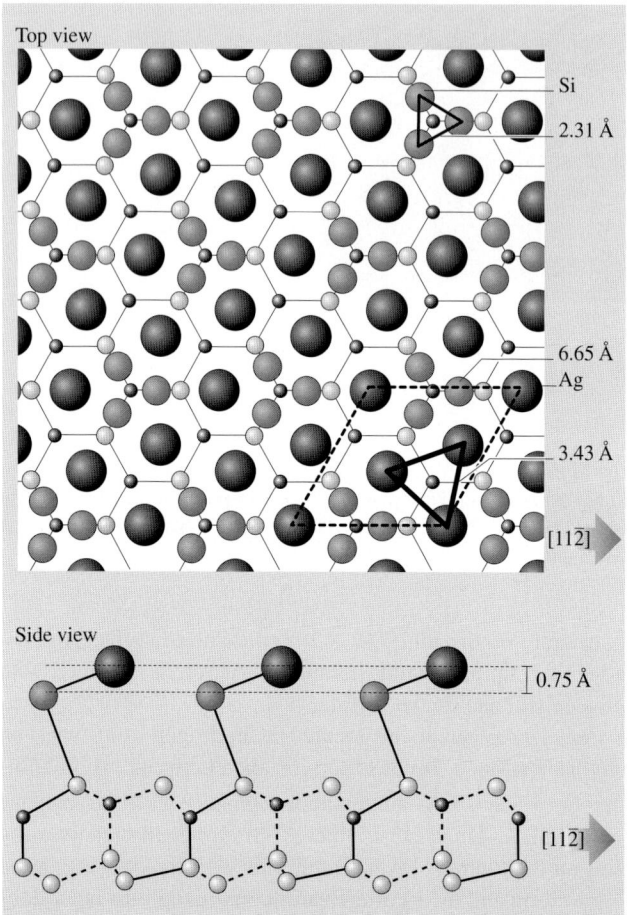

Fig. 4.14. HCT model for the structure of the Si(111) $\sqrt{3} \times \sqrt{3}$-Ag surface. *Black closed circle, gray closed circle, open circle,* and *closed circle* with *horizontal line* indicate Ag atom at the topmost layer, Si atom at the second layer, Si atom at the third layer, and Si atom at the fourth layer, respectively. *Rhombus* indicates $\sqrt{3} \times \sqrt{3}$ unit cell. *Thick, large solid triangle* indicates an Ag trimer. *Thin, small, solid triangle* indicates a Si trimer

distances between the nearest neighbor Ag atoms forming Ag trimer and between lower Si atoms forming Si trimer are 3.43 Å and 2.31 Å, respectively. The apexes of Si trimers and Ag trimers face the [11$\bar{2}$] direction and the direction tilted a little to the [$\bar{1}\bar{1}$2] direction, respectively.

In Fig. 4.15, we show the noncontact AFM images measured using a normal Si tip at the frequency shift of (a) −37 Hz, (b) −43 Hz, and (c) −51 Hz, respectively. These frequency shifts correspond to the tip-sample distance of about 0–3 Å. We defined the zero position of tip-sample distance, i.e., the contact point, where the

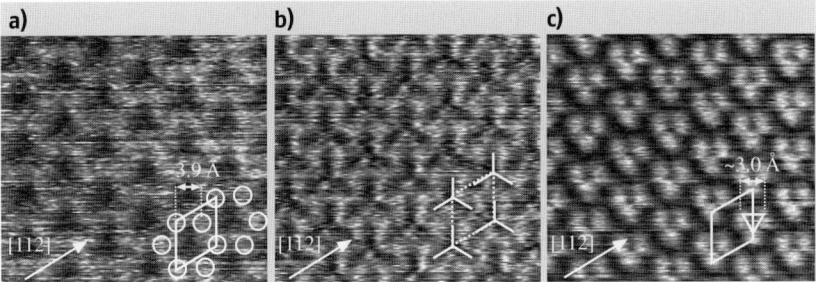

Fig. 4.15. Noncontact AFM images obtained at the frequency shifts of (**a**) −37 Hz, (**b**) −43 Hz, and (**c**) −51 Hz on a Si(111) $\sqrt{3} \times \sqrt{3}$-Ag surface. This distance dependence was obtained with the Si tip. Scan area is 38 Å × 34 Å. A *rhombus* indicates $\sqrt{3} \times \sqrt{3}$ unit cell

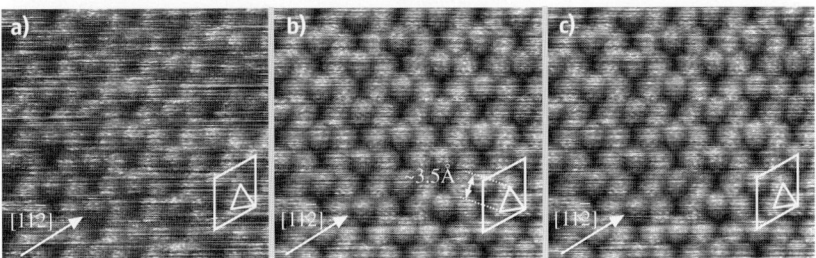

Fig. 4.16. Noncontact AFM images obtained at the frequency shifts of (**a**) −4.4 Hz, (**b**) −6.9 Hz, and (**c**) −9.4 Hz on a Si(111) $\sqrt{3} \times \sqrt{3}$-Ag surface. This distance dependence was obtained with the Ag adsorbed tip. Scan area is 38 Å × 34 Å

vibration amplitude began to decrease. The rhombus indicates the $\sqrt{3} \times \sqrt{3}$ unit cell. When the tip approached the surface, the contrast of the noncontact AFM images became strong and the pattern remarkably changed. That is, by moving the tip toward the sample surface, the hexagonal pattern, the trefoil-like pattern composed of three dark lines, and the triangle pattern can be observed sequentially. In Fig. 4.15a, the distance between the bright spots is 3.9 ± 0.2 Å. In Fig. 4.15c, the distance between the bright spots is 3.0 ± 0.2 Å, and the direction of the apex of all triangles composed of three bright spots is [11$\bar{2}$].

In Fig. 4.16, we show the noncontact AFM images measured by using an Ag adsorbed tip at the frequency shift of (a) −4.4 Hz, (b) −6.9 Hz, and (c) −9.4 Hz, respectively. The tip-sample distances Z are roughly estimated to be Z = 1.9 Å, 0.6 Å, and ∼ 0 Å (in noncontact region), respectively. When the tip approached the surface, the pattern of the noncontact AFM images didn't change, although the contrasts became clearer. A triangle pattern can be observed. The distance between the bright spots is 3.5 ± 0.2 Å. The direction of the apex of all triangles composed of three bright spots is tilted a little from the [$\bar{1}\bar{1}2$] direction.

Thus, noncontact AFM images measured on a Si(111) $\sqrt{3} \times \sqrt{3}$-Ag surface showed two types of distance dependence of the image patterns, depending on the atom species on the tip apex.

By using the normal Si tip with dangling bond, as in Fig. 4.15a, the measured distance between the bright spot of 3.9 ± 0.2 Å agrees with the distance of 3.84 Å between the center of the Ag trimer in the HCT model within the experimental error. Furthermore, the hexagonal pattern composed of six bright spots also agrees with the honeycomb structure of the Ag trimer in the HCT model. So the most appropriate site corresponding to the bright spots in Fig. 4.15a is the site of the center of the Ag trimers. In Fig. 4.15c, the measured distance of 3.0 ± 0.2 Å between the bright spots forming the triangle pattern does not agree with either the distance between the Si trimer of 2.31 Å or the distance between the Ag trimer of 3.43 Å in the HCT model, while the direction of the apex of triangles composed of three bright spots agrees with the [11$\bar{2}$] direction of the apex of the Si trimer in the HCT model. So the most appropriate site corresponding to the bright spots in Fig. 4.15c is the intermediate site between Si atoms and Ag atoms. On the other hand, by using the Ag adsorbed tip, the measured distance of 3.5 ± 0.2 Å between the bright spots in Fig. 4.16 agrees with the distance of 3.43 Å between the nearest neighbor Ag atoms forming the Ag trimer at the topmost layer in the HCT model within the experimental error. In addition, the direction of the apex of the triangles composed of three bright spots also agrees with the direction of the apex of the Ag trimer, i.e., tilted [$\bar{1}\bar{1}2$], in the HCT model. So the most appropriate site corresponding to the bright spots in Fig. 4.16 is the site of individual Ag atoms forming the Ag trimer at the topmost layer.

It should be noted that by using the noncontact AFM with an Ag adsorbed tip the individual Ag atom on the $\sqrt{3}$-Ag surface could be resolved in real space for the first time, although it could not be resolved by using the noncontact AFM with a Si tip. So far, the $\sqrt{3}$-Ag surface has been observed by the scanning tunneling microscope (STM) with atomic resolution. However, the STM can measure the local charge density of state near Fermi level on the surface. From the first principle calculation, it was proven that the unoccupied surface state is densely distributed around the center of the Ag trimer. As a result, bright contrast is obtained at the center of the Ag trimer with the STM.

Finally, we consider the origin of the atomic resolution imaging of the individual Ag atoms on the $\sqrt{3}$-Ag surface. Here, we discuss the difference of the force interactions between the Si and the Ag adsorbed tips. As shown in Fig. 4.17a, by using the Si tip, there is the dangling bond out of the topmost Si atom on the Si tip apex. As a result, the force interaction is dominated by the physical bonding interaction such as Coulomb force far from the surface and by chemical bonding interaction very close to the surface. In other words, if the reactive Si tip with dangling bond approaches the surface, at the distance far from the surface, the Coulomb force acts between the electron localized at the dangling bond out of the topmost Si atom on the tip apex and the positive charge distributed around the center of the Ag trimer. At the distance very close to the surface, the chemical bonding interaction will occur due to theonset of the orbital hybridization between the dangling bond out of the topmost Si

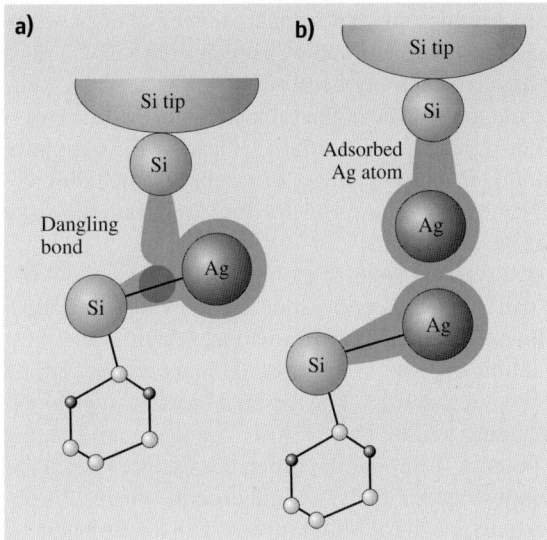

Fig. 4.17. Schematic illustration of (**a**) the Si atom with dangling bond and (**b**) the Ag adsorbed tip above the Si-Ag covalent bond on Si(111) $\sqrt{3} \times \sqrt{3}$-Ag surface

atom on the Si tip apex and the Si-Ag covalent bond on the surface. Hence, the individual Ag atoms will not be resolved, the image pattern will change depending on the tip-sample distance. On the other hand, as shown in Fig. 4.17b, by using an Ag adsorbed tip, the dangling bond localized out of the topmost Si atom on the Si tip apex is terminated by the adsorbed Ag atom. As a result, even at very close tip-sample distance, the force interaction is dominated by the physical bonding interaction such as the vdW force interaction. If the Ag adsorbed tip approaches the surface, the vdW force acts between the Ag atom on the tip apex and Ag or Si atom on the surface. Ag atoms at the topmost layer on the $\sqrt{3}$-Ag surface are located higher than the Si atoms at the lower layer. Hence, the individual Ag atoms (or nearly true topography) will be resolved, and the image pattern will not change even at very close tip-sample distance. It should be emphasized that there is a possibility to identify or recognize atom species on a sample surface using the noncontact AFM if we can control an atomic species at the tip apex.

4.4 Applications to Insulators

Insulators such as alkali halides, fluorides, and metal oxides are key materials in many applications, including optics, microelectronics, catalysis, and so on. Surface properties are important in these technologies, but they are usually poorly understood. This is due to their low conductivity, which makes it difficult to investigate them using electron- and ion-based measurement techniques such as low-energy electron diffraction, ion scattering spectroscopy, and scanning tunneling microscopy (STM). Surface imaging by noncontact atomic force microscopy (NC-AFM) does not require the sample to exhibit high conductivity, because NC-AFM detects a force between the tip on the cantilever and the surface of the sample. Since the first report of atomically resolved NC-AFM on a Si(111) 7×7 surface [33], several groups have succeeded in obtaining "true" atomic resolution images of insulators, including defects, and it has been shown that NC-AFM is a powerful new tool for the atomic-scale surface investigation of insulators.

In this section, we will describe typical results of atomically resolved NC-AFM imaging of insulators such as alkali halides, fluorides, and metal oxides. For alkali halides and fluorides, we will focus on the contrast formation, which is the most important issue for interpreting atomically resolved images of binary compounds on the basis of experimental and theoretical results. For the metal oxides, typical examples of atomically resolved imaging will be shown. Also, simultaneous imaging using STM and NC-AFM will be described, and the difference between the STM and NC-AFM images will be demonstrated. Finally, we will describe the results obtained when we imaged an antiferromagnetic NiO(001) surface with a ferromagnetic Fe-coated tip in order to explore the possibility of detecting short-range magnetic interaction using the NC-AFM.

4.4.1 Alkali Halides, Fluorides, and Metal Oxides

The surfaces of alkali halides were the first insulating materials to be imaged by NC-AFM with true atomic resolution [53]. Up until now, there have been reports on atomically resolved images of (001) cleaved surfaces for single crystal NaF, RbBr, LiF, KI, NaCl [54], KBr [55], and thin films of NaCl(001) on Cu(111) [56]. In this section, we describe the contrast formation of alkali halides surface on the basis of experimental and theoretical results.

Alkali Halides

In experiments of alkali halides, the symmetry of the observed topographic images indicates that the protrusions exhibit only one type of ions, either positively or negatively charged ions. This leads to the conclusion that the atomic contrast is dominantly caused by electrostatic interactions between a charged atom at the tip apex and the surface ions, i.e., the long-range forces between the macroscopic tip and the sample such as van der Waals force is modulated by an alternating short-range electrostatic interaction of the surface ions. Theoretical work employing the atomistic simulation technique has revealed the mechanism for contrast formation on an ionic surface [57]. A significant part of the contrast is due to the displacement of ions in the force field, not only enhancing the atomic corrugations, but also contributing to the electrostatic potential by forming dipoles at the surface. The experimentally observed atomic corrugation height is determined by the interplay of the long- and short-range forces. In the case of NaCl, it has been experimentally demonstrated that a more blunt tip produces a larger corrugation when the tip-sample distance is shorter [54].

This result shows that the increased long-range forces induced by a blunt tip allow for more stable imaging closer to the surface. The stronger electrostatic short-range interaction and the larger ion displacement produce a more pronounced atomic corrugation.

At steps and kinks of an NaCl thin film on Cu(111), the amplitude of the atomic corrugations has been observed to increase by up to a factor of two, as large as that of atomically flat terraces [56]. This increase in the contrast is due to the low coordination of the step-edge and the kinked ions. The low coordination of the ions results in an enhancement of the electrostatic potential over the site and an increase in the displacement induced by the interaction with the tip. These results demonstrate that the short-range interaction dominates the atomic contrast and leads to a considerable displacement of the surface ions from the ideal position.

Theoretical study predicts that the image contrast depends on the chemical species of the tip apex. *Bennewitz* et al. [56] have performed the calculations using an MgO tip terminated by oxygen, an Mg ion, and an OH. The magnitude of the atomic contrast for the Mg-terminated tip shows a slight increase in comparison with an oxygen-terminated tip. The atomic contrast with the oxygen-terminated tip is dominated by the attractive electrostatic interaction between the oxygen on the tip apex and the Na ion, but the Mg-terminated tip attractively interacts with the Cl ion. Also, it has been demonstrated that the magnitude of the contrast for an OH-terminated tip is almost an order of magnitude less than for the oxygen- and Mg-terminated tips.

In order to explore the imaging mechanism for the alkali halides, the heterogeneous surface of NaCl crystal containing $(OCN)^-$ molecules as impurities has been observed by NC-AFM [58]. In this case, the atomic periodicity of the NaCl lattice is imaged. However, high-resolution investigations of impurities are limited by the convolution between the tip apex and the molecules at the surface. *Bennewitz* et al. [59] reported on the heterogeneous surfaces of alkali halide crystals. In this experiment, which was designed to observe the chemically nonhomogeneous alkali halide surface, a mixed crystal composing of 60% KCl and 40% KBr was used as the sample. In this crystal, which is a so-called solid solution, the Cl and Br ions are mixed randomly. The image of the cleaved $KCl_{0.6}Br_{0.4}(001)$ surface indicates that only one type of ion is imaged as protrusions as if it were a pure alkali halide crystal. However, there is a significant difference in comparison with the surface of a pure crystal. The amplitude of atomic corrugation varies strongly between the positions of the ions imaged as depressions. This variation in the corrugations corresponds to the constituents of the crystal, i.e., the Cl and Br ions, and it is concluded that the tip apex is negatively charged. Also, it is demonstrated that a comparison between the distribution of the deep depressions and the composition of the mixed crystal assigns the deep depressions to Br ions. Although an evaluation of the effect of the surface buckling induced by a random distribution of anions is needed in order to understand the contrast mechanism, this result demonstrates that NC-AFM measurement exhibits chemical sensitivity on an atomic scale. For applications to alkali halide surfaces, real space imaging with NC-AFM presents the possibility for chemical discrimination at the surface.

Fluorides

Fluorides are important materials for the progress of atomic-scale resolution NC-AFM imaging on insulators. There are reports in the literature of surface images for single crystal BaF_2, SrF_2 [60], and CaF_2 [61–63]. Surfaces of fluorite-type crystals are prepared by cleaving along the (111) planes. Their structure is more complex than the structure of alkali halides with the NaCl structure. The complexity is of great interest for atomic-resolution imaging using an NC-AFM and also for theoretical predictions of the interpretation of atomic-scale contrast information.

The first atomically resolved images of a $CaF_2(111)$ surface have been obtained in topographic mode [62] in which the tip-sample distance is regulated by maintaining a constant frequency shift. In this imaging mode, the surface ions mostly appear as spherical caps. Similar measurements have been performed on the (111) surfaces of SrF_2 and BaF_2.

Barth et al. [64] have found that the $CaF_2(111)$ surface images obtained using the constant height mode, in which the frequency shift is recorded with a very low loop gain, can be categorized into two contrast patterns. In the first, the ions appear as triangles, and in the other, they have the appearance of circles, similar to the contrast obtained in a topographic image. These differences are revealed by comparing the cross sections along the [121] direction. Theoretical studies demonstrated that these two different contrast patterns could be explained as a result of imaging with tips of different polarity [64–66]. When imaging with a positively charged (cation-terminated) tip, the triangular pattern appears. In this case, the contrast is dominated by the strong short-range electrostatic attraction between the positive tip and the negative F ions. The cross section along the [121] direction of the triangular image shows two maxima: one is a larger peak over the F ions located in the topmost layer, and the other is a smaller peak at the position of the F ions in the third layer. The minima appear at the position of the Ca ions in the second layer. When imaging with a negatively charged (anion-terminated) tip, the spherical image contrast appears, and the main periodicity is created by the Ca ions in the second layer between the topmost and the third F ion layers. In the cross section along the [121] direction, the large maxima correspond to the Ca sites, because of the strong attraction of the negative tip, and the minima appear at the sites of maximum repulsion over the F ions in the first layer. At a position between two F ions, there are smaller maxima. This reflects the weaker repulsion over the third layer F ion sites compared to the protruding topmost F ion sites and a slower decay in the contrast on one side of the Ca ions. These results demonstrate that the image contrast of a $CaF_2(111)$ surface depends on the state of the tip apex.

The triangular pattern obtained with a positively charged tip appears at relatively large tip-sample distance, as shown in Fig. 4.18a. On the cross section along the [121] direction, experimental results and theoretical studies both demonstrate the large peak and small shoulder characteristic for the triangular pattern image (Fig. 4.18d). When the tip approaches the surface more closely, the triangular pattern of the experimental images is more vivid (Fig. 4.18b), just as predicted in the theoretical works. On the cross section along the [121] direction, as the tip approaches, the amplitude of

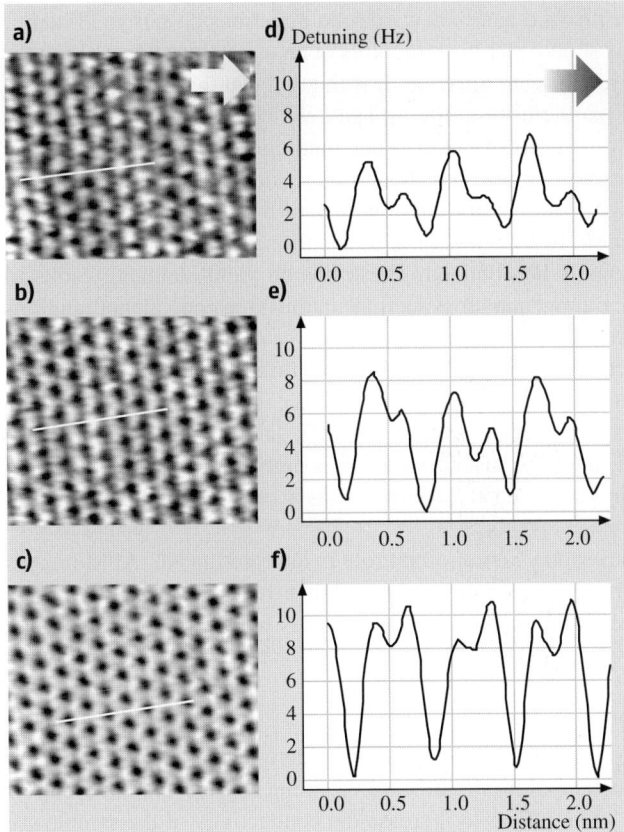

Fig. 4.18. (a)–(c) CaF$_2$(111) surface images obtained by using the constant height mode. From (a) to (c) the frequency shift was lowered. The *white lines* represent the positions of the cross section. (d)–(f) The cross section extracted from the Fourier-filtered images of (a)–(c). The *white* and *black arrows* represent the scanning direction. The images and the cross sections are from [64]

the shoulder increases until it is equal to that of the main peak, and this feature gives rise to the honeycomb pattern image, as shown Fig. 4.18c. These results demonstrate that there is a more complex process of image contrast formation. Also, from detailed theoretical analysis of the electrostatic potential [67], it is suggested that the change in displacement of the ions due to the proximity of the tip plays an important role in the formation of the image contrast. Such a drastic change in image contrast, depending on both the polarity of the terminated tip atom and the tip-sample distance, is inherent in the fluoride (111) surface, and this image contrast feature cannot be seen on the (001) surface of alkali halides with a simple crystal structure.

These results have demonstrated that there is good agreement between experiment and theory. However, the results of careful experiments show another feature:

The cross sections in the forward and backward scan directions do not yield identical results [64]. Also, it is shown that the structures of cross sections taken along the three equivalent [121] directions are different. It is thought that this can be attributed to the asymmetry of the nano-cluster at the tip apex, which leads to different interactions in the equivalent directions. A better understanding of the asymmetric image contrast may require a more complicated modeling of the tip structure. However, the most desirable solution would be the development of the suitable techniques for well-defined tip preparation. In fact, it should be mentioned that perfect tips on an atomic scale can occasionally be obtained. These tips do yield identical results in forward and backward scanning, and cross sections in the three equivalent directions taken with this tip are almost identical [68].

The fluoride (111) surface is an excellent standard surface for calibrating tips on an atomic scale. The polarity of the tip-terminated atom can be determined from the image contrast pattern (spherical or triangular pattern). The irregularities in the tip structure can be detected, since the surface structure is highly symmetric. Therefore, once such a tip has been prepared, it can be used as a calibrated tip for imaging unknown surfaces.

The states of the tip apex play an important role in interpreting NC-AFM images of alkali halide and fluoride surfaces. It is expected that the achievement of a good correlation between experimental and theoretical studies will help advance surface imaging for insulators by NC-AFM.

Metal Oxides

Most of the metal oxides that have attracted strong interest for their technological importance are insulative. Therefore, in the case of atomically resolved imaging of metal oxide surfaces by STM, efforts to increase the conductivity of the sample are needed, for example, the introduction of defects of anions or cations, doping with other atoms, and surface observations during heating of the sample. However, in principle, NC-AFM provides the possibility of observing nonconductive metal oxides without these efforts. In cases where the conductivity of the metal oxides is high enough for a tunneling current to flow, it should be noted that most of the surface images obtained by NC-AFM and STM are not identical.

Since the first report of atomically resolved images on a $TiO_2(110)$ surface with oxygen point defects [69], they have also been reported on $TiO_2(100)$ [70–72], $SnO_2(110)$ [73], $NiO(001)$ [74, 75], $SrTiO_3(100)$ [76], and $CeO_2(111)$ [77] surfaces. Also, *Barth* et al. [78] have succeeded in obtaining atomically resolved NC-AFM images of a clean $\alpha\text{-}Al_2O_3(0001)$ surface, which is impossible to investigate using STM. In this section, we describe typical results of the imaging of metal oxides by NC-AFM.

The $\alpha\text{-}Al_2O_3(0001)$ surface exists in several ordered phases that can be reversibly transformed into each other by thermal treatments and oxygen exposure. It is known that the high temperature phase has a large ($\sqrt{31} \times \sqrt{31})R\pm9°$ unit cell. However, the details of the atomic structure of this surface have not been revealed, and two models are proposed from the results of low-energy electron diffraction and X-ray diffraction studies. Recently, *Barth* et al. [78] have directly observed this reconstructed

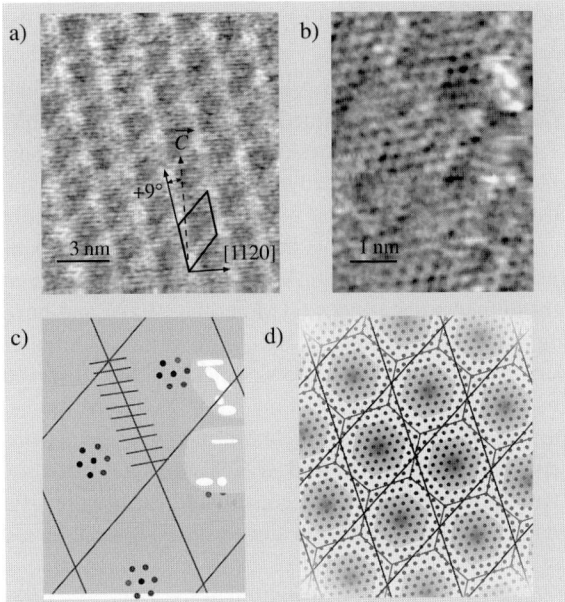

Fig. 4.19. (a) Image of the high temperature, reconstructed clean α-Al$_2$O$_3$ surface obtained by using the constant height mode. The *rhombus* represents the unit cell of the ($\sqrt{31} \times \sqrt{31}$)$R+9°$ reconstructed surface. (b) Higher magnification image of (a). Imaging was performed at a reduced tip-sample distance. (c) Schematic representation of the indicating regions of hexagonal order in the center of reconstructed rhombi. (d) Superposition of the hexagonal domain with reconstruction rhombi found by NC-AFM imaging. Atoms in *gray shaded regions* are well-ordered. The images and the schematic representations are from [78]

α-Al$_2$O$_3$(0001) surface by NC-AFM. They confirmed that the dominant contrast of the low magnification image corresponds to a rhombic grid representing a unit cell of ($\sqrt{31} \times \sqrt{31}$)$R+9°$, as shown in Fig. 4.19a. Also, more details of atomic structures are determined from the higher magnification image (Fig. 4.19b), which was taken at a reduced tip-sample distance. In this atomically resolved image, it is revealed that each side of the rhombus is intersected by 10 atomic rows, and that a hexagonal arrangement of atoms exists in the center of the rhombi (Fig. 4.19c). This feature agrees with the proposed surface structure that predicts order in the center of hexagonal surface domains and disorder at the domain boundaries. Their result is an excellent demonstration of the capabilities of the NC-AFM as a powerful tool for the atomic-scale surface investigation of insulators.

In addition, *Barth* et al. [78] have performed observations of the Al$_2$O$_3$(0001) surface when it is exposed to water and hydrogen. It is suggested that the region of atomic disorder has an extremely high reactivity. Such investigations of the surface structure of adsorbed insulator substrates can only be provided by NC-AFM. The NC-AFM imaging of metal growth on oxides [79] has also been reported. A description of the adsorbed molecular arrangement on a TiO_2 surface can be found in the next section.

The atomic structure of $SrTiO_3(100)$-($\sqrt{5} \times \sqrt{5}$) $R26.6°$ surface, as well as that of $Al_2O_3(0001)$-($\sqrt{31} \times \sqrt{31}$)$R \pm 9°$ can be determined on the basis of the results of NC-AFM imaging [76]. $SrTiO_3$ is one of the perovskite oxides and its (100) surface exhibits the many different kinds of reconstructed structures. In the case of the ($\sqrt{5} \times \sqrt{5}$)$R26.6°$ reconstruction, the oxygen vacancy-Ti^{3+}-oxygen model (where the terminated surface is TiO_2 and the observed spots are related to oxygen vacancies) was proposed from the results of STM imaging. As shown in Fig. 4.20, *Kubo et al.* [76] have performed measurements using both STM and NC-AFM and have found that the size of the bright spots, as observed by NC-AFM, is always smaller than that for STM measurement and that the dark spots, which are not observed by STM, are arranged along the [001] and [010] directions in the NC-AFM image. A theoretical simulation of the NC-AFM image using first-principles calculations shows that the bright and dark spots correspond to Sr and oxygen atoms, respectively. It has been proposed that the structural model of the reconstructed surface consists of an ordered Sr adatom located at the oxygen fourfold site on the TiO_2-terminated layer (Fig. 4.20c).

Because STM images are related to the spatial distribution of the wave functions near the Fermi level, atoms, which do not have a local density of states near the Fermi level, are generally invisible even on the conductive materials. On the other hand, the NC-AFM image reflects the strength of the tip-sample interaction force originating from chemical, electrostatic interactions, and so on. Therefore, even STM and NC-AFM images obtained using an identical tip and sample will generally differ.

Simultaneous imaging of a $TiO_2(110)$ surface with STM and NC-AFM shows a typical result that differs in terms of the atoms that are visualized [72]. For simultaneous imaging, the tip-sample distance is controlled by keeping a constant frequency shift, and the tunneling current between the oscillating cantilever and the surface is measured. The two images show different atomic contrasts, i.e., the atomic corrugation of the STM image is in antiphase in the (1×1) and in phase in the (1×2) with respect to the NC-AFM image. In the (1×1), the STM image shows that the dangling bond states at the tip apex overlap with the dangling bonds of the $3d$ states that are protruding from the Ti atom, while the NC-AFM primarily imaged the uppermost oxygen atom. The results of $SrTiO_3$ and TiO_2 surface imaging show that the information obtained by STM and NC-AFM differ, and that the simultaneous imaging of a metal oxide surface enables the investigation of a more detailed surface structure.

4.4.2 Atomically Resolved Imaging of a NiO(001) Surface

The transition metal oxides such as NiO, CoO, and FeO feature the simultaneous existence of an energy gap and unpaired electrons, which gives rise to magnetic properties. Such magnetic insulators are widely used for the exchange biasing in spin valve devices. The NC-AFM enables the direct surface imaging of magnetic insulators on an atomic scale. The forces detected by NC-AFM originate from several kinds of interaction between the surface and the tip, including magnetic interactions in some

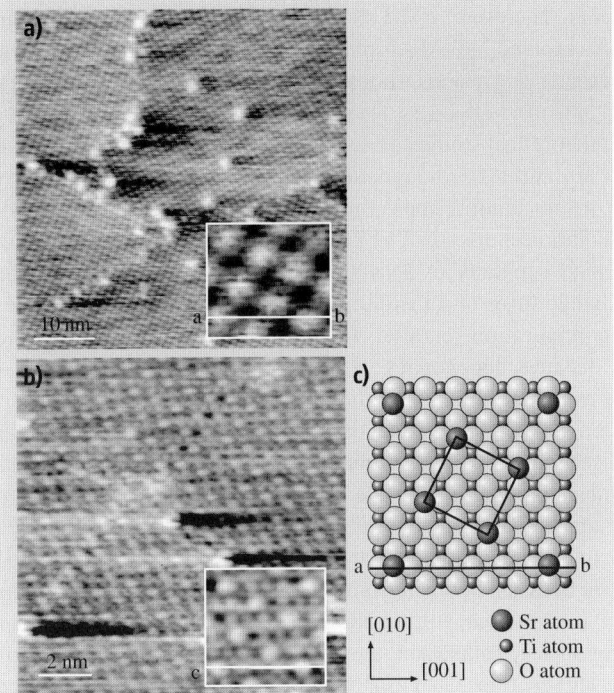

Fig. 4.20. (a) STM and (b) NC-AFM images of a SrTiO$_3$(100) surface. (c) A proposed model of the SrTiO$_3$(100)-($\sqrt{5} \times \sqrt{5}$)$R26.6°$ surface reconstruction. The images and the schematic representations are from [76]

cases. Theoretical studies predict that short-range magnetic interactions such as the exchange interaction should enable the NC-AFM to image magnetic moments on an atomic scale. In this section, we will describe an imaging of the antiferromagnetic NiO(001) surface using a ferromagnetic Fe-coated tip in order to explore the detection of the short-range magnetic interaction. Also, theoretical studies of the exchange force interaction between a magnetic tip and a sample will be described.

Theoretical Studies of the Exchange Force

The NC-AFM measures various kinds of interaction between tip and sample such as van der Waals, and electrostatic and short-range chemical interactions. In the case of a magnetic tip and sample, the interaction detected by the NC-AFM includes the long-range magnetic dipole interaction and the short-range magnetic interaction. The short-range magnetic interaction between the atoms at the tip apex and the sample surface is governed by a combination of the Coulomb interaction and the Pauli exclusion principle and may be included among the short-range chemical interactions. The magnetic interaction energy depends on the electron spin state of the atoms, and the energy difference in spin alignments (parallel or anti-parallel) is referred to as the exchange interaction energy. Therefore, if the short-range magnetic interaction can be detected, an atomically resolved NC-AFM image, in which the short-range interaction contributes to the atomic protrusions, should reveal the local energy difference in the spin alignment.

In the past, extensive theoretical studies on the short-range magnetic interaction between a ferromagnetic tip and a ferromagnetic sample have been performed by a simple calculation [80], a tight-binding approximation [81], and first-principles calculations [82]. In the calculations performed by *Nakamura* et al. [82], three-atomic-layer Fe(001) films, which are separated by a vacuum gap, are used as a model for the tip and sample. The exchange force (F_{ex}) is defined as the difference of the forces (F_P, F_{AP}) in each spin configuration (parallel: P, anti-parallel: AP) of the tip and sample. The tip-sample distance dependency of the exchange force demonstrates that the amplitude of the exchange force is on the order of 10^{-10} N or above and is large enough to be detected by NC-AFM. Also, the site dependency of the exchange force indicates that the discrimination of the exchange force enables the direct imaging of the magnetic moments on an atomic scale. Recently, the interaction between a spin-polarized H atom and a Ni atom on a NiO(001) surface has been investigated theoretically [83]. *Foster* and *Shluger* demonstrated that the difference in magnitude of the exchange interaction between opposite spin Ni ions in a NiO surface could be sufficient to be measured in a low temperature NC-AFM experiment. Also, in order to investigate the effect of surface relaxation, they calculated the interaction of different metal tips with the MgO surface used as a model system. The results indicated that the interaction over the oxygen site is stronger than that over the Mg site, and it was pointed out that the tips should be coated in metals that interact weakly with the surface for the detection of the exchange interaction between the atom at the tip apex and the Ni atom on the NiO surface. However, we must take into account another possibility, that is, that a metal atom at the ferromagnetic tip apex may interact with a Ni atom on the second layer through a magnetic interaction mediated by the electrons in an oxygen atom on the surface.

4.4.3 Atomically Resolved Imaging Using Noncoated and Fe-Coated Si Tips

It is thought that the magnetic insulator NiO is one of the best candidates for detecting short-range magnetic interactions such as the exchange interaction, because an experimental configuration of a ferromagnetic tip and an antiferromagnetic sample minimizes the long-range magnetic dipole interaction that might hinder the detection of the short-range interaction.

Below the Néel temperature of 525 K, NiO is antiferromagnetic and has a rhombohedral structure contracted along the $\langle 111 \rangle$ axes. The direction of the spin on nickel atoms in the same $\{111\}$ plane is parallel, a so-called ferromagnetic sheet. The direction of spin in the adjacent $\{111\}$ planes is anti-parallel. As a result, the spins at the Ni atoms on the (001) surface are arranged in a checkerboard pattern, i.e., the spin alignment along the $\langle 111 \rangle$ direction is antiferromagnetic in order and that along the $\langle 1-1\,0 \rangle$ direction is ferromagnetic in order.

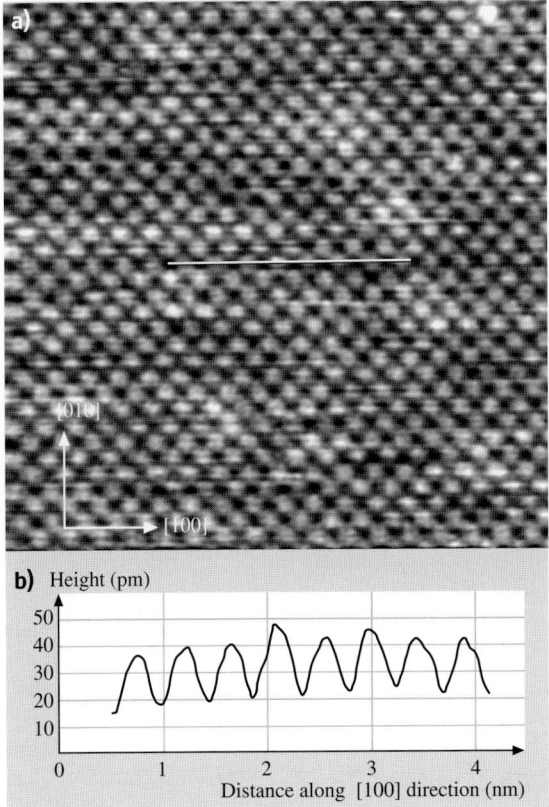

Fig. 4.21. (a) Atomically resolved image obtained with an Fe-coated tip. (b) shows the cross sections of the middle part in (a). Their corrugations are about 30 pm

Figure 4.21a shows an atomically resolved image of a NiO(001) surface with a ferromagnetic Fe-coated tip [84]. The bright protrusions correspond to atoms spaced about 0.42 nm apart, consistent with the expected periodic arrangement of the NiO(001) surface. The corrugation amplitude is typically 30 pm, which is comparable to the value previously reported [74, 75], as shown in Fig. 4.21b. Also, because of the symmetry of the image, it was concluded that only one type of atom is visualized by NC-AFM. However, from this image, we think that it is difficult to distinguish which of the atoms are observed, oxygen or nickel. The theoretical works indicate that a metal tip interacts strongly with the oxygen atoms on the MgO(001) surface [83]. From this result, it is presumed that the bright protrusions correspond to the oxygen atoms. However, it is still questionable which of the atoms are visible with a Fe-coated tip.

If the short-range magnetic interaction is included in the interactions detected by NC-AFM, the corrugation amplitude of the atoms should depend on the direction of the spin at the atom-site. The results of a comparison of the corrugation amplitudes of the same species of atoms with opposite spin can reveal the difference in the short-range magnetic interaction, because the nonmagnetic interactions peculiar to a particular atomic species should be identical. From the results of first-principles calculations [82], the contribution of the short-range magnetic interaction to the measured corrugation amplitude is expected to be about a few percent of the total interaction. Discrimination of such small perturbations is therefore needed. In order to reduce the noise, the corrugation amplitude was added on the basis of the periodicity of the NC-AFM image. In addition, the topographical asymmetry, which is the index characterizing the difference in atomic corrugation amplitude, has been defined [85]. The result shows that the value of the topographical asymmetry calculated from the image obtained with an Fe-coated Si tip depends on the direction of summing of the corrugation amplitude, and that the dependency corresponds to the antiferromagnetic spin ordering of the NiO(001) surface. With a paramagnetic Si tip, no significant indication of a directional dependency of the topographical asymmetry can be seen. Therefore, it is believed that the results obtained using a Fe-coated tip support the conclusion that the dependency of the topographical asymmetry originates in the short-range magnetic interaction.

The measurements presented here demonstrate the possibility of imaging magnetic structures on an atomic scale by NC-AFM. However, the origin of the detected short-range magnetic interaction is under discussion. Therefore, further experiments such as the measurement of force-distance curves and a theoretical study of the interaction between the ferromagnetic tip and an antiferromagnetic sample are needed.

4.5 Applications to Molecules

In the future, it is expected that electronic, chemical, and medical devices will be downsized to a nanoscale. To achieve this, visualizing and assembling individual molecular components is of fundamental importance. Topographic imaging of nonconductive materials is a challenge for atomic force microscopy (AFM). Nanometer-sized domains of surfactants terminated with different functional groups have been identified by lateral force microscopy (LFM) [86] and chemical force microscopy (CFM) [87] as extensions of AFM. At a higher resolution, a periodic array of molecules, Langmuir–Blodgett films [88] for example, was recognized by AFM. However, it remains difficult to visualize an isolated molecule, molecule vacancy, or the boundary of different periodic domains with a microscope with the tip in contact.

4.5.1 Why Molecules and What Molecules?

The access to individual molecules has not been a trivial task even for noncontact atomic force microscopy (NC-AFM). The force pulling the tip into the surface is less sensitive to the gap width (r), especially when chemically stable molecules cover the surface. The attractive potential between two stable molecules is shallow and exhibits r^{-6} decay [13].

Fig. 4.22. The constant frequency-shift topography of domain boundaries on a C_{60} multilayered film deposited on a Si(111) surface based on [91]. Image size: 35×35 nm^2

High-resolution topography of formate (HCOO$^-$) [89] was first reported in 1997 as a molecular adsorbate. The number of imaged molecules is now increasing because of the technological importance of molecular interfaces. To date, the following studies on molecular topography have been published: C_{60} [90, 91], DNAs [92, 93], adenine and thymine [94], alkanethiols [94, 95], a perylene-derivative (PTCDA) [96], a metal porphyrin (Cu-TBPP) [97], glycine sulfate [98], polypropylene [99], vinylidene fluoride [100], and a series of carboxylates (RCOO$^-$) [101–107]. Two of these are presented in Figs. 4.22 and 4.23 to demonstrate the current stage of achievement. The proceedings of the annual NC-AFM conference represent a convenient opportunity for us to update the list of molecules imaged.

4.5.2 Mechanism of Molecular Imaging

A systematic study on carboxylate (RCOO$^-$) with R = H, CH$_3$, C(CH$_3$)$_3$, C ≡ CH, and CF$_3$ revealed that the van der Waals force is responsible for the molecule-dependent microscope topography despite its long-range (r^{-6}) nature. Carboxylates adsorbed on the (110) surface of rutile TiO$_2$ have been extensively studied as a prototype of organic materials interfaced with an inorganic metal oxide [108]. A carboxylic acid molecule (RCOOH) dissociates on this surface to a carboxylate (RCOO$^-$) and a proton (H$^+$) at room temperature, as illustrated in Fig. 4.24. The pair of negatively charged oxygen atoms in RCOO$^-$ coordinate two positively charged Ti atoms at the surface. The adsorbed carboxylates create a long-range ordered monolayer. The lateral distances of the adsorbates in the ordered monolayer are regulated at 0.65 and 0.59 nm along the [1$\bar{1}$0] and [001] directions. By scanning a mixed monolayer containing different carboxylates, the microscope topography of the terminal groups can be quantitatively compared while minimizing tip-dependent artifacts.

Figure 4.25 presents the observed constant frequency-shift topography of four carboxylates terminated by different alkyl groups. On the formate-covered surface of panel (a) individual formates (R = H) were resolved as protrusions of a uniform

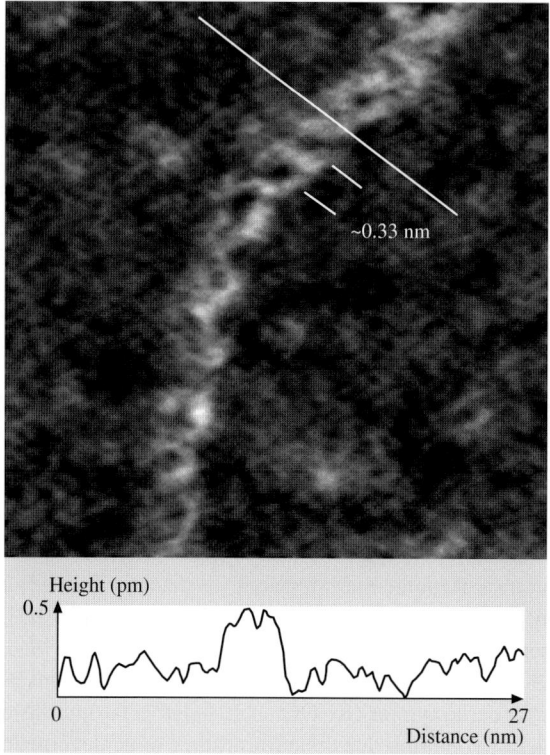

Fig. 4.23. The constant frequency-shift topography of a DNA helix on a mica surface based on [93]. Image size: 43 × 43 nm². The image revealed features with a spacing of 3.3 nm, consistent with the helix turn of B-DNA

brightness. The dark holes represent unoccupied surface sites. The cross section in the lower panel shows that the accuracy of the height measurement was 0.01 nm or better. Brighter particles appeared in the image when the formate monolayer was exposed to acetic acid (CH_3COOH), as shown in panel (b). Some formates were exchanged with acetates (R = CH_3) impinging from the gas phase [110]. Because the number of brighter spots increased with exposure time to acetic acid, the brighter particle was assigned to the acetate [102]. Twenty-nine acetates and 188 formates were identified in the topography. An isolated acetate and its surrounding formates exhibited an image height difference of 0.06 nm. Pivalate is terminated by bulky R = $C(CH_3)_3$. Nine bright pivalates were surrounded by formates of ordinary brightness in the image of panel (c) [104]. The image height difference of an isolated pivalate over the formates was 0.11 nm. Propiolate with C≡CH is a needle-like adsorbate of single-atom diameter. That molecule exhibited in panel (d) a microscope topography 0.20 nm higher than that of the formate [106].

The image topography of formate, acetate, pivalate, and propiolate followed the order of the size of the alkyl groups. Their physical topography can be assumed on

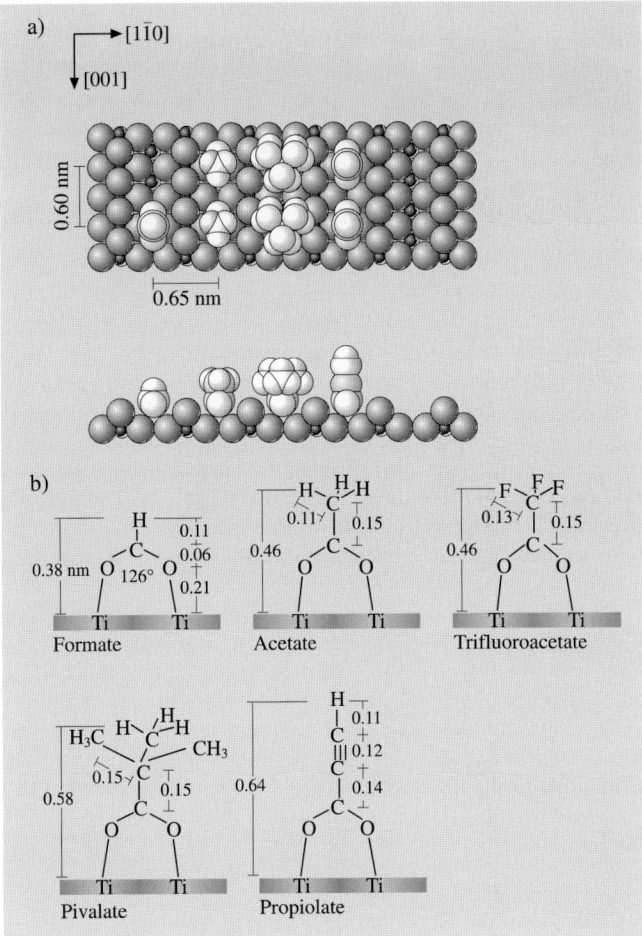

Fig. 4.24. The carboxylates and TiO₂ substrate. (**a**) Top and side view ball model. *Small shaded* and *large shaded balls* represent Ti and O atoms in the substrate. Protons yielded in the dissociation reaction are not shown. (**b**) Atom geometry of formate, acetate, pivalate, propiolate, and trifluoroacetate adsorbed on the TiO$_2$(110) surface. The O-Ti distance and O – C – O angle of the formate were determined in the quantitative analysis of photoelectron diffraction [109]

the C−C and C−H bond lengths in the corresponding RCOOH molecules in the gas phase [111] and is illustrated in Fig. 4.24. The top hydrogen atom of the formate is located 0.38 nm above the surface plane containing the Ti atom pair, while three equivalent hydrogen atoms of the acetate are more elevated at 0.46 nm. The uppermost H atoms in the pivalate are raised by 0.58 nm relative to the Ti plane. The H atom terminating the triple-bonded carbon chain in the propiolate is at 0.64 nm. Figure 4.26 summarizes the observed image heights relative to the formate as a function

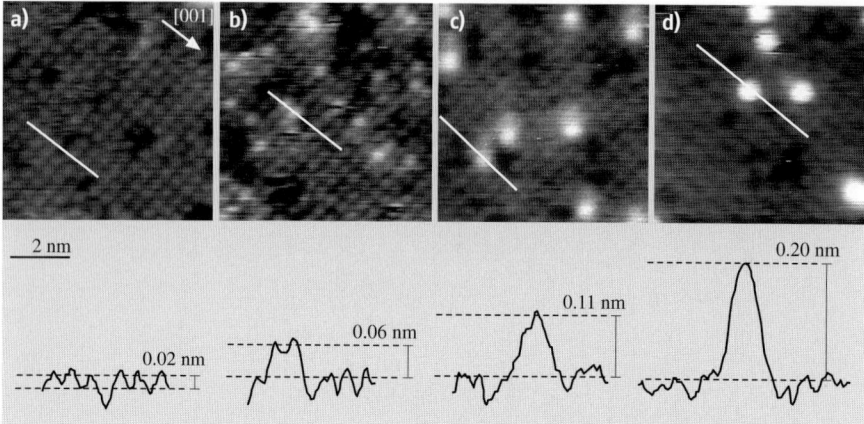

Fig. 4.25. The constant frequency-shift topography of carboxylate monolayers prepared on the TiO$_2$(110) surface based on [102, 104, 106]. Image size: 10×10 nm^2. (**a**) Pure formate monolayer; (**b**) formate-acetate mixed layer; (**c**) formate-pivalate mixed layer; (**d**) formate-propiolate mixed layer. Cross sections determined on the *lines* are shown in the lower panel

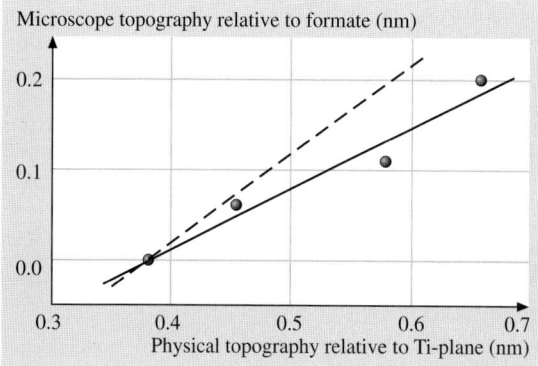

Fig. 4.26. The constant frequency-shift topography of the alkyl-substituted carboxylates as a function of their physical topography given in the model of Fig. 4.4 based on [104]

of the physical height of the topmost H atoms given in the model. The straight line fitted the four observations [104]. When the horizontal axis was scaled with other properties (molecular weight, the number of atoms in a molecule, or the number of electrons in valence states), the correlation became poor.

On the other hand, if the tip apex traced the contour of a molecule composed of hard-sphere atoms, the image topography would reproduce the physical topography in a one-to-one ratio, as shown by the broken line in Fig. 4.26. However, the slope of the fitted line was 0.7. A slope of less than unity is interpreted with the long-range nature of the tip-molecule force. The observable frequency shift reflects the sum of the forces between the tip apex and individual molecules. When the tip passes above a tall molecule embedded in short molecules, it is pulled up to compensate for the increased force originating from the tall molecule. Forces between the lifted tip and the short molecules are reduced due to the increased tip-surface distance. Feedback regulation pushes down the probe to restore the lost forces.

This picture predicts that microscope topography is sensitive to the lateral distribution of the molecules, and that was in fact the case. Two-dimensionally clustered acetates exhibited enhanced image height over an isolated acetate [102]. The tip-molecule force, therefore, remained non-zero at distances over the lateral separation of the carboxylates on this surface (0.59–0.65 nm). Chemical bond interactions cannot be important across such a wide tip-molecule gap, whereas atom-scale images of Si(111)-(7 × 7) are interpreted with the fractional formation of tip-surface chemical bonds [24, 45, 49]. Instead, the attractive component of the van der Waals force is probable for the observed molecule-dependent topography. The absence of the tip-surface chemical bond is reasonable on the carboxylate-covered surface terminated with stable C–H bonds.

The attractive component of the van der Waals force contains electrostatic terms caused by permanent-dipole/permanent-dipole coupling, permanent-dipole/induced-dipole coupling, and induced-dipole/induced-dipole coupling (dispersion force). The four carboxylates examined are equivalent in terms of their permanent electric dipole, because the alkyl groups are nonpolar. The image contrast of one carboxylate relative to another is thus ascribed to the dispersion force and/or the force created by the coupling between the permanent dipole on the tip and the induced dipole on the molecule. If we further assume that the Si tip used exhibits the smallest permanent dipole, the dispersion force remains dominant to create the NC-AFM topography dependent on the nonpolar groups of atoms. A numerical simulation on this assumption [106] successfully reproduced the propiolate topography of Fig. 4.25d. A calculation that does not include quantum chemical treatments is expected to work, unless the tip approaches the surface too closely, or the molecule possesses a dangling bond.

In addition to the contribution of the dispersion force, the permanent dipole moment of molecules may perturb the microscope topography through electrostatic coupling with the tip. Its possible role was demonstrated by imaging a fluorine-substituted acetate. The strongly polarized C–F bonds were expected to perturb the electrostatic field over the molecule. The constant frequency-shift topography of acetate (R = CH_3) and trifluoroacetate (R = CF_3) was indeed sensitive to the fluorine substitution. The acetate was observed 0.05 nm higher than the trifluoroacetate [103], although the F atoms in the trifluoroacetate, as well as the H atoms in the acetate were lifted by 0.46 nm from the surface plane, as illustrated in Fig. 4.24.

4.5.3 Perspectives

Experimental results summarized in this section prove the feasibility of NC-AFM in identifying individual molecules. A systematic study on the constant frequency-shift topography of carboxylates with R = H, CH_3, $C(CH_3)_3$, C≡CH, and CF_3 has revealed the mechanism behind the high-resolution imaging of the chemically stable molecules. The dispersion force is primarily responsible for the molecule-dependent topography. The permanent dipole moment of the imaged molecule, if it exists, perturbs the topography through the electrostatic coupling with the tip. A tiny calculation containing empirical force fields works in simulating the microscope topography.

These results make us optimistic about analyzing physical and chemical properties of nanoscale supramolecular assemblies constructed on a solid surface. If the accuracy of topographic measurement is developed by one more order of magnitude, which is not an unrealistic concept, it may be possible to identify structural isomers, chiral isomers, and conformational isomers of a molecule. Kelvin probe force microscopy (KPFM), an extension of NC-AFM, provides a nanoscale analysis of molecular electronic properties [99, 100]. Force spectroscopy with chemically modified tips seems promising for the detection of a selected chemical force. Operation in a liquid atmosphere [112] is demanded to observe biochemical materials in the raw.

Acknowledgement

Thanks to Tom Albrecht, Alexis Baratoff, Hartmut Bielefeldt, Gerd Binnig, Dominik Brändlin, Peter van Dongen, Urs Dürig, Christoph Gerber, Stefan Hembacher, Markus Herz, Lukas Howald, Christian Laschinger, Ulrich Mair, Jochen Mannhart, Thomas Ottenthal, Calvin Quate, Marco Tortonese, and the BMBF for funding under project no. 13N6918.

References

1. G. Binnig. Atomic force microscope and method for imaging surfaces with atomic resolution. *US Patent 4,724,318*, 1986.
2. G. Binnig, C. F. Quate, and C. Gerber. Atomic force microscope. *Phys. Rev. Lett.*, 56:930–933, 1986.
3. G. Binnig, H. Rohrer, C. Gerber, and E. Weibel. Surface studies by scanning tunneling microscopy. *Phys. Rev. Lett.*, 49:57–61, 1982.
4. G. Binnig and H. Rohrer. The scanning tunneling microscope. *Sci. Am.*, 253:50–56, 1985.
5. G. Binnig and H. Rohrer. In touch with atoms. *Rev. Mod. Phys.*, 71:S324–S320, 1999.
6. C. J. Chen. *Introduction to Scanning Tunneling Microscopy*. Oxford Univ. Press, 1993.
7. H.-J. Güntherodt and R. Wiesendanger, editors. *Scanning Tunneling Microscopy I-III*. Springer, 1991.
8. J. A. Stroscio and W. J. Kaiser, editors. *Scanning Tunneling Microscopy*. Academic, 1993.
9. R. Wiesendanger. *Scanning Probe Microscopy and Spectroscopy: Methods and Applications*. Cambridge Univ. Press, 1994.
10. S. Morita, R. Wiesendanger, and E. Meyer, editors. *Noncontact Atomic Force Microscopy*. Springer, 2002.
11. R. Garcia and R. Perez. Dynamic atomic force microscopy methods. *Surf. Sci. Rep.*, 47:197–301, 2002.

12. F. J. Giessibl. Advances in atomic force microscopy. *Rev. Mod. Phys.*, 75:949–983, 2003.
13. J. Israelachvili. *Intermolecular and Surface Forces*. Academic, 2nd edition, 1992.
14. L. Olsson, N. Lin, V. Yakimov, and R. Erlandsson. A method for *in situ* characterization of tip shape in ac-mode atomic force microscopy using electrostatic interaction. *J. Appl. Phys.*, 84:4060–4064, 1998.
15. S. Akamine, R. C. Barrett, and C. F. Quate. Improved atomic force microscopy images using cantilevers with sharp tips. *Appl. Phys. Lett.*, 57:316–318, 1990.
16. T. R. Albrecht, S. Akamine, T. E. Carver, and C. F. Quate. Microfabrication of cantilever styli for the atomic force microscope. *J. Vac. Sci. Technol. A*, 8:3386–3396, 1990.
17. M. Tortonese, R. C. Barrett, and C. Quate. Atomic resolution with an atomic force microscope using piezoresistive detection. *Appl. Phys. Lett.*, 62:834–836, 1993.
18. O. Wolter, T. Bayer, and J. Greschner. Micromachined silicon sensors for scanning force microscopy. *J. Vac. Sci. Technol.*, 9:1353–1357, 1991.
19. D. Sarid. *Scanning Force Microscopy*. Oxford Univ. Press, 2nd edition, 1994.
20. F. J. Giessibl and B. M. Trafas. Piezoresistive cantilevers utilized for scanning tunneling and scanning force microscope in ultrahigh vacuum. *Rev. Sci. Instrum.*, 65:1923–1929, 1994.
21. P. Güthner, U. C. Fischer, and K. Dransfeld. Scanning near-field acoustic microscopy. *Appl. Phys. B*, 48:89–92, 1989.
22. K. Karrai and R. D. Grober. Piezoelectric tip-sample distance control for near field optical microscopes. *Appl. Phys. Lett.*, 66:1842–1844, 1995.
23. F. J. Giessibl. High-speed force sensor for force microscopy and profilometry utilizing a quartz tuning fork. *Appl. Phys. Lett.*, 73:3956–3958, 1998.
24. F. J. Giessibl, S. Hembacher, H. Bielefeldt, and J. Mannhart. Subatomic features on the silicon (111)-(7 × 7) surface observed by atomic force microscopy. *Science*, 289:422–425, 2000.
25. F. Giessibl, C. Gerber, and G. Binnig. A low-temperature atomic force/scanning tunneling microscope for ultrahigh vacuum. *J. Vac. Sci. Technol., B*, 9:984–988, 1991.
26. F. Ohnesorge and G. Binnig. True atomic resolution by atomic force microscopy through repulsive and attractive forces. *Science*, 260:1451–1456, 1993.
27. F. J. Giessibl and G. Binnig. True atomic resolution on KBr with a low-temperature atomic force microscope in ultrahigh vacuum. *Ultramicroscopy*, 42-44:281–286, 1992.
28. S. P. Jarvis, H. Yamada, H. Tokumoto, and J. B. Pethica. Direct mechanical measurement of interatomic potentials. *Nature*, 384:247–249, 1996.
29. L. Howald, R. Lüthi, E. Meyer, P. Guthner, and H.-J. Güntherodt. Scanning force microscopy on the Si(111)7 × 7 surface reconstruction. *Z. Phys. B*, 93:267–268, 1994.
30. L. Howald, R. Lüthi, E. Meyer, and H.-J. Güntherodt. Atomic-force microscopy on the si(111)7 × 7 surface. *Phys. Rev. B*, 51:5484–5487, 1995.
31. Y. Martin, C. C. Williams, and H. K. Wickramasinghe. Atomic force microscope – force mapping and profiling on a sub 100 Å scale. *J. Appl. Phys.*, 61:4723–4729, 1987.
32. T. R. Albrecht, P. Grutter, H. K. Horne, and D. Rugar. Frequency modulation detection using high-q cantilevers for enhanced force microscope sensitivity. *J. Appl. Phys.*, 69:668–673, 1991.
33. F. J. Giessibl. Atomic resolution of the silicon (111)-(7 × 7) surface by atomic force microscopy. *Science*, 267:68–71, 1995.
34. S. Kitamura and M. Iwatsuki. Observation of silicon surfaces using ultrahigh-vacuum noncontact atomic force microscopy. *Jpn. J. Appl. Phys.*, 35:L668–L671, 1995.

35. R. Erlandsson, L. Olsson, and P. Martensson. Inequivalent atoms and imaging mechanisms in ac-mode atomic-force microscopy of Si(111)7 × 7. *Phys. Rev. B*, 54:R8309–R8312, 1996.
36. N. Burnham and R. J. Colton. Measuring the nanomechanical and surface forces of materials using an atomic force microscope. *J. Vac. Sci. Technol. A*, 7:2906–2913, 1989.
37. D. Tabor and R. H. S. Winterton. Direct measurement of normal and retarded van der waals forces. *Proc. R. Soc. London A*, 312:435, 1969.
38. F. J. Giessibl. Forces and frequency shifts in atomic resolution dynamic force microscopy. *Phys. Rev. B*, 56:16010–16015, 1997.
39. G. Binnig, H. Rohrer, C. Gerber, and E. Weibel. 7×7 reconstruction on Si(111) resolved in real space. *Phys. Rev. Lett.*, 50:120–123, 1983.
40. H. Goldstein. *Classical Mechanics*. Addison Wesley, 1980.
41. U. Dürig. Interaction sensing in dynamic force microscopy. *New J. Phys.*, 2:5.1–5.12, 2000.
42. F. J. Giessibl. A direct method to calculate tip-sample forces from frequency shifts in frequency-modulation atomic force microscopy. *Appl. Phys. Lett.*, 78:123–125, 2001.
43. U. Dürig, H. P. Steinauer, and N. Blanc. Dynamic force microscopy by means of the phase-controlled oscillator method. *J. Appl. Phys.*, 82:3641–3651, 1997.
44. F. J. Giessibl, H. Bielefeldt, S. Hembacher, and J. Mannhart. Calculation of the optimal imaging parameters for frequency modulation atomic force microscopy. *Appl. Surf. Sci.*, 140:352–357, 1999.
45. M. A. Lantz, H. J. Hug, R. Hoffmann, P. J. A. van Schendel, P. Kappenberger, S. Martin, A. Baratoff, and H.-J. Güntherodt. Quantitative measurement of short-range chemical bonding forces. *Science*, 291:2580–2583, 2001.
46. F. J. Giessibl, M. Herz, and J. Mannhart. Friction traced to the single atom. *Proc. Nat. Acad. Sci. USA*, 99:12006–12010, 2002.
47. N. Nakagiri, M. Suzuki, K. Oguchi, and H. Sugimura. Site discrimination of adatoms in Si(111)-7 × 7 by noncontact atomic force microscopy. *Surf. Sci. Lett.*, 373:L329–L332, 1997.
48. Y. Sugawara, M. Ohta, H. Ueyama, and S. Morita. Defect motion on an InP(110) surface observed with noncontact atomic force microscopy. *Science*, 270:1646–1648, 1995.
49. T. Uchihashi, Y. Sugawara, T. Tsukamoto, M. Ohta, and S. Morita. Role of a covalent bonding interaction in noncontact-mode atomic-force microscopy on Si(111)7×7. *Phys. Rev. B*, 56:9834–9840, 1997.
50. K. Yokoyama, T. Ochi, A. Yoshimoto, Y. Sugawara, and S. Morita. Atomic resolution imaging on si(100)2 × 1 and si(100)2 × 1-h surfaces using a noncontact atomic force microscope. *Jpn. J. Appl. Phys.*, 39:L113–L115, 2000.
51. Y. Sugawara, T. Minobe, S. Orisaka, T. Uchihashi, T. Tsukamoto, and S. Morita. Non-contact afm images measured on si(111) $\sqrt{3} \times \sqrt{3}$-ag and ag(111) surfaces. *Surf. Interface Anal.*, 27:456–461, 1999.
52. K. Yokoyama, T. Ochi, Y. Sugawara, and S. Morita. Atomically resolved ag imaging on si(111) $\sqrt{3} \times \sqrt{3}$-ag surface with noncontact atomic force microscope. *Phys. Rev. Lett.*, 83:5023–5026, 1999.
53. M. Bammerlin, R. Lüthi, E. Meyer, A. Baratoff, J. Lü, M. Guggisberg, Ch. Gerber, L. Howald, and H.-J. Güntherodt. True atomic resolution on the surface of an insulator via ultrahigh vacuum dynamic force microscopy. *Probe Microsc.*, 1:3–7, 1997.
54. M. Bammerlin, R. Lüthi, E. Meyer, A. Baratoff, J. Lü, M. Guggisberg, C. Loppacher, Ch. Gerber, and H.-J. Güntherodt. Dynamic sfm with true atomic resolution on alkali halide surfaces. *Appl. Phys. A*, 66:S293–S294, 1998.

55. R. Hoffmann, M. A. Lantz, H. J. Hug, P. J. A. van Schendel, P. Kappenberger, S. Martin, A. Baratoff, and H.-J. Güntherodt. Atomic resolution imaging and force versus distance measurements on KBr(001) using low temperature scanning force microscopy. *Appl. Surf. Sci.*, 188:238–244, 2002.
56. R. Bennewitz, A. S. Foster, L. N. Kantorovich, M. Bammerlin, Ch. Loppacher, S. Schär, M. Guggisberg, E. Meyer, and A. L. Shluger. Atomically resolved edges and kinks of NaCl islands on Cu(111): experiment and theory. *Phys. Rev. B*, 62:2074–2084, 2000.
57. A. I. Livshits, A. L. Shluger, A. L. Rohl, and A. S. Foster. Model of noncontact scanning force microscopy on ionic surfaces. *Phys. Rev. B*, 59:2436–2448, 1999.
58. R. Bennewitz, M. Bammerlin, Ch. Loppacher, M. Guggisberg, L. Eng, E. Meyer, H.-J. Güntherodt, C. P. An, and F. Luty. Molecular impurities at the NaCl(100) surface observed by scanning force microscopy. *Rad. Effects Defects Solids*, 150:321–326, 1998.
59. R. Bennewitz, O. Pfeiffer, S. Schär, V. Barwich, E. Meyer, and L. N. Kantorovich. Atomic corrugation in nc-afm of alkali halides. *Appl. Surf. Sci.*, 188:232–237, 2002.
60. C. Barth and M. Reichling. Resolving ions and vacancies at step edges on insulating surfaces. *Surf. Sci.*, 470:L99–L103, 2000.
61. R. Bennewitz, M. Reichling, and E. Matthias. Force microscopy of cleaved and electron-irradiated CaF_2(111) surfaces in ultra-high vacuum. *Surf. Sci.*, 387:69–77, 1997.
62. M. Reichling and C. Barth. Scanning force imaging of atomic size defects on the CaF_2(111) surface. *Phys. Rev. Lett.*, 83:768–771, 1999.
63. M. Reichling, M. Huisinga, S. Gogoll, and C. Barth. Degradation of the CaF_2(111) surface by air exposure. *Surf. Sci.*, 439:181–190, 1999.
64. C. Barth, A. S. Foster, M. Reichling, and A. L. Shluger. Contrast formation in atomic resolution scanning force microscopy of CaF_2(111): experiment and theory. *J. Phys. Condens. Matter*, 13:2061–2079, 2001.
65. A. S. Foster, C. Barth, A. L. Shluger, and M. Reichling. Unambiguous interpretation of atomically resolved force microscopy images of an insulator. *Phys. Rev. Lett.*, 86:2373–2376, 2001.
66. A. S. Foster, A. L. Rohl, and A. L. Shluger. Imaging problems on insulators: what can be learnt from nc-afm modeling on CaF_2? *Appl. Phys. A*, 72:S31–S34, 2001.
67. A. S. Foster, A. L. Shluger, and R. M. Nieminen. Quantitative modeling in scanning force microscopy on insulators. *Appl. Surf. Sci.*, 188:306–318, 2002.
68. M. Reichling and C. Barth. *Atomically resoluiton imaging on fluorides*. Springer, 2002.
69. K. Fukui, H. Ohnishi, and Y. Iwasawa. Atom-resolved image of the TiO_2(110) surface by noncontact atomic force microscopy. *Phys. Rev. Lett.*, 79:4202–4205, 1997.
70. H. Raza, C. L. Pang, S. A. Haycock, and G. Thornton. Non-contact atomic force microscopy imaging of TiO_2(100) surfaces. *Appl. Surf. Sci.*, 140:271–275, 1999.
71. C. L. Pang, H. Raza, S. A. Haycock, and G. Thornton. Imaging reconstructed TiO_2(100) surfaces with non-contact atomic force microscopy. *Appl. Surf. Sci.*, 157:223–238, 2000.
72. M. Ashino, T. Uchihashi, K. Yokoyama, Y. Sugawara, S. Morita, and M. Ishikawa. Stm and atomic-resolution noncontact afm of an oxygen-deficient TiO_2(110) surface. *Phys. Rev. B*, 61:13955–13959, 2000.
73. C. L. Pang, S. A. Haycock, H. Raza, P. J. Møller, and G. Thornton. Structures of the 4×1 and 1×2 reconstructions of SnO_2(110). *Phys. Rev. B*, 62:R7775–R7778, 2000.
74. H. Hosoi, K. Sueoka, K. Hayakawa, and K. Mukasa. Atomic resolved imaging of cleaved NiO(100) surfaces by nc-afm. *Appl. Surf. Sci.*, 157:218–221, 2000.
75. W. Allers, S. Langkat, and R. Wiesendanger. Dynamic low-temperature scanning force microscopy on nickel oxide (001). *Appl. Phys. A*, 72:S27–S30, 2001.
76. T. Kubo and H. Nozoye. Surface structure of $srtio_3$(100)-($\sqrt{5} \times \sqrt{5}$) – $R26.6°$. *Phys. Rev. Lett.*, 86:1801–1804, 2001.

77. K. Fukui, Y. Namai, and Y. Iwasawa. Imaging of surface oxygen atoms and their defect structures on $CeO_2(111)$ by noncontact atomic force microscopy. *Appl. Surf. Sci.*, 188:252–256, 2002.
78. C. Barth and M. Reichling. Imaging the atomic arrangements on the high-temperature reconstructed α-Al_2O_3 surface. *Nature*, 414:54–57, 2001.
79. C. L. Pang, H. Raza, S. A. Haycock, and G. Thornton. Growth of copper and palladium α-$Al_2O_3(0001)$. *Surf. Sci.*, 460:L510–L514, 2000.
80. K. Mukasa, H. Hasegawa, Y. Tazuke, K. Sueoka, M. Sasaki, and K. Hayakawa. Exchange interaction between magnetic moments of ferromagnetic sample and tip: Possibility of atomic-resolution images of exchange interactions using exchange force microscopy. *Jpn. J. Appl. Phys.*, 33:2692–2695, 1994.
81. H. Ness and F. Gautier. Theoretical study of the interaction between a magnetic nanotip and a magnetic surface. *Phys. Rev. B*, 52:7352–7362, 1995.
82. K. Nakamura, H. Hasegawa, T. Ohuchi, K. Sueoka, K. Hayakawa, and K. Mukasa. First-principles calculation of the exchange interaction and the exchange force between magnetic Fe films. *Phys. Rev. B*, 56:3218–3221, 1997.
83. A. S. Foster and A. L. Shluger. Spin-contrast in non-contact sfm on oxide surfaces: theoretical modeling of NiO(001) surface. *Surf. Sci.*, 490:211–219, 2001.
84. H. Hosoi, K. Kimura, K. Sueoka, K. Hayakawa, and K. Mukasa. Non-contact atomic force microscopy of an antiferromagnetic NiO(100) surface using a ferromagnetic tip. *Appl. Phys. A*, 72:S23–S26, 2001.
85. H. Hosoi, K. Sueoka, K. Hayakawa, and K. Mukasa. *Atomically resolved imaging of a NiO(001) surface*, pages 125–134. Springer, 2002.
86. R. M. Overney, E. Meyer, J. Frommer, D. Brodbeck, R. Lüthi, L. Howald, H.-J. Güntherodt, M. Fujihira, H. Takano, and Y. Gotoh. Friction measurements on phase-separated thin films with a modified atomic force microscope. *Nature*, 359:133–135, 1992.
87. D. Frisbie, L. F. Rozsnyai, A. Noy, M. S. Wrighton, and C. M. Lieber. Functional group imaging by chemical force microscopy. *Science*, 265:2071–2074, 1994.
88. E. Meyer, L. Howald, R. M. Overney, H. Heinzelmann, J. Frommer, H.-J. Güntherodt, T. Wagner, H. Schier, and S. Roth. Molecular-resolution images of langmuir–blodgett films using atomic force microscopy. *Nature*, 349:398–400, 1992.
89. K. Fukui, H. Onishi, and Y. Iwasawa. Imaging of individual formate ions adsorbed on $TiO_2(110)$ surface by non-contact atomic force microscopy. *Chem. Phys. Lett.*, 280:296–301, 1997.
90. K. Kobayashi, H. Yamada, T. Horiuchi, and K. Matsushige. Investigations of C_60 molecules deposited on Si(111) by noncontact atomic force microscopy. *Appl. Surf. Sci.*, 140:281–286, 1999.
91. K. Kobayashi, H. Yamada, T. Horiuchi, and K. Matsushige. Structures and electrical properties of fullerene thin films on Si(111)-7 × 7 surface investigated by noncontact atomic force microscopy. *Jpn. J. Appl. Phys.*, 39:3821–3829, 2000.
92. Y. Maeda, T. Matsumoto, and T. Kawai. Observation of single- and double-strand dna using non-contact atomic force microscopy. *Appl. Surf. Sci.*, 140:400–405, 1999.
93. T. Uchihashi, M. Tanigawa, M. Ashino, Y. Sugawara, K. Yokoyama, S. Morita, and M. Ishikawa. Identification of b-form dna in an ultrahigh vacuum by noncontact-mode atomic force microscopy. *Langmuir*, 16:1349–1353, 2000.
94. T. Uchihashi, T. Ishida, M. Komiyama, M. Ashino, Y. Sugawara, W. Mizutani, K. Yokoyama, S. Morita, H. Tokumoto, and M. Ishikawa. High-resolution imaging of organic monolayers using noncontact afm. *Appl. Surf. Sci*, 157:244–250, 2000.

95. T. Fukuma, K. Kobayashi, T. Horiuchi, H. Yamada, and K. Matsushige. Alkanethiol self-assembled monolayers on Au(111) surfaces investigated by non-contact afm. *Appl. Phys. A*, 72:S109–S112, 2001.
96. B. Gotsmann, C. Schmidt, C. Seidel, and H. Fuchs. Molecular resolution of an organic monolayer by dynamic afm. *Europ. Phys. J. B*, 4:267–268, 1998.
97. Ch. Loppacher, M. Bammerlin, M. Guggisberg, E. Meyer, H.-J. Güntherodt, R. Lüthi, R. Schlittler, and J. K. Gimzewski. Forces with submolecular resolution between the probing tip and Cu-tbpp molecules on Cu(100) observed with a combined afm/stm. *Appl. Phys. A*, 72:S105–S108, 2001.
98. L. M. Eng, M. Bammerlin, Ch. Loppacher, M. Guggisberg, R. Bennewitz, R. Lüthi, E. Meyer, and H.-J. Güntherodt. Surface morphology, chemical contrast, and ferroelectric domains in tgs bulk single crystals differentiated with uhv non-contact force microscopy. *Appl. Surf. Sci.*, 140:253–258, 1999.
99. S. Kitamura, K. Suzuki, and M. Iwatsuki. High resolution imaging of contact potential difference using a novel ultrahigh vacuum non-contact atomic force microscope technique. *Appl. Surf. Sci.*, 140:265–270, 1999.
100. H. Yamada, T. Fukuma, K. Umeda, K. Kobayashi, and K. Matsushige. Local structures and electrical properties of organic molecular films investigated by non-contact atomic force microscopy. *Appl. Surf. Sci*, 188:391–398, 2000.
101. K. Fukui and Y. Iwasawa. Fluctuation of acetate ions in the (2 × 1)-acetate overlayer on TiO_2(110)-(1 × 1) observed by noncontact atomic force microscopy. *Surf. Sci.*, 464:L719–L726, 2000.
102. A. Sasahara, H. Uetsuka, and H. Onishi. Single-molecule analysis by non-contact atomic force microscopy. *J. Phys. Chem. B*, 105:1–4, 2001.
103. A. Sasahara, H. Uetsuka, and H. Onishi. Nc-afm topography of HCOO and CH_3COO molecules co-adsorbed on TiO_2(110). *Appl. Phys. A*, 72:S101–S103, 2001.
104. A. Sasahara, H. Uetsuka, and H. Onishi. Image topography of alkyl-substituted carboxylates observed by noncontact atomic force microscopy. *Surf. Sci.*, 481:L437–L442, 2001.
105. A. Sasahara, H. Uetsuka, and H. Onishi. Noncontact atomic force microscope topography dependent on permanent dipole of individual molecules. *Phys. Rev. B*, 64:121406(R), 2001.
106. A. Sasahara, H. Uetsuka, T. Ishibashi, and H. Onishi. A needle-like organic molecule imaged by noncontact atomic force microscopy. *Appl. Surf. Sci.*, 188:265–271, 2002.
107. H. Onishi, A. Sasahara, H. Uetsuka, and T. Ishibashi. Molecule-dependent topography determined by noncontact atomic force microscopy: Carboxylates on TiO_2(110). *Appl. Surf. Sci.*, 188:257–264, 2002.
108. H. Onishi. *Carboxylates adsorbed on* $TiO_2(110)$, pages 75–89. Springer, 2002.
109. S. Thevuthasan, G. S. Herman, Y. J. Kim, S. A. Chambers, C. H. F. Peden, Z. Wang, R. X. Ynzunza, E. D. Tober, J. Morais, and C. S. Fadley. The structure of formate on tio_2(110) by scanned-energy and scanned-angle photoelectron diffraction. *Surf. Sci.*, 401:261–268, 1998.
110. H. Uetsuka, A. Sasahara, A. Yamakata, and H. Onishi. Microscopic identification of a bimolecular reaction intermediate. *J. Phys. Chem. B*, 106:11549–11552, 2002.
111. D. R. Lide. *Handbook of Chemistry and Physics*. CRC, 81st edition, 2000.
112. K. Kobayashi, H. Yamada, and K. Matsushige. Dynamic force microscopy using fm detection in various environments. *Appl. Surf. Sci.*, 188:430–434, 2002.

5

Low Temperature Scanning Probe Microscopy

Markus Morgenstern, Alexander Schwarz, and Udo D. Schwarz

Summary. This chapter is dedicated to scanning probe microscopy, one of the most important techniques in nanotechnology. In general, scanning probe techniques allow the measurement of physical properties down to the nanometer scale. Some techniques, such as the scanning tunneling microscope and the scanning force microscope even go down to the atomic scale. The properties that are accessible are various. Most importantly, one can image the arrangement of atoms on conducting surfaces by scanning tunneling microscopy and on insulating substrates by scanning force microscopy. But also the arrangement of electrons (scanning tunneling spectroscopy), the force interaction between different atoms (scanning force spectroscopy), magnetic domains (magnetic force microscopy), the local capacitance (scanning capacitance microscopy), the local temperature (scanning thermo microscopy), and local light-induced excitations (scanning near-field microscopy) can be measured with high spatial resolution. In addition, some techniques even allow the manipulation of atomic configurations.

Probably the most important advantage of the low-temperature operation of scanning probe techniques is that they lead to a significantly better signal-to-noise ratio than measuring at room temperature. This is why many researchers work below 100 K. However, there are also physical reasons to use low-temperature equipment. For example, the manipulation of atoms or scanning tunneling spectroscopy with high energy resolution can only be realized at low temperatures. Moreover, some physical effects such as superconductivity or the Kondo effect are restricted to low temperatures. Here, we describe the design criteria of low-temperature scanning probe equipment and summarize some of the most spectacular results achieved since the invention of the method about 20 years ago. We first focus on the scanning tunneling microscope, giving examples of atomic manipulation and the analysis of electronic properties in different material arrangements. Afterwards, we describeresults obtained by scanning force microscopy, showing atomic-scale imaging on insulators, as well as force spectroscopy analysis. Finally, the magnetic force microscope, which images domain patterns in ferromagnets and vortex patterns in superconductors, is discussed. Although this list is far from complete, we feel that it gives an adequate impression of the fascinating possibilities of low-temperature scanning probe instruments.

In this chapter low temperatures are defined as lower than about 100 K and are normally achieved by cooling with liquid nitrogen or liquid helium. Applications in which SPMs are operated close to 0 °C are not covered in this chapter.

5.1 Introduction

More than two decades ago, the first design of an experimental setup was presented where a sharp tip was systematically scanned over a sample surface in order to obtain local information on the tip-sample interaction down to the atomic scale. This original instrument used the tunneling current between a conducting tip and a conducting sample as a feedback signal and was named *scanning tunneling microscope* accordingly [1]. Soon after this historic breakthrough, it became widely recognized that virtually any type of tip-sample interaction can be used to obtain local information on the sample by applying the same general principle, provided that the selected interaction was reasonably short-ranged. Thus, a whole variety of new methods has been introduced, which are denoted collectively as *scanning probe methods*. An overview is given by *Wiesendanger* [2].

The various methods, especially the above mentioned scanning tunneling microscopy (STM) and scanning force microscopy (SFM) – which is often further classified into subdisciplines such as the topography-reflecting atomic force microscopy (AFM), the magnetic force microscopy (MFM), or the electrostatic force microscopy (EFM) – have been established as standard methods for surface characterization on the nanometer scale. The reason is that they feature extremely high resolution (often down to the atomic scale for STM and AFM), despite a principally simple, compact, and comparatively inexpensive design.

A side effect of the simple working principle and the compact design of many scanning probe microscopes (SPMs) is that they can be adapted to different environments such as air, all kinds of gaseous atmospheres, liquids, or vacuum with reasonable effort. Another advantage is their ability to work within a wide temperature range. A microscope operation at higher temperatures is chosen to study surface diffusion, surface reactivity, surface reconstructions that only manifest at elevated temperatures, high-temperature phase transitions, or to simulate conditions as they occur, e.g., in engines, catalytic converters, or reactors. Ultimately, the upper limit for the operation of an SPM is determined by the stability of the sample, but thermal drift, which limits the ability to move the tip in a controlled manner over the sample, as well as the depolarization temperature of the piezoelectric positioning elements might further restrict successful measurements.

On the other hand, low-temperature (LT) application of SPMs is much more widespread than operation at high temperatures. Essentially five reasons make researchers adapt their experimental setups to low-temperature compatibility. These are: (1) the reduced thermal drift, (2) lower noise levels, (3) enhanced stability of tip and sample, (4) the reduction in piezo hysteresis/creep, and (5) probably the most obvious, the fact that many physical effects are restricted to low temperature. Reasons (1) to (4) only apply unconditionally if the whole microscope body is kept at low temperature (typically in or attached to a bath cryostat, see Sect. 5.3). Setups in which only the sample is cooled may show considerably less favorable operating characteristics. As a result of (1) to (4), ultrahigh resolution and long-term stability can be achieved on a level that significantly exceeds what can be accomplished at room temperature even under the most favorable circumstances. Typical examples for (5) are superconductivity [3] and the Kondo effect [4].

5.2 Microscope Operation at Low Temperatures

Nevertheless, before we devote ourselves to a small overview of experimental LT-SPM work, we will take a closer look at the specifics of microscope operation at low temperatures, including a discussion of the corresponding instrumentation.

5.2.1 Drift

Thermal drift originates from thermally activated movements of the individual atoms, which are reflected by the thermal expansion coefficient. At room temperature, typical values for solids are on the order of $(1-50) \times 10^{-6}\,\text{K}^{-1}$. If the temperature could be kept precisely constant, any thermal drift would vanish, regardless of the absolute temperature of the system. The close coupling of the microscope to a large temperature bath that keeps a constant temperature ensures a significant reduction in thermal drift and allows for distortion-free, long-term measurements. Microscopes that are efficiently attached to sufficiently large bath cryostats, therefore, show a one- to two-order-of-magnitude increase in thermal stability compared with non-stabilized setups operated at room temperature.

A second effect also helps suppress thermally induced drift of the probing tip relative to a specific location on the sample surface. The thermal expansion coefficients are at liquid helium temperature two or more orders of magnitude smaller than at room temperature. Consequently, the thermal drift during low-temperature operation decreases accordingly.

For some specific scanning probe methods, there may be additional ways a change in temperature can affect the quality of the data. In *frequency-modulation SFM* (FM-SFM), for example, the measurement principle relies on the accurate determination of the eigenfrequency of the cantilever, which is determined by its spring constant and its effective mass. However, the spring constant changes with temperature due to both thermal expansion (i.e., the resulting change in the cantilever dimensions) and the variation of Young's modulus with temperature. Assuming drift rates of about 2 mK/min, as is typical for room temperature measurements, this effect might have a significant influence on the obtained data.

5.2.2 Noise

The theoretically achievable resolution in SPMs often increases with decreasing temperature due to a decrease in thermally induced noise. An example is the thermal noise in SFMs, which is proportional to the square root of the temperature [5, 6]. Lowering the temperature from $T = 300\,\text{K}$ to $T = 10\,\text{K}$ thus results in a reduction of the thermal frequency noise of more than a factor of five. Graphite, e.g., has been imaged atomically resolved only at low temperatures due to its extremely low corrugation, which was below the room temperature noise level [7, 8].

Another, even more striking, example is the spectroscopic resolution in *scanning tunneling spectroscopy* (STS). It depends linearly on the temperature [2] and is consequently reduced even more at LT than the thermal noise in AFM. This provides the opportunity to study structures or physical effects not accessible at room temperature such as spin and Landau levels in semiconductors [9].

Finally, it might be worth mentioning that the enhanced stiffness of most materials at low temperatures (increased Young's modulus) leads to a reduced coupling to external noise. Even though this effect is considered small [6], it should not be ignored.

5.2.3 Stability

There are two major stability issues that considerably improve at low temperature. First, low temperatures close to the temperature of liquid helium inhibit most of the thermally activated diffusion processes. As a consequence, the sample surfaces show a significantly increased long-term stability, since defect motion or adatom diffusion is massively suppressed. Most strikingly, even single xenon atoms deposited on suitable substrates can be successfully imaged [10, 11], or even manipulated [12]. In the same way, the low temperatures also stabilize the atomic configuration at the tip end by preventing sudden jumps of the most loosely bound foremost tip atom(s). Second, the large cryostat that usually surrounds the microscope acts as an effective cryo pump. Thus samples can be kept clean for several weeks, which is a multiple of the corresponding time at room temperature (about 3–4 h).

5.2.4 Piezo Relaxation and Hysteresis

The last important benefit from low-temperature operation of SPMs is that artifacts given by the response of the piezoelectric scanners are substantially reduced. After applying a voltage ramp to one electrode of a piezoelectric scanner, its immediate initial deflection, l_0, is followed by a much slower relaxation, Δl, with a logarithmic time dependence. This effect, known as piezo relaxation or "creep", diminishes substantially at low temperatures, typically by a factor of ten or more. As a consequence, piezo nonlinearities and piezo hysteresis decrease accordingly. Additional information is given by *Hug* et al. [13].

5.3 Instrumentation

The two main design criteria for all vacuum-based scanning probe microscope systems are (1) to provide an efficient decoupling of the microscope from the vacuum system and other sources of external vibrations, and (2) to avoid most internal noise sources through a high mechanical rigidity of the microscope body itself. In vacuum systems designed for low-temperature applications, a significant degree of complexity is added, since, on the one hand, close thermal contact of the SPM and cryogen is necessary to ensure the (approximately) drift-free conditions described above, while, on the other hand, a good vibration isolation (both from the outside world, as well as from the boiling or flowing cryogen) has to be maintained.

Plenty of microscope designs have been presented in the last ten to 15 years, predominantly in the field of STM. Due to the variety of the different approaches, we will, somewhat arbitrarily, give two examples on different levels of complexity that might serve as illustrative model designs.

5.3.1 A Simple Design for a Variable Temperature STM

A simple design for a variable temperature STM system is presented in Fig. 5.1 (Similar systems are also offered by Omicron (Germany) or Jeol (Japan)). It should give an impression of what the minimum requirements are, if samples are to be investigated successfully at low temperatures. It features a single ultrahigh vacuum (UHV) chamber that houses the microscope in its center. The general idea to keep the setup simple is that only the sample is cooled by means of a flow cryostat that ends in the small liquid nitrogen (LN) reservoir. This reservoir is connected to the sample holder with copper braids. The role of the copper braids is to thermally attach the LN reservoir to the sample located on the sample holder in an effective manner, while vibrations due to the flow of the cryogen should be blocked as much as possible. In this way, a sample temperature of about 100 K is reached. Alternatively, with liquid helium operation, a base temperature of below 30 K can be achieved, while a heater that is integrated into the sample stage enables high-temperature operation up to 1000 K.

A typical experiment would run as follows: First, the sample is brought into the system by placing it in the so-called *load-lock*. This small part of the chamber can be separated from the rest of the system by a valve, so that the main part of the system can remain under vacuum at all times (i.e., even if the load-lock is opened for introducing the sample). After vacuum is re-established, the sample is transferred to the main chamber using the transfer arm. A linear motion feedthrough enables the storage of sample holders or, alternatively, specialized holders that carry replacement tips for the STM. Extending the transfer arm further, the sample can be placed on the sample stage and subsequently cooled down to the desired temperature. The scan head, which carries the STM tip, is then lowered with the scan head manipulator onto the sample holder (see Fig. 5.2). The special design of the scan head (see [14] for details) allows not only a flexible positioning of the tip on any desired location on the sample surface, but also compensates to a certain degree for thermal drift that inevitably occurs in such a design due to temperature gradients.

In fact, thermal drift is often much more prominent in LT-SPM designs, where only the sample is cooled, than in room temperature designs. Therefore, to fully benefit from the high stability conditions described in the introduction, it is mandatory to keep the whole microscope at the exact same temperature. This is mostly realized by using bath cryostats, which add a certain degree of complexity.

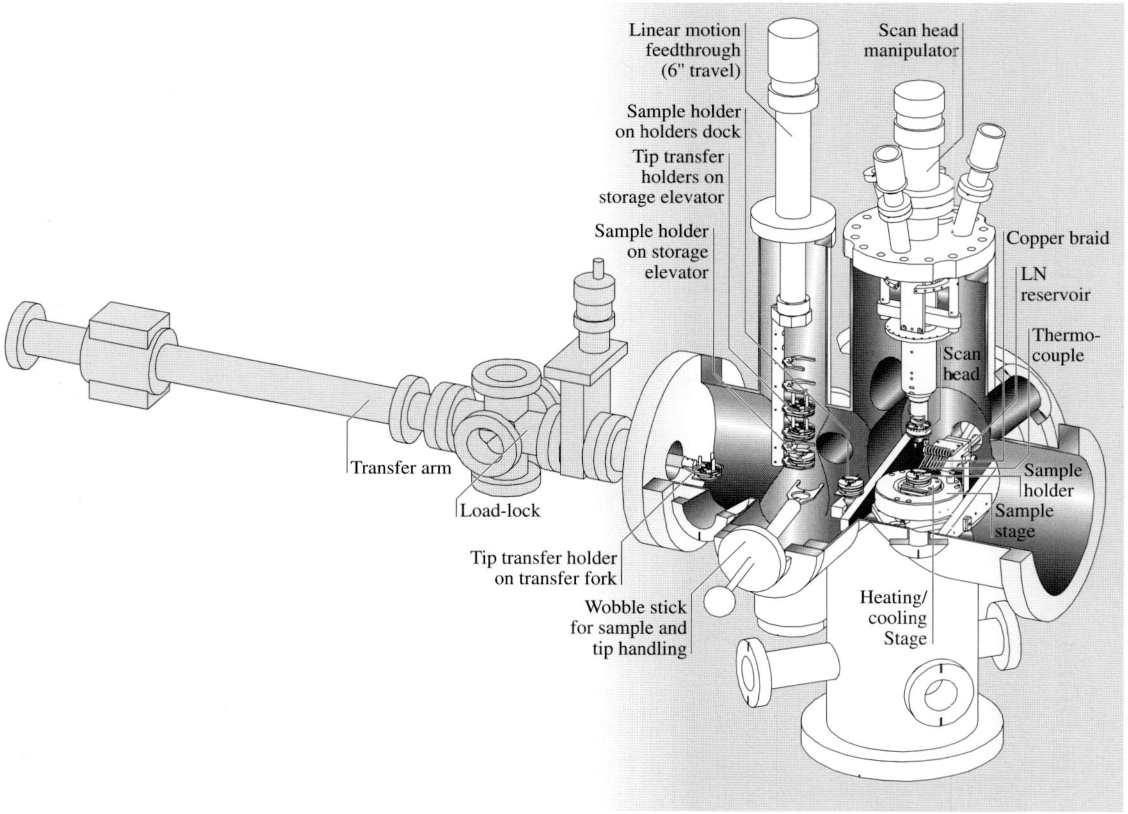

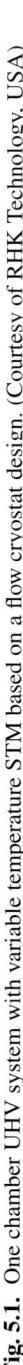

Fig. 5.1. One chamber UHV system with variable temperature STM based on a flow cryostat design. (Courtesy of RHK Technology, USA)

Fig. 5.2. Photograph of the STM located inside the system sketched in Fig. 5.1. After the scan head has been lowered onto the sample holder, it is fully decoupled from the scan head manipulator and can be moved laterally using the three piezo legs it stands on. (Courtesy of RHK Technology, USA)

5.3.2 A Low Temperature SFM Based on a Bath Cryostat

As an example for an LT-SPM setup based on a bath cryostat, let us take a closer look at the LT-SFM system sketched in Fig. 5.3, which has been used to acquire the images on graphite, xenon, NiO, and InAs presented in Sect. 5.5. The force microscope is built into a UHV system that comprises three vacuum chambers: one for cantilever and sample preparation, which also serves as a transfer chamber, one for analysis purposes, and a main chamber that houses the microscope. A specially designed vertical transfer mechanism based on a double chain allows the lowering of the microscope into a UHV compatible bath cryostat attached underneath the main chamber. To damp the system, it is mounted on a table carried by pneumatic damping legs, which, in turn, stand on a separate foundation to decouple it from building vibrations. The cryostat and dewar are separated from the rest of the UHV system by a bellow. In addition, the dewar is surrounded by sand for acoustic isolation.

In this design, tip and sample are exchanged at room temperature in the main chamber. After the transfer into the cryostat, the SFM can be cooled by either liquid nitrogen or liquid helium, reaching temperatures down to 10 K. An all-fiber interferometer as the detection mechanism for the cantilever deflection ensures high resolution, while simultaneously allowing the construction of a comparatively small, rigid, and symmetric microscope.

Figure 5.4 highlights the layout of the SFM body itself. Along with the careful choice of materials, the symmetric design eliminates most of the problems with drift inside the microscope encountered when cooling or warming it up. The microscope body has an overall cylindrical shape with a 13 cm height and a 6 cm diameter and exact mirror symmetry along the cantilever axis. The main body is made of a single

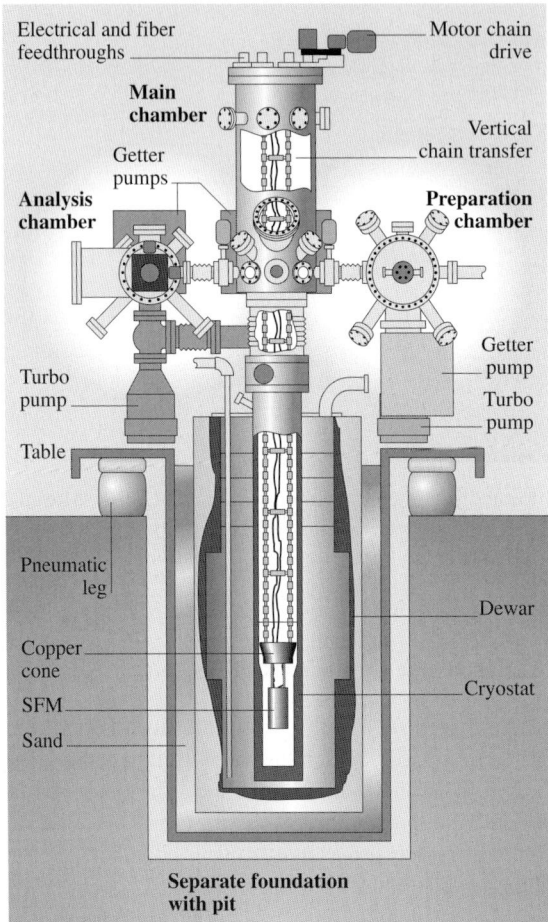

Fig. 5.3. Three-chamber UHV and bath cryostat system for scanning force microscopy, front view

block of macor, a machinable glass ceramic, which ensures a rigid and stable design. For most of the metallic parts titanium was used, which has a temperature coefficient similar to macor. The controlled but stable accomplishment of movements, such as coarse approach and lateral positioning in other microscope designs, is a difficult task at low temperatures. The present design uses a special type of piezo motor that moves a sapphire prism (see the "fiber approach" and the "sample approach" signs in Fig. 5.3). It is described in detail in [15]. More information regarding this design is given in [16].

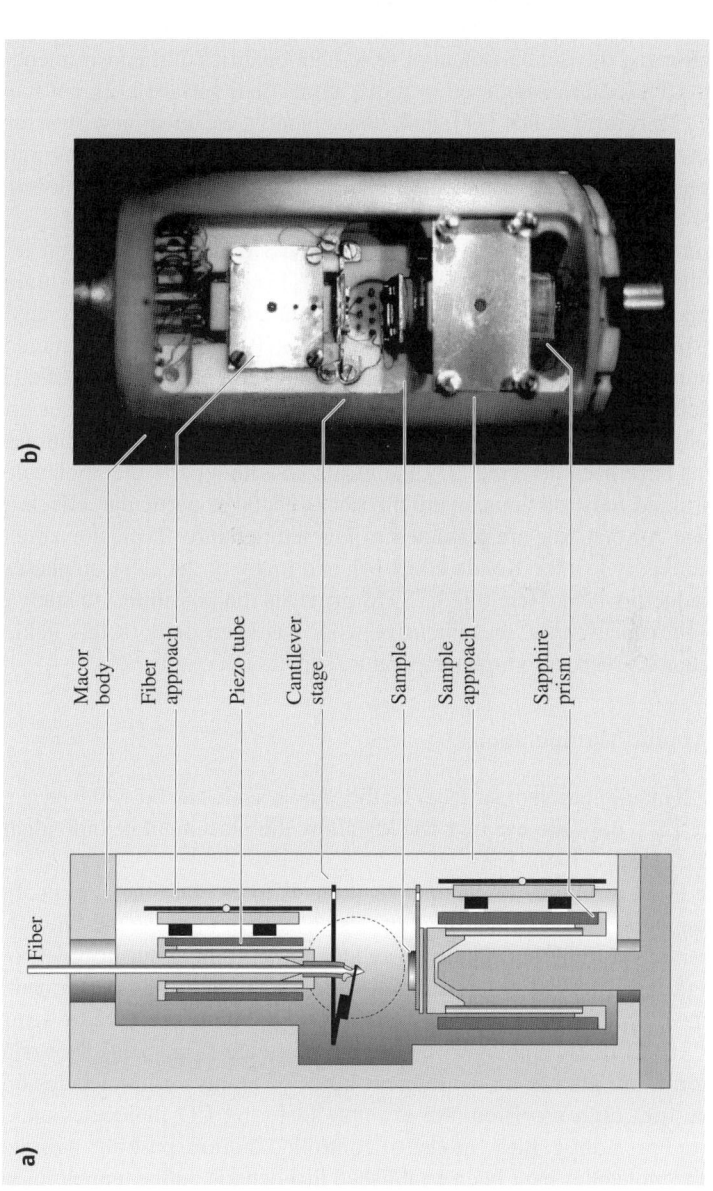

Fig. 5.4. The scanning force microscope incorporated into the system presented in Fig. 5.3. (**a**) Section along plane of symmetry. (**b**) Photo from the front

5.4 Scanning Tunneling Microscopy and Spectroscopy

In this section, we review some of the most important results achieved by LTSTM. After summarizing the results with emphasis on the necessity of LT equipment, we turn to details of the different experiments and the physical meaning of the obtained results.

As described in Sect. 5.2, the LT equipment has basically three advantages for scanning tunneling microscopy (STM) and spectroscopy (STS). First, the instruments are much more stable with respect to thermal drift and the coupling to external noise, allowing the establishment of new functionalities of the instrument. In particular, the LTSTM has been used to move atoms on a surface [12], cut molecules in pieces [17], reform bonds [18], and, consequently, establish new structures on the nanometer scale. Also, the detection of light resulting from tunneling into a particular atom [19] and the visualization of thermally induced atomic movements [20] partly require LT instrumentation.

Second, the spectroscopic resolution in STS depends linearly on temperature and is, therefore, considerably reduced at LT. This provides the opportunity to study physical effects unaccessible at room temperature. Obvious examples are the resolution of spin and Landau levels in semiconductors [9], or the investigation of lifetime broadening effects of particular electronic states on the nanometer scale [21]. More spectacular, electronic wave functions have been imaged for the first time in real space using an LTSTM [22], and vibrational levels giving rise to additional inelastic tunneling have been detected [23] and localized within particular molecules [24].

Third and most obvious, many physical effects, in particular, effects guided by electronic correlations, are restricted to low temperature. Typical examples are superconductivity [3], the Kondo effect [4], and many of the electron phases found in semiconductors [25]. Here, the LTSTM provides the possibility to study electronic effects on a local scale, and intensive work has been done in this field, the most elaborate with respect to high temperature superconductivity [26].

5.4.1 Atomic Manipulation

Although manipulation of surfaces on the atomic scale can be achieved at room temperature [27], only the use of LTSTMs allow the placement of individual atoms at desired atomic positions [28].

The usual technique to manipulate atoms is to increase the current above a certain atom, which reduces the tip-atom distance, then to move the tip with the atom to a desired position, and finally to reduce the current again in order to decouple atom and tip. The first demonstration of this technique was performed by Eigler and Schweizer (1990), who used Xe atoms on a Ni(110) surface to write the three letters "IBM" (their employer) on the atomic scale (Fig. 5.5a). Nowadays, many laboratories are able to move different kinds of atoms and molecules on different surfaces with high precision. An example featuring CO molecules on Cu(110) is shown in Fig. 5.5b–g. Basic modes of controlled motion, pushing, pulling, and sliding of the molecules have been established that depend on the tunneling current, i.e., the distance and the particular molecule-substrate combination [29]. It is believed that the electric field between tip and molecule is the strongest force moving the molecules, but other mechanisms such as electromigration caused by the high current density [28] or modifications of the surface potential due to the presence of the tip [30] have been put forth as important for some of the manipulation modes.

Meanwhile, other types of manipulation on the atomic scale have been developed. Some of them require inelastic tunneling into vibrational or rotational modes of molecules or atoms. They lead to controlled desorption [31], diffusion [32], pick-up of molecules by the tip [18], or rotation of individual entities [33, 34]. Also, dissociation of molecules by voltage pulses [17], conformational changes induced by dramatic change of tip-molecule distance [35], and association of pieces into larger molecules by reducing their lateral distance [18] have been shown. Fig. 5.5h–m shows the production of biphenyl from two iodobenzene molecules [36]. The iodine is abstracted by voltage pulses (Fig. 5.5i,j), then the iodine is moved to the terrace by the pulling mode (Fig. 5.5k,l), and finally the two phenyl parts are slid along the step edge until they are close enough to react (Fig. 5.5m). The chemical identification of the components is not deduced straightforwardly from STM images and partly requires detailed calculations of their apparent shape.

Low temperatures are not always required in these experiments, but they increase the reproducibility because of the higher stability of the instrument, as discussed in Sect. 5.2. Moreover, rotation or diffusion of entities could be excited at higher temperatures, making the intentionally produced configurations unstable.

5.4.2 Imaging Atomic Motion

Since individual manipulation processes last seconds to minutes, they probably cannot be used to manufacture large and repetitive structures. A possibility to construct such structures is self-assembled growth. It partly relies on the temperature dependence of different diffusion processes on the surface. A detailed knowledge of the diffusion parameters is required, which can be deduced from sequences of STM images measured at temperatures close to the onset of the process of interest [37]. Since many diffusion processes set on at LT, LT are partly required [20]. Consecutive images of so-called HtBDC molecules on Cu(110) recorded at $T = 194$ K are shown in Fig. 5.6a–c [38]. As indicated by the arrows, the position of the molecules changes with time, implying diffusion. Diffusion parameters are obtained from Arrhenius plots of the determined hopping rate h, as shown in Fig. 5.6d. Of course, one must make sure that the diffusion process is not influenced by the presence of the tip, since it is known from manipulation experiments that the presence of the tip can move a molecule. However, particularly at low tunneling voltages these conditions can be fulfilled.

Besides the determination of diffusion parameters, studies of the diffusion of individual molecules showed the importance of mutual interactions in diffusion, which can lead to concerted motion of several molecules [20], or, very interestingly, the influence of quantum tunneling [39]. The latter is deduced from the Arrhenius plot of hopping rates of H and D on Cu(001), as shown in Fig. 5.6e. The hopping rate of

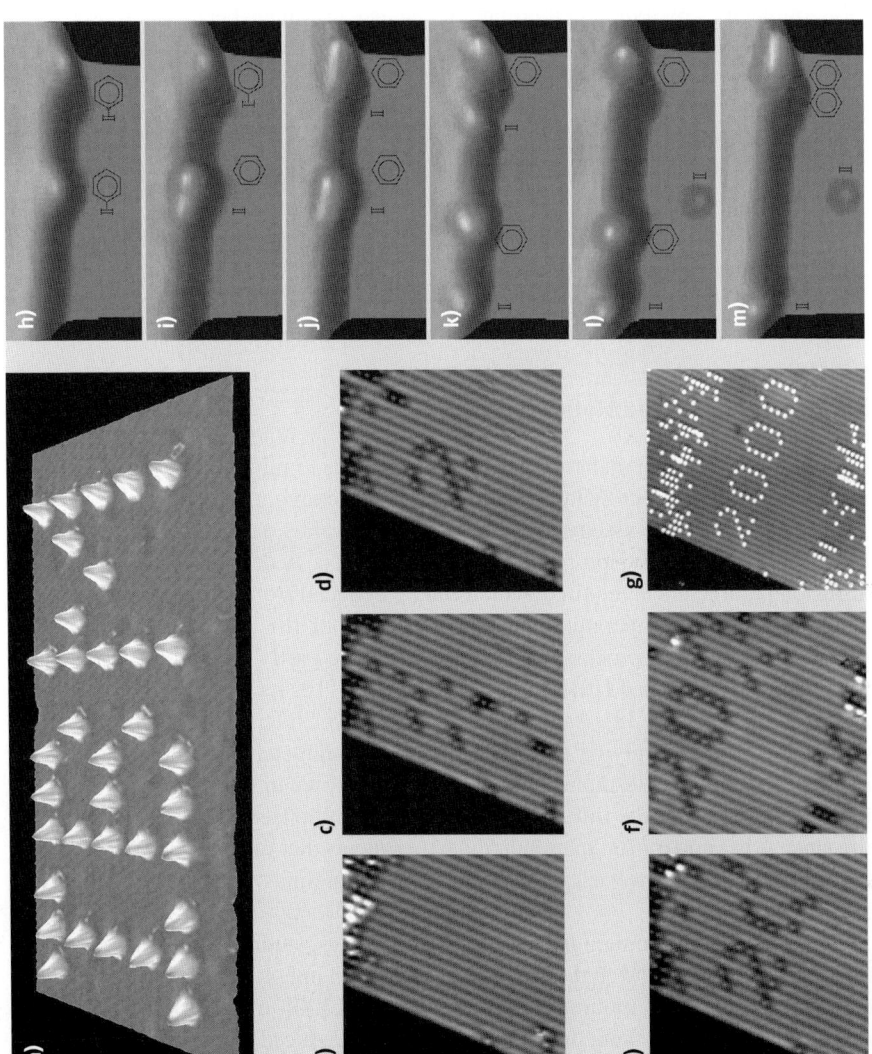

Fig. 5.5. (a) STM image of single Xe atoms positioned on a Ni(110) surface in order to realize the letters IBM on the atomic scale (courtesy of Eigler, IBM); (b)–(f) STM images recorded after different positioning processes of CO molecules on a Cu(110) surface; (g) final artwork greeting the new millennium on the atomic scale ((b)–(g) courtesy of Meyer, Berlin). (h)–(m) Synthesis of biphenyl from two iodobenzene molecules on Cu(111): First, iodine is abstracted from both molecules (i), (j), then the iodine between the two phenyl groups is removed from the step (k), and finally one of the phenyls is slid along the Cu-step (l) until it reacts with the other phenyl (m); the line drawings symbolize the actual status of the molecules ((h)–(m) courtesy of Hla and Rieder, Ohio)

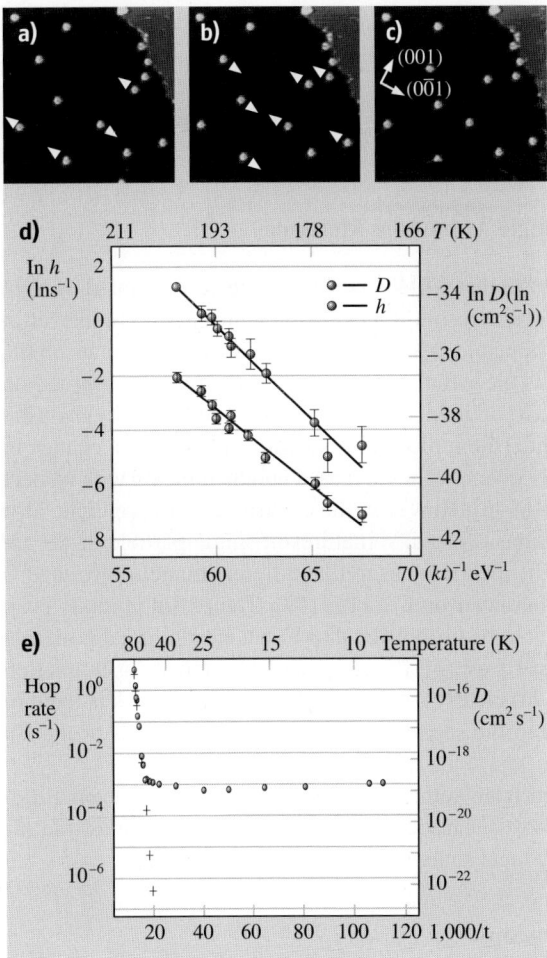

Fig. 5.6. (a)–(c) Consecutive STM images of hexa-tert-butyl decacyclene molecules on Cu(110) imaged at $T = 194$ K; *arrows* indicate the direction of motion of the molecules between two images. (d) Arrhenius plot of the hopping rate h determined from images like (a)–(c) as a function of inverse temperature (*grey symbols*); the *brown symbols* show the corresponding diffusion constant D; *lines* are fit results revealing an energy barrier of 570 meV for molecular diffusion ((a)–(d) courtesy of M. Schuhnack and F. Besenbacher, Aarhus). (e) Arrhenius plot for D (*crosses*) and H (*circles*) on Cu(001). The constant hopping rate of H below 65 K indicates a non-thermal diffusion process, probably tunneling (courtesy of Ho, Irvine)

H levels off at about 65 K, while the hopping rate of the heavier D atom goes down to nearly zero, as expected from thermally induced hopping.

Other diffusion processes such as the movement of surface vacancies [40] or of bulk interstitials close to the surface [41], and the Brownian motion of vacancy islands [42] have also been displayed.

5.4.3 Detecting Light from Single Atoms and Molecules

It had already been realized in 1988 that STM experiments are accompanied by light emission [43]. The fact that molecular resolution in the light intensity was achieved at LT (Fig. 5.7a,b) [19] raised the hope of performing quasi-optical experiments on the molecular scale. Meanwhile, it is clear that the basic emission process observed on metals is the decay of a local plasmon induced in the area around the tip by inelastic tunneling processes [44, 45]. Thus, the molecular resolution is basically a change in the plasmon environment, largely given by the increased height of the tip with respect to the surface above the molecule [46]. However, the electron can, in principle, also decay via single-particle excitations. Indeed, signatures of single-particle levels are observed. Figure 5.7c shows light spectra measured at different tunneling voltage V above a nearly complete Na monolayer on Cu(111) [47]. The plasmon mode peak energy (arrow) is found, as usual, to be proportional to V, but an additional peak that does not move with V appears at 1.6 eV (p). Plotting the light intensity as a function of photon energy and V (Fig. 5.7d) clearly shows that this additional peak is fixed in photon energy and corresponds to the separation of quantum well levels of the Na ($E_n - E_m$).

Light has also been detected from semiconductors [48], including heterostructures [49]. There, the light is mostly caused by single-particle relaxation of the injected electrons, allowing a very local source of electron injection.

5.4.4 High Resolution Spectroscopy

One of the most important modes of LTSTM is STS. It detects the differential conductivity dI/dV as a function of the applied voltage V and the position (x, y). The dI/dV-signal is basically proportional to the local density of states (LDOS) of the sample, the sum over squared single-particle wave functions Ψ_i [2]

$$\frac{dI}{dV}(V, x, y) \propto \text{LDOS}(E, x, y) = \sum_{\Delta E} |\Psi_i(E, x, y)|^2 , \qquad (5.1)$$

where ΔE is the energy resolution of the experiment. In simple terms, each state corresponds to a tunneling channel if it is located between the Fermi levels (E_F) of the tip and the sample. Thus, all states located in this energy interval contribute to I, while $dI/dV(V)$ detects only the states at the energy E corresponding to V. The local intensity of each channel depends further on the LDOS of the state at

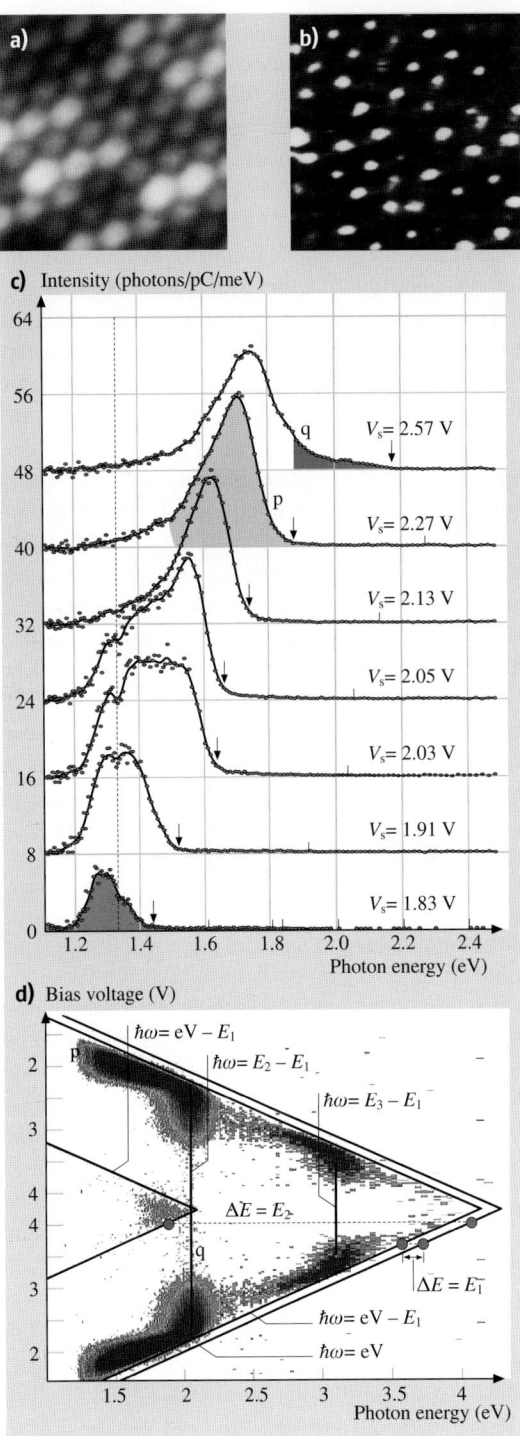

Fig. 5.7. (a) STM image of C_{60} molecules on Au(110) imaged at $T = 50$ K. (b) STM-induced photon intensity map of the same area; all photons from 1.5 eV to 2.8 eV contribute to the image, tunneling voltage $V = -2.8$ V ((a),(b) courtesy of Berndt, Kiel). (c) STM-induced photon spectrum measured on 0.6 monolayer of Na on Cu(111) at different tunneling voltages as indicated. Besides the shifting plasmon mode marked by an *arrow*, an energetically constant part named p is recognizable. (d) Greyscale map of photon intensity as a function of tunnelling voltage and photon energy measured on 2.0 monolayer Na on Cu(111). The energetically constant photons are identified with intersubband transitions of the Na quantum well, as marked by $E_n - E_m$ ((c), (d) courtesy of Hoffmann, Hamburg)

the corresponding surface position and its decay length into vacuum. For s-like tip states, *Tersoff* and *Hamann* have shown that it is simply proportional to the LDOS at the position of the tip [50]. Therefore, as long as the decay length is spatially constant, one measures the LDOS at the surface (5.1). Note that the contributing states are not only surface states, but also bulk states. However, surface states usually dominate if present. *Chen* has shown that higher orbital tip states lead to the so-called derivation rule [51]: p_z-type tip states detect $d(LDOS)/dz$, d_z^2-states detect $d^2(LDOS)/dz^2$, and so on. As long as the decay into vacuum is exponential and spatially constant, this leads only to an additional factor in dI/dV. Thus, it is still the LDOS that is measured (5.1). The requirement of a spatially constant decay is usually fulfilled on larger length scales, but not on the atomic scale [51]. There, states located close to the atoms show a stronger decay into vacuum than the less localized states in the interstitial region. This effect can lead to corrugations that are larger than the real LDOS corrugations [52].

The voltage dependence of dI/dV is sensitive to a changing decay length with V, which increases with V. Additionally, $dI/dV(V)$-curves might be influenced by possible structures in the DOS of the tip, which also contributes to the number of tunneling channels [53]. However, these structures can usually be identified, and only tips free of characteristic DOS structures are used for quantitative experiments.

Importantly, the energy resolution ΔE is largely determined by temperature. It is defined as the smallest energy distance of two δ-peaks in the LDOS that can still be resolved as two individual peaks in $dI/dV(V)$-curves and is $\Delta E = 3.3\,kT$ [2]. The temperature dependence is nicely demonstrated in Fig. 5.8, where the tunneling gap of the superconductor Nb is measured at different temperatures [54]. The peaks at the rim of the gap get wider at temperatures well below the critical temperature of the superconductor ($T_c = 9.2$ K).

Lifetime Broadening

Besides ΔE, intrinsic properties of the sample lead to a broadening of spectroscopic features. Basically, the finite lifetime of the electron or hole in the corresponding state broadens its energetic width. Any kind of interaction such as electron-electron interaction can be responsible. Lifetime broadening has usually been measured by photoemission spectroscopy (PES), but it turned out that lifetimes of surface states on noble metal surfaces determined by STS (Fig. 5.9a,b) are up to a factor of three larger than the ones measured by PES [55]. The reason is probably that defects broaden the PES spectrum. Defects are unavoidable in a spatially integrating technique such as PES, thus, STS has the advantage of choosing a particularly clean area for lifetime measurements. The STS results can be successfully compared to theory, highlighting the dominating influence of intraband transitions for the surface state lifetime on Au(111) and Cu(111), at least close to the onset of the surface band [21].

With respect to band electrons, the analysis of the width of the band onset in $dI/dV(V)$-curves has the disadvantage of being restricted to the onset energy. Another method circumvents this problem by measuring the decay of standing electron waves scattered from a step edge as a function of energy [56]. Figure 5.9c,d shows

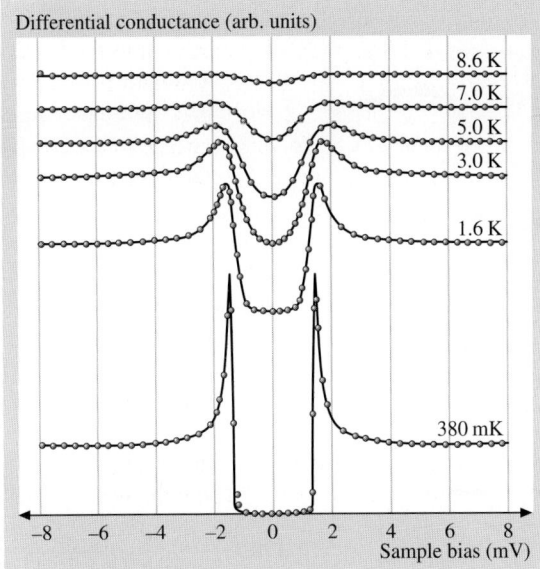

Fig. 5.8. Differential conductivity curve $dI/dV(V)$ measured on a Au surface by a Nb tip (*symbols*). Different temperatures are indicated; the lines are fits according to the superconducting gap of Nb folded with the temperature broadened Fermi distribution of the Au (courtesy of Pan, Houston)

the resulting oscillating dI/dV-signal measured for two different energies. To deduce the coherence length L_ϕ, which is inversely proportional to the lifetime τ_ϕ, one has to consider that the finite ΔE in the experiment also leads to a decay of the standing wave away from the step edge. The dotted fit line using $L_\phi = \infty$ indicates this effect and, more importantly, shows a discrepancy with the measured curve. Only including a finite coherence length of 6.2 nm results in good agreement, which, in turn, determines L_ϕ and thus τ_ϕ, as displayed in Fig. 5.9c. The found $1/E^2$-dependence of τ_ϕ points to a dominating influence of electron-electron interactions at higher energies in the surface band.

Landau and Spin Levels

Moreover, the increased energy resolution at LT allows the resolution of electronic states that are not resolvable at RT. For example, Landau and spin quantization appearing in magnetic field B have been probed on InAs(110) [9, 57]. The corresponding quantizationenergies are given by $E_{Landau} = \hbar eB/m_{eff}$ and $E_{spin} = g\mu B$. Thus InAs is a good choice, since it exhibits a low effective mass $m_{eff}/m_e = 0.023$ and a high g-factor of 14 in the bulk conduction band. The values in metals are $m_{eff}/m_e \approx 1$ and $g \approx 2$, resulting in energy splittings of only 1.25 meV and 1.2 meV at $B = 10$ T. This is obviously lower than the typical lifetime broadenings discussed in the previous section and also close to $\Delta E = 1.1$ meV achievable at $T = 4$ K.

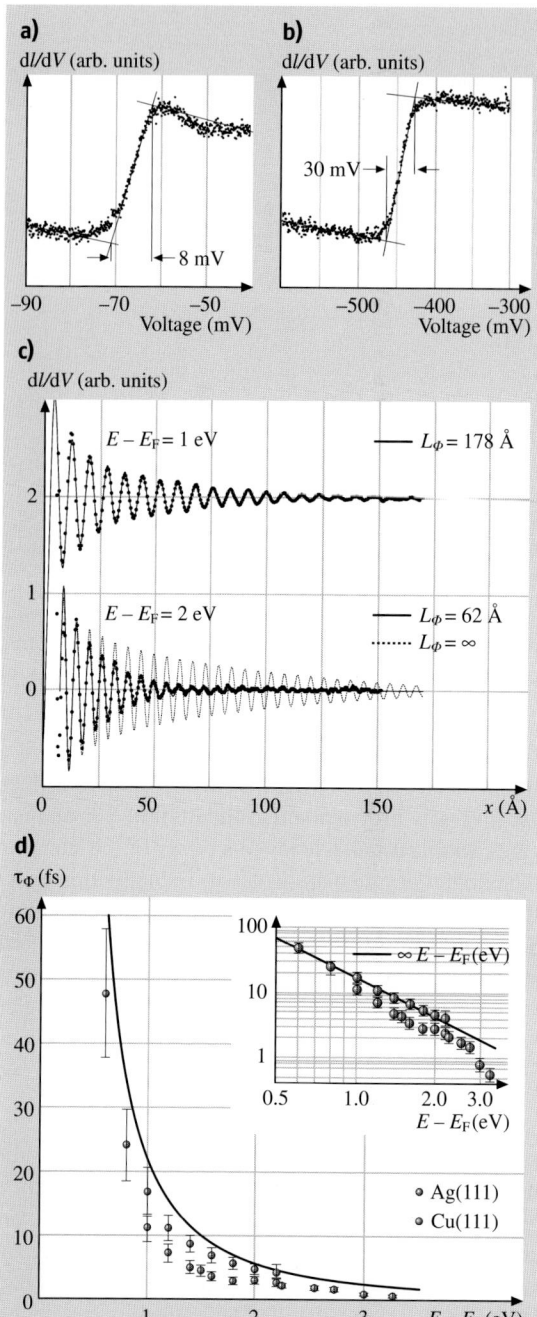

Fig. 5.9. (**a**), (**b**) spatially averaged $dI/dV(V)$-curves of Ag(111) and Cu(111); both surfaces exhibit a surface state with parabolic dispersion starting at -65 meV and -430 meV, respectively. The lines are drawn to determine the energetic width of the onset of these surface bands ((**a**), (**b**) courtesy of Berndt, Kiel); (**c**) dI/dV-intensity as a function of position away from a step edge of Cu(111) measured at the voltages ($E - E_F$), as indicated (*points*); the lines are fits assuming standing electron waves with a phase coherence length L_Φ as marked; (**d**) resulting phase coherence time as a function of energy for Ag(111) and Cu(111). *Inset* shows the same data on a double logarithmic scale evidencing the E^{-2}-dependence (*line*) ((**c**), (**d**) courtesy of Brune, Lausanne)

Fortunately, the electron density in doped semiconductors is much lower, and thus the lifetime increases significantly. Figure 5.10a shows a set of spectroscopy curves obtained on InAs(110) in different magnetic fields [9]. Above E_F, oscillations with increasing intensity and energy distance are observed. They show the separation expected from Landau quantization. In turn, they can be used to deduce m_{eff} from the peak separation (Fig. 5.10b). An increase of m_{eff} with increasing E has been found as expected from theory. Also, at high fields spin quantization is observed (Fig. 5.10c). It is larger than expected from the bare g-factor due to contributions from exchange enhancement [58]. A detailed discussion of the peaks revealed that they belong to the so-called tip-induced quantum dot resulting from the work function difference between tip and sample.

Vibrational Levels

As discussed with respect to light emission in STM, inelastic tunneling processes contribute to the tunneling current. The coupling of electronic states to vibrational levels is one source of inelastic tunneling [23]. It provides additional channels contributing to $dI/dV(V)$ with final states at energies different from V. The final energy is simply shifted by the energy of the vibrational level. If only discrete vibrational energy levels couple to a smooth electronic DOS, one expects a peak in d^2I/dV^2 at the vibrational energy. This situation appears for molecules on noble metal surfaces. As usual, the isotope effect on the vibrational energy can be used to verify the vibrational origin of the peak. First indications of vibrational levels have been found for H_2O and D_2O on TiO_2 [59], and completely convincing work has been performed for C_2H_2 and C_2D_2 on Cu(001) [23] (Fig. 5.11a). The technique can be used to identify individual molecules on the surface by their characteristic vibrational levels. In particular, surface reactions, as described in Fig. 5.5h–m, can be directly verified. Moreover, the orientation of complexes with respect to the surface can be determined to a certain extent, since the vibrational excitation depends on the position of the tunneling current within the molecule. Finally, the excitation of certain molecular levels can induce such corresponding motions as hopping [32], rotation [34] (Fig. 5.11b–e), or desorption [31], leading to additional possibilities for manipulation on the atomic scale.

Kondo Resonance

A rather intricate interaction effect is the Kondo effect. It results from a second order scattering process between itinerate states and a localized state [60]. The two states exchange some degree of freedom back and forth, leading to a divergence of the scattering probability at the Fermi level of the itinerate states. Due to the divergence, the effect strongly modifies sample properties. For example, it leads to an unexpected increase in resistance with decreasing temperature for metals containing magnetic impurities [4]. Here, the exchanged degree of freedom is the spin. A spectroscopic signature of the Kondo effect is a narrow peak in the DOS at the Fermi level disappearing above a characteristic temperature (Kondo temperature). STS provides the opportunity to study this effect on the local scale [61, 62].

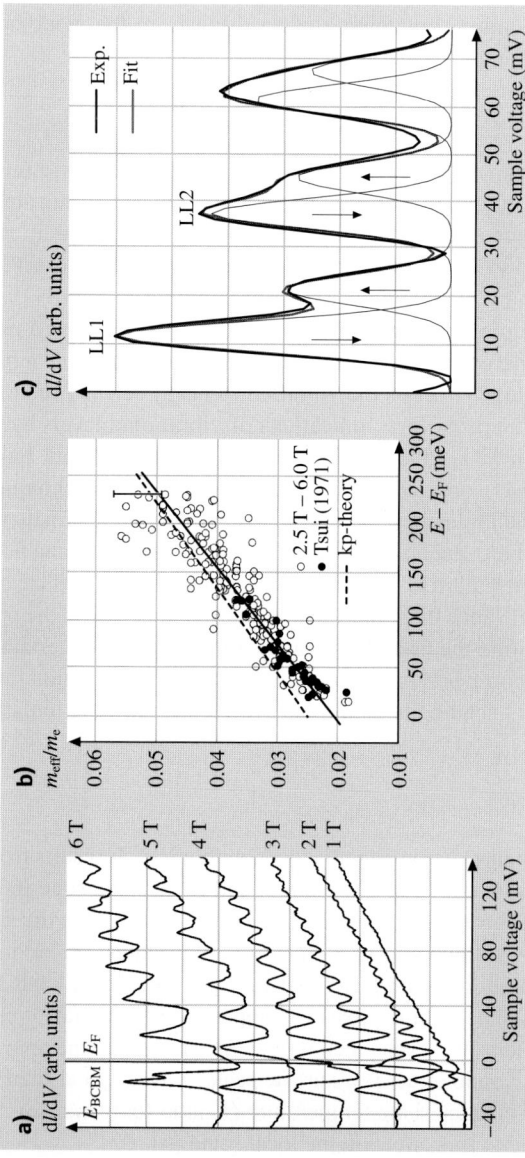

Fig. 5.10. (a) dI/dV-curves of n-InAs(110) at different magnetic fields as indicated; E_{BCBM} marks the bulk conduction band minimum; oscillations above E_{BCBM} are caused by Landau quantization; the double peaks at $B = 6$ T are caused by spin quantization. (b) Effective mass data deduced from the distance of adjacent Landau peaks ΔE according to $\Delta E = \hbar eB/m_{\rm eff}$ (open symbols); filled symbols are data from planar tunnel junctions (Tsui); the solid line is a mean-sqare fit of the data and the dashed line is the expected effective mass of InAs according to kp-theory. (c) Magnification of a dI/dV-curve at $B = 6$ T exhibiting spin splitting; the Gaussian curves marked by arrows are the fitted spin levels

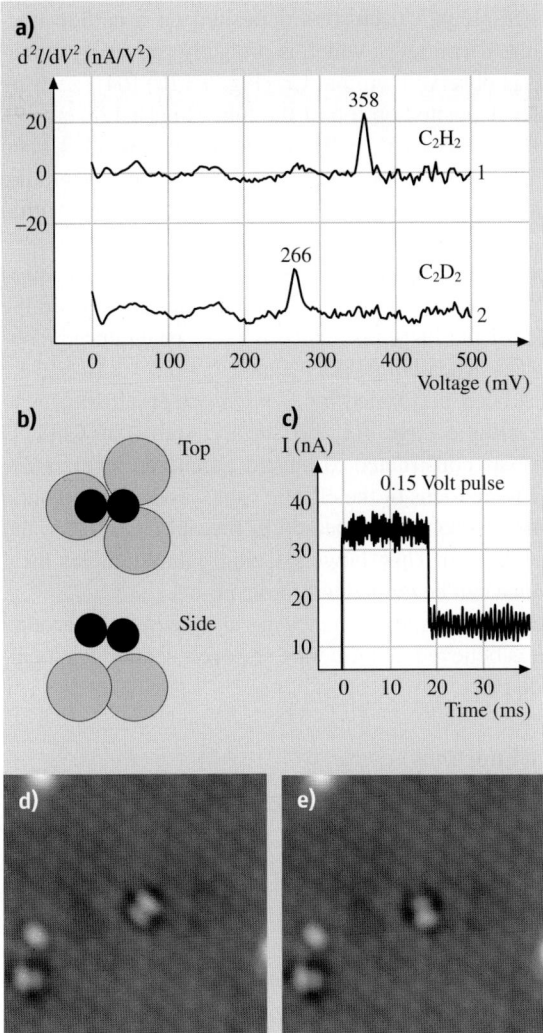

Fig. 5.11. (a) d^2I/dV^2-curves taken above a C_2H_2 and a C_2D_2 molecule on Cu(100); the peaks correspond to the C-H, respectively, C-D stretch mode energy of the molecule. (b) Sketch of O_2 molecule on Pt(111). (c) Tunnelling current above an O_2 molecule on Pt(111) during a voltage pulse of 0.15 V; the jump in current indicates rotation of the molecule. (d), (e) STM image of an O_2 molecule on Pt(111) ($V = 0.05$ V) prior and after rotation induced by a voltage pulse to 0.15 V ((**a**)–(**e**) courtesy of Ho, Irvine)

Figure 5.12a–d shows an example of Co clusters deposited on a carbon nanotube [63]. While a small dip at the Fermi level, which is probably caused by curvature influences on the π-orbitals, is observed without Co (Fig. 5.12b) [64], a strong peak is found around a Co cluster deposited on top of the tube (Fig. 5.12a, arrow). The peak is slightly shifted with respect to $V = 0$ mV due to the so-called Fano resonance [65], which results from interference of the tunneling processes into the localized Co-level and the itinerant nanotube levels. The resonance disappears within a several nanometer distance from the cluster, as shown in Fig. 5.12d.

The Kondo effect has also been detected for different magnetic atoms deposited on noble metal surfaces [61, 62]. There, it disappears at about 1 nm away from the magnetic impurity, and the effect of the Fano resonance is more pronounced, contributing to dips in $dI/dV(V)$-curves instead of peaks.

A fascinating experiment has been performed by *Manoharan* et al. [66], who used manipulation to form an elliptic cage for the surface states of Cu(111) (Fig. 5.12e, bottom). This cage was constructed to have a quantized level at E_F. Then, a cobalt atom was placed in one focus of the elliptic cage, producing a Kondo resonance. Surprisingly, the same resonance reappeared in the opposite focus, but not away from the focus (Fig. 5.12e, top). This shows amazingly that complex local effects such as the Kondo resonance can be guided to remote points.

Orbital scattering as a source of the Kondo resonance has also been found around a defect on Cr(001) [67]. Here, it is believed that itinerate sp-levels scatter at a localized d-level to produce the Kondo peak.

5.4.5 Imaging Electronic Wave Functions

Since STS measures the sum of squared wave functions (5.1), it is an obvious task to measure the local appearance of the most simple wave functions in solids, namely, the Bloch waves.

Bloch Waves

The atomically periodic part of the Bloch wave is always measured if atomic resolution is achieved (inset of Fig. 5.14a). However, the long-range, wavy part requires the presence of scatterers. The electron wave impinges on the scatterer and gets reflected, leading to self-interference. In other words, the phase of the Bloch wave gets fixed by the scatterer.

Such self-interference patterns were first found on Graphite(0001) [68] and later on noble metal surfaces, where adsorbates or step edges scatter the surface states (Fig. 5.13a) [22]. Fourier transforms of the real-space images reveal the k-space distribution of the corresponding states [69], which may include additional contributions besides the surface state [70]. Using particular geometries as the so-called quantum corrals to form a cage for the electron wave, the scattering state can be rather complex (Fig. 5.13b). Anyway, it can usually be reproduced by simple calculations involving single-particle states [71].

Bloch waves in semiconductors scattered at charged dopants (Fig. 5.13c,d) [72], Bloch states confined in semiconductor quantum dots (Fig. 5.13e–g) [73], and Bloch

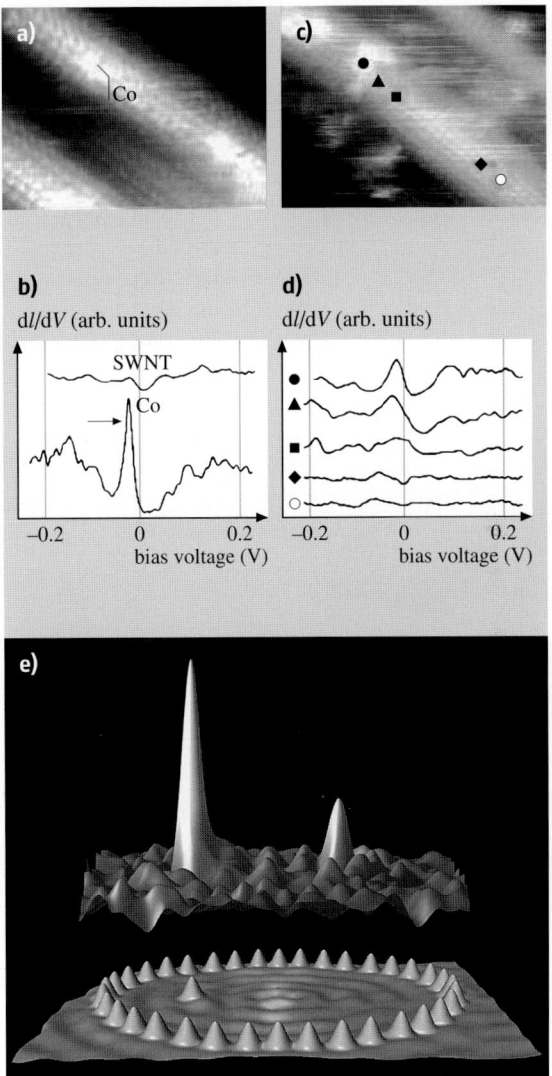

Fig. 5.12. (a) STM image of a Co cluster on a single-wall carbon nanotube (SWNT). (b) dI/dV-curves taken directly above the Co cluster (Co) and far away from the Co cluster (SWNT); the *arrow* marks the Kondo peak. (c) STM image of another Co cluster on a SWNT with *symbols* marking the positions where the dI/dV-curves displayed in (d) are taken. (d) dI/dV-curves taken at the positions marked in (c) ((a)–(d) courtesy of Lieber, Cambridge). (e) *Lower part:* STM image of a quantum corral of elliptic shape made from Co atoms on Cu(111); one Co atom is placed in one of the focii of the ellipse. *Upper part:* map of the strength of the Kondo signal in the corral; note that there is also a Kondo signal in the focus, which is not covered by a Co atom ((e) courtesy of Eigler, Almaden)

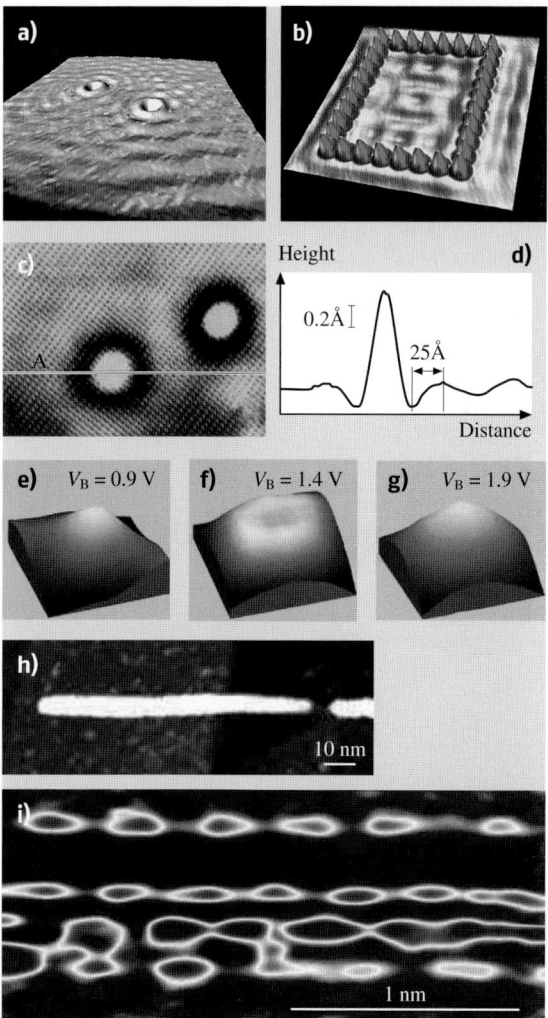

Fig. 5.13. (a) Low voltage STM image of Cu(111), including two defect atoms; the waves are electronic Bloch waves scattered at the defects. (b) Low voltage STM image of a rectangular quantum corral made from single atoms on Cu(111); the pattern inside the corral is the confined state of the corral close to E_F; ((a), (b) courtesy of Eigler, Almaden). (c) STM image of GaAs(110) around a Si donor, $V = -2.5$ V; the line scan along A shown in (d) exhibits an additional oscillation around the donor caused by a standing Bloch wave; the grid like pattern corresponds to the atomic corrugation of the Bloch wave ((c), (d) courtesy of van Kempen, Nijmegen). (e)–(g) STM images of an InAs/ZnSe-core/shell-nanocluster at different V. The image is measured in the so-called constant-height mode, i.e., the images display the tunneling current at constant height above the surface; the hill in (e) corresponds to the s-state of the cluster, the ring in (f) to the degenerate p_x- and p_y-state and the hill in (g) to the p_z-state ((e)–(g) courtesy of Millo, Jerusalem). (h) STM-image of a short-cut carbon nanotube. (i) Colour plot of the dI/dV intensity inside the short-cut nanotube as a function of position and tunneling voltage; four wavy patterns of different wavelength are visible in the voltage range from −0.1 to 0.15 V ((h), (i) courtesy of Dekker, Delft)

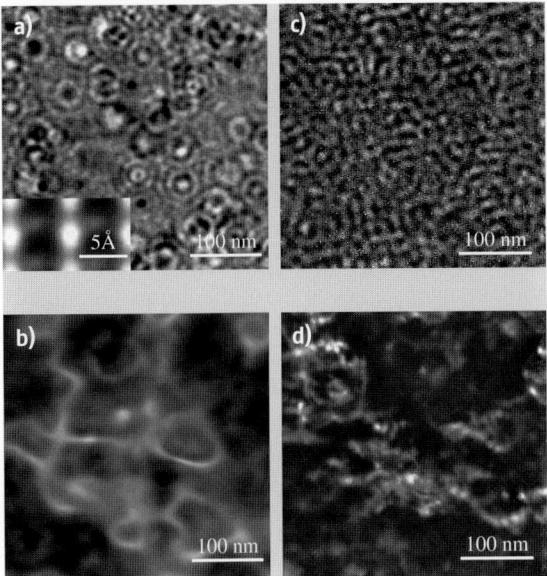

Fig. 5.14. (a) dI/dV-image of InAs(110) at $V = 50\,\text{mV}$, $B = 0\,\text{T}$; circular wave patterns corresponding to standing Bloch waves around each S donor are visible; inset shows a magnification revealing the atomically periodic part of the Bloch wave. (**b**) Same as (**a**), but at $B = 6\,\text{T}$; the stripe structures are drift states. (**c**) dI/dV-image of a 2-D electron system on InAs(110) induced by the deposition of Fe, $B = 0\,\text{T}$. (**d**) Same as (**c**) but at $B = 6\,\text{T}$; note that the contrast in (**a**) is increased by a factor of ten with respect to (**b**)–(**d**)

waves confined in short-cut carbon nanotubes (Fig. 5.13h,i) [74, 75] have been visualized.

Drift States

More complex wave functions result from interactions. A nice playground to study such interactions are doped semiconductors. The reduced electron density with respect to metals increases the importance of electron interactions with potential disorder and other electrons. Applying a magnetic field quenches the kinetic energy, further enhancing the importance of interactions. A dramatic effect can be observed on InAs(110), where 3-D bulk states are displayed. While the usual scattering states around individual dopants are observed at $B = 0\,\text{T}$ (Fig. 5.14a) [76], stripe structures are found in high magnetic field (Fig. 5.14b) [77]. They run along equipotential lines of the disorder potential. This can be understood by recalling that the electron tries to move in a cyclotron orbit, which is accelerated and decelerated in electrostatic potential, leading to a drift motion along an equipotential line [78].

The same effect has been found in 2-D electron systems (2-DES) of the same substrate, where the scattering states at $B = 0\,\text{T}$ are, however, found to be more complex (Fig. 5.14c) [79]. The reason is the tendency of a 2-DES to exhibit closed scattering paths [80]. Consequently, the self-interference does not result from scattering at individual scatterers, but from complicated self-interference paths involving many scatterers. But drift states are also observed in the 2-DES at high magnetic fields (Fig. 5.14d) [81].

Charge Density Waves

Another interaction modifying the LDOS is the electron-phonon interaction. Phonons scatter electrons between different Fermi points. If the wave vectors connecting Fermi points exhibit a preferential orientation, a so-called Peierls instability occurs [82]. The corresponding phonon energy goes to zero, the atoms are slightly displaced with the periodicity of the corresponding wave vector, and a charge density wave (CDW) with the same periodicity appears. Essentially, the CDW increases the overlap of the electronic states with the phonon by phase fixing with respect to the atomic lattice. The Peierls transition naturally occurs in 1-D systems, where only two Fermi points are present, and, hence, preferential orientation is pathological. It can also occur in 2-D systems if large areas of the Fermi line run in parallel.

STS studies of CDWs are numerous (e.g., [83, 84]). Examples of a 1-D CDW on a quasi 1-D-like bulk material and of a 2-D CDW are shown in Fig. 5.15a–d and Fig. 5.15e, respectively [85, 86]. In contrast to usual scattering states where LDOS corrugations are only found close to the scatterer, the corrugations of CDWs are continous across the surface. Heating the substrate toward the transition temperature leads to a melting of the CDW lattice, as shown in Fig. 5.15f–h.

CDWs have also been found on monolayers of adsorbates such as a monolayer of Pb on Ge(111) [87]. These authors performed a nice temperature-dependent study revealing that the CDW is nucleated by scattering states around defects as one might expect [88]. 1-D systems have also been prepared on surfaces showing Peierls transitions [89, 90]. Finally, the energy gap occurring at the transition has been studied by measuring $dI/dV(V)$-curves [91].

Superconductors

An intriguing effect resulting from electron-phonon interaction is superconductivity. Here, the attractive part of the electron-phonon interaction leads to the coupling of electronic states with opposite wave vector and mostly opposite spin [92]. Since the resulting Cooper pairs are bosons, they can condense at LT, forming a coherent many-particle phase, which can carry current without resistance. Interestingly, defect scattering does not influence the condensate if the coupling along the Fermi surface is homogeneous (s-wave superconductor). The reason is that the symmetry of the scattering of the two components of a Cooper pair effectively leads to a scattering from one Cooper pair state to another without affecting the condensate. This is different if the scatterer is magnetic, since the different spin components of the pair are scattered differently, leading to an effective pair breaking visible as a single-particle excitation within the superconducting gap. On a local scale, the effect has first been demonstrated by putting Mn, Gd, and Ag atoms on a Nb(110) surface [93]. While the nonmagnetic Ag does not modify the gap shown in Fig. 5.16a, it is modified in an asymmetric fashion close to Mn or Gd adsorbates, as shown in Fig. 5.16b. The

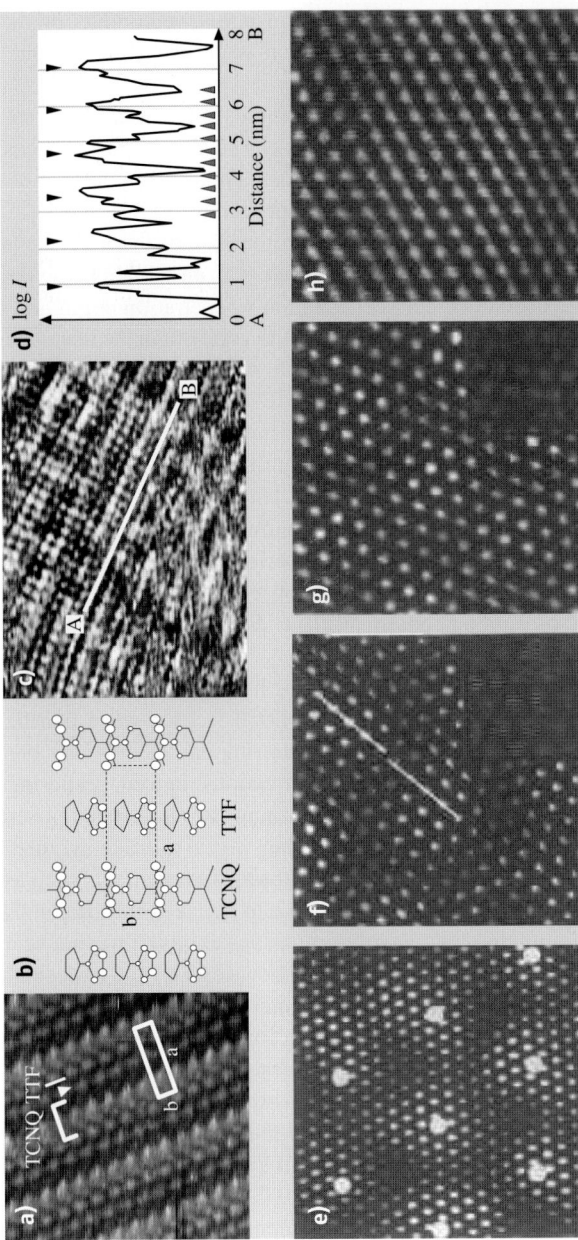

Fig. 5.15. (a) STM image of the ab-plane of the organic quasi 1-D conductor TTF-TCNQ, $T = 300$ K; while the TCNQ chains are conducting, the TTF chains are insulating. (b) Stick and ball model of the ab-plane of TTF-TCNQ. (c) STM image taken at $T = 61$ K, the additional modulation due to the Peierls-transition is visible in the profile along line A shown in (d); the *open triangles* mark the atomic periodicity and the *filled triangles* the expected CDW periodicity ((a)–(d) courtesy of Kageshima, Kanagawa). (e)–(h) Low voltage STM-images of the two-dimensional CDW-system 1T-TaS$_2$ at $T = 242$ K (e), 298 K (f), 349 K (g), 357 K (h). A long-range, hexagonal modulation is visible besides the atomic spots; its periodicity is highlighted by *big white dots* in (e); the additional modulation obviously weakens with increasing T, but is still apparent in (f) and (g), as evidenced in the lower magnification images in the insets ((e)–(h) courtesy of Lieber, Cambridge)

asymmetry of the additional intensity is caused by the breaking of the particle-hole symmetry due to the exchange interaction between the localized Mn state and the itinerate Nb states.

Another important local effect is caused by the relatively large coherence length of the condensate. At a material interface, the condensate wave function cannot stop abruptly, but laps into the surrounding material (proximity effect). Consequently, a superconducting gap can be measured in areas of non-superconducting material. Several studies have shown this effect on the local scale using metals and doped semiconductors as surrounding materials [94, 95].

While the classical type-I superconductors are ideal diamagnets, the so-called type-II superconductors can contain magnetic flux. The flux forms vortices, each containing one flux quantum. These vortices are accompanied by a disappearance of the superconducting gap and, therefore, can be probed by STS [96]. LDOS maps measured inside the gap lead to bright features in the area of the vortex core. Importantly, the length scale of these features is different from the length scale of the magnetic flux due to the difference between London's penetration depth and coherence length. Thus, STS probes another property of the vortex than the usual magnetic imaging techniques (see Sect. 5.5.4). Surprisingly, first measurements of the vortices on $NbSe_2$ revealed vortices shaped as a sixfold star [97] (Fig. 5.16c). With increasing voltage inside the gap, the orientation of the star rotates by 30° (Fig. 5.16d,e). The shape of these stars could finally be reproduced by theory, assuming an anisotropic pairing of electrons in the superconductor (Fig. 5.16f–h) [98]. Additionally, bound states inside the vortex core, which result from confinement by the surrounding superconducting material, are found [97]. Further experiments investigated the arrangement of the vortex lattice, including transitions between hexagonal and quadratic lattices [99], the influence of pinning centers [100], and the vortex motion induced by current [101].

An important topic is still the understanding of high temperature superconductivity (HTCS). An almost accepted property of HTCS is its d-wave pairing symmetry. In contrast to s-wave superconductors, scattering can lead to pair breaking, since the Cooper pair density vanishes in certain directions. Indeed, scattering states (bound states in the gap) around nonmagnetic Zn impurities have been observed in $Bi_2Sr_2CaCu_2O_{8+\delta}$ (BSCCO) (Fig. 5.16i,j) [26]. They reveal a d-like symmetry, but not the one expected from simple Cooper pair scattering. Other effects such as magnetic polarization in the environment probably have to be taken into account [102]. Moreover, it has been found that magnetic Ni impurities exhibit a weaker scattering structure than Zn impurities [103]. Thus, BSCCO shows exactly the opposite behavior of the Nb discussed above (Fig. 5.16a,b). An interesting topic is the importance of inhomogeneities in HTCS materials. Evidence for inhomogeneities has indeed been found in underdoped materials, where puddles of the superconducting phase are shown to be embedded in non-superconducting areas [104].

Of course, vortices have also been investigated in HTCS materials [105]. Bound states are found, but at energies that are in disagreement with simple models, assuming a BCS-like d-wave superconductor [106, 107]. Theory predicts, instead, that the bound states are magnetic-field-induced spin-density waves stressing the com-

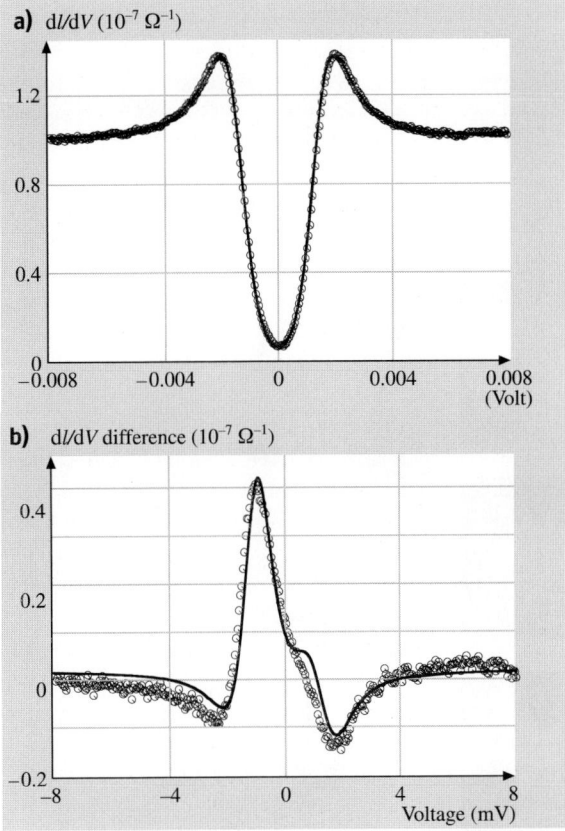

Fig. 5.16. (a) dI/dV-curve of Nb(110) at $T = 3.8$ K (*symbols*) in comparison with a BCS-fit of the superconducting gap of Nb (*line*). (b) Difference between the dI/dV-curve taken directly above a Mn-atom on Nb(110) and the dI/dV-curve taken above the clean Nb(110) (*symbols*) in comparison with a fit using the Bogulubov-de Gennes equations (*line*) ((**a**), (**b**) courtesy of Eigler, Almaden). (**c**)–(**e**) dI/dV-images of a vortex core in the type-II superconductor 2H-NbSe$_2$ at 0 mV (**c**), 0.24 mV (**d**), and 0.48 mV (**e**) ((**c**)–(**e**) courtesy of H. F. Hess). (**f**)–(**h**) Corresponding calculated LDOS images within the Eilenberger framework ((**f**)–(**h**) courtesy of Machida, Okayama). (**i**) Overlap of an STM image at $V = -100$ mV (backround 2-D image) and a dI/dV-image at $V = 0$ mV (overlapped 3-D image) of optimally doped Bi$_2$Sr$_2$CaCu$_2$O$_{8+\delta}$ containing 0.6% Zn impurities. The STM image shows the atomic structure of the cleavage plane, while the dI/dV-image shows a bound state within the superconducting gap, which is located around a single Zn impurity. The fourfold symmetry of the bound state reflects the d-like symmetry of the superconducting pairing function; (**j**) dI/dV-curves taken at different positions across the Zn impurity; the bound state close to 0 mV is visible close to the Zn atom; (**k**) LDOS in the vortex core of slightly overdoped Bi$_2$Sr$_2$CaCu$_2$O$_{8+\delta}$, $B = 5$ T; the dI/dV-image taken at $B = 5$ T is integrated over $V = 1$–12 mV, and the corresponding dI/dV-image at $B = 0$ T is subtracted to highlight the LDOS induced by the magnetic field. The checkerboard pattern within the seven vortex cores exhibits a periodicity, which is fourfold with respect to the atomic lattice shown in (**i**) and is thus assumed to be a CDW ((**i**)–(**k**) courtesy of S. Davis, Cornell and S. Uchida, Tokyo)

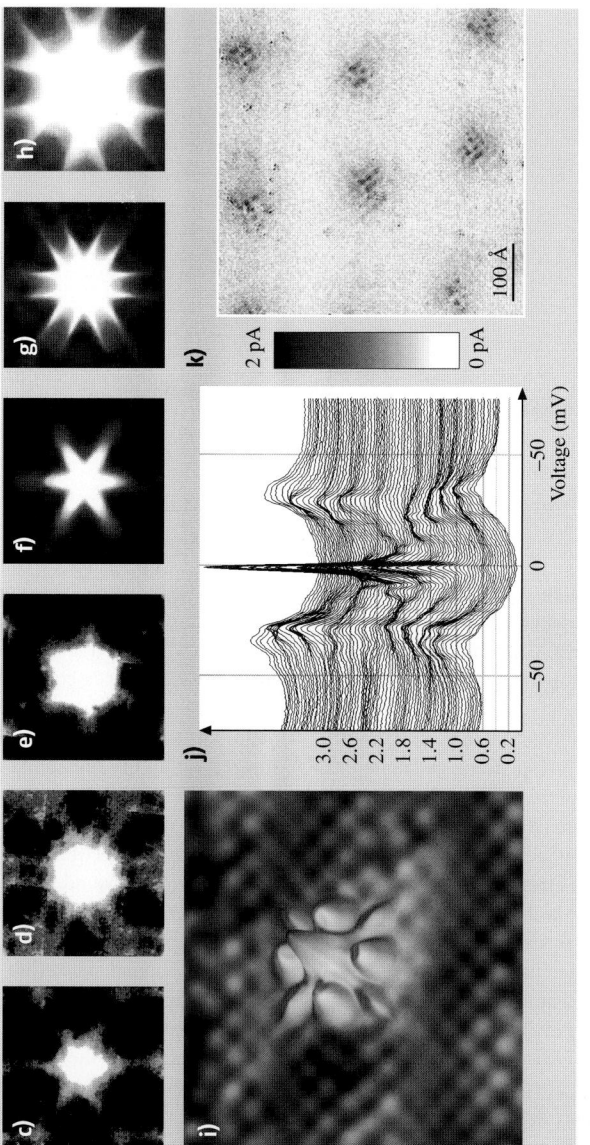

Fig. 5.16. continue

petition between anti-ferromagnetic order and superconductivity in HTCS materials [108]. Since the spin density wave is accompanied by a charge density wave of half wavelength, it can be probed by STS [109]. Indeed, a checkerboard pattern of the right periodicity has been found in and around vortex cores in BSCCO (Fig. 5.16k). It exceeds the width of an individual vortex core, implying that the superconducting coherence length is different from the anti-ferromagnetic one.

Complex Systems (Manganites)

Complex phase diagrams are not restricted to HTCS materials (cuprates). They exist with similar complexity for other doped oxides such as manganites. Only a few studies of these materials have been performed by STS, mainly showing the inhomogeneous evolution of metallic and insulating phases [110, 111]. Similarities to the granular case of an underdoped HTCS material are obvious. Since inhomogeneities seem to be crucial in many of these materials, a local method such as STS might continue to be important for the understanding of their complex properties.

5.4.6 Imaging Spin Polarization: Nanomagnetism

Conventional STS couples to the LDOS, i.e., the charge distribution of the electronic states. Since electrons also have spin, it is desirable to also probe the spin distribution of the states. This can be achieved if the tunneling tip is covered by a ferromagnetic material [112]. The coating acts as a spin filter or, more precisely, the tunneling current depends on the relative angle α_{ij} between the spins of the tip and the sample according to $\cos(\alpha_{ij})$. In ferromagnets, the spins mostly have one preferential orientation along the so-called easy axis, i.e., a particular tip is not sensitive to spin orientations of the sample perpendicular to the spin orientation of the tip. Different tips have to be prepared to detect different spin orientations of the sample. Moreover, the magnetic stray field of the tip can perturb the spin orientation of the sample. To avoid this, a technique using anti-ferromagnetic Cr as a tip coating material has been developed [113]. This avoids stray fields, but still provides a preferential spin orientation of the few atoms at the tip apex that dominate the tunneling current. Depending on the thickness of the Cr coating, spin orientations perpendicular or parallel to the sample surface are prepared.

So far, the described technique has been used to image the evolution of magnetic domains with increasing B-field(Fig. 5.17a–d) [114], the anti-ferromagnetic order of a Mn monolayer on W(110) (Fig. 5.17e,f) [115], and the out-of-plane orientation predicted for a magnetic vortex core as it exists in the center of a Fe island exhibiting four domains in the flux closure configuration (Fig. 5.17g,h) [116].

Besides the obvious strong impact on nanomagnetism, the technique might also be used to investigate other electronic phases such as the proposed spin density wave around a HTCS vortex core.

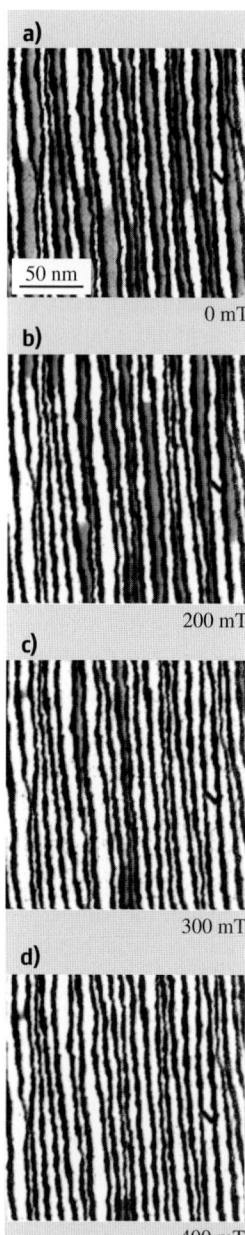

Fig. 5.17. (**a**)–(**d**) Spin-polarized STM images of 1.65 monolayer of Fe deposited on a stepped W(110) surface measured at different *B*-fields, as indicated. Double layer and monolayer Fe stripes are formed on the W substrate; only the double layer stripes exhibit magnetic contrast with an out-of-plane sensitive tip as used here. *White* and *grey areas* correspond to different domains. Note that more white areas appear with increasing field. (**e**) STM image of an antiferromagnetic Mn monolayer on W(110). (**f**) Spin-polarized STM-image of the same surface (in-plane tip). The insets in (**e**) and (**f**) show the calculated STM and spin-polarized STM images, respectively, and the stick and ball models symbolize the atomic and the magnetic unit cell ((**a**)–(**f**) courtesy of M. Bode, Hamburg). (**g**) Spin-polarized STM image of a 6-nm-high Fe island on W(110) (in-plane tip). Four different areas are identified as four different domains with domain orientations as indicated by the *arrows*. (**h**) Spin-polarized STM image of the central area of an island; the size of the area is indicated by the rectangle in (**g**); the measurement is performed with an out-of-plane sensitive tip showing that the magnetization turns out-of-plane in the center of the island

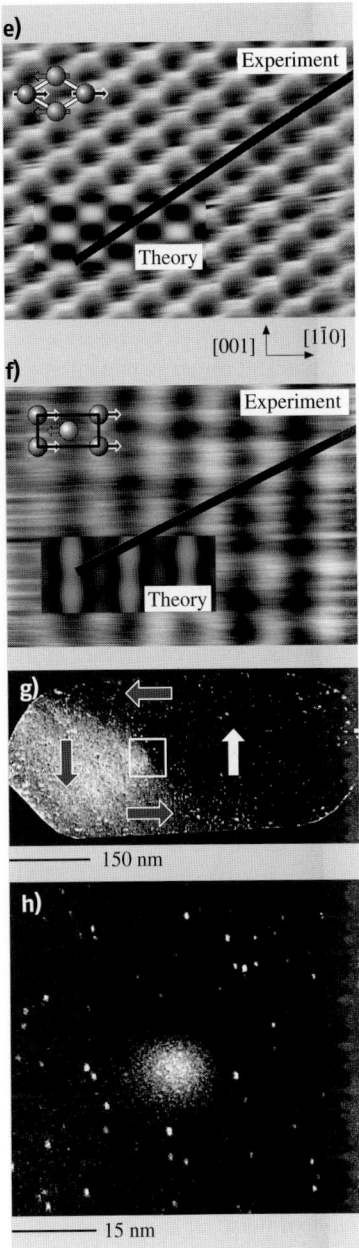

Fig. 5.17. continue

5.5 Scanning Force Microscopy and Spectroscopy

SFSThe examples discussed in the previous section show the wide variety of physical questions that have been tackled with the help of LTSTM. Here, we turn to the other prominent scanning probe method that is applied at low temperatures, namely, SFM, which gives complementary information on sample properties on the atomic scale.

The ability to sensitively detect *forces* with spatial resolution down to the atomic scale is of great interest, since force is one of the most fundamental quantities in physics. Mechanical force probes usually consist of a cantilever with a tip at its free end that is brought close to the sample surface. The cantilever can be mounted parallel or perpendicular to the surface (general aspects of force probe designs are described in Chap. 2.4.4). Basically, two methods exist to detect forces with cantilever-based probes – the *static* and the *dynamic* mode (see Chap. I). They can be used to generate a laterally resolved image (*microscopy* mode) or determine its distance dependence (*spectroscopy* mode). One can argue about the terminology, since spectroscopy is usually related to energies and not to distance dependences. Nevertheless, we will use it throughout the text, because it avoids lenghty paraphrases and is established in this sense throughout the literature.

In the static mode, a force that acts on the tip bends the cantilever. By measuring its deflection Δz the tip-sample force F_{ts} can be directly calculated with Hooke's law: $F_{ts} = c_z \cdot \Delta z$, where c_z denotes the spring constant of the cantilever. In the various dynamic modes, the cantilever is oscillated with amplitude A at or near its eigenfrequency f_0, but in some applications also off resonance. At ambient pressures or in liquids, amplitude modulation (AM-SFM) is used to detect amplitude changes or the phase shift between driving force and cantilever oscillation. In vacuum, the frequency shift Δf of the cantilever due to a tip-sample interaction is measured by frequency modulation technique (FM-SFM). The nomenclature is not standardized. Terms like tapping mode or intermittent contact mode are used instead of AM-SFM, and NC-AFM (noncontact atomic force microscopy) or DFM (dynamic force microscopy) instead of FM-SFM or FM-AFM. However, all these modes are *dynamic*, i.e., they involve an oscillating cantilever and can be used in the noncontact, as well as in the contact regime. Therefore, we believe that the best and most consistent way is to distinguish them by their different detection schemes. Converting the measured quantity (amplitude, phase, or frequency shift) into a physically meaningful quantity, e.g., the tip-sample interaction force F_{ts} or the force gradient $\partial F_{ts}/\partial z$, is not always straightforward and requires an analysis of the equation of motion of the oscillating tip (see Chaps. 5.5.4, 3.3.2).

Whatever method is used, the resolution of a cantilever-based force detection is fundamentally limited by its intrinsic *thermomechanical* noise. If the cantilever is in thermal equilibrium at a temperature T, the equipartition theorem predicts a thermally induced *root mean square* (rms) motion of the cantilever in z direction of $z_{rms} = (k_B T/c_{eff})^{1/2}$, where k_B is the Boltzmann constant and $c_{eff} = c_z + \partial F_{ts}/\partial z$. Note that usually $dF_{ts}/dz \gg c_z$ in contact and $dF_{ts}/dz < c_z$ in noncontact. Evidently, this fundamentally limits the force resolution in the static mode, particularly if operated in noncontact. Of course, the same is true for the different dynamic modes, because the thermal energy $k_B T$ excites the eigenfrequency f_0 of the cantilever. Thermal noise is *white* noise, i.e., its spectral density is flat. However, if the cantilever transfer function is taken into account, one can see that the thermal energy mainly excites f_0. This explains the term "thermo" in thermomechanical noise, but what is the "mechanical" part?

A more detailed analysis reveals that the thermally induced cantilever motion is given by

$$z_{rms} = \sqrt{\frac{2k_B T B}{\pi c_z f_0 Q}}, \qquad (5.2)$$

where B is the measurement bandwidth and Q is the quality factor of the cantilever. Analog expressions can be obtained for all quantities measured in dynamic modes, because the deflection noise translates, e.g., into frequency noise [5]. Note that f_0 and c_z are correlated with each other via $2\pi f_0 = (c_z/m_{eff})^{1/2}$, where the effective mass m_{eff} depends on the geometry, density, and elasticity of the material. The Q-factor of the cantilever is related to the external damping of the cantilever motion in a medium and on the intrinsic damping within the material. This is the "mechanical" part of the fundamental cantilever noise.

It is possible to operate a low temperature force microscope directly immersed in the cryogen [117, 118] or in the cooling gas [119], whereby the cooling is simple and very effective. However, it is evident from (5.2) that the smallest fundamental noise is achievable in vacuum, where the Q-factors are more than 100 times larger than in air, and at low temperatures.

The best force resolution up to now, which is better than 1×10^{-18} N/Hz$^{1/2}$, has been achieved by *Mamin* et al. [120] in vacuum at a temperature below 300 mK. Due to the reduced thermal noise and the lower thermal drift, which results in a higher stability of the tip-sample gap and a better signal-to-noise ratio, the highest resolution is possible at low temperatures in ultrahigh vacuum with FM-SFM. A vertical rms-noise below 2 pm [121, 122] and a force resolution below 1 aN [120] have been reported.

Besides the reduced noise, the application of force detection at low temperatures is motivated by the increased stability and the possibility to observe phenomena, which appear below a certain critical temperature T_c, as outlined at page 187. The experiments, which have been performed at low temperatures up until now, were motivated by at least one of these reasons and can be roughly divided into four groups: (i) atomic-scale imaging, (ii) force spectroscopy, (iii) investigation of quantum phenomena by measuring electrostatic forces, and (iv) utilizing magnetic probes to study ferromagnets, superconductors, and single spins. In the following, we describe some exemplary results.

5.5.1 Atomic-Scale Imaging

In a simplified picture, the dimensions of the tip end and its distance to the surface limit the lateral resolution of force microscopy, since it is a near-field technique.

Consequently, atomic resolution requires a stable single atom at the tip apex that has to be brought within a distance of some tenths of a nanometer to an atomically flat surface. Both conditions can be fulfilled by FM-AFM, and nowadays *true* atomic resolution is routinely obtained (see Chap. 3.3.2). However, the above statement says nothing about the nature and the underlying processes of imaging with atomic resolution. Si(111)-(7 × 7) was the first surface on which true atomic resolution was achieved [123], and several studies have been performed at low temperatures on this well-known material [124–126]. First principle simulations performed on semiconductors with a silicon tip revealed that *chemical* interactions, i.e., a significant charge redistribution between the dangling bonds of tip and sample, dominate the atomic-scale contrast [127–129]. On V-III semiconductors, it was found that only one atomic species, the group V atoms, is imaged as protrusions with a silicon tip [128, 129]. Furthermore, these simulations revealed that the sample, as well as the tip atoms are noticeably displaced from their equilibrium position due to the interaction forces. At low temperatures, both aspects could be observed with silicon tips on indium arsenide [121, 130].

Chemical Sensitivity of Force Microscopy

The (110) surface of the III-V semiconductor indium arsenide exhibits both atomic species in the top layer (see Fig. 5.18a). Therefore, this sample is well suited to study the chemical sensitivity of force microscopy [121]. In Fig. 5.18b, the usually observed atomic-scale contrast on InAs(110) is displayed. As predicted, the arsenic atoms, which are shifted by 80 pm above the indium layer due to the (1×1) relaxation, are imaged as protrusions. While this general appearance was similar for most tips, two other distinctively different contrasts were also observed: a second protrusion (c) and a sharp depression (d). The arrangement of these two features corresponds well to the zigzag configuration of the indium and arsenic atoms along the [1$\bar{1}$0]-direction. A sound explanation would be as follows: The contrast usually obtained with one feature per surface unit cell corresponds to a silicon-terminated tip, as predicted by simulations. A different atomic species at the tip apex, however, can result in a very different charge redistribution. Since the atomic-scale contrast is due to a chemical interaction, the two other contrasts would then correspond to a tip that has been accidentally contaminated with sample material (arsenic or indium-terminated tip apex). Nevertheless, this explanation has not yet been verified by simulations for this material.

Tip-Induced Atomic Relaxation

Schwarz et al. [121] were able to directly visualize the predicted tip-induced relaxation during atomic-scale imaging near a point defect. Figure 5.19 shows two FM-AFM images of the same point defect recorded with different constant frequency shifts on InAs(110), i.e., the tip was closer to the surface in (b) compared to (a). The arsenic atoms are imaged as protrusions with the used silicon tip. From the symmetry of the defect, an indium-site defect can be inferred, since the distance-dependent

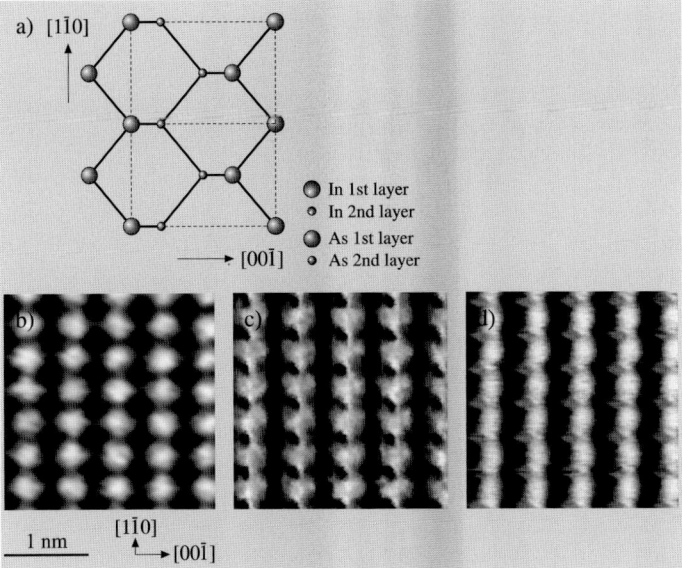

Fig. 5.18. The structure of InAs(110) as seen from top (**a**) and three FM-AFM images of this surface obtained with different tips at 14 K (**b**)–(**d**). In (**b**), only the arsenic atoms are imaged as protrusions, as predicted for a silicon tip. The two features in (**c**) and (**d**) corresponds to the zigzag arrangement of the indium and arsenic atoms. Since force microscopy is sensitive to short-range chemical forces, the appearance of the indium atoms can be associated with a chemically different tip apex

contrast is consistent with what has to be expected for an indium vacancy. This expectation is based on calculations performed for the similar III-V semiconductor GaP(110), where the two surface gallium atoms around a P-vacancy were found to relax downward [131]. This corresponds to the situation in Fig. 5.19a, where the tip is relatively far away and an inward relaxation of the two arsenic atoms is observed. The considerably larger attractive force in Fig. 5.19b, however, pulls the two arsenic atoms toward the tip. All other arsenic atoms are also pulled, but they are less displaced, because they have three bonds to the bulk, while the two arsenic atoms in the neighborhood of an indium vacancy have only two bonds. This direct experimental proof of the presence of tip-induced relaxations is also relevant for STM measurements, because the tip-sample distances are similar during atomic resolution imaging. Moreover, the result demonstrates that FM-AFM can probe elastic properties on an atomic level.

Imaging of Weakly Interacting van der Waals Surfaces

For weakly interacting van der Waals surfaces like the (0001) surface of graphite, much smaller atomic corrugation amplitudes are expected compared to strongly interacting surfaces such as semiconductors. In graphite, a layered material, the carbon atoms are covalently bonded and arranged in a honeycomb structure within the

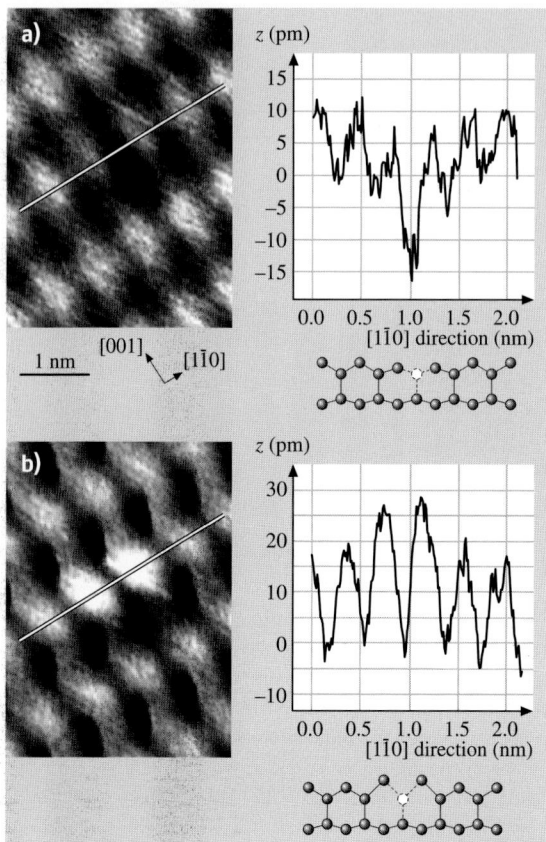

Fig. 5.19. Two FM-AFM images of the identical indium-site point defect (presumably an indium vacancy) recorded at 14 K. If the tip is relatively far away, the theoretically predicted inward relaxation of two arsenic atoms adjacent to an indium vacancy is visible (**a**). At a closer tip-sample distance (**b**), the two arsenic atoms are pulled farther toward the tip compared to the other arsenic atoms, since they have only two instead of three bonds

(0001) plane. Individual layers of (0001) planes (*ABA*... stacking) stick together only by weak van der Waals forces. Three distinctive sites exist on the surface: carbon atoms with (*A*-type) and without (*B*-type) a neighbor in the next graphite layer and the *hollow site* (H-site) in the hexagon center. Neither static contact force microscopy nor STM can resolve the three different sites. For both methods, the contrast is well understood and exhibits protrusions with a sixfold symmetry and a periodicity of 246 pm. In high resolution FM-AFM images acquired at low temperatures, a large maximum and two different minima have been resolved (see Fig. 5.20a). A simulation using the Lennard-Jones potential, in which the attractive part is given by the short-range interatomic van der Waals force, reproduced these three features very well. By comparison with the experimental data (cf. the solid and the dotted line in

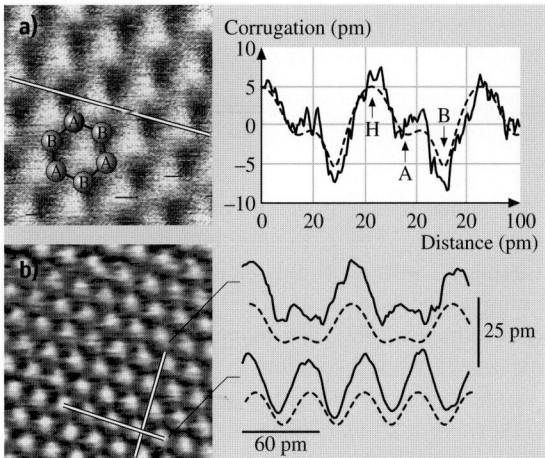

Fig. 5.20. FM-AFM images of graphite(0001) (**a**) and Xe(111) (**b**) recorded at 22 K. On the right side, line sections taken from the experimental data (*solid lines*) are compared to simulations (*dotted lines*). A- and B-type carbon atoms, as well as the hollow site (H-site) on graphite can be distinguished, but are imaged with inverted contrast, i.e., the carbon sites are displayed as minima. Such an inversion does not occur on Xe(111)

the section of Fig. 5.20a), the large maximum could be assigned to the hollow site, while the two different minima represent A- and B-type carbon atoms [132]. Note that the image contrast is inverted with respect the arrangement of the atoms, i.e., the minima correspond to the position of the carbon atoms. The carbon-carbon distance of only 142 pm is the smallest interatomic distance that has been resolved with FM-AFM so far. Moreover, the weak tip-sample interaction results in typical corrugation amplitudes around 10 pm peak-to-peak, i.e., just above the noise level [7].

While experiments on graphite basically gain advantages from the increased stability and signal-to-noise ratio at low temperatures, solid xenon (melting temperature $T_m = 161$ K) can only be observed at sufficiently low temperatures [8]. In addition, xenon is a pure van der Waals crystal, and since it is an insulator, FM-AFM is the only real space method available today that allows the study of solid xenon on the atomic scale.

Allers et al. [8] adsorbed a well-ordered xenon film on cold graphite(0001) ($T < 55$ K) and subsequently studied it at 22 K by FM-AFM (see Fig. 5.20b). The sixfold symmetry and the distance between the protrusions correspond well with the nearest neighbor distance in the closed, packed (111) plane of bulk xenon, which crystallizes in the face-centered cubic structure. A comparison between experiment and simulation confirmed that the protrusions correspond to the position of the xenon atoms [133]. However, the simulated corrugation amplitudes do not fit as well as for graphite (see sections in Fig. 5.20b). A possible reason is that tip-induced relaxations, which were not considered in the simulations, are more important for this pure van der Waals crystal xenon than they are for graphite, because in-plane graphite exhibits strong covalent bonds. Nevertheless, the results demonstrated that a weakly bonded van der Waals crystal could be imaged nondestructively on the atomic scale for the first time.

5.5.2 Force Spectroscopy

A wealth of information about the nature of the tip-sample interaction can be obtained by measuring its distance dependence. This is usually done by recording the measured quantity (deflection, frequency shift, amplitude change, phase shift) and applying an appropriate voltage ramp to the z-electrode of the scanner piezo, while the z-feedback is switched off. According to (5.2), low temperatures and high Q-factors (vacuum) considerably increase the force resolution. In the static mode, long-range forces and contact forces can be examined. Force measurements at small tip-sample distances are inhibited by the *jump-to-contact* phenomenon: If the force gradient $\partial F_{ts}/\partial z$ becomes larger than the spring constant c_z, the cantilever cannot resist the attractive tip-sample forces and the tip snaps onto the surface. Sufficiently large spring constants prevent this effect, but reduce the force resolution. In the dynamic modes, the jump-to-contact can be avoided due to the additional restoring force ($c_z A$) at the lower turnaround point. The highest sensitivity can be achieved in vacuum by using the FM technique, i.e., by recording $\Delta f(z)$-curves. An alternative FM spectroscopy method, the recording of $\Delta f(A)$-curves, has been suggested by *Hölscher* et al. [134]. Note that if the amplitude is much larger than the characteristic decay length of the tip-sample force, the frequency shift cannot simply be converted into force gradients by using $\partial F_{ts}/\partial z = 2c_z \cdot \Delta f/f_0$ [135]. Several methods have been published to convert $\Delta f(z)$ data into the tip-sample potential $V_{ts}(z)$ and tip-sample force $F_{ts}(z)$ (see, e.g., [136, 137]).

Measurement of Interatomic Forces at Specific Atomic Sites

FM force spectroscopy has been successfully used to measure and determine quantitatively the short-range chemical force between the foremost tip atom and specific surface atoms [109, 138, 139]. Figure 5.21 displays an example for the quantitative determination of the short-range force. Figure 5.21a shows two $\Delta f(z)$-curves measured with a silicon tip above a corner hole and above an adatom. Their position is indicated by arrows in the inset, which displays the atomically resolved Si(111)-(7×7) surface. The two curves differ from each other only for small tip-sample distances, because the long-range forces do not contribute to the atomic-scale contrast. The low, thermally induced lateral drift and the high stability at low temperatures were required to precisely address the two specific sites. To extract the short-range force, the long-range van der Waals and/or electrostatic forces can be subtracted from the total force. The grey curve in Fig. 5.21b has been reconstructed from the $\Delta f(z)$-curve recorded above an adatom and represents the total force. After removing the long-range contribution from the data, the much steeper black line is obtained, which corresponds to the short-range force between the adatom and the atom at the tip apex. The measured maximum attractive force (-2.1 nN) agrees well with first principle calculations (-2.25 nN).

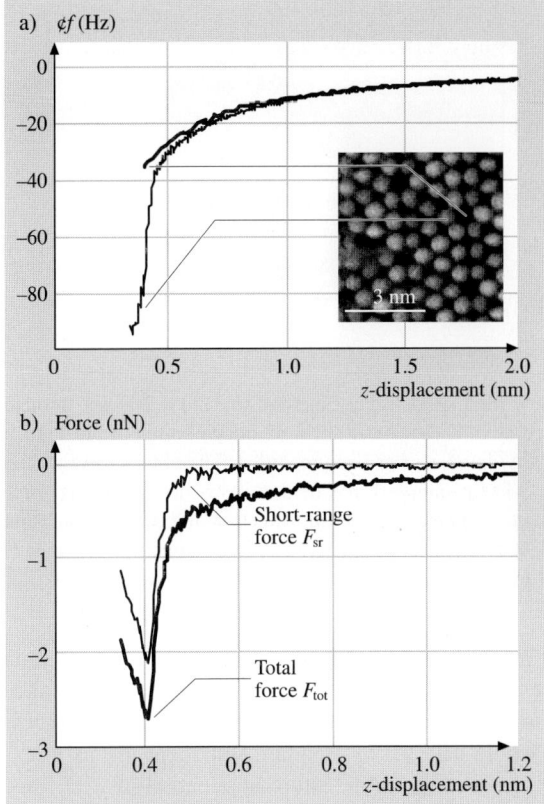

Fig. 5.21. FM force spectroscopy on specific atomic sites at 7.2 K. In (**a**), an FM-SFM image of the Si(111)-(7 × 7) surface is displayed together with two $\Delta f(z)$-curves, which have been recorded at the positions indicated by the *arrows*, i.e., above the cornerhole (*thin*) and above an adatom (*thick*). In (**b**), the total force above an adatom (*thin line*) has been recovered from the $\Delta f(z)$-curve. After subtraction of the long-range part, the short-range force can be determined (*thick line*) (courtesy of H. J. Hug; cf. [138])

Three-dimensional Force Field Spectroscopy

Further progress with the FM technique has been made by *Hölscher* et al. [140]. They acquired a complete 3-D force field on NiO(001) with atomic resolution (*3-D force field spectroscopy*). In Fig. 5.22a, the atomically resolved FM-AFM image of NiO(001) is shown together with the used coordinate system and the tip to illustrate the measurement principle. NiO(001) crystallizes in the rock salt structure. The distance between the protrusions corresponds to the lattice constant of 417 pm, i.e., only one type of atom (most likely the oxygen) is imaged as protrusion. In an area of 1 nm × 1 nm, 32 × 32 individual $\Delta f(z)$-curves have been recorded at every (x, y) image point and converted into $F_{ts}(z)$-curves. The $\Delta f(x, y, z)$ data set is thereby converted into the 3-D force field $F_{ts}(x, y, z)$. Figure 5.22b, where a specific *x-z*-plane

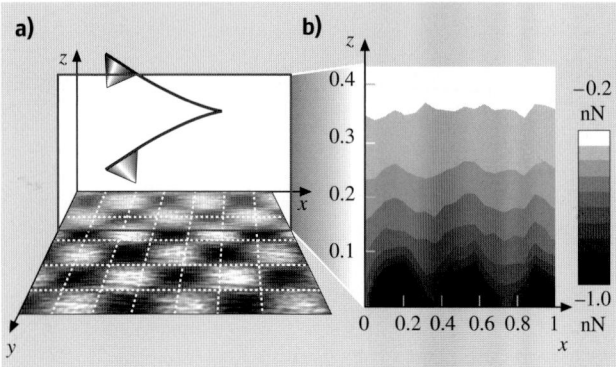

Fig. 5.22. Principle of the 3-D force filed spectroscopy method (**a**) and a 2-D cut through the 3-D force field $F_{ts}(x,y,z)$ recorded at 14 K (**b**). At all 32 × 32 image points of the 1 nm × 1 nm large scan area on NiO(001), a $\Delta f(z)$-curve has been recorded. The obtained $\Delta f(x,y,z)$ data set is then converted into the 3-D tip-sample force field $F_{ts}(x,y,z)$. The *shaded slice* $F_{ts}(x, y = $ const$, z)$ in (**a**) corresponds to a cut along the [100]-direction and demonstrates that atomic resolution has been obtained, because the distance between the protrusions corresponds well to the lattice constant of nickel oxide

is displayed, demonstrates that atomic resolution is achieved. It represents a 2-D cut $F_{ts}(x, y = $ const$, z)$ along the [100]-direction (corresponding to the shaded slice marked in Fig. 5.22a). Since a large number of curves have been recorded, *Langkat* et al. [139] could evaluate the whole data set by standard statistical means to extract the long- and short-range forces. A possible future application of 3-D force field spectroscopy could be to map the short-range forces of complex molecules with functionalized tips in order to locally resolve their chemical reactivity.

Noncontact Friction

Another approach to achieve small tip-sample distances in combination with high force sensitivity is to use soft springs in a perpendicular configuration. The much higher cantilever stiffness along the cantilever axis prevents the jump-to-contact, but the lateral resolution is limited by the magnitude of the oscillation amplitude. However, with such a setup at low temperatures, *Stipe* et al. [141] measured the distance dependence of the very small force due to noncontact friction between tip and sample in vacuum. The effect was attributed to electric charges, which are moved parallel to the surface by the oscillating tip. Since the topography was not recorded in situ, the influence of contaminants or surface steps remained unknown.

5.5.3 Electrostatic Force Microscopy

Electrostatic forces are readily detectable by a force microscope, because tip and sample can be regarded as two electrodes of a capacitor. If they are electrically connected via their back sides and have different work functions, electrons will flow between tip and sample until their Fermi levels are equalized. As a result, an electric field and, consequently, an attractive electrostatic force exists between them at zero bias. This *contact potential difference* can be balanced by applying an appropriate bias voltage. It has been demonstrated that individual doping atoms in semiconducting materials can be detected by electrostatic interactions due to the local variation of the surface potential around them [142, 143].

Detection of Edge Channels in the Quantum Hall Regime

At low temperatures, electrostatic force microscopy has been used to measure the electrostatic potential in the quantum Hall regime of a *two-dimensional electron gas* (2-DEG) buried in epitaxially grown GaAs/AlGaAs heterostructures [144–147]. In the 2-DEG, electrons can move freely in the x-y-plane, but they cannot move in z-direction. Electrical transport properties of a 2-DEG are very different compared to normal metallic conduction. Particularly, the Hall resistance $R_H = h/ne^2$ (where h represents Planck's constant, e is the electron charge, and $n = 1, 2, \ldots$) is quantized in the quantum Hall regime, i.e., at sufficiently low temperatures ($T < 4$ K) and high magnetic fields (up to 20 T). Under these conditions, theoretical calculations predict the existence of *edge channels* in a Hall bar. A Hall bar is a strip conductor that is contacted in a specific way to allow longitudinal and transversal transport measurements in a perpendicular magnetic field. The current is not evenly distributed over the cross section of the bar, but passes mainly along rather thin paths close to the edges. This prediction has been verified by measuring profiles of the electrostatic potential across a Hall bar in different perpendicular external magnetic fields [144–146].

Figure 5.23a shows the experimental setup used to observe these edge channels on top of a Hall bar with a force microscope. The tip is positioned above the surface of a Hall bar under which the 2-DEG is buried. The direction of the magnetic field is oriented perpendicular to the 2-DEG. Note that although the 2-DEG is located several tens of nanometers below the surface, its influence on the electrostatic surface potential can be detected. In Fig. 5.23b, the results of scans perpendicular to the Hall bar are plotted against the magnitude of the external magnetic field. The value of the electrostatic potential is grey-coded in arbitrary units. In certain field ranges, the potential changes linearly across the Hall bar, while in other field ranges the potential drop is confined to the edges of the Hall bar. The predicted edge channels can explain this behavior. The periodicity of the phenomenon is related to the filling factor ν, i.e., the number of Landau levels that are filled with electrons (see also Sect. 5.4.4). Its value depends on $1/B$ and is proportional to the electron concentration n_e in the 2-DEG ($\nu = n_e h/eB$, where h represents Planck's constant and e the electron charge).

5.5.4 Magnetic Force Microscopy

To detect magnetostatic tip-sample interactions with magnetic force microscopy (MFM), a ferromagnetic probe has to be used. Such probes are readily prepared by evaporating a thin magnetic layer, e.g., 10 nm iron, onto the tip. Due to the in-plane

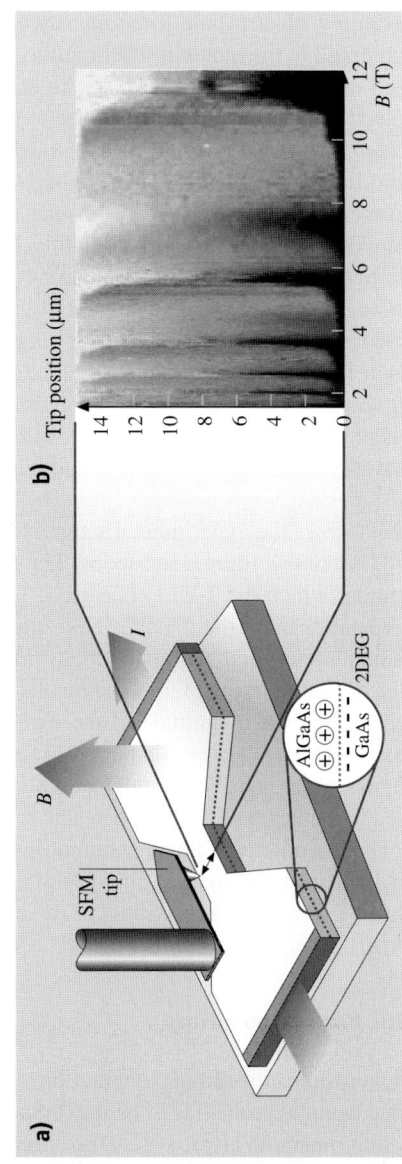

Fig. 5.23. Configuration of the Hall bar within a low temperature ($T < 1$ K) force microscope (**a**) and profiles (y-axis) at different magnetic field (x-axis) of the electrostatic potential across a 14-μm-wide Hall bar in the quantum Hall regime (**b**). The external magnetic field is oriented perpendicular to the 2-DEG, which is buried below the surface. *Bright* and *dark regions* reflect the characteristic changes of the electrostatic potential across the Hall bar at different magnetic fields and can be explained by the existence of the theoretically predicted edge channels (courtesy of E. Ahlswede; cf. [146])

shape anisotropy of thin films, the magnetization of such tips lies predominantly along the tip axis, i.e., perpendicular to the surface. Since magnetostatic interactions are long-range, they can be separated from the topography by scanning in a certain constant height (typically around 20 nm) above the surface, where the z-component of the sample stray field is probed (see Fig. 5.24a). Therefore, MFM is always operated in noncontact. The signal from the cantilever is directly recorded while the z-feedback is switched off. MFM can be operated in the static mode or in the dynamic modes (AM-MFM at ambient pressures and FM-MFM in vacuum). A lateral resolution below 50 nm can be routinely obtained.

Observation of Domain Patterns

MFM is widely used to visualize domain patterns of ferromagnetic materials. At low temperatures, *Moloni* et al. [148] observed the domain structure of magnetite below its Verwey transition temperature ($T_V = 122$ K), but most of the work concentrated on thin films of $La_{1-x}Ca_xMnO_3$ [149–151]. Below T_V, the conductivity decreases by two orders of magnitude and a small structural distortion is observed. The domain structure of this mixed valence manganite is of great interest, because its resistivity strongly depends on the external magnetic field, i.e., it exhibits a large colossal magneto resistive effect. To investigate the field dependence of domain patterns in ambient conditions, electromagnets have to be used. They can cause severe thermal drift problems due to Joule-heating of the coils by large currents. Field on the order of 100 mT can be achieved. In contrast, much larger fields (more than 10 T) can be rather easily produced by implementing a superconducting magnet in low temperature setups. With such a design, *Liebmann* et al. [151] recorded the domain structure along the major hysteresis loop of $La_{0.7}Ca_{0.3}MnO_3$ epitaxially grown on $LaAlO_3$ (see Fig. 5.24b–f). The film geometry (thickness is 100 nm) favors an in-plane magnetization, but the lattice mismatch with the substrate induces an out-of-plane anisotropy. Thereby, an irregular pattern of strip domains appears in zero field. If the external magnetic field is increased, the domains with anti-parallel orientation shrink and finally disappear in saturation (see Fig. 5.24b,c). The residual contrast in saturation (d) reflects topographic features. If the field is decreased after saturation (see Fig. 5.24e,f), cylindrical domains first nucleate and then start to grow. At zero field, the maze-type domain pattern has been evolved again. Such data sets can be used to analyze domain nucleation and the domain growth mode. Moreover, due to the negligible drift, domain structure and surface morphology can be directly compared, because every MFM can be used as a regular topography-imaging force microscope.

Detection of Individual Vortices in Superconductors

Numerous low temperature MFM experiments have been performed on superconductors [152–159]. Some basic features of superconductors have been mentioned already in Sect. 5.4.5. The main difference of STM/STS compared to MFM is the high sensitivity to the electronic properties of the surface. Therefore, careful sample preparation is a prerequisite. This is not so important for MFM experiments, since the tip is scanned at a certain distance above the surface.

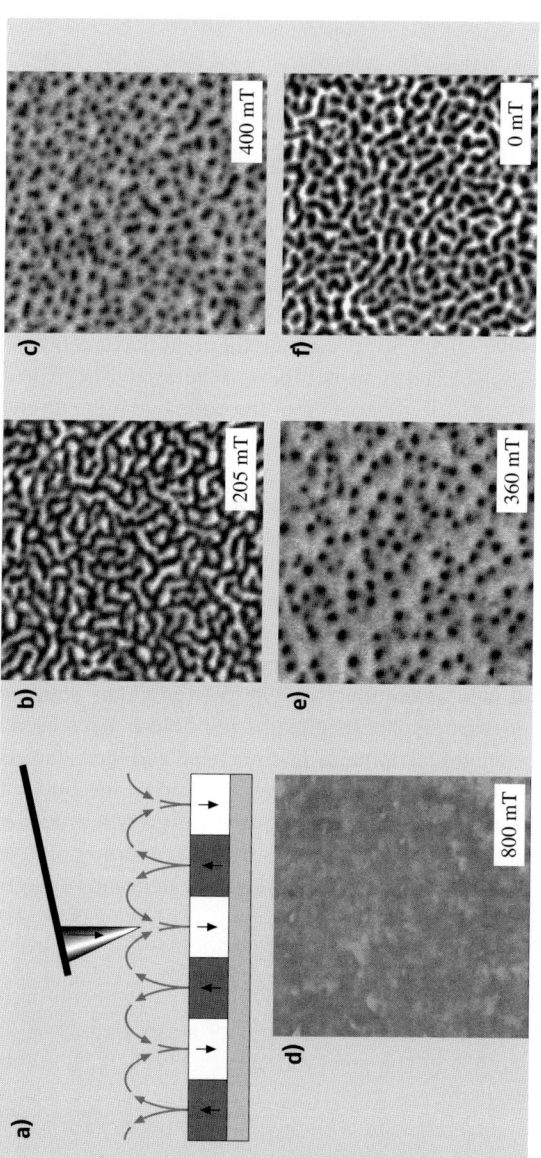

Fig. 5.24. Principle of MFM operation (**a**) and field-dependent domain structure of a ferromagnetic thin film (**b**)–(**f**) recorded at 5.2 K with FM-MFM. All images were recorded on the same 4 μm × 4 μm large scan area. The $La_{0.7}Ca_{0.3}MnO_3/LaAlO_3$ system exhibits a substrate-induced out-of-plane anisotropy. *Bright* and *dark areas* are visible and correspond to attractive and repulsive magnetostatic interactions, respectively. The series shows how the domain pattern evolves along the major hysteresis loop from, i.e., zero field to saturation at 600 mT and back to zero field

Superconductors can be divided into two classes with respect to their behavior in an external magnetic field. For type-I superconductors, any magnetic flux is entirely excluded below their critical temperature T_c (Meissner effect), while for type-II superconductors, cylindrical inclusions (*vortices*) of normal material exist in a superconducting matrix (*vortex* state). The radius of the vortex *core*, where the Cooper pair density decreases to zero, is on the order of the coherence length ξ. Since the superconducting gap vanishes in the core, they can be detected by STS (see Sect. 5.4.5). Additionally, each vortex contains one magnetic quantum flux $\Phi = h/2e$ (where h represents Planck's constant and e the electron charge). Circular supercurrents around the core screen the magnetic field associated with a vortex; their radius is given by the London penetration depth λ of the material. This magnetic field of the vortices can be detected by MFM. Investigations have been performed on the two most popular copper oxide high-T_c superconductors, $YBa_2Cu_3O_7$ [152, 153, 155] and $Bi_2Sr_2CaCu_2O_8$ [153, 159], on the only elemental conventional type-II superconductor Nb [156, 157], and on the layered compound crystal $NbSe_2$ [154, 156].

Most often, vortices have been generated by cooling the sample from the normal state to below T_c in an external magnetic field. After such a *field cooling* procedure, the most energetically favorable vortex arrangement is a regular triangular Abrikosov lattice. *Volodin* et al. [154] were able to observe such an Abrikosov lattice on $NbSe_2$. The intervortex distance d is related to the external field during B cool down via $d = (4/3)^{1/4} (\Phi/B)^{1/2}$. Another way to introduce vortices into a type-II superconductor is vortex penetration from the edge by applying a magnetic field at temperatures below T_c. According to the Bean model, a vortex density gradient exists under such conditions within the superconducting material. *Pi* et al. [159] slowly increased the external magnetic field until the vortex front approaching from the edge reached the scanning area.

If the vortex configuration is dominated by the *pinning* of vortices at randomly distributed structural defects, no Abrikosov lattice emerges. The influence of pinning centers can be studied easily by MFM, because every MFM can be used to scan the topography in its AFM mode. This has been done for natural growth defects by *Moser* et al. [155] on $YBa_2Cu_3O_7$ and for $YBa_2Cu_3O_7$ and niobium thin films, respectively, by *Volodin* et al. [158]. *Roseman* et al. [160] investigated the formation of vortices in the presence of an artificial structure on niobium films, while *Pi* et al. [159] produced columnar defects by heavy ion bombardment in a $Bi_2Sr_2CaCu_2O_8$ single crystal to study the strong pinning at these defects.

Figure 5.25 demonstrates that MFM is sensitive to the polarity of vortices. In Fig. 5.25a, six vortices have been produced in a niobium film by field cooling in + 0.5 mT. External magnetic field and tip magnetization are parallel, and, therefore, the tip-vortex interaction is attractive (bright contrast). To remove the vortices, the niobium was heated above T_c ($\approx$ 9 K). Thereafter, vortices of opposite polarity were produced by field cooling in − 0.5 mT, which appear dark in Fig. 5.25b. The vortices are probably bound to strong pinning sites, because the vortex positions are identical in both images of Fig. 5.25. By imaging the vortices at different scanning heights, *Roseman* et al. [157] tried to extract values for the London penetration depth from the scan height dependence of their profiles. While good qualitative agreement with

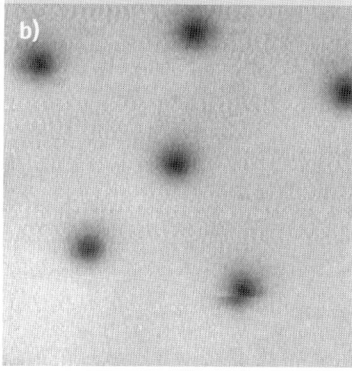

Fig. 5.25. Two 5 μm × 5 μm FM-MFM images of vortices in a niobium thin film after field-cooling at 0.5 mT (**a**) and −0.5 mT (**b**), respectively. Since the external magnetic field was parallel in (**a**) and antiparallel in (**b**) with respect to the tip magnetization, the vortices exhibit reversed contrast. Strong pinning dominates the position of the vortices, since they appear at identical locations in (**a**) and (**b**) and are not arranged in a regular Abrikosov lattice (courtesy of P. Grütter; cf. [157])

theoretical predictions has been found, the absolute values do not agree with published literature values. The disagreement was attributed to the convolution between tip and vortex stray field. Better values might be obtained with calibrated tips.

Toward Single Spin Detection

So far, only collective magnetic phenomena like ferromagnetic domains have been observed via magnetostatic tip-sample interactions detected by MFM. However, magnetic ordering exists due to the exchange interaction between the electron spins of neighboring atoms in a solid. The most energetically favorable situation can be either ferromagnetic (parallel orientation) or anti-ferromagnetic (anti-parallel orientation) ordering. It has been predicted that the exchange force between an individual spin of a magnetically ordered sample and the spin of the foremost atom of a magnetic tip can be detected at sufficiently small tip-sample distances [161, 162].

The experimental realization, however, is very difficult, because the exchange force is about a factor of ten weaker and of even shorter range than the chemical interactions that are responsible for the atomic-scale contrast. FM-AFM experiments with a ferromagnetic tip have been performed on the anti-ferromagnetic NiO(001) surface at room temperature [163] and with a considerable better signal-to-noise ratio at low temperatures [122]. Although it was possible to achieve atomic resolution, a periodic contrast that could be attributed to the anti-ferromagnetically ordered spins of the nickel atoms could not be observed.

Even more ambitious is the proposed detection of individual nuclear spins by magnetic resonance force microscopy (MRFM) using a magnetic tip [164, 165]. The interest in this technique is driven by the possibility of reaching subsurface true 3-D imaging with atomic resolution and chemical specificity. The idea of MRFM is to combine aspects of force microscopy (atomic resolution capability) with nuclear magnetic resonance (3-D imaging and elemental selectivity). Low temperatures are required, because the forces are extremely small.

Up until now, no individual spins have been detected by MRFM. However, electron spin resonance [166, 167], nuclear magnetic resonance [168], and ferromagnetic resonance [169] experiments of spin ensembles have been performed with micrometer resolution. On the way toward single spin detection, the design of ultrasensitive cantilevers made considerable progress, and the detection of forces below 1×10^{-18} N has been achieved [120].

References

1. G. Binnig, H. Rohrer, Ch. Gerber, and E. Weibel. Surface studies by scanning tunneling microscopy. *Phys. Rev. Lett.*, 49:57–61, 1982.
2. R. Wiesendanger. *Scanning Probe Microscopy and Spectroscopy*. Cambridge Univ. Press, 1994.
3. M. Tinkham. *Introduction to Superconductivity*. McGraw-Hill, 1996.
4. J. Kondo. Theory of dilute magnetic alloys. *Solid State Phys.*, 23:183–281, 1969.
5. T. R. Albrecht, P. Grütter, H. K. Horne, and D. Rugar. Frequency modulation detection using high-q cantilevers for enhanced force microscope sensitivity. *J. Appl. Phys.*, 69:668–673, 1991.
6. F. J. Giessibl, H. Bielefeld, S. Hembacher, and J. Mannhart. Calculation of the optimal imaging parameters for frequency modulation atomic force microscopy. *Appl. Surf. Sci.*, 140:352–357, 1999.
7. W. Allers, A. Schwarz, U. D. Schwarz, and R. Wiesendanger. Dynamic scanning force microscopy at low temperatures on a van der waals surface: graphite(0001). *Appl. Surf. Sci.*, 140:247–252, 1999.
8. W. Allers, A. Schwarz, U. D. Schwarz, and R. Wiesendanger. Dynamic scanning force microscopy at low temperatures on a noble-gas crystal: atomic resolution on the xenon(111) surface. *Europhys. Lett.*, 48:276–279, 1999.
9. M. Morgenstern, D. Haude, V. Gudmundsson, C. Wittneven, R. Dombrowski, and R. Wiesendanger. Origin of landau oscillations observed in scanning tunneling spectroscopy on n-InAs(110). *Phys. Rev. B*, 62:7257–7263, 2000.
10. D. M. Eigler, P. S. Weiss, E. K. Schweizer, and N. D. Lang. Imaging Xe with a low-temperature scanning tunneling microscope. *Phys. Rev. Lett.*, 66:1189–1192, 1991.
11. P. S. Weiss and D. M. Eigler. Site dependence of the apparent shape of a molecule in scanning tunneling micoscope images: Benzene on Pt{111}. *Phys. Rev. Lett.*, 71:3139–3142, 1992.
12. D. M. Eigler and E. K. Schweizer. Positioning single atoms with a scanning tunneling microscope. *Nature*, 344:524–526, 1990.

13. H. Hug, B. Stiefel, P.J.A. van Schendel, A. Moser, S. Martin, and H.-J. Güntherodt. A low temperature ultrahigh vacuum scanning force microscope. *Rev. Sci. Instrum.*, 70:3627–3640, 1999.
14. S. Behler, M.K. Rose, D.F. Ogletree, and F. Salmeron. Method to characterize the vibrational response of a beetle type scanning tunneling microscope. *Rev. Sci. Instrum.*, 68:124–128, 1997.
15. C. Wittneven, R. Dombrowski, S.H. Pan, and R. Wiesendanger. A low-temperature ultrahigh-vacuum scanning tunneling microscope with rotatable magnetic field. *Rev. Sci. Instrum.*, 68:3806–3810, 1997.
16. W. Allers, A. Schwarz, U.D. Schwarz, and R. Wiesendanger. A scanning force microscope with atomic resolution in ultrahigh vacuum and at low temperatures. *Rev. Sci. Instrum.*, 69:221–225, 1998.
17. G. Dujardin, R.E. Walkup, and Ph. Avouris. Dissociation of individual molecules with electrons from the tip of a scanning tunneling microscope. *Science*, 255:1232–1235, 1992.
18. H.J. Lee and W. Ho. Single-bond formation and characterization with a scanning tunneling microscope. *Science*, 286:1719–1722, 1999.
19. R. Berndt, R. Gaisch, J.K. Gimzewski, B. Reihl, R.R. Schlittler, W.D. Schneider, and M. Tschudy. Photon emission at molecular resolution induced by a scanning tunneling microscope. *Science*, 262:1425–1427, 1993.
20. B.G. Briner, M. Doering, H.P. Rust, and A.M. Bradshaw. Microscopic diffusion enhanced by adsorbate interaction. *Science*, 278:257–260, 1997.
21. J. Kliewer, R. Berndt, E.V. Chulkov, V.M. Silkin, P.M. Echenique, and S. Crampin. Dimensionality effects in the lifetime of surface states. *Science*, 288:1399–1401, 2000.
22. M.F. Crommie, C.P. Lutz, and D.M. Eigler. Imaging standing waves in a two-dimensional electron gas. *Nature*, 363:524–527, 1993.
23. B.C. Stipe, M.A. Rezaei, and W. Ho. Single-molecule vibrational spectroscopy and microscopy. *Science*, 280:1732–1735, 1998.
24. H.J. Lee and W. Ho. Structural determination by single-molecule vibrational spectroscopy and microscopy: Contrast between copper and iron carbonyls. *Phys. Rev. B*, 61:R16347–R16350, 2000.
25. C.W.J. Beenakker and H. van Houten. Quantum transport in semiconductor nanostructures. *Solid State Phys.*, 44:1–228, 1991.
26. S.H. Pan, E.W. Hudson, K.M. Lang, H. Eisaki, S. Uchida, and J.C. Davis. Imaging the effects of individual zinc impurity atoms on superconductivity in $Bi_2Sr_2CaCu_2O_{8+\delta}$. *Nature*, 403:746–750, 2000.
27. R.S. Becker, J.A. Golovchenko, and B.S. Swartzentruber. Atomic-scale surface modifications using a tunneling microscope. *Nature*, 325:419–42, 1987.
28. J.A. Stroscio and D.M. Eigler. Atomic and molecular manipulation with the scanning tunneling microscope. *Science*, 254:1319–1326, 1991.
29. L. Bartels, G. Meyer, and K.H. Rieder. Basic steps of lateral manipulation of single atoms and diatomic clusters with a scanning tunneling microscope. *Phys. Rev. Lett.*, 79:697–700, 1997.
30. J.J. Schulz, R. Koch, and K.H. Rieder. New mechanism for single atom manipulation. *Phys. Rev. Lett.*, 84:4597–4600, 2000.
31. T.C. Shen, C. Wang, G.C. Abeln, J.R. Tucker, J.W. Lyding, Ph. Avouris, and R.E. Walkup. Atomic-scale desorption through electronic and vibrational excitation mechanisms. *Science*, 268:1590–1592, 1995.
32. T. Komeda, Y. Kim, M. Kawai, B.N.J. Persson, and H. Ueba. Lateral hopping of molecules induced by excitations of internal vibration mode. *Science*, 295:2055–2058, 2002.

33. Y. W. Mo. Reversible rotation of antimony dimers on the silicon(001) surface with a scanning tunneling microscope. *Science*, 261:886–888, 1993.
34. B. C. Stipe, M. A. Rezaei, and W. Ho. Inducing and viewing the rotational motion of a single molecule. *Science*, 279:1907–1909, 1998.
35. F. Moresco, G. Meyer, K. H. Rieder, H. Tang, A. Gourdon, and C. Joachim. Conformational changes of single molecules by scanning tunneling microscopy manipulation: a route to molecular switching. *Phys. Rev. Lett.*, 86:672–675, 2001.
36. S. W. Hla, L. Bartels, G. Meyer, and K. H. Rieder. Inducing all steps of a chemical reaction with the scanning tunneling microscope tip: Towards single molecule engineering. *Phys. Rev. Lett.*, 85:2777–2780, 2000.
37. E. Ganz, S. K. Theiss, I. S. Hwang, and J. Golovchenko. Direct measurement of diffusion by hot tunneling microscopy: Activations energy, anisotropy, and long jumps. *Phys. Rev. Lett.*, 68:1567–1570, 1992.
38. M. Schuhnack, T. R. Linderoth, F. Rosei, E. Laegsgaard, I. Stensgaard, and F. Besenbacher. Long jumps in the surface diffusion of large molecules. *Phys. Rev. Lett.*, 88:156102, 1–4, 2002.
39. L. J. Lauhon and W. Ho. Direct observation of the quantum tunneling of single hydrogen atoms with a scanning tunneling microscope. *Phys. Rev. Lett.*, 85:4566–4569, 2000.
40. N. Kitamura, M. Lagally, and M. B. Webb. Real-time observation of vacancy diffusion on Si(001)-(2×1) by scanning tunneling microscopy. *Phys. Rev. Lett.*, 71:2082–2085, 1993.
41. M. Morgenstern, T. Michely, and G. Comsa. Onset of interstitial diffusion determined by scanning tunneling microscopy. *Phys. Rev. Lett.*, 79:1305–1308, 1997.
42. K. Morgenstern, G. Rosenfeld, B. Poelsema, and G. Comsa. Brownian motion of vacancy islands on Ag(111). *Phys. Rev. Lett.*, 74:2058–2061, 1995.
43. B. Reihl, J. H. Coombs, and J. K. Gimzewski. Local inverse photoemission with the scanning tunneling microscope. *Surf. Sci.*, 211–212:156–164, 1989.
44. R. Berndt, J. K. Gimzewski, and P. Johansson. Inelastic tunneling excitation of tip-induced plasmon modes on noble-metal surfaces. *Phys. Rev. Lett.*, 67:3796–3799, 1991.
45. P. Johansson, R. Monreal, and P. Apell. Theory for light emission from a scanning tunneling microscope. *Phys. Rev. B*, 42:9210–9213, 1990.
46. J. Aizpurua, G. Hoffmann, S. P. Apell, and R. Berndt. Electromagnetic coupling on an atomic scale. *Phys. Rev. Lett.*, 89:156803, 1–4, 2002.
47. G. Hoffmann, J. Kliewer, and R. Berndt. Luminescence from metallic quantum wells in a scanning tunneling microscope. *Phys. Rev. Lett.*, 78:176803, 1–4, 2001.
48. A. Downes and M. E. Welland. Photon emission from Si(111)-(7×7) induced by scanning tunneling microscopy: atomic scale and material contrast. *Phys. Rev. Lett.*, 81:1857–1860, 1998.
49. M. Kemerink, K. Sauthoff, P. M. Koenraad, J. W. Geritsen, H. van Kempen, and J. H. Wolter. Optical detection of ballistic electrons injected by a scanning-tunneling microscope. *Phys. Rev. Lett.*, 86:2404–2407, 2001.
50. J. Tersoff and D. R. Hamann. Theory and application for the scanning tunneling microscope. *Phys. Rev. Lett.*, 50:1998–2001, 1983.
51. C. J. Chen. *Introduction to Scanning Tunneling Microscopy*. Oxford Univ. Press, 1993.
52. J. Winterlin, J. Wiechers, H. Brune, T. Gritsch, H. Hofer, and R. J. Behm. Atomic-resolution imaging of close-packed metal surfaces by scanning tunneling microscopy. *Phys. Rev. Lett.*, 62:59–62, 1989.
53. A. L. Vazquez de Parga, O. S. Hernan, R. Miranda, A. Levy Yeyati, N. Mingo, A. Martin-Rodero, and F. Flores. Electron resonances in sharp tips and their role in tunneling spectroscopy. *Phys. Rev. Lett.*, 80:357–360, 1998.

54. S. H. Pan, E. W. Hudson, and J. C. Davis. Vacuum tunneling of superconducting quasiparticles from atomically sharp scanning tunneling microscope tips. *Appl. Phys. Lett.*, 73:2992–2994, 1998.
55. J. T. Li, W. D. Schneider, R. Berndt, O. R. Bryant, and S. Crampin. Surface-state lifetime measured by scanning tunneling spectroscopy. *Phys. Rev. Lett.*, 81:4464–4467, 1998.
56. L. Bürgi, O. Jeandupeux, H. Brune, and K. Kern. Probing hot-electron dynamics with a cold scanning tunneling microscope. *Phys. Rev. Lett.*, 82:4516–4519, 1999.
57. J. W. G. Wildoer, C. J. P. M. Harmans, and H. van Kempen. Observation of landau levels at the InAs(110) surface by scanning tunneling spectroscopy. *Phys. Rev. B*, 55:R16013–R16016, 1997.
58. M. Morgenstern, V. Gudmundsson, C. Wittneven, R. Dombrowski, and R. Wiesendanger. Nonlocality of the exchange interaction probed by scanning tunneling spectroscopy. *Phys. Rev. B*, 63:201301(R), 1–4, 2001.
59. M. V. Grishin, F. I. Dalidchik, S. A. Kovalevskii, N. N. Kolchenko, and B. R. Shub. Isotope effect in the vibrational spectra of water measured in experiments with a scanning tunneling microscope. *JETP Lett.*, 66:37–40, 1997.
60. A. Hewson. *From the Kondo Effect to Heavy Fermions*. Cambridge Univ. Press, 1993.
61. V. Madhavan, W. Chen, T. Jamneala, M. F. Crommie, and N. S. Wingreen. Tunneling into a single magnetic atom: Spectroscopic evidence of the kondo resonance. *Science*, 280:567–569, 1998.
62. J. Li, W. D. Schneider, R. Berndt, and B. Delley. Kondo scattering observed at a single magnetic impurity. *Phys. Rev. Lett.*, 80:2893–2896, 1998.
63. T. W. Odom, J. L. Huang, C. L. Cheung, and C. M. Lieber. Magnetic clusters on single-walled carbon nanotubes: the kondo effect in a one-dimensional host. *Science*, 290:1549–1552, 2000.
64. M. Ouyang, J. L. Huang, C. L. Cheung, and C. M. Lieber. Energy gaps in metallic single-walled carbon nanotubes. *Science*, 292:702–705, 2001.
65. U. Fano. Effects of configuration interaction on intensities and phase shifts. *Phys. Rev.*, 124:1866–1878, 1961.
66. H. C. Manoharan, C. P. Lutz, and D. M. Eigler. Quantum mirages formed by coherent projection of electronic structure. *Nature*, 403:512–515, 2000.
67. O. Y. Kolesnychenko, R. de Kort, M. I. Katsnelson, A. I. Lichtenstein, and H. van Kempen. Real-space observation of an orbital kondo resonance on the Cr(001) surface. *Nature*, 415:507–509, 2002.
68. H. A. Mizes and J. S. Foster. Long-range electronic perturbations caused by defects using scanning tunneling microscopy. *Science*, 244:559–562, 1989.
69. P. T. Sprunger, L. Petersen, E. W. Plummer, E. Laegsgaard, and F. Besenbacher. Giant friedel oscillations on beryllium (0001) surface. *Science*, 275:1764–1767, 1997.
70. P. Hofmann, B. G. Briner, M. Doering, H. P. Rust, E. W. Plummer, and A. M. Bradshaw. Anisotropic two-dimensional friedel oscillations. *Phys. Rev. Lett.*, 79:265–268, 1997.
71. E. J. Heller, M. F. Crommie, C. P. Lutz, and D. M. Eigler. Scattering and adsorption of surface electron waves in quantum corrals. *Nature*, 369:464–466, 1994.
72. M. C. M. M. van der Wielen, A. J. A. van Roij, and H. van Kempen. Direct observation of friedel oscillations around incorporated Si_{Ga} dopants in GaAs by low-temperature scanning tunneling microscopy. *Phys. Rev. Lett.*, 76:1075–1078, 1996.
73. O. Millo, D. Katz, Y. W. Cao, and U. Banin. Imaging and spectroscopy of artificial-atom states in core/shell nanocrystal quantum dots. *Phys. Rev. Lett.*, 86:5751–5754, 2001.
74. L. C. Venema, J. W. G. Wildoer, J. W. Janssen, S. J. Tans, L. J. T. Tuinstra, L. P. Kouwenhoven, and C. Dekker. Imaging electron wave functions of quantized energy levels in carbon nanotubes. *Nature*, 283:52–55, 1999.

75. S. G. Lemay, J. W. Jannsen, M. van den Hout, M. Mooij, M. J. Bronikowski, P. A. Willis, R. E. Smalley, L. P. Kouwenhoven, and C. Dekker. Two-dimensional imaging of electronic wavefunctions in carbon nanotubes. *Nature*, 412:617–620, 2001.
76. C. Wittneven, R. Dombrowski, M. Morgenstern, and R. Wiesendanger. Scattering states of ionized dopants probed by low temperature scanning tunneling spectroscopy. *Phys. Rev. Lett.*, 81:5616–5619, 1998.
77. D. Haude, M. Morgenstern, I. Meinel, and R. Wiesendanger. Local density of states of a three-dimensional conductor in the extreme quantum limit. *Phys. Rev. Lett.*, 86:1582–1585, 2001.
78. R. Joynt and R. E. Prange. Conditions for the quantum hall effect. *Phys. Rev. B*, 29:3303–3317, 1984.
79. M. Morgenstern, J. Klijn, C. Meyer, M. Getzlaff, R. Adelung, R. A. Römer, K. Rossnagel, L. Kipp, M. Skibowski, and R. Wiesendanger. Direct comparison between potential landscape and local density of states in a disordered two-dimensional electron system. *Phys. Rev. Lett.*, 89:136806, 1–4, 2002.
80. E. Abrahams, P. W. Anderson, D. C. Licciardello, and T. V. Ramakrishnan. Scaling theory of localization: absence of quantum diffusion in two dimensions. *Phys. Rev. Lett.*, 42:673–676, 1979.
81. M. Morgenstern, J. Klijn, and R. Wiesendanger. Real space observation of drift states in a two-dimensional electron system at high magnetic fields. *Phys. Rev. Lett.*, 90:056804, 1–4, 2003.
82. R. E. Peierls. *Quantum Theory of Solids*. Clarendon, 1955.
83. C. G. Slough, W. W. McNairy, R. V. Coleman, B. Drake, and P. K. Hansma. Charge-density waves studied with the use of a scanning tunneling microscope. *Phys. Rev. B*, 34:994–1005, 1986.
84. X. L. Wu and C. M. Lieber. Hexagonal domain-like charge-density wave of TaS_2 determined by scanning tunneling microscopy. *Science*, 243:1703–1705, 1989.
85. T. Nishiguchi, M. Kageshima, N. Ara-Kato, and A. Kawazu. Behaviour of charge density waves in a one-dimensional organic conductor visualized by scanning tunneling microscopy. *Phys. Rev. Lett.*, 81:3187–3190, 1998.
86. X. L. Wu and C. M. Lieber. Direct observation of growth and melting of the hexagonal-domain charge-density-wave phase in 1 T-TaS_2 by scanning tunneling microscopy. *Phys. Rev. Lett.*, 64:1150–1153, 1990.
87. J. M. Carpinelli, H. H. Weitering, E. W. Plummer, and R. Stumpf. Direct observation of a surface charge density wave. *Nature*, 381:398–400, 1996.
88. H. H. Weitering, J. M. Carpinelli, A. V. Melechenko, J. Zhang, M. Bartkowiak, and E. W. Plummer. Defect-mediated condensation of a charge density wave. *Science*, 285:2107–2110, 1999.
89. H. W. Yeom, S. Takeda, E. Rotenberg, I. Matsuda, K. Horikoshi, J. Schäfer, C. M. Lee, S. D. Kevan, T. Ohta, T. Nagao, and S. Hasegawa. Instability and charge density wave of metallic quantum chains on a silicon surface. *Phys. Rev. Lett.*, 82:4898–4901, 1999.
90. K. Swamy, A. Menzel, R. Beer, and E. Bertel. Charge-density waves in self-assembled halogen-bridged metal chains. *Phys. Rev. Lett.*, 86:1299–1302, 2001.
91. J. J. Kim, W. Yamaguchi, T. Hasegawa, and K. Kitazawa. Observation of mott localization gap using low temperature scanning tunneling spectroscopy in commensurate 1 T-$TaSe_2$. *Phys. Rev. Lett.*, 73:2103–2106, 1994.
92. J. Bardeen, L. N. Cooper, and J. R. Schrieffer. Theory of superconductivity. *Phys. Rev.*, 108:1175–1204, 1957.
93. A. Yazdani, B. A. Jones, C. P. Lutz, M. F. Crommie, and D. M. Eigler. Probing the local effects of magnetic impurities on superconductivity. *Science*, 275:1767–1770, 1997.

94. S. H. Tessmer, M. B. Tarlie, D. J. van Harlingen, D. L. Maslov, and P. M. Goldbart. Probing the superconducting proximity effect in $NbSe_2$ by scanning tunneling micrsocopy. *Phys. Rev. Lett*, 77:924–927, 1996.
95. K. Inoue and H. Takayanagi. Local tunneling spectroscopy of Nb/InAs/Nb superconducting proximity system with a scanning tunneling microscope. *Phys. Rev. B*, 43:6214–6215, 1991.
96. H. F. Hess, R. B. Robinson, R. C. Dynes, J. M. Valles, and J. V. Waszczak. Scanning-tunneling-microscope observation of the abrikosov flux lattice and the density of states near and inside a fluxoid. *Phys. Rev. Lett.*, 62:214–217, 1989.
97. H. F. Hess, R. B. Robinson, and J. V. Waszczak. Vortex-core structure observed with a scanning tunneling microscope. *Phys. Rev. Lett.*, 64:2711–2714, 1990.
98. N. Hayashi, M. Ichioka, and K. Machida. Star-shaped local density of states around vortices in a type-ii superconductor. *Phys. Rev. Lett.*, 77:4074–4077, 1996.
99. H. Sakata, M. Oosawa, K. Matsuba, and N. Nishida. Imaging of vortex lattice transition in YNi_2B_2C by scanning tunneling spectroscopy. *Phys. Rev. Lett.*, 84:1583–1586, 2000.
100. S. Behler, S. H. Pan, P. Jess, A. Baratoff, H.-J. Güntherodt, F. Levy, G. Wirth, and J. Wiesner. Vortex pinning in ion-irrediated $NbSe_2$ studied by scanning tunneling microscopy. *Phys. Rev. Lett.*, 72:1750–1753, 1994.
101. R. Berthe, U. Hartmann, and C. Heiden. Influence of a transport current on the abrikosov flux lattice observed with a low-temperature scanning tunneling microscope. *Ultramicroscopy*, 42–44:696–698, 1992.
102. A. Polkovnikov, S. Sachdev, and M. Vojta. Impurity in a d-wave superconductor: Kondo effect and stm spectra. *Phys. Rev. Lett.*, 86:296–299, 2001.
103. E. W. Hudson, K. M. Lang, V. Madhavan, S. H. Pan, S. Uchida, and J. C. Davis. Interplay of magnetism and high-t_c superconductivity at individual Ni impurity atoms in $Bi_2Sr_2CaCu_2O_{8+\delta}$. *Nature*, 411:920–924, 2001.
104. K. M. Lang, V. Madhavan, J. E. Hoffman, E. W. Hudson, H. Eisaki, S. Uchida, and J. C. Davis. Imaging the granular structure of high-t_c superconductivity in underdoped $Bi_2Sr_2CaCu_2O_{8+\delta}$. *Nature*, 415:412–416, 2002.
105. I. Maggio-Aprile, C. Renner, E. Erb, E. Walker, and Ø. Fischer. Direct vortex lattice imaging and tunneling spectroscopy of flux lines on $YBa_2Cu_3O_{7-\delta}$. *Phys. Rev. Lett.*, 75:2754–2757, 1995.
106. C. Renner, B. Revaz, K. Kadowaki, I. Maggio-Aprile, and Ø. Fischer. Observation of the low temperature pseudogap in the vortex cores of $Bi_2Sr_2CaCu_2O_{8+\delta}$. *Phys. Rev. Lett.*, 80:3606–3609, 1998.
107. S. H. Pan, E. W. Hudson, A. K. Gupta, K. W. Ng, H. Eisaki, S. Uchida, and J. C. Davis. Stm studies of the electronic structure of vortex cores in $Bi_2Sr_2CaCu_2O_{8+\delta}$. *Phys. Rev. Lett.*, 85:1536–1539, 2000.
108. D. P. Arovas, A. J. Berlinsky, C. Kallin, and S. C. Zhang. Superconducting vortex with antiferromagnetic core. *Phys. Rev. Lett.*, 79:2871–2874, 1997.
109. J. E. Hoffmann, E. W. Hudson, K. M. Lang, V. Madhavan, H. Eisaki, S. Uchida, and J. C. Davis. A four unit cell periodic pattern of quasi-particle states surrounding vortex cores in $Bi_2Sr_2CaCu_2O_{8+\delta}$. *Science*, 295:466–469, 2002.
110. M. Fäth, S. Freisem, A. A. Menovsky, Y. Tomioka, J. Aaarts, and J. A. Mydosh. Spatially inhomogeneous metal–insulator transition in doped manganites. *Science*, 285:1540–1542, 1999.
111. C. Renner, G. Aeppli, B. G. Kim, Y. A. Soh, and S. W. Cheong. Atomic-scale images of charge ordering in a mixed-valence manganite. *Nature*, 416:518–521, 2000.
112. M. Bode, M. Getzlaff, and R. Wiesendanger. Spin-polarized vacuum tunneling into the exchange-split surface state of Gd(0001). *Phys. Rev. Lett.*, 81:4256–4259, 1998.

113. A. Kubetzka, M. Bode, O. Pietzsch, and R. Wiesendanger. Spin-polarized scanning tunneling microscopy with antiferromagnetic probe tips. *Phys. Rev. Lett.*, 88:057201, 1–4, 2002.
114. O. Pietzsch, A. Kubetzka, M. Bode, and R. Wiesendanger. Observation of magnetic hysteresis at the nanometer scale by spin-polarized scanning tunneling spectroscopy. *Science*, 292:2053–2056, 2001.
115. S. Heinze, M. Bode, A. Kubetzka, O. Pietzsch, X. Xie, S. Blügel, and R. Wiesendanger. Real-space imaging of two-dimensional antiferromagnetism on the atomic scale. *Science*, 288:1805–1808, 2000.
116. A. Wachowiak, J. Wiebe, M. Bode, O. Pietzsch, M. Morgenstern, and R. Wiesendanger. Internal spin-structure of magnetic vortex cores observed by spin-polarized scanning tunneling microscopy. *Science*, 298:577–580, 2002.
117. M. D. Kirk, T. R. Albrecht, and C. F. Quate. Low-temperature atomic force microscopy. *Rev. Sci. Instrum.*, 59:833–835, 1988.
118. D. Pelekhov, J. Becker, and J. G. Nunes. Atomic force microscope for operation in high magnetic fields at millikelvin temperatures. *Rev. Sci. Instrum.*, 70:114–120, 1999.
119. J. Mou, Y. Jie, and Z. Shao. An optical detection low temperature atomic force microscope at ambient pressure for biological research. *Rev. Sci. Instrum.*, 64:1483–1488, 1993.
120. H. J. Mamin and D. Rugar. Sub-attonewton force detection at millikelvin temperatures. *Appl. Phys. Lett.*, 79:3358–3360, 2001.
121. A. Schwarz, W. Allers, U. D. Schwarz, and R. Wiesendanger. Dynamic mode scanning force microscopy of n-InAs(110)-(1 × 1) at low temperatures. *Phys. Rev. B*, 61:2837–2845, 2000.
122. W. Allers, S. Langkat, and R. Wiesendanger. Dynamic low-temperature scanning force microscopy on nickel oxide(001). *Appl. Phys. A*, 72:S27–S30, 2001.
123. F. J. Giessibl. Atomic resolution of the silicon(111)-(7 × 7) surface by atomic force microscopy. *Science*, 267:68–71, 1995.
124. M. A. Lantz, H. J. Hug, P. J. A. van Schendel, R. Hoffmann, S. Martin, A. Baratoff, A. Abdurixit, and H.-J. Güntherodt. Low temperature scanning force microscopy of the Si(111)-(7 × 7) surface. *Phys. Rev. Lett.*, 84:2642–2465, 2000.
125. K. Suzuki, H. Iwatsuki, S. Kitamura, and C. B. Mooney. Development of low temperature ultrahigh vacuum force microscope/scanning tunneling microscope. *Jpn. J. Appl. Phys.*, 39:3750–3752, 2000.
126. N. Suehira, Y. Sugawara, and S. Morita. Artifact and fact of Si(111)-(7 × 7) surface images observed with a low temperature noncontact atomic force microscope (lt-nc-afm). *Jpn. J. Appl. Phys.*, 40:292–294, 2001.
127. R. Peréz, M. C. Payne, I. Štich, and K. Terakura. Role of covalent tip-surface interactions in noncontact atomic force microscopy on reactive surfaces. *Phys. Rev. Lett.*, 78:678–681, 1997.
128. S. H. Ke, T. Uda, R. Peréz, I. Štich, and K. Terakura. First principles investigation of tip-surface interaction on GaAs(110): Implication for atomic force and tunneling microscopies. *Phys. Rev. B*, 60:11631–11638, 1999.
129. J. Tobik, I. Štich, R. Peréz, and K. Terakura. Simulation of tip-surface interactions in atomic force microscopy of an InP(110) surface with a Si tip. *Phys. Rev. B*, 60:11639–11644, 1999.
130. A. Schwarz, W. Allers, U. D. Schwarz, and R. Wiesendanger. Simultaneous imaging of the In and As sublattice on InAs(110)-(1 × 1) with dynamic scanning force microscopy. *Appl. Surf. Sci.*, 140:293–297, 1999.

131. G. Schwarz, A. Kley, J. Neugebauer, and M. Scheffler. Electronic and structural properties of vacancies on and below the GaP(110) surface. *Phys. Rev. B*, 58:1392–1499, 1998.
132. H. Hölscher, W. Allers, U. D. Schwarz, A. Schwarz, and R. Wiesendanger. Interpretation of 'true atomic resolution' images of graphite (0001) in noncontact atomic force microscopy. *Phys. Rev. B*, 62:6967–6970, 2000.
133. H. Hölscher, W. Allers, U. D. Schwarz, A. Schwarz, and R. Wiesendanger. Simulation of nc-afm images of xenon(111). *Appl. Phys. A*, 72:S35–S38, 2001.
134. H. Hölscher, W. Allers, U. D. Schwarz, A. Schwarz, and R. Wiesendanger. Determination of tip-sample interaction potentials by dynamic force spectroscopy. *Phys. Rev. Lett.*, 83:4780–4783, 1999.
135. H. Hölscher, U. D. Schwarz, and R. Wiesendanger. Calculation of the frequency shift in dynamic force microscopy. *Appl. Surf. Sci.*, 140:344–351, 1999.
136. B. Gotsman, B. Anczykowski, C. Seidel, and H. Fuchs. Determination of tip-sample interaction forces from measured dynamic force spectroscopy curves. *Appl. Surf. Sci.*, 140:314–319, 1999.
137. U. Dürig. Extracting interaction forces and complementary observables in dynamic probe microscopy. *Appl. Phys. Lett.*, 76:1203–1205, 2000.
138. M. A. Lantz, H. J. Hug, R. Hoffmann, P. J. A. van Schendel, P. Kappenberger, S. Martin, A. Baratoff, and H.-J. Güntherodt. Quantitative measurement of short-range chemical bonding forces. *Science*, 291:2580–2583, 2001.
139. S. M. Langkat, H. Hölscher, A. Schwarz, and R. Wiesendanger. Determination of site specific forces between an iron coated tip and the NiO(001) surface by force field spectroscopy. *Surf. Sci.*, 2002.
140. H. Hölscher, S. M. Langkat, A. Schwarz, and R. Wiesendanger. Measurement of three-dimensional force fields with atomic resolution using dynamic force spectroscopy. *Appl. Phys. Lett.*, 2002.
141. B. C. Stipe, H. J. Mamin, T. D. Stowe, T. W. Kenny, and D. Rugar. Noncontact friction and force fluctuations between closely spaced bodies. *Phys. Rev. Lett.*, 87, 2001.
142. C. Sommerhalter, T. W. Matthes, T. Glatzel, A. Jäger-Waldau, and M. C. Lux-Steiner. High-sensitivity quantitative kelvin probe microscopy by noncontact ultra-high-vacuum atomic force microscopy. *Appl. Phys. Lett.*, 75:286–288, 1999.
143. A. Schwarz, W. Allers, U. D. Schwarz, and R. Wiesendanger. Dynamic mode scanning force microscopy of n-InAs(110)-(1 × 1) at low temperatures. *Phys. Rev. B*, 62:13617–13622, 2000.
144. K. L. McCormick, M. T. Woodside, M. Huang, M. Wu, P. L. McEuen, C. Duruoz, and J. S. Harris. Scanned potential microscopy of edge and bulk currents in the quantum hall regime. *Phys. Rev. B*, 59:4656–4657, 1999.
145. P. Weitz, E. Ahlswede, J. Weis, K. v. Klitzing, and K. Eberl. Hall-potential investigations under quantum hall conditions using scanning force microscopy. *Physica E*, 6:247–250, 2000.
146. E. Ahlswede, P. Weitz, J. Weis, K. v. Klitzing, and K. Eberl. Hall potential profiles in the quantum hall regime measured by a scanning force microscope. *Physica B*, 298:562–566, 2001.
147. M. T. Woodside, C. Vale, P. L. McEuen, C. Kadow, K. D. Maranowski, and A. C. Gossard. Imaging interedge-state scattering centers in the quantum hall regime. *Phys. Rev. B*, 64, 2001.
148. K. Moloni, B. M. Moskowitz, and E. D. Dahlberg. Domain structures in single crystal magnetite below the verwey transition as observed with a low-temperature magnetic force microscope. *Geophys. Res. Lett.*, 23:2851–2854, 1996.

149. Q. Lu, C. C. Chen, and A. de Lozanne. Observation of magnetic domain behavior in colossal magnetoresistive materials with a magnetic force microscope. *Science*, 276:2006–2008, 1997.
150. G. Xiao, J. H. Ross, A. Parasiris, K. D. D. Rathnayaka, and D. G. Naugle. Low-temperature mfm studies of cmr manganites. *Physica C*, 341–348:769–770, 2000.
151. M. Liebmann, U. Kaiser, A. Schwarz, R. Wiesendanger, U. H. Pi, T. W. Noh, Z. G. Khim, and D. W. Kim. Domain nucleation and growth of $La_{07}Ca_{0.3}MnO_{3-\delta}$/$LaAlO_3$ films studied by low temperature mfm. *J. Appl. Phys.*, 93:8319–8321, 2003.
152. A. Moser, H. J. Hug, I. Parashikov, B. Stiefel, O. Fritz, H. Thomas, A. Baratoff, H. J. Güntherodt, and P. Chaudhari. Observation of single vortices condensed into a vortex-glass phase by magnetic force microscopy. *Phys. Rev. Lett.*, 74:1847–1850, 1995.
153. C. W. Yuan, Z. Zheng, A. L. de Lozanne, M. Tortonese, D. A. Rudman, and J. N. Eckstein. Vortex images in thin films of $YBa_2Cu_3O_{7-x}$ and $Bi_2Sr_2Ca_1Cu_2O_{8-x}$ obtained by low-temperature magnetic force microscopy. *J. Vac. Sci. Technol. B*, 14:1210–1213, 1996.
154. A. Volodin, K. Temst, C. van Haesendonck, and Y. Bruynseraede. Observation of the abrikosov vortex lattice in $NbSe_2$ with magnetic force microscopy. *Appl. Phys. Lett.*, 73:1134–1136, 1998.
155. A. Moser, H. J. Hug, B. Stiefel, and H. J. Güntherodt. Low temperature magnetic force microscopy on $YBa_2Cu_3O_{7-\delta}$ thin films. *J. Magn. Magn. Mater.*, 190:114–123, 1998.
156. A. Volodin, K. Temst, C. van Haesendonck, and Y. Bruynseraede. Imaging of vortices in conventional superconductors by magnetic force microscopy images. *Physica C*, 332:156–159, 2000.
157. M. Roseman and P. Grütter. Estimating the magnetic penetration depth using constant-height magnetic force microscopy images of vortices. *New J. Phys.*, 3:24.1–24.8, 2001.
158. A. Volodin, K. Temst, C. van Haesendonck, Y. Bruynseraede, M. I. Montero, and I. K. Schuller. Magnetic force microscopy of vortices in thin niobium films: Correlation between the vortex distribution and the thickness-dependent film morphology. *Europhys. Lett.*, 58:582–588, 2002.
159. U. H. Pi, T. W. Noh, Z. G. Khim, U. Kaiser, M. Liebmann, A. Schwarz, and R. Wiesendanger. Vortex dynamics in $Bi_2Sr_2CaCu_2O_8$ single crystal with low density columnar defects studied by magnetic force microscopy. *J. Low Temp. Phys.*, 131:993–1002, 2003.
160. M. Roseman, P. Grütter, A. Badia, and V. Metlushko. Flux lattice imaging of a patterned niobium thin film. *J. Appl. Phys.*, 89:6787–6789, 2001.
161. K. Nakamura, H. Hasegawa, T. Oguchi, K. Sueoka, K. Hayakawa, and K. Mukasa. First-principles calculation of the exchange interaction and the exchange force between magnetic Fe films. *Phys. Rev. B*, 56:3218–3221, 1997.
162. A. S. Foster and A. L. Shluger. Spin-contrast in non-contact afm on oxide surfaces: Theoretical modeling of NiO(001) surface. *Surf. Sci.*, 490:211–219, 2001.
163. H. Hoisoi, M. Kimura, K. Hayakawa, K. Sueoka, and K. Mukasa. Non-contact atomic force microscopy of an antiferromagnetic NiO(100) surface using a ferromagnetic tip. *Appl. Phys. A*, 72:S23–S26, 2001.
164. J. A. Sidles, J. L. Garbini, and G. P. Drobny. The theory of oscillator-coupled magnetic resonance with potential applications to molecular imaging. *Rev. Sci. Instrum.*, 63:3881–3899, 1992.
165. J. A. Sidles, J. L. Garbini, K. J. Bruland, D. Rugar, O. Züger, S. Hoen, and C. S. Yannoni. Magnetic resonance force microscopy. *Rev. Mod. Phys.*, 67:249–265, 1995.
166. D. Rugar, C. S. Yannoni, and J. A. Sidles. Mechanical detection of magnetic resonance. *Nature*, 360:563–566, 1992.

167. K. Wago, D. Botkin, O. Züger, R. Kendrick, C. S. Yannoni, and D. Rugar. Force-detected electron spin resonance: Adiabatic inversion, nutation and spin echo. *Phys. Rev. B*, 57:1108–1114, 1998.
168. D. Rugar, O. Züger, S. Hoen, C. S. Yannoni, H. M. Vieth, and R. D. Kendrick. Force detection of nuclear magnetic resonance. *Science*, 264:1560–1563, 1994.
169. Z. Zhang, P. C. Hammel, and P. E. Wigen. Observation of ferromagnetic resonance in a microscopic sample using magnetic resonance force microscopy. *Appl. Phys. Lett.*, 68:2005–2007, 1996.

6

Dynamic Force Microscopy

André Schirmeisen, Boris Anczykowski, and Harald Fuchs

Summary. This chapter presents an introduction to the concept of the dynamic operational mode of the atomic force microscope (dynamic AFM). While the static, or contact mode AFM is a widespread technique to obtain nanometer resolution images on a wide variety of surfaces, true atomic resolution imaging is routinely observed only in the dynamic mode. We will explain the jump-to-contact phenomenon encountered in static AFM and present the dynamic operational mode as a solution to overcome this effect. The dynamic force microscope is modeled as a harmonic oscillator to gain a basic understanding of the underlying physics in this mode.

Dynamic AFM comprises a whole family of operational modes. A systematic overview of the different modes typically encountered in force microscopy is presented, and special care is taken to explain the distinct features of each mode. Two modes of operation dominate the application of dynamic AFM. First, the amplitude modulation mode (also called tapping mode) is shown to exhibit an instability, which separates the purely attractive force interaction regime from the attractive–repulsive regime. Second, the self-excitation mode is derived and its experimental realization is outlined. While the first is primarily used for imaging in air and liquid, the second dominates imaging in UHV (ultrahigh vacuum) for atomic resolution imaging. In particular, we explain the influence of different forces on spectroscopy curves obtained in dynamic force microscopy. A quantitative link between the measurement values and the interaction forces is established.

Force microscopy in air suffers from small quality factors of the force sensor (i.e., the cantilever beam), which are shown to limit the achievable resolution. Also, the above mentioned instability in amplitude modulation mode often hinders imaging of soft and fragile samples. A combination of the amplitude modulation with the self-excitation mode is shown to increase the quality, or Q-factor, and extend the regime of stable operation, making the so-called Q-control module a valuable tool. Apart from the advantages of dynamic force microscopy as a nondestructive, high-resolution imaging method, it can also be used to obtain information about energy dissipation phenomena at the nanometer scale. This measurement channel can provide crucial information on electric and magnetic surface properties. Even atomic resolution imaging has been obtained in the dissipation mode. Therefore, in the last section, the quantitative relation between the experimental measurement values and the dissipated power is derived.

6.1 Motivation: Measurement of a Single Atomic Bond

The direct measurement of the force interaction between two distinct molecules has been the motivation for scientists for many years now. The fundamental forces responsible for the solid state of matter can be directly investigated, ultimately between defined single molecules. But it has not been until very recently that the chemical forces could be quantitatively measured for a single atomic bond [1]. How can we reliably measure forces that may be as small as one billionth of 1 N? How can we identify one single pair of atoms as the source of the force interaction?

The same mechanical principle that is used to measure the gravitational force exerted by your body weight (e.g., with the scale in your bathroom) can be employed to measure the forces between single atoms. A spring with a defined elasticity is compressed by an arbitrary force (e.g., your weight). The compression Δz of the spring (with spring constant k) is a direct measure of the force F exerted, which in the regime of elastic deformation obeys Hooke's law:

$$F = k\Delta z \:. \tag{6.1}$$

The only difference with regard to your bathroom scale is the sensitivity of the measurement. Typically springs with a stiffness of 0.1 N/m to 10 N/m are used, which will be deflected by 0.1 nm to 10 nm upon application of an interatomic force of some nN. Experimentally, a laser deflection technique is used to measure the movement of the spring. The spring is a bendable cantilever microfabricated from silicon wafers. If a sufficiently sharp tip, usually directly attached to the cantilever, is approached toward a surface within some nanometers, we can measure the interacting forces through changes in the deflected laser beam. This is a static measurement; hence it is called *static AFM*. Alternatively, the cantilever can be excited to vibrate at its resonant frequency. Under the influence of tip-sample forces the resonant frequency (and consequently also amplitude and phase) of the cantilever will change and serve as measurement parameters. This is called the *dynamic AFM*. Due to the multitude of possible operational modes, expressions like non-contact mode, intermittent-contact mode, tapping mode, FM-mode, AM-mode, self-excitation, constant-excitation, or constant-amplitude mode AFM are found in the literature, which will be systematically categorized in the following paragraphs.

In fact, the first AFMs were operated in the dynamic mode. In 1986, *Binnig, Quate* and *Gerber* presented the concept of the atomic force microscope [2]. The deflection of the cantilever with the tip was measured with sub-angstrom precision by an additional scanning tunneling microscope (STM). While the cantilever was externally oscillated close to its resonant frequency, the amplitude and phase of the oscillation were measured. If the tip is approached toward the surface, the oscillation parameters, amplitude and phase, are influenced by the tip-surface interaction, and can, therefore, be used as feedback channels. A certain set-point, for example, the amplitude, is given, and the feedback loop will adjust the tip-sample distance such that the amplitude remains constant. The controller parameter is recorded as a function of the lateral position of the tip with respect to the sample, and the scanned image essentially represents the surface topography.

What then is the difference between the static and the dynamic mode of operation for the AFM? The static deflection AFM directly gives the interaction force between tip and sample using (6.1). In the dynamic mode, we find that the resonant frequency, amplitude, and phase of the oscillation change as a consequence of the interaction forces (and also dissipative processes, as discussed in the last section).

In order to get a basic understanding of the underlying physics, it is instructive to consider a very simplified case. Assume that the vibration amplitude is small compared to the range of force interaction. Since van der Waals forces range over typical distances of 10 nm, the vibration amplitude should be less than 1 nm. Furthermore, we require that the force gradient $\partial F_{ts}/\partial z$ does not vary significantly over one oscillation cycle.

We can view the AFM setup as a coupling of two springs (see Fig. 6.1). Whereas the cantilever is represented by a spring with spring constant k, the force interaction between tip and surface can be modeled by a second spring. The derivative of the force with respect to tip-sample distance is the force gradient and represents the spring constant k_{ts} of the interaction spring. This spring constant k_{ts} is constant only with respect to one oscillation cycle, but varies with the average tip-sample distance as the probe is approached to the sample. The two springs are coupled in series. Therefore, we can write for the total spring constant of the AFM system:

$$k_{\text{total}} = k + k_{ts} = k - \frac{\partial F_{ts}}{\partial z}. \tag{6.2}$$

From the simple harmonic oscillator (neglecting any damping effects) we find that the resonant frequency ω of the system is shifted by $\Delta\omega$ from the free resonant frequency ω_0 due to the force interaction:

$$\omega^2 = (\omega_0 + \Delta\omega)^2$$
$$= k_{\text{total}}/m^* = \left(k + \frac{\partial F_{ts}}{\partial z}\right)/m^*. \tag{6.3}$$

Here m^* represents the effective mass of the cantilever. A detailed analysis of how m^* is related to the geometry and total mass of the cantilever can be found in the literature [3]. In the approximation that $\Delta\omega$ is much smaller than ω_0, we can write:

$$\frac{\Delta\omega}{\omega_0} \cong -\frac{1}{2k}\frac{\partial F_{ts}}{\partial z}. \tag{6.4}$$

Therefore, we find that the frequency shift of the cantilever resonance is proportional to the force gradient of the tip-sample interaction.

Although the above consideration is based on a very simplified model, it shows qualitatively that in dynamic force microscopy we will find that the oscillation frequency depends on the force gradient, while static force microscopy measures the force itself. In principle, we can calculate the force curve from the force gradient and vice versa (neglecting a constant offset). It seems, therefore, that the two methods are equivalent, and our choice will depend on whether we can measure the beam deflection or the frequency shift with better precision at the cost of technical effort.

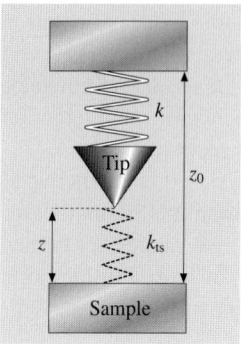

Fig. 6.1. Model of the AFM tip while experiencing tip-sample forces. The tip is attached to a cantilever with spring constant k, and the force interaction is modeled by a spring with a stiffness k_{ts} equal to the force gradient. Note that the force interaction spring is not constant, but depends on the tip-sample distance z

However, we have neglected one important issue for the operation of the AFM thus far: the mechanical stability of the measurement. In static AFM, the tip is slowly approached toward the surface. The force between the tip and the surface will always be counteracted by the restoring force of the cantilever. In Fig. 6.2, you can see a typical force-distance curve. Upon approach of the tip toward the sample, the negative attractive forces, representing, e.g., van der Waals or chemical interaction forces, increase until a maximum is reached. This turnaround point is due to the onset of repulsive forces caused by coulomb repulsion, which will start to dominate upon further approach. The spring constant of the cantilever is represented by the slope of the straight line. The position of the z-transducer (typically a piezo element), which moves the probe, is at the intersection of the line with the horizontal axis. The position of the tip, shifted from the probe's base due to the lever bending, can be found at the intersection of the cantilever line with the force curve. Hence, the total force is zero, i.e., the cantilever is in its equilibrium position (note that the spring constant line here shows attractive forces, although in reality the forces are repulsive, i.e., pulling back the tip from the surface). As soon as the position A in Fig. 6.2 is reached, we find two possible intersection points, and upon further approach there are even three force equilibrium points. However, between point A to B the tip is at a local energy minimum and, therefore, will still follow the force curve. But at point B, when the adhesion force upon further approach would become larger than the spring restoring force, the tip will suddenly jump to point C. We can then probe the predominantly repulsive force interaction by further reducing the tip-sample distance. Then, while retracting the tip, we will pass point C, because the tip is still in a local energy minimum. At position D the tip will jump suddenly to point A again, since the restoring force now exceeds the adhesion. From Fig. 6.2 we can see that the sudden instability will happen at exactly the point where the slope of the adhesion force exceeds the slope of the spring constant. Therefore, if the negative force gradient of the tip-sample interaction will at any point exceed the spring con-

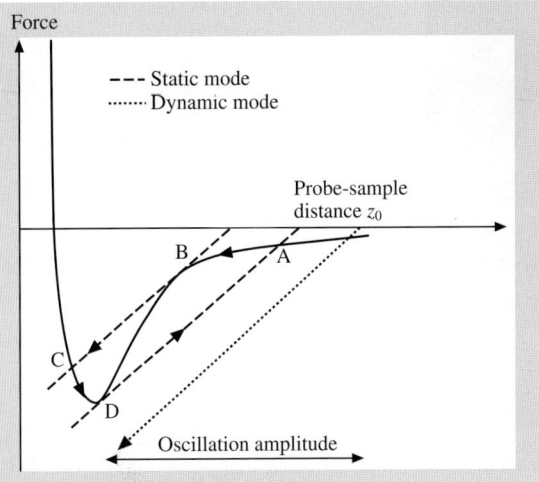

Fig. 6.2. Force-distance curve of a typical tip-sample interaction. In static mode AFM, the tip would follow the force curve until point B is reached. If the slope of the force curve becomes larger than the spring constant of the cantilever (*dashed line*), the tip will suddenly jump to position C. Upon retraction, a different path will be followed along D and A again. In dynamic AFM, the cantilever oscillates with amplitude A. Although the equilibrium position of the oscillation is far from the surface, the tip will experience the maximum attractive force at point D during some parts of the oscillation cycle. However, the total force is always pointing away from the surface, therefore, avoiding an instability

stant, a mechanical instability occurs. Mathematically speaking, we demand that for a stable measurement:

$$\left.\frac{\partial F_{ts}}{\partial z}\right|_z > k \quad \text{for all points } z \,. \tag{6.5}$$

The phenomenon of mechanical instability is often referred to as the "jump-to-contact".

Looking at Fig. 6.2, we realize that large parts of the force curve cannot be measured if the jump-to-contact phenomenon occurs. We will not be able to measure the point at which the attractive forces reach their maximum, representing the temporary chemical bonding of the tip and the surface atoms. Secondly, the sudden instability, the jump-to-contact, will often cause the tip to change the very last tip or surface atoms. A smooth, careful approach needed to measure the full force curve does not seem feasible. Our goal of measuring the chemical interaction forces of two single molecules may become impossible.

There are several solutions to the jump-to-contact problem: On the one hand, we can simply choose a sufficiently stiff spring, so that (6.5) is fulfilled at all points of the force curve. On the other hand, we can resort to a trick to enhance the counteracting force of the cantilever: We can oscillate the cantilever with large amplitude, thereby making it virtually stiffer at the point of strong force interaction.

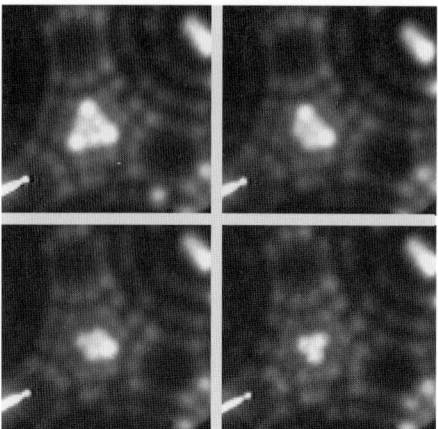

Fig. 6.3. Manipulation of the apex atoms of an AFM tip using field ion microscopy (FIM). Images are obtained at a tip bias of 4.5 kV. The last six atoms of the tip can be inspected in this example. Field evaporation to remove single atoms is performed by increasing the bias voltage for a short time to 5.2 kV. Each of the outer three atoms can be consecutively removed, eventually leaving a trimer tip apex

Consider the first solution, which seems simpler at first glance. Chemical bonding forces extend over a distance range of about 0.1 nm. Typical binding energies of a couple of eV will lead to adhesion forces on the order of some nN. Force gradients will, therefore, reach values of some 10 N/m. A spring for stable force measurements will have to be as stiff as 100 N/m to ensure that no instability occurs (a safety factor of ten seems to be a minimum requirement, since usually one cannot be sure a priori that only one atom will dominate the interaction). In order to measure the nN interaction force, a static cantilever deflection of 0.01 nm has to be detected. With standard beam deflection AFM setups this becomes a challenging task.

This problem was solved [4, 5] using an in situ optical interferometer measuring the beam deflection at liquid nitrogen temperature in a UHV environment. In order to ensure that the force gradients are smaller than the lever spring constant (50 N/m), the tips were fabricated to terminate in only three atoms, therefore, minimizing the total force interaction. The only tool known to be able to engineer SPM tips down to atomic dimensions is the field ion microscope (FIM). This technique not only allows imaging of the tip apex with atomic precision, but also can be used to manipulate the tip atoms by field evaporation [6], as shown in Fig. 6.3. Atomic interaction forces were measured with sub-nanonewton precision, revealing force curves of only a few atoms interacting without mechanical hysteresis. However, the technical effort to achieve this type of measurement is considerable, and most researchers today have resorted to the second solution.

The alternative solution can be visualized in Fig. 6.2. The straight, dashed line now represents the force values of the oscillating cantilever, with amplitude A assuming Hooke's law is valid. This is the tensile force of the cantilever spring pulling the tip away from the sample. The restoring force of the cantilever is at all points stronger than the adhesion force. For example, the total force at point D is still pointing away from the sample, although the spring has the same stiffness as before. Mathematically speaking, the measurement is stable as long as the cantilever spring force $F_{cb} = kA$ is larger than the attractive tip-sample force F_{ts} [7]. In the static mode we would already experience an instability at that point. However, in the dynamic mode, the spring is pre-loaded with a force stronger than the attractive tip-sample force. The equilibrium point of the oscillation is still far away from the point of closest contact of the tip and surface atoms. The total force curve can be probed by varying the equilibrium point of the oscillation, i.e., by adjusting the z-piezo.

The diagram also shows that the oscillation amplitude has to be quite large if fairly soft cantilevers are to be used. With lever spring constants of 10 N/m, the amplitude must be at least 1 nm to ensure forces of 1 nN can be reliably measured. In practical applications, amplitudes of 10–100 nm are used to stay on the safe side. This means that the oscillation amplitude is much larger than the force interaction range. The above simplification, that the force gradient remains constant within one oscillation cycle, does not hold anymore. Measurement stability is gained at the cost of a simple quantitative analysis of the experiments. In fact, dynamic AFM was first used to obtain atomic resolution images of clean surfaces [8], and it took another six years [1] before quantitative measurements of single bond forces were obtained.

The technical realization of dynamic mode AFMs is based on the same key components as a DC-AFM setup. The most common principle is the method of laser deflection sensing (see, e.g., Fig. 6.4). A laser beam is focused on the back side of a microfabricated cantilever. The reflected laser spot is detected with a positional sensitive diode (PSD). This photodiode is sectioned into two parts that are read-out separately (usually even a four-quadrant diode is used to detect torsional movements of the cantilever for lateral friction measurements). With the cantilever at equilibrium, the spot is adjusted such that the two sections show the same intensity. If the cantilever bends up or down, the spot moves, and the difference signal between the upper and lower sections is a measure of the bending.

In order to enhance sensitivity, several groups have adopted an interferometer system to measure the cantilever deflection. A thorough comparison of different measurement methods with analysis of sensitivity and noise level is given by [3].

The cantilever is mounted on a device that allows oscillation of the beam. Often a piezo-element serves this purpose. The reflected laser beam is analyzed for oscillation amplitude, frequency, and phase difference between excitation and vibration. Depending on the mode of operation, a feedback mechanism will adjust oscillation parameters and/or tip-sample distance during the scanning. The setup can be operated in air, UHV, and even in fluids. This allows measurement of a wide range of surface properties from atomic resolution imaging [8] up to studying biological processes in liquid [9, 10].

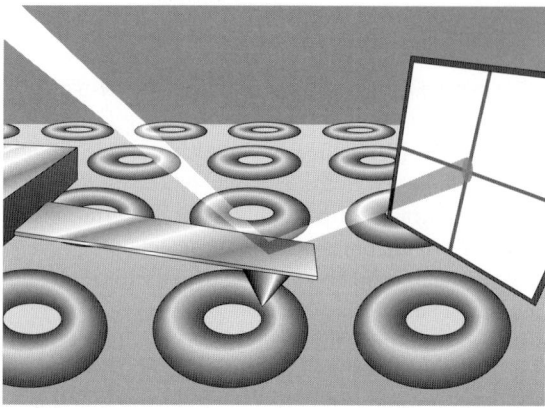

Fig. 6.4. Representation of an AFM setup with the laser beam deflection method. Cantilever and tip are microfabricated from silicon wafers. A laser beam is deflected from the back side of the cantilever and again focused on a photosensitive diode. The diode is segmented into four quadrants, which allows the measurement of vertical and torsional bending of the cantilever (artwork from *Jörg Heimel* rendered with POV-Ray 3.0)

6.2 Harmonic Oscillator: A Model System for Dynamic AFM

The oscillating cantilever has three degrees of freedom: the amplitude, the frequency, and the phase difference between excitation and oscillation. Let us consider the damped driven harmonic oscillator. The cantilever is mounted on a piezoelectric element that is oscillating with amplitude A_d at frequency ω:

$$z_d(t) = A_d \cos(\omega t) \,. \tag{6.6}$$

We assume that the cantilever spring obeys Hooke's law. Secondly, we introduce a friction force that is proportional to the speed of the cantilever motion, whereas α denotes the damping coefficient (Amontons's law). With Newton's first law we find for the oscillating system the following equation of motion for the position $z(t)$ of the cantilever tip (see also Fig. 6.1):

$$m\ddot{z}(t) = -\alpha \dot{z}(t) - kz(t) - kz_d(t) \,. \tag{6.7}$$

We define $\omega_0^2 = k/m^*$, which turns out to be the resonant frequency of the free (undamped, i.e., $\alpha = 0$) oscillating beam. We further define the dimensionless quality factor $Q = m^*\omega_0/\alpha$, antiproportional to the damping coefficient. The quality factor describes the number of oscillation cycles, after which the damped oscillation amplitude decays to $1/e$ of the initial amplitude with no external excitation ($A_d = 0$). After some basic math, this results in the following differential equation:

$$\ddot{z}(t) + \frac{\omega_0}{Q}\dot{z}(t) + \omega_0^2 z(t) = A_d \omega_0^2 \cos(\omega t) \,. \tag{6.8}$$

The solution is a linear combination of two regimes [11]. Starting from rest and switching on the piezo-excitation at $t = 0$, the amplitude will increase from zero to the final magnitude and reach a steady state, where amplitude, phase, and frequency of the oscillation stay constant over time. The steady-state solution $z_1(t)$ is reached after $2Q$ oscillation cycles and follows the external excitation with amplitude A_0 and phase difference φ:

$$z_1(t) = A_0 \cos(\omega t + \varphi) \,. \tag{6.9}$$

The oscillation amplitude in the transient regime during the first $2Q$ cycles follows:

$$z_2(t) = A_t e^{-\omega_0 t/2Q} \sin(\omega_0 t + \varphi_t) \,. \tag{6.10}$$

We emphasize the important fact that the exponential term causes $z_2(t)$ to diminish exponentially with time constant τ:

$$\tau = 2Q/\omega_0 \,. \tag{6.11}$$

In vacuum conditions, only the internal dissipation due to bending of the cantilever is present, and Q reaches values of 10 000 at typical resonant frequencies of 100 000 Hz. This results in a relatively long transient regime of $\tau \cong 30$ ms, which limits the possible operational modes for dynamic AFM (detailed analysis by [11]). Changes in the measured amplitude, which reflect a change of atomic forces, will have a time lag of 30 ms, which is very slow considering one wants to scan a 200×200 point image within a few minutes. In air, however, viscous damping due to air friction dominates and Q goes down to less than 1000, resulting in a time constant below the millisecond level. This response time is fast enough to use the amplitude as a measurement parameter.

If we evaluate the steady state solution $z_1(t)$ in the differential equation, we find the following well-known solution for amplitude and phase of the oscillation as a function of the excitation frequency ω:

$$A_0 = \frac{A_d Q \omega_0^2}{\sqrt{\omega^2 \omega_0^2 + Q^2 \left(\omega_0^2 - \omega^2\right)^2}} \tag{6.12}$$

$$\varphi = \arctan\left(\frac{\omega \omega_0}{Q\left(\omega_0^2 - \omega^2\right)}\right) \,. \tag{6.13}$$

Amplitude and phase diagrams are depicted in Fig. 6.9. As can be seen from (6.12), the amplitude will reach its maximum at a frequency different from ω_0, if Q has a finite value. The damping term of the harmonic oscillator causes the resonant frequency to shift from ω_0 to ω_0^*:

$$\omega_0^* = \omega_0 \sqrt{1 - \frac{1}{2Q^2}} \,. \tag{6.14}$$

The shift is negligible for Q-factors of 100 and above, which is the case for most applications in vacuum or air. However, for measurements in liquids, Q can be smaller than 10 and ω_0 differs significantly from ω_0^*. As we will discuss later, it is also possible to enhance Q by using a special excitation method.

In the case that the excitation frequency is equal to the resonant frequency of the undamped cantilever, $\omega = \omega_0$, we find the useful relation:

$$A_0 = QA_d \quad \text{for } \omega = \omega_0 . \tag{6.15}$$

Since $\omega_0^* \approx \omega_0$ for most cases, we find that (6.15) holds true for exciting the cantilever at its resonance. From a similar argument, the phase becomes approximately 90 deg for the resonance case. We also see that in order to reach vibration amplitudes of some 10 nm, the excitation only has to be as small as 1 pm for typical cantilevers operated in vacuum.

So far, we have not considered an additional force term, describing the interaction between the probing tip and the sample. For typical, large vibration amplitudes of 10–100 nm no general solution for this analytical problem has been found yet. The cantilever tip will experience a whole range of force interactions during one single oscillation cycle, rather than one defined tip-sample force. Only in the special case of a self-excited cantilever oscillation has the problem been solved semi-analytically, as we will see later.

6.3 Dynamic AFM Operational Modes

While the quantitative interpretation of force curves in contact AFM is straightforward using (6.1), we explained in the previous paragraphs that its application to assess short-range attractive interatomic forces is rather limited. The dynamic mode of operation seems to open a viable direction toward achieving this task. However interpretation of the measurements generally appears to be more difficult. Different operational modes are employed in dynamic AFM, and the following paragraphs are intended to distinguish these modes and categorize them in a systematic way.

The oscillation trajectory of a dynamically driven cantilever is determined by three parameters: the amplitude, the phase, and the frequency. Tip-sample interactions can influence all three parameters, in the following, termed the internal parameters. The oscillation is driven externally, with excitation amplitude A_d and excitation frequency ω. These variables will be referred to as the external parameters. The external parameters are set by the experimentalist, whereas the internal parameters are measured and contain the crucial information about the force interaction. In scanning probe applications, it is common to control the probe-surface distance z_0 in order to keep an internal parameter constant (i.e., the tunneling current in STM or the beam deflection in contact AFM), which represents a certain tip-sample interaction. In z-spectroscopy mode, the distance is varied in a certain range, and the change of the internal parameters are measured as a fingerprint of the tip-sample interactions.

In dynamic AFM the situation is rather complex. Any of the internal parameters can be used for feedback of the tip-sample distance z_0. However, we already realized that, in general, the tip-sample forces could only be fully assessed by measuring all three parameters. This makes it difficult to obtain images where the distance z_0 is representative of the surface at one force set-point. A solution to this problem is to establish additional feedback loops, which keep the internal parameters constant by adjusting the external variables.

In the simplest setup, the excitation frequency is set to a predefined value, and the excitation amplitude remains constant by a feedback loop. This is called the AM mode (amplitude modulation), or tapping mode (TM). As stated before, in principle, any of the internal parameters can be used for feedback to the tip-sample distance – in AM mode the amplitude signal is used. A certain amplitude (smaller than the free oscillation amplitude) at a frequency close to the resonance of the cantilever is chosen, the tip is approached toward the surface under investigation, and the approach is stopped as soon as the set-point amplitude is reached. The oscillation-phase is usually recorded during the scan, however, the shift of the resonant frequency of the cantilever cannot be directly accessed, since this degree of freedom is blocked by the external excitation at a fixed frequency. It turns out that this mode is simple to operate from a technical perspective, but quantitative information about the tip-sample interaction forces has so far not been reliably extracted from AM-mode AFM. Still, it is one of the most commonly used modes in dynamic AFM operated in air, and even in liquid. The strength of this mode is the qualitative imaging of a large variety of surfaces.

It is interesting to discuss the AM mode in the extreme situation that the external excitation frequency is much lower than the resonant frequency [12, 13]. This results in a quasi-static measurement, although a dynamic oscillation force is applied, and, therefore, this mode can be viewed as a hybrid between static and dynamic AFM. Unfortunately, it has the drawbacks of the static mode, namely, that stiff spring constants must be used and, therefore, the sensitivity of the deflection measurement must be very good, typically employing a high resolution interferometer. Still, it has the advantage of the static measurement in quantitative interpretation, since in the regime of small amplitudes (< 0.1 nm) a direct interpretation of the experiments is possible. In particular, the force gradient at tip-sample distance z_0 is given by the change of the amplitude A and the phase angle φ:

$$\left.\frac{\partial F_{ts}}{\partial z}\right|_{z_0} = k\left(1 - \frac{A_0}{A}\cos\varphi\right) . \tag{6.16}$$

In effect, the modulated AFM, in contrast to the purely static AFM, can enhance sensitivity due to the use of lock-in techniques, which allows the measurement of the amplitude and phase of the oscillation signal with high precision.

As stated before, the internal parameters can be fed back to the external excitation variables. One of the most useful applications in this direction is the self-excitation system. Here the resonant frequency of the cantilever is detected and selected againas the excitation frequency. In a typical setup, this is done with a phase shift of 90 deg by feeding back the detector signal to the excitation piezo, i.e., the cantilever is always excited in resonance. Influences of the tip-sample interaction forces on the resonant frequency do not change the two other parameters of the oscillation, i.e., amplitude and phase; only the oscillation frequency is shifted. Therefore, it is sufficient to measure the frequency shift between the free oscillation and the oscillation with tip-sample interaction. Since the phase remains at a fixed value, the oscillating system is much better defined than before, and the degrees of freedom for the oscillation are reduced. To even reduce the last degree of freedom, the oscillation amplitude, another feedback loop can be established to keep the oscillation amplitude A constant by varying the excitation amplitude A_d. Now, all internal parameters have a fixed relation to the external excitation variables, the system is well-defined, and all parameters can be assessed during the measurement. As it turns out, this mode is the only dynamic mode in which a quantitative relation between tip-sample forces and the change of the resonant frequency can be established.

In the following section we want to discuss the two most popular operational modes, tapping mode and self-excitation mode, in more detail.

6.3.1 Amplitude-Modulation/ Tapping-Mode AFMs

In tapping mode, or AM-AFM, the cantilever is excited externally at a constant frequency close to its resonance. Oscillation amplitude and phase during approach of tip and sample serve as the experimental observation channels. Figure 6.5 shows a diagram of a typical tapping-mode AFM setup. The oscillation amplitude and the phase (not shown in diagram) detected with the photodiode are analyzed with a lock-in amplifier. The amplitude is compared to the set-point, and the difference or error signal is used to adjust the z-piezo, i.e., the probe-sample distance. The external modulation unit supplies the signal for the excitation piezo, and, at the same time, the oscillation signal serves as the reference for the lock-in amplifier.

During one oscillation cycle with amplitudes of 10–100 nm, the tip-sample interaction will range over a wide distribution of forces, including attractive, as well as repulsive forces. We will, therefore, measure a convolution of the force-distance curve with the oscillation trajectory. This complicates the interpretation of AM-AFM measurements appreciably.

At the same time, the resonant frequency of the cantilever will change due to the appearing force gradients, as could already be seen in the simplified model from (6.4). If the cantilever is excited exactly at its resonant frequency before, it will be excited off-resonance after interaction forces are encountered. This, in turn, changes amplitude and phase (6.12, 6.13), which serve as the measurement signals. Consequently, a different amplitude will cause a change in the encountered effective force. We can see already from this simple *gedanken*-experiment that the interpretation of force curves will be highly complicated. In fact, there is no quantitative theory for AM-AFM available, which allows the experimentalist to unambiguously convert the experimental data to a force-distance relationship.

The qualitative behavior for amplitude versus z_0-position curves is depicted in Fig. 6.6. At large distances, where the forces between tip and sample are negligible, the cantilever oscillates with its free oscillation amplitude. Upon approach of the probe toward the surface the interaction forces cause the amplitude to change, resulting typically in an amplitude getting smaller with continuously decreased tip-sample

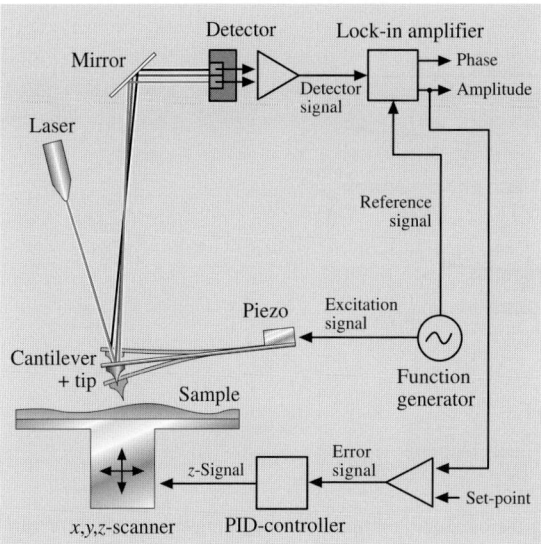

Fig. 6.5. Setup of a dynamic force microscope operated in the AM or tapping mode. A laser beam is deflected by the back side of the cantilever, and the deflection is detected by a split photodiode. The excitation frequency is chosen externally with a modulation unit, which drives the excitation piezo. A lock-in amplifier analyzes phase and amplitude of the cantilever oscillation. The amplitude is used as the feedback signal for the probe-sample distance control

distance. This is expected, since the force-distance curve will eventually reach the repulsive part and the tip is hindered from indenting further into the sample, resulting in smaller oscillation amplitudes.

However, in order to gain some qualitative insight into the complex relationship between forces and oscillation parameters, we resort to numerical simulations. *Anczykowski* et al. [14, 15] have calculated the oscillation trajectory of the cantilever under the influence of a given force model. Van der Waals interactions were considered the only effective, attractive forces, and the total interaction resembled a Lennard-Jones-type potential. Mechanical relaxations of the tip and sample surface were treated in the limits of continuum theory with the numerical MYD/BHW [16, 17] approach, which allows the simulations to be compared to corresponding experiments. Figure 6.7 shows the force-distance curves for different tip radii underlying the dynamic AFM simulations.

The cantilever trajectory was analyzed by solving the differential equation (6.7) extended by the force-distance relations from Fig. 6.7 using the numerical Verlet algorithm [18, 19]. The results of the simulation for the amplitude and phase of the tip oscillation as a function of z-position of the probe are presented in Fig. 6.8. One has to keep in mind that the z-position of the probe is not equivalent to the real tip-sample distance at equilibrium position, since the cantilever might bend statically due to the interaction forces. The behavior of the cantilever can be subdivided into three different regimes. We distinguish the cases in which the beam is oscillated below its

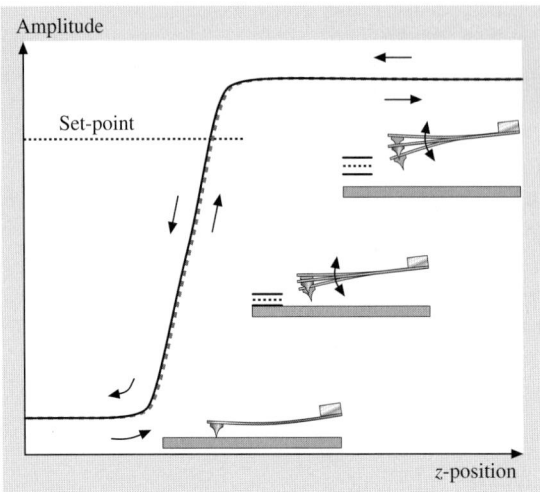

Fig. 6.6. Simplified model showing the oscillation amplitude in tapping-mode AFM for various probe-sample distances

resonant frequency ω_0, exactly at ω_0, and above ω_0. In the following, we will refer to ω_0 as the resonant frequency, although the correct resonant frequency is ω_0^* if taking into account the finite Q-value.

Clearly, Fig. 6.8 exhibits more features than were anticipated from the initial, simple arguments. Amplitude and phase seem to change rather abruptly at certain points when the z_0-position is decreased. Besides, the amplitude or phase-distance curves don't resemble the force-distance curves from Fig. 6.7 in a simple, direct manner. Additionally, we find a hysteresis between approach and retraction.

As an example, let us start by discussing the discontinuous features in the AFM spectroscopy curves of the first case, where the excitation frequency is smaller than ω_0. Consider the oscillation amplitude as a function of excitation frequency in Fig. 6.9 in conjunction with a typical force curve, as depicted in Fig. 6.7. Upon approach of probe and sample, attractive forces will lower the effective resonant frequency of the oscillator. Therefore, the excitation frequency will now be closer to the resonant frequency, causing the vibration amplitude to increase. This, in turn, reduces the tip-sample distance, which again gives rise to a stronger attractive force. The system becomes unstable until the point $z_0 = d_{\mathrm{app}}$ is reached, where repulsive forces stop the self-enhancing instability. This can be clearly observed in Fig. 6.8. Large parts of the force-distance curve cannot be measured due to this instability.

In the second case, where the excitation equals the free resonant frequency, only a small discontinuity is observed upon reduction of the z-position. Here, a shift of the resonant frequency toward smaller values, induced by the attractive force interaction, will reduce the oscillation amplitude. The distance between tip and sample is, therefore, reduced as well, and the self-amplifying effect with the sudden instability does not occur as long as repulsive forces are not encountered. However, at closer

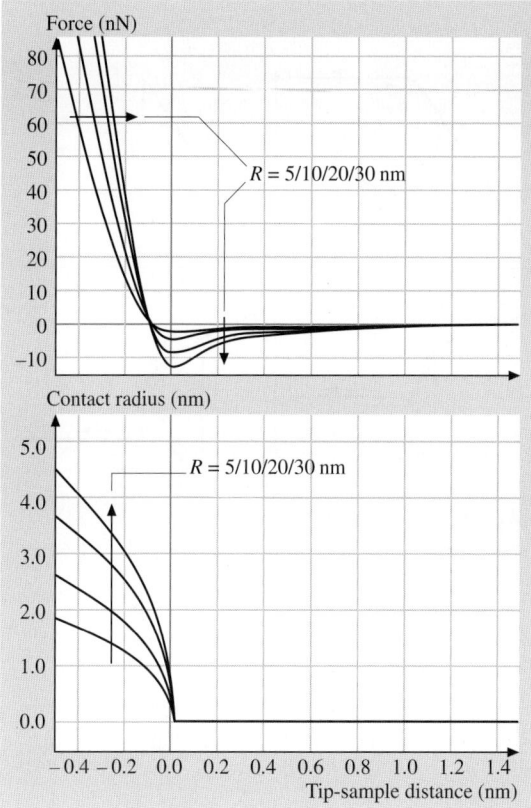

Fig. 6.7. Force curves and corresponding contact radius calculated with the MYD/BHW-model as a function of tip radius for a Si–Si contact. These force curves are used for the tapping-mode AFM simulations

tip-sample distances, repulsive forces will cause the resonant frequency to shift again toward higher values, increasing the amplitude with decreasing tip-sample distance. Therefore, a self-enhancing instability will also occur in this case, but at the crossover from purely attractive forces to the regime where repulsive forces occur. Correspondingly, a small kink in the amplitude curve can be observed in Fig. 6.8. An even clearer indication of this effect is manifested by the sudden change in the phase signal at d_{app}.

In the last case, with $\omega > \omega_0$, the effect of amplitude reduction due to the resonant frequency shift is even larger. Again, we find no instability in the amplitude signal during approach in the attractive force regime. However, as soon as the repulsive force regime is reached, the instability occurs due to the induced positive frequency shift. Consequently, a large jump in the phase curve from values smaller than 90 deg to values larger than 90 deg is observed. The small change in the amplitude curve is not resolved in the simulated curves in Fig. 6.8, however, it can be clearly seen in the experimental curves in Fig. 6.10.

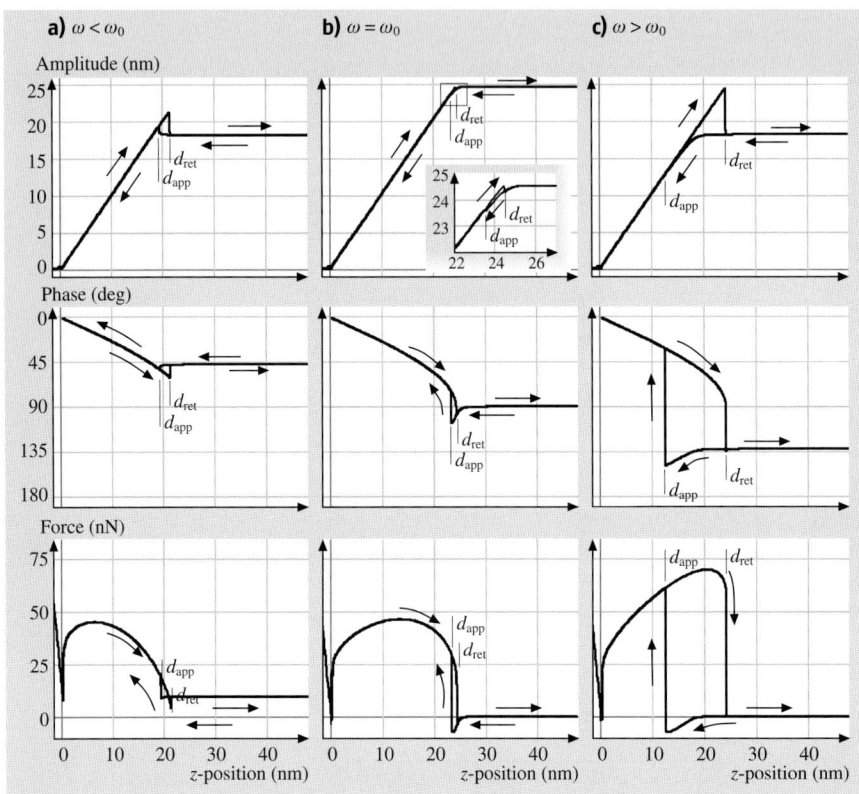

Fig. 6.8. Amplitude and phase diagrams with excitation frequency (**a**) below, (**b**) exactly at, and (**c**) above the resonant frequency for tapping-mode AFM from the numerical simulations. Additionally, the bottom diagrams show the interaction forces at the point of closest tip–sample distance, i.e., the lower turnaround point of the oscillation

Figure 6.10 depicts the corresponding experimental amplitude and phase curves. The measurements were performed in air with a Si cantilever approaching a Si wafer, with a cantilever resonant frequency of 299.95 kHz. Qualitatively, all prominent features of the simulated curves can also be found in the experimental data sets. Hence, the above model seems to capture the important factors necessary for an appropriate description of the experimental situation.

But what is the reason for this unexpected behavior? We have to turn to the numerical simulations again, where we have access to all physical parameters, in order to understand the underlying processes. The lower part of Fig. 6.8 also shows the interaction force between the tip and the sample at the point of closest approach, i.e., the sample-sided turnaround point of the oscillation. We see that exactly at the points of the discontinuities the total interaction force changes from the net-attractive regime to the attractive-repulsive regime, also termed the intermittent contact regime. The term net-attractive is used to emphasize that the total force is attractive, despite

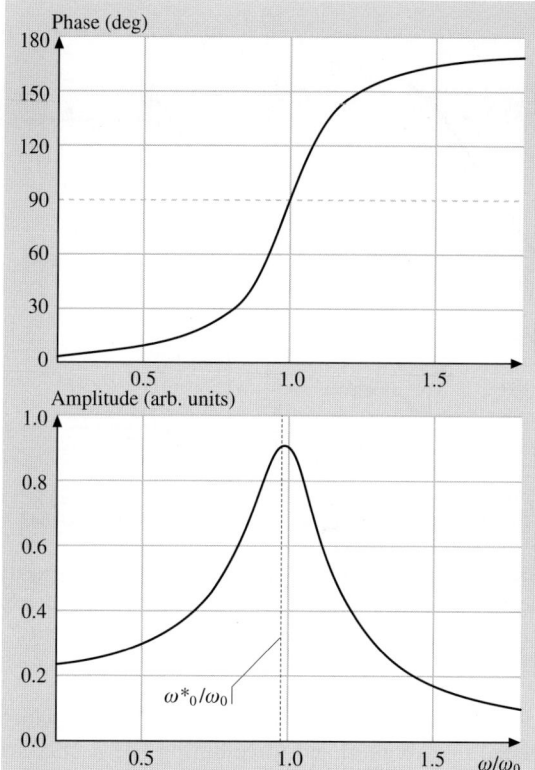

Fig. 6.9. Amplitude and phase versus excitation frequency curves for the damped harmonic oscillator with a quality factor of $Q = 4$

the fact that some minor contributions might still originate from repulsive forces. As soon as a minimum distance is reached, the tip also starts to experience repulsive forces, which completely changes the oscillation behavior. In other words, the dynamic system switches between two oscillatory states.

Directly related to this fact is the second phenomenon: the hysteresis effect. We find separate curves for the approach of the probe toward the surface and the retraction. This seems to be somewhat counterintuitive, since the tip is constantly approaching and retracting from the surface, and the average values of amplitude and phase should be independent of the direction of the average tip-sample distance movement. A hysteresis between approach and retraction within one oscillation due to dissipative processes should directly influence amplitude and phase. However, no dissipation models were included in the simulation. In this case, the hysteresis in Fig. 6.8 is due to the fact that the oscillation jumps into different modes, the system exhibits bistability. This effect is often observed in oscillators under the influence of nonlinear forces [20].

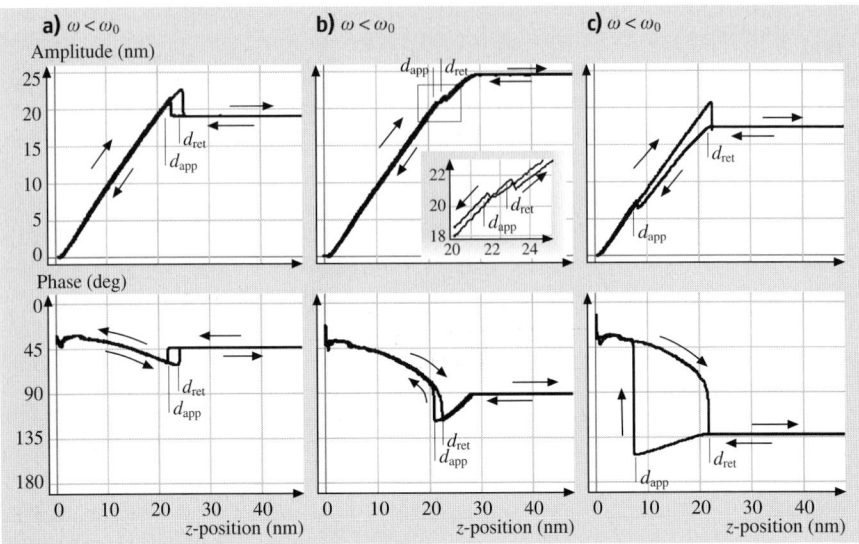

Fig. 6.10. Amplitude and phase diagrams with excitation frequency (**a**) below, (**b**) exactly at, and (**c**) above the resonant frequency for tapping-mode AFM from experiments with a Si cantilever on a Si wafer in air

For the interpretation of these effects it is helpful to look at Fig. 6.11, which shows the behavior of the simulated tip vibration and force during one oscillation cycle over time. The data is recorded at the same z-position at a point where hysteresis is observed, while a) was taken during the approach and b) during the retraction. Excitation is in resonance, representing case 2, where the amplitude shows a small hysteresis. Also note that the amplitude is almost exactly the same in a) and b). We see that the oscillation at the same z-position exhibits two different modes: While in a) the experienced force is net-attractive, in b) the tip is exposed to attractive and repulsive interactions. Experimental and simulated data show that the change between the net-attractive and intermittent contact mode takes place at different z-positions (d_{app} and d_{ret}) for approach and retraction. Between d_{app} and d_{ret} the system is in a bistable mode. Depending on the history of the measurement, e.g., whether the position d_{app} during the approach (or d_{ret} during retraction) has been reached, the system flips to the other oscillation mode. While the amplitude might not be influenced strongly (case 2), the phase is a clear indicator of the mode switch. On the other hand, if point d_{app} is never reached during the approach, the system will stay in the net-attractive regime and no hysteresis is observed, i.e., the system remains stable.

In conclusion, we find that although a qualitative interpretation of the interaction forces is possible, the AM- nor tapping-mode AFM is not suitable to gain direct quantitative knowledge of tip-sample force interactions. However, it is a very useful tool for imaging nanometer-sized structures in a wide variety of setups, in air or even in liquid. We find that two distinct modes exist for the externally excited oscillation

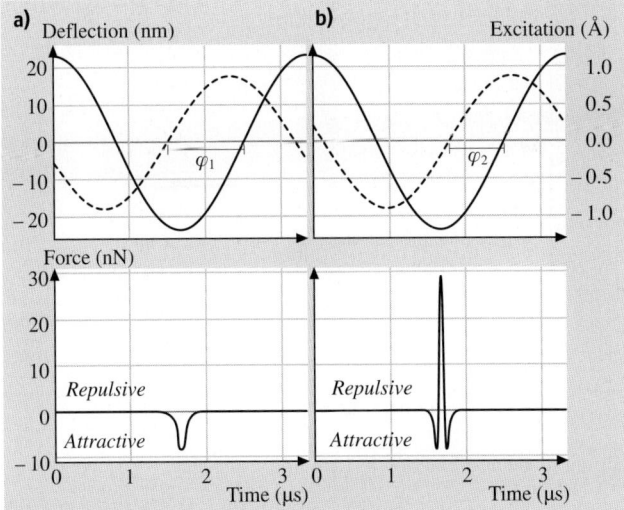

Fig. 6.11. Simulation of the tapping-mode cantilever oscillation in the (**a**) net-attractive and (**b**) the intermittent contact regime. The *dashed line* represents the excitation amplitude and the *solid line* is the oscillation amplitude

– the net-attractive and the intermittent contact mode – which describe what kind of forces the tip-sample interaction is governed by. The phase can be used as an indicator of which mode the system is running in.

In particular, it can be easily seen that if the free resonant frequency of the cantilever is higher than the excitation frequency, the system cannot stay in the net-attractive regime due to a self-enhancing instability. Since in many applications involving soft and delicate biological samples strong repulsive forces should be avoided, the tapping-mode AFM should be operated at frequencies equal to or above the free resonant frequency [21]. Even then, statistical changes of tip-sample forces during the scan might induce a sudden jump into the intermittent contact mode, and the previously explained hysteresis will tend to keep the system in this mode. It is, therefore, of high importance to tune the oscillation parameters in such a way that the AFM stays in the net-attractive regime [22]. A concept that achieves this task is the Q-control system, which will be discussed in some detail in the forthcoming paragraphs.

A last word concerning the overlap of simulation and experimental data: While the qualitative agreement down to the detailed shape of hysteresis and instabilities is rather striking, we still find some quantitative discrepancies between the positions of the instabilities d_{app} and d_{ret}. This is probably due to the simplified force model, which only takes into account van der Waals and repulsive forces. Especially at ambient conditions, an omnipresent water meniscus between tip and sample will give rise to much stronger attractive and also dissipative forces than considered in the model. A very interesting feature is that the simulated phase curves in the intermittent con-

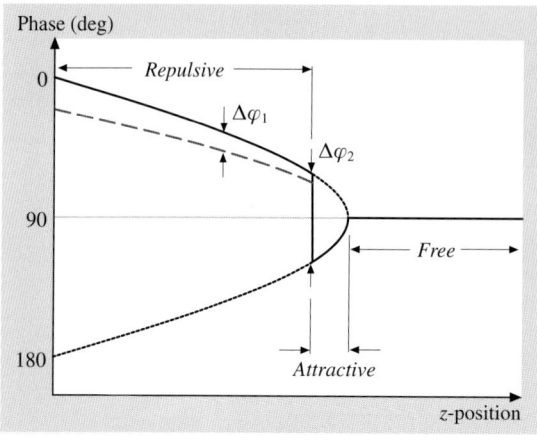

Fig. 6.12. Phase shift in tapping mode as a function of tip-sample distance

tact regime tend to have a steeper slope in the simulation than in the experiments (see also Fig. 6.12). We will later show that this effect is a fingerprint of an effect that had not been included in the above simulation at all: dissipative processes during the oscillation, giving rise to an additional loss of oscillation energy.

6.3.2 Self-Excitation Modes

Despite the wide range of technical applications of the AM or tapping mode of dynamic AFM, it has been found unsuitable for measurements in an environment extremely useful for scientific research: vacuum or ultrahigh vacuum (UHV) with pressures reaching 1×10^{-10} mbar. The STM has already shown how much insight can be gained from some highly defined experiments under those conditions. Consider (6.11) from the above section. The time constant τ for the amplitude to adjust to a different tip-sample force scales with $1/Q$. In vacuum applications, Q of the cantilever is on the order of 10 000, which means that τ is in the range of some 10 ms. This is clearly too long for a scan of at least (100×100) data points. The temperature-induced drift of the sample will render useful interpretation of images an impossible task. However, the resonant frequency of the system will react instantaneously to tip-sample forces. This has led *Albrecht* et al. [11] to use a modified excitation scheme.

The system is always oscillated at its resonant frequency. This is achieved by feeding back the oscillation signal from the cantilever into the excitation piezoelement. Figure 6.13 pictures the method in a block diagram. The signal from the PSD is phase shifted by 90 deg (and, therefore, always exciting in resonance) and used as the excitation signal of the cantilever. An additional feedback loop adjusts the excitation amplitude in such a way that the oscillation amplitude remains constant. This ensures that the tip-sample distance is not influenced by changes in the oscillation amplitude. The only degree of freedom of the oscillation system that can react to the tip-sample forces is the change of the resonant frequency. This shift of the

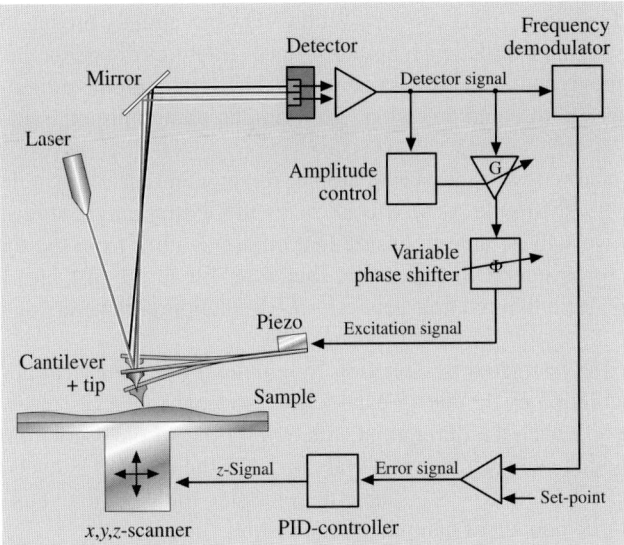

Fig. 6.13. Dynamic AFM operated in the self-excitation mode, where the oscillation signal is directly feedbacked to the excitation piezo. The detector signal is amplified with the variable gain G and phase shifted by phase ϕ. The frequency demodulator detects the frequency shift due to tip-sample interactions, which serves as the control signal for the probe-sample distance

frequency is detected and used as the set-point signal for surface scans. Therefore, this mode is also called the FM (frequency modulated) mode.

Let us take a look at the sensitivity of the dynamic AFM. If electronic noise, laser noise, and thermal drift can be neglected, the main noise contribution will come from thermal excitation of the cantilever. A detailed analysis of a dynamic system yields for the minimum detectable force gradient the following relation [11]:

$$\left.\frac{\partial F}{\partial z}\right|_{MIN} = \sqrt{\frac{4k_B T B \cdot k}{\omega_0 Q \langle z_{osc}^2 \rangle}} \;. \tag{6.17}$$

Here B is the bandwidth of the measurement, T the temperature, and $\langle z_{osc}^2 \rangle$ is the mean-square amplitude of the oscillation. Please note that this sensitivity limit was deliberately calculated for the FM mode. A similar analysis of the AM mode, however, yields virtually the same result [23]. We find that the minimum detectable force gradient, i.e., the measurement sensitivity, is inversely proportional to the square root of the Q-factor of the cantilever. This means that it should be possible to achieve very high-resolution imaging in vacuum conditions. In contrast, AM- or tapping-mode AFM cannot be usefully pursued with large Q. Only the FM mode makes it possible to take practical advantage of (6.17).

A breakthrough in high-resolution AFM application was the atomic resolution imaging of the Si(111)-(7×7) surface reconstruction by Giessibl [8] under UHV conditions. Today, atomic resolution imaging has become a standard feature guaranteed by industrially produced dynamic AFM systems. While STM has already proven to be an indispensable tool to gain detailed insight into surface structures of conductors with atomic resolution, the dynamic AFM has opened up the avenue into investigating nonconductive surfaces with equal precision (for example, on aluminium oxide by *Barth* et al. [24]).

However, we are concerned with measuring atomic force potentials of a single pair of molecules. Clearly, FM-mode AFM will allow us to identify single atoms, and with sufficient care we will be able to ensure that only one atom from the tip contributes to the total force interaction. Can we, therefore, fill in the last bit of information and find a quantitative relation between the oscillation parameters and the force?

Gotsmann et al. [25] investigated this relation by employing a numerical approach. During each oscillation cycle the tip experiences a whole range of forces. For each step during the approach the differential equation for the whole oscillation loop (including also the feedback system) was evaluated and the relation between force and frequency shift revealed. It was possible to determine the quantitative interaction forces of a metallic contact of nanometer dimensions.

However, it was shown that there also exists an analytical relationship if some approximations are accepted [7, 26, 27]. Here, we will follow the route indicated by *Dürig* [27], although different alternative ways have also proven successful. Consider the tip oscillation trajectory reaching over a large part of the force gradient curve in Fig. 6.2. We model the tip-sample interaction as a spring constant of stiffness $k_{ts}(z) = \partial F/\partial z|_{z_0}$, as in Fig. 6.1. For small oscillation amplitudes, we already found that the frequency shift is proportional to the force gradient in (6.4). For large amplitudes, we can calculate an effective force gradient k_{eff} as a convolution of the force and the fraction of time the tip spends between the positions x and $x + dx$:

$$k_{eff}(z) = \frac{2}{\pi A^2} \int_z^{z+2A} F(x) g\left(\frac{x-z}{A} - 1\right) dx \tag{6.18}$$

$$\text{with } g(u) = -\frac{u}{\sqrt{1-u^2}}.$$

In the approximation that the vibration amplitude is much larger than the range of the tip-sample forces, the above equation can be simplified to:

$$k_{eff}(z) = \frac{\sqrt{2}}{\pi} A^{3/2} \int_z^\infty \frac{F(x)}{\sqrt{x-z}} dx. \tag{6.19}$$

This effective force gradient can now be used in (6.4), the relation between frequency shift and force gradient. We find:

$$\Delta f = \frac{f_0}{\sqrt{2}\pi k A^{3/2}} \int_z^\infty \frac{F(x)}{\sqrt{x-z}} dx. \tag{6.20}$$

If we separate the integral from other parameters, we can define:

$$\Delta f = \frac{f_0}{k A^{3/2}} \gamma(z)$$

$$\text{with } \gamma(z) = \frac{1}{\sqrt{2\pi}} \int_z^\infty \frac{F(x)}{\sqrt{x-z}} \, dx \, . \tag{6.21}$$

This means we can define $\gamma(z)$, which is only dependent on the shape of the force curve $F(z)$ but independent of the external parameters of the oscillation. The function $\gamma(z)$ is also referred to as the "normalized frequency shift" [7], a very useful parameter, which allows us to compare measurements independent of resonant frequency, amplitude and spring constant of the cantilever.

The dependence of the frequency shift on the vibration amplitude is an especially useful relation, since this parameter can be easily varied during one experiment. A nice example is depicted in Fig. 6.14, where frequency shift curves for different amplitudes were found to coincide very well in the $\gamma(z)$ diagrams [28].

This relationship has been nicely exploited for the calibration of the vibration amplitude by *Guggisberg* [29], which is a problem often encountered in dynamic AFM operation and worthwhile discussing. One approaches tip and sample and takes frequency shift curves versus distance, which show a reproducible shape. Then, the z-feedback is permanently turned off, and several curves with different amplitudes are taken. The amplitudes are typically chosen by adjusting the amplitude set-point in volts. One has to take care that drift in the z-direction is negligible. An analysis of the corresponding $\gamma(z)$-curves will show the same curves (as in Fig. 6.14), but the curves will be shifted in the horizontal axis. These shifts correspond to the change in amplitude, allowing one to correlate the voltage values with the z-distances.

For the oft-encountered force contributions from electrostatic, van der Waals, and chemical binding forces the frequency shift has been calculated from the force laws. In the approximation that the tip radius R is larger than the tip-sample distance z, an electrostatic potential V will yield a normalized frequency shift of (adapted from [30]):

$$\gamma(z) = \frac{\pi \varepsilon_0 R V^2}{\sqrt{2}} z^{-1/2} \, . \tag{6.22}$$

For van der Waals forces with Hamaker constant H and with R larger than z, we find accordingly:

$$\gamma(z) = \frac{HR}{12\sqrt{2}} z^{-3/2} \, . \tag{6.23}$$

Finally, short-range chemical forces represented by the well-known Morse potential (with the parameters binding energy U_0, decay length λ, and equilibrium distance z_{equ}) yield:

$$\gamma(z) = \frac{U_0 \sqrt{2}}{\sqrt{\pi \lambda}} \exp\left(-\frac{(z - z_{\text{equ}})}{\lambda}\right) \, . \tag{6.24}$$

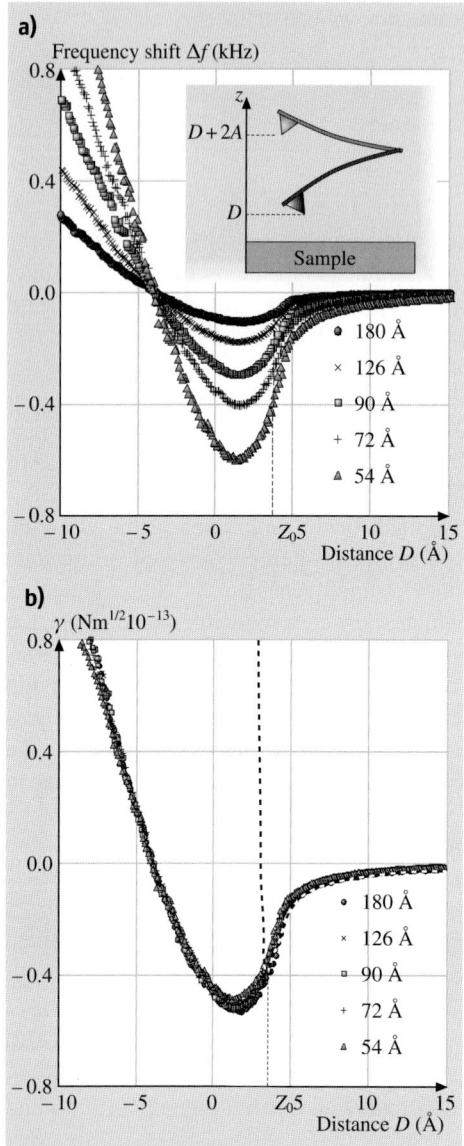

Fig. 6.14. (a) Frequency-shift curves for different oscillation amplitudes for a silicon tip on a graphite surface in UHV, (b) γ-curves calculated from the Δf-curves in (a). (After [28] with permission, Copyright (2000) by The American Physical Society)

These equations allow the experimentalist to directly interpret the spectroscopic measurements. For example, the contributions of the electrostatic and van der Waals forces can be easily distinguished by their slope in a log-log plot (for an example, see [30]).

Alternatively, if the force law is not known beforehand, the experimentalist wants to analyze the experimental frequency-shift data curves and extract the force or energy potential curves. We, therefore, have to invert the integral in (6.21) to find the tip-sample interaction potential V_{ts} from the $\gamma(z)$ curves [27]:

$$V_{ts}(z) = \sqrt{2} \int_z^\infty \frac{\gamma(x)}{\sqrt{x-z}} \, dx. \tag{6.25}$$

Using this method, quantitative force curves were extracted from Δf-spectroscopy measurements on different, atomically resolved sites of the Si(111)-(7 × 7) reconstruction [1]. Comparison to theoretical MD simulations showed good quantitative agreement with theory and confirmed the assumption that force interactions were governed by a single atom at the tip apex. Our initially formulated goal seems to be achieved: With FM-AFM we have found a powerful method that allows us to measure the chemical bond formation of single molecules. The last uncertainty, the exact shape and identity of the tip apex atom, can possibly be resolved by employing the FIM technique to characterize the tip surface in combination with FM-AFM.

All the above equations are strictly valid only in the approximation that the oscillation amplitudes are much larger than the distance range of the encountered forces. For amplitudes of 10 nm and long-range forces like electrostatic interactions this approximation may not be valid. An iterative approach has been presented by *Dürig* [31] to overcome this problem. The interaction force is calculated from the frequency shift curves [gradient of (6.25)]. This force curve again is used to calculate the frequency shift with the exact relation from (6.18). The difference between the experimental Δf-curve and the reconstructed Δf-curve can now be used to calculate a first-order correction term to the interaction force. This procedure can be iteratively followed until sufficient agreement is found.

In this context it is worthwhile to point out a slightly different dynamic AFM method. While in the typical FM-AFM setup the oscillation amplitude is controlled to stay constant by a dedicated feedback circuit, one could simply keep the excitation amplitude constant (this has been termed *CA = constant amplitude*, as opposed to the *cem = constant excitation* mode). It is expected that this mode is more gentle to the surface, because any dissipative interaction will reduce the amplitude and therefore increase the tip-sample distance. The tip is prevented from deeply indenting the surface. This mode has been employed to image soft biological molecules like DNA or thiols in UHV [32]. However, quantitative interpretation of the obtained frequency spectra is more complicated, since the amplitude and tip-sample distance are altered during the measurement. Until now, we have always associated the self-excitation scheme with vacuum applications. Although it is difficult to operate the FM-AFM in constant amplitude mode in air, since large dissipative effects make it difficult to ensure a constant amplitude, it is indeed possible to use the constant-excitation FM-AFM in air or even in liquid. Still, only a few applications of FM-AFM under ambient or liquid conditions have been reported so far. In fact, a low budget construction set (employing a tuning-fork force sensor) for a cem-mode dynamic AFM setup has been published on the Internet (*http://www.sxm4.uni-muenster.de*).

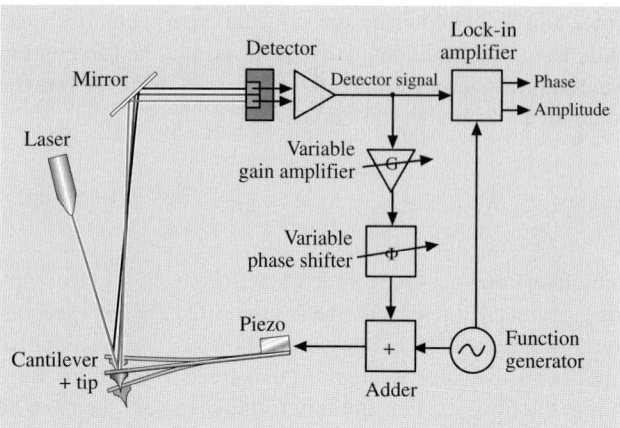

Fig. 6.15. Schematic diagram of operating a Q-Control feedback circuit with an externally driven dynamic AFM. The tapping mode setup is in effect extended by an additional feedback loop

6.4 Q-Control

We have already discussed the virtues of a high Q value for high sensitivity measurements: The minimum detectable force gradient was inversely proportional to the square root of Q. In vacuum, Q mainly represents the internal dissipation of the cantilever during oscillation, an internal damping factor. Little damping is obtained by using high quality cantilevers, which are cut (or etched) from defect-free, single-crystal silicon wafers. Under ambient or liquid conditions, the quality factor is dominated by dissipative interactions between the cantilever and the surrounding medium, and Q values can be as low as 100 for air or even 5 in liquid. Still, we ask if it is somehow possible to compensate for the damping effect by exciting the cantilever in a sophisticated way?

It turns out that the shape of the resonance curves in Fig. 6.16 can be influenced toward higher (or lower) Q values by an amplitude feedback loop. In principle, there are several mechanisms to couple the amplitude signal back to the cantilever, by the photothermal effect [33] or capacitive forces [34]. Figure 6.15 shows a method in which the amplitude feedback is mediated directly by the excitation piezo [35]. This has the advantage that no additional mechanical setups are necessary.

The working principle of the feedback loop can be understood by analyzing the equation of motion of the modified dynamic system:

$$m^* \ddot{z}(t) + \alpha \dot{z}(t) + kz(t) - F_{ts}(z_0 + z(t))$$
$$= F_{ext} \cos(\omega t) + G e^{i\phi} z(t) . \quad (6.26)$$

This ansatz takes into account the feedback of the detector signal through a phase shifter, amplifier, and adder as an additional force, which is linked to the cantilever

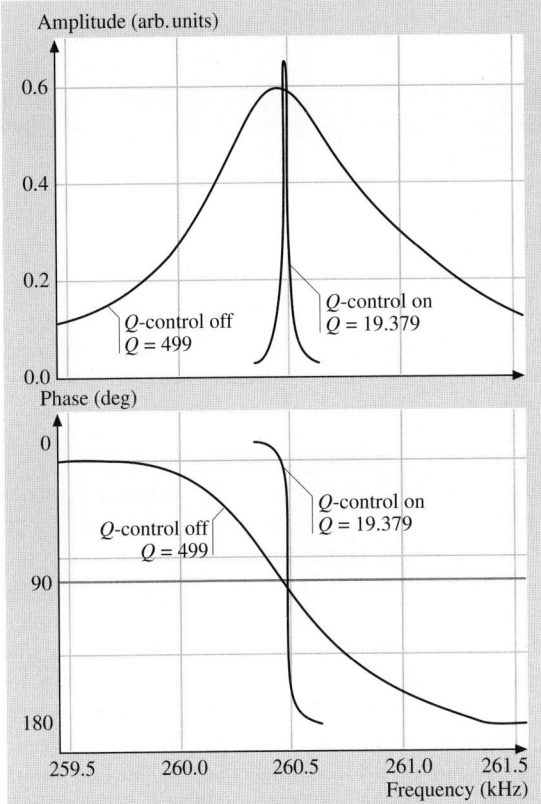

Fig. 6.16. Amplitude and phase diagrams measured in air with a Si cantilever far away from the sample. The quality factor can be increased from 450 to 20 000 by using the Q-Control feedback method

deflection $z(t)$ through the gain G and the phase shift $e^{i\phi}$. To a good approximation, we assume that the oscillation can be described by a harmonic trajectory. With a phase shift of $\phi = \pm\pi/2$ we find:

$$e^{\pm i\pi/2} z(t) = \pm \frac{1}{\omega} \dot{z}(t) \ . \tag{6.27}$$

This means that the additional feedback force signal $Ge^{i\phi}z(t)$ is proportional to the velocity of the cantilever, just like the damping term in the equation of motion. We can define an effective damping constant α_{eff}, which combines the two terms:

$$m^* \ddot{z}(t) + \alpha_{\text{eff}} \dot{z}(t) + kz(t) - F_{\text{ts}}(z_0 + z(t))$$
$$= F_{\text{ext}} \cos(\omega t) \tag{6.28}$$

$$\text{with } \alpha_{\text{eff}} = \alpha \mp \frac{1}{\omega} G \quad \text{for } \phi = \pm \frac{\pi}{2} \ .$$

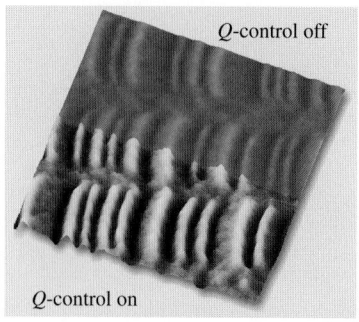

Fig. 6.17. Enhancement of the contrast in the phase channel due to Q-Control on a magnetic hard disk measured with a magnetic tip in tapping-mode AFM in air. Scan size $5 \times 5\,\mu m$, phase range 10 deg (www.nanoanalytics.com)

Equation (6.28) shows that the dampening of the oscillator can be enhanced or weakened by the choice of G, $\phi = +\pi/2$ or $\phi = -\pi/2$, respectively. The feedback loop, therefore, allows us to vary the effective quality factor $Q_{\text{eff}} = m\omega_0/\alpha_{\text{eff}}$ of the complete dynamic system. Hence, we term this system Q-Control. Figure 6.16 shows experimental data on the effect of Q-Control on the amplitude and phase as a function of the external excitation frequency [35]. While $Q = 449$, without Q-Control the quality factor of the system operated in air is enhanced by a factor of more than 40 by applying the feedback loop.

The effect of improved image contrast is demonstrated in Fig. 6.17. Here, a computer hard disk was analyzed with a magnetic tip in tapping mode, where the magnetic contrast is observed in the phase image. The upper part shows the recorded magnetic data structures in standard mode, whereas in the lower part of the image Q-Control feedback was activated, giving rise to an improved signal, i.e., magnetic contrast. A more detailed analysis of measurements on a magnetic tape shows that the signal amplitude (upper diagrams in Fig. 6.18), i.e., the image contrast, was increased by a factor of 12.4 by the Q-Control feedback. The lower image shows a noise analysis of the signal, indicating an increase of the signal-to-noise ratio by a factor of 2.3.

Note that the diagrams represent measurements in air with an AFM operated in AM mode. Only then can we make a distinction between excitation and vibration frequency, since in the FM mode these two frequencies are equal by definition. Although the relation between sensitivity and Q-factor in (6.17) is the same for AM and FM mode, it must be critically investigated to see whether the enhanced quality factor by Q-Control can be inserted in the equation for FM-mode AFM. In vacuum applications, Q is already very high, which makes it oblique to operate an additional Q-Control module.

As stated before, we can also use Q-Control to enhance the dampening in the oscillating system. This would decrease the sensitivity of the system. But on the other hand, the response time of the amplitude change is decreased as well. For tapping-mode applications, where high speed scanning is the goal, Q-Control will be able to reduce the relaxation time [36].

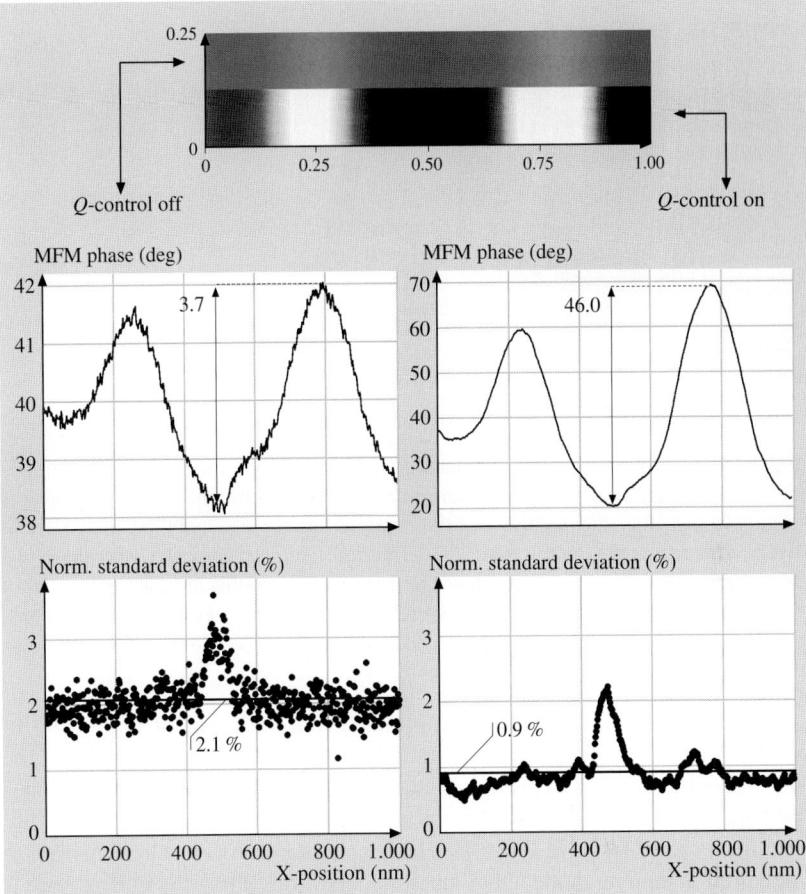

Fig. 6.18. Signal-to-noise analysis with a magnetic tip in tapping-mode AFM on a magnetic tape sample with Q-Control

A large quality factor Q does not only have the virtue of increasing the force sensitivity of the instrument. It also has the advantage of increasing the parameter space of stable AFM operation in tapping- or AM-mode AFM. Consider the resonance curve of Fig. 6.9. When approaching the tip toward the surface there are two competing mechanisms: On the one hand, we bring the tip closer to the sample, which results in an increase in attractive forces (see Fig. 6.2). On the other hand, for the case $\omega > \omega_0$, the resonant frequency of the cantilever is shifted toward smaller values due to the attractive forces, which causes the amplitude to become smaller, preventing a tip-sample contact. This is the desirable regime, where stable operation of the AFM is possible in the net-attractive regime. But as explained before, below a certain tip-sample separation d_{app}, the system switches suddenly into intermittent contact mode, where surface modifications are likely due to the onset of

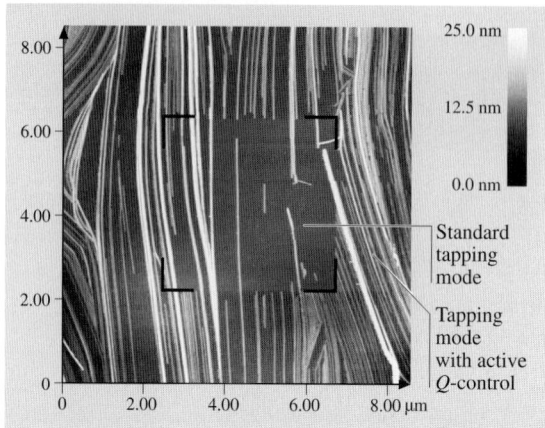

Fig. 6.19. Imaging of a delicate organic surface with Q-Control. Sample was a Langmuir-Blodgett film (ethyl-2,3-dihydroxyoctadecanoate) on a mica substrate. The topographical image clearly shows that the highly sensitive sample surface can only be imaged non-destructively with active Q-Control, whereas the periodic repulsive contact with the probe in standard operation without Q-Control leads to a significant modification or destruction of the surface structure. (Data courtesy of *Lifeng Chi* and coworkers, Westfälische Wilhelms-Universität, Münster, Germany)

strong repulsive forces. The steeper the amplitude curve the larger the regime of stable, net-attractive AFM operation. Looking at Fig. 6.9 we find that the slope of the amplitude curve is governed by the quality factor Q. A high Q, therefore, facilitates stable operation of the AM-AFM in the net-attractive regime.

An example can be found in Fig. 6.19. Here, a surface scan of an ultrathin organic film is acquired in tapping mode under ambient conditions. First, the inner square is scanned without the Q enhancement, and then a wider surface area was scanned with applied Q-Control. The high quality factor provides a larger parameter space for operating the AFM in the net-attractive regime, allowing good resolution of the delicate organic surface structure. Without the Q-Control the surface structures are deformed and even destroyed due to the strong repulsive tip-sample interactions [37–39]. This also allowed imaging of DNA structures without predominantly depressing the soft material during imaging. It was then possible to observe a DNA diameter close to the theoretical value with the Q-Control feedback [40].

In conclusion, we have shown that by applying an additional feedback circuit to the dynamic AFM system it is possible to influence the quality factor Q of the oscillator system. High-resolution, high-speed, or low-force scanning is then possible.

6.5 Dissipation Processes Measured with Dynamic AFM

Dynamic AFM methods have proven their great potential for imaging surface structures at the nanoscale, and we have also discussed methods that allow the assessment of forces between distinct single molecules. However, there is another physical mechanism that can be analyzed with the dynamic mode and has been mentioned in some previous paragraphs: energy dissipation. In Fig. 6.12, we have already shown an example, where the phase signal in tapping mode cannot be explained by conservative forces alone; dissipative processes must also play a role. In constant-amplitude FM mode, where the quantitative interpretation of experiments has proven to be less difficult, an intuitive distinction between conservative and dissipative tip-sample interaction is possible. While we have shown the correlation between forces and frequency shifts of the oscillating system, we have neglected one experimental input channel. The excitation amplitude, which is necessary to keep the oscillation amplitude constant, is a direct indication of the energy dissipated during one oscillation cycle. *Dürig* [41] has shown that in self-excitation mode (with an excitation-oscillation phase difference of 90 deg), conservative and dissipative interactions can be strictly separated. Part of this energy is dissipated in the cantilever itself, another part is due to external viscous forces in the surrounding medium. But more interestingly, some energy is dissipated at the tip-sample junction. This is the focus of the following paragraphs.

In contrast to conservative forces acting at the tip-sample junction, which at least in vacuum can be understood in terms of van der Waals, electrostatic, and chemical interactions, the dissipative processes are poorly understood. *Stowe* et al. [42] have shown that if a voltage potential is applied between tip and sample, charges are induced in the sample surface, which will follow the tip motion (here the oscillation is parallel to the surface). Due to the finite resistance of the sample material, energy will be dissipated during the charge movement. This effect has been exploited to image the doping level of semiconductors. Energy dissipation has also been observed in imaging magnetic materials. *Liu* et al. [43] found that energy dissipation due to magnetic interactions was enhanced at the boundaries of magnetic domains, which was attributed to domain wall oscillations. But also in the absence of external electromagnetic fields, energy dissipation was observed in close proximity of tip and sample, within 1 nm. Clearly, mechanical surface relaxations must give rise to energy losses. One could model the AFM tip as a small hammer, hitting the surface at high frequency, possibly resulting in phonon excitations. From a contiuum mechanics point of view, we assume that the mechanical relaxation of the surface is not only governed by elastic responses. Viscoelastic effects of soft surfaces will also render a significant contribution to energy dissipation. The whole area of phase imaging in tapping mode is concerned with those effects [44–47].

In the atomistic view, the last tip atom can be envisaged to change position while yielding to the tip-sample force field. A strictly reversible change of position would not result in a loss of energy. Still, it has been pointed out by *Sasaki* et al. [48] that a change in atom position would result in a change in the force interaction itself. Therefore, it is possible that the tip atom changes position at different tip-surface distances during approach and retraction, effectively causing an atomic-scale hysteresis to develop. In fact, *Hoffmann* et al. [13] have measured short-range energy dissipation for a tungsten tip on silicon in UHV. A theoretical model explaining this effect on the basis of a two-energy-state system was developed. However, a clear understanding of the underlying physical mechanism is still lacking.

Nonetheless, the dissipation channel has been used to image surfaces with atomic resolution [49]. Instead of feedbacking the distance on the frequency shift, the excitation amplitude in FM mode has been used as the control signal. The Si(111)-(7×7) reconstruction was successfully imaged in this mode. The step edges of monoatomic NaCl islands on single-crystalline copper have also rendered atomic resolution contrast in the dissipation channel [50]. The dissipation processes discussed so far are mostly in the configuration in which the tip is oscillated perpendicular to the surface. Friction is usually referred to as the energy loss due to lateral movement of solid bodies in contact. It is interesting to note in this context that *Israelachivili* [51] has pointed out a quantitative relationship between lateral and vertical (with respect to the surface) dissipation. He states that the hysteresis in vertical force-distance curves should equal the energy loss in lateral friction. An experimental confirmation of this conjecture at the molecular level is still missing.

Physical interpretation of energy dissipation processes at the atomic scale seems to be a daunting task at this point. Notwithstanding, we can find a quantitative relation between the energy loss per oscillation cycle and the experimental parameters in dynamic AFM, as will be shown in the following section.

In static AFM it was found that permanent changes of the sample surface by indentations can cause a hysteresis between approach and retraction (e.g., [5]). The area between the approach and retraction curves in a force-distance diagram represents the lost or dissipated energy caused by the irreversible change of the surface structure. In dynamic-mode AFM, the oscillation parameters like amplitude, frequency, and phase must contain the information about the dissipated energy per cycle. So far, we have resorted to a treatment of the equation of motion of the cantilever vibration in order to find a quantitative correlation between forces and the experimental parameters. For the dissipation it is useful to treat the system from the energy conservation point of view.

Assuming a dynamic system is in equilibrium, the average energy input must equal the average energy output or dissipation. Applying this rule to an AFM running in a dynamic mode means that the average power fed into the cantilever oscillation by an external driver, denoted $\bar{P}_{in}$, must equal the average power dissipated by the motion of the cantilever beam $\bar{P}_0$ and by tip-sample interaction $\bar{P}_{tip}$:

$$\bar{P}_{in} = \bar{P}_0 + \bar{P}_{tip} . \tag{6.29}$$

The term $\bar{P}_{tip}$ is what we are interested in, since it gives us a direct physical quantity to characterize the tip-sample interaction. Therefore, we have first to calculate and then measure the two other terms in (6.29) in order to determine the power dissipated when the tip periodically probes the sample surface. This requires an appropriate rheological model to describe the dynamic system. Although there are investigations in which the complete flexural motion of the cantilever beam has been considered [52], a simplified model, comprising a spring and two dashpots (Fig. 6.20), represents a good approximation in this case [53].

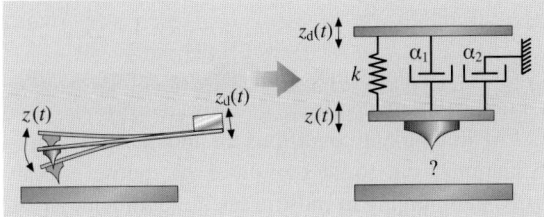

Fig. 6.20. Rheological model applied to describe the dynamic AFM system, comprising the oscillating cantilever and tip interacting with the sample surface. The movement of the cantilever base and the tip is denoted as $z_d(t)$ and $z(t)$, respectively. The cantilever is characterized by the spring constant k and the damping constant α. In a first approach, damping is broken into two pieces, α_1 and α_2. First, intrinsic damping caused by the movement of the cantilever's tip relative to its base. Second, damping related to the movement of the cantilever body in a surrounding medium, e.g., air damping

The spring, characterized by the constant k according to Hooke's law, represents the only channel through which power P_{in} can be delivered to the oscillating tip $z(t)$ by the external driver $z_d(t)$. Therefore, the instantaneous power fed into the dynamic system is equal to the force exerted by the driver times the velocity of the driver (the force that is necessary to move the base side of the dashpot can be neglected, since this power is directly dissipated and, therefore, does not contribute to the power delivered to the oscillating tip):

$$P_{in}(t) = F_d(t)\dot{z}_d(t) = k\left[z(t) - z_d(t)\right]\dot{z}_d(t) . \tag{6.30}$$

Assuming a sinusoidal steady state response and that the base of the cantilever is driven sinusoidally (see (6.6)) with amplitude A_d and frequency ω, the deflection from equilibrium of the end of the cantilever follows (6.9), where A and $0 \leq \varphi \leq \pi$ are the oscillation amplitude and phase shift, respectively. This allows us to calculate the average power input per oscillation cycle by integrating (6.30) over one period $T = 2\pi/\omega$:

$$\bar{P}_{in} = \frac{1}{T}\int_0^T P_{in}(t)\,dt - \frac{1}{2}k\omega A_d A \sin\varphi . \tag{6.31}$$

This contains the familiar result that the maximum power is delivered to an oscillator when the response is 90 deg out of phase with the drive.

The simplified rheological model, as it is depicted in Fig. 6.20, exhibits two major contributions to the damping term $\bar{P}_0$. Both are related to the motion of the cantilever body and assumed to be well modeled by viscous damping with coefficients α_1 and α_2. The dominant damping mechanism in UHV conditions is intrinsic damping caused by the deflection of the cantilever beam, i.e., the motion of the tip relative to the cantilever base. Therefore, the instantaneous power dissipated by such a mechanism is given by

$$P_{01}(t) = |F_{01}(t)\dot{z}(t)| = |\alpha_1 [\dot{z}(t) - \dot{z}_d(t)] \dot{z}(t)| \;. \tag{6.32}$$

Note that the absolute value has to be calculated, since all dissipated power is "lost" and, therefore, cannot be returned to the dynamic system.

However, when running an AFM in ambient conditions an additional damping mechanism has to be considered. Damping due to the motion of the cantilever body in the surrounding medium, e.g., air damping, is in most cases the dominant effect. The corresponding instantaneous power dissipation is given by

$$P_{02}(t) = |F_{02}(t)\dot{z}(t)| = \alpha_2 \dot{z}^2(t) \;. \tag{6.33}$$

In order to calculate the average power dissipation, (6.32) and (6.33) have to be integrated over one complete oscillation cycle. This yields

$$\bar{P}_{01} = \frac{1}{T} \int_0^T P_{01}(t) \, dt$$

$$= \frac{1}{\pi} \alpha_1 \omega^2 A \left[(A - A_d \cos\varphi) \right.$$

$$\times \arcsin\left(\frac{A - A_d \cos\varphi}{\sqrt{A^2 + A_d^2 - 2AA_d \cos\varphi}} \right)$$

$$\left. + A_d \sin\varphi \right] , \tag{6.34}$$

and

$$\bar{P}_{02}(t) = \frac{1}{T} \int_0^T P_{02}(t) \, dt = \frac{1}{2} \alpha_2 \omega^2 A^2 \;. \tag{6.35}$$

Considering the fact that commonly used cantilevers exhibit a quality factor of at least several hundreds (in UHV several tens of thousands), we can assume that the oscillation amplitude is significantly larger than the drive amplitude when the dynamic system is driven at or near its resonance frequency: $A \gg A_d$. Therefore, (6.34) can be simplified in first order approximation to an expression similar to (6.35). Combining the two equations yields the total average power dissipated by the oscillating cantilever

$$\bar{P}_0 = \frac{1}{2} \alpha \omega^2 A^2 \quad \text{with } \alpha = \alpha_1 + \alpha_2 \;, \tag{6.36}$$

where α denotes the overall effective damping constant.

We can now solve (6.29) for the power dissipation localized to the small interaction volume of the probing tip with the sample surface, represented by the question mark in Fig. 6.20. Furthermore, by expressing the damping constant α in terms of experimentally accessible quantities such as the spring constant k, the quality factor Q, and the natural resonant frequency ω_0 of the free oscillating cantilever, $\alpha = \frac{k}{Q\omega_0}$, we obtain:

$$\bar{P}_{\text{tip}} = \bar{P}_{\text{in}} - \bar{P}_0$$
$$= \frac{1}{2}\frac{k\omega}{Q}\left(Q_{\text{cant}}A_{\text{d}}A \sin\varphi - A^2\frac{\omega}{\omega_0}\right). \quad (6.37)$$

Note that so far no assumptions have been made regarding how the AFM is operated, except that the motion of the oscillating cantilever has to remain sinusoidal to a good approximation. Therefore, (6.37) is applicable to a variety of different dynamic AFM modes.

For example, in FM-mode AFM the oscillation frequency ω changes due to tip-sample interaction, while at the same time the oscillation amplitude A is kept constant by adjusting the drive amplitude A_{d}. By measuring these quantities, one can apply (6.37) to determine the average power dissipation related to tip-sample interaction. In spectroscopy applications, usually $A_{\text{d}}(d)$ is not measured directly, but a signal $G(d)$ proportional to $A_{\text{d}}(d)$ is acquired representing the gain factor applied to the excitation piezo. With the help of (6.15) we can write:

$$A_{\text{d}}(d) = \frac{A_0 G(d)}{Q G_0}, \quad (6.38)$$

while A_0 and G_0 are the amplitude and gain at large tip-sample distances, where the tip-sample interactions are negligible.

Now let us consider the tapping-mode, or AM-AFM. In this case, the cantilever is driven at a fixed frequency and with constant drive amplitude, while the oscillation amplitude and phase shift may change when the probing tip interacts with the sample surface. Assuming that the oscillation frequency is chosen to be ω_0, (6.37) can be further simplified by again employing (6.15) for the free oscillation amplitude A_0. This yields

$$\bar{P}_{\text{tip}} = \frac{1}{2}\frac{k\omega_0}{Q}\left(A_0 A \sin\varphi - A^2\right). \quad (6.39)$$

Equation (6.39) implies that if the oscillation amplitude A is kept constant by a feedback loop, like it is commonly done in tapping mode, simultaneously acquired phase data can be interpreted in terms of energy dissipation [45, 47, 54, 55]. When analyzing such phase images [56–58], one has also to consider the fact that the phase may also change due to the transition from net-attractive ($\varphi > 90°$) to intermittent contact ($\varphi < 90°$) interaction between the tip and the sample [15, 35, 59, 60]. For example, consider the phase shift in tapping mode as a function of z-position, Fig. 6.12. If phase measurements are performed close to the point where the oscillation switches from the net-attractive to the intermittent contact regime, a large contrast in the phase

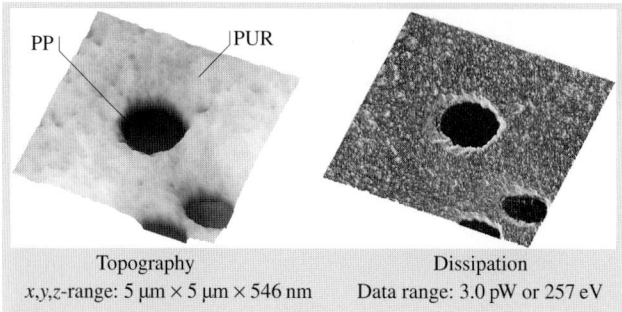

Fig. 6.21. Topography and phase image in tapping-mode AFM of a polymer blend composed of polypropylene (PP) particles embedded in a polyurethane (PUR) matrix. The dissipation image shows a strong contrast between the harder PP (little dissipation, *dark*) to the softer PUR (large dissipation, *bright*) surface

channel is observed. However, this contrast is not due to dissipative processes. Only a variation of the phase signal within the intermittent contact regime will give information of the tip-sample dissipative processes.

An example of a dissipation measurement is depicted in Fig. 6.21. The surface of a polymer blend was imaged in air, simultaneously acquiring the topography and dissipation. The dissipation on the softer polyurethane-matrix is significantly larger than on the embedded, mechanically stiffer polypropylene particles.

6.6 Conclusion

Dynamic force microscopy is a powerful tool that is capable of imaging surfaces with atomic precision. It also allows us to look at surface dynamics, and it can operate in vacuum, air, or even in liquid. However, the oscillating cantilever system introduces a level of complexity that disallows a straightforward interpretation of acquired images. An exception is the self-excitation mode, where tip-sample forces can be successfully extracted from spectroscopic experiments. However, not only conservative forces can be investigated with dynamic AFM; energy dissipation also influences the cantilever oscillation and can, therefore, serve as a new information channel.

Open questions still exist concerning the exact geometric and chemical identity of the probing tip, which significantly influences the imaging and spectroscopic results. Using predefined tips like single-walled nanotubes or atomic resolution techniques like field ion microscopy to image the tip itself are possible approaches to addressing this issue. Furthermore, little is known about the dissipative processes at the scale where only a few atoms between tip and sample interact. It is also desirable to learn more about the interpretation of images acquired in tapping mode.

References

1. M. A. Lantz, H. J. Hug, R. Hoffmann, P. J. A. van Schendel, P. Kappenberger, S. Martin, A. Baratoff, and H.-J. Güntherodt. Quantitative measurement of short-range chemical bonding forces. *Science*, 291:2580–2583, 2001.
2. G. Binnig, C. F. Quate, and Ch. Gerber. Atomic force microscope. *Phys. Rev. Lett.*, 56:930–933, 1986.
3. O. Marti. *AFM instrumentation and tips*, pages 81–144. CRC, 2nd edition, 1999.
4. G. Cross, A. Schirmeisen, A. Stalder, P. Grütter, M. Tschudy, and U. Dürig. Adhesion interaction between atoically defined tip and sample. *Phys. Rev. Lett.*, 80:4685–4688, 1998.
5. A. Schirmeisen, G. Cross, A. Stalder, P. Grütter, and U. Dürig. Metallic adhesion and tunneling at the atomic scale. *New J. Phys.*, 2:29.1–29.10, 2000.
6. A. Schirmeisen. Metallic adhesion and tunneling at the atomic scale, 1999.
7. F. J. Giessibl. Forces and frequency shifts in atomic-resolution dynamic-force microscopy. *Phys. Rev. B*, 56:16010–16015, 1997.
8. F. J. Giessibl. Atomic resolution of the silicon (111)-(7 × 7) surface by atomic force microscopy. *Science*, 267:68–71, 1995.
9. M. Bezanilaa, B. Drake, E. Nudler, M. Kashlev, P. K. Hansma, and H. G. Hansma. Motion and enzymatic degradation of dna in the atomic force microscope. *Biophys. J.*, 67:2454–2459, 1994.
10. Y. Jiao, D. I. Cherny, G. Heim, T. M. Jovin, and T. E. Schäffer. Dynamic interactions of p53 with dna in solution by time-lapse atomic force microscopy. *J. Mol. Biol.*, 314:233–243, 2001.
11. T. R. Albrecht, P. Grütter, D. Horne, and D. Rugar. Frequency modulation detection using high-q cantilevers for enhanced force microscopy sensitivity. *J. Appl. Phys.*, 69:668–673, 1991.
12. S. P. Jarvis, M. A. Lantz, U. Dürig, and H. Tokumoto. Off resonance ac mode force spectroscopy and imaging with an atomic force microscope. *Appl. Surf. Sci.*, 140:309–313, 1999.
13. P. M. Hoffmann, S. Jeffery, J. B. Pethica, H. Ö. Özer, and A. Oral. Energy dissipation in atomic force microscopy and atomic loss processes. *Phys. Rev. Lett.*, 87:265502–265505, 2001.
14. B. Anczykowski, D. Krüger, and H. Fuchs. Cantilever dynamics in quasinoncontact force microscopy: Spectroscopic aspects. *Phys. Rev. B*, 53:15485–15488, 1996.
15. B. Anczykowski, D. Krüger, K. L. Babcock, and H. Fuchs. Basic properties of dynamic force spectroscopy with the scanning force microscope in experiment and simulation. *Ultramicroscopy*, 66:251–259, 1996.
16. V. M. Muller, V. S. Yushchenko, and B. V. Derjaguin. On the influence of molecular forces on the deformation of an elastic sphere and its sticking to a rigid plane. *J. Coll. Interf. Sci.*, 77:91–101, 1980.
17. B. D. Hughes and L. R. White. 'Soft' contact problems in linear elasticity. *Quart. J. Mech. Appl. Math.*, 32:445–471, 1979.
18. L. Verlet. Computer "experiments" on classical fluids. i. thermodynamical properties of lennard–jones molecules. *Phys. Rev.*, 159:98–103, 1967.
19. L. Verlet. Computer "experiments" on classical fluids. ii. equilibrium correlation functions. *Phys. Rev.*, 165:201–214, 1968.
20. P. Gleyzes, P. K. Kuo, and A. C. Boccara. Bistable behavior of a vibrating tip near a solid surface. *Appl. Phys. Lett.*, 58:2989–2991, 1991.

21. A. San Paulo and R. García. High-resolution imaging of antibodies by tapping-mode atomic force microscopy: Attractive and repulsive tip-sample interaction regimes. *Biophys. J.*, 78:1599–1605, 2000.
22. D. Krüger, B. Anczykowski, and H. Fuchs. Physical properties of dynamic force microscopies in contact and noncontact operation. *Ann. Phys.*, 6:341–363, 1997.
23. Y. Martin, C. C. Williams, and H. K. Wickramasinghe. Atomic force microscope – force mapping and profiling on a sub 100-Å scale. *J. Appl. Phys.*, 61:4723–4729, 1987.
24. C. Barth and M. Reichling. Imaging the atomic arrangement on the high-temperature reconstructed α-Al_2O_3(0001) surface. *Nature*, 414:54–57, 2001.
25. B. Gotsmann and H. Fuchs. Dynamic force spectroscopy of conservative and dissipative forces in an Al-Au(111) tip-sample system. *Phys. Rev. Lett.*, 86:2597–2600, 2001.
26. H. Hölscher, W. Allers, U. D. Schwarz, A. Schwarz, and R. Wiesendanger. Determination of tip-sample interaction potentials by dynamic force spectroscopy. *Phys. Rev. Lett.*, 83:4780–4783, 1999.
27. U. Dürig. Relations between interaction force and frequency shift in large-amplitude dynamic force microscopy. *Appl. Phys. Lett.*, 75:433–435, 1999.
28. H. Hölscher, A. Schwarz, W. Allers, U. D. Schwarz, and R. Wiesendanger. Quantitative analysis of dynamic-force-spectroscopy data on graphite(0001) in the contact and noncontact regime. *Phys. Rev. B*, 61:12678–12681, 2000.
29. M. Guggisberg. Lokale messung von atomaren kräften, 2000.
30. M. Guggisberg, M. Bammerlin, E. Meyer, and H.-J. Güntherodt. Separation of interactions by noncontact force microscopy. *Phys. Rev. B*, 61:11151–11155, 2000.
31. U. Dürig. Extracting interaction forces and complementary observables in dynamic probe microscopy. *Appl. Phys. Lett.*, 76:1203–1205, 2000.
32. T. Uchihasi, T. Ishida, M. Komiyama, M. Ashino, Y. Sugawara, W. Mizutani, K. Yokoyama, S. Morita, H. Tokumoto, and M. Ishikawa. High-resolution imaging of organic monolayers using noncontact afm. *Appl. Surf. Sci.*, 157:244–250, 2000.
33. J. Mertz, O. Marti, and J. Mlynek. Regulation of a microcantilever response by force feedback. *Appl. Phys. Lett.*, 62:2344–2346, 1993.
34. D. Rugar and P. Grütter. Mechanical parametric amplification and thermomechanical noise squeezing. *Phys. Rev. Lett.*, 67:699–702, 1991.
35. B. Anczykowski, J. P. Cleveland, D. Krüger, V. B. Elings, and H. Fuchs. Analysis of the interaction mechanisms in dynamic mode sfm by means of experimental data and computer simulation. *Appl. Phys. A*, 66:885–889, 1998.
36. T. Sulchek, G. G. Yaralioglu, C. F. Quate, and S. C. Minne. Characterization and optimisation of scan speed for tapping-mode atomic force microscopy. *Rev. Sci. Instrum.*, 73:2928–2936, 2002.
37. L. F. Chi, S. Jacobi, B. Anczykowski, M. Overs, H.-J. Schäfer, and H. Fuchs. Supermolecular periodic structures in monolayers. *Adv. Mater.*, 12:25–30, 2000.
38. S. Gao, L. F. Chi, S. Lenhert, B. Anczykowski, C. Niemeyer, M. Adler, and H. Fuchs. High-quality mapping of dna-protein complexes by dynamic scanning force microscopy. *ChemPhysChem*, 6:384–388, 2001.
39. B. Zou, M. Wang, D. Qiu, X. Zhang, L. F. Chi, and H. Fuchs. Confined supramolecular nanostructures of mesogen-bearing amphiphiles. *Chem. Commun.*, 9:1008–1009, 2002.
40. B. Pignataro, L. F. Chi, S. Gao, B. Anczykowski, C. Niemeyer, M. Adler, and H. Fuchs. Dynamic scanning force microscopy study of self-assembled dna-protein nanostructures. *Appl. Phys. A*, 74:447–452, 2002.
41. U. Dürig. Interaction sensing in dynamic force microscopy. *New J. Phys.*, 2:5.1–5.2, 2000.

42. T. D. Stowe, T. W. Kenny, D. J. Thomson, and D. Rugar. Silicon dopant imaging by dissipation force microscopy. *Appl. Phys. Lett.*, 75:2785–2787, 1999.
43. Y. Liu and P. Grütter. Magnetic dissipation force microscopy studies of magnetic materials. *J. Appl. Phys.*, 83:7333–7338, 1998.
44. J. Tamayo and R. García. Effects of elastic and inelastic interactions on phase contrast images in tapping-mode scanning force microscopy. *Appl. Phys. Lett.*, 71:2394–2396, 1997.
45. J. P. Cleveland, B. Anczykowski, A. E. Schmid, and V. B. Elings. Energy dissipation in tapping-mode atomic force microscopy. *Appl. Phys. Lett.*, 72:2613–2615, 1998.
46. R. García, J. Tamayo, M. Calleja, and F. García. Phase contrast in tapping-mode scanning force microscopy. *Appl. Phys. A*, 66:309–312, 1998.
47. B. Anczykowski, B. Gotsmann, H. Fuchs, J. P. Cleveland, and V. B. Elings. How to measure energy dissipation in dynamic mode atomic force microscopy. *Appl. Surf. Sci.*, 140:376–382, 1999.
48. N. Sasaki and M. Tsukada. Effect of microscopic nonconservative process on noncontact atomic force microscopy. *Jpn. J. Appl. Phys.*, 39:L1334–L1337, 2000.
49. R. Lüthi, E. Meyer, M. Bammerlin, A. Baratoff, L. Howald, C. Gerber, and H.-J Güntherodt. Ultrahigh vacuum atomic force microscopy: true atomic resolution. *Surf. Rev. Lett.*, 4:1025–1029, 1997.
50. R. Bennewitz, A. S. Foster, L. N. Kantorovich, M. Bammerlin, Ch. Loppacher, S. Schär, M. Guggisberg, E. Meyer, and A. L. Shluger. Atomically resolved edges and kinks of NaCl islands on Cu(111): experiment and theory. *Phys. Rev. B*, 62:2074–2084, 2000.
51. J. Israelachvili. *Intermolecular and Surface Forces*. Academic, 1992.
52. U. Rabe, J. Turner, and W. Arnold. Analysis of the high-frequency response of atomic force microscope cantilevers. *Appl. Phys. A*, 66:277–282, 1998.
53. T. R. Rodríguez and R. García. Tip motion in amplitude modulation (tapping-mode) atomic-force microscopy: Comparison between continuous and point-mass models. *Appl. Phys. Lett.*, 80:1646–1648, 2002.
54. J. Tamayo and R. García. Relationship between phase shift and energy dissipation in tapping-mode scanning force microscopy. *Appl. Phys. Lett.*, 73:2926–2928, 1998.
55. R. García, J. Tamayo, and A. San Paulo. Phase contrast and surface energy hysteresis in tapping mode scanning force microscopy. *Surf. Interface Anal.*, 27:312–316, 1999.
56. S. N. Magonov, V. B. Elings, and M. H. Whangbo. Phase imaging and stiffness in tapping-mode atomic force microscopy. *Surf. Sci.*, 375:L385–L391, 1997.
57. J. P. Pickering and G. J. Vancso. Apparent contrast reversal in tapping mode atomic force microscope images on films of polystyrene-b-polyisoprene-b-polystyrene. *Polymer Bull.*, 40:549–554, 1998.
58. X. Chen, S. L. McGurk, M. C. Davies, C. J. Roberts, K. M. Shakesheff, S. J. B. Tendler, P. M. Williams, J. Davies, A. C. Dwakes, and A. Domb. Chemical and morphological analysis of surface enrichment in a biodegradable polymer blend by phase-detection imaging atomic force microscopy. *Macromolecules*, 31:2278–2283, 1998.
59. A. Kühle, A. H. Sørensen, and J. Bohr. Role of attractive forces in tapping tip force microscopy. *J. Appl. Phys.*, 81:6562–6569, 1997.
60. A. Kühle, A. H. Sørensen, J. B. Zandbergen, and J. Bohr. Contrast artifacts in tapping tip atomic force microscopy. *Appl. Phys. A*, 66:329–332, 1998.

7
Molecular Recognition Force Microscopy

Peter Hinterdorfer

Summary. Atomic force microscopy (AFM), developed in the late eighties to explore atomic details on hard material surfaces, has evolved to an imaging method capable of achieving fine structural details on biological samples. Its particular advantage in biology is that the measurements can be carried out in aqueous and physiological environment, which opens the possibility to study the dynamics of biological processes in vivo. The additional potential of the AFM to measure ultra-low forces at high lateral resolution has paved the way for measuring inter- and intra-molecular forces of bio-molecules on the single molecule level. Molecular recognition studies using AFM open the possibility to detect specific ligand–receptor interaction forces and to observe molecular recognition of a single ligand–receptor pair. Applications include biotin–avidin, antibody–antigen, NTA nitrilotriacetate–hexahistidine 6, and cellular proteins, either isolated or in cell membranes.

The general strategy is to bind ligands to AFM tips and receptors to probe surfaces (or vice versa), respectively. In a force–distance cycle, the tip is first approached towards the surface whereupon a single receptor–ligand complex is formed, due to the specific ligand receptor recognition. During subsequent tip–surface retraction a temporarily increasing force is exerted to the ligand–receptor connection thus reducing its lifetime until the interaction bond breaks at a critical force (unbinding force). Such experiments allow for estimation of affinity, rate constants, and structural data of the binding pocket. Comparing them with values obtained from ensemble-average techniques and binding energies is of particular interest. The dependences of unbinding force on the rate of load increase exerted to the receptor–ligand bond reveal details of the molecular dynamics of the recognition process and energy landscapes. Similar experimental strategies were also used for studying intra-molecular force properties of polymers and unfolding–refolding kinetics of filamentous proteins. Recognition imaging, developed by combing dynamic force microscopy with force spectroscopy, allows for localization of receptor sites on surfaces with nanometer positional accuracy.

7.1 Introduction

Macromolecular interactions play a key role in cellular regulation and biological function. Molecular recognition is one of the most important regulatory elements, as it is often the initiating step in reaction pathways and cascades. Since the immune system has evolved recognition molecules that are both highly specific and versatile, antibody-antigen interaction is often used as a paradigm for molecular recognition. However, molecular recognition also plays fundamental roles in the regulation of gene expression, cellular metabolism, and drug design.

Molecular recognition studies emphasize specific biological interactions between receptors and their cognitive ligands. Despite a growing literature on the structure and function of receptor-ligand complexes, it is still not possible to predict reaction kinetics or energetics for any given complex formation, even when the structures are known. Additional insights, in particular about the molecular dynamics within the complex during the association and dissociation process, are needed. The high-end strategy is to probe quantitatively the science that underlies a specific biological recognition in which the chemical structures can be defined at atomic resolution, and genetic or chemical means are available to vary the structure of the interacting partners.

Receptor-ligand complexes are usually formed by a few tens of parallel weak interactions of contacting chemical groups in complementary determining regions supported by framework residues providing structurally conserved scaffolding. Both the complementary determining regions and the framework have a considerable amount of plasticity and flexibility resulting in conformational changes upon association and dissociation (induced fit). In addition to what is known about structure, energies, and kinetic constants, information about conformational movements is required for the understanding of the recognition process. It is likely that insight into the temporal and spatial action of the many weak interactions, in particular the cooparativity of bond formation, is the key to understanding receptor-ligand recognition.

For this, experiments on the single-molecule level at time scales typical for receptor-ligand complex formation and dissociation appear to be required. The potential of the atomic force microscope (AFM) [1] to measure ultralow forces at high lateral resolution has paved the way for single-molecule recognition force microscopy studies. The particular advantage of AFM in biology is that the measurements can be carried out in aqueous and physiological environments, which opens the possibility for studying biological processes in vivo. The methodology for investigating the molecular dynamics of receptor-ligand interactions described in this chapter, *Molecular Recognition Force Microscopy* (MRFM) [2–4], was developed from scanning probe microscopy (SPM) [1]. A force is exerted on a receptor-ligand complex, and the dissociation process is followed over time. Dynamic aspects of recognition are addressed in force spectroscopy (FS) experiments, in which distinct force-time profiles are applied to give insight into the changes in conformations and states during receptor-ligand dissociation. It will be shown that MRFM is a versatile tool to explore kinetic and structural details of receptor-ligand recognition.

7.2 Ligand Tip Chemistry

In MRFM experiments, the binding of ligands on AFM tips to surface-bound receptors (or vice versa) is studied by applying a force to the receptor-ligand complex that reduces its lifetime until the bond breaks at a measurable unbinding force.

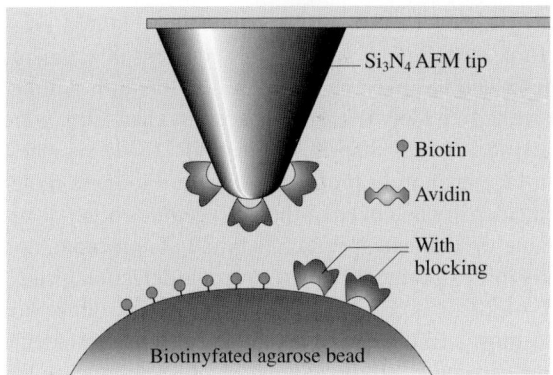

Fig. 7.1. Avidin-functionalized AFM tip. A dense layer of biotinylated BSA was adsorbed to the tip and subsequently saturated with avidin. The biotinylated agarose bead opposing the tip also contained a high surface density of reactive sites, which were, therefore, partly blocked with avidin to achieve single-molecule binding events (after [3])

This requires a careful AFM tip sensor design, including tight attachment of the ligands to the tip surface. In the first pioneering demonstrations of single-molecule recognition force measurements [2, 3], strong physical adsorption of bovine serum albumin (BSA) was used to directly coat the tip [3] or, alternatively, a glass bead glued to it [2]. This physisorbed protein layer may then serve as a functional matrix for the biochemical modification with chemically active ligands (Fig. 7.1). In spite of the large number of probe molecules on the tip (E3–$10^4/\mu m^2$), the low fraction of properly oriented molecules, or internal blocks of most reactive sites (see Fig. 7.1), allowed the measurement of single receptor-ligand unbinding forces. Nevertheless, parallel breakage of multiple bonds was predominately observed with this configuration.

For measuring interactions between isolated receptor-ligand pairs, strictly defined conditions need to be fulfilled. Covalently coupling ligands to gold-coated tip surfaces via freely accessible SH-groups guarantees a sufficiently stable attachment, because these bonds are about ten times stronger than typical ligand-receptor interactions [5]. This chemistry has been used to detect the forces between complementary DNA strands [6] and single nucleotides [7]. Self-assembled monolayers of dithiobis(succinimidylundecanoate) were formed to enable covalent coupling of biomolecules via amines [8], and were used to study the binding strength between cell adhesion proteoglycans [9] and between biotin-directed IgG antibodies and biotin [10]. A vectorial orientation of Fab molecules on gold tips was achieved by a firsthand site-directed chemical binding via their SH-groups [11], without the need of additional linkers. For this, antibodies were digested with papain and subsequently purified to generate Fab fragments with freely accessible SH-groups in the hinge region.

Gold surfaces provide a unique and selective affinity for thiols, although the adhesion strength of the formed bond is weaker than that of other covalent bonds [5]. Since all commercially available AFM tips are etched of silicon nitride or silicon oxide material, deposition of a gold layer onto the tip surface is required prior to using this chemistry. Therefore, designing a sensor with covalent attachments of the biomolecules to the silicon surface may be more straightforward. Amine-functionalization procedures, a strategy widely used in surface bio-chemistry, were applied using ethanolamine [4, 12] and various silanization methods [13–15] as a first step in thoroughly developed surface anchoring protocols suitable for single molecule experiments. Since the amine density achieved in these procedures often determines the number of ligands on the tip that can specifically bind to the receptors on the surface, it has to be sufficiently low to guarantee single molecule detection [4, 12]. One molecule per tip apex ($\sim 5 - -20$ nm) corresponds to a macroscopic density of about 500 molecules per μm^2. A most striking example of a single ligand molecule tip was realized by gluing a single nanotube to the cantilever, thus minimizing the contact between tip and surface with a chemically well-defined and extremely small tip apex [16]. The ligand molecule was reacted to an activated chemical site at the open end of the nanotube.

In a number of laboratories, a distensible and flexible linker was used to space the ligand molecule from the tip surface (e.g., [4, 14]) (Fig. 7.2). At a given low number of spacer molecules per tip, the ligand can freely orient and diffuse within a certain volume provided by the length of the tether to achieve unconstrained binding to its receptor. The unbinding process occurs with little torque, and the ligand molecule escapes the danger of being squeezed between the tip and the surface. It also opens the possibility of site-directed coupling for a defined orientation of the ligand relative to the receptor at receptor-ligand unbinding. As a crosslinking element, poly(ethylene glycol) (PEG), a water soluble, nontoxic polymer with a wide range of applications in surface technology and clinical research, was often used [17]. PEG is known to prevent surface adsorption of proteins and lipid structures and therefore appears ideally suited for this purpose. Gluteraldehyde [13] and DNA [6] were also successfully applied in recognition force studies. Crosslinker lengths, ideally arriving at a good compromise between high tip molecule mobility and narrow lateral resolution of the target recognition site, varied from 2 to 100 nm.

For coupling to the tip surface and to the ligands, respectively, the crosslinker is mostly equipped with two different functional ends, e.g., an N-hydroxysuccinimidyl (NHS) residue on the one end is reactive to amines on the tip and a 2-pyridyldithiopropionyl (PDP) [18] or a vinyl sulfon [19] residue on the other end, can be covalently bound to thiols of ligands (Fig. 7.2). This sulfur chemistry is highly advantageous, since it is very reactive and renders site-directed coupling possible. However, free thiols are hardly available on native ligands and must, therefore, be generated.

Different strategies were used to achieve this goal. Lysins of protein-ligands were derived from the short heterobifunctional linker N-succinnimidyl-3-(S-acethylthio)propionate (SATP) and subsequent deprotection with NH_2OH led to reactive SH groups [18]. Since it is very delicate to react distinct lysins with this method, the coupling to the crosslinker is often not specifically site directed. Several protocols are commercially available (Pierce, Rockford, IL) to generate active antibody fragments with free cysteines. Half-antibodies are produced by cleaving the

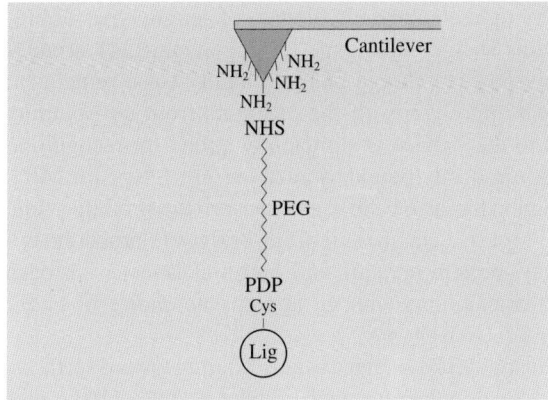

Fig. 7.2. Linkage of ligands to AFM tips. Ligands were covalently coupled to AFM tips via a heterobifunctional polyethylene glycol (PEG) derivative of 8 nm length. Silicon tips were first functionalized with ethanolamine ($NH_2 - C_2H_4OH \cdot HCl$). Then, the NHS (N-hydroxysuccinimide) end of the PEG linker was covalently bound to amines on the tip surface before ligands were attached to the PDP (Pyridoyldithiopropionate) end via a free thiol or cysteine

two disulfides in the central region of the heavy chain using 2-mercaptoethylamine HCl [20] and Fab fragments are generated from digestion using papain [11]. The most elegant methods are to mutate a cysteine into the primary sequence of proteins and to append a thiol to the end of synthesized DNA strands [6], because they allow for a defined, sequence-specific coupling of the ligand to the crosslinker.

A nice alternative for a most common, strong but noncovalent site-directed anchor on the spacer has been introduced recently. The binding strength of the NTA (nitrilotriacetate) - His_6 (histidine 6) system, routinely used on chromatographic and biosensor matrices for the binding of recombinant proteins to which a His_6 tag is appended to the primary sequence, was found to be significantly larger than typical values of other ligand-receptor systems [21–23]. Therefore, a crosslinker containing an NTA residue is ideally suited for coupling a recombinant ligand carrying a His_6 in its sequence to the AFM tip. This general, site-directed, and oriented coupling strategy also allows rigid and fast control of the specificity of ligand-receptor recognition by using Ni^{++} as a molecular switch of the NTA-His_6 bond.

7.3 Fixation of Receptors to Probe Surfaces

For the recognition by ligands on the AFM tip, receptors should be tightly attached to probe surfaces. Loose receptor fixation could lead to a pull-off of the receptor from the surface by the ligand on the tip, which would consequently block ligand-receptor recognition and obscure the recognition force experiments.

Freshly cleaved muscovite mica is a perfectly pure and atomically flat surface and, therefore, ideally suited for MRFM studies. Moreover, the strong negative charge of mica accomplishes very tight electrostatic binding of certain types of biomolecules. Some receptor proteins such as lysozyme [20] or avidin [24] strongly adhere to mica due to the strong positive charge of these highly basic proteins at pH < 8. In this case, it is safe to purely adsorb the receptors from the solution, since the unspecific attachment to the surface is sufficiently strong for recognition force experiments. Nucleic acids are firmly bound to mica via Zn^{2+}, Ni^{2+}, or Mg^{2+} bridges [25]. Electrostatic interaction has also been used to adsorb the strongly acidic sarcoplasmic domain of the Ca^{2+} release channel via Ca^{2+}-bridges to probe the cytoplasmic surface [26]. Similarly, protein crystals and bacterial layers have been deposited onto mica in defined orientations with an appropriate choice of buffer conditions [27, 28].

In many cases, however, one cannot rely on electrostatic binding toward surfaces. Ideally, water-soluble receptors like globular antigenic proteins or extracellular protein chimeras are then anchored covalently. When glass, silicon, or mica is used as the probe surface, exactly the same surface chemistry described for the silicon AFM tips applies (see above). The number of reactive SiOH groups of the relatively chemically inert mica can be optionally increased by water plasma treatment [29]. Frequently, crosslinkers were also used on probe surfaces to provide the receptors with motional freedom, too [4]. Protein binding to photo-crosslinkers thereby enables rigid temporal control of the reaction [30].

A major limitation of the silicon chemistry is the impossibility of reaching a high surface density of functional sites (< $1000/\mu m^2$). By comparison, a monolayer of streptavidin would consist of 60 000 molecules per μm^2 and a phospholipid monolayer would consist of 1.7×10^6 molecules per μm^2. The latter density is also achieved by chemisorption of alkanethiols to gold. Tightly bound, ordered, self-assembled monolayers are formed on ultraflat gold surfaces and display excellent probes for AFM [10]. Alkanethiols can carry reactive groups at the free end that allow for covalent attachment of biomolecules [10, 31] (Fig. 7.3).

A recently developed anchoring strategy to gold uses dithio-phospholipids consisting of a propyldithio group at their hydrophobic end and either a phosphocholine head group (as host lipid) or a phosphethanolamine head group [32]. The latter head group is chemically reactive and was derived from a long chain biotin for the molecular recognition of streptavidin molecules in an initial study [32]. This phospholipid layer closely mimics a cell surface and has been optimized by nature to afford little nonspecific adsorption. Additionally, it can be spread as an insoluble monolayer at an air-water interface. Thereby, the ratio of functionalized thio-lipids in host thio-lipids accurately defines the surface density of bio-reactive sites in the monolayer. Subsequent transfer onto gold substrates leads to covalent and, therefore, tight attachment of the monolayer, and thus it is suited for recognition force studies.

Immobilization of cells mainly depends on the type of cell. Cells with adherent growth are readily usable for MRFM, whereas cells that grow in solution without contact to a surface have to be adsorbed. Various protocols for a tight cell anchoring are available. The easiest way is to grow the cells directly on glass or other surfaces in their cell culture medium [33]. Various adhesive coatings like Cell-Tak [34], gelatin, and poly-lysin increase the strength of adhesion and/or display appropriate surfaces

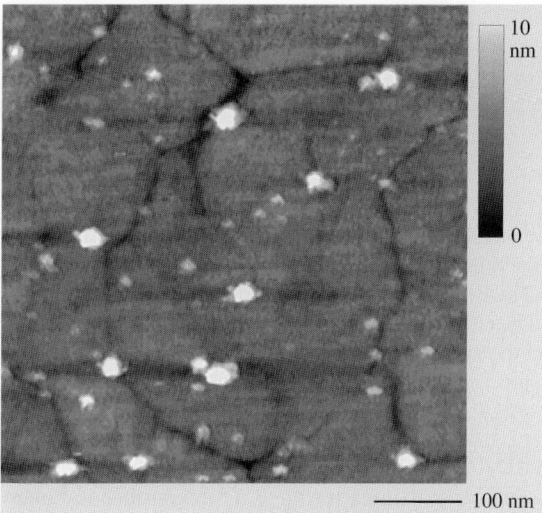

Fig. 7.3. AFM image of hisRNAP molecules specifically bound to nickel-NTA domains on a functionalized gold surface. Alkanethiols terminated with ethylene glycol groups to resist unspecific protein adsorption served as host matrices and were doped with 10% nickel-NTA alkenthiols. The sample was prepared to achieve full monolayer coverage. Ten individual his-RNAP molecules can be clearly seen bound to the surface. The more abundant, smaller, lower features are NTA islands with no bound molecules. The underlying morphology of the gold can also be distinguished (after [31])

to immobilize spherical cells. Other hydrophic surfaces like gold or carbon are suitable matrices as well [35]. Covalent binding of cells to surfaces can be accomplished by using PEG crosslinkers as the one described for the tip chemistry, since they react with free thiols on the cell surface [34]. Alternatively, PEG crosslinkers carrying a fatty acid penetrate into the interior of the cell membrane, which guarantees a sufficiently strong fixation without interference with membrane proteins [34].

7.4 Single-Molecule Recognition Force Detection

Quantification of recognition forces goes back to ensemble techniques such as shear flow detachment measurements (SFD) [36] and the surface force apparatus (SFA) [37]. In SFD, receptors are fixed to a surface to which beads or cells containing ligands attach via their specific bonds. A fluid shear stress is applied to the surface-bound particles by a flow of given velocity that disrupts the ligand-receptor-bonds at a critical force. However, the unbinding force per single bond can only be estimated, because the calculation of the net force on the particle is complicated by the stress distribution and the number of bonds per particle is assumed from the geometry of the contact area.

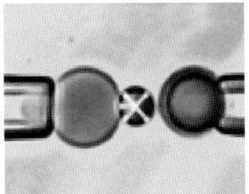

Fig. 7.4. Experimental setup of the biomembrane force probe (BFP). The spring in the BFP is a pressurized membrane capsule. Its spring constant is set by the membrane tension, which is controlled by micropipette suction. The BFP tip, a glass microbead of 1–2 μm diameter, was chemically glued to the membrane. Operated on the stage of an inverted microscope, the BFP (on the *left*) was kept stationary and the microbead test surface (on the *right*) was translated to/from contact with the BFP tip by precision piezo control (after [38])

SFA directly measures the vertical forces between two macroscopic surfaces containing receptors and ligands, respectively, using a spring as a force prober and interferometry for detection. Although SFA does not have a great sensitivity in measuring absolute forces and requires assumptions to be made for down-scaling to single bonds, its z-resolution is highly accurate (better than 1 Å) over a long range. Adhesion and compression forces, as well as rapid transient effects can be followed in real time.

The bio force probe (BFP) technique (for a recent paper see [38]) (Fig. 7.4) uses soft membranes as the force transducer, rather than a mechanical spring. A single cell or giant vesicle is held by suction at the tip of a glass micropipette and its spring constants are adjusted by the aspiration pressure and can be varied over several orders of magnitude. Interaction forces with opposing cells, vesicles, or functionalized beads are obtained from the micromechanical analysis of the deformations of the membranes and/or via interference contrast microscopy. A force range from 0.1 pN to 1 nN is accessible with high force sensitivity down to the single-bond level.

In optical tweezers (OT) small particles are manipulated by optical traps [39]. Three-dimensional light intensity gradients of a focused laser beam are used to pull or push the particles against another one. Particle displacements can be followed with nanometer resolution. Movements of single molecular motors [40] and force-extension profiles of single extensible molecules such as titin [41] and DNA [42] were measured. Defined, force-controlled twisting of DNA using rotating, magnetically manipulated particles gave even further insights into DNA viscoelastic properties [43].

The atomic force microscope (AFM) is the force measuring method with the smallest-sized force sensor and therefore achieves the highest lateral resolution. Radii of commercially available AFM tips vary between 2 and 50 nm. In contrast, the sensing particles in SFD, BFP, and OT are in the 1 to 10 μm range, and the surfaces used in SFA exceed millimeter extensions. The small apex of the AFM tip allows for the probing of single biomolecules. Originally designed to explore the topography of rigorous surfaces with atomic resolution [1] in area scans, the AFM was used in biology to image the secondary structures of nucleotides [25], as well as

7 Molecular Recognition Force Microscopy 291

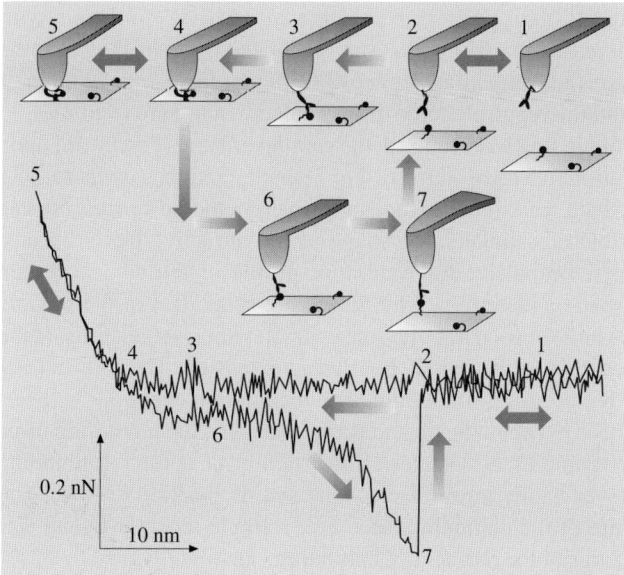

Fig. 7.5. Single-molecule recognition event detected with AFM. Raw data from a force-distance cycle with 100 nm z-amplitude at 1 Hz sweep frequency measured in PBS. Binding of the antibody on the tip to the antigen on the surface during approach (trace points 1 to 5) physically connects tip to probe. This causes a distinct force signal of distinct shape (points 6 to 7) during tip retraction, reflecting extension of the distensible crosslinker-antibody-antigen connection. The force increases until unbinding occurs at an unbinding force of 268 pN (points 7 to 2). (After [4])

polar lipid head groups and subunits of proteins embedded in membranes in liquid environments [27, 28].

The additional potential of the AFM to detect ultralow forces has been successfully applied to the detection of interaction forces of single receptor-ligand pairs [2–4]. These forces are measured in so called force-distance cycles using a ligand-carrying tip mounted to a cantilever and a probe surface with receptors firmly attached. A force-distance cycle, where a distensible tether was used to couple ligands to tips and receptors to surfaces is shown in Fig. 7.5 (from [4]). At a fixed lateral position the tip approaches the probe surface and is subsequently retracted, and the cantilever deflection Δx is measured in dependence of the tip-surface separation Δz. Cantilever deflections are detected by reflecting a laser beam off the back of the cantilever onto a split photodiode. Bending of the cantilever causes a change in the position where the laser beam strikes the photodiode and, therefore, a change in the output voltage of the photodiode.

The force F acting on the cantilever directly relates to the cantilever deflection Δx according to Hook's law $F = k\Delta x$, where k is the cantilever spring constant. During tip-surface approach (trace, and line 1 to 5), the cantilever deflection remains zero far away from the surface (1 to 4), because there is no detectable tip-surface interaction.

Upon tip-surface contact (4), the cantilever bends upward (4 to 5), consistent with a repulsive force that increases linearly with z the harder the tip is pushed into the surface. Subsequent tip-surface retraction (retrace, 5 to 7) first leads to relaxation of the cantilever bending and the repulsive force drops back down to zero (5 to 4).

When ligand-receptor binding has occurred, the cantilever bends downward upon retraction, reflecting an attractive force (retrace, 4 to 7) that increases with increasing tip-surface separation. Since the cognitive molecules were tethered to the surfaces via flexible and distensible crosslinkers, the shape of the attractive force-distance profile is, in contrast to the repulsive force-distance profile of the contact region (4 to 7), nonlinear. Its shape is determined by the elastic properties of the flexible PEG crosslinker [17, 44] and shows parabolic-like characteristics, which reflects the increase of the spring constant of the crosslinker during extension. The force attractive profile contains the features of a single-molecule recognition event.

The physical connection between the tip and surface sustains the increasing force until the ligand-receptor complex dissociates at a certain critical force (unbinding force, 7), and the cantilever finally jumps back to the resting position (7 to 2). The quantitative force measure of the unbinding force f of a single receptor-ligand pair is directly given by the force at the moment of unbinding (7).

If the ligand on the tip does not form a specific bond with one of the receptors on the probe surface, the recognition event is missing and retrace looks like trace. Also, the specificity of the ligand-receptor binding is usually demonstrated in block experiments. Free ligands are injected in solution so as to block receptor sites on the surfaces. As a further consequence, recognition signals completely disappear, because the receptor sites on the surface are blocked by the ligand of the solution and, thus, prevent recognition by the ligand on the tip.

The force resolution of the AFM is limited by the thermal noise of the cantilever that is determined by its spring constant. According to the equipartition theorem, the cantilever has on average a thermal energy of $0.5\,k_B T$, therefore, $0.5\,k\Delta z^2 = 0.5\,k_B T$, where k_B is Boltzmann's constant and T the absolute temperature. Since $F = k\Delta x$, the force sensitivity is given by $\Delta F = (k_B T k)^{1/2}$ and is, therefore, better the softer the cantilever. The smallest force that can be detected with commercially available cantilevers is in the few picoNewton range. For thinner levers with low spring constants, the force resolution decreases by an order of magnitude [45]. Due to their smaller mass, they have a higher resonance frequency (30–100 kHz in liquid). When deflection is recorded at a typical bandwidth of 10 kHz, only a fraction of the thermal noise is included and fast movements of the tip can still be detected [45]. An alternative route for improving the force resolution was found by electronically increasing the apparent damping constant, which results in the reduction of thermal noise [46].

Besides the detection of intermolecular forces between receptors and ligands, the AFM also shows great potential in measuring intramolecular force profiles of single molecules. The molecule is clamped in between the tip and the probe surface and its visco-elastic properties are studied in force-distance cycles. The detailed force extension profile of the recognition event shown in Fig. 7.5 contains, for example, the elastic features of the crosslinker (PEG in this particular case) [44] with which the ligand is tethered to the tip of the AFM.

7.5 Principles of Molecular Recognition Force Spectroscopy

Specific interactions underlying molecular recognition are a distinct class of highly complementary, noncovalent bonds between biological macromolecules composed typically of a few tens of weak interactions such as electrostatic, polar, van der Waals, hydrophobic, and hydrogen bonds. The energy of each single bond is just slightly higher than the thermal energy $k_B T$. Due to the power law dependence of each weak interaction potential and the directionality of the hydrogen bonds, the attractive forces between the receptors and their cognitive ligands are extremely short-range. Therefore, a close geometrical and chemical fit within the binding pocket is a prerequisite for molecular recognition.

Conformational flexibility of at least one of the two binding partners is required for their structural adaptation. The weak interaction bonds are believed to be formed in a spatially and temporarily correlated fashion, thus creating a strong and highly specific molecular interaction. Recognition binding sites are structurally and chemically unique and selective for distinct receptor/ligand combinations. The overall affinity or binding energy of a receptor-ligand bond of a few tens of $k_B T$ is typically sufficient to effect chemical or physical changes of the binding partners, which often initiate biochemical reaction cascades or cellular metabolic pathways.

The binding energy E_B, given by the free energy difference between the bound and the free state, is the common parameter to describe the strength of a bond and can be determined in ensemble average calorimetric experiments. E_B determines the ratio of bound complexes, [RL], to the product of free reactants, [R] [L], at equilibrium in solution and is related to the equilibrium dissociation constant K_D, the affinity parameter, according to $E_B \alpha - \log(K_D)$. At thermodynamic equilibrium, the number of complexes that form per unit of time, k_{on} [R] [L], equals the number of complexes that dissociate per time unit, k_{off} [RL]. The empirical kinetic rate constants, on-rate k_{on} and off-rate k_{off}, are related to the equilibrium dissociation constant K_D through $K_D = k_{off}/k_{on}$.

In order to get an estimate for the interaction forces f from binding energies E_B, the dimension of the binding pocket may be used as the characteristic length dimension l; thus $f = E_B/l$. Typical values of $E_B = 20 k_B T$ and $l = 0.5$ nm yields $f \sim 170$ pN as an order of magnitude feeling for the strength of a single receptor-ligand bond. Classical mechanics describes the force required to separate interacting molecules as the gradient in energy along the interaction potential. The complexed molecules would, therefore, dissociate when force exceeds the steepest gradient in energy. However, activation barriers, temperature, time scales, and the detailed characteristics of the energy landscape, all known to be essential for the understanding of receptor-ligand dissociation, are not considered in this purely mechanical picture.

Ligand-receptor binding is generally a reversible reaction. Viewed on the single-molecule level, the average lifetime of a ligand-receptor bond, $\tau(0)$, is given by the inverse of the kinetic off-rate, k_{off}, so $\tau(0) = k_{off}^{-1}$. Therefore, ligands will dissociate from receptors without any force applied to the bond at times larger than $\tau(0)$ (covering a range from milliseconds to days for different receptor-ligand combinations). In contrast, if molecules are pulled faster than $\tau(0)$, the bond will resist and require a force for detachment. The size of the unbinding force may vary over more than one order of magnitude, and even exceed the adiabatic limit given by the steepest energy gradient of the interaction potential if bond breakage occurs faster than diffusive relaxation (nanosecond range for biomolecules in viscous aqueous medium) and friction effects become dominant [47]. Therefore, unbinding forces do not resemble unitary values, and the dynamics of the experiment defines the physics underlying the unbinding process.

At the millisecond to second time scale of AFM experiments, thermal impulses govern the unbinding process. In the thermal activation model, the lifetime of a complex in solution is described by a Boltzmann ansatz, $\tau(0) = \tau_{osc} \exp(E_b/k_B T)$ [48], where τ_{osc} is the inverse of the natural oscillation frequency and E_b the energy barrier for dissociation. Hence, due to the thermal energy there is a finite probability of overcoming the energy barrier E_b, which leads to the separation of the receptor-ligand complex. The mean complex lifetime in solution is, therefore, determined by the ratio of E_b over the thermal energy.

A force acting on a binding complex deforms the interaction energy landscape and lowers the activation energy barrier, very similar to the mode of operation of an enzyme (Fig. 7.6). Thus the input of thermal energy by forces acting in this time regime reduces the bond lifetime. The lifetime $\tau(f)$ of a bond loaded with a constant force f is given as $\tau(f) = \tau_{osc} \exp[(E_b - x_\beta f)/k_B T]$ [48], x_β being interpreted as the distance of the energy barrier E_b from the energy minimum along the direction of the applied force. In this model, the energy barrier decreases linearly with the force applied. A detailed physical basis for Bell's theory has been derived in [49], where the strength of weak non-covalent bonds was studied in liquid using Kramer's theory for reaction kinetics in liquids under the influence of force.

The lifetime $\tau(f)$ under constant force f compares to the lifetime at zero force, $\tau(0)$, according to $\tau(f) = \tau(0) \exp(-x_\beta f/k_B T)$ [4] for a single sharp energy barrier for which a mono-exponential dependence is characteristic. Using AFM, receptor-ligand unbinding is commonly measured in force-distance cycles (cf. previous chapter), during which the applied force to the complex does not remain constant. Rather, it increases in a complex fashion at a nonlinear rate determined by the pulling velocity, the spring constant of the cantilever, and the force-distance profile of the complexed biomolecules. The main contribution of the thermal activation, however, comes from the part of the force curve, which is close to unbinding. Therefore, an effective force increase or loading rate r can be deduced from $r = df/dt =$ pulling velocity times effective spring constant at the end of the force curve just before unbinding occurs. The approximation of a constant loading rate using force-distance cycles with AFM is sound with a properly defined effective loading rate [51].

The dependence of the unbinding force on a linear loading rate in the thermally activated regime was first derived in [49] and further described in [50]. Force-induced dissociation of receptor-ligand complexes using AFM or BFP can be regarded as an

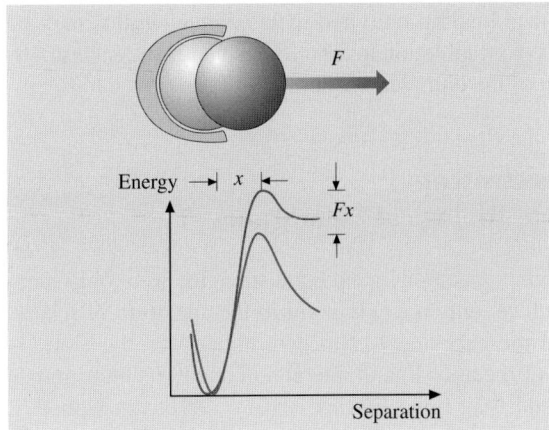

Fig. 7.6. Dissociation over a single sharp energy barrier. Under a constant force, the barrier is linearly decreased by the applied force F, giving rise to a characteristic length scale x that is interpreted as the distance of the energy barrier from the energy minimum along the projection of the force (after [50])

irreversible process, because the two binding partners are further separated after dissociation has occurred. Therefore, rebinding and back-reactions are negligible. The unbinding process itself is of a stochastic nature, and the likelihood of bond survival is expressed in the master equation as time-dependent probability $N(t)$ to be in the bound state under a linearly increasing force $f = rt$, thus $dN(t) = -k_{off}(rt)N(t)$ [50].

Rewriting Bell's formula using $\tau(f) = k_{off}^{-1}(f)$ relates the kinetic off-rate at a given force f, $k_{off}(f)$, to the off-rate at zero force, $k_{off}(0)$: $k_{off}(f) = k_{off}(0) \exp(l_r f / k_B T)$. The master equation combined with Bell's formula results in a spectrum of Gaussian-like distributions of unbinding forces (frequency of occurrence versus force) parameterized by the loading rate r [50]. Thus the maximum of each force distribution, $f^*(r)$, reflects the most probable force of unbinding for the respective loading rate r. f^* is related to r through $f^* = f_\beta \log_e(rk_{off}^{-1}/f_\beta)$, where the slope is governed by f_β, a force scale set by the ratio of the thermal energy $k_B T$ to the length scale x_β, which marks the thermally averaged position of the energy barrier along the direction of the force [49, 50]. Apparently, the unbinding force f^* scales linearly with the logarithm of the loading rate. For a single barrier, this would give rise to a simple, linear dependence of the force versus log loading rate. In cases where more barriers are involved along the escape path, the curve will follow a sequence of linear regimes, each of which marks a particular barrier [38]. Hence, transition from one regime to the other is associated with an abrupt change of slope determined by the characteristic barrier length scale.

In force spectroscopy experiments, the dynamics of pulling on specific receptor-ligand bonds is varied, which leads to detailed structural and kinetic information of the bond breakage, as discussed above and shown in the following chapter on explicit examples. Length scales of energy barriers are obtained from the slope of the spectroscopy plot (force versus loading rate) and their relative heights may be gained from the force shifts [38]. Extrapolation to zero forces yields the kinetic off-rate constant for the dissociation of the complex in solution [52].

7.6 Recognition Force Spectroscopy: From Isolated Molecules to Biological Membranes

In the early pioneering AFM studies receptor-ligand bond strengths were only measured at one loading rate. Therefore, only a single point in the spectrum of forces, that depend on the dynamics of the experiment, was determined. Avidin–biotin is often regarded as the prototype of receptor-ligand interaction due to its enormously high affinity ($K_D = 10^{-13}$ M) and long bond lifetime ($\tau(0) = 80$ days). Hence, it may not be too surprising that the first realizations of single-molecule recognition force detections were made with biotin and its cognitive receptors, streptavidin [2], the succinilated form of avidin, and avidin [3]. Unbinding forces of 250–300 pN and 160 pN were found for streptavidin and avidin, respectively. It was also shown [53] that the forces vary with the spring constant of the cantilever using the same pulling velocity, which is consistent with the loading rate dependency of the unbinding force discussed above. The recognition forces of various biotin analogs with avidin and streptavidin [53], as well as with site-directed streptavidin mutants [54] were related to their energies and found to be correlated with the equilibrium binding enthalpy [53] and the enthalpic activation barrier [54] but independent of free energy parameters. It was suggested that internal energies of the bond breakage, rather than entropic changes were probed by the force measurements [54].

7.6.1 Forces, Energies, and Kinetic Rates

In first measurements of inter- and intramolecular forces of the DNA double helix [6], complementary single-strand DNA oligonucleotides covalently attached to the AFM tip and the probe surface, respectively, were probed. The multimodal force distributions obtained (three distinct, separated force peaks) were associated with interchain interactions of single pairs of 12, 16, and 20 base sequence length. A long strain single-strand DNA with cohesive ends was sandwiched in between tip and substrate and its intramolecular force pattern revealed elastic properties comparable to polymeric molecules [6]. Single molecule interactions between all possible 16 combinations of the four bases were also studied [7] and yielded specific binding only when complemetary bases were present on the tip and the probe surface, indicating that AFM is capable of following Watson-Crick base pairing.

Anibody-antigen interactions are of key importance for the function of the immune system. Their affinities are known to vary over orders of magnitudes. For single-molecule recognition AFM studies, the molecules were surface-coupled via flexible and distensible crosslinkers [4, 10, 13, 14] to provide them with sufficient motional freedom, so that problems of misorientation and steric hindrance that can obscure specific recognition are avoided. In [4, 12], the tips were functionalized with a low antibody density, so that on the average only a single antibody on the tip end could access the antigens on the surface. Hence, isolated single molecular antibody-antigen complexes could be examined. It was observed that the interaction sites of the two Fab fragments of the antibody (the antibody consists of two antigen-active Fab-fragments and one Fc-portion) are able to bind simultaneously and independently with equal binding probability. Single antibody-antigen binding events were also studied with tip-bound antigens [10] using an antibody biosensor probe surface [13], or single chain antibody-fragments attached to gold via freely accessible sulfhidryl groups [14]. In the latter study, the unbinding forces of two mutant single-chain antibodies, varying by an order of magnitude in their affinity, were compared. The mutant of ten times lower affinity (consistent with 12% lower free energy due to the logarithmic dependence of the affinity on free energy) showed 20% lower unbinding force.

All these studies describe detection of intermolecular forces between single receptors and their cognitive ligands. However, the AFM can also be used to characterize single-molecule intramolecular forces, e.g., interactions responsible for stabilizing the conformations and tertiary structures of polymers and proteins. The force-extension profile (cf. also stretching of PEG in Fig. 7.5) of polysaccharide dextran revealed multiple elastic regimes dominated by entropic effects, internal bond angle twistings, and conformational changes [55, 56]. The complex quantities of elastic and viscous force contributions upon stretching a single dextran molecule were resolved by resonant energy tracking using dynamic force spectroscopy [57]. Titin, a fibrillose muscle protein, contains a large number of repetitive IgG domains that shows a characteristic sawtooth pattern in the force profile, reflecting their sequential unfolding (Fig. 7.7) [58]. Unfolding and refolding kinetics and energetics of proteins were studied in pull-hold-release cycles [59] by carefully handling the molecule between tip and surface. These studies yield crucial information on the relation between structure and function of single molecules.

Besides the investigation of forces, single-molecule recognition force microscopy studies allow for the estimation of association and dissociation rates [4, 12, 22, 52, 60, 61], energies [38], and structural parameters of the binding pocket [4, 12, 15, 60, 61]. For single molecule studies, quantification of the on-rate constant k_{on} for the association of the ligand on the tip to a receptor on the surface requires determination of the interaction time $t_{0.5}$ needed for half-maximal probability of binding. This value is obtained from an experiment in which the encounter duration of the ligand-coated tip on the receptor-containing surface is varied until a range of the unbinding activity from zero to saturation is achieved [60]. With the knowledge of the effective ligand concentration c_{eff} on the tip available for receptor interaction, k_{on} is given by $k_{on} = t_{0.5}^{-1} c_{eff}^{-1}$. The effective concentration c_{eff} is described by the effective volume V_{eff} the tip-tethered ligand diffuses about the tip, which yields $c_{eff} = N_A^{-1} V_{eff}^{-1}$, where N_A is the Avogadro's number. V_{eff} is essentially a half-sphere with a radius of the effective tether length, and the kinetic on-rate can, therefore, be calculated from $k_{on} = t_{0.5}^{-1} N_A 2\pi/3r^3$. However, since k_{on} critically depends on the tether length r, only order of magnitude estimates are gained [60].

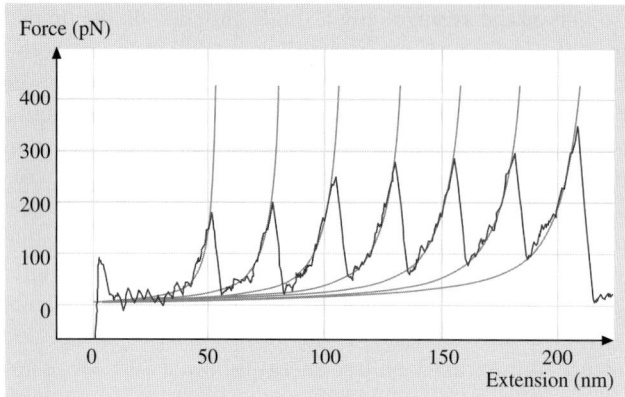

Fig. 7.7. Force extension profile of a single titin molecule. The periodic sawtooth-like pattern reflects domain unfolding in the peptide chain consistent with its modular construction and can be explained as a stepwise increase of the molecule's length upon each unfolding event (after [58])

Additional information about the unbinding process is contained in the unbinding force distributions. Not only the maximum f^*, but also its intrinsic width σ increases with increasing loading rate [22, 38]. Apparently, at slower loading rates the systems adjust closer to equilibrium, which leads to smaller values of both the force f^* and its variation σ. This is another indication that dynamics is an important issue when studying recognition forces. An effective lifetime $\tau(f)$ of the bond under an applied force f was estimated by the time the cantilever spends in the force window spanned by the standard deviation σ of the f distribution [4]. The time the force increases from $f^* - \sigma$ to $f^* + \sigma$ is thus given by $\tau(f) \approx 2\sigma/df/dt$ [4]. In an example of the Ni^{2+}-chelating ligand/receptor interaction nitrotrilacetate/hexa-histidin [22], the lifetime $\tau(f)$ decreased with increasing pulling force f from 17 ms at 150 pN to 2.5 ms at 194 pN. The data were fitted with Bell's formula, confirming the theoretically predicted exponential lifetime-force relation for the reduction of the lifetime $\tau(f)$ by the applied force f and, thus yielding a lifetime at zero force, $\tau_0 = 15$ s. Direct measurements of lifetimes are only possible by using a force clamp, where a constant adjustable force is applied to the complex via a feedback loop and the time duration of bond survival is detected. This configuration has been developed recently to study the force dependence of the unfolding probability of the filament muscle protein titin [62].

Theory predicts (cf. previous chapter) that the unbinding force of specific and reversible ligand-receptor bonds is dependent on the rate of the increasing force [47, 49, 63] during force-distance cycles. Indeed, in experiments unbinding forces were found to assume not a unitary value, but rather were dependent on both the pulling velocity [52, 60] and the cantilever spring constant [2]. The theoretical findings were confirmed by experimental studies and revealed a logarithmic dependence of the unbinding force on the loading rate [15, 22, 38, 52, 60, 61], consistent with the exponential lifetime-force relation described above. On a half-logarithmic plot, a single linear slope is characteristic for a single energy barrier probed in the thermally activated regime [38]. The same relation was also found for unfolding forces of proteins [55]. The slope of the force spectroscopy curve contains the length scale of the energy barrier, which may be connected to the length dimension of the interaction binding pocket [60]. The kinetic off-rate k_{off} at zero force is gained from the extrapolation to zero force [52, 60]. Combining k_{off} with k_{on} leads to a value for the equilibrium dissociation constant K_D, according to $K_D = k_{off}/k_{on}$.

A simple correlation between unbinding force and solution kinetics was found for nine different single-chain Fv fragments constructed from point mutations of three unrelated anti-fluorescein antibodies [61]. These molecules provide a good model system because they are monovalent and follow a 1:1 binding model without allosteric effects, and the ligand (fluoerscein) is inert and rigid. For each mutant, the affinity K_D, the kinetic rate constants k_{on} and k_{off}, and the temperature dependence of k_{off} yielding the activation barrier E_b, were determined in solution. In a series of single-molecule AFM recognition events, the loading rate (r) dependence of the unbinding forces (f) was measured. The maxima of the force distributions f^* scaled linearly with $\log r$. Therefore, unbinding was dictated by a single energy barrier along the reaction path (Fig. 7.8). Since the measured unbinding forces of all mutants precisely correlated with k_{off}, the unbinding path followed and the transition state crossed cannot be too dissimilar in spontaneous and forced unbinding. Interestingly, the distance of the energy barrier along the unbinding pathway determined from the force spectroscopy curve was found to be proportional to the barrier height and, thus, most likely includes elastic stretching of the antibody construct during the unbinding process.

7.6.2 Complex Bonds and Energy Landscapes

More complex force spectra were found from unbinding force measurements between complementary DNA strands of different lengths [15]. DNA duplexes with 10, 20, and 30 base pairs, were investigated to reveal unbinding forces from 20 to 50 pN of single duplexes, depending on the loading rate and sequence length. The unbinding force of the respective duplexes scaled with the logarithm of the loading rate, however, each of them with a different slope (Fig. 7.9). The distance of the energy barrier along the separation path and the logarithm of the kinetic off-rate, both obtained from the force spectroscopy plot as parameters describing the energy landscape, were found to be proportional to the number of DNA complexes per duplex. This defined scaling leads to a model of fully cooperative bonds in a series for the unbinding of the base pairs in the duplex, described by one distinct length and time scale for each duplex length, where each single bond contributes a defined increment [15]. A temperature study revealed that entropic contributions also play a role in the unbinding between complementary DNA strands [64].

Energy landscapes of receptor-ligand bonds in biology can be rugged terrains with more than one prominent energy barrier. The inner barriers are undetectable

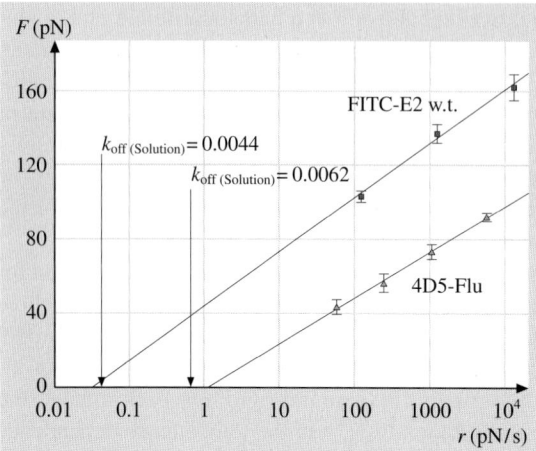

Fig. 7.8. Loading rate dependence of the unbinding force dependence of two anti-fluoresceins. For both FITC-E2 w.t. and 4D5-Flu, a strictly mono-logarithmic dependence was found in the range accessed, indicating that only a single energy barrier was probed. The same energy barrier dominates dissociation without forces applied, because extrapolation to zero force matches kinetic off-rates determined in solution (indicated by *arrow*) (after [61])

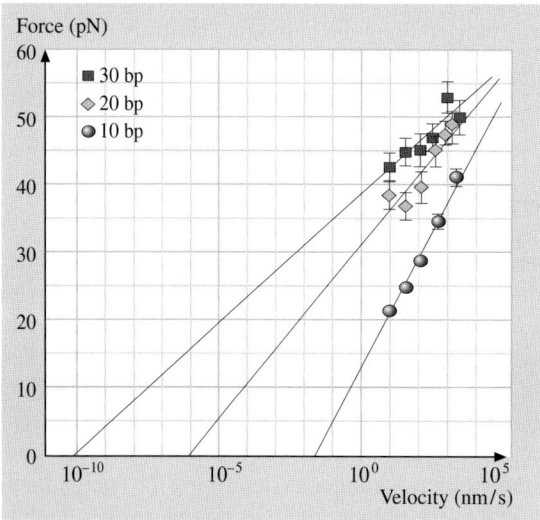

Fig. 7.9. Dependence of the unbinding force between DNA single-strand duplexes on the retract velocity. Besides the mono-logarithmic behavior, the unbinding force scales with the length of the strands, increasing from the 10- to 20- to 30-base-pair duplexes (after [15])

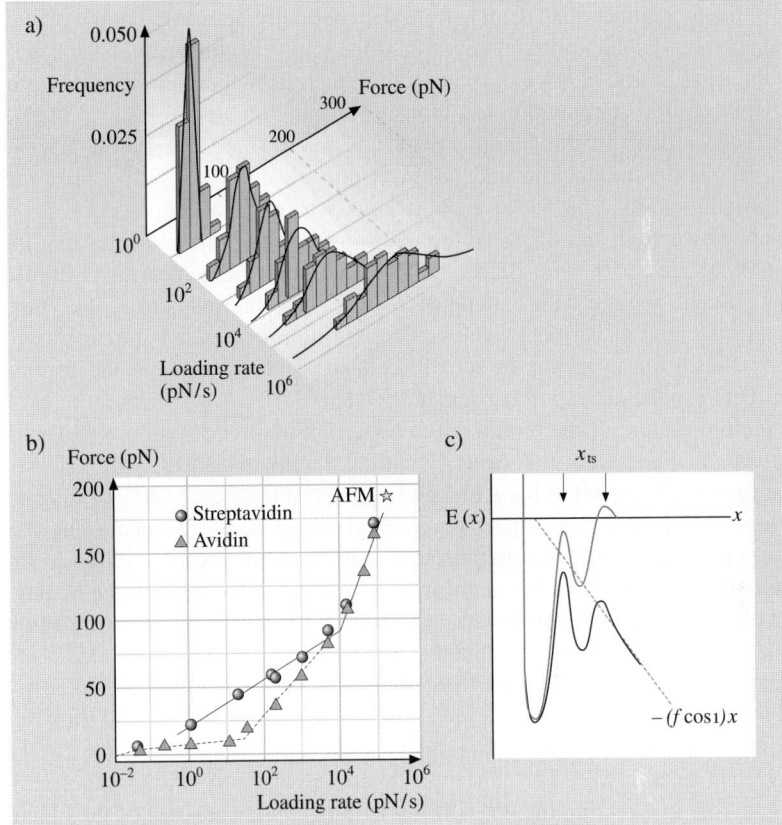

Fig. 7.10. Unbinding force distributions and energy landscape of a complex molecular bond. (**a**) Force histograms of single biotin-streptavidin bonds recorded at different loading rates. The shift in peak location and the increase in width with increasing loading rate is clearly demonstrated. (**b**) Dynamic force spectra for biotin-streptavidin (*circles*) and biotin-avidin (*triangles*). The slopes of the linear regimes mark distinct activation barriers along the direction of force. (**c**) Conceptual energy landscape traversed along a reaction coordinate under force under an angle θ to the molecular coordinate. The external force f adds a mechanical potential that tilts the energy landscape and lowers the barriers. The inner barrier starts to dominate when the outer has fallen below it due to the applied force (after [38])

in solution assays, but can become predominant for certain time scales in force-induced unbinding. Therefore, force spectroscopy appears to be a proper tool to discover hidden activation barriers and explore the energy landscape in greater detail. In initial AFM studies [53, 54], it was concluded that the interaction strength of biotin-streptavidin is in the range of 200–300 pN and the strength of biotin-avidin is somewhat lower (~ 150 pN). However, using BFP, the loading rate was varied over eight orders of magnitude (from 10^{-2} to 10^6 pN/s) and gained a detailed picture of the force spectroscopy curve with forces increasing from 5 to 200 pN (Fig. 7.10) [38].

Distinct linear regimes that demonstrate the thermally activated nature of the bond breakage are visible, abrupt changes in slope imply a number of sharp barriers along the dissociation pathway. Above 85 pN, there is a common high strength regime for both biotin-streptavidin and biotin-avidin with a slope of $f_\beta = 34$ pN, locating a barrier deep in the binding pocket at $x_\beta = 0.12$ nm. Below 85 pN, the slopes for biotin-streptavidin and biotin-avidin are different. $f_\beta = 8$ pN for biotin-strepavidin marks a barrier at $x_\beta = 0.5$ nm, whereas the steeper slope $f_\beta = 13 - -14$ pN between 38 and 85 pN in the biotin-avidin spectrum indicates that its next barrier is located at 0.3 nm. Below 11 pN, the biotin-avidin spectrum exhibits a very low strength regime with a slope of $f_\beta = 1.4$ pN that maps to $x_\beta = 3$ nm. In addition to the map of barrier locations, the logarithmic intercepts found by extrapolation of each linear regime to zero force also yield estimates of the energy differences between the activation barriers (Fig. 7.10).

A meticulous picture of the biotin-avidin bond rupture process with amino acid resolution was obtained with molecular dynamics simulation (MDS) [47, 63]. Although the time scales of MDS for AFM pulling experiments are orders of magnitudes faster, forces in laboratory speeds were approximated. By reconstructing the interaction potential, the experimentally found transition states were readily identified. Simulations of antibody-antigen unbinding, however, revealed a large heterogeneity of enforced unbinding pathways and a correspondingly large flexibility of the binding pocket region, which exhibited significant induced fit motions [65].

The velocity dependence of rupture forces and adhesion probability of the receptor/ligand system P-selectin/P-selectin glycoprotein ligand-1 (PSGL-1) and their relation to kinetic and thermodynamic parameters obtained in ensemble average assays were analyzed in detail in [52]. The adhesive interactions of this system play a central immunogenic role in directing the PSGLT-1-containing leukocytes out of the blood stream and into sites of inflammation on cell layers via P-selectin. The high molecular elasticity of the P-selectin/PSGL-1 complex was determined from the force extension profile gained in force-distance cycles and described by the freely jointed chain, a polymer model that yields the Contour length (overall maximum extension), the persistence length (effective segment dimension), and the molecular spring constant. Rupture forces of single complex unbinding increased logarithmically with the loading rate (spring constant times pulling velocity), as expected from theory, and yielded $k_{off} = 0.02$ s^{-1} and $l = 0.25$ nm.

Another characteristic of the P-selectin/PSGL-1 relevant to its biological function was found by investigating the adhesion probability, defined as the ratio between the force-distance cycles showing an adhesion event and the total number of cycles, on the pulling velocity. Counterintuitively and in contrast to experiments with antibody-antigen [4], cell adhesion proteins [9] and cadherins [60], the adhesion probability increased with increasing velocities [52]. Since the adhesion probability approaches 1.0 at fast velocities, P-selectin/PSGL-1 complex formation must occur as soon as the tip reaches the surface, thus reflecting a fast kinetic on-rate. Upon the increase in force during pulling, the complexes only dissociate under a measurable force when pulled at fast velocities. No force will be detected at low velocities because the complex has already dissociated by the pure thermal energy and/or a force beneath the noise limit. A quantitative analysis revealed a fast kinetic off-rate of $k_{\text{off}} = 15 \pm 2\,\text{s}^{-1}$. The fast kinetic rates and high chain-like elasticity of the PSGL-1 system guarantee a rapid exchange of the leukocytes on endothelial cell surfaces and therefore provide the basis of the physiological leukocyte *roling* process [52].

PSGL-1 recognition to another member of the selectin family, L-selectin, was studied using BFP [66]. The loading rate was varied over three orders of magnitude, thereby gaining rupture forces from a few to 200 pN. Plotted on a logarithmic scale of loading rates, the rupture forces reveal two prominent energy barriers along the unbinding pathway. Strengths above 75 pN arise from rapid detachment (< 0.01 s) impeded by an inner barrier that requires Ca^{2+}, whereas strengths below 75 pN occur under slow detachment (> 0.01 s) impeded by an outer barrier, most likely constituted by an array of weak hydrogen bonds. It was speculated that a complex hierarchy of inner and outer barriers is significant in selectin-mediated function [66].

Single molecule recognition was also used to investigate the mechanism of the cadherin-mediated adhesion process of vascular endothelial cells. A monolayer of these cells constitutes the major barrier of the body that separates the blood compartment from the extracellular space of tissues. Homophilic cadherin adhesion acting within the cell junctions allows fast dynamic cellular remodeling to change the barrier properties of the cell layers. Recordings of force-distance cycles showed specific recognition events between tip- and surface-bound chimeric VE-cadherin dimers. Measuring the binding activity in dependence of the free Ca^{2+} concentration [60] revealed an apparent K_D of 1.15 mM with a Hill coefficient of $n_H = 5.04$, indicating high cooperativity and steep dependency. It might be of physiological relevance that a local drop of free Ca^{2+} in the narrow intercellular cleft weakens intercellular adhesion and is, therefore, involved in facilitating cellular remodeling.

The rather low trans-interaction force of a single bond between two cadherin strand dimers was found to be amplified by complexes of cumulative binding strength, visible as distinct force quanta in the adhesion force distributions. These associates were formed by time-controlled lateral and trans-oligomerization, suggesting that cadherins from opposing cells associate to form complexes and, thus, increase the intercellular binding strength. The rather low values for the kinetic rates and the adhesive binding activity determined as described above, make cadherins ideal candidates for adhesion regulation by cytoskeletal tethering, thereby setting the kinetic rate of lateral self-association [60].

7.6.3 Live Cells and Membranes

So far, there have been only a few attempts to apply recognition force microscopy to living cells. Recently, *Lehenkari* et al. [67] measured the binding forces between integrin receptors in intact cells and Arg-Gly-Asp (RGD) amino acid sequence containing extracellular matrix protein ligands. Ligands, which had a higher predicted affinity for the receptor, gave greater unbinding forces, and the amino acid sequence/pH/divalent cation dependency of the receptor-ligand interaction examined by AFM was similar to the one observed under bulk measurement conditions using other techniques. In another study [68], the AFM was used to detect the adhesive strength between concavalin A (con A) coupled to an AFM tip and Con A receptors on the surface of NIH3T3 fibroblast cells. Crosslinking of receptors led to an increase in adhesion that could be attributed to enhanced cooperativity among adhesion complexes, which has been proposed as a physiological mechanism for modulating cell adhesion.

The sidedness and accessability of protein epitopes of the Na^{2+}/D-glucose cotransporter in intact brush border membrane vesicles was sensed with an AFM tip containing an antibody directed against an a sequence close to the glucose binding site [35]. Orientations and conformationsof the transporter molecule were found to change during glucose binding and transmembrane transport, as evident from the observation of three distinct states discernable in interaction strength and binding activity. Force spectroscopy experiments between leukocyte function-associated antigen-1 (LFA-1) on the tip and intercellular adhesion molecule-1 (ICAM-1) expressed on cell surfaces [69] revealed a bilinear behavior in the force vs. log loading rate curve, marking a steep inner activation barrier and a wide outer activation barrier. Since addition of Mg^{2+}, a co-factor that stabilizes the LFA-1/ICAM-1 interaction, elevated the unbinding force of the complex in the slow loading regime and EDTA suppressed the inner barrier, it was suggested that the equilibrium dissociation constant is regulated by the energetics of the outer barrier, whereas the inner barrier determines the ability to resist an external force.

Forces about ten times higher (3 nN) were achieved when whole cells were used to modify the tip instead of single molecules, as shown in measurements of inter-cell force detections between trophoplasts and uterine pithelium cells [70]. In a similar configuration but with defined control of the interaction, de-adhesion forces at the resolution of individual cell adhesion molecules in cell membranes were detected and the expression of a gene was quantitatively linked to the function of its product in cell adhesion [71].

7.7 Recognition Imaging

Besides studying ligand-receptor recognition processes, the identification and localization of receptor binding sites on bio-surfaces such as cells and membranes are of particular interest. For this, high resolution imaging must by combined with force detection, so that binding sites can be assigned to structures.

A method for imaging functional groups using chemical force microscopy was developed in [72]. Both tip and surface were chemically modified, and the adhesive interactions between tip and surface directly correlated with friction images of the patterned sample surfaces. Simultaneous information for topography and recognition forces on a µm-sized streptavidin pattern was obtained by lateral force mapping on a µm-sized streptavidin pattern with a biotinylated AFM tip. An approach-retract cycle was performed in every pixel of the image, and the information for topography was gained from the contact region, whereas a recognition image was constructed from the recognition forces of the force curve [72]. This method was also used to map functional receptors on living cells [73] and differentiate red blood cells of different blood groups within a mixed population using AFM tips conjugated with a blood group A specific lectin [74].

A first identification and localization of single-molecule antigenic sites was realized by laterally scanning an AFM tip containing antibodies in one dimension across a surface containing a low density of antigens during recording force-distance cycles. The binding probability was determined in dependence of the lateral position, resulting in Gaussian-like profiles [4, 12]. The position of the antigen could be determined from the position of the maximum, yielding a positional accuracy of 1.5 nm. The width of the binding profile of 6 nm reflects the dynamical reach of the antibody on the tip, provided by the crosslinker used for coupling the antibody to the tip, and can be considered the lateral resolution obtained by this technique [4]. With a similar configuration, height and adhesion force images were simultaneously obtained with resolution approaching the single-molecule level as well [75].

The strategies of force mapping, however, lack high lateral resolution [72] and/or are much slower in data acquisition [4, 12, 75] than topography imaging, since the frequency of the retract-approach cycles performed in every pixel is limited by hydrodynamic forces in the aqueous solution. In addition, obtaining the force image requires the ligand to be disrupted from the receptor in each retract-approach cycle. For this, the z-amplitude of the retract-approach cycle must be at least 50 nm and, therefore, the ligand on the tip is without access to the receptor on the surface for most of the experiment. The development of shorter cantilevers [45], however, increases the speed for force mapping, because the hydrodynamic forces are significantly reduced and the resonance frequency is higher than that of commercially available cantilevers. The short cantilevers were also recently applied to follow the association and dissociation of individual chaperonin proteins GroES to GroEL in real time using dynamic force microscopy topography imaging [76].

An imaging method for the mapping of antigenic sites on surfaces was recently developed [20] by combining molecular recognition force spectroscopy [4] with dynamic force microscopy (DFM) [25, 77]. In DFM, the AFM tip is oscillated across a surface and the amplitude reduction arising from tip-surface interactions is held constant by a feedback loop that lifts or lowers the tip according to the detected amplitude signal. Since the tip only intermittently contacts the surface, this technique provides very gentle tip-surface interactions, and the specific interaction of the antibody on the tip with the antigen on the surface can be used to localize antigenic sites for recording recognition images. The principles of this method are described in detail below and in Fig. 7.11. A magnetically coated tip is oscillated by an alternating magnetic field at an amplitude of 5 nm while being scanned along the surface. Since the tether has a length of 8 nm, the antibody on the tip always has a chance of recognition when passing an antigenic site, which increases the binding probability enormously. Since the oscillation frequency is more than 100 times faster than typical frequencies in conventional force mapping, the data acquisition rate is much higher.

Half-antibodies were used to provide a monovalent ligand on the tip. The antigen, lysozyme, was tightly adsorbed to mica in conditions that yielded a low surface coverage (for details see [20]). A topographic image of this preparation was first

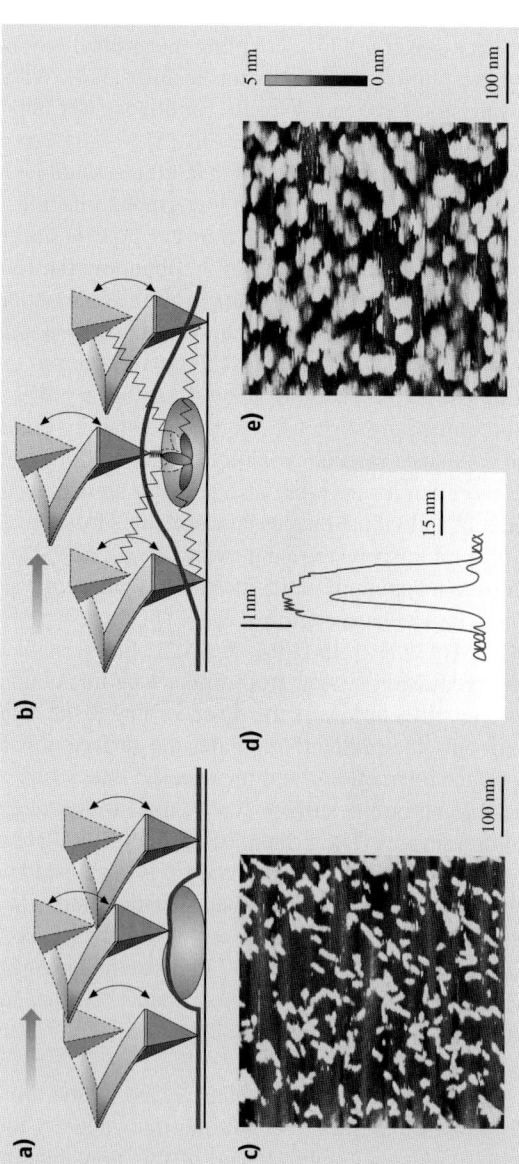

Fig. 7.11. Topography and recognition image. (**a**) AFM tip-lysozyme interaction during topography imaging. (**b**) AFM tip-lysozyme interaction during recognition imaging. Half-antibodies are bound to the AFM tip via a flexible tether (*brown line*) for the recognition of lysozyme on the surface. Imaging results in a height profile, as indicated (*brown line*). (**c**) Topography image. Single lysozyme molecules can be clearly resolved. Sometimes small lysozyme aggregates are observed. Image size was 500 nm. False color bar for heights from 0 (*dark*) to 5 (*bright*) nm. (**d**) Height profiles. Cross-sectional profiles of single lysozyme molecules obtained from the topography (*black line*) and the recognition (*brown line*) image. (**e**) Recognition image. The *bright dots* represent recognition profiles of single lysozyme molecules. Imaging was performed using an AFM tip carrying one half-antibody with access to the antigens on the surface. Conditions were exactly the same as in (**c**). (After [20])

recorded in buffer using a bare AFM tip as a control (Fig. 7.11c). Single lysozyme molecules were clearly resolved (Fig. 7.11c). A cross-sectional analysis (Fig. 7.11d, profile in black) reveals that the molecules appear 8 to 12 nm in diameter and 2.0 to 2.5 nm in height. Imaging with a half-antibody tethered to the tip in conditions identical to those used to obtain the topographical image gave strikingly different images (Fig. 7.11e). They differ significantly both in height and diameter compared to the topography image (Fig. 7.11c). Cross-sectional analysis (Fig. 7.11d, trace in red) reveals a height of 3.0 to 3.5 nm and a diameter of 20 to 25 nm. Therefore, profiles obtained from the recognition image appear at least 1 nm higher and 10 nm broader than profiles from the topography image.

The antibody-antigen recognition process during imaging is depicted in Fig. 7.11b. Approaching the antigen in a lateral scan from the left, the antibody on the tip binds to the antigen about 10 nm before the tip end is above the antigenic site (Fig. 7.11b, left tip), due to the flexible tethering provided by the crosslinker. In the bound state, the z-oscillation of the cantilever is further reduced by the attractive force of the crosslinker-antibody-antigen connection, which is acting as a nonlinear spring. Since the AFM detects the z-projection of the force, the amount of the attractive force measured increases when the tip moves further to the right and reaches its maximum just above the position of the antigenic site (Fig. 7.11b, tip in middle). At lateral distances comparable to the length of the antibody-crosslinker connection, the antibody on the tip dissociates from the antibody on the surface and the attractive force goes to zero.

The diameter of cross-sectional profiles obtained from the recognition image (Fig. 7.11d, brown trace) corresponds to about twice the length of the crosslinker (6 nm) plus antibody (6 nm). Increased heights detected in comparison to profiles of the topography image (Fig. 7.11d, black trace) reflect the amplitude reduction owing to antibody-antigen recognition. The maxima of the cross-sectional profiles of the recognition image, as shown in Fig. 7.11d (brown trace), indicate the position of the antigenic site. The accuracy of maximum determination, which in turn reflects the positional accuracy of determining the position of the antigenic site, was 3 nm. The specific nature of the antibody-antigen interaction was tested by injecting free antibody into the liquid cell, so as to block the antigenic sites on the surface, and subsequent images showed a reduction in apparent height.

With this methodology, topography and recognition images can be obtained at the same time and distinct receptor sites in the recognition image can be assigned to structures from the topography image. The described methodology [20] is applicable to any ligand and, therefore, it should prove possible to recognize many types of proteins or protein layers and carry out epitope mapping on a nanometer scale on membranes and cells.

7.8 Concluding Remarks

Atomic force microscopy has evolved into the imaging method that yields the greatest structural details on live, biological samples like proteins, nucleotides, membranes and cells in their native, aqueous environment at ambient conditions. Due to its high lateral resolution and sensitive force detection capability, the exciting possibility of measuring inter- and intramolecular forces of biomolecules on the single-molecule level has become possible. The proof-of-principle stage of the pioneering experiments has already developed into a high-end analysis method for exploring kinetic and structural details of interactions underlying protein folding and molecular recognition by taking the advantages of single molecule studies. The information obtained from force spectroscopy includes physical parameters unachievable by other methods and opens new perspectives in exploring the regulation of the dynamics of biological processes such as molecular recognition. With ongoing instrumental developments to increase sensitivity and speed, exciting new fields may also open up in drug and cell screening technologies.

References

1. G. Binnig, C. F. Quate, and Ch. Gerber. Atomic force microscope. *Phys. Rev. Lett.*, 56:930–933, 1986.
2. G. U. Lee, D. A. Kidwell, and R. J. Colton. Sensing discrete streptavidin-biotin interactions with atomic force microscopy. *Langmuir*, 10:354–357, 1994.
3. E. L. Florin, V. T. Moy, and H. E. Gaub. Adhesion forces between individual ligand receptor pairs. *Science*, 264:415–417, 1994.
4. P. Hinterdorfer, W. Baumgartner, H. J. Gruber, K. Schilcher, and H. Schindler. Detection and localization of individual antibody-antigen recognition events by atomic force microscopy. *Proc. Nat. Acad. Sci. USA*, 93:3477–3481, 1996.
5. M. Grandbois, W. Dettmann, M. Benoit, and H. E. Gaub. How strong is a covalent bond? *Science*, 283:1727–1730, 1999.
6. G. U. Lee, A. C. Chrisey, and J. C. Colton. Direct measurement of the forces between complementary strands of dna. *Science*, 266:771–773, 1994.
7. T. Boland and B. D. Ratner. Direct measurement of hydrogen bonding in dna nucleotide bases by atomic force microscopy. *Proc. Nat. Acad. Sci. USA*, 92:5297–5301, 1995.
8. P. Wagner, M. Hegner, P. Kernen, F. Zaugg, and G. Semenza. Covalent immobilization of native biomolecules onto au(111) via n-hydroxysuccinimide ester functionalized self assembled monolayers for scanning probe microscopy. *Biophys. J*, 70:2052–2066, 1996.
9. U. Dammer, O. Popescu, P. Wagner, D. Anselmetti, H.-J. Güntherodt, and G. M. Misevic. Binding strength between cell adhesion proteoglycans measured by atomic force microscopy. *Science*, 267:1173–1175, 1995.
10. U. Dammer, M. Hegner, D. Anselmetti, P. Wagner, M. Dreier, W. Huber, and H.-J. Güntherodt. Specific antigen/antibody interactions measured by force microscopy. *Biophys. J.*, 70:2437–2441, 1996.
11. Y. Harada, M. Kuroda, and A. Ishida. Specific and quantized antibody-antigen interaction by atomic force microscopy. *Langmuir*, 16:708–715, 2000.
12. P. Hinterdorfer, K. Schilcher, W. Baumgartner, H. J. Gruber, and H. Schindler. A mechanistic study of the dissociation of individual antibody-antigen pairs by atomic force microscopy. *Nanobiology*, 4:39–50, 1998.
13. S. Allen, X. Chen, J. Davies, M. C. Davies, A. C. Dawkes, J. C. Edwards, C. J. Roberts, J. Sefton, S. J. B. Tendler, and P. M. Williams. Spatial mapping of specific molecular recognition sites by atomic force microscopy. *Biochemistry*, 36:7457–7463, 1997.

14. R. Ros, F. Schwesinger, D. Anselmetti, M. Kubon, R. Schäfer, A. Plückthun, and L. Tiefenauer. Antigen binding forces of individually addressed single-chain fv antibody molecules. *Proc. Nat. Acad. Sci. USA*, 95:7402–7405, 1998.
15. T. Strunz, K. Oroszlan, R. Schäfer, and H.-J. Güntherodt. Dynamic force spectroscopy of single dna molecules. *Proc. Nat. Acad. Sci. USA*, 96:11 277–11 282, 1999.
16. S. S. Wong, E. Joselevich, A. T. Woolley, C. L. Cheung, and C. M. Lieber. Covalently functionalyzed nanotubes as nanometre-sized probes in chemistry and biology. *Nature*, 394:52–55, 1998.
17. P. Hinterdorfer, F. Kienberger, A. Raab, H. J. Gruber, W. Baumgartner, G. Kada, C. Riener, S. Wielert-Badt, C. Borken, and H. Schindler. Poly(ethylene glycol): an ideal spacer for molecular recognition force microscopy/spectroscopy. *Single Mol.*, 1:99–103, 2000.
18. Th. Haselgrübler, A. Amerstorfer, H. Schindler, and H. J. Gruber. Synthesis and applications of a new poly(ethylene glycol) derivative for the crosslinking of amines with thiols. *Bioconj. Chem.*, 6:242–248, 1995.
19. C. K. Riener, G. Kada, C. Borken, F. Kienberger, P. Hinterdorfer, H. Schindler, G. J. Schütz, T. Schmidt, C. D. Hahn, and H. J. Gruber. Bioconjugation for biospecific detection of single molecules in atomic force microscopy (afm) and in single dye tracing (sdt). *Recent Res. Devel. Bioconj. Chem.*, 1:133–149, 2002.
20. A. Raab, W. Han, D. Badt, S. J. Smith-Gill, S. M. Lindsay, H. Schindler, and P. Hinterdorfer. Antibody recognition imaging by force microscopy. *Nature Biotech.*, 17:902–905, 1999.
21. M. Conti, G. Falini, and B. Samori. How strong is the coordination bond between a histidine tag and ni-nitriloacetate? an experiment of mechanochemistry on single molecules. *Angew. Chem.*, 112:221–224, 2000.
22. F. Kienberger, G. Kada, H. J. Gruber, V. Ph. Pastushenko, C. Riener, M. Trieb, H.-G. Knaus, H. Schindler, and P. Hinterdorfer. Recognition force spectroscopy studies of the nta-his6 bond. *Single Mol.*, 1:59–65, 2000.
23. L. Schmitt, M. Ludwig, H. E. Gaub, and R. Tampe. A metal-chelating microscopy tip as a new toolbox for single-molecule experiments by atomic force microscopy. *Biophys. J.*, 78:3275–3285, 2000.
24. C. Yuan, A. Chen, P. Kolb, and V. T. Moy. Energy landscape of avidin-biotin complexes measured by atomic force microscopy. *Biochemistry*, 39:10 219–10 223, 2000.
25. W. Han, S. M. Lindsay, M. Dlakic, and R. E. Harrington. Kinked dna. *Nature*, 386:563, 1997.
26. G. Kada, L. Blaney, L. H. Jeyakumar, F. Kienberger, V. Ph. Pastushenko, S. Fleischer, H. Schindler, F. A. Lai, and P. Hinterdorfer. Recognition force microscopy/spectroscopy of ion channels: applications to the skeletal muscle ca^{2+} release channel (ryr1). *Ultramicroscopy*, 86:129–137, 2001.
27. D. J. Müller, W. Baumeister, and A. Engel. Controlled unzipping of a bacterial surface layer atomic force microscopy. *Proc. Nat. Acad. Sci. USA*, 96:13170–13174, 1999.
28. F. Oesterhelt, D. Oesterhelt, M. Pfeiffer, A. Engle, H. E. Gaub, and D. J. Müller. Unfolding pathways of individual bacteriorhodopsins. *Science*, 288:143–146, 2000.
29. E. Kiss and C.-G. Gölander. Chemical derivatization of muscovite mica surfaces. *Coll. Surf.*, 49:335–342, 1990.
30. S. Karrasch, M. Dolder, F. Schabert, J. Ramsden, and A. Engel. Covalent binding of biological samples to solid supports for scanning probe microscopy in buffer solution. *Biophys. J.*, 65:2437–2446, 1993.

31. N. H. Thomson, B. L. Smith, N. Almquist, L. Schmitt, M. Kashlev, E. T. Kool, and P. K. Hansma. Oriented, active *escherichia coli* rna polymerase: an atomic force microscopy study. *Biophys. J.*, 76:1024–1033, 1999.
32. G. Kada, C. K. Riener, P. Hinterdorfer, F. Kienberger, C. M. Stroh, and H. J. Gruber. Dithio-phospholipids for biospecific immobilization of proteins on gold surfaces. *Single Mol.*, 3:119–125, 2002.
33. C. Le Grimellec, E. Lesniewska, M. C. Giocondi, E. Finot, V. Vie, and J. P. Goudonnet. Imaging of the surface of living cells by low-force contact-mode atomic force microscopy. *Biophys. J.*, 75(2):695–703, 1998.
34. K. Schilcher, P. Hinterdorfer, H. J. Gruber, and H. Schindler. A non-invasive method for the tight anchoring of cells for scanning force microscopy. *Cell. Biol. Int.*, 21:769–778, 1997.
35. S. Wielert-Badt, P. Hinterdorfer, H. J. Gruber, J.-T. Lin, D. Badt, H. Schindler, and R. K.-H. Kinne. Single molecule recognition of protein binding epitopes in brush border membranes by force microscopy. *Biophys. J.*, 82:2767–2774, 2002.
36. P. Bongrand, C. Capo, J.-L. Mege, and A.-M. Benoliel. *Use of hydrodynamic flows to study cell adhesion*, pages 125–156. CRC Press, 1988.
37. J. N. Israelachvili. *Intermolecular and Surface Forces*. Academic Press, 2nd edition, 1991.
38. R. Merkel, P. Nassoy, A. Leung, K. Ritchie, and E. Evans. Energy landscapes of receptor–ligand bonds explored with dynamic force spectroscopy. *Nature*, 397:50–53, 1999.
39. A. Askin. Optical trapping and manipulation of neutral particles using lasers. *Proc. Nat. Acad. Sci. USA*, 94:4853–4860, 1997.
40. K. Svoboda, C. F. Schmidt, B. J. Schnapp, and S. M. Block. Direct observation of kinesin stepping by optical trapping interferometry. *Nature*, 365:721–727, 1993.
41. M. S. Z. Kellermayer, S. B. Smith, H. L. Granzier, and C. Bustamante. Folding-unfolding transitions in single titin molecules characterized with laser tweezers. *Science*, 276:1112–1216, 1997.
42. S. Smith, Y. Cui, and C. Bustamante. Overstretching b-dna: the elastic response of individual double-stranded and single-stranded dna molecules. *Science*, 271:795–799, 1996.
43. T. R. Strick, J. F. Allemend, D. Bensimon, A. Bensimon, and V. Croquette. The elasticity of a single supercoiled dna molecule. *Biophys. J.*, 271:1835–1837, 1996.
44. F. Kienberger, V. Ph. Pastushenko, G. Kada, H. J. Gruber, C. Riener, H. Schindler, and P. Hinterdorfer. Static and dynamical properties of single poly(ethylene glycol) molecules investigated by force spectroscopy. *Single Mol.*, 1:123–128, 2000.
45. B. V. Viani, T. E. Schäffer, A. Chand, M. Rief, H. E. Gaub, and P. K. Hansma. Small cantilevers for force spectroscopy of single molecules. *J. Appl. Phys.*, 86:2258–2262, 1999.
46. S. Liang, D. Medich, D. M. Czajkowsky, S. Sheng, J.-Y. Yuan, and Z. Shao. Thermal noise reduction of mechanical oscillators by actively controlled external dissipative forces. *Ultramicroscopy*, 84:119–125, 2000.
47. H. Grubmüller, B. Heymann, and P. Tavan. Ligand binding: molecular mechanics calculation of the streptavidin-biotin rupture force. *Science*, 271:997–999, 1996.
48. G. I. Bell. Models for the specific adhesion of cells to cells. *Science*, 200:618–627, 1978.
49. E. Evans and K. Ritchie. Dynamic strength of molecular adhesion bonds. *Biophys. J.*, 72:1541–1555, 1997.
50. T. Strunz, K. Oroszlan, I. Schumakovitch. H.-J. Güntherodt, and M. Hegner. Model energy landscapes and the force-induced dissociation of ligand-receptor bonds. *Biophys. J.*, 79:1206–1212, 2000.

51. E. Evans and K. Ritchie. Strength of weak bondconnecting flexible polymer chains. *Biophys. J.*, 76:2439–2447, 1999.
52. J. Fritz, A. G. Katopidis, F. Kolbinger, and D. Anselmetti. Force-mediated kinetics of single p-selectin/ligand complexes observed by atomic force microscopy. *Proc. Nat. Acad. Sci. USA*, 95:12 283–12 288, 1998.
53. V. T. Moy, E.-L. Florin, and H. E. Gaub. Adhesive forces between ligand and receptor measured by afm. *Science*, 266:257–259, 1994.
54. A. Chilkoti, T. Boland, B. Ratner, and P. S. Stayton. The relationship between ligand-binding thermodynamics and protein-ligand interaction forces measured by atomic force microscopy. *Biophys. J.*, 69:2125–2130, 1995.
55. M. Rief, F. Oesterhelt, B. Heymann, and H. E. Gaub. Single molecule force spectroscopy on polysaccharides by atomic force microscopy. *Science*, 275:1295–1297, 1997.
56. P. E. Marzsalek, A. F. Oberhauser, Y.-P. Pang, and J. M. Fernandez. Polysaccharide elasticity governed by chair-boat transitions of the glucopyranose ring. *Nature*, 396:661–664, 1998.
57. A. D. L. Humphries, J. Tamayo, and M. J. Miles. Active quality factor control in liquids for force spectroscopy. *Langmuir*, 16:7891–7894, 2000.
58. M. Rief, M. Gautel, F. Oesterhelt, J. M. Fernandez, and H. E. Gaub. Reversible unfolding of individual titin immunoglobulin domains by afm. *Science*, 276:1109–1112, 1997.
59. A. F. Oberhauser, P. E. Marzsalek, H. P. Erickson, and J. M. Fernandez. The molecular elasticity of the extracellular matrix tenascin. *Nature*, 393:181–185, 1998.
60. W. Baumgartner, P. Hinterdorfer, W. Ness, A. Raab, D. Vestweber, H. Schindler, and D. Drenckhahn. Cadherin interaction probed by atomic force microscopy. *Proc. Nat. Acad. Sci. USA*, 97:4005–4010, 2000.
61. F. Schwesinger, R. Ros, T. Strunz, D. Anselmetti, H.-J. Güntherodt, A. Honegger, L. Jermutus, L. Tiefenauer, and A. Plückthun. Unbinding forces of single antibody-antigen complexes correlate with their thermal dissociation rates. *Proc. Nat. Acad. Sci. USA*, 29:9972–9977, 2000.
62. A. F. Oberhauser, P. K. Hansma, M. Carrion-Vazquez, and J. M. Fernandez. Stepwise unfolding of titin under force-clamp atomic force microscopy. *Proc. Nat. Acad. Sci. USA*, 98:468–472, 2001.
63. S. Izraelev, S. Stepaniants, M. Balsera, Y. Oono, and K. Schulten. Molecular dynamics study of unbinding of the avidin-biotin complex. *Biophys. J.*, 72:1568–1581, 1997.
64. I. Schumakovitch, W. Grange, T. Strunz, P. Bertoncini, H.-J. Güntherodt, and M. Hegner. Temperature dependence of unbinding forces between complementary dna strands. *Biophys. J.*, 82:517–521, 2002.
65. B. Heymann and H. Grubmüller. Molecular dynamics force probe simulations of antibody/antigen unbinding: entropic control and non additivity of unbinding forces. *Biophys. J.*, 81:1295–1313, 2001.
66. E. Evans, E. Leung, D. Hammer, and S. Simon. Chemically distinct transition states govern rapid dissociation of single l-selectin bonds under force. *Proc. Nat. Acad. Sci. USA*, 98:3784–3789, 2001.
67. P. P. Lehenkari and M. A. Horton. Single integrin molecule adhesion forces in intact cells measured by atomic force microscopy. *Biochem. Biophys. Res. Com.*, 259:645–650, 1999.
68. A. Chen and V. T. Moy. Cross-linking of cell surface receptors enhances cooperativity of molecular adhesion. *Biophys. J.*, 78:2814–2820, 2000.
69. X. Zhang, E. Woijcikiewicz, and V. T. Moy. Force spectroscopy of the leukocyte function-associated antigen-1/intercellular adhesion molecule-1 interaction. *Biophys. J.*, 83:2270–2279, 2002.

70. M. Thie, R. Rospel, W. Dettmann, M. Benoit, M. Ludwig, H. E. Gaub, and H. W. Denker. Interactions between trophoblasts and uterine epithelium: monitoring of adhesive forces. *Hum. Reprod.*, 13:3211–3219, 1998.
71. M. Benoit, D. Gabriel, G. Gerisch, and H. E. Gaub. Discrete interactions in cell adhesion measured by single molecule force spectroscopy. *Nature Cell Biol.*, 2:313–317, 2000.
72. M. Ludwig, W. Dettmann, and H. E. Gaub. Atomic force microscopy imaging contrast based on molecular recognition. *Biophys. J.*, 72:445–448, 1997.
73. P. P. Lehenkari, G. T. Charras, G. T. Nykänen, and M. A. Horton. Adapting force microscopy for cell biology. *Ultramicroscopy*, 82:289–295, 2000.
74. M. Grandbois, M. Beyer, M. Rief, H. Clausen-Schaumann, and H. E. Gaub. Affinity imaging of red blood cells using an atomic force microscope. *J. Histochem. Cytochem.*, 48:719–724, 2000.
75. O. H. Willemsen, M. M. E. Snel, K. O. van der Werf, B. G. de Grooth, J. Greve, P. Hinterdorfer, H. J. Gruber, H. Schindler, Y. van Kyook, and C. G. Figdor. Simultaneous height and adhesion imaging of antibody antigen interactions by atomic force microscopy. *Biophys. J.*, 57:2220–2228, 1998.
76. B. V. Viani, L. I. Pietrasanta, J. B. Thompson, A. Chand, I. C. Gebeshuber, J. H. Kindt, M. Richter, H. G. Hansma, and P. K. Hansma. Probing protein-protein interactions in real time. *Nature Struct. Biol.*, 7:644–647, 2000.
77. W. Han, S. M. Lindsay, and T. Jing. A magnetically driven oscillating probe microscope for operation in liquid. *Appl. Phys. Lett.*, 69:1–3, 1996.

Part II

Nanotribology and Nanomechanics

8

Micro/Nanotribology and Materials Characterization Studies Using Scanning Probe Microscopy

Bharat Bhushan

Summary. A sharp AFM/FFM tip sliding on a surface simulates just one asperity contact. However, asperities come in all shapes and sizes. The effect of radius of a single asperity (tip) on the friction/adhesion performance can be studied using tips of different radii. AFM/FFM are used to study the various tribological phenomena, which include surface/roughness, adhesion, friction, scratching, wear, indentation, detection of material transfer, and boundary lubrication.

Directionality in the friction is observed on both micro- and macroscales and results from the surface roughness and surface preparation. Microscale friction is generally found to be smaller than the macrofriction, as there is less plowing contribution in microscale measurements. The mechanism of material removal on the microscale is studied. Evolution of wear has also been studied using AFM. Wear is found to be initiated at nanoscratches. For a sliding interface requiring near-zero friction and wear, contact stresses should be below the hardness of the softer material to minimize plastic deformation, and surfaces should be free of nanoscratches. Wear precursors can be detected at early stages of wear by using surface potential measurements. Detection of material transfer on a nanoscale is possible with AFM. In situ surface characterization of local deformation of materials and thin coatings can be carried out using a tensile stage inside an AFM.Boundary lubrication studies can be conducted using AFM. Chemically bonded lubricant films and self-assembled monolayers are superior in friction and wear resistance. For chemically bonded lubricant films, the adsorption of water, the formation of meniscus and its change during sliding, viscosity, and surface properties play an important role on the friction, adhesion, and durability of these films. For SAMs, their friction mechanism is explained by a so-called "molecular spring" model.

8.1 Introduction

The mechanisms and dynamics of the interactions of two contacting solids during relative motion ranging from atomic- to microscale need to be understood in order to develop a fundamental understanding of adhesion, friction, wear, indentation, and lubrication processes. At most solid-solid interfaces of technological relevance contact occurs at many asperities. Consequently, the importance of investigating single asperity contacts in studies of the fundamental micro/nanomechanical and micro/nanotribological properties of surfaces and interfaces has long been recognized. The recent emergence and proliferation of proximal probes, including scanning probe microscopies (the scanning tunneling microscope and the atomic force microscope), the surface force apparatus, and computational techniques for simulating tip-surface interactions and interfacial properties, have allowed systematic investigations of interfacial problems with high resolution, as well as ways and means of modifying and manipulating nanoscale structures. These advances have led to the appearance of the new field of micro/nanotribology, which pertains to experimental and theoretical investigations of interfacial processes on scales ranging from the atomic- and molecular- to the microscale, occurring during adhesion, friction, scratching, wear, indentation, and thin-film lubrication at sliding surfaces [1–12].

Micro/nanotribological studies are needed to develop a fundamental understanding of interfacial phenomena on a small scale and study interfacial phenomena in micro/nanostructures used in magnetic storage systems, micro/nanoelectromechanical systems (MEMS/NEMS), and other applications [3, 7–9, 13, 14]. Friction and wear of lightly loaded micro/nanocomponents are highly dependent on the surface interactions (few atomic layers). These structures are generally lubricated with molecularly thin films. Micro/nanotribological studies are also valuable in understanding interfacial phenomena in macrostructures and provide a bridge between science and engineering.

The surface force apparatus (SFA), the scanning tunneling microscopes (STM), atomic force and friction force microscopes (AFM and FFM) are widely used in micro/nanotribological studies. Typical operating parameters are compared in Table 8.1. The SFA was developed in 1968 and is commonly employed to study both static and dynamic properties of molecularly thin films sandwiched between two molecularly smooth surfaces. The STM, developed in 1981, allows imaging of electrically conducting surfaces with atomic resolution and has been used for the imaging of clean surfaces, as well as of lubricant molecules. The introduction of the atomic force microscope in 1985 provided a method for measuring ultra-small forces between a probe tip and an engineering (electrically conducting or insulating) surface and has been used for topographical measurements of surfaces on the nanoscale, as well as for adhesion and electrostatic force measurements. Subsequent modifications of the AFM led to the development of the friction force microscope (FFM), designed for atomic-scale and microscale studies of friction. This instrument measures forces in the scanning direction. The AFM is also used in investigations of scratching, wear, indentation, detection of transfer of material, boundary lubrication, and fabrication and machining. Meanwhile, significant progress in understanding the fundamental nature of bonding and interactions in materials, combined with advances in computer-based modeling and simulation methods, have allowed theoretical studies of complex interfacial phenomena with high resolution in space and time. Such simulations provide insights into atomic-scale energetics, structure, dynamics, thermodynamics, transport and rheological aspects of tribological processes.

The nature of interactions between two surfaces brought close together and those between two surfaces in contact as they are separated have been studied experimentally with the surface force apparatus. This has led to a basic understanding of the normal forces between surfaces and the way in which these are modified by the pres-

Table 8.1. Comparison of typical operating parameters in SFA, STM, and AFM/FFM used for micro/nanotribological studies

Operating parameter	SFA	STM[a]	AFM/FFM
Radius of mating surface/tip	~ 10 mm	5–100 nm	5–100 nm
Radius of contact area	10–40 μm	N/A	0.05–0.5 nm
Normal load	10–100 mN	N/A	< 0.1 nN – 500 nN
Sliding velocity	0.001–100 μm/s	0.02–200 μm/s (scan size ~ 1 nm ×1 nm to 125 μm ×125 μm; scan rate < 1 – –122 Hz)	0.02–200 μm/s (scan size ~ 1 nm ×1 nm to 125 μm ×125 μm; scan rate < 1 – –122 Hz)
Sample limitations	Typically atomically smooth, optically transparent mica; opaque ceramic, smooth surfaces can also be used	Electrically conducting samples	None

[a] Can only be used for atomic-scale imaging

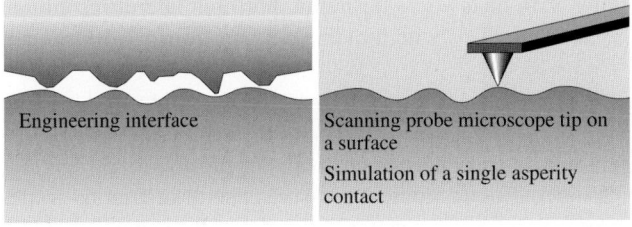

Fig. 8.1. Schematics of an engineering interface and scanning probe microscope tip in contact with an engineering interface

ence of a thin liquid or a polymer film. The frictional properties of such systems have been studied by moving the surfaces laterally, and such experiments have provided insights into the molecular-scale operation of lubricants such as thin liquid or polymer films. Complementary to these studies are those in which the AFM or FFM is used to provide a model asperity in contact with a solid or lubricated surface, Fig. 8.1. These experiments have demonstrated that the relationship between friction and surface roughness is not always simple or obvious. AFM studies have also revealed much about the nanoscale nature of intimate contact during wear and indentation.

In this chapter, we present a review of significant aspects of micro/nanotribological studies conducted using AFM/FFM.

8.2 Description of AFM/FFM and Various Measurement Techniques

An atomic force microscope (AFM) was developed by *Binnig* et al. in 1985. It is capable of investigating surfaces of scientific and engineering interest on an atomic scale [15, 16]. The AFM relies on a scanning technique to produce very high resolution, 3-D images of sample surfaces. It measures ultrasmall forces (less than 1 nN) present between the AFM tip surface mounted on a flexible cantilever beam and a sample surface. These small forces are determined by measuring the motion of a very flexible cantilever beam with an ultra-small mass by a variety of measurement techniques including optical deflection, optical interference, capacitance, and tunneling current. The deflection can be measured to within 0.02 nm, so for a typical cantilever spring constant of 10 N/m, a force as low as 0.2 nN can be detected. To put these numbers in perspective, individual atoms and human hair are typically a fraction of a nanometer and about 75 μm in diameter, respectively, and a drop of water and an eyelash have a mass of about 10 μN and 100 nN, respectively. In the operation of high resolution AFM, the sample is generally scanned, rather than the tip, because any cantilever movement would add vibrations. AFMs are available for measurement of large samples where the tip is scanned and the sample is stationary. To obtain atomic resolution with the AFM, the spring constant of the cantilever should be weaker than the equivalent spring between atoms. A cantilever beam with a spring constant of about 1 N/m or lower is desirable. For high lateral resolution, tips should be as sharp as possible. Tips with a radius ranging from 10–100 nm are commonly available.

A modification to AFM, providing a sensor to measure the lateral force, led to the development of the friction force microscope (FFM), or the lateral force microscope (LFM), designed for atomic-scale and microscale studies of friction [3–5, 7, 8, 11, 17–25] and lubrication [26–30]. This instrument measures lateral or friction forces (in the plane of sample surface and in the scanning direction). By using a standard or a sharp diamond tip mounted on a stiff cantilever beam, AFM is also used in investigations of scratching and wear [6, 9, 11, 22, 31–33], indentation [6, 9, 11, 22, 34, 35], and fabrication/machining [4, 11, 22].

Surface roughness, including atomic-scale imaging, is routinely measured using the AFM, Fig. 8.2. Adhesion, friction, wear, and boundary lubrication at the interface between two solids with and without liquid films are studied using AFM and FFM. Nanomechanical properties are also measured using an AFM. In situ surface characterization of local deformation of materials and thin coatings has been carried out by imaging the sample surfaces using an AFM during tensile deformation using a tensile stage.

8.2.1 Surface Roughness and Friction Force Measurements

Simultaneous measurements of surface roughness and friction force can be made with a commercial AFM/FFM. These instruments are available for the measurement of small samples and large samples. In a small sample AFM, shown in Fig. 8.2a, the

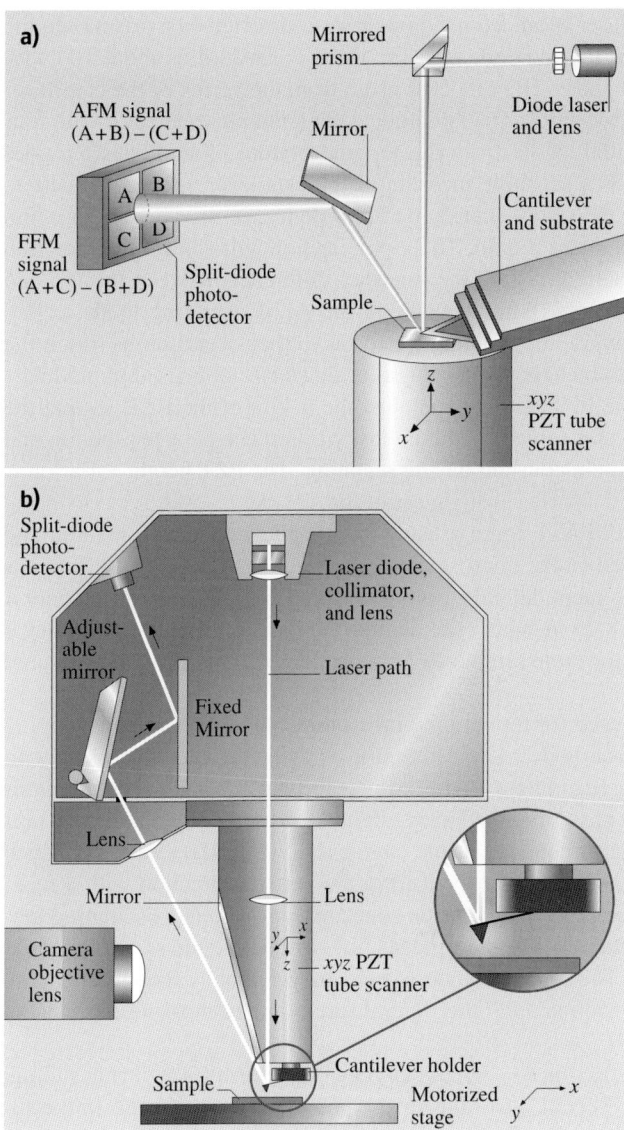

Fig. 8.2. Schematics (**a**) of a commercial small sample atomic force microscope/friction force microscope (AFM/FFM) and (**b**) a large sample AFM/FFM

sample, generally no larger than 10 mm × 10 mm, is mounted on a PZT tube scanner that consists of separate electrodes to scan precisely the sample in the x-y plane in a raster pattern and move it in the vertical (z) direction. A sharp tip at the free end of a flexible cantilever is brought in contact with the sample. Normal and frictional forces being applied at the tip-sample interface are measured using a laser beam deflection technique. A laser beam from a diode laser is directed by a prism onto the back of a cantilever near its free end, which is tilted downward at about 10° with respect to the horizontal plane. The reflected beam from the vertex of the cantilever is directed through a mirror onto a quad photodetector (split photodetector with four quadrants). The differential signal from the top and bottom photodiodes provides the AFM signal, which is a sensitive measure of the cantilever vertical deflection. Topographic features of the sample cause the tip to deflect in the vertical direction as the sample is scanned under the tip. This tip deflection will change the direction of the reflected laser beam, changing the intensity difference between the top and bottom sets of photodetectors (AFM signal). In the AFM operating mode, called the height mode, for topographic imaging, or for any other operation in which the applied normal force is to be kept a constant, a feedback circuit is used to modulate the voltage applied to the PZT scanner to adjust the height of the PZT, so that the cantilever vertical deflection (given by the intensity difference between the top and bottom detector) will remain constant during scanning. The PZT height variation is thus a direct measure of the surface roughness of the sample.

In a large sample AFM both force sensors using optical deflection method and scanning unit are mounted on the microscope head, Fig. 8.2b. Because of vibrations added by cantilever movement, lateral resolution of this design is somewhat poorer than the design in Fig. 8.2a in which the sample is scanned instead of cantilever beam. The advantage of the large sample AFM is that large samples can be measured readily.

Most AFMs can be used for topography measurements in the so-called tapping mode (intermittent contact mode), also referred to as dynamic force microscopy. In the tapping mode, during scanning over the surface, the cantilever/tip assembly is sinusoidally vibrated by a piezo mounted above it, and the oscillating tip slightly taps the surface at the resonant frequency of the cantilever (70–400 Hz) with a constant (20–100 nm) oscillating amplitude introduced in the vertical direction with a feedback loop keeping the average normal force constant, Fig. 8.3. The oscillating amplitude is kept large enough so that the tip does not get stuck to the sample because of adhesive attractions. The tapping mode is used in topography measurements to minimize the effects of friction and other lateral forces and to measure topography of soft surfaces.

For measurement of friction force being applied at the tip surface during sliding, left-hand and right-hand sets of quadrants of the photodetector are used. In the so-called friction mode, the sample is scanned back and forth in a direction orthogonal to the long axis of the cantilever beam. A friction force between the sample and the tip will produce a twisting of the cantilever. As a result, the laser beam will be reflected out of the plane defined by the incident beam and the beam reflected vertically from an untwisted cantilever. This produces an intensity difference of the laser beam received in the left-hand and right-hand sets of quadrants of the photodetector. The intensity difference between the two sets of detectors (FFM signal) is directly related to the degree of twisting and, hence, to the magnitude of the friction force. One problem associated with this method is that any misalignment between the laser beam and the photodetector axis would introduce error in the measurement. However, by fol-

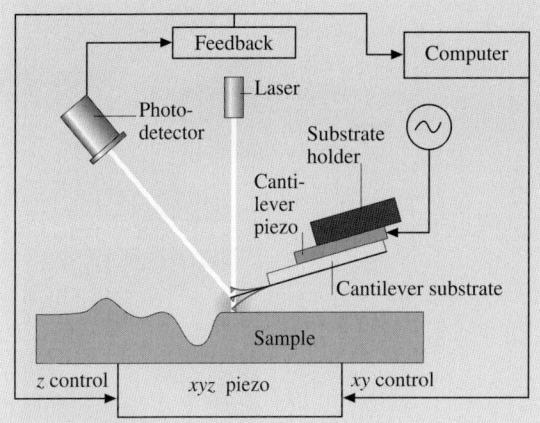

Fig. 8.3. Schematic of tapping mode used for surface topography measurements

lowing the procedures developed by *Ruan* and *Bhushan* [19], in which the average FFM signal for the sample scanned in two opposite directions is subtracted from the friction profiles of each of the two scans, the misalignment effect is eliminated. This method provides three-dimensional maps of friction force. By following the friction force calibration procedures developed by *Ruan* and *Bhushan* [19], voltages corresponding to friction forces can be converted to force units. The coefficient of friction is obtained from the slope of friction force data measured as a function of normal loads typically ranging from 10–150 nN. This approach eliminates any contributions due to the adhesive forces [22]. For calculation of the coefficient of friction based on a single point measurement, friction force should be divided by the sum of applied normal load and intrinsic adhesive force. Furthermore, it should be pointed out that for a single asperity contact, the coefficient of friction is not independent of load. This will be discussed in greater detail later in the chapter.

Topographic measurements in the contact mode are typically made using a sharp, microfabricated square-pyramidal Si_3N_4 tip with a radius of 30–50 nm on a triangular cantilever beam (Fig. 8.4a) with normal stiffness on the order of 0.5 N/m at a normal load of about 10 nN, and friction measurements are carried out in the load range of 10–150 nN. Topography measurements in the tapping mode utilize a stiff cantilever with high resonant frequency; typically, a square-pyramidal, etched, single-crystal silicon tip with a tip radius of 5–10 nm mounted on a stiff rectangular silicon cantilever beam (Fig. 8.4a) with a normal stiffness on the order of 50 N/m is used. Carbon nanotube tips with small diameters (few nm) and high aspect ratios attached on the single-crystal silicon, square pyramidal tips are used for high resolution imaging of surfaces and deep trenches in the tapping mode or noncontact mode (Fig. 8.4b). To study the effect of the radius of a single asperity (tip) on adhesion and friction, microspheres of silica with radii ranging from about 4–15 μm are attached with epoxy at the ends of tips of Si_3N_4 cantilever beams. Optical micrographs of two of the microspheres mounted at the ends of triangular cantilever beams are shown in Fig. 8.4c.

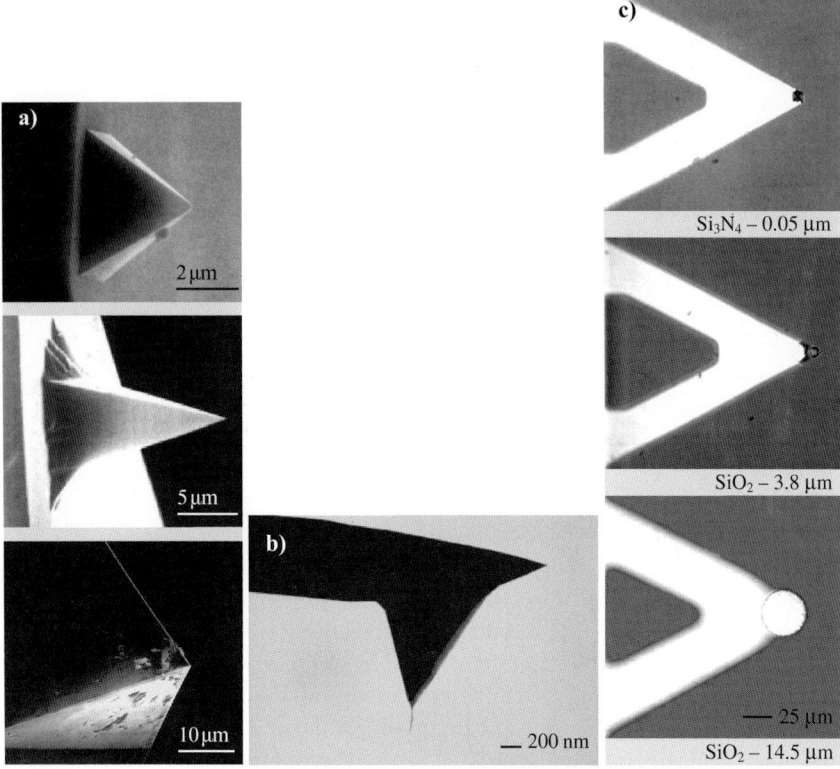

Fig. 8.4. (a) SEM micrographs of a square-pyramidal PECVD Si_3N_4 tip with a triangular cantilever beam (*top*), a square-pyramidal etched single-crystal silicon tip with a rectangular silicon cantilever beam (*middle*), and a three-sided pyramidal natural diamond tip with a square stainless steel cantilever beam (*bottom*), (b) SEM micrograph of a multiwalled carbon nanotube (MWNT) physically attached on the single-crystal silicon, square-pyramidal tip, and (c) optical micrographs of a commercial Si_3N_4 tip and two modified tips showing SiO_2 spheres mounted over the sharp tip and at the end of the triangular Si_3N_4 cantilever beams. (Radii of the tips are given in the figure)

The tip is scanned in such a way that its trajectory on the sample forms a triangular pattern, Fig. 8.5. Scanning speeds in the fast and slow scan directions depend on the scan area and scan frequency. Scan sizes ranging from less than 1 nm × 1 nm to 125 µm × 125 µm and scan rates from less than 0.5 to 122 Hz typically can be used. Higher scan rates are used for smaller scan lengths. For example, scan rates in the fast and slow scan directions for an area of 10 µm × 10 µm scanned at 0.5 Hz are 10 µm/s and 20 nm/s, respectively.

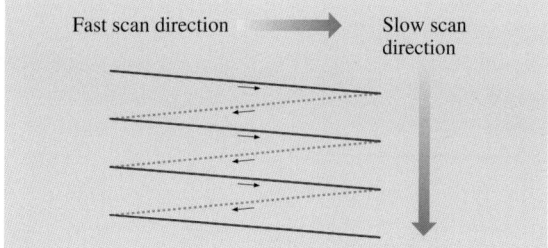

Fig. 8.5. Schematic of triangular pattern trajectory of the tip as the sample (or the tip) is scanned in two dimensions. During scanning, data are recorded only during scans along the solid scan lines

8.2.2 Adhesion Measurements

Adhesive force measurements are performed in the so-called force calibration mode. In this mode, force-distance curves are obtained, for an example, see Fig. 8.6. The horizontal axis gives the distance the piezo (and hence the sample) travels, and the vertical axis gives the tip deflection. As the piezo extends, it approaches the tip, which is at this point in free air and hence shows no deflection. This is indicated by the flat portion of the curve. As the tip approaches within a few nanometers of the sample (point A), an attractive force exists between the atoms of the tip surface and the atoms of the sample surface. The tip is pulled toward the sample and contact occurs at point B on the graph. From this point on, the tip is in contact with the surface, and as the piezo extends farther, the tip gets deflected farther. This is represented by the sloped portion of the curve. As the piezo retracts, the tip goes beyond the zero deflection (flat) line, due to attractive forces (van der Waals forces and long-range meniscus forces), into the adhesive regime. At point C in the graph, the tip snaps free of the adhesive forces and is again in free air. The horizontal distance between points B and C along the retrace line gives the distance the tip moved in the adhesive regime. This distance multiplied by the stiffness of the cantilever gives the adhesive force. Incidentally, the horizontal shift between the loading and unloading curves results from the hysteresis in the PZT tube [4, 36].

8.2.3 Scratching, Wear and Fabrication/Machining

For microscale scratching, microscale wear, nanofabrication/nanomachining, and nanoindentation hardness measurements, an extremely hard tip is required. A three-sided pyramidal single-crystal natural diamond tip with an apex angle of 80° and a radius of about 100 nm mounted on a stainless steel cantilever beam with normal stiffness of about 25 N/m is used at relatively higher loads (1–150 µN), Fig. 8.4a. For scratching and wear studies, the sample is generally scanned in a direction orthogonal to the long axis of the cantilever beam (typically at a rate of 0.5 Hz), so that friction can be measured during scratching and wear. The tip is mounted on the cantilever such that one of its edges is orthogonal to the long axis of the beam. Therefore, wear during scanning along the beam axis is higher (about two to three times)

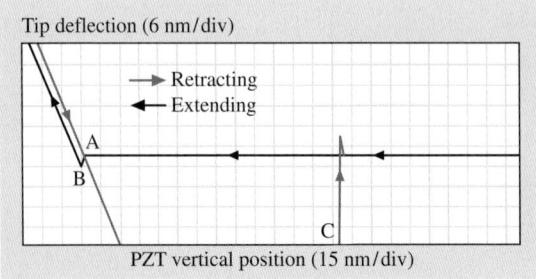

Fig. 8.6. Typical force-distance curve for a contact between Si_3N_4 tip and single-crystal silicon surface in measurements made in the ambient environment. Contact between the tip and silicon occurs at point B; tip breaks free of adhesive forces at point C as the sample moves away from the tip

than during scanning orthogonal to the beam axis. For wear studies, an area on the order of $2\,\mu m \times 2\,\mu m$ is scanned at various normal loads (ranging from 1 to $100\,\mu N$) for a selected number of cycles [4, 22].

Scratching can also be performed at ramped loads and the coefficient of friction can be measured during scratching [33]. A linear increase in the normal load approximated by a large number of normal load increments of small magnitude is applied using a software interface (lithography module in Nanoscope III) that allows the user to generate controlled movement of the tip with respect to the sample. The friction signal is tapped out of the AFM and is recorded on a computer. A scratch length on the order of $25\,\mu m$ and a velocity on the order of $0.5\,\mu m/s$ are used and the number of loading steps is usually 50.

Nanofabrication/nanomachining is conducted by scratching the sample surface with a diamond tip at specified locations and scratching angles. The normal load used for scratching (writing) is on the order of $1-100\,\mu N$ with a writing speed on the order of $0.1-200\,\mu m/s$ [4, 5, 11, 22, 37].

8.2.4 Surface Potential Measurements

To detect wear precursors and study the early stages of localized wear, the multimode AFM can be used to measure the potential difference between the tip and the sample by applying a DC bias potential and an oscillating (AC) potential to a conducting tip over a grounded substrate in a Kelvin probe microscopy, or so-called "nano-Kelvin probe" technique [38–40].

Mapping of the surface potential is made in the so-called "lift mode" (Fig. 8.7). These measurements are made simultaneously with the topography scan in the tapping mode using an electrically conducting (nickel-coated single-crystal silicon) tip. After each line of the topography scan is completed, the feedback loop controlling the vertical piezo is turned off, and the tip is lifted from the surface and traced over the same topography at a constant distance of 100 nm. During the lift mode,

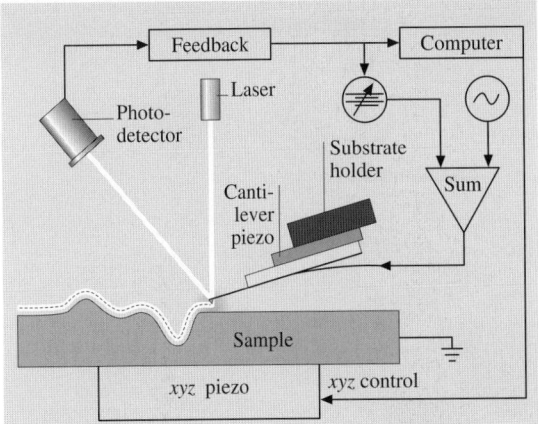

Fig. 8.7. Schematic of lift mode used to make surface potential measurement. The topography is collected in tapping mode in the primary scan. The cantilever piezo is de-activated. Using topography information of the primary scan, the cantilever is scanned across the surface at a constant height above the sample. An oscillating voltage at the resonant frequency is applied to the tip and a feedback loop adjusts the DC bias of the tip to maintain the cantilever amplitude at zero. The output of the feedback loop is recorded by the computer and becomes the surface potential map [4]

a DC bias potential and an oscillating potential (3–7 V) is applied to the tip. The frequency of oscillation is chosen to be equal to the resonant frequency of the cantilever (~ 80 kHz). When a DC bias potential equal to the negative value of surface potential of the sample (on the order of ±2 V) is applied to the tip, it does not vibrate. During scanning, a difference between the DC bias potential applied to the tip and the potential of the surface will create DC electric fields that interact with the oscillating charges (as a result of the AC potential), causing the cantilever to oscillate at its resonant frequency, as in tapping mode. However, a feedback loop is used to adjust the DC bias on the tip to exactly nullify the electric field and, thus, the vibrations of the cantilever. The required bias voltage follows the localized potential of the surface. The surface potential was obtained by reversing the sign of the bias potential provided by the electronics [39, 40]. Surface and subsurface changes of structure and/or chemistry can cause changes in the measured potential of a surface. Thus, the mapping of the surface potential after sliding can be used for detecting wear precursors and studying the early stages of localized wear.

8.2.5 In Situ Characterization of Local Deformation Studies

In situ characterization of local deformation of materials can be carried out by performing tensile, bending, or compression experiments inside an AFM and by observing nanoscale changes during the deformation experiment [6]. In these experiments, small deformation stages are used to deform the samples inside an AFM. In tensile testing of the polymeric films carried out by *Bobji* and *Bhushan* [41, 42], a tensile

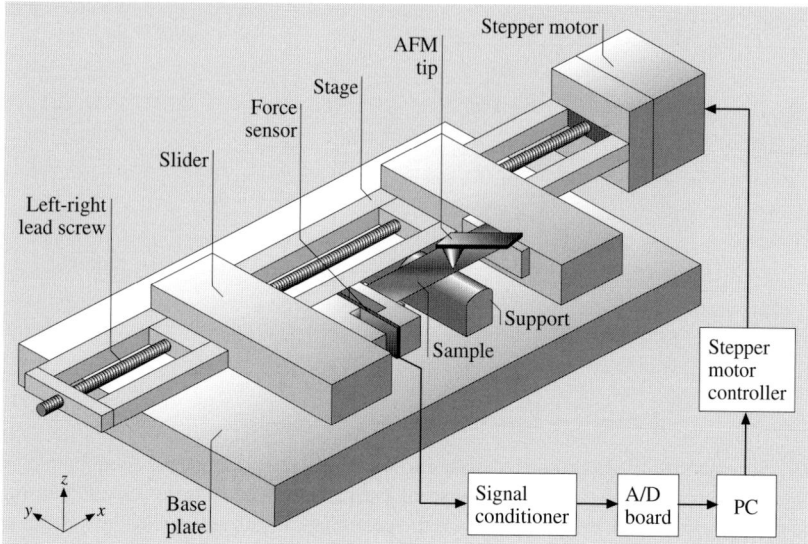

Fig. 8.8. Schematic of the tensile stage to conduct in situ tensile testing of the polymeric films in AFM [42]

stage was used (Fig. 8.8). The stage with a left-right combination lead screw (which helps to move the slider in the opposite direction) was used to stretch the sample and minimize the movement of the scanning area, which was kept close to the center of the tensile specimen. One end of the sample was mounted on the slider via a force sensor to monitor the tensile load. The samples were stretched for various strains using a stepper motor, and the same control area at different strains was imaged. In order to better locate the control area for imaging, a set of four markers was created at the corners of a 30 μm × 30 μm square at the center of the sample by scratching the sample with a sharp silicon tip. The scratching depth was controlled such that it did not affect cracking behavior of the coating. A minimum displacement of 1.6 μm could be obtained. This corresponded to a strain increment of $8 \times 10^{-3}\%$ for a sample length of 38 mm. The maximum travel was about 100 mm. The resolution of the force sensor was 10 mN with a capacity of 45 N. During stretching, a stress-strain curve was obtained during the experiment to study any correlation between the degree of plastic strain and propensity of cracks.

8.2.6 Nanoindentation Measurements

For nanoindentation hardness measurements, the scan size is set to zero, and a normal load is then applied to make the indents using the diamond tip. During this procedure, the tip is continuously pressed against the sample surface for about two seconds at various indentation loads. The sample surface is scanned before and after the scratching, wear, or indentation to obtain the initial and final surface topographies at a low normal load of about 0.3 μN using the same diamond tip. An area larger than the indentation region is scanned to observe the indentation marks. Nanohardness is calculated by dividing the indentation load by the projected residual area of the indents [34].

Direct imaging of the indent allows one to quantify piling up of ductile material around the indenter. However, it becomes difficult to identify the boundary of the indentation mark with great accuracy. This makes the direct measurement of contact area somewhat inaccurate. A technique with the dual capability of depth-sensing, as well as in situ imaging, which is most appropriate in nanomechanical property studies, is used for accurate measurement of hardness with shallow depths [4, 35]. This indentation system is used to make load-displacement measurement and, subsequently, carry out in situ imaging of the indent, if required. The indentation system consists of a three-plate transducer with electrostatic actuation hardware used for direct application of normal load and a capacitive sensor used for measurement of vertical displacement. The AFM head is replaced with this transducer assembly while the specimen is mounted on the PZT scanner, which remains stationary during indentation experiments. Indent area and, consequently, hardness value can be obtained from the load-displacement data. The Young's modulus of elasticity is obtained from the slope of the unloading curve.

8.2.7 Localized Surface Elasticity and Viscoelasticity Mapping

Indentation experiments provide a single-point measurement of the Young's modulus of elasticity calculated from the slope of the indentation curve during unloading. Localized surface elasticity maps can be obtained using dynamic force microscopy in which an oscillating tip is scanned over the sample surface in contact under steady and oscillating load. Lower frequency operation mode in the kHz range such as force modulation microscopy [43–45] and pulsed force microscopy [46] are well suited for soft samples such as polymers. However, if the tip-sample contact stiffness becomes significantly higher than the cantilever stiffness, the sensitivity of these techniques strongly decreases. In this case, sensitivity of the measurement of stiff materials can be improved by using high-frequency operation modes in the MHz range such as acoustic (ultrasonic) force microscopy [47]. We only describe here the force modulation technique.

In the force modulation technique, the oscillation is applied to the cantilever substrate with a cantilever piezo (bimorph), Fig. 8.9a. For measurements, an etched silicon tip is first brought in contact with a sample under a static load of 50–300 nN. In addition to the static load applied by the sample piezo, a small oscillating (modulating) load is applied by a bimorph, generally at a frequency (about 8 kHz) far below that of the natural resonance of the cantilever (70–400 kHz). When the tip is brought into contact with the sample, the surface resists the oscillations of the tip, and the cantilever deflects. Under the same applied load, a stiff area on the sample would deform less than a soft one; i.e., stiffer areas cause greater deflection amplitudes of the cantilever, Fig. 8.9b. The variations in the deflection amplitudes provide a measure of the relative stiffness of the surface. Contact analyses can be used to obtain a quantitative measure of localized elasticity of soft surfaces [44]. The elasticity data are

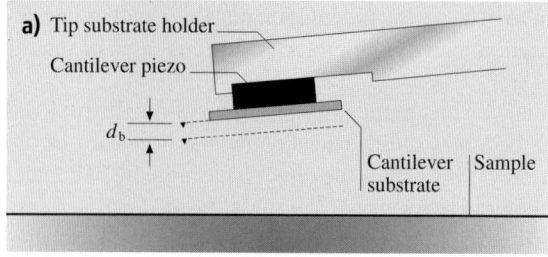

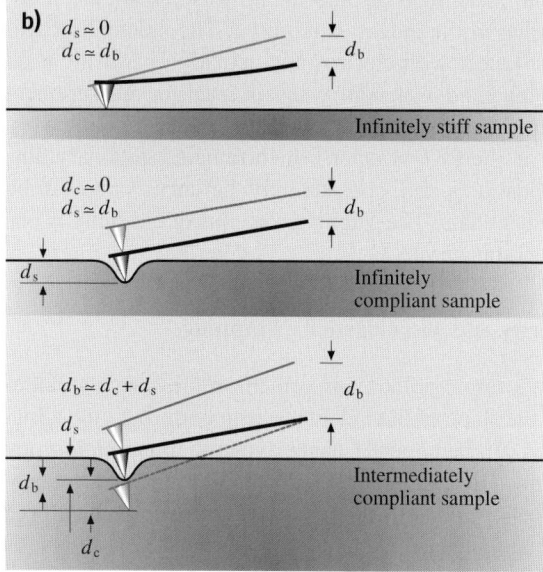

Fig. 8.9. Schematics (**a**) of the bimorph assembly and (**b**) of the motion of the cantilever and tip as a result of the oscillations of the bimorph for an infinitely stiff sample, an infinitely compliant sample, and an intermediate sample. The *thin line* represents the cantilever at the top cycle, and the *thick line* corresponds to the bottom of the cycle. The *dashed line* represents the position of the tip if the sample was not present or was infinitely compliant. d_c, d_s and d_b are the oscillating (AC) deflection amplitude of the cantilever, penetration depth, and oscillating (AC) amplitude of the bimorph, respectively [44]

collected simultaneously with the surface height data using a so-called negative lift mode technique. In this mode, each scan line of each topography image (obtained in tapping mode) is replaced with the tapping action disabled and the tip lowered into steady contact with the surface.

Another form of dynamic force microscopy, phase contrast microscopy, is used to detect the contrast in viscoelastic (viscous energy dissipation) properties of the different materials across the surface [48–52]. In this technique, both deflection amplitude and phase angle contrasts are measured, which are the measures of the relative stiffness and viscoelastic properties, respectively. Tapping mode is utilized. For tapping mode during scanning, the cantilever/tip assembly is sinusoidally vibrated at its resonant frequency, and the sample X-Y-Z piezo is adjusted using feedback control in the z-direction to maintain a constant set point. As shown in Fig. 8.10, the cantilever/tip assembly is vibrated at some amplitude, here referred to as the tapping amplitude, by a cantilever piezo before the tip engages the sample. The tip engages the sample at some set point, which may be thought of as the amplitude of the cantilever as influenced by contact with the sample. A lower set point gives a reduced amplitude and closer mean tip-to-sample distance. This is also depicted in Fig. 8.10. The feedback signal to the z-direction sample piezo (to keep the set point constant) is a measure of topography. This height data is recorded on a computer. The extender electronics is used to measure the phase angle lag between the cantilever piezo drive signal and the cantilever response during sample engagement. As illustrated in Fig. 8.10, the phase angle lag is (at least partially) a function of the viscoelastic properties of the sample material. A range of tapping amplitudes and set points can be used for measurements. A commercially etched single-crystal silicon tip (DI TESP) used for tapping mode, with a radius of 5–10 nm, a stiffness of 20–100 N/m, and a natural frequency of 350–400 kHz is normally used. Scanning is normally set to a rate of 1 Hz along the fast axis.

8.2.8 Boundary Lubrication Measurements

To study nanoscale boundary lubrication, adhesive forces are measured in the force calibration mode, as previously described. The adhesive forces are also calculated from the horizontal intercept of friction versus normal load curves at a zero value of friction force. For friction measurements, the samples are typically scanned using an Si_3N_4 tip over an area of 2×2 µm at the normal load ranging from 5 to 130 nN. The samples are generally scanned with a scan rate of 0.5 Hz resulting in a scanning speed of 2 µm/s. Velocity effects on friction are studied by changing the scan frequency from 0.1 to 60 Hz while the scan size is maintained at 2×2 µm, which allows the velocity to vary from 0.4 to 240 µm/s. To study the durability properties, the friction force and coefficient of friction are monitored during scanning at a normal load of 70 nN and a scanning speed of 0.8 µm/s for a desired number of cycles [27, 28, 30].

8.3 Friction and Adhesion

8.3.1 Atomic-Scale Friction

To study friction mechanisms on an atomic scale, a well characterized, freshly cleaved surface of highly oriented pyrolytic graphite (HOPG) has been used by *Mate* et al. [17] and *Ruan* and *Bhushan* [20].

The atomic-scale friction force of HOPG exhibited the same periodicity as that of the corresponding topography (Fig. 8.11a), but the peaks in friction and those in topography are displaced relative to each other (Fig. 8.11b). A Fourier expansion

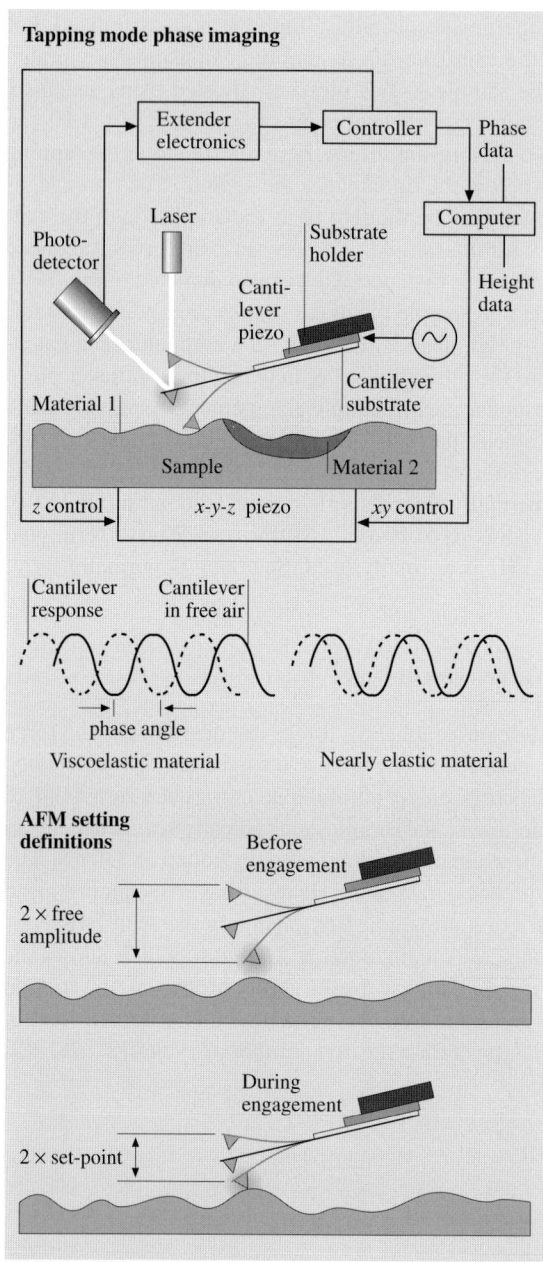

Fig. 8.10. Schematic of tapping mode used to obtain height and phase data and definitions of tapping amplitude and set point. During scanning the cantilever is vibrated at its resonant frequency and the sample x-y-z piezo is adjusted by feedback control in the z-direction to maintain a constant set point. The computer records height and phase angle (which is a function of the viscoelastic properties of the sample materials) data

of the interatomic potential was used to calculate the conservative interatomic forces between atoms of the FFM tip and those of the graphite surface. Maxima in the interatomic forces in the normal and lateral directions do not occur at the same location, which explains the observed shift between the peaks in the lateral force and those in the corresponding topography. Furthermore, the observed local variations in friction force were explained by variation in the intrinsic lateral force between the sample and the FFM tip resulting from an atomic-scale stick-slip process [17, 20].

8.3.2 Microscale Friction

Local variations in the microscale friction of cleaved graphite have been observed (Fig. 8.12). These arise from structural changes that occur during the cleaving process [21]. The cleaved HOPG surface is largely atomically smooth, but exhibits line-shaped regions in which the coefficient of friction is more than an order of magnitude larger. Transmission electron microscopy indicates that the line-shaped regions consist of graphite planes of different orientation, as well as of amorphous carbon. Differences in friction have also been observed for multiphase ceramic materials [53]. Figure 8.13 shows the surface roughness and friction force maps of Al_2O_3-TiC (70–30 wt%). TiC grains have a Knoop hardness of about $2800\,kg/mm^2$. Therefore, they do not polish as much and result in a slightly higher elevation (about 2–3 nm higher than that of Al_2O_3 grains). TiC grains exhibit higher friction force than Al_2O_3 grains. The coefficients of friction of TiC and Al_2O_3 grains are 0.034 and 0.026, respectively, and the coefficient of friction of Al_2O_3-TiC composite is 0.03. Local variation in friction force also arises from the scratches present on the Al_2O_3-TiC surface. *Meyer* et al. [54] also used FFM to measure structural variations of organic mono- and multilayer films. All of these measurements suggest that the FFM can be used for structural mapping of the surfaces. FFM measurements can be used to map chemical variations, as indicated by the use of the FFM with a modified probe tip to map the spatial arrangement of chemical functional groups in mixed organic monolayer films [55]. Here, sample regions that had stronger interactions with the functionalized probe tip exhibited larger friction.

Local variations in the microscale friction of nominally rough, homogeneous surfaces can be significant and are seen to depend on the local surface slope, rather than the surface height distribution, Fig. 8.14. This dependence was first reported by *Bhushan* and *Ruan* [18], *Bhushan* et al. [22], and *Bhushan* [37] and later discussed in more detail in [56] and [57]. In order to show elegantly any correlation between local values of friction and surface roughness, surface roughness and friction force maps of a gold-coated ruling with somewhat rectangular grids and a silicon grid with square pits were obtained, see Fig. 8.15 [57]. Figures 8.14 and 8.15 show the surface roughness map, the slopes of the roughness map taken along the sliding direction (surface slope map), and the friction force map for various samples. There is a strong correlation between the surface slopes and friction forces. For example, in Fig. 8.15, friction force is high locally at the edge of the grids and pits with a positive slope and is low at the edges with a negative slope.

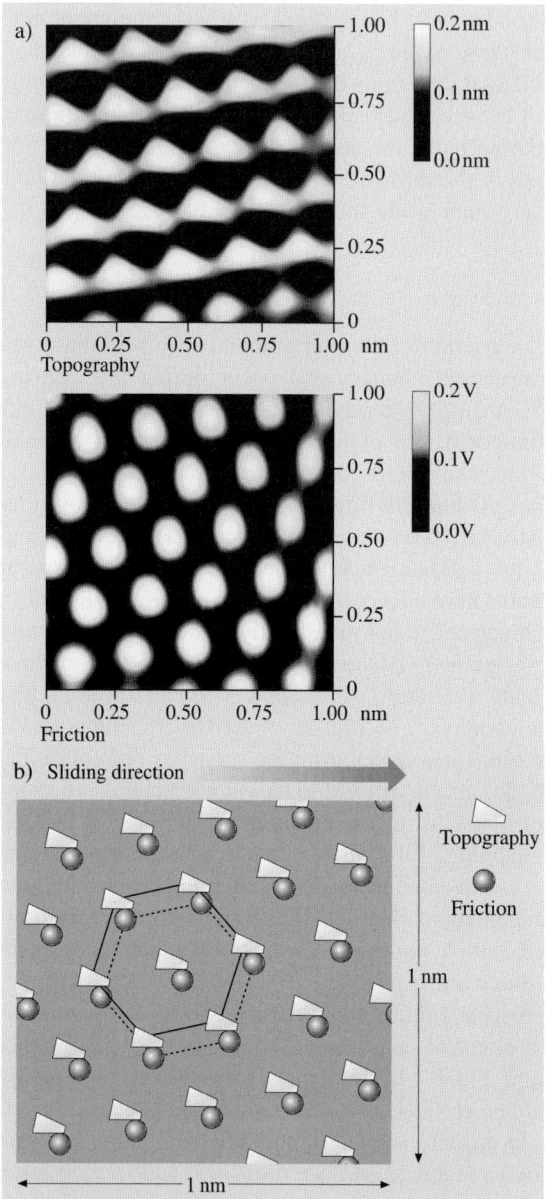

Fig. 8.11. (a) Gray-scale plots of surface topography and friction force maps of a 1 nm × 1 nm area of freshly cleaved HOPG showing the atomic-scale variation of topography and friction and (b) schematic of superimposed topography and friction maps from (a); the symbols correspond to maxima. Note the spatial shift between the two plots [20]

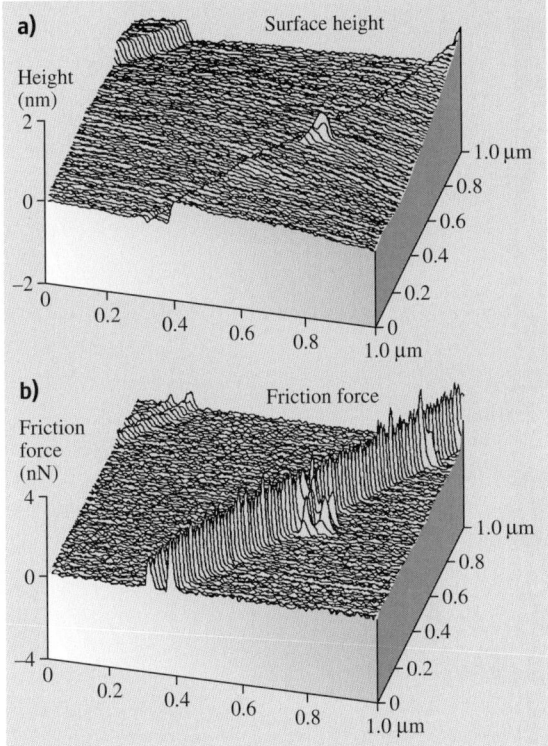

Fig. 8.12. (a) Surface roughness and (b) friction force maps at a normal load of 42 nN of freshly cleaved HOPG surface against an Si_3N_4 FFM tip. Friction in the *line-shaped region* is over an order of magnitude larger than the *smooth areas* [20]

We now examine the mechanism of microscale friction, which may explain the resemblance between the slope of surface roughness maps and the corresponding friction force maps [4, 5, 18, 20–22, 56, 57]. There are three dominant mechanisms of friction: adhesive, adhesive and roughness (ratchet), and plowing [10, 57]. First, we may assume these to be additive. The adhesive mechanism cannot explain the local variation in friction. Next, we consider the ratchet mechanism. We consider a small tip sliding over an asperity making an angle θ with the horizontal plane, Fig. 8.16. The normal force W (normal to the general surface) applied by the tip to the sample surface is constant. The friction force F on the sample would be a constant for a smooth surface if the friction mechanism does not change. For a rough surface, as shown in Fig. 8.16, if the adhesive mechanism does not change during sliding, the local value of coefficient of friction remains constant,

$$\mu_0 = S/N , \tag{8.1}$$

where S is the local friction force and N is the local normal force. However, the friction and normal forces are measured with respect to global horizontal and normal

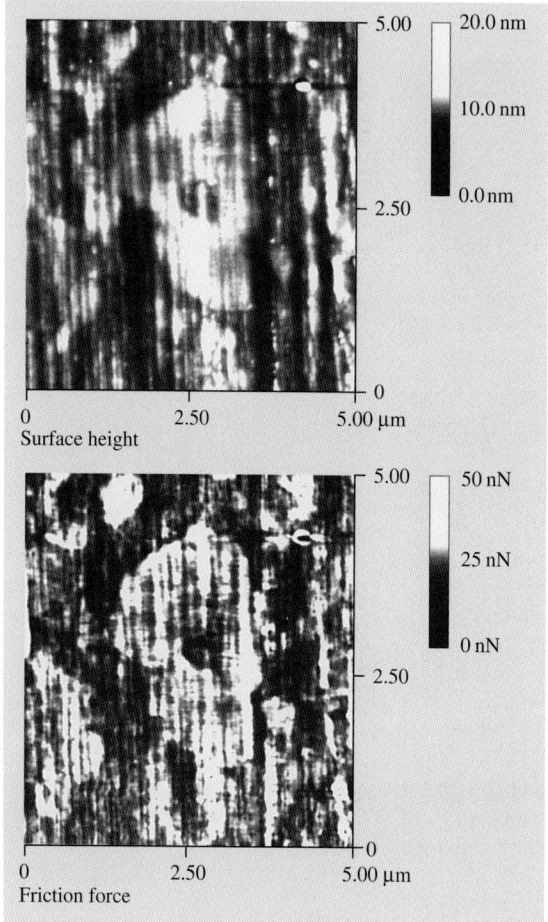

Fig. 8.13. Gray-scale surface roughness ($\sigma = 0.80$ nm) and friction force maps (mean = 7.0 nN, $\sigma = 0.90$ nN) for Al_2O_3-TiC (70–30 wt%) at a normal load of 138 nN [53]

axes, respectively. The measured local coefficient of friction μ_1 in the ascending part is

$$\mu_1 = F/W$$
$$= (\mu_0 + \tan\theta) / (1 - \mu_0 \tan\theta) \sim \mu_0 + \tan\theta,$$
for small $\mu_0 \tan\theta$, \hfill (8.2)

indicating that in ascending part of the asperity one may simply add the friction force and the asperity slope to one another. Similarly, on the right-hand side (descending part) of the asperity,

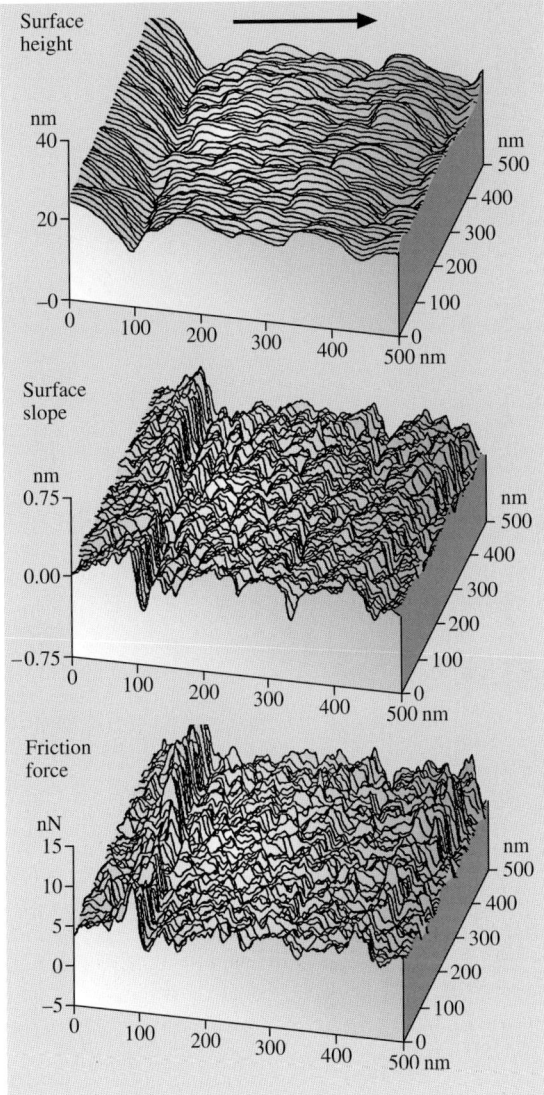

Fig. 8.14. Surface roughness map ($\sigma = 4.4$ nm); surface slope map taken in the sample sliding direction, the horizontal axis, (mean $= 0.023$, $\sigma = 0.197$); and friction force map (mean $= 6.2$ nN, $\sigma = 2.1$ nN) for a lubricated thin-film magnetic rigid disk for a normal load of 160 nN [22]

$$\mu_2 = (\mu_0 - \tan\theta) / (1 + \mu_0 \tan\theta) \sim \mu_0 - \tan\theta , \quad (8.3)$$
for small $\mu_0 \tan\theta$.

For a symmetrical asperity, the average coefficient of friction experienced by the FFM tip traveling across the whole asperity is

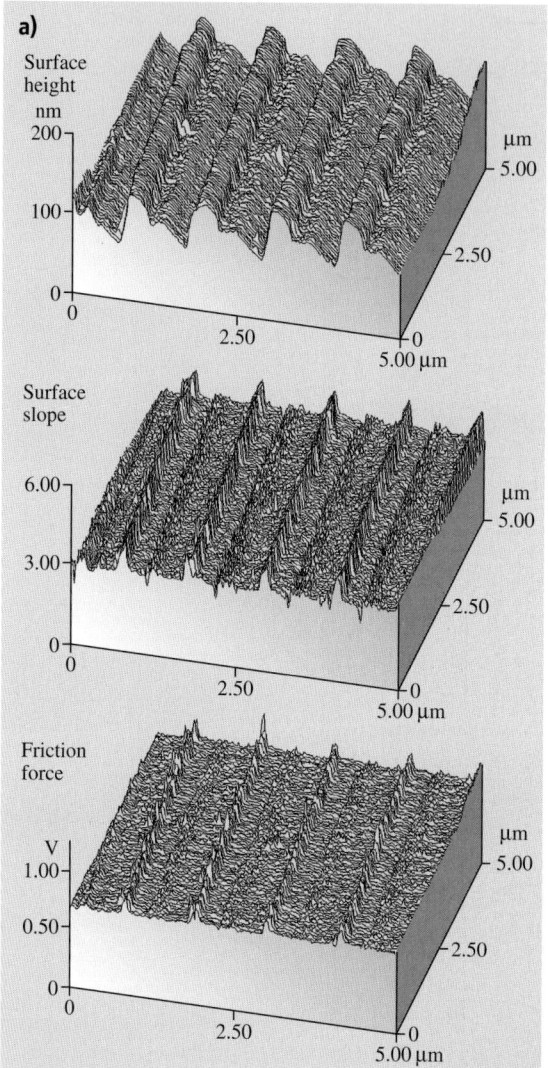

Fig. 8.15. Surface roughness map, surface slope map taken in the sample sliding direction (the horizontal axis), and friction force map for (**a**) a gold-coated ruling (with somewhat rectangular grids with a pitch of 1 μm and a ruling step height of about 70 nm) at a normal load of 25 nN and (**b**) a silicon grid (with 5 μm square pits of 180-nm depth and a pitch of 10 μm) [57]

$$\begin{aligned}\mu_{\text{ave}} &= (\mu_1 + \mu_2)/2 \\ &= \mu_0 \left(1 + \tan^2 \theta\right) / \left(1 - \mu_0^2 \tan^2 \theta\right) \\ &\sim \mu_0 \left(1 + \tan^2 \theta\right), \\ &\text{for small } \mu_0 \tan \theta.\end{aligned} \qquad (8.4)$$

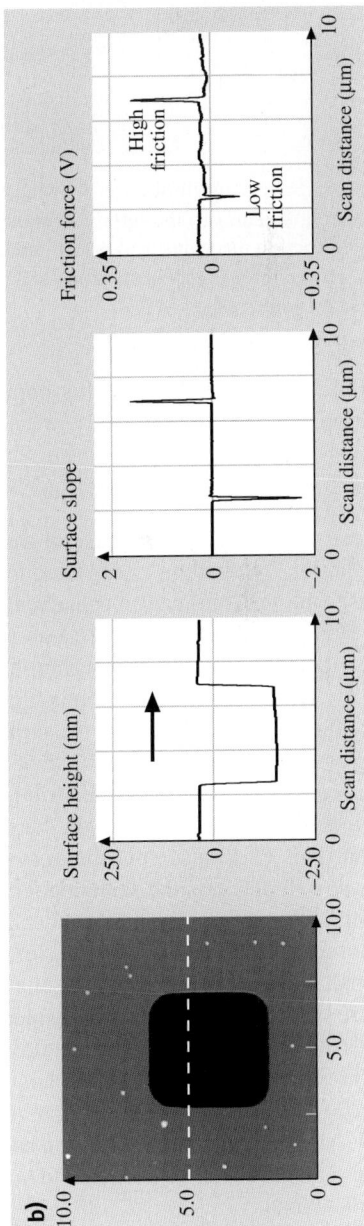

Fig. 8.15. continue

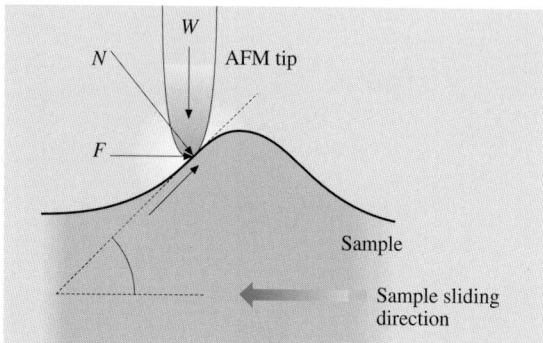

Fig. 8.16. Schematic illustration showing the effect of an asperity (making an angle θ with the horizontal plane) on the surface in contact with the tip on local friction in the presence of adhesive friction mechanism. W and F are the normal and friction forces, respectively, and S and N are the force components along and perpendicular to the local surface of the sample at the contact point, respectively

Finally, we consider the plowing component of friction with tip sliding in either direction, which is [10, 58]

$$\mu_p \sim \tan \theta \,. \tag{8.5}$$

Because we notice little damage of the sample surface in FFM measurements, the contribution by plowing is expected to be small and the ratchet mechanism is believed to be the dominant mechanism for the local variations in the friction force map. With the tip sliding over the leading (ascending) edge of an asperity, the surface slope is positive; it is negative during sliding over the trailing (descending) edge of an asperity. Thus, measured friction is high at the leading edge of asperities and low at the trailing edge. In addition to the slope effect, the collision of the tip when encountering an asperity with a positive slope produces additional torsion of the cantilever beam, leading to higher measured friction force. When encountering an asperity with the same negative slope, however, there is no collision effect and hence no effect on torsion. This effect also contributes to the difference in friction forces when the tip scans up and down on the same topography feature. The ratchet mechanism and the collision effects thus explain semiquantitatively the correlation between the slopes of the roughness maps and friction force maps observed in Figs. 8.14 and 8.15. We note that in the ratchet mechanism, the FFM tip is assumed to be small compared to the size of asperities. This is valid since the typical radius of curvature of the tips is about 10–50 nm. The radii of curvature of samples, the asperities measured here (the asperities that produce most of the friction variation) are found to typically be about 100–200 nm, which is larger than that of the FFM tip [59]. It is important to note that the measured local values of friction and normal forces are measured with respect to global (and not local) horizontal and vertical axes, which are believed to be relevant in applications.

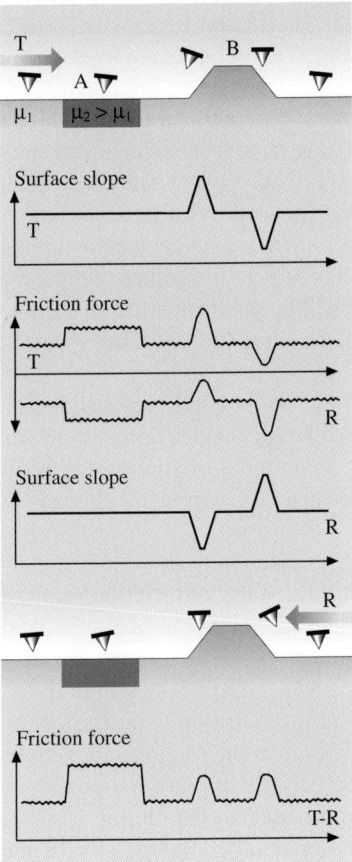

Fig. 8.17. Schematic of friction forces expected when a tip traverses a sample that is composed of different materials and sharp changes in topography. A schematic of surface slope is also shown (T = Trace, R = Retrace)

8.3.3 Directionality Effect on Microfriction

During friction measurements, the friction force data from both the forward (trace) and backward (retrace) scans are useful in understanding the origins of the observed friction forces. Magnitudes of material-induced effects are independent of the scanning direction, whereas topography-induced effects are different between forward and backward scanning directions. Since the sign of the friction force changes as the scanning direction is reversed (because of the reversal of torque applied to the end of the tip), addition of the friction force data of the forward and backward scan eliminates the material-induced effects, while topography-induced effects still remain. Subtraction of the data between forward and backward scans does not eliminate either effect, see Fig. 8.17 [57].

Owing to the reversal of the sign of the retrace (R) friction force with respect to the trace (T) data, the friction force variations due to topography are in the same direction (peaks in trace correspond to peaks in retrace). However, the magnitudes of the peaks in trace and retrace at a given location are different. An increase in the friction force experienced by the tip when scanning up a sharp change in topography is more than the decrease in the friction force experienced when scanning down the same topography change, partly due to the collision effects discussed earlier. Asperities on engineering surfaces are asymmetrical, which also affect the magnitude of friction force in the two directions. In addition, asymmetry in tip shape may have an effect on the directionality effect of friction. We will later note that the magnitude of surface slopes are virtually identical. Therefore, the tip shape asymmetry should not have much effect.

Figure 8.18 shows surface height and friction force data for gold ruler and a silicon grid in the trace and retrace directions. Subtraction of two friction data yields a residual peak, because of the differences in the magnitudes of friction forces in the two directions. This effect is observed at all locations of significant changes in topography.

In order to facilitate comparison of directionality effect on friction, it is important to take into account the sign change of the surface slope and friction force in the trace and retrace directions. Figure 8.19 shows surface height, surface slope, and friction force data for the two samples in the trace and retrace directions. The correlation between surface slope and friction forces is clear. The third column in the figures shows retrace slope and friction data with an inverted sign (−retrace). Now we can compare trace data with −retrace data. It is clear that the friction experienced by the tip is dependent upon the scanning direction because of surface topography. In addition to the effect of topographical changes discussed earlier, during surface-finishing processes, material can be transferred preferentially onto one side of the asperities, which also causes asymmetry and direction dependence. Reduction in local variations and in directionality of friction properties requires careful optimization of surface roughness distributions and of surface-finishing processes.

The directionality as a result of surface asperities effect will also be manifested in macroscopic friction data, i.e., the coefficient of friction may be different in one sliding direction than in the other direction. Asymmetrical shape of asperities accentuates this effect. The frictional directionality can also exist in materials with particles with a preferred orientation. The directionality effect in friction on a macroscale is observed in some magnetic tapes. In a macroscale test, a 12.7-mm wide polymeric magnetic tape was wrapped over an aluminium drum and slid in a reciprocating motion with a normal load of 0.5 N and a sliding speed of about 60 mm/s [3]. The coefficient of friction as a function of sliding distance in either direction is shown in Fig. 8.20. We note that the coefficient of friction on a macroscale for this tape varies in different directions. Directionality in friction is sometimes observed on the macroscale; on the microscale this is the norm [4, 13]. On the macroscale, the effect of surface asperities is normally averaged out over a large number of contacting asperities.

8.3.4 Velocity Dependence on Microfriction

AFM/FFM experiments are generally conducted at relative velocities as high as a few tens of μm/s. To simulate applications, it is of interest to conduct friction experiments at higher velocities. High velocities can be achieved by mounting either the sample or the cantilever beam on a shear wave transducer (ultrasonic transducer) to produce surface oscillations at MHz frequencies [24, 25, 60]. The velocities on the order of a few mm/s can thus be achieved. The effect of in-plane and out-of-plane sample vibration amplitude on the coefficient of friction is shown in Fig. 8.21. Vibration of a sample at ultrasonic frequencies (> 20 kHz) can substantially reduce the coefficient of friction, known as ultrasonic lubrication or sonolubrication. When the surface is vibrated in-plane, classical hydrodynamic lubrication develops hydrodynamic pressure, which supports the tip and reduces friction. When the surface is vibrated out-of-plane, a lift-off caused by the squeeze-film lubrication (a form of hydrodynamic lubrication), reduces friction.

8.3.5 Effect of Tip Radii and Humidity on Adhesion and Friction

The tip radius and relative humidity affect adhesion and friction for dry and lubricated surfaces [23, 36].

Experimental Observations

Figure 8.22 shows the variation of single-point adhesive-force measurements as a function of tip radius on a Si(100) sample for several humidities. The adhesive force data are also plotted as a function of relative humidity for several tip radii. The general trend at humidities up to the ambient is that a 50-nm radius Si_3N_4 tip exhibits a slightly lower adhesive force compared to the other microtips of larger radii; in the latter case, values are similar. Thus, for the microtips there is no appreciable variation in adhesive force with tip radius at a given humidity up to the ambient. The adhesive force increases as relative humidity increases for all tips.

Sources of adhesive force between a tip and a sample surface are van der Waals attraction and meniscus formation [10, 58]. Relative magnitudes of the forces from the two sources are dependent upon various factors, including the distance between the tip and the sample surface, their surface roughness, their hydrophobicity, and relative humidity [61]. For most rough surfaces, meniscus contribution dominates at moderate to high humidities, which arise from capillary condensation of water vapor from the environment. If enough liquid is present to form a meniscus bridge, the meniscus force should increase with an increase in tip radius (proportional to tip radius for a spherical tip). In addition, an increase in tip radius results in increased contact area leading to higher values of van der Waals forces. However, if nanoasperities on the tip and the sample are considered, then the number of contacting and near-contacting asperities forming meniscus bridges increases with an increase of humidity leading to an increase in meniscus forces. These explain the trends observed in Fig. 8.22. From the data, the tip radius has little effect on the

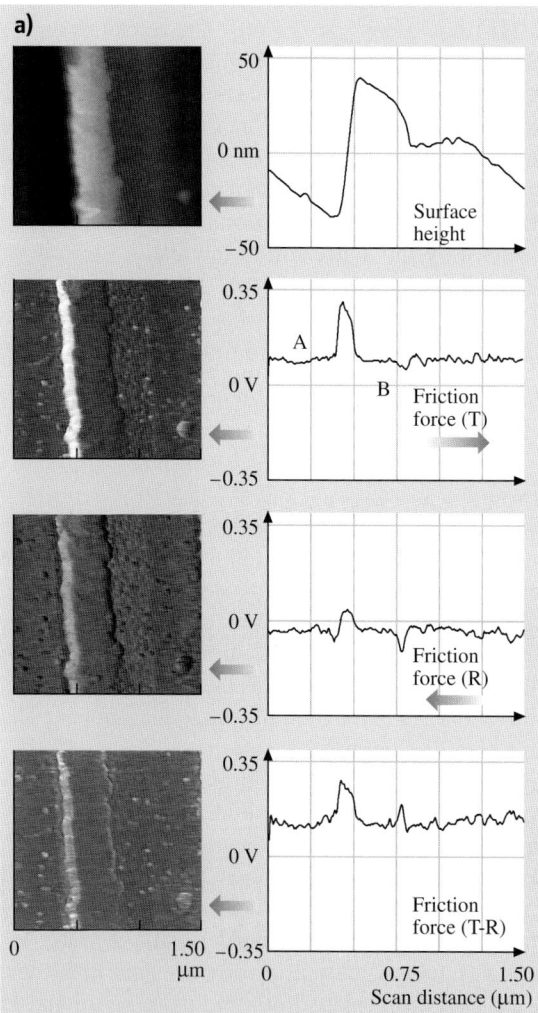

Fig. 8.18. (a) Gray-scale images and two-dimensional profiles of surface height and friction forces across a single ruling of the gold-coated ruling

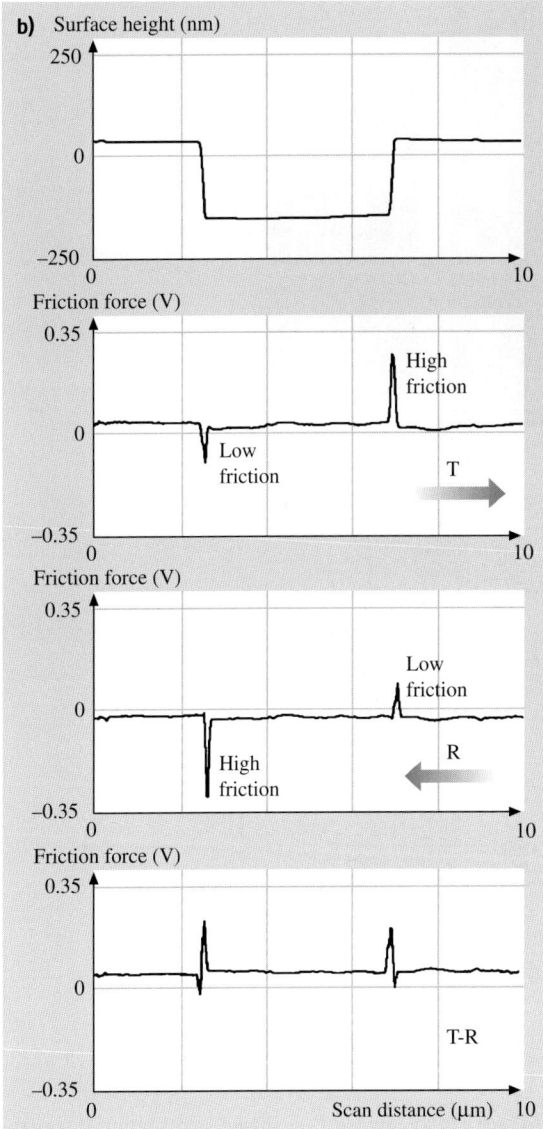

Fig. 8.18. (b) two-dimensional profiles of surface height and friction forces across a silicon grid pit. Friction force data in trace (T) and retrace (R) directions and substrated force data are presented

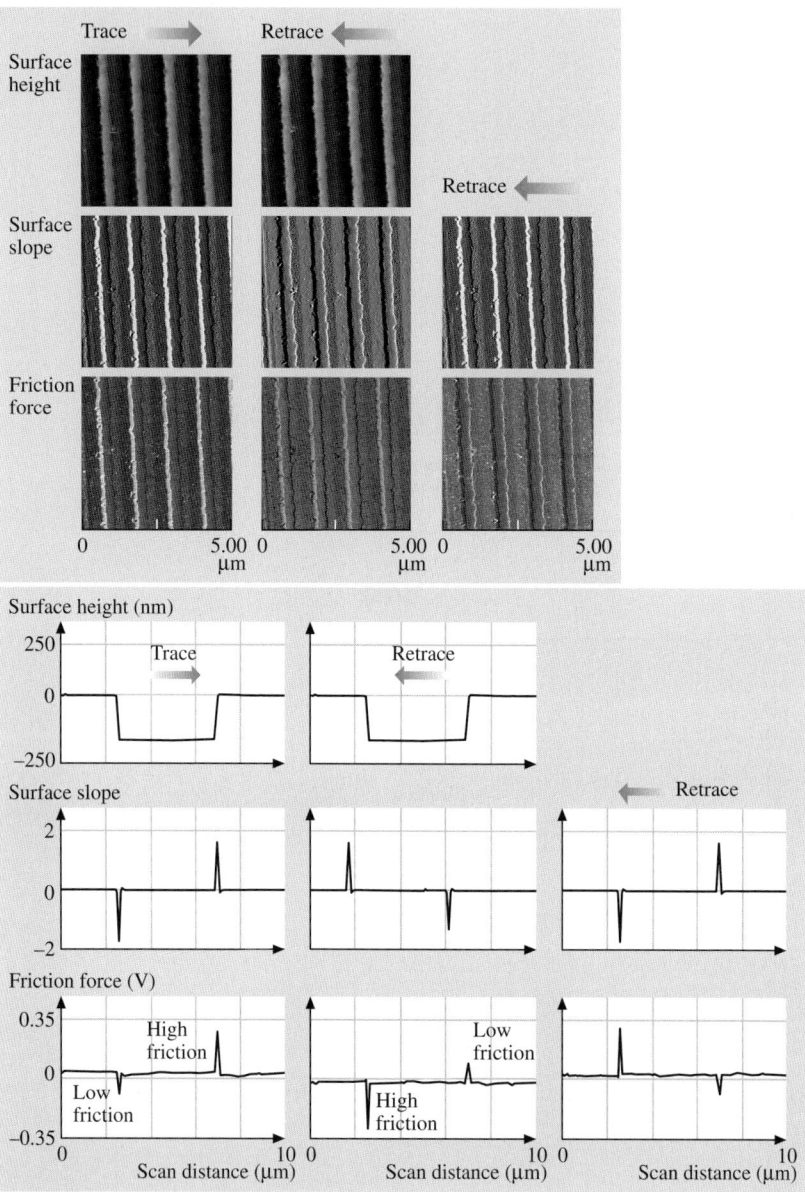

Fig. 8.19. (a) Gray-scale images of surface heights, surface slopes, and friction forces for scans across a gold-coated ruling, and (b) two-dimensional profiles of surface heights, surface slopes, and friction forces for scans across the silicon grid pit. *Arrows* indicate the tip sliding direction [57]

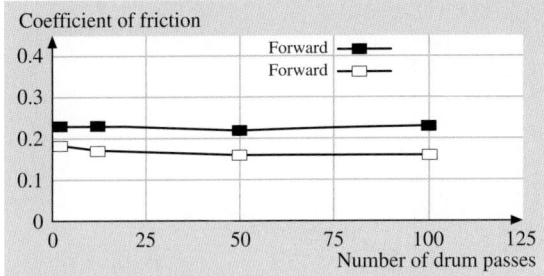

Fig. 8.20. Coefficient of macroscale friction as a function of drum passes for a polymeric magnetic tape sliding over an aluminium drum in a reciprocating mode in both directions. Normal load = 0.5 N over 12.7-mm wide tape, sliding speed = 60 mm/s [37]

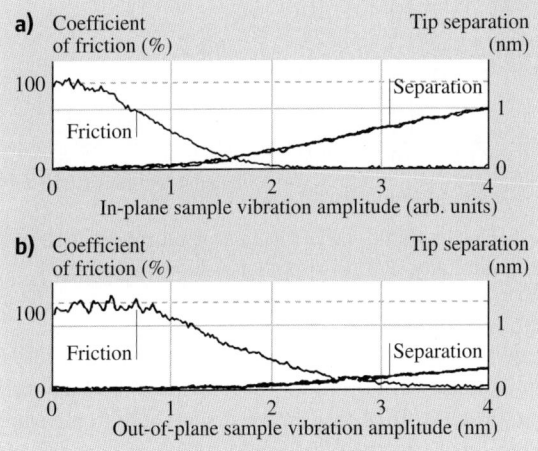

Fig. 8.21. Reduction of coefficient of friction measured at a normal load of 100 nN and average tip separation as a function of surface amplitude on a single-crystal silicon subjected to (a) in-plane and (b) out-of-plane vibrations at about 1 MHz against a silicon nitride tip [24]

adhesive forces at low humidities, but increases with tip radius at high humidity. Adhesive force also increases with an increase in humidity for all tips. This observation suggests that thickness of the liquid film at low humidities is insufficient to form continuous meniscus bridges to affect adhesive forces in the case of all tips.

Figure 8.22 also shows the variation in coefficient of friction as a function of tip radius at a given humidity and as a function of relative humidity for a given tip radius for Si(100). It can be observed that for 0% RH, the coefficient of friction is about the same for the tip radii except for the largest tip, which shows a higher value. At all other humidities, the trend consistently shows that the coefficient of friction increases with tip radius. An increase in friction with tip radius at low to moderate humidities arises from increased contact area (higher van der Waals forces) and

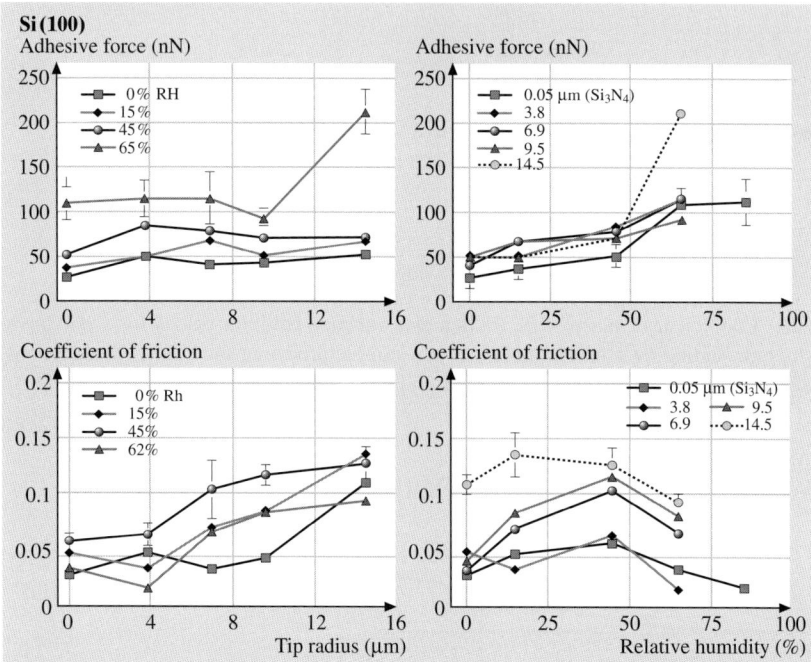

Fig. 8.22. Adhesive force and coefficient of friction as a function of tip radius at several humidities and as a function of relative humidity (RH) at several tip radii on Si(100) [23]

higher values of shear forces required for larger contact area. At high humidities, similar to adhesive force data, an increase with tip radius occurs due to both contact area and meniscus effects. Although AFM/FFM measurements are able to measure the combined effect of the contribution of van der Waals and meniscus forces toward friction force or adhesive force, it is difficult to measure their individual contributions separately. It can be seen that for all tips, the coefficient of friction increases with humidity to about ambient, beyond which it starts to decrease. The initial increase in the coefficient of friction with humidity arises from the fact that the thickness of the water film increases with an increase in the humidity, which results in a larger number of nanoasperities forming meniscus bridges and leads to higher friction (larger shear force). The same trend is expected with the microtips beyond 65% RH. This is attributed to the fact that at higher humidities, the adsorbed water film on the surface acts as a lubricant between the two surfaces. Thus the interface is changed at higher humidities, resulting in lower shear strength and, hence, lower friction force and coefficient of friction.

Adhesion and Friction Force Expressions for a Single-Asperity Contact

We now obtain the expressions for the adhesive force and coefficient of friction for a single-asperity contact with a meniscus formed at the interface. For a spherical asperity of radius R in contact with a flat and smooth surface with the composite modulus of elasticity E^* and with a concave meniscus, the attractive meniscus force (adhesive force) W_{ad} is given as [10, 58]

$$W_{ad} = 2\pi R \gamma_1 (\cos\theta_1 + \cos\theta_2) . \tag{8.6}$$

For an elastic contact for both extrinsic (W) and intrinsic (W_{ad}) normal load, the friction force is given as

$$F_e = \pi\tau \left(\frac{3(W + W_{ad})R}{4E^*}\right)^{2/3} , \tag{8.7}$$

where γ_1 is the surface tension of the liquid in air, θ_1 and θ_2 are the contact angles between the liquid and the two surfaces, W is the external load, and τ is the average shear strength of the contacts. (Surface energy effects are not considered here.) Note that adhesive force increases linearly with an increase in the tip radius, and the friction force increases with an increase in tip radius as $R^{2/3}$ and with normal load as $(W + W_{ad})^{2/3}$. The experimental data in support of $W^{2/3}$ dependence on the friction force can be found in various references (see, for example, [62]). The coefficient of friction μ_e is obtained from (8.7) as

$$\mu_e = \frac{F_e}{(W + W_{ad})}$$
$$= \pi\tau \left(\frac{3R}{4E^*}\right)^{2/3} \frac{1}{(W + W_{ad})^{1/3}} . \tag{8.8}$$

In the plastic contact regime [58], the coefficient of friction μ_p is obtained as

$$\mu_p = \frac{F_p}{(W + W_{ad})} = \frac{\tau}{H_s} , \tag{8.9}$$

where H_s is the hardness of the softer material. Note that in the plastic contact regime, the coefficient of friction is independent of external load, adhesive contributions, and surface geometry.

In comparison, for multiple-asperity contacts in the elastic contact regime, the total adhesive force W_{ad} is the summation of adhesive forces at n individual contacts,

$$W_{ad} = \sum_{i=1}^{n}(W_{ad})_i \quad \text{and}$$
$$\mu_e \approx \frac{3.2\tau}{E^*\left(\sigma_p/R_p\right)^{1/2} + W_{ad}/W} , \tag{8.10}$$

where σ_p and R_p are the standard deviation of summit heights and average summit radius, respectively. Note that the coefficient of friction depends upon the surface roughness. In the plastic contact regime, the expression for μ_p in (8.9) does not change.

The source of the adhesive force, in a wet contact in the AFM experiments being performed in an ambient environment, includes mainly attractive meniscus force due to capillary condensation of water vapor from the environment. The meniscus force for a single contact increases with an increase in tip radius. A sharp AFM tip in contact with a smooth surface at low loads (on the order of a few nN) for most materials can be simulated as a single-asperity contact. At higher loads, for rough and soft surfaces, multiple contacts would occur. Furthermore, at low loads (nN range) for most materials, the local deformation would be primarily elastic. Assuming that shear strength of contacts does not change, the adhesive force for smooth and hard surfaces at low normal load (on the order of few nN) (for a single-asperity contact in the elastic contact regime) would increase with an increase in tip radius, and the coefficient of friction would decrease with an increase in total normal load as $(W + W_{ad})^{-1/3}$ and would increase with an increase of tip radius as $R^{2/3}$. In this case, the Amontons law of friction, which states that the coefficient of friction is independent of normal load and is independent of apparent area of contact, does not hold. For a single-asperity plastic contact and multiple-asperity plastic contacts, neither the normal load nor tip radius come into play in the calculation of the coefficient of friction. In the case of multiple-asperity contacts, the number of contacts increase with an increase of normal load. Therefore, adhesive force increases with an increase in load.

In the data presented earlier in this section, the effect of tip radius and humidity on the adhesive forces and coefficient of friction is investigated for experiments with Si(100) surface at loads in the range of 10–100 nN. The multiple-asperity elastic contact regime is relevant for this study. An increase in humidity generally results in an increase in the number of meniscus bridges that would increase the adhesive force. As suggested earlier, that increase in humidity may also decrease the shear strength of contacts. A combination of an increase in adhesive force and a decrease in shear strength would affect the coefficient of friction. An increase in tip radius would increase the meniscus force (adhesive force). A substantial increase in the tip radius may also increase interatomic forces. These effects influence the coefficient of friction with an increase in the tip radius.

8.3.6 Scale Dependence on Friction

Table 8.2 shows the coefficient of friction measured for various materials on nano- and macroscales [19]. It is clearly observed that friction values are scale-dependent. The values on the nano/microscale are much lower than those on the macroscale. There can be the following four, and possibly more, differences in the operating conditions responsible for these differences. First, the contact stresses at AFM conditions, in spite of small tip radii, generally do not exceed the sample hardness that minimizes plastic deformation. Average contact stresses in macrocontacts are generally lower than in AFM contacts, however, a large number of asperities come into contact that go through some plastic deformation. Second, when measured for the small contact areas and very low loads used in microscale studies, indentation hardness is higher than at the macroscale, as will be discussed later [11, 34, 35]. Lack of

Table 8.2. Surface roughness (standard deviation of surface heights or σ) and coefficients of friction on nano- and macroscales of various samples in air

Material	σ(nm)	Coefficient of nanoscale friction versus Si_3N_4 tip[a]	Coefficient of macroscale friction versus Si_3N_4 ball[b]
Graphite (HOPG)	0.09	0.006	0.1
Natural diamond	2.3	0.04	0.2
Si(100)	0.14	0.07	0.4

[a] Tip radius of about 50 nm in the load range of 10–150 nN (2.5–6.1 GPa) and a scanning speed of 0.5 nm/s and scan area of 1 nm × 1 nm for HOPG and a scanning speed of 4 µm/s and scan area of 1 µm × 1 µm for diamond and Si(100)
[b] Ball radius of 3 mm at a normal load of 1 N (0.6 GPa) and average sliding speed of 0.8 mm/s

plastic deformation and improved mechanical properties reduce the degree of wear and friction. Third, the smallapparent area of contact reduces the number of particles trapped at the interface, and thus minimizes the third-body plowing contribution to the friction force [11]. As a fourth and final difference, we have seen in the previous section that coefficient of friction increases with an increase in the AFM tip radius. AFM data are taken with a sharp tip, whereas asperities coming in contact in macroscale tests range from nanoasperities to much larger asperities, which may be responsible for larger values of friction force on the macroscale.

To demonstrate the load dependence on the coefficient of friction, stiff cantilevers were used to conduct friction experiments at high loads, see Fig. 8.23 [63]. At higher loads (with contact stresses exceeding the hardness of the softer material), as anticipated, the coefficient of friction for microscale measurements increases toward values comparable to those obtained from macroscale measurements, and surface damage also increases. At high loads, plowing is a dominant contributor to the friction force. Based on these results, Amontons rule of friction, which states that the coefficient of friction is independent of apparent area and normal load, does not hold for microscale measurements. These findings also suggest that microcomponents sliding under lightly loaded conditions should experience ultralow friction and near-zero wear [11].

8.4 Scratching, Wear, Local Deformation, and Fabrication/Machining

8.4.1 Nanoscale Wear

Bhushan and *Ruan* [18] conducted nanoscale wear tests on polymeric magnetic tapes using conventional silicon nitride tips at two different loads of 10 and 100 nN (Fig. 8.24). For a low normal load of 10 nN, measurements were made twice. There was no discernible difference between consecutive measurements for this load. However, as the load was increased from 10 nN to 100 nN, topographical changes were

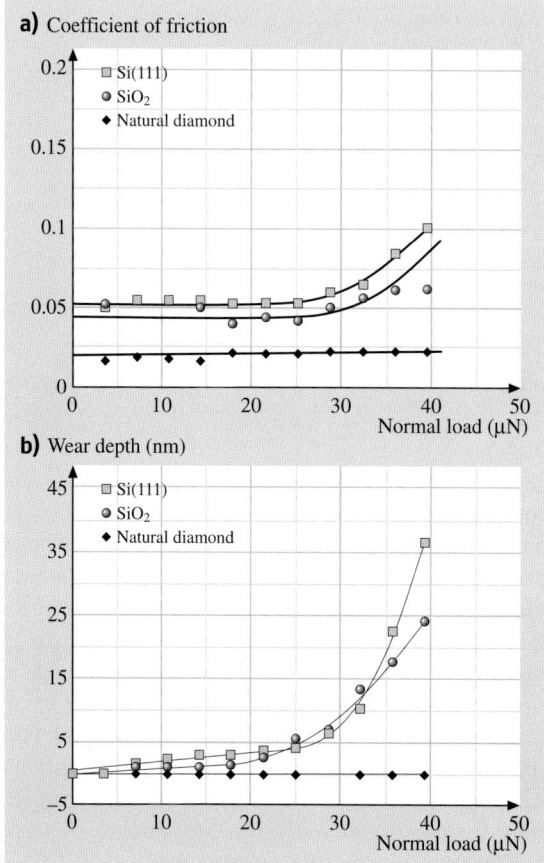

Fig. 8.23. (a) Coefficient of friction as a function of normal load and (b) corresponding wear depth as a function of normal load for silicon, SiO_2 coating, and natural diamond. Inflections in the curves for silicon and SiO_2 correspond to the contact stresses equal to the hardnesses of these materials [63]

observed during subsequent scanning at a normal load of 10 nN; material was pushed in the sliding direction of the AFM tip relative to the sample. The material movement is believed to occur as a result of plastic deformation of the tape surface. Thus, deformation and movement of the soft materials on a nanoscale can be observed.

8.4.2 Microscale Scratching

The AFM can be used to investigate how surface materials can be moved or removed on micro- to nanoscales, for example, in scratching and wear [4] (where these things are undesirable), and nanofabrication/nanomachining (where they are desirable). Figure 8.25a shows microscratches made on Si(111) at various loads and

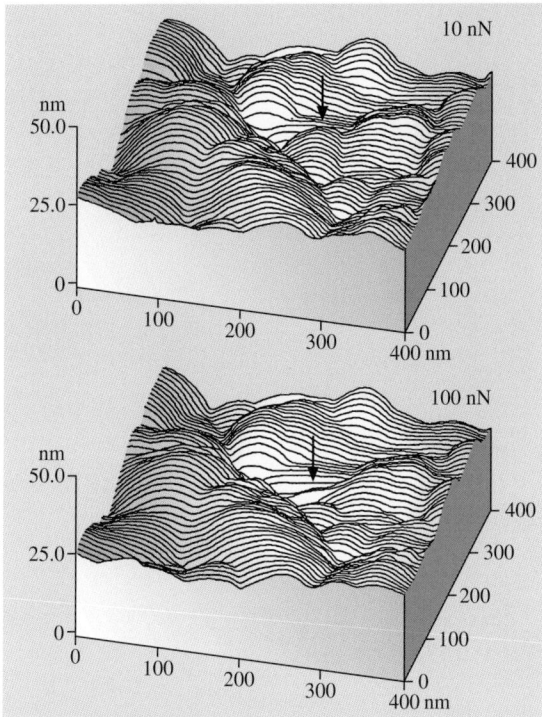

Fig. 8.24. Surface roughness maps of a polymeric magnetic tape at the applied normal load of 10 nN and 100 nN. Location of the change in surface topography as a result of nanowear is indicated by *arrows* [18]

scanning velocity of 2 μm/s after 10 cycles [22]. As expected, the scratch depth increases linearly with load. Such microscratching measurements can be used to study failure mechanisms on the microscale and evaluate the mechanical integrity (scratch resistance) of ultrathin films at low loads.

To study the effect of scanning velocity, unidirectional scratches, 5 μm in length, were generated at scanning velocities ranging from 1–100 μm/s at various normal loads ranging from 40–140 μN. There is no effect of scanning velocity obtained at a given normal load. (For representative scratch profiles at 80 μN, see Fig. 8.25b.) This may be because of a small effect of frictional heating with the change in scanning velocity used here. Furthermore, for a small change in interface temperature, there is a large underlying volume to dissipate the heat generated during scratching.

Scratching can be performed under ramped loading to determine the scratch resistance of materials and coatings. The coefficient of friction is measured during scratching, and the load at which the coefficient of friction increases rapidly is known as the "critical load", which is a measure of scratch resistance. In addition, postscratch imaging can be performed in situ with the AFM in tapping mode to study failure mechanisms. Figure 8.26 shows data from a scratch test on Si(100) with

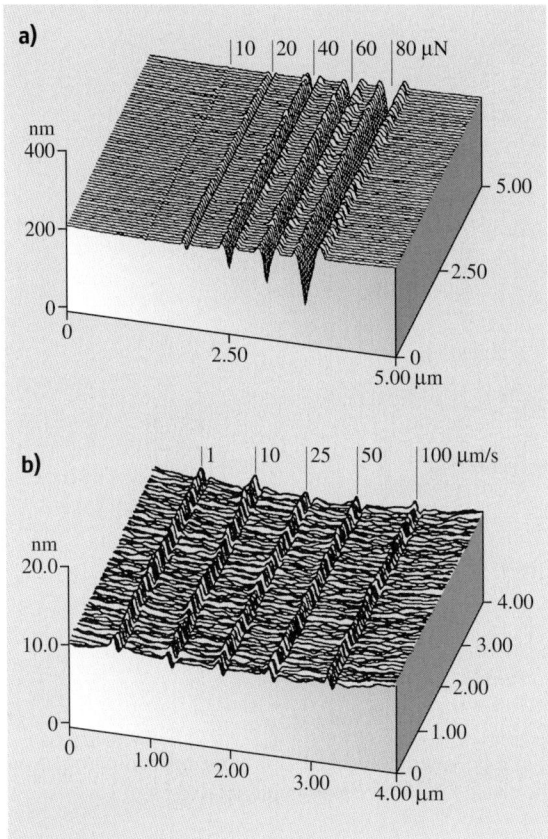

Fig. 8.25. Surface plots of (**a**) Si(111) scratched for 10 cycles at various loads and a scanning velocity of 2 μm/s. Note that x and y axes are in μm and z axis is in nm. (**b**) Si(100) scratched in one unidirectional scan cycle at a normal force of 80 μN and different scanning velocities

a scratch length of 25 μm and a scratching velocity of 0.5 μm/s. At the beginning of the scratch, the coefficient of friction is 0.04, which indicates a typical value for silicon. At about 35 μN (indicated by the arrow in the figure), there is a sharp increase in the coefficient of friction, which indicates the critical load. Beyond the critical load, the coefficient of friction continues to increase steadily. In the post-scratch image, we note that at the critical load, a clear groove starts to form. This implies that Si(100) was damaged by plowing at the critical load, associated with the plastic flow of the material. At and after the critical load, small and uniform debris is observed and the amount of debris increases with increasing normal load. *Sundararajan* and *Bhushan* [33] have also used this technique to measure the scratch resistance of diamond-like carbon coatings ranging in thickness from 3.5–20 nm.

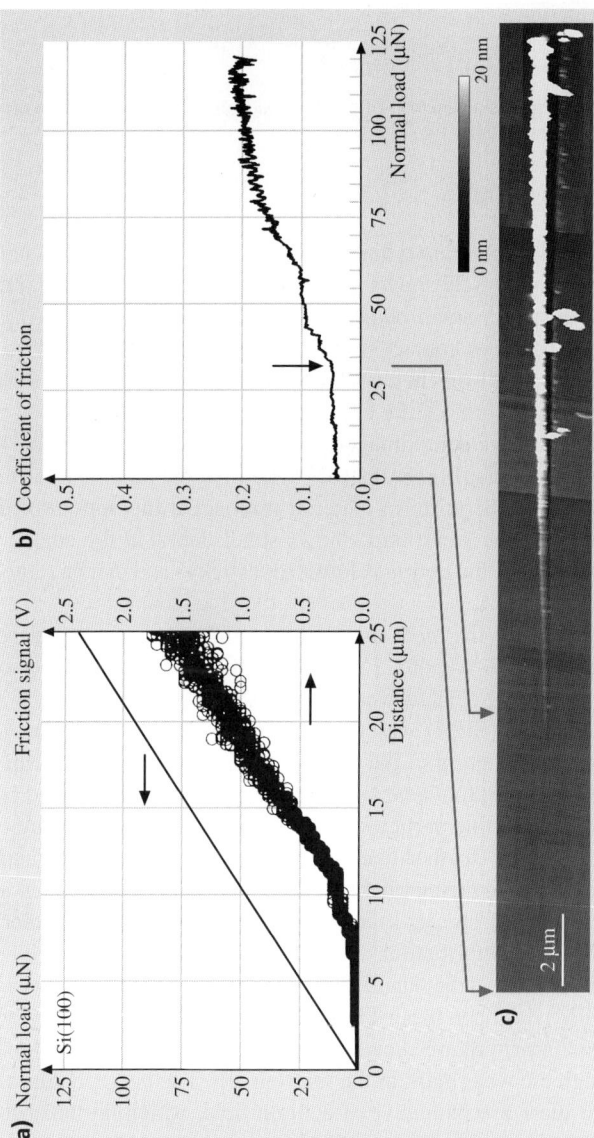

Fig. 8.26. (a) Applied normal load and friction signal measured during the microscratch experiment on Si(100) as a function of scratch distance, (b) friction data plotted in the form of coefficient of friction as a function of normal load, and (c) AFM surface height image of scratch obtained in tapping mode [33]

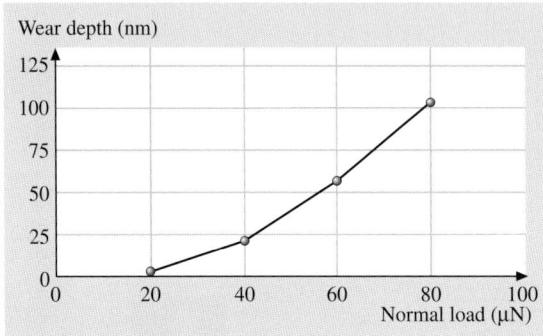

Fig. 8.27. Wear depth as a function of normal load for Si(100) after onecycle [64]

8.4.3 Microscale Wear

By scanning the sample in two dimensions with the AFM, wear scars are generated on the surface. Figure 8.27 shows the effect of normal load on wear depth. We note that wear depth is very small below 20 μN of normal load [64, 65]. A normal load of 20 μN corresponds to contact stresses comparable to the hardness of the silicon. Elastic deformation at loads below 20 μN is primarily responsible for low wear [63].

A typical wear mark, of the size 2 μm × 2 μm, generated at a normal load of 40 μN for one scan cycle and imaged using AFM with scan size of 4 μm × 4 μm at 300 nN load is shown in Fig. 8.28a [65]. The inverted map of wear marks shown in Fig. 8.28b indicates the uniform material removal at the bottom of the wear mark. An AFM image of the wear mark shows small debris at the edges, swiped during AFM scanning. Thus, the debris is loose (not sticky) and can be removed during the AFM scanning.

Next, we examine the mechanism of material removal on a microscale in AFM wear experiments [23, 64, 65]. Figure 8.29 shows a secondary electron image of the wear mark and associated wear particles. The specimen used for the scanning electron microscope (SEM) was not scanned with the AFM after initial wear, in order to retain wear debris in the wear region. Wear debris is clearly observed. In the SEM micrographs, the wear debris appear to be agglomerated because of high surface energy of the fine particles. Particles appear to be a mixture of rounded and so-called cutting type (feather-like or ribbon-like material). *Zhao* and *Bhushan* [64] reported an increase in the number and size of cutting-type particles with the normal load. The presence of cutting-type particles indicates that the material is removed primarily by plastic deformation.

To better understand the material removal mechanisms, transmission electron microscopy (TEM) has been used. The TEM micrograph of the worn region and associated diffraction pattern are shown in Fig. 8.30a,b. The bend contours are observed to pass through the wear mark in the micrograph. The bend contours around and inside the wear mark are indicative of a strain field, which, in the absence of applied stresses, can be interpreted as plastic deformation and/or elastic residual stresses.

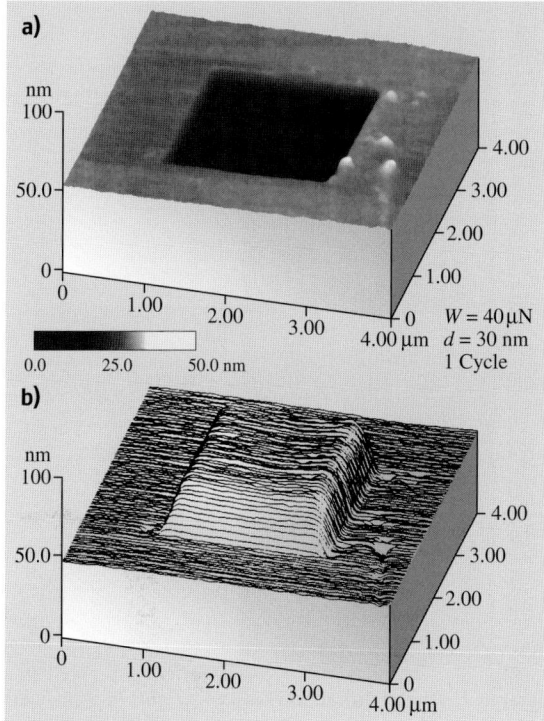

Fig. 8.28. (a) Typical gray-scale and (b) inverted AFM images of wear mark created using a diamond tip at a normal load of 40 µN and one scan cycle on Si(100) surface

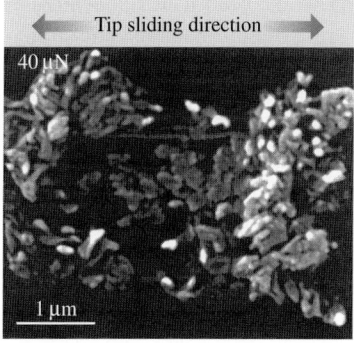

Fig. 8.29. Secondary electron image of wear mark and debris for Si(100) produced at a normal load of 40 µN and one scan cycle

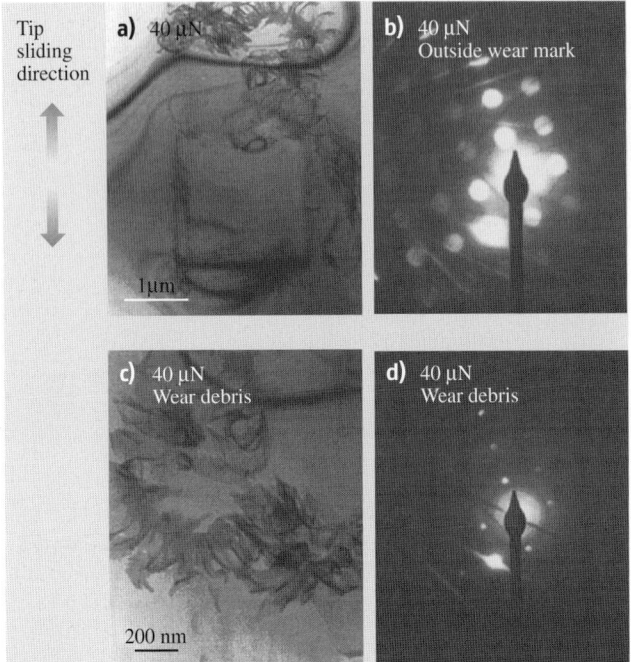

Fig. 8.30. Bright field TEM micrographs (*left*) and diffraction patterns (*right*) of wear mark (**a**), (**b**) and wear debris (**c**), (**d**) in Si(100) produced at a normal load of 40 μN and one scan cycle. Bend contours around and inside wear mark are observed

Often, localized plastic deformation during loading would lead to residual stresses during unloading; therefore, bend contours reflect a mix of elastic and plastic strains. The wear debris is observed outside the wear mark. The enlarged view of the wear debris in Fig. 8.30c shows that much of the debris is ribbon-like, indicating that material is removed by a cutting process via plastic deformation, which is consistent with the SEM observations. The diffraction pattern from inside the wear mark is similar to that of virgin silicon, showing no evidence of any phase transformation (amorphization) during wear. A selected area diffraction pattern of the wear debris shows some diffuse rings, which indicates the existence of amorphous material in the wear debris, confirmed as silicon oxide products from chemical analysis. It is known that plastic deformation occurs by generation and propagation of dislocations. No dislocation activity or cracking was observed at 40 μN. However, dislocation arrays could be observed at 80 μN. Figure 8.31 shows the TEM micrographs of the worn region at 80 μN; for better observation of the worn surface, wear debris was moved out of the wear mark using AFM with a large area scan at 300 nN after the wear test. The existence of dislocation arrays confirms that material removal occurs by plastic deformation. This corroborates the observations made in scratch tests at ramped load in the previous section. It is concluded that the material on microscale at high loads is removed by plastic deformation with a small contribution from elastic fracture [64].

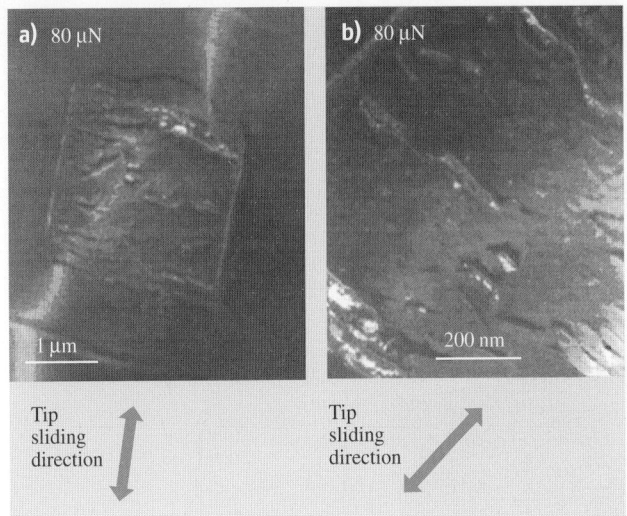

Fig. 8.31. (a) Bright field and (b) weak beam TEM micrographs of wear mark in Si(100) produced at a normal load of 80 μN and one scan cycle showing bend contours and dislocations [64]

To understand wear mechanisms, evolution of wear can be studied using AFM. Figure 8.32 shows evolution of wear marks of a DLC-coated disk sample. The data illustrate how the microwear profile for a load of 20 μN develops as a function of the number of scanning cycles [22]. Wear is not uniform, but is initiated at the nanoscratches. Surface defects (with high surface energy) present at nanoscratches act as initiation sites for wear. Coating deposition also may not be uniform on and near nanoscratches, which may lead to coating delamination. Thus, scratch-free surfaces will be relatively resistant to wear.

Wear precursors (precursors to measurable wear) can be studied by making surface potential measurements [38–40]. The contact potential difference, or simply surface potential between two surfaces, depends on a variety of parameters such as electronic work function, adsorption, and oxide layers. The surface potential map of an interface gives a measure of changes in the work function, which is sensitive to both physical and chemical conditions of the surfaces, including structural and chemical changes. Before material is actually removed in a wear process, the surface experiences stresses that result in surface and subsurface changes of structure and/or chemistry. These can cause changes in the measured potential of a surface. An AFM tip allows mapping of surface potential with nanoscale resolution. Surface height and change in surface potential maps of a polished single-crystal aluminium (100) sample abraded using a diamond tip at loads of 1 μN and 9 μN are shown in Fig. 8.33a. (Note that the sign of the change in surface potential is reversed here from that in [38].) It is evident that both abraded regions show a large potential contrast (~ 0.17 V) with respect to the non-abraded area. The black region in the lower right-hand part of the

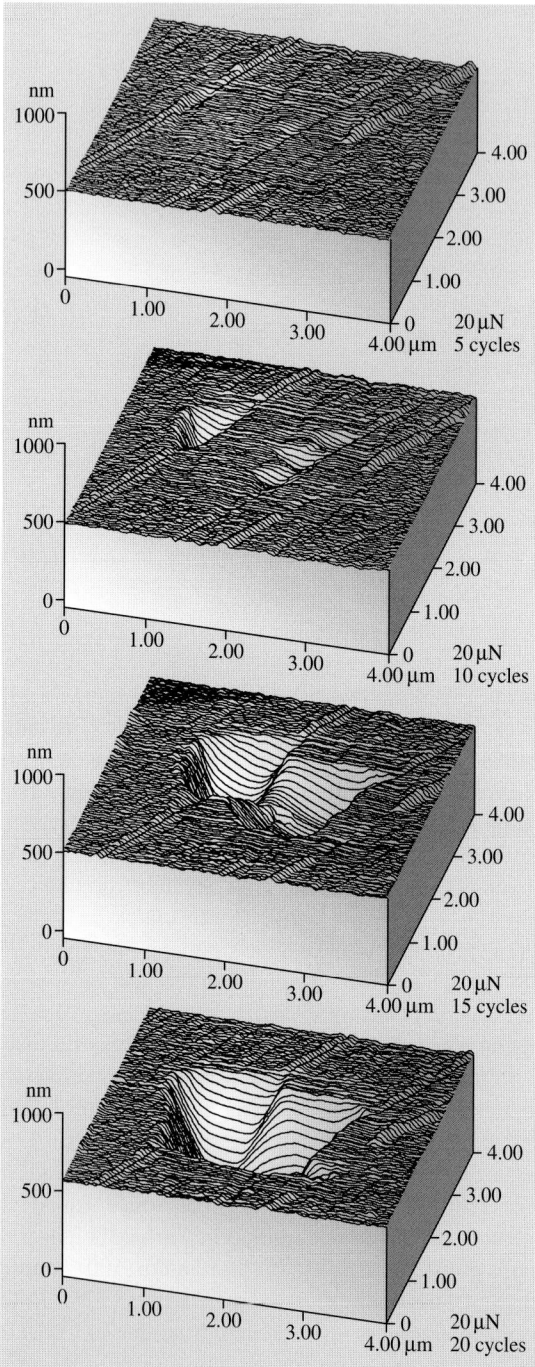

Fig. 8.32. Surface plots of diamond-like carbon-coated thin-film disk showing the worn region; the normal load and number of test cycles are indicated [22]

topography scan shows a step that was created during the polishing phase. There is no potential contrast between the high region and the low region of the sample, indicating that the technique is independent of surface height. Figure 8.33b shows a close-up scan of the upper (low load) wear region in Fig. 8.33a. Notice that while there is no detectable change in the surface topography, there is, nonetheless, a large change in the potential of the surface in the worn region. Indeed, the wear mark of Fig. 8.33b might not have been visible at all in the topography map were it not for the noted absence of wear debris generated nearby and then swept off during the low load scan. Thus, even in the case of zero wear (no measurable deformation of the surface using AFM), there can be a significant change in the surface potential inside the wear mark, which is useful for the study of wear precursors. It is believed that the removal of the thin contaminant layer including the natural oxide layer gives rise to the initial change in surface potential. The structural changes, which precede generation of wear debris and/or measurable wear scars, occur under ultralow loads in the top few nanometers of the sample and are primarily responsible for the subsequent changes in surface potential.

8.4.4 In Situ Characterization of Local Deformation

In situ surface characterization of local deformation of materials and thin films is carried out using a tensile stage inside an AFM. Failure mechanisms of polymeric thin films were studied by *Bobji* and *Bhushan* [41, 42]. The specimens were strained at a rate of $4 \times 10^{-3}\%$ per second, and AFM images were captured at different strains up to about 10% to monitor generation and propagation of cracks and deformation bands.

Bobji and *Bhushan* [41, 42] studied three magnetic tapes with thickness ranging from 7 to 8.5 µm. One of these was with acicular-shaped metal particle (MP) coating, and the other two with metal-evaporated (ME) coating and with and without a thin diamond-like carbon (DLC) overcoat, both on a polymeric substrate and all with particulate back coating [13]. They also studied the polyethylene terephthalate (PET) substrate with 6 µm thickness. They reported that cracking of the coatings started at about 1% strain for all tapes, much before the substrate starts to yield at about 2% strain. Figure 8.34 shows the topographical images of the MP tape at different strains. At 0.83% strain, a crack can be seen originating at the marked point. As the tape is further stretched along the direction, as shown in Fig. 8.34, the crack propagates along the shorter boundary of the ellipsoidal particle. However, the general direction of the crack propagation remains perpendicular to the direction of the stretching. The length, width, and depth of the cracks increase with strain, and at the same time, newer cracks keep on nucleating and propagating with reduced crack spacing. At 3.75% strain, another crack can be seen nucleating. This crack continues to grow parallel to the first. When the tape is unloaded after stretching up to a strain of about 2%, i.e., within the elastic limit of the substrate, the cracks rejoin perfectly and it is impossible to determine the difference from the unstrained tape.

Figure 8.35 shows topographical images of the three magnetic tapes and the PET substrate after being strained to 3.75%, which is well beyond the elastic limit of the

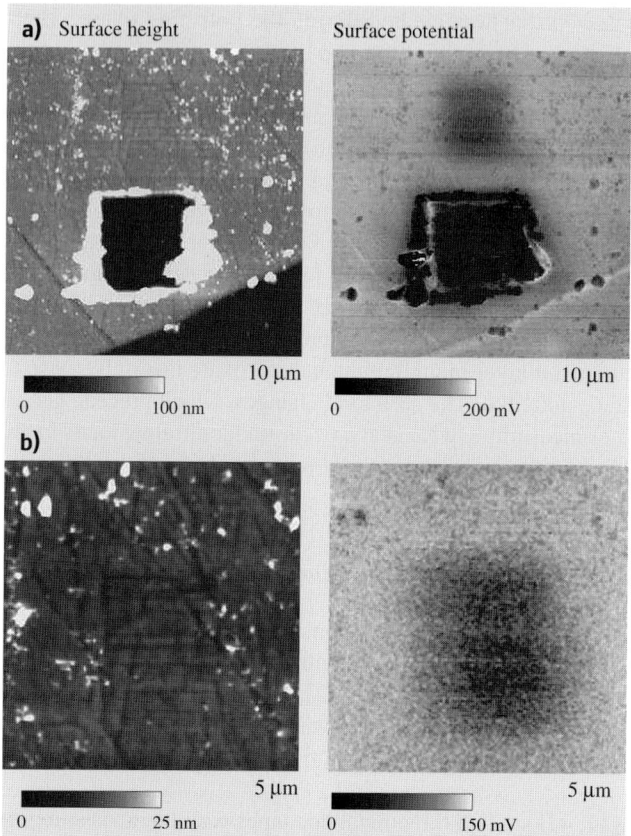

Fig. 8.33. (a) Surface height and change in surface potential maps of wear regions generated at 1 μN (*top*) and 9 μN (*bottom*) on a single-crystal aluminium sample showing bright contrast in the surface potential map on the worn regions. (**b**) Close up of upper (low load) wear region [38]

substrate. MP tape develops short and numerous cracks perpendicular to the direction of loading. In tapes with metallic coating, the cracks extend throughout the tape width. In ME tape with DLC coating, there is a bulge in the coating around the primary cracks that are initiated when the substrate is still elastic, like crack A in the figure. The white band on the right-hand side of the figure is the bulge of another crack. The secondary cracks like B and C are generated at higher strains and are straighter compared to the primary cracks. In ME tape, which has a Co-O film on PET substrate, with a thickness ratio of 0.03, both with and without DLC coating, no difference is observed in the rate of growth between primary and secondary cracks. The failure is cohesive with no bulging of the coating. This seems to suggest that the DLC coating has residual stresses that relax when the coating cracks, causing delamination. Since the stresses are already relaxed, the secondary crack does not

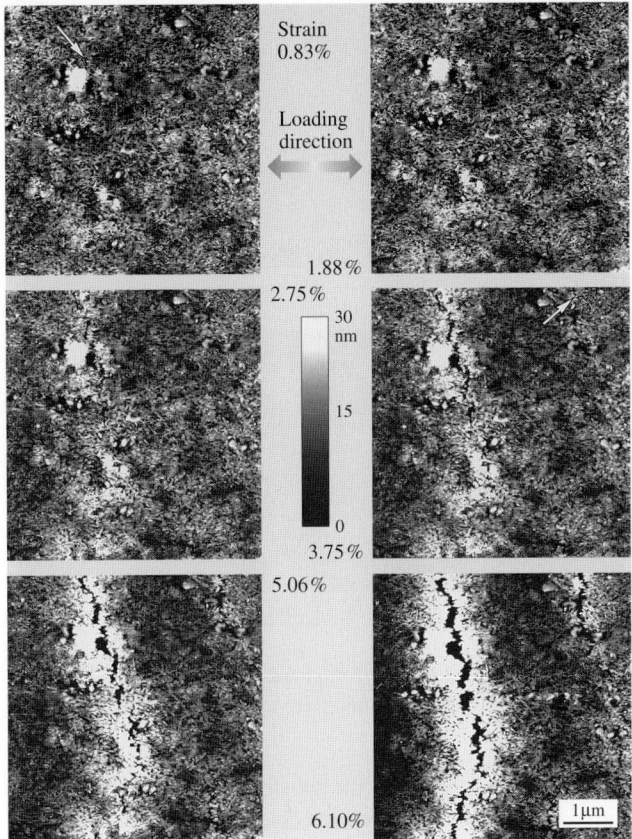

Fig. 8.34. Topographical images of the MP magnetic tape at different strains [41]

result in delamination. The presence of the residual stress is confirmed by the fact that a freestanding ME tape curls up (in a cylindrical form with its axis perpendicular to the tape length) with a radius of curvature of about 6 mm; the ME tape without the DLC does not curl. The magnetic coating side of PET substrate is much smoother at smaller scan lengths. However, in 20 μm scans it has a lot of bulging out, which appears as white spots in the figure. These spots change shape even while scanning the samples in tapping mode at very low contact forces.

The variation of average crack width and average crack spacing with strain is plotted in Fig. 8.36. The crack width is measured at a spot along a given crack over a distance of 1 μm in the 5 μm scan image at different strains. The crack spacing is obtained by averaging the inter-crack distance measured in five separate 50 μm scans at each strain. It can be seen that the cracks nucleate at a strain of about 0.7–1.0%, well within the elastic limit of the substrate. There is a definite change in the slope of the load-displacement curve at the strain where cracks nucleate, and the slope after

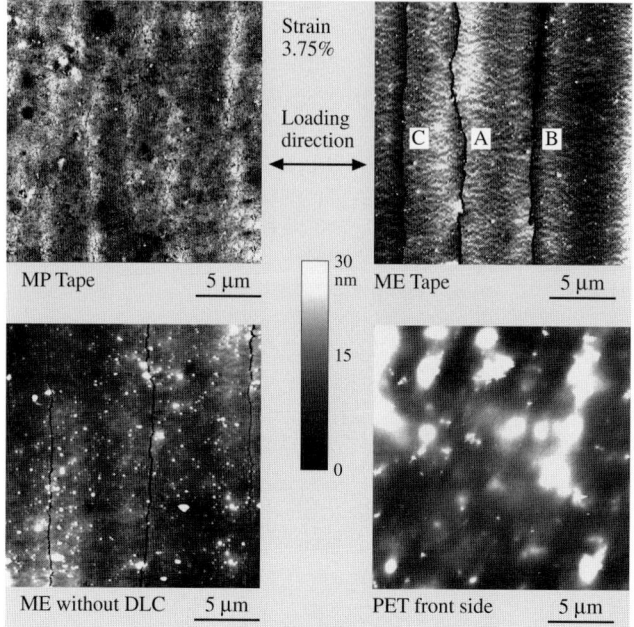

Fig. 8.35. Comparison of crack morphologies at 3.75% strain in three magnetic tapes and PET substrate. Cracks B and C, nucleated at higher strains, are more linear than crack A [42]

that is closer to the slope of the elastic portion of the substrate. This would mean that most of the load is supported by the substrate once the coating fails by cracking.

In situ surface characterization of unstretched and stretched films has been used to measure Poisson's ratio of polymeric thin films [66]. Uniaxial tension is applied by the tensile stage. Surface height profiles obtained from the AFM images of unstretched and stretched samples are used to monitor simultaneously the changes in displacements of the polymer films in the longitudinal and lateral directions.

8.4.5 Nanofabrication/Nanomachining

An AFM can be used for nanofabrication/nanomachining by extending the microscale scratching operation [4, 11, 22, 37]. Figure 8.37 shows two examples of nanofabrication. The patterns were created on a single-crystal silicon (100) wafer by scratching the sample surface with a diamond tip at specified locations and scratching angles. Each line is scribed manually at a normal load of 15 µN and a writing speed of 0.5 µm/s. The separation between lines is about 50 nm, and the variation in line width is due to the tip asymmetry. Nanofabrication parameters – normal load, scanning speed, and tip geometry – can be controlled precisely to control the depth and length of the devices.

Nanofabrication using mechanical scratching has several advantages over other techniques. Better control over the applied normal load, scan size, and scanning

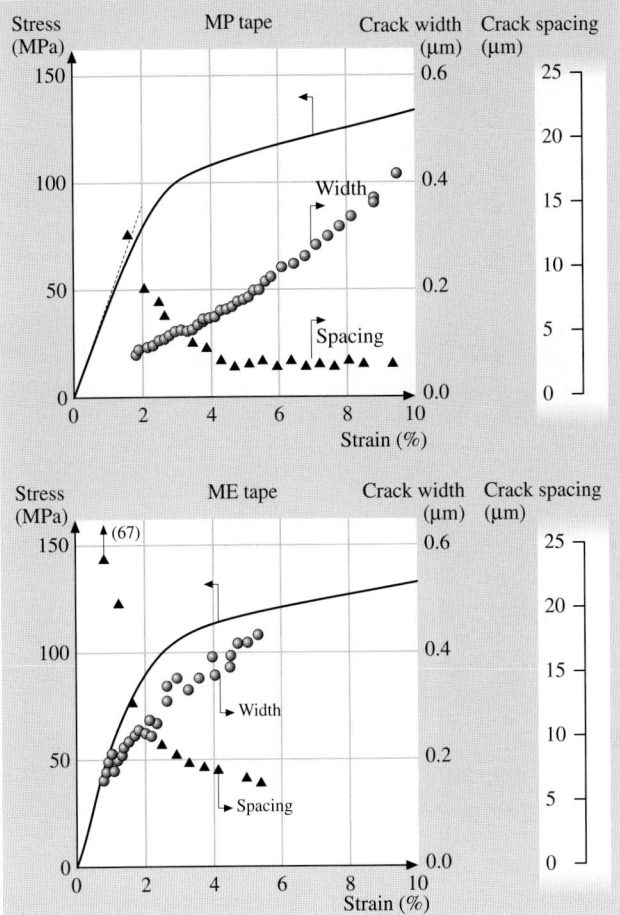

Fig. 8.36. Variation of stress, crack width, and crack spacing with strain in three magnetic tapes and PET substrate [41]

speed can be used for nanofabrication of devices. Using the technique, nanofabrication can be performed on any engineering surface. Use of chemical etching or reactions is not required, and this dry nanofabrication process can be used where use of chemicals and electric field is prohibited. One disadvantage of this technique is the formation of debris during scratching. At light loads, debris formation is not a problem compared to high-load scratching. However, debris can be removed easily from the scan area at light loads during scanning.

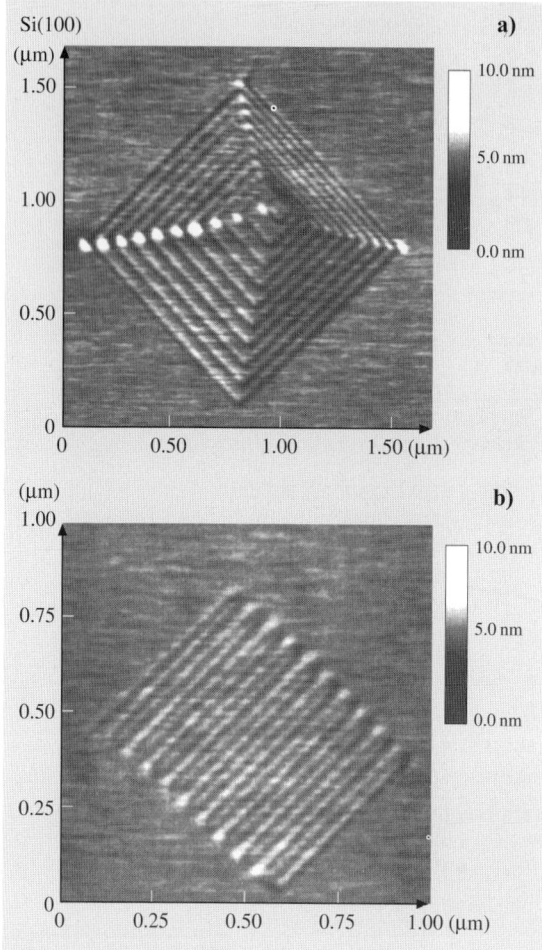

Fig. 8.37. (a) Trim and (b) spiral patterns generated by scratching a Si(100) surface using a diamond tip at a normal load of 15 μN and writing speed of 0.5 μm/s

8.5 Indentation

Mechanical properties such as hardness and Young's modulus of elasticity can be determined on micro- to picoscales using the AFM [18, 22, 31, 34] and a depth-sensing indentation system used in conjunction with an AFM [35, 67–69].

8.5.1 Picoindentation

Indentability on the scale of subnanometers of soft samples can be studied in the force calibration mode (Fig. 8.6) by monitoring the slope of cantilever deflection as a function of sample traveling distance after the tip is engaged and the sample is

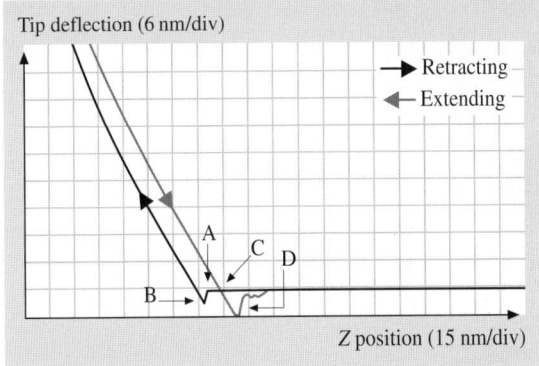

Fig. 8.38. Tip deflection (normal load) as a function of the Z (separation distance) curve for a polymeric magnetic tape [18]

pushed against the tip. For a rigid sample, cantilever deflection equals the sample traveling distance, but the former quantity is smaller if the tip indents the sample. In an example of a polymeric magnetic tape shown in Fig. 8.38, the line in the left portion of the figure is curved with a slope of less than 1 shortly after the sample touches the tip, which suggests that the tip has indented the sample [18]. Later, the slope is equal to 1, suggesting that the tip no longer indents the sample. This observation indicates that the tape surface is soft locally (polymer-rich) but hard (as a result of magnetic particles) underneath. Since the curves in extending and retracting modes are identical, the indentation is elastic up to a maximum load of about 22 nN used in the measurements.

Detection of transfer of material on a nanoscale is possible with the AFM. Indentation of C_{60}-rich fullerene films with an AFM tip has been shown [70] to result in the transfer of fullerene molecules to the AFM tip, as indicated by discontinuities in the cantilever deflection as a function of sample traveling distance in subsequent indentation studies.

8.5.2 Nanoscale Indentation

The indentation hardness of surface films with an indentation depth as small as about 1 nm can be measured using an AFM [11, 34, 35]. Figure 8.39 shows the gray-scale plots of indentation marks made on Si(111) at normal loads of 60, 65, 70, and 100 µN. Triangular indents can be clearly observed with very shallow depths. At a normal load of 60 µN, indents are observed and the depth of penetration is about 1 nm. As the normal load is increased, the indents become clearer and indentation depth increases. For the case of hardness measurements at shallow depths on the same order as variations in surface roughness, it is desirable to subtract the original (unindented) map from the indent map for accurate measurement of the indentation size and depth [22].

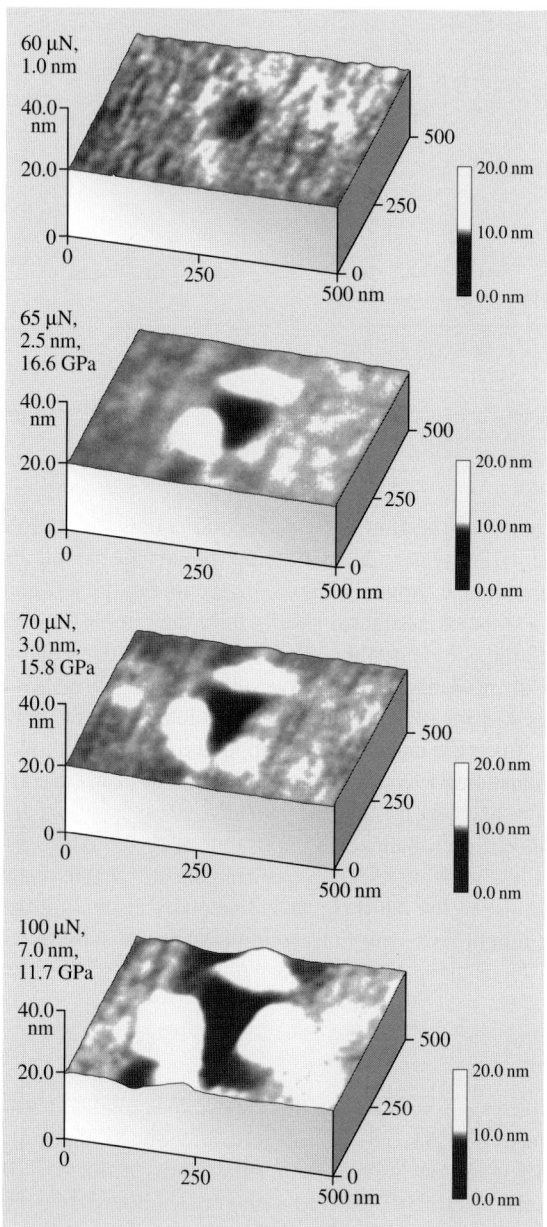

Fig. 8.39. Gray-scale plots of indentation marks on the Si(111) sample at various indentation loads. Loads, indentation depths, and hardness values are listed in the figure [34]

To make accurate measurements of hardness at shallow depths, a depth-sensing indentation system is used [35]. Figure 8.40 shows the load-displacement curves at

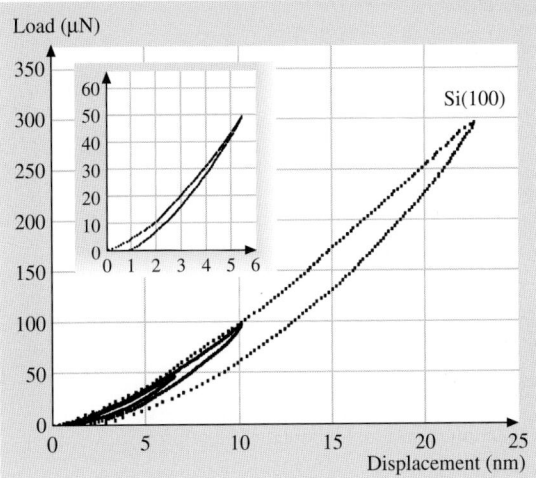

Fig. 8.40. Load-displacement curves at various peak loads for Si(100) [35]

different peak loads for Si(100). Loading/unloading curves often exhibit sharp discontinuities, particularly at high loads. Discontinuities, referred to as pop-ins, during the initial loading part of the curve mark a sharp transition from pure elastic loading to a plastic deformation of the specimen surface, thus corresponding to an initial yield point. The sharp discontinuities in the unloading part of the curves are believed to be due to the formation of lateral cracks that form at the base of the median crack, which results in the surface of the specimen being thrust upward. Load-displacement data at residual depths as low as about 1 nm can be obtained, and the indentation hardness of surface films has been measured for Si(100) [35, 67–69]. The hardness of silicon on a nanoscale is found to be higher than on a microscale (Fig. 8.41). Microhardness has also been reported to be higher than on the millimeter scale by several investigators. The data reported to date show that hardness exhibits size effect. According to the strain gradient plasticity theory advanced by *Fleck* et al. [71], large strain gradients inherent in small indentations lead to the accumulation of geometrically necessary dislocations for strain compatibility reasons that cause enhanced hardening. In addition, the decrease in hardness with an increase in indentation depth can possibly be rationalized on the basis that as the volume of deformed material increases, there is a higher probability of encountering material defects.

Bhushan and *Koinkar* [31] have used AFM measurements to show that ion implantation of silicon surfaces increases their hardness and thus their wear resistance. Formation of surface alloy films with improved mechanical properties by ion implantation is of growing technological importance as a means of improving the mechanical properties of materials. Hardness of 20 nm-thick diamond-like carbon films has been measured by *Kulkarni* and *Bhushan* [69].

The creep and strain-rate effects (viscoelastic effects) of ceramics can be studied using a depth-sensing indentation system. *Bhushan* et al. [35] and *Kulkarni* and

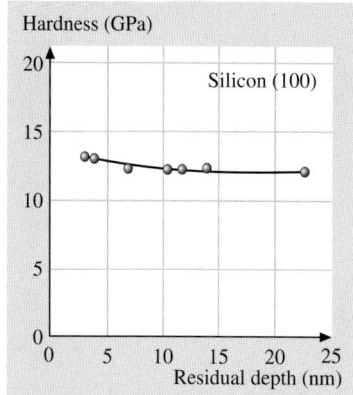

Fig. 8.41. Indentation hardness as a function of residual indentation depth for Si(100) [35]

Bhushan [67–69] have reported that ceramics exhibit significant plasticity and creep on a nanoscale. Figure 8.42a shows the load-displacement curves for single-crystal silicon at various peak loads held at 180 s. To demonstrate the creep effects, the load-displacement curves for a 500 µN peak load held at 0 and 30 s are also shown as an inset. Note that significant creep occurs at room temperature. Nanoindenter experiments conducted by *Li* et al. [72] exhibited significant creep only at high temperatures (greater than or equal to 0.25 times the melting point of silicon). The mechanism of dislocation glide plasticity is believed to dominate the indentation creep process on the macroscale. To study the strain-rate sensitivity of silicon, data at two different (constant) rates of loading are presented in Fig. 8.42b. Note that a change in the loading rate by a factor of about five results in a significant change in the load-displacement data. The viscoelastic effects observed here for silicon at ambient temperature could arise from size effects mentioned earlier. Most likely, creep and strain rate experiments are being conducted on the hydrated films present on the silicon surface in ambient environment, and these films are expected to be viscoelastic.

8.5.3 Localized Surface Elasticity and Viscoelasticity Mapping

The Young's modulus of elasticity is calculated from the slope of the indentation curve during unloading. However, these measurements provide a single-point measurement. By using the force modulation technique, it is possible to get localized elasticity maps of soft and compliant materials with penetration depths of less than 100 nm. This technique has been successfully used for polymeric magnetic tapes, which consist of magnetic and nonmagnetic ceramic particles in a polymeric matrix. Elasticity maps of a tape can be used to identify relative distribution of hard magnetic and nonmagnetic ceramic particles on the tape surface, which has an effect on friction and stiction at the head-tape interface [13]. Figure 8.43 shows surface height and elasticity maps on a polymeric magnetic tape [44]. The elasticity image reveals sharp variations in surface elasticity due to the composite nature of the film. As can

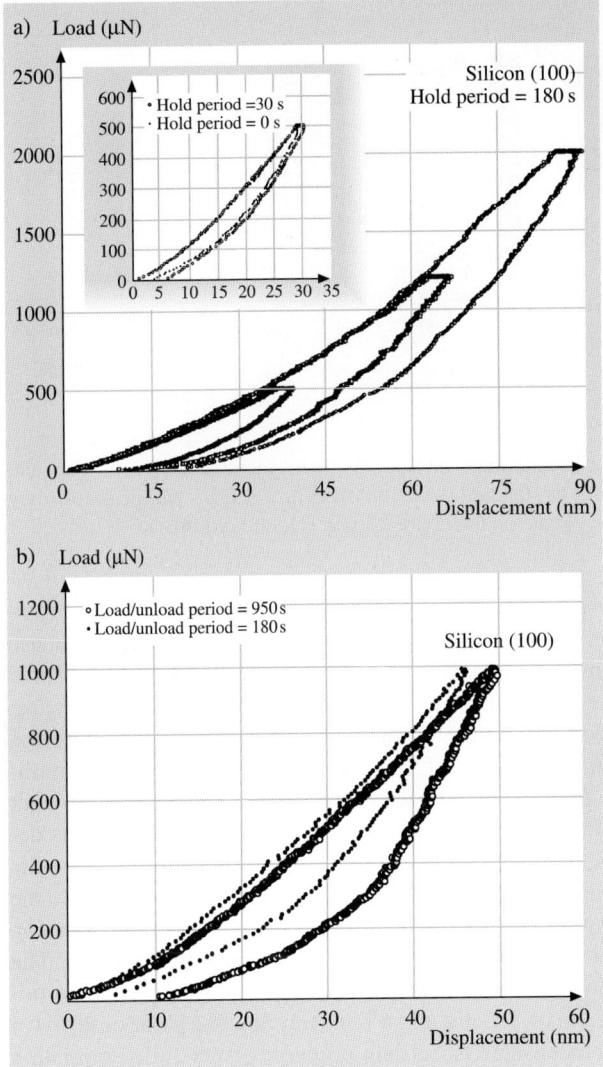

Fig. 8.42. (a) Creep behavior and (b) strain-rate sensitivity of Si(100) [35]

be clearly seen, regions of high elasticity do not always correspond to high or low topography. Based on a Hertzian elastic-contact analysis, the static indentation depth of these samples during the force modulation scan is estimated to be about 1 nm. We conclude that the contrast seen is influenced most strongly by material properties in the top few nanometers, independent of the composite structure beneath the surface layer.

By using phase contrast microscopy, it is possible to get phase contrast maps or the contrast in viscoelastic properties of near surface regions. This technique has

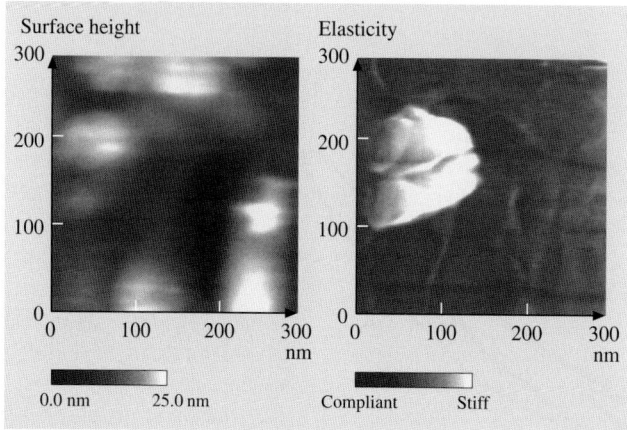

Fig. 8.43. Surface height and elasticity maps on a polymeric magnetic tape ($\sigma = 6.72$ nm and P-V $= 31.7$ nm; σ and P-V refer to standard deviation of surface heights and peak-to-valley distance, respectively). The gray scale on the elasticity map is arbitrary [44]

been used successfully for polymeric films and magnetic tapes that consist of ceramic particles in a polymeric matrix [51, 52]. Figure 8.44 shows topography, phase angle, and friction force images for polyethylene terephthalate (PET) film. Three particles, which show up as approximately circular, lightly shaded (high points) regions, are seen in the topography image. In the phase angle images, the particles show up dark against the film background. A combination of high tapping amplitude and low set point was found to maximize contrast in phase images. This result may be explained by the degree to which the tip is able to penetrate and thus deform the sample. A high tapping amplitude and low set point should maximize the force with which the tip strikes the sample. This should give morepenetration. So the material or viscoelastic contribution, as opposed to the surface force and adhesion hysteresis contributions, should be more dominant for this tapping condition. The phase angle image shown in Fig. 8.44 was acquired using tapping mode with a relatively high tapping amplitude of 190 nm and a relatively low set point of 40% (of the tapping amplitude) to emphasize viscoelastic properties. Very little correlation is found between the phase angle and friction force images. Friction force images were obtained using an etched single-crystal silicon tip (DI MESP) with a radius of 20–50 nm, a stiffness of 1–5 N/m, and a natural frequency of 60–70 kHz. A lack of correlation may be a further indication that something other than adhesion, which should correlate to friction force, causes the phase angle contrast; this may imply that viscoelastic properties may be dominant. These results also indicate that phase angle imaging yields information that cannot be obtained using other, more conventional AFM modes.

8.6 Boundary Lubrication

8.6.1 Perfluoropolyether Lubricants

The classical approach to lubrication uses freely supported multimolecular layers of liquid lubricants [10, 13, 58, 73]. The liquid lubricants are sometimes chemically bonded to improve their wear resistance [10, 13, 58]. Partially chemically bonded, molecularly thick perfluoropolyether (PFPE) films are used for lubrication of magnetic storage media [13]. These are considered potential candidate lubricants for micro/nanoelectromechanical systems (MEMS/NEMS). Molecularly thick PFPEs are well suited because of the following properties: low surface tension and low contact angle, which allow easy spreading on surfaces and provide hydrophobic properties; chemical and thermal stability, which minimize degradation under use; low vapor pressure, which provides low out-gassing; high adhesion to substrate via organic functional bonds; and good lubricity, which reduces contact surface wear.

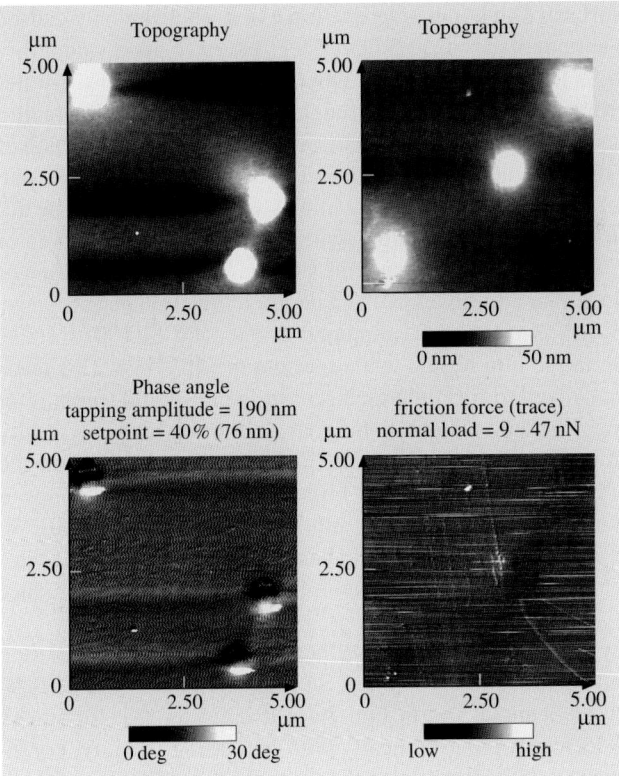

Fig. 8.44. Tapping mode height and phase angle images acquired using a TESP tip at a tapping amplitude of 190 nm and a setpoint of 40% (74 nm), and contact mode height and friction force (trace) images using an MESP tip with a normal load between 9 and 47 nN for PET film with embedded ceramic particles [51]

For boundary lubrication studies, friction, adhesion, and durability experiments have been performed on virgin Si(100) surfaces and silicon surfaces lubricated with two PFPE lubricants – Z-15 (with $-CF_3$ nonpolar end groups) and Z-DOL (with $-OH$ polar end groups) [27, 28, 30]. Z-DOL film was thermally bonded at 150 °C for 30 minutes and an unbonded fraction was removed by a solvent (BW) [13]. The thicknesses of Z-15 and Z-DOL films were 2.8 nm and 2.3 nm, respectively.

The adhesive forces of Si(100), Z-15, and Z-DOL (BW) measured by force calibration plot and friction force versus normal load plot are summarized in Fig. 8.45. The results measured by these two methods are in good agreement. Figure 8.45 shows that the presence of mobile Z-15 lubricant film increases the adhesive force as compared to that of Si(100) by meniscus formation. The presence of solid phase Z-DOL (BW) film reduces the adhesive force as compared to that of Si(100) because of the absence of mobile liquid. The schematic (bottom) in Fig. 8.45 shows relative size and sources of meniscus. It is well-known that the native oxide layer (SiO_2) on the top of a Si(100) wafer exhibits hydrophilic properties, and some water molecules can be adsorbed on this surface. The condensed water will form meniscus as the tip approaches the sample surface. The larger adhesive force in Z-15 is not only caused by the Z-15 meniscus; the nonpolarized Z-15 liquid does not have good wettability and strong bonding with Si(100). In the ambient environment, the condensed water molecules from the environment will permeate through the liquid Z-15 lubricant film and compete with the lubricant molecules present on the substrate. The interaction of the liquid lubricant with the substrate is weakened, and a boundary layer of the liquid lubricant forms puddles [27, 28]. This de-wetting allows water molecules to be adsorbed on the Si(100) surface as aggregates along with Z-15 molecules. And both of them can form meniscus while the tip approaches the surface. Thus, the de-wetting of liquid Z-15 film results in higher adhesive force and poorer lubrication performance. In addition, as the Z-15 film is pretty soft compared to the solid Si(100) surface, and penetration of the tip in the film occurs while pushing the tip down. This leads the large area of the tip involved to form the meniscus at the tip-liquid (mixture of Z-15 and water) interface. It should also be noted that Z-15 has a higher viscosity compared to water and, therefore, Z-15 film provides higher resistance to motion and coefficient of friction. In the case of Z-DOL (BW) film, both of the active groups of Z-DOL molecules are mostly bonded on Si(100) substrate. Thus, the Z-DOL (BW) film has low free surface energy and cannot be displaced readily by water molecules or readily adsorb water molecules. Therefore, the use of Z-DOL (BW) can reduce the adhesive force.

To study the velocity effect on friction and adhesion, the variation of friction force, adhesive force, and coefficient of friction of Si(100), Z-15 and Z-DOL (BW) as a function of velocity are summarized in Fig. 8.46. It indicates that for silicon wafer the friction force decreases logarithmically with increasing velocity. For Z-15, the friction force decreases with increasing velocity up to $10\,\mu m/s$, after which it remains almost constant. The velocity has very small effect on the friction force of Z-DOL (BW): It reduced slightly only at very high velocity. Figure 8.46 also indicates that the adhesive force of Si(100) increases when the velocity is higher than $10\,\mu m/s$. The adhesive force of Z-15 is reduced dramatically with a velocity increase up to

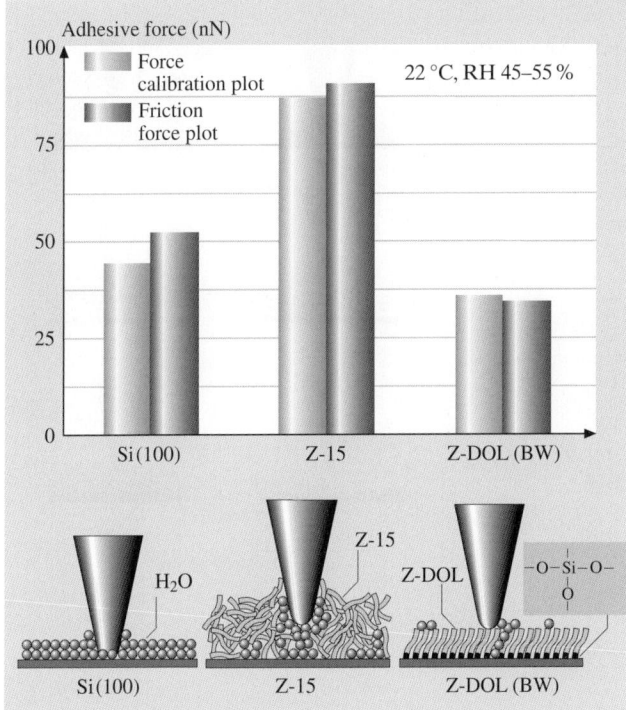

Fig. 8.45. Summary of the adhesive forces of Si(100) and Z-15 and Z-DOL (BW) films measured by force calibration plots and friction force versus normal load plots in ambient air. The schematic (*bottom*) showing the effect of meniscus formed between the AFM tip and surface sample on the adhesive and friction forces [30]

20 μm/s, after which it is reduced slightly. And the adhesive force of Z-DOL (BW) also decreases at high velocity. In the testing range of velocity, only the coefficient of friction of Si(100) decreases with velocity, but the coefficients of friction of Z-15 and Z-DOL (BW) almost remain constant. This implies that the friction mechanisms of Z-15 and Z-DOL (BW) do not change with the variation of velocity.

The mechanisms of the effect of velocity on the adhesion and friction are explained based on schematics shown in Fig. 8.46 (right). For Si(100), tribochemical reaction plays a major role. Although at high velocity the meniscus is broken and does not have enough time to rebuild, the contact stresses and high velocity lead to tribochemical reactions of Si(100) wafer (which has native oxide (SiO_2)) and a Si_3N_4 tip with water molecules and form $Si(OH)_4$. The $Si(OH)_4$ is removed and continuously replenished during sliding. The $Si(OH)_4$ layer between the tip and Si(100) surface is known to be of low shear strength and causes a decrease in friction force and coefficient of friction [10, 58]. The chemical bonds of Si−OH between the tip and Si(100) surface induce large adhesive force. For Z-15 film, at high velocity the meniscus formed by condensed water and Z-15 molecules is broken and does not

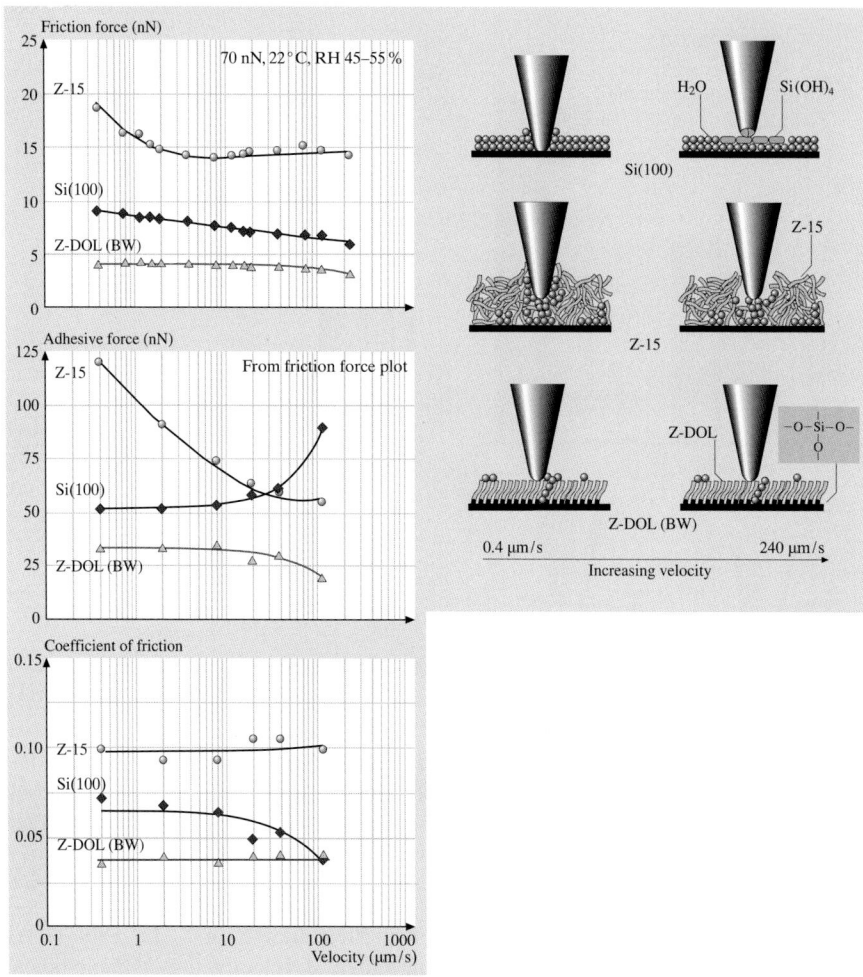

Fig. 8.46. The influence of velocity on the friction force, adhesive force, and coefficient of friction of Si(100) and Z-15 and Z-DOL (BW) films at 70 nN, in ambient air. The schematic (*right*) shows the change of surface composition (by tribochemical reaction) and formation of meniscus while increasing the velocity [30]

have enough time to rebuild. Therefore, the adhesive force and, consequently, friction force are reduced. The friction mechanism for Z-15 film still is shearing the same viscous liquid even at high velocity range, thus, the coefficient of friction of Z-15 does not change with velocity. For Z-DOL (BW) film, the surface can adsorb few water molecules in ambient condition, and at high velocity these molecules are displaced, which causes a slight decrease in friction and adhesive forces. *Koinkar* and *Bhushan* [27, 28] have suggested that in the case of samples with mobile films such as condensed water and Z-15 films, alignment of liquid molecules (shear thinning) is responsible for the drop in friction force with an increase in scanning velocity. This could be another reason for the decrease in friction force for Si(100) and Z-15 film with velocity in this study.

To study the relative humidity effect on friction and adhesion, the variations of friction force, adhesive force, and coefficient of friction of Si(100), Z-15, and Z-DOL (BW) as a function of relative humidity are shown in Fig. 8.47. It shows that for Si(100) and Z-15 film, the friction force increases with a relative humidity increase of up to 45%, and then slightly decreases with a further increase in the relative humidity. Z-DOL (BW) has a smaller friction force than Si(100) and Z-15 in the whole testing range. And its friction force shows a relative apparent increase when the relative humidity is higher than 45%. For Si(100), Z-15, and Z-DOL (BW), their adhesive forces increase with relative humidity, and their coefficients of friction increase with relative humidity up to 45%, after which they decrease with further increasing of the relative humidity. It is also observed that the humidity effect on Si(100) really depends on the history of the Si(100) sample. As the surface of Si(100) wafer readily adsorbs water in air, without any pre-treatment the Si(100) used in our study almost reaches to its saturate stage of adsorbed water and is responsible for less effect during increasing relative humidity. However, once the Si(100) wafer was thermally treated by baking at 150 °C for 1 hour, a bigger effect was observed.

The schematic in Fig. 8.47 (right) shows that Si(100), because its high free surface energy, can adsorb more water molecules with increasing relative humidity. As discussed earlier, for Z-15 film in the humid environment, the condensed water from the humid environment competes with the lubricant film present on the sample surface, interaction of the liquid lubricant film with the silicon substrate is weakened, and a boundary layer of the liquid lubricant forms puddles. This dewetting allows water molecules to be adsorbed on the Si(100) substrate mixed with Z-15 molecules [27, 28]. Obviously, more water molecules can be adsorbed on Z-15 surface while increasing relative humidity. The more adsorbed water molecules in the case of Si(100), along with lubricant molecules in Z-15 film case, form bigger water meniscus which leads to an increase of friction force, adhesive force, and coefficient of friction of Si(100) and Z-15 with humidity. But at a very high humidity of 70%, large quantities of adsorbed water can form a water layer that separates the tip and sample surface and act as a kind of lubricant, which causes a decrease in the friction force and coefficient of friction. For Z-DOL (BW) film, because of its hydrophobic surface properties, water molecules can be adsorbed at a humidity higher than 45% and causes an increase in the adhesive and friction forces.

To study the temperature effect on friction and adhesion, the variations of friction force, adhesive force, and coefficient of friction of Si(100), Z-15, and Z-DOL (BW) as a function of temperature are summarized in Fig. 8.48. It shows that the increasing temperature causes a decrease in friction force, adhesive force, and coefficient of friction of Si(100), Z-15, and Z-DOL (BW). The schematic (right) in Fig. 8.48 indicates that at high temperature, desorption of water leads to the decrease of friction force, adhesive forces, and coefficient of friction for all of the samples. For Z-15 film, the reduction of viscosity at high temperature also contributes to the decrease of friction force and coefficient of friction. In the case of Z-DOL (BW) film, molecules

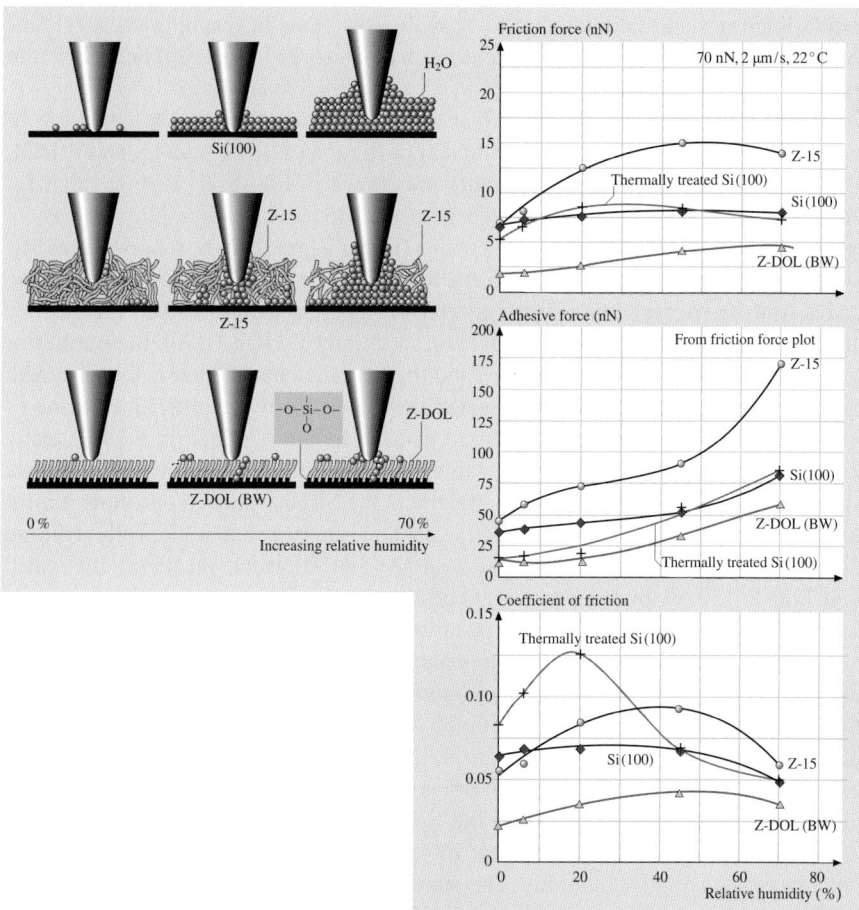

Fig. 8.47. The influence of relative humidity (RH) on the friction force, adhesive force, and coefficient of friction of Si(100) and Z-15 and Z-DOL (BW) films at 70 nN, 2 μm/s, and in 22 °C air. Schematic (*right*) shows the change of meniscus while increasing the relative humidity. In this figure, the thermally treated Si(100) represents the Si(100) wafer that was baked at 150 °C for 1 hour in an oven (in order to remove the adsorbed water) just before it was placed in the 0% RH chamber [30]

are more easily oriented at high temperature, which may be partly responsible for the low friction force and coefficient of friction.

To study the durability of lubricant films at the nanoscale, the friction of Si(100), Z-15, and Z-DOL (BW) as a function of the number of scanning cycles is shown in Fig. 8.49. As observed earlier, friction force and coefficient of friction of Z-15 is higher than that of Si(100) with the lowest values for Z-DOL (BW). During cycling, friction force and coefficient of friction of Si(100) show a slight decrease during the initial few cycles then remain constant. This is related to the removal of the top ad-

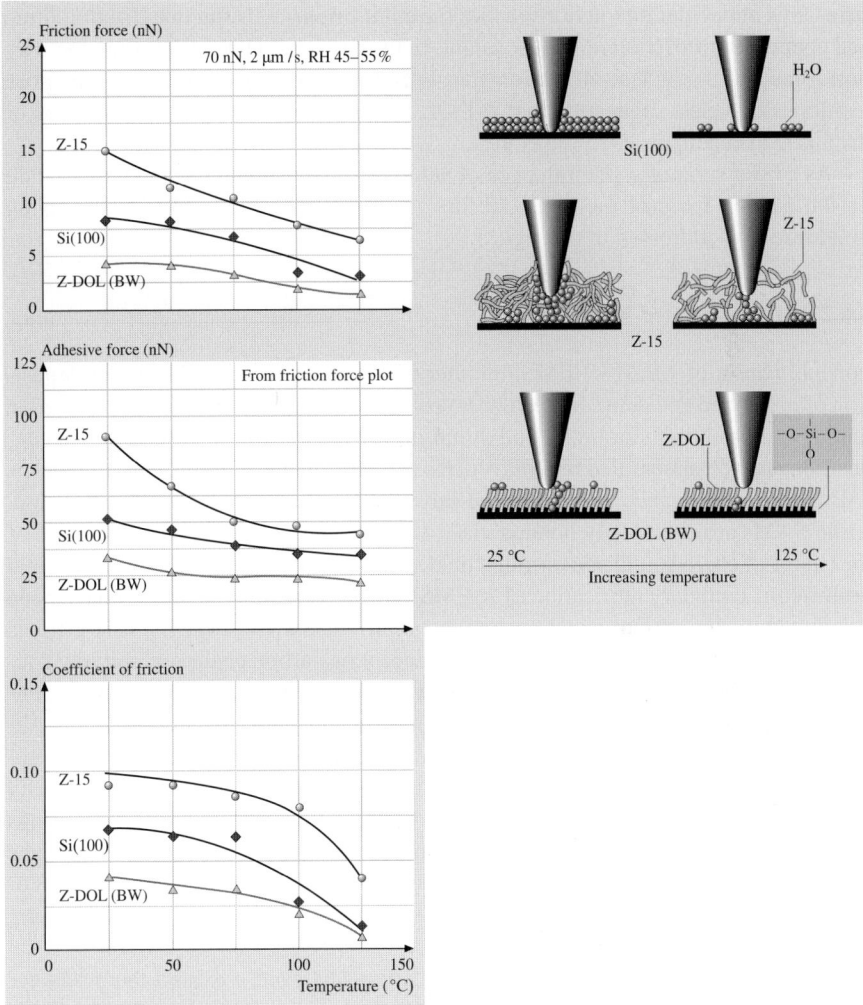

Fig. 8.48. The influence of temperature on the friction force, adhesive force, and coefficient of friction of Si(100) and Z-15 and Z-DOL (BW) films at 70 nN, at 2 µm/s, and in RH 40–50% air. The schematic (*right*) shows that at high temperature desorption of water decreases the adhesive forces. And the reduced viscosity of Z-15 leads to the decrease of the coefficient of friction. High temperature facilitates orientation of molecules in Z-DOL (BW) film, which results in lower coefficient of friction [30]

sorbed layer. In the case of Z-15 film, the friction force and coefficient of friction show an increase during the initial few cycles, and then approach higher and stable values. This is believed to be caused by the attachment of the Z-15 molecules to the tip. The molecular interaction between these attached molecules to the tip and molecules on the film surface is responsible for an increase in friction. But after several scans, this molecular interaction reaches the equilibrium, and after that, friction force and coefficient of friction remain constant. In the case of Z-DOL (BW) film, the friction force and coefficient of friction start out low and remain low during the entire test for 100 cycles, suggesting that Z-DOL (BW) molecules do not get attached or displaced as readily as Z-15 molecules.

As a brief summary, the influence of velocity, relative humidity, and temperature on the friction force of mobile Z-15 film is presented in Fig. 8.50. The changing trends are also addressed in this figure.

8.6.2 Self-Assembled Monolayers

For lubrication of MEMS/NEMS, another effective approach involves the deposition of organized and dense molecular layers of long-chain molecules. Two common methods to produce monolayers and thin films are the Langmuir–Blodgett (L-B) deposition and self-assembled monolayers (SAMs) by chemical grafting of molecules. L-B films are physically bonded to the substrate by weak van der Waals attraction, while SAMs are chemically bonded via covalent bonds to the substrate. Because of the choice of chain length and terminal linking group that SAMs offer, they hold great promise for boundary lubrication of MEMS/NEMS. A number of studies have been conducted to study tribological properties of various SAMs [26, 29, 74–76]. It has been reported that SAMs with high-compliance long carbon chains exhibit low friction; chain compliance is desirable for low friction. Based on [29], the friction mechanism of SAMs is explained by a so-called "molecular spring" model (Fig. 8.51). According to this model, the chemically adsorbed self-assembled molecules on a substrate are just like assembled molecular springs anchored to the substrate. An asperity sliding on the surface of SAMs is like a tip sliding on the top of "molecular springs or brush". The molecular spring assembly has compliant features and can experience orientation and compression under load. The orientation of the molecular springs or brush under normal load reduces the shearing force at the interface, which, in turn, reduces the friction force.

The SAMs with high-compliance long carbon chains also exhibit the best wear resistance [29, 75]. In wear experiments, the wear depth as a function of normal load curves shows critical normal loads (Fig. 8.52). Below the critical normal load SAMs undergo orientation, at the critical load SAMs wear away from the substrate due to weak interface bond strengths, while above the critical normal load severe wear takes place on the substrate.

8.6.3 Liquid Film Thickness Measurements

Liquid film thickness measurement of thin lubricant films (on the order of 10 nm or thicker) with nanometer lateral resolution can be made with the AFM [4, 59]. The lubricant thickness is obtained by measuring the force on the tip as it approaches, contacts, and pushes through the liquid film and ultimately contacts the substrate.

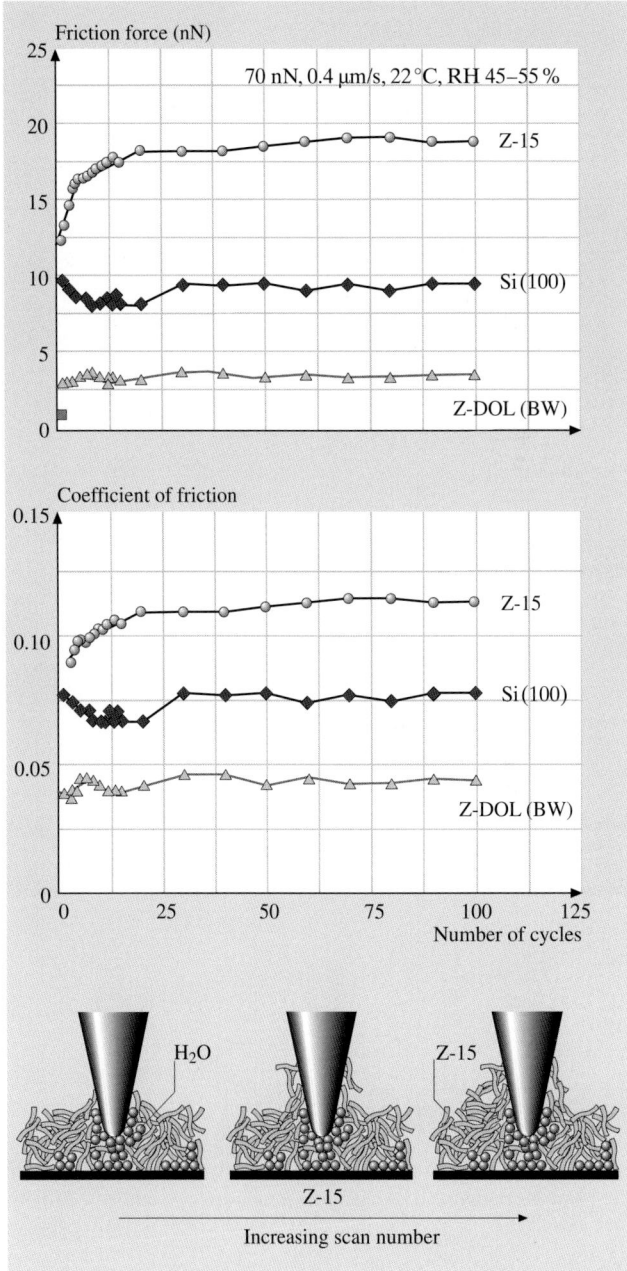

Fig. 8.49. Friction force and coefficient of friction versus number of sliding cycles for Si(100) and Z-15 and Z-DOL (BW) films at 70 nN, 0.8 μm/s, and in ambient air. Schematic (*bottom*) shows that some liquid Z-15 molecules can be attached to the tip. The molecular interaction between the attached molecules to the tip with the Z-15 molecules in the film results in an increase of the friction force with multiple scanning [30]

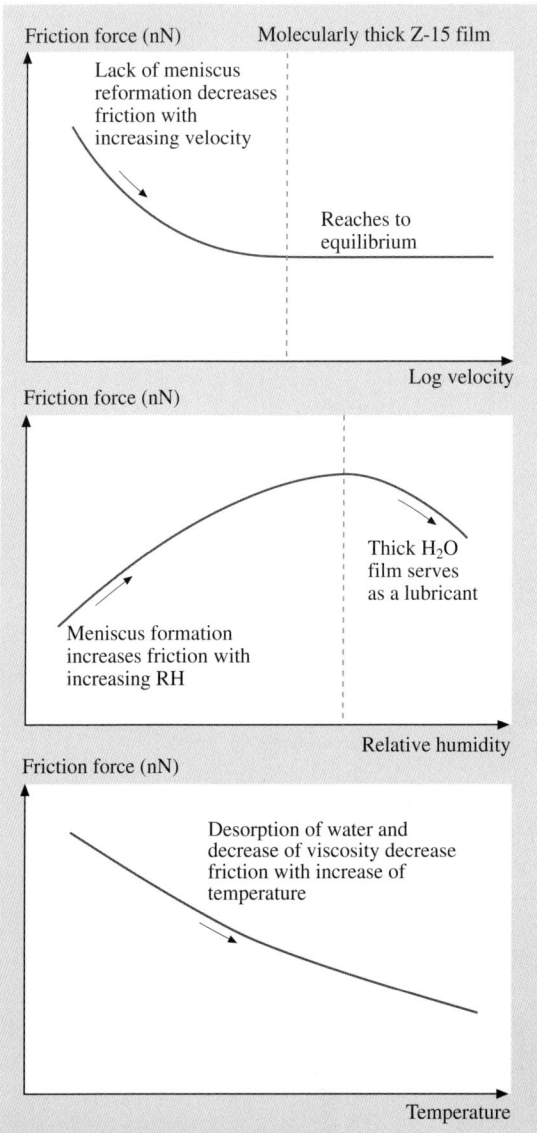

Fig. 8.50. Schematic shows the change of friction force of molecularly thick Z-15 films with log velocity, relative humidity, and temperature. The changing trends are also addressed in this figure [30]

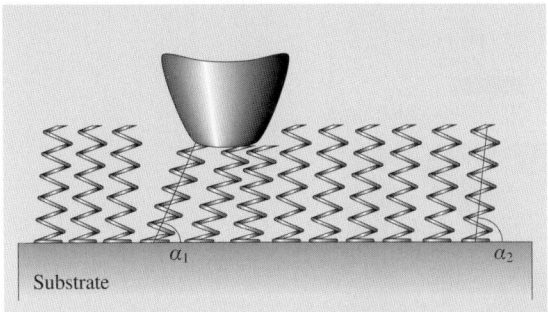

Fig. 8.51. Molecular spring model of SAMs. In this figure, $\alpha_1 < \alpha_2$, which is caused by the further orientation under the normal load applied by an asperity tip [29]

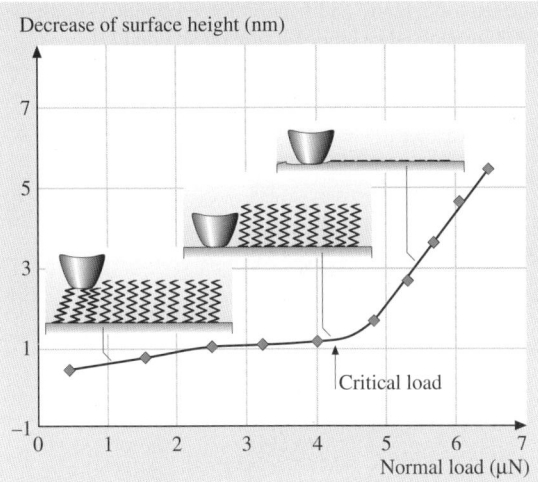

Fig. 8.52. Illustration of the wear mechanism of SAMs with increasing normal load [75]

The distance between the sharp snap-in (owing to the formation of a liquid meniscus between the film and the tip) at the liquid surface and the hard repulsion at the substrate surface is a measure of the liquid film thickness.

Lubricant film thickness mapping of ultrathin films (on the order of 2 nm) can be obtained using friction force microscopy [27] and adhesive force mapping [36]. Figure 8.53 shows gray-scale plots of the surface topography and friction force obtained simultaneously for unbonded Demnum-type PFPE lubricant film on silicon. The friction force plot shows well distinguished low and high friction regions roughly corresponding to high and low regions in surface topography (thick and thin lubricant regions). A uniformly lubricated sample does not show such a variation in the friction. Friction force imaging can thus be used to measurethe lubricant uniformity on the sample surface, which cannot be identified by surface topography alone. Fig-

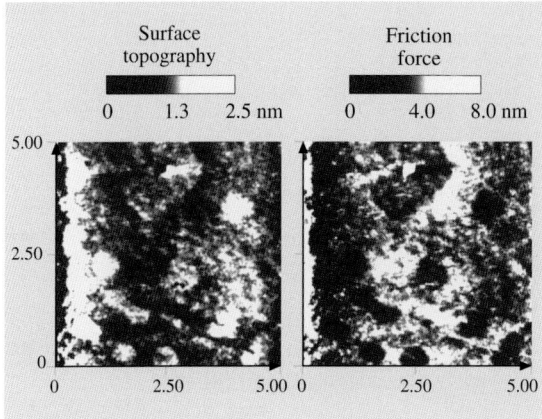

Fig. 8.53. Gray-scale plots of the surface topography and friction force obtained simultaneously for unbonded Demnum-type perfluoropolyether lubricant film on silicon [27]

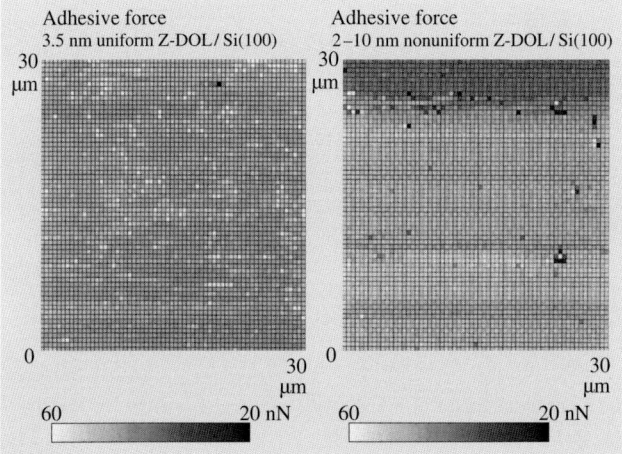

Fig. 8.54. Gray-scale plots of the adhesive force distribution of a uniformly coated, 3.5-nm thick unbonded Z-DOL film on silicon and 3- to 10-nm thick unbonded Z-DOL film on silicon that was deliberately coated nonuniformly by vibrating the sample during the coating process [36]

ure 8.54 shows the gray-scale plots of the adhesive force distribution for silicon samples coated uniformly and nonuniformly with Z-DOL type PFPE lubricant. It can be clearly seen that a region exists that has an adhesive force distinctly different from the other region for the nonuniformly coated sample. This implies that the liquid film thickness is nonuniform, giving rise to a difference in the meniscus forces.

8.7 Closure

At most solid-solid interfaces of technological relevance, contact occurs at many asperities. A sharp AFM/FFM tip sliding on a surface simulates just one such contact. However, asperities come in all shapes and sizes. The effect of radius of a single asperity (tip) on the friction/adhesion performance can be studied using tips of different radii. AFM/FFM are used to study various tribological phenomena, including surface roughness, adhesion, friction, scratching, wear, indentation, detection of material transfer, and boundary lubrication. Measurement of atomic-scale friction of a freshly cleaved, highly oriented pyrolytic graphite exhibits the same periodicity as that of the corresponding topography. However, the peaks in friction and those in the corresponding topography are displaced relative to one another. Variations in atomic-scale friction and the observed displacement can be explained by the variation in interatomic forces in the normal and lateral directions. Local variations in microscale friction occur and are found to correspond to the local slopes, suggesting that a ratchet mechanism and collision effects are responsible for this variation. Directionality in the friction is observed on both micro- and macroscales, resulting from the surface roughness and surface preparation. Anisotropy in surface roughness accentuates this effect. Microscale friction is generally found to be smaller than the macrofriction, as there is less plowing contribution in microscale measurements. Microscale friction is load-dependent, and friction values increase with an increase in the normal load, approaching the macrofriction at contact stresses higher than the hardness of the softer material. The tip radius also affects adhesion and friction.

Mechanism of material removal on the microscale is studied. Wear rate for single-crystal silicon is negligible below 20 μN and is much higher and remains approximately constant at higher loads. Elastic deformation at low loads is responsible for negligible wear. Most of the wear debris is loose. SEM and TEM studies of the wear region suggest that the material on the microscale is removed by plastic deformation with a small contribution from elastic fracture; this observation corroborates the scratch data. Evolution of wear has also been studied using AFM. Wear is found to be initiated at nanoscratches. For a sliding interface requiring near-zero friction and wear, contact stresses should be below the hardness of the softer material to minimize plastic deformation, and surfaces should be free of nanoscratches. Furthermore, wear precursors can be detected at early stages of wear by using surface potential measurements. It is found that even in the case of zero wear (no measurable deformation of the surface using AFM), there can be a significant change in the surface potential inside the wear mark, which is useful for study of wear precursors. Detection of material transfer on a nanoscale is possible with AFM.

In situ surface characterization of local deformation of materials and thin coatings can be carried out using a tensile stage inside an AFM. An AFM can also be used for nanofabrication/nanomachining.

By using the force modulation technique, localized surface elasticity maps of composite materials with penetrating depths of less than 100 nm can be obtained. It is possible using phase contrast microscopy to get phase contrast maps or the contrast in viscoelastic properties of near surface regions. Scratching and indentation on nanoscales are powerful ways to screen for adhesion and resistance to deformation of ultrathin films. Modified AFM can be used to obtain load-displacement curves and for measurement of nanoindentation hardness and Young's modulus of elasticity, with depth of indentation as low as 1 nm. Hardness of ceramics on the nanoscale is found to be higher than that on the microscale. Ceramics exhibit significant plasticity and creep on a nanoscale.

Boundary lubrication studies and measurement of lubricant-film thickness with a lateral resolution on a nanoscale can be conducted using AFM. Chemically bonded lubricant films and self-assembled monolayers are superior in friction and wear resistance. For chemically bonded lubricant films, the adsorption of water, the formation of meniscus, and its change during sliding, viscosity, and surface properties play an important role on the friction, adhesion, and durability of these films. For SAMs, their friction mechanism is explained by a so-called "molecular spring" model. The films with high-compliance long carbon chains exhibit low friction and wear.

Investigations of wear, scratching, and indentation on the nanoscale using the AFM can provide insights into failure mechanisms of materials. Coefficients of friction, wear rates, and mechanical properties such as hardness have been found to be different on the nanoscale than on the macroscale; generally, coefficients of friction and wear rates on micro- and nanoscales are smaller, whereas hardness is greater. Therefore, micro/nanotribological studies may help define the regimes for ultralow friction and near-zero wear. These studies also provide insight into the atomic origins of adhesion, friction, wear, and lubrication mechanisms.

References

1. I. L. Singer and H. M. Pollock. *Fundamentals of Friction: Macroscopic and Microscopic Processes*, volume 220 of *Nato Sci. Ser. E*. Kluwer, 1992.
2. B. N. J. Persson and E. Tosatti. *Physics of Sliding Friction*, volume 311 of *Nato Sci. Ser. E*. Kluwer, 1996.
3. B. Bhushan. *Micro/Nanotribology and Its Applications*, volume 330 of *Nato Sci. Ser. E*. Kluwer, 1997.
4. B. Bhushan. *Handbook of Micro/Nanotribology*. CRC, 2nd edition, 1999.
5. B. Bhushan. Nanoscale tribophysics and tribomechanics. *Wear*, 225-229:465–492, 1999.
6. B. Bhushan. Wear and mechanical characterisation on micro- to picoscales using afm. *Int. Mat. Rev.*, 44:105–117, 1999.
7. B. Bhushan. *Modern Tribology Handbook, Vol. 1: Principles of Tribology*. CRC, 2001.
8. B. Bhushan. *Fundamentals of tribology and bridging the gap between the macro- and micro/nanoscales*, volume 10 of *NATO Sci. Ser. E*. Kluwer, 2001.
9. B. Bhushan. Nano- to microscale wear and mechanical characterization studies using scanning probe microscopy. *Wear*, 251:1105–1123, 2001.
10. B. Bhushan. *Introduction to Tribology*. Wiley, 2002.
11. B. Bhushan, J. N. Israelachvili, and U. Landman. Nanotribology: Friction, wear and lubrication at the atomic scale. *Nature*, 374:607–616, 1995.
12. H. J. Guntherodt, D. Anselmetti, and E. Meyer. *Forces in Scanning Probe Methods*, volume 286 of *Nato Sci. Ser. E*. Kluwer, 1995.

13. B. Bhushan. *Tribology and Mechanics of Magnetic Storage Devices*. Springer, 2nd edition, 1996.
14. B. Bhushan. *Tribology Issues and Opportunities in MEMS*. Kluwer, 1998.
15. G. Binnig, C. F. Quate, and Ch. Gerber. Atomic force microscopy. *Phys. Rev. Lett.*, 56:930–933, 1986.
16. G. Binnig, Ch. Gerber, E. Stoll, T. R. Albrecht, and C. F. Quate. Atomic resolution with atomic force microscope. *Europhys. Lett.*, 3:1281–1286, 1987.
17. C. M. Mate, G. M. McClelland, R. Erlandsson, and S. Chiang. Atomic-scale friction of a tungsten tip on a graphite surface. *Phys. Rev. Lett.*, 59:1942–1945, 1987.
18. B. Bhushan and J. Ruan. Atomic-scale friction measurements using friction force microscopy: part ii – application to magnetic media. *ASME J. Tribol.*, 116:389–396, 1994.
19. J. Ruan and B. Bhushan. Atomic-scale friction measurements using friction force microscopy: part i – general principles and new measurement techniques. *ASME J. Tribol.*, 116:378–388, 1994.
20. J. Ruan and B. Bhushan. Atomic-scale and microscale friction of graphite and diamond using friction force microscopy. *J. Appl. Phys.*, 76:5022–5035, 1994.
21. J. Ruan and B. Bhushan. Frictional behavior of highly oriented pyrolytic graphite. *J. Appl. Phys.*, 76:8117–8120, 1994.
22. B. Bhushan, V. N. Koinkar, and J. Ruan. Microtribology of magnetic media. *Proc. Inst. Mech. Eng., Part J: J. Eng. Tribol.*, 208:17–29, 1994.
23. B. Bhushan and S. Sundararajan. Micro/nanoscale friction and wear mechanisms of thin films using atomic force and friction force microscopy. *Acta Mater.*, 46:3793–3804, 1998.
24. V. Scherer, W. Arnold, and B. Bhushan. *Active friction control using ultrasonic vibration*, pages 463–469. Kluwer, 1998.
25. V. Scherer, W. Arnold, and B. Bhushan. Lateral force microscopy using acoustic friction force microscopy. *Surf. Interface Anal.*, 27:578–587, 1999.
26. B. Bhushan, A. V. Kulkarni, V. N. Koinkar, M. Boehm, L. Odoni, C. Martelet, and M. Belin. Microtribological characterization of self-assembled and langmuir–blodgett monolayers by atomic and friction force microscopy. *Langmuir*, 11:3189–3198, 1995.
27. V. N. Koinkar and B. Bhushan. Micro/nanoscale studies of boundary layers of liquid lubricants for magnetic disks. *J. Appl. Phys.*, 79:8071–8075, 1996.
28. V. N. Koinkar and B. Bhushan. Microtribological studies of unlubricated and lubricated surfaces using atomic force/friction force microscopy. *J. Vac. Sci. Technol. A.*, 14:2378–2391, 1996.
29. B. Bhushan and H. Liu. Nanotribological properties and mechanisms of alkylthiol and biphenyl thiol self-assembled monolayers studied by afm. *Phys. Rev. B*, 63:245412–1–245412–11, 2001.
30. H. Liu and B. Bhushan. Nanotribological characterization of molecularly-thick lubricant films for applications to mems/nems by afm. *Ultramicroscopy*, 97:321–340, 2003.
31. B. Bhushan and V. N. Koinkar. Tribological studies of silicon for magnetic recording applications. *J. Appl. Phys.*, 75:5741–5746, 1994.
32. V. N. Koinkar and B. Bhushan. Microtribological properties of hard amorphous carbon protective coatings for thin film magnetic disks and heads. *Proc. Inst. Mech. Eng. Part J: J. Eng. Tribol.*, 211:365–372, 1997.
33. S. Sundararajan and B. Bhushan. Development of a continuous microscratch technique in an atomic force microscope and its application to study scratch resistance of ultra-thin hard amorphous carbon coatings. *J. Mater. Res.*, 16:75–84, 2001.
34. B. Bhushan and V. N. Koinkar. Nanoindentation hardness measurements using atomic force microscopy. *Appl. Phys. Lett.*, 64:1653–1655, 1994.

35. B. Bhushan, A. V. Kulkarni, W. Bonin, and J. T. Wyrobek. Nano/picoindentation measurement using a capacitance transducer system in atomic force microscopy. *Philos. Mag.*, 74:1117–1128, 1996.
36. B. Bhushan and C. Dandavate. Thin-film friction and adhesion studies using atomic force microscopy. *J. Appl. Phys.*, 87:1201–1210, 2000.
37. B. Bhushan. Micro/nanotribology and its applications to magnetic storage devices and mems. *Tribol. Int.*, 28:85–95, 1995.
38. D. DeVecchio and B. Bhushan. Use of a nanoscale kelvin probe for detecting wear precursors. *Rev. Sci. Instrum.*, 69:3618–3624, 1998.
39. B. Bhushan and A. V. Goldade. Measurements and analysis of surface potential change during wear of single crystal silicon (100) at ultralow loads using kelvin probe microscopy. *Appl. Surf. Sci*, 157:373–381, 2000.
40. B. Bhushan and A. V. Goldade. Kelvin probe microscopy measurements of surface potential change under wear at low loads. *Wear*, 244:104–117, 2000.
41. M. S. Bobji and B. Bhushan. Atomic force microscopic study of the micro-cracking of magnetic thin films under tension. *Scripta Mater.*, 44:37–42, 2001.
42. M. S. Bobji and B. Bhushan. In-situ microscopic surface characterization studies of polymeric thin films during tensile deformation using atomic force microscopy. *J. Mater. Res.*, 16:844–855, 2001.
43. P. Maivald, H. J. Butt, S. A. C. Gould, C. B. Prater, B. Drake, J. A. Gurley, V. B. Elings, and P. K. Hansma. Using force modulation to image surface elasticities with the atomic force microscope. *Nanotechnol.*, 2:103–106, 1991.
44. D. DeVecchio and B. Bhushan. Localized surface elasticity measurements using an atomic force microscope. *Rev. Sci. Instrum.*, 68:4498–4505, 1997.
45. V. Scherer, B. Bhushan, U. Rabe, and W. Arnold. Local elasticity and lubrication measurements using atomic force and friction force microscopy at ultrasonic frequencies. *IEEE Trans. Mag.*, 33:4077–4079, 1997.
46. H. U. Krotil, T. Stifter, H. Waschipky, K. Weishaupt, S. Hild, and O. Marti. Pulse force mode: A new method for the investigation of surface properties. *Surf. Interface Anal.*, 27:336–340, 1999.
47. S. Amelio, A. V. Goldade, U. Rabe, V. Scherer, B. Bhushan, and W. Arnold. Measurements of elastic properties of ultra-thin diamond-like carbon coatings using atomic force acoustic microscopy. *Thin Solid Films*, 392:75–84, 2001.
48. B. Anczykowski, D. Kruger, K. L. Babcock, and H. Fuchs. Basic properties of dynamic force microscopy with the scanning force microscope in experiment and simulation. *Ultramicroscopy*, 66:251–259, 1996.
49. J. Tamayo and R. Garcia. Deformation, contact time, and phase contrast in tapping mode scanning force microscopy. *Longmuir*, 12:4430–4435, 1996.
50. R. Garcia, J. Tamayo, M. Calleja, and F. Garcia. Phase contrast in tapping-mode scanning force microscopy. *Appl. Phys. A*, 66:309–312, 1998.
51. W. W. Scott and B. Bhushan. Use of phase imaging in atomic force microscopy for measurement of viscoelastic contrast in polymer nanocomposites and molecularly-thick lubricant films. *Ultramicroscopy*, 97:151–169, 2003.
52. B. Bhushan and J. Qi. Phase contrast imaging of nanocomposites and molecularly- thick lubricant films in magnetic media. *Nanotechnol.*, 14:886–895, 2003.
53. V. N. Koinkar and B. Bhushan. Microtribological studies of Al_2O_3-TiC, polycrystalline and single-crystal Mn-Zn ferrite and SiC head slider materials. *Wear*, 202:110–122, 1996.
54. E. Meyer, R. Overney, R. Luthi, D. Brodbeck, L. Howald, J. Frommer, H. J. Guntherodt, O. Wolter, M. Fujihira, T. Takano, and Y. Gotoh. Friction force microscopy of mixed langmuir–blodgett films. *Thin Solid Films*, 220:132–137, 1992.

55. C. D. Frisbie, L. F. Rozsnyai, A. Noy, M. S. Wrighton, and C. M. Lieber. Functional group imaging by chemical force microscopy. *Science*, 265:2071–2074, 1994.
56. V. N. Koinkar and B. Bhushan. Effect of scan size and surface roughness on microscale friction measurements. *J. Appl. Phys.*, 81:2472–2479, 1997.
57. S. Sundararajan and B. Bhushan. Topography-induced contributions to friction forces measured using an atomic force/friction force microscope. *J. Appl. Phys.*, 88:4825–4831, 2000.
58. B. Bhushan. *Principles and Applications of Tribology*. Wiley, 1999.
59. B. Bhushan and G. S. Blackman. Atomic force microscopy of magnetic rigid disks and sliders and its applications to tribology. *ASME J. Tribol.*, 113:452–458, 1991.
60. M. Reinstaedtler, U. Rabe, V. Scherer, U. Hartmann, A. Goldade, B. Bhushan, and W. Arnold. On the nanoscale measurement of friction using atomic-force microscope cantilever torsional resonances. *Appl. Phys. Lett.*, 82:2604–2606, 2003.
61. T. Stifter, O. Marti, and B. Bhushan. Theoretical investigation of the distance dependence of capillary and van der waals forces in scanning probe microscopy. *Phys. Rev. B*, 62:13667–13673, 2000.
62. U. D. Schwarz, O. Zwoerner, P. Koester, and R. Wiesendanger. *Friction force spectroscopy in the low-load regime with well-defined tips*, pages 233–238. Kluwer, 1997.
63. B. Bhushan and A. V. Kulkarni. Effect of normal load on microscale friction measurements. *Thin Solid Films*, 278:49–56, 1996.
64. X. Zhao and B. Bhushan. Material removal mechanism of single-crystal silicon on nanoscale and at ultralow loads. *Wear*, 223:66–78, 1998.
65. V. N. Koinkar and B. Bhushan. Scanning and transmission electron microscopies of single-crystal silicon microworn/machined using atomic force microscopy. *J. Mater. Res.*, 12:3219–3224, 1997.
66. B. Bhushan, P. S. Mokashi, and T. Ma. A new technique to measure poisson's ratio of ultrathin polymeric films using atomic force microscopy. *Rev. Sci. Instrumen.*, 74:1043–1047, 2003.
67. A. V. Kulkarni and B. Bhushan. Nanoscale mechanical property measurements using modified atomic force microscopy. *Thin Solid Films*, 290-291:206–210, 1996.
68. A. V. Kulkarni and B. Bhushan. Nano/picoindentation measurements on single-crystal aluminum using modified atomic force microscopy. *Mater. Lett.*, 29:221–227, 1996.
69. A. V. Kulkarni and B. Bhushan. Nanoindentation measurement of amorphous carbon coatings. *J. Mater. Res.*, 12:2707–2714, 1997.
70. J. Ruan and B. Bhushan. Nanoindentation studies of fullerene films using atomic force microscopy. *J. Mater. Res.*, 8:3019–3022, 1993.
71. N. A. Fleck, G. M. Muller, M. F. Ashby, and J. W. Hutchinson. Strain gradient plasticity: Theory and experiment. *Acta Metall. Mater.*, 42:475–487, 1994.
72. W. B. Li, J. L. Henshall, R. M. Hooper, and K. E. Easterling. The mechanism of indentation creep. *Acta Metall. Mater.*, 39:3099–3110, 1991.
73. F. P. Bowden and D. Tabor. *The Friction and Lubrication of Solids*, volume 1. Clarendon, 1950.
74. H. Liu and B. Bhushan. Investigation of the adhesion, friction, and wear properties of biphenyl thiol self-assembled monolayers by atomic force microscopy. *J. Vac. Sci. Technol. A*, 19:1234–1240, 2001.
75. H. Liu and B. Bhushan. Investigation of nanotribological properties of self-assembled monolayers with alkyl and biphenyl spacer chains. *Ultramicroscopy*, 91:185–202, 2002.
76. B. Bhushan. *Self-assembled monolayers for controlling hydrophobicity and/or friction and wear*, pages 909–929. CRC, 2001.

9

Surface Forces and Nanorheology of Molecularly Thin Films

Marina Ruths, Alan D. Berman, and Jacob N. Israelachvili

Summary. In this chapter, we describe the static and dynamic normal forces that occur between surfaces in vacuum or liquids and the different modes of friction that can be observed between (i) bare surfaces in contact (dry or interfacial friction), (ii) surfaces separated by a thin liquid film (lubricated friction), and (iii) surfaces coated with organic monolayers (boundary friction).

Experimental methods suitable for measuring normal surface forces, adhesion and friction (lateral or shear) forces of different magnitude at the molecular level are described. We explain the molecular origin of van der Waals, electrostatic, solvation and polymer mediated interactions, and basic models for the contact mechanics of adhesive and nonadhesive elastically deforming bodies. The effects of interaction forces, molecular shape, surface structure and roughness on adhesion and friction are discussed.

Simple models for the contributions of the adhesion force and external load to interfacial friction are illustrated with experimental data on both unlubricated and lubricated systems, as measured with the surface forces apparatus. We discuss rate-dependent adhesion (adhesion hysteresis) and how this is related to friction. Some examples of the transition from wearless friction to friction with wear are shown.

Lubrication in different lubricant thickness regimes is described together with explanations of nanorheological concepts. The occurrence of and transitions between smooth and stick–slip sliding in various types of dry (unlubricated and solid boundary lubricated) and liquid lubricated systems are discussed based on recent experimental results and models for stick–slip involving memory distance and dilatancy.

9.1 Introduction: Types of Surface Forces

In this chapter, we discuss the most important types of surface forces and the relevant equations for the force and friction laws. Several different attractive and repulsive forces operate between surfaces and particles. Some forces occur in vacuum, for example, attractive van der Waals and repulsive hard-core interactions. Other types of forces can arise only when the interacting surfaces are separated by another condensed phase, which is usually a liquid. The most common types of surface forces and their main characteristics are listed in Table 9.1.

Type of force	Subclasses or alternative names	Main characteristics
van der Waals	Debye induced dipole force (v & s) London dispersion force (v & s) Casimir force (v & s)	Ubiquitous, occurs both in vacuum and in liquids
Electrostatic	Ionic bond (v) Coulombic force (v & s) Hydrogen bond (v) Charge-exchange interaction (v & s) Acid–base interaction (s) "Harpooning" interaction (v)	Strong, long-range, arises in polar solvents; requires surface charging or charge-separation mechanism
Ion correlation	van der Waals force of polarizable ions (s)	Requires mobile charges on surfaces in a polar solvent
Quantum mechanical	Covalent bond (v) Metallic bond (v) Exchange interaction (v)	Strong, short-range, responsible for contact binding of crystalline surfaces
Solvation	Oscillatory force (s) Depletion force (s)	Mainly entropic in origin, the oscillatory force alternates between attraction and repulsion
Hydrophobic	Attractive hydration force (s)	Strong, apparently long-range; origin not yet understood
Specific binding	"Lock-and-key" or complementary binding (v & s) Receptor–ligand interaction (s) Antibody–antigen interaction (s)	Subtle combination of different noncovalent forces giving rise to highly specific binding; main recognition mechanism of biological systems
Repulsive forces		
van der Waals	van der Waals disjoining pressure (s)	Arises only between dissimilar bodies interacting in a medium
Electrostatic	Coulombic force (v & s)	Arises only for certain constrained surface charge distributions
Quantum mechanical	Hard-core or steric repulsion (v) Born repulsion (v)	Short-range, stabilizing attractive covalent and ionic binding forces, effectively determine molecular size and shape
Solvation	Oscillatory solvation force (s) Structural force (s) Hydration force (s)	Monotonically repulsive forces, believed to arise when solvent molecules bind strongly to surfaces
Entropic	Osmotic repulsion (s) Double-layer force (s) Thermal fluctuation force (s) Steric polymer repulsion (s) Undulation force (s) Protrusion force (s)	Due to confinement of molecular or ionic species; requires mechanism that keeps trapped species between the surfaces
Dynamic interactions		
Nonequilibrium	Hydrodynamic forces (s) Viscous forces (s) Friction forces (v & s) Lubrication forces (s)	Energy-dissipating forces occurring during relative motion of surfaces or bodies

Table 9.0. continue

Note: (v) applies only to interactions in vacuum, (s) applies only to interactions in solution (or to surfaces separated by a liquid), and (v & s) applies to interactions occurring both in vacuum and in solution.

In *vacuum*, the two main long-range interactions are the attractive van der Waals and electrostatic (Coulomb) forces. At smaller surface separations (corresponding to molecular contact at surface separations of $D \approx 0.2$ nm), additional attractive interactions can be found such as covalent or metallic bonding forces. These attractive forces are stabilized by the hard-core repulsion. Together they determine the surface and interfacial energies of planar surfaces, as well as the strengths of materials and adhesive junctions. Adhesion forces are often strong enough to elastically or plastically deform bodies or particles when they come into contact.

In *vapors* (e.g., atmospheric air containing water and organic molecules), solid surfaces in or close to contact will generally have a surface layer of chemisorbed or physisorbed molecules, or a capillary condensed liquid bridge between them. A surface layer usually causes the adhesion to decrease, but in the case of capillary condensation, the additional Laplace pressure or attractive "capillary" force may make the adhesion between the surfaces stronger than in inert gas or vacuum.

When totally immersed in a *liquid*, the force between particles or surfaces is completely modified from that in vacuum or air (vapor). The van der Waals attraction is generally reduced, but other forces can now arise that can qualitatively change both the range and even the sign of the interaction. The attractive force in such a system can be either stronger or weaker than in the absence of the intervening liquid. For example, the overall attraction can be stronger in the case of two hydrophobic surfaces separated by water, but weaker for two hydrophilic surfaces. Depending on the different forces that may be operating simultaneously in solution, the overall force law is not generally monotonically attractive even at long range; it can be repulsive, or the force can change sign at some finite surface separation. In such cases, the potential energy minimum, which determines the adhesion force or energy, occurs not at true molecular contact between the surfaces, but at some small distance farther out.

The forces between surfaces in a liquid medium can be particularly complex at *short range*, i.e., at surface separations below a few nanometers or 4–10 molecular diameters. This is partly because with increasing confinement, a liquid ceases to behave as a structureless continuum with bulk properties; instead, the size and shape of its molecules begin to determine the overall interaction. In addition, the surfaces themselves can no longer be treated as inert and structureless walls (i.e., mathematically flat) and their physical and chemical properties at the atomic scale must now be taken into account. The force laws will then depend on whether the surfaces are amorphous or crystalline (and whether the lattices of crystalline surfaces are matched or not), rough or smooth, rigid or soft (fluid-like), and hydrophobic or hydrophilic.

It is also important to distinguish between *static* (i.e., equilibrium) interactions and *dynamic* (i.e., nonequilibrium) forces such as viscous and friction forces. For example, certain liquid films confined between two contacting surfaces may take a surprisingly long time to equilibrate, as may the surfaces themselves, so that the short-range and adhesion forces appear to be time-dependent, resulting in "aging" effects.

9.2 Methods Used to Study Surface Forces

9.2.1 Force Laws

The full force law $F(D)$ between two surfaces, i.e., the force F as a function of surface separation D, can be measured in a number of ways [1–5]. The simplest is to move the base of a spring by a known amount, ΔD_0. Figure 9.1 illustrates this method when applied to the interaction of two magnets. However, the method is applicable also at the microscopic or molecular level, and it forms the basis of all direct force-measuring apparatuses such as the surface forces apparatus (SFA; [2, 6]) and the atomic force microscope (AFM; [7–9]). If there is a detectable force between the surfaces, this will cause the force-measuring spring to deflect by ΔD_s, while the surface separation changes by ΔD. These three displacements are related by

$$\Delta D_s = \Delta D_0 - \Delta D . \tag{9.1}$$

The difference in force, ΔF, between the initial and final separations is given by

$$\Delta F = k_s \Delta D_s , \tag{9.2}$$

where k_s is the spring constant. The equations above provide the basis for measurements of the force difference between any two surface separations. For example, if a force-measuring apparatus with a known k_s can measure D (and thus ΔD), ΔD_0, and ΔD_s, the force difference ΔF can be measured between a large initial or reference separation D, where the force is zero ($F = 0$), and another separation $D - \Delta D$. By working one's way in increasing increments of $\Delta D = \Delta D_0 - \Delta D_s$, the full force law $F(D)$ can be constructed over any desired distance regime.

In order to measure an equilibrium force law, it is essential to establish that the two surfaces have stopped moving before the displacements are measured. When displacements are measured while two surfaces are still in relative motion, one also measures a viscous or frictional contribution to the total force. Such dynamic force measurements have enabled the viscosities of liquids near surfaces and in thin films to be accurately determined [11–13].

In practice, it is difficult to measure the forces between two perfectly flat surfaces, because of the stringent requirement of perfect alignment for making reliable measurements at distances of a few tenths of a nanometer. It is far easier to measure the forces between curved surfaces, e.g., two spheres, a sphere and a flat surface, or two crossed cylinders. Furthermore, the force $F(D)$ measured between two curved surfaces can be directly related to the energy per unit area $E(D)$ between two flat surfaces at the same separation, D, by the so-called Derjaguin approximation [14]:

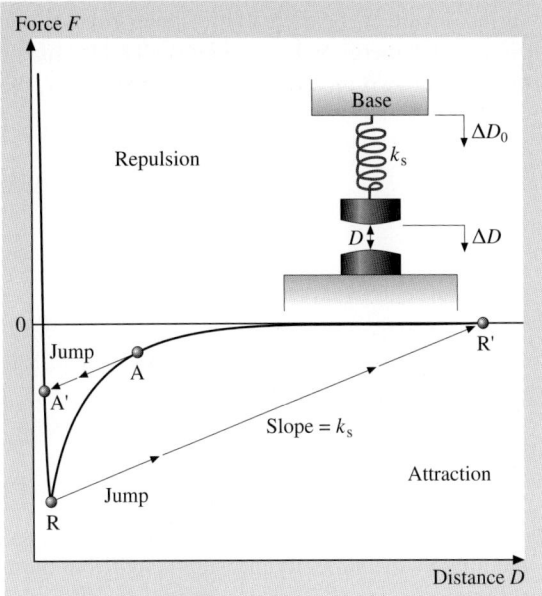

Fig. 9.1. Schematic attractive force law between two macroscopic objects such as two magnets, or between two microscopic objects such as the van der Waals force between a metal tip and a surface. On lowering the base supporting the spring, the latter will expand or contract such that at any equilibrium separation D the attractive force balances the elastic spring restoring force. If the gradient of the attractive force dF/dD exceeds the gradient of the spring's restoring force (defined by the spring constant k_s), the upper surface will jump from A into contact at A' (A for "advancing"). On separating the surfaces by raising the base, the two surfaces will jump apart from R to R' (R for "receding"). The distance R–R' multiplied by k_s gives the adhesion force, i.e., the value of F at the point R (after [10] with permission)

$$E(D) = \frac{F(D)}{2\pi R}, \tag{9.3}$$

where R is the radius of the sphere (for a sphere and a flat surface) or the radii of the cylinders (for two crossed cylinders).

9.2.2 Adhesion Forces

The most direct way to measure the adhesion of two solid surfaces (such as two spheres or a sphere on a flat) is to suspend one of them on a spring and measure the adhesion or "pull-off" force needed to separate the two bodies from its deflection. If k_s is the stiffness of the force-measuring spring and ΔD the distance the two surfaces jump apart when they separate, then the adhesion force F_s is given by

$$F_s = F_{\max} = k_s \Delta D, \tag{9.4}$$

where we note that in liquids, the maximum or minimum in the force may occur at some nonzero surface separation (see Fig. 9.7). From F_s and a known surface geometry, and assuming that the surfaces were everywhere in molecular contact, one may also calculate the surface or interfacial energy γ. For an elastically deformable sphere of radius R on a flat surface, or for two crossed cylinders of radius R, we have [2, 15]

$$\gamma = \frac{F_s}{3\pi R}, \tag{9.5}$$

while for two spheres of radii R_1 and R_2

$$\gamma = \frac{F_s}{3\pi}\left(\frac{1}{R_1} + \frac{1}{R_2}\right), \tag{9.6}$$

where γ is in units of $J\,m^{-2}$ (see Sect. 9.5.2).

9.2.3 The SFA and AFM

In a typical force-measuring experiment, at least two of the above displacement parameters – ΔD_0, ΔD, and ΔD_s – are directly or indirectly measured, and from these the third displacement and the resulting force law $F(D)$ are deduced using (9.1) and (9.2) together with a measured value of k_s. For example, in SFA experiments, ΔD_0 is changed by expanding or contracting a piezoelectric crystal by a known amount or by moving the base of the spring with sensitive motor-driven mechanical stages. The resulting change in surface separation ΔD is measured optically, and the spring deflection ΔD_s can then be obtained according to (9.1). In AFM experiments, ΔD_0 and ΔD_s are measured using a combination of piezoelectric, optical, capacitance or magnetic techniques, from which the change in surface separation ΔD is deduced. Once a force law is established, the geometry of the two surfaces (e.g., their radii) must also be known before the results can be compared with theory or with other experiments.

The SFA (Fig. 9.2) is used for measurements of adhesion and force laws between two curved molecularly smooth surfaces immersed in liquids or controlled vapors [2, 6, 16]. The surface separation is measured by multiple beam interferometry with an accuracy of ± 0.1 nm. From the shape of the interference fringes one also obtains the radius of the surfaces, R, and any surface deformation that arises during an interaction [17–19]. The resolution in the lateral direction is about 1 µm. The surface separation can be independently controlled to within 0.1 nm, and the force sensitivity is about 10^{-8} N. For a typical surface radius of $R \approx 1$ cm, γ values can be measured to an accuracy of about 10^{-3} mJ m^{-2}.

Several different materials have been used to form the surfaces in the SFA, including mica [20, 21], silica [22], sapphire [23], and polymer sheets [24]. These materials can also be used as supporting substrates in experiments on the forces between adsorbed or chemically bound polymer layers [13, 25–30], surfactant and lipid monolayers and bilayers [31–34], and metal and metal oxide layers [35–42]. The range of liquids and vapors that can be used is almost endless, and have thus far included aqueous solutions, organic liquids and solvents, polymer melts, various petroleum oils and lubricant liquids, and liquid crystals.

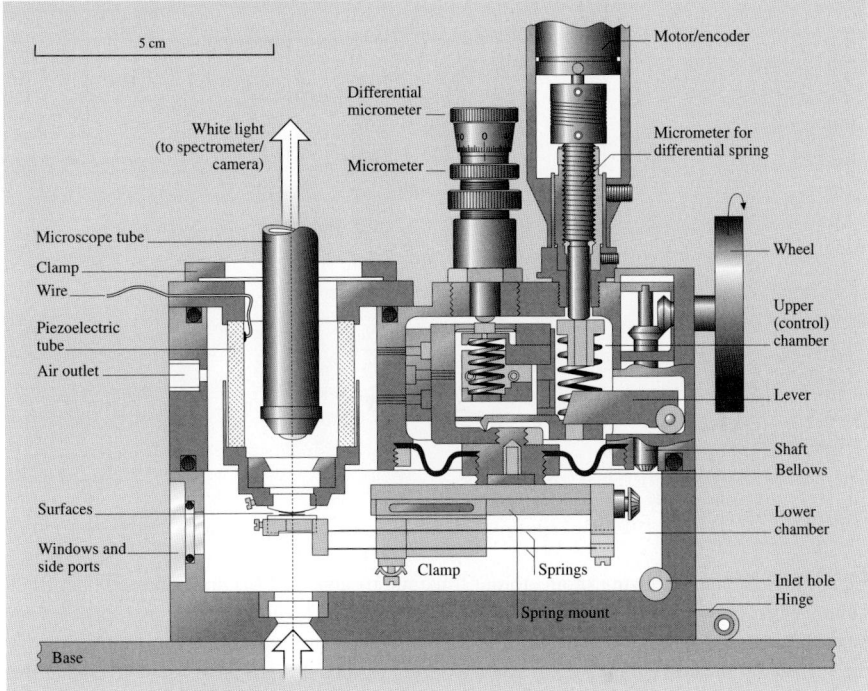

Fig. 9.2. A surface forces apparatus (SFA) where the intermolecular forces between two macroscopic, cylindrical surfaces of local radius R can be directly measured as a function of surface separation over a large distance regime from tenths of a nanometer to micrometers. Local or transient surface deformations can be detected optically. Various attachments for moving one surface laterally with respect to the other have been developed for friction measurements in different regimes of sliding velocity and sliding distance (after [16] with permission)

Friction attachments for the SFA [43–47] allow for the two surfaces to be sheared laterally past each other at varying sliding speeds or oscillating frequencies, while simultaneously measuring both the transverse (frictional or shear) force and the normal force (load) between them. The ranges of friction forces and sliding speeds that can be studied with such methods are currently 10^{-7}–10^{-1} N and 10^{-13}–10^{-2} m s^{-1}, respectively [48]. The externally applied load, L, can be varied continuously, and both positive and negative loads can be applied. The distance between the surfaces, D, their true molecular contact area, their elastic (or viscoelastic or elastohydrodynamic) deformation, and their lateral motion can all be monitored simultaneously by recording the moving interference fringe pattern.

In the atomic force microscope (Fig. 9.3), the force is measured by monitoring the deflection of a soft cantilever supporting a sub-microscopic tip ($R \approx 10$–200 nm) as this is interacting with a flat, macroscopic surface [7, 49, 50]. The measurements can be done in a vapor or liquid. The normal (bending) spring stiffness of the cantilever can be as small as 0.01 N m^{-1}, allowing measurements of normal forces as

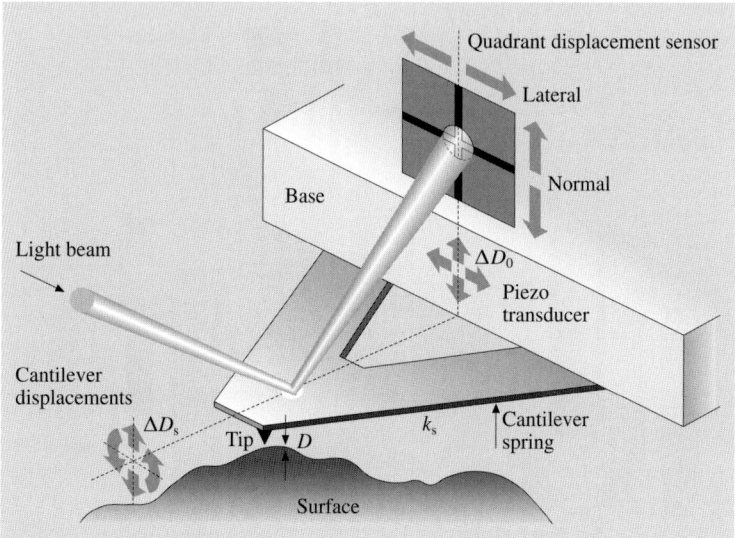

Fig. 9.3. Schematic drawing of an atomic force microscope (AFM) tip supported on a triangular cantilever interacting with an arbitrary solid surface. The normal force and topology are measured by monitoring the calibrated deflection of the cantilever as the tip is moved across the surface by means of a piezoelectric transducer. Various designs have been developed that move either the sample or the cantilever during the scan. Friction forces can be measured from the torsion of the cantilever when the scanning is in the direction perpendicular to its long axis (after [53] with permission)

small as 1 pN (10^{-12} N), which corresponds to the bond strength of single molecules [51, 52]. Distances can be inferred with an accuracy of about 1 nm, and changes in distance can be measured to about 0.1 nm. Since the contact area is small, different interaction regimes can be resolved on samples with a heterogeneous composition on lateral scales of about ten nanometers. Height differences and the roughness of the sample can be measured directly from the cantilever deflection or, alternatively, by using a feedback system to raise or lower the sample so that the deflection (the normal force) is kept constant during a scan over the area of interest. Weak interaction forces and larger (microscopic) interaction areas can be investigated by replacing the tip with a micrometer-sized sphere to form a "colloidal probe" [8].

The atomic force microscope can also be used for friction measurements (lateral force microscopy, LFM) by monitoring the torsion of the cantilever as the sample is scanned in the direction perpendicular to the long axis of the cantilever [9, 50, 54, 55]. Typically, the stiffness of the cantilever to lateral bending is much larger than to bending in the normal direction and to torsion, so that these signals are decoupled and height and friction can be detected simultaneously. The torsional spring constant can be as low as 0.1 N m^{-1}, giving a lateral (friction) force sensitivity of 10^{-11} N.

Rapid technical developments have facilitated the calibrations of the normal [56, 57] and lateral spring constants [55, 58], as well as in situ measurements of the macroscopic tip radius [59]. Cantilevers of different shapes with a large range of spring constants, tip radii, and surface treatments (inorganic or organic coatings) are commercially available. The flat surface, and also the particle in the colloidal probe technique, can be any material of interest. However, remaining difficulties with this technique are that the distance between the tip and the substrate, D, and the deformations of the tip and sample are not directly measurable. Another important difference between the AFM/LFM and SFA techniques is the different size of the contact area, and the related observation that even when a cantilever with a very low spring constant is used in the AFM, the pressure in the contact zone is typically much higher than in the SFA. Hydrodynamic effects in liquids also affect the measurements of normal forces differently on certain time scales [60–62].

9.2.4 Some Other Force-Measuring Techniques

A large number of other techniques are available for the measurements of the normal forces between solid or fluid surfaces (see [4, 53]). The techniques discussed in this section are not used for lateral (friction) force measurements, but are commonly used to study normal forces, particularly in biological systems.

Micropipette aspiration is used to measure the forces between cells or vesicles, or between a cell or vesicle and another surface [63–65]. The cell or vesicle is held by suction at the tip of a glass micropipette and deforms elastically in response to the net interactions with another surface and to the applied suction. The shape of the deformed surface (cell membrane) is measured and used to deduce the force between the surfaces and the membrane tension [64]. The membrane tension, and thus the stiffness of the cell or vesicle, is regulated by applying different hydrostatic pressures. Forces can be measured in the range of 0.1 pN to 1 nN, and the distance resolution is a few nanometers. The interactions between a colloidal particle and another surface can be studied by attaching the particle to the cell membrane [66].

In the osmotic stress technique, pressures are measured between colloidal particles in aqueous solution, membranes or bilayers, or other ordered colloidal structures (viruses, DNA). The separation between the particle surfaces and the magnitude of membrane undulations are measured by X-ray or neutron scattering techniques. This is combined with a measurement of the osmotic pressure of the solution [67–70]. The technique has been used to measure repulsive forces, such as DLVO interactions, steric forces, and hydration forces [71]. The sensitivity in pressure is $0.1 \, \text{mN m}^{-2}$, and distances can be resolved to 0.1 nm.

The optical tweezers technique is based on the trapping of dielectric particles at the center of a focused laser beam by restoring forces arising from radiation pressure and light intensity gradients [72, 73]. The forces experienced by particles as they are moved toward or away from one another can be measured with a sensitivity in the pN range. Small biological molecules are typically attached to a larger bead of a material with suitable refractive properties. Recent development allows determinations of position with nanometer resolution [74], which makes this technique useful for studying the forces during extension of single molecules.

In total internal reflection microscopy (TIRM), the potential energy between a micrometer-sized colloidal particle and flat surface in aqueous solution is deduced from the average equilibrium height of the particle above the surface, measured from the intensity of scattered light. The average height ($D \approx 10\text{--}100$ nm) results from a balance of gravitational force, radiation pressure from a laser beam focused at the particle from below, and intermolecular forces [75]. The technique is particularly suitable for measuring weak forces (sensitivity ca. 10^{-14} N), but is moredifficult to use for systems with strong interactions. A related technique is reflection interference contrast microscopy (RICM), where optical interference is used to also monitor changes in the shape of the approaching colloidal particle or vesicle [76].

An estimate of bond strengths can be obtained from the hydrodynamic shear force exerted by a fluid on particles or cells attached to a substrate [77, 78]. At a critical force, the bonds are broken and the particle or cell will be detached and move with the velocity of the fluid. This method requires knowledge of the contact area and the flow velocity profile of the fluid. Furthermore, a uniform stress distribution in the contact area is generally assumed. At low bond density, this technique can be used to determine the strength of single bonds (1 pN).

9.3 Normal Forces Between Dry (Unlubricated) Surfaces

9.3.1 Van der Waals Forces in Vacuum and Inert Vapors

Forces between macroscopic bodies (such as colloidal particles) across vacuum arise from interactions between the constituent atoms or molecules of each body across the gap separating them. These intermolecular interactions are electromagnetic forces between permanent or induced dipoles (van der Waals forces), and between ions (electrostatic forces). In this section, we describe the van der Waals forces, which occur between all atoms and molecules and between all macroscopic bodies (see [2]).

The interaction between two permanent dipoles with a fixed relative orientation can be attractive or repulsive. For the specific case of two freely rotating permanent dipoles in a liquid or vapor (orientational or Keesom interaction), and for a permanent dipole and an induced dipole in an atom or polar or nonpolar molecule (induction or Debye interaction), the interaction is on average always attractive. The third type of van der Waals interaction, the fluctuation or London dispersion interaction, arises from instantaneous polarization of one nonpolar or polar molecule due to fluctuations in the charge distribution of a neighboring nonpolar or polar molecule (Fig. 9.4a). Correlation between these fluctuating induced dipole moments gives an attraction that is present between any two molecules or surfaces across vacuum. At very small separations, the interaction will ultimately be repulsive as the electron clouds of atoms and molecules begin to overlap. The total interaction is thus a combination of a short-range repulsion and a relatively long-range attraction.

Except for in highly polar materials such as water, London dispersion interactions give the largest contribution (70–100 %) to the van der Waals attraction. The

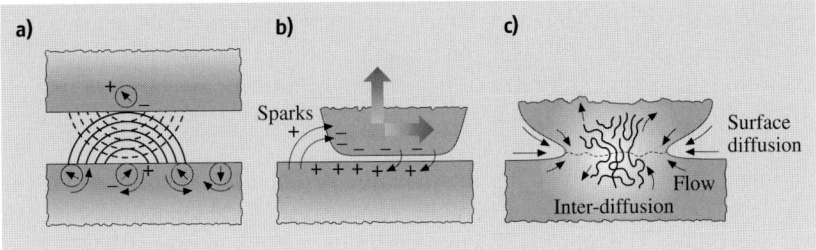

Fig. 9.4. Schematic representation of (**a**) van der Waals interaction (dipole–induced dipole interaction), (**b**) charge exchange, which acts to increase adhesion and friction forces, and (**c**) sintering between two surfaces

interaction energy of the van der Waals force between atoms or molecules depends on the separation r as

$$E(D) = \frac{-C_{vdW}}{r^6}, \tag{9.7}$$

where the constant C_{vdW} depends on the dipole moments and polarizabilities of the molecules. At large separations (> 10 nm), the London interaction is weakened by a randomizing effect caused by the rapid fluctuations. That is, the induced temporary dipole moment of one molecule may have changed during the time needed for the transmission of the electromagnetic wave (photon) generated by its fluctuating charge density to another molecule and the return of the photon generated by the induced fluctuation in this second molecule. This phenomenon is called retardation and causes the interaction energy to decay as r^{-7} at large separations [79].

Dispersion interactions are to a first approximation additive, and their contribution to the interaction energy between two macroscopic bodies (such as colloidal particles) across vacuum can be found by summing the pair-wise interactions [80]. The interaction is generally described in terms of the Hamaker constant, A_H. Another approach is to treat the interacting bodies and an intervening medium as continuous phases and determine the strength of the interaction from bulk dielectric properties of the materials [81, 82]. Unlike the pair-wise summation, this method takes into account the screening of the interactions between molecules inside the bodies by the molecules closer to the surfaces and the effects of the intervening medium. For the interaction between material 1 and material 3 across material 2, the non-retarded Hamaker constant given by the Lifshitz theory is approximately [2]:

$$A_{H,123} = A_{H,\nu=0} + A_{H,\nu>0}$$

$$\approx \tfrac{3}{4}k_B T \left(\tfrac{\varepsilon_1-\varepsilon_2}{\varepsilon_1+\varepsilon_2}\right)\left(\tfrac{\varepsilon_3-\varepsilon_2}{\varepsilon_3+\varepsilon_2}\right)$$

$$+ \tfrac{3h\nu_e}{8\sqrt{2}} \tfrac{(n_1^2-n_2^2)(n_3^2-n_2^2)}{\sqrt{(n_1^2+n_2^2)}\sqrt{(n_3^2+n_2^2)}\left(\sqrt{(n_1^2+n_2^2)}+\sqrt{(n_3^2+n_2^2)}\right)}, \tag{9.8}$$

where the first term ($\nu = 0$) represents the permanent dipole and dipole–induced dipole interactions and the second ($\nu > 0$) the London (dispersion) interaction. ε_i and n_i

are the static dielectric constants and refractive indexes of the materials, respectively. ν_e is the frequency of the lowest electron transition (around 3×10^{15} s^{-1}). Either one of the materials 1, 2, or 3 in (9.8) can be vacuum or air ($\varepsilon = n = 1$). A_H is typically $10^{-20} - 10^{-19}$ J (the higher values are found for metals) for interactions between solids and liquids across vacuum or air.

The interaction energy between two macroscopic bodies is dependent on the geometry and is always attractive between two bodies of the same material [A_H positive, see (9.8)]. The van der Waals interaction energy and force laws ($F = -dE(D)/dD$) for some common geometries are given in Table 9.1. Because of the retardation effect, the equations in Table 9.1 will lead to an overestimation of the dispersion force at large separations. It is, however, apparent that the interaction energy between macroscopic bodies decays more slowly with separation (i.e., has a longer range) than between two molecules.

For inert nonpolar surfaces, e.g., consisting of hydrocarbons or van der Waals solids and liquids, the Lifshitz theory has been found to apply even at molecular contact, where it can be used to predict the surface energies (surface tensions) of such solids and liquids. For example, for hydrocarbon surfaces, $A_H = 5 \times 10^{-20}$ J. Inserting this value into the equation for two flat surfaces (Table 9.1) and using a "cut-off" distance of $D_0 \approx 0.165$ nm as an effective separation when the surfaces are in contact [2], we obtain for the surface energy γ (which is defined as half the interaction energy)

$$\gamma = \frac{E}{2} = \frac{A_H}{24\pi D_0^2} \approx 24 \text{ mJ m}^{-2}, \qquad (9.9)$$

a value that is typical for hydrocarbon solids and liquids [83].

If the adhesion force is measured between a spherical surface of radius $R = 1$ cm and a flat surface using an SFA, we expect the adhesion force to be (see Table 9.1) $F = A_H R/(6D_0^2) = 4\pi R\gamma \approx 3.0$ mN. Using a spring constant of $k_s = 100$ N m^{-1}, such an adhesive force will cause the two surfaces to jump apart by $\Delta D = F/k_s = 30$ μm, which can be accurately measured. (For elastic bodies that deform in adhesive contact, R changes during the interaction and the measured adhesion force is 25% lower, see Sect. 9.5.2). Surface energies of solids can thus be directly measured with the SFA and, in principle, with the AFM if the contact geometry can be quantified. The measured values are in good agreement with calculated values based on the known surface energies γ of the materials, and for nonpolar low-energy solids they are well accounted for by the Lifshitz theory [2].

9.3.2 Charge Exchange Interactions

Electrostatic interactions are present between ions (Coulomb interactions), between ions and permanent dipoles, and between ions and nonpolar molecules in which a charge induces a dipole moment. The interaction energy between ions or between a charge and a fixed permanent dipole can be attractive or repulsive. For an induced dipole or a freely rotating permanent dipole in vacuum or air, the interaction energy with a charge is always attractive.

Table 9.1. Van der Waals interaction energy and force between macroscopic bodies of different geometries

Geometry of bodies with surfaces D apart ($D \ll R$)		van der Waals interaction Energy, E	van der Waals interaction Force, F
Two atoms or small molecules	$r \geqslant \sigma$	$\dfrac{-C_{vdW}}{r^6}$	$\dfrac{-6C_{vdW}}{r^7}$
Two flat surfaces (per unit area)	$r \gg D$	$\dfrac{-A_H}{12\pi D^2}$	$\dfrac{-A_H}{6\pi D^3}$
Two spheres or macromolecules of radii R_1 and R_2	$R_1 \gg D$ $R_2 \gg D$	$\dfrac{-A_H}{6D}\left(\dfrac{R_1 R_2}{R_1 + R_2}\right)$	$\dfrac{-A_H}{6D^2}\left(\dfrac{R_1 R_2}{R_1 + R_2}\right)$
Sphere or macromolecule of radius R near a flat surface	$R \gg D$	$\dfrac{-A_H R}{6D}$	$\dfrac{-A_H R}{6D^2}$
Two parallel cylinders or rods of radii R_1 and R_2 (per unit length)	$R_1 \gg D$ $R_2 \gg D$	$\dfrac{-A_H}{12\sqrt{2}\, D^{3/2}} \times \left(\dfrac{R_1 R_2}{R_1 + R_2}\right)^{1/2}$	$\dfrac{-A_H}{8\sqrt{2}\, D^{5/2}} \times \left(\dfrac{R_1 R_2}{R_1 + R_2}\right)^{1/2}$
Cylinder of radius R near a flat surface (per unit length)	$R \gg D$	$\dfrac{-A_H \sqrt{R}}{12\sqrt{2}\, D^{3/2}}$	$\dfrac{-A_H \sqrt{R}}{8\sqrt{2}\, D^{5/2}}$
Two cylinders or filaments of radii R_1 and R_2 crossed at 90°	$R_1 \gg D$ $R_2 \gg D$	$\dfrac{-A_H \sqrt{R_1 R_2}}{6D}$	$\dfrac{-A_H \sqrt{R_1 R_2}}{6D^2}$

A negative force (A_H positive) implies attraction, a positive force means repulsion (A_H negative) (after [53] with permission)

Spontaneous charge transfer may occur between two dissimilar materials in contact [84]. The phenomenon is especially prominent in contacts between a metal, for example, mercury, and a material with low conductivity, but is also observed, for example, between two different polymer layers. During separation, rolling or sliding of one body over the other, the surfaces experience both charge transition from one surface to the other and charge transfer (conductance) along each surface (Fig. 9.4b). The latter process is typically slower, and, as a result, charges remain on the surfaces as they are separated in vacuum or dry nitrogen gas. The charging gives rise to a strong adhesion with adhesion energies of over $1000\,\text{mJ}\,\text{m}^{-2}$, similar to fracture or cohesion energies of the solid bodies themselves [84, 85]. Upon separating the surfaces farther apart, a strong, long-range electrostatic attraction is observed. The charging can be decreased through discharges across the gap between the surfaces (which requires a high charging) or through conducting in the solids. It has been suggested that charge exchange interactions are particularly important in rolling friction between dry surfaces (which can simplistically be thought of as an adhesion–separation process), where the distance dependence of forces acting normally to the surfaces plays a larger role than in sliding friction. Recent experiments on the sliding friction between metal–insulator surfaces indicate that stick–slip would be accompanied by charge transfer events [86].

Photo-induced charge transfer, or harpooning, involves the transfer of an electron between an atom in a molecular beam or at a solid surface (typically alkali or transition metal) to an atom or molecule in a gas (typically halides) to form a negatively charged molecular ion in a highly excited vibrational state. This transfer process can occur at atomic distances of 0.5–0.7 nm, which is far from molecular contact. The formed molecular ion is attracted to the surface and chemisorbs onto it. Photo-induced charge transfer processes also occur in the photosynthesis in green plants and in photoelectrochemical cells (solar cells) at the junction between two semiconductors or between a semiconductor and an electrolyte solution [87].

9.3.3 Sintering and Cold Welding

When macroscopic particles in a powder or in a suspension come into molecular contact, they can bond together to form a network or solid body with very different density and shear strength compared to the powder (a typical example is porcelain). The rate of bonding is dependent on the surface energy (causing a stress at the edge of the contact) and the atomic mobility (diffusion rate) of the contacting materials. To increase the diffusion rate, objects formed from powders are heated to about one-half of the melting temperature of the components in a process called sintering, which can be done in different atmospheres or in a liquid.

In the sintering process, the surface energy of the system is lowered due to the reduction of total surface area (Fig. 9.4c). In metal and ceramic systems, the most important mechanism is solid-state diffusion, initially surface diffusion. As the surface area decreases and the grain boundaries increase at the contacts, grain boundary diffusion and diffusion through the crystal lattice become more important. The grain boundaries will eventually migrate, so that larger particles are formed (coarsening).

Mass can also be transferred through evaporation and condensation, and through viscous and plastic flow. In liquid-phase sintering, the materials can melt, which increases the mass transport. Amorphous materials like polymers and glasses do not have real grain boundaries and sinter by viscous flow [88].

Some of these mechanisms (surface diffusion and evaporation–condensation) reduce the surface area and increase the grain size (coarsening) without densification, in contrast to bulk transport mechanisms like grain boundary diffusion and plastic and viscous flow. As the material becomes denser, elongated pores collapse to form smaller, spherical pores with a lower surface energy. Models for sintering typically consider the size and growth rate of the grain boundary (the "neck") formed between two spherical particles. At a high stage of densification, the sintering stress σ at the curved neck between two particles is given by [88]

$$\sigma = \frac{2\gamma_{ss}}{G} + \frac{2\gamma_{sv}}{r_p}, \tag{9.10}$$

where γ_{ss} is the solid–solid grain boundary energy, γ_{sv} is the solid–vapor surface energy, G is the grain size, and r_p is the radius of the pore.

A related phenomenon is cold welding, which is the spontaneous formation of strong junctions between clean (unoxidized) metal surfaces with a mutual solubility when they are brought in contact with or without an applied pressure. The plastic deformations accompanying the formation and breaking of such contacts on a molecular scale during motion of one surface normally (see Fig. 9.10c,d) or laterally (shearing) with respect to the other have been studied both experimentally [89] and theoretically [90–95]. The breaking of a cold welded contact is generally associated with damage or deformation of the surface structure.

9.4 Normal Forces Between Surfaces in Liquids

9.4.1 Van der Waals Forces in Liquids

The dispersion interaction in a medium will be significantly lower than in vacuum, since the attractive interaction between two solute molecules in a medium (solvent) involves displacement and reorientation of the nearest-neighbor solvent molecules. Even though the surrounding medium may change the dipole moment and polarizability from that in vacuum, the interaction between two identical molecules remains attractive in a binary mixture. The extension of the interactions to the case of two macroscopic bodies is the same as described in Sect. 9.3.1. Typically, the Hamaker constants for interactions in a medium are an order of magnitude lower than in vacuum. Between macroscopic surfaces in liquids, van der Waals forces become important at distances below 10–15 nm and may at these distances start to dominate interactions of different origin that have been observed at larger separations.

Figure 9.5 shows the measured van der Waals forces between two crossed cylindrical mica surfaces in water and various salt solutions. Good agreement is obtained between experiment and theory. At larger surface separations, above about 5 nm,

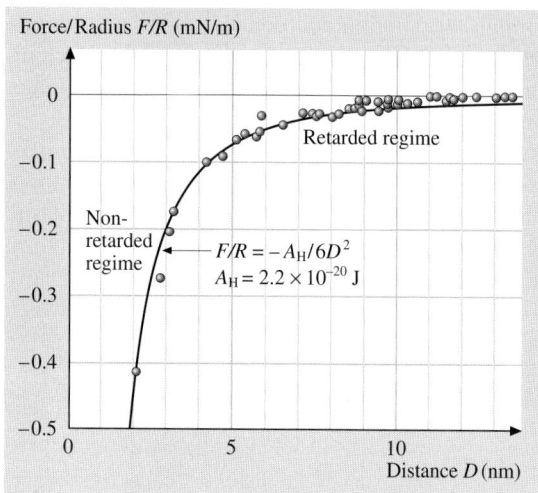

Fig. 9.5. Attractive van der Waals force F between two curved mica surfaces of radius $R \approx$ 1 cm measured in water and various aqueous electrolyte solutions. The electrostatic interaction has been subtracted from the total measured force. The measured non-retarded Hamaker constant is $A_H = 2.2 \times 10^{-20}$ J. Retardation effects are apparent at distances larger than 5 nm, as expected theoretically. (After [2]. Copyright 1991, with permission from Elsevier Science)

the measured forces fall off more rapidly than as D^{-2}. This retardation effect (see Sect. 9.3.1) is also predicted by Lifshitz theory and is due to the time needed for propagation of the induced dipole moments over large distances.

From Fig. 9.5, we may conclude that at separations above about 2 nm, or 8 molecular diameters of water, the *continuum* Lifshitz theory is valid. This would mean that water films as thin as 2 nm may be expected to have bulk-like properties, at least as far as their interaction forces are concerned. Similar results have been obtained with other liquids, where in general continuum properties are manifested, both as regards their interactions and other properties such as viscosity, at a film thickness larger than 5 or 10 molecular diameters. In the absence of a solvent (in vacuum), the agreement of measured van der Waals forces with the continuum Lifshitz theory is generally good at all separations down to molecular contact ($D = D_0$).

Van der Waals interactions in a system of three or more different materials (see (9.8)) can be attractive or repulsive, depending on their dielectric properties. Numerous experimental studies show the attractive van der Waals forces in various systems [2], and also repulsive van der Waals forces have been measured directly [96]. A practical consequence of the repulsive interaction obtain ed across a medium with intermediate dielectric properties is that the van der Waals forces will give rise to preferential, nonspecific adsorption of molecules with an intermediate dielectric constant. This is commonly seen as adsorption of vapors or solutes to a solid surface. It is also possible to diminish the attractive interaction between dispersed colloidal particles by adsorption of a thin layer of material with dielectric properties close to those of the surrounding medium (matching of refractive index), or by adsorption of a polymer that gives a steric repulsive force that keeps the particles separated at a distance where the magnitude of the van der Waals attraction is negligible. Thermal motion will then keep the particles dispersed.

9.4.2 Electrostatic and Ion Correlation Forces

Most surfaces in contact with a highly polar liquid (such as water) acquire a surface charge, either by dissociation of ions from the surface into the solution or by preferential adsorption of certain ions from the solution. The surface charge is balanced by a layer of oppositely charged ions (counterions) in the solution at some small distance from the surface (see [2]). In dilute solution, this distance is the Debye length, κ^{-1}, which is purely a property of the electrolyte *solution*. The Debye length falls with increasing ionic strength (i.e., with the molar concentration M_i and valency z_i) of the ions in solution:

$$\kappa^{-1} = \left(\frac{\varepsilon \varepsilon_0 k_B T}{e^2 N_A \sum_i z_i^2 M_i} \right)^{1/2}, \tag{9.11}$$

where e is the electronic charge. For example, for 1:1 electrolytes at 25 °C, $\kappa^{-1} = 0.304$ nm$/\sqrt{M_{1:1}}$, where M_i is given in M (mol dm^{-3}). κ^{-1} is thus ca. 10 nm in a 1 mM NaCl solution and 0.3 nm in a 1 M solution. In totally pure water at pH 7, where $M_i = 10^{-7}$ M, κ^{-1} is 960 nm, or about 1 µm. The Debye length also relates the surface charge density σ of a surface to the electrostatic surface potential ψ_0 via the Grahame equation, which for 1:1 electrolytes can be expressed as:

$$\sigma = \sqrt{8\varepsilon\varepsilon_0 k_B T} \sinh(e\psi_0/2k_B T) \times \sqrt{M_{1:1}}. \tag{9.12}$$

Since the Debye length is a measure of the thickness of the diffuse atmosphere of counterions near a charged surface, it also determines the range of the electrostatic "double-layer" interaction between two charged surfaces. The electrostatic double-layer interaction is an entropic effect that arises upon decreasing the thickness of the liquid film containing the dissolved ions. Because of the attractive force between the dissolved ions and opposite charges on the surfaces, the ions stay between the surfaces, but an osmotic repulsion arises as their concentration increases. The long-range electrostatic interaction energy at large separations (weak overlap) between two similarly charged molecules or surfaces is typically repulsive and is roughly an exponentially decaying function of D:

$$E(D) \approx +C_{ES} e^{-\kappa D}, \tag{9.13}$$

where C_{ES} is a constant that depends on the geometry of the interacting surfaces, on their surface charge density, and the solution conditions (Table 9.2). We see that the Debye length is the decay length of the interaction energy between two surfaces (and of the mean potential away from one surface). C_{ES} can be determined by solving the so-called Poisson–Boltzmann equation or by using other theories [97, 98].The equations in Table 9.2 are expressed in terms of a constant, Z, defined as

$$Z = 64\pi\varepsilon\varepsilon_0(k_B T/e)^2 \tanh^2\left[ze\psi_0/(4k_B\S T)\right],\qquad(9.14)$$

which depends only on the properties of the *surfaces*.

The above approximate expressions are accurate only for surface separations larger than about one Debye length. At smaller separations one must use numerical solutions of the Poisson–Boltzmann equation to obtain the exact interaction potential, for which there are no simple expressions. In the limit of small D, it can be shown that the interaction energy depends on whether the surfaces remain at constant potential ψ_0 (as assumed in the above equations) or at constant charge σ (when the repulsion exceeds that predicted by the above equations), or somewhere between these limits. In the "constant charge limit" the total *number* of counterions in the compressed film does not change as D is decreased, whereas at constant potential, the *concentration* of counterions is constant.The limiting pressure (or force per unit area) at constant charge is the osmotic pressure of the confined ions:

$$\begin{aligned}F &= k_B T \times \text{ion number density}\\ &= 2\sigma k_B T/(zeD),\quad \text{for } D \ll \kappa^{-1}.\end{aligned}\qquad(9.15)$$

That is, as $D \to 0$ the double-layer pressure at constant surface charge becomes infinitely repulsive and independent of the salt concentration (at constant potential the force instead becomes a constant at small D). However, at small separations, the van der Waals attraction (which goes as D^{-2} between two spheres or as D^{-3} between two planar surfaces, see Table 9.1) wins out over the double-layer repulsion, unless some other short-range interaction becomes dominant (see Sect. 9.4.4). This is the theoretical prediction that forms the basis of the so-called Derjaguin–Landau–Verwey–Overbeek (DLVO) theory [97, 99], illustrated in Fig. 9.6.

Because of the different distance dependence of the van der Waals and electrostatic interactions, the total force law, as described by the DLVO theory, can show several minima and maxima. Typically, the depth of the outer (secondary) minimum is a few $k_B T$, enough to cause reversible flocculation of particles from an aqueous dispersion. If the force barrier between the secondary and primary minimum is lowered, for example, by increasing the electrolyte concentration, particles can be irreversibly coagulated in the primary minimum. In practice, other forces (describedin the following sections) often appear at very small separations, so that the full force law between two surfaces or colloidal particles in solution can be more complex than might be expected from the DLVO theory.

There are situations when the double-layer interaction can be attractive at short range even between surfaces of similar charge, especially in systems with charge regulation due to dissociation of chargeable groups on the surfaces [100]; ion condensation [101], which may lower the effective surface charge density in systems containing di-and trivalent counterions; or ion correlation, which is an additional van der Waals-like attraction due to mobile and therefore highly polarizable counterions

Table 9.2. Electrical "double-layer" interaction energy $E(D)$ and force ($F = -dE/dD$) between macroscopic bodies

Geometry of bodies with surfaces D apart ($D \ll R$)		Electric "double-layer" interaction Energy, E	Force, F
Two ions or small molecules	$r > \sigma$	$\dfrac{+z_1 z_2 e^2}{4\pi\varepsilon\varepsilon_0 r} \dfrac{e^{-\kappa(r-\sigma)}}{(1+\kappa\sigma)}$	$\dfrac{+z_1 z_2 e^2}{4\pi\varepsilon\varepsilon_0 r^2} \dfrac{(1+\kappa r)}{(1+\kappa\sigma)} e^{-\kappa(r-\sigma)}$
Two flat surfaces (per unit area)	$r \gg D$	$(\kappa/2\pi) Z e^{-\kappa D}$	$(\kappa^2/2\pi) Z e^{-\kappa D}$
Two spheres or macromolecules of radii R_1 and R_2	$R_1 \gg D$ $R_2 \gg D$	$\left(\dfrac{R_1 R_2}{R_1+R_2}\right) Z e^{-\kappa D}$	$\kappa \left(\dfrac{R_1 R_2}{R_1+R_2}\right) Z e^{-\kappa D}$
Sphere or macro molecule of radius R near a flat surface	$R \gg D$	$R Z e^{-\kappa D}$	$\kappa R Z e^{-\kappa D}$
Two parallel cylinders or rods of radii R_1 and R_2 (per unit length)	$R_1 \gg D$ $R_2 \gg D$	$\dfrac{\kappa^{1/2}}{\sqrt{2\pi}} \left(\dfrac{R_1 R_2}{R_1+R_2}\right)^{1/2} Z e^{-\kappa D}$	$\dfrac{\kappa^{3/2}}{\sqrt{2\pi}} \left(\dfrac{R_1 R_2}{R_1+R_2}\right)^{1/2} Z e^{-\kappa D}$
Cylinder of radius R near a flat surface (per unit length)	$R \gg D$	$\kappa^{1/2} \sqrt{\dfrac{R}{2\pi}} Z e^{-\kappa D}$	$\kappa^{3/2} \sqrt{\dfrac{R}{2\pi}} Z e^{-\kappa D}$
Two cylinders or filaments of radii R_1 and R_2 crossed at 90°	$R_1 \gg D$ $R_2 \gg D$	$\sqrt{R_1 R_2}\, Z e^{-\kappa D}$	$\kappa \sqrt{R_1 R_2}\, Z e^{-\kappa D}$

The interaction energy and force for bodies of different geometries is based on the Poisson–Boltzmann equation (a continuum, mean-field theory). Equation (9.14) gives the interaction constant Z (in terms of surface potential ψ_0) for the interaction between similarly charged (ionized) surfaces in aqueous solutions of monovalent electrolyte. It can also be expressed in terms of the surface charge density σ by applying the Grahame equation (9.12) (after [53] with permission)

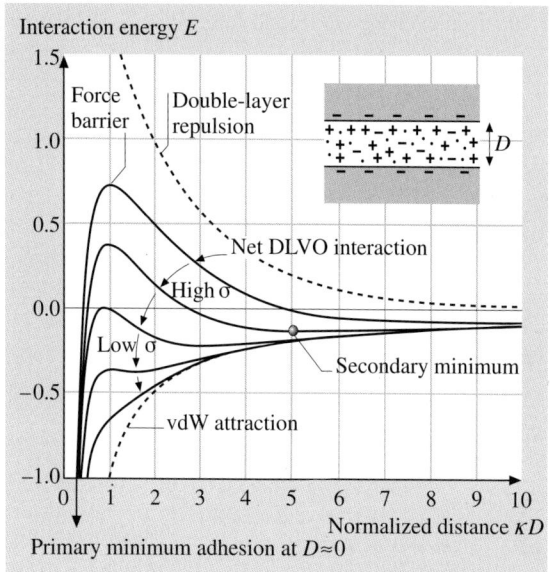

Fig. 9.6. Schematic plots of the DLVO interaction potential energy E between two flat, charged surfaces (or, according to the Derjaguin approximation, (9.3), the force F between two curved surfaces) as a function of the surface separation normalized by the Debye length, κ^{-1}. The van der Waals attraction (inverse power-law dependence on D) together with the repulsive electrostatic "double-layer" force (roughly exponential) at different surface charge σ (or potential, see (9.12)) determine the net interaction potential in aqueous electrolyte solution (after [53] with permission)

located at the surface [102]. The ion correlation (or charge fluctuation) force becomes significant at separations below 4 nm and increases with the surface charge density σ and the valency z of the counterions. Computer simulations have shown that at high charge density and monovalent counterions, the ion correlation force can reduce the effective double-layer repulsion by 10–15 %. With divalent counterions, the ion correlation force was found to exceed the double-layer repulsion and the total force then became attractive at a separation below 2 nm even in dilute electrolyte solution [103]. Experimentally, such short-range attractive forces have been found in charged bilayer systems [104, 105].

9.4.3 Solvation and Structural Forces

When a liquid is confined within a restricted space, for example, a very thin film between two surfaces, it ceases to behave as a structureless continuum. At small surface separations (below about 10 molecular diameters), the van der Waals force between two surfaces or even two solute molecules in a liquid (solvent) is no longer a smoothly varying attraction. Instead, there arises an additional "solvation" force that generally oscillates between attraction and repulsion with distance, with a periodicity equal to some mean dimension σ of the liquid molecules [106]. Figure 9.7a shows the force law between two smooth mica surfaces across the hydrocarbon liquid tetradecane, whose inert, chain-like molecules have a width of $\sigma \approx 0.4$ nm.

The short-range oscillatory force law is related to the "density distribution function" and "potential of mean force" characteristic of intermolecular interactions in liquids. These forces arise from the confining effects two surfaces have on liquid molecules, forcing them to order into quasi-discrete layers. Such layers are energetically or entropically favored and correspond to the minima in the free energy, whereas fractional layers are disfavored (energy maxima). This effect is quite general and arises in all simple liquids when they are confined between two smooth, rigid surfaces, both flat and curved. Oscillatory forces do not require any attractive liquid–liquid or liquid–wall interaction, only two hard walls confining molecules whose shape is not too irregular and that are free to exchange with molecules in a bulk liquid reservoir. In the absence of any attractive pressure between the molecules, the bulk liquid density could be maintained by an external hydrostatic pressure – in real liquids attractive van der Waals forces play the role of such an external pressure.

Oscillatory forces are now well understood theoretically, at least for simple liquids, and a number of theoretical studies and computer simulations of various confined liquids (including water) that interact via some form of Lennard-Jones potential have invariably led to an oscillatory solvation force at surface separations below a few molecular diameters [110–117]. In a first approximation, the oscillatory force law may be described by an exponentially decaying cosine function of the form

$$E \approx E_0 \cos(2\pi D/\sigma) e^{-D/\sigma} , \qquad (9.16)$$

where both theory and experiments show that the oscillatory period and the characteristic decay length of the envelope are close to σ.

Once the solvation zones of the two surfaces overlap, the mean liquid density in the gap is no longer the same as in the bulk liquid. Since the van der Waals interaction depends on the optical properties of the liquid, which in turn depends on the density, the van der Waals and the oscillatory solvation forces are not strictly additive. It is more correct to think of the solvation force as *the* van der Waals force at small separations with the molecular properties and density variations of the medium taken into account. It is also important to appreciate that solvation forces do not arise simply because liquid molecules tend to structure into semi-ordered layers at surfaces. They arise because of the disruption or *change* of this ordering during the approach of a second surface. The two effects are related; the greater the tendency toward structuring at an isolated surface the greater the solvation force between two such surfaces, but there is a real distinction between the two phenomena that should be borne in mind.

Oscillatory forces lead to different adhesion values depending on the energy minimum from which two surfaces are being separated. For an interaction energy described by (9.16), "quantized" adhesion energies will be E_0 at $D = 0$ (primary minimum), E_0/e at $D = \sigma$, E_0/e^2 at $D = 2\sigma$, etc. E_0 can be thought of as a depletion force (see Sect. 9.4.5) that is approximately given by the osmotic limit $E_0 \approx -k_B T/\sigma^2$, which can exceed the contribution to the adhesion energy

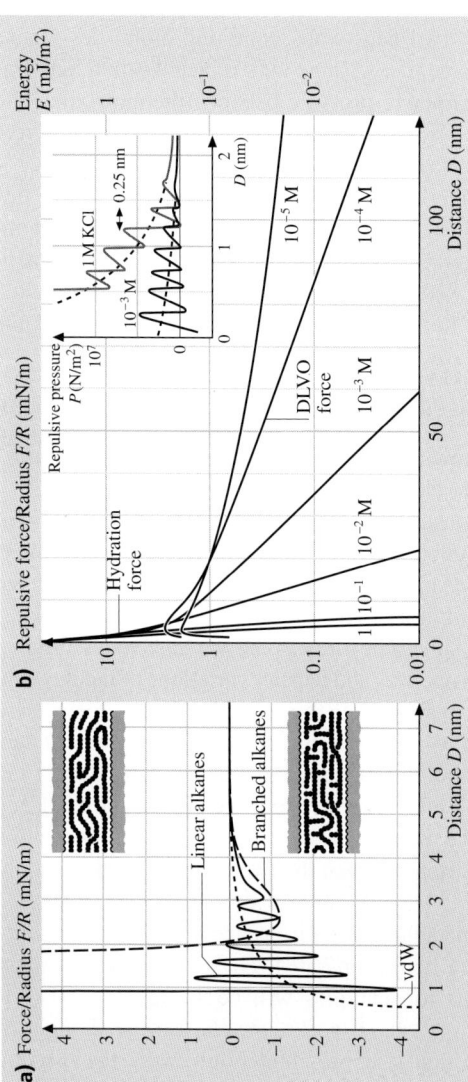

Fig. 9.7. (**a**) *Solid curve*: Forces measured between two mica surfaces across saturated linear chain alkanes such as *n*-tetradecane and *n*-hexadecane [107, 108]. The 0.4 nm periodicity of the oscillations indicates that the molecules are preferentially oriented parallel to the surfaces, as shown schematically in the *upper insert*. The theoretical continuum van der Waals attraction is shown as a *dotted curve*. *Dashed curve*: Smooth, non-oscillatory force law exhibited by irregularly shaped alkanes (such as 2-methyloctadecane) that cannot order into well-defined layers (*lower insert*) [108, 109]. Similar non-oscillatory forces are also observed between "rough" surfaces, even when these interact across a saturated linear chain liquid. This is because the irregularly shaped surfaces (rather than the liquid) now prevent the liquid molecules from ordering in the gap. (**b**) Forces measured between charged mica surfaces in KCl solutions of varying concentrations [20]. In dilute solutions (10^{-5} and 10^{-4} M), the measured forces are excellently described by the DLVO theory, based on exact solutions to the nonlinear Poisson–Boltzmann equation for the electrostatic forces and the Lifshitz theory for the van der Waals forces (using a Hamaker constant of $A_H = 2.2 \times 10^{-20}$ J). At higher concentrations, as more hydrated K^+ cations adsorb onto the negatively charged surfaces, an additional hydration force appears superimposed on the DLVO interaction at distances below 3–4 nm. This force has both an oscillatory and a monotonic component. *Insert*: Short-range hydration forces between mica surfaces shown as pressure versus distance. The lower and upper curves show surfaces 40% and 95% saturated with K^+ ions. At larger separations, the forces are in good agreement with the DLVO theory. (After [2]. Copyright 1991, with permission from Elsevier Science)

in contact from the van der Waals forces (at $D_0 \approx 0.15$–0.20 nm, as discussed in Sect. 9.3.1, keeping in mind that the Lifshitz theory fails to describe the force law at *intermediate* distances). Such multivalued adhesion forces have been observed in a number of systems, including the interactions of fibers.

Measurements of oscillatory forces between different surfaces across both aqueous and nonaqueous liquids have revealed their richness of properties [118–121], for example, their great sensitivity to the shape and rigidity of the solvent molecules, to the presence of other components, and to the structure of the confining surfaces (see Sects. 9.5.3 and 9.9). In particular, the oscillations can be smeared out if the molecules are irregularly shaped (e.g., branched) and therefore unable to pack into ordered layers, or when the interacting surfaces are rough or fluid-like (see Sect. 9.4.6).

It is easy to understand how oscillatory forces arise between two flat, plane parallel surfaces. Between two curved surfaces, e.g., two spheres, one might imagine the molecular ordering and oscillatory forces to be smeared out in the same way that they are smeared out between two randomly rough surfaces (see Sect. 9.5.3); however, this is not the case. Ordering can occur so long as the curvature or roughness is itself regular or uniform, i.e., not random. This is due to the Derjaguin approximation (9.3). If the energy between two flat surfaces is given by a decaying oscillatory function (for example, a cosine function as in (9.16)), then the force (and energy) between two curved surfaces will also be an oscillatory function of distance with some phase shift. Likewise, two surfaces with regularly curved regions will also retain their oscillatory force profile, albeit modified, as long as the corrugations are truly regular, i.e., periodic. On the other hand, surface roughness, even on the nanometer scale, can smear out oscillations if the roughness is random and the confined molecules are smaller than the size of the surface asperities [122, 123]. If an organic liquid contains small amounts of water, the expected oscillatory force can be replaced by a strongly attractive capillary force (see Sect. 9.5.1).

9.4.4 Hydration and Hydrophobic Forces

The forces occurring in water and electrolyte solutions are more complex than those occurring in nonpolar liquids. According to continuum theories, the attractive van der Waals force is always expected to win over the repulsive electrostatic "double-layer" force at small surface separations (Fig. 9.6). However, certain surfaces (usually oxide or hydroxide surfaces such as clays or silica) swell spontaneously or repel each other in aqueous solution, even at high salt concentrations. Yet in all these systems one would expect the surfaces or particles to remain in strong adhesive contact or coagulate in a primary minimum if the only forces operating were DLVO forces.

There are many other aqueous systems in which the DLVO theory fails and where there is an additional short-range force that is not oscillatory but monotonic. Between hydrophilic surfaces this force is exponentially repulsive and is commonly referred to as the *hydration*, or *structural*, force. The origin and nature of this force has long been controversial, especially in the colloidal and biological literature. Repulsive hydration forces are believed to arise from strongly hydrogen-bonding surface groups, such as hydrated ions or hydroxyl (–OH) groups, which modify the hydrogen-bonding network of liquid water adjacent to them. Because this network is quite extensive in range [124], the resulting interaction force is also of relatively long range.

Repulsive hydration forces were first extensively studied between clay surfaces [125]. More recently, they have been measured in detail between mica and silica surfaces [20–22], where they have been found to decay exponentially with decay lengths of about 1 nm. Their effective range is about 3 to 5 nm, which is about twice the range of the oscillatory solvation force in water. Empirically, the hydration repulsion between two hydrophilic surfaces appears to follow the simple equation

$$E = E_0 e^{-D/\lambda_0} , \qquad (9.17)$$

where $\lambda_0 \approx 0.6$–1.1 nm for 1:1 electrolytes and $E_0 = 3$–30 mJ m^{-2} depending on the hydration (hydrophilicity) of the surfaces, higher E_0 values generally being associated with lower λ_0 values.

The interactions between molecularly smooth mica surfaces in dilute electrolyte solutions obey the DLVO theory (Fig. 9.7b). However, at higher salt concentrations, specific to each electrolyte, hydrated cations bind to the negatively charged surfaces and give rise to a repulsive hydration force [20, 21]. This is believed to be due to the energy needed to dehydrate the bound cations, which presumably retain some of their water of hydration on binding. This conclusion was arrived at after noting that the strength and range of the hydration forces increase with the known hydration numbers of the electrolyte cations in the order: $Mg^{2+} > Ca^{2+} > Li^+ \sim Na^+ > K^+ > Cs^+$. Similar trends are observed with other negatively charged colloidal surfaces.

While the hydration force between two mica surfaces is overall repulsive below a distance of 4 nm, it is not always monotonic below about 1.5 nm but exhibits oscillations of mean periodicity of 0.25 ± 0.03 nm, roughly equal to the diameter of the water molecule. This is shown in the insert in Fig. 9.7b, where we may note that the first three minima at $D = 0, 0.28$, and 0.56 nm occur at negative energies, a result that rationalizes observations on certain colloidal systems. For example, clay platelets such as montmorillonite often repel each other increasingly strongly as they come closer together, but they are also known to stack into stable aggregates with water interlayers of typical thickness 0.25 and 0.55 nm between them [126, 127], suggestive of a turnabout in the force law from a monotonic repulsion to discretized attraction. In chemistry we would refer to such structures as stable hydrates of fixed stoichiometry, whereas in physics we may think of them as experiencing an oscillatory force.

Both surface force and clay swelling experiments have shown that hydration forces can be modified or "regulated" by exchanging ions of different hydration on surfaces, an effect that has important practical applications in controlling the stability of colloidal dispersions. It has long been known that colloidal particles can be precipitated (coagulated or flocculated) by increasing the electrolyte concentration, an effect that was traditionally attributed to the reduced screening of the electrostatic double-layer repulsion between the particles due to the reduced Debye length. However, there are many examples where colloids are stabilized at high salt concentrations, not at low concentrations. This effect is now recognized as being due to the increased hydration repulsion experienced by certain surfaces when they bind highly hydrated ions at higher salt concentrations. Hydration regulation of adhesion and interparticle forces is an important practical method for controlling various processes such as clay swelling [126, 127], ceramic processing and rheology [128, 129], material fracture [128], and colloidal particle and bubble coalescence [130].

Water appears to be unique in having a solvation (hydration) force that exhibits both a monotonic and an oscillatory component. Between hydrophilic surfaces the monotonic component is repulsive (Fig. 9.7b), but between hydrophobic surfaces it is attractive and the final adhesion is much greater than expected from the Lifshitz theory.

A hydrophobic surface is one that is inert to water in the sense that it cannot bind to water molecules via ionic or hydrogen bonds. Hydrocarbons and fluorocarbons are hydrophobic, as is air, and the strongly attractive hydrophobic force has many important manifestations and consequences such as the low solubility or miscibility of water and oil molecules, micellization, protein folding, strong adhesion and rapid coagulation of hydrophobic surfaces, non-wetting of water on hydrophobic surfaces, and hydrophobic particle attachment to rising air bubbles (the basic principle of froth flotation).

In recent years, there has been a steady accumulation of experimental data on the force laws between various hydrophobic surfaces in aqueous solution [131–143]. These studies have found that the hydrophobic force law between two macroscopic surfaces is of surprisingly long range, decaying exponentially with a characteristic decay length of 1 to 2 nm in the separation range of 0 to 10 nm, and then more gradually further out. The hydrophobic force can be far stronger than the van der Waals attraction, especially between hydrocarbon surfaces in water, for which the Hamaker constant is quite small. The magnitude of the hydrophobic attraction has been found to decrease with the decreasing hydrophobicity (increasing hydrophilicity) of lecithin lipid bilayer surfaces [31] and silanated surfaces [139], whereas examples of the opposite trend have been shown for some Langmuir–Blodgett-deposited monolayers [144].

For two surfaces in water the purely hydrophobic interaction energy (ignoring DLVO and oscillatory forces) in the range of 0 to 10 nm is given by

$$E = -2\gamma e^{-D/\lambda_0} , \qquad (9.18)$$

where typically $\lambda_0 = 1$–2 nm, and $\gamma = 10$–50 mJ m^{-2}. The higher value corresponds to the interfacial energy of a pure hydrocarbon–water interface.

At a separation below 10 nm, the hydrophobic force appears to be insensitive or only weakly sensitive to changes in the type and concentration of electrolyte ions in the solution. The absence of a "screening" effect by ions attests to the nonelectrostatic origin of this interaction. In contrast, some experiments have shown that at separations greater than 10 nm, the attraction does depend on the intervening electrolyte, and that in dilute solutions, or solutions containing divalent ions, it can continue to exceed the van der Waals attraction out to separations of 80 nm [136, 145].

The long-range nature of the hydrophobic interaction has a number of important consequences. It accounts for the rapid coagulation of hydrophobic particles in water and may also account for the rapid folding of proteins. It also explains the ease with which water films rupture on hydrophobic surfaces. In this case, the van der Waals force across the water film is repulsive and therefore favors wetting, but this is more than offset by the attractive hydrophobic interaction acting between the two hydrophobic phases across water. Hydrophobic forces are increasingly being implicated in the adhesion and fusion of biological membranes and cells. It is known that both osmotic and electric field stresses enhance membrane fusion, an effect that may be due to the concomitant increase in the hydrophobic area exposed between two adjacent surfaces.

From the previous discussion we can infer that hydration and hydrophobic forces are not of a simple nature. These interactions are probably the most important, yet the least understood of all the forces in aqueous solutions. The unusual properties of water and the nature of the surfaces (including their homogeneity and stability) appear to be equally important. Some particle surfaces can have their hydration forces regulated, for example, by ion exchange. Others appear to be intrinsically hydrophilic (e.g., silica) and cannot be coagulated by changing the ionic condition, but can be rendered hydrophobic by chemically modifying their surface groups. For example, on heating silica to above 600 °C, two adjacent surface silanol (–OH) groups release a water molecule and form a hydrophobic siloxane (–O–) group, whence the repulsive hydration force changes into an attractive hydrophobic force.

How do these exponentially decaying repulsive or attractive forces arise? Theoretical work and computer simulations [112, 114, 146, 147] suggest that the solvation forces in water should be purely oscillatory, whereas other theoretical studies [148–153] suggest a monotonically exponential repulsion or attraction, possibly superimposed on an oscillatory force. The latter is consistent with experimental findings, as shown in the inset to Fig. 9.7b, where it appears that the oscillatory force is simply additive with the monotonic hydration and DLVO forces, suggesting that these arise from essentially different mechanisms.

It is probable that the short-range hydration force between all smooth, rigid, or crystalline surfaces (e.g., mineral surfaces such as mica) has an oscillatory component. This may or may not be superimposed on a monotonic force due to image interactions [150] and/or structural or hydrogen bonding interactions [148, 149].

Like the repulsive hydration force, the origin of the hydrophobic force is still unknown. *Luzar* et al. [152] carried out a Monte Carlo simulation of the interaction between two hydrophobic surfaces across water at separations below 1.5 nm. They obtained a decaying oscillatory force superimposed on a monotonically attractive curve. In more recent work [154, 155], it has been suggested that hydrophobic surfaces generate a depleted region of water around them, and that a long-range attractive force due to depletion arises between two such surfaces.

It is questionable whether the hydration or hydrophobic force should be viewed as an ordinary type of solvation or structural force that is reflecting the packing of water molecules. The energy (or entropy) associated with the hydrogen bonding network, which extends over a much larger region of space than the molecular correlations, is probably at the root of the long-range interactions of water. The situation in water appears to be governed by much more than the molecular packing effects that dominate the interactions in simpler liquids.

9.4.5 Polymer-Mediated Forces

Polymers or macromolecules are chain-like molecules consisting of many identical segments (monomers or repeating units) held together by covalent bonds. The size of a polymer coil in solution or in the melt is determined by a balance between van der Waals attraction (and hydrogen bonding, if present) between polymer segments, and the entropy of mixing, which causes the polymer coil to expand. In polymer melts above the glass transition temperature, and at certain conditions in solution, the attraction between polymer segments is exactly balanced by the entropy effect. The polymer solution will then behave virtually ideally, and the density distribution of segments in the coil is Gaussian. This is called the theta (θ) condition, and it occurs at the theta or Flory temperature for a particular combination of polymer and solvent or solvent mixture. At lower temperatures (in a poor or bad solvent), the polymer–polymer interactions dominate over the entropic, and the coil will shrink or precipitate. At higher temperatures (good solvent conditions), the polymer coil will be expanded.

High molecular weight polymers form large coils, which significantly affect the properties of a solution even when the total mass of polymer is very low. The radius of the polymer coil is proportional to the segment length, a, and the number of segments, n. At theta conditions, the hydrodynamic radius of the polymer coil (the root-mean-square separation of the ends of one polymer chain) is theoretically given by $R_h = a\,n^{1/2}$, and the unperturbed radius of gyration (the average root-mean-square distance of a segment from the center of mass of the molecule) is $R_g = a\,(n/6)^{1/2}$. In a good solvent the perturbed size of the polymer coil, the Flory radius R_F, is proportional to $n^{3/5}$.

Polymers interact with surfaces mainly through van der Waals and electrostatic interactions. The physisorption of polymers containing only one type of segment is reversible and highly dynamic, but the rate of exchange of adsorbed chains with free chains in the solution is low, since the polymer remains bound to the surface as long as one segment along the chain is adsorbed. The adsorption energy per segment is on the order of $k_B T$. In a good solvent, the conformation of a polymer on a surface is very different from the coil conformation in bulk solution. Polymers adsorb in "trains", separated by "loops" extending into solution and dangling "tails" (the ends of the chain). Compared to adsorption at lower temperatures, good solvent conditions favor more of the polymer chain being in the solvent, where it can attain its optimum conformation. As a result, the extension of the polymer is longer, even though the total amount of adsorbed polymer is lower. In a good solvent, the polymer chains can also be effectively repelled from a surface, if the loss in conformational entropy close to the surface is not compensated for by a gain in enthalpy from adsorption of segments. In this case, there will be a layer of solution (thickness $\approx R_g$) close to the surfaces that is depleted of polymer.

The interaction forces between two surfaces across a polymer solution will depend on whether the polymer is adsorbing onto the surfaces or is repelled from them, and also on whether the interaction occurs at "true" or "restricted" thermodynamic equilibrium. At true or full equilibrium, the polymer between the surfaces can equilibrate (exchange) with polymer in the bulk solution at all surface separations. Some theories [156, 157] predict that at full equilibrium, the polymer chains would move from the confined gap into the bulk solution where they could attain entropically more favorable conformations, and that a monotonic attraction at all distances would result from bridging and depletion interactions (which will be discussed below). Other theories suggest that the interaction at small separations would be ultimately repulsive, since some polymer chains would remain in the gap due to their attractive interactions with many sites on the surface (enthalpic) – more sites would be available to the remaining polymer chains if some others desorbed and diffused out from the gap [64, 158, 159].

At restricted equilibrium, the polymer is kinetically trapped, and the adsorbed amount is thus constant as the surfaces are brought toward each other, but the chains can still rearrange on the surfaces and in the gap. Experimentally, the true equilibrium situation is very difficult to attain, and most experiments are done at restricted equilibrium conditions. Even the equilibration of conformations assumed in theoretical models for restricted equilibrium conditions can be so slow that this condition is difficult to reach.

In systems of adsorbing polymer, bridging of chains from one surface to the other can give rise to a long-range attraction, since the bridging chains would gain conformational entropy if the surfaces were closer together. In poor solvents, both bridging and intersegment interactions contribute to an attraction [26]. However, regardless of solvent and equilibrium conditions, a strong repulsion due to the osmotic interactions is seen as small surface separations in systems of adsorbing polymers at restricted equilibrium.

In systems containing high concentrations of non-adsorbing polymer, the difference in solute concentration in the bulk and between the surfaces at separations smaller than the approximate polymer coil diameter ($2R_g$, i.e., when the polymer has been squeezed out from the gap between the surfaces) may give rise to an attractive osmotic force ("depletion attraction") [160–164]. In addition, if the polymer coils become initially compressed as the surfaces approach each other, this can give rise to a repulsion ("depletion stabilization") at large separations [163]. For a system of two cylindrical surfaces or radius R, the maximum depletion force, F_{dep}, is expected to occur when the surfaces are in contact and is given by multiplying the depletion (osmotic) pressure, $P_{dep} = \rho k_B T$, by the contact area πr^2, where r is given by the chord theorem: $r^2 = (2R - R_g)R_g \approx 2RR_g$ [2]:

$$F_{dep}/R = -2\pi R_g \rho k_B T, \qquad (9.19)$$

where ρ is the number density of the polymer in the bulk solution.

If a part of the polymer (typically an end group) is different from the rest of the chain, this part may preferentially adsorb to the surface. End-adsorbed polymer is attached to the surface at only one point, and the extension of the chain is dependent on the grafting density, i.e., the average distance, s, between adsorbed end groups on the surface (Fig. 9.8). One distinguishes between different regions of increasing overlap of the chains (stretching) called pancake, mushroom, and brush regimes [165]. In the mushroom regime, where the coverage is sufficiently low so that there is no overlap between neighboring chains, the thickness of the adsorbed layer is proportional to $n^{1/2}$ (i.e., to R_g) at theta conditions and to $n^{3/5}$ in a good solvent.

Several models [165, 170–174] have been developed for the extension and interactions between two brushes (strongly stretched grafted chains). They are based on a balance between osmotic pressure within the brush layers (uncompressed and compressed) and elastic energy of the chains and differ mainly in the assumptions of the segment density profile, which can be a step function or parabolic. At high coverage (in the brush regime), where the chains will avoid overlapping each other, the thickness of the layer is proportional to n.

Experimental work on both monodisperse [27, 28, 166, 175] and polydisperse [30] systems at different solvent conditions has confirmed the expected range and magnitude of the repulsive interactions resulting from compression of densely packed grafted layers.

9.4.6 Thermal Fluctuation Forces

If a surface is not rigid but very soft or even fluid-like, this can act to smear out any oscillatory solvation force. This is because the thermal fluctuations of such interfaces make them dynamically "rough" at any instant, even though they may be perfectly smooth on a time average. The types of surfaces that fall into this category are fluid-like amphiphilic surfaces of micelles, bilayers, emulsions, soap films, etc., but also solid colloidal particle surfaces that are coated with surfactant monolayers, as occurs in lubricating oils, paints, toners, etc.

These thermal fluctuation forces (also called entropic or steric forces) are usually short-range and repulsive and are very effective at stabilizing the attractive van der Waals forces at some small but finite separation. This can reduce the adhesion energy or force by up to three orders of magnitude. It is mainly for this reason that fluid-like micelles and bilayers, biological membranes, emulsion droplets, or gas bubbles adhere to each other only very weakly.

Because of their short range it was, and still is, commonly believed that these forces arise from water ordering or "structuring" effects at surfaces, and that they reflect some unique or characteristic property of water. However, it is now known that these repulsive forces also exist in other liquids. Moreover, they appear to become stronger with increasing temperature, which is unlikely if the force would originate from molecular ordering effects at surfaces. Recent experiments, theory, and computer simulations [176–178] have shown that these repulsive forces have an entropic origin arising from the osmotic repulsion between exposed thermally mobile surface groups once these overlap in a liquid. These phenomena include undulating and peristaltic forces between membranes or bilayers, and, on the molecular scale, protrusion and head group overlap forces where the interactions also are influenced by hydration forces.

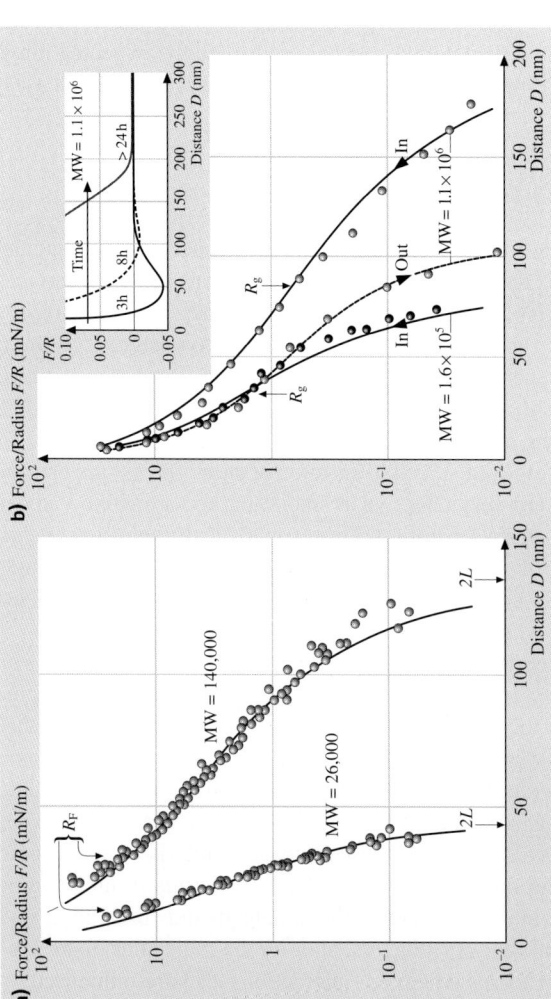

Fig. 9.8. Experimentally determined forces in systems of two interacting polymer brushes: (**a**) Polystyrene brush layers grafted via an adsorbing chain-end group onto mica surfaces in toluene (a good solvent for polystyrene). *Left curve*: MW = 26 000 g/mol, R_F = 12 nm. *Right curve*: MW = 140 000 g/mol, R_F = 32 nm. Both force curves were reversible on approach and separation. The *solid curves* are theoretical fits using the Alexander–de Gennes theory with the following measured parameters: spacing between attachments sites: s = 8.5 nm, brush thickness: L = 22.5 nm and 65 nm, respectively. (Adapted from [166]). (**b**) Polyethylene oxide layers physisorbed onto mica from 150 μg/ml solution in aqueous 0.1 M KNO$_3$ (a good solvent for polyethylene oxide). *Main figure*: Equilibrium forces at full coverage after ∼ 16 h adsorption time. *Left curve*: MW = 160 000 g/mol, R_g = 32 nm. *Right curve*: MW = 1 100 000 g/mol, R_g = 86 nm. Note the hysteresis (irreversibility) on approach and separation for this *physisorbed* layer, in contrast to the absence of hysteresis with *grafted* chains in case (**a**). The *solid curves* are based on a modified form of the Alexander–de Gennes theory. *Inset* in (**b**): Evolution of the forces with the time allowed for the higher MW polymer to adsorb from solution. Note the gradual reduction in the attractive bridging component. (Adapted from [167–169]. After [2]. Copyright 1991, with permission from Elsevier Science)

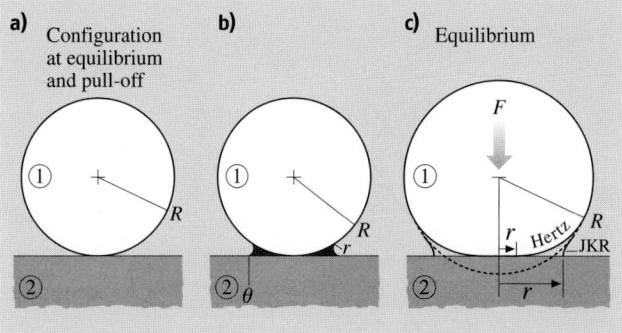

Fig. 9.9. Adhesion and capillary forces: (**a**) a non-deforming sphere on a rigid, flat surface in an inert atmosphere and (**b**) in a vapor that can "capillary condense" around the contact zone. At equilibrium, the concave radius, r, of the liquid meniscus is given by the Kelvin equation. For a concave meniscus to form, the contact angle θ has to be less than 90°. In the case of hydrophobic surfaces surrounded by water, a vapor cavity can form between the surfaces. As long as the surfaces are perfectly smooth, the contribution of the meniscus to the adhesion force is independent of r. (After [10] with permission.) (**c**) Elastically deformable sphere on a rigid flat surface in the absence (Hertz) and presence (JKR) of adhesion [(**a**) and (**c**) after [2]. Copyright 1991, with permission from Elsevier Science]

9.5 Adhesion and Capillary Forces

9.5.1 Capillary Forces

When considering the adhesion of two solid surfaces or particles in air or in a liquid, it is easy to overlook or underestimate the important role of capillary forces, i.e., forces arising from the Laplace pressure of curved menisci formed by condensation of a liquid between and around two adhering surfaces (Fig. 9.9).

The adhesion force between a non-deformable spherical particle of radius R and a flat surface in an inert atmosphere (Fig. 9.9a) is

$$F_s = 4\pi R \gamma_{SV} . \tag{9.20}$$

But in an atmosphere containing a condensable vapor, the expression above is replaced by

$$F_s = 4\pi R (\gamma_{LV} \cos\theta + \gamma_{SL}) , \tag{9.21}$$

where the first term is due to the Laplace pressure of the meniscus and the second is due to the direct adhesion of the two contacting solids within the liquid. Note that the above equation does not contain the radius of curvature, r, of the liquid meniscus (see Fig. 9.9b). This is because for smaller r the Laplace pressure γ_{LV}/r increases, but the area over which it acts decreases by the same amount, so the two effects cancel out. Experiments with inert liquids, such as hydrocarbons, condensing between two mica surfaces indicate that (9.21) is valid for values of r as small as 1–2 nm,

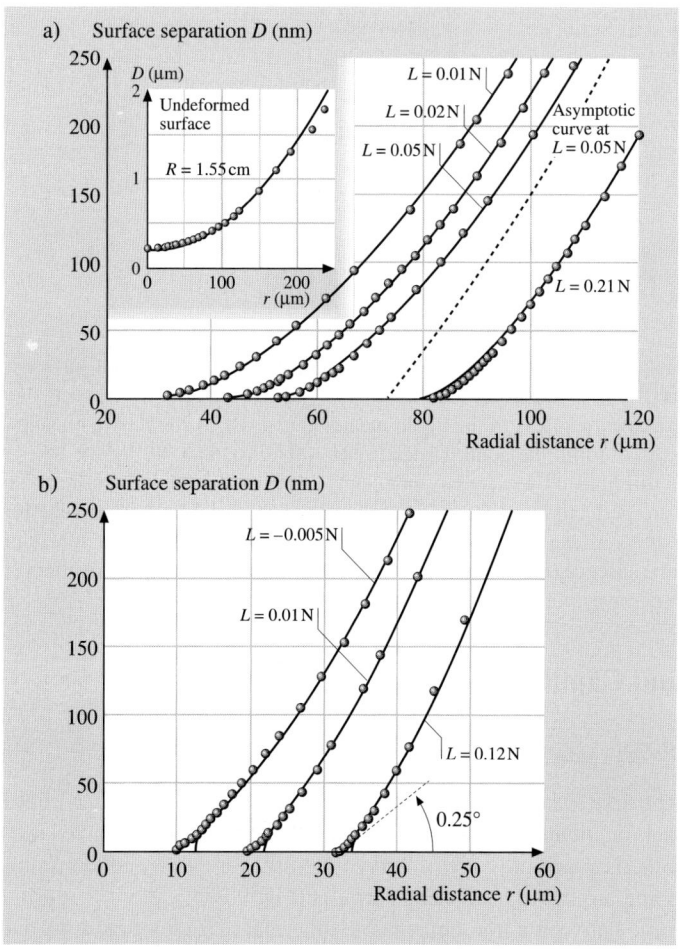

Fig. 9.10. Experimental and computer simulation data on contact mechanics for ideal Hertz and JKR contacts. (**a**) Measured profiles of surfaces in nonadhesive contact (*circles*) compared with Hertz profiles (*continuous curves*). The system was mica surfaces in a concentrated KCl solution in which they do not adhere. When not in contact, the surface shape is accurately described by a sphere of radius $R = 1.55$ cm (*inset*). The applied loads were 0.01, 0.02, 0.05, and 0.21 N. The last profile was measured in a different region of the surfaces where the local radius of curvature was 1.45 cm. The Hertz profiles correspond to central displacements of $\delta = 66.5$, 124, 173, and 441 nm. The *dashed line* shows the shape of the undeformed sphere corresponding to the curve at a load of 0.05 N; it fits the experimental points at larger distances (not shown). (**b**) Surface profiles measured with adhesive contact (mica surfaces adhering in dry nitrogen gas) at applied loads of -0.005, 0.01, and 0.12 N. The continuous lines are JKR profiles obtained by adjusting the central displacement in each case to get the best fit to points at larger distances. The values are $\delta = -4.2$, 75.6, and 256 nm. Note that the scales of this figure exaggerate the apparent angle at the junction of the surfaces. This angle, which is insensitive to load, is only about $0.25°$

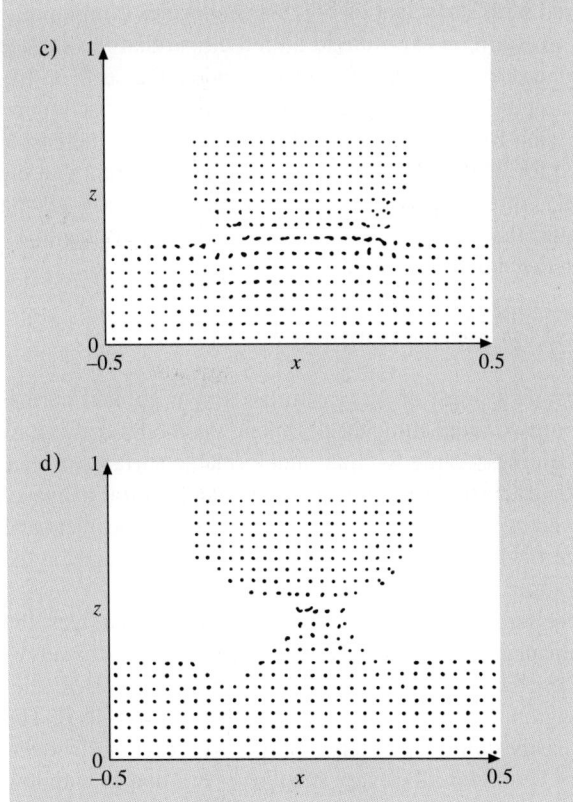

Fig. 9.10. (c) and (d) Molecular dynamics simulation illustrating the formation of a connective neck between an Ni tip (topmost eight layers) and an Au substrate. The figures show the atomic configuration in a slice through the system at indentation (c) and during separation (d). Note the crystalline structure of the neck. Distances are given in units of x and z, where $x = 1$ and $z = 1$ correspond to 6.12 nm. [(a) and (b) after [179]. Copyright 1987, with permission from Elsevier Science. (c) and (d) after [91], with kind permission from Kluwer Academic Publishers]

corresponding to vapor pressures as low as 40% of saturation [121, 180, 181]. Capillary condensation also occurs in binary liquid systems, e.g., when water dissolved in hydrocarbon liquids condense around two contacting hydrophilic surfaces or when a vapor cavity forms in water around two hydrophobic surfaces. In the case of water condensing from vapor or from oil, it also appears that the bulk value of γ_{LV} is applicable for meniscus radii as small as 2 nm.

The capillary condensation of liquids, especially water, from vapor can have additional effects on the physical state of the contact zone. For example, if the surfaces contain ions, these will diffuse and build up within the liquid bridge, thereby changing the chemical composition of the contact zone, as well as influencing the adhesion.

In the case of surfaces covered with surfactant or polymer molecules (amphiphilic surfaces), the molecules can overturn on exposure to humid air, so that the surface nonpolar groups become replaced by polar groups, which renders the surfaces hydrophilic. When two such surfaces come into contact, water will condense around the contact zone and the adhesion force will also be affected – generally increasing well above the value expected for inert hydrophobic surfaces. It is apparent that the adhesion in vapor or a solvent is often largely determined by capillary forces arising from the condensation of liquid that may be present only in very small quantities, e.g., 10–20 % of saturation in the vapor, or 20 ppm in the solvent.

9.5.2 Adhesion Mechanics

Two bodies in contact deform as a result of surface forces and/or applied normal forces. For the simplest case of two interacting elastic spheres (a model that is easily extended to an elastic sphere interacting with an undeformable surface (or vice versa)) and in the absence of attractive surface forces, the vertical central displacement (compression) was derived by *Hertz* [182]. In this model, the displacement and the contact area are equal to zero when no external force (load) is applied, i.e., at the points of contact and of separation.

In systems where attractive surface forces are present between the surfaces, the deformations are more complicated. Modern theories of the adhesion mechanics of two contacting solid surfaces are based on the Johnson–Kendall–Roberts (JKR) theory [15, 183], or on the Derjaguin–Muller–Toporov (DMT) theory [184–186]. The JKR theory is applicable to easily deformable, large bodies with high surface energy, whereas the DMT theory better describes very small and hard bodies with low surface energy [187].

In the JKR theory, two spheres of radii R_1 and R_2, bulk elastic modulus K, and surface energy γ will flatten due to attractive surface forces when in contact at no external load. The contact area will increase under an external load or normal force F, such that at mechanical equilibrium the radius of the contact area, r, is given by

$$r^3 = \frac{R}{K}\left[F + 6\pi R\gamma + \sqrt{12\pi R\gamma F + (6\pi R\gamma)^2}\right], \quad (9.22)$$

where $R = R_1 R_2/(R_1 + R_2)$. In the absence of surface energy, γ, equation (9.22) is reduced to the expression for the radius of the contact area in the Hertz model. Another important result of the JKR theory gives the adhesion force or "pull-off" force:

$$F_S = -3\pi R\gamma_S, \quad (9.23)$$

where the surface energy, γ_S, is defined as $W = 2\gamma_S$, where W is the reversible work of adhesion. Note that according totheJKR theory a finite elastic modulus, K, while having an effect on the load–area curve, has no effect on the adhesion force, an interesting and unexpected result that has nevertheless been verified experimentally [15, 179, 188, 189].

Equation (9.22) and (9.23) provide the framework for analyzing results of adhesion measurements of contacting solids, known as contact mechanics [183, 190], and for studying the effects of surface conditions and time on adhesion energy hysteresis (see Sect. 9.5.4).

9.5.3 Effects of Surface Structure, Roughness, and Lattice Mismatch

In a contact between two rough surfaces, the real area of contact varies with the applied load in a different manner than between smooth surfaces [191, 192]. For a Hertzian contact, it has been shown that the contact area for rough surfaces increases approximately linearly with the applied normal force (load), F, instead of as $F^{2/3}$ for smooth surfaces. In systems with attractive surface forces, there is a competition between this attraction and repulsive forces arising from compression of high asperities. As a result, the adhesion in such systems can be very low, especially if the surfaces are not easily deformed [193–195]. The opposite is possible for soft (viscoelastic) surfaces where the real (molecular) contact area might be larger than for two perfectly smooth surfaces.

Adhesion forces may also vary depending on the commensurability of the crystallographic lattices of the interacting surfaces. *McGuiggan* and *Israelachvili* [196] measured the adhesion between two mica surfaces as a function of the orientation (twist angle) of their surface lattices. The forces were measured in air, water, and an aqueous salt solution where oscillatory structural forces were present. In humid air, the adhesion was found to be relatively independent of the twist angle θ due to the adsorption of a 0.4-nm-thick amorphous layer of organics and water at the interface. In contrast, in water, sharp adhesion peaks (energy minima) occurred at $\theta = 0°$, $\pm 60°$, $\pm 120°$ and $180°$, corresponding to the "coincidence" angles of the surface lattices (Fig. 9.11). As little as $\pm 1°$ away from these peaks, the energy decreased by 50%. In aqueous KCl solution, due to potassium ion adsorption the water between the surfaces becomes ordered, resulting in an oscillatory force profile where the adhesive minima occur at discrete separations of about 0.25 nm, corresponding to integral numbers of water layers. The whole interaction potential was now found to depend on the orientation of the surface lattices, and the effect extended at least four molecular layers.

It has also been appreciated that the structure of the confining surfaces is just as important as the nature of the liquid for determining the solvation forces [90, 122, 123, 198–202]. Between two surfaces that are completely flat but "unstructured", the liquid molecules will order into layers, but there will be no lateral ordering within the layers. In other words, there will be positional ordering normal but not parallel to the surfaces. If the surfaces have a crystalline (periodic) lattice, this may induce ordering parallel to the surfaces, as well, and the oscillatory force then also depends on the structure of the surface lattices. Further, if the two lattices have different dimensions ("mismatched" or "incommensurate" lattices), or if the lattices are similar but are not in register relative to each other, the oscillatory force law is further modified [196, 203] and the tribological properties of the film are also influenced, as discussed in Sect. 9.9 [203, 204].

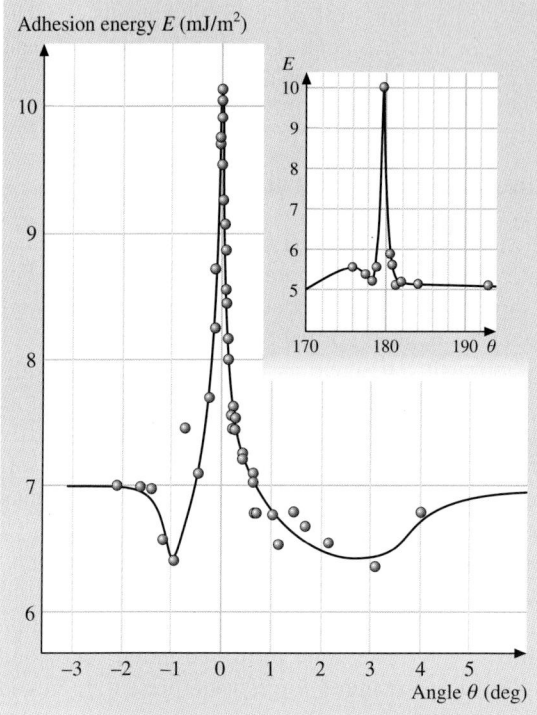

Fig. 9.11. Adhesion energy for two mica surfaces in contact in water (in the primary minimum of an oscillatory force curve) as a function of the mismatch angle θ about $\theta = 0°$ and $180°$ between the mica surface lattices. (After [197] with permission)

As shown by the experiments, these effects can alter the magnitude of the adhesive minima found at a given separation within the last one or two nanometers of a thin film by a factor of two. The force barriers (maxima) may also depend on orientation. This could be even more important than the effects on the minima. A high barrier could prevent two surfaces from coming closer together into a much deeper adhesive well. Thus the maxima can effectively contribute to determining not only the final separation of two surfaces, but also their final adhesion. Such considerations should be particularly important for determining the thickness and strength of intergranular spaces in ceramics, the adhesion forces between colloidal particles in concentrated electrolyte solution, and the forces between two surfaces in a crack containing capillary condensed water.

For surfaces that are *randomly* rough, oscillatory forces become smoothed out and disappear altogether, to be replaced by a purely monotonic solvation force [108, 122, 123]. This occurs even if the liquid molecules themselves are perfectly capable of ordering into layers. The situation of *symmetric* liquid molecules confined between *rough* surfaces is therefore not unlike that of *asymmetric* molecules between *smooth* surfaces (see Sect. 9.4.3 and Fig. 9.7a). To summarize, for there to be an oscillatory solvation force, the liquid molecules must be able to be correlated over a reasonably long range. This requires that both the liquid molecules and the surfaces have a high degree of order or symmetry. If either is missing, so will the oscillations. A roughness of only a few tenths of a nanometer is often sufficient to eliminate any oscillatory component of the force law.

9.5.4 Nonequilibrium and Rate-Dependent Interactions: Adhesion Hysteresis

Under ideal conditions the adhesion energy is a well-defined thermodynamic quantity. It is normally denoted by E or W (the work of adhesion) or γ (the surface tension, where $W = 2\gamma$) and gives the reversible work done on bringing two surfaces together or the work needed to separate two surfaces from contact. Under ideal, equilibrium conditions these two quantities are the same, but under most realistic conditions they are not: The work needed to separate two surfaces is always greater than that originally gained by bringing them together. An understanding of the molecular mechanisms underlying this phenomenon is essential for understanding many adhesion phenomena, energy dissipation during loading–unloading cycles, contact angle hysteresis, and the molecular mechanisms associated with many frictional processes.

It is wrong to think that hysteresis arises because of some imperfection in the system such as rough or chemically heterogeneous surfaces, or because the supporting material is viscoelastic. Adhesion hysteresis can arise even between perfectly smooth and chemically homogenous surfaces supported by perfectly elastic materials. It can be responsible for such phenomena as rolling friction and elastoplastic adhesive contacts [190, 205–208] during loading–unloading and adhesion–decohesion cycles.

Adhesion hysteresis may be thought of as being due to mechanical effects such as instabilities, or chemical effects such as interdiffusion, interdigitation, molecular reorientations and exchange processes occurring at an interface after contact, as illustrated in Fig. 9.12. Such processes induce roughness and chemical heterogeneity even though initially (and after separation and re-equilibration) both surfaces are perfectly smooth and chemically homogeneous. In general, if the energy change, or work done, on separating two surfaces from adhesive contact is not fully recoverable on bringing the two surfaces back into contact again, the adhesion hysteresis may be expressed as

$$W_R > W_A$$
Receding   Advancing

or

$$\Delta W = (W_R - W_A) > 0, \qquad (9.24)$$

where W_R and W_A are the adhesion or surface energies for receding (separating) and advancing (approaching) two solid surfaces, respectively.

Hysteresis effects are also commonly observed in wetting/dewetting phenomena [213]. For example, when a liquid spreads and then retracts from a surface the

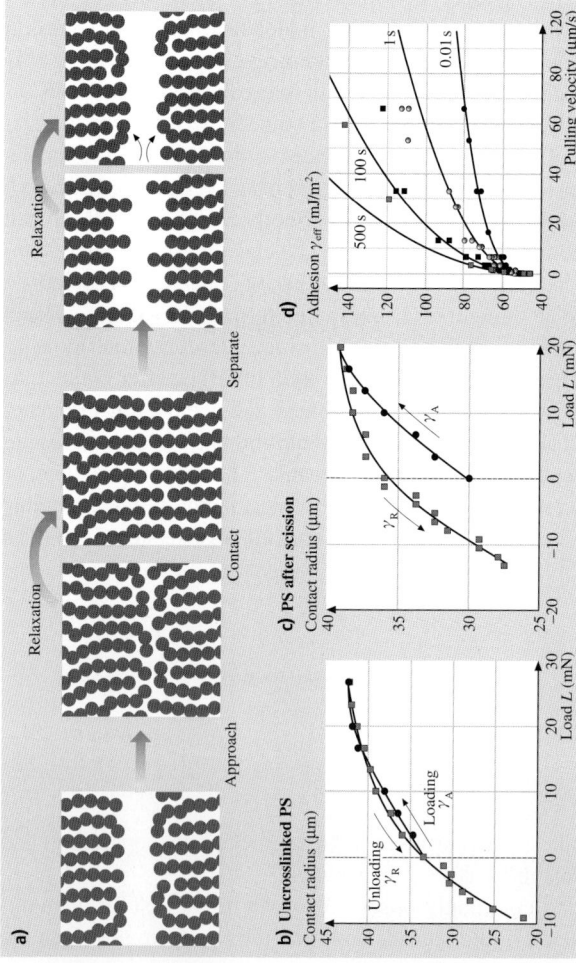

Fig. 9.12. (**a**) Schematic representation of interpenetrating chains. (**b**) and (**c**): JKR plots (contact radius r as a function of applied load L) showing small adhesion hysteresis for uncrosslinked polystyrene and larger adhesion hysteresis after chain scission at the surfaces after 18 h irradiation with ultraviolet light in an oxygen atmosphere. The adhesion hysteresis continues to increase with the irradiation time. (**d**) Rate-dependent adhesion of CTAB surfactant monolayers. The *solid curves* [209] are fits to experimental data on CTAB adhesion after different contact times [210] using an approximate analytical solution for a JKR model, including crack tip dissipation. Due to the limited range of validity of the approximation, the fits rely on the low effective adhesion energy part of the experimental data only. From the fits one can determine the thermodynamic adhesion energy, the characteristic dissipation velocity, and the intrinsic dissipation exponent of the model. [(**a**) after [211]. Copyright 1993 American Chemical Society. (**b**) and (**c**) after [212]. Copyright 2002 American Association for the Advancement of Science. (**d**) after [209]. Copyright 2000 American Chemical Society]

advancing contact angle θ_A is generally larger than the receding angle θ_R. Since the contact angle, θ, is related to the liquid–vapor surface tension, γ_L, and the solid–liquid adhesion energy, W, by the Dupré equation,

$$(1 + \cos\theta)\gamma_L = W, \qquad (9.25)$$

we see that *wetting hysteresis* or *contact angle hysteresis* ($\theta_A > \theta_R$) actually implies adhesion hysteresis, $W_R > W_A$, as given by (9.24).

Energy dissipating processes such as adhesion and contact angle hysteresis arise because of practical constraints of the *finite time* of measurements and the *finite elasticity* of materials. This prevents many loading–unloading or approach–separation cycles from being thermodynamically reversible, even though if carried out infinitely slowly they would be. By thermodynamically irreversible one simply means that one cannot go through the approach–separation cycle via a continuous series of equilibrium states, because some of these are connected via spontaneous – and therefore thermodynamically irreversible – instabilities or transitions where energy is liberated and therefore "lost" via heat or phonon release [214]. This is an area of much current interest and activity, especially regarding the fundamental molecular origins of adhesion and friction in polymer and surfactant systems, and the relationships between them [190, 206, 209, 210, 212, 215–218].

9.6 Introduction: Different Modes of Friction and the Limits of Continuum Models

Most frictional processes occur with the sliding surfaces becoming damaged in one form or another [205]. This may be referred to as "normal" friction. In the case of brittle materials, the damaged surfaces slide past each other while separated by relatively large, micron-sized wear particles. With more ductile surfaces, the damage remains localized to nanometer-sized, plastically deformed asperities. Some features of the friction between damaged surfaces will be described in Sect. 9.7.4.

There are also situations in which sliding can occur between two perfectly smooth, undamaged surfaces. This may be referred to as "interfacial" sliding or "boundary" friction and is the focus of the following sections. The term "boundary lubrication" is more commonly used to denote the friction of surfaces that contain a thin protective lubricating layer such as a surfactant monolayer, but here we shall use the term more broadly to include any molecularly thin solid, liquid, surfactant, or polymer film.

Experiments have shown that as a liquid film becomes progressively thinner, its physical properties change, at first quantitatively and then qualitatively [44, 47, 219–222]. The quantitative changes are manifested by an increased viscosity, non-Newtonian flow behavior, and the replacement of normal melting by a glass transition, but the film remains recognizable as a liquid (Fig. 9.13). In tribology, this regime is commonly known as the "mixed lubrication" regime, where the rheological properties of a film are intermediate between the bulk and boundary properties. One may also refer to it as the "intermediate" regime (Table 9.3).

Table 9.3. The three main tribological regimes characterizing the changing properties of liquids subjected to an increasing confinement between two solid surfaces[a]

Regime	Conditions for getting into this regime	Static/equilibrium properties[b]	Dynamic properties[c]
Bulk	• Thick films (> 10 molecular diameters, $\gg R_g$ for polymers) • Low or zero loads • High shear rates	Bulk (continuum) properties: • Bulk liquid density • No long-range order	Bulk (continuum) properties: • Newtonian viscosity • Fast relaxation times • No glass temperature • No yield point • Elastohydrodynamic lubrication
Intermediate mixed	• Intermediately thick films (4–10 molecular diameters, $\sim R_g$ for polymers) • Low loads or pressure	Modified fluid properties include: • Modified positional and orientational order[a] • Medium to long-range molecular correlations • Highly entangled states	Modified rheological properties include: • Non-Newtonian flow • Glassy states • Long relaxation times • Mixed lubrication
Boundary	• Molecularly thin films (< 4 molecular diameters) • High loads or pressure • Low shear rates • Smooth surfaces or asperities	Onset of non-fluidlike properties: • Liquid-like to solid-like phase transitions • Appearance of new liquid-crystalline states • Epitaxially induced long-range ordering	Onset of tribological properties: • No flow until yield point or critical shear stress reached • Solid-like film behavior characterized by defect diffusion, dislocation motion, shear melting • Boundary lubrication

Based on work by *Granick* [220], *Hu* and *Granick* [222], and others [38, 211, 223] on the dynamic properties of short chain molecules such as alkanes and polymer melts confined between surfaces

[a] Confinement can lead to an increased or decreased order in a film, depending both on the surface lattice structure and the geometry of the confining cavity

[b] In each regime both the static and dynamic properties change. The static properties include the film density, the density distribution function, the potential of mean force, and various positional and orientational order parameters

[c] Dynamic properties include viscosity, viscoelastic constants, and tribological yield points such as the friction coefficient and critical shear stress

For even thinner films, the changes in behavior are more dramatic, resulting in a qualitative change in properties. Thus first-order phase transitions can now occur to solid or liquid-crystalline phases [46, 201, 211, 223–227], whose properties can no longer be characterized even qualitatively in terms of bulk or continuum liquid

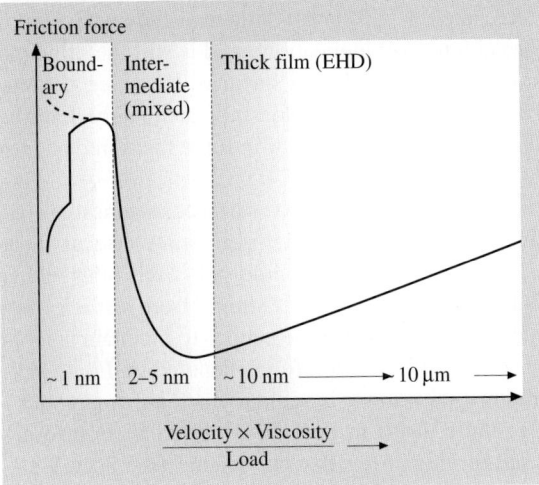

Fig. 9.13. Stribeck curve: an empirical curve giving the trend generally observed in the friction forces or friction coefficients as a function of sliding velocity, the bulk viscosity of the lubricating fluid, and the applied load. The three friction/lubrication regimes are known as the boundary lubrication regime (see Sect. 9.7), the intermediate or mixed lubrication regime (Sect. 9.8.2), and thick film or elastohydrodynamic (EHD) lubrication regime (Sect. 9.8.1). The film thicknesses believed to correspond to each of these regimes are also shown. For thick films, the "friction" force is purely viscous, e.g., Couette flow at low shear rates, but may become complicated at higher shear rates where EHD deformations of surfaces can occur during sliding. (After [10] with permission)

properties such as viscosity. These films now exhibit yield points (characteristic of fracture in solids) and their molecular diffusion and relaxation times can be ten orders of magnitude longer than in the bulk liquid or even in films that are just slightly thicker. The three friction regimes are summarized in Table 9.3.

9.7 Relationship Between Adhesion and Friction Between Dry (Unlubricated and Solid Boundary Lubricated) Surfaces

9.7.1 Amontons' Law and Deviations from It Due to Adhesion: The Cobblestone Model

Early theories and mechanisms for the dependence of friction on the applied normal force or load, L, were developed by *da Vinci*, *Amontons*, *Coulomb* and *Euler* [228]. For the macroscopic objects investigated, the friction was found to be directly proportional to the load, with no dependence on the contact area. This is described by the so-called Amontons' law:

$$F = \mu L, \tag{9.26}$$

where F is the shear or friction force and μ is a constant defined as the coefficient of friction. This friction law has a broad range of applicability and is still the principal means of quantitatively describing the friction between surfaces. However, particularly in the case of adhering surfaces, Amontons' law does not adequately describe the friction behavior with load, because of the finite friction force measured at zero and even negative applied loads.

When a lateral force, or shear stress, is applied to two surfaces in adhesive contact, the surfaces initially remain "pinned" to each other until some critical shear force is reached. At this point, the surfaces begin to slide past each other either smoothly or in jerks. The frictional force needed to initiate sliding from rest is known as the *static* friction force, denoted by F_s, while the force needed to maintain smooth sliding is referred to as the *kinetic* or *dynamic* friction force, denoted by F_k. In general, $F_s > F_k$. Two sliding surfaces may also move in regular jerks, known as stick–slip sliding, which is discussed in more detail in Sect. 9.8.3. Such friction forces cannot be described by models used for thick films that are viscous (see Sect. 9.8.1) and, therefore, shear as soon as the smallest shear force is applied.

Experimentally, it has been found that during both smooth and stick–slip sliding the local geometry of the contact zone remains largely unchanged from the static geometry. In an adhesive contact, the contact area as a function of load is thus generally well described by the JKR equation, (9.22). The friction force between two molecularly smooth surfaces sliding in adhesive contact is not simply proportional to the applied load, L, as might be expected from Amontons' law. There is an additional adhesion contribution that is proportional to the area of contact, A. Thus, in general, the interfacial friction force of dry, unlubricated surfaces sliding smoothly past each other in adhesive contact is given by

$$F = F_k = S_c A + \mu L, \tag{9.27}$$

where S_c is the "critical shear stress" (assumed to be constant), $A = \pi r^2$ is the contact area of radius r given by (9.22), and μ is the coefficient of friction. For low loads we have:

$$\begin{aligned} F &= S_c A = S_c \pi r^2 \\ &= S_c \pi \left[\frac{R}{K} \left(L + 6\pi R\gamma + \sqrt{12\pi R\gamma L + (6\pi R\gamma)^2} \right) \right]^{2/3}, \end{aligned} \tag{9.28}$$

whereas for high loads (or high μ), (9.27) reduces to Amontons' law: $F = \mu L$. Depending on whether the friction force in (9.27) is dominated by the first or second term, one may refer to the friction as *adhesion-controlled* or *load-controlled*, respectively.

The following friction model, first proposed by *Tabor* [229] and developed further by *Sutcliffe* et al. [230], *McClelland* [231], and *Homola* et al. [45], has been quite successful at explaining the interfacial and boundary friction of two solid crystalline surfaces sliding past each other in the absence of wear. The surfaces may be unlubricated, or they may be separated by a monolayer or more of some boundary lubricant or liquid molecules. In this model, the values of the critical shear stress S_c, and coefficient of friction μ, of (9.27) are calculated in terms of the energy needed to overcome the attractive intermolecular forces and compressive externally applied load as one surface is raised and then slid across the molecular-sized asperities of the other.

This model (variously referred to as the *interlocking asperity model*, *Coulomb friction*, or the *cobblestone model*) is similar to pushing a cart over a road of cobblestones where the cartwheels (which represent the molecules of the upper surface or film) must be made to roll over the cobblestones (representing the molecules of the lower surface) before the cart can move. In the case of the cart, the downward force of gravity replaces the attractive intermolecular forces between two material surfaces. When at rest, the cartwheels find grooves between the cobblestones where they sit in potential energy minima, and so the cart is at some stable mechanical equilibrium. A certain lateral force (the "push") is required to raise the cartwheels against the force of gravity in order to initiate motion. Motion will continue as long as the cart is pushed, and rapidly stops once it is no longer pushed. Energy is dissipated by the liberation of heat (phonons, acoustic waves, etc.) every time a wheel hits the next cobblestone. The cobblestone model is not unlike the *Coulomb* and *interlocking asperity* models of friction [228] except that it is being applied at the molecular level and for a situation where the external load is augmented by attractive intermolecular forces.

There are thus two contributions to the force pulling two surfaces together: the externally applied load or pressure, and the (internal) attractive intermolecular forces that determine the adhesion between the two surfaces. Each of these contributions affects the friction force in a different way, which we will discuss in more detail below.

9.7.2 Adhesion Force and Load Contribution to Interfacial Friction

Adhesion Force Contribution

Consider the case of two surfaces sliding past each other, as shown in Fig. 9.14. When the two surfaces are initially in adhesive contact, the surface molecules will adjust themselves to fit snugly together [232], in an analogous manner to the self-positioning of the cartwheels on the cobblestone road. A small tangential force applied to one surface will therefore not result in the sliding of that surface relative to the other. The attractive van der Waals forces between the surfaces must first be overcome by having the surfaces separate by a small amount. To initiate motion, let the separation between the two surfaces increase by a small amount ΔD, while the lateral distance moved is $\Delta \sigma$. These two values will be related via the geometry of the two surface lattices. The energy put into the system by the force F acting over a lateral distance $\Delta \sigma$ is

Input energy: $F \times \Delta \sigma$. (9.29)

This energy may be equated with the change in interfacial or surface energy associated with separating the surfaces by ΔD, i.e., from the equilibrium separation

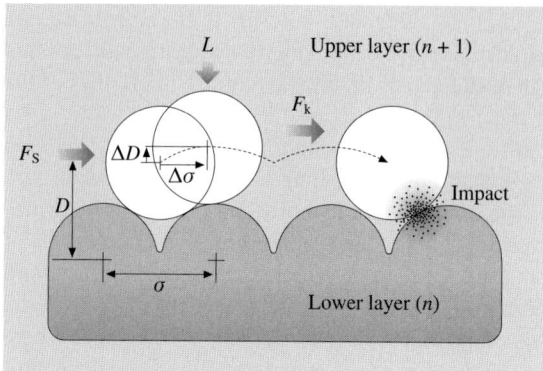

Fig. 9.14. Schematic illustration of how one molecularly smooth surface moves over another when a lateral force F is applied (the "cobblestone model"). As the upper surface moves laterally by some fraction of the lattice dimension $\Delta\sigma$, it must also move up by some fraction of an atomic or molecular dimension ΔD before it can slide across the lower surface. On impact, some fraction ε of the kinetic energy is "transmitted" to the lower surface, the rest being "reflected" back to the colliding molecule (upper surface). (After [233] with permission)

$D = D_0$ to $D = (D_0 + \Delta D)$. Since $\gamma \propto D^{-2}$ for two flat surfaces, the energy cost may be approximated by:

Surface energy change × area:

$$2\gamma A \left[1 - D_0^2/(D_0 + \Delta D)^2\right] \approx 4\gamma A (\Delta D/D_0), \qquad (9.30)$$

where γ is the surface energy, A the contact area, and D_0 the surface separation at equilibrium. During steady state sliding (kinetic friction), not all of this energy will be "lost" or absorbed by the lattice every time the surface molecules move by one lattice spacing: Some fraction will be reflected during each impact of the "cartwheel" molecules [231]. Assuming that a fraction ε of the above surface energy is "lost" every time the surfaces move across the characteristic length $\Delta\sigma$ (Fig. 9.14), we obtain after equating (9.29) and (9.30)

$$S_c = \frac{F}{A} = \frac{4\gamma\varepsilon\Delta D}{D_0 \Delta\sigma}. \qquad (9.31)$$

For a typical hydrocarbon or a van der Waals surface, $\gamma \approx 25$ mJ m^{-2}. Other typical values would be: $\Delta D \approx 0.05$ nm, $D_0 \approx 0.2$ nm, $\Delta\sigma \approx 0.1$ nm, and $\varepsilon \approx 0.1$–0.5. Using the above parameters, (9.31) predicts $S_c \approx (2.5$–$12.5) \times 10^7$ N m^{-2} for van der Waals surfaces. This range of values compares very well with typical experimental values of 2×10^7 N m^{-2} for hydrocarbon or mica surfaces sliding in air (see Fig. 9.16) or separated by one molecular layer of cyclohexane [45].

The above model suggests that all interfaces, whether dry or lubricated, dilate just before they shear or slip. This is a small but important effect: The dilation provides the crucial extra space needed for the molecules to slide across each other or

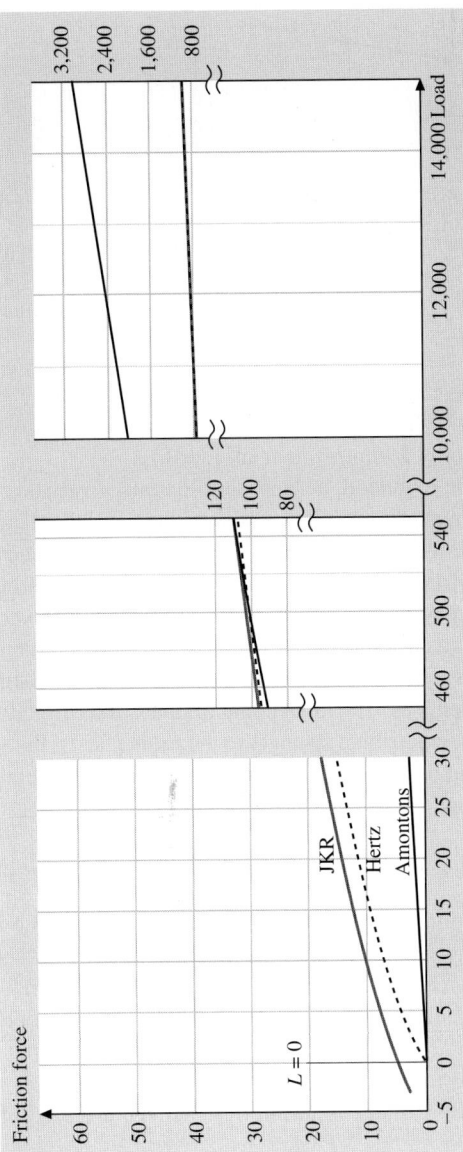

Fig. 9.15. Friction as a function of load for smooth surfaces. At low loads, the friction is dominated by the C_1A term of (9.37). The adhesion contribution (JKR curve) is most prominent near zero load where the Hertzian and Amontons' contributions to the friction are minimal. As the load increases, the adhesion contribution becomes smaller as the JKR and Hertz curves converge. In this range of loads, the linear C_2L contribution surpasses the area contribution to the friction. At much higher loads the explicit load dependence of the friction dominates the interactions, and the observed behavior approaches Amontons' law. It is interesting to note that for smooth surfaces the pressure over the contact area does not increase as rapidly as the load. This is because as the load is increased, the surfaces deform to increase the surface area and thus moderate the contact pressure. (After [234] with permission of Kluwer Academic Publishers)

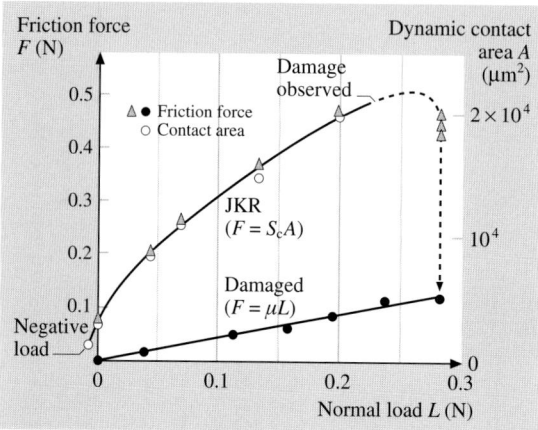

Fig. 9.16. Friction force F and contact area A vs. load L for two mica surfaces sliding in adhesive contact in dry air. The contact area is well described by the JKR theory, (9.22), even during sliding, and the friction force is found to be directly proportional to this area, (9.28). The *vertical dashed line* and *arrow* show the transition from interfacial to normal friction with the onset of wear (*lower curve*). Sliding velocity is 0.2 μm s^{-1}. (After [45] with permission, copyright 1989 American Society of Mechanical Engineers)

flow. This dilation has been computed by *Thompson* and *Robbins* [201] and *Zaloj* et al. [235] and measured by *Dhinojwala* et al. [236].

This model may be extended, at least semi-quantitatively, to lubricated sliding, where a thin liquid film is present between the surfaces. With an increase in the number of liquid layers between the surfaces, D_0 increases while ΔD decreases, hence the lower the friction force. This is precisely what is observed, but with more than one liquid layer between two surfaces the situation becomes too complex to analyze analytically (actually, even with one or no interfacial layers, the calculation of the fraction of energy dissipated per molecular collision ε is not a simple matter). Furthermore, even in systems as simple as linear alkanes, interdigitation and interdiffusion have been found to contribute strongly to the properties of the system [117, 237]. Sophisticated modeling based on computer simulations is now required, as described in the following section.

Relation Between Boundary Friction and Adhesion Energy Hysteresis

While the above equations suggest that there is a direct correlation between friction and adhesion, this is not the case. The correlation is really between friction and adhesion hysteresis, described in Sect. 9.5.4. In the case of friction, this subtle point is hidden in the factor ε, which is a measure of the amount of energy absorbed (dissipated, transferred, or "lost") by the lower surface when it is impacted by a molecule from the upper surface. If $\varepsilon = 0$, all the energy is reflected, and there will be no kinetic friction force nor any adhesion hysteresis, but the absolute magnitude of the adhesion force or energy will remain finite and unchanged. This is illustrated in Figs. 9.17 and 9.19.

The following simple model shows how adhesion hysteresis and friction may be quantitatively related. Let $\Delta\gamma = \gamma_R - \gamma_A$ be the adhesion energy hysteresis per unit area, as measured during a typical loading–unloading cycle (see Figs. 9.17a and 9.19c,d). Now consider the same two surfaces sliding past each other and assume that frictional energy dissipation occurs through the same mechanism as adhesion energy dissipation, and that both occur over the same characteristic molecular length scale σ. Thus, when the two surfaces (of contact area $A = \pi r^2$) move a distance σ, equating the frictional energy ($F \times \sigma$) to the dissipated adhesion energy ($A \times \Delta\gamma$), we obtain

$$\text{Friction force: } F = \frac{A \times \Delta\gamma}{\sigma} = \frac{\pi r^2}{\sigma}(\gamma_R - \gamma_A) \;, \tag{9.32}$$

or Critical shear stress: $S_c = F/A = \Delta\gamma/\sigma$, (9.33)

which is the desired expression and has been found to give order of magnitude agreement between measured friction forces and adhesion energy hysteresis [211]. If we equate (9.33) with (9.31), since $4\Delta D/(D_0\Delta\sigma) \approx 1/\sigma$, we obtain the intuitive relation

$$\varepsilon = \frac{\Delta\gamma}{\gamma} \;. \tag{9.34}$$

External Load Contribution to Interfacial Friction

When there is no interfacial adhesion, S_c is zero. Thus, in the absence of any adhesive forces between two surfaces, the only "attractive" force that needs to be overcome for sliding to occur is the externally applied load or pressure.

For a preliminary discussion of this question, it is instructive to compare the magnitudes of the *externally* applied pressure to the *internal* van der Waals pressure between two smooth surfaces. The internal van der Waals pressure between two flat surfaces is given (see Table 9.1) by $P = A_H/6\pi D_0^3 \approx 1$ GPa (10^4 atm), using a typical Hamaker constant of $A_H = 10^{-19}$ J, and assuming $D_0 \approx 2$ nm for the equilibrium interatomic spacing. This implies that we should not expect the externally applied load to affect the interfacial friction force F, as defined by (9.27), until the externally applied pressure L/A begins to exceed ~ 100 MPa (10^3 atm). This is in agreement with experimental data [239] where the effect of load became dominant at pressures in excess of 10^3 atm.

For a more general semiquantitative analysis, again consider the cobblestone model used to derive (9.31), but now include an additional contribution to the surface energy change of (9.30) due to the work done against the external load or pressure, $L\Delta D = P_{\text{ext}} A\Delta D$ (this is equivalent to the work done against gravity in the case of a cart being pushed over cobblestones). Thus:

$$S_c = \frac{F}{A} = \frac{4\gamma\varepsilon\Delta D}{D_0\Delta\sigma} + \frac{P_{\text{ext}}\varepsilon\Delta D}{\Delta\sigma} \;, \tag{9.35}$$

which gives the more general relation

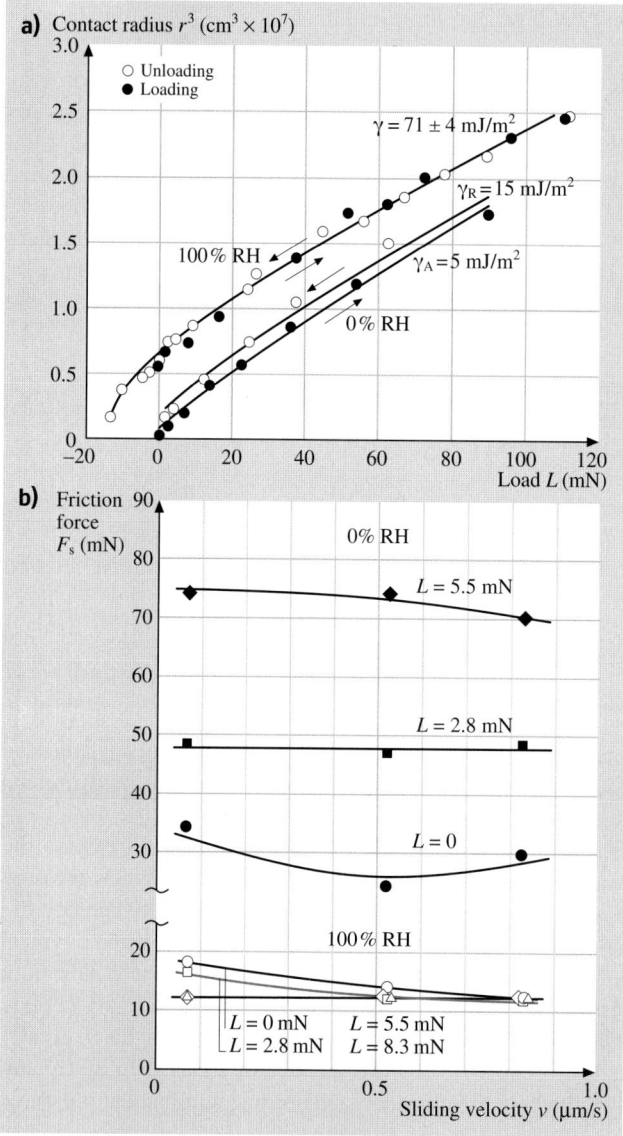

Fig. 9.17. (a) Contact radius r versus externally applied load L for loading and unloading of two hydrophilic silica surfaces exposed to dry and humid atmospheres. Note that while the adhesion is higher in humid air, the *hysteresis* in the adhesion is higher in dry air. (b) Effect of velocity on the static friction force F_s for hydrophobic (heat-treated electron beam evaporated) silica in dry and humid air. The effects of humidity, load, and sliding velocity on the friction forces, as well as the stick–slip friction of the hydrophobic surfaces, are qualitatively consistent with a "friction" phase diagram representation as in Fig. 9.28. (After [41]. Copyright 1994, with permission from Elsevier Science)

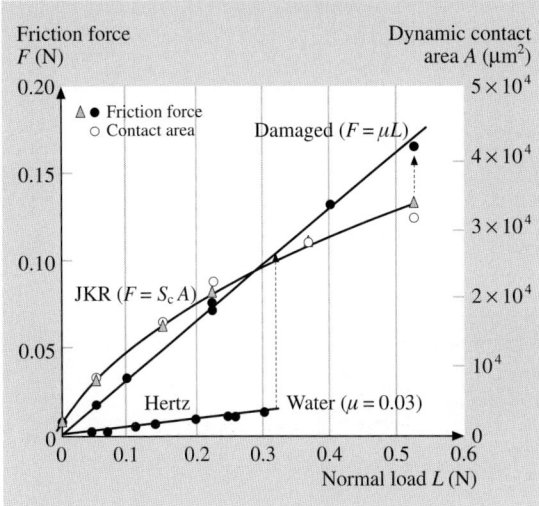

Fig. 9.18. Sliding of mica surfaces, each coated with a 2.5-nm-thick monolayer of calcium stearate surfactant, in the absence of damage (obeying JKR type boundary friction) and in the presence of damage (obeying Amontons' type normal friction). Note that both for this system and for the bare mica in Figs. 9.16 and 9.20, the friction force obeys Amontons' law with a friction coefficient of $\mu \approx 0.3$ after damage occurs. At much higher applied loads, the undamaged surfaces also follow Amontons' type sliding, but for a different reason: The dependence on adhesion becomes smaller. *Lower line:* Interfacial sliding with a monolayer of water between the mica surfaces, shown for comparison. (After [238]. Copyright 1990, with permission from Elsevier Science)

$$S_c = F/A = C_1 + C_2 P_{\text{ext}}, \tag{9.36}$$

where $P_{\text{ext}} = L/A$ and C_1 and C_2 are constants characteristic of the surfaces and sliding conditions. The constant $C_1 = 4\gamma\varepsilon\Delta D/(D_0\Delta\sigma)$ depends on the mutual adhesion of the two surfaces, while both C_1 and $C_2 = \varepsilon\Delta D/\Delta\sigma$ depend on the topography or atomic bumpiness of the surface groups (Fig. 9.14). The smoother the surface groups the smaller the ratio $\Delta D/\Delta\sigma$ and hence the lower the value of C_2. In addition, both C_1 and C_2 depend on ε (the fraction of energy dissipated per collision), which depends on the relative masses of the shearing molecules, the sliding velocity, the temperature, and the characteristic molecular relaxation processes of the surfaces. This is by far the most difficult parameter to compute, and yet it is the most important since it represents the energy transfer mechanism in any friction process, and since ε can vary between 1 and 0, it determines whether a particular friction force will be large or close to zero. Molecular simulations offer the best way to understand and predict the magnitude of ε, but the complex multi-body nature of the problem makes simple conclusions difficult to draw [240–242]. Some of the basic physics of the energy transfer and dissipation of the molecular collisions can be drawn from simplified models such as a 1-D three-body system [214].

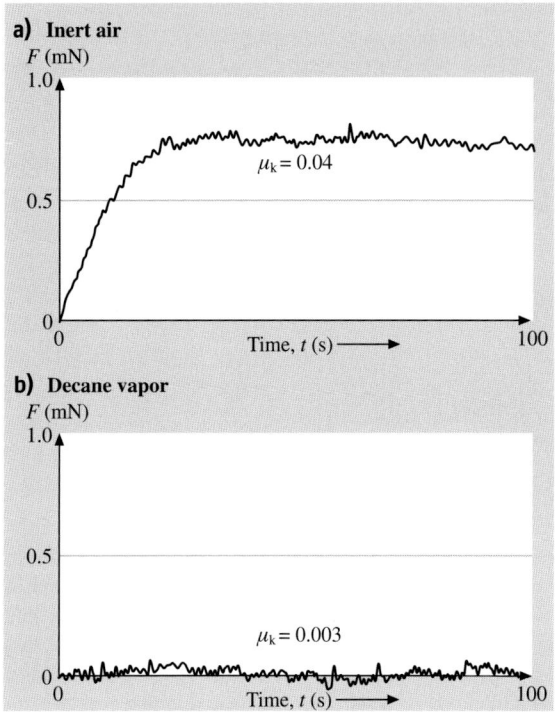

Fig. 9.19. *Top:* Friction traces for two fluid-like calcium alkylbenzene sulfonate monolayer-coated surfaces at 25 °C showing that the friction force is much higher between dry monolayers (**a**) than between monolayers whose fluidity has been enhanced by hydrocarbon penetration from vapor (**b**). *Bottom:* Contact radius vs. load (r^3 vs. L) data measured for the same two surfaces as above and fitted to the JKR equation (9.22), shown by the *solid curves*. For dry monolayers (**c**) the adhesion energy on unloading ($\gamma_R = 40$ mJ m^{-2}) is greater than that on loading ($\gamma_R = 28$ mJ m^{-2}), indicative of an adhesion energy hysteresis of $\Delta\gamma = \gamma_R - \gamma_A = 12$ mJ m^{-2}. For monolayers exposed to saturated decane vapor (**d**) their adhesion hysteresis is zero ($\gamma_A = \gamma_R$), and both the loading and unloading data are well fitted by the thermodynamic value of the surface energy of fluid hydrocarbon chains, $\gamma = 24$ mJ m^{-2}. (After [211] with permission. Copyright 1993 American Chemical Society)

Finally, the above equation may also be expressed in terms of the friction force F:

$$F = S_c A = C_1 A + C_2 L \,. \tag{9.37}$$

Equations similar to (9.36) and (9.37) were previously derived by *Derjaguin* [243, 244] and by *Briscoe* and *Evans* [245], where the constants C_1 and C_2 were interpreted somewhat differently than in this model.

In the absence of any attractive interfacial force, we have $C_1 \approx 0$, and the second term in (9.36) and (9.37) should dominate (Fig. 9.15). Such situations typically arise when surfaces repel each other across the lubricating liquid film. In such cases, the total frictional force should be low and should increase *linearly* with the external

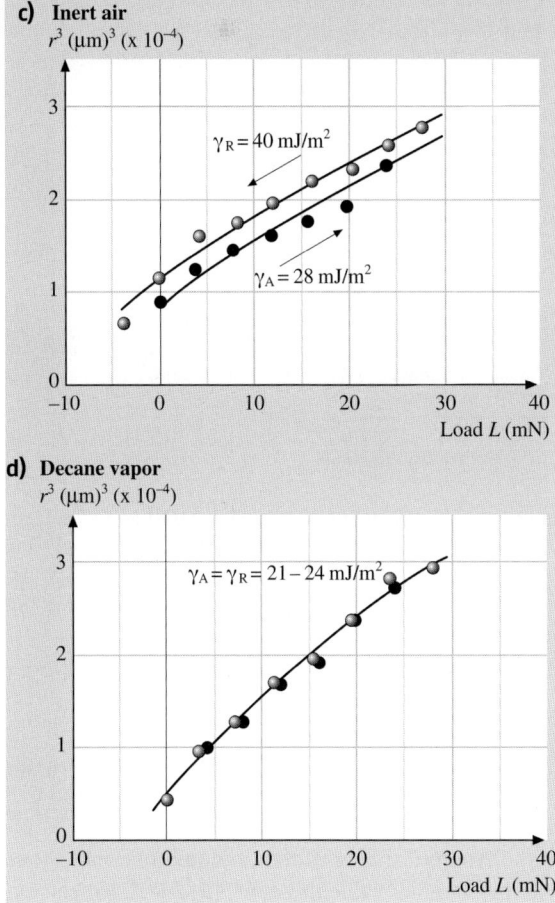

Fig. 9.19. continue

load according to

$$F = C_2 L. \tag{9.38}$$

An example of such lubricated sliding occurs when two mica surfaces slide in water or in salt solution (see Fig. 9.20), where the short-range "hydration" forces between the surfaces are repulsive. Thus, for sliding in 0.5 M KCl it was found that $C_2 = 0.015$ [246]. Another case where repulsive surfaces eliminate the adhesive contribution to friction is for polymer chains attached to surfaces at one end and swollen by a good solvent [175]. For this class of systems, $C_2 < 0.001$ for a finite range of polymer layer compressions (normal loads, L). The low friction between the surfaces in this regime is attributed to the entropic repulsion between the opposing brush layers with a minimum of entanglement between the two layers. However, with higher normal loads, the brush layers become compressed and begin to entangle, which results in higher friction (see [247]).

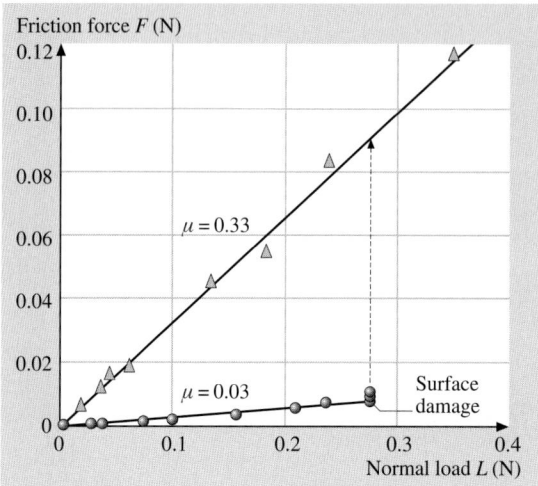

Fig. 9.20. Two mica surfaces sliding past each other while immersed in a 0.01 M KCl salt solution (nonadhesive conditions). The water film is molecularly thin, 0.25 to 0.5 nm thick, and the interfacial friction force is very low: $S_c \approx 5 \times 10^5 \, \text{N m}^{-2}$, $\mu \approx 0.03$ (before damage occurs). After the surfaces have become damaged, the friction coefficient is ca. 0.3. (After [238]. Copyright 1990, with permission from Elsevier Science)

It is important to note that (9.38) has exactly the same form as Amontons' Law

$$F = \mu L, \tag{9.39}$$

where μ is the coefficient of friction.

At the molecular level a thermodynamic analog of the Coulomb or cobblestone models (see Sect. 9.7.1) based on the "contact value theorem" [2, 234, 246] can explain why $F \propto L$ also holds at the microscopic or molecular level. In this analysis we consider the surface molecular groups as being momentarily compressed and decompressed as the surfaces move along. Under irreversible conditions, which always occur when a cycle is completed in a finite amount of time, the energy "lost" in the compression–decompression cycle is dissipated as heat. For two non-adhering surfaces, the stabilizing pressure P_i acting locally between any two elemental contact points i of the surfaces may be expressed by the contact value theorem [2]:

$$P_i = \rho_i k_B T = k_B T / V_i, \tag{9.40}$$

where $\rho_i = V_i^{-1}$ is the local number density (per unit volume) or activity of the interacting entities, be they molecules, atoms, ions or the electron clouds of atoms. This equation is essentially the osmotic or entropic pressure of a gas of confined molecules. As one surface moves across the other, local regions become compressed and decompressed by a volume ΔV_i. The work done per cycle can be written as $\varepsilon P_i \Delta V_i$, where ε ($\varepsilon \leq 1$) is the fraction of energy per cycle "lost" as heat, as defined earlier. The energy balance shows that for each compression–decompression cycle, the dissipated energy is related to the friction force by

$$F_i x_i = \varepsilon P_i \Delta V_i ,\qquad(9.41)$$

where x_i is the lateral distance moved per cycle, which can be the distance between asperities or the distance between surface lattice sites. The pressure at each contact junction can be expressed in terms of the local normal load L_i and local area of contact A_i as $P_i = L_i/A_i$. The volume change over a cycle can thus be expressed as $\Delta V_i = A_i z_i$, where z_i is the vertical distance of confinement. Inserting these into (9.41), we get

$$F_i = \varepsilon L_i (z_i/x_i) ,\qquad(9.42)$$

which is independent of the local contact area A_i. The total friction force is thus

$$\begin{aligned}F = \sum F_i &= \sum \varepsilon L_i (z_i/x_i)\\ &= \varepsilon \langle z_i/x_i \rangle \sum L_i = \mu L ,\end{aligned}\qquad(9.43)$$

where it is assumed that on average the local values of L_i and P_i are independent of the local "slope" z_i/x_i. Therefore, the friction coefficient μ is a function only of the average surface topography and the sliding velocity, but is independent of the local (real) or macroscopic (apparent) contact areas.

While this analysis explains non-adhering surfaces, there is still an additional explicit contact area contribution for the case of adhering surfaces, as in (9.37). The distinction between the two cases arises because the initial assumption of the contact value theorem, (9.40), is incomplete for adhering systems. A more appropriate starting equation would reflect the full intermolecular interaction potential, including the attractive interactions, in addition to the purely repulsive contributions of (9.40), much as the van der Waals equation of state modifies the ideal gas law.

9.7.3 Examples of Experimentally Observed Friction of Dry Surfaces

Numerous model systems have been studied with a surface forces apparatus (SFA) modified for friction experiments (see Sect. 9.2.3). The apparatus allows for control of load (normal force) and sliding speed, and simultaneous measurement of surface separation, surface shape, true (molecular) area of contact between smooth surfaces, and friction forces. A variety of both unlubricated and solid and liquid lubricated surfaces have been studied both as smooth single-asperity contacts and after they have been roughened by shear-induced damage.

Figure 9.16 shows the contact area, A, and friction force, F, both plotted against the applied load, L, in an experiment in which two molecularly smooth surfaces of mica in adhesive contact were slid past each other in an atmosphere of dry nitrogen gas. This is an example of the low load, adhesion controlled limit, which is excellently described by (9.28). In a number of different experiments, S_c was measured to be 2.5×10^7 N m^{-2} and to be independent of the sliding velocity [45, 238]. Note that there is a friction force even at negative loads, where the surfaces are still sliding in adhesive contact.

Figure 9.17 shows the correlation between adhesion hysteresis and friction for two surfaces consisting of silica films deposited on mica substrates [41]. The friction between undamaged hydrophobic silica surfaces showed stick–slip both at dry conditions and at 100% relative humidity. Similar to the mica surfaces in Figs. 9.16, 9.18, and 9.20, the friction of damaged silica surfaces obeyed Amontons' law with a friction coefficient of 0.25 both at dry conditions and at 55% relative humidity.

The high friction force of unlubricated sliding can often be reduced by treating the solid surface with a boundary layer of some other solid material that exhibits lower friction such as a surfactant monolayer, or by ensuring that during sliding, a thin liquid film remains between the surfaces (as will be discussed in Sect. 9.8). The effectiveness of a solid boundary lubricant layer on reducing the forces of friction is illustrated in Fig. 9.18. Comparing this with the friction of the unlubricated/untreated surfaces (Fig. 9.16) shows that the critical shear stress has been reduced by a factor of about ten: from 2.5×10^7 to 3.5×10^6 N m^{-2}. At much higher applied loads or pressures, the friction force is proportional to the load, rather than the area of contact [239], as expected from (9.27).

Yamada and *Israelachvili* [248] studied the adhesion and friction of fluorocarbon surfaces (surfactant-coated boundary lubricant layers), which were compared to those of hydrocarbon surfaces. They concluded that well-ordered fluorocarbon surfaces have a high friction, in spite of their lower adhesion energy (in agreement with previous findings). The low friction coefficient of PTFE (Teflon) must, therefore, be due to some effect other than a low adhesion. For example, the softness of PTFE, which allows material to flow at the interface, which thus behaves like a fluid lubricant. On a related issue, *Luengo* et al. [249] found that C_{60} surfaces also exhibited low adhesion but high friction. In both cases the high friction appears to arise from the bulky surface groups – fluorocarbon compared to hydrocarbon groups in the former, large fullerene spheres in the latter. Apparently, the fact that C_{60} molecules rotate *in their lattice* does not make them a good lubricant: The molecules of the opposing surface must still climb over them in order to slide, and this requires energy that is independent of whether the surface molecules are fixed or freely rotating. Larger particles such as ~ 25 nm-sized nanoparticles (also known as "inorganic fullerenes") do appear to produce low friction by behaving like molecular ball bearings, but the potential of this promising new class of solid lubricant has still to be explored [250].

Figure 9.19 illustrates the relationship between adhesion hysteresis and friction for surfactant-coated surfaces under different conditions. This effect, however, is much more general and has been shown to hold for other surfaces as well [41, 233, 251].

Direct comparisons between absolute adhesion energies and friction forces show little correlation. In some cases, higher adhesion energies for the same system under different conditions correspond with lower friction forces. For example, for hydrophilic silica surfaces (Fig. 9.17) it was found that with increasing relative humidity the adhesion energy *increases*, but the adhesion energy hysteresis measured in a loading–unloading cycle *decreases*, as does the friction force [41]. For hydrophobic silica surfaces under dry conditions, the friction at load $L = 5.5$ mN was $F = 75$ mN. For the same sample, the adhesion energy hysteresis was $\Delta\gamma = 10$ mJ m^{-2}, with a contact area of $A \approx 10^{-8}$ m^2 at the same load. Assuming a value for the characteristic distance σ on the order of one lattice spacing, $\sigma \approx 1$ nm, and inserting these values into (9.32), the friction force is predicted to be $F \approx 100$ mN for the kinetic friction force, which is close to the measured value of 75 mN. Alternatively, we may conclude that the dissipation factor is $\varepsilon = 0.75$, i.e., that almost all the energy is dissipated as heat at each molecular collision.

A liquid lubricant film (Sect. 9.8.3) is usually much more effective at lowering the friction of two surfaces than a solid boundary lubricant layer. However, to successfully use a liquid lubricant, it must "wet" the surfaces, that is, it should have a high affinity for the surfaces, so that not all the liquid molecules become squeezed out when the surfaces come close together, even under a large compressive load. Another important requirement is that the liquid film remains a liquid under tribological conditions, i.e., that it does not epitaxially solidify between the surfaces.

Effective lubrication usually requires that the lubricant be injected between the surfaces, but in some cases the liquid can be made to condense from the vapor. This is illustrated in Fig. 9.20 for two untreated mica surfaces sliding with a thin layer of water between them. A monomolecular film of water (of thickness 0.25 nm per surface) has reduced S_c from its value for dry surfaces (Fig. 9.16) by a factor of more than 30, which may be compared with the factor of ten attained with the boundary lubricant layer (of thickness 2.5 nm per surface) in Fig. 9.18. Water appears to have unusual lubricating properties and usually gives wearless friction with no stick–slip.

The effectiveness of a water film only 0.25 nm thick to lower the friction force by more than an order of magnitude is attributed to the "hydrophilicity" of the mica surface (mica is "wetted" by water) and to the existence of a strongly repulsive short-range hydration force between such surfaces in aqueous solutions, which effectively removes the adhesion-controlled contribution to the friction force [246]. It is also interesting that a 0.25-nm-thick water film between two mica surfaces is sufficient to bring the coefficient of friction down to 0.01–0.02, a value that corresponds to the unusually low friction of ice. Clearly, a single monolayer of water can be a very good lubricant – much better than most other monomolecular liquid films – for reasons that will be discussed in Sect. 9.9.

Dry polymer layers (Fig. 9.21) typically show a high initial static friction ("stiction") as sliding commences from rest in adhesive contact. The development of the friction force after a change in sliding direction, a gradual transition from stick–slip to smooth sliding, is shown in Fig. 9.21. A correlation between adhesion hysteresis and friction similar to the one observed for silica surfaces in Fig. 9.17 can be seen for dry polymer layers below their glass transition temperature. As shown in Fig. 9.12b,c, the adhesion hysteresis for polystyrene surfaces can be increased by irradiation to induce scission of chains, and it has been found that the steady state friction force (kinetic friction) shows a similar increase with irradiation time [212].

Figure 9.22 shows an example of a computer simulation of the sliding of two unlubricated silicon surfaces (modeled as a tip sliding over a planar surface) [91].

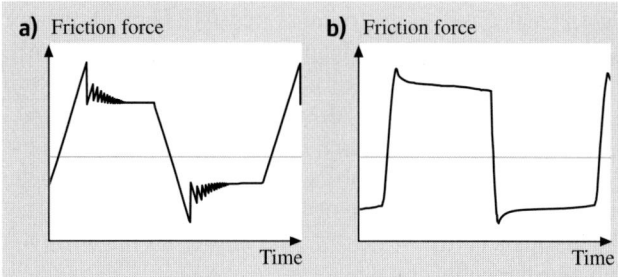

Fig. 9.21. Typical friction traces showing how the friction force varies with the sliding time for two symmetric, glassy polymer films at dry conditions. Qualitative features that are common to both polystyrene and polyvinyl benzyl chloride: (**a**) Decaying stick–slip motion is observed until smooth sliding is attained if the motion continues for a sufficiently long distance. (**b**) Smooth sliding observed at sufficiently high speeds. Similar observations have been made by *Berthoud* et al. [252] in measurements on polymethyl methacrylate. (After [212] with permission. Copyright 2002 American Association for the Advancement of Science)

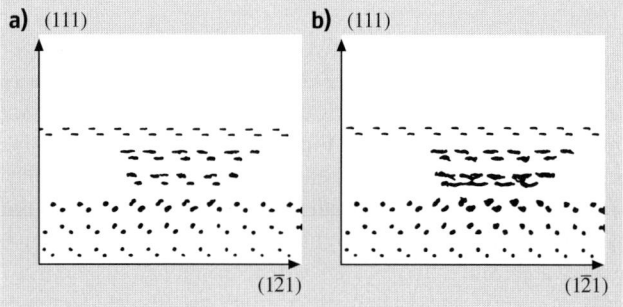

Fig. 9.22. Computer simulation of the sliding of two contacting Si surfaces (a tip and a flat surface). Shown are particle trajectories in a constant-force simulation, $F_{z,\text{external}} = -2.15 \times 10^{-8}$ N, viewed along the $(10\bar{1})$ direction just before (**a**) and after (**b**) a stick–slip event for a large, initally ordered, dynamic tip. (After [91] with permission of Kluwer Academic Publishers)

The sliding proceeds through a series of stick–slip events, and information on the friction force and the local order of the initially crystalline surfaces can be obtained. Similar studies for cold welding systems [91] have demonstrated the occurrence of shear or friction damage within the sliding surface (tip) as the lowest layer of it adheres to the bottom surface.

9.7.4 Transition from Interfacial to Normal Friction with Wear

Frictional damage can have many causes such as adhesive tearing at high loads or overheating at high sliding speeds. Once damage occurs, there is a transition from "interfacial" to "normal" or load-controlled friction as the surfaces become forced apart by the torn-out asperities (wear particles). For low loads, the friction changes from obeying $F = S_c A$ to obeying Amontons' law, $F = \mu L$, as shown in Figs. 9.16 and 9.18, and sliding now proceeds smoothly with the surfaces separated by a 10–100 nm forest of wear debris (in this case, mica flakes). The wear particles keep the surfaces apart over an area that is much greater than their size, so that even one submicroscopic particle or asperity can cause a significant reduction in the area of contact and, therefore, in the friction [238]. For this type of frictional sliding, one can no longer talk of the molecular contact area of the two surfaces, although the macroscopic or "apparent" area is still a useful parameter.

One remarkable feature of the transition from interfacial to normal friction of brittle surfaces is that while the strength of interfacial friction, as reflected in the values of S_c, is very dependent on the type of surface and on the liquid film between the surfaces, this is not the case once the transition to normal friction has occurred. At the onset of damage, the material properties of the underlying substrates control the friction. In Figs. 9.16, 9.18, and 9.20 the friction for the damaged surfaces is that of any damaged mica–mica system, $\mu \approx 0.3$, *independent of the initial surface coatings or liquid films between the surfaces*. A similar friction coefficient was found for damaged silica surfaces [41].

In order to practically modify the frictional behavior of such brittle materials, it is important to use coatings that will both alter the interfacial tribological character and remain intact and protect the surfaces from damage during sliding. *Berman et al.* [253] found that the friction of a strongly bound octadecyl phosphonic acid monolayer on alumina surfaces was higher than for untreated, undamaged α-alumina surfaces, but the bare surfaces easily became damaged upon sliding, resulting in an ultimately higher friction system with greater wear rates than the more robust monolayer-coated surfaces.

Clearly, the mechanism and factors that determine *normal* friction are quite different from those that govern *interfacial* friction (Sects. 9.7.1–9.7.2). This effect is not general and may only apply to brittle materials. For example, the friction of ductile surfaces is totally different and involves the continuous plastic deformation of contacting surface asperities during sliding, rather than the rolling of two surfaces on hard wear particles [205]. Furthermore, in the case of ductile surfaces, water and other surface-active components do have an effect on the friction coefficients under "normal" sliding conditions.

9.8 Liquid Lubricated Surfaces

9.8.1 Viscous Forces and Friction of Thick Films: Continuum Regime

Experimentally, it is usually difficult to unambiguously establish which type of sliding mode is occurring, but an empirical criterion, based on the Stribeck curve (Fig. 9.13), is often used as an indicator. This curve shows how the friction force or the coefficient of friction is expected to vary with sliding speed, depending on which type of friction regime is operating. For thick liquid lubricant films whose behavior can be described by bulk (continuum) properties, the friction forces are essentially the hydrodynamic or viscous drag forces. For example, for two plane parallel surfaces of area A separated by a distance D and moving laterally relative to each other with velocity v, if the intervening liquid is *Newtonian*, i.e., if its viscosity η is independent of the shear rate, the frictional force experienced by the surfaces is given by

$$F = \frac{\eta A v}{D}, \tag{9.44}$$

where the shear rate $\dot{\gamma}$ is defined by

$$\dot{\gamma} = \frac{v}{D}. \tag{9.45}$$

At higher shear rates, two additional effects often come into play. First, certain properties of liquids may change at high $\dot{\gamma}$ values. In particular, the effective viscosity may become non-Newtonian, one form given by

$$\eta = \dot{\gamma}^n, \tag{9.46}$$

where $n = 0$ (i.e., $\eta_{\text{eff}} = $ constant) for Newtonian fluids, $n > 0$ for shear thickening (dilatent) fluids, and $n < 0$ for shear thinning (pseudoplastic) fluids (the latter become less viscous, i.e., flow more easily, with increasing shear rate). An additional effect on η can arise from the higher local stresses (pressures) experienced by the liquid film as $\dot{\gamma}$ increases. Since the viscosity is generally also sensitive to the pressure (usually increasing with P), this effect also acts to increase η_{eff} and thus the friction force.

A second effect that occurs at high shear rates is surface deformation, arising from the large hydrodynamic forces acting on the sliding surfaces. For example, Fig. 9.23 shows how two surfaces deform elastically when the sliding speed increases to a high value. These deformations alter the hydrodynamic friction forces, and this type of friction is often referred to as *elastohydrodynamic lubrication* (EHD or EHL), as mentioned in Table 9.3.

How thin can a liquid film be before its dynamic, e.g., viscous flow, behavior ceases to be described by bulk properties and continuum models? Concerning the static properties, we have already seen in Sect. 9.4.3 that films composed of simple liquids display continuum behavior down to thicknesses of 4–10 molecular diameters. Similar effects have been found to apply to the dynamic properties, such as the viscosity, of simple liquids in thin films. Concerning viscosity measurements, a number of dynamic techniques were recently developed [11–13, 43, 254, 255] for directly measuring the viscosity as a function of film thickness and shear rate across very thin liquid films between two surfaces. By comparing the results with theoretical predictions of fluid flow in thin films, one can determine the effective positions of the shear planes and the onset of non-Newtonian behavior in very thin films.

The results show that for simple liquids, including linear chain molecules such as alkanes, the viscosity in thin films is the same, within 10%, as the bulk even for

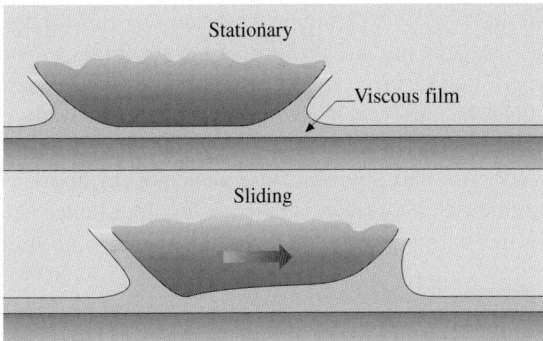

Fig. 9.23. *Top:* Stationary surfaces (one more deformable and one rigid) separated by a thick liquid film. *Bottom:* Elastohydrodynamic deformation of the upper surface during sliding. (After [10] with permission)

films as thin as 10 molecular diameters (or segment widths) [11–13, 254, 255]. This implies that the shear plane is effectively located within one molecular diameter of the solid–liquid interface, and these conclusions were found to remain valid even at the highest shear rates studied (of $\sim 2 \times 10^5\,\text{s}^{-1}$). With water between two mica or silica surfaces [22, 254–256] this has been found to be the case (to within $\pm 10\%$) down to surface separations as small as 2 nm, implying that the shear planes must also be within a few tenths of a nanometer of the solid–liquid interfaces. These results appear to be independent of the existence of electrostatic "double-layer" or "hydration" forces. For the case of the simple liquid toluene confined between surfaces with adsorbed layers of C_{60} molecules, this type of viscosity measurement has shown that the traditional no-slip assumption for flow at a solid interface does not always hold [257]. The C_{60} layer at the mica–toluene interface results in a "full slip" boundary, which dramatically lowers the viscous drag or effective viscosity for regular Couette or Poiseuille flow.

With polymeric liquids (polymer melts) such as polydimethylsiloxanes (PDMS) and polybutadienes (PBD), or with polystyrene (PS) adsorbed onto surfaces from solution, the far-field viscosity is again equal to the bulk value, but with the non-slip plane (hydrodynamic layer thickness) being located at $D = 1-2R_g$ away from each surface [11, 47], or at $D = L$ or less for polymer brush layers of thickness L per surface [13, 258]. In contrast, the same technique was used to show that for non-adsorbing polymers in solution, there is actually a depletion layer of nearly pure solvent that exists at the surfaces that affects the confined solution flow properties [256]. These effects are observed from near contact to surface separations in excess of 200 nm.

Further experiments with surfaces closer than a few molecular diameters ($D < 2-4$ nm for simple liquids, or $D < 2-4R_g$ for polymer fluids) indicate that large deviations occur for thinner films, described below. One important conclusion from these studies is, therefore, that the dynamic properties of simple liquids, including water, near an *isolated* surface are similar to those of the bulk liquid *already within the first layer of molecules adjacent to the surface*, only changing when another surface approaches the first. In other words, the viscosity and position of the shear plane near a surface are not simply a property of that surface, but of how far that surface is from another surface. The reason for this is because when two surfaces are close together, the constraining effects on the liquid molecules between them are much more severe than when there is only one surface. Another obvious consequence of the above is that one should not make measurements on a single, isolated solid–liquid interface and then draw conclusions about the state of the liquid or its interactions in a thin film *between* two surfaces.

9.8.2 Friction of Intermediate Thickness Films

For liquid films in the thickness range between 4 and 10 molecular diameters, the properties can be significantly different from those of bulk films. Still, the fluids remain recognizable as fluids; in other words, they do not undergo a phase transition into a solid or liquid-crystalline phase. This regime has recently been studied by *Granick* et al. [44, 219–222], who used a friction attachment [43, 44] to the SFA where a sinusoidal input signal to a piezoelectric device makes the two surfaces slide back and forth laterally past each other at small amplitudes. This method provides information on the real and imaginary parts (elastic and dissipative components, respectively) of the shear modulus of thin films at different shear rates and film thickness. *Granick* [220] and *Hu* et al. [221] found that films of simple liquids become non-Newtonian in the 2.5–5 nm regime (about 10 molecular diameters, see Fig. 9.24). Polymer melts become non-Newtonian at much larger film thicknesses, depending on their molecular weight [47].

Klein and *Kumacheva* [46, 226, 260] studied the interaction forces and friction of small quasi-spherical liquid molecules such as cyclohexane between molecularly smooth mica surfaces. They concluded that surface epitaxial effects can cause the liquid film to already solidify at 6 molecular diameters, resulting in a sudden (discontinuous) jump to high friction at low shear rates. Such dynamic first-order transitions, however, may depend on the shear rate.

A generalized friction map (Fig. 9.24c,d) has been proposed by *Luengo* et al. [259] that illustrates the changes in η_{eff} from bulk Newtonian behavior ($n = 0$, $\eta_{\text{eff}} = \eta_{\text{bulk}}$) through the transition regime where n reaches a minimum of -1 with decreasing shear rate to the solid-like creep regime at very low $\dot{\gamma}$ where n returns to 0. A number of results from experimental, theoretical, and computer simulation work have shown values of n from $-1/2$ to -1 for this transition regime for a variety of systems and assumptions [220, 222, 240, 261–267].

The intermediate regime appears to extend over a narrow range of film thickness, from about 4 to 10 molecular diameters or polymer radii of gyration. Thinner films begin to adopt "boundary" or "interfacial" friction properties (described below, see also Table 9.4). Note that the intermediate regime is actually a very narrow one when defined in terms of film thickness, for example, varying from about $D = 2$ to $4\,\text{nm}$ for hexadecane films [220].

Table 9.4. Effect of molecular shape and short-range forces on tribological properties[a]

Liquid	Short-range force	Type of friction	Friction coefficient	Bulk liquid viscosity (cP)
Organic (water-free)				
Cyclohexane	Oscillatory	Quantized stick–slip	$\gg 1$	0.6
OMCTS[b]	Oscillatory	Quantized stick–slip	$\gg 1$	2.3
Octane	Oscillatory	Quantized stick–slip	1.5	0.5
Tetradecane	Oscillatory ↔ smooth	stick–slip ↔ smooth	1.0	2.3
Octadecane (branched)	Oscillatory ↔ smooth	stick–slip ↔ smooth	0.3	5.5
PDMS[b] ($M = 3700$ g mol^{-1}, melt)	Oscillatory ↔ smooth	Smooth	0.4	50
PBD[b] ($M = 3500$ g mol^{-1}, branched)	Smooth	Smooth	0.03	800
Water				
Water (KCl solution)	Smooth	Smooth	0.01–0.03	1.0

[a] For molecularly thin liquid films between two shearing mica surfaces at 20 °C
[b] OMCTS: Octamethylcyclotetrasiloxane, PDMS: Polydimethylsiloxane, PBD: Polybutadiene

A fluid's effective viscosity η_{eff} in the intermediate regime is usually higher than in the bulk, but η_{eff} usually *decreases* with increasing sliding velocity, v (known as *shear thinning*). When two surfaces slide in the intermediate regime, the motion tends to thicken the film (dilatancy). This sends the system into the bulk EHL regime where, as indicated by (9.44), the friction force now *increases* with velocity. This initial decrease, followed by an increase, in the frictional forces of many lubricant systems is the basis for the empirical Stribeck curve of Fig. 9.13. In the transition from bulk to boundary behavior there is first a quantitative change in the material properties (viscosity and elasticity), which can be continuous, to discontinuous qualitative changes that result in yield stresses and non-liquidlike behavior.

The rest of this section is devoted to friction in the boundary regime. Boundary friction may be thought of as applying to the case where a lubricant film is present, but where this film is of molecular dimensions – a few molecular layers or less.

9.8.3 Boundary Lubrication of Molecularly Thin Films: Nanorheology

When a liquid is confined between two surfaces or within any narrow space whose dimensions are less than 4 to 10 molecular diameters, both the static (equilibrium) and dynamic properties of the liquid, such as its compressibility and viscosity, can no

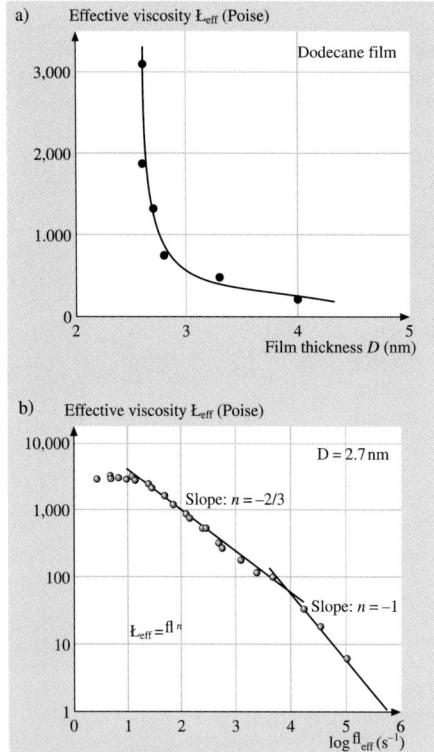

Fig. 9.24. Typical rheological behavior of liquid films in the mixed lubrication regime. (**a**) Increase in effective viscosity of dodecane film between two mica surfaces with decreasing film thickness. At distances larger than 4–5 nm, the effective viscosity η_{eff} approaches the bulk value η_{bulk} and does not depend on the shear rate $\dot{\gamma}$. (After [220]. Copyright 1991 American Association for the Advancement of Science.) (**b**) Non-Newtonian variation of η_{eff} with shear rate of a 2.7-nm-thick dodecane film at a net normal pressure of 0.12 MPa and at 28 °C. The effective viscosity decays as a power law, as in (9.46). In this example, $n = 0$ at the lowest $\dot{\gamma}$ and changes to $n = -2/3$ and -1 at higher $\dot{\gamma}$. For films of bulk thickness, dodecane is a low viscosity Newtonian fluid ($n = 0$). (**c**) Proposed general friction map of effective viscosity η_{eff} (arbitrary units) as a function of effective shear rate $\dot{\gamma}$ (arbitrary units) on logarithmic scales. Three main classes of behavior emerge: (i) Thick films: elastohydrodynamic sliding. At $L = 0$, approximating bulk conditions, η_{eff} is independent of shear rate except when shear thinning might occur at sufficiently large $\dot{\gamma}$. (ii) Boundary layer films, intermediate regime. A Newtonian regime is again observed [η_{eff} = constant, $n = 0$ in (9.46)] at low loads and low shear rates, but η_{eff} is much higher than the bulk value. As the shear rate $\dot{\gamma}$ increases beyond $\dot{\gamma}_{\text{min}}$, the effective viscosity starts to drop with a power-law dependence on the shear rate [see panel (**b**)], with n in the range $-1/2$ to -1 most commonly observed. As the shear rate $\dot{\gamma}$ increases still more, beyond $\dot{\gamma}_{\text{max}}$, a second Newtonian plateau is encountered. (iii) Boundary layer films, high load. The η_{eff} continues to grow with load and to be Newtonian provided that the shear rate is sufficiently low. Transition to sliding at high velocity is discontinuous ($n < -1$) and usually of the stick–slip variety. *See next page*

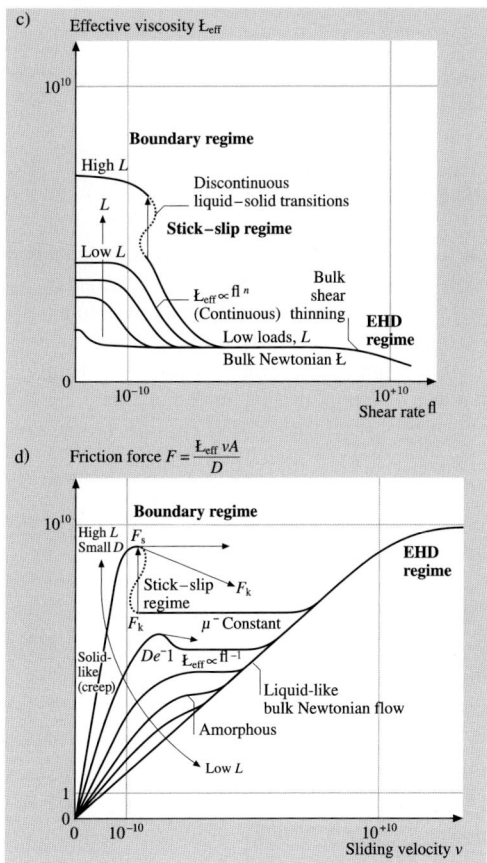

Fig. 9.24. (d) Proposed friction map of friction force as a function of sliding velocity in various tribological regimes. With increasing load, Newtonian flow in the elastohydrodynamic (EHD) regimes crosses into the boundary regime of lubrication. Note that even EHD lubrication changes, at the highest velocities, to limiting shear stress response. At the highest loads (L) and smallest film thickness (D), the friction force goes through a maximum (the static friction, F_s), followed by a regime where the friction coefficient (μ) is roughly constant with increasing velocity (meaning that the kinetic friction, F_k, is roughly constant). Non-Newtonian shear thinning is observed at somewhat smaller load and larger film thickness; the friction force passes through a maximum at the point where $De = 1$. De, the Deborah number, is the point at which the applied shear rate exceeds the natural relaxation time of the boundary layer film. The velocity axis from 10^{-10} to 10^{10} (arbitrary units) indicates a large span. (Panels (**b**)–(**d**) after [259]. Copyright 1996, with permission from Elsevier Science)

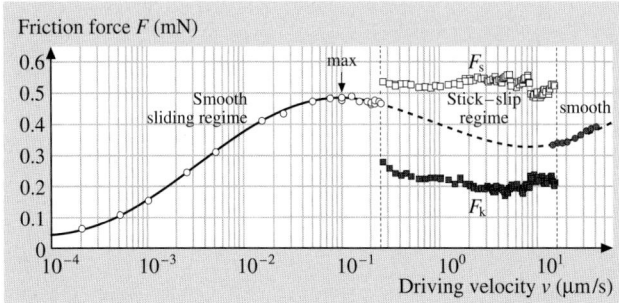

Fig. 9.25. Transition from smooth sliding to "inverted" stick–slip and to a second smooth sliding regime with increasing driving velocity during shear of two adsorbed surfactant monolayers in aqueous solution at a load of $L = 4.5$ mN and $T = 20\,°\mathrm{C}$. The smooth sliding (*open circles*) to inverted stick–slip (*squares*) transition occurs at $v_\mathrm{c} \sim 0.3\,\mu\mathrm{m/s}$. Prior to the transition, the kinetic stress levels off at after a logarithmic dependence on velocity. The quasi-smooth regime persists up to the transition at v_c. At high driving velocities (*filled circles*), a new transition to a smooth sliding regime is observed between 14 and 17 μm/s. (After [268] with permission)

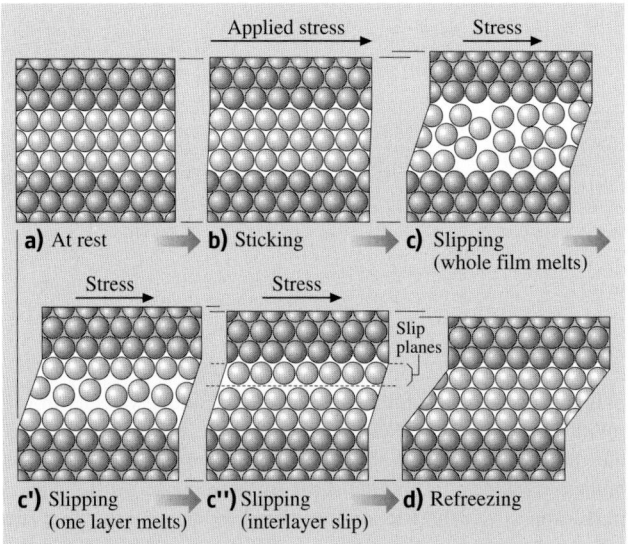

Fig. 9.26. Idealized schematic illustration of molecular rearrangements occurring in a molecularly thin film of spherical or simple chain molecules between two solid surfaces during shear. Depending on the system, a number of different molecular configurations within the film are possible during slipping and sliding, shown here as stages (**c**): total disorder as the whole film melts; (**c′**): partial disorder; and (**c″**): order persists even during sliding with slip occurring at a single slip plane either within the film or at the walls. A dilation is predicted in the direction normal to the surfaces. (After [224] with permission)

longer be described even qualitatively in terms of the bulk properties. The molecules confined within such molecularly thin films become ordered into layers ("out-of-plane" ordering), and within each layer they can also have lateral order ("in-plane" ordering). Such films may be thought of as behaving more like a liquid crystal or a solid than a liquid.

As described in Sect. 9.4.3, the measured *normal* forces between two solid surfaces across molecularly thin films exhibit exponentially decaying oscillations, varying between attraction and repulsion with a periodicity equal to some molecular dimension of the solvent molecules. Thus most liquid films can sustain a finite normal stress, and the adhesion force between two surfaces across such films is "quantized", depending on the thickness (or number of liquid layers) between the surfaces. The structuring of molecules in thin films and the oscillatory forces it gives rise to are now reasonably well understood, both experimentally and theoretically, at least for simple liquids.

Work has also recently been done on the dynamic, e.g., viscous or shear, forces associated with molecularly thin films. Both experiments [38, 46, 203, 223, 226, 227, 269, 270] and theory [200, 201, 261, 271] indicate that even when two surfaces are in steady state sliding they still prefer to remain in one of their stable potential energy minima, i.e., a sheared film of liquid can retain its basic layered structure. Thus even during motion the film does not become totally liquid-like. Indeed, if there is some "in-plane" ordering within a film, it will exhibit a yield-point before it begins to flow. Such films can therefore sustain a finite shear stress, in addition to a finite normal stress. The value of the yield stress depends on the number of layers comprising the film and represents another "quantized" property of molecularly thin films [200].

The dynamic properties of a liquid film undergoing shear are very complex. Depending on whether the film is more liquid-like or solid-like, the motion will be smooth or of the stick–slip type. During sliding, transitions can occur between n layers and $(n-1)$ or $(n+1)$ layers (see Fig. 9.27). The details of the motion depend critically on the externally applied load, the temperature, the sliding velocity, the twist angle between the two surface lattices, and the sliding direction relative to the lattices.

Smooth and Stick–Slip Sliding

Recent advances in friction-measuring techniques have enabled the interfacial friction of molecularly thin films to be measured with great accuracy. Some of these advances have involved the surface forces apparatus technique [38, 45, 47, 223, 238, 269, 270], while others have involved the atomic force microscope [9, 54, 231, 272]. In addition, computer simulations [90, 200, 201, 271, 273, 274] have become sufficiently sophisticated to enable fairly complex tribological systems to be studied. All these advances are necessary if one is to probe such subtle effects as smooth or stick–slip friction, transient and memory effects, and ultralow friction mechanisms at the molecular level.

The theoretical models presented in this section will be concerned with a situation commonly observed experimentally: Stick–slip occurs between a static state

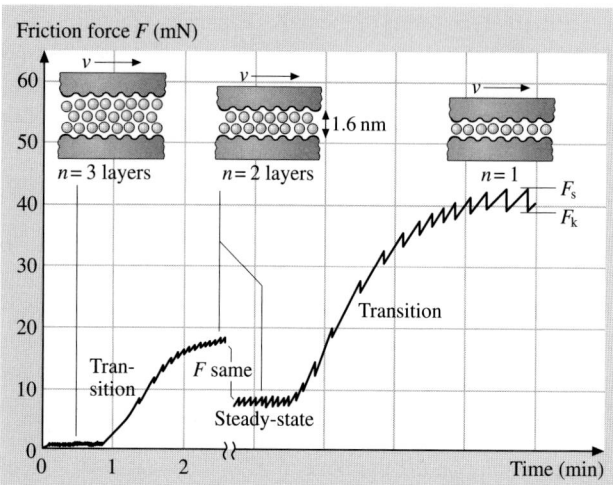

Fig. 9.27. Measured change in friction during interlayer transitions of the silicone liquid octamethylcyclotetrasiloxane (OMCTS), an inert liquid whose quasispherical molecules have a diameter of 0.8 nm. In this system, the shear stress $S_c = F/A$ was found to be constant as long as the number of layers, n, remained constant. Qualitatively similar results have been obtained with other quasi-spherical molecules such as cyclohexane [269]. The shear stresses are only weakly dependent on the sliding velocity v. However, for sliding velocities above some critical value, v_c, the stick–slip disappears and sliding proceeds smoothly at the kinetic value. (After [223] with permission)

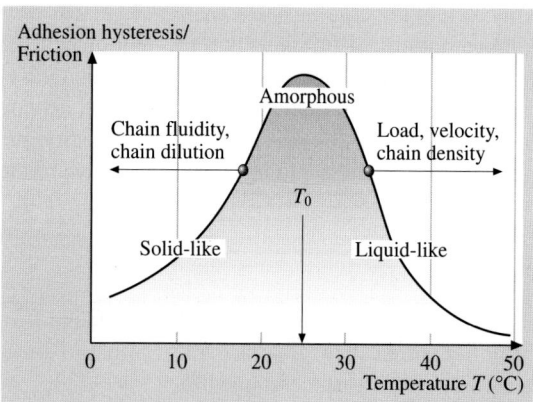

Fig. 9.28. Schematic friction phase diagram representing the trends observed in the boundary friction of a variety of different surfactant monolayers. The characteristic bell-shaped curve also correlates with the adhesion energy hysteresis of the monolayers. The *arrows* indicate the direction in which the whole curve is dragged when the load, velocity, etc., is increased. (After [233] with permission)

with high friction and a low-friction kinetic state, and a transition from this sliding regime to smooth sliding can be induced by an increase in velocity. Experimental data on various systems showing this behavior are shown in Figs. 9.27, 9.30b, and 9.31a. Recent studies on adhesive systems have revealed the possibility of other dynamic responses such as "inverted" stick–slip between two kinetic states of higher and lower friction and with a transition from smooth sliding to stick–slip with increasing velocity, as shown in Fig. 9.25 [268].

With the added insights provided by computer simulations, a number of distinct molecular processes have been identified during smooth and stick–slip sliding in model systems for the more familiar static-to-kinetic stick–slip and transition from stick–slip to smooth sliding. These are shown schematically in Fig. 9.26 for the case of spherical liquid molecules between two solid crystalline surfaces. The following regimes may be identified:

Surfaces at rest (Fig. 9.26a): Even with no externally applied load, solvent–surface epitaxial interactions can cause the liquid molecules in the film to attain a solid-like state. Thus at rest the surfaces are stuck to each other through the film.

Sticking regime (frozen, solid-like film) (Fig. 9.26b): A progressively increasing lateral shear stress is applied. The solid-like film responds elastically with a small lateral displacement and a small increase or "dilatancy" in film thickness (less than a lattice spacing or molecular dimension, σ). In this regime the film retains its frozen, solid-like state: All the strains are elastic and reversible, and the surfaces remain effectively stuck to each other. However, slow creep may occur over long time periods.

Slipping and sliding regimes (molten, liquid-like film) (Fig. 9.26c,c',c''): When the applied shear stress or force has reached a certain critical value, the *static* friction force, F_s, the film suddenly melts (known as "shear melting") or rearranges to allow for wall-slip or slip within the film to occur at which point the two surfaces begin to slip rapidly past each other. If the applied stress is kept at a high value, the upper surface will continue to slide indefinitely.

Refreezing regime (resolidification of film) (Fig. 9.26d): In many practical cases, the rapid slip of the upper surface relieves some of the applied force, which eventually falls below another critical value, the *kinetic* friction force F_k, at which point the film resolidifies and the whole stick–slip cycle is repeated. On the other hand, if the slip rate is smaller than the rate at which the external stress is applied, the surfaces will continue to slide smoothly in the kinetic state and there will be no more stick–slip. The critical velocity at which stick–slip disappears is discussed in more detail in Sect. 9.8.3.

Experiments with linear chain (alkane) molecules show that the film thickness remains quantized during sliding, so that the structure of such films is probably more like that of a nematic liquid crystal where the liquid molecules have become shear aligned in some direction, enabling shear motion to occur while retaining some order within the film. Experiments on the friction of two molecularly smooth mica surfaces separated by three molecular layers of the liquid OMCTS (Fig. 9.27) show how the friction increases to higher values in a quantized way when the number of layers falls from $n = 3$ to $n = 2$ and then to $n = 1$.

Computer simulations for simple spherical molecules [201] further indicate that during the slip the film thickness is roughly 15% higher than at rest (i.e., the film density falls), and that the order parameter within the film drops from 0.85 to about 0.25. Such dilatancy has been investigated both experimentally [236] and in further computer simulations [235]. The changes in thickness and in order parameter are consistent with a disorganized state for the whole film during the slip [271], as illustrated schematically in Fig. 9.26c. At this stage, we can only speculate on other possible configurations of molecules in the slipping and sliding regimes. This probably depends on the shapes of the molecules (e.g., whether spherical or linear or branched), on the atomic structure of the surfaces, on the sliding velocity, etc. [275]. Figure 9.26c,c′,c″ shows three possible sliding modes wherein the shearing film either totally melts, or where the molecules retain their layered structure and where slip occurs between two or more layers. Other sliding modes, for example, involving the movement of dislocations or disclinations are also possible, and it is unlikely that one single mechanism applies in all cases.

Both friction and adhesion hysteresis vary non-linearly with temperature, often peaking at some particular temperature, T_0. The temperature dependence of these forces can, therefore, be represented on a friction phase diagram such as the one shown in Fig. 9.28. Experiments have shown that T_0, and the whole bell-shaped curve, are shifted along the temperature-axis (as well as in the vertical direction) in a systematic way when the load, sliding velocity, etc., are varied. These shifts also appear to be highly correlated with one another, for example, an increase in temperature producing effects that are similar to *decreasing* the sliding speed or load.

Such effects are also commonly observed in other energy dissipating phenomena such as polymer viscoelasticity [276], and it is likely that a similar physical mechanism is at the heart all of such phenomena. A possible molecular process underlying the energy dissipation of chain molecules during boundary layer sliding is illustrated in Fig. 9.29, which shows the three main dynamic phase states of boundary monolayers.

In contrast to the characteristic relaxation time associated with fluid lubricants, it has been established that for unlubricated (dry, solid, rough) surfaces, there is a characteristic memory distance that must be exceeded before the system looses all memory of its initial state (original surface topography). The underlying mechanism for a characteristic distance was first used to successfully explain rock mechanics and earthquake faults [278] and, more recently, the tribological behavior of unlubricated surfaces of ceramics, paper and elastomeric polymers [252, 279]. Recent experiments [275, 280, 281] suggest that fluid lubricants composed of complex branched-chained or polymer molecules may also have characteristic distances (in addition to characteristic relaxation times) associated with their tribological behavior – the characteristic distance being the total sliding distance that must be exceeded before the system reaches its steady-state tribological conditions (see Sect. 9.8.3).

Abrupt vs. Continuous Transitions Between Smooth and Stick–Slip Sliding

An understanding of stick–slip is of great practical importance in tribology [282], since these spikes are the major cause of damage and wear of moving parts. Stick–

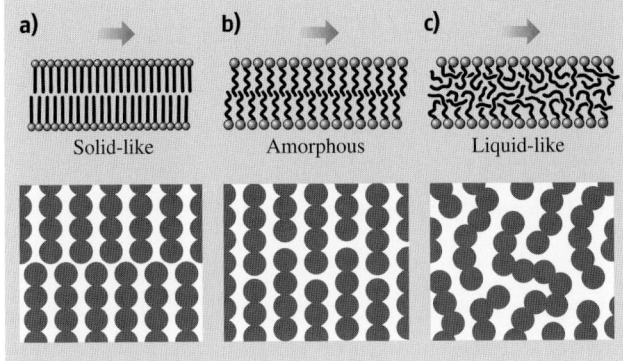

Fig. 9.29. Different dynamic phase states of boundary monolayers during adhesive contact and/or frictional sliding. Solid-like (**a**) and liquid-like monolayers (**c**) exhibit low adhesion hysteresis and friction. Increasing the temperature generally shifts a system from the left to the right. Changing the load, sliding velocity, or other experimental conditions can also change the dynamic phase state of surface layers, as shown in Fig. 9.28. (After [233] with permission)

slip motion is a very common phenomenon and is also the cause of sound generation (the sound of a violin string, a squeaking door, or the chatter of machinery), sensory perception (taste texture and feel), earthquakes, granular flow, nonuniform fluid flow such as the spurting flow of polymeric liquids, etc. In the previous section, the stick–slip motion arising from freezing–melting transitions in thin interfacial films was described. There are other mechanisms that can give rise to stick–slip friction, which will now be considered. However, before proceeding with this, it is important to clarify exactly what one is measuring during a friction experiment.

Most tribological systems and experiments can be described in terms of an equivalent mechanical circuit with certain characteristics. The friction force F_0, which is generated at the surfaces, is generally measured as F at some other place in the setup. The mechanical coupling between the two may be described in terms of a simple elastic stiffness or compliance, K, or as more complex nonelastic coefficients, depending on the system. The distinction between F and F_0 is important because in almost all practical cases, the applied, measured, or detected force, F, is *not* the same as the "real" or "intrinsic" friction force, F_0, generated at the surfaces. F and F_0 are coupled in a way that depends on the mechanical construction of the system, for example, the axle of a car wheel that connects it to the engine. This coupling can be modeled as an elastic spring of stiffness K and mass m. This is the simplest type of mechanical coupling and is also the same as in SFA and AFM type experiments. More complicated real systems can be reduced to a system of springs and dashpots, as described by *Peachey* et al. [283] and *Luengo* et al. [47].

We now consider four different models of stick–slip friction, where the mechanical couplings are assumed to be of the simple elastic spring type. The first three mechanisms may be considered traditional or classical mechanisms or models [282],

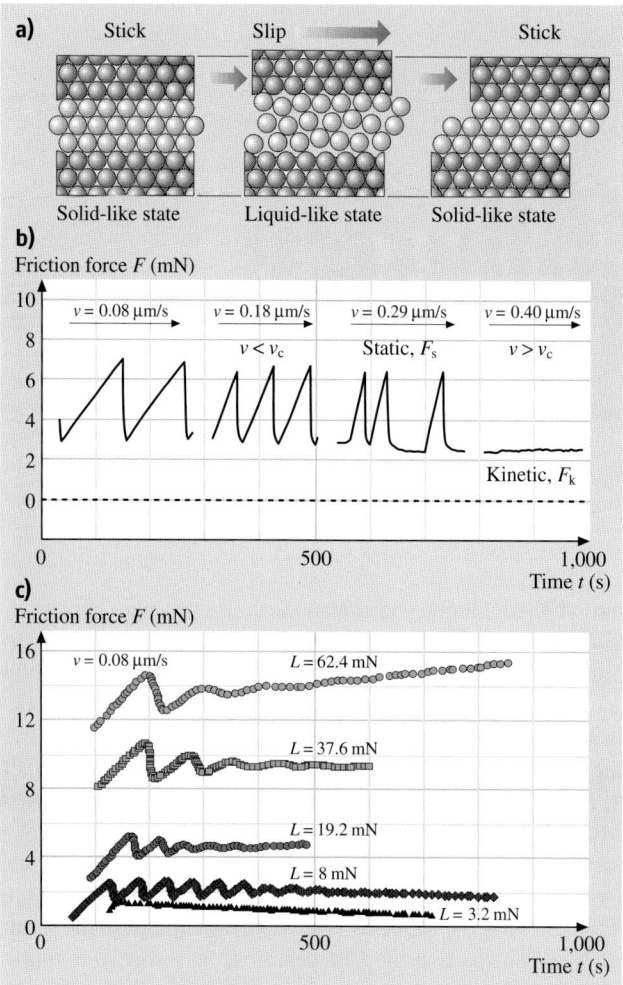

Fig. 9.30. (a) "Phase transitions" model of stick–slip where a thin liquid film alternately freezes and melts as it shears, shown here for 22 spherical molecules confined between two solid crystalline surfaces. In contrast to the velocity-dependent friction model, the intrinsic friction force is assumed to change abruptly (at the transitions), rather than smooth or continuously. The resulting stick–slip is also different, for example, the peaks are sharper and the stick–slip disappears above some critical velocity v_c. Note that while the slip displacement is here shown to be only two lattice spacings, in most practical situations it is much larger, and that freezing and melting transitions at surfaces or in thin films may not be the same as freezing or melting transitions between the bulk solid and liquid phases. (**b**) Exact reproduction of a chart-recorder trace of the friction force for hexadecane between two untreated mica surfaces at increasing sliding velocity v, plotted as a function of time. In general, with increasing sliding speed, the stick–slip spikes increase in frequency and decrease in magnitude. As the critical sliding velocity v_c is approached, the spikes become erratic, eventually disappearing altogether at $v = v_c$.

Fig. 9.30. At higher sliding velocities the sliding continues smoothly in the kinetic state. Such friction traces are fairly typical for simple liquid lubricants and dry boundary lubricant systems (see Fig. 9.31a) and may be referred to as the "conventional" type of static–kinetic friction (in contrast to Fig. 9.25). Experimental conditions: contact area $A = 4 \times 10^{-9}\,\mathrm{m}^2$, load $L = 10\,\mathrm{mN}$, film thickness $D = 0.4$–$0.8\,\mathrm{nm}$, $v = 0.08$–$0.4\,\mathrm{\mu m\,s}^{-1}$, $v_c \approx 0.3\,\mathrm{\mu m\,s}^{-1}$, atmosphere: dry N_2 gas, $T = 18\,°\mathrm{C}$. [(**a**) and (**b**) after [277] with permission. Copyright 1993 American Chemical Society.] (**c**) Friction response of a thin squalane (a branched hydrocarbon) film at different loads and a constant sliding velocity $v = 0.08\,\mathrm{\mu m\,s}^{-1}$, slightly above the critical velocity for this system at low loads. Initally, with increasing load, the stick–slip amplitude and the mean friction force decrease with sliding time or sliding distance. However, at high loads or pressures, the mean friction force increases with time, and the stick–slip takes on a more symmetrical, sinusoidal shape. At all loads investigated, the stick–slip component gradually decayed as the friction proceeded towards smooth sliding. (*Gourdon and Israelachvili*, unpublished data)

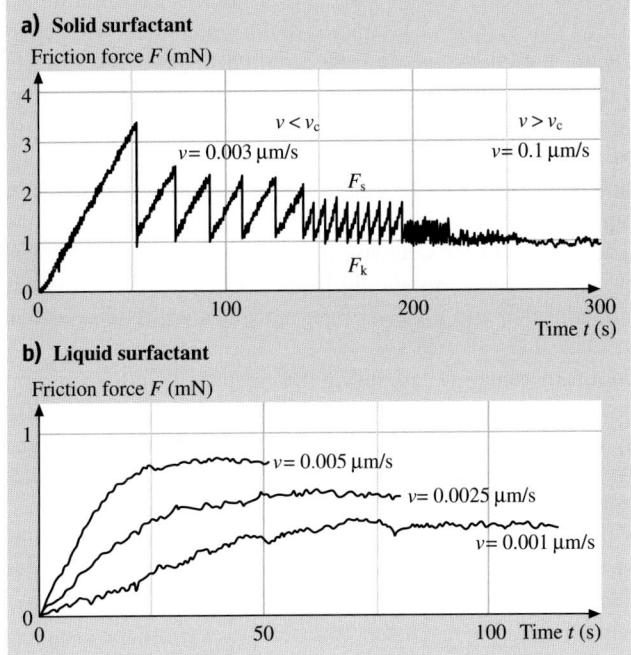

Fig. 9.31. (**a**) Exact reproduction of chart-recorder trace for the friction of closely packed surfactant monolayers (L-α-dimirystoyl-phosphatidyl-ethanolamine, DMPE) on mica (dry boundary friction) showing qualitatively similar behavior to that obtained with a liquid hexadecane film (Fig. 9.30b). In this case, $L = 0$, $v_c \approx 0.1\,\mathrm{\mu m\,s}^{-1}$, atmosphere: dry N_2 gas, $T = 25\,°\mathrm{C}$. (**b**) Sliding typical of liquid-like monolayers, here shown for calcium alkylbenzene sulfonate in dry N_2 gas at $T = 25\,°\mathrm{C}$ and $L = 0$. (After [211] with permission. Copyright 1993 American Chemical Society)

the fourth is essentially the same as the freezing–melting phase-transition model described in Sect. 9.8.3.

Rough Surfaces or Surface Topology Model Rapid slips can occur whenever an asperity on one surface goes over the top of an asperity on the other surface. The extent of the slip will depend on asperity heights and slopes, on the speed of sliding, and on the elastic compliance of the surfaces and the moving stage. As in all cases of stick–slip motion, the driving velocity v may be constant, but the resulting motion at the surfaces v_0 will display large slips. This type of stick–slip has been described by *Rabinowicz* [282]. It will not be of much concern here since it is essentially a noise-type fluctuation, resulting from surface imperfections rather than from the intrinsic interaction between two surfaces. Actually, at the atomic level, the regular atomic-scale corrugations of surfaces can lead to periodic stick–slip motion of the type shown here. This is what is sometimes measured by AFM tips [9, 54, 231, 272].

Distance-Dependent or Creep Model Another theory of stick–slip, observed in solid-on-solid sliding, is one that involves a characteristic *distance*, but also a characteristic time, τ_s, this being the characteristic time required for two asperities to increase their adhesion strength after coming into contact. Originally proposed by *Rabinowicz* [282, 284], this model suggests that two rough macroscopic surfaces adhere through their microscopic asperities of characteristic length. During shearing, each surface must first creep this distance (the size of the contacting junctions) after which the surfaces continue to slide, but with a lower (kinetic) friction force than the original (static) value. The reason for the decrease in the friction force is that even though, on average, new asperity junctions should form as rapidly as the old ones break, the time-dependent adhesion and friction of the new ones will be lower than the old ones.

The friction force, therefore, remains high during the creep stage of the slip. But once the surfaces have moved the characteristic distance, the friction rapidly drops to the kinetic value. For any system where the kinetic friction is less than the static force (or one that has a negative slope over some part of its F_0 versus v_0 curve) will exhibit regular stick–slip sliding motion for certain values of K, m, and driving velocity, v.

This type of friction has been observed in a variety of dry (unlubricated) systems such as paper-on-paper [285, 286] and steel-on-steel [284, 287, 288]. This model is also used extensively in geologic systems to analyze rock-on-rock sliding [289, 290]. While originally described for adhering macroscopic asperity junctions, the distance-dependent model may also apply to molecularly smooth surfaces. For example, for polymer lubricant films, the characteristic length would now be the chain–chain entanglement length, which could be much larger in a confined geometry than in the bulk.

Velocity-Dependent Friction Model In contrast to the two friction models mentioned above, which apply mainly to unlubricated, solid-on-solid contacts, the stick–slip of surfaces with thin liquid films between them is better described by other mechanisms. The velocity-dependent friction model is the most studied mechanism of stick–slip and, until recently, was considered to be the only cause of intrinsic stick–slip. If a friction force decreases with increasing sliding velocity, as occurs with boundary films exhibiting shear thinning, the force (F_s) needed to initiate motion will be higher than the force (F_k) needed to maintain motion.

A decreasing intrinsic friction force F_0 with sliding velocity v_0 results in the sliding surface or stage moving in a periodic fashion, where during each cycle rapid acceleration is followed by rapid deceleration. So long as the drive continues to move at a fixed velocity v, the surfaces will continue to move in a periodic fashion punctuated by abrupt stops and starts whose frequency and amplitude depend not only on the function $F_0(v_0)$, but also on the stiffness K and mass m of the moving stage, and on the starting conditions at $t = 0$.

More precisely, the motion of the sliding surface or stage can be determined by solving the following differential equation:

$$m\ddot{x} = (F_0 - F) = F_0 - (x_0 - x)K$$
$$\text{or} \quad m\ddot{x} + (x_0 - x)K - F_0 = 0, \quad (9.47)$$

where $F_0 = F_0(x_0, v_0, t)$ is the intrinsic or "real" friction force at the shearing surfaces, F is the force on the spring (the externally applied or measured force), and $(F_0 - F)$ is the force on the stage. To fully solve (9.47), one must also know the initial (starting) conditions at $t = 0$, and the driving or steady-state conditions at finite t. Commonly, the driving condition is: $x = 0$ for $t < 0$ and $x = vt$ for $t > 0$, where v = constant. In other systems, the appropriate driving condition may be F = constant.

Various, mainly phenomenological, forms for $F_0 = F_0(x_0, v_0, t)$ have been proposed to explain various kinds of stick–slip phenomena. These models generally assume a particular functional form for the friction as a function of velocity only, $F_0 = F_0(v_0)$, and they may also contain a number of mechanically coupled elements comprising the stage [291, 292]. One version is a two-state model characterized by two friction forces, F_s and F_k, which is a simplified version of the phase transitions model described below. More complicated versions can have a rich F–v spectrum, as proposed by *Persson* [293]. Unless the experimental data is very detailed and extensive, these models cannot generally distinguish between different types of mechanisms. Neither do they address the basic question of the *origin* of the friction force, since this is assumed to begin with.

Experimental data has been used to calculate the friction force as a function of velocity *within* an individual stick–slip cycle [294]. For a macroscopic granular material confined between solid surfaces, the data shows a velocity-weakening friction force during the first half of the slip. However, the data also shows a hysteresis loop in the friction–velocity plot, with a different behavior in the deceleration half of the slip phase. Similar results were observed for a 1–2 nm liquid lubricant film between mica surfaces [275]. These results indicate that a purely velocity-dependent friction law is insufficient to describe such systems, and an additional element such as the *state* of the confined material must be considered.

Phase Transitions Model In recent molecular dynamics computer simulations it has been found that thin interfacial films undergo first-order phase transitions between solid-like and liquid-like states during sliding [201, 273]. It has been suggested that this is responsible for the observed stick–slip behavior of simple isotropic liquids between two solid crystalline surfaces. With this interpretation, stick–slip is seen to arise because of the abrupt change in the flow properties of a film at a transition [224, 225, 261], rather than the gradual or continuous change, as occurs in the previous example. Other computer simulations indicate that it is the stick–slip that induces a disorder ("shear-melting") in the film, not the other way around [295].

An interpretation of the well-known phenomenon of decreasing coefficient of friction with increasing sliding velocity has been proposed by *Thompson* and *Robbins* [201] based on their computer simulation. This postulates that it is not the friction that changes with sliding speed v, but rather the time various parts of the system spend in the sticking and sliding modes. In other words, at any instant during sliding, the friction at any local region is always F_s or F_k, corresponding to the "static" or "kinetic" values. The measured frictional force, however, is the sum of all these discrete values averaged over the whole contact area. Since as v increases each local region spends more time in the sliding regime (F_k) and less in the sticking regime (F_s), the overall friction coefficient falls. One may note that this interpretation reverses the traditional way that stick–slip has been explained. Rather than invoking a decreasing friction with velocity to explain stick–slip, it is now the more fundamental stick–slip phenomenon that is producing the apparent decrease in the friction force with increasing sliding velocity. This approach has been studied analytically by *Carlson* and *Batista* [296], with a comprehensive rate- and state- dependent friction force law. This model includes an analytic description of the freezing–melting transitions of a film, resulting in a friction force that is a function of sliding velocity in a natural way. This model predicts a full range of stick–slip behavior observed experimentally.

An example of the rate- and state-dependent model is observed when shearing thin films of OMCTS between mica surfaces [297, 298]. In this case, the static friction between the surfaces is dependent on the time that the surfaces are at rest with respect to each other, while the intrinsic kinetic friction $F_{k,0}$ is relatively constant over the range of velocities. At slow driving velocities, the system responds with stick–slip sliding with the surfaces reaching maximum static friction before each slip event, and the amplitude of the stick–slip, $F_s - F_k$, is relatively constant. As the driving velocity increases, the static friction decreases as the time at relative rest becomes shorter with respect to the characteristic time of the lubricant film. As the static friction decreases with increasing drive velocity, it eventually equals the intrinsic kinetic friction $F_{k,0}$, which defines the critical velocity v_c, above which the surfaces slide smoothly without the jerky stick–slip motion.

The above classifications of stick–slip are not exclusive, and molecular mechanisms of real systems may exhibit aspects of different models simultaneously. They do, however, provide a convenient classification of existing models and indicate which experimental parameters should be varied to test the different models.

Critical Velocity for Stick–Slip For any given set of conditions, stick–slip disappears above some critical sliding velocity v_c, above which the motion continues smoothly in the liquid-like or kinetic state. The critical velocity is well described by two simple equations. Both are based on the phase transition model, and both include some parameter associated with the inertia of the measuring instrument. The first equation is based on both experiments and simple theoretical modeling [277]:

$$v_c \approx \frac{(F_s - F_k)}{5K\tau_0}, \tag{9.48}$$

where τ_0 is the *characteristic nucleation time* or freezing time of the film. For example, inserting the following typically measured values for a $\sim$ 1-nm-thick hexadecane film between mica: $(F_s - F_k) \approx 5$ mN, spring constant $K \approx 500$ N m^{-1}, and nucleation time [277] $\tau_0 \approx 5$ s, we obtain $v_c \approx 0.4\,\mu\text{m s}^{-1}$, which is close to typically measured values (Fig. 9.30b).

The second equation is based on computer simulations [273]:

$$v_c \approx 0.1\sqrt{\frac{F_s \sigma}{m}}, \tag{9.49}$$

where σ is a molecular dimension and m is the mass of the stage. Again, inserting typical experimental values into this equation, viz., $m \approx 20$ g, $\sigma \approx 0.5$ nm, and $(F_s - F_k) \approx 5$ mN as before, we obtain $v_c \approx 0.3\,\mu\text{m s}^{-1}$, which is also close to measured values.

Stick–slip also disappears above some critical temperature T_c, which is not the same as the melting temperature of the bulk fluid. Certain correlations have been found between v_c and T_c and between various other tribological parameters that appear to be consistent with the principle of time–temperature superposition (see Sect. 9.8.3), similar to that occurring in viscoelastic polymer fluids [276, 299, 300].

Recent work on the coupling between the mechanical resonances of the sliding system and molecular-scale relaxations [235, 301–303] has resulted in a better understanding of a phenomenon previously noted in various engineering applications: the vibrating of one of the sliding surfaces perpendicularly to the sliding direction can lead to a significant reduction of the friction. At certain oscillation amplitudes and a frequency higher than the molecular-scale relaxation frequency, stick–slip friction can be eliminated and replaced by an ultralow kinetic friction state.

9.9 Role of Molecular Shape and Surface Structure in Friction

The above scenario is already quite complicated, and yet this is the situation for the simplest type of experimental system. The factors that appear to determine the critical velocity v_c depend on the type of liquid between the surfaces (as well as on the surface lattice structure). Small spherical molecules such as cyclohexane and OMCTS have been found to have very high v_c, which indicates that these molecules can rearrange relatively quickly in thin films. Chain molecules and especially branched chain molecules have been found to have much lower v_c, which is to be expected, and such liquids tend to slide smoothly or with erratic stick–slip [275], rather than in a stick–slip fashion (see Table 9.4). With highly asymmetric molecules, such as multiply branched isoparaffins and polymer melts, no regular spikes or stick–slip behavior occurs at any speed, since these molecules can never order themselves sufficiently to solidify. Examples of such liquids are some perfluoropolyethers and polydimethylsiloxanes (PDMS).

Recent computer simulations [274, 304, 305] of the structure, interaction forces, and tribological behavior of chain molecules between two shearing surfaces indicate that both linear *and* singly or doubly branched chain molecules order between two flat surfaces by aligning into discrete layers parallel to the surfaces. However, in the case of the weakly branched molecules, the expected oscillatory forces do not appear because of a complex cancellation of entropic and enthalpic contributions to the interaction free energy, which results in a monotonically smooth interaction, exhibiting a weak energy minimum rather than the oscillatory force profile that is characteristic of linear molecules. During sliding, however, these molecules can be induced to further align, which can result in a transition from smooth to stick–slip sliding.

Table 9.4 shows the trends observed with some organic and polymeric liquid between smooth mica surfaces. Also listed are the bulk viscosities of the liquids. From the data of Table 9.4 it appears that there is a direct correlation between the shapes of molecules and their coefficient of friction or effectiveness as lubricants (at least at low shear rates). Small spherical or chain molecules have high friction with stick–slip, because they can pack into ordered solid-like layers. In contrast, longer chained and irregularly shaped molecules remain in an entangled, disordered, fluid-like state even in very thin films, and these give low friction and smoother sliding. It is probably for this reason that irregularly shaped branched chain molecules are usually better lubricants. It is interesting to note that the friction coefficient generally decreases as the bulk viscosity of the liquids *increases*. This unexpected trend occurs because the factors that are conducive to low friction are generally conducive to high viscosity. Thus molecules with side groups such as branched alkanes and polymer melts usually have higher bulk viscosities than their linear homologues for obvious reasons. However, in thin films the linear molecules have higher shear stresses, because of their ability to become ordered. The only exception to the above correlations is water, which has been found to exhibit both low viscosity *and* low friction (see Fig. 9.20, and Sect. 9.7.3). In addition, the presence of water can drastically lower the friction and eliminate the stick–slip of hydrocarbon liquids when the sliding surfaces are hydrophilic.

If an "effective" viscosity, η_{eff}, were to be calculated for the liquids of Table 9.4, the values would be many orders of magnitude higher than those of the bulk liquids. This can be demonstrated by the following simple calculation based on the usual equation for Couette flow (see (9.44)):

$$\eta_{\text{eff}} = F_k D / A v , \qquad (9.50)$$

where F_k is the kinetic friction force, D is the film thickness, A the contact area, and v the sliding velocity. Using typical values for experiments with hexadecane [277]: $F_k = 5\,\text{mN}$, $D = 1\,\text{nm}$, $A = 3 \times 10^{-9}\,\text{m}^2$, and $v = 1\,\mu\text{m s}^{-1}$, one gets $\eta_{\text{eff}} \approx 2000\,\text{Ns m}^{-2}$, or 20 000 Poise, which is $\sim 10^6$ times higher than the bulk viscosity, η_{bulk}, of the liquid. It is instructive to consider that this very high effective

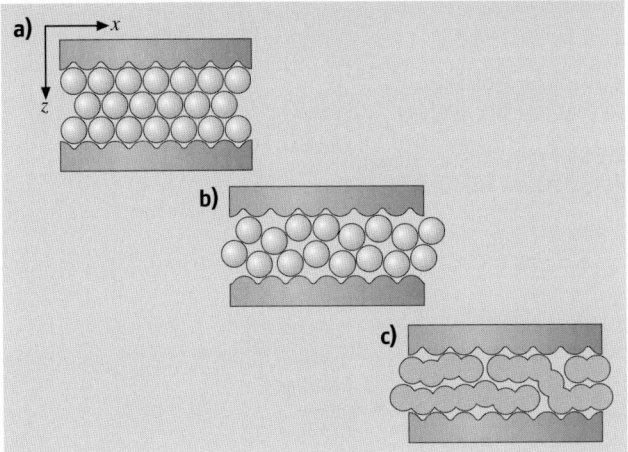

Fig. 9.32. Schematic view of interfacial film composed of spherical molecules under a compressive pressure between two solid crystalline surfaces. (**a**) If the two surface lattices are free to move in the x-y-z directions, so as to attain the lowest energy state, they could equilibrate at values of x, y, and z, which induce the trapped molecules to become "epitaxially" ordered into a "solid-like" film. (**b**) Similar view of trapped molecules between two solid surfaces that are not free to adjust their positions, for example, as occurs in capillary pores or in brittle cracks. (**c**) Similar to (**a**), but with chain molecules replacing the spherical molecules in the gap. These may not be able to order as easily as do spherical molecules even if x, y, and z can adjust, resulting in a situation that is more akin to (**b**). (After [277] with permission. Copyright 1993 American Chemical Society)

viscosity nevertheless still produces a low friction force or friction coefficient μ of about 0.25. It is interesting to speculate that if a 1 nm film were to exhibit bulk viscous behavior, the friction coefficient under the same sliding conditions would be as low as 0.000 001. While such a low value has never been reported for any tribological system, one may consider it a theoretical lower limit that could conceivably be attained under certain experimental conditions.

Various studies [44, 219, 220, 222] have shown that confinement and load generally increase the effective viscosity and/or relaxation times of molecules, suggestive of an increased glassiness or solid-like behavior (Figs. 9.32 and 9.33). This is in marked contrast to studies of liquids in small confining capillaries where the opposite effects have been observed [306, 307]. The reason for this is probably because the two modes of confinement are different. In the former case (confinement of molecules between two structured solid surfaces), there is generally little opposition to any lateral or vertical displacement of the two surface lattices relative to each other. This means that the two lattices can shift in the x-y-z plane (Fig. 9.32a) to accommodate the trapped molecules in the most crystallographically commensurate or epitaxial way, which would favor an ordered, solid-like state. In contrast, the walls of capillaries are rigid and cannot easily move or adjust to accommodate the confined molecules (Fig. 9.32b), which will therefore be forced into a more disordered, liquid-

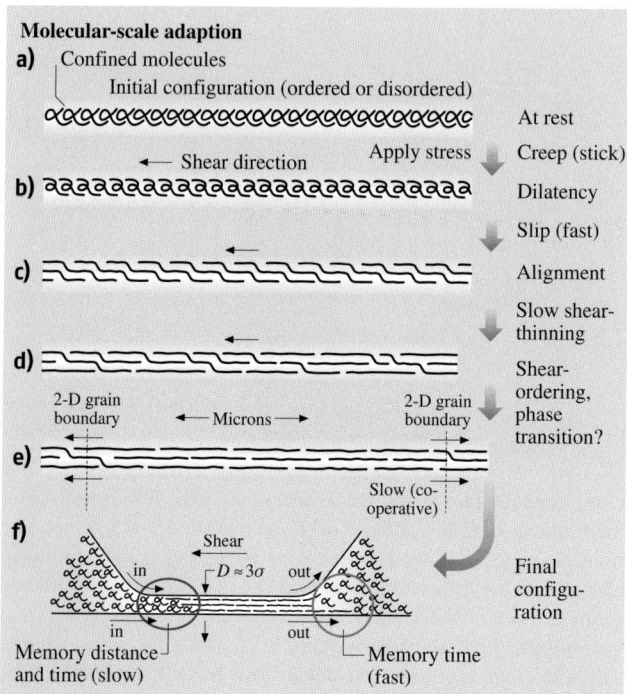

Fig. 9.33. Schematic representation of the film under shear. (**a**) The lubricant molecules are just confined, but not oriented in any particular direction. Because of the need to shear, the film dilates (**b**). The molecules disentangle (**c**) and get oriented in a certain direction related to the shear direction (**d**). (**e**) Slowly evolving domains grow inside the contact region. These macroscopic domains are responsible for the long relaxation times. (**f**) At the steady-state, a continuous gradient of confinement time and molecular order is established in the contact region, which is different for molecules adsorbed on the upper and lower surfaces. Molecules entering into the contact are not oriented or ordered. The required sliding distance to modify their state defines a characteristic distance. Molecules leaving the contact region need some (short) characteristic time to regain their bulk, unconfined configuration. (After [281] with permission. Copyright 2000 American Chemical Society)

like state (unless the capillary wall geometry and lattice are *exactly* commensurate with the liquid molecules, as occurs in certain zeolites [307]).

Experiments have demonstrated the effects of surface lattice mismatch on the friction between surfaces [203, 204, 308]. Similar to the effects of lattice mismatch on adhesion (Fig. 9.11), the static friction of a confined liquid film is maximum when the lattices of the confining surfaces are aligned. For OMCTS confined between mica surfaces [204] the static friction was found to vary by more than a factor of 4, while for bare mica surfaces the variation was by a factor of 3.5 [308]. In contrast to the sharp variations in adhesion energy over small twist angles, the variations in friction as a function of twist angle were much more broad both in magnitude and angular spread. Similar variations in friction as a function of twist or misfit angles have also been observed in computer simulations [309].

Robbins et al. [274] computed the friction forces of two clean crystalline surfaces as a function of the angle between their surface lattices. They found that for all non-zero angles (finite "twist" angles) the friction forces fell to zero due to incommensurability effects. They further found that submonolayer amounts of organic or other impurities trapped between two incommensurate surfaces can generate a finite friction force. They, therefore, concluded that any finite friction force measured between incommensurate surfaces is probably due to such "third body" effects.

With rough surfaces, i.e., those that have *random* protrusions rather than being periodically structured, we expect a smearing out of the correlated intermolecular interactions that are involved in film freezing and melting (and in phase transitions in general). This should effectively eliminate the highly regular stick–slip and may also affect the location of the slipping planes [123, 252, 279]. The stick–slip friction of "real" surfaces, which are generally rough, may, therefore, be quite different from those of perfectly smooth surfaces composed of the same material. We should note, however, that even between rough surfaces, most of the contacts occur between the tips of microscopic asperities, which may be smooth over their microscopic contact area.

References

1. K. B. Lodge. Techniques for the measurement of forces between solids. *Adv. Colloid Interface Sci.*, 19:27–73, 1983.
2. J. N. Israelachvili. *Intermolecular and Surface Forces*. Academic Press, 2nd edition, 1991.
3. P. F. Luckham and B. A. de L. Costello. Recent developments in the measurement of interparticle forces. *Adv. Colloid Interface Sci.*, 44:183–240, 1993.
4. P. M. Claesson, T. Ederth, V. Bergeron, and M. W. Rutland. Techniques for measuring surface forces. *Adv. Colloid Interface Sci.*, 67:119–183, 1996.
5. V. S. J. Craig. An historical review of surface force measurement techniques. *Colloids Surf. A*, 129–130:75–94, 1997.
6. J. N. Israelachvili and G. E. Adams. Measurements of forces between two mica surfaces in aqueous electrolyte solutions in the range 0–100 nm. *J. Chem. Soc. Faraday Trans. I*, 74:975–1001, 1978.
7. G. Binnig, C. F. Quate, and C. Gerber. Atomic force microscope. *Phys. Rev. Lett.*, 56:930–933, 1986.
8. W. A. Ducker, T. J. Senden, and R. M. Pashley. Direct measurement of colloidal forces using an atomic force microscope. *Nature*, 353:239–241, 1991.
9. E. Meyer, R. M. Overney, K. Dransfeld, and T. Gyalog. *Nanoscience: Friction and Rheology on the Nanometer Scale*. World Scientific, 1998.
10. J. N. Israelachvili. *Surface Forces and Microrheology of Molecularly Thin Liquid Films*, pages 267–319. CRC Press, 1995.
11. J. N. Israelachvili. Measurements of the viscosity of thin fluid films between two surfaces with and without adsorbed polymers. *Colloid Polym. Sci.*, 264:1060–1065, 1986.

12. J. P. Montfort and G. Hadziioannou. Equilibrium and dynamic behavior of thin films of a perfluorinated polyether. *J. Chem. Phys.*, 88:7187–7196, 1988.
13. A. Dhinojwala and S. Granick. Surface forces in the tapping mode: Solvent permeability and hydrodynamic thickness of adsorbed polymer brushes. *Macromolecules*, 30:1079–1085, 1997.
14. B. V. Derjaguin. Untersuchungen über die reibung und adhäsion, iv. theorie des anhaftens kleiner tcilchen. *Kolloid Z.*, 69:155–164, 1934.
15. K. L. Johnson, K. Kendall, and A. D. Roberts. Surface energy and the contact of elastic solids. *Proc. R. Soc. London A*, 324:301–313, 1971.
16. J. N. Israelachvili and P. M. McGuiggan. Adhesion and short-range forces between surfaces. part 1: New apparatus for surface force measurements. *J. Mater. Res.*, 5:2223–2231, 1990.
17. J. N. Israelachvili. Thin film studies using multiple-beam interferometry. *J. Colloid Interface Sci.*, 44:259–272, 1973.
18. Y. L. Chen, T. Kuhl, and J. Israelachvili. Mechanism of cavitation damage in thin liquid films: Collapse damage vs. inception damage. *Wear*, 153:31–51, 1992.
19. M. Heuberger, G. Luengo, and J. Israelachvili. Topographic information from multiple beam interferometry in the surface forces apparatus. *Langmuir*, 13:3839–3848, 1997.
20. R. M. Pashley. Dlvo and hydration forces between mica surfaces in Li^+, Na^+, K^+, and Cs^+ electrolyte solutions: A correlation of double-layer and hydration forces with surface cation exchange properties. *J. Colloid Interface Sci.*, 83:531–546, 1981.
21. R. M. Pashley. Hydration forces between mica surfaces in electrolyte solution. *Adv. Colloid Interface Sci.*, 16:57–62, 1982.
22. R. G. Horn, D. T. Smith, and W. Haller. Surface forces and viscosity of water measured between silica sheets. *Chem. Phys. Lett.*, 162:404–408, 1989.
23. R. G. Horn, D. R. Clarke, and M. T. Clarkson. Direct measurements of surface forces between sapphire crystals in aqueous solutions. *J. Mater. Res.*, 3:413–416, 1988.
24. W. W. Merrill, A. V. Pocius, B. V. Thakker, and M. Tirrell. Direct measurement of molecular level adhesion forces between biaxially oriented solid polymer films. *Langmuir*, 7:1975–1980, 1991.
25. J. Klein. Forces between mica surfaces bearing adsorbed macromolecules in liquid media. *J. Chem. Soc. Faraday Trans. I*, 79:99–118, 1983.
26. S. S. Patel and M. Tirrell. Measurement of forces between surfaces in polymer fluids. *Annu. Rev. Phys. Chem.*, 40:597–635, 1989.
27. H. Watanabe and M. Tirrell. Measurements of forces in symmetric and asymmetric interactions between diblock copolymer layers adsorbed on mica. *Macromolecules*, 26:6455–6466, 1993.
28. T. L. Kuhl, D. E. Leckband, D. D. Lasic, and J. N. Israelachvili. *Modulation and modeling of interaction forces between lipid bilayers exposing terminally grafted polymer chains*, pages 73–91. CRC Press, 1995.
29. J. Klein. Shear, friction, and lubrication forces between polymer-bearing surfaces. *Annu. Rev. Mater. Sci.*, 26:581–612, 1996.
30. M. Ruths, D. Johannsmann, J. Rühe, and W. Knoll. Repulsive forces and relaxation on compression of entangled, polydisperse polystyrene brushes. *Macromolecules*, 33:3860–3870, 2000.
31. C. A. Helm, J. N. Israelachvili, and P. M. McGuiggan. Molecular mechanisms and forces involved in the adhesion and fusion of amphiphilic bilayers. *Science*, 246:919–922, 1989.

32. Y. L. Chen, C. A. Helm, and J. N. Israelachvili. Molecular mechanisms associated with adhesion and contact angle hysteresis of monolayer surfaces. *J. Phys. Chem.*, 95:10736–10747, 1991.
33. D. E. Leckband, J. N. Israelachvili, F.-J. Schmitt, and W. Knoll. Long-range attraction and molecular rearrangements in receptor-ligand interactions. *Science*, 255:1419–1421, 1992.
34. J. Peanasky, H. M. Schneider, S. Granick, and C. R. Kessel. Self-assembled monolayers on mica for experiments utilizing the surface forces apparatus. *Langmuir*, 11:953–962, 1995.
35. C. J. Coakley and D. Tabor. Direct measurement of van der waals forces between solids in air. *J. Phys. D*, 11:L77–L82, 1978.
36. J. L. Parker and H. K. Christenson. Measurements of the forces between a metal surface and mica across liquids. *J. Chem. Phys.*, 88:8013–8014, 1988.
37. C. P. Smith, M. Maeda, L. Atanasoska, H. S. White, and D. J. McClure. Ultrathin platinum films on mica and the measurement of forces at the platinum/water interface. *J. Phys. Chem.*, 92:199–205, 1988.
38. S. J. Hirz, A. M. Homola, G. Hadziioannou, and C. W. Frank. Effect of substrate on shearing properties of ultrathin polymer films. *Langmuir*, 8:328–333, 1992.
39. J. M. Levins and T. K. Vanderlick. Reduction of the roughness of silver films by the controlled application of surface forces. *J. Phys. Chem.*, 96:10405–10411, 1992.
40. S. Steinberg, W. Ducker, G. Vigil, C. Hyukjin, C. Frank, M. Z. Tseng, D. R. Clarke, and J. N. Israelachvili. Van der waals epitaxial growth of α-alumina nanocrystals on mica. *Science*, 260:656–659, 1993.
41. G. Vigil, Z. Xu, S. Steinberg, and J. Israelachvili. Interactions of silica surfaces. *J. Colloid Interface Sci.*, 165:367–385, 1994.
42. M. Ruths, M. Heuberger, V. Scheumann, J. Hu, and W. Knoll. Confinement-induced film thickness transitions in liquid crystals between two alkanethiol monolayers on gold. *Langmuir*, 17:6213–6219, 2001.
43. J. Van Alsten and S. Granick. Molecular tribometry of ultrathin liquid films. *Phys. Rev. Lett.*, 61:2570–2573, 1988.
44. J. Van Alsten and S. Granick. Shear rheology in a confined geometry: Polysiloxane melts. *Macromolecules*, 23:4856–4862, 1990.
45. A. M. Homola, J. N. Israelachvili, M. L. Gee, and P. M. McGuiggan. Measurements of and relation between the adhesion and friction of two surfaces separated by molecularly thin liquid films. *J. Tribol.*, 111:675–682, 1989.
46. J. Klein and E. Kumacheva. Simple liquids confined to molecularly thin layers. i. confinement-induced liquid-to-solid phase transitions. *J. Chem. Phys.*, 108:6996–7009, 1998.
47. G. Luengo, F.-J. Schmitt, R. Hill, and J. Israelachvili. Thin film rheology and tribology of confined polymer melts: Contrasts with bulk properties. *Macromolecules*, 30:2482–2494, 1997.
48. E. Kumacheva. Interfacial friction measurements in surface force apparatus. *Prog. Surf. Sci.*, 58:75–120, 1998.
49. A. L. Weisenhorn, P. K. Hansma, T. R. Albrecht, and C. F. Quate. Forces in atomic force microscopy in air and water. *Appl. Phys. Lett.*, 54:2651–2653, 1989.
50. G. Meyer and N. M. Amer. Simultaneous measurement of lateral and normal forces with an optical-beam-deflection atomic force microscope. *Appl. Phys. Lett.*, 57:2089–2091, 1990.
51. E. L. Florin, V. T. Moy, and H. E. Gaub. Adhesion forces between individual ligand–receptor pairs. *Science*, 264:415–417, 1994.

52. G. U. Lee, D. A. Kidwell, and R. J. Colton. Sensing discrete streptavidin–biotin interactions with atomic force microscopy. *Langmuir*, 10:354–357, 1994.
53. D. Leckband and J. Israelachvili. Intermolecular forces in biology. *Quart. Rev. Biophys.*, 34:105–267, 2001.
54. C. M. Mate, G. M. McClelland, R. Erlandsson, and S. Chiang. Atomic-scale friction of a tungsten tip on a graphite surface. *Phys. Rev. Lett.*, 59:1942–1945, 1987.
55. R. W. Carpick and M. Salmeron. Scratching the surface: Fundamental investigations of tribology with atomic force microscopy. *Chem. Rev.*, 97:1163–1194, 1997.
56. J. P. Cleveland, S. Manne, D. Bocek, and P. K. Hansma. A nondestructive method for determining the spring constant of cantilevers for scanning force microscopy. *Rev. Sci. Instrum.*, 64:403–405, 1993.
57. J. E. Sader, J. W. M. Chon, and P. Mulvaney. Calibration of rectangular atomic force microscope cantilevers. *Rev. Sci. Instrum.*, 70:3967–3969, 1999.
58. Y. Liu, T. Wu, and D. F. Evans. Lateral force microscopy study on the shear properties of self-assembled monolayers of dialkylammonium surfactant on mica. *Langmuir*, 10:2241–2245, 1994.
59. C. Neto and V. S. J. Craig. Colloid probe characterization: Radius and roughness determination. *Langmuir*, 17:2097–2099, 2001.
60. R. G. Horn and J. N. Israelachvili. Molecular organization and viscosity of a thin film of molten polymer between two surfaces as probed by force measurements. *Macromolecules*, 21:2836–2841, 1988.
61. R. G. Horn, S. J. Hirz, G. Hadziioannou, C. W. Frank, and J. M. Catala. A reevaluation of forces measured across thin polymer films: Nonequilibrium and pinning effects. *J. Chem. Phys.*, 90:6767–6774, 1989.
62. O. I. Vinogradova, H.-J. Butt, G. E. Yakubov, and F. Feuillebois. Dynamic effects on force measurements. i. viscous drag on the atomic force microscope cantilever. *Rev. Sci. Instrum.*, 72:2330–2339, 2001.
63. E. Evans and D. Needham. Physical properties of surfactant bilayer membranes: Thermal transitions, elasticity, rigidity, cohesion, and colloidal interactions. *J. Phys. Chem.*, 91:4219–4228, 1987.
64. E. Evans and D. Needham. Attraction between lipid bilayer membranes in concentrated solutions of nonadsorbing polymers: Comparison of mean-field theory with measurements of adhesion energy. *Macromolecules*, 21:1822–1831, 1988.
65. S. E. Chesla, P. Selvaraj, and C. Zhu. Measuring two-dimensional receptor-ligand binding kinetics by micropipette. *Biophys. J.*, 75:1553–1572, 1998.
66. E. Evans, K. Ritchie, and R. Merkel. Sensitive force technique to probe molecular adhesion and structural linkages at biological interfaces. *Biophys. J.*, 68:2580–2587, 1995.
67. D. M. LeNeveu, R. P. Rand, and V. A. Parsegian. Measurements of forces between lecithin bilayers. *Nature*, 259:601–603, 1976.
68. A. Homola and A. A. Robertson. A compression method for measuring forces between colloidal particles. *J. Colloid Interface Sci.*, 54:286–297, 1976.
69. V. A. Parsegian, N. Fuller, and R. P. Rand. Measured work of deformation and repulsion of lecithin bilayers. *Proc. Nat. Acad. Sci. USA*, 76:2750–2754, 1979.
70. R. P. Rand and V. A. Parsegian. Hydration forces between phospholipid bilayers. *Biochim. Biophys. Acta*, 988:351–376, 1989.
71. S. Leikin, V. A. Parsegian, D. C. Rau, and R. P. Rand. Hydration forces. *Annu. Rev. Phys. Chem.*, 44:369–395, 1993.
72. S. Chu, J. E. Bjorkholm, A. Ashkin, and A. Cable. Experimental observation of optically trapped atoms. *Phys. Rev. Lett.*, 57:314–317, 1986.

73. A. Ashkin. Optical trapping and manipulation of neutral particles using lasers. *Proc. Nat. Acad. Sci. USA*, 94:4853–4860, 1997.
74. K. Visscher, S. P. Gross, and S. M. Block. Construction of multiple-beam optical traps with nanometer-resolution positioning sensing. *IEEE J. Sel. Top. Quantum Electron.*, 2:1066–1076, 1996.
75. D. C. Prieve and N. A. Frej. Total internal reflection microscopy: A quantitative tool for the measurement of colloidal forces. *Langmuir*, 6:396–403, 1990.
76. J. Rädler and E. Sackmann. On the measurement of weak repulsive and frictional colloidal forces by reflection interference contrast microscopy. *Langmuir*, 8:848–853, 1992.
77. P. Bongrand, C. Capo, J.-L. Mege, and A.-M. Benoliel. *Use of hydrodynamic flows to study cell adhesion*, pages 125–156. CRC Press, 1988.
78. G. Kaplanski, C. Farnarier, O. Tissot, A. Pierres, A.-M. Benoliel, A.-C. Alessi, S. Kaplanski, and P. Bongrand. Granulocyte–endothelium initial adhesion: Analysis of transient binding events mediated by E-selectin in a laminar shear flow. *Biophys. J.*, 64:1922–1933, 1993.
79. H. B. G. Casimir and D. Polder. The influence of retardation on the london–van der waals forces. *Phys. Rev.*, 73:360–372, 1948.
80. H. C. Hamaker. The london–van der waals attraction between spherical particles. *Physica*, 4:1058–1072, 1937.
81. E. M. Lifshitz. The theory of molecular attraction forces between solid bodies. *Sov. Phys. JETP (English translation)*, 2:73–83, 1956.
82. I. E. Dzyaloshinskii, E. M. Lifshitz, and L. P. Pitaevskii. The general theory of van der waals forces. *Adv. Phys.*, 10:165–209, 1961.
83. H. W. Fox and W. A. Zisman. The spreading of liquids on low-energy surfaces. iii. hydrocarbon surfaces. *J. Colloid Sci.*, 7:428–442, 1952.
84. B. V. Derjaguin and V. P. Smilga. Electrostatic component of the rolling friction force moment. *Wear*, 7:270–281, 1964.
85. R. G. Horn and D. T. Smith. Contact electrification and adhesion between dissimilar materials. *Science*, 256:362–364, 1992.
86. R. Budakian and S. J. Putterman. Correlation between charge transfer and stick–slip friction at a metal–insulator interface. *Phys. Rev. Lett.*, 85:1000–1003, 2000.
87. M. Grätzel. Photoelectrochemical cells. *Nature*, 414:338–344, 2001.
88. R. M. German. *Sintering Theory and Practice*. Wiley, 1996.
89. R. Budakian and S. J. Putterman. Time scales for cold welding and the origins of stick–slip friction. *Phys. Rev. B*, 65:235429/1–5, 2002.
90. U. Landman, W. D. Luedtke, N. A. Burnham, and R. J. Colton. Atomistic mechanisms and dynamics of adhesion, nanoindentation, and fracture. *Science*, 248:454–461, 1990.
91. U. Landman, W. D. Luedtke, and E. M. Ringer. Molecular dynamics simulations of adhesive contact formation and friction. *NATO Science Ser. E*, 220:463–510, 1992.
92. W. D. Luedtke and U. Landman. Solid and liquid junctions. *Comput. Mater. Sci.*, 1:1–24, 1992.
93. U. Landman and W. D. Luedtke. Interfacial junctions and cavitation. *MRS Bull.*, 18:36–44, 1993.
94. B. Bhushan, J. N. Israelachvili, and U. Landman. Nanotribology: Friction, wear and lubrication at the atomic scale. *Nature*, 374:607–616, 1995.
95. M. R. Sørensen, K. W. Jacobsen, and P. Stoltze. Simulations of atomic-scale sliding friction. *Phys. Rev. B*, 53:2101–2113, 1996.
96. A. Meurk, P. F. Luckham, and L. Bergström. Direct measurement of repulsive and attractive van der waals forces between inorganic materials. *Langmuir*, 13:3896–3899, 1997.

97. E. J. W. Verwey and J. T. G. Overbeek. *Theory of the Stability of Lyophobic Colloids.* Elsevier, 1st edition, 1948.
98. D. Y. C. Chan, R. M. Pashley, and L. R. White. A simple algorithm for the calculation of the electrostatic repulsion between identical charged surfaces in electrolyte. *J. Colloid Interface Sci.*, 77:283–285, 1980.
99. B. Derjaguin and L. Landau. Theory of the stability of strongly charged lyophobic sols and of the adhesion of strongly charged particles in solutions of electrolytes. *Acta Physicochim. URSS*, 14:633–662, 1941.
100. D. Chan, T. W. Healy, and L. R. White. Electrical double layer interactions under regulation by surface ionization equilibriums – dissimilar amphoteric surfaces. *J. Chem. Soc. Faraday Trans. 1*, 72:2844–2865, 1976.
101. G. S. Manning. Limiting laws and counterion condensation in polyelectrolyte solutions. i. colligative properties. *J. Chem. Phys.*, 51:924–933, 1969.
102. L. Guldbrand, V. Jönsson, H. Wennerström, and P. Linse. Electrical double-layer forces: A monte carlo study. *J. Chem. Phys.*, 80:2221–2228, 1984.
103. H. Wennerström, B. Jönsson, and P. Linse. The cell model for polyelectrolyte systems. exact statistical mechanical relations, monte carlo simulations, and the poisson–boltzmann approximation. *J. Chem. Phys.*, 76:4665–4670, 1982.
104. J. Marra. Effects of counterion specificity on the interactions between quaternary ammonium surfactants in monolayers and bilayers. *J. Phys. Chem.*, 90:2145–2150, 1986.
105. J. Marra. Direct measurement of the interaction between phosphatidylglycerol bilayers in aqueous-electrolyte solutions. *Biophys. J.*, 50:815–825, 1986.
106. R. G. Horn and J. N. Israelachvili. Direct measurement of structural forces between two surfaces in a nonpolar liquid. *J. Chem. Phys.*, 75:1400–1411, 1981.
107. H. K. Christenson, D. W. R. Gruen, R. G. Horn, and J. N. Israelachvili. Structuring in liquid alkanes between solid surfaces: Force measurements and mean-field theory. *J. Chem. Phys.*, 87:1834–1841, 1987.
108. M. L. Gee and J. N. Israelachvili. Interactions of surfactant monolayers across hydrocarbon liquids. *J. Chem. Soc. Faraday Trans.*, 86:4049–4058, 1990.
109. J. N. Israelachvili, S. J. Kott, M. L. Gee, and T. A. Witten. Forces between mica surfaces across hydrocarbon liquids: Effects of branching and polydispersity. *Macromolecules*, 22:4247–4253, 1989.
110. I. K. Snook and W. van Megen. Solvation forces in simple dense fluids i. *J. Chem. Phys.*, 72:2907–2913, 1980.
111. W. J. van Megen and I. K. Snook. Solvation forces in simple dense fluids ii. effect of chemical potential. *J. Chem. Phys.*, 74:1409–1411, 1981.
112. R. Kjellander and S. Marcelja. Perturbation of hydrogen bonding in water near polar surfaces. *Chem. Phys. Lett.*, 120:393–396, 1985.
113. P. Tarazona and L. Vicente. A model for the density oscillations in liquids between solid walls. *Mol. Phys.*, 56:557–572, 1985.
114. D. Henderson and M. Lozada-Cassou. A simple theory for the force between spheres immersed in a fluid. *J. Colloid Interface Sci.*, 114:180–183, 1986.
115. J. E. Curry and J. H. Cushman. Structure in confined fluids: Phase separation of binary simple liquid mixtures. *Tribol. Lett.*, 4:129–136, 1998.
116. M. Schoen, T. Gruhn, and D. J. Diestler. Solvation forces in thin films confined between macroscopically curved substrates. *J. Chem. Phys.*, 109:301–311, 1998.
117. F. Porcheron, B. Rousseau, M. Schoen, and A. H. Fuchs. Structure and solvation forces in confined alkane films. *Phys. Chem. Chem. Phys.*, 3:1155–1159, 2001.
118. H. K. Christenson. Forces between solid surfaces in a binary mixture of non-polar liquids. *Chem. Phys. Lett.*, 118:455–458, 1985.

119. H. K. Christenson and R. G. Horn. Solvation forces measured in non-aqueous liquids. *Chem. Scr.*, 25:37–41, 1985.
120. J. Israelachvili. Solvation forces and liquid structure, as probed by direct force measurements. *Acc. Chem. Res.*, 20:415–421, 1987.
121. H. K. Christenson. Non-dlvo forces between surfaces – solvation, hydration and capillary effects. *J. Disp. Sci. Technol.*, 9:171–206, 1988.
122. L. J. D. Frink and F. van Swol. Solvation forces between rough surfaces. *J. Chem. Phys.*, 108:5588–5598, 1998.
123. J. Gao, W. D. Luedtke, and U. Landman. Structures, solvation forces and shear of molecular films in a rough nano-confinement. *Tribol. Lett.*, 9:3–13, 2000.
124. H. E. Stanley and J. Teixeira. Interpretation of the unusual behavior of H_2O and D_2O at low temperatures: Tests of a percolation model. *J. Chem. Phys.*, 73:3404–3422, 1980.
125. H. van Olphen. *An Introduction to Clay Colloid Chemistry*. Wiley, 2nd edition, 1977.
126. U. Del Pennino, E. Mazzega, S. Valeri, A. Alietti, M. F. Brigatti, and L. Poppi. Interlayer water and swelling properties of monoionic montmorillonites. *J. Colloid Interface Sci.*, 84:301–309, 1981.
127. B. E. Viani, P. F. Low, and C. B. Roth. Direct measurement of the relation between interlayer force and interlayer distance in the swelling of montmorillonite. *J. Colloid Interface Sci.*, 96:229–244, 1983.
128. R. G. Horn. Surface forces and their action in ceramic materials. *J. Am. Ceram. Soc.*, 73:1117–1135, 1990.
129. B. V. Velamakanni, J. C. Chang, F. F. Lange, and D. S. Pearson. New method for efficient colloidal particle packing via modulation of repulsive lubricating hydration forces. *Langmuir*, 6:1323–1325, 1990.
130. R. R. Lessard and S. A. Zieminski. Bubble coalescence and gas transfer in aqueous electrolytic solutions. *Ind. Eng. Chem. Fundam.*, 10:260–269, 1971.
131. J. Israelachvili and R. Pashley. The hydrophobic interaction is long range, decaying exponentially with distance. *Nature*, 300:341–342, 1982.
132. R. M. Pashley, P. M. McGuiggan, B. W. Ninham, and D. F. Evans. Attractive forces between uncharged hydrophobic surfaces: Direct measurements in aqueous solutions. *Science*, 229:1088–1089, 1985.
133. P. M. Claesson, C. E. Blom, P. C. Herder, and B. W. Ninham. Interactions between water-stable hydrophobic langmuir–blodgett monolayers on mica. *J. Colloid Interface Sci.*, 114:234–242, 1986.
134. Ya. I. Rabinovich and B. V. Derjaguin. Interaction of hydrophobized filaments in aqueous electrolyte solutions. *Colloids Surf.*, 30:243–251, 1988.
135. J. L. Parker, D. L. Cho, and P. M. Claesson. Plasma modification of mica: Forces between fluorocarbon surfaces in water and a nonpolar liquid. *J. Phys. Chem.*, 93:6121–6125, 1989.
136. H. K. Christenson, J. Fang, B. W. Ninham, and J. L. Parker. Effect of divalent electrolyte on the hydrophobic attraction. *J. Phys. Chem.*, 94:8004–8006, 1990.
137. K. Kurihara, S. Kato, and T. Kunitake. Very strong long range attractive forces between stable hydrophobic monolayers of a polymerized ammonium surfactant. *Chem. Lett.*, pages 1555–1558, 1990.
138. Y. H. Tsao, D. F. Evans, and H. Wennerstrom. Long-range attractive force between hydrophobic surfaces observed by atomic force microscopy. *Science*, 262:547–550, 1993.
139. Ya. I. Rabinovich and R.-H. Yoon. Use of atomic force microscope for the measurements of hydrophobic forces between silanated silica plate and glass sphere. *Langmuir*, 10:1903–1909, 1994.

140. V. S. J. Craig, B. W. Ninham, and R. M. Pashley. Study of the long-range hydrophobic attraction in concentrated salt solutions and its implications for electrostatic models. *Langmuir*, 14:3326–3332, 1998.
141. P. Kékicheff and O. Spalla. Long-range electrostatic attraction between similar, charge-neutral walls. *Phys. Rev. Lett.*, 75:1851–1854, 1995.
142. H. K. Christenson and P. M. Claesson. Direct measurements of the force between hydrophobic surfaces in water. *Adv. Colloid Interface Sci.*, 91:391–436, 2001.
143. P. Attard. Nanobubbles and the hydrophobic attraction. *Adv. Colloid Interface Sci.*, 104:75–91, 2003.
144. M. Hato. Attractive forces between surfaces of controlled "hydrophobicity" across water: A possible range of "hydrophobic interactions" between macroscopic hydrophobic surfaces across water. *J. Phys. Chem.*, 100:18530–18538, 1996.
145. H. K. Christenson, P. M. Claesson, J. Berg, and P. C. Herder. Forces between fluorocarbon surfactant monolayers: Salt effects on the hydrophobic interaction. *J. Phys. Chem.*, 93:1472–1478, 1989.
146. N. I. Christou, J. S. Whitehouse, D. Nicholson, and N. G. Parsonage. A monte carlo study of fluid water in contact with structureless walls. *Faraday Symp. Chem. Soc.*, 16:139–149, 1981.
147. B. Jönsson. Monte carlo simulations of liquid water between two rigid walls. *Chem. Phys. Lett.*, 82:520–525, 1981.
148. S. Marcelja, D. J. Mitchell, B. W. Ninham, and M. J. Sculley. Role of solvent structure in solution theory. *J. Chem. Soc. Faraday Trans. II*, 73:630–648, 1977.
149. D. W. R. Gruen and S. Marcelja. Spatially varying polarization in water: A model for the electric double layer and the hydration force. *J. Chem. Soc. Faraday Trans. 2*, 79:225–242, 1983.
150. B. Jönsson and H. Wennerström. Image-charge forces in phospholipid bilayer systems. *J. Chem. Soc. Faraday Trans. 2*, 79:19–35, 1983.
151. D. Schiby and E. Ruckenstein. The role of the polarization layers in hydration forces. *Chem. Phys. Lett.*, 95:435–438, 1983.
152. A. Luzar, D. Bratko, and L. Blum. Monte carlo simulation of hydrophobic interaction. *J. Chem. Phys.*, 86:2955–2959, 1987.
153. P. Attard and M. T. Batchelor. A mechanism for the hydration force demonstrated in a model system. *Chem. Phys. Lett.*, 149:206–211, 1988.
154. K. Leung and A. Luzar. Dynamics of capillary evaporation. ii. free energy barriers. *J. Chem. Phys.*, 113:5845–5852, 2000.
155. D. Bratko, R. A. Curtis, H. W. Blanch, and J. M. Prausnitz. Interaction between hydrophobic surfaces with metastable intervening liquid. *J. Chem. Phys.*, 115:3873–3877, 2001.
156. P. G. de Gennes. Polymers at an interface. 2. interaction between two plates carrying adsorbed polymer layers. *Macromolecules*, 15:492–500, 1982.
157. J. M. H. M. Scheutjens and G. J. Fleer. Interaction between two adsorbed polymer layers. *Macromolecules*, 18:1882–1900, 1985.
158. E. A. Evans. Force between surfaces that confine a polymer solution: Derivation from self-consistent field theories. *Macromolecules*, 22:2277–2286, 1989.
159. H. J. Ploehn. Compression of polymer interphases. *Macromolecules*, 27:1627–1636, 1994.
160. S. Asakura and F. Oosawa. Interaction between particles suspended in solutions of macromolecules. *J. Polym. Sci.*, 33:183–192, 1958.
161. J. F. Joanny, L. Leibler, and P. G. de Gennes. Effects of polymer solutions on colloid stability. *J. Polym. Sci. Polym. Phys.*, 17:1073–1084, 1979.

162. B. Vincent, P. F. Luckham, and F. A. Waite. The effect of free polymer on the stability of sterically stabilized dispersions. *J. Colloid Interface Sci.*, 73:508–521, 1980.
163. R. I. Feigin and D. H. Napper. Stabilization of colloids by free polymer. *J. Colloid Interface Sci.*, 74:567–571, 1980.
164. P. G. de Gennes. Polymer solutions near an interface. 1. adsorption and depletion layers. *Macromolecules*, 14:1637–1644, 1981.
165. P. G. de Gennes. Polymers at an interface; a simplified view. *Adv. Colloid Interface Sci.*, 27:189–209, 1987.
166. H. J. Taunton, C. Toprakcioglu, L. J. Fetters, and J. Klein. Interactions between surfaces bearing end-adsorbed chains in a good solvent. *Macromolecules*, 23:571–580, 1990.
167. J. Klein and P. Luckham. Forces between two adsorbed poly(ethylene oxide) layers immersed in a good aqueous solvent. *Nature*, 300:429–431, 1982.
168. J. Klein and P. F. Luckham. Long-range attractive forces between two mica surfaces in an aqueous polymer solution. *Nature*, 308:836–837, 1984.
169. P. F. Luckham and J. Klein. Forces between mica surfaces bearing adsorbed homopolymers in good solvents. *J. Chem. Soc. Faraday Trans.*, 86:1363–1368, 1990.
170. S. Alexander. Adsorption of chain molecules with a polar head. a scaling description. *J. Phys. (France)*, 38:983–987, 1977.
171. P. G. de Gennes. Conformations of polymers attached to an interface. *Macromolecules*, 13:1069–1075, 1980.
172. S. T. Milner, T. A. Witten, and M. E. Cates. Theory of the grafted polymer brush. *Macromolecules*, 21:2610–2619, 1988.
173. S. T. Milner, T. A. Witten, and M. E. Cates. Effects of polydispersity in the end-grafted polymer brush. *Macromolecules*, 22:853–861, 1989.
174. E. B. Zhulina, O. V. Borisov, and V. A. Priamitsyn. Theory of steric stabilization of colloid dispersions by grafted polymers. *J. Colloid Interface Sci.*, 137:495–511, 1990.
175. J. Klein, E. Kumacheva, D. Mahalu, D. Perahia, and L. J. Fetters. Reduction of frictional forces between solid surfaces bearing polymer brushes. *Nature*, 370:634–636, 1994.
176. J. N. Israelachvili and H. Wennerström. Hydration or steric forces between amphiphilic surfaces? *Langmuir*, 6:873–876, 1990.
177. J. N. Israelachvili and H. Wennerström. Entropic forces between amphiphilic surfaces in liquids. *J. Phys. Chem.*, 96:520–531, 1992.
178. M. K. Granfeldt and S. J. Miklavic. A simulation study of flexible zwitterionic monolayers. interlayer interaction and headgroup conformation. *J. Phys. Chem.*, 95:6351–6360, 1991.
179. R. G. Horn, J. N. Israelachvili, and F. Pribac. Measurement of the deformation and adhesion of solids in contact. *J. Colloid Interface Sci.*, 115:480–492, 1987.
180. L. R. Fisher and J. N. Israelachvili. Direct measurements of the effect of meniscus forces on adhesion: A study of the applicability of macroscopic thermodynamics to microscopic liquid interfaces. *Colloids Surf.*, 3:303–319, 1981.
181. H. K. Christenson. Adhesion between surfaces in unsaturated vapors – a reexamination of the influence of meniscus curvature and surface forces. *J. Colloid Interface Sci.*, 121:170–178, 1988.
182. H. Hertz. Über die berührung fester elastischer körper. *J. Reine Angew. Math.*, 92:156–171, 1881.
183. H. M. Pollock, D. Maugis, and M. Barquins. The force of adhesion between solid surfaces in contact. *Appl. Phys. Lett.*, 33:798–799, 1978.
184. B. V. Derjaguin, V. M. Muller, and Yu. P. Toporov. Effect of contact deformations on the adhesion of particles. *J. Colloid Interface Sci.*, 53:314–326, 1975.

185. V. M. Muller, V. S. Yushchenko, and B. V. Derjaguin. On the influence of molecular forces on the deformation of an elastic sphere and its sticking to a rigid plane. *J. Colloid Interface Sci.*, 77:91–101, 1980.
186. V. M. Muller, B. V. Derjaguin, and Y. P. Toporov. On 2 methods of calculation of the force of sticking of an elastic sphere to a rigid plane. *Colloids Surf.*, 7:251–259, 1983.
187. D. Maugis. Adhesion of spheres: The jkr–dmt transition using a dugdale model. *J. Colloid Interface Sci.*, 150:243–269, 1992.
188. V. Mangipudi, M. Tirrell, and A. V. Pocius. Direct measurement of molecular level adhesion between poly(ethylene terephthalate) and polyethylene films: Determination of surface and interfacial energies. *J. Adh. Sci. Technol.*, 8:1251–1270, 1994.
189. H. K. Christenson. Surface deformations in direct force measurements. *Langmuir*, 12:1404–1405, 1996.
190. M. Barquins and D. Maugis. Fracture mechanics and the adherence of viscoelastic bodies. *J. Phys. D: Appl Phys.*, 11:1989–2023, 1978.
191. J. A. Greenwood and J. B. P. Williamson. Contact of nominally flat surfaces. *Proc. R. Soc. London A*, 295:300–319, 1966.
192. B. N. J. Persson. Elastoplastic contact between randomly rough surfaces. *Phys. Rev. Lett.*, 87:116101/1–4, 2001.
193. K. N. G. Fuller and D. Tabor. The effect of surface roughness on the adhesion of elastic solids. *Proc. R. Soc. London A*, 345:327–342, 1975.
194. D. Maugis. On the contact and adhesion of rough surfaces. *J. Adh. Sci. Technol.*, 10:161–175, 1996.
195. B. N. J. Persson and E. Tosatti. The effect of surface roughness on the adhesion of elastic solids. *J. Chem. Phys.*, 115:5597–5610, 2001.
196. P. M. McGuiggan and J. N. Israelachvili. Adhesion and short-range forces between surfaces. part ii: Effects of surface lattice mismatch. *J. Mater. Res.*, 5:2232–2243, 1990.
197. P. M. McGuiggan and J. Israelachvili. Measurements of the effects of angular lattice mismatch on the adhesion energy between two mica surfaces in water. *Mat. Res. Soc. Symp. Proc.*, 138:349–360, 1989.
198. C. L. Rhykerd, Jr., M. Schoen, D. J. Diestler, and J. H. Cushman. Epitaxy in simple classical fluids in micropores and near-solid surfaces. *Nature*, 330:461–463, 1987.
199. M. Schoen, D. J. Diestler, and J. H. Cushman. Fluids in micropores. i. structure of a simple classical fluid in a slit-pore. *J. Chem. Phys.*, 87:5464–5476, 1987.
200. M. Schoen, C. L. Rhykerd, Jr., D. J. Diestler, and J. H. Cushman. Shear forces in molecularly thin films. *Science*, 245:1223–1225, 1989.
201. P. A. Thompson and M. O. Robbins. Origin of stick–slip motion in boundary lubrication. *Science*, 250:792–794, 1990.
202. K. K. Han, J. H. Cushman, and D. J. Diestler. Grand canonical monte carlo simulations of a stockmayer fluid in a slit micropore. *Mol. Phys.*, 79:537–545, 1993.
203. M. Ruths and S. Granick. Influence of alignment of crystalline confining surfaces on static forces and shear in a liquid crystal, 4'-n-pentyl-4-cyanobiphenyl (5cb). *Langmuir*, 16:8368–8376, 2000.
204. A. D. Berman. Dynamics of molecules at surfaces, 1996.
205. F. P. Bowden and D. Tabor. *The Friction and Lubrication of Solids*. Clarendon, 1971.
206. J. A. Greenwood and K. L. Johnson. The mechanics of adhesion of viscoelastic solids. *Phil. Mag. A*, 43:697–711, 1981.
207. D. Maugis. Subcritical crack growth, surface energy, fracture toughness, stick–slip and embrittlement. *J. Mater. Sci.*, 20:3041–3073, 1985.
208. F. Michel and M. E. R. Shanahan. Kinetics of the jkr experiment. *C. R. Acad. Sci. II (Paris)*, 310:17–20, 1990.

209. E. Barthel and S. Roux. Velocity dependent adherence: An analytical approach for the jkr and dmt models. *Langmuir*, 16:8134–8138, 2000.
210. M. Ruths and S. Granick. Rate-dependent adhesion between polymer and surfactant monolayers on elastic substrates. *Langmuir*, 14:1804–1814, 1998.
211. H. Yoshizawa, Y. L. Chen, and J. Israelachvili. Fundamental mechanisms of interfacial friction. 1: relation between adhesion and friction. *J. Phys. Chem.*, 97:4128–4140, 1993.
212. N. Maeda, N. Chen, M. Tirrell, and J. N. Israelachvili. Adhesion and friction mechanisms of polymer-on-polymer surfaces. *Science*, 297:379–382, 2002.
213. C. A. Miller and P. Neogi. *Interfacial Phenomena: Equilibrium and Dynamic Effects*. Dekker, 1985.
214. J. Israelachvili and A. Berman. Irreversibility, energy dissipation, and time effects in intermolecular and surface interactions. *Israel J. Chem.*, 35:85–91, 1995.
215. A. N. Gent and A. J. Kinloch. Adhesion of viscoelastic materials to rigid substrates. iii. energy criterion for failure. *J. Polym. Sci. A-2*, 9:659–668, 1971.
216. A. N. Gent. Adhesion and strength of viscoelastic solids. is there a relationship between adhesion and bulk properties? *Langmuir*, 12:4492–4496, 1996.
217. H. R. Brown. The adhesion between polymers. *Annu. Rev. Mater. Sci.*, 21:463–489, 1991.
218. M. Deruelle, M. Tirrell, Y. Marciano, H. Hervet, and L. Léger. Adhesion energy between polymer networks and solid surfaces modified by polymer attachment. *Faraday Discuss.*, 98:55–65, 1995.
219. J. Van Alsten and S. Granick. The origin of static friction in ultrathin liquid films. *Langmuir*, 6:876–880, 1990.
220. S. Granick. Motions and relaxations of confined liquids. *Science*, 253:1374–1379, 1991.
221. H.-W. Hu, G. A. Carson, and S. Granick. Relaxation time of confined liquids under shear. *Phys. Rev. Lett.*, 66:2758–2761, 1991.
222. H. W. Hu and S. Granick. Viscoelastic dynamics of confined polymer melts. *Science*, 258:1339–1342, 1992.
223. M. L. Gee, P. M. McGuiggan, J. N. Israelachvili, and A. M. Homola. Liquid to solidlike transitions of molecularly thin films under shear. *J. Chem. Phys.*, 93:1895–1906, 1990.
224. J. Israelachvili, M. Gee, P. McGuiggan, P. Thompson, and M. Robbins. Melting–freezing transitions in molecularly thin liquid films during shear, 1990.
225. J. Israelachvili, P. McGuiggan, M. Gee, A. Homola, M. Robbins, and P. Thompson. Liquid dynamics in molecularly thin films. *J. Phys.: Condens. Matter*, 2:SA89–SA98, 1990.
226. J. Klein and E. Kumacheva. Confinement-induced phase transitions in simple liquids. *Science*, 269:816–819, 1995.
227. A. L. Demirel and S. Granick. Glasslike transition of a confined simple fluid. *Phys. Rev. Lett.*, 77:2261–2264, 1996.
228. D. Dowson. *History of Tribology*. Professional Engineering Publishing, 2nd edition, 1998.
229. D. Tabor. The role of surface and intermolecular forces in thin film lubrication. *Tribol. Ser.*, 7:651–682, 1982.
230. M. J. Sutcliffe, S. R. Taylor, and A. Cameron. Molecular asperity theory of boundary friction. *Wear*, 51:181–192, 1978.
231. G. M. McClelland. *Friction at weakly interacting interfaces*, pages 1–16. Springer, 1989.
232. D. H. Buckley. The metal-to-metal interface and its effect on adhesion and friction. *J. Colloid Interface Sci.*, 58:36–53, 1977.

233. J. N. Israelachvili, Y.-L. Chen, and H. Yoshizawa. Relationship between adhesion and friction forces. *J. Adh. Sci. Technol.*, 8:1231–1249, 1994.
234. A. Berman and J. Israelachvili. Control and minimization of friction via surface modification. *NATO ASI Ser. E: Appl. Sci.*, 330:317–329, 1997.
235. V. Zaloj, M. Urbakh, and J. Klafter. Modifying friction by manipulating normal response to lateral motion. *Phys. Rev. Lett.*, 82:4823–4826, 1999.
236. A. Dhinojwala, S. C. Bae, and S. Granick. Shear-induced dilation of confined liquid films. *Tribol. Lett.*, 9:55–62, 2000.
237. L. M. Qian, G. Luengo, and E. Perez. Thermally activated lubrication with alkanes: The effect of chain length. *Europhys. Lett.*, 61:268–274, 2003.
238. A. M. Homola, J. N. Israelachvili, P. M. McGuiggan, and M. L. Gee. Fundamental experimental studies in tribology: The transition from "interfacial" friction of undamaged molecularly smooth surfaces to "normal" friction with wear. *Wear*, 136:65–83, 1990.
239. B. J. Briscoe, D. C. B. Evans, and D. Tabor. The influence of contact pressure and saponification on the sliding behavior of steric acid monolayers. *J. Colloid Interface Sci.*, 61:9–13, 1977.
240. M. Urbakh, L. Daikhin, and J. Klafter. Dynamics of confined liquids under shear. *Phys. Rev. E*, 51:2137–2141, 1995.
241. M. G. Rozman, M. Urbakh, and J. Klafter. Origin of stick–slip motion in a driven two-wave potential. *Phys. Rev. E*, 54:6485–6494, 1996.
242. M. G. Rozman, M. Urbakh, and J. Klafter. Stick–slip dynamics as a probe of frictional forces. *Europhys. Lett.*, 39:183–188, 1997.
243. B. V. Derjaguin. Molekulartheorie der äußeren reibung. *Z. Physik*, 88:661–675, 1934.
244. B. V. Derjaguin. Mechanical properties of the boundary lubrication layer. *Wear*, 128:19–27, 1988.
245. B. J. Briscoe and D. C. B. Evans. The shear properties of langmuir–blodgett layers. *Proc. R. Soc. London A*, 380:389–407, 1982.
246. A. Berman, C. Drummond, and J. Israelachvili. Amontons' law at the molecular level. *Tribol. Lett.*, 4:95–101, 1998.
247. P. F. Luckham and S. Manimaaran. Investigating adsorbed polymer layer behaviour using dynamic surface force apparatuses – a review. *Adv. Colloid Interface Sci.*, 73:1–46, 1997.
248. S. Yamada and J. Israelachvili. Friction and adhesion hysteresis of fluorocarbon surfactant monolayer-coated surfaces measured with the surface forces apparatus. *J. Phys. Chem. B*, 102:234–244, 1998.
249. G. Luengo, S. E. Campbell, V. I. Srdanov, F. Wudl, and J. N. Israelachvili. Direct measurement of the adhesion and friction of smooth C_{60} surfaces. *Chem. Mater.*, 9:1166–1171, 1997.
250. L. Rapoport, Y. Bilik, Y. Feldman, M. Homyonfer, S. R. Cohen, and R. Tenne. Hollow nanoparticles of WS_2 as potential solid-state lubricants. *Nature*, 387:791–793, 1997.
251. J. Israelachvili, Y.-L. Chen, and H. Yoshizawa. *Relationship between adhesion and friction forces*, pages 261–279. VSP, 1995.
252. P. Berthoud, T. Baumberger, C. G'Sell, and J. M. Hiver. Physical analysis of the state- and rate-dependent friction law: Static friction. *Phys. Rev. B*, 59:14313–14327, 1999.
253. A. Berman, S. Steinberg, S. Campbell, A. Ulman, and J. Israelachvili. Controlled microtribology of a metal oxide surface. *Tribol. Lett.*, 4:43–48, 1998.
254. D. Y. C. Chan and R. G. Horn. The drainage of thin liquid films between solid surfaces. *J. Chem. Phys.*, 83:5311–5324, 1985.

255. J. N. Israelachvili and S. J. Kott. Shear properties and structure of simple liquids in molecularly thin films: The transition from bulk (continuum) to molecular behavior with decreasing film thickness. *J. Colloid Interface Sci.*, 129:461–467, 1989.
256. T. L. Kuhl, A. D. Berman, S. W. Hui, and J. N. Israelachvili. Part 1: Direct measurement of depletion attraction and thin film viscosity between lipid bilayers in aqueous polyethylene glycol solutions. *Macromolecules*, 31:8250–8257, 1998.
257. S. E. Campbell, G. Luengo, V. I. Srdanov, F. Wudl, and J. N. Israelachvili. Very low viscosity at the solid–liquid interface induced by adsorbed C_{60} monolayers. *Nature*, 382:520–522, 1996.
258. J. Klein, Y. Kamiyama, H. Yoshizawa, J. N. Israelachvili, G. H. Fredrickson, P. Pincus, and L. J. Fetters. Lubrication forces between surfaces bearing polymer brushes. *Macromolecules*, 26:5552–5560, 1993.
259. G. Luengo, J. Israelachvili, A. Dhinojwala, and S. Granick. Generalized effects in confined fluids: New friction map for boundary lubrication. *Wear*, 200:328–335, 1996.
260. E. Kumacheva and J. Klein. Simple liquids confined to molecularly thin layers. ii. shear and frictional behavior of solidified films. *J. Chem. Phys.*, 108:7010–7022, 1998.
261. P. A. Thompson, G. S. Grest, and M. O. Robbins. Phase transitions and universal dynamics in confined films. *Phys. Rev. Lett.*, 68:3448–3451, 1992.
262. Y. Rabin and I. Hersht. Thin liquid layers in shear: Non-newtonian effects. *Physica A*, 200:708–712, 1993.
263. P. A. Thompson, M. O. Robbins, and G. S. Grest. Structure and shear response in nanometer-thick films. *Israel J. Chem.*, 35:93–106, 1995.
264. M. Urbakh, L. Daikhin, and J. Klafter. Sheared liquids in the nanoscale range. *J. Chem. Phys.*, 103:10707–10713, 1995.
265. A. Subbotin, A. Semenov, E. Manias, G. Hadziioannou, and G. ten Brinke. Rheology of confined polymer melts under shear flow: Strong adsorption limit. *Macromolecules*, 28:1511–1515, 1995.
266. A. Subbotin, A. Semenov, E. Manias, G. Hadziioannou, and G. ten Brinke. Rheology of confined polymer melts under shear flow: Weak adsorption limit. *Macromolecules*, 28:3901–3903, 1995.
267. J. Huh and A. Balazs. Behavior of confined telechelic chains under shear. *J. Chem. Phys.*, 113:2025–2031, 2000.
268. C. Drummond, J. Elezgaray, and P. Richetti. Behavior of adhesive boundary lubricated surfaces under shear: A new dynamic transition. *Europhys. Lett.*, 58:503–509, 2002.
269. J. N. Israelachvili, P. M. McGuiggan, and A. M. Homola. Dynamic properties of molecularly thin liquid films. *Science*, 240:189–191, 1988.
270. A. M. Homola, H. V. Nguyen, and G. Hadziioannou. Influence of monomer architecture on the shear properties of molecularly thin polymer melts. *J. Chem. Phys.*, 94:2346–2351, 1991.
271. M. Schoen, S. Hess, and D. J. Diestler. Rheological properties of confined thin films. *Phys. Rev. E*, 52:2587–2602, 1995.
272. G. M. McClelland and S. R. Cohen. *Chemistry and Physics of Solid Surfaces VIII*. Springer, 1990.
273. M. O. Robbins and P. A. Thompson. Critical velocity of stick–slip motion. *Science*, 253:916, 1991.
274. G. He, M. Müser, and M. Robbins. Adsorbed layers and the origin of static friction. *Science*, 284:1650–1652, 1999.
275. C. Drummond and J. Israelachvili. Dynamic phase transitions in confined lubricant fluids under shear. *Phys. Rev. E.*, 63:041506/1–11, 2001.

276. J. D. Ferry. *Viscoelastic Properties of Polymers*. Wiley, 3rd edition, 1980.
277. H. Yoshizawa and J. Israelachvili. Fundamental mechanisms of interfacial friction. 2: stick–slip friction of spherical and chain molecules. *J. Phys. Chem.*, 97:11300–11313, 1993.
278. A. Ruina. Slip instability and state variable friction laws. *J. Geophys. Res.*, 88:10359–10370, 1983.
279. T. Baumberger, P. Berthoud, and C. Caroli. Physical analysis of the state- and rate-dependent friction law. ii. dynamic friction. *Phys. Rev. B*, 60:3928–3939, 1999.
280. J. Israelachvili, S. Giasson, T. Kuhl, C. Drummond, A. Berman, G. Luengo, J.-M. Pan, M. Heuberger, W. Ducker, and N. Alcantar. Some fundamental differences in the adhesion and friction of rough versus smooth surfaces. *Tribol. Ser.*, 38:3–12, 2000.
281. C. Drummond and J. Israelachvili. Dynamic behavior of confined branched hydrocarbon lubricant fluids under shear. *Macromolecules*, 33:4910–4920, 2000.
282. E. Rabinowicz. *Friction and Wear of Materials*. Wiley, 2nd edition, 1995.
283. J. Peachey, J. Van Alsten, and S. Granick. Design of an apparatus to measure the shear response of ultrathin liquid films. *Rev. Sci. Instrum.*, 62:463–473, 1991.
284. E. Rabinowicz. The intrinsic variables affecting the stick–slip process. *Proc. Phys. Soc.*, 71:668–675, 1958.
285. T. Baumberger, F. Heslot, and B. Perrin. Crossover from creep to inertial motion in friction dynamics. *Nature*, 367:544–546, 1994.
286. F. Heslot, T. Baumberger, B. Perrin, B. Caroli, and C. Caroli. Creep, stick–slip, and dry-friction dynamics: Experiments and a heuristic model. *Phys. Rev. E*, 49:4973–4988, 1994.
287. J. Sampson, F. Morgan, D. Reed, and M. Muskat. Friction behavior during the slip portion of the stick–slip process. *J. Appl. Phys.*, 14:689–700, 1943.
288. F. Heymann, E. Rabinowicz, and B. Rightmire. Friction apparatus for very low-speed sliding studies. *Rev. Sci. Instrum.*, 26:56–58, 1954.
289. J. H. Dieterich. Time-dependent friction and the mechanics of stick–slip. *Pure Appl. Geophys.*, 116:790–806, 1978.
290. J. H. Dieterich. Modeling of rock friction. 1. experimental results and constitutive equations. *J. Geophys. Res.*, 84:2162–2168, 1979.
291. G. A. Tomlinson. A molecular theory of friction. *Phil. Mag.*, 7:905–939, 1929.
292. J. M. Carlson and J. S. Langer. Mechanical model of an earthquake fault. *Phys. Rev. A*, 40:6470–6484, 1989.
293. B. N. J. Persson. Theory of friction: The role of elasticity in boundary lubrication. *Phys. Rev. B*, 50:4771–4786, 1994.
294. S. Nasuno, A. Kudrolli, and J. P. Gollub. Friction in granular layers: Hysteresis and precursors. *Phys. Rev. Lett.*, 79:949–952, 1997.
295. P. Bordarier, M. Schoen, and A. Fuchs. Stick–slip phase transitions in confined solidlike films from an equilibrium perspective. *Phys. Rev. E*, 57:1621–1635, 1998.
296. J. M. Carlson and A. A. Batista. Constitutive relation for the friction between lubricated surfaces. *Phys. Rev. E*, 53:4153–4165, 1996.
297. A. D. Berman, W. A. Ducker, and J. N. Israelachvili. Origin and characterization of different stick–slip friction mechanisms. *Langmuir*, 12:4559–4563, 1996.
298. A. D. Berman, W. A. Ducker, and J. N. Israelachvili. Experimental and theoretical investigations of stick–slip friction mechanisms. *NATO ASI Ser. E: Appl. Sci.*, 311:51–67, 1996.
299. K G. McLaren and D. Tabor. Viscoelastic properties and the friction of solids. friction of polymers and influence of speed and temperature. *Nature*, 197:856–858, 1963.

300. K. A. Grosch. Viscoelastic properties and friction of solids. relation between friction and viscoelastic properties of rubber. *Nature*, 197:858–859, 1963.
301. J. Gao, W. D. Luedtke, and U. Landman. Friction control in thin-film lubrication. *J. Phys. Chem. B*, 102:5033–5037, 1998.
302. M. Heuberger, C. Drummond, and J. Israelachvili. Coupling of normal and transverse motions during frictional sliding. *J. Phys. Chem. B*, 102:5038–5041, 1998.
303. L. Bureau, T. Baumberger, and C. Caroli. Shear response of a frictional interface to a normal load modulation. *Phys. Rev. E*, 62:6810–6820, 2000.
304. J. P. Gao, W. D. Luedtke, and U. Landman. Structure and solvation forces in confined films: Linear and branched alkanes. *J. Chem. Phys.*, 106:4309–4318, 1997.
305. J. Gao, W. D. Luedtke, and U. Landman. Layering transitions and dynamics of confined liquid films. *Phys. Rev. Lett.*, 79:705–708, 1997.
306. J. Warnock, D. D. Awschalom, and M. W. Shafer. Orientational behavior of molecular liquids in restricted geometries. *Phys. Rev. B*, 34:475–478, 1986.
307. D. D. Awschalom and J. Warnock. Supercooled liquids and solids in porous glass. *Phys. Rev. B*, 35:6779–6785, 1987.
308. M. Hirano, K. Shinjo, R. Kaneko, and Y. Murata. Anisotropy of frictional forces in muscovite mica. *Phys. Rev. Lett.*, 67:2642–2645, 1991.
309. T. Gyalog and H. Thomas. Friction between atomically flat surfaces. *Europhys. Lett.*, 37:195–200, 1997.

10

Friction and Wear on the Atomic Scale

Enrico Gnecco, Roland Bennewitz, Oliver Pfeiffer, Anisoara Socoliuc, and Ernst Meyer

Summary. Friction is an old subject of research: the empirical da Vinci–Amontons laws are common knowledge. Macroscopic experiments systematically performed by the school of *Bowden* and *Tabor* have revealed that macroscopic friction can be related to the collective action of small asperities. During the last 15 years, experiments performed with the atomic force microscope gave new insight into the physics of single asperities sliding over surfaces. This development, together with complementary experiments by means of surface force apparatus and quartz microbalance, established the new field of nanotribology. At the same time, increasing computing power allowed for the simulation of the processes in sliding contacts consisting of several hundred atoms. It became clear that atomic processes cannot be neglected in the interpretation of nanotribology experiments. Experiments on even well-defined surfaces directly revealed atomic structures in friction forces. This chapter will describe friction force microscopy experiments that reveal, more or less directly, atomic processes in the sliding contact.

We will begin by introducing friction force microscopy, including the calibration of cantilever force sensors and special aspects of the ultra-high vacuum environment. The empirical Tomlinson model largely describes atomic stick-slip results and is therefore presented in detail. We review experimental results regarding atomic friction. These include thermal activation, velocity dependence, as well as temperature dependence. The geometry of the contact plays a crucial role in the interpretation of experimental results, as we will demonstrate, for example, for the calculation of the lateral contact stiffness. The onset of wear on atomic scale has recently come into the scope of experimental studies and is described here. In order to compare the respective results, we present molecular dynamics simulations that are directly related to atomic friction experiments. We close the chapter with a discussion of dissipation measurements performed in noncontact forcemicroscopy, which may become an important complementary tool for the study of mechanical dissipation in nanoscopic devices.

10.1 Friction Force Microscopy in Ultra-High Vacuum

The *friction force microscope* (FFM, also called lateral force microscope, LFM) exploits the interaction of a sharp tip sliding on a surface to quantify dissipative processes down to the atomic scale (Fig. 10.1).

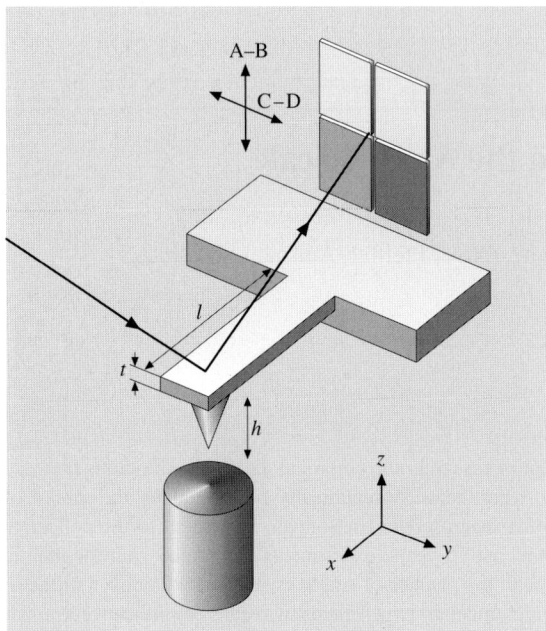

Fig. 10.1. Schematic diagram of a beam-deflection friction force microscope

10.1.1 Friction Force Microscopy

The relative motion of tip and surface is realized by a *scanner* formed by piezoelectric elements, which moves the surface perpendicularly to the tip with a certain periodicity. The scanner can be also extended or retracted in order to vary the normal force, F_N, which is applied on the surface. This force is responsible for the deflection of the *cantilever* that supports the tip. If the normal force F_N increases while scanning due to the local slope of the surface, the scanner is retracted by a feedback loop. On the other hand, if F_N decreases, the surface is brought closer to the tip by extending the scanner. In such way, the surface topography can be determined line by line from the vertical displacement of the scanner. An accurate control of such vertical movement is made possible by a light beam, which is reflected from the rear of the lever into a photodetector. When the bending of the cantilever changes, the light spot on the detector moves up or down and causes a variation of the photocurrent that corresponds to the normal force F_N to be controlled.

Usually the relative sliding of tip and surface is also accompanied by *friction*. A lateral force, F_L, with the opposite direction of the scan velocity, v, hinders the motion of the tip. This force provokes the torsion of the cantilever, and it can be observed with the topography if the photodetector can reveal not only the normal deflection but also the lateral movement of the lever while scanning. In practice this is realized by four-quadrants photodetectors, as shown in Fig. 10.1. We should notice that friction forces also cause the lateral bending of the cantilever, but this effect is negligible if the thickness of the lever is much less than the width.

The FFM was first used by *Mate* et al. in 1987 to reveal friction with atomic features [1], i.e. just one year after *Binnig, Quate,* and *Gerber* introduced the atomic force microscope [2]. In their experiment, Mate used a tungsten wire and a slightly different technique to detect lateral forces (non-fiber interferometry). The optical beam deflection was introduced later by *Marti* et al., and *Meyer* et al. [3, 4]. Other methods to measure

the forces between tip and surface are given by capacitance detection [5], dual fiber interferometry [6], and piezolevers [7]. In the first method, two plates close to the cantilever reveal capacitance while scanning. The second technique uses two optical fibers to detect the cantilever deflection along two orthogonal directions angled 45° with respect to the surface normal. Finally, in the third method, cantilevers with two Wheatstone bridges at their base reveal normal and lateral forces, which are respectively proportional to the sum and the difference of both bridge signals.

10.1.2 Force Calibration

The force calibration is relatively simple if rectangular cantilevers are used. Due to possible discrepancies with the geometric values provided by manufacturers, one should use optical and electron microscopes to determine the width, thickness, and length of cantilever, w, t, l, the tip height, h, and the position of the tip with respect to the cantilever. The thickness of the cantilever can also be determined from the resonance frequency of the lever, f_0, using the relation [8]:

$$t = \frac{2\sqrt{12}\pi}{1.875^2} \sqrt{\frac{\rho}{E}} f_0 l^2 , \qquad (10.1)$$

In (10.1) ρ is the density of the cantilever and E is its Young's modulus. The normal spring constant, c_N, and the lateral spring constant, c_L, of the lever are given by

$$c_N = \frac{Ewt^3}{4l^3} , \quad c_L = \frac{Gwt^3}{3h^2 l} , \qquad (10.2)$$

where G is the shear modulus. Figure 10.2 shows some SEM images of rectangular silicon cantilevers used for FFM. In the case of silicon, $\rho = 2.33 \times 10^3 \text{ kg/m}^3$, $E = 1.69 \times 10^{11} \text{ N/m}^2$ and $G = 0.5 \times 10^{11} \text{ N/m}^2$. Thus, for the cantilever in Fig. 10.2, $c_N = 1.9 \text{ N/m}$ and $c_L = 675 \text{ N/m}$.

The next step in force calibration consists of measuring the sensitivity of the photodetector, S_z (nm/V). For beam-deflection FFMs, the sensitivity S_z can be determined by force vs. distance curves measured on hard surfaces (e.g., Al_2O_3), where elastic deformations are negligible and the vertical movement of the scanner equals the deflection of the cantilever. A typical relation between the difference of the vertical signals on the four-quadrant detector, V_N, and the distance from the surface, z, is

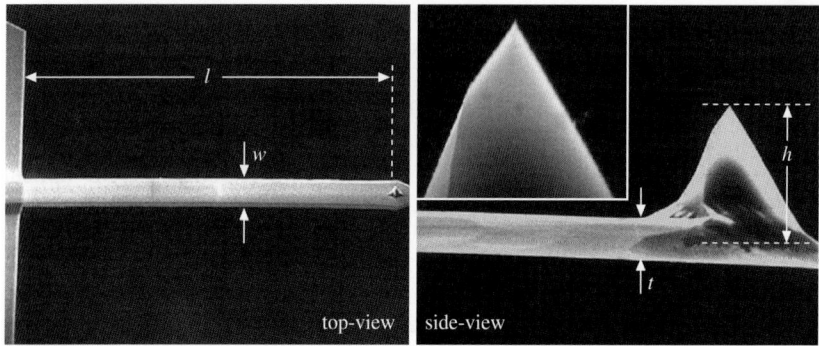

Fig. 10.2. SEM images of a rectangular cantilever. The relevant dimensions are $l = 445\,\mu\text{m}$, $w = 43\,\mu\text{m}$, $t = 4.5\,\mu\text{m}$, $h = 14.75\,\mu\text{m}$. Note that h is given by the sum of the tip height and half of the cantilever thickness. (After [9])

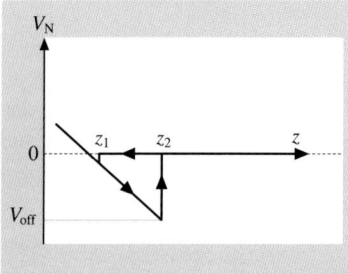

Fig. 10.3. Sketch of a typical force vs. distance curve

sketched in Fig. 10.3. When the tip is approached, no signal is revealed until the tip jumps into contact at $z = z_1$. Further extension or retraction of the scanner results in an elastic behavior until the tip jumps again out of contact at a distance $z_2 > z_1$. The slope of the elastic part of the curve gives the required sensitivity, S_z.

The normal and lateral forces are related to the voltage V_N, and the difference between the horizontal signals, V_L, as follows:

$$F_N = c_N S_z V_N, \quad F_L = \frac{3}{2} c_L \frac{h}{l} S_z V_L, \tag{10.3}$$

in which it is assumed that the light beam is positioned above the probing tip.

The normal spring constant c_N can also be calibrated with alternative methods. *Cleveland* et al. [10] attached tungsten spheres to the tip, which changes the resonance frequency f_0 according to the formula

$$f_0 = \frac{1}{2\pi} \sqrt{\frac{c_N}{M + m^*}}. \tag{10.4}$$

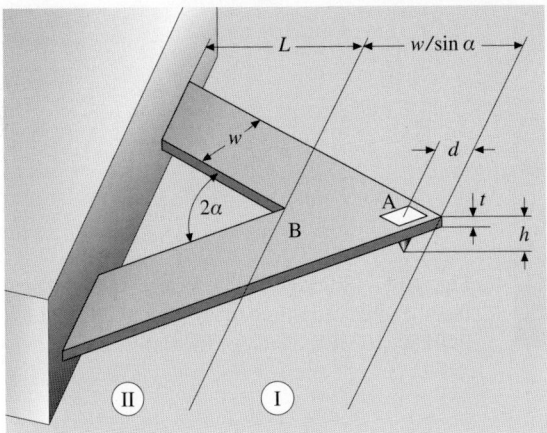

Fig. 10.4. Geometry of a V-shaped cantilever. (After [13])

M is the mass of the added object, and m^* is an effective mass of the cantilever, which depends on its geometry [10]. The spring constant can be extrapolated from the frequency shifts corresponding to the different masses attached.

As an alternative, *Hutter* et al. observed that the spring constant c_N can be related to the area of the power spectrum of the thermal fluctuations of the cantilever, P [11]. The correct relation is $c_N = 4k_B T/3P$, where $k_B = 1.38 \times 10^{-3}$ J/K is the Boltzmann's constant and T is the temperature [12].

Cantilevers with different shape require finite element analysis, although in few cases analytical formulas can be derived. For V-shaped cantilevers, *Neumeister* et al. found the following approximation for the lateral spring constant c_L [13]:

$$c_L = \frac{Et^3}{3(1+\nu)h^2} \tag{10.5}$$
$$\times \left(\frac{1}{\tan \alpha} \ln \frac{w}{d \sin \alpha} + \frac{L \cos \alpha}{w} - \frac{3 \sin 2\alpha}{8} \right)^{-1},$$

where the geometrical quantities L, w, α, d, t, and h are defined in Fig. 10.4. The expression for the normal constant is more complex and can be found in the cited reference.

Surfaces with well-defined profiles provide an alternative in situ calibration of lateral forces [14]. We present a slightly modified version of the method [15]. Figure 10.5 shows a commercial grating formed by alternated faces with opposite inclinations with respect to the scan direction. When the tip slides on the inclined planes, the normal force, F_N, and the lateral force, F_L, with respect to the surface, are different from the two components $F_\perp$ and $F_\parallel$, which are separated by the photodiode (see Fig. 10.6a).

If the linear relation $F_L = \mu F_N$ holds (see Sect. 10.5), the component $F_\parallel$ can be expressed in terms of $F_\perp$:

Fig. 10.5. Silicon grating formed by alternated faces angled ±55° (courtesy Silicon-MDT Ltd., POB 50, 103305, Moscow, Russia)

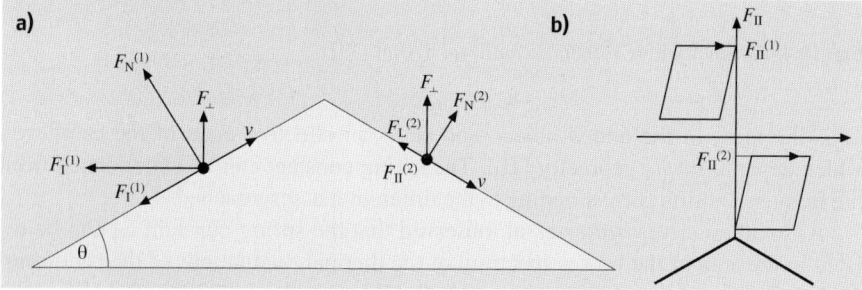

Fig. 10.6. (a) Forces acting on a FFM tip sliding on the grating in Fig. 10.5; (b) Friction loops acquired on the two faces

$$F_\parallel = \frac{\mu + \tan\theta}{1 - \mu\tan\theta} F_\perp .\tag{10.6}$$

The component $F_\perp$ is kept constant by the feedback loop. The sum and the difference of $F_\parallel$ on the two planes (1) and (2) are given by

$$\begin{aligned}F_+ \equiv F_\parallel^{(1)} + F_\parallel^{(2)} &= \frac{2\mu\left(1 + \tan^2\theta\right)}{1 - \mu^2\tan^2\theta} F_\perp \\ F_- \equiv F_\parallel^{(1)} - F_\parallel^{(2)} &= \frac{2\left(1 + \mu^2\right)\tan\theta}{1 - \mu^2\tan^2\theta} F_\perp\end{aligned}\tag{10.7}$$

The values of F_+ and F_- (in volts) can be measured by scanning the profile back and forth (Fig. 10.6b). If F_+ and F_- are recorded with different values of $F_\perp$, one can determine the conversion ratio between volts and nanonewtons as well as the coefficient of friction, μ.

An accurate analysis of the errors in lateral force calibration was given by *Schwarz* et al., who revealed the important role of the cantilever's oscillations induced by the feedback loop and of the so-called pull-off force (Sect. 10.5) in friction measurements, aside from the geometrical setting of cantilevers and laser beams [16].

For some applications, an adequate estimation of the radius of curvature of the tip, R, is important (Sect. 10.5.2). This quantity can be evaluated with a scanning electron microscope. Otherwise, well-defined structures as step sites [17, 18] or whiskers [19] can be imaged. The image of these high-aspect ratio structures is a convolution with the tip structure. A deconvolution algorithm that allows for the extraction of the probing tip's radius of curvature was suggested by *Villarrubia* [20].

10.1.3 The Ultra-high Vacuum Environment

Atomic friction studies require well-defined surfaces and – whenever possible – tips. For the surfaces, the established methods of surface science can be employed working in ultra-high vacuum (UHV). Ionic crystals such as NaCl have become standard materials for friction force microscopy on atomic scale. Atomically clean and flat surfaces can be prepared by cleavage in UHV. The crystal has to be heated to about 150 °C for 1 h in order to remove charging after the cleavage process. Metal surfaces can be cleaned and flattened by cycles of sputtering with argon ions and annealing. Even surfaces prepared in air or liquids like self-assembled molecular monolayers can be transferred into the vacuum and studied after careful heating procedures, which remove water layers.

Tip preparation in UHV is a more difficult subject. Most force sensors for friction studies have silicon nitride or pure silicon tips. Tips can be cleaned and oxide layers removed by sputtering with argon ions. But the sharpness of the tip is normally reduced by sputtering. As an alternative, tips can be etched in fluoric acid directly before transfer to the UHV. The significance of tip preparation is limited by the fact that the chemical and geometrical structure of the tip can undergo significant changes when sliding over the surface.

The implementation of the friction force microscope into UHV requires some additional efforts. First of all, only materials with a low vapor pressure can be used, thereby excluding most plastics and lubricants. Beam-deflection type force microscopes employ either a light source in the vacuum chamber or an optical fiber guiding the light into the chamber. The positioning of the light beam onto the cantilever and of the reflected beam onto the position-sensitive detector is performed by motorized mirrors [21] or by moving light source or detector [22]. Furthermore, a motorized sample approach has to be realized.

The quality force sensor's electrical signal can seriously deteriorate on the way out of the vacuum chamber. Low noise and high bandwidth can be preserved by implementing a pre-amplifier into the vacuum. Again the choice of materials for print and devices is limited by the condition of low vapor pressure. Stronger heating of the electrical circuitry in vacuum, therefore, must be considered.

10.1.4 A Typical Microscope in UHV

A typical AFM for UHV application is shown in Fig. 10.7. The housing (1) contains the light source and a set of lenses that focus the light onto the cantilever. Alternatively, the light can be guided via an optical fiber into the vacuum. By using light

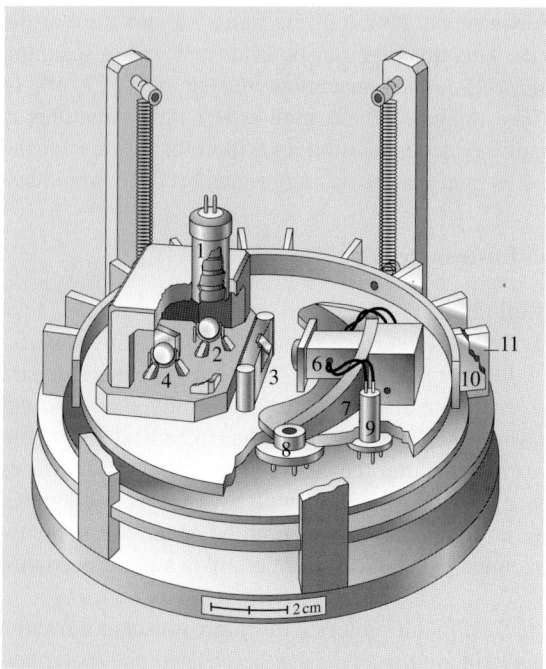

Fig. 10.7. Schematic view of the UHV-AVM realized at the University of Basel. (After [21])

emitting diodes with low coherency you avoid disturbing interference effects often found in instruments equipped with lasers as light source. A plane mirror fixed on the spherical rotor of a first stepping motor (2) can be rotated around a vertical and horizontal axis to guide the light beam onto the rear of the cantilever, mounted on a removable carrier plate (3). The light is reflected off the cantilever toward a second motorized mirror (4) that guides the beam to the center of the quadrant photodiode (5), where the light is then converted into four photo-currents. Four preamplifiers in close vicinity to the photodiode allow low-noise measurements with 3 MHz bandwidth.

The two motors with spherical rotors, used to realign the light path after exchange of the cantilever, work as *inertial stepping motors*: the sphere is resting on three piezoelectric legs that can be moved by a small amount tangentially to the sphere. Each step of the motor consists of a slow forward motion of two legs followed by an abrupt jump backwards. During the slow forward motion the sphere follows the legs due to friction, whereas it cannot follow the sudden jump due to its inertia. A series of such tiny steps rotates the sphere macroscopically.

The sample, also on an exchangeable carrier plate, is mounted on the end of a tube scanner (6), which can move the sample in three dimensions over several micrometers. The whole scanning head (7) is the slider of a third inertial stepping motor for coarse positioning of the sample. It rests with its flat and polished bottom on three supports. Two of them are symmetrically placed piezoelectric legs (8), whereas the third central support is passive. The slider (7) can be translated in two dimensions and rotated about a vertical axis by several millimeters (rotation is achieved by antiparallel operation of the two legs). The slider is held down by two magnets, close to the active supports, and its travel is limited by two fixed posts (9) that also serve as cable attachments. The whole platform is suspended by four springs. A ring of radial copper lamellae (10), floating between a ring of permanent magnets (11) on the base flange, acts as efficient eddy current damping.

10.2 The Tomlinson Model

In Sect. 10.3, we will show that the FFM can reveal friction forces down to the atomic scale, which are characterized by a typical sawtooth pattern. This phenomenon can be seen as a consequence of a *stick-slip* mechanism, first discussed by *Tomlinson* in 1929 [23].

10.2.1 One-dimensional Tomlinson Model

In the Tomlinson model the motion of the tip is influenced by both the interaction with the atomic lattice of the surface and the elastic deformations of the cantilever. The shape of the tip-surface potential, $V(\mathbf{r})$, depends on several factors, such as the chemical composition of the materials in contact and the atomic arrangement at the tip end. For the sake of simplicity, we will start the analysis in the one-dimensional case considering a sinusoidal profile with the periodicity of the atomic lattice, a, and a peak-to-peak amplitude E_0. In Sect. 10.5, we will show how the elasticity of the cantilever and of the contact area are described in a unique framework by introducing an effective lateral spring constant, k_{eff}. If the cantilever moves with a constant velocity v along the x direction, the total energy of the system is

$$E_{\text{tot}}(x,t) = -\frac{E_0}{2}\cos\frac{2\pi x}{a} + \frac{1}{2}k_{\text{eff}}(vt-x)^2 . \tag{10.8}$$

Figure 10.8 shows the energy profile $E_{\text{tot}}(x,t)$ at two different instants. When $t = 0$ the tip is localized in the absolute minimum of E_{tot}. This minimum increases with time due to the cantilever motion, until the tip position becomes unstable when $t = t^*$.

At a given time t the position of the tip can be determined by equating to zero the first derivative of $E_{\text{tot}}(x,t)$ with respect to x:

$$\frac{\partial E_{\text{tot}}}{\partial x} = \frac{\pi E_0}{a}\sin\frac{2\pi x}{a} - k_{\text{eff}}(vt-x) = 0 . \tag{10.9}$$

The critical position x^* corresponding to $t = t^*$ is determined by equating to zero the second derivative $\partial^2 E_{\text{tot}}(x,t)/\partial x^2$, which gives

$$x^* = \frac{a}{4}\arccos\left(-\frac{1}{\gamma}\right), \quad \gamma = \frac{2\pi^2 E_0}{k_{\text{eff}} a^2} . \tag{10.10}$$

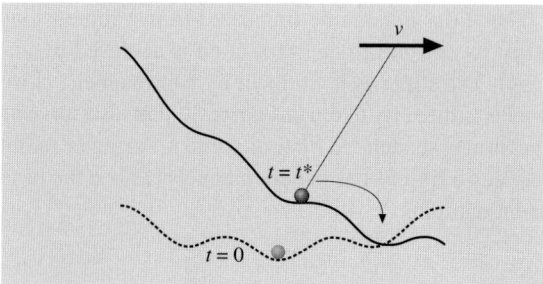

Fig. 10.8. Energy profile experienced by the FFM tip (*dark circle*) at $t = 0$ (*dotted line*) and $t = t^*$ (*continuous line*)

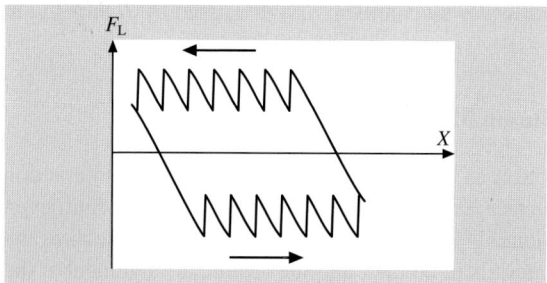

Fig. 10.9. Friction loop obtained by scanning back and forth in the 1-D Tomlinson model. The effective spring constant k_{eff} is the slope of the sticking part of the loop (if $\gamma \gg 1$)

The coefficient γ compares the strength of interaction between tip and surface with the stiffness of the system. When $t = t^*$ the tip suddenly *jumps* into the next minimum of the potential profile. The lateral force $F^* = k_{\text{eff}}(vt - x^*)$, which induces the jump, can be evaluated from (10.9) and (10.10):

$$F^* = \frac{k_{\text{eff}}\, a}{2\pi} \sqrt{\gamma^2 - 1}. \tag{10.11}$$

Thus the stick-slip is observed only if $\gamma > 1$, i.e. when the system is not too stiff or the tip–surface interaction is strong enough. Figure 10.9 shows the lateral force F_L as a function of the cantilever position, X. When the cantilever is moved rightward, the lower part of the curve in Fig. 10.9 is obtained. If, at a certain point, the cantilever's direction of motion is suddenly inverted, the force has the profile in the upper part of the curve. The area of the *friction loop* obtained by scanning back and forth gives the total energy dissipated.

10.2.2 Two-dimensional Tomlinson Model

In two dimensions, the energy of our system is given by

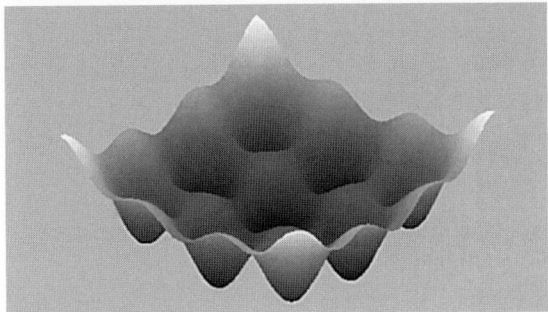

Fig. 10.10. Energy landscape experienced by the FFM tip in 2-D

$$E_{tot}(\mathbf{r}, t) = U(\mathbf{r}) + \frac{k_{eff}}{2}(\mathbf{v}t - \mathbf{r})^2, \qquad (10.12)$$

where $\mathbf{r} \equiv (x, y)$ and $\mathbf{v}$ is arbitrarily oriented on the surface (note that $\mathbf{v} \neq d\mathbf{r}/dt$!).
Figure 10.10 shows the total energy corresponding to a periodic potential of the form

$$U(x, y, t) = -\frac{E_0}{2}\left(\cos\frac{2\pi x}{a} + \cos\frac{2\pi y}{a}\right) + E_1 \cos\frac{2\pi x}{a} \cos\frac{2\pi y}{a}. \qquad (10.13)$$

The equilibrium condition becomes

$$\nabla E_{tot}(\mathbf{r}, t) = \nabla U(\mathbf{r}) + k_{eff}(\mathbf{r} - \mathbf{v}t) = 0. \qquad (10.14)$$

The stability of the equilibrium can be discussed introducing the Hessian matrix

$$H = \begin{pmatrix} \dfrac{\partial^2 U}{\partial x^2} + k_{eff} & \dfrac{\partial^2 U}{\partial x \partial y} \\ \dfrac{\partial^2 U}{\partial y \partial x} & \dfrac{\partial^2 U}{\partial y^2} + k_{eff} \end{pmatrix}. \qquad (10.15)$$

When both eigenvalues $\lambda_{1,2}$ of the Hessian are positive, the position of the tip is stable. Figure 10.11 shows such regions for a potential of the form (10.13). The tip follows the cantilever adiabatically as long as it remains in the (++)-region. When the tip is dragged to the border of the region, it suddenly jumps into the next (++)-region. A comparison between a theoretical friction map deduced from the 2-D Tomlinson model and an experimental map acquired by UHV-FFM is given in the next section.

10.2.3 Friction Between Atomically Flat Surfaces

So far we have implicitly assumed that the tip is terminated by only one atom. It is also instructive to consider the case of a periodic surface sliding on another periodic

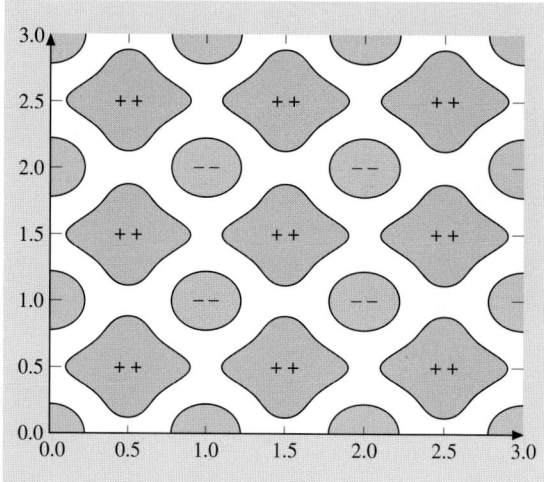

Fig. 10.11. Regions on the tip plane labeled according to the signs of the eigenvalues of the Hessian matrix. (After [24])

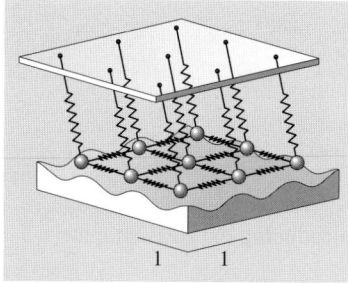

Fig. 10.12. The FKT model in 2-D. (After [25])

surface. In the Frenkel–Kontorova–Tomlinson (FKT) model, the atoms of one surface are harmonically coupled with their nearest neighbors. We will restrict ourselves to the case of quadratic symmetries, with lattice constants a_1 and a_2 for the upper and lower surface, respectively (Fig. 10.12). In such context the role of *commensurability* is essential. It is well known that any real number z can be represented as a continued fraction:

$$z = N_0 + \cfrac{1}{N_1 + \cfrac{1}{N_2 + \cdots}} . \tag{10.16}$$

The sequence that converges most slowly is obtained when all $N_i = 1$, which corresponds to the *golden mean* $\bar{z} = (\sqrt{5} - 1)/2$. In 1-D *Weiss* and *Elmer* predicted that friction should decrease with decreasing commensurability, the minimum of friction being reached when $a_1/a_2 = \bar{z}$ [26].

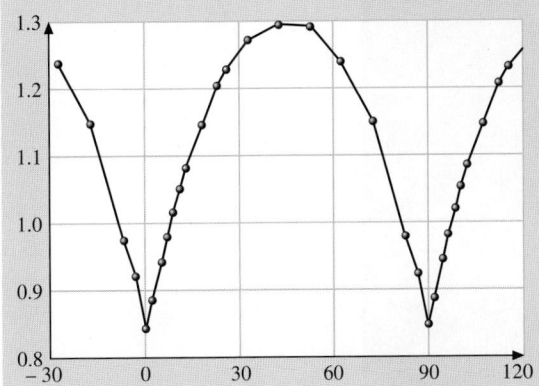

Fig. 10.13. Friction as a function of the sliding angle φ in the 2-D FKT model. (After [25])

In 2-D *Gyalog* and *Thomas* studied the case $a_1 = a_2$, with a misalignment between the two lattices given by an angle θ [25]. When the sliding direction changes, friction also varies from a minimum value corresponding to the sliding angle $\varphi = \theta/2$ to a maximum value, which is reached when $\varphi = \theta/2 + \pi/4$ (Fig. 10.13). The misfit angle θ is related with commensurability. Since the misfit angles giving rise to commensurate structure form a dense subset, the dependence of friction on θ should be discontinuous. The numerical simulations performed by Gyalog are in agreement with this conclusion.

10.3 Friction Experiments on Atomic Scale

Figure 10.14 shows the first friction map observed by Mate on atomic scale. The periodicity of the lateral force is the same as of the atomic lattice of graphite. The series of friction loops in Fig. 10.15 reveals the stick-slip effect discussed in the previous section. The applied loads are in the range of tens of µN. According to the continuum models discussed in Sect. 10.5 these values correspond to contact diameters of 100 nm. A possible explanation for the atomic features observed at such high loads is that graphite flakes might have been detached from the surface and adhered on the tip [27]. Another explanation is that the contact between tip and surface consisted of few nm-scale asperities and the corrugation was not entirely averaged out while sliding. The load dependence of friction found by Mate is rather linear, with a small friction coefficient $\mu = 0.01$ (Fig. 10.16).

The UHV environment reduces the influence of contaminants on the surface and leads to more precise and reproducible results. *Meyer* et al. [28] obtained a series of interesting results on ionic crystals using the UHV-FFM apparatus described in Sect. 10.1.4. In Fig. 10.17 a friction map recorded on KBr(100) is compared with a theoretical map, which was obtained with the 2-D-Tomlinson model. The periodicity $a = 0.47$ nm corresponds to the spacing between equally charged ions. No

Fig. 10.14. First atomic friction map acquired on graphite with a normal force $F_N = 56\,\mu N$. Frame size: 2 nm. (After [1])

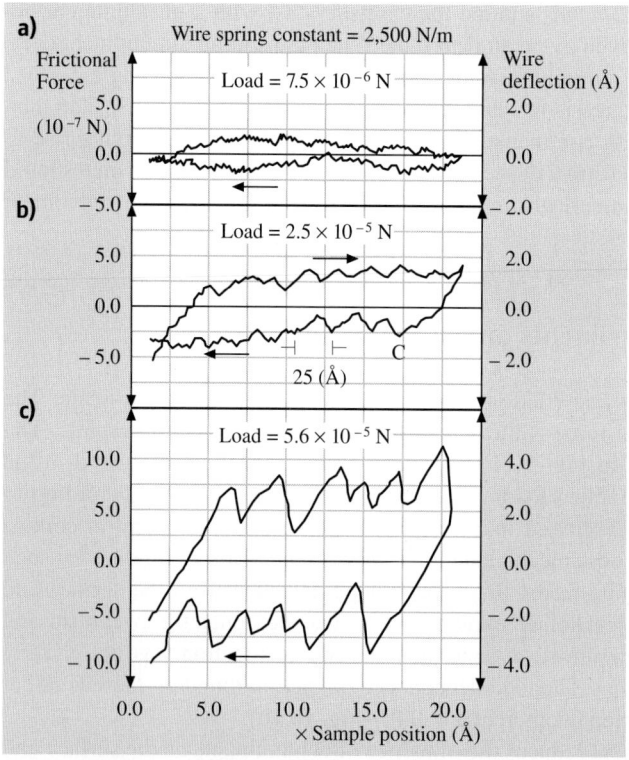

Fig. 10.15. Friction loops on graphite acquired with (**a**) $F_N = 7.5\,\mu N$, (**b**) $24\,\mu N$ and (**c**) $75\,\mu N$. (After [1])

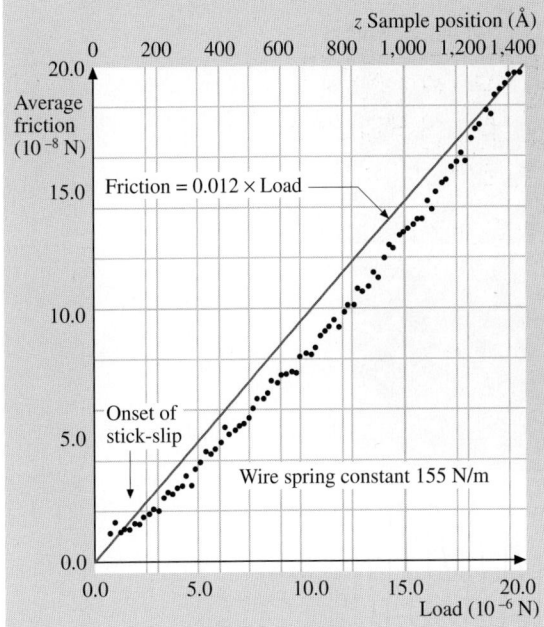

Fig. 10.16. Load dependence of friction on graphite. (After [1])

individual defects were observed. One possible reason is that the contact realized by the FFM tip is always formed by many atoms, which superimpose and average their effects. Molecular dynamics (MD) calculations (Sect. 10.7) show that even a single-atom contact may cause rather large stresses in the sample, which lead to the motion of defects far away from the contact area. In a picturesque frame, we can say that "defects behave like dolphins that swim away in front of an ocean cruiser" [28].

Lüthi et al. [30] detected atomic-scale friction even on the reconstructed Si(111)7 × 7 surface. But uncoated Si-tips and tips coated with Pt, Au, Ag, Cr, Pt/C damaged the sample irreversibly, and the observation of atomic features was achieved only after coating the tips with polytetrafluoroethylene (PTFE), which has lubricant properties and does not react with the dangling bonds of Si(111)7 × 7 (Fig. 10.18).

Recently friction could be resolved on atomic scale even on metallic surfaces in UHV [31]. In Fig. 10.19a a reproducible stick-slip process on Cu(111) is shown. Sliding on the (100) surface of copper produced irregular patterns, although atomic features were recognized even in this case (Fig. 10.19b). Molecular dynamics suggests that wear should occur more easily on the Cu(100) surface than on the closed packed Cu(111) (Sect. 10.7). This conclusion was achieved by adopting copper tips in computer simulations. The assumption that the FFM tip used in the experiments was covered by copper is supported by current measurements performed at the same time.

Atomic stick-slip on diamond was observed by *Germann* et al. with an apposite diamond tip prepared by chemical vapor deposition [33] and, a few years later, by

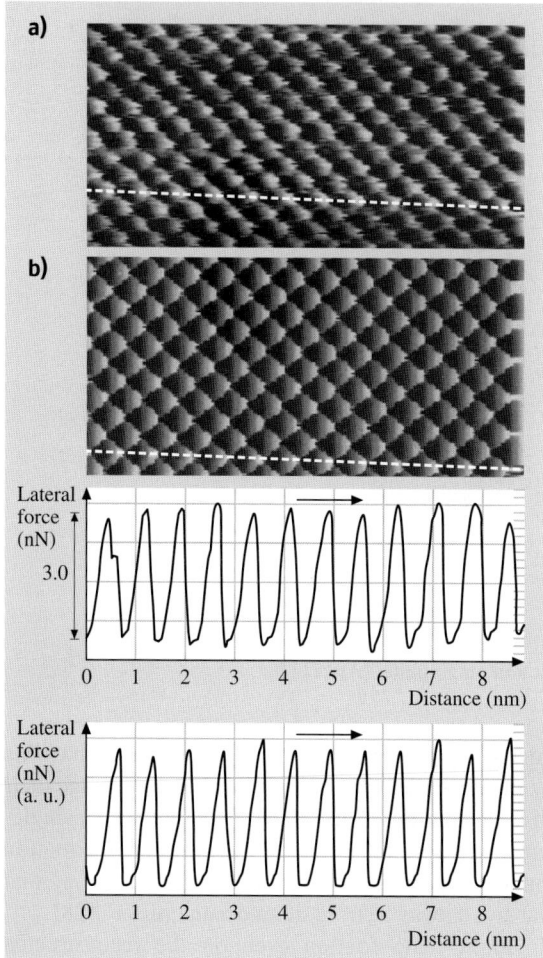

Fig. 10.17. (a) Measured and (b) theoretical friction map on KBr(100). (After [29])

van der Oetelaar et al. [34] with standard silicon tips. The values of friction have huge discrepancies corresponding to the presence or absence of hydrogen on the surface.

Fujisawa et al. [35] measured friction on mica and on MoS_2 with a 2-D-FFM apparatus, which could also reveal forces perpendicular to the scan direction. The features in Fig. 10.20 correspond to a zig-zag walk of the tip, which is predicted by the 2-D-Tomlinson model [36]. The 2-D stick-slip on NaF was detected with normal forces below 14 nN, whereas loads up to 10 μN could be applied on layered materials. The contact between tip and NaF was thus formed by one or a few atoms. The zig-zag walk on mica was also observed by *Kawakatsu* et al. using an original 2-D-FFM with two laser beams and two quadrant photodetectors [37].

Fig. 10.18. (a) Topography and (b) friction image of Si(111)7 × 7 measured with a PTFE coated Si-tip. (After [30])

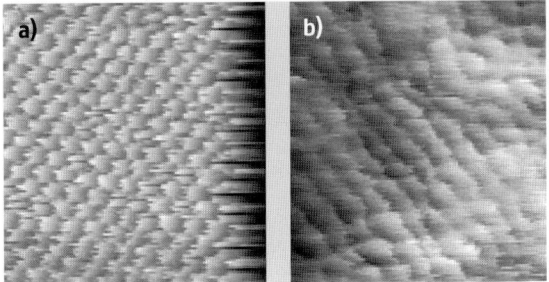

Fig. 10.19. Friction images of (a) Cu(111) and (b) Cu(100). Frame size: 3 nm. (After [32])

10.3.1 Anisotropy of Friction

The importance of the misfit angle in the reciprocal sliding of two flat surfaces was first observed experimentally by *Hirano* et al. in the contact of two mica sheets [38]. Friction increased when the two surfaces formed commensurate structures, in agreement with the discussion in Sect. 10.2.3. In more recent measurements with a monocrystalline tungsten tip on Si(001), *Hirano* et al. observed *superlubricity* in the incommensurate case [39].

Overney et al. [40] studied the effects of friction anisotropy on a bilayer lipid film. In such case different molecular alignments resulted in significant variation of friction. Other measurements of friction anisotropy on stearic acid single crystals were reported by *Takano* and *Fujihira* [41].

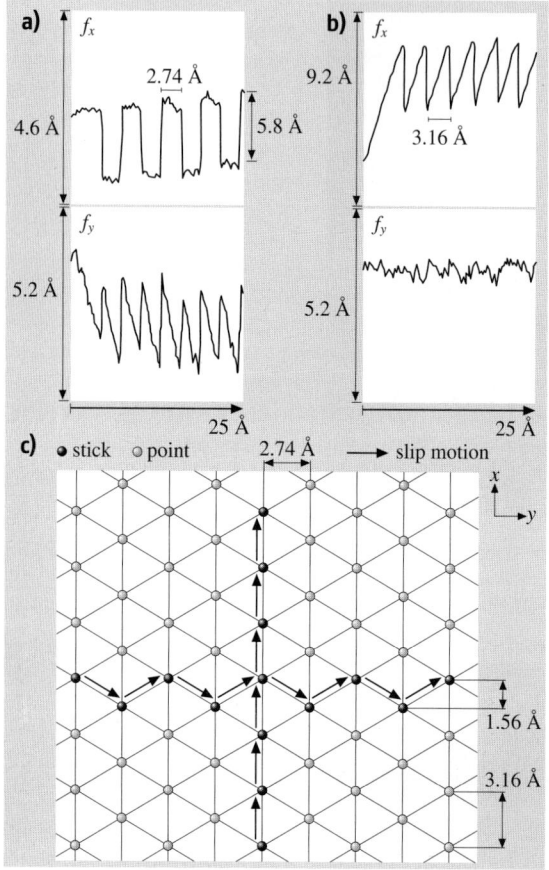

Fig. 10.20. (a) Friction force on MoS$_2$ acquired by scanning along the cantilever and (b) across the cantilever. (c) Motion of the tip on the sample. (After [35])

Liley et al. [42] observed flower-shaped islands of a lipid monolayer on mica, which consisted of domains with different molecular orientation (Fig. 10.21). The angular dependence of friction reflects the tilt direction of the alkyl chains of the monolayer, which was revealed by other techniques.

Lüthi et al. [43] used the FFM tip to move C$_{60}$ islands, which slid on sodium chloride in UHV without disruption (Fig. 10.22). In this experiment friction was found to be independent of the sliding direction. This was not the case in other experiments performed by *Sheehan* and *Lieber*, who observed that the misfit angle is relevant when MoO$_3$ islands are dragged on the MoS$_2$ surface [44]. In these experiments, sliding was possible only along low index directions. The weak orientation dependence found by *Lüthi* et al. [43] is probably due to the large mismatch of C$_{60}$ on NaCl.

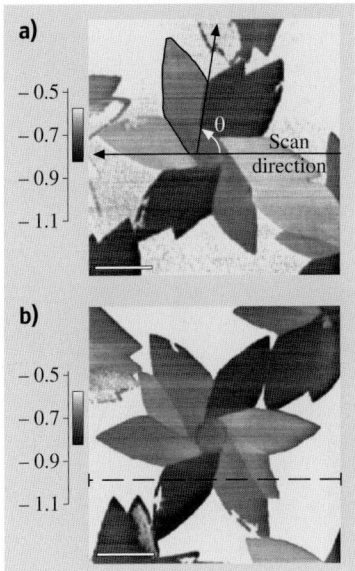

Fig. 10.21. Friction images of a thiolipid monolayer on a mica surface. (After [42])

A recent example of friction anisotropy is related to carbon nanotubes. *Falvo* et al. [45] manipulated nanotubes on graphite using a FFM tip (Fig. 10.23). A dramatic increase of the lateral force was found in the directions corresponding to commensurate contact. At the same time, the nanotube motion changed from sliding/rotating to stick-roll.

10.4 Thermal Effects on Atomic Friction

Although the Tomlinson model gives a good interpretation of the basic mechanism of the atomic stick-slip discussed in Sect. 10.2, it cannot explain some minor features observed in the atomic friction. For example, Fig. 10.24 shows a friction loop acquired on NaCl(100). The peaks in the sawtooth profile have different heights, which is in contrast to the result in Fig. 10.9. Another effect is observed if the scan velocity v is varied: the mean friction force increases with the logarithm of v (Fig. 10.25). This effect cannot be interpreted within the mechanical approach in Sect. 10.2 without further assumptions.

10.4.1 The Tomlinson Model at Finite Temperature

Let us focus again on the energy profile discussed in Sect. 10.2.1. For sake of simplicity, we will assume that $\gamma \gg 1$. At a given time $t < t^*$, the tip jump is prevented by the energy barrier $\Delta E = E(x_{max}, t) - E(x_{min}, t)$, where x_{max} corresponds to the

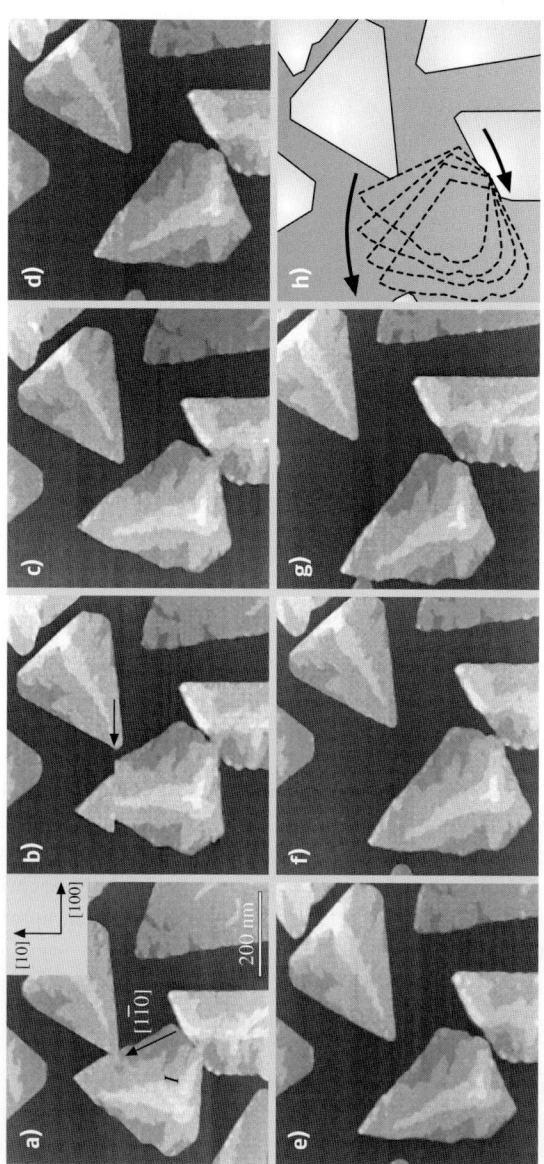

Fig. 10.22. Sequence of topography images of C_{60} islands on NaCl(100). (After [43])

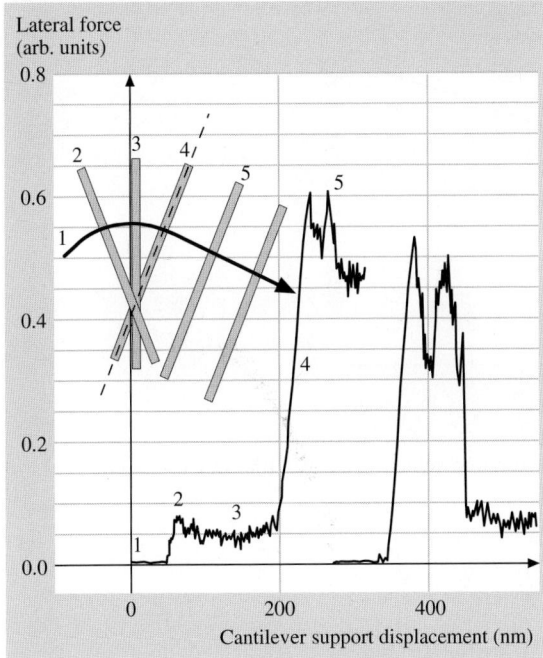

Fig. 10.23. Friction force experienced as a carbon nanotube is rotated into (*left trace*) and out of (*right trace*) commensurate contact. (After [45])

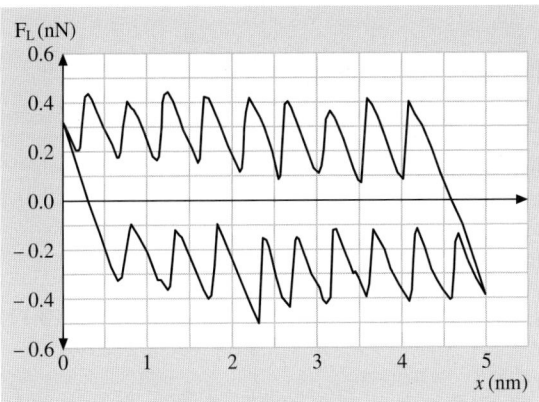

Fig. 10.24. Friction loop on NaCl(100). (After [46])

first maximum observed in the energy profile and x_{min} is the actual position of the tip (Fig. 10.26). The quantity ΔE decreases with time or, equivalently, with the frictional force F_L until it vanishes when $F_L = F^*$ (Fig. 10.27). Close to the critical point the

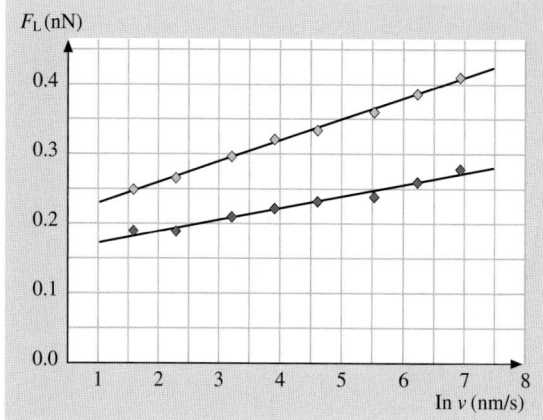

Fig. 10.25. Mean friction force vs. scanning velocity on NaCl(100) at $F_N = 0.44$nN (+) and $F_N = 0.65$nN (×). (After [46])

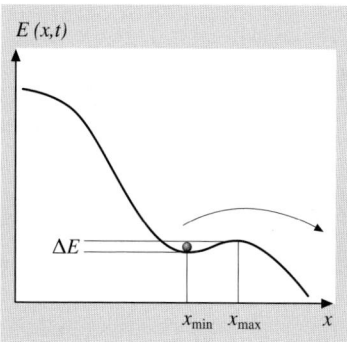

Fig. 10.26. Energy barrier that hinders the tip jump in the Tomlinson model

energy barrier can be written approximately as

$$\Delta E = \lambda(\tilde{F} - F_L), \tag{10.17}$$

where $\tilde{F}$ is close to the critical value $F^* = \pi E_0/a$.

At finite temperature T, the lateral force required to induce a jump is lower than F^*. To estimate the most probable value of F_L at this point, we first consider the probability p that the tip does *not* jump. The probability p changes with time t according to the master equation

$$\frac{dp(t)}{dt} = -f_0 \exp\left(-\frac{\Delta E(t)}{k_B T}\right) p(t), \tag{10.18}$$

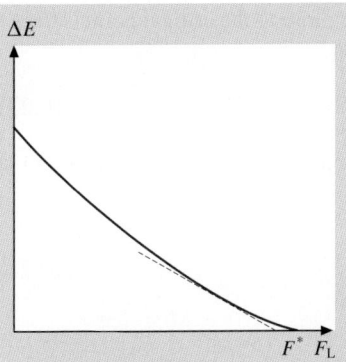

Fig. 10.27. Energy barrier ΔE as a function of the lateral force F_L. The *dashed line* close to the critical value corresponds to the linear approximation (10.17)

where f_0 is a characteristic frequency of the system. The physical meaning of this frequency is discussed in Sect. 10.4.2. We should note that the probability of a reverse jump is neglected, since in this case the energy barrier to overcome is much higher than ΔE. If time is replaced by the corresponding lateral force, the master equation becomes

$$\frac{dp(F_L)}{dF_L} = -f_0 \exp\left(-\frac{\Delta E(F_L)}{k_B T}\right)\left(\frac{dF_L}{dt}\right)^{-1} p(F_L). \tag{10.19}$$

At this point, we substitute

$$\frac{dF_L}{dt} = \frac{dF_L}{dX}\frac{dX}{dt} = k_{\text{eff}} v \tag{10.20}$$

and use the approximation (10.17). The maximum probability transition condition $d^2 p(\S F)/dF^2 = 0$ then yields

$$F_L(v) = F^* - \frac{k_B T}{\lambda} \ln \frac{v_c}{v} \tag{10.21}$$

with

$$v_c = \frac{f_0 k_B T}{k_{\text{eff}} \lambda}. \tag{10.22}$$

Thus the lateral force depends logarithmically on the sliding velocity, as observed experimentally. But the approximation (10.17) does not hold when the tip jump occurs very close to the critical point $x = x^*$, which is the case at high velocities. In this instance, the factor $(dF_L/dt)^{-1}$ in (10.19) is small and, consequently, the probability $p(t)$ does not change significantly until it suddenly approaches 1 when $t \to t^*$. Thus friction is constant at high velocities, in agreement with the classical Coulomb's law of friction [28].

Sang et al. [47] observed that the energy barrier close to the critical point is better approximated by a relation like

$$\Delta E = \mu (F^* - F_L)^{3/2} . \tag{10.23}$$

The same analysis performed using the approximation (10.23) instead of (10.17) leads to the expression [48]

$$\frac{\mu (F^* - F_L)^{3/2}}{k_B T} = \ln \frac{v_c}{v} - \ln \sqrt{1 - \frac{F^*}{F_L}} , \tag{10.24}$$

where the critical velocity v_c is now

$$v_c = \frac{\pi \sqrt{2}}{2} \frac{f_0 k_B T}{k_{\text{eff}} a} . \tag{10.25}$$

The velocity v_c discriminates between two different regimes. If $v \ll v_c$ the second logarithm in (10.24) can be neglected, which leads to the logarithmic dependence

$$F_L(v) = F^* - \left(\frac{k_B T}{\mu} \right)^{2/3} \left(\ln \frac{v_c}{v} \right)^{2/3} . \tag{10.26}$$

In the opposite case, $v \gg v_c$, the term on the left in (10.23) is negligible and

$$F_L(v) = F^* \left(1 - \frac{v_c}{v} \right)^2 . \tag{10.27}$$

In such a case, the lateral force F_L tends to F^*, as expected.

10.4.2 Velocity Dependence of Friction

The velocity dependence of friction was studied by FFM only recently. *Zwörner* et al. observed that friction between silicon tips and diamond, graphite or amorphous carbon is constant with scan velocities of few μm/s [49]. Friction decreased when v is reduced below 1 μm/s. In their experiment on lipid films on mica (Sect. 10.3.1), *Gourdon* et al. [50] explored a range of velocities from 0.01 to 50 μm/s and found a critical velocity $v_c = 3.5$ μm/s, which discriminates between an increasing friction and a constant friction regime (Fig. 10.28). Although these results were not explained with thermal activation, we argue that the previous theoretical discussion gives the correct interpretative key. A clear observation of a logarithmic dependence of friction on the micrometer scale was reported by *Bouhacina* et al., who studied friction on triethoxysilane molecules and polymers grafted on silica with sliding velocity up to $v = 300$ μm/s [51]. The result was explained with a thermally activated Eyring model, which does not differ significantly from the model discussed in the previous subsection [52, 53].

The first measurements on the atomic scale were performed by *Bennewitz* et al. on copper and sodium chloride [31, 46]; in both cases a logarithmic dependence

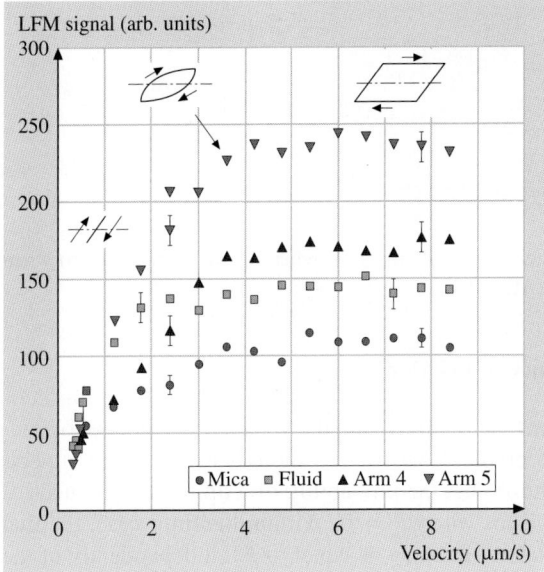

Fig. 10.28. Velocity dependence of friction on mica and on lipid films with different orientation (arms 4 and 5) and in a fluid phase. (After [50])

of friction was revealed up to $v < 1\,\mu\text{m/s}$ (Fig. 10.25), in agreement with (10.21). Higher values of velocities were not explored, due to the limited range of the scan frequencies applicable by FFM on atomic scale. The same limitation does not allow a clear distinction between (10.21) and (10.26) in the interpretation of the experimental results.

At this point we would like to discuss the physical meaning of the characteristic frequency f_0. With lattice constants a of a few angstroms and effective spring constants k_{eff} of about 1 N/m, which is typical in FFM experiments, (10.25) gives values of few hundreds of kHz for f_0. This is the characteristic range in which the torsional eigenfrequencies of the cantilevers are located in both contact and noncontact modes (Fig. 10.29). Future work might clarify whether or not f_0 has to be identified with these frequencies.

To conclude this paragraph, we should emphasize that the increase of friction with increasing velocity is ultimately related to the materials and the environment in which the measurements are realized. In humid environment, *Riedo* et al. observed that the surface wettability has an important role [55]. Friction *decreases* with increasing velocity on hydrophilic surfaces, and the rate of such decrease depends drastically on humidity. On partially hydrophobic surfaces, a logarithmic increase is found again (Fig. 10.30). These results were interpreted considering the thermally activated nucleation of water bridges between tip and sample asperities, as discussed in the cited reference.

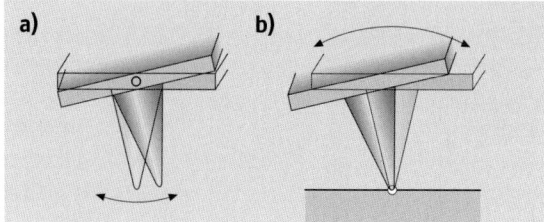

Fig. 10.29. Torsional modes of cantilever oscillations (**a**) when the tip is free and (**b**) when the tip is in contact with a surface. (After [54])

10.4.3 Temperature Dependence of Friction

Thus far we have used thermal activation to explain the velocity dependence of friction. The same mechanism also predicts that friction should change with temperature. Master equation (10.18) shows that the probability of a tip jump is reduced at low temperatures T until it vanishes when $T = 0$. Within this limit case, thermal activation is excluded, and the lateral force F_L is equal to F^*, independently of the scanning velocity v.

To our knowledge stick-slip processes at low temperatures have not been reported. A significant increase of the mean friction with decreasing temperature was recentlymeasured by *He* et al. [56] by FFM (Fig. 10.31). Neglecting the logarithmic contributions (10.21) and (10.26) predict $(F^* - F_L) \sim T$ and $(F^* - F_L) \sim T^{2/3}$ respectively for the temperature dependence of friction. Although He et al. applied a linear fit to their data, the range of 30 K they considered is again not large enough to prove that a $T^{2/3}$ fit would be preferable.

10.5 Geometry Effects in Nanocontacts

Friction is ultimately related to the real shape of the contact between the sliding surfaces. On macroscopic scale the contact between two bodies is studied within continuum mechanics, which are all based on the elasticity theory developed by Hertz in 19th century. Various FFM experiments showed that continuum mechanics is still valid down to contact areas a few nanometers in size. Only when contact is reduced to few atoms does the continuum frame become unsuitable, and other approaches like molecular dynamics are necessary. This section will deal with continuum mechanics theory; molecular dynamics will be discussed in Sect. 10.7.

10.5.1 Continuum Mechanics of Single Asperities

The lateral force F_L between two surfaces in reciprocal motion depends on the size of the real area of contact, A, which can be a few orders of magnitude smaller than the apparent area of contact. The simplest assumption is that friction is proportional to A; the proportionality factor is called *shear strength* σ [57]:

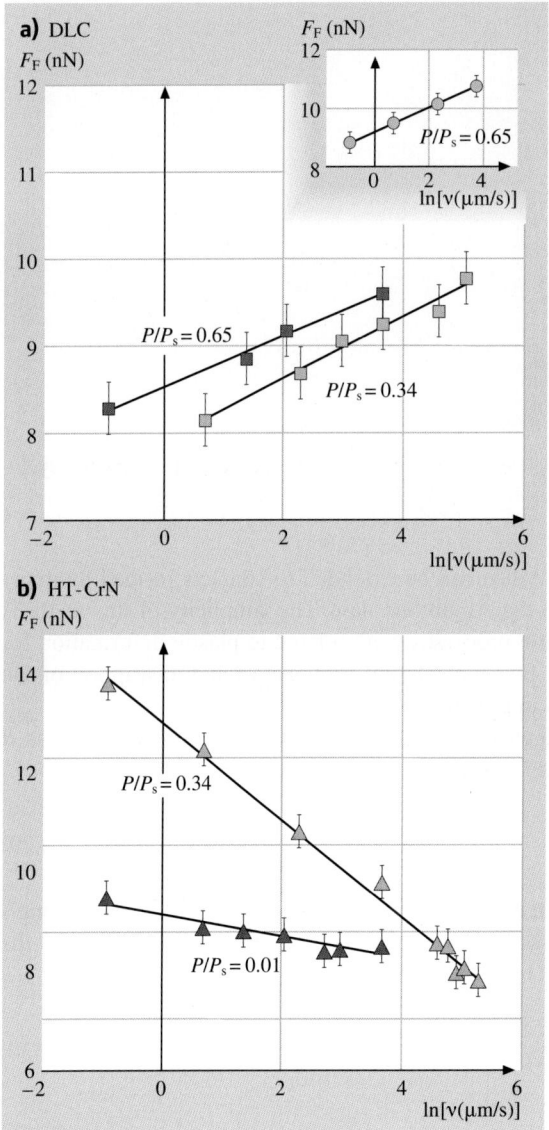

Fig. 10.30. Friction vs. sliding velocity (**a**) on hydrophobic surfaces and (**b**) on hydrophilic surfaces (After [55])

$$F_L = \sigma A \ . \tag{10.28}$$

In case of plastic deformation, the asperities are compressed until the pressure, p, equals a certain yield value, p^*. The resulting contact area is thus $A = F_N/p^*$, and the well-known Amontons' law is obtained: $F_L = \mu F_N$, where $\mu = \sigma/p^*$ is the

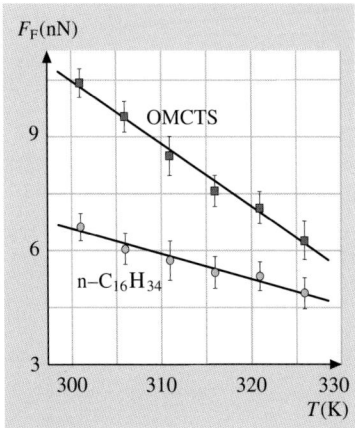

Fig. 10.31. Temperature dependence of friction on *n*-hexadecane and octamethylcyclotetrasiloxane. (After [56])

coefficient of friction. The same idea can be extended to contacts formed by many asperities, and it leads again to the Amontons' law. The simplicity of this analysis explains why most of the friction processes were related to plastic deformation for a long time. Such mechanism, however, should provoke a quick disruption of the surfaces, which is not observed in practice.

Elastic deformation can be easily studied in the case of a sphere of radius R pressed against a flat surface. In such case the contact area is

$$A(F_N) = \pi \left(\frac{R}{K}\right)^{2/3} F_N^{2/3}, \qquad (10.29)$$

where $K = 3E^*/4$ and E^* is an effective Young's modulus, related to the Young's moduli, E_1 and E_2, and the Poisson's numbers, ν_1 and ν_2, of sphere and plane, by the following relation [58]:

$$\frac{1}{E^*} = \frac{1-\nu_1^2}{E_1} + \frac{1-\nu_2^2}{E_2}. \qquad (10.30)$$

The result $A \propto F_N^{2/3}$ is in contrast with the Amontons' law. But a linear relation between F_L and F_N can be obtained for contacts formed by several asperities in particular cases. For example, the area of contact between a flat surface and a set of asperities with an exponential height distribution and the same radius of curvature R depends linearly on the normal force F_N [59]. The same conclusion holds approximately even for a Gaussian height distribution. But the hypothesis that the radii of curvature are the same for all asperities is not realistic. A general model was recently proposed by *Persson*, who derived analytically the proportionality between contact area and load for a large variety of elastoplastic contacts formed by surfaces with arbitrary roughness [60]. The discussion is not straightforward and goes beyond the purposes of this section.

Further effects are observed if adhesive forces between the asperities are taken into account. If the range of action of these forces is smaller than the elastic deformation, (10.29) is extended to the Johnson–Kendall–Roberts (JKR) relation

$$A(F_N) = \pi\left(\frac{R}{K}\right)^{2/3} \times \left(F_N + 3\pi\gamma R + \sqrt{6\pi\gamma R F_N + (3\pi\gamma R)^2}\right)^{2/3}, \quad (10.31)$$

where γ is the surface tension of sphere and plane [61]. The real contact area at zero load is finite and the sphere can be detached only by pulling it away with a certain force. This is also true in the opposite case, in which the range of action of adhesive forces is larger than the elastic deformation. In such a case, the relation between contact area and load has the simple form

$$A(F_N) = \pi\left(\frac{R}{K}\right)^{2/3} (F_N - F_{\text{off}})^{2/3}, \quad (10.32)$$

where F_{off} is the negative load required to break the contact. The Hertz-plus-offset relation (10.32) can be derived from the Derjaguin–Muller–Toporov (DMT) model [62]. To discriminate between the JKR or DMT models *Tabor* introduced a nondimensional parameter

$$\Phi = \left(\frac{9R\gamma^2}{4K^2 z_0^3}\right)^{1/3}, \quad (10.33)$$

where z_0 is the equilibrium distance in contact. The JKR model can be applied if $\Phi > 5$; the DMT model holds when $\Phi < 0.1$ [63]. For intermediate values of Φ, the Maugis–Dugdale model [64] could reasonably explain experimental results (Sect. 10.5.3).

10.5.2 Load Dependence of Friction

The FFM tip represents a single asperity sliding on a surface. The previous discussion suggests a nonlinear dependence of friction on the applied load, provided that continuum mechanics is applicable. Schwarz et al. observed the Hertz-plus-offset relation (10.32) on graphite, diamond, amorphous carbon and C_{60} in argon atmosphere (Fig. 10.32). In their measurements they used well-defined spherical tips with radii of curvature of tens of nanometers, obtained by contaminating silicon tips with amorphous carbon in a transmission electron microscope. In order to compare the tribological behavior of different materials, *Schwarz* et al. suggested the introduction of an effective coefficient of friction, $\tilde{C}$, which is independent of the tip curvature [65].

Meyer et al., *Carpick* et al., and *Polaczyc* et al. performed friction measurements in UHV in agreement with JKR theory [17, 66, 67]. In these experiments different

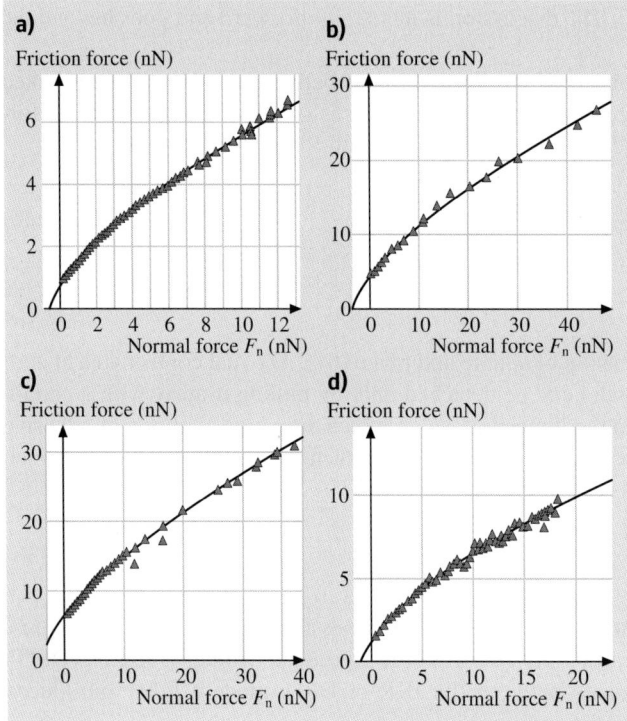

Fig. 10.32. Friction vs. load curve on amorphous carbon in argon atmosphere. Curves (**a**)–(**d**) refer to tips with different radii of curvature. (After [65])

materials were considered, i.e. ionic crystals, mica and metals. In order to correlate lateral and normal forces with improved statistics, Meyer et al. applied an original 2-D-histogram technique (Fig. 10.33). Carpick et al. extended the JKR relation (10.32) to include nonspherical tips. In the case of axisymmetric tip profile $z \propto r^{2n}$ ($n > 1$), it can be proven analytically that the increase of friction becomes less pronounced with increasing n (Fig. 10.34).

10.5.3 Estimation of the Contact Area

In contrast to other tribological instruments, such as the surface force apparatus [68], the contact area cannot be directly measured by FFM. Indirect methods are provided by contact stiffness measurements. The contact between the FFM tip and the sample can be modeled by a series of two springs (Fig. 10.35). The effective constant k^z_{eff} of the series is given by

$$\frac{1}{k^z_{\text{eff}}} = \frac{1}{k^z_{\text{contact}}} + \frac{1}{c_N}, \tag{10.34}$$

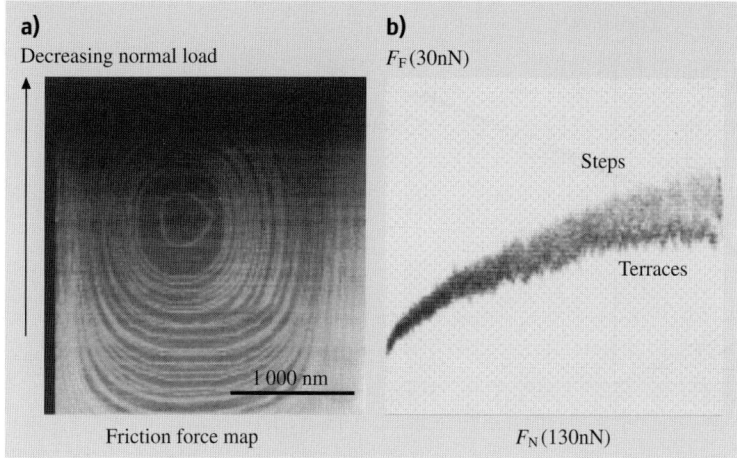

Fig. 10.33. (a) Friction force map on NaCl(100). The load is decreased from 140 to 0 nN (jump-off point) during imaging. (b) 2-D-hystogram of (a). (After [17])

where c_N is the normal spring constant of the cantilever and $k_{contact}^z$ is the normal stiffness of the contact. This quantity is related to the radius of the contact area a, by the simple relation

$$k_{contact}^z = 2aE^*, \qquad (10.35)$$

where E^* is the effective Young's modulus previously introduced [69]. Typical values of $k_{contact}^z$ are order of magnitude larger than c_N, however, and a practical application of (10.34) is not possible.

Carpick et al. suggested an alternative method independently [70, 71]. According to various models, the *lateral* contact stiffness of the contact between a sphere and a flat surface is [72]

$$k_{contact}^x = 8aG^*, \qquad (10.36)$$

where the effective shear stress G^* is defined by

$$\frac{1}{G^*} = \frac{2-v_1^2}{G_1} + \frac{2-v_2^2}{G_2} \qquad (10.37)$$

(G_1, G_2 are the shear moduli of sphere and plane). The contact between the FFM tip and the sample can be modeled again by a series of springs (Fig. 10.35). The effective constant k_{eff}^x of the series is given by

$$\frac{1}{k_{eff}^x} = \frac{1}{k_{contact}^x} + \frac{1}{k_{tip}^x} + \frac{1}{c_L}, \qquad (10.38)$$

where c_L is the lateral spring constant of the cantilever and $k_{contact}^x$ is the lateral stiffness of the contact. As suggested by Lantz, (10.38) includes also the lateral stiffness

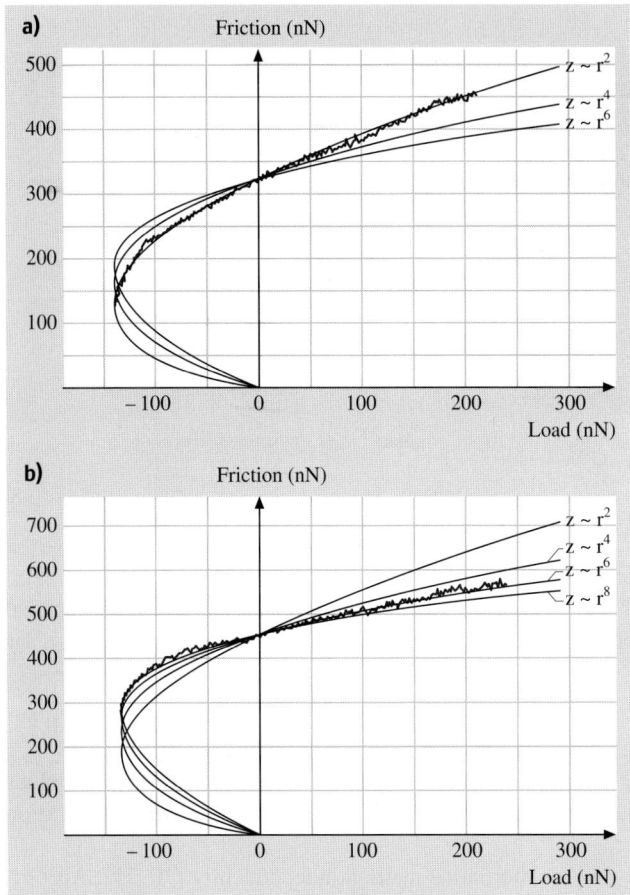

Fig. 10.34. Friction vs. load curves (**a**) for a spherical tip and (**b**) for a blunted tip. The *solid curves* are determined with the JKR theory. (After [66])

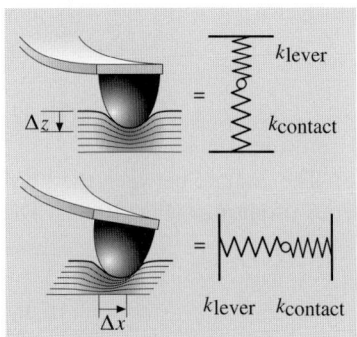

Fig. 10.35. Sketch of normal and lateral stiffness of the contact between tip and surface. (After [70])

of the tip, k_{tip}^x, which can be comparable to the lateral spring constant. The effective spring constant k_{eff}^x is simply given by the slope dF_L/dx of the friction loops (Sect. 10.2.1). Once k_{contact}^x is determined, the contact radius a is easily estimated by (10.36).

The lateral stiffness method was applied to contacts between silicon nitride and muscovite mica in air and between $NbSe_2$ and graphite in UHV. Both the spring constant k_{eff}^x and the lateral force F_L dependence on the load F_N were explained within the same models (JKR and Maugis–Dugdale, respectively), which confirms that friction is proportional to the contact area in the applied range of loads (up to $F_N = 40$ nN in both experiments).

Enachescu et al. estimated the contact area by measuring the contact conductance on diamond as a function of the applied load [73, 74]. Their experimental data were fitted with the DMT model, which was also used to explain the dependence of friction vs. load. Since the contact conductance is proportional to the contact area, the validity of the hypothesis (10.28) was confirmed again.

10.6 Wear on the Atomic Scale

If the normal force F_N applied with the FFM exceeds a critical value, which depends on the tip shape and on the material under investigation, the surface topography is permanently modified. In some cases wear is exploited to create patterns with well-defined shape. Here we will focus on the mechanisms acting on the nanometer scale, where recent experiments have proven the unique capability of the FFM in both scratching and imaging surfaces down to the atomic scale.

10.6.1 Abrasive Wear on the Atomic Scale

In the case of ionic crystals, *Lüthi* et al. observed the appearance of wear at very low loads, i.e. $F_N = 3$ nN [29]. Atomically resolved images of the damage produced by scratching the FFM tip areas on potassium bromide were obtained very recently by *Gnecco* et al. [75]. In Fig. 10.36 a small mound grown up at the end of a groove on KBr(100) is shown under different magnifications. The groove was created a few minutes before imaging by repeated scanning with the normal force $F_N = 21$ nN. The image shows a lateral force map acquired with a load of about 1 nN; no atomic features were observed in the corresponding topography signal. Figure 10.36a,b shows that the debris extracted from the groove recrystallized with the same atomic arrangement of the undamaged surface, which suggests that the wear process occurred like an epitaxial growth assisted by the microscope tip.

Although it is not straightforward to understand how wear is initiated and how the tip transports the debris, important indications are given by the profile of the lateral force F_L recorded while scratching. Figure 10.37 shows some friction loops acquired when the tip was scanned laterally on 5×5 nm^2 areas. The mean lateral force multiplied by the scanned length gives the total energy dissipated in the process. The result of the tip movement are the pits in Fig. 10.38a. Thanks to the pseudo-atomic

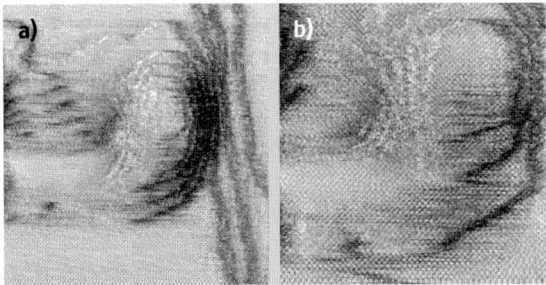

Fig. 10.36. Lateral force images acquired at the end of a groove scratched 256 times with a normal force $F_N = 21$ nN. Frame sizes: (**a**) 39 nm, (**b**) 25 nm

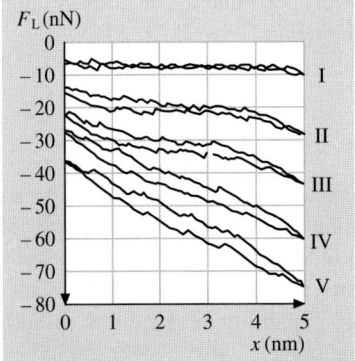

Fig. 10.37. Friction loops acquired while scratching the KBr surface on 5 nm long lines with different loads $F_N = 5.7$ to 22.8 nN. (After [75])

resolution obtained by FFM (Fig. 10.38b), the number of removed atoms could be determined from the lateral force images, which allowed estimating that 70% of the dissipated energy went into wear-less friction [75]. Figures 10.37 and 10.38 clearly show that the damage increases with increasing load. On the other hand, changing the scan velocity v between 25 and 100 nm/s did not produce any significant variation in the wear process.

A different kind of wear was observed on layered materials. *Kopta* et al. [76] could remove layers from a muscovite mica surface by scratching with the normal force $F_N = 230$ nN (Fig. 10.39a). Fourier filtered images acquired on very small areas revealed the different periodicities of the underlying layers, which reflect the complex structure of the muscovite mica (Fig. 10.39b,c).

10.6.2 Wear Contribution to Friction

In Fig. 10.40 the mean lateral force detected while scratching the KBr(100) surface with a fixed load $F_N = 11$ nN is shown. A rather continuous increase of "friction"

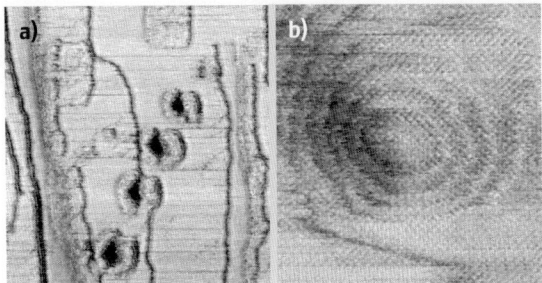

Fig. 10.38. (a) Lateral force images of the pits produced with $F_N = 5.7$ to 22.8 nN. Frame size: 150 nm; (b) Detailed image of the fourth pit from the top with psuedo-atomic resolution. Frame size: 20 nm

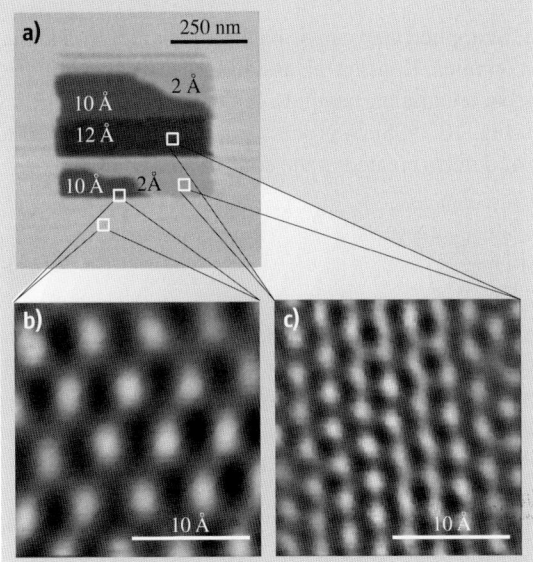

Fig. 10.39. (a) Topography image of an area scratched on muscovite mica with $F_N = 230$ nN; (b), (c) Fourier filtered images of different regions. (After [76])

with the number of scratches N is observed, which can be approximated with the following exponential law:

$$F_L = F_0\, e^{-N/N_0} + F_\infty \left(1 - e^{-N/N_0}\right) . \tag{10.39}$$

Equation (10.39) is easily interpreted assuming that friction is proportional to the contact area $A(\S N)$, and that time evolution of $A(\S N)$ can be described by

$$\frac{dA}{dN} = \frac{A_\infty - A(\S N)}{N_0}, \tag{10.40}$$

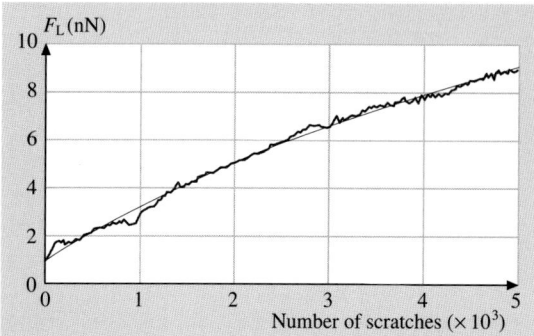

Fig. 10.40. Mean value of the lateral force during repeated scratching with $F_N = 11$ nN on a 500 nm line. (After [75])

A_∞ being the limit area in which the applied load can be balanced without scratching.

To interpret their experiment on mica, Kopta et al. assumed that wear is initiated by atomic defects. When the defects accumulate beyond a critical concentration, they grow to form the scars shown in Fig. 10.39. Such process was once again related to thermal activation. The number of defects created in the contact area $A(F_N)$ is

$$N_{\text{def}}(F_N) = t_{\text{res}} n_0 A(F_N) f_0 \exp\left(-\frac{\Delta E}{k_B T}\right), \tag{10.41}$$

where t_{res} is the residence time of the tip, n_0 is the surface density of atoms, and f_0 is the attempt frequency to overcome the energy barrier ΔE to break a Si–O bond, which depends on the applied load. When the defect density reaches a critical value, a hole is nucleated. The friction force during the creation of a hole could also be estimated with thermal activation by Kopta et al., who derived the formula

$$F_L = c(F_N - F_{\text{off}})^{2/3} + \gamma F_N^{2/3} \exp\left(B_0 F_N^{2/3}\right). \tag{10.42}$$

The first term on the right gives the wearless dependence of friction in the Hertz-plus-offset model (Sect. 10.5.1); the second term is the contribution of the defect production. The agreement between (10.42) and the experiment can be observed in Fig. 10.41.

10.7 Molecular Dynamics Simulations of Atomic Friction and Wear

Section 10.5 has introduced how small sliding contacts can be modeled by continuum mechanics. This modeling has several limitations. The first, most obvious, is that continuum mechanics cannot account for atomic-scale processes like atomic stick-slip. While this limit can be overcome by semiclassical descriptions like the

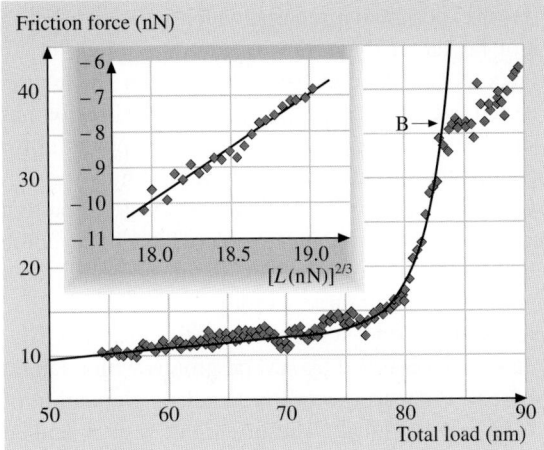

Fig. 10.41. Friction vs. load curve during the creation of a hole in the muscovite mica. (After [76])

Tomlinson model, a definite limit is the determination of contact stiffness for contacts with a radius of a few nanometers. Interpretation of experimental results with the methods introduced in Sect. 10.5.3 regularly yields contact radii of atomic or even smaller size, in clear contradiction to the minimal contact size given by adhesion forces. Such macroscopic quantities as shear modulus or pressure fail to describe the mechanical behavior of these contacts. A microscopic modeling including the atomic structure of the contact is therefore required. This is usually done by a *molecular dynamics* (MD) simulation of the contact. In such simulations, the sliding contact is set up by boundaries of fixed atoms in relative motion and the atoms of the contact, which are allowed to relax their positions according to interactions between each pair of atoms. The methods of computer simulations in tribology are discussed elsewhere in this book. In Sect. 10.7.1 we will discuss such simulations, which can be directly compared to experimental results showing atomic friction processes. The major outcome of the simulations beyond the inclusion of the atomic structure is the importance of displacement of atoms for a correct prediction of forces. In the following Sect. 10.7.2, we will present simulation studies including wear of tip or surface.

10.7.1 Molecular Dynamics Simulation of Friction Processes

The first experiments that exhibited atomic friction features were performed on layered materials, often graphite. A theoretical study of forces between an atomically sharp diamond tip and the graphite surface has been reported by *Tang* et al. [77]. The authors found a significant distance dependence of forces, in which the strongest contrast appeared at different distances for normal and lateral forces due to the strong displacement of surface atoms. The order of magnitude found in this study was one nanonewton, much less than in most experimental reports, which indicated that in such experiments a contact area of far larger dimension was realized. Tang et al. already determined that the distance dependence of forces could even change the symmetry appearance of the observed lateral forces. The experimental situation has also been studied in numerical simulations using simplified one-atom potential for the tip–surface interaction but including the spring potential of the probing force sensor [36]. The motivation for these studies was the observation of a hexagonal pattern in the friction force, while the surface atoms of graphite are ordered in a honeycomb structure. The simulations revealed how the jump path of the tip under lateral force is dependent on the force constant of the probing force sensor.

Surfaces of ionic crystals have become model systems for studies in atomic friction. Atomic stick-slip behavior has been observed by several groups with a lateral force modulation of the order of 1 nN. Pioneering work in atomistic simulation of sliding contacts has been done by *Landman* et al.. The first ionic system studied was a CaF_2 tip sliding over a $CaF_2(111)$ surface [78]. In MD calculation with controlled temperature, the tip was first approached toward the surface up to the point at which an attractive normal force of −3 nN acted on the tip. Then the tip was laterally moved, and the lateral force determined. An oscillation with the atomic periodicity of the surface and with an amplitude decreasing from 8 nN was found. Inspection of the atomic positions revealed a wear process by shear cleavage of the tip. Such transfer of atoms between tip and surface plays a crucial role in atomic friction studies as was shown by *Shluger* et al. [79]. These authors simulated a MgO tip scanning laterally over a LiF(100) surface. Initially an irregular oscillation of the system's energy is found together with transfer of atoms between surface and tip. After a while, the tip apex structure is changed by adsorption of Li and F ions in such a way that nondestructive sliding with perfectly regular oscillation of the energy with the periodicity of the surface could be observed. The authors called this effect self-lubrication and speculate that generally a dynamic self-organization of the surface material on the tip might promote the observation of periodic forces. In a less costly approach of a molecular mechanics study, in which the forces are calculated for each fixed tip–sample configuration, *Tang* et al. produced lateral and normal force maps for a diamond tip over a NaCl(100) surface including such defects as vacancies and a step [80]. In accordance with the studies mentioned before, they found that significant atomic force contrast can be expected for tip–sample distances below 0.35 nm, while distances below 0.15 nm result in destructive forces. For the idealized conditions of scanning at constant height in this regime, the authors predict that even atomic-sized defects could be imaged. Experimentally, these conditions cannot be stabilized in static modes, which so far have been used in lateral force measurements. But dynamic modes of force microscopy have proven atomic resolution of defects when entering the distance regime between 0.2 and 0.4 nm [81]. Recent experimental progress in atomic friction studies of surfaces of ionic crystals include the velocity dependence of lateral forces and atomic-scale wear processes. Such phenomena are not yet accessible by MD studies: the experimentally realized scanning time scale is too far from the atomic relaxation time scales that govern MD simulations. Furthermore, the

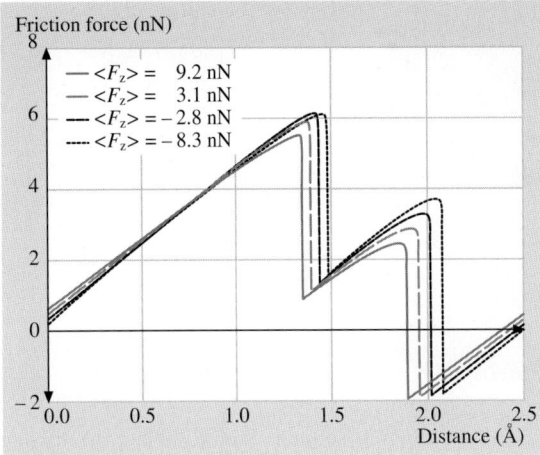

Fig. 10.42. Lateral force acting on a Cu(111) tip in matching contact with a Cu(111) substrate as a function of the sliding distance at different loads. (After [83])

number of freely transferable atoms that can be included in a simulation is simply limited by meaningful calculation time.

Landman et al. also simulated a system of high reactivity, namely a silicon tip sliding over a silicon surface [82]. A clear stick-slip variation of the lateral force was observed for this situation. Strong atom displacements created an interstitial atom under the influence of the tip, which was annealed as the tip moved on. Permanent damage was predicted, however, in the case that the tip enters the repulsive force regime. Although the simulated Si(111) surface is experimentally not accessible, it should be mentioned that on the Si(111)7 × 7 reconstructed surface the tip had to be passivated by a Teflon layer before nondestructive contact-mode measurements became possible (Sect. 10.3). It is worth noting that the simulations for the Cu(111) surface revealed a linear relation between contact area and mean lateral force, similar to classical macroscopic laws.

For metallic sliding over metallic surfaces, wear processes are predicted by several MD studies, which will be discussed in the following section. For a (111)-terminated copper tip over the Cu(111) surface, however, Sørensen et al. found that nondestructive sliding is possible while the lateral force exhibits the sawtooth-like shape characteristic of atomic stick-slip (Fig. 10.42). In contrast, a Cu(100) surface would be disordered by a sliding contact (Fig. 10.43). This difference between the (100) and (111) plane as well as the absolute lateral forces has been confirmed experimentally (Sect. 10.3).

10.7.2 Molecular Dynamics Simulations of Abrasive Wear

The long time scales characteristic of the wear processes and the large amount of material involved make any attempt to simulate these mechanisms on a computer

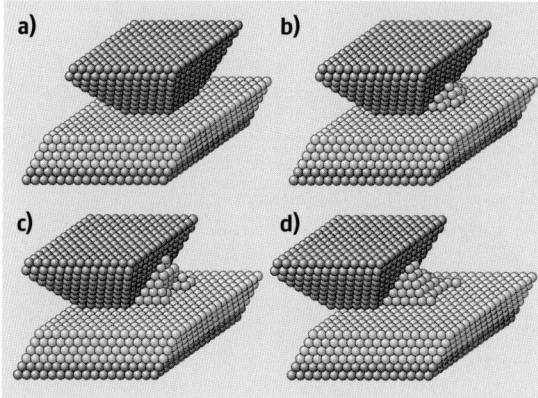

Fig. 10.43. Snapshot of a Cu(100) tip on a Cu(100) substrate during sliding. (**a**) Starting configuration; (**b**)–(**d**) snapshots after 2, 4, and 6 slips. (After [83])

a tremendous challenge. Despite these premises, MD can provide useful insight on the mechanism of removal and deposition of single atoms operated by the FFM tip, which is a kind of information not directly observable in the experiments. Complex processes like abrasive wear and nanolithography can be investigated only within approximate classical mechanics.

The observation made by *Livshits* and *Shluger* that the FFM tip undergoes a process of self-lubrication when scanning ionic surfaces (Sect. 10.7.1) proves that friction and wear are strictly related phenomena. In their MD simulations on copper, *Sørensen* et al. considered not only ordered (111) and (100) terminated tips but even amorphous structures obtained by "heating" the tip at high temperatures [83]. The lateral motion of the neck thus formed revealed a stick-slip behavior due to combined sliding and stretching, and also ruptures, accompanied by deposition of debris on the surface (Fig. 10.44).

To our knowledge, only a few examples of abrasive wear simulations on the atomic scale have been reported. *Buldum* and *Ciraci*, for instance, studied nanoindentation and sliding of a sharp Ni(111) tip on Cu(110) and of a blunt Ni(001) tip on Cu(100) [84]. In the case of the sharp tip quasiperiodic variations of the lateral force were observed, due to stick-slip involving phase-transition. One layer of the asperity was deformed to match the substrate during the first slip and then two asperity layers merged into one through structural transition during the second slip. In the case of the blunt tip, the stick-slip was less regular.

Different results have been reported in which the tip is harder than the underlying sample. *Komanduri* et al. considered an infinitely hard Ni indenter scratching single crystal aluminium at extremely low depths (Fig. 10.45) [85]. In this case a linear relation between friction and load was found, with a high coefficient of friction $\mu = 0.6$, independent of the scratch depth. Nanolithography simulations were recently performed by *Fang* et al. [86], who investigated the role of the displacement of the FFM tip along the direction of slow motion between a scan line and the next one.

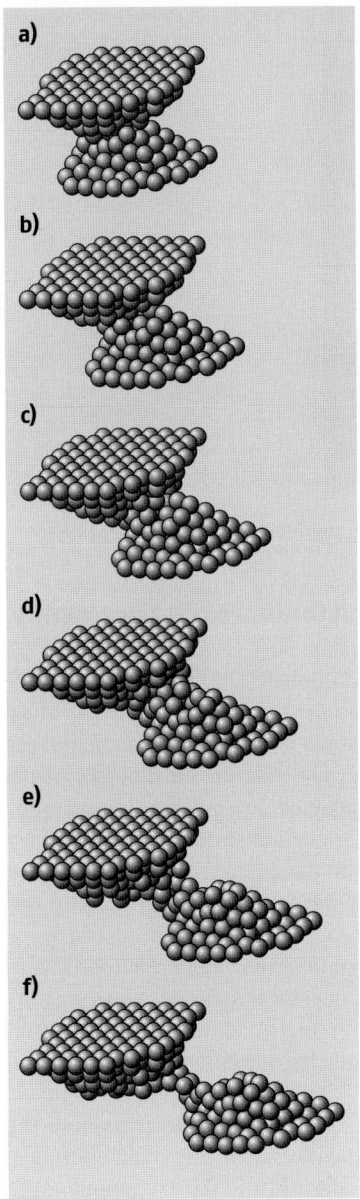

Fig. 10.44. Snapshot of a Cu(100) neck during shearing starting from the configuration (**a**). The upper substrate has been displaced 4.2 Å between subsequent pictures. (After [83])

They found a certain correlation with FFM experiments on silicon films coated with aluminium.

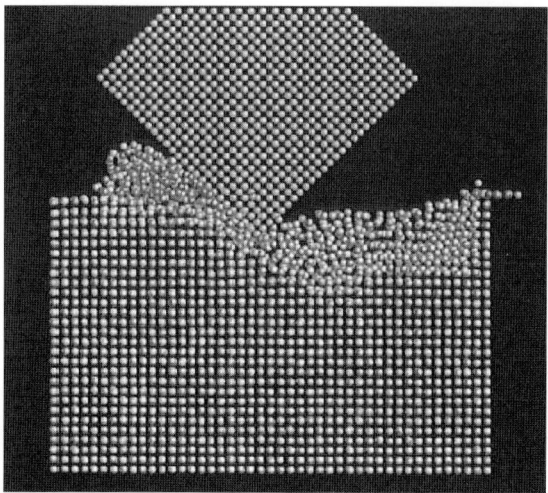

Fig. 10.45. MD simulation of a scratch realized with an infinitely hard tool. (After [85])

10.8 Energy Dissipation in Noncontact Atomic Force Microscopy

Historically, the measurement of energy dissipation induced by tip–sample interaction has been the domain of friction-force microscopy, where the sharp AFM-tip is sliding in gentle contact over a sample. The origins of dissipation in friction are related to phonon excitation, electronic excitation, and irreversible changes of the surface. In a typical stick-slip experiment, the energy dissipated in a single atomic slip event is of the order of 1 eV.

But the lateral resolution of force microscopy in the contact mode is limited by a minimum contact area containing several atoms due to adhesion between tip and sample.

This problem has been overcome in noncontact dynamic force microscopy. In the dynamic mode, the tip is oscillating with a constant amplitude A of typically 1–20 nm at the eigenfrequency f of the cantilever, which shifts by Δf due to interaction forces between tip and sample. This technique has been described in detail in Part B of this book.

Dissipation also occurs in noncontact mode of force microscopy, where the atomic structure of tip and sample are reliably preserved. In this dynamic mode, the damping of the cantilever oscillation can be deduced from the excitation amplitude A_{exc} required to maintain the constant tip oscillation amplitude on resonance.

Compared to friction force microscopy, the interpretation of noncontact-AFM (nc-AFM) experiments is complicated due to the perpendicular oscillation of the tip, typically with an amplitude that is large compared to the minimum tip–sample separation. Another problem is to relate the measured damping of the cantilever to the different origins of dissipation.

In all dynamic force microscopy measurements a power dissipation P_0 caused by internal friction of the freely oscillating cantilever is observed, which is proportional to the eigenfrequency ω_0 and to the square of the amplitude A and inverse proportional to the known Q-value of the cantilever. When the tip–sample distance is reduced, the tip interacts with the sample and therefore an additional damping of the oscillation is encountered. This extra dissipation P_{ts} caused by tip–sample interaction can be calculated from the excitation signal A_{exc} [87].

The observed energy losses per oscillation cycle (100 meV) [88] are roughly similar to the 1 eV energy loss in the contact slip-process. When estimating the contact area in the contact-mode to a few atoms, the energy dissipation per atom that can be associated with a bond being broken and reformed is also around 100 meV.

The idea to relate additional damping of the tip oscillation to dissipative tip–sample interaction has recently attracted much attention [89]. The origin of this additional dissipation are manifold: one may distinguish between apparent energy dissipation (for example from an inharmonic cantilever-motion, artifacts from the phase-controller, or slow fluctuations round the steady-state solution [89, 90]), velocity dependent dissipation (for example electric- and magnetic-field mediated Joule dissipation [91, 92]) and hysteresis-related dissipation (due to atomic instabilities [93, 94] or hysteresis due to adhesion [95]).

By recording Δf and A_{exc} simultaneously during a typical AFM-experiment, forces and dissipation can be measured. Many experiments show true atomic contrast in topography (controlled by Δf) and in the dissipation-signal A_{exc} [96]; however, the origin of the atomic energy dissipation process is still not completely resolved.

To prove that the observed atomic-scale variation of the damping is indeed due to atomic-scale energy dissipation and not an artefact of the distance feedback, *Loppacher* et al. [88] carried out an nc-AFM experiment on Si(111)-7 × 7 at constant height, i.e. with stopped distance feedback. Frequency-shift and dissipation exhibit an atomic-scale contrast, demonstrating a true atomic-scale variation of force and dissipation.

A strong atomic-scale dissipation contrast at step edges has been demonstrated in a few experiments (NaCl on Cu [81]; or measurements on KBr [97]). In Fig. 10.46 ultrathin NaCl islands grown on Cu(111) are shown. As it is shown in Fig. 10.46a, the island edges have a higher contrast compared to the NaCl-terrace and show an atomically resolved corrugation. The strongly enhanced contrast of the step edges and kink sites could be attributed to a slower decay of the electric field and to easier relaxation of theses positions of the ions at these locations. The dissipation image shown in Fig. 10.46b was recorded simultaneously. To establish a direct spatial correspondence between the excitation and the topography-signal, the match between topography and A_{exc} on many images has been studied. Sometimes topography and A_{exc} are in phase, sometimes they are shifted a little bit, and sometimes A_{exc} is at a minimum when topography is at a maximum. The local contrast formation thus depends strongly on the atomic tip structure. In fact, the strong dependence of the dissipation contrast on the atomic state of the tip apex is impressively confirmed by the tip change observed in the experimental images shown in Fig. 10.46b. The dissipation contrast is seriously enhanced, while the topography contrast remains nearly un-

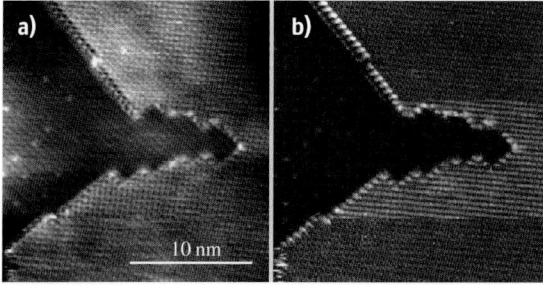

Fig. 10.46. (a) Topography and (b) A_{exc} images of a NaCl island on Cu(111). The tip changes after 1/4 of the scan, thereby changing the contrast in the topography and enhancing the contrast in A_{exc}. After 2/3 of the scan, the contrast from the *lower part* of the image is reproduced, indicating that the tip change was reversible. (After [81])

changed. Clearly, the dissipation depends strongly on the state of the tip and exhibits more short-range character than the frequency shift.

More directly related to friction measurements, where the tip is sliding in contact over the sample, are nc-AFM experiments, where the tip is oscillating parallel to the surface. *Stowe* et al. [98] oriented cantilever beams with in-plane tips perpendicular to the surface, so that the tip motion was approximately parallel to the surface. The noncontact damping of the lever was used to measure localized electrical Joule dissipation. They were able to image the dopant density for n- and p-type silicon samples with 150 nm spatial resolution. An U_{ts}^2 dependence on the tip–sample voltage for the dissipation was found, as proposed by *Denk* and *Pohl* [91] for electric-field Joule-dissipation.

Stipe et al. [99] measured the noncontact friction between a Au(111)-surface and a gold-coated tip with the same setup. They observed the same U_{ts}^2 dependence of the Bias-voltage and a distance dependence that follows a power law $1/d^n$, where n is between 1.3 and 3 [99, 100]. Even when the external Bias-voltage is zero, a substantial electric-field is present. The presence of inhomogeneous tip–sample electric fields is difficult to avoid, even under the best experimental conditions. Although this dissipation is electrical in origin, the detailed mechanism is not totally clear. The most straightforward mechanism is to assume that inhomogeneous fields emanating from the tip and the sample induce surface charges in the nearby metallic sample. When the tip moves, currents are induced, causing Ohmic dissipation [91, 98]. But in metals with good electrical conductivity, Ohmic dissipation is insufficient to account for the observed effect [101]. Thus the tip–sample electric field must have an additional effect, such as driving the motion of adsorbates and surface defects.

When exciting the torsional oscillation of commercial, rectangular AFM cantilevers, the tip is oscillating approximately parallel to the surface. In this mode, lateral forces acting on the tip at step edges and near impurities could be measured quantitatively [54]. An enhanced energy-dissipation was observed at the impurities as well. When the tip is further approached toward the sample, contact formation transforms the nearly free torsional oscillation of the cantilever into a different mode with the tip–sample contact acting as a hinge. When this contact is formed, a rapid increase of the power required to maintain a constant tip oscillation amplitude and a positive frequency shift are found. The onset of the simultaneously recorded damping and positive frequency shift are sharp and essentially coincide. It is assumed that these changes indicate the formation of a tip–sample contact. Two recent studies [102, 103] report on the use of the torsional eigenmode for measuring the elastic properties of the tip–sample contact, where the tip is in contact with the sample and the shear stiffness depends on the normal load.

Kawagishi et al. [104] scanned with lateral amplitudes in the order of 10 pm to 3 nm; their imaging technique gave contrast between graphite terraces, silicon and silicon dioxide, graphite and mica. Torsional self-excitation showed nanometric features of self-assembled monolayer islands due to different lateral dissipation.

Recently *Giessibl* et al. [105] established true atomic resolution of both conservative and dissipative forces by lateral force microscopy. The interaction between a single-tip atom oscillated parallel to an Si(111)-7 × 7 surface is measured. A dissipation energy of up to 4 eV per oscillation cycle is found, which is explained by a plucking action of one atom on to the other as described by *Tomlinson* in 1929 [23].

A detailed review of dissipation phenomena in noncontact force microscopy has been given by *Hug* [106].

10.9 Conclusion

Over the last 15 years, two instrumental developments have stimulated scientific activities in the field of nanotribology. On the one hand, the invention and development of friction force microscopy allows us to quantitatively study friction of a single asperity. As we have discussed in this chapter, atomic processes are observed by forces on the order of 1 nN, i.e. forces related to single chemical bonds. On the other hand, the enormous increase of achievable computing power provides the basis for molecular dynamics simulations of systems containing several hundred atoms. Therewithin, the development of the atomic structure in a sliding contact can be analyzed and forces predicted.

The most prominent observation of atomic friction is a stick-slip behavior with the periodicity of the surface atomic lattice. Semiclassical models can explain the experimental findings including the velocity dependence, which is a consequence of thermal activation of slip events. Classical continuum mechanics can also describe the load dependence of friction in contacts with an extension of several ten nanometers. But when trying to apply continuum mechanics to contacts formed at just ten atoms, obviously wrong numbers result, such as, for example, for the contact radius. Only a comparison with atomistic simulations can provide a full meaningful picture of the physical parameters of such sliding contacts. These simulations predict a close connection between wear and friction, in particular a transfer of atoms between surface and tip, which in some case can even lower friction in a process of self-lubrication.

First experiments have succeeded in studying the onset of wear with atomic resolution. The research in microscopic wear processes will certainly grow in importance as nanostructures are produced and their mechanical properties exploited. Simulation of such processes involving transfer of thousands of atoms will become feasible with further increase of computing power. Another perspective of nanotribology is the expansion of atomic friction experiments toward surfaces with a well-defined roughness. In general, the problem of bridging the gap between single asperity experiments on well-defined surfaces and macroscopic friction should be approached, both experimentally and in modeling.

References

1. C. M. Mate, G. M. McClelland, R. Erlandsson, and S. Chiang. Atomic-scale friction of a tungsten tip on a graphite surface. *Phys. Rev. Lett.*, 59:1942–1945, 1987.
2. G. Binnig, C. F. Quate, and Ch. Gerber. Atomic force microscope. *Phys. Rev. Lett.*, 56:930–933, 1986.
3. O. Marti, J. Colchero, and J. Mlynek. Combined scanning force and friction microscopy of mica. *Nanotechnology*, 1:141–144, 1990.
4. G. Meyer and N. Amer. Simultaneous measurement of lateral and normal forces with an optical-beam-deflection atomic force microscope. *Appl. Phys. Lett.*, 57:2089–2091, 1990.
5. G. Neubauer, S. R. Cohen, G. M. McClelland, D. E. Horn, and C. M. Mate. Force microscopy with a bidirectional capacitance sensor. *Rev. Sci. Instrum.*, 61:2296–2308, 1990.
6. G. M. McClelland and J. N. Glosli. *Friction at the atomic scale*, volume 220, pages 405–425. Kluwer, 1992.
7. R. Linnemann, T. Gotszalk, I. W. Rangelow, P. Dumania, and E. Oesterschulze. Atomic force microscopy and lateral force microscopy using piezoresistive cantilevers. *J. Vac. Sci. Technol. B*, 14:856–860, 1996.
8. M. Nonnenmacher, J. Greschner, O. Wolter, and R. Kassing. Scanning force microscopy with micromachined silicon sensors. *J. Vac. Sci. Technol. B*, 9:1358–1362, 1991.
9. R. Lüthi. Untersuchungen zur nanotribologie und zur auflösungsgrenze im ultrahochvakuum mittels rasterkraftmikroskopie, 1996.
10. J. Cleveland, S. Manne, D. Bocek, and P. K. Hansma. A nondestructive method for determining the spring constant of cantilevers for scanning force microscopy. *Rev. Sci. Instrum.*, 64:403–405, 1993.
11. J. L. Hutter and J. Bechhoefer. Calibration of atomic-force microscope tips. *Rev. Sci. Instrum.*, 64:1868–1873, 1993.
12. H. J. Butt and M. Jaschke. Calculation of thermal noise in atomic-force microscopy. *Nanotechnology*, 6:1–7, 1995.
13. J. M. Neumeister and W. A. Ducker. Lateral, normal, and longitudinal spring constants of atomic-force microscopy cantilevers. *Rev. Sci. Instrum.*, 65:2527–2531, 1994.
14. D. F. Ogletree, R. W. Carpick, and M. Salmeron. Calibration of frictional forces in atomic force microscopy. *Rev. Sci. Instrum.*, 67:3298–3306, 1996.
15. E. Gnecco. Afm study of friction phenomena on the nanometer scale, 2001.
16. U. D. Schwarz, P. Köster, and R. Wiesendanger. Quantitative analysis of lateral force microscopy experiments. *Rev. Sci. Instrum.*, 67:2560–2567, 1996.

17. E. Meyer, R. Lüthi, L. Howald, M. Bammerlin, M. Guggisberg, and H.-J. Güntherodt. Site-specific friction force spectroscopy. *J. Vac. Sci. Technol. B*, 14:1285–1288, 1996.
18. S. S. Sheiko, M. Möller, E. M. C. M. Reuvekamp, and H. W. Zandberger. Calibration and evaluation of scanning-force-microscopy probes. *Phys. Rev B*, 48:5675, 1993.
19. F. Atamny and A. Baiker. Direct imaging of the tip shape by afm. *Surf. Sci.*, 323:L314, 1995.
20. J. S. Villarrubia. Algorithms for scanned probe microscope image simulation, surface reconstruction, and tip estimation. *J. Res. Natl. Inst. Stand. Technol.*, 102:425–454, 1997.
21. L. Howald, E. Meyer, R. Lüthi, H. Haefke, R. Overney, H. Rudin, and H.-J. Güntherodt. Multifunctional probe microscope for facile operation in ultrahigh vacuum. *Appl. Phys. Lett.*, 63:117–119, 1993.
22. Q. Dai, R. Vollmer, R. W. Carpick, D. F. Ogletree, and M. Salmeron. A variable temperature ultrahigh vacuum atomic force microscope. *Rev. Sci. Instrum.*, 66:5266–5271, 1995.
23. G. A. Tomlinson. A molecular theory of friction. *Philos. Mag. Ser.*, 7:905, 1929.
24. T. Gyalog, M. Bammerlin, R. Lüthi, E. Meyer, and H. Thomas. Mechanism of atomic friction. *Europhys. Lett.*, 31:269–274, 1995.
25. T. Gyalog and H. Thomas. Friction between atomically flat surfaces. *Europhys. Lett.*, 37:195–200, 1997.
26. M. Weiss and F. J. Elmer. Dry friction in the frenkel–kontorova–tomlinson model: Static properties. *Phys. Rev. B*, 53:7539–7549, 1996.
27. J. B. Pethica. Comment on "Interatomic forces in scanning tunneling microscopy: Giant corrugations of the graphite surface". *Phys. Rev. Lett.*, 57:3235, 1986.
28. E. Meyer, R. M. Overney, K. Dransfeld, and T. Gyalog. *Nanoscience, Friction and Rheology on the Nanometer Scale*. World Scientific, 1998.
29. R. Lüthi, E. Meyer, M. Bammerlin, L. Howald, H. Haefke, T. Lehmann, C. Loppacher, H.-J. Güntherodt, T. Gyalog, and H. Thomas. Friction on the atomic scale: An ultrahigh vacuum atomic force microscopy study on ionic crystals. *J. Vac. Sci. Technol. B*, 14:1280–1284, 1996.
30. L. Howald, R. Lüthi, E. Meyer, and H.-J. Güntherodt. Atomic-force microscopy on the Si(111)7 × 7 surface. *Phys. Rev. B*, 51:5484–5487, 1995.
31. R. Bennewitz, T. Gyalog, M. Guggisberg, M. Bammerlin, E. Meyer, and H.-J. Güntherodt. Atomic-scale stick-slip processes on Cu(111). *Phys. Rev. B*, 60:R11301–R11304, 1999.
32. R. Bennewitz, E. Gnecco, T. Gyalog, and E. Meyer. Atomic friction studies on well-defined surfaces. *Tribol. Lett.*, 10:51–56, 2001.
33. G. J. Germann, S. R. Cohen, G. Neubauer, G. M. McClelland, and H. Seki. Atomic scale friction of a diamond tip on diamond (100) and (111) surfaces. *J. Appl. Phys.*, 73:163–167, 1993.
34. R. J. A. van den Oetelaar and C. F. J. Flipse. Atomic-scale friction on diamond(111) studied by ultra-high vacuum atomic force microscopy. *Surf. Sci.*, 384:L828–L835, 1997.
35. S. Fujisawa, E. Kishi, Y. Sugawara, and S. Morita. Atomic-scale friction observed with a two-dimensional frictional-force microscope. *Phys. Rev. B*, 51:7849–7857, 1995.
36. N. Sasaki, M. Kobayashi, and M. Tsukada. Atomic-scale friction image of graphite in atomic-force microscopy. *Phys. Rev. B*, 54:2138–2149, 1996.
37. H. Kawakatsu and T. Saito. Scanning force microscopy with two optical levers for detection of deformations of the cantilever. *J. Vac. Sci. Technol. B*, 14:872–876, 1996.
38. M. Hirano, K. Shinjo, R. Kaneko, and Y. Murata. Anisotropy of frictional forces in muscovite mica. *Phys. Rev. Lett.*, 67:2642–2645, 1991.

39. M. Hirano, K. Shinjo, R. Kaneko, and Y. Murata. Observation of superlubricity by scanning tunneling microscopy. *Phys. Rev. Lett.*, 78:1448–1451, 1997.
40. R. M. Overney, H. Takano, M. Fujihira, W. Paulus, and H. Ringsdorf. Anisotropy in friction and molecular stick-slip motion. *Phys. Rev. Lett.*, 72:3546–3549, 1994.
41. H. Takano and M. Fujihira. Study of molecular scale friction on stearic acid crystals by friction force microscopy. *J. Vac. Sci. Technol. B*, 14:1272–1275, 1996.
42. M. Liley, D. Gourdon, D. Stamou, U. Meseth, T. M. Fischer, C. Lautz, H. Stahlberg, H. Vogel, N. A. Burnham, and C. Duschl. Friction anisotropy and asymmetry of a compliant monolayer induced by a small molecular tilt. *Science*, 280:273–275, 1998.
43. R. Lüthi, E. Meyer, H. Haefke, L. Howald, W. Gutmannsbauer, and H.-J. Güntherodt. Sled-type motion on the nanometer scale: Determination of dissipation and cohesive energies of C_{60}. *Science*, 266:1979–1981, 1994.
44. P. E. Sheehan and C. M. Lieber. Nanotribology and nanofabrication of MoO_3 structures by atomic force microscopy. *Science*, 272:1158–1161, 1996.
45. M. R. Falvo, J. Steele, R. M. Taylor, and R. Superfine. Evidence of commensurate contact and rolling motion: Afm manipulation studies of carbon nanotubes on hopg. *Tribol. Lett.*, 9:73–76, 2000.
46. E. Gnecco, R. Bennewitz, T. Gyalog, Ch. Loppacher, M. Bammerlin, E. Meyer, and H.-J. Güntherodt. Velocity dependence of atomic friction. *Phys. Rev. Lett.*, 84:1172–1175, 2000.
47. Y. Sang, M. Dubé, and M. Grant. Thermal effects on atomic friction. *Phys. Rev. Lett.*, 87:174301, 2001.
48. E. Riedo, E. Gnecco, R. Bennewitz, E. Meyer, and H. Brune. Interaction potential and hopping dynamics governing sliding friction. *Phys. Rev. Lett.*, 91:084502, 2003.
49. O. Zwörner, H. Hölscher, U. D. Schwarz, and R. Wiesendanger. The velocity dependence of frictional forces in point-contact friction. *Appl. Phys. A*, 66:263–267, 1998.
50. D. Gourdon, N. A. Burnham, A. Kulik, E. Dupas, F. Oulevey, G. Gremaud, D. Stamou, M. Liley, Z. Dienes, H. Vogel, and C. Duschl. The dependence of friction anisotropies on the molecular organization of lb films as observed by afm. *Tribol. Lett.*, 3:317–324, 1997.
51. T. Bouhacina, J. P. Aimé, S. Gauthier, D. Michel, and V. Heroguez. Tribological behavior of a polymer grafted on silanized silica probed with a nanotip. *Phys. Rev. B*, 56:7694–7703, 1997.
52. H. J. Eyring. The activated complex in chemical reactions. *J. Chem. Phys.*, 3:107, 1937.
53. J. N. Glosli and G. M. McClelland. Molecular dynamics study of sliding friction of ordered organic monolayers. *Phys. Rev. Lett.*, 70:1960–1963, 1993.
54. O. Pfeiffer, R. Bennewitz, A. Baratoff, E. Meyer, and P. Grütter. Lateral-force measurements in dynamic force microscopy. *Phys. Rev. B*, 65:161403, 2002.
55. E. Riedo, F. Lévy, and H. Brune. Kinetics of capillary condensation in nanoscopic sliding friction. *Phys. Rev. Lett.*, 88:185505, 2002.
56. M. He, A. S. Blum, G. Overney, and R. M. Overney. Effect of interfacial liquid structuring on the coherence length in nanolubrucation. *Phys. Rev. Lett.*, 88:154302, 2002.
57. F. P. Bowden and F. P. Tabor. *The Friction and Lubrication of Solids*. Oxford Univ. Press, 1950.
58. L. D. Landau and E. M. Lifshitz. *Introduction to Theoretical Physics*. Nauka, 1998.
59. J. A. Greenwood and J. B. P. Williamson. Contact of nominally flat surfaces. *Proc. R. Soc. Lond. A*, 295:300, 1966.
60. B. N. J. Persson. Elastoplastic contact between randomly rough surfaces. *Phys. Rev. Lett.*, 87:116101, 2001.

61. K. L. Johnson, K. Kendall, and A. D. Roberts. Surface energy and contact of elastic solids. *Proc. R. Soc. Lond. A*, 324:301, 1971.
62. B. V. Derjaguin, V. M. Muller, and Y. P. Toporov. Effect of contact deformations on adhesion of particles. *J. Colloid Interface Sci.*, 53:314–326, 1975.
63. D. Tabor. Surface forces and surface interactions. *J. Colloid Interface Sci.*, 58:2–13, 1977.
64. D. Maugis. Adhesion of spheres: the jkr-dmt transition using a dugdale model. *J. Colloid Interface Sci.*, 150:243–269, 1992.
65. U. D. Schwarz, O. Zwörner, P. Köster, and R. Wiesendanger. Quantitative analysis of the frictional properties of solid materials at low loads. *Phys. Rev. B*, 56:6987–6996, 1997.
66. R. W. Carpick, N. Agraït, D. F. Ogletree, and M. Salmeron. Measurement of interfacial shear (friction) with an ultrahigh vacuum atomic force microscope. *J. Vac. Sci. Technol. B*, 14:1289–1295, 1996.
67. C. Polaczyk, T. Schneider, J. Schöfer, and E. Santner. Microtribological behavior of Au(001) studied by afm/ffm. *Surf. Sci.*, 402:454–458, 1998.
68. J. N. Israelachvili and D. Tabor. Measurement of van der waals dispersion forces in range 1.5 to 130 nm. *Proc. R. Soc. Lond. A*, 331:19, 1972.
69. S. P. Jarvis, A. Oral, T. P. Weihs, and J. B. Pethica. A novel force microscope and point-contact probe. *Rev. Sci. Instrum.*, 64:3515–3520, 1993.
70. R. W. Carpick, D. F. Ogletree, and M. Salmeron. Lateral stiffness: A new nanomechanical measurement for the determination of shear strengths with friction force microscopy. *Appl. Phys. Lett.*, 70:1548–1550, 1997.
71. M. A. Lantz, S. J. O'Shea, M. E. Welland, and K. L. Johnson. Atomic-force-microscope study of contact area and friction on NbSe$_2$. *Phys. Rev. B*, 55:10776–10785, 1997.
72. K. L. Johnson. *Contact Mechanics*. Cambridge Univ. Press, 1985.
73. M. Enachescu, R. J. A. van den Oetelaar, R. W. Carpick, D. F. Ogletree, C. F. J. Flipse, and M. Salmeron. Atomic force microscopy study of an ideally hard contact: the diamond(111)/tungsten carbide interface. *Phys. Rev. Lett.*, 81:1877–1880, 1998.
74. M. Enachescu, R. J. A. van den Oetelaar, R. W. Carpick, D. F. Ogletree, C. F. J. Flipse, and M. Salmeron. Observation of proportionality between friction and contact area at the nanometer scale. *Tribol. Lett.*, 7:73–78, 1999.
75. E. Gnecco, R. Bennewitz, and E. Meyer. Abrasive wear on the atomic scale. *Phys. Rev. Lett.*, 88:215501, 2002.
76. S. Kopta and M. Salmeron. The atomic scale origin of wear on mica and its contribution to friction. *J. Chem. Phys.*, 113:8249–8252, 2000.
77. H. Tang, C. Joachim, and J. Devillers. Interpretation of afm images – the graphite surface with a diamond tip. *Surf. Sci.*, 291:439–450, 1993.
78. U. Landman, W. D. Luedtke, and E. M. Ringer. Atomistic mechanisms of adhesive contact formation and interfacial processes. *Wear*, 153:3–30, 1992.
79. A. I. Livshits and A. L. Shluger. Self-lubrication in scanning-force-microscope image formation on ionic surfaces. *Phys. Rev. B*, 56:12482–12489, 1997.
80. H. Tang, X. Bouju, C. Joachim, C. Girard, and J. Devillers. Theoretical study of the atomic-force-microscopy imaging process on the NaCl(100) surface. *J. Chem. Phys.*, 108:359–367, 1998.
81. R. Bennewitz, A. S. Foster, L. N. Kantorovich, M. Bammerlin, Ch. Loppacher, S. Schär, M. Guggisberg, E. Meyer, and A. L. Shluger. Atomically resolved edges and kinks of NaCl islands on Cu(111): Experiment and theory. *Phys. Rev. B*, 62:2074–2084, 2000.
82. U. Landman, W. D. Luetke, and M. W. Ribarsky. Structural and dynamical consequences of interactions in interfacial systems. *J. Vac. Sci. Technol. A*, 7:2829–2839, 1989.

83. M. R. Sørensen, K. W. Jacobsen, and P. Stoltze. Simulations of atomic-scale sliding friction. *Phys. Rev. B*, 53:2101–2113, 1996.
84. A. Buldum and C. Ciraci. Contact, nanoindentation, and sliding friction. *Phys. Rev. B*, 57:2468–2476, 1998.
85. R. Komanduri, N. Chandrasekaran, and L. M. Raff. Molecular dynamics simulation of atomic-scale friction. *Phys. Rev. B*, 61:14007–14019, 2000.
86. T. H. Fang, C. I. Weng, and J. G. Chang. Molecular dynamics simulation of nanolithography process using atomic force microscopy. *Surf. Sci.*, 501:138–147, 2002.
87. B. Gotsmann, C. Seidel, B. Anczykowski, and H. Fuchs. Conservative and dissipative tip-sample interaction forces probed with dynamic afm. *Phys. Rev. B*, 60:11051–11061, 1999.
88. C. Loppacher, R. Bennewitz, O. Pfeiffer, M. Guggisberg, M. Bammerlin, S. Schär, V. Barwich, A. Baratoff, and E. Meyer. Experimental aspects of dissipation force microscopy. *Phys. Rev. B*, 62:13674–13679, 2000.
89. M. Gauthier and M. Tsukada. Theory of noncontact dissipation force microscopy. *Phys. Rev. B*, 60:11716–11722, 1999.
90. J. P. Aimé, R. Boisgard, L. Nony, and G. Couturier. Nonlinear dynamic behavior of an oscillating tip-microlever system and contrast at the atomic scale. *Phys. Rev. Lett.*, 82:3388–3391, 1999.
91. W. Denk and D. W. Pohl. Local electrical dissipation imaged by scanning force microscopy. *Appl. Phys. Lett.*, 59:2171–2173, 1991.
92. S. Hirsekorn, U. Rabe, A. Boub, and W. Arnold. On the contrast in eddy current microscopy using atomic force microscopes. *Surf. Interface Anal.*, 27:474–481, 1999.
93. U. Dürig. *Forces in scanning probe methods*, volume 286 of *Ser. E*, pages 353–366. Kluwer, 1995.
94. N. Sasaki and M. Tsukada. Effect of microscopic nonconservative process on noncontact atomic force microscopy. *Jpn. J. of Appl. Phys.*, 39:L1334–L1337, 2000.
95. B. Gotsmann and H. Fuchs. The measurement of hysteretic forces by dynamic afm. *Appl. Phys. A*, 72:55–58, 2001.
96. M. Guggisberg, M. Bammerlin, A. Baratoff, R. Lüthi, C. Loppacher, F. M. Battiston, J. Lü, R. Bennewitz, E. Meyer, and H. J. Güntherodt. Dynamic force microscopy across steps on the Si(111)-(7 × 7) surface. *Surf. Sci.*, 461:255–265, 2000.
97. R. Bennewitz, S. Schär, V. Barwich, O. Pfeiffer, E. Meyer, F. Krok, B. Such, J. Kolodzej, and M. Szymonski. Atomic-resolution images of radiation damage in KBr. *Surf. Sci.*, 474:197–202, 2001.
98. T. D. Stowe, T. W. Kenny, J. Thomson, and D. Rugar. Silicon dopant imaging by dissipation force microscopy. *Appl. Phys. Lett*, 75:2785–2787, 1999.
99. B. C. Stipe, H. J. Mamin, T. D. Stowe, T. W. Kenny, and D. Rugar. Noncontact friction and force fluctuations between closely spaced bodies. *Phys. Rev. Lett*, 87:96801, 2001.
100. B. Gotsmann and H. Fuchs. Dynamic force spectroscopy of conservative and dissipative forces in an Al-Au(111) tip-sample system. *Phys. Rev. Lett.*, 86:2597–2600, 2001.
101. B. N. J. Persson and A. I. Volokitin. Comment on "Brownian motion of microscopic solids under the action of fluctuating electromagnetic fields". *Phys. Rev. Lett.*, 84:3504, 2000.
102. K. Yamanaka, A. Noguchi, T. Tsuji, T. Koike, and T. Goto. Quantitative material characterization by ultrasonic afm. *Surf. Interface Anal.*, 27:600–606, 1999.
103. T. Drobek, R. W. Stark, and W. M. Heckl. Determination of shear stiffness based on thermal noise analysis in atomic force microscopy: Passive overtone microscopy. *Phys. Rev. B*, 64:045401, 2001.

104. T. Kawagishi, A. Kato, Y. Hoshi, and H. Kawakatsu. Mapping of lateral vibration of the tip in atomic force microscopy at the torsional resonance of the cantilever. *Ultramicroscopy*, 91:37–48, 2002.
105. F. J. Giessibl, M. Herz, and J. Mannhart. Friction traced to the single atom. *PNAS*, 99:12006–12010, 2002.
106. H.-J. Hug and A. Baratoff. *Measurement of dissipation induced by tip-sample interactions*, page 395. Springer, 2002.

11

Nanoscale Mechanical Properties – Measuring Techniques and Applications

Andrzej J. Kulik, András Kis, Gérard Gremaud, Stefan Hengsberger,
Philippe K. Zysset, and Lásló Forró

Summary. The first part of this chapter describes local (at the scale of nanometers) measurements of mechanical properties. It includes detailed state-of-the-art presentation and in-depth analysis of experimental techniques, results, and interpretations.

After a short introduction, the second part describes local mechanical spectroscopy using coupled Atomic Force Microscopy and ultrasound. This technique allows us to map quickly not only spatial distribution of the elasticity but anelastic properties as well. At one point in the sample, semi-quantitative measurements can be made as a function of the temperature. On the nanometer scale, results have close similitudes to bulk measurements and interpretable differences. Local elasticity and damping were measured during phase transition of polymer samples and shape-memory alloys.

The third part describes the "nano-Swiss cheese" method of measuring the elastic properties of such tubular nanometer size objects as carbon nanotubes and microtubules. It is probably the only experiment in which properties of single-wall nanotube ropes were measured as a function of the rope diameter. We extended this idea to biological objects, microtubules, and successfully solved major experimental difficulties. We not only measured the temperature dependency of microtubule modulus in pseudo-physiological conditions but also estimated shear modulus using the same microtubule with several lengths of suspended segments.

The fourth section demonstrates the scanning nanoindentation technique as applied to human bone tissue. This instrument allows performing topography scans and indentation tests using the identical tip. The available surface scan allows a high positioning precision of the indenter tip on the structure of interest. For very inhomogeneous samples, such as bone tissue, this tool provides a probe to detect local variations of the mechanical properties. The indentation test supplies quantitative parameters like elastic modulus and hardness on the submicron level. Local mechanical properties of compact and trabecular bone lamellae were tested under both dry and pseudo-physiological conditions.

Finally, last part is given to a discussion of future prospects and conclusions.

11.1 Introduction

Experiments measuring the mechanical properties at nanoscale are key to understanding mesoscopic materials and inhomogeneous materials.

The most prominent technique for such local measurements implements an Atomic Force Microscopy in static mode to obtain force–distance curves (f-d). Ideally this provides the force applied to the tip and the tip–sample distance allowing the local reduced Young's modulus to be determined. In practice, however, these values are not measured directly. Instead, the applied force is offset by adhesive and capillary forces. Furthermore, the tip–sample distance must be obtained by subtraction of the sample (Z-piezo) displacement and the deflection of the cantilever, neither of which are calibrated on most commercial instruments. Usually only the voltage applied to nonlinear and hysteretic Z-piezo and the uncalibrated cantilever deflection is available. These available data must be tediously calibrated and converted to real displacements and forces. The calibration procedure given by *Radmacher* [1] works well but only when applied to very compliant materials such as biological or polymer samples. An additional difficulty is that since one end of the cantilever is fixed, neither the orientation of the tip nor its position on the sample surface remains constant during f-d curve acquisition. Movement of the tip on the sample surface, therefore occurs not only vertically but also along the cantilever axis. As a result, applied contact mechanics may not be valid. Finally, uncertainty about the exact tip shape and radius complicate the procedure even further. Instrumental challenges are so significant that applying static AFM to obtain absolute values of the elastic modulus using commercial instruments is difficult at best, especially on stiff surfaces. And this, of course, is the motivation for developing of other techniques.

Although absolute values of mechanical properties are difficult to measure, relative measurements or maps (images of physical properties) are still very interesting. Using acoustic vibrations of the AFM sample surface one can access local elastic and anelastic properties. The first part of this chapter describes these techniques and applications.

Studies of mesoscopic specimens are even more challenging: it is difficult and time consuming to locate a sample, specially properly positioned. In the second part of the chapter, we used 'nano Swiss cheese' technique to get insight into the axial and shear modulus of several species of carbon nanotubes. The method was further extended to biological microtubules, where The experiment was performed in liquid and as a function of the temperature.

Reliable mechanical measurements can be obtained by combining the best of two worlds, AFM and nanoindenter. Classical nanoindenters use optical tip positioning, which are inadequate in the nanoworld. Scanning nanoindenter uses the same diamond tip for imaging (with microNewton forces) and measuring. Its application to human bones in liquid and 37 °C is described in the third part of this chapter.

One should underline the importance of AFM linearity and stability for such experiments: only good designs may limit frustration and experiment time of courageous PhD students, and lot of improvements are still lacking.

11.2 Local Mechanical Spectroscopy by Contact AFM

Mechanical properties of solids (elasticity, anelasticity, plasticity) are generally measured on macroscopic samples. But many phenomena in materials science require measurements of mechanical properties at the surface of a material or at the interfaces between thin layers deposited on a surface. High spatial resolution is also important, for example in the cases of multiphased materials or composite materials, phase transitions, lattice softening in shape-memory alloys, precipitation in light alloys, and glass transition of amorphous materials.

In the case of multiphased materials, such as nanomaterials, composites, alloys, or polymer blends, the location of the dissipative mechanism in one phase must be determined either through modeling or by separately studying each phase, when possible, without changing its behavior. The latter is only possible in a limited number of cases due to the interactions between the different phases within a material. To give an example, the global behavior of a composite is mostly driven by the stress transfer properties between the reinforcement and the matrix, which are controlled by the local mechanical properties in the interface region and in particular the dynamics of the structural defects in this area. It is obviously impossible to prepare a sample only composed of interface regions. Therefore a method for locally studying the dynamics of the structural defects will thus provide important steps in the understanding and the improvement of such materials.

Different techniques based on Scanning Probe Microscopies (SPM) have been developed to probe the elastic and anelastic properties of surfaces, interfaces, or phases of inhomogeneous materials at the micrometer and the nanometer scales. Scanning Acoustic Microscopy (SAMscanning acoustic microscopy), first developped in the mid-1970s, allows one to study the materials properties at the micrometer scale.

Among the different ways explored to study local mechanical properties of materials, several groups have recently used techniques based on Scanning Force Microscopy (SFM) [2]. For most of these, the focus has been placed on "elasticity," using the so-called Force Modulation Mode (FMM) at low frequencies [3]. Force modulation mode generally uses a large amplitude (greater than 10 nm), low frequency (some kHz) vibration of the sample underneath the SFM tip. The component of the tip motion at the excitation frequency and the tip mean position are simultaneously recorded, giving several images of the sample surface. In particular, the in-phase and out-of-phase components of force modulation mode at room temperatures have been interpreted in terms of stiffness ("elasticity") and damping ("viscoelasticity") [4, 5]. But, it has been recently shown by *Mazeran* et al. [6] that the contrast of force modulation mode is dominated by friction properties, giving only little information on the elasticity. Consequently, some care has to be taken in the interpretation of these low-frequency studies. A way to suppress the influence of friction on the contrast is to use smaller amplitudes (some Å) at higher frequencies [7]. Scanning Local-Acceleration Microscopy (SLAM) implements this idea [8]: SLAM is a modification of contact-mode scanning force microscopy. Its principle is to vibrate the sample at a frequency just above the resonance of the tip–sample system. In this case, the inertia of the tip prevents it from completely following the imposed high frequency displacement, inducing nonnegligible forces and giving rise to elastic deformation of the sample. Contact stiffness is obtained from the measure of the residual displacement of the tip. Mapping the contact stiffness at different temperatures with SLAM [9] has allowed local mechanical spectroscopy. Some other techniques also use high frequencies but with different approaches to image elasticity at room temperature [7, 10, 11]. Each of these high-frequency techniques is capable of mapping properties such as stiffness or adhesion at constant temperature with a very high lateral resolution.

11.2.1 The Variable-Temperature SLAM (T-SLAM)

The technique described here combines the lateral resolution of scanning force microscopy with the physical information available from temperature ramps. It is the variable-temperature SLAM (T-SLAM) [12, 13]. Ramping the temperature of the sample during a SLAM measurement and acquiring both the amplitude and the phase of the tip's motion allows one to obtain local mechanical spectroscopy data. A simple model enables the interpretation of the measurement in terms of material properties.

T-SLAM variable-temperature SLAMis based on SLAM [8, 14], scanning local-acceleration microscopyusing a variable temperature sample holder [9, 15]. An ultrasonic transducer is placed beneath the sample in a commercial scanning force microscope (Fig. 11.1) and excited by means of a function generator. The ultrasound is transmitted through the sample, forcing periodic local deformation of the sample's surface underneath the tip. The motion of the scanning force microscopy tip is detected optically by laser beam deflection from the backside of the cantilever. The detection signal is then fed to a lock-in amplifier, which extracts the tip's amplitude (related to elasticity) and phase (related to viscoelasticity) relative to the transducer's motion. The transducer's typical frequency is 825 kHz. This installation can operate in two ways, either by mapping the amplitude and phase of the signal as a function of position at fixed temperature or by recording the amplitude and phase as a function of temperature at a "fixed" location. The first method is known as "SLAM imaging," where the output signal of the lock-in is fed into an auxiliary data acquisition channel of the microscope. An extra computer is used to store local mechanical spectroscopy data as a function of the temperature at a fixed location and to control temperature. The heat is produced with a small resistive heater below the transducer, and the temperature is measured with a thermocouple. The sample must be prepared with a sufficiently low surface roughness in order to avoid artifacts in the measurements due to the sample's topography. Due to the geometry of the contact, SLAM measures only a small volume near the surface. Viewing SLAM as a very fast indentation measurement, the probed volume is approximately a half-sphere with a radius of $10a$, where a is the contact radius between tip and sample [16]. The typical value of a is some nm. So even if the sample is a thick film, T-SLAM gives access to the near-surface mechanical properties that may differ from bulk.

The mechanical properties of the deformed region are obtained from the measure of the residual displacement of the tip. The tip vibration amplitude d_1 is related to the contact stiffness, proportional to the dynamic elastic modulus, while the phase

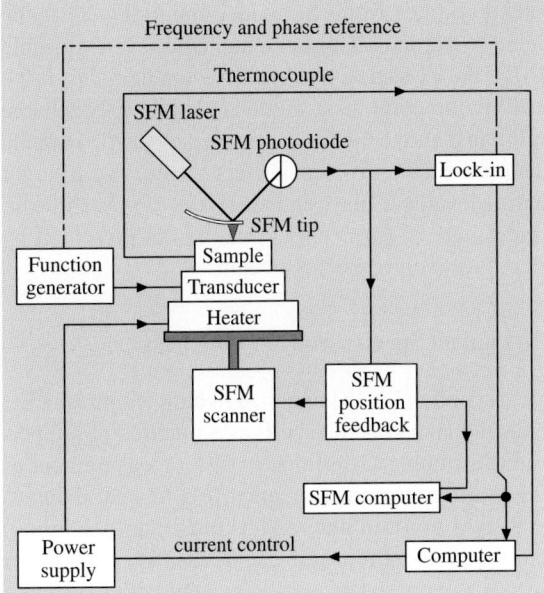

Fig. 11.1. *Schematic diagram of T-SLAM.* Ultrasound is generated with a transducer connected to a function generator and placed beneath the sample. The motion of the scanning force microscope tip is detected optically. The detection signal is fed to a lock-in amplifier, which extracts amplitude and phase relative to the transducer's motion. The temperature is controlled with a small resistive heater and measured with a thermocouple. The rest of the microscope head, to which the tip, the laser and the photodiode are attached, is not represented in this diagram

lag φ between the tip motion and the surface motions is related to the internal friction (energy dissipation inside the deformed volume). Mapping d_1 and φ at different temperatures with T-SLAM allows the study of the homogeneity of mechanical relaxations or of phase transitions. By recording d_1 and φ as a function of temperature at a fixed location, local mechanical spectroscopy can be performed.

When used just above the first resonance of the tip–sample system, the SLAM system can be described and analyzed using a point mass rheological model [9, 12], which allows one to obtain the equation relating the damping (loss factor η') to the measured parameters

$$\eta' = \left(\frac{1}{2}\frac{k_e}{k_c - m\omega^2}\right)\frac{\sin\varphi}{d_1/z_1}, \qquad (11.1)$$

where ω is the measurement frequency, k_c is a parameter related to the elastic modulus of the AFM cantilever and k_e to the elastic modulus of the sample, z_1 is the transducer vibration amplitude, and $m = k_c/\omega_c^2$ is the equivalent point mass of the tip, where ω_c is the free resonant frequency of the cantilever.

The principal limitation of (11.1) comes from the model assumption that the cantilever is a point mass restricted to vertical displacements. Due to the possible lateral displacements of the tip [6] or the existence of other vibrational modes of the cantilever, which are not described by the point mass model, (11.1) is only reliable for the measurement of the damping just above the first resonance of the tip–sample system. In order to obtain a quantitative measurement of this quantity in a larger frequency spectrum, it is necessary to develop a more realistic model of the cantilever interacting with the sample surface such as, for example, the model presented by *Dupas* et al. [17], in which the cantilever is modeled as a beam.

11.2.2 Example One: Local Mechanical Spectroscopy of Polymers

Figure 11.2 shows local mechanical spectra [13, 18] obtained in bulk technical PVC (including plasticizer and pigments and taken off-the-shelf) as a function of temperature (Fig. 11.2a) and a Differential Scanning Calorimetry (DSC) measurement of the same sample (Fig. 11.2b). The mechanical measurement (Fig. 11.2a) displays the amplitude of vibration of the SLAM tip (thin line) and its phase lag (thick line) as a function of temperature. Four temperature domains can be identified. At lower temperatures (1), the first domain shows a phase lag peak associated with a decrease of amplitude. The second temperature (2) domain corresponds to a zone where the vibration amplitude increases, without any variation in the phase lag. The third temperature domain (3) is characterized by a large phase lag peak and a significant decrease of vibration amplitude. The last temperature domain (4) shows a slow increase in the phase lag without variation of the vibration amplitude.

The calorimetry curves (b) display the heat flow as a function of temperature for the same material but at a larger size scale. For clarity, the same temperature domains have been reported on the graph. The dashed line displays the first heating, where both reversible and irreversible events are present; the solid line shows the second heating, where only reversible events are still present. On the solid line, an endothermic event (labeled A) can be observed near the border between temperature domains 2 and 3. An irreversible endothermic relaxation is superimposed with (A) on the first heating. A reversible endothermic peak is observed around 380 K, in domain 4. The endothermic event (A) has the characteristics of a glass transition: the specific heat goes from one value to another without a peak [19]. The glass transition temperature of this PVC is, therefore, approximately 340 K. The irreversible relaxation occurring in the same temperature range is certainly associated with physical aging or structural relaxation. The reversible peak at higher temperature is associated with the melting of the small crystalline volume fraction.

These local mechanical spectroscopy results are in good qualitative agreement with macroscopic global measurements [13, 18]. They yield the same information as the macroscopic global measurements, but from a much smaller volume, and allow as a consequence the location of the different mechanisms. Both measurements (local and global) show a peak in phase for the primary and secondary relaxations, connected with a large decrease of the stiffness for the primary relaxation and a much smaller one for the secondary relaxation. Plasticity induces an increase of the phase

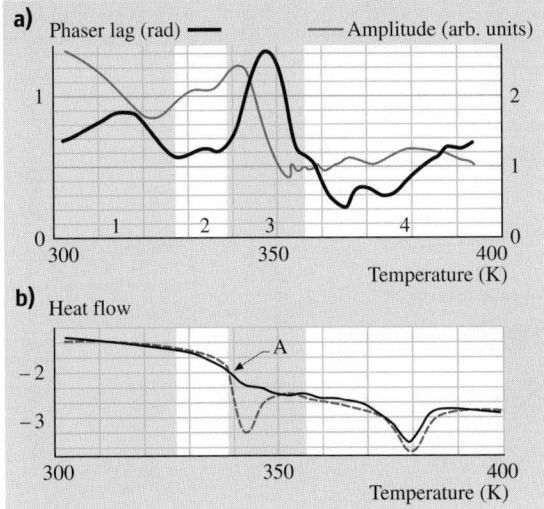

Fig. 11.2. (**a**) *Local mechanical spectroscopy spectra as a function of temperature of a technical PVC*: The vibration amplitude is displayed as a *thin line* and the phase lag as a *thick line*. Four temperature domains can be distinguished: From left to right, a small "vibration amplitude" decrease is associated with a first "phase lag" peak (1), in the next domain, "vibration amplitude" increases (2), then a large decrease of "vibration amplitude" is associated with a large "phase lag" peak (3) and finally "phase lag" increases slowly (4). (**b**) *DSC measurements of the same sample*. For clarity, the temperature domains observed on Fig. 11.5a have been reported. The graph displays the first (*dashed line*) and the second (*solid line*) heating. The glass transition can easily be recognized around 340 K (labeled *A*), slightly below the temperature range of domain 3. An irreversible endothermic relaxation takes place in the same temperature range (only visible on the first run). A reversible endothermic event occurs around 380 K

lag in both cases. The temperatures observed for the relaxation related to the glass transition compare well with the calorimetric data. Based on all these examples, there is no doubt that the amplitude and phase lag of T-SLAM are functions of the stiffness and damping. In this regard, T-SLAM provides an extension of the global method, allowing location of dissipative phenomena and study of the spatial homogeneity of phase transitions or of relaxations. This will bring new insight into the field of inhomogeneous or composite materials. In addition, T-SLAM measures only a small volume near the surface. So even if the sample is a thick film, T-SLAM gives access to the near-surface mechanical properties, which may differ from bulk properties. This will bring interesting perspectives to the still debated field of surface dynamics.

11.2.3 Example Two: Local Mechanical Spectroscopy of NiTi

Near-stoichiometric NiTi alloys exhibit a martensitic phase transformation between a low-temperature monoclinic phase, called martensite, and a high temperature cubic

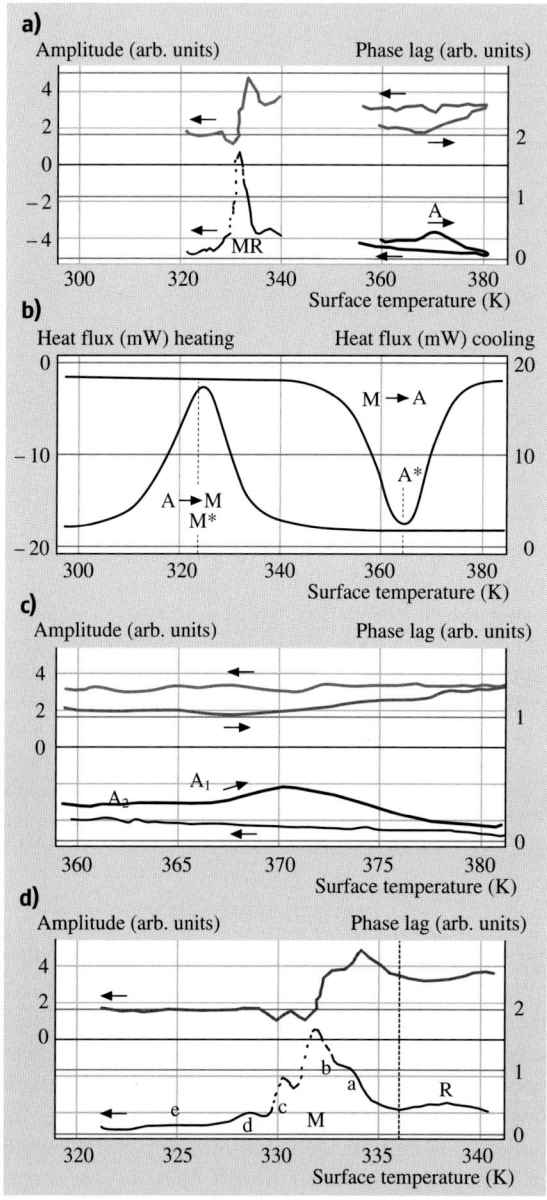

Fig. 11.3. (**a**)–(**b**) *Comparison of the local mechanical spectroscopy measurements of* NiTi *with calorimetric experiment*. (**a**) Both the reverse and direct transformation are associated with a phase lag peak and a modulus variation. (**b**) Calorimetric measurement of the same sample. (**c**)–(**d**) *Zoom on the reverse (c) and direct (d) transformation's temperature ranges.* The peak (A1) may exhibit a shoulder on the low temperature side (A2). The direct transformation is characterized by a recovery of the original vibration amplitude and a complex phase lag spectrum, formed of two main maxima, noted R and M. Peaks a,b,c,d,e are a substructure of the M peak

phase with B2 structure, called austenite (1). This transformation is responsible for the shape memory and pseudo-elastic effects in deformed NiTi alloys. Upon transformation to the martensitic phase, an intermediate rhombohedral (R) phase can be formed [20, 21]. Although *Bataillard* [21] tends to demonstrate that the R phase is finely dispersed inside the material, a controversy remains over the spatial scale at which the decomposition of the transformation occurs. Another puzzling question concerns the transformation itself. Optical microscopy observation suggests that the transformation occurs very suddenly inside an austenite grain. This has led to the concept of "military transformation." The width of the globally measured transformation would then be a sum of different narrow contributions coming from different places inside the sample. This image is, however, not universally accepted.

A measurement inside one single grain of a polycrystal would be a way to address these questions. Both the spatial scale of the R phase distribution and the "military" character of the transformation will have an effect on the result of such a measurement. This is the reason for which local mechanical spectroscopy measurements of the martensitic phase transition of NiTi by T-SLAM have been performed [12, 22]. They are presented in Fig. 11.3. The global transformation behavior of NiTi is defined by calorimetry spectra (Fig. 11.3b, temperature scanning rate 10 K/min).

The vibration amplitude ("elasticity") and the phase lag ("internal friction") are shown in the top and bottom curves, respectively. The presented data is incomplete due to experimental limitations. A phase lag peak can be observed upon heating at approximately 370 K, associated with a change in the vibration amplitude level. This event can be correlated with the phase transformation from martensite to austenite observed by calorimetry (Fig. 11.3b). Upon cooling, no event is observed around 370 K: the vibration amplitude is stable and the phase lag curve does not exhibit any peak. But, an intense phase lag peak can be observed upon cooling near 330 K, associated with a restoration of the vibration amplitude level characteristic of the martensite. This event correlates with the phase transformation from austenite to martensite observed by calorimetry. A change in the vibration amplitude spectrum is correlated with each event on the phase lag spectrum. Two main features should be noted. First, the transformation peaks observed with the SLAM method are narrower than those observed with the (global) calorimetry measurement. Second, the measured temperatures for the local peak are located on the high temperature side of the peaks measured by calorimetry. Details of the two transformation peaks are displayed in Figs. 11.3c and 11.3d. The reverse transformation (Fig. 11.3c) is characterized by a change in the vibration amplitude level and a phase lag peak. This peak (A1) may have a shoulder on the low temperature side (A2). A small decrease of the vibration amplitude may precede this increase. But, the magnitude of this softening and the very low intensity of peak A2 are too small to exclude experimental artifacts. The direct transformation (Fig. 11.3d) displays a recovery of the original vibration amplitude and a complicated phase lag maximum.

The correlation of the transition temperatures measured by calorimetry and the local mechanical spectroscopy lead to the conclusion that the peaks observed in phase lag and the change in vibration amplitude originate from the martensitic phase transition. At the scale of the observation, the martensite transforms into austenite around 370 K (A1 peak) and the austenite transforms into martensite close to 332 K (M peak), with the formation of the rhombohedral R phase at 337 K (R peak). The peaks A2 and e could be linked to the presence of two different types of martensite as already observed elsewhere [20, 21]. It is striking to note that all the events observed in macroscopic experiments seem to be reproducible at the scale of observation, namely less than 10^{-3} μm^3. In particular, this would confirm that the R phase precipitates are very finely distributed in the austenite matrix as observed by Bataillard [21]. Otherwise the R peak could not be observed reproducibly using such a technique. The local measurement differs from global ones in at least two aspects: first, the width of the transformation temperature range is smaller; second, the M peak exhibits a substructure.

Global internal friction measurements on these materials have given spectra with peaks having a similar breadth as the calorimetry peaks [23], whereas the peaks measured locally are narrower. This is easy to understand if the martensitic transformation occurs inhomogeneously inside the sample. Grains tend to transform at different temperatures depending on their stress state. As the local measurement has sufficient spatial resolution to probe a single grain inside the material, it is logical that the transformation occurs over a narrower temperature range than in a global experiment probing a large number of grains.

The substructure in the M peak could have three origins: (i) the technique is sensitive to the mechanical relaxation inside newly formed plates, such as stress relaxation by twin motion; (ii) the probed volume contains several martensite plates that grow one by one, each with its own, distinct "elementary" peak (a, b, etc.); (iii) the analyzed region of the sample surface changes due to thermal drifts, therefore probing the transformation of several growing martensite plates, leading to multiple "elementary" peaks.

11.3 Static Methods – Mesoscopic Samples

11.3.1 Carbon Nanotubes –
Introduction to Basic Morphologies and Production Methods

Carbon nanotubes (CNTs) are the newest form of carbon, found in 1991 by *Iijima* [24]. Because of their remarkable properties, this discovery has opened whole new fields of study in physics, chemistry, and material science. They possess a unique combination of small size (diameters ranging from ~ 1 to 50 nm with lengths up to ~ 10 μm), low density (similar to that of graphite), high stiffness, high strength, and a broad range of electronic properties from metallic to p- and n-doped semiconducting. Their potential field of application is immense and includes reinforcing elements in high strength composites, electron sources in field emission displays and small X-ray sources, ultra-sharp and resistant AFM tips with high aspect ratios, gas sensors, and components of future, nanoscale electronics. In addition, they represent a widely used system for studying fundamental physical phenomena on the mesoscopic scale.

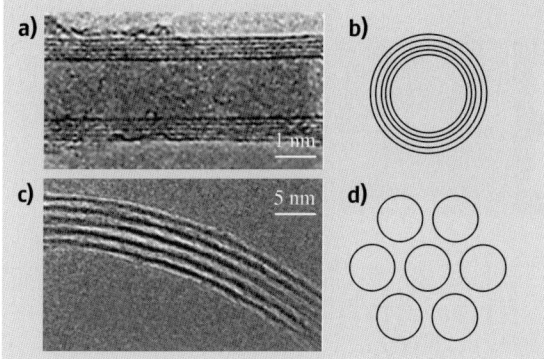

Fig. 11.4. TEM images and schematic drawings of cross sections for different morphologies of carbon nanotubes. (**a**) and (**b**) Multiwalled carbon nanotubes, consisting of concentric, nested tubes (MWNT). (**c**) and (**d**) ropes of single-walled carbon nanotubes, consisting of single carbon nanotubes bundled up in ropes and held together by van der Waals interaction

Following advances in manufacturing and processing they are likely to be integral to many devices we use in our everyday life.

From the structural point of view, carbon nanotubes can be thought of as rolled up single sheets of graphite, *graphene*. They can be divided in two distinct groups. The first, multiwalled carbon nanotubes (MWNT), exhibit a Russian doll-like structure of nested, concentric tubes, Fig. 11.4a and b. They were also the first CNT to be discovered experimentally. The interlayer spacing can range from 0.342 to 0.375 nm, depending on the diameter and number of shells comprising the tube [25]. For comparison, the interlayer spacing in graphite is 0.335 nm, suggesting relatively weak interaction between individual shells. This fact has been corroborated by studies of mechanical and electronic properties of CNTs.

The second type of carbon nanotubes is in the basic form of a rolled-up graphitic sheet: a single-walled CNT (SWNT). During the production, their diameter distribution is relatively narrow so they often bundle in the form of crystalline "ropes" [27], Fig. 11.4c and d, in which the single tubes are held together by van der Waals interaction.

There are several distinct classes of production methods. The earliest is based on the cooling of carbon plasma. When voltage is applied between two graphitic electrodes in an inert atmosphere, they gradually evaporate and form plasma. On cooling, soot containing multiwalled nanotubes is formed [24]. If the anode is filled with catalysts such as, for example, cobalt of iron particles, SWNT ropes form. Another way of producing CNTs from carbon plasma is by laser ablation of a graphitic target [27]. It is considered that, in general, these methods produce CNTs of higher quality albeit in very small quantities and without the possibility of scale-up to industrial production. Other methods are based on chemical vapor deposition (CVD), a catalytic decomposition of various hydrocarbons, e.g. methane or acetylene mixed with nitrogen or hydrogen in the presence of catalysts [28]. This method offers the

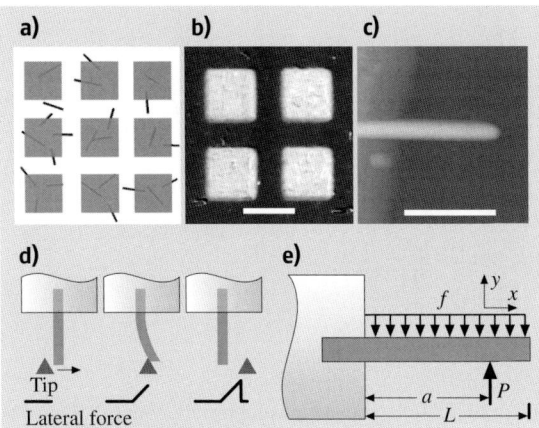

Fig. 11.5. Schematic of the experiment performed by *Wong* et al. [26] (**a**) nanotubes were dispersed on a substrate and pinned down by a deposition of SiO pads. (**b**) Optical micrograph showing the pads (*light*) and MoS$_2$ substrate (*dark*). The scale bar is 8 μm long. (**c**) AFM image of a SiC nanorod protruding from the pad. The scale bar is 500 nm long. The same experimental setup was used for elastically deforming MWNTs. (**d**) The tip shown as a triangle moves in the direction of the arrow. The lateral force is indicated by the trace at the bottom. Before the tip contacts the beam, the lateral force remains constant and equal to the friction force. During the bending a linear increase in the lateral force with deflection is measured. After the tip has passed over the beam the lateral force drops to its initial value and the beam snaps back to its equilibrium position. (**e**) Schematic of a pinned beam with a free end. The beam of length L is subjected to a point load P at $x = a$ and friction force f (abstracted with permission from [26] © American Association for the Advancement of Science)

possibility of controlling the growth of nanotubes by patterning the catalyst [29] and is therefore more suitable for producing nanoscale structures with integrated CNTs. This method is also capable of producing CNTs in industrial quantities. Their main disadvantage is the higher concentration of defects, which as a consequence diminish their mechanical properties.

11.3.2 Measurements of the Mechanical Properties of Carbon Nanotubes by SPM

Mechanical measurements on CNTs performed with the AFM have confirmed theoretical expectations [30] of their superior mechanical properties. They involve measurements of deformations under controlled forces by bending immobilized carbon nanotubes either in the lateral [26] or normal direction [31] and also tensile stretching of CNTs fixed on their two ends to AFM tips observed in an electron microscope [32].

It is not obvious that continuum mechanics and its concepts like Young's, shear moduli, and tensile strength should work on the nanoscale. In order to apply them, one should also define the "thickness" of the nanotube's walls, a graphene sheet, in the frame of the continuous beam approximation. Most scientists working in this field are using the value 0.34 nm, close to the interlayer separation in graphite as the thickness of a nanotube. When comparing different results, however, one has to bear in mind that to convert relatively precise force–displacement measurements into macroscopic quantities like Young's or shear modulus, one has to introduce various geometrical factors, including diameter and length. Even a small imprecision in their determination is very unforgiving because the diameter d enters into equations for beam deformation as d^4 and length l as l^3, leading to large uncertainty in final results.

The first quantitative measurement of the Young's modulus of carbon nanotubes was reported by *Wong* et al. [26] in which they laterally bent MWNTs and SiC nanorods (similar in dimensions to MWNTs) deposited on flat surfaces. MWNTs were first randomly dispersed on a flat surface of MoS_2 single crystals that were used because of their low friction coefficient and exceedingly flat surface. Friction between the tubes and substrate was further reduced by performing the measurements in water. Tubes were then pinned on one side to this substrate by a deposition of an array of square pads through a shadow mask, Fig. 11.5a–c. AFM was used to locate and characterize the dimensions of protruding tubes. The beam was deformed laterally by the AFM tip, until at a certain deformation the tip would pass over the tube, allowing the tube to snap back to its relaxed position. During measurements, the force–distance curves were acquired at different positions along the chosen beam, Fig. 11.5d–e. Maximum deflection of the nanobeam can be controlled to a certain degree by the applied normal load, and in this way tube breaking can be avoided or achieved in a controlled manner. The applied lateral load P in terms of lateral displacement y at the position x along the beam is given by the equation:

$$P(x,y) = 3EI\frac{y}{x^3} + \frac{f}{8}\left(x - 4L - 6\frac{L^2}{x}\right), \tag{11.2}$$

where E is the Young's modulus of the beam, I the second moment of the cross section, equal to $\pi d^4/64$ for a solid cylinder of diameter d and f the unknown friction force, presumably small due to the experimental design. The lateral force, Fig. 11.6a, is known only up to a factor of proportionality because the AFM lever's lateral force constant wasn't calibrated for these measurements. This uncertainty and the effect of friction were eliminated by calculating the nanobeam lateral force constant:

$$\frac{dP}{dy} \equiv k = \frac{3\pi d^4}{64x^3}E \tag{11.3}$$

shown in Fig. 11.6b.

The mean value for the Young's modulus of MWNTs was $E = 1.3 \pm 0.6$ TPa, similar to that of diamond ($E = 1.2$ TPa). For larger deformations, discontinuities in bending curves were also observed, attributed to elastic buckling of nanotubes [35].

In another series of experiments, *Salvetat* et al. measured the Young's modulus of isolated SWNTs and SWNT ropes [31], MWNTs produced using different methods [34], and the shear modulus of SWNT ropes [33]. The experimental setup that enabled them to perform measurements on such a wide range of CNT morphologies

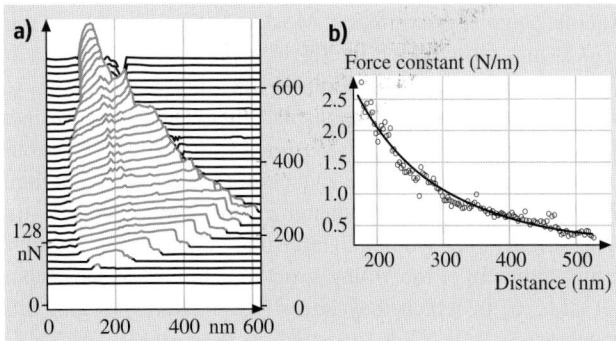

Fig. 11.6. (a) A series of lateral force distance curves acquired by *Wong* et al. [26] for different positions along a MWNT. (b) The lateral spring constant as a function of position on the beam. The curve is a fit to (11.3) (abstracted with permission from [26] © American Association for the Advancement of Science)

involved measuring the vertical deflection of nanotubes bridging holes in a porous membrane.

In their measurement method, they suspended CNTs in ethanol and deposited them on the surface of a well-polished alumina Al_2O_3 ultrafiltration membrane. Tubes adhered to the surface due to the van der Waals interaction occasionally spanning holes, Fig. 11.7a. After a suitable nanotube had been found, a series of AFM images was taken under different loads, in which every image would thus correspond to the surface (and the tube) under a given normal load. Extracted linescans across the tube revealed the vertical deformation, Fig. 11.7b. For the range of applied normal loads, the deflection of a thin, long nanotube at the midpoint δ as a function of the normal force F can be fitted using the clamped beam formula [36]:

$$\delta = \frac{FL^3}{192EI}, \tag{11.4}$$

where L is the suspended length.

The fitted line does not pass through the origin because the force acting on the nanotube is not equal to the nominal force alone; it contains a constant term arising from the attractive force between the AFM tip and the tube. The tube's deflection should also contain a term corresponding to the interaction between the tube and the substrate, but this is generally regarded as negligible.

This variable load imaging technique is advantageous for obtaining quantitative information as one is assured that the AFM tip is in the desired location when deforming the tube. Equation (11.4) is valid only if the tubes adhere well to the substrate, confirmed by the fact that the images reveal no displacement of the parts of the tube in contact with the membrane.

Using this technique, Young's modulus of 1 TPa was found for SWNTs. Values for the MWNTs show a strong dependence on the amount of disorder in the graphitic layers: an average value of $E = 870\,\text{GPa}$ was found for the arc-discharge grown

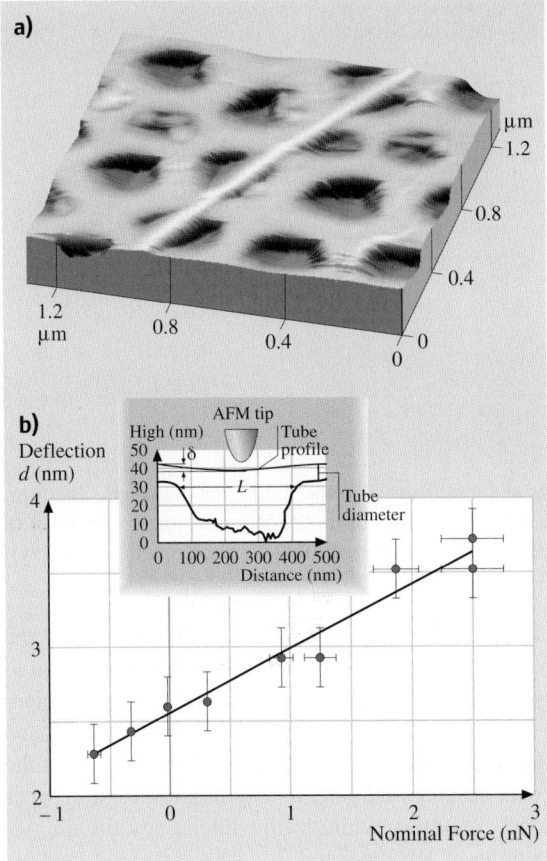

Fig. 11.7. (a) An AFM image of a CNT lying on a porous alumina [33] filter. (b) Measured dependence of vertical deflection on the applied nominal force. The inset shows a comparison between linescans taken on the tube and over a hole [34] (abstracted with permission (a) from [33] © 1999 American Physical Society, (b) from [34] © 1999 Wiley)

tubes, while the catalytically grown MWNTs, known to include a high concentration of deffects, can have a Young's modulus as low as 12 GPa.

For the deflection of SWNT ropes, an additional term in the bending formula has to be taken into account because of the influence of shearing between the tubes comprising the rope. Single CNTs are held together in the tube only by weak, van der Waals interactions. Ropes, therefore, behave as an assembly of individual tubes rather than as a thick beam. The deflection can be modeled as a sum of deflections due to bending and shearing [36]:

$$\delta = \delta_{\text{bending}} + \delta_{\text{shearing}}$$
$$= \frac{FL^3}{192EI} + f_s \frac{FL}{4GA} = \frac{FL^3}{192E_{\text{bending}}I}, \quad (11.5)$$

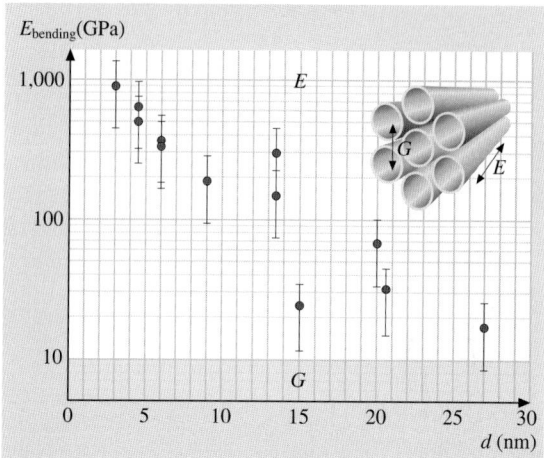

Fig. 11.8. Values of the shear modulus for 12 SWNT ropes of different diameters. The measured $E_{bending}$ of thin ropes corresponds to E, while for thick ropes one obtains the value of shear modulus G (abstracted with permission from [33] © 1999 American Physical Society)

where f_s is a shape factor, equal to 10/9 for a cylinder, G the shear modulus and A the area of the beam's cross section. $E_{bending}$ is the effective, bending modulus, equal to Young's modulus in the case in which the influence of shearing can be neglected (for thin, long ropes).

The Young's and the shear modulus can thus be extrapolated by measuring the $E_{bending}$ of an ensemble of ropes with different diameter to length ratios: for thin ropes one obtains the value of the Young's modulus, while for the thick ones, the $E_{bending}$ approaches the value of G on the order of 10 GPa (see Fig. 11.8).

Walters et al. pinned ropes of single-wall nanotubes beneath metal pads on an oxidized silicon surface, then released them by wet etching, Fig. 11.9a. The SWNT rope was deflected in the lateral direction using an AFM tip, Fig. 11.9b. As the suspended length is on the order of μm, the SWNT rope can be modeled as an elastic string stretched between the pads. Upon deformation all of the strain goes into stretching. In the simple case of a tube lying perpendicular to the trench and the AFM tip deforming the tube in the middle, the force F exerted on the tube by the AFM tip is given by the expression:

$$F = 2T \sin\theta = 2T \frac{2x}{L} \approx \frac{8kx^3}{L_0}, \qquad (11.6)$$

where T is the tension in the string, L_0 its equilibrium length, k the spring constant and x the lateral deflection in the middle. Using this setup, they deformed SNWT ropes to the maximal strain of $5.8 \pm 0.9\%$ and determined a lower bound of 45 ± 7 GPa on the tensile strength, assuming a value of 1.2 TPa for the Young's modulus.

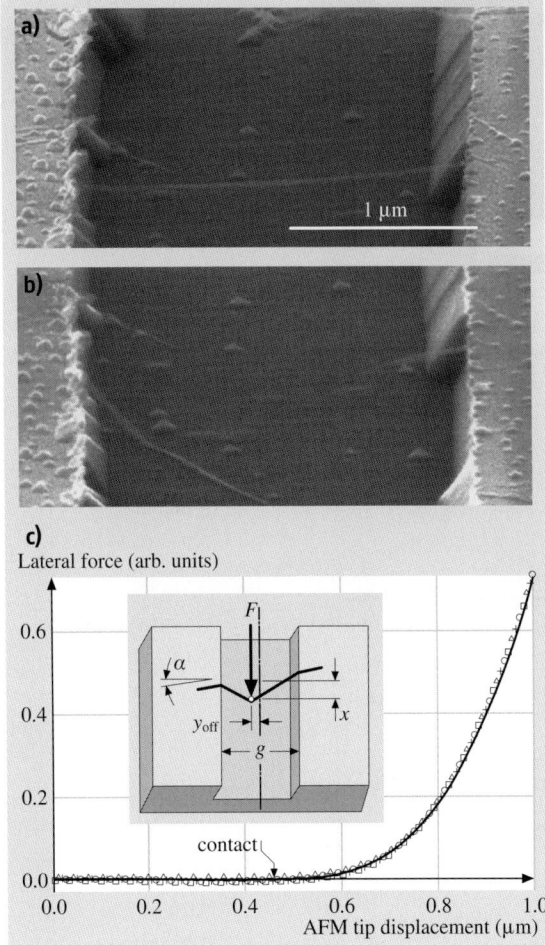

Fig. 11.9. (a) SEM image of a SWNT rope suspended over trench in silicon before and (b) after being deformed past its elastic limit. (c) Lateral force on a single-wall nanotube rope as a function of AFM tip displacement (abstracted with permission from [37] © American Institute of Physics)

Kim et al. used a setup in which the SWNT rope was embedded in metallic electrodes deposited on a silicon substrate coated with poly(methylmethacrylate) (PMMA). In their experiment the tube can also be modeled as an elastic string. Using an AFM tip, they vertically deformed the rope and obtained an estimate of $E = 0.4$ TPa for the Young's modulus of an SWNT rope.

Finally, the first direct measurements of the elastic properties of CNTs that haven't relied on the beam or stretching string setup involved deforming MWNTs [38] and SWNT ropes [32] under axial strain. This was achieved by identifying and attaching opposite ends of MWNTs or SWNT ropes to two AFM tips,

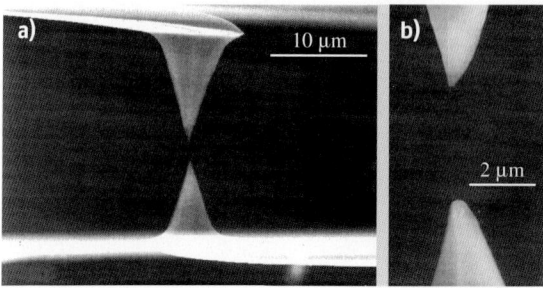

Fig. 11.10. (a) A SEM image of a MWNT mounted between two opposing AFM tips. (b) A close-up of the region indicated by a rectangle in (a) (abstracted with permission from [38] © 1999 American Association for the Advancement of Science)

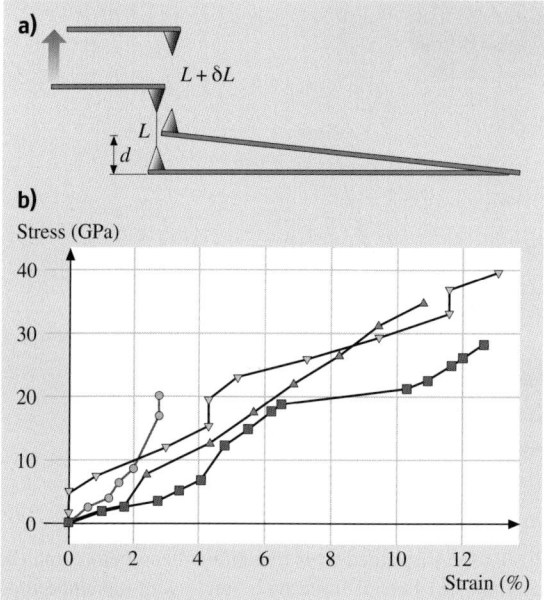

Fig. 11.11. (a) The principle of the experiment performed by *Yu* et al. As the rigid cantilever is driven upward, the lower, compliant cantilever bends by the amount d and the nanotube is stretched by δL. As a result, the nanotube is strained by $\delta L/L$ under the action of force $F = kd$, where k is the elastic constant of the lower AFM lever. (b) Plot of stress vs. strain curves for different individual MWNTs (abstracted with permission from [38] © 1999 American Association for the Advancement of Science)

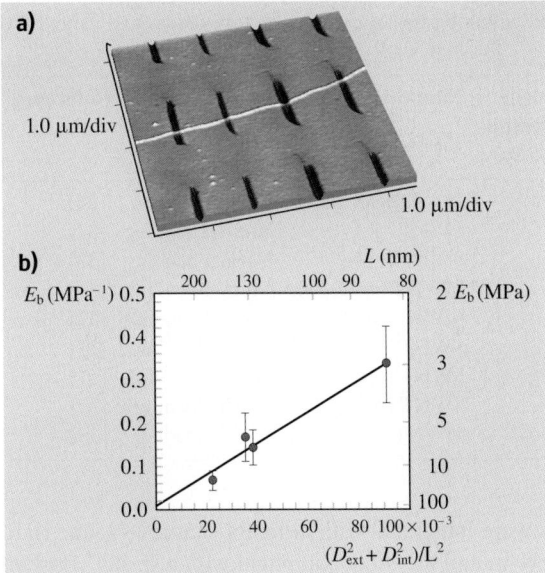

Fig. 11.12. (a) A pseudo 3-D rendering of a single microtubule suspended over four different-sized holes in PMMA. (b) From the variation of the E_{bending} with varying length, the shear and the Young's modulus have been determined for the microtubule displayed in (a) (abtracted with permission from [39] © 2002 American Physical Society)

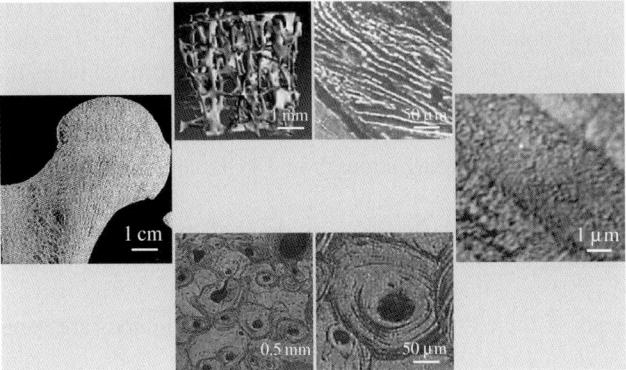

Fig. 11.13. Hierarchy of the human femoral neck (from left to the right) the far left figure shows a cut through a frontal plane of the femoral neck. The outer shell is constituted of compact bone while the inside is made of the spongier trabecular bone. The left pair of images shows the trabecular structure (*top*) and a transverse cut of the compact shell that shows vascular canals (*bottom*). The right pair of images shows packets of trabecular bone lamellae (*top*) and a single osteon (*bottom*), a vascular canal surrounded by concentric lamellae. Packets of trabecular bone lamellae and osteons represent a structural unit (BSU). Note the alternation of bright (*thick*) and dark (*thin*) lamellae. The figure far right shows three bone lamellae, structures that are similar for trabecular and compact bone. The bone matrix within the lamellae are mainly constituted of collagen fibers and hydroxyapatite crystals

Table 11.1. Summary of the mechanical properties of carbon nanotubes measured using SPM methods

Young's modulus E (GPa)	Shear modulus G (GPa)	Tensile strength σ (GPa)	Nanotube type	Deformation method	Reference
1300±600	–	–	MWNTs arc grown	Lateral bending	[26]
1000±600	–	–	SWNTs	Normal bending	[31]
1000±600	~ 1 GPa	–	SWNT ropes	Normal bending	[33]
870±400	–	–	MWNT arc-grown	Normal bending	[31]
12±6	–	–	MWNT catalytic	Normal bending	[31]
–	–	45±7	SWNT rope	Lateral bending	[37]
400	–	–	SWNT rope	Normal bending	[41]
1020	–	30	SWNT ropes	Tensile loading	[32]
270–950	–	11–63	MWNTs arc-grown	Tensile loading	[38]

all inside a SEM. The AFM tips were integrated with different cantilevers, one rigid with a spring constant above 20 N/m and the other compliant with a spring constant of 0.1 N/m, Fig. 11.10. The rigid lever was then driven using a linear piezomotor. On the other end, the compliant lever bent under the applied tensile load. The deflection of the compliant cantilever – corresponding to the force applied on the nanotube – and the strain of the nanotube were simultaneously measured. The force F is calculated as $F = kd$ where k is the spring constant of the flexible AFM lever and d its displacement in the vertical direction. The strain of the nanotube is $\delta L/L$, Fig. 11.11. From the stress–strain curves obtained in this fashion, Fig. 11.11b, Young's moduli ranging from 270–950 GPa were found. Examinations of the same broken tubes inside a TEM revealed that nanotubes break with the "sword in sheath" mechanism, where only the outer layer appears to have carried the load. After it breaks, pullout of inner shells follows. An average bending strength of 14 GPa and axial strengths up to 63 GPa were found.

Firm attachment of nanotubes to AFM tips was ensured by a deposition of carbonaceous material induced by the electron beam concentrated in the contact area [40].

All these measurements of the elastic properties of carbon nanotubes are summarized in Table 11.1.

Before comparing them, it should be noted that the absolute values of these mechanical constants have relatively large uncertainties because of the huge influence of the precision of determining tube diameters and lengths on the final result. Also, the cited values represent mean values of results obtained on several tubes, with the exceptions of the lowest value given for catalytically grown MWNTs by Salvetat et al. [31], a single value for an individual SWNT rope from *Kim* et al. [41], and the lowest and highest values from *Yu* et al. [38]. One also has to bear in mind that concepts like Young's and shear moduli and tensile strength were introduced for describing macroscopic and continuous structures. Their application for describing mesoscopic objects like nanotubes, therefore, has its limitations.

The methods presented above are all in their nature "single-molecule" methods, in the sense that they measure properties of individual objects. A result of a single experiment, therefore, represents properties of a particular object and differs from case to case because of, for example, defects coming from production and purification, or for more prosaic reasons such as experimental errors. In order to perform comparisons, therefore, it is more practical to deal with values averaged for multiple tubes. The average values for the Young's modulus of high quality tubes are, within the experimental error, all on the order of 1 TPa, close to that of diamond (1.2 TPa), while the tensile strength can be 30 times higher than that of steel (1.9 GPa). Catalytically grown MWNTs with the Young's modulus that can be as low as 12±6 GPa are definitely disappointing and show that the production method plays an important role in the quality of carbon nanotubes from the point of view of their mechanical properties.

Future improvements in large scale production and processing are therefore necessary before applying them as the ultimate reinforcement fiber. Even so, the progress in measuring the mechanical properties of CNTs will continue to be closely related to, and often motivate, the progress in nanoscale manipulation in general.

11.3.3 Microtubules and Their Elastic Properties

Microtubules are a vital biological nanostructure, similar in dimension and shape to carbon nanotubes. In fact, the first name given to carbon nanotubes by their discoverer Iijima was "microtubules of graphitic carbon." From the structural point of view, they are much more complicated, as is common for biological structures. They self-assemble in buffers maintained at the physiological temperature of 37 °C out of protein subunits, α and β-tubulin, each having a molecular weight of 40 kDa. These protein subunits are bound laterally into protofilaments, which in turn are arranged in a shape of a hollow cylinder with an external diameter of 25 nm and an internal diameter of 15 nm. Microtubules are a remarkable material: inside living cells, their length incessantly fluctuates; they can even completely disassemble and consume energy in the form of GTP.

Together with actin and intermediate filaments, microtubules constitute the cellular cytoskeleton. In addition, they perform various unique vital functions: they act as tracks along which molecular motors move, they help pull apart chromosomes during cell division and form bundles that propel sperm and some bacteria. All these roles are determined by their structure and mechanical properties. Yet, after more than a decade, there is still a large discrepancy in the values of their Young's modulus reported in the literature. Several methods have been applied, such as bending or buckling microtubules using optical tweezers [42], hydrodynamic flow [43], thermally induced vibrations or shape fluctuations [44], buckling in vesicles [45], or squashing with an AFM tip [46]. These methods yielded results ranging from 1 MPa [46] to several GPa [42].

Since microtubules are geometrically similar to nanotubes, *Kis* et al. [39] used the suspended tube configuration. Microtubules were deposited on porous membranes, and AFM images were acquired under different nominal normal loads. All the measurements were performed in liquid and at controlled temperatures, in order to prevent the degradation of proteins, so the substrate had to be functionalized in order to ensure good adhesion between the tubes and the support. They used silicon with a layer of PMMA as a support. Slits were prepared in PMMA using electron-beam lithography, providing the possibility to measure the elastic response of individual microtubules lying over four different-sized holes ranging from 80–200 nm across and 400 nm deep, Fig. 11.12.

Clearly the results are dependent on the hole diameter, which could only result from a shear component within the microtubules. Simplifying (11.5), the deformation of microtubules can be modeled as:

$$\frac{1}{E_{bending}} = \frac{1}{E} + \frac{10}{3}\frac{D_{ext}^2 + D_{int}^2}{L^2}\frac{1}{G}, \quad (11.7)$$

where and D_{int} represent the external and internal MT's diameter. From the plot of $1/E_{bending}$ vs. the $\left(D_{ext}^2 + D_{int}^2\right)/L^2$ the shear modulus can be calculated from the slope of the fit, and the inverse Young's modulus will be equal to the intercept on the y-axis. In this way the shear modulus of $G = 1.4 \pm 0.3$ GPa and a lower limit of $E = 150$ MPa were simultaneously obtained from a measurement on an individual microtubule.

This anisotropy comes from the fact that microtubules are built of monomers that are strongly bound in the longitudinal direction, along single protofilaments. The link between neighboring protofilaments is much weaker. Microtubules, therefore, have to be considered as anisotropic beams, having an $E_{bending}$ that depends on the scale on which they are deformed [39]. This is in all respects analogous to the situation in SWNT ropes that are built of stiff individual SWNTs, held together in bundles by a weak van der Waals interaction. This conclusion may explain the large discrepancy in the value of Young's moduli reported in the literature over the past ten years using inadequate modeling.

As in the case of carbon nanotubes, measuring the mechanical properties of microtubules can provide valuable insight into their structure and provide deeper understanding about the functioning of these remarkable structures.

11.4 Scanning Nanoindentation: An Application to Bone Tissue

In the following section we discuss nanoindentation as a tool to determine nanomechanical properties. This method was developed in the early 1980s evolving from traditional Vicker hardness testing devices. The latter is based on the concept that a pyramidal tip is loaded into a material applying a known force. After the test, the size of the remaining imprint is measured under an optical microscope. The ratio of the employed force and the imprint area after load removal was defined as hardness that represents a mean pressure the material can resist. Unfortunately this mechanical parameter is a complex combination of elastic and postyield properties and can't be easily explained on the level of continuum mechanics. This point raised important concerns for determining elastic constants such as the Young modulus from an indentation test. Further improvements on the transducer sensitivity were necessary to provide continuous acquisition of the employed load and the resulting indentation depth. Nanoindentation represents the state of the art of this development allowing mechanical tests on the nanometer scale.

11.4.1 Scanning Nanoindentation

Based on this continuous force–displacement data, an elastic modulus of volumes in the submicron regime can be quantified. To investigate features down to the micrometer regime, classical nanoindentation tools are combined with an optical microscope for positioning the indenter tip on the region of interest. But to defeat the limits of optical resolution, indentation and scanning probe microscopy were combined using the same tip to allow nanoscale control of the indenting tip position [47]. This instrument allows for scanning the material topography in an AFM mode and performing nanoindentation tests employing the same tip. The mechanical tests are restricted to the scanned area (100 μm × 100 μm maximum) where the indenter can be positioned with a precision of better than 100 nm. The surface roughness can be measured, which is helpful for choosing the area to be indented. Furthermore the available in situ scan of the indented region can provide information about the piling-up or sink-in behavior of the material. *Kulkarni* and *Bhushan* [48, 49] introduced this device and demonstrated that measurements on aluminium and silicon were similar to results of non-scanning indentation systems.

11.4.2 Application of Scanning Nanoindentation

The development of the scanning nanoindentation (SN) technique has led to a variety of studies, primarily characterizing thin layers. Applications range from the mechanical characterization of corrosion-free film apposition on single-crystalline iron [50] to indentation and microscratch tests on Fe-N/Ti-N multilayers [51]. One author included nanoindentation to discuss two different electrochemical deposition techniques of thin Ni-P layers on pure Ni [52]. *Rar* et al. [53] reported studies on the growth of thin gold layer on native oxides of silicium while other investigations focussed on wear-resistant $TiB_2(N)$ coatings [54].

Studies of heterogeneous materials clearly demonstrate the advantage of the available surface scanning. *Malkow* and *Bull* [55] investigated the elastic/plastic behaviour of carbon-nitride films deposited on silicon. Therefore they determined the load-on and load-off hardness, the latter being accessible by scanning the remaining indent impression. *Shima* et al. [56] studied silicon oxynitride films on pure silicon and demonstrated hardness as a function of the employed deposition temperature. *Göken* et al. [57] used the high positioning precision to characterize individual lamellae of a TiAl alloy that consists of a two-phase structure and also to study the mechanical properties of nanometer-size precipitates of nickel-based superalloys [58]. Performing in situ electrochemical treatments of iron single crystals, *Seo* et al. [59] used SN to study the variation of the remaining imprint shape with time.

In the field of biomechanics, work was done on bone tissue in wild-type and gene-mutated zebrafish. The characterization of the residual indentations in AFM mode supported the statement that gene mutation can change bone brittleness [60]. Other investigations focussed on the demineralising effects of soft drinks on tooth enamel by studying changes of elastic properties and surface topography [61]. *Habelitz* et al. [62] and *Marshall* et al. [63] characterized the junction between human tooth enamel and the mechanically different dentin.

Hengsberger et al. [64] took advantage of the nanometer positioning capability of the SN to investigate the elastic and plastic properties of individual human bone lamellae. Related work is presented in greater detail below to demonstrate the use and benefits of the SN technique at this example.

11.4.3 Example: Study of Mechanical Properties of Bone Lamellae Using SN

Before we discuss nanomechanical properties of bone lamellae, it is useful to know the structure on the macroscopic level. Figure 11.13 presents the hierarchy of human bone tissue at the example of the femoral neck. The outer shell is constituted of cortical bone while the porous trabecular bone structure gives inner support. On the next lower level, both bone types show structural units (BSU, after [65]) constituted of some tens of lamellae. For compact bone the lamellae have a concentric organization while for trabecular bone these lamellae are parallel. The nomenclature "BSU" was motivated by its underlying cellular process. The formation of such an individual structural unit occurs within a single cellular process. The optical contrast shows two alternating types of lamellae, thin lamellae that appear dark, and thick lamellae that appear bright. These bone lamellae show widths ranging from 1 to 3 µm for thin and 2 to 4 µm for thick lamellae [64]. On the next lower level the bone matrix is constituted of a complex collagen and mineral structure. It is still not entirely understood what type of variations in the collagen/mineral meshwork are responsible for the lamellar structure.

The mechanical properties on these different levels of bone hierarchy are increasingly well understood. But, little is known about how the macroscopic mechanical properties of the whole bone relate to its nanomechanical properties. The possibility that factors such as fracture risk might be better understood with an analysis of small bone volumes motivates the application of nanoindentation in this field.

A set of nanoindentation studies is available (mainly employing the device combined with an optical microscope) that presents the intrinsic mechanical properties of bone tissue. Among recent work, differences were reported between donors, anatomical sites, BSU, and thin and thick lamellae [64, 66–69]. Due to technical constraints of this sensitive nanomechanical device, the majority of these studies present dehydrated or dried tissue properties measured at ambient temperature [70–73]. But, removal of the water content may lead to anisotropic shrinking of the matrix that creates microcracks and alters the mechanical properties of the bone constituents. For an accurate characterization of the in vivo properties, the nanomechanical tests should therefore be done under physiological conditions.

There have been a few attempts to characterize the in vivo bone properties [66, 68, 74]. This was achieved by studying fresh bone samples that are kept moist with a thin layer of liquid (less than a hundred microns) on the surface or with subsurface water irrigation. But, local evaporation of the thin liquid layer may have led to indents on areas that were partially dried during the test. One possible solution to such local drying might be to conduct measurements in which the indentation tip and tissue sample are both fully immersed in a liquid cell and simultaneously heated to body temperature.

Practical limitations on temperature stability and inaccuracy in contact force detection due to liquid on the surface make such measurements extraordinarily difficult, however. For statistically powerful studies it seems unavoidable to dry the samples.

The objective of this study was to use SN to determine the effect of drying on the stiffness of single bone lamellae. The goal was to determine a conversion factor that allows dry tissue properties to be recalculated into their in vivo properties.

For this purpose, we measured the identical set of lamellae selected from human trabecular and compact bone at first under physiological and then under dry conditions.

Experimental Setup and Technical Features:

Figure 11.14 demonstrates an optical picture and a sketch of the combined AFM and nanoindenter device (Hysitron Inc. Minneapolis, MN). The transducer consists of a three-plate capacitor on whose central plate a Berkovich (three-sided pyramid) diamond tip is mounted. The transducer provides a contact force feedback between the tip and the sample surface. The sample is mounted on a piezoelectric scanner that allows moving in the x,y and z-directions. During x,y-surface scanning, the piezoelectric scanner keeps this feedback signal constant by correcting the z-height. The movement of the piezoelectric scanner, therefore, describes the negative surface of constant contact force, commonly called an AFM-scan. For the liquid cell tests, the sample was glued in a plexiglass cup for the addition of a several millimeter high liquid layer. A commercially available liquid cell tip (Hysitron Inc. Minneapolis, MN) was used, which contains an additional shaft of approximately 300 μm diameter and 5 mm length to protect the transducer from the fluid. A small layer of latex was placed between the sample holder and the magnetic stick of the piezoelectric scanner to protect the latter from liquid. The additional shaft of the indenter tip and the latex layer represent elastic components that increased the machine compliance (C_m = 7 nm/mN instead of 3 nm/mN). This variable corrects the deformation of all the machine components while indentation data are recorded and can be determined from the tip shape calibration curve (Hysitron, Minneapolis). The nanoindentation device was installed in a custom-made thermal chamber to allow sample heating to 37 °C. Note that the increased temperature and humidity also changed the value of the electrostatic force constant (EFC). The latter corrects the force due to the springs that support the central plate on which the indenter tip is mounted. The EFC can be calibrated doing out-of-contact indents and varying this value until a zero-line in the force–displacement curve is measured.

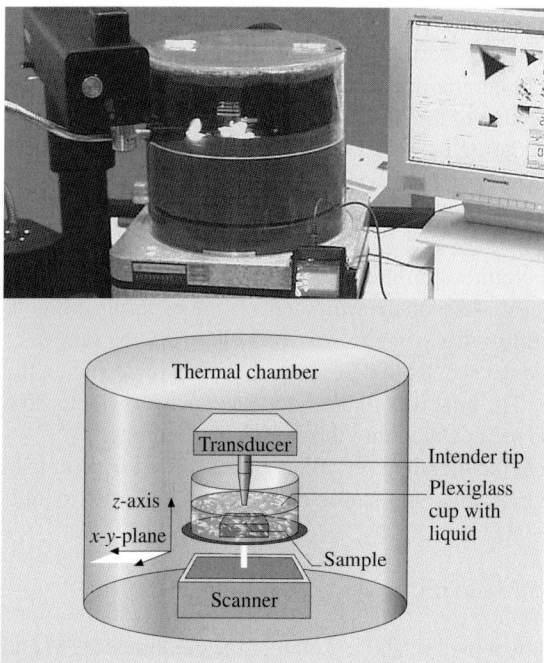

Fig. 11.14. Optical picture (*top*) and sketch (*bottom*) of the scanning nanoindentation device for measurements under physiological conditions. For this purpose the sample was installed in a plexiglass cup for addition of liquid. The entire instrument was heated in a custom-made thermal chamber

Based on an image acquired in AFM mode, the indentation area can be selected with a high spatial resolution (< 0.1 μm). Figure 11.15 shows the force–displacement data of a typical nanoindentation curve. In the first step, the tip is pressed into the material, resulting in indistinguishable elastic and plastic deformation. Then the tip is held at maximum force resulting in creep of the material under the tip. In a third step, unloading is done that leads to elastic recovery of the material. Typically each indent requires between 15 seconds to several minutes, representing a compromise between a desired quasistatic strain rate and the thermal drift of the instrument (possibly below 0.1 nm/second). This device is load-controlled, and linearly increasing and decreasing loading protocols were therefore applied. The loading rate corresponds to

$$\frac{dP}{dt} = \frac{P_{max}}{T} \qquad (11.8)$$

with P_{max} as the maximum load and T as the (un)loading time.

The elastic constants of the sample were determined using the unloading part (step 3) of the nanoindentation curve. Based on the analytical work by *Sneddon* [75], *Oliver* and *Pharr* [76] derived the following equation for an indenter of revolution pressed into an isotropic elastic material:

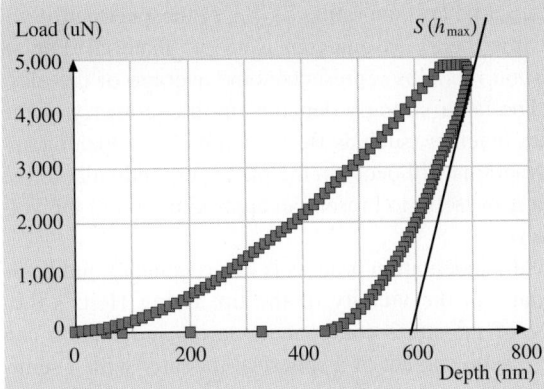

Fig. 11.15. Force–displacement curve obtained during a nanoindentation test that consists of three parts (see text). Hardness is determined at maximum load. The elastic indentation modulus is calculated with the slope at the point of initial unloading $S(h_{max})$

$$S(h_{max}) = \frac{dP}{dh}(h_{max}) = \frac{2}{\sqrt{\pi}} E_r \sqrt{A_c(h_{max})}. \tag{11.9}$$

Pharr et al. [77] showed that this equation is a good approximation for a Berkovich indenter tip. P represents the applied load, $S(h_{max})$ is the derivative of the unloading curve at the point of initial unloading h_{max}. This variable is determined by fitting typically 40% to 95% of the unloading curve to avoid the influence of viscoelastic effects at initial unloading when a singularity in the strain rate occurs. $A_c(h)$ is the contact area over which the material and the indenter are in instantaneous contact. The latter function has to be calibrated by performing indents with increasing depth in a standard material, typically fused silica. The latter has a known reduced modulus of $E_r = 69.9$ GPa that allows calibration of the contact area $A_c(h)$ using the measured contact stiffness $S(h)$ and (11.9). The reduced modulus E_r combines the deformation of the material and the diamond tip as follows [78]:

$$\frac{1}{E_r} = \frac{1 - v_{specimen}^2}{E_{specimen}} + \frac{1 - v_{tip}^2}{E_{tip}}. \tag{11.10}$$

We use the indentation modulus

$$E_{ind} = \left(\frac{1}{E_r} - \frac{1 - v_{tip}^2}{E_{tip}}\right)^{-1} \tag{11.11}$$

that can be calculated with the reduced modulus and the elastic properties of the diamond indenter tip $v_{tip} = 0.07$ and $E_{tip} = 1140$ GPa. This variable combines with

$$E_{ind} = \frac{E_{specimen}}{1 - v_{specimen}^2} \tag{11.12}$$

the local Young's modulus E_{specimen} and Poisson ratio v_{specimen} of thespecimen.

It is important to note that this theory assumes an isotropic material. For an anisotropic material the indentation modulus represents some average of the elastic properties in all directions. The latter strongly depends on the geometry of the indenter. For indents with a blunt indenter (such as Berkovich with an opening angle of 143°) nonnegligible deformations will occur in the plane perpendicular to the loading direction. In this context it is useful to present an approximation of the volume ganged by the indentation test.

The stress field generated by the indentation process is heterogeneous and leads to plastic deformation and damage in the vicinity of the tip. Using Hertz's theory [78], the spatial dependence of the stress components during indentation can be estimated by considering the elastic contact of a spherical indenter with a semi-infinite half space. In the direction of loading (z-axis), the stress component below the indenter decreases according to

$$\frac{\sigma_{zz}}{p_0} = -\left(1 + \frac{z^2}{a^2}\right), \tag{11.13}$$

where p_0 indicates the maximum pressure below the indenter and a the contact radius.

In a horizontal plane at the surface ($z = 0$), the radial and circumferential components of the stress field next to the contact area obey

$$\frac{\sigma_{rr}}{p_0} = -\frac{\sigma_{\theta\theta}}{p_0} = \frac{(1-2v)a^2}{3r^2}, \tag{11.14}$$

where θ and r are the cylindrical coordinates of the periphery of the indenter and v is the Poisson ratio. For $v = 0.3$ the stress field components reach their 10%-boundary defined by $\sigma_{zz} = -0.1p_0$ and $\sigma_{rr} = -\sigma_{\theta\theta} = 0.1p_0$ at a depth of $z \cong 3a$ or in lateral direction at $r \cong 2/\sqrt{3}a$. For a Berkovich tip, the ratio between maximum indentation depth and contact radius is approximately $a \cong 3h_{\text{max}}$. The mechanical properties measured by nanoindentation, therefore, correspond to a semi-ellipsoidal volume extending to about nine times the employed indentation depth ($z \cong 3a \cong 9h_{\text{max}}$) in the vertical direction (z) and about seven times this same depth ($2r \cong 4/\sqrt{3}a \cong 7h_{\text{max}}$) in the radial direction (r).

The load-on hardness is determined with

$$H = \frac{P(h_{\text{max}})}{A_c(h_{\text{max}})} \tag{11.15}$$

as the ratio of the maximum load and the ("on load") contact area at maximum load. This is different from Vicker hardness in which the contact area is characterized by the remaining imprint after load removal. Differences between nanoindentation and Vicker, therefore, occur for materials with nonnegligible elastic recovery. After unloading, such materials may expose an imprint smaller than the area of contact at maximum load. Since the SN tool can be used to image the remaining imprint after unloading, hardness values using both definitions may thus be compared.

For fused silica the literature provides a hardness range between 8.3 to 9.5 GPa while an intercomparison of SN-users showed an average of 8.96 GPa (Surface, Hückelhoven). Conveniently, the hardness value also represents a possibility to check the area function of the tip based on the reduced modulus of the calibration material.

Tests of Bone Tissue Under Pseudo-physiological Conditions

Samples containing trabecular and compact bone lamellae were dissected from the medial part of the femoral neck of an 86-year-old female. After embedding in PMMA, the samples were polished with successive grades of carbide paper and finished with 0.05 μm alumina solution. Polishing represents an important preparation step since the here-employed theory assumes an infinitely flat surface. The mean surface roughness of the indented area should, therefore, be far below the employed indentation depth. Unfortunately no objective criteria that determines the maximum allowable surface roughness as a function of the indentation depth has been formulated so far.

Thin and thick lamellae of trabecular and compact bone were first characterized under physiological conditions, i.e. fully wet and at 37 °C. Then the specimens were dried for 24 h at 50 °C and identical tests were carried out again but under dry conditions.

In both cases 16 indentation tests were performed to maximum depths of 100 nm and 500 nm. Each test consisted of 10 s loading, 10 s holding, and 10 s unloading. The maximum allowable thermal drift was set 0.1 nm/s. To avoid proximity effects of neighbouring indents, an adjacent testing area in the identical lamellae was chosen after changing the testing conditions.

The tests under fully wet conditions at body temperature were found to be very sensitive with respect to thermal stability. The nanoindentation device was heated for several days to reach stable thermal conditions of the instrument components before the sample could be installed. The electrostatic force constant (EFC) was checked daily before beginning data acquisition. The approach of the indenter tip in the liquid environment was performed employing a contact force of 7 μN. This value is offset by approximately 1 μN per mm of water penetration, however, due to Archimedes force acting on the fluid cell tip. Such additional effects as water surface tension may also have a repulsive force on the tip.

AFM-scans as in Fig. 11.16, which shows the topography of trabecular bone lamellae under dry conditions, allowed the two lamella types to be identified. Thick lamellae correspond to the tops (bright contrast) and thin lamellae to the valleys (dark contrast) where the surface relief results from preferential polishing.

Indentation Modulus

Trabecular bone showed under wet conditions a mean indentation modulus of 12.5 ± 4.0 GPa when the data of both lamella types and indentation depths were combined. Under dry conditions the results showed a mean of 19.6 ± 2.6 GPa, an increase of 57%. Lamellae of compact bone increased their mean stiffness by 76.5% from 14.9 ± 4.5 GPa to 26.4 ± 3.8 GPa.

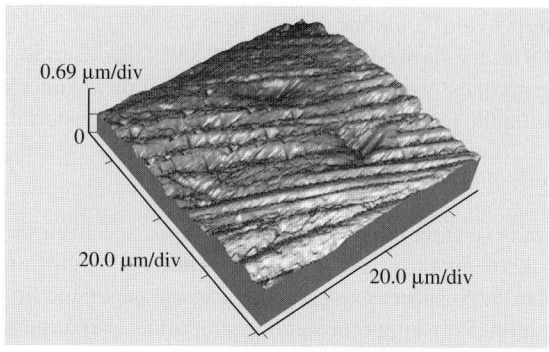

Fig. 11.16. Surface topography of a trabecular bone structural unit that shows the lamellar structure. Thick (*bright*) lamellae correspond to the tops in the topography. Note, the two holes are not remaining indents but are ellipsoidal lacunae that embed bone cells

These determined conversion factors between dried and fully wet tissue properties should be compared with other studies. *Rho* [74] has reported an increase of indentation modulus of bovine compact bone by 15.8%. Our study demonstrates a change of +76.5% for human compact bone. This high discrepancy may be attributed to different preparation and testing protocols. Rho tested the bone samples at ambient temperature while kept moist by a thin film of deionized water. In our study the tissue properties were determined under fully wet conditions and at body temperature. Furthermore, Rho dried for 14 days at ambient temperature while in our study the drying process occurred during 24 h at 50 °C. These points may explain the different relative change of mechanical properties we detected.

Figure 11.17 presents the indentation moduli (combining both depths) normalized with respect to their initial wet values under physiological conditions. It is interesting that the increase of this elastic parameter after drying was significantly ($p < 0.00001$) higher for thin than for thick lamellae. The relative change of stiffness due to drying was +44% for thick lamellae and +109% for thin lamellae of compact bone. For trabecular bone the corresponding values were +37% and +78% for thick and thin lamellae, respectively.

Table 11.2 shows the results for both indentation depths. The differences are likely related to the volume sampled during the indentation, a semi-ellipsoid following the approximations of (11.13) and (11.14) with 0.7 μm diameter and 0.9 μm height for 100 nm indents (and 3.5 μm×4.5 μm for 500 nm indents). Given the typical lamellae dimensions of 1–4 μm, the shallow indentation measurements, therefore, represent properties of single lamellae whereas the deeper indents include neighboring lamellae. It is also worth noting that thin lamellae showed a greater effect of drying when only the shallow indents were considered.

This important result should be discussed in the frame of published models that address the phenomenon of bone lamellation. *Marotti* [79] proposed that bone lamellae are the result of alternating changes of the collagen fiber density. According to his SEM-results the density of collagen is higher in thin lamellae than in thick lamellae.

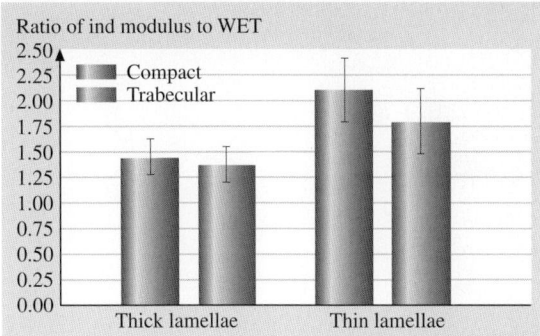

Fig. 11.17. Relative increase of indentation modulus for compact and trabecular bone lamellae after drying. The results are normalized with respect to their initial value under wet conditions. Note that thin lamellae are more affected by drying for both types of bone

Table 11.2. Absolute and relative changes of the indentation modulus with respect to the initial values under wet conditions. Note the greater increase for the thin lamellae after drying. Indents to 100 nm depth represent properties of single lamellae while 500 nm indents are also influenced by neighboring lamellae

Indentation modulus	Lamella	Ind. depth (nm)	Wet (GPa)	Dry (GPa)	Rel. change (%)
Trabecular bone	Thin	100	11.6 ± 4.1	20.5 ± 4.3	+76
		500	9.8 ± 1.9	18.1 ± 3.2	+84
	Thick	100	12.4±5	19.1 ± 1.11	+54
		500	15.6 ± 1.3	19.0 ± 1.1	+22
Compact bone	Thin	100	10.4 ± 0.2	23.4 ± 3.6	+124
		500	13.9 ± 2.0	27.1 ± 3.1	+95
	Thick	100	19.9 ± 4.8	27.9 ± 4.6	+40
		500	15.5 ± 2.6	23.0 ± 0.8	+49

Collagen fibers are long chains of proteins and contain adsorption sites for polar water molecules. A higher density of the collagen fibers in the thinner lamellae results in a higher water binding capacity that may explain the higher relative influence of drying.

Giraud-Guille [80], on the other hand, proposed a nested-arc structure for bone lamellae with smooth orientation changes between adjacent collagen fibers. According to this model indents on thick lamellae load into the longitudinal direction of the fibers, whereas the load is perpendicular to the long axis for thin lamellae. Removal of the liquid phase leads to packaging of the collagen fibers, an effect that is intuitively anisotropic and that may explain why the effect of drying was different for both lamella types. In addition, drying leads to microfracture initiation within thick lamellae [64], possibly explaining the diminished effect of drying on these structure.

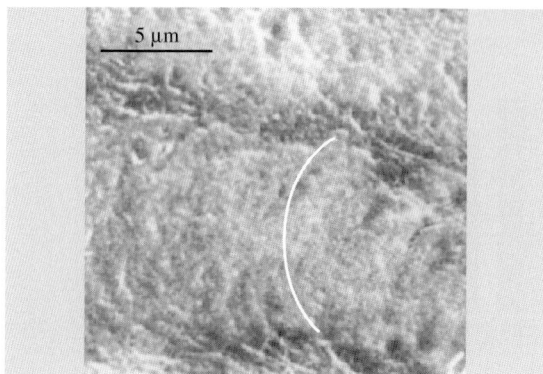

Fig. 11.18. SEM-scan of bone lamellae adjacent to the spot characterized by nanoindentation. This SEM-scan gives support for the bone lamellation model that is based on a smooth orientation change of collagen fibers [80]

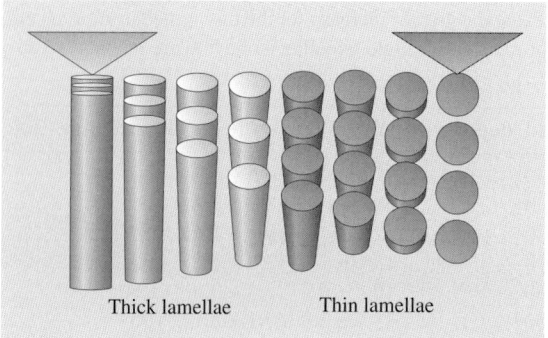

Fig. 11.19. Bone lamellation model proposed by this study. This model represents a combination of a smooth orientation change and density variations of the collagen fibers. Note also the spots where the indentations were done. For thin lamellae a greater change of mechanical properties was detected after drying

Figure 11.18 presents a SEM-image from a bone sample used in this study. This scan confirms the nested-arc structure Giraud-Guille observed with TEM.

Our results are, therefore, in agreement with both the Marotti and the Giraud-Guille models. This implies a model that combines smooth orientation changes of the collagen fibers with density changes. Such an architecture is sketched in Fig. 11.19. Note also the spots where the indentations were done.

Hardness

Compact bone revealed a mean hardness of 0.46 GPa under wet conditions that increased by 74% to 0.80 GPa after drying. Due to drying trabecular bone lamellae showed a mean hardness change of 76% from 0.41 up to 0.72 GPa.

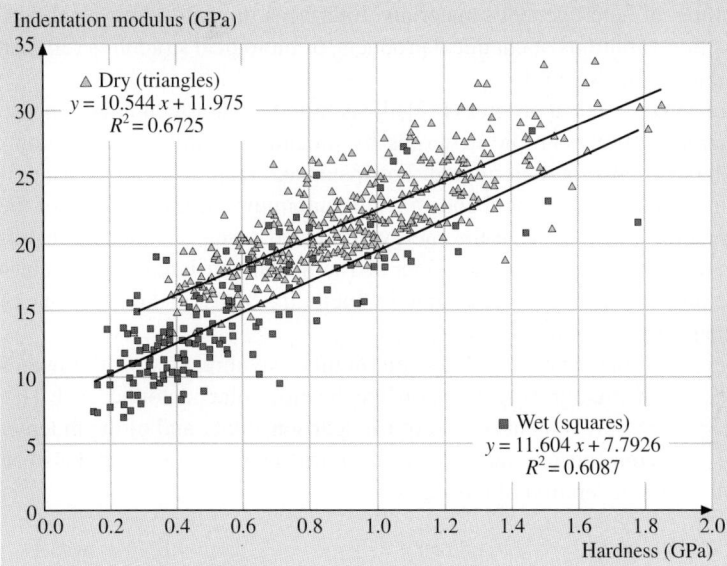

Fig. 11.20. Indentation modulus and hardness under dry and wet conditions using raw data from *Hengsberger* et al. [64]. Drying does not change the correlation coefficient and changes slightly but significantly the slope of the regression curve

For compact bone the effect of drying was again significantly ($p = 0.0006$) greater for thin lamellae (+108%) than for thick lamellae (+44%). Similar results were found for trabecular bone lamellae with a +99% increase for thin and a +56% increase for thick lamellae.

The results for hardness, therefore, appear to similar trends as the indentation modulus. This motivates us to discuss a further point, the correlation between indentation modulus and hardness. These mechanical parameters show an empirically proven relationship [81] that may change as a result of drying. For this purpose we related the data of the indentation modulus of Hengsberger et al. [64] with the corresponding unpublished hardness results. The correlation between hardness and indentation modulus is similar under wet ($R^2 = 0.61$) and under dry ($R^2 = 0.67$) conditions (Fig. 11.20). The slopes of the linear regression, however, show close but significantly different values ($p < 0.0001$) for the wet and dry samples. Hardness and indentation modulus show a similar but significantly different relative shift as a result of drying.

11.4.4 Conclusion

Modern scanning nanoindentation has clearly solved two problems: indentations in the nanometer scale can be performed with high force and displacement resolution, and nanometer lateral control of the indentation position allows the characterization of small structures of heterogeneous materials. Interfaces in composites, local density or composition variations of chemical products, or biological structures can then be investigated using the SN tool.

Based on recent theoretical progress [82], the nanoindentation technique can also be used to characterize anisotropic materials by indenting in different directions. Other parameters can also be determined. For instance, creep behavior is accessible when the indentation load is held constant at a maximum load. Hysteresis of the force–displacement curve represents the energy dissipated during an indentation providing information about post-elastic behavior of a sample. Such dynamical variables as viscoelastic properties are also accessible by including sinusoidal oscillations of different frequencies to the loading history.

Future interests of the nanotechnology community may direct the work towards tests on lower levels of structural organization like the molecular or atomic level. This would require strong improvements of the transducer sensitivity and of the indenter tip machining. New concepts to reduce the thermal drifting properties would also be necessary for the next generation of this device.

11.5 Conclusions and Perspectives

We have demonstrated that AFM is an ideal tool for investigating variations in local properties of bulk materials, like bone, and also for performing physical measurements on individual nanometer scale objects such as, for example, carbon nanotubes and protein polymers. In the latter case, the technique provides previously inaccessible quantities in living matter such as, for instance, the shear modulus. It should be emphasized that shear is omnipresent since biostructures are "composite" materials, with strong anisotropic interactions between their constituents. In the case of proteins, we have explored the mechanical properties as a function of temperature, as well. Interactions can vary remarkably even over a range of a few tens of degrees, providing deeper insight in the functioning of these structures. One task, which seems very difficult for the time being, is the frequency dependent response of nano and biostructures. In the latter case it is limited to a few kilohertz, mainly because of the surrounding liquid, but for a meaningful analysis one requires several decades in frequency.

Regarding the scanning probe tips themselves, improvements are necessary to provide better resolution, longer lifetimes, and easy functionalization for sensing different chemical environments. This might be achieved by carbon nanotube tips (Fig. 11.21), which have a Young's modulus of 1 TPa, a very well-defined and sharp tip structure, and pentagon "defects" at the apex providing a site for functionalization. It is generally agreed upon by the scanning probe community that carbon nanotubes will open new avenues for the study of living matter.

Finally the development of the photonic force microscope in Heidelberg has allowed imaging of bio-structures with unprecedented three-dimensional resolution, even including features that are otherwise inaccessible to AFM tips. In the future, this instrument will certainly provide radical new insight in biological functioning.

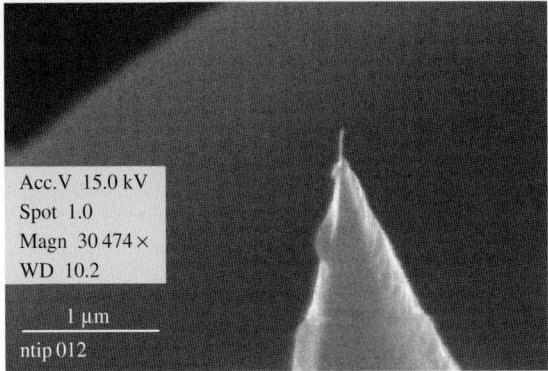

Fig. 11.21. SEM picture of an AFM tip with attached multiwall Carbon nanotube

References

1. M. Radmacher. Measuring the elastic properties of biological samples with the afm. *IEEE Eng. Med. Biol. Mag.*, 16(2):47–57, 1997.
2. G. Binnig, C. F. Quate, and C. Gerber. Atomic force microscope. *Phys. Rev. Lett.*, 56:930–933, 1986.
3. P. Maivald, H. J. Butt, S. A. C. Gould, C. B. Prater, B. Drake, J. A. Gurley, V. B. Elings, and P. K. Hansma. Using force modulation to image surface elasticities with the atomic force microscope. *Nanotechnology*, 2:103–106, 1991.
4. T. Kajiyama, K. Takata, I. Ohki, S.-R. Ge, J.-S. Yoon, and A. Takahara. Imaging of dynamic viscoelstic properties of a phase-separated polymer by forced oscillation atomic force microscopy. *Macromolecules*, 27:7932–7934, 1994.
5. B. Nysten, R. Legras, and J.-L. Costa. Atomic force microscopy imaging of viscoelastic properties in toughened polypropylene resins. *J. Appl. Phys.*, 78:5953–5958, 1995.
6. P.-E. Mazeran and J.-L. Loubet. Normal and lateral modulation with a scanning force microscope, an analysis: Implication in quantitative elastic and friction imaging. *Tribol. Lett.*, 3:125–129, 1997.
7. B. Cretin and F. Sthal. Scanning microdeformation microscopy. *Appl. Phys. Lett.*, 62:829–831, 1993.
8. N. A. Burnham, A. J. Kulik, G. Gremaud, P.-J. Gallo, and F. Oulevey. Scanning local-acceleration microscopy. *J. Vac. Sci. Technol. B*, 14:794–799, 1996.
9. F. Oulevey, N. A. Burnham, A. J. Kulik, G. Gremaud, and W. Benoit. Mechanical properties studied at the nanoscale using scanning local-acceleration microscopy (slam). *J. Phys.*, IV:C8–731–734, 1996.
10. U. Rabe and W. Arnold. Acoustic microscopy by atomic force microscopy. *Appl.Phys. Lett*, 64:1493–1495, 1994.
11. O. Kolosov and K. Yamanaka. Nonlinear detection of ultrasonic vibrations in an atomic force microscope. *Jpn. J. Appl. Phys.*, 32:22–25, 1993.
12. F. Oulevey. Cartographie et spectrométrie des propriétés mécaniques à l'échelle nanométrique par microscopie acoustique en champ proche, 1999.
13. F. Oulevey, G. Gremaud, A. Semoroz, A. J. Kulik, N. A. Burnham, E. Dupas, and D. Gourdon. Local mechanical spectroscopy with nanometer-scale lateral resolution. *Rev. Sci. Instrum.*, 69:2085–2094, 1998.

14. N. A. Burnham, G. Gremaud, A. J. Kulik, P.-J. Gallo, and F. Oulevey. Materials' properties measurements: Choosing the optimal scanning probe microscope configuration. *J. Vac. Sci. Technol. B*, 14:1308–1312, 1996.
15. F. Oulevey, G. Gremaud, A. J. Kulik, and B. Guisolan. Simple low-drift heating stage for scanning probe microscopes. *Rev. Sci. Instrum.*, 70:1889–1890, 1999.
16. H. M. Pollock. *Nanoindentation*, volume 18 of *AMS Handbook*, page 419. AMS, 1992.
17. E. Dupas, G. Gremaud, A. Kulik, and J.-L. Loubet. High-frequency mechanical spectroscopy with an atomic force microscope. *Rev. Sci. Instrum.*, 72(10):3891–3897, 2001.
18. F. Oulevey, N. A. Burnham, G. Gremaud, A. J. Kulik, H. M. Pollock, A. Hammiche, M. Reading, M. Song, and D. J. Hourston. Dynamic mechanical analysis at the submicron scale. *Polymer*, 41:3087–3092, 2000.
19. W. Wm. Wendlandt. *Thermal Analysis in Chemical Analysis*, volume 19. Wiley, 3rd edition, 1996.
20. D. Mari and D. C. Dunand. NiTi and niTi-TiC composites: Part i. transformation and thermal cycling behavior. *Metall. Mater. Trans. A*, 26A:2833–2847, 1995.
21. L. Bataillard. Transformation martensitique multiple dans un alliage à mémoire de forme Ni-Ti, 1996.
22. F. Oulevey, G. Gremaud, D. Mari, A. J. Kulik, N. A. Burnham, and W. Benoit. Martensitic transformation of NiTi studied at the nanometer scale by local mechanical spectroscopy. *Scripta Mater.*, 42:31–36, 2000.
23. D. Mari, L. Bataillard, D. C. Dunand, and R. Gotthardt. Martensitic transformation of NiTi and NiTi-TiC composites. *J. Phys. (France)*, IV:C8–659–664, 1995.
24. S. Iijima. Helical microtubules of graphitic carbon. *Nature*, 354:56–58, 1991.
25. C.-H. Kiang, M. Endo, P. M. Ajayan, G. Dresselhaus, and M. S. Dresselhaus. Size effects in carbon nanotubes. *Phys. Rev. Lett.*, 81:1869–1872, 1998.
26. E. W. Wong, P. E. Sheehan, and C. M. Lieber. Nanobeam mechanics: Elasticity, strength and toughness of nanorods and nanotubes. *Science*, 277:1971–1975, 1997.
27. A. Thess, R. Lee, P. Nikolaev, H. Dai, P. Petit, J. Robert, C. Xu, Y. H. Lee, S. G. Kim, A. G. Rinzler, Daniel T. Colbert, G. U. Scuseria, D. Tománek, J. E. Fischer, and R. E. Smalley. Crystalline ropes of metallic carbon nanotubes. *Science*, 273:483–487, 1996.
28. W. Z. Li, S. S. Xie, L. X. Qian, B. H. Chang, B. S. Zou, W. Y. Zhou, R. A. Zhao, and G. Wang. Large-scale synthesis of aligned carbon nanotubes. *Science*, 274:1701–1703, 1996.
29. J. Kong, H. T. Soh, A. M. Cassell, C. F. Quate, and H. J. Dai. Synthesis of individual single-walled carbon nanotubes on patterned silicon wafers. *Nature*, 395:878–881, 1998.
30. J. P. Lu. Elastic properties of carbon nanotubes and nanoropes. *Phys. Rev. Lett.*, 79:1297–1300, 1997.
31. J.-P. Salvetat, J. M. Bonard, N. H. Thomson, A. J. Kulik, L. Forró, W. Benoit, and L. Zuppiroli. Mechanical properties of carbon nanotubes. *Appl. Phys. A*, 69:255–260, 1999.
32. M.-F. Yu, B. S. Files, S. Arepalli, and R. S. Ruoff. Tensile loading of ropes of single wall carbon nanotubes and their mechanical properties. *Phys. Rev. Lett.*, 84:5552–5555, 2000.
33. J.-P. Salvetat, G. A. D. Briggs, J.-M. Bonard, R. R. Bacsa, A. J. Kulik, T. Stöckli, N. Burnham, and L. Forró. Elastic and shear moduli of single-walled carbon nanotube ropes. *Phys. Rev. Lett.*, 82:944–947, 1999.
34. J.-P. Salvetat, A. J. Kulik, J.-M. Bonard, G. A. D. Briggs, T. Stöckli, K. Méténier, S. Bonnamy, F. Béguin, N. A. Burnham, and L. Forró. Elastic modulus of ordered and disordered multiwalled carbon nanotubes. *Adv. Mater.*, 11:161–165, 1999.
35. S. Iijima, C. Brabec, A. Maiti, and J. Bernholc. Structural flexibility of carbon nanotubes. *J. Chem. Phys.*, 104:2089–2092, 1996.

36. J. M. Gere and S. P. Timoshenko. *Mechanics of Materials*. PWS-Kent, 1990.
37. D. A. Walters, L. M. Ericson, M. J. Casavant, J. Liu, D. T. Colbert, K. A. Smith, and R. E. Smalley. Elastic strain of freely suspended single-wall carbon nanotube ropes. *Appl. Phys. Lett.*, 74:3803–3805, 1999.
38. M.-F. Yu, O. Lourie, M. J. Dyer, K. Moloni, T. F. Kelly, and R. S. Ruoff. Strength and breaking mechanism of multiwalled carbon nanotubes under tensile load. *Science*, 287:637–640, 2000.
39. A. S. K. Kis, B. Babic, A. J. Kulik, W. Benoît, G. A. D. Briggs, C. Schönenberger, S. Catsicas, and L. Forró. Nanomechanics of microtubules. *Phys. Rev. Lett.*, 89:248101, 2002.
40. T. Fujii, M. Suzuki, M. Miyashita, M. Yamaguchi, T. Onuki, H. Nakamura, T. Matsubara, H. Yamada, and K. Nakayama. Micropattern measurement with an atomic force microscope. *J. Vac. Sci. Technol. B*, 9:666–669, 1991.
41. G.-T. Kim, G. Gu, U. Waizmann, and S. Roth. Simple method to prepare individual suspended nanofibers. *Appl. Phys. Lett.*, 80:1815–1817, 2002.
42. M. Kurachi, M. Hoshi, and H. Tashiro. Buckling of a single microtubule by optical trapping forces: Direct measurement of microtubule rigidity. *Cell. Motil. Cyt.*, 30:221–228, 1995.
43. R. B. Dye, S. P. Fink, and R. C. Williams. Taxol-induced flexibillity of microtubules and its reversal by map-2 and tau. *J. Biol. Chem.*, 268:6847–6850, 1993.
44. F. Gittes, B. Mickey, J. Nettleton, and J. Howard. Flexural rigidity of microtubules and actin filaments measured from thermal fluctuations in shape. *J. Cell Biol.*, 120:923–934, 1993.
45. M. Elbaum, D. K. Fygenson, and A. Libchaber. Buckling microtubules in vesicles. *Phys. Rev. Lett.*, 76:4078–4081, 1996.
46. A. Vinckier, C. Dumortier, Y. Engelborghs, and L. Hellemans. Dynamical and mechanical study of immobilized microtubules with atomic force microscopy. *J. Vac. Sci. Technol. B*, 14:1427–1431, 1996.
47. Hysitron, incorporated 2010 east hennepin avenue minneapolis, mn 55413.
48. A. V. Kulkarni and B. Bhushan. Nanoscale mechanical property measurements using modified atomic force microscopy. *Thin Solid Films*, 290-291:206–210, 1996.
49. A. V. Kulkarni and B. Bhushan. Nano/picoindentation measurements on single-crystal aluminum using modified atomic force microscopy. *Mater. Lett.*, 29:221–227, 1996.
50. M. Chiba and M. Seo. Effects of dichromate treatment on mechanical properties of passivated single crystal iron (100) and (110) surfaces. *Corros. Sci.*, 44:2379–2391, 2002.
51. X. C. Lu, B. Shi, L. K. Y. Li, J. Luo, X. Chang, Z. Tian, and J. I. Mou. Nanoindentation and microtribological behavior of Fe-N/Ti-N multilayers with different thickness of Ti-N layers. *Wear*, 251:1144–1149, 2001.
52. P. Peeters, Gvd. Horn, T. Daenen, A. Kurowski, and G. Staikov. Properties of electroless and electroplated Ni-P and its application in microgalvanics. *Electrochim. Acta*, 47:161–169, 2001.
53. A. Rar, J. N. Zhou, W. J. Liu, J. A. Barnard, A. Bennett, and S. C. Street. Dendrimer-mediated growth of very flat ultrathin Au films. *Appl. Surf. Sci.*, 175-176:134–139, 2001.
54. R. D. Ott, C. Ruby, F. Huang, M. L. Weaver, and J. A. Barnard. Nanotribology and surface chemistry of reactively sputtered Ti-B-N hard coatings. *Thin Solid Films*, 377-378:602–606, 2000.
55. T. Malkow and S. J. Bull. Hardness measurements on thin ibad CN_x films – a comparative study. *Surf. Coat. Technol.*, 137:197–204, 2001.
56. Y. Shima, H. Hasuyama, T. Kondoh, Y. Imaoka, T. Watari, K. Baba, and R. Hatada. Mechanical properties of silicon oxynitride thin films prepared by low energy ion beam assisted deposition. *Nucl. Instrum. Methods Phys. Res., Sect. B*, 148:599–603, 1999.

57. M. Göken, M. Kempf, and W. D. Nix. Hardness and modulus of the lamellar microstructure in pst-TiAl studied by nanoindentations and afm. *Acta Mater.*, 49:903–911, 2001.
58. M. Göken and M. Kempf. Microstructural properties of superalloys investigated by nanoindentations in an atomic force microscope. *Acta Mater.*, 47:1043–1052, 1999.
59. M. Seo, M. Chiba, and K. Suzuki. Nano-mechano-electrochemistry of the iron (100) surface in solution. *J. Electroanal. Chem.*, 473:49–53, 1999.
60. Y. Zhang, F. Z. Cui, X. M. Wang, Q. L. Feng, and X. D. Zhu. Mechanical properties of skeletal bone in gene-mutated *stöpsel^{dtl28d}* and wild-type zebrafish (*Danio rerio*) measured by atomic force microscopy-based nanoindentation. *Bone*, 30:541–546, 2002.
61. M. Finke, J. A. Hughes, D. M. Parker, and K. D. Jandt. Mechanical properties of in situ demineralised human enamel measured by afm nanoindentation. *Surf. Sci.*, 491:456–467, 2001.
62. S. Habelitz, S. J. Marshall, G. W. Marshall, J. R. Balooch, and M. Balooch. The functional width of the dentino–enamel junction determined by afm-based nanoscratching. *J. Struct. Biol.*, 135:294–301, 2001.
63. G. W. Marshall Jr., M. Balooch, R. R. Gallagher, S. A. Gansky, and S. J. Marshall. Mechanical properties of the dentinoenamel junction: Afm studies of nanohardness, elastic modulus, and fracture. *J. Biomed. Mater. Res.*, 54:87–95, 2001.
64. S. Hengsberger and A. Kulik. Nanoindentation discriminates the elastic properties of individual human bone lamellae under dry and physiological conditions. *Bone*, 30:178–184, 2002.
65. E. F. Eriksen, D. W. Axelrod, and F. M. Melsen. *Bone Histomorphometry*. Raven, 1st edition, 1994.
66. C. E. Hoffler, K. E. Moore, K. Kozloff, P. K. Zysset, M. B. Brown, and S. A. Goldstein. Heterogeneity of bone lamellar-level elastic moduli. *Bone*, 26:603–609, 2000.
67. C. E. Hoffler, K. E. Moore, K. Kozloff, P. K. Zysset, and S. A. Goldstein. Age, gender, and bone lamellae elastic moduli. *J. Orth. Res.*, 18:432–437, 2000.
68. P. K. Zysset, X. E. Guo, C. E. Hoffler, K. E. Moore, and S. A. Goldstein. Elastic modulus and hardness of cortical and trabecular bone lamellae measured by nanoindentation in the human femur. *J. Biomech.*, 32:1005–1012, 1999.
69. J. Y. Rho, P. Zioupos, J. D. Currey, and G. M. Pharr. Variations in the individual thick lamellar properties within osteons by nanoindentation. *Bone*, 25:295–300, 1999.
70. J. Y. Rho, M. E. Roy, T. Y. Tsui, and G. M. Pharr. Elastic properties of microstructural components of human bone tissue as measured by nanoindentation. *J. Biomed. Mater. Res.*, 45:48–54, 1999.
71. S. Hengsberger, A. Kulik, and P. Zysset. A combined atomic force microscopy and nanoindentation technique to investigate the elastic properties of bone structural units. *Europ. Cells Mater.*, 1:12–16, 2001.
72. C. H. Turner, J. Y. Rho, Y. Takano, T. Y. Tsui, and G. M. Pharr. The elastic properties of trabecular and cortical bone tissues are similar: Results from two microscopic measurement techniques. *J. Biomech.*, 32:437–441, 1999.
73. M. E. Roy, J. Y. Rho, T. Y. Tsui, N. S. Evans, and G. M. Pharr. Mechanical and morphological variation of the human lumbar vertebral cortical and trabecular bone. *J. Biomed. Mater. Res.*, 44:191–199, 1999.
74. J. Y. Rho and G. M. Pharr. Effects of drying on the mechanical properties of bovine femur measured by nanoindentation. *J. Mater. Sci.: Mater. Med.*, 10:485–488, 1999.
75. I. N. Sneddon. The relation between load and penetration in the axisymmetric boussinesq problem for a punch of arbitrary profile. *Int. J. Eng. Sci.*, 3:47–57, 1965.

76. W. C. Oliver and G. M. Pharr. An improved technique for determining hardness and elastic modulus using load and displacement sensing indentation experiments. *Mat. Res. Soc.*, 7/6:1564–1583, 1992.
77. G. M. Pharr, W. C. Oliver, and F. R. Brotzen. On the generality of the relationship among contact stiffness, contact area, and elastic modulus during indentation. *J. Mater. Res.*, 7/3:613–617, 1992.
78. K. L. Johnson. *Contact Mechanics*. Cambridge Univ. Press, 1st edition, 1985.
79. G. Marotti. A new theory of bone lamellation. *Calcif Tissue Int.*, 53:47–56, 1993.
80. M. M. Giraud-Guille. Plywood structures in nature. *Curr. Op. Solid State Mater. Sci.*, 3:221–227, 1998.
81. G. P. Evans, J. C. Behiri, J. D. Currey, and W. Bonfield. Microhardness and young's modulus in cortical bone exhibiting a wide range of mineral volume fractions, and in a bone analogue. *J. Mater. Sci.: Mater. Med.*, 1:38–43, 1990.
82. J. G. Swadener and G. M. Pharr. Indentation of elastically anisotropic half-spaces by cones and parabolae of revolution. *Philos. Mag. A*, 81:447–466, 2001.

12

Nanomechanical Properties of Solid Surfaces and Thin Films

Adrian B. Mann

Summary. Instrumentation for the testing of mechanical properties on the submicron scale has developed enormously in recent years. This has enabled the mechanical behavior of surfaces, thin films, and coatings to be studied with unprecedented accuracy. In this chapter, the various techniques available for studying nanomechanical properties are reviewed with particular emphasis on nanoindentation. The standard methods for analyzing the raw data obtained using these techniques are described, along with the main sources of error. These include residual stresses, environmental effects, elastic anisotropy, and substrate effects. The methods that have been developed for extracting thin-film mechanical properties from the often convoluted mix of film and substrate properties measured by nanoindentation are discussed. Interpreting the data is frequently difficult, as residual stresses can modify the contact geometry and, hence, invalidate the standard analysis routines. Work hardening in the deformed region can also result in variations in mechanical behavior with indentation depth. A further unavoidable complication stems from the ratio of film to substrate mechanical properties and the depth of indentation in comparison to film thickness. Even very shallow indentations may be influenced by substrate properties if the film is hard and very elastic but the substrate is compliant. Under these circumstances nonstandard methods of analysis must be used. For multilayered systems many different mechanisms affect the nanomechanical behavior, including Orowan strengthening, Hall–Petch behavior, image force effects, coherency and thermal stresses, and composition modulation.

The application of nanoindentation to the study of phase transformations in semiconductors, fracture in brittle materials, and mechanical properties in biological materials are described. Recent developments such as the testing of viscoelasticity using nanoindentation methods are likely to be particularly important in future studies of polymers and biological materials. The importance of using a range of complementary methods such as electron microscopy, in situ AFM imaging, acoustic monitoring, and electrical contact measurements is emphasized. These are especially important on the nanoscale because so many different physical and chemical processes can affect the measured mechanical properties.

12.1 Introduction

When two bodies come into contact their surfaces experience the first and usually largest mechanical loads. Hence, characterizing and understanding the mechanical properties of surfaces is of paramount importance in a wide range of engineering applications. Obvious examples of where surface mechanical properties are important are in wear-resistant coatings on reciprocating surfaces and hard coatings for machine tool bits. This chapter details the current methods for measuring the mechanical properties of surfaces and highlights some of the key experimental results that have been obtained.

The experimental technique that is highlighted in this chapter is nanoindentation. This is for the simple reason that it is now recognized as the preferred method for testing thin film and surface mechanical properties. Despite this recognition, there are still many pitfalls for the unwary researcher when performing nanoindentation tests. The commercial instruments that are currently available all have attractive, user-friendly software, which makes the performance and analysis of nanoindentation tests easy. Hidden within the software, however, are a myriad of assumptions regarding the tests that are being performed and the material that is being examined. Unless the researcher is aware of these, there is a real danger that the results obtained will say more about the analysis routines than they do about the material being tested.

12.2 Instrumentation

The instruments used to examine nanomechanical properties of surfaces and thin films can be split into those based on point probes and those complimentary methods that can be used separately or in conjunction with point probes. The complimentary methods include a wide variety of techniques ranging from optical tests such as micro-Raman spectroscopy to high-energy diffraction studies using X-rays, neutrons, or electrons to mechanical tests such as bulge or blister testing.

Point-probe methods have developed from two historically different methodologies, namely, scanning probe microscopy [1] and microindentation [2]. The two converge at a length scale between 10–1000 nm. Point-probe mechanical tests in this range are often referred to as nanoindentation.

12.2.1 AFM and Scanning Probe Microscopy

Atomic force microscopy (AFM) and other scanning probe microscopies are covered in detail elsewhere in this volume, but it is worth briefly highlighting the main features in order to demonstrate the similarities to nanoindentation. There are now a myriad of different variants on the basic scanning probe microscope. All use piezoelectric stacks to move either a probe tip or the sample with subnanometer precision in the lateral and vertical planes. The probe itself can be as simple as a tungsten wire electrochemically polished to give a single atom at the tip, or as complex as an AFM tip that is bio-active with, for instance, antigens attached. A range of scanning probes have been developed with the intention of measuring specific physical properties such as magnetism and heat capacity.

To measure mechanical properties with an AFM, the standard configuration is a hard probe tip (such as silicon nitride or diamond) mounted on a cantilever (see

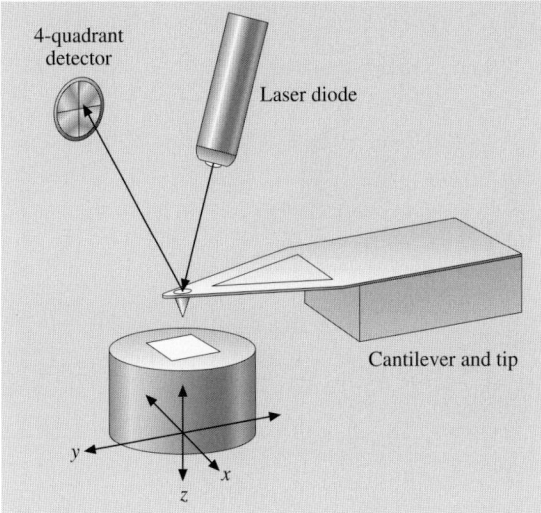

Fig. 12.1. Diagram of a commercial AFM. The AFM tip is mounted on a compliant cantilever, and a laser light is reflected off the back of the cantilever onto a position-sensitive detector (4-quadrant detector). Any movements of the cantilever beam cause a deflection of the laser light that the detector senses. The sample is moved using piezo-electric stack, and forces are calculated from the cantilever's stiffness and the measured deflection

Fig. 12.1). The elastic deflection of the cantilever is monitored either directly or via a feedback mechanism to measure the forces acting on the probe. In general, the forces experienced by the probe tip split into attractive or repulsive forces. As the tip approaches the surface, it experiences intermolecular forces that are attractive, although they can be repulsive under certain circumstances [3]. Once in contact with the surface the tip usually experiences a combination of attractive intermolecular forces and repulsive elastic forces. Two schools of thought exist regarding the attractive forces when the tip is in contact with the surface. The first is often referred to as the DMT or Bradley model. It holds that attractive forces only act outside the region of contact [4–7]. The second theory, usually called the JKR model, assumes that all the forces experienced by the tip, whether attractive or repulsive, act in the region of contact [8]. Most real nanoscale contacts lie somewhere between these two theoretical extremes.

12.2.2 Nanoindentation

The fundamental difference between AFM and nanoindentation is that during a nanoindentation experiment an external load is applied to the indenter tip. This load enables the tip to be pushed into the sample, creating a nanoscale impression on the surface, otherwise referred to as a nanoindentation or nanoindent.

Conventional indentation or microindentation tests involve pushing a hard tip of known geometry into the sample surface using a fixed peak load. The area of

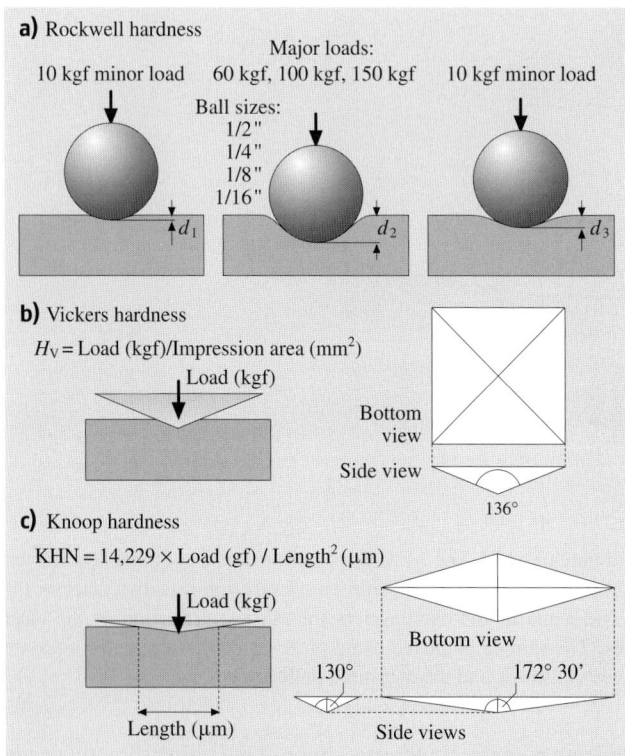

Fig. 12.2. (a) The standard Rockwell hardness test involves pushing a ball into the sample with a minor load, recording the depth, d_1, then applying a major load and recording the depth, d_2, then returning to the lower load and recording the depth, d_3. Using the depths, the hardness is calculated. (b) Vickers hardness testing uses a four-sided pyramid pushed into the sample with a known load. The area of the resulting indentation is measured optically and the hardness calculated as the load is divided by area. (c) Knoop indentation uses the same definition of hardness as the Vickers test, load divided by area, but the indenter geometry has one long diagonal and one short diagonal

indentation that is created is then measured, and the mechanical properties of the sample, in particular its hardness, is calculated from the peak load and the indentation area. Various types of indentation testing are used in measuring hardness, including Rockwell, Vickers, and Knoop tests. The geometries and definitions of hardness used in these tests are shown by Fig. 12.2.

When indentations are performed on the nanoscale there is a basic problem in measuring the size of the indents. Standard optical techniques cannot easily be used to image anything smaller than a micron, while electron microscopy is simply impractical due to the time involved in finding and imaging small indents. To overcome these difficulties, nanoindentation methods have been developed that continuously record the load, displacement, time, and contact stiffness throughout the indentation

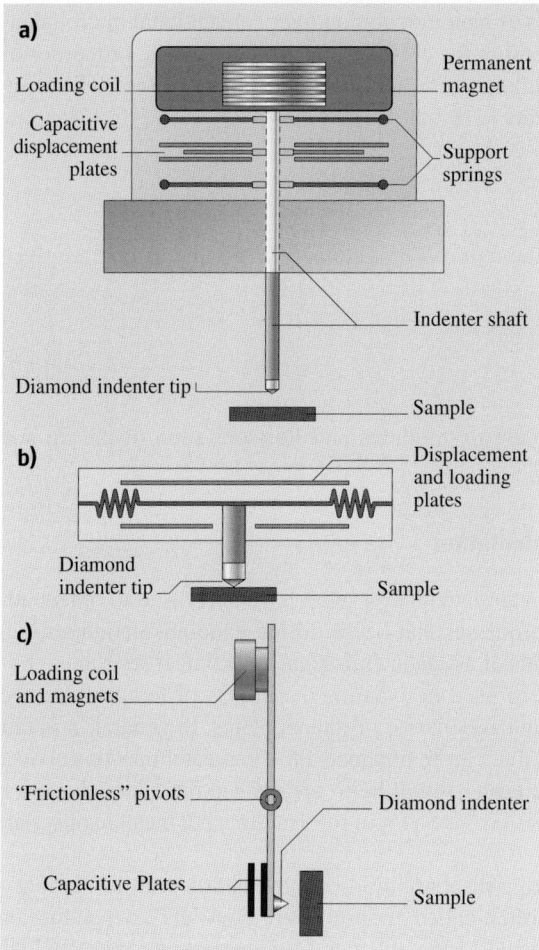

Fig. 12.3. Schematics of three commercial nanoindentation devices made by (**a**) MTS Nanoinstruments, Oak Ridge, Tennessee, (**b**) Hysitron Inc., Minneapolis, Minnesota, (**c**) Micro Materials Limited, Wrexham, UK. Instruments (**a**) and (**c**) use electromagnetic loading, while (**b**) uses electrostatic loads

process. This type of continuously recording indentation testing was originally developed in the former Soviet Union [9–13] as an extension of microindentation tests. It was applied to nanoscale indentation testing in the early 1980s [14, 15], hence, giving rise to the field of nanoindentation testing.

In general, nanoindentation instruments include a loading system that may be electrostatic, electromagnetic, or mechanical, along with a displacement measuring system that may be capacitive or optical. Schematics of several commercial nanoindentation instruments are shown in Fig. 12.3a–c.

Among the many advantages of nanoindentation over conventional microindentation testing is the ability to measure the elastic, as well as the plastic properties of the test sample. The elastic modulus is obtained from the contact stiffness (S) using the following equation that appears to be valid for all elastic contacts [16, 17]:

$$S = \frac{2}{\sqrt{\pi}} E_r \sqrt{A} . \tag{12.1}$$

A is the contact area and E_r is the reduced modulus of the tip and sample as given by:

$$\frac{1}{E_r} = \frac{(1-v_t^2)}{E_t} + \frac{(1-v_s^2)}{E_s} , \tag{12.2}$$

where E_t, v_t and E_s, v_s are the elastic modulus and Poissons ratio of the tip and sample, respectively.

12.2.3 Adaptations of Nanoindentation

Several adaptations to the basic nanoindentation setup have been used to obtain additional information about the processes that occur during nanoindentation testing, for example, in situ measurements of acoustic emissions and contact resistance. Environmental control has also been used to examine the effects of temperature and surface chemistry on the mechanical behavior of nanocontacts. In general, it is fair to say that the more information that can be obtained and the greater the control over the experimental parameters the easier it will be to understand the nanoindentation results. Load-displacement curves provide a lot of information, but they are only part of the story.

During nanoindentation testing discontinuities are frequently seen in the load–displacement curve. These are often called "pop-ins" or "pop-outs", depending on their direction. These sudden changes in the indenter displacement, at a constant load (see Fig. 12.4), can be caused by a wide range of events, including fracture, delamination, dislocation multiplication, or nucleation and phase transformations. To help distinguish between the various sources of discontinuities, acoustic transducers have been placed either in contact with the sample or immediately behind the indenter tip. For example, the results of nanoindentation tests that monitor acoustic emissions have shown that the phase transformations seen in silicon during nanoindentation are not the sudden events that they would appear to be from the load–displacement curve. There is no acoustic emission associated with the pop-out seen in the unloading curve of silicon [18]. An acoustic emission would be expected if there were a very rapid phase transformation causing a sudden change in volume. Fracture and delamination of films, however, give very strong acoustic signals [19], but the exact form of the signal appears to be more closely related to the sample geometry than to the event [20].

Additional information about the nature of the deformed region under the nanoindentation can be obtained by performing in situ measurements of contact resistance.

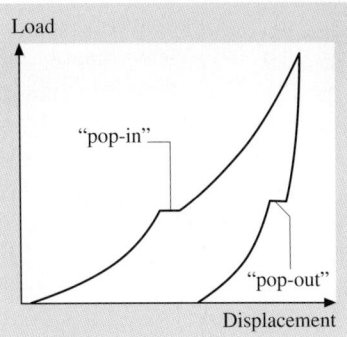

Fig. 12.4. Sketch of a load/displacement curve showing a pop-in and a pop-out

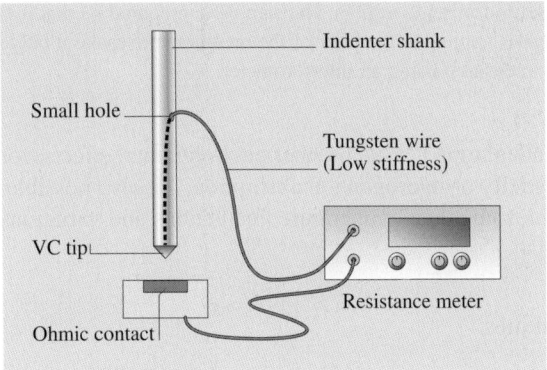

Fig. 12.5. Schematic of the basic setup for making contact resistance measurements during nanoindentation testing

The basic setup for this type of testing is shown in Fig. 12.5. An electrically conductive tip is needed to study contact resistance. Consequently, a conventional diamond tip is of limited use. Elastic, hard, and metallically conductive materials such as vanadium carbide can be used as substitutes for diamond [21, 22], or a thin conductive film (e.g., Ag) can be deposited on the diamond's surface (such a film is easily transferred to the indented surface so great care must be taken if multiple indents are performed). Measurements of contact resistance have been most useful for examining phase transformations in semiconductors [21, 22] and the dielectric breakdown of oxide films under mechanical loading [23].

One factor that is all too frequently neglected during nanoindentation testing is the effect of the experimental environment. Two obvious ways in which the environment can affect the results of nanoindentation tests are increases in temperature, which give elevated creep rates, and condensation of water vapor, which modifies the tip-sample interactions. Both of these environmental effects have been shown to significantly affect the measured mechanical properties and the modes of defor-

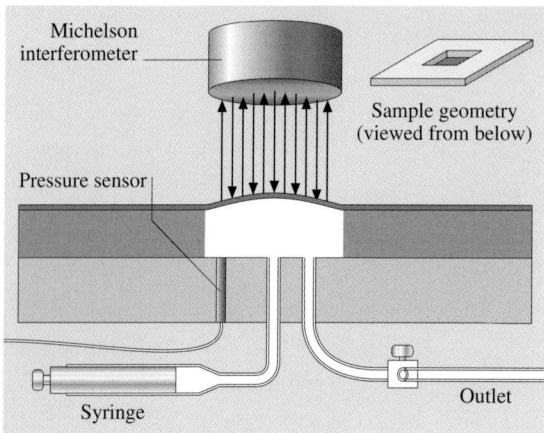

Fig. 12.6. Schematic of the basic setup for bulge testing. The sample is prepared so that it is a thin membrane, and then a pressure is applied to the back of the membrane to make it bulge upwards. The height of the bulge is measured using an interferometer

mation that occur during nanoindentation [24–27]. Other environmental effects, for instance, those due to photoplasticity or hydrogen ion absorption, are also possible, but they are generally less troublesome than temperature fluctuations and variations in atmospheric humidity.

12.2.4 Complimentary Techniques

Nanoindentation testing is probably the most important technique for characterizing the mechanical properties of thin films and surfaces, but there are many alternative or additional techniques that can be used. One of the most important alternative methods for measuring the mechanical properties of thin films uses bulge or blister testing [28]. Bulge tests are performed on thin films mounted on supporting substrates. A small area of the substrate is removed to give a window of unsupported film. A pressure is then applied to one side of the window causing it to bulge. By measuring the height of the bulge, the stress-strain curve and the residual stress are obtained. The basic configuration for bulge testing is shown in Fig. 12.6.

12.2.5 Bulge Tests

The original bulge tests used circular windows because they are easier to analyze mathematically, but now square and rectangular windows have become common [29]. These geometries tend to be easier to fabricate. Unfortunately, there are several sources of errors in bulge testing that can potentially lead to large errors in the measured mechanical properties. These errors at one time led to the belief that multilayer films can show a "super modulus" effect, where the elastic modulus of the multilayer is several times that of its constituent layers [30]. It is now accepted that any enhancement to the elastic modulus in multilayer films is small, on the order of 15% [31]. The main sources of error stem from compressive stresses in the film (tensile stress is not a problem), small variations in the dimensions of the window, and uncertainty in the exact height of the bulge. Despite these difficulties, one advantage of bulge testing over nanoindentation testing is that the stress state is biaxial, so that only properties in the plane of the film are measured. In contrast, nanoindentation testing measures a combination of in-plane and out-of-plane properties.

12.2.6 Acoustic Methods

Acoustic and ultrasonic techniques have been used for many years to study the elastic properties of materials. Essentially, these techniques take advantage of the fact that the velocity of sound in a material is dependent on the inter-atomic or inter-molecular forces in the material. These, of course, are directly related to the material's elastic constants. In fact, any nonlinearity of inter-atomic forces enables slight variations in acoustic signals to be used as a measure of residual stress.

An acoustic method ideally suited to studying surfaces is scanning acoustic microscopy (SAMscanning acoustic microscopy) [32]. There are also several other techniques that have been used to study surface films and multilayers, but we will first consider SAMscanning acoustic microscopy in detail. In a SAMscanning acoustic microscopy, a lens made of sapphire is used to bring acoustic waves to focus via a coupling fluid on the surface. A small piezoelectric transducer at the top of the lens generates the acoustic signal. The same transducer can be used to detect the signal when the SAMscanning acoustic microscopy is used in reflection mode. The use of a transducer as both generator and detector, a common imaging mode, necessitates the use of a pulsed rather than a continuous acoustic signal. Continuous waves can be used if phase changes are used to build up the image. The transducer lens generates two types of acoustic waves in the material: longitudinal and shear. The ability of a solid to sustain both types of wave (liquids can only sustain longitudinal waves) gives rise to a third type of acoustic wave called a Rayleigh, or surface, wave. These waves are generated as a result of superposition of the shear and longitudinal waves with a common phase velocity. The stresses and displacements associated with a Rayleigh wave are only of significance to a depth of ≈ 0.6 Rayleigh wavelengths below the solid surface. Hence, using SAMscanning acoustic microscopy to examine Rayleigh waves in a material is a true surface characterization technique.

Using a SAMscanning acoustic microscopy in reflection mode gives an image where the contrast is directly related to the Rayleigh wave velocity, which is in turn a function of the material's elastic constants. The resolution of the image depends on the frequency of the transducer used, i.e., for a 2 GHz signal a resolution better than 1 μm is achievable. The contrast in the image results from the interference of two different waves in the coupling fluid. Rayleigh waves that are excited in the surface "leak" into the coupling fluid and interfere with the acoustic signal that is directly reflected back from the surface. It is usually assumed that the properties of the coupling fluid are well characterized. The interference of the two waves gives a characteristic $V(z)$ curve, as illustrated by Fig. 12.7, where z is the separation between the lens and

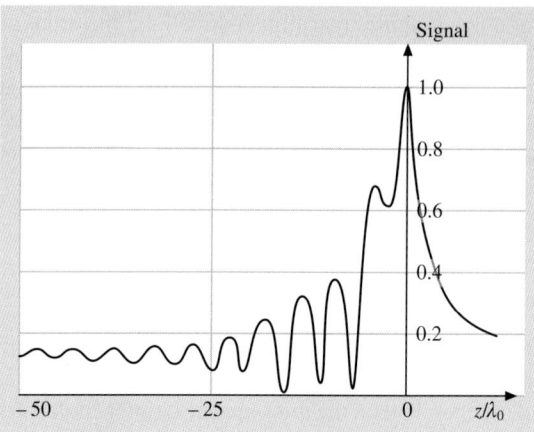

Fig. 12.7. A typical $V(z)$ curve obtained with a SAM when testing fused silica

the surface. Analyzing the periodicity of the $V(z)$ curves provides information on the Rayleigh wave velocity. As with other acoustic waves, the Rayleigh velocity is related to the elastic constants of the material. When using the SAMscanning acoustic microscopy for a material's characterization, the lens is usually held in a fixed position on the surface. By using a lens designed specifically to give a line-focus beam, rather than the standard spherical lens, it is possible to use SAMscanning acoustic microscopy to look at anisotropy in the wave velocity [33] and hence in elastic properties by producing waves with a specific direction.

One advantage of using SAMscanning acoustic microscopy in conjunction with nanoindentation to characterize a surface is that the measurements obtained with the two methods have a slightly different dependence on the test material's elastic properties, E_s and v_s (the elastic modulus and Poisson's ratio). As a result, it is possible to use SAMscanning acoustic microscopy and nanoindentation combined to find both E_s and v_s, as illustrated by Fig. 12.8 [34]. This is not possible when using only one of the techniques alone.

In addition to measuring surface properties, SAMscanning acoustic microscopy has been used to study thin films on a surface. However, the Rayleigh wave velocity can be dependent on a complex mix of the film and substrate properties. Other acoustic methods have been utilized to study freestanding films. A freestanding film can be regarded as a plate, and, therefore, it is possible to excite Lamb waves in the film. Using a pulsed laser to generate the waves and a heterodyne interferometer to detect the arrival of the Lamb wave, it is possible to measure the flexural modulus of the film [35]. This has been successfully demonstrated for multilayer films with a total thickness $< 10\,\mu$m. In the plate configuration, due to the nonlinearity of elastic properties, it is also possible to measure stress. This has been demonstrated for horizontally polarized shear waves in plates [36], but thin plates require very high frequency transducers or laser sources.

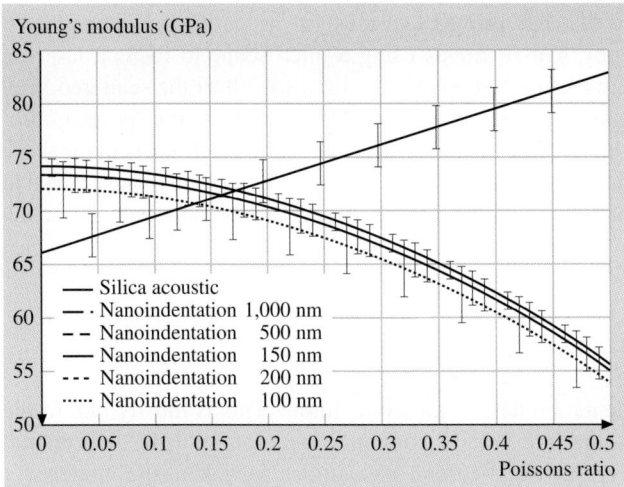

Fig. 12.8. Because SAM and nanoindentation have different dependencies on Young's modulus, E, and Poissons ratio, v, it is possible to use the two techniques in combination to find E and v [34]. On the graph, the intersection of the curves gives E and v

12.2.7 Imaging Methods

When measuring the mechanical properties of a surface or thin film using nanoindentation, it is not always easy to visualize what is happening. In many instances there is a risk that the mechanical data can be completely misinterpreted if the geometry of the test is not as expected. To expedite the correct interpretation of the mechanical data, it is generally worthwhile to use optical, electron, or atomic force microscopy to image the nanoindentations. Obviously, optical techniques are only of use for larger indentations, but they will often reveal the presence of median or lateral cracks [37]. Electron microscopy and AFM, however, can be used to examine even the smallest nanoindentations. The principle problem with these microscopy techniques is the difficulty in finding the nanoindentations. It is usually necessary to make large, "marker" indentations in the vicinity of the nanoindentations to be examined in order to find them [38].

It is possible to see features such as extrusions with a scanning electron microscope (SEM) [39], as well as pile-up and sink-in around the nanoindents, though AFM is generally better for this. Transmission electron microscopy (TEM) is useful for examining what has happened subsurface, for instance, the indentation induced dislocations in a metal [40] or the phases present under a nanoindent in silicon [21]. However, with TEM there is the added difficulty of sample preparation and the associated risk of observing artifacts. Recently, there has been considerable interest in the use of focused ion beams to cut cross sections through nanoindents [41]. When used in conjunction with SEM or TEM this provides an excellent means to see what has happened in the subsurface region.

One other technique that has proved to be useful in studying nanoindents is micro-Raman spectroscopy. This involves using a microscope to focus a laser on the sample surface. The same microscope is also used to collect the scattered laser light, which is then fed into a spectroscope. The Raman peaks in the spectrum provide information on the bonding present in a material, while small shifts in the wave number of the peaks can be used as a measure of strain. Micro-Raman has proven to be particularly useful for examining the phases present around nanoindentations in silicon [42].

12.3 Data Analysis

The analysis of nanoindentation data is far from simple. This is mostly due to the lack of effective models that are able to combine elastic and plastic deformation under a contact. However, provided certain precautions are taken, the models for perfectly elastic deformation and ideal plastic materials can be used in the analysis of nanoindentation data. For this reason, it is worth briefly reviewing the models for perfect contacts.

12.3.1 Elastic Contacts

The theoretical modeling of elastic contacts can be traced back many years, at least to the late nineteenth century and the work of *Hertz* (1882) [43] and *Boussinesq* (1885) [44]. These models, which are still widely used today, consider two axisymmetric curved surfaces in contact over an elliptical region (see Fig. 12.9). The contact region is taken to be small in comparison to the radius of curvature of the contacting surfaces, which are treated as elastic half-spaces. For an elastic sphere, radius R, in contact with a flat, elastic half-space, the contact region will be circular and the Hertz model gives the following relationships:

$$a = \sqrt[3]{\frac{3PR}{4E_r}}, \tag{12.3}$$

$$\delta = \sqrt[3]{\frac{9P^2}{16RE_r^2}}, \tag{12.4}$$

$$P_0 = \sqrt[3]{\frac{6PE_r^2}{\pi^3 R^2}}, \tag{12.5}$$

where a is the radius of the contact region, E_r is given by (12.2), δ is the displacement of the sphere into the surface, P is the applied load and P_0 is the maximum pressure under the contact (in this case at the center of the contact).

The work of *Hertz* and *Boussinesq* was extended by *Love* [45, 46] and later by *Sneddon* [47], who simplified the analysis using Hankel transforms. *Love* showed

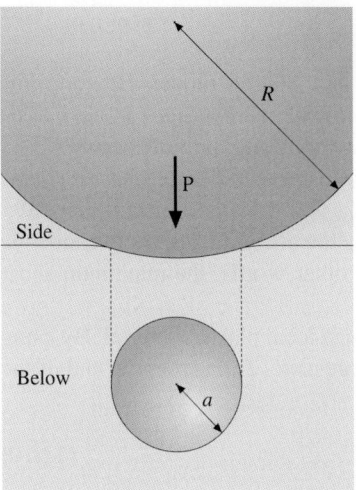

Fig. 12.9. Hertzian contact of a sphere, radius R, on a semi-infinite, flat surface. The contact in this case is a circular region of radius a

how *Boussinesq*'s model could be used for a flat-ended cylinder and a conical indenter, while *Sneddon* produced a generalized relationship for any rigid axisymmetric punch pushed into an elastic half-space. *Sneddon* applied his new analysis to punches of various shapes and derived the following relationships between the applied load, P, and displacement, δ, into the elastic half-space for, respectively, a flat-ended cylinder, a cone of semi-vertical angle ϕ, and a parabola of revolution where $a^2 = 2k\delta$:

$$P = \frac{4\mu a \delta}{1-\nu}, \tag{12.6}$$

$$P = \frac{4\mu \cot \phi}{\pi (1-\nu)} \delta^2, \tag{12.7}$$

$$P = \frac{8\mu}{3(1-\nu)} \left(2k\delta^3\right)^{1/2}, \tag{12.8}$$

where μ and ν are the shear modulus and Poisson's ratio of the elastic half-space, respectively.

The key point to note about (12.6), (12.7), and (12.8) is that they all have the same basic form, namely:

$$P = \alpha \delta^m, \tag{12.9}$$

where α and m are constants for each geometry.

Equation (12.9) and the relationships developed by Hertz and his successors, (12.3–12.8), form the foundation for much of the current nanoindentation data analysis routines.

12.3.2 Indentation of Ideal Plastic Materials

Plastic deformation during indentation testing is not easy to model. However, the indentation response of ideal plastic metals was considered by *Tabor* in his classic text, "The Hardness of Metals" [48]. An ideal plastic material (or more accurately an ideal elastic-plastic material) has a linear stress-strain curve until it reaches its elastic limit and then yields plastically at a yield stress, Y_0, that remains constant even after deformation has commenced. In a 2-D problem, the yielding occurs because the Huber-Mises [49] criterion has been reached. In other words, the maximum shear stress acting on the material is around $1.15Y_0/2$.

First, we consider a 2-D flat punch pushed into an ideal plastic material. By using the method of slip lines it is found that the mean pressure, P_m, across the end of the punch is related to the yield stress by:

$$P_m = 3Y_0 . \tag{12.10}$$

If the *Tresca* criterion [50] is used, then P_m is closer to $2.6Y_0$. In general, for both 2-D and three-dimensional punches pushed into ideal plastic materials, full plasticity across the entire contact region can be expected when $P_m = 2.6$ to $3.0Y_0$. However, significant deviations from this range can be seen if, for instance, the material undergoes work-hardening during indentation, or the material is a ceramic, or there is friction between the indenter and the surface.

The apparently straightforward relationship between P_m and Y_0 makes the mean pressure a very useful quantity to measure. In fact, P_m is very similar to the Vickers hardness, H_V, of a material:

$$H_V = 0.927 P_m . \tag{12.11}$$

During nanoindentation testing it is the convention to take the mean pressure as the nanohardness. Thus, the "nanohardness", H, is defined as the peak load, P, applied during a nanoindentation divided by the projected area, A, of the nanoindentation in the plane of the surface, hence:

$$H = \frac{P}{A} . \tag{12.12}$$

12.3.3 Adhesive Contacts

During microindentation testing and even most nanoindentation testing the effects of intermolecular and surface forces can be neglected. Very small nanoindentations, however, can be influenced by the effects of intermolecular forces between the sample and the tip. These adhesive effects are most readily seen when testing soft polymers, but there is some evidence that forces between the tip and sample may be important in even relatively strong materials [51, 52].

Contact adhesion is usually described by either the JKR or DMT model, as discussed earlier in this chapter. Both the models consider totally elastic spherical contacts under the influence of attractive surface forces. The JKR model considers the

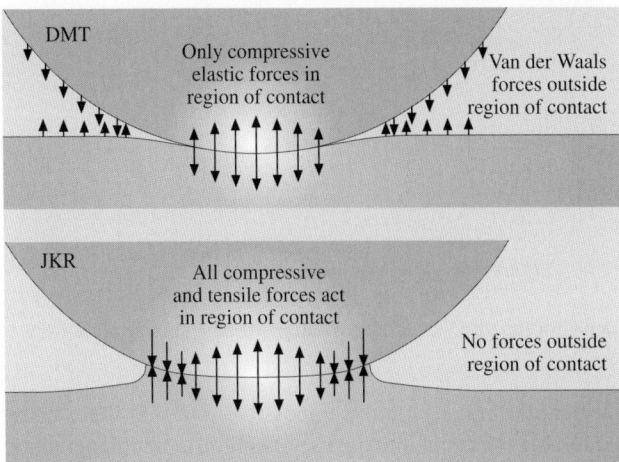

Fig. 12.10. The contact geometry for the DMT and JKR models for adhesive contact. Both models are based on the Hertzian model. In the DMT, model van der Waals forces outside the region of contact introduce an additional load in the Hertz model. But for the JKR model, it is assumed that tensile, as well as compressive stresses can be sustained within the region of contact

surface forces in terms of the associated surface energy, whereas the DMT model considers the effects of adding van der Waals forces to the Hertzian contact model. The differences between the two models are illustrated by Fig. 12.10.

For nanoindentation tests conducted in air the condensation of water vapor at the tip-sample interface usually determines the size of the adhesive force acting during unloading. The effects of water vapor on a single nanoasperity contact have been studied using force-controlled AFM techniques [53] and, more recently, nanoindentation methods [26]. Unsurprisingly, it has also been found that water vapor can affect the deformation of surfaces during nanoindentation testing [27].

In addition to water vapor, other surface adsorbates can cause dramatic changes in the nanoscale mechanical behavior. For instance, oxygen on a clean metal surface can cause an increase in the apparent strength of the metal [54]. These effects are likely to be related to, firstly, changes in the surface and intermolecular forces acting between the tip and the sample and, secondly, changes in the mechanical stability of surface nanoasperities and ledges. Adsorbates can help stabilize atomic-scale variations in surface morphology, thereby making defect generation at the surface more difficult.

12.3.4 Indenter Geometry

All of the indenter geomctries considered up to this point have been axisymmetric, largely because they are easier to deal with theoretically. Unfortunately, fabricating axisymmetric nanoindentation tips is extremely difficult, because shaping a hard

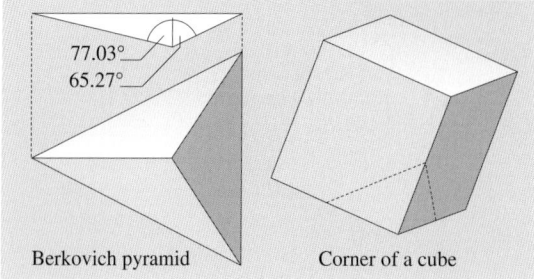

Fig. 12.11. The ideal geometry for the three-sided Berkovich pyramid and cube corner tips

tip on the scale of a few nanometers is virtually impossible. Despite these problems, there has been considerable effort put into the use of spherical nanoindentation tips [55]. This clearly demonstrates that the spherical geometry can be useful at larger indentation depths.

Because of the problems associated with creating axisymmetric nanoindentation tips, pyramidal indenter geometries have now become standard during nanoindentation testing. The most common geometries are the three-sided Berkovich pyramid and cube-corner (see Fig. 12.11). The Berkovich pyramid is based on the four-sided Vickers pyramid, the opposite sides of which make an 136° angle. For both the Vickers and Berkovich pyramids the cross-sectional area of the pyramid's base, A, is related to the pyramid's height, D, by:

$$A = 24.5D^2 . \qquad (12.13)$$

The cube-corner geometry is now widely used for making very small nanoindentations, because it is much sharper than the Berkovich pyramid. This makes it easier to initiate plastic deformation at very light loads, but great care should be taken when using the cube-corner geometry. Sharp cube-corners can wear down quickly and become blunt, hence the cross-sectional area as a function of depth can change over the course of several indentations. There is also a potential problem with the standard analysis routines [56], which were developed for much blunter geometries and are based on the elastic contact models outlined earlier. The elastic contact models all assume the displacement into the surface is small compared to the tip radius. For the cube-corner geometry this is probably only the case for nanoindentations that are no more than a few nanometers deep.

12.3.5 Analyzing Load/Displacement Curves

The load/displacement curves obtained during nanoindentation testing are deceptively simple. Most newcomers to the area will see the curves as being somewhat akin to the stress/strain curves obtained during tensile testing. There is also a real temptation just to use the values of hardness, H, and elastic modulus, E, obtained from standard analysis software packages as the "true" values. This may be the case

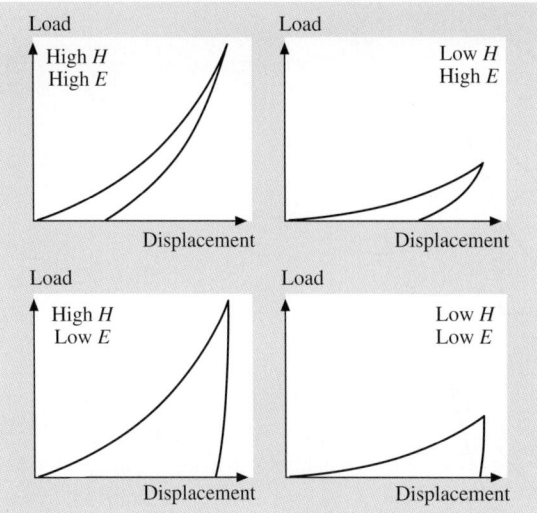

Fig. 12.12. Examples of load/displacement curves for idealized materials with a range of hardness and elastic properties

in many instances, but for very shallow nanoindents and tests on thin films the geometry of the contact can differ significantly from the geometry assumed in the analysis routines. Consequently, experimentalists should think very carefully about the test itself before concluding that the values of H and E are correct.

The basic shape of a load/displacement curve can reveal a great deal about the type of material being tested. Figure 12.12 shows some examples of ideal curves for materials with different elastic moduli and yield stresses. Discontinuities in the load/displacement curve can also provide information on such processes as fracture, dislocation nucleation, and phase transformations. Initially, though, we will consider ideal situations such as those illustrated by Fig. 12.12.

The loading section of the load/displacement curve approximates a parabola [57] whose width depends on a combination of the material's elastic and plastic properties. The unloading curve, however, has been shown to follow a more general relationship [56] of the form:

$$P = \alpha (\delta - \delta_i)^m , \tag{12.14}$$

where δ is the total displacement and δ_i is the intercept of the unloading curve with the displacement axis shown in Fig. 12.13.

Equation (12.14) is essentially the same as (12.9) but with the origin displaced. Since (12.9) is obtained by considering purely elastic deformation, it follows that the unloading curve is exhibiting purely elastic behavior. Since the shape of the unloading curve is determined by the elastic recovery of the indented region, it is not entirely surprising that its shape resembles that found for purely elastic deformation. What is fortuitous is that the elastic analysis used for an elastic half-space seems to

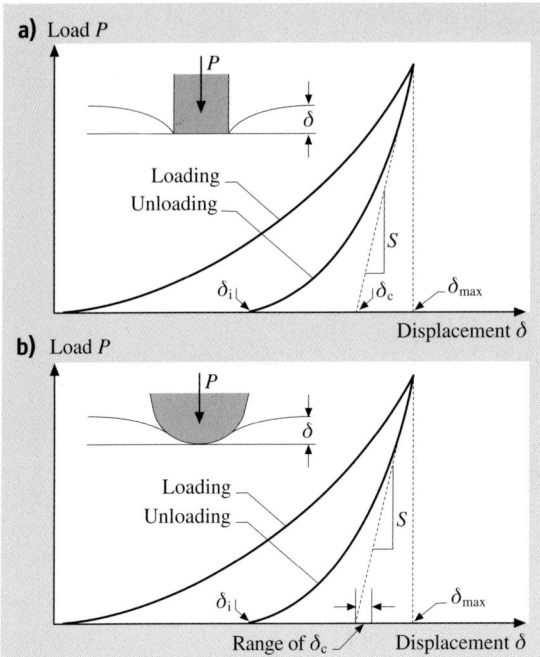

Fig. 12.13. Analysis of the load/displacement curve gives the contact stiffness, S, and the contact depth, δ_c. These can then be used to find the hardness, H, and elastic, or Young's modulus, E. (**a**) The first method of analysis [58–60] assumed the unloading curve could be approximated by a flat punch on an elastic half-space. (**b**) A more refined analysis [56] uses a paraboloid on an elastic half-space

be valid for a surface where there is a plastically formed indentation crater present under the contact. However, the validity of this analysis may only hold when the crater is relatively shallow and the geometry of the surface does not differ significantly from that of a flat, elastic half-space. For nanoindentations with a Berkovich pyramid, this is generally the case.

Before *Oliver* and *Pharr* [56] proposed their now standard method for analyzing nanoindentation data, the analysis had been based on the observation that the initial part of the unloading curve is almost linear. A linear unloading curve, equivalent to $m = 1$ in (12.14), is expected when a flat punch is used on an elastic half-space. The flat punch approximation for the unloading curve was used in [58–60] to analyze nanoindentation data. When Oliver and Pharr looked at a range of materials they found m was typically larger than 1, and that $m = 1.5$, or a paraboloid, was a better approximation than a flat punch. Oliver and Pharr used (12.1) and (12.12) to obtain the values for a material's elastic modulus and hardness. Equation (12.1) relates the contact stiffness during the initial part of the unloading curve (see Fig. 12.13) to the reduced elastic modulus and the contact area at the peak load. Equation (12.12) gives the hardness as the peak load divided by the contact area. It is immediately obvious

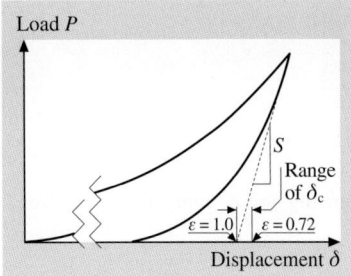

Fig. 12.14. Load-displacement curve showing how δ_c varies with ε

that the key to measuring the mechanical properties of a material is knowing the contact area at the peak load. This is the single most important factor in analyzing nanoindentation data. Most mistakes in the analysis come from incorrect assumptions about the contact area.

To find the contact area, a function relating the contact area, A_c, to the contact depth, δ_c, is needed. For a perfect Berkovich pyramid this would be the same as (12.13). But since making a perfect nanoindenter tip is impossible, an expanded equation is used:

$$A_c(\delta_c) = 24.5\delta_c^2 + \sum_{j=1}^{7} C_j \sqrt[2j]{\delta_c}, \qquad (12.15)$$

where C_j are calibration constants of the tip.

There is a crucial step in the analysis before A_c can be calculated, namely, finding δ_c. The contact depth is not the same as the indentation depth, because the surface around the indentation will be elastically deflected during loading, as illustrated by Fig. 12.15. Sneddon's analysis [47] provides a way to calculate the deflection of the surface at the edge of an axisymmetric contact. Subtracting the deflection from the total indentation depth at peak load gives the contact depth. For a paraboloid, as used by *Oliver* and *Pharr* [56] in their analysis, the elastic deflection at the edge of the contact is given by:

$$\delta_s = \varepsilon \frac{P}{S} = 0.75 \frac{P}{S}, \qquad (12.16)$$

where S is the contact stiffness and P the peak load. The constant ε is 0.75 for a paraboloid, but ranges between 0.72 (conic indenter) and 1 (flat punch). Figure 12.14 shows how the contact depth depends on the value of ε. The contact depth at the peak load is, therefore:

$$\delta_c = \delta - \delta_s. \qquad (12.17)$$

Using the load/displacement data from the unloading curve and (12.1), (12.2), (12.12), (12.14–12.17), the hardness and reduced elastic modulus for the test sample

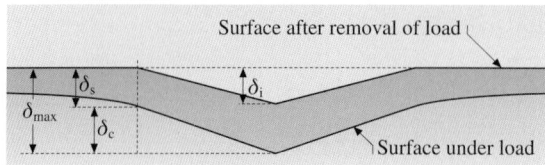

Fig. 12.15. Profile of surface under load and unloaded showing how δ_c compares to δ_i and δ_{max}

can be calculated. To find the elastic modulus of the sample, E_s, it is also necessary to know Poisson's ratio, ν_s, for the sample, as well as the elastic modulus, E_t, and Poisson's ratio, ν_t, of the indenter tip. For diamond these are 1141 GPa and 0.07, respectively.

There also remains the issue of calibrating the tip shape, or finding the values for C_j in (12.15). Knowing the exact expansion of $A_c(\delta_c)$ is vital if the values for E_s and H are to be accurate. Several methods for calibrating the tip shape have been used, including imaging the tip with an electron microscope, measuring the size of nanoindentations using SEM or TEM of negative replicas, and using scanning probes to examine either the tip itself or the nanoindentations made with the tip. There are strengths and weaknesses to each of these methods. In general, however, the accuracy and usefulness of the methods depends largely on how patient and rigorous the experimentalist is in performing the calibration.

Because of the experimental difficulties and time involved in calibrating the tip shape by these methods, *Oliver* and *Pharr* [56] developed a method for calibration based on standard specimens. With a standard specimen that is mechanically isotropic and has a known E and H that does not vary with indentation depth, it should be possible to perform nanoindentations to a range of depths, and then use the analysis routines in reverse to deduce the tip area function, $A_c(\delta_c)$. In other words, if you perform a nanoindentation test, you can find the contact stiffness, S, at the peak load, P, and the contact depth, δ_c, from the unloading curve. Then if you know E a priori, (12.1) can be used to calculate the contact area, A, and, hence, you have a value for A_c at a depth δ_c. Repeating this procedure for a range of depths will give a numerical version of the function $A_c(\delta_c)$. Then, it is simply a case of fitting (12.15) to the numerical data. If the hardness, H, is known and not a function of depth, and the calibration specimen was fully plastic during testing, then essentially the same approach could be used but based on (12.12). Situations where a constant H is used to calibrate the tip are extremely rare.

In addition to the tip shape function, the machine compliance must be calibrated. Basic Newtonian mechanics tells us that for the tip to be pushed into a surface the tip must be pushing off of another body. During nanoindentation testing the other body is the machine frame. As a result, during a nanoindentation test it is not just the sample, but the machine frame that is being loaded. Consequently, a very small elastic deformation of the machine frame contributes to the total stiffness obtained from the unloading curve. The machine frame is usually very stiff, $> 10^6$ N/m, so the effect is only important at relatively large loads.

To calibrate the machine frame stiffness or compliance, large nanoindentations are made in a soft material such as aluminium with a known, isotropic elastic modulus. For very deep nanoindentations made with a Berkovich pyramid, the contact area, $A_c(\delta_c)$, can be reasonably approximated to $24.5\delta_c^2$, thus (12.1) can be used to find the expected contact stiffness for the material. Any difference between the expected value of S and the value measured from the unloading curve will be due to the compliance of the machine frame. Performing a number of deep nanoindentations enables an accurate value for the machine frame compliance to be obtained.

Currently, because of its ready availability and predictable mechanical properties, the most popular calibration material is fused silica ($E = 72$ GPa, $v = 0.17$), though aluminium is still used occasionally.

12.3.6 Modifications to the Analysis

Since the development of the analysis routines in the early 1990s, it has become apparent that the standard analysis of nanoindentation data is not applicable in all situations, usually because errors occur in the calculated contact depth or contact area. *Pharr* et al. [61–64] have used finite element modeling (FEM) to help understand and overcome the limitations of the standard analysis. Two important sources of errors have been identified in this way. The first is residual stress at the sample surface. The second is the change in the shape of nanoindents after elastic recovery.

The effect of residual stresses at a surface on the indentation properties has been the subject of debate for many years [65–67]. The perceived effect was that compressive stresses increased hardness, while tensile stresses decreased hardness. Using FEM it is possible to model a pointed nanoindenter being pushed into a model material that is in residual tension or compression. An FEM model of nanoindentation into aluminium alloy 8009 [62] has confirmed earlier experimental observations [68] indicating that the contact area calculated from the unloading curve is incorrect if there are residual stresses. In the FEM model of an aluminium alloy the mechanical behavior of the material is modeled using a stress-strain curve, which resembles that of an elastic-perfectly-plastic metal with a flow stress of 425.6 MPa. Yielding starts at 353.1 MPa and includes a small amount of work hardening. The FEM model was used to find the contact area directly and using the simulated unloading curve in conjunction with Oliver and Pharr's method. The results as a function of residual stress are illustrated in Fig. 12.16. Note that the differences between the two measured contact areas lead to miscalculations of E and H.

Errors in the calculated contact area stem from incorrect assumptions about the pile-up and sink-in at the edge of the contact, as illustrated by Fig. 12.17. The Oliver and Pharr analysis assumes the geometry of the sample surface is the same as that given by *Sneddon* [47] in his analytical model for the indentation of elastic surfaces. Clearly, for materials where there is significant plastic deformation, it is possible that there will be large deviations from the surface geometry found using Sneddon's elastic model. In reality, the error in the contact area depends on how much the geometry of the test sample surface differs from that of the calibration material (typically fused

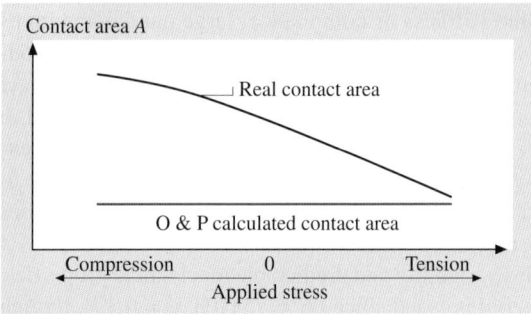

Fig. 12.16. When a surface is in a state of stress there is a significant difference between the contact area calculated using the Oliver and Pharr method and the actual contact area [62]. For an aluminium alloy this can lead to significant errors in the calculated hardness and elastic modulus

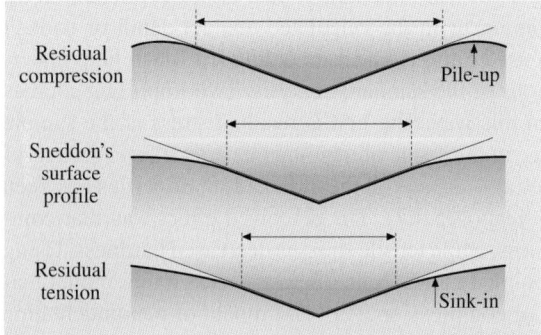

Fig. 12.17. Pile-up and sink-in are affected by residual stresses, and, hence, errors are introduced into standard Oliver and Pharr analysis

silica). It is possible that a test sample, even without a residual stress, will have a different surface geometry and, hence, contact area at a given depth, when compared to the calibration material. This is often seen for thin films on a substrate (e.g., *Tsui* et al. [69, 70]). Residual stresses increase the likelihood that the contact area calculated using Oliver and Pharr's method will be incorrect.

The issue of sink-in and pile-up is always a factor in nanoindentation testing. However, there is still no effective way to deal with these phenomena other than reverting to imaging of the indentations to identify the true contact area. Even this is difficult, as the edge of an indentation is not easy to identify using AFM or electron microscopy. One approach that has been used [71] with some success is measuring the ratio E_r^2/H, rather than E_r and H separately. Because E_r is proportional to $1/\sqrt{A}$ and H is proportional to $1/A$, E_r^2/H should be independent of A and, hence, unaffected by pile-up or sink-in. While this does not provide quantitative values for mechanical properties, it does provide a way to identify any variations in mechanical

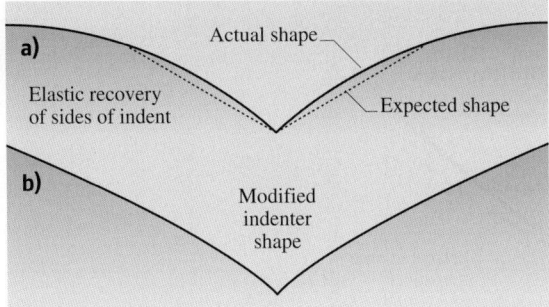

Fig. 12.18. (**a**) *Hay* et al. [63] found from experiments and FEM simulations that the actual shape of an indentation after unloading is not as expected. (**b**) They introduced a γ term to correct for this effect. This assumes the indenter has slightly concave sides

properties with indentation depth or between similar samples with different residual stresses.

Another source of error in the Oliver and Pharr analysis is due to incorrect assumptions about the nanoindentation geometry after unloading [63]. Once again, this is due to differences between the test sample and the calibration material. The exact shape of an unloaded nanoindentation on a material exhibiting elastic recovery is not simply an impression of the tip shape; rather, there is some elastic recovery of the nanoindentation sides giving them a slightly convex shape (see Fig. 12.18). The shape actually depends on Poisson's ratio, so the standard Oliver and Pharr analysis will only be valid for a material where $\nu = 0.17$, the value for fused silica, assuming it is used for the calibration.

To deal with the variations in the recovered nanoindentation shape, it has been suggested [63] that a modified nanoindenter geometry with a slightly concave side be used in the analysis (see Fig. 12.18). This requires a modification to (12.1):

$$S = \gamma 2 E_r \sqrt{\frac{A}{\pi}}, \qquad (12.18)$$

where γ is a correction term dependent on the tip geometry. For a Berkovich pyramid the best value is:

$$\gamma = \frac{\frac{\pi}{4} + 0.15483073 \cot \Phi \left(\frac{(1-2\nu_s)}{4(1-\nu_s)} \right)}{\left[\frac{\pi}{2} - 0.83119312 \cot \Phi \left(\frac{(1-2\nu_s)}{4(1-\nu_s)} \right) \right]^2}, \qquad (12.19)$$

where $\Phi = 70.32°$. For a cube corner the correction can be even larger and γ is given by:

$$\gamma = 1 + \left(\frac{(1-2\nu_s)}{4(1-\nu_s) \tan \Phi} \right), \qquad (12.20)$$

where $\Phi = 42.28°$. Figure 12.19 shows how the modified contact area varies with depth for a real diamond Berkovich pyramid.

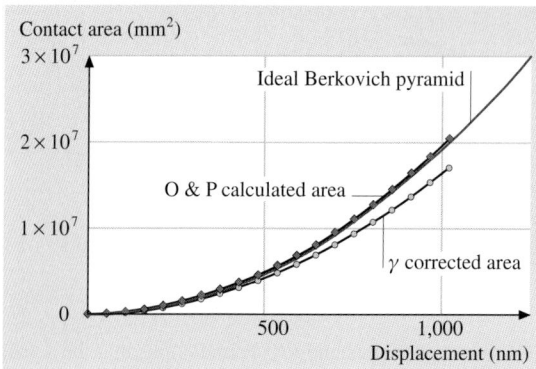

Fig. 12.19. For a real Berkovich tip the γ corrected area [63] is less at a given depth than the area calculated using the Oliver and Pharr method

The validity of the γ-modified geometry is questionable from the perspective of contact mechanics since it relies on assuming an incorrect geometry for the nanoindenter tip to correct for an error in the geometry of the nanoindentation impression. The values for E and H obtained using the γ-modification are, however, good and can be significantly different to the values obtained with the standard Oliver and Pharr analysis.

12.3.7 Alternative Methods of Analysis

All of the preceding discussion on the analysis of nanoindentation curves has focused on the unloading curve, virtually ignoring the loading curve data. This is for the simple reason that the unloading curve can in many cases be regarded as purely elastic, whereas the shape of the loading curve is determined by a complex mix of elastic and plastic properties.

It is clear that there is substantially more data in the loading curve if it can be extracted. *Page* et al. [57, 72] have explored the possibility of curve fitting to the loading data using a combination of elastic and plastic properties. By a combination of analysis and empirical fitting to experimental data, it was suggested that the loading curve is of the following form:

$$P = E\left(\psi\sqrt{\frac{H}{E}} + \phi\sqrt{\frac{E}{H}}\right)^{-2} \delta^2 , \qquad (12.21)$$

where ψ and ϕ are determined experimentally to be 0.930 and 0.194, respectively. For homogenous samples this equation gives a linear relationship between P and δ^2. Coatings, thin film systems, and samples that strain-harden can give significant deviations from linearity. Analysis of the loading curve has yet to gain popularity as a standard method for examining nanoindentation data, but it should certainly be regarded as a prime area for further investigation.

Another alternative method of analysis is based on the work involved in making an indentation. In essence, the nanoindentation curve is a plot of force against distance indicating integration under the loading curve will give the total work of indentation, or the sum of the elastic strain energy and the plastic work of indentation. Integrating under the unloading curve should give only the elastic strain energy. Thus, the work involved in both elastic and plastic deformation during nanoindentation can be found. *Cheng* and *Cheng* [73] combined measurements of the work of indentation with a dimensional analysis that deals with the effects of scaling in a material that work-hardens to estimate H/E_r. They subsequently evaluated H and E using the Oliver and Pharr approach to find the contact area.

12.3.8 Measuring Contact Stiffness

As discussed earlier, it is possible to add a small AC load on top of the DC load used during nanoindentation testing, providing a way to measure the contact stiffness throughout the entire loading and unloading cycle [74, 75]. The AC load is typically at a frequency of $\approx 60\,\text{Hz}$ and creates a dynamic system, with the sample acting as a spring with stiffness S (the contact stiffness), and the nanoindentation system acting as a series of springs and dampers. Figure 12.20 illustrates how the small AC load is added to the DC load. Figure 12.21 shows how the resulting dynamic system can be modeled. An analysis of the dynamic system gives the following relationships for S based on the amplitude of the AC displacement oscillation and the phase difference between the AC load and displacement signals:

$$\left|\frac{P_{os}}{\delta(\omega)}\right| = \sqrt{\left[(S^{-1} + C_f)^{-1} + K_s - m\omega^2\right]^2 + \omega^2 D^2}, \tag{12.22}$$

$$\tan(\chi) = \frac{\omega D}{(S^{-1} + C_f)^{-1} + K_s - m\omega^2}, \tag{12.23}$$

where C_f is the load frame compliance (the reciprocal of the load frame stiffness), K_s is the stiffness of the support springs (typically in the region of 50–100 N/m), D is the damping coefficient, P_{os} is the magnitude of the load oscillation, $\delta(\omega)$ is the magnitude of the displacement oscillation, ω is the oscillation frequency, m is the mass of the indenter, and χ is the phase angle between the force and the displacement.

In order to find S using either (12.22) or (12.23), it is necessary to calibrate the dynamic response of the system when the tip is not in contact with a sample ($S^{-1} = 0$). This calibration combined with the standard DC calibrations will provide the values for all of the constants in the two equations. All that needs to be measured in order to obtain S is either $\delta(\omega)$ or χ, both of which are measured by the lock-in amplifier used to generate the AC signal. Since the S obtained is the same as the S in (12.1), it follows that the Oliver and Pharr analysis can be applied to obtain E_r and H throughout the entire nanoindentation cycle.

The dynamic analysis detailed here was developed for the MTS Nanoindenter™ (Oakridge, Tennessee), but a similar analysis has been applied to other commercial instruments such as the Hysitron Triboscope™ (Minneapolis, Minnesota) [76]. For

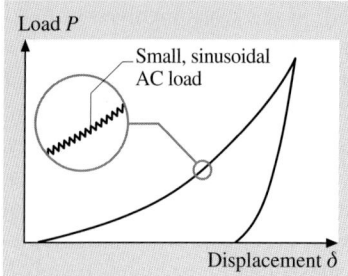

Fig. 12.20. A small AC load can be added to the DC load. This enables the contact stiffness, S, to be calculated throughout the indentation cycle

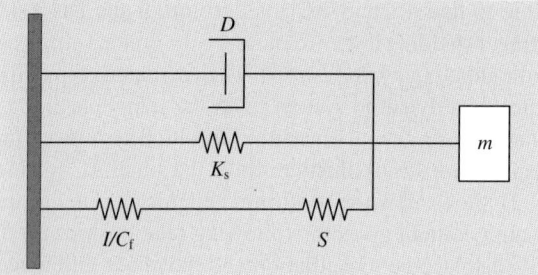

Fig. 12.21. The dynamic model used in the analysis of the AC response of a nanoindentation device

all instruments, an AC oscillation is used in addition to the DC voltage, and a dynamic model is used to analyze the response.

12.3.9 Measuring Viscoelasticity

Using an AC oscillation in addition to the DC load introduces the possibility of measuring viscoelastic properties during nanoindentation testing. This has recently been the subject of considerable interest with researchers looking at the loss modulus, storage modulus, and loss tangent of various polymeric materials [25, 77]. Recording the displacement response to the AC force oscillation enables the complex modulus (including the loss and storage modulus) to be found. If the modulus is complex, it is clear from (12.1) that the stiffness also becomes complex. In fact, the stiffness will have two components: S', the component in phase with the AC force and S'', the component out of phase with the AC force.

The dynamic model illustrated in Fig. 12.21 is no longer appropriate for this situation, as the contact on the test sample also includes a damping term, shown in Fig. 12.22. Equations (12.22) and (12.23) must also be revised. Neglecting the load frame compliance, C_f, which in most real situations is negligible, (12.22) and (12.23) when the sample damping, D_s, is included become:

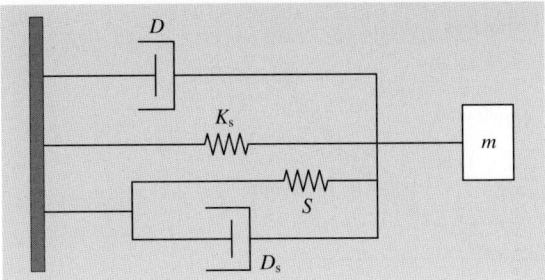

Fig. 12.22. The simplified dynamic model used when the sample is viscoelastic. It is assumed that the load frame compliance is negligible

$$\left|\frac{P_{os}}{\delta(\omega)}\right| = \sqrt{\{S + K_s - m\omega^2\}^2 + \omega^2 (D + D_s)^2} , \qquad (12.24)$$

$$\tan(\chi) = \frac{\omega(D + D_s)}{S + K_s - m\omega^2} . \qquad (12.25)$$

In order to find the loss modulus and storage modulus, (12.1) is used to relate S' (storage component) and S'' (loss component) to the complex modulus.

This method for measuring viscoelastic properties using nanoindentation has now been proven in principal, but has still only been applied to a very small range of polymers and remains an area of future growth.

12.4 Modes of Deformation

As described earlier, the analysis of nanoindentation data is based firmly on the results of elastic continuum mechanics. In reality, this idealized, purely elastic situation rarely occurs. For very shallow contacts on metals with thin surface films such as oxides, carbon layers, or organic layers [78, 79], the contact can initially be very similar to that modeled by Hertz and, later, Sneddon. It is very important to realize that this in itself does not constitute proof that the contact is purely elastic, because in many cases a small number of defects are present. These may be preexisting defects that move in the strain-field generated beneath the contact. Alternatively, defects can be generated either when the contact is first made or during the initial loading [52, 80]. When defects such as short lengths of dislocation are present the curves may still appear to be elastic even though inelastic processes like dislocation glide and cross-slip are taking place.

12.4.1 Defect Nucleation

Nucleation of defects during nanoindentation testing has been the subject of many experimental [81, 82] and theoretical studies [83, 84]. This is probably because

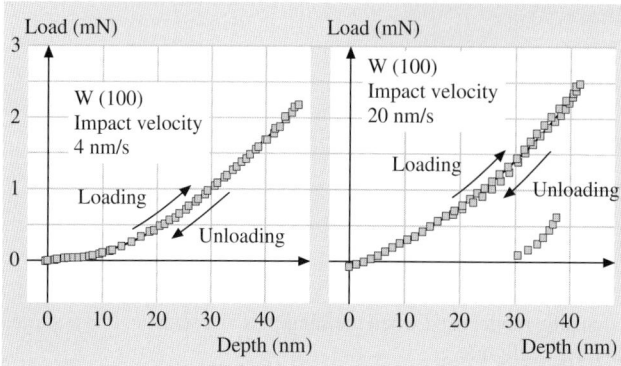

Fig. 12.23. Load-displacement curves for W(100) showing how changes in the impact velocity can cause a transition from perfectly elastic behavior to yielding during unloading

nanoindentation is seen as a way to deform a small, defect-free volume of material to its elastic limit and beyond in a highly controlled geometry. There are, unfortunately, problems in comparing experimental results with theoretical predictions, largely because the kinetic processes involved in defect nucleation are difficult to model. Simulations conducted at 0 K do not permit kinetic processes, and molecular dynamics simulations are too fast (nanoseconds or picoseconds). Real nanoindentation experiments take place at ≈ 293 K and last for seconds or even minutes.

Kinetic effects appear in many forms, for instance, during the initial contact between the indenter tip and the surface when defects can be generated by the combined action of the impact velocity and surface forces [51]. A second example of a kinetic effect occurs during hold cycles at large loads when what appears to be an elastic contact can suddenly exhibit a large discontinuity in the displacement data [80]. Figure 12.23 shows how these kinetic effects can affect the nanoindentation data and the apparent yield point load.

During the initial formation of a contact, the deformation of surface asperities [51] and ledges [85] can create either point defects or short lengths of dislocation line. During the subsequent loading, the defects can help in the nucleation and multiplication of dislocations. The large strains present in the region surrounding the contact, coupled with the existence of defects generated on contact, can result in the extremely rapid multiplication of dislocations and, hence, pronounced discontinuities in the load-displacement curve. It is important to realize that the discontinuities are due to the rapid multiplication of dislocations, which may or may not occur at the same time that the first dislocation is nucleated. Dislocations may have been present for some time with the discontinuity only occurring when the existing defects are configured appropriately, as a Frank-Read source, for instance. Even under large strains, the time taken for a dislocation source to form from preexisting defects may be long. It is, therefore, not surprising that large discontinuities can be seen during hold cycles or unloading.

The generation of defects at the surface and the initiation of yielding is a complex process that is extremely dependent on surface asperities and surface forces. These, in turn, are closely related to the surface chemistry. It is not only the magnitude of surface forces, but also their range in comparison to the height of surface asperities that determines whether defects are generated on contact. Small changes in the surface chemistry or the velocity of the indenter tip when it first contacts the surface, can cause a transition from a situation in which defects are generated on contact to one where the contact is purely elastic [52].

When the generation of defects during the initial contact is avoided and the deformed region under the contact is truly defect free, then the yielding of the sample should occur at the yield stress of a perfect crystal lattice. The load at which plastic deformation commences under these circumstances becomes very reproducible [86]. Unfortunately, nanoindenter tips on the near-atomic scale are not perfectly smooth or axisymmetric. As a result, accurately measuring the yield stress is very difficult. In fact, a slight rotation in the plane of the surface of either the sample or the tip can give a substantial change in the observed yield point load. Coating the surface in a cushioning self-assembled monolayer [87] can alleviate some of these variations, but it also introduces a large uncertainty in the contact area. Surface oxide layers, which may be several nanometers thick, have also been found to enhance the elastic behavior seen for very shallow nanoindentations on metallic surfaces [78]. Removal of the oxide has been shown to alleviate the initial elastic response.

While nanoindentation testing is ideal for examining the mechanical properties of defect-free volumes and looking at the generation of defects in perfect crystal lattices, it should be clear from the preceding discussion that great care must be taken in examining how the surface properties and the loading rate affect the results, particularly when comparisons are being made to theoretical models for defect generation.

12.4.2 Variations with Depth

Ideal elastic-plastic behavior, as described by *Tabor* [48], can be seen during indentation testing, provided the sample has been work-hardened so that the flow stress is a constant. However, it is often the case that the mechanical properties appear to change as the load (or depth) is increased. This apparent change can be a result of several processes, including work-hardening during the test. This is a particularly important effect for soft metals like copper. These metals usually have a high hardness at shallow depths, but it decreases asymptotically with increasing indentation depth to a hardness value that may be less than half that observed at shallow depths. This type of behavior is due to the increasing density of geometrically necessary dislocations at shallow depths [88]. Hence the effects of work-hardening are most pronounced at shallow depths. For hard materials the effect is less obvious.

Work-hardening is one of the factors that contribute to the so-called indentation size effect (ISE), whereby at shallow indentation depths the material appears to be harder. The ISE has been widely observed during microindentation testing, with at least part of the effect appearing to result from the increased difficulty in optically measuring the area of an indentation when it is small. During nanoindentation testing the ISE can also be observed, but it is often due to the tip area function, $A_c(\delta_c)$, being incorrectly calibrated. However, there are physical reasons other than work-hardening for expecting an increase in mechanical strength in small volumes. As described in the previous section, small volumes of crystalline materials can have either no defects or only a small number of defects present, making plastic yielding more difficult. Also, because of dislocation line tension, the shear stress required to make a dislocation bow out increases as the radius of the bow decreases. Thus, the shear stress needed to make a dislocation bow out in a small volume is greater than it is in a large volume. These physical reasons for small volumes appearing stronger than large volumes are particularly important in thin film systems, as will be discussed later. Note, however, that these physical reasons for increased hardness do not apply for an amorphous material such as fused silica, which partially explains its value as a calibration material.

12.4.3 Anisotropic Materials

The analysis methods detailed earlier were concerned primarily with the interpretation of data from nanoindentations in isotropic materials where the elastic modulus is assumed to be either independent of direction or a polycrystalline average of a material's elastic constants. Many crystalline materials exhibit considerable anisotropy in their elastic constants, hence, these analysis techniques may not always be appropriate. The theoretical problem of a rigid indenter pressed into an elastic, anisotropic half-space has been considered by *Vlassak* and *Nix* [89]. Their aim was to identify the feasibility of interpreting data from a depth-sensing indentation apparatus for samples with elastic constants that are anisotropic. Nanoindentation experiments [90] have shown the validity of the elastic analysis for crystalline zinc, copper, and beta-brass. The observed indentation modulus for zinc, as predicted, varied by as much as a factor of 2 between different orientations. The variations in the observed hardness values for the same materials were smaller, with a maximum variation with orientation of 20% detected in zinc. While these variations are clearly detectable with nanoindentation techniques, the variations are small in comparison to the actual anisotropy of the test material's elastic properties. This is because the indentation modulus is a weighted average of the stiffness in all directions.

At this time the effects of anisotropy on the hardness measured using nanoindentation have not been fully explored. For materials with many active slip planes it is likely that the small anisotropy observed by *Vlassak* and *Nix* is correct once plastic flow has been initiated. It is possible, however, that for defect-free crystalline specimens with a limited number of active slip planes that very shallow nanoindentations may show a much larger anisotropy in the observed hardness and initial yield point load.

12.4.4 Fracture and Delamination

Indentation testing has been widely used to study fracture in brittle materials [91], but the lower loads and smaller deformation regions of nanoindentation tests make it harder to initiate cracks and, hence, less useful as a way to evaluate fracture toughness. To overcome these problems the cube corner geometry, which generates larger shear stresses than the Berkovich pyramid, has been used with nanoindentation testing to study fracture [92]. These studies have had mixed success, because the cube corner geometry blunts very quickly when used on hard materials. In many cases, brittle materials are very hard.

Depth sensing indentation is better suited to studying delamination of thin films. Recent work extends the research conducted by *Marshall* et al. [93, 94], who examined the deformation of residually stressed films by indentation. A schematic of their analysis is given by Fig. 12.24. Their indentations were several microns deep, but the basic analysis is valid for nanoindentations. The analysis has been extended to multilayers [95], which is important since it enables a quantitative assessment of adhesion energy when an additional stressed film has been deposited on top of the film and substrate of interest. The additional film limits the plastic deformation of the film of interest and also applies extra stress that aids in the delamination. After indentation, the area of the delaminated film is measured optically or with an AFM to assess the extent of the delamination. This measurement, coupled with the load-displacement data, enables quantitative assessment of the adhesion energy to be made for metals [96] and polymers [97].

12.4.5 Phase Transformations

The pressure applied to the surface of a material during indentation testing can be very high. Equation (12.10) indicates that the pressure during plastic yielding is about three times the yield stress. For many materials, high hydrostatic pressures can cause phase transformations, and provided the transformation pressure is less than the pressure required to cause plastic yielding, it is possible during indentation testing to induce a phase transformation. This was first reported for silicon [98], but it has also been speculated [99] that many other materials may show the same effects. Most studies still focus on silicon because of its enormous technological importance, although there is some evidence that germanium also undergoes a phase transformation during the nanoindentation testing [100].

Recent results [21, 22, 41, 101, 102] indicate that there are actually multiple phase transformations during the nanoindentation of silicon. TEM of nanoindentations in diamond cubic silicon have shown the presence of amorphous-Si and the body-centered cubic BC-8 phase (see Fig. 12.25). Micro-Raman spectroscopy has indicated the presence of a further phase, the rhombhedral R-8 (see Fig. 12.26). For many nanoindentations on silicon there is a characteristic discontinuity in the unloading curve (see Fig. 12.27), which seems to correlate with a phase transformation. The exact sequence in which the phases form is still highly controversial with some [42], suggesting that the sequence during loading and unloading is:

Increasing load $\rightarrow$
Diamond cubic Si $\rightarrow$ β-Sn Si

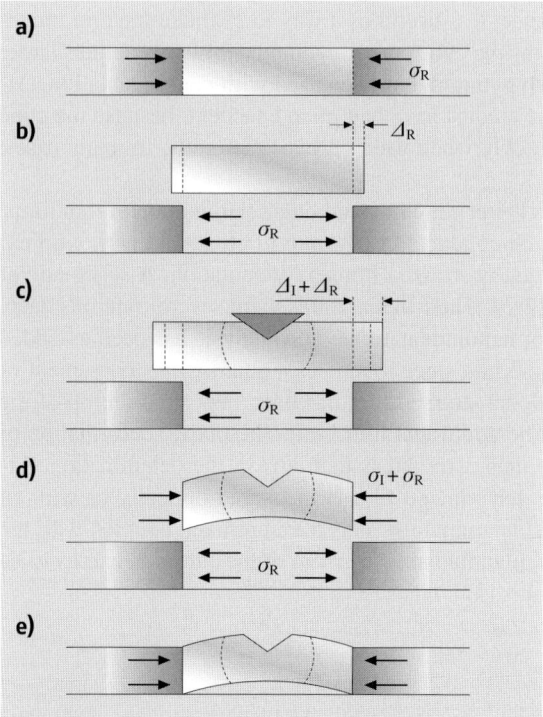

Fig. 12.24. To model delamination *Kriese* et al. [95] adapted the model developed by *Marshall* and *Evans* [93]. The model considers a segment of removed stressed film that is allowed to expand and then indented, thereby expanding it further. Replacing the segment in its original position requires an additional stress, and the segment bulges upwards

$\leftarrow$ **Decreasing load**
BC-8 Si and R-8 Si $\leftarrow$ β-Sn Si

Other groups [21, 22] suggest that the above sequence is only valid for shallow nanoindentations that do not exhibit an unloading discontinuity. For large nanoindentations that show an unloading discontinuity, they suggest the sequence will be:

Increasing load $\rightarrow$
Diamond cubic Si $\rightarrow$ β-Sn Si

$\leftarrow$ **Decreasing load**
α-Si $\leftarrow$ BC-8 Si and R-8 Si $\leftarrow$ β-Sn Si

The disagreement is over the origin of the unloading discontinuity. *Mann* et al. [21] suggest it is due to the formation of amorphous silicon, while *Gogotsi* et al. [42] believe it is the β-Sn Si to BC-8 or R-8 transformation. Mann et al. argue that the high contact resistance before the discontinuity and the low contact resistance afterwards rule out the discontinuity being the metallic β-Sn Si transforming

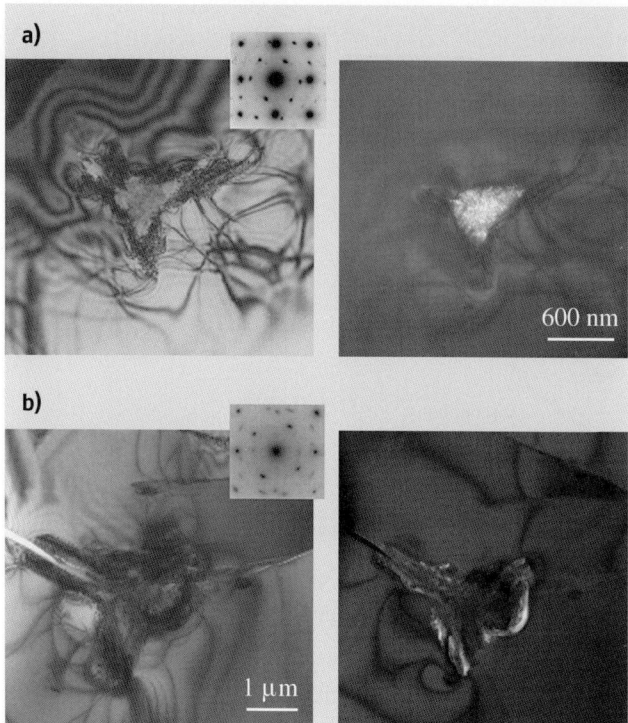

Fig. 12.25. Bright-field and dark-field TEM of (**a**) small and (**b**) large nanoindentations in Si. In small nanoindents the metastable phase BC-8 is seen in the center, but for large nanoindents BC-8 is confined to the edge of the indent, while the center is amorphous

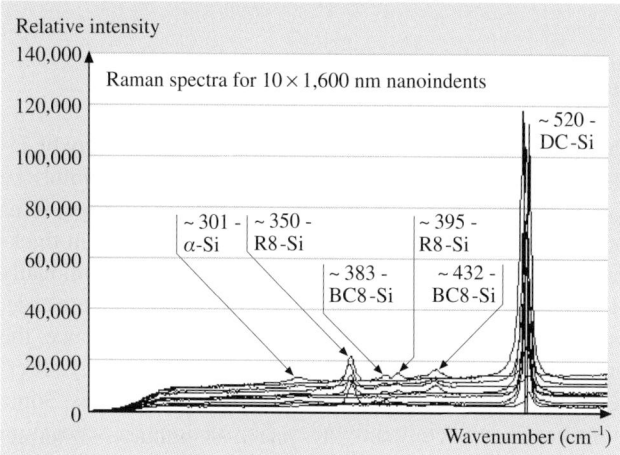

Fig. 12.26. Micro-Raman generally shows the BC-8 and R-8 Si phases that are at the edge of the nanoindents, but the amorphous phase in the center is not easy to detect, as it is often subsurface and the Raman peak is broad

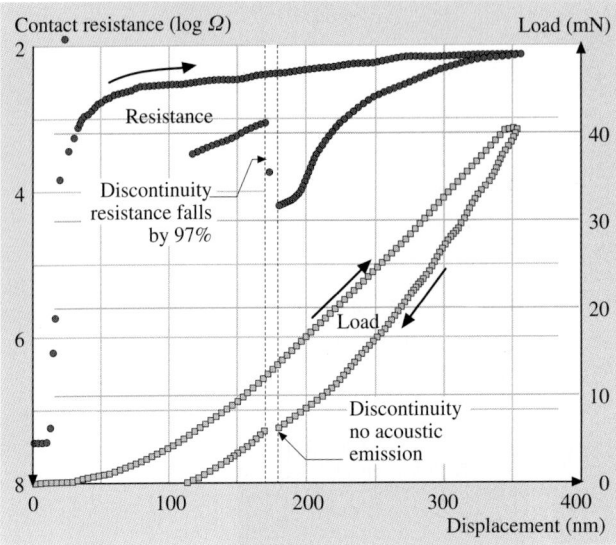

Fig. 12.27. Nanoindentation curves for deep indents on Si show a discontinuity during unloading and simultaneously a large drop in contact resistance

to the more resistive BC-8 or R-8. The counterargument is that amorphous Si is only seen with micro-Raman spectroscopy when the unloading is very rapid or there is a large nanoindentation with no unloading discontinuity. The importance of unloading rate and cracking in determining the phases present are further complications. The controversy will remain until in situ characterization of the phases present is undertaken.

12.5 Thin Films and Multilayers

In almost all real applications, surfaces are coated with thin films. These may be intentionally added such as hard carbide coatings on a tool bit, or they may simply be native films such as an oxide layer. It is also likely that there will be adsorbed films of water and organic contaminants that can range from a single molecule in thickness up to several nanometers. All of these films, whether native or intentionally placed on the surface, will affect the surface's mechanical behavior on the nanoscale. Adsorbates can have a significant impact on the surface forces [3] and, hence, the geometry and stability of asperity contacts. Oxide films can have dramatically different mechanical properties to the bulk and will also modify the surface forces. Some of the effects of native films have been detailed in the earlier sections on dislocation nucleation and adhesive contacts.

The importance of thin films in enhancing the mechanical behavior of surfaces is illustrated by the abundance of publications on thin film mechanical properties (see for instance *Nix* [88] or *Cammarata* [31] or *Was* and *Foecke* [103]). In the following sections, the mechanical properties of films intentionally deposited on the surface will be discussed.

12.5.1 Thin Films

Measuring the mechanical properties of a single thin surface film has always been difficult. Any measurement performed on the whole sample will inevitably be dominated by the bulk substrate. Nanoindentation, since it looks at the mechanical properties of a very small region close to the surface, offers a possible solution to the problem of measuring thin film mechanical properties. However, there are certain inherent problems in using nanoindentation testing to examine the properties of thin films. The problems stem in part from the presence of an interface between the film and substrate. The quality of the interface can be affected by many variables, resulting in a range of effects on the apparent elastic and plastic properties of the film. In particular, when the deformation region around the indent approaches the interface, the indentation curve may exhibit features due to the thin film, the bulk, the interface, or a combination of all three. As a direct consequence of these complications, models for thin-film behavior must attempt to take into account not only the properties of the film and substrate, but also the interface between them.

If, initially, the effect of the interface is neglected, it is possible to divide thin-coated systems into a number of categories that depend on the values of E (elastic modulus) and Y (the yield stress) of the film and substrate. These categories are typically [104, 105]:

1. coatings with high E and high Y, substrates with high E and high Y;
2. coatings with high E and high Y, substrates with high or low E and low Y;
3. coatings with high or low E and low Y, substrates with high E and high Y;
4. coatings with high or low E and low Y, substrates with high or low E and low Y.

The reasons for splitting thin film systems into these different categories have been amply demonstrated experimentally by *Whitehead* and *Page* [104, 105] and theoretically by *Fabes* et al. [106]. Essentially, hard, elastic materials (high E and Y) will possess smaller plastic zones than soft, inelastic (low E and Y) materials. Thus, when different combinations of materials are used as film and substrate, the overall plastic zone will differ significantly. In some cases, the plasticity is confined to the film, and in other cases, it is in both the film and substrate, as shown by Fig. 12.28. If the standard nanoindentation analysis routines are to be used, it is essential that the plastic zone and the elastic strain field are both confined to the film and do not reach the substrate. Clearly, this is difficult to achieve unless extremely shallow nanoindentations are used. There is an often quoted 10% rule, that says nanoindents in a film must have a depth of less than 10% of the film's thickness if only the film properties are to be measured. This has no real validity [107]. There are film/substrate combinations for which 10% is very conservative, while for other combinations even 5% may be too deep. The effect of the substrate for different combinations of film and substrate properties has been studied using FEM [108], which has shown that the

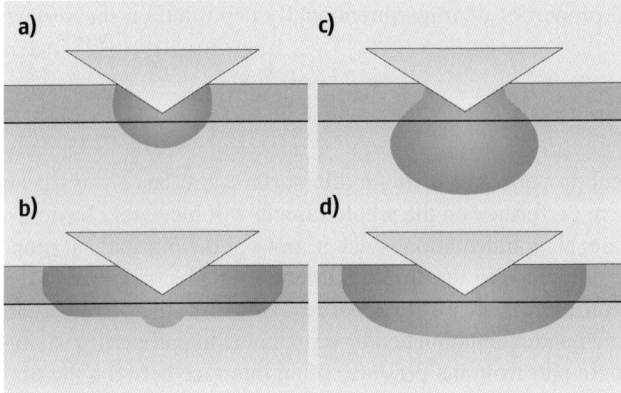

Fig. 12.28. Variations in the plastic zone for indents on films and substrates of different properties. (**a**) Film and substrate have high E and Y, (**b**) film has a high E and Y, substrate has a high or low E and low Y, (**c**) film has a high or low E and a low Y, and substrate has a high E and Y, (**d**) film has a high or low E and a low Y, and substrate has a high or low E and a low Y

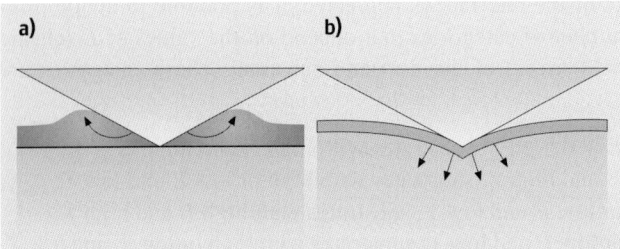

Fig. 12.29. Two different modes of deformation during nanoindentation of films. In (**a**) materials move upwards and outwards, while in (**b**) the film acts like a membrane and the substrate deforms

maximum nanoindentation depth to measure film only properties decreases in moving from soft on hard to hard on soft combinations. For a very soft film on a hard substrate, nanoindentations of 50% of the film thickness are alright, but this drops to < 10% for a hard film on a soft substrate. For a very strong film on a soft substrate, the surface film behaves like an elastic membrane or a bending plate.

Theoretical analysis of thin-film mechanical behavior is difficult. One theoretical approach that has been adopted uses the volumes of plastically deformed material in the film and substrate to predict the overall hardness of the system. However, it should be noted that this method is only really appropriate for soft coatings and indentation depths below the thickness of the coating (see cases c and d of Fig. 12.28), otherwise the behavior will be closer to that detailed later and shown by Fig. 12.29.

The technique of combining the mechanical properties of the film and substrate to evaluate the overall hardness of the system is generally referred to as the rule of mixtures. It stems from work by *Burnett* et al. [109–111] and *Sargent* [112], who derived a weighted average to relate the "composite" hardness (H) to the volumes of plastically deformed material in the film (V_f) and substrate (V_s) and their respective values of hardness, H_f and H_s. Thus,

$$H = \frac{H_f V_f + H_s V_s}{V_{total}}, \qquad (12.26)$$

where V_{total} is $V_f + V_s$.

Equation (12.26) was further developed by *Burnett* and *Page* [109] to take into account the indentation size effect. They replaced H_s with $K\delta_c^{n-2}$, where K and n are experimentally determined constants dependent on the indenter and sample, and δ_c is the contact depth. This expression is derived directly from Meyer's law for spherical indentations, which gives the relationship $P = Kd^n$ between load, P, and the indentation dimension, d. Burnett and Page also employed a further refinement to enable the theory to fit experimental results from a specific sample, ion-implanted silicon. This particular modification essentially took into account the different sizes of the plastic zones in the two materials by multiplying H_s by a dimensionless factor (V_s/V_{total}). While this seems to be a sensible approach, it is mostly empirical, and the physical justification for using this particular factor is not entirely clear. Later, *Burnett* and *Rickerby* [110, 111] took this idea further and tried to generalize the equations to take into account all of the possible scenarios. Thus, the following equations were suggested:

$$H = \frac{H_f(\Omega^3) V_f + H_s V_s}{V_{total}}, \qquad (12.27)$$

$$H = \frac{H_f V_f + H_s(\Omega^3) V_s}{V_{total}}. \qquad (12.28)$$

The first of these, (12.27), deals with the case of a soft film on a hard substrate, and the second, (12.28), with a hard film on a soft substrate. The Ω term expresses the variation of the total plastic zone from the ideal hemispherical shape. This was taken still further by *Bull* and *Rickerby* [113], who derived an approximation for Ω based on the film and substrate zone radii being related to their respective hardness and elastic modulii [114, 115]. Hence:

$$\Omega = (E_f H_s / E_s H_f)^l, \qquad (12.29)$$

where l is determined empirically. E_f and H_f and E_s and H_s are the elastic modulus and hardness of the film and substrate, respectively.

Experimental data [116] indicate that the effect of the substrate on the elastic modulus of the film can be quite

different than the effect on hardness, due to the zones of the elastic and plastic strain fields being different sizes.

Chechechin et al. [117] have recently studied the behavior of Al_2O_3 films of various thicknesses on different substrates. Their results indicate that many of the models correctly predict the transition between the properties of the film and those of the substrate, but do not always fit the observed hardness against depth curves. This group have also studied the pop-in behavior of Al_2O_3 films [118] and have attempted to model the range of loads and depths at which they occur via a Weibull-type distribution, as utilized in fracture analysis.

A point raised by Burnett and Rickerby should be emphasized. They state that there are two very distinct modes of deformation during nanoindentation testing. The first, referred to as *Tabor*'s [48] model for low Y/E materials, involves the buildup of material at the side of the indenter through movement of material at slip lines. The second, for materials with large Y/E does not result in surface pile-up. The displaced material is then accommodated by radial displacements [115]. The point is that a thin, strong, and well-bonded surface film can cause a substrate that would normally deform by Tabor's method to behave more like a material with high Y/E (see Fig. 12.29). It should be noted that this only applies as long as the film does not fail.

In recent theoretical and experimental studies the importance of material pile-up and sink-in has been investigated extensively. As discussed in an earlier section, pile-up can be increased by residual compressive stresses, but even in the absence of residual stresses pile-up can introduce a significant error in the calculated contact area. This is most pronounced in materials that do not work-harden [61]. For these materials using the Oliver and Pharr method fails to account for the pile-up and results in a large error in the values for E and H. For thin films *Tsui* et al. have used a focused ion beam to section through Knoop indentations in both soft films on hard substrates [69] and hard films on soft substrates [70]. The soft films, as expected, exhibit pile-up, while the hard film acts more like a membrane and the indentation exhibits sink-in with most of the plasticity in the substrate. Thus, there are three clearly identifiable factors affecting the pile-up and sink-in around nanoindents during testing of thin films:

1. Residual stresses
2. Degree of work-hardening
3. Ratio of film and substrate mechanical properties.

The bonding or adhesion between the film and substrate could also be added to this list. And it should not be forgotten that the depth of the nanoindentation relative to the film thickness also affects pile-up. For a very deep nanoindentation into a thin, soft film on a hard substrate pile-up is reduced, due to the combined constraints on the film of the tip and substrate [119]. Due to all of these complications, using nanoindentation to study thin film mechanical properties is fraught with danger. Many unprepared researchers have misguidedly taken the values of E and H obtained during nanoindentation testing to be absolute values only to find out later that the values contain significant errors.

Many of the problems associated with nanoindentation testing are related to incorrectly calculating the contact area, A. The *Joslin* and *Oliver* method [71] is one

Table 12.1. Results for some experimental studies of multilayer hardness

Study	Multilayer	Maximum hardness and multilayer repeat length	Reference hardness value	Range of hardness values for multilayers
Isostructural Knoop hardness [124]	Cu/Ni	524 at 11.6 nm	284 (interdiffused)	295–524
Non-isostructural Nanoindentation [125]	Mo/NbN	33 GPa at 2 nm	NbN – 17 GPa Mo – 2.7 GPa Wo – 7 GPa	12–33 GPa
	W/NbN	29 GPa at 3 nm	(individual layer materials)	23–29 GPa

way that A can be removed from the calculations. This approach has been used with some success to look at strained epitaxial II/VI semiconductor films [120], but there is evidence that the lattice mismatch in these films can cause dramatic changes in the mechanical properties of the films [121]. This may be due to image forces and the film/substrate interface acting as a barrier to dislocation motion. Recently, it has been shown that using films and substrates with known matching elastic moduli, it is possible to use the assumption of constant elastic modulus with depth to evaluate H [122]. In effect, this is using (12.1) to evaluate A from the contact stiffness data, and then substituting the value for A into (12.12). The value of E is measured independently, for instance, using acoustic techniques.

12.5.2 Multilayers

Multilayered materials with individual layers that are a micron or less in thickness, sometimes referred to as superlattices, can exhibit substantial enhancements in hardness or strength. This should be distinguished from the super modulus effect discussed earlier, which has been shown to be largely an artifact. The enhancements in hardness can be as much as 100% when compared to the value expected from the rule of mixtures, which is essentially a weighted average of the hardness for the constituents of the two layers [123]. Table 12.1 shows how the properties of isostructural multilayers can show a substantial increase in hardness over that for fully interdiffused layers. The table also shows how there can be a substantial enhancement in hardness for non-isostructural multilayers compared to the values for the same materials when they are homogeneous.

There are many factors that contribute to enhanced hardness in multilayers. These can be summarized as [103]:

1. Hall-Petch behavior
2. Orowan strengthening
3. Image effects
4. Coherency and thermal stresses
5. Composition modulation

Hall-Petch behavior is related to dislocations piling-up at grain boundaries. (Note that pile-up is used to describe two distinct effects: One is material building up at the side of an indentation, the other is an accumulation of dislocations on a slip-plane.) The dislocation pile-up at grain boundaries impedes the motion of dislocations. For materials with a fine grain structure there are many grain boundaries, and, hence, dislocations find it hard to move. In polycrystalline multilayers, it is often the case that the size of the grains within a layer scales with the layer thickness so that reducing the layer thickness reduces the grain size. Thus, the Hall-Petch relationship (below) should be applicable to polycrystalline multilayer films with the grain size, d_g, replaced by the layer thickness.

$$Y = Y_0 + k_{HP} d_g^{-0.5}, \qquad (12.30)$$

where Y is the enhanced yield stress, Y_0 is the yield stress for a single crystal, and k_{HP} is a constant.

There is an ongoing argument about whether Hall-Petch behavior really takes place in nanostructured multilayers. The basic model assumes many dislocations are present in the pile-up, but such large dislocation pile-ups are not seen in small grains [126] and are unlikely to be present in multilayers. As a direct consequence, studies have found a range of values, between 0 and -1, for the exponent in (12.30), rather than the -0.5 predicted for Hall-Petch behavior.

Orowan strengthening is due to dislocations in layered materials being effectively pinned at the interfaces. As a result, the dislocations are forced to bow out along the layers. In narrow films, dislocations are pinned at both the top and bottom interfaces of a layer and bow out parallel to the plane of the interface [127, 128]. Forcing a dislocation to bow out in a layered material requires an increase in the applied shear stress beyond that required to bow out a dislocation in a homogeneous sample. This additional shear stress would be expected to increase as the film thickness is reduced.

Image effects were suggested by *Koehler* [129] as a possible source of enhanced yield stress in multilayered materials. If two metals, A and B, are used to make a laminate and one of them, A, has a high dislocation line energy, but the other, B, has a low dislocation line energy, then there will be an increased resistance to dislocation motion due to image forces. However, if the individual layers are thick enough that there may be a dislocation source present within the layer, then dislocations could pile-up at the interface. This will create a local stress concentration point and the enhancement to the strength will be very limited. If the layers are thin enough that there will be no dislocation source present, the enhanced mechanical strength may be substantial. In Koehler's model only nearest neighbor layers were taken to contribute to the image forces. However, this was extended to include more layers [130] without substantial changes in the results. The consequence on image effects of reducing the thickness of the individual layers in a multilayer is that it prevents dislocation sources from being active within the layer.

For many multilayer systems there is an increase in strength as the bilayer repeat length is reduced, but there is often a critical repeat length (e.g., 3 nm for the W/NbN multilayer of Table 12.1) below which the strength falls. One explanation for the fall in strength involves the effects of coherency and thermal stresses on dislocation energy. Unlike image effects where the energy of dislocations are a maximum or minimum in the center of layers, the energy maxima and minima are at the interfaces for coherency stresses. Combining the effects of varying moduli and coherency stresses shows that the dependence of strength on layer thickness has a peak near the repeat period where coherency strains begin decreasing [131].

Another source of deviations in behavior at very small repeat periods is the imperfect nature of interfaces. With the exception of atomically perfect epitaxial films, interfaces are generally not atomically flat and there is some interdiffusion. For the Cu/Ni film of Table 12.1, the effects of interdiffusion on hardness were examined [124] by annealing the multilayers. The results were in agreement with a model by *Krzanowski* [132] that predicted the variations in hardness would be proportional to the amplitude of the composition modulation.

It is interesting to note that the explanations for enhanced mechanical properties in multilayered materials are all based on dislocation mechanisms. So it would seem natural to assume that multilayered materials that do not contain dislocations will show no enhanced hardness over their rule of mixtures values. This has been verified by studies on amorphous metal multilayers [133], which shows that the hardness of the multilayers, firstly, lies between that of the two individual materials and, secondly, has almost no variation with repeat period.

12.6 Developing Areas

Over the past 20 to 30 years, the driving force for studying nanomechanical behavior of surfaces and thin films has been largely, though not exclusively, the microelectronics industry. The importance of electronics to the modern world is only likely to grow in the foreseeable future, but other technological areas may overtake microelectronics as the driving force for research, including the broad fields of biomaterials and nanotechnology. In several places in this chapter, a number of developing areas have been mentioned. These include nanoscale measurements of viscoelasticity and the study of environmental effects (temperature and surface chemistry) on nanomechanical properties. Both of these topics will be vital in the study of biological systems and, as a result, will be increasingly important from a research point of view.

We still have a relatively rudimentary understanding of the nanomechanics of complex biological systems such as bone cells (osteoblasts and osteoclasts) and skin cells (fibroblasts), or even, for that matter, simpler biological structures such as dental enamel. For example, Fig. 12.30 shows how the mechanical properties of dental enamel can vary within a single tooth [134]. But this is still a relatively large-scale measure of mechanical behavior. The prismatic structure of enamel means that there are variations in mechanical properties on a range of scales from millimeters down to nanometers.

In terms of data analysis, there remains much to be done. If an analysis method can be developed that deals with the problems of pile-up and sink-in, the utility of nanoindentation testing will be greatly enhanced.

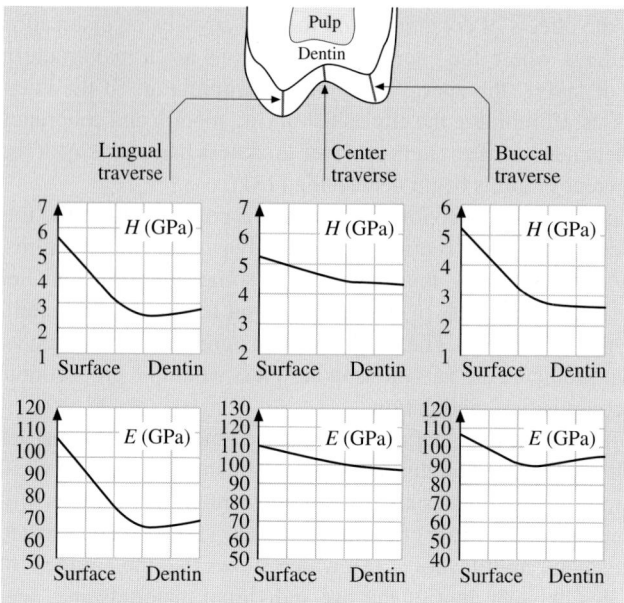

Fig. 12.30. Variations in E and H across human dental enamel. The sample is an upper 2nd molar cut in cross section from the lingual to the buccal side. Nanoindentations are performed across the surface to examine how the mechanical properties vary

References

1. G. Binnig, C. F. Quate, and C. Gerber. Atomic force microscope. *Phys. Rev. Lett.*, 56:930–933, 1986.
2. P. J. Blau and B. R. Lawn, editors. *Microindentation Techniques in Materials Science and Engineering*. ASTM, 1986.
3. J. N. Israelachvili. *Intermolecular and Surface Forces*. Academic, 1992.
4. R. S. Bradley. The cohesive force between solid surfaces and the surface energy of solids. *Philos. Mag.*, 13:853–862, 1932.
5. B. V. Derjaguin, V. M. Muller, and Yu. P. Toporov. Effect of contact deformations on the adhesion of particles. *J. Coll. Interface Sci.*, 53:314–326, 1975.
6. V. M. Muller, V. S. Yuschenko, and B. V. Derjaguin. On the influence of molecular forces on the deformation of an elastic sphere and its sticking to a rigid plane. *J. Coll. Interface Sci.*, 77:91–101, 1980.
7. V. M. Muller, B. V. Derjaguin, and Yu. P. Toporov. On two methods of calculation of the force of sticking of an elastic sphere to a rigid plane. *Coll. Surf.*, 7:251–259, 1983.
8. K. L. Johnson, K. Kendal, and A. D. Roberts. Surface energy and the contact of elastic solids. *Proc. R. Soc. A*, 324:301–320, 1971.
9. A. P. Ternovskii, V. P. Alekhin, M. Kh. Shorshorov, M. M. Khrushchov, and V. N. Skvortsov. *Zavod Lab.*, 39:1242, 1973.
10. S. I. Bulychev, V. P. Alekhin, M. Kh. Shorshorov, A. P. Ternovskii, and G. D. Shnyrev. Determining young's modulus from the indentor penetration diagram. *Zavod Lab.*, 41:1137, 1975.

11. S. I. Bulychev, V. P. Alekhin, M. Kh. Shorshorov, and A. P. Ternovskii. Mechanical properties of materials studied from kinetic diagrams of load versus depth of impression during microimpression. *Prob. Prochn.*, 9:79, 1976.
12. S. I. Bulychev and V. P. Alekhin. *Zavod Lab.*, 53:76, 1987.
13. M. Kh. Shorshorov, S. I. Bulychev, and V. P. Alekhin. *Sov. Phys. Doklady*, 26:769, 1982.
14. J. B. Pethica. *Microhardness tests with penetration depths less than ion implanted layer thickness*, page 147. Pergamon, 1982.
15. D. Newey, M. A. Wilkens, and H. M. Pollock. An ultra-low-load penetration hardness tester. *J. Phys. E: Sci. Instrum.*, 15:119, 1982.
16. D. Kendall and D. Tabor. An ultrasonic study of the area of contact between stationary and sliding surfaces. *Proc. R. Soc. A*, 323:321–340, 1971.
17. G. M. Pharr, W. C. Oliver, and F. R. Brotzen. On the generality of the relationship among contact stiffness, contact area and elastic-modulus during indentation. *J. Mater. Res.*, 7:613, 1992.
18. T. P. Weihs, C. W. Lawrence, B. Derby, C. B. Scruby, and J. B. Pethica. Acoustic emissions during indentation tests. *MRS Symp. Proc.*, 239:361–366, 1992.
19. D. F. Bahr, J. W. Hoehn, N. R. Moody, and W. W. Gerberich. Adhesion and acoustic emission analysis of failures in nitride films with 14 metal interlayer. *Acta Mater.*, 45:5163, 1997.
20. D. F. Bahr and W. W. Gerberich. Relationships between acoustic emission signals and physical phemomena during indentation. *J. Mat. Res.*, 13:1065, 1998.
21. A. B. Mann, D. van Heerden, J. B. Pethica, and T. P. Weihs. Size-dependent phase transformations during point-loading of silicon. *J. Mater. Res.*, 15:1754, 2000.
22. A. B. Mann, D. van Heerden, J. B. Pethica, P. Bowes, and T. P. Weihs. Contact resistance and phase transformations during nanoindentation of silicon. *Philos. Mag. A*, 82:1921, 2002.
23. S. Jeffery, C. J. Sofield, and J. B. Pethica. The influence of mechanical stress on the dielectric breakdown field strength of SiO_2 films. *Appl. Phys. Lett.*, 73:172, 1998.
24. B. N. Lucas and W. C. Oliver. Indentation power-law creep of high-purity indium. *Metall. Trans. A*, 30:601, 1999.
25. S. A. Syed Asif. Time dependent micro deformation of materials, 1997.
26. S. A. Syed Asif, R. J. Colton, and K. J. Wahl. *Nanoscale Surface Mechanical Property Measurements: Force Modulation Techniques Applied to Nanoindentation*. ACS Books, 2000.
27. A. B. Mann and J. B. Pethica. Nanoindentation studies in a liquid environment. *Langmuir*, 12:4583, 1996.
28. J. W. Beams. *Mechanical properties of thin films of gold and silver*, pages 183–192. Wiley, 1959.
29. J. J. Vlassak and W. D. Nix. A new bulge test technique for the determination of youngs modulus and poissons ratio of thin-films. *J. Mater. Res.*, 7:3242, 1992.
30. W. M. C. Yang, T. Tsakalakos, and J. E. Hilliard. Enhanced elastic modulus in composition modulated gold-nickel and copper-palladium foils. *J. Appl. Phys.*, 48:876, 1977.
31. R. C. Cammarata. Mechanical properties of nanocomposite thin-films. *Thin Solid Films*, 240:82, 1994.
32. G. A. D. Briggs. *Acoustic Microscopy*. Clarendon, 1992.
33. J. Kushibiki and N. Chubachi. Material characterization by line-focus-beam acoustic microscope. *IEEE Trans. Sonics Ultrasonics*, 32:189–212, 1985.
34. M. J. Bamber, K. E. Cooke, A. B. Mann, and B. Derby. Accurate determination of young's modulus and poisson's ratio of thin films by a combination of acoustic microscopy and nanoindentation. *Thin Solid Films*, 398:299–305, 2001.

35. S. E. Bobbin, R. C. Cammarata, and J. W. Wagner. Determination of the flexural modulus of thin-films from measurement of the 1st arrival of the symmetrical lamb wave. *Appl. Phys. Lett.*, 59:1544–1546, 1991.
36. R. B. King and C. M. Fortunko. Determination of inplane residual-stress state in plates using horizontally polarized shear waves. *J. Appl. Phys.*, 54:3027–3035, 1983.
37. R. F. Cook and G. M. Pharr. Direct observation and analysis of indentation cracking in glasses and ceramics. *J. Am. Ceram. Soc.*, 73:787–817, 1990.
38. T. F. Page, W. C. Oliver, and C. J. McHargue. The deformation-behavior of ceramic crystals subjected to very low load (nano)indentations. *J. Mater. Res.*, 7:450–473, 1992.
39. G. M. Pharr, W. C. Oliver, and D. S. Harding. New evidence for a pressure-induced phase-transformation during the indentation of silicon. *J. Mater. Res.*, 6:1129–1130, 1991.
40. C. F. Robertson and M. C. Fivel. The study of submicron indent-induced plastic deformation. *J. Mater. Res.*, 14:2251–2258, 1999.
41. J. E. Bradby, J. S. Williams, J. Wong-Leung, M. V. Swain, and P. Munroe. Transmission electron microscopy observation of deformation microstructure under spherical indentation in silicon. *Appl. Phys. Lett.*, 77:3749–3751, 2000.
42. Y. G. Gogotsi, V. Domnich, S. N. Dub, A. Kailer, and K. G. Nickel. Cyclic nanoindentation and raman microspectroscopy study of phase transformations in semiconductors. *J. Mater. Res.*, 15:871–879, 2000.
43. H. Hertz. Über die berührung fester elastischer körper. *J. reine angew. Math.*, 92:156–171, 1882.
44. J. Boussinesq. *Application des potentiels à l'étude de l'équilibre et du mouvement des solides élastiques.* Blanchard, 1885.
45. A. E. H. Love. The stress produced in a semi-infinite solid by pressure on part of the boundary. *Philos. Trans. R. Soc.*, 228:377–420, 1929.
46. A. E. H. Love. Boussinesq's problem for a rigid cone. *Quarter. J. Math.*, 10:161, 1939.
47. I. N. Sneddon. The relationship between load and penetration in the axisymmetric boussinesq problem for a punch of arbitrary profile. *Int. J. Eng. Sci.*, 3:47–57, 1965.
48. D. Tabor. *Hardness of Metals.* Oxford Univ. Press, 1951.
49. R. von Mises. Mechanik der festen körper in plastisch deformablen zustand. *Goettinger Nachr. Math.-Phys.*, K1:582–592, 1913.
50. H. Tresca. Sur l'ecoulement des corps solids soumis s fortes pression. *Compt. Rend.*, 59:754, 1864.
51. A. B. Mann and J. B. Pethica. The role of atomic-size asperities in the mechanical deformation of nanocontacts. *Appl. Phys. Lett.*, 69:907–909, 1996.
52. A. B. Mann and J. B. Pethica. The effect of tip momentum on the contact stiffness and yielding during nanoindentation testing. *Philos. Mag. A*, 79:577–592, 1999.
53. S. P. Jarvis. Atomic force microscopy and tip-surface interactions, 1993.
54. J. B. Pethica and D. Tabor. Contact of characterised metal surfaces at very low loads: Deformation and adhesion. *Surf. Sci.*, 89:182, 1979.
55. J. S. Field and M. V. Swain. Determining the mechanical-properties of small volumes of materials from submicrometer spherical indentations. *J. Mater. Res.*, 10:101–112, 1995.
56. W. C. Oliver and G. M. Pharr. An improved technique for determining hardness and elastic-modulus using load and displacement sensing indentation experiments. *J. Mater. Res.*, 7:1564–1583, 1992.
57. S. V. Hainsworth, H. W. Chandler, and T. F. Page. Analysis of nanoindentation load-displacement loading curves. *J. Mater. Res.*, 11:1987–1995, 1996.
58. J. L. Loubet, J. M. Georges, O. Marchesini, and G. Meille. Vickers indentation curves of magnesium oxide (MgO). *Mech. Eng.*, 105:91–92, 1983.

59. J. L. Loubet, J. M. Georges, O. Marchesini, and G. Meille. Vickers indentation curves of magnesium oxide (mgo). *J. Tribol. Trans. ASME*, 106:43–48, 1984.
60. M. F. Doerner and W. D. Nix. A method for interpreting the data from depth sensing indentation experiments. *J. Mater. Res.*, 1:601–609, 1986.
61. A. Bolshakov and G. M. Pharr. Influences of pileup on the measurement of mechanical properties by load and depth sensing instruments. *J. Mater. Res.*, 13:1049–1058, 1998.
62. A. Bolshakov, W. C. Oliver, and G. M. Pharr. Influences of stress on the measurement of mechanical properties using nanoindentation. 2. finite element simulations. *J. Mater. Res.*, 11:760–768, 1996.
63. J. C. Hay, A. Bolshakov, and G. M. Pharr. A critical examination of the fundamental relations used in the analysis of nanoindentation data. *J. Mater. Res.*, 14:2296–2305, 1999.
64. G. M. Pharr, T. Y. Tsui, A. Bolshakov, and W. C. Oliver. Effects of residual-stress on the measurement of hardness and elastic-modulus using nanoindentation. *MRS Symp. Proc.*, 338:127–134, 1994.
65. T. R. Simes, S. G. Mellor, and D. A. Hills. A note on the influence of residual-stress on measured hardness. *J. Strain Anal. Eng. Des.*, 19:135–137, 1984.
66. W. R. Lafontaine, B. Yost, and C. Y. Li. Effect of residual-stress and adhesion on the hardness of copper-films deposited on silicon. *J. Mater. Res.*, 5:776–783, 1990.
67. W. R. Lafontaine, C. A. Paszkiet, M. A. Korhonen, and C. Y. Li. Residual stress measurements of thin aluminum metallizations by continuous indentation and x-ray stress measurement techniques. *J. Mater. Res.*, 6:2084–2090, 1991.
68. T. Y. Tsui, W. C. Oliver, and G. M. Pharr. Influences of stress on the measurement of mechanical properties using nanoindentation. 1. experimental studies in an aluminum alloy. *J. Mater. Res.*, 11:752–759, 1996.
69. T. Y. Tsui, J. Vlassak, and W. D. Nix. Indentation plastic displacement field: Part i. the case of soft films on hard substrates. *J. Mater. Res.*, 14:2196–2203, 1999.
70. T. Y. Tsui, J. Vlassak, and W. D. Nix. Indentation plastic displacement field: Part ii. the case of hard films on soft substrates. *J. Mater. Res.*, 14:2204–2209, 1999.
71. D. L. Joslin and W. C. Oliver. A new method for analyzing data from continuous depth-sensing microindentation tests. *J. Mater. Res.*, 5:123–126, 1990.
72. M. R. McGurk and T. F. Page. Using the p-delta(2) analysis to deconvolute the nanoindentation response of hard-coated systems. *J. Mater. Res.*, 14:2283–2295, 1999.
73. Y. T. Cheng and C. M. Cheng. Relationships between hardness, elastic modulus, and the work of indentation. *Appl. Phys. Lett.*, 73:614–616, 1998.
74. J. B. Pethica and W. C. Oliver. Mechanical properties of nanometer volumes of material: Use of the elastic response of small area indentations. *MRS Symp. Proc.*, 130:13–23, 1989.
75. W. C. Oliver and J. B. Pethica. Method for continuous determination of the elastic stiffness of contact between two bodies, united states patent number 4,848,141. 1989.
76. S. A. S. Asif, K. J. Wahl, and R. J. Colton. Nanoindentation and contact stiffness measurement using force modulation with a capacitive load-displacement transducer. *Rev. Sci. Instrum.*, 70:2408–2413, 1999.
77. J. L. Loubet, W. C. Oliver, and B. N. Lucas. Measurement of the loss tangent of low-density polyethylene with a nanoindentation technique. *J. Mater. Res.*, 15:1195–1198, 2000.
78. W. W. Gerberich, J. C. Nelson, E. T. Lilleodden, P. Anderson, and J. T. Wyrobek. Indentation induced dislocation nucleation: The initial yield point. *Acta Mater.*, 44:3585–3598, 1996.

79. J. D. Kiely and J. E. Houston. Nanomechanical properties of Au(111), (001), and (110) surfaces. *Phys. Rev. B*, 57:12588–12594, 1998.
80. D. F. Bahr, D. E. Wilson, and D. A. Crowson. Energy considerations regarding yield points during indentation. *J. Mater. Res.*, 14:2269–2275, 1999.
81. D. E. Kramer, K. B. Yoder, and W. W. Gerberich. Surface constrained plasticity: Oxide rupture and the yield point process. *Philos. Mag. A*, 81:2033–2058, 2001.
82. S. G. Corcoran, R. J. Colton, E. T. Lilleodden, and W. W. Gerberich. Anomalous plastic deformation at surfaces: Nanoindentation of gold single crystals. *Phys. Rev. B*, 55:16057–16060, 1997.
83. E. B. Tadmor, R. Miller, R. Phillips, and M. Ortiz. Nanoindentation and incipient plasticity. *J. Mater. Res.*, 14:2233–2250, 1999.
84. J. A. Zimmerman, C. L. Kelchner, P. A. Klein, J. C. Hamilton, and S. M. Foiles. Surface step effects on nanoindentation. *Phys. Rev. Lett.*, 87:article 165507 (1–4), 2001.
85. J. D. Kiely, R. Q. Hwang, and J. E. Houston. Effect of surface steps on the plastic threshold in nanoindentation. *Phys. Rev. Lett.*, 81:4424–4427, 1998.
86. A. B. Mann, P. C. Searson, J. B. Pethica, and T. P. Weihs. The relationship between near-surface mechanical properties, loading rate and surface chemistry. *Mater. Res. Soc. Symp. Proc.*, 505:307–318, 1998.
87. R. C. Thomas, J. E. Houston, T. A. Michalske, and R. M. Crooks. The mechanical response of gold substrates passivated by self-assembling monolayer films. *Science*, 259:1883–1885, 1993.
88. W. D. Nix. Elastic and plastic properties of thin films on substrates: Nanoindentation techniques. *Mater. Sci. Eng. A*, 234:37–44, 1997.
89. J. J. Vlassak and W. D. Nix. Indentation modulus of elastically anisotropic half-spaces. *Philos. Mag. A*, 67:1045–1056, 1993.
90. J. J. Vlassak and W. D. Nix. Measuring the elastic properties of anisotropic materials by means of indentation experiments. *J. Mech. Phys. Solids*, 42:1223–1245, 1994.
91. B. R. Lawn. *Fracture of Brittle Solids*. Cambridge Univ. Press, 1993.
92. G. M. Pharr. Measurement of mechanical properties by ultra-low load indentation. *Mater. Sci. Eng. A*, 253:151–159, 1998.
93. D. B. Marshall and A. G. Evans. Measurement of adherence of residually stressed thin-films by indentation. 1. mechanics of interface delamination. *J. Appl. Phys.*, 56:2632–2638, 1984.
94. C. Rossington, A. G. Evans, D. B. Marshall, and B. T. Khuriyakub. Measurement of adherence of residually stressed thin-films by indentation. 2. experiments with ZnO/Si. *J. Appl. Phys.*, 56:2639–2644, 1984.
95. M. D. Kriese, W. W. Gerberich, and N. R. Moody. Quantitative adhesion measures of multilayer films: Part i. indentation mechanics. *J. Mater. Res.*, 14:3007–3018, 1999.
96. M. D. Kriese, W. W. Gerberich, and N. R. Moody. Quantitative adhesion measures of multilayer films: Part ii. indentation of W/Cu, W/W, Cr/W. *J. Mater. Res.*, 14:3019–3026, 1999.
97. M. Li, C. B. Carter, M. A. Hillmyer, and W. W. Gerberich. Adhesion of polymer-inorganic interfaces by nanoindentation. *J. Mater. Res.*, 16:3378–3388, 2001.
98. D. R. Clarke, M. C. Kroll, P. D. Kirchner, R. F. Cook, and B. J. Hockey. Amorphization and conductivity of silicon and germanium induced by indentation. *Phys. Rev. Lett.*, 60:2156–2159, 1988.
99. J. J. Gilman. Insulator-metal transitions at microindentation. *J. Mater. Res.*, 7:535–538, 1992.

100. G. M. Pharr, W. C. Oliver, R. F. Cook, P. D. Kirchner, M. C. Kroll, T. R. Dinger, and D. R. Clarke. Electrical-resistance of metallic contacts on silicon and germanium during indentation. *J. Mater. Res.*, 7:961–972, 1992.
101. A. Kailer, Y. G. Gogotsi, and K. G. Nickel. Phase transformations of silicon caused by contact loading. *J. Appl. Phys.*, 81:3057–3063, 1997.
102. J. E. Bradby, J. S. Williams, J. Wong-Leung, M. V. Swain, and P. Munroe. Mechanical deformation in silicon by micro-indentation. *J. Mater. Res.*, 16:1500–1507, 2000.
103. G. S. Was and T. Foecke. Deformation and fracture in microlaminates. *Thin Solid Films*, 286:1–31, 1996.
104. A. J. Whitehead and T. F. Page. Nanoindentation studies of thin-film coated systems. *Thin Solid Films*, 220:277–283, 1992.
105. A. J. Whitehead and T. F. Page. Nanoindentation studies of thin-coated systems. *NATO ASI Ser. E*, 233:481–488, 1993.
106. B. D. Fabes, W. C. Oliver, R. A. McKee, and F. J. Walker. The determination of film hardness from the composite response of film and substrate to nanometer scale indentations. *J. Mater. Res.*, 7:3056–3064, 1992.
107. T. F. Page and S. V. Hainsworth. Using nanoindentation techniques for the characterization of coated systems – a critique. *Surface Coat. Technol.*, 61:201–208, 1993.
108. X. Chen and J. J. Vlassak. Numerical study on the measurement of thin film mechanical properties by means of nanoindentation. *J. Mater. Res.*, 16:2974–2982, 2001.
109. P. J. Burnett and T. F. Page. Surface softening in silicon by ion-implantation. *J. Mater. Sci.*, 19:845–860, 1984.
110. P. J. Burnett and D. S. Rickerby. The mechanical-properties of wear resistant coatings. 1. modeling of hardness behavior. *Thin Solid Films*, 148:41–50, 1987.
111. P. J. Burnett and D. S. Rickerby. The mechanical-properties of wear resistant coatings. 2. experimental studies and interpretation of hardness. *Thin Solid Films*, 148:51–65, 1987.
112. P. M. Sargent. A better way to present results from a least-squares fit to experimental-data – an example from microhardness testing. *J. Test. Eval.*, 14:122–127, 1986.
113. S. J. Bull and D. S. Rickerby. Evaluation of coatings. *Brit. Ceram. Trans. J.*, 88:177–183, 1989.
114. B. R. Lawn, A. G. Evans, and D. B. Marshall. Elastic/plastic indentation damage in ceramics: The median/radial crack system. *J. Am. Ceram. Soc.*, 63:574–581, 1980.
115. R. Hill. *The Mathematical Theory of Plasticity*. Clarendon, 1950.
116. W. C. Oliver, C. J. McHargue, and S. J. Zinkle. Thin-film characterization using a mechanical-properties microprobe. *Thin Solid Films*, 153:185–196, 1987.
117. N. G. Chechechin, J. Bottiger, and J. P. Krog. Nanoindentation of amorphous aluminum oxide films. 1. influence of the substrate on the plastic properties. *Thin Solid Films*, 261:219–227, 1995.
118. N. G. Chechechin, J. Bottiger, and J. P. Krog. Nanoindentation of amorphous aluminum oxide films. 2. critical parameters for the breakthrough and a membrane effect in thin hard films on soft substrates. *Thin Solid Films*, 261:228–235, 1995.
119. D. E. Kramer, A. A. Volinsky, N. R. Moody, and W. W. Gerberich. Substrate effects on indentation plastic zone development in thin soft films. *J. Mater. Res.*, 16:3150–3157, 2001.
120. A. B. Mann. Nanomechanical measurements: Surface and environmental effects, 1995.
121. A. B. Mann, J. B. Pethica, W. D. Nix, and S. Tomiya. Nanoindentation of epitaxial films: A study of pop-in events. *Mater. Res. Soc. Symp. Proc.*, 356:271–276, 1995.
122. R. Saha and W. D. Nix. Effects of the substrate on the determination of thin film mechanical properties by nanoindentation. *Acta Mater.*, 50:23–38, 2002.

123. S. A. Barnett. *Deposition and mechanical properties of superlattice thin films.* Academic, 1993.
124. R. R. Oberle and R. C. Cammarata. Dependence of hardness on modulation amplitude in electrodeposited cu-ni compositionally modulated thin-films. *Scripta Metall.*, 32:583–588, 1995.
125. A. Madan, Y. Y. Wang, S. A. Barnett, C. Engstrom, H. Ljungcrantz, L. Hultman, and M. Grimsditch. Enhanced mechanical hardness in epitaxial nonisostructural Mo/NbN and W/NbN superlattices. *J. Appl. Phys.*, 84:776–785, 1998.
126. R. Venkatraman and J. C. Bravman. Separation of film thickness and grain-boundary strengthening effects in Al thin-films on Si. *J. Mater. Res.*, 7:2040–2048, 1992.
127. J. D. Embury and J. P. Hirth. On dislocation storage and the mechanical response of fine-scale microstructures. *Acta Mater.*, 42:2051–2056, 1994.
128. D. J. Srolovitz, S. M. Yalisove, and J. C. Bilello. Design of multiscalar metallic multilayer composites for high-strength, high toughness, and low cte mismatch. *Metall. Trans. A*, 26:1805–1813, 1995.
129. J. S. Koehler. Attempt to design a strong solid. *Phys. Rev. B*, 2:547–551, 1970.
130. S. V. Kamat, J. P. Hirth, and B. Carnahan. Image forces on screw dislocations in multilayer structures. *Scripta Metall.*, 21:1587–1592, 1987.
131. M. Shinn, L. Hultman, and S. A. Barnett. Growth, structure, and microhardness of epitaxial TiN/NbN superlattices. *J. Mater. Res.*, 7:901–911, 1992.
132. J. E. Krzanowski. The effect of composition profile on the strength of metallic multilayer structures. *Scripta Metall.*, 25:1465–1470, 1991.
133. J. B. Vella, R. C. Cammarata, T. P. Weihs, C. L. Chien, A. B. Mann, and H. Kung. Nanoindentation study of amorphous metal multilayered thin films. *MRS Symp. Proc.*, 594:25–29, 2000.
134. J. L. Cuy, A. B. Mann, K. J. Livi, M. F. Teaford, and T. P. Weihs. Nanoindentation mapping of the mechanical properties of human molar tooth enamel. *Arch. Oral Biol.*, 47:281–291, 2002.

13

Computational Modeling of Nanometer-Scale Tribology

Seong-Jun Heo, Susan B. Sinnott, Donald W. Brenner, and Judith A. Harrison

Summary. Friction and wear have long been acknowledged as limiting factors to numerous applications and many areas of technology, which has lead to significant interest in understanding and controlling these processes. Current interest in microscale and nanoscale machines with moving parts add to this interest, especially as the mechanisms that lead to friction at the atomic-scale can sometimes be quite distinct from the mechanisms that dominate at the macroscale.

This chapter presents a review of the applications of computational modeling methods to atomic-scale and nanometer-scale tribology. It includes a discussion of computational modeling methods frequently employed in these studies, with some analysis of the conditions under which these methods are best applied. This is followed by a review of the findings of computational studies of nanometer-scale indentation, friction, and lubrication.

In this chapter, a relatively complete discussion of the contribution that molecular dynamics and related simulations are making in the area of nanotribology is presented. The examples discussed above make it clear that these approaches are providing exciting insights into friction, wear, and related processes at the atomic scale that could not have been obtained in any other way. Furthermore, the synergy between these simulations and new and experimental techniques such as the surface force apparatus and proximal probe microscopes is producing a revolution in our understanding of the origin of friction at its most fundamental atomic level.

13.1 Introduction

Friction and wear have long been recognized as limiting factors to numerous applications and many areas of technology. Thus, there has been significant interest in understanding and controlling these processes. Some historical examples include the ancient Egyptians, who invented new technologies to move the stones used to build the pyramids [1]; Coulomb, whose studies of friction were motivated by the need to move ships easily and without wear from land to the water [1]; and *Johnson* et al. [2], whose study of automobile windshield wipers led to a better understanding of contact mechanics and surface energies. Today, research and development is focused on microscale and nanoscale machines with moving parts, which continue to challenge our fundamental understanding of friction and wear. This has motivated researchers to study friction on ever-smaller scales to determine its fundamental cause.

The study of atomic-scale friction has also paved the way for new innovations, such as the development of self-lubricating surfaces and wear-resistant materials. As will be shown over the course of this article, the mechanisms that lead to friction at the atomic scale can sometimes be quite distinct from the mechanisms that dominate at the macroscale. This has implications for nanometer-regime devices such as magnetic storage disks, which have been shrinking in size steadily over the last few years, and microelectromechanical systems (MEMS) where atomic-scale friction, adhesion and wear are the dominant processes.

The scientific study of atomic-scale friction has been on the rise since the late 1980's [3–14]. This is due to the simultaneous development of sophisticated new experimental tools to measure friction over nanometer-scale distances at low loads, the rapid increase in computer power and processing speed, and the maturation of theoretical methods needed to study material processes in a realistic manner. For example, the surface force apparatus (SFA) has provided new information related to friction and lubrication for many liquid and solid systems with unprecedented resolution [15]. The friction-force and atomic-force microscopes (FFM and AFM) allow the frictional properties of solids to be characterized with atomic-scale resolution under single asperity contact conditions [16–19]. Other methods, such as the quartz crystal microbalance (QCM), also can be used to provide insight into the origin of friction [3, 20]. With the experimental apparatus currently available, it is possible to study sliding surfaces at the atomic-scale, and ultimately relate observed behavior to macroscopically-observed phenomena.

Theoretical models and simulations assist in the interpretation of experimental data and provide predictions of phenomena that the experiments can subsequently confirm or refute. These include analytic models and large-scale molecular dynamics (MD) simulations, among others [21]. Analytic models have long been used to study friction. These include the early studies by *Tomlinson* [22] and *Frenkel* and *Kontorova* [23] and more recent studies by *Mate* et al. [17], *Sokoloff* [12, 24–26], and others [27–29]. The strength of these idealized models is that they can be used to divide the complex motions that create friction into basic components defined by quantities such as spring constants, the curvature and magnitude of potential wells, and bulk phonon frequencies. The main weakness of these approaches is that simplifying assumptions must be made to apply these models to study friction that may lead to incorrect results.

Molecular dynamics computer simulations represent a compromise between analytic models and experimental conditions. For instance, MD simulations rely on approximate interatomic forces and classical dynamics, which is similar to what is needed for analytic models. In addition, simulations can reveal unanticipated phenomena that require further exploration, which is similar to what occurs in experiments. Moreover, a poor choice of simulation conditions can result in meaningless result, which is also the case in experimental studies. Thus, a thorough understanding of the strengths and weaknesses of MD simulations is crucial to both successfully implementing this approach and understanding its results.

Atomistic computer simulations appear, on the surface, to be rather straightforward to carry out: given a set of initial conditions and a way of describing in-

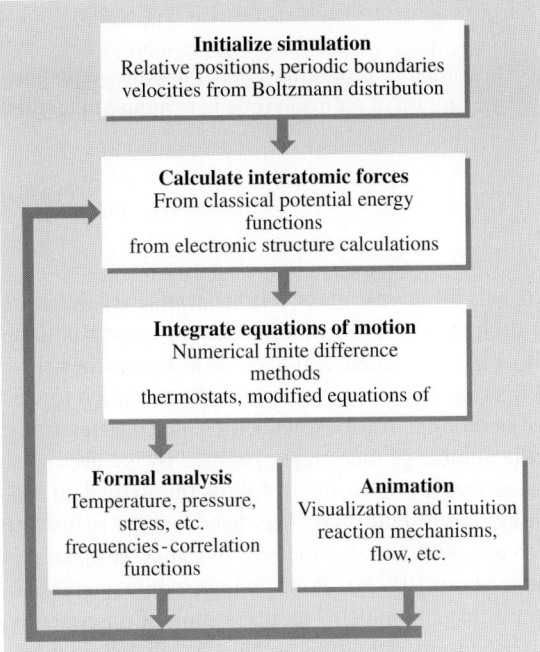

Fig. 13.1. Flow chart of an MD simulation (After [11])

teratomic forces, one simply integrates Newton's classical equation of motion using one of several standard methods [30]. The simulations yield new relative atomic positions, velocities and forces and the atomic responses can be followed using animated movies, as outlined in Fig. 13.1. However, the effective use of MD simulations to study atomic-scale friction requires an understanding of many details not apparent in this simple analysis. For this reason, the next section provides a brief review of MD techniques. The reader is referred to a number of other more comprehensive sources [31–34] for a more detailed overview of MD simulations, including computer algorithms.

13.2 Computational Molecular Dynamics Simulations

This section begins with a brief review of several interatomic potentials that have been used in MD simulations of atomic-scale friction, indentation, and wear. The reader is referred to [35, 36] for a broader discussion of potentials. This section is followed by a brief discussion of thermodynamic ensembles, and their use in different types of simulations. The section then closes with a description of several of the thermostatting techniques used to regulate temperature during an MD simulation. This topic is particularly relevant for tribological simulations because friction and indentation do work on the system, raising its kinetic energy.

13.2.1 Interatomic Potentials

As stated earlier, MD simulations are computer programs where Newton's equations of motion are integrated to monitor the motion of atoms in time in response to applied forces

$$F = ma \tag{13.1a}$$

$$-\nabla E = m\left(\partial^2 r/\partial t^2\right), \tag{13.1b}$$

where F is the force on each atom, m is the atomic mass, a is the atomic acceleration, E is the potential energy felt by each atom, r is the atomic position, and t is time. The forces acting on any given atom are calculated, and then the atoms move a short increment ∂t (called a time step) forward in time in response to these applied forces. This is accompanied by a change in atomic positions, velocities, and accelerations. The process is then repeated for the number of time steps specified by the user.

The advantage of this method is that it is possible to follow the individual motions of all the atoms in a given system in real time. The disadvantage is that the time-scales are very limited (ps to ns). Additionally, the size of the system that can be considered is presently limited to about 10^6–10^8 atoms, which, while impressive, is still far removed from real systems that contain 10^{23} atoms or more. Consequently, while MD simulations have been used to model atomic-scale friction and do a reasonably good job of providing insight into the atomic-scale mechanisms of friction, it is still limited to length and time scales that are significantly smaller than experimental values.

To model a material atomistically, mathematical expressions are needed to characterize the potential energy of the material [E in (13.1b)]. Currently, there are two approaches that are widely used to do this. In the first, one assumes that the potential energy of the atoms can be represented as a function of only their relative atomic positions. These functions, called empirical potentials, are typically based on simplified interpretations of more general quantum mechanical principles and usually contain some number of free parameters. The parameters are then chosen to closely reproduce a set of physical properties of the system of interest. While this may sound uncomplicated, there are many intricacies involved in developing a useful potential energy function. For example, the parameters used to fit the potential energy function are determined from a limited set of known system properties. For a metal, the properties to which a potential energy function is fit might include the lattice constant, cohesive energy, elastic constants, and vacancy formation energy. A consequence of this is that other properties, including those that might be crucial to determining the outcome of a given simulation, are determined solely by the assumed functional form. Predicted properties for a metal might then include surface reconstructions, energetics of interstitial defects, and response (both elastic and plastic) to applied load. The form of the potential is, therefore, crucial if the simulation is to sufficiently capture the relevant physics of the system.

The second approach, which has become more common with the ready availability of powerful computers, is the calculation of interatomic forces directly from first-principles [37] or semiempirical [38, 39] calculations that explicitly include electrons. An advantage of this approach is that, in general the number of unknown parameters is smaller than in empirical potentials and, because, the calculated forces are based on quantum principles, the interatomic interactions may be reasonably transferable between different atomic environments. However, this does not guarantee that forces from a semiempirical electronic structure calculation are accurate. Poorly chosen parameterizations and functional forms can still yield nonphysical results. The disadvantage of this approach is that the potentials involved require significantly more computational effort than those used in empirical potential functions. Longer simulation times require that both the system size and the timescale studied be smaller than when using empirical potentials. For example, the largest systems that may currently be considered using first principles and semi-empirical MD simulations are a few hundred atoms and a few thousand atoms, respectively. Thus, while these approaches have been used to study the forces responsible for friction [21], they have not yet found widespread application for the type of large-scale modeling discussed here.

The simplest approach for developing a continuous potential energy function is to assume that the binding energy E_b, can be written as a sum over pairs of atoms,

$$E_b = \sum_i \sum_{j>i} V_{\text{pair}}(r_{ij}) , \qquad (13.2)$$

where the indices i and j are atom labels, r_{ij} is the scalar distance between atoms i and j, and $V_{\text{pair}}(r_{ij})$ is an assumed functional form for the energy. Some traditional forms for the pair term are given by

$$V_{\text{pair}}(r_{ij}) = D \times X(X - 1) , \qquad (13.3)$$

where the parameter D determines the minimum energy for pairs of atoms. Two common forms of this expression are the Morse potential ($X = e_{ij}^{-\beta r}$), and the Lennard-Jones (LJ) "12-6" potential ($X = (\sigma/r_{ij})^6$), where β and σ are arbitrary parameters that are used to fit the potential to observed properties. The short-range exponential form for the Morse function provides a reasonable description of repulsive forces between atomic cores, while the $1/r^6$ term of the LJ potential describes the leading term in long-range dispersion forces.

For systems with significant Coulombic interactions, a fractional point charge q_i can be assigned to each atom. These point charges then interact according to the following expression:

$$V_{\text{pair}}(r_{ij}) = \frac{q_i q_j}{r_{ij}} . \qquad (13.4)$$

Because the $1/r$ Coulomb interactions act over distances that are long relative to atomic dimensions, simulations that include them require special attention to boundary conditions [40, 41].

There are numerous other pair potentials that have been used, and each has its strengths and weaknesses. However, the approximation of a pairwise-additive binding energy is so severe that in most cases no form of pair potential will adequately describe every property of a given system except for systems where the dominant interactions are van der Waals, such as occurs in systems of rare gas atoms. This is not meant to imply that pair potentials are not useful. In fact, just the opposite is true. Numerous general principles of many-body dynamics have been gleaned from simulations that have used pair potentials, and they will continue to find a central role in MD simulations. As discussed below, this is especially true for simulations of systems of confined fluids.

A rational extension of the pair potential is to assume that the binding energy can be written as a many-body expansion of the relative positions of the atoms, as shown in the following expression

$$E_b = \frac{1}{2} \sum_i \sum_j V_{\text{2-body}} + \frac{1}{3!} \sum_i \sum_j \sum_k V_{\text{3-body}} + \frac{1}{4!} \sum_i \sum_j \sum_k \sum_l V_{\text{4-body}} + \cdots . \quad (13.5)$$

Normally, it is assumed that this series converges rapidly enough that four-body and higher terms can be safely ignored. Several functional forms of this type have had considerable success in simulations. The most prominent of these is the potential by *Stillinger* and *Weber* [42] for silicon, that has found widespread use. Another example is the work of *Murrell* and coworkers [43] who have developed a number of potentials of this type for different gas-phase and condensed-phase systems.

The valence force field method is a common form of the many-body expansion (13.5). In this technique, interatomic interactions are included via a Taylor series expansion over bond lengths, bond angles, and torsional angles. Some sort of non-bonded interaction is typically included as well. Accounting for electrostatic induction interactions is one way of including many-body effects in systems where Coulombic interactions are important. This way, each point charge will give rise to an electric field, and will induce a dipole moment on neighboring atoms. This phenomenon can be modeled by including terms for the atomic or molecular dipoles in the interaction potential, and solving for the values of the dipoles at each step in the MD simulation.

Numerous potentials beyond the many-body expansion have been successfully developed and are widely used in MD simulations. For metals, the embedded-atom (EAM) and related methods have been quite successful in reproducing a host of properties [44–47]. In this potential, the energy of an atom interacting with the atoms that surround it is modeled as the energy of the atom interacting with a homogeneous electron gas with a compensating positive background. The density of the electron gas is approximated as the sum of electron densities of the surrounding atoms. The functional form of the EAM also includes a repulsive term to account for core–core interactions. Within this set of approximations, the total binding energy is given as a sum over atomic sites

$$E_b = \sum_i E_i , \quad (13.6)$$

where each site energy is given as a pair sum plus a contribution from a functional (called an embedding function) that, in turn, depends on the sum of electron densities at that site

$$E_i = \frac{1}{2}\sum_j \Phi(r_{ij}) + F\left[\sum_j \rho(r_{ij})\right]. \tag{13.7}$$

The function $\Phi(r_{ij})$ is the pairwise core–core repulsion term, F is the embedding function, and $\rho(r_{ij})$ is the contribution to the electron density at site i from atom j. For practical applications, functional forms are assumed for the core–core repulsion, the embedding function, and the contribution of the electron densities from surrounding atoms [48].

A potential based on bond orders has been developed [49] to model covalently bound materials. Originally adapted by *Tersoff* [50] to model silicon, the approach has found use in computer simulations of a wide range of materials [51–54]. Like the EAM potentials, Tersoff potentials begin by approximating the binding energy of a system of atoms as a sum over the atomic sites

$$E_b = \sum_i E_i. \tag{13.8}$$

Unlike the EAM potential, however, each site energy is given by an expression that resembles a pair potential

$$E_i = \sum_{j>i}\left[V_R(r_{ij}) + B_{ij} \times V_A(r_{ij})\right]. \tag{13.9}$$

The functions $V_A(r_{ij})$ and $V_R(r_{ij})$ are pairwise-additive attractive and repulsive terms, respectively, and B_{ij} is an empirical bond-order term that modulates the attractive pair potential. Thus, this bond-order term is where the many-body effects are introduced. As is the case for other empirical potentials, functional forms for the bond order and the pair terms are fit to a range of properties for the systems of interest. Applications of this approach have used functional forms for the bond order that decrease in value as the number of nearest-neighbors of a given pair of bonded atoms increases. Physically, the attractive pair term can be envisioned as bonding due to valence electrons, with the bond order destabilizing the bond as the valence electrons are shared among more neighbors. If properly parameterized, the potential yields Pauling's bond order relations [49, 51, 54] and correctly reproduces many of the properties of covalently bound materials. These properties make this expression quite adaptable at modeling the properties of a range of different structures, including molecules, clusters, surfaces, and solids, and therefore, useful for predicting new phenomena in MD simulations.

Although they are based on different principles, the EAM and Tersoff expressions are fundamentally similar to one another. In the EAM, binding energy is defined by electron densities through the embedding function. The electron densities are, in turn, defined by the arrangement of neighboring atoms. In the same way, in the Tersoff expression the binding energy is defined directly by the number and arrangement of neighbors through the bond-order expression. As a matter of fact, the EAM and Tersoff approaches can both be derived from tight binding theory, and it has been shown that the two are identical for simplified expressions provided that angular interactions are not used [55].

In simulations of ionic materials, the ions have well-defined charges and typically interact via a Coulombic potential. Their short ranged repulsion is frequently described by an empirical potential, typically the Buckingham potential (an exponential repulsion term plus a van der Waals attraction term). The development of the shell model [56] to describe the atomic polarizability adds a further level of materials specificity.

Typically simulations involving friction between different classes of materials use simple pairwise functional forms to describe the (usually repulsive) interactions between, e.g., a diamond tip and a metal surface. Frequently in these types of simulations the tip is held perfectly rigid to remove the need for realistic interfacial interactions. In recent years there has been some development of empirical potentials that model heterogeneous material interactions in a realistic manner. For instance, *Streitz* and *Mintmire* [57] developed a method to simulate the interface between Al and Al_2O_3. This combines a scheme for self-consistent charge determination with electrostatic interactions and an EAM potential. This approach has been subsequently extended to Ti/TiO_2 [58] but, despite its elegance, has not been widely used. In another approach, the modified EAM (MEAM) of Baskes, originally developed for Si and for metals, [59–61] has been successfully used to describe ionic materials [62], obviating the Coulombic interactions entirely. The justification for this approach lies in the frequently-made observation that due to the near balance between attractive and repulsive Coulombic interactions, the effective range of ionic forces is actually quite short [63, 64]. The weakness in this approach is that it is not evident that there is transferability of parameters between different charge states of the ions. Lastly, *Yasukawa* [65] has extended the bond-order method of Tersoff by including self-consistent charge determination and an electrostatic term in the spirit of Streitz and Mintmire. Yasukawa applied his approach to the Si/SiO_2 system and *Iwasaki* and *Miura* [66] have explicitly shown its ability to simulate metal/ceramic interfaces. While these developments have increased the applicability of empirical potentials to heterogeneous systems, none of them have yet been applied to model friction between different classes of materials.

This section provides a brief introduction to the potentials that have found the most use in the simulations of friction, wear, and related phenomena in MD simulations, and discusses the development of new potentials that are likely to play an important role in future simulations, especially of heterogeneous systems.

13.2.2 Temperature Regulation

Constant energy trajectories are not optimal to study friction and indentation because these processes require work to be performed on the system which raises the system energy and temperature without bound. In contrast, in actual macroscopic systems, heat is dissipated through the surroundings and a fairly constant temperature is maintained. If the simulation systems were as large as the experimental systems, i.e. on the order of 10^{23} atoms, this dissipation process would occur computationally as well. However, as discussed above, atomistic computer simulations are limited to much smaller systems that are many orders of magnitude smaller than systems that are studied experimentally. For this reason, a thermodynamic ensemble where the temperature (T), rather than the energy, is held constant more closely mimics reality. Conditions with constant-NVT correspond to the canonical ensemble.

A constant temperature is maintained in a canonical ensemble by using any of a large number of thermostats, several of which are described below. Typically in simulations of indentation or friction, the thermostat is applied to a region of the simulation cell that is well removed from the interface where friction and indentation is taking place. In this way, local heating of the interface as work is done on the system, but excess heat is efficiently dissipated for the system as a whole. These simulations can be thought of as operating under "hybrid" NVE/NVT conditions, which, although not rigorously a member of any true thermodynamic ensemble, are useful for mimicking reality and are commonly used.

The simplest approach for controlling system temperature is to simply intermittently rescale the atomic velocities to yield desired temperature [67]. This approach was widely used in early MD simulations, and is often effective at maintaining a given temperature during the course of a simulation. Nevertheless, it has several disadvantages that have spurred the development of more sophisticated methods. For instance, there is little theoretical justification for the rescaling of system velocities. For typical MD simulation system sizes, averaged quantities, such as pressure, do not correspond to those obtained from any particular thermodynamic ensemble. Additionally, the dynamics produced are not time reversible, which is inconsistent with classical mechanics. Lastly, the rate and mode of heat dissipation are not determined by system properties, but rather depend on how frequently the atomic velocities are rescaled. This may have an effect on the dynamics of the system.

A more sophisticated technique for maintaining system temperature is with Langevin dynamics [33]. This method was originally used to describe Brownian motion and has found widespread use in MD simulations. In this technique, additional terms are added to the equations of motion that correspond to a frictional term and a random force [31, 68, 69]. The equation of motion for atoms subjected to Langevin thermostats is given by the following term [rather than (13.1)]

$$m\mathbf{a} = \mathbf{F} - m\xi\mathbf{v} + R(t) , \qquad (13.10)$$

where $\mathbf{F}$ are the forces due to the interatomic potential, the quantities m and $\mathbf{v}$ are the particle's mass and velocity, respectively, ξ is a friction coefficient, and $R(t)$ represents a random force that acts as "white noise". The friction term is defined in terms of a memory function in formal applications; terms developed for harmonic solids have been used successfully in MD simulations [70–72].

It is important to keep in mind that, as is the case with any thermostat, the atomic velocities are altered in the process of controlling the temperature with Langevin thermostats. This has the potential to perturb any dynamical properties of the system being studied. One approach that is effective at minimizing this problem is to add Langevin forces only to those atoms in a region away from where the dynamics of interest occurs. Thus, coupling to a heat bath occurs away from the processes of interest, and simplified approximations for the friction term can be used without unduly influencing the dynamics produced by the interatomic forces.

The random force in (13.10) is typically given by a Gaussian distribution where the width, which is chosen to satisfy the fluctuation–dissipation theorem, is determined from the equation

$$\langle R(0) \times R(t) \rangle = 2mkT\xi\delta(t) . \tag{13.11}$$

The function R is the random force in (13.10), m is the particle mass, T is the desired temperature, k is Boltzmann's constant, t is time, and ξ is the friction coefficient. Note that the random forces are uncoupled from those at the previous steps (as denoted by the delta function), and the width of the Gaussian distribution, from which the random force are obtained, varies with temperature.

The Langevin approach outlined above does not require any feedback from the current temperature of the system; rather, the random forces are determined solely from (13.11). A slightly different approach, called the Berendsen thermostat, has been developed that eliminates the random forces and replaces the constant friction coefficient with one that depends on the ratio of the desired temperature to current kinetic energy of the system (measured as a temperature) [73]. The resulting equation of motion with this approach is

$$m\mathbf{a} = \mathbf{F} + m\xi\left(\frac{T_0}{T} - 1\right)\mathbf{v} , \tag{13.12}$$

where $\mathbf{F}$ is the force due to the potential energy (as given by the empirical potential in classical MD simulations), T_0 and T are the desired and actual temperatures, and $\mathbf{v}$ is again the particle velocity. This approach does not require the evaluation of random forces, which can be expensive for a large number of thermostat atoms. In practice, however, if the system is not pre-equilibrated to properly populate the atomic vibrational modes, or if nonrandom external forces are applied to the system, the system can be slow to reach equilibrium with this approach. Conversely, the Langevin approach using a random force on each atom does not require feedback from the system, and thermostats each atom individually. Consequently, the random forces are much more efficient in eliminating these nonphysical reflecting waves.

Nonequilibrium equations of motion have also been developed to maintain a constant temperature [31]. Like the Berendsen thermostat, this approach adds a frictional term to the interatomic forces. However, it is derived from Gauss' Principle of Least Constraint, which maintains that the sum of the squares of any constraining forces on a system should be as small as possible. Using a Lagrange multiplier, a frictional force on each atom i of the form

$$\mathbf{F}_i^{\text{friction}} = -\varsigma m_i \mathbf{v}_i , \tag{13.13}$$

where

$$\varsigma = \sum_i (F_i \times v_i) / \sum_i (m v_i^2) \;, \tag{13.14}$$

can be derived that maintains a constant temperature. The quantity m_i is the mass of atom i, v_i is its velocity, and F_i is the total force on atom i due to the interatomic potential. It should be noted that there is no target temperature in this approach. Rather, the temperature of the system when the constraint is initiated is maintained as the simulation moves forward in time. There are several obvious advantages to this approach. The first is that it does not rely on an approximated input such as the Debye frequency as in the simplified Langevin or Berendsen thermostats. As a result, heat loss and gain are determined only by implicit system properties. Secondly, because a random force is not required, it does not significantly increase computational time. Thirdly, the equations of motion are time reversible. Lastly, by differentiating energy with respect to time, the heat loss (or gain) due to the thermostat can be calculated directly (which is also true of the Langevin thermostat). As with the Berendsen thermostat, however, coupling of the friction to global properties of the system is typically slow to randomize nonphysical vibrational disturbances.

Nosé developed a thermostat that corresponds rigorously to a canonical ensemble [74, 75]. This represents a significant advance from the methods described so far. This approach also adds a friction term to the equation of motion, but maintains the correct distribution of vibrational modes. This is achieved by adding a new dimensionless variable to the standard classical equations of motion that corresponds to a virtual, large heat bath that couples to each of the physical degrees of freedom. The consequence of this variable, however, is to scale the coordinates of either time or mass in the system. Thus, while the dynamics of the expanded system correspond to the microcanonical ensemble, when they are projected onto only the physical degrees of freedom they generate a trajectory in the canonical ensemble. Sampling problems associated with very small or very stiff systems are overcome by attaching a series of Nosé–Hoover thermostats to the system [76]. The resulting equations of motion are time reversible and the resulting trajectories can be analyzed exactly with well-established statistical mechanical methods [76]. The reader is referred to an article by *Hoover* [31] for a more complete description of the Nosé thermostat and a comparison of the dynamics generated with this approach and the others that have been reviewed here.

A unique alternative to the grand canonical ensemble is that chosen by *Cushman* and coworkers [77, 78] who performed a series of grand canonical Monte Carlo simulations [33, 79] at various points along a hypothetical sliding trajectory. These simulations were used to calculate the correct particle numbers at a fixed chemical potential, which were then used as inputs to non-sliding, constant-NVE MD simulations at each of the chosen trajectory points. Because the system was fully equilibrated at each step along the sliding trajectory, the sliding speed could be assumed to be infinitely slow. This offers a useful alternative to continuous MD simulations that are currently restricted to sliding speeds of roughly 1 m/s or greater, which are orders of magnitude larger than most experimental studies.

Each of the components used in an atomistic computer simulation of dynamical friction, indentation, or wear described above, including potential energy functions, thermodynamic ensembles, and thermostats, have their own advantages and disadvantages. The optimum choice of components consequently depends strongly on the system of interest and the process being simulated. It also depends strongly on the type of information in which one is interested. For example, general principles related to liquid lubrication in confined areas may be most easily understood and generalized from simulations that use pair potentials and may not require a thermostat. On the other hand, if one wants to study the wear or indentation of a surface of a particular metal, then EAM or other semiempirical potentials, together with a thermostat, would be expected to yield more reliable results. Even more detailed studies, including the evaluation of electronic degrees of freedom, require interatomic forces derived from electronic structure calculations. In short, the best way to set up or interpret the results of an MD simulation of friction and related processes is to fully understand the pluses and minuses of each of these approaches, decide what one wishes to learn from the simulation, and form conclusions based on careful interpretation of the results.

13.3 Indentation

To develop the fundamental ideas needed to design new coatings with tailor-made friction and wear properties, it is crucial to understand material properties at the nanometer scale. One of the ways in which these properties are being characterized is through the use of the AFM, which has proved to be a versatile tool that provides a rich variety of atomic-scale information pertaining to a given tip/sample interaction [80]. In these AFM indentation experiments, the microscope tip has a radius in the range of 1–100 nm and is pressed against the surface under ultra-high vacuum (UHV) conditions, in air (ambient conditions) or in a liquid. The tip moves normal to the surface to indent (AFM) and/or across the surface (FFM) at sliding speeds of 1 nm/s–1 μm/s. These indentation or sliding speeds are in contrast to the speeds used in MD simulations of sliding or indentation, where much higher sliding speeds of 1–100 m/s are used due to the high computational costs of slower speeds.

When the AFM tip is rastered across the sample substrate, the force on the tip perpendicular to the substrate is measured at each point and a force map of the surface is obtained that can be related to the actual surface topography [81]. Rastering the AFM tip across a substrate while measuring the deflection of the tip in the lateral direction produces a friction map of the surface [18]. In addition, by moving the tip perpendicular to the surface of the substrate, AFMs can be used as nanoindenters that probe the mechanical properties of various substrates and thin films [82, 83]. This indentation is reflected in a dramatic increase in force as the tip is moved further into the substrate (Fig. 13.2b); this region of the force curve is known as the repulsive wall region [80], or, when considered without the rest of the force curve, an indentation curve. Retraction of the tip following indentation results in enhanced adhesion between the tip and surface, which is evidenced by hysteresis in the force curve. The origin of this enhanced adhesion is discussed below.

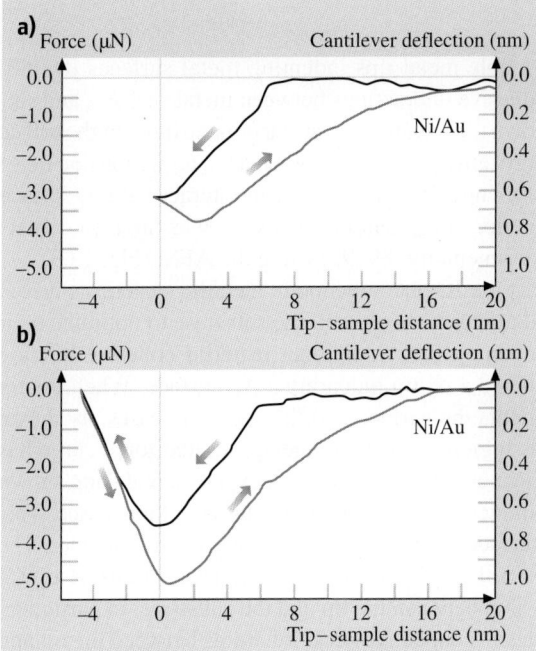

Fig. 13.2. Experimentally measured forces versus tip-to-sample distances curves for a Ni tip interacting with an Au substrate for contact followed by separation in (**a**) and contact, indentation, then separation in (**b**). These curves were derived from AFM measurements taken in dry nitrogen (After [84])

Many types of tip–substrate interface adhesion are possible. For example, adhesion might result from the formation of covalent chemical bonds between the tip and the sample. Additionally, real surfaces typically have a layer of liquid contamination on the surface that can lead to capillary formation between the surface and the AFM tip that leads to adhesion. In metallic systems a different sort of wetting is possible: that is, the sample can be wet by the tip (or vice versa). This wetting results in the formation of an adhesive "connective neck" of metallic atoms between the tip and the sample and a consequent adhesion. Lastly, entanglement of molecules that are anchored on the tip with molecules anchored on the sample can be responsible for force curve hysteresis.

MD simulations of nanoindentation were first employed in an effort to shed light on the physical phenomena that are responsible for the qualitative shape of AFM force curves. Not only have they succeeded at this task, MD simulations have also revealed a wealth of atomic-scale phenomena that occur during the nanoindentation process. The rest of this section discusses some important findings from MD simulations of nanoindentation.

13.3.1 Metals

Many MD simulations of deformable metal tips indenting metal surfaces [84–88] have clarified the nature of the adhesive interactions between metal surfaces and single metal asperities. It is well-known that clean metal surfaces have quite high surface energies and, therefore, are strongly attracted to each other. This attraction can be so strong that when the tip gets close enough to the surface to interact with it, surface atoms "jump" upwards to wet the tip. This phenomena is known as jump-to-contact (JC) and has been confirmed experimentally [89–92] using the AFM (Fig. 13.3).

Landman et al. [84] noted that the JC phenomenon in metallic systems is driven by the tendency of the interfacial atoms of the tip and the substrate to optimize their embedding energies while maintaining their individual material cohesive binding energies. His initial study considered a Ni tip indenting a Au surface. When the tip advances past the JC point it indents the surface and the force increases, as shown in Fig. 13.4, points D to M. This region of the computer-generated force curve has a maximum not present in the force curve generated from experimental data, shown in Fig. 13.4, point L, due to tip-induced flow of the metal atoms in the surface. This flow causes "piling-up" of the surface atoms around the edges of the indenter. The force curve is completed by reversal of the tip motion (Fig. 13.4, points M to X). Hysteresis is present due to adhesion between the tip and the substrate. In particular, as the tip retracts from the sample, a "connective neck" of atoms between the tip and the substrate forms (Fig. 13.5). This connective-neck of atoms is largely composed of metal atoms from the surface; however, some atoms from the metal indenter also diffuse into the structure. As a result, continued retraction of the tip causes the magnitude of the force to increase (i.e., become more negative) until, at a critical force, the atoms in adjacent layers of the connective neck rearrange so that an additional row of atoms form in the neck. These rearrangement events are the essence of the elongation process and are responsible for the fine structure (apparent as a series of maxima) present in the retraction portion of the force curve. These elongation and rearrangement steps are repeated until the connective neck of atoms is severed.

Constant-energy simulations by *Tomagnini* et al. [94] of Au tips indenting Pb substrates predict that a JC is initiated by Pb atoms wetting the Au tip. The energy released due to the wetting of the tip is predicted to cause an increase in the temperature of the tip of approximately 15 K when the simulation takes place at room temperature. Extensive structural rearrangements in the tip are also predicted to occur when the tip/sample distance is decremented further. Increasing the substrate temperature to 600 K causes the formation of a liquid Pb layer that is approximately 4 layers thick to form on the surface of the substrate. As a result, during the indentation the distance at which the JC occurs increases by about 1.5 Å. The high diffusivity of the Pb surface atoms at this temperature also causes the contact area to increase. Eventually the Au tip dissolves in the liquid Pb and this liquid-like connective neck of atoms follows the tip upon retraction. Consequently, the liquid-solid interface moves further back into the bulk Pb substrate, increasing the length of the connective neck. Similar elongation events have been observed experimentally. For example, scanning

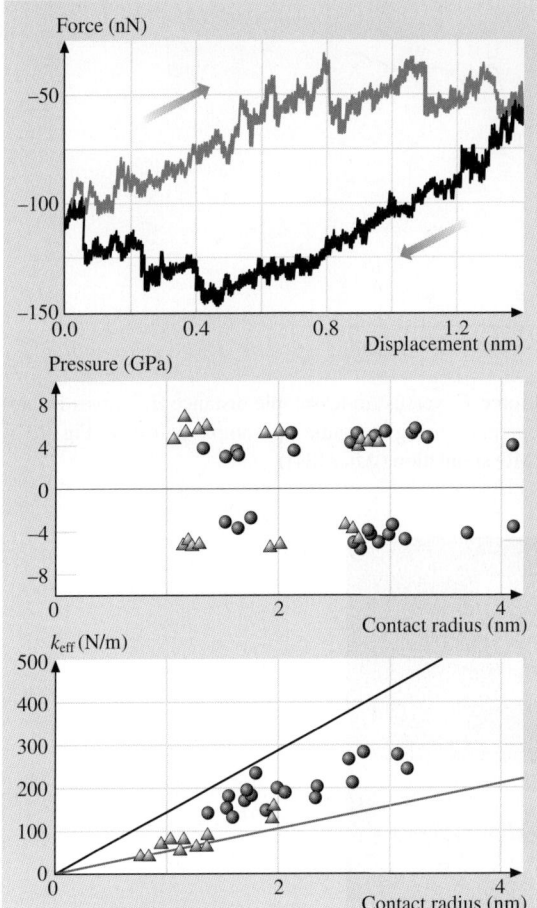

Fig. 13.3. *Top*: The experimental values for the force between a tip and a surface that have a connective neck between them. The neck contracts and extends without breaking on the scales shown. *Bottom*: The effective spring constant k_{eff} determined experimentally for the connective necks and corresponding maximum pressures, versus contact radius of the tip. The *triangles* indicate measurements taken at room temperature; the *circles* are the measurements taken at liquid He temperatures (After [93])

tunneling microscopy (STM) experiments on the same surface demonstrate that the neck can elongate approximately 2500 Å without breaking [95].

When nanoindentation occurs with a tip that is significantly stiffer than the surface, it causes pile-up of surface atoms around the tip to relieve the stresses induced by the indentation. In contrast, damage is predominately done to the tip when a soft tip is used to indent a hard substrate. Simulations using perfectly rigid tips by *Belak* and *Stowers* [96] illustrate the mechanism by which the surface yields plastically after its elastic threshold is exceeded. This occurs by popping atoms out onto the

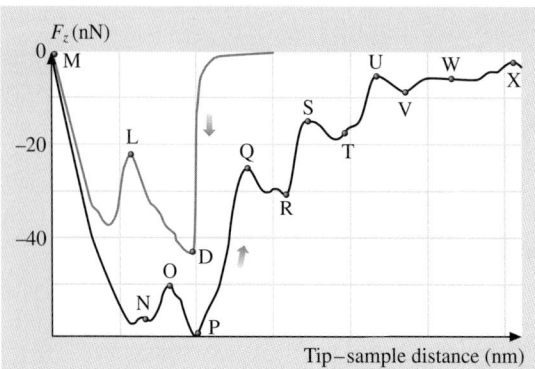

Fig. 13.4. Computationally derived force F_z versus tip-to-sample distance d_{hs} curves for approach, contact, indentation, then separation using the same tip–sample pair as in Fig. 13.5. These data were calculated from an MD simulation (After [84])

Fig. 13.5. Illustration of atoms in the MD simulation of a Ni tip being pulled back from an Au substrate. This causes the formation of a connective neck of atoms between the tip and the surface. (After [84])

surface under the tip, which leads to atomic pileup. In this study, variations in the indentation rate from 1 m/s to 100 m/s reveal that point defects created as a result of nanoindentation relax by moving through the surface only at the slower rate; at the higher indentation rate, there is no time for the point defects to relax and move away from the indentation area and so strain builds up more rapidly. Rigid indenter simulations are analogous to experiments that use surface passivation to prevent JC between the tip and the surface [97, 98]. These experiments show the predicted results of pile-up and crater formation, as illustrated in Fig. 13.6 [97].

It is possible to model the indentation of a surface with a hard-sphere indenter in a manner that is independent of the indentation speed. In one such study by *Kelchner*

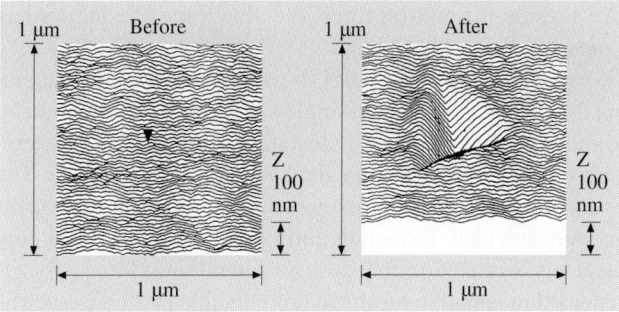

Fig. 13.6. Images of a Au surface before and after being indented with a pyramidal shaped diamond tip in air. The indentation created a surface crater. Note the pileup around the crater edges (After [97])

et al. [99], rather than use MD, the indenter is pushed against the substrate and then the system is allowed to relax using standard energy minimization methods. When the system is fully relaxed, the tip is moved closer and the process is repeated. Dislocations at the surface are generated as a result. When the tip is pulled back after indenting less than a critical value, the atoms in the dislocations are able to slide back to their original positions and the surface is healed. However, if the tip is indented past the critical depth, more dislocations are created that interfere with the healing process resulting in a surface crater.

Thus, MD simulations are able to explain the atomic scale mechanisms behind measured experimental force curves obtained when metal tips indent homogeneous metal surfaces to nanometer-scale depths. There is currently interest in using the process of JC to generate metal nanowires [100–102].

MD simulations have also been used to study the indentation of heterogeneous metal substrates, such as an Ir tip used to indent an Ir substrate covered with a single layer of Pb in work by *Raffi-Tabar* and *Kawazoe* [103]. In this case, the disruption of the Pb monolayer caused by the JC also results in local deformation of the Ir substrate beneath the monolayer. As a result, further indentation leads to penetration of the Pb film at only one atomic site and no Ir–Ir adhesive bonds are formed. Because Ir–Ir adhesive bonds are stronger than the Ir–Pb adhesive bonds, the separation force is significantly less in this case than in the absence of the Pb monolayer [104]. Additionally, the Ir substrate is not deformed as a result of the indentation because the Ir tip does not wet the substrate.

In contrast, when a Pb tip is used to indent an Ir-covered Pb substrate, the Pb tip atoms wet the Ir monolayer during the JC [103]. As the contact area between the tip and the Ir monolayer is large, there is no discernible crystal structure present in the Pb tip. Rather, the tip has the structure and properties of a liquid drop wetting a surface. Because of the presence of the Ir monolayer, continued indentation of the tip does not result in the formation of any new, adhesive Pb–Pb bonds. During pull off, the Pb tip forms a connective neck that decreases in width as it separates from the surface. The radius of this connective neck of atoms is smaller than it is in the absence of the Ir layer. As a result, the pull-off force, (i.e., the force of adhesion), is less when the Ir layer is present.

MD simulations by *Komvopoulos* and *Yan* [105] show the atomic-scale material responses of dynamic metal-like substrates due to single and repeated indentation by metallic or covalently bound, rigid tips. A single indentation even produces hysteresis in the force curve due to inelastic deformation and heating of the substrate. The repulsive force decreases abruptly during the permanent deformation caused by penetration of the tip. The connective neck formation during unloading observed only in indentation with a metal tip is related to the stronger interatomic forces associated with this system. Repeated indentation results in reproducible processes such as cyclic work hardening and softening by annealing observed at the macroscale.

To summarize these findings, a reduction in the force of adhesion is observed when a metal monolayer with different interactions with the metal tip than the underlying surface is placed between the tip and the substrate. However, the mechanism responsible for this reduction differs for the two systems discussed here. In the Ir/Pb/Ir case, formation of strong Ir−Ir bonds is prevented by the presence of the Pb film; therefore, the pull-off force is reduced. In the Pb/Ir/Pb case, the smaller radius of the connective neck between the tip and substrate is responsible for the reduction in the force of adhesion.

Molecular dynamics simulations have also been used to examine the relationship between nanoindentation and surface structure. *Zimmerman* et al., for example, modeled indentation of a single crystal gold substrate both near and far from a surface step [106]. Their simulations showed that the onset of plastic deformation is strongly influenced by the distance of indentation from the step, and whether the indentation is on the plane above or below the step. In related simulations, *Shenderova*, *Mewkill* and *Brenner* explored the issue of whether very shallow elastic nanoindentation can be used to nondestructively probe surface stress distributions associated with surface structures such as a trench and a dislocation intersecting a surface. The simulations, which carried out the nanoindentation to a constant depth, showed maximum loads that reflect the in-plane stresses at the point of contact between the indenter and the substrate.

Studies performed since the 1930's using hardness measurement techniques [107–110] and indentation methods [111, 112] suggest that the hardness of a material depends on applied in-plane uni- and bi-axial strain. In general, increases in hardness under compressive in-plane strain were reported, while tensile strain appeared to decrease hardness. This behavior had traditionally been attributed to the contribution of stresses from the in-plane strain and the local strain from the indentation to the resolved shear stresses [108, 110]. In 1996, however, *Pharr* and co-workers pointed out that changes in elastic modulus determined from unloading curves of strained substrates using contact areas estimated via an elastic model appear too large to have physical significance, a result that called into question the interpretation of prior hardness data [111, 112]. Based on experimental nanoindentation studies of a strained polycrystalline aluminium alloy and finite element calculations on an isotropic solid [111, 112] they hypothesized that any apparent change in modulus (and hardness) with in-plane strain is primarily due to changes in contact area that are not typically taken into consideration in elastic half-space models. In particular they suggested that in-plane compression increases pile-up around the indentor, which when not taken into account in the analysis of unloading curves, implies a (nonphysical) increase in modulus. Similarly, they suggested that in-plane tensile strain reduces the amount of material that is pile-up around an indenter, leading to a corresponding reduced (nonphysical) modulus when interpreting unloading curves using elastic models.

To further explore the issue of pile-up and its influence on the interpretation of loading curves, *Schall* and *Brenner* used molecular simulations and EAM potentials to model the plastic nanoindentation of a single-crystal gold surface under an applied in-plane strain [113]. These simulations showed that the mean pressure calculated from true contact areas, which take into account plastic pile-up around the indenter, varies only slightly with applied pre-stress, with higher values in compression than in tension, and that the modulus calculated from the true contact area is essentially independent of the pre-stress level in the substrate. On the other hand, if the contact area is estimated from approximate elastic formulae, the contact area is underestimated, leading to a strong, incorrect dependence of apparent modulus on the pre-stress level. In agreement with the Pharr model, the simulations showed larger pile-up in compression than in tension, with both regimes producing contact areas larger than those typically assumed in elastic analyses.

13.3.2 Ceramics

Nanometer-scale indentation of ceramic systems has also been investigated with MD simulations. In the bulk, ceramics are stiffer and more brittle than metals. The study of nanoindentation can provide insight into the manner by which cracks and defects form in covalent and ionic materials.

In the case of ionically bound materials, *Landman* et al. [87, 114] considered the interaction of a CaF_2 tip with a CaF_2 substrate in constant-temperature MD simulations. The attractive force between the tip and the substrate increased gradually as the tip approached the substrate. This attractive force increased dramatically at the critical distance of 2.3 Å, as the interlayer spacing of the tip increased (i.e., the tip elongated). This process is similar to the JC phenomenon observed in metals; however, the amount of elongation (0.35 Å) is much smaller in this case than the amount predicted to occur in metallic tips (several Å). Decreasing the distance between the tip-holder and the substrate causes a further increase in the attractive energy until a maximum value is reached. Indentation beyond this point results in a repulsive tip/substrate interaction, compression of the tip, and ionic bonding between the tip and substrate. It is these bonds that are responsible for the hysteresis predicted to occur in the force curve. Retraction from the substrate ultimately leads to plastic deformation of the tip and its fracture.

Kallman et al. [115] were among the first to use MD simulations to study the nanometer-scale indentation of covalently bound materials when they examined the microstructure of amorphous and crystalline silicon before, during, and after indentation. The goal was to determine if and how Si directly beneath the indenter undergoes two different pressure-induced, solid–solid phase transformations during indentation. The first involves the high-pressure β-Sn structure (above 100 kbar) and the second involves a thermodynamically unstable amorphous phase. At the highest indentation rate and the lowest temperature, *Kallman* et al. [115] determined that amorphous and crystalline Si have similar yield strengths of 138 and 179 kbar, respectively. However, at temperatures near the melting temperature and at the slowest indentation rate, lower yield strengths of 30 kbar are predicted for both amorphous and crystalline Si. This study shows how the simulated yield strength of Si at the nanometer scale depends on structure, rate of deformation, and sample temperature. Interestingly, amorphous Si does not show any sign of crystallization upon indentation, while indentation of crystalline Si at temperatures near the melting point reveals a tendency to transform to the amorphous phase near the indenter surface. In addition, no evidence of the transformation to the β-Sn structure under warm indentation is found. These results agree with the outcomes of scratching experiments [116] that showed that amorphous silicon emerged from room-temperature scratching of pure silicon.

The indentation of diamond (111) substrates beyond the elastic limit, with and without hydrogen-termination, using a hydrogen-terminated sp^3-bonded tip, was investigated by *Harrison* et al. [117]. The simulations indicated the depth at which the diamond (111) substrate incurrs a plastic deformation due to indentation (see Figs. 13.7a–c). No hysteresis in the potential energy versus distance curve is observed (Fig. 13.7a) when the maximum normal force on the tip holder is 200 nN or less. At this point the indentation is non-adhesive and elastic, and therefore, the tip/substrate system does not sustain any permanent damage as a result. This information is also apparent from a comparison of initial and final tip/substrate geometries.

In contrast, increasing the maximum normal force on the tip holder to 250 nN prior to retraction causes plastic deformation of both the tip and the substrate. This finding is apparent from the marked hysteresis in the plot of potential energy versus distance (Fig. 13.7b) and from a comparison of the initial and final tip/substrate geometries (Figs. 13.8a–d). Specific, atomic-scale motions associated with certain features in the plot of potential energy as a function of distance are apparent from the simulations. As the tip is withdrawn from the substrate, connective strings of atoms are formed between the two, as illustrated in Fig. 13.8c. These strings break as the distance between the tip and crystal increases. Each break is accompanied by a sudden drop in the potential energy at large positive values of tip/substrate separation (Fig. 13.7b).

Other phenomena that are predicted by the simulations is that the end of the tip twists to minimize interatomic repulsions between the hydrogen atoms on the tip and the substrate. This twisting leads to the formation of new covalent bonds between the tip and carbon atoms below the first layer of the surface. For this reason the indentation process in this case is disordered and ultimately leads to the formation of connective strings of atoms between the substrate and tip on tip retraction.

When the Si atoms on the surface are not terminated with hydrogen atoms, indentation with the hydrogen-terminated Si tip leads to approximately the same value of maximum indentation force; plastic deformation of the tip and surface is again

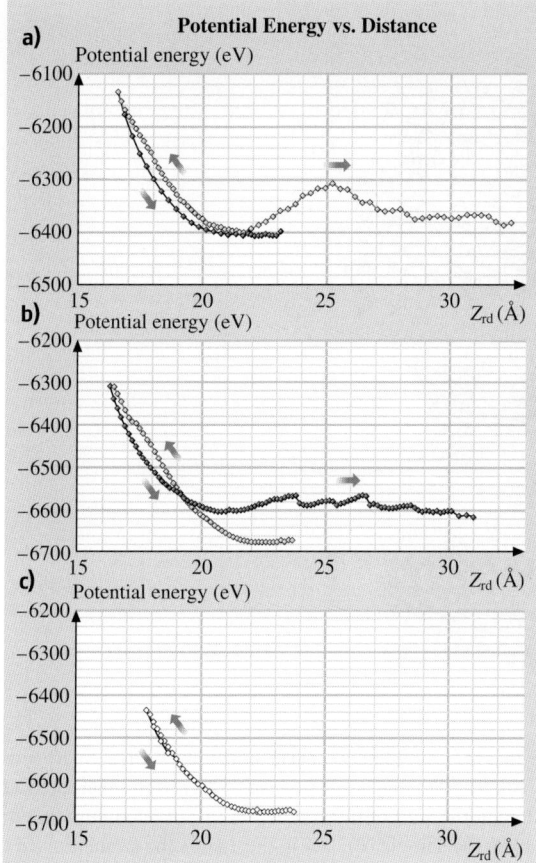

Fig. 13.7. Potential energy of rigid-layer separation generated from an MD simulation of an elastic (nonadhesive) indentation (**a**) shows the results of the same tip indenting a diamond (111) surface after the hydrogen termination layer was removed (**b**) of a hydrogen-terminated diamond (111) surface using a hydrogen-terminated, sp^3-hyrbized tip. (**a**) and plastic (adhesive) indentation. (After [117])

observed. However, the atomic-scale details, including the degree of damage, differ greatly from the case described above where both the surface and tip are hydrogen-terminated. In particular, the absence of hydrogen on the surface of the substrate minimizes repulsive interactions during indentation, and causes the tip to indent the substrate without twisting [117]. Because C–C bonds are formed between the tip and the first layer of substrate, the indentation is ordered (i.e., the surface is not disrupted as much through interacting with the tip) and the eventual fracture of the tip during retraction results in minimal damage to the substrate (Figs. 13.9a,b). The concerted fracture of all bonds in the tip gave rise to the single maximum in the potential versus distance curve at large distance (Fig. 13.7c).

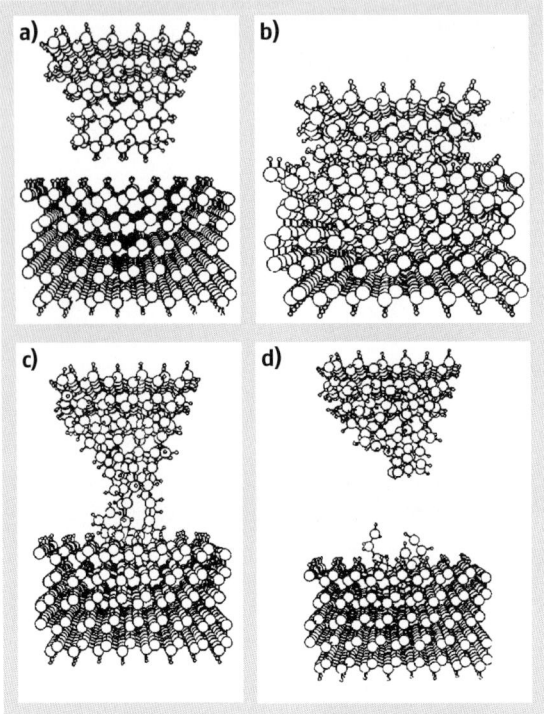

Fig. 13.8. Illustration of atoms in the MD simulation of the indentation of a hydrogen-terminated diamond (111) substrate with a hydrogen-terminated, sp^3-hybridized tip at selected time intervals. The tip–substrate system at the start of the simulation (**a**), at maximum indentation (**b**), as the tip was withdrawn from the sample (**c**), and at the end of the simulation (**d**). Large and small spheres represent carbon and hydrogen atoms, respectively (After [117])

When atomically-sharp tips are used to indent thin films where there is a large mismatch in the mechanical properties of the film and the substrate, there is difficulty determining the true contact area. In the case of soft films on hard substrates, pileups can occur around the tip increasing the contact area between the film and the tip. In contrast, with hard films on soft substrates, "sink-in" is experienced around the tip decreasing the true contact area.

A commonly used thin film material is diamond-like carbon (DLC) that is amorphous. Although nearly has hard as crystalline diamond, DLC coatings have very low friction coefficients (<0.01) [118–120]. There have been several MD simulation studies to determine the mechanical and atomic-scale frictional properties of DLC coatings that have provided much information about them. For example, a study by *Sinnott* et al. [121] explored the reason why the indentation behavior by a diamond tip is very different on a bare, single-crystal diamond surface and a diamond surface covered with about 20 layers of amorphous carbon. In the former case, the tip goes through shear and twist deformations at low loads that change to plastic deformation

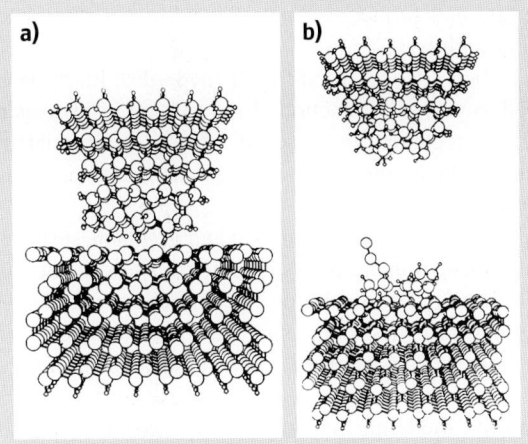

Fig. 13.9. Illustration of atoms in the MD simulation of the indentation of a non-hydrogen-terminated diamond (111) substrate with a hydrogen-terminated, sp^3-hybridized tip. The tip–substrate system prior to indentation (**a**) and subsequent to indentation (**b**). The *large spheres* represent carbon atoms and the *small spheres* represent hydrogen atoms (After [117])

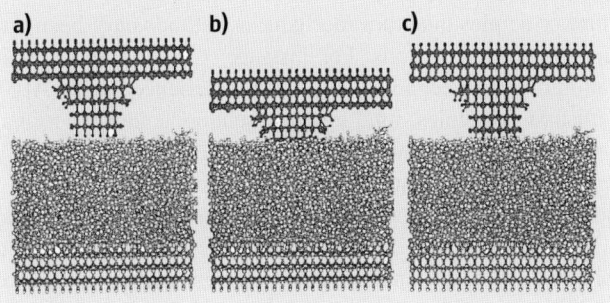

Fig. 13.10. Snapshots from a simulation where a pyramidal diamond tip indented an amorphous carbon thin film that is 20 layers thick. The simulation took place at room temperature and the carbon atoms in the film were 21% sp^3-hydridized and 58% sp^2-hydridized (remaining atoms were on the surface and were not counted) (After [121])

and adhesion with the surface at high loads. When the surface is covered with a thin film, the tip easily penetrates the film, which "heals" easily when the tip is retracted so that no crater or other evidence of the indentation is left behind (Fig. 13.10).

Other simulations by *Glosli* et al. [122] of the indentation of DLC films that are about 20 nm thick give similar results. In this case a larger, rigid diamond tip was used in the indentations and was also slid across the surface. During sliding the tip plows the surface and causes some changes to the film not seen in the simple indent.

However, adhesion between the film and surface was not allowed in this study and this might have affected the results slightly.

The hardness of amorphous carbon coatings used for magnetic disk lubrication has been experimentally observed to vary as a function of the amount of hydrogen within the film [123]. As the hydrogen content increases, the film is more resistant to scratching up to a critical value. If the amount of hydrogen is increased past this critical value, the film degrades more easily.

Garg et al. [124] considered the indentation of diamond and graphene surfaces with carbon nanotube proximal probe tips using MD simulations. The simulations reveal that nanotubes do not plastically deform during tip crashes on unreactive surfaces. Instead, they elastically deform, buckle, and slip. However, in the case of highly reactive surfaces strong adhesion can occur between the CNT and the surface that destroys the CNT.

Some success has been achieved modeling nanometer-scale indentation with methods other than purely atomistic approaches. For example, the interactions of a STM tip with a surface of were modeled by *Shluger* et al. [125] using an atomistic treatment for the interaction between the tip and the surface (including the van der Waals interactions) and a macroscopic parameter of cantilever deflection, (Fig. 13.11). The results show that if the tip is charged and in hard contact with the surface, tip and surface distortions are possible that can lead to motion of the surface ions within the surface plane and the transfer of some of the ions onto the tip.

Additionally, some first principles quantum mechanical methods have been applied to the study of tip–surface interactions. For instance, *Cho* and *Joamopoulos* [126] examined the atomic-scale mechanical hysteresis experienced by an AFM tip indenting Si(100) with first principles, total energy pseudopotential functional methods. The calculations predict that at low rates it is possible to cycle repeatedly between two buckled configurations of the surface without adhesion.

The study of ceramic materials under contact loading carried out with spherical indenters by *Licht* et al. [127] shows that ceramic materials are primarily damaged through two main mechanisms. The first is unstable crack growth and material failure that occurs in regions under tension. The second mechanism is the nucleation and growth of microcracks, which occurs in regions under pressure.

In short, MD simulations reveal the properties of ceramic tips and surfaces with covalent or ionic bonding that are most important for nanometer-scale indentation. Brittle fracture of the tip and strong adhesion with the substrate is possible in some cases, while in others neither the tip nor the surface is affected by the process. The insight gained from these simulations helps in the interpretation of experimental data as well as revealing the nanometer-scale mechanisms by which, for example, tip buckling and permanent modification of the surface occur.

13.3.3 Molecular Systems

MD simulations have been used to study the indentation of a metal surface covered with a liquid n-hexadecane film with a metal AFM tip, as shown in Fig. 13.12. As the tip touches the film, some of the molecules from the surface attach to the tip,

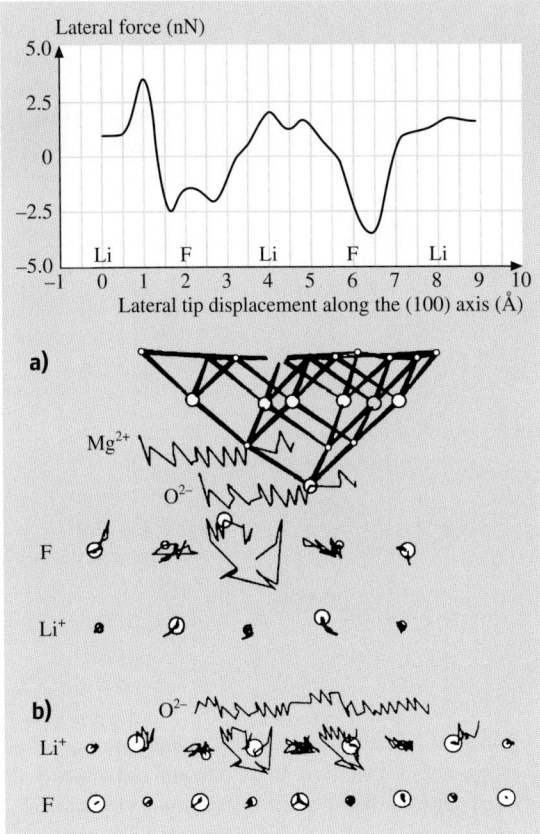

Fig. 13.11. *Top*: The lateral force calculated for a MgO tip scanning in the ⟨001⟩ direction on LiF(001). *Bottom*: A view of the side of the surface plane along the scan direction. The surface Li^+ and F^- atoms are seen to relax to relieve the frictional energy and this relaxation motions is indicated in the figure by the *category lines*. (**a**) How a F^- ion on the surface can be moved into an interstitial site by the tip and then returned to its original position. (**b**) How the relaxation of the surface atoms is reversible (After [125])

which causes the film to "swell". As the tip continues to push against the surface, the hydrocarbon film moves to engulf the tip and wet its sides. In contrast to the strong attractive interactions between clean metal surfaces and tips discussed above, the tip–surface interactions are purely repulsive in the presence of the hydrocarbon film, resulting in surface passivation.

Tupper and *Brenner* used MDsimulations to model the compression of a thiol self-assembled monolayer on a rigid gold surface using both a smooth compressing surface [128] and a compressing surface with an asperity [129]. Compression with the smooth surface results in an apparent compression-induced structural change that leads to a change in slope of the simulated force versus compression curve. This transition, which is reversible, involves a change in the ordered arrangement of the

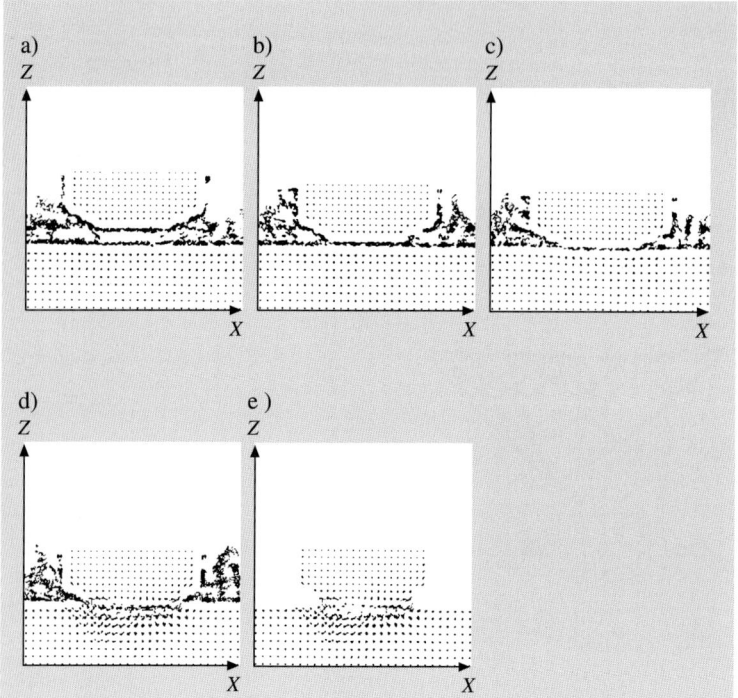

Fig. 13.12. Cutaways of the side view from molecular dynamics simulations of a Ni tip indenting a Au(001) surface covered with a hexadecane film. In (e) only the metal atoms are shown. Note how the hexadecane is forced out from between the metal surfaces (After [87])

sulfur head-groups on the gold surface. Tupper and Brenner noted a similar change in slope in experimental indentation curves by *Houston* and coworkers [130] that had not been apparent prior to the simulations. In the simulations with the asperity it was shown that the asperity is able to penetrate the tail groups of the self-assembled monolayer before an appreciable load is apparent on the compressing surface. This result suggests that it is possible to image the head groups of a thiol self-assembled monolayer that is adsorbed onto the surface of a gold substrate using STM, and therefore, ordered images of these systems may not be indicative of the arrangement of the tail groups.

The end groups on polymer lubricants have a significant influence on the lubrication properties of the polymers [131]. For instance, fluorinated end groups are less reactive than regular alcohol end groups. When fluorinated films are indented, the normal force becomes more attractive as the distance between the tip and film decreases until the hard wall limit is reached and the interactions become repulsive. On the other hand, when AFM tips indent hydrogenated films, the forces become increasingly repulsive as the distance between them decreases, as shown in Figs. 13.13 and 13.14. This is due to the compression of the end group beneath the tip. For the

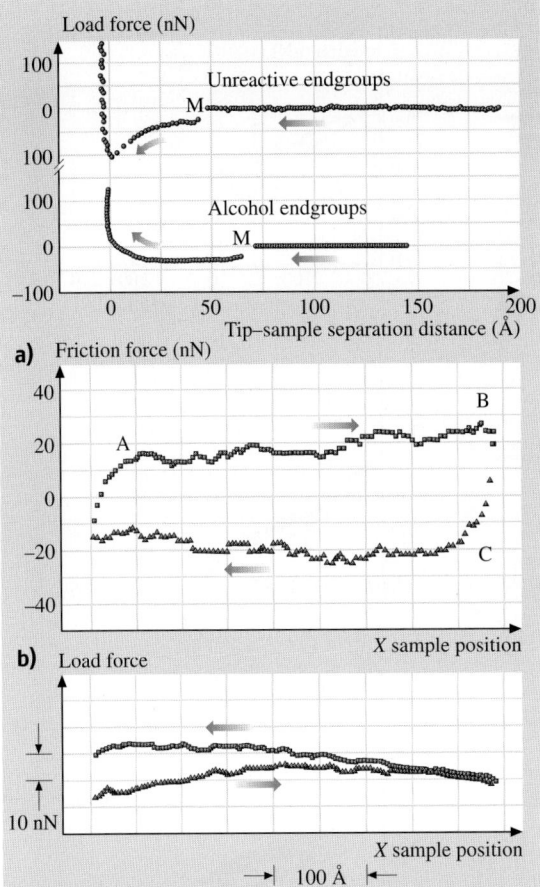

Fig. 13.13. *Top*: The force versus distance curve (indentation part only) for unbonded perfluoropolyether on Si(100). The unreactive end groups were from a 10 Å thick film; the reactive alcohol end groups were from a 30 Å thick film. The negative forces represent attractive interactions between the tip and the surface. *Bottom*: Measured plots of friction (**a**) and load (**b**) forces of the tip as it slides over the sample with the alcohol end groups (After [131])

lubricant molecules to be squeezed out from between the tip and the surface, the hydrogen-bonding between the two must first be broken, which increases the force needed to indent the system. Therefore, a major effect of the alcohol end groups is to dramatically increase the load that a liquid lubricant can withstand before failure (solid–solid contact) occurs.

It has also been shown that the polarity of the lubricant can affect atomic-scale friction [132]. Non-polar polymer lubricants on Si are depleted from the wear track within a few wear cycles, while similar polar lubricants show resistance to degra-

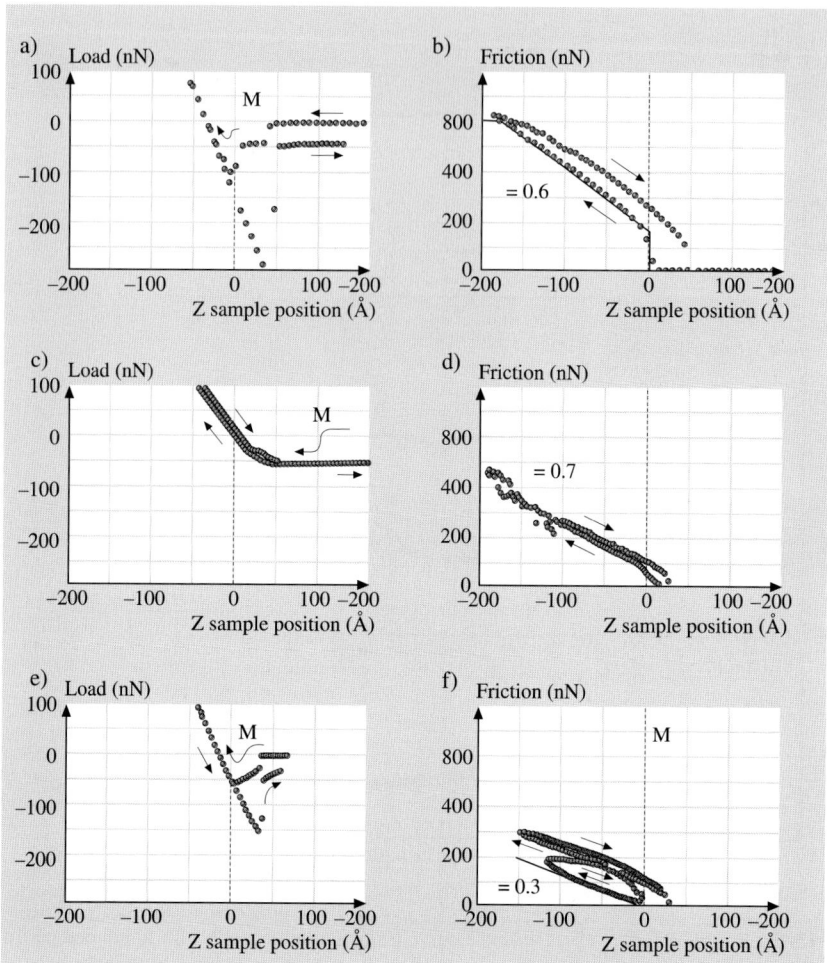

Fig. 13.14. Measured values for friction and load as an atomic-force microscope tip is scanned across a 30 Å thick sample of perfluoropolyether on Si(100). (**a**), (**b**) The unbonded polymer with unreactive end groups. (**c**), (**d**) The unbonded polymer with alcohol end groups. (**e**), (**f**) A bonded polymer (After [131])

dation. In each case, however, thicker films are more durable and behave like soft polymer solids.

Classical MD simulations by *Tutein* et al. [133] have also been used to examine the indentation of monolayers composed of linear hydrocarbon chains that are chemically bound (or anchored) to a diamond substrate. Indentation is accomplished using both a flexible and rigid single-wall, capped nanotubes as the tip. Regardless of the type of nanotube considered, the simulations show that indentation of the hydrocarbon monolayers causes a disruption of the original ordering of the monolayer, the pinning of selected hydrocarbon chains beneath the tube, and the formation of gauche defects within the monolayer. Because nanotubes are stiff along their axial direction, the flexible nanotube is distorted only slightly by its interaction with the softer monolayers. However, interaction with the hard diamond substrate causes the tube to buckle. Severe indents with a rigid nanotube tip result in rupture of chemical bonds within the hydrocarbon monolayer.

This section shows that the repulsive interaction between surfaces covered molecular films and tips result in surface passivation. The lubrication properties of polymers can vary with the nature of the end groupsand the polarity of polymers. In some cases, indentations can disrupt the initial ordering of the polymers and break the bonds within the hydrocarbon monolayer.

13.4 Friction

When the work of sliding is converted into some less ordered form, friction will occur at sliding solid interfaces. In the case of strongly adhering systems, the work of sliding may be converted into damage within the bulk. As the adhesive forces between solid surfaces decreases, the conversion of work changes to mechanical damage at or near the surface, which can lead to the formation of wear debris and transfer films [134, 135].

It should be pointed out that while the thermodynamic principles of the conversion of work to heat are well-known, many of the detailed mechanisms that take place at sliding interfaces are just now being explored in any detail. This is despite the fact that a detailed understanding of these processes is important for producing interfaces with specific friction (and wear) properties.

Atomic-scale friction has been investigated using MD simulations for systems composed of a variety of materials, in a number of geometries. For instance, the atomic-scale friction and wear of diamond surfaces, both atomically flat and rough, has been examined this way by *Harrison* et al. [136–138]. Additionally, atomic-scale friction between monolayers of alkane chains bound to rigid substrates [139], between perfluorocarboxylic acid and hydrocarboxylic Langmuir–Blodgett (LB) monolayers [140], between contacting Cu surfaces [141, 142], between a Si tip and a Si substrate [114, 143], and between contacting diamond surfaces in the presence of third-body molecules [144] have all been examined using MD simulations. These and other studies are discussed below.

13.4.1 Friction Between Bare Sliding Surfaces

When there is no lubricant between two surfaces in sliding contact, the friction is labeled 'dry' even if it is taking place in air. Dry sliding friction can be readily modeled by considering the motion of a single atom over a monoatomic chain, as shown in Fig. 13.15 [145]. The result provides insight into the effect of elastic deformation of the substrate caused by the sliding atom on energy dissipation, as indicated in Fig. 13.16a. This model also shows how the average frictional force varies with

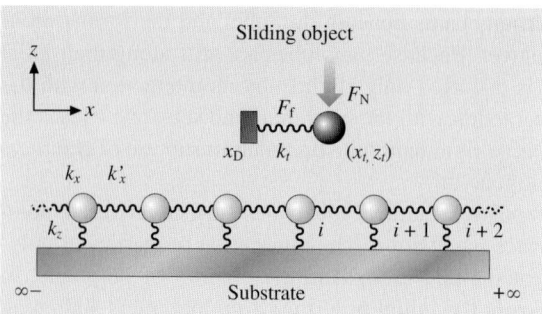

Fig. 13.15. Model of a single atom moving over a monotomic chain. x_D is the location of the sliding object and k_t, k_x, k'_x, and k_z are spring constants (After [145])

the changes in the force constant of the substrate in the direction normal to the scan direction, as indicated in Fig. 13.16b. Interestingly, much of the correct behavior involved in dry sliding friction is captured by simple models such as this. However, more complex MD simulations provide even more information on the process, as discussed below.

MD simulations have been used extensively to study the sliding of metal tips across clean metal surfaces by numerous groups [142, 146–150]. An illustrative case is shown in Figs. 13.17 and 13.18 for a Cu tip indenting Cu(111) and Cu(100), respectively, by *Sorensen* et al. [142]. Adhesion and wear occurs when the attractive force between the atoms on the tip and the atoms at the surface becomes greater than the attractive forces within the tip itself. Atomic-scale stick and slip can occur through nucleation and subsequent motion of dislocations between the close-packed (111) surfaces, leading to kinetic friction. Wear also occurs if part of the tip gets left behind on the surface, as shown in Fig. 13.18. The required forces and other conditions necessary to produce adhesion and wear are clearly indicated in the simulations.

MD simulations can further provide data on how the characteristic 'stick–slip' friction motion can depend on the area of contact, the rate of sliding, and the sliding direction (Fig. 13.19). For example, the simulations show that during the 'slip' state, friction energy is dissipated by the system through an increase in temperature at the interface between the sliding surfaces [142].

In general, the macroscopic scale results of experimental studies show good agreement with the results of these simulations. However, the simulations discussed above assume that the electronic contributions to friction on metals are negligible and experiments have measured a non-negligible contribution of conduction electrons to friction [151]. Thus, future simulations of metal tip/metal substrate interactions using more sophisticated methods that include electronic effects are encouraged.

Layered ceramics, which have weak van der Waals bonds between the layers, have long been known to have good lubricating properties because of the ease with which the layers slide over one another. It is, therefore, not surprising that some of the earliest experimental studies of nanometer-scale friction [17, 152] were on

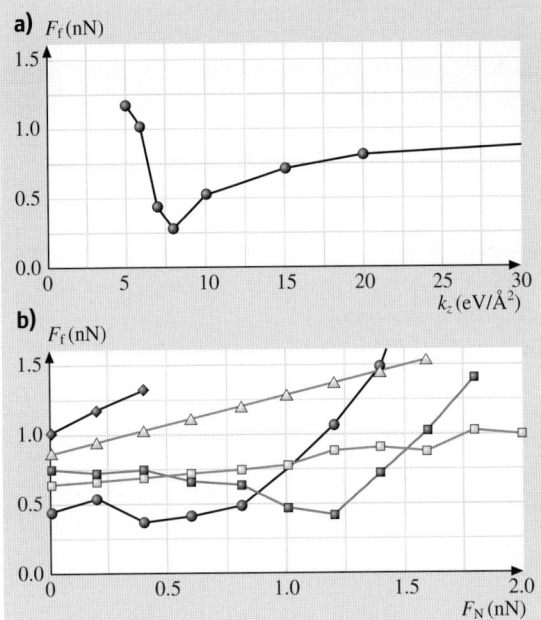

Fig. 13.16. (a) Results for the model shown in Fig. 13.15 for the change in the average frictional force as a function of the perpendicular force constant k_z. The normal force applied was 0.2 nN. (b) The average frictional force versus normal force for various values of k_z. Symbols: *filled diamonds*, $k_z = 5\,\text{eV/Å}^2$, *triangles*, $k_z = \infty$; *filled squares*, $k_z = 15\,\text{eV/Å}^2$, *filled circles*, $k_z = 10\,\text{eV/Å}^2$, *empty squares*, an anharmonic potential of substrate atoms in the perpendicular direction (After [145])

layered compounds such as mica, graphite and MoS$_2$. In these early studies, it was hypothesized that at high loads measured friction forces were related to "incipient sliding" [153, 154] caused by a small flake from the surface becoming attached to the end of the tip. If this were true, then all measured interactions would be between the flake and the surface, which would have a larger contact area than the clean tip. Subsequent simulations of constant force AFM images of graphite by *Tang* et al. [155] show that there is no need for the assumption of a graphite flake under the tip to reproduce the experimental images of a graphite surface.

Layered ceramics show surprisingly strong localized fluctuations in atomic-scale friction [156–159]. For example, square-well signals with sub-Å lateral width are obtained in FFM scans on MoS$_2$(001) in the direction across the scan direction, while sawtooth signals are detected along the scan direction, as shown in Fig. 13.20. A stick–slip model by *Mate* et al. [17] and *Erlandsson* et al. [152], where the tip does a zigzag walk along the scan, is able to explain this finding. Variations in the frictional force with the periodicity of cleavage planes, shown in Figs. 13.21 and 13.22 [152], are consistent with the results of this simple model. However, additional experiments indicate a more complex tip–surface interaction, such as changes

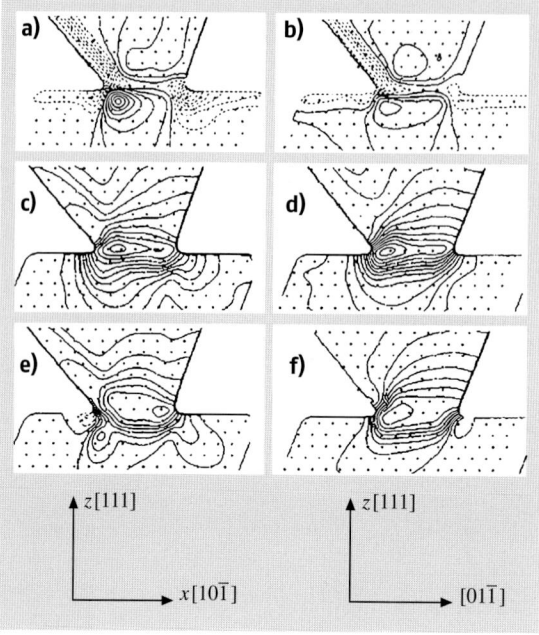

Fig. 13.17. Contour plots of the stresses at the interface between a Cu tip and a Cu(111) surface as the tip slides across the surface. The *dashed lines* indicate negative stresses, the *solid lines* are positive stresses (After [142])

in the intrinsic lateral force between the substrate and the AFM tip (Fig. 13.23) [160] or sliding-induced chemistry between the tip and the surface [93, 161].

Crystalline ceramics differ from layered ceramics in that they are held together by relatively strong covalent or ionic bonds. One crystalline ceramic for which there is an extensive literature related to friction is diamond [162] because, while it is one of the hardest materials known, it also exhibits relatively low friction. It is speculated that energy dissipation during friction on the macroscale might occur by the "ratchet mechanism", where energy is released by the transfer of normal force from one surface asperity to another, or the elastic mechanism, where the released energy comes from elastic strain in an asperity. Decreasing friction coefficients through implantation with, for example, nitrogen, is a well-established method of decreasing the friction coefficient because the implantation processes decreases the shear stress in the diamond. However, when the same surface is scratched with an AFM tip, the wear is larger than prior to implantation [163], which could be due to increased atomic-scale surface roughness caused by the implantation process.

Atomic-scale friction has been measured experimentally [18] for diamond tips with near atomic-scale radii sliding over hydrogen-terminated diamond (111) and (100) surfaces that are able to detect the 2×1 reconstruction on the (100) surface. In addition, the average friction coefficient determined with an AFM on H-terminated diamond (111) surfaces is about two orders of magnitude smaller than the value

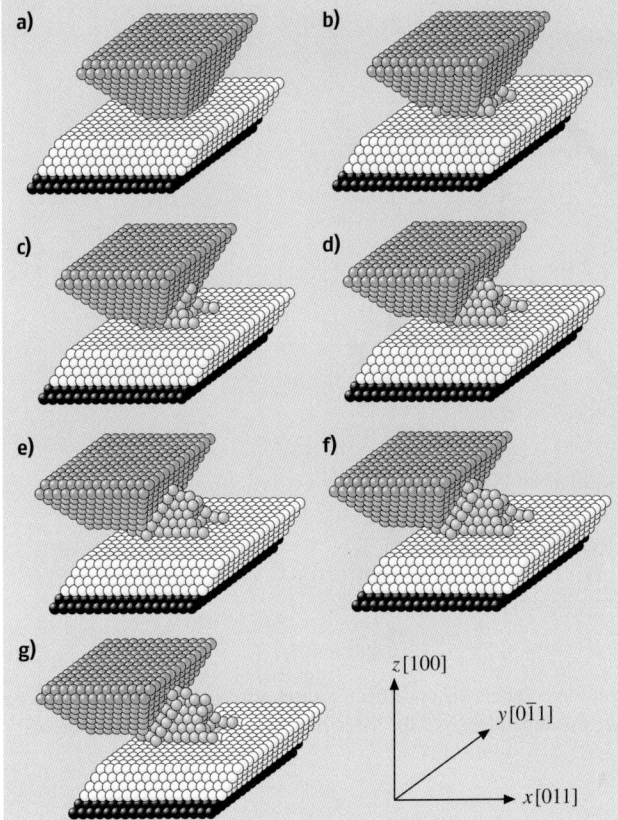

Fig. 13.18. Snapshots from a MD simulation of a Cu tip sliding across a Cu(100) surface. Note the connective neck between the two that is sheared during the sliding, leading to wear of the tip. The simulation was performed at a temperature of 0 K (After [142])

measured on bare, 2 × 1 diamond (111) surfaces, indicating greater adhesion in the latter case [164].

Extensive MD simulations of friction on sliding hydrogen-terminated diamond (111) surfaces have been carried out by *Harrison* et al. [136]. These simulations reveal that the slip–stick atomic-scale motion of the terminating *H* atoms changes depending on the direction of rotation. Sliding in the $[11\bar{2}]$ direction allows the H atoms to "revolve" around one another, thus decreasing the repulsive interaction between the sliding surfaces since the hydrogen atoms are not forced to pass directly over one another [136]. In the [110] sliding direction, the hydrogen atoms are located at diagonals relative to one another and as they begin to interact, they are "pushed" apart in the plane of the surface in response to the repulsive forces between them. Since they are never in contact as they are during sliding in the $[11\bar{2}]$ direction, the overall

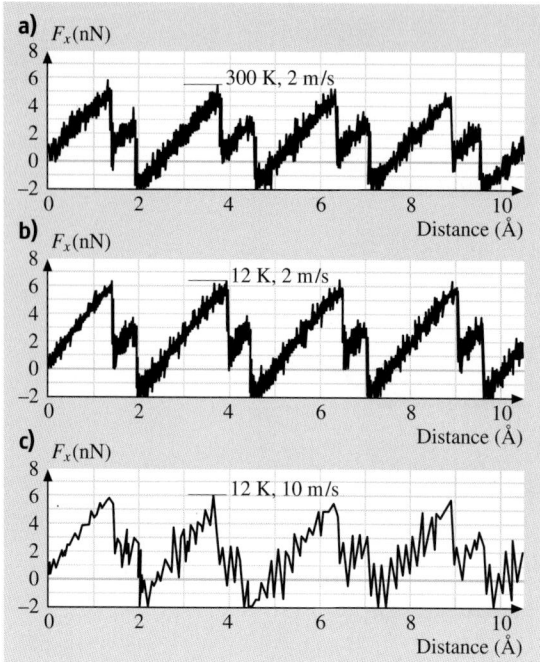

Fig. 13.19. Plots of the lateral force versus distance from a simulation similar to that shown in Fig. 13.18. The plots illustrate the dependence of the force on temperature and sliding velocity (After [142])

friction coefficient is lower and the atomic-scale mechanisms of sliding friction are different. Figures 13.24 and 13.25 summarize these results.

MD simulations by *Mulliah* et al. [165] have been used to study the atomic-scale stick–slip phenomenon of a pyramidal diamond tip inserted into the Ag(010) surface by considering various sliding rates and vertical support displacements. Simulations show that dislocations are related to the stick events emitting a dislocation in the substrate near the tip. In the case of small vertical displacements, the scratch in the substrate is discrete because of tip jumping over the surface. A continuous scratch, however, is formed at high displacements, such as 15 Å. The dynamic friction coefficient and the static friction coefficient increase with increasing depth. The tip moves continuously through a stick and slip motion at large depths, whereas it comes to a halt for shallow indents. Although the sliding rate can change the exact points of stick and slip, the distribution of sliding rates over the range of values considered in this study (1.0 to 5.0 m/s) have no influence on the damage to the substrate, the atomistic stick–slip mechanisms, or the friction coefficients.

Finally, diamond tips are frequently used in friction measurements due to the high mechanical strength of diamond and the belief that such tips are wear resistant. However, diamond tips that were used to scratch diamond and Si surfaces and then imaged showed significant wear that increased with the increasing hardness of the

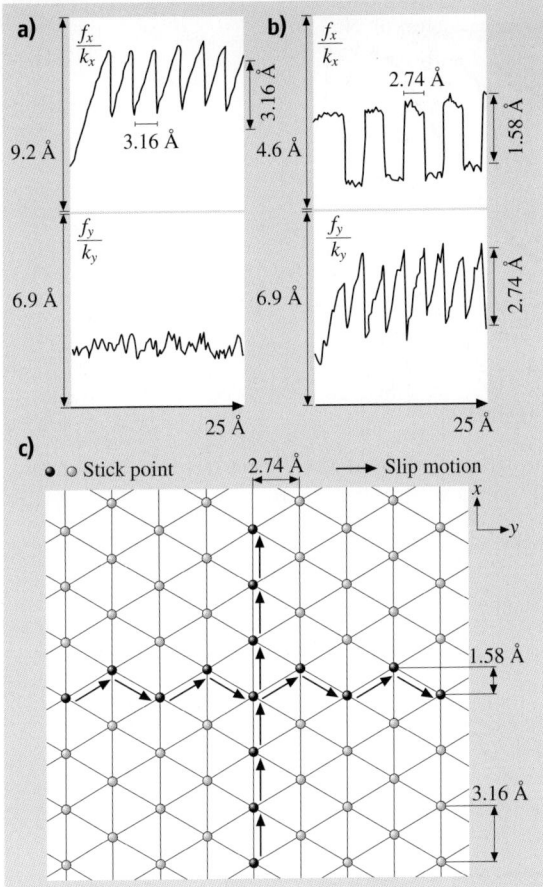

Fig. 13.20. Displacement data from a scan across a MoS$_2$(001) surface. The data in (**a**) and (**b**) are form scans along the *x*- and *y*-directions, respectively, on the surface shown in (**c**) (After [159])

tested material [166, 167]. This wear affected the shape of the tip, which could affect the contact area and measured friction values.

In short, MD simulations provide insight into dry sliding friction and the sliding of metal tips across clean metal surfaces. Stick–slip friction or wear can occur depending on the sliding conditions. In the case of ceramics, layered ceramics have good lubricating properties due to their weak interlayer bonds and show the localized fluctuations in atomic-scale friction. Crystalline ceramics, such as diamond, exhibit relatively low friction and their friction coefficients can be reduced by the implantation. MD simulations also show that the slip–stick atomic-scale motion of the hydrogen-terminated diamond (111) surfaces changes with the sliding direction and the applied load.

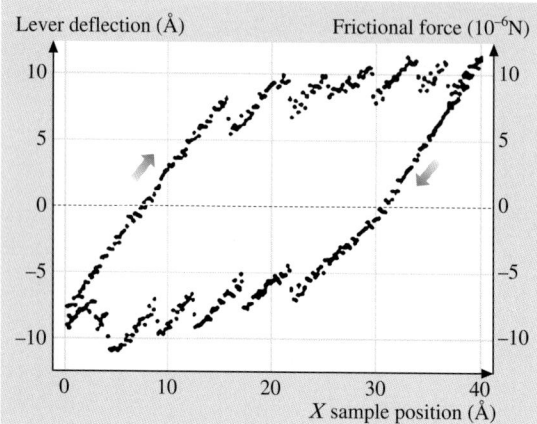

Fig. 13.21. Experimental data for the deflection of an AFM lever and, therefore, friction force for a mica sample scanned repeatedly with a W tip (After [152])

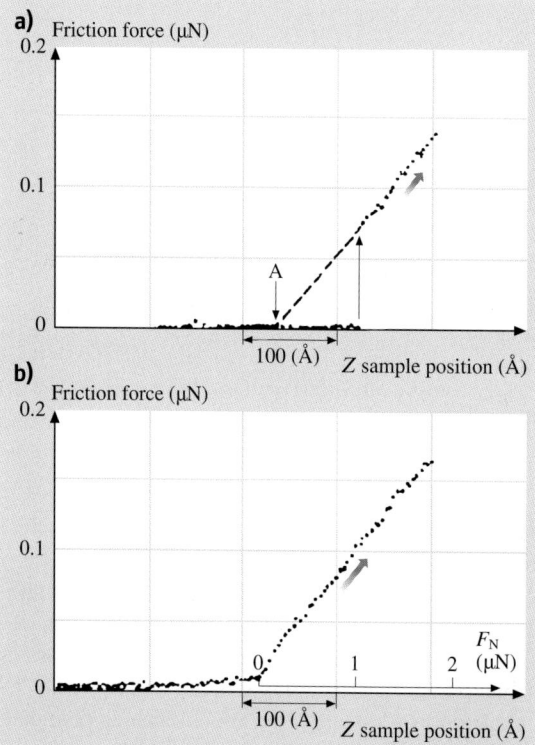

Fig. 13.22. Experimentally measured frictional force on mica as a friction of sample position as the sample is moved closer to the tip. (**a**) The sample approaches from far away. (**b**) The sample approaches after being retracted. The point marked "A" shows the point where the extrapolated data in (**a**) intersect the x-axis (After [152])

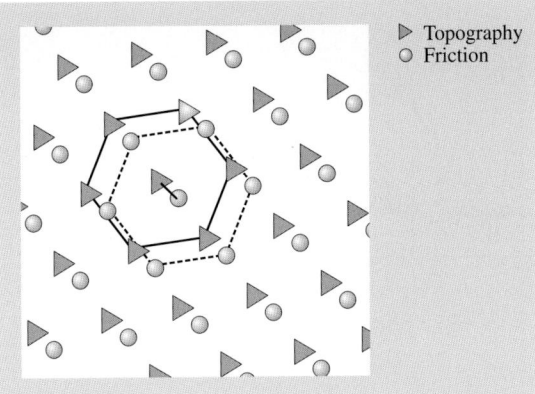

Fig. 13.23. The relationship between measured topography and friction on a freshly cleaved, highly oriented pyrolytic graphite surface. The figure shows that, due to the lateral forces, only every other carbon atom would be seen with a normal force profile (After [160])

13.4.2 Friction in the Presence of a Third Body

While dry sliding friction assumes that ambient gas particles have no direct effect on friction study results, MD simulations show that free particles at an interface can influence friction to a surprisingly large degree. This is exemplified by the results of *Perry* and *Harrison* [144, 168, 169] who used MD simulations to investigate the effect of trapping small hydrocarbon molecules on the atomic-scale fiction of two (111) crystal faces of diamond. These trapped molecules, sometimes referred to as third-body molecules, might represent hydrocarbon contamination trapped between contacting surfaces prior to a "dry" sliding experiment, or hydrocarbon debris formed by wear. In separate studies, Perry et al. considered the effects of methane (CH_4), ethane (C_2H_6), and isobutane $(CH_3)_3CH$ molecules (Fig. 13.26). Examination of the average frictional force as a function of load reveals that, for all three systems, the frictional force generally increases as the load increases, as shown in Fig. 13.27. However, the presence of each of the third-body molecules markedly reduces the average frictional force compared to the results predicted for pristine hydrogen-terminated surfaces.

It is found that the presence of the molecules significantly reduces friction, especially at high loads where the molecules act as a boundary layer (Fig. 13.28). The simulations also show the manner in which the frictional forces at a given load vary as the larger ethane and isobutene molecules change orientation during sliding. Molecular configurations that lead to increased interactions between the diamond surfaces result in higher average frictional force. Despite the fact that ethane and isobutane cause the two sliding surfaces to be further apart during sliding than the methane molecules, friction is predicted to be larger than with methane because the larger molecules interact more with the diamond surfaces while sliding.

Different behavior is observed by *Harrison* et al. [137, 138, 170, 171] when similar hydrocarbon molecules (methyl, ethyl, and *n*-propyl groups) are chemisorbed to

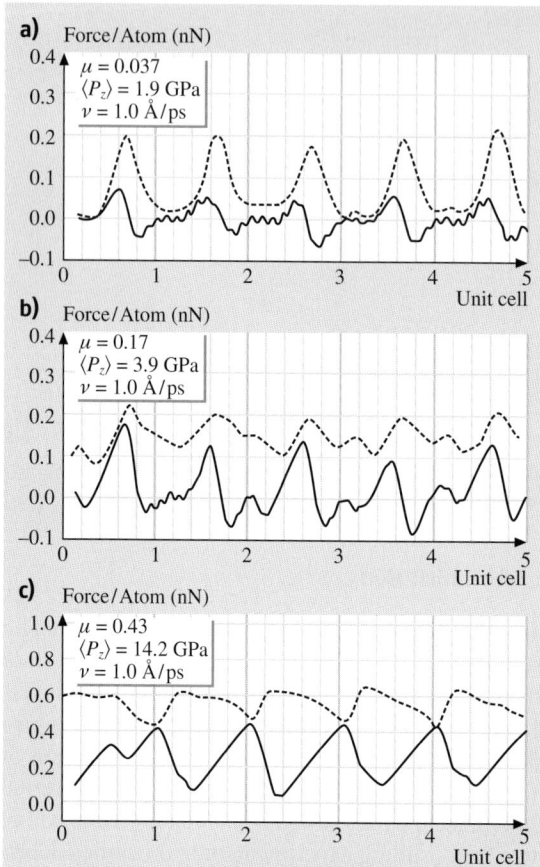

Fig. 13.24. Calculated frictional force (*solid lines*) and normal force (*dashed lines*) felt by a hydrogen-terminated diamond (111) surface as it slides against another hydrogen-terminated diamond (111) surface in a MD simulation. The sliding direction is the [11$\bar{2}$] direction at a speed of 1.0 Å/ps at room temperature. The *three plots* show how the simulated stick–slip motion changes as a function of the applied load (After [136])

one of the sliding diamond surfaces. For example, methyl-termination does not decrease friction to a significant degree but does result in frictional forces that are approximately the same as in the case of hydrogen-terminated diamond surfaces [172]. The trapped methane molecules lower the frictional force more than the chemisorbed methyl groups, but the frictional force versus normal force data are comparable for the ethyl-terminated and ethane systems, with the former giving slightly higher frictional forces. The attached groups have less freedom to move around, which decreases their mechanical interaction with the surface. Finally, the simulations provide insight into the rich, nonequilibrium tribochemistry that can occur between lubricants and sliding surfaces, summarized in Fig. 13.29 [173]. This simulation is unique in

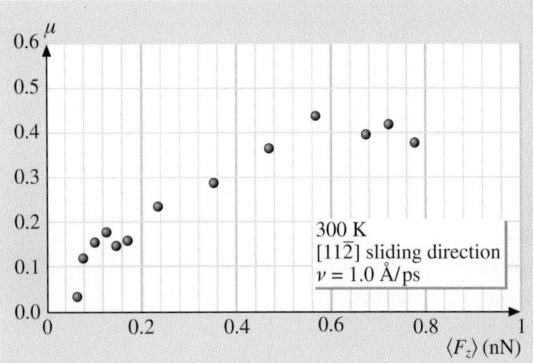

Fig. 13.25. A plot of the friction coefficient versus applied load determined from the MD simulations described for Fig. 13.24 (After [136])

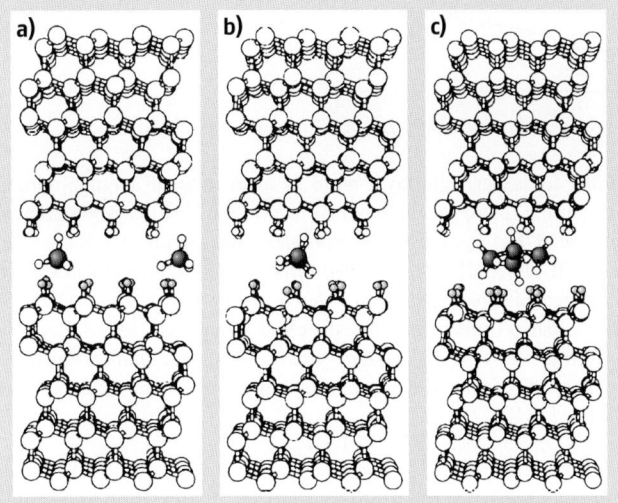

Fig. 13.26. Initial configuration at low load for the diamond plus third-body molecules systems. These systems are composed of two diamond surfaces, viewed along the [110] direction, and two methane molecules in (**a**), one ethane molecule in (**b**), and one isobutane molecule in (**c**). *Large white* and *dark gray spheres* represent carbon atoms of the diamond surfaces and the third-body molecules, respectively. *Small gray spheres* represent hydrogen atoms of the lower diamond surface. Hydrogen atoms of the upper diamond surface and the third body molecules are both represented by *small white spheres*. Sliding is achieved by moving the rigid layers of the upper surface from *left* to *right* in the figure (After [144])

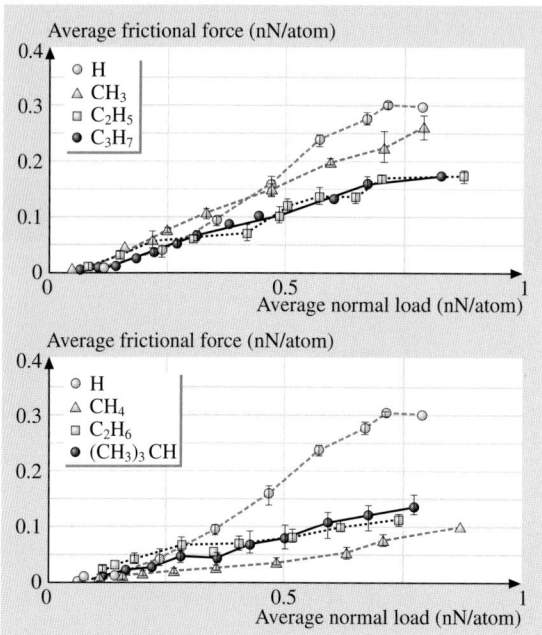

Fig. 13.27. Average frictional force per rigid-layer atom as a function of average normal load per rigid-layer atom for sliding the upper diamond surface in the [11$\bar{2}$] crystallographic direction. Data for the methane (CH_4) system (*open triangles*), the ethane (C_2H_6) system (*open squares*), the isobutane [$(CH_3)_3CH$] system (*filled circles*), and diamond surfaces in the absence of third-body molecules (*open circles*) are shown in the *lower panel*. Data for the methyl-terminated ($-CH_3$) system (*open triangles*), the ethyl-terminated ($-C_2H_5$) system (*open squares*), the n-propyl-terminated ($-C_3H_7$) system (*filled circles*), and diamond surfaces in the absence of third-body molecules (*open circles*) are shown in the *upper panel*. *Lines* have been drawn to aid the eye (After [144])

that it shows atomic-scale mechanisms for the degradation of lubricant molecules due to friction. Similar types of debris molecules have been observed in macroscopic experiments that have examined the friction between diamond surfaces [174]. The presence of small impurity molecules or atoms on thin metal films can be expected to affect the film's properties. Calculations have shown that resistivity changes in the metal are strongly dependent on the nature of the adsorption bond [175]. When this result is applied to the atomic-scale friction results obtained with the QCM, the sliding of adsorbate structures on metal surfaces can be understood to be a combination of electron excitation and lattice vibrations. When the adsorbate is very different chemically from the surface on which it is sliding, other interesting quantum effects can come into play. For example, the electronic frictional forces acting on small, inert atoms and molecules, such as C_2H_6 and Xe, sliding on metal surfaces have been calculated by *Persson* [176], where the metal surface is approximated by a jellium (electron gas) model. It is found that the Pauli repulsive and attractive van der Waals

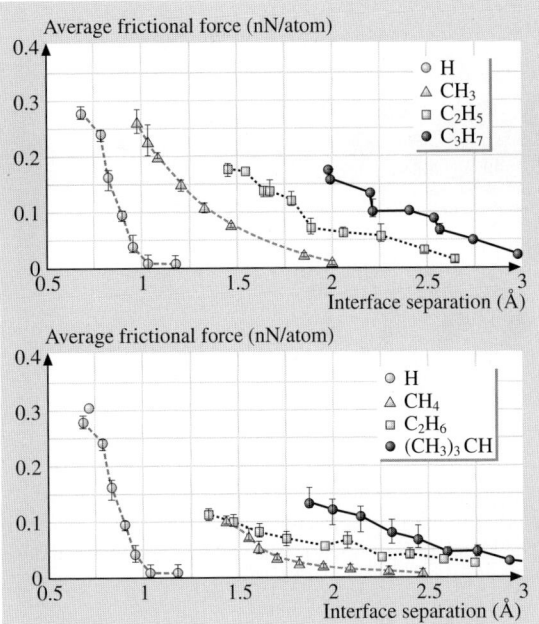

Fig. 13.28. Average frictional force per rigid-layer atom as a function of interface separation for sliding the upper diamond surface in the [11$\bar{2}$] crystallographic direction. Data for the third-body and the chemisorbed systems are shown in the *lower* and *upper* panels, respectively. The *symbols* are the same as in Fig. 13.27. *Lines* have been drawn to aid the eye (After [144])

forces, and therefore, the dissipative forces are of similar magnitudes. The calculated electronic friction agrees well with the values derived from surface resistivity by *Grabhorn* et al. [29, 177] and QCM measurements. Therefore, it was concluded that parallel friction is mainly due to electronic effects while perpendicular friction is phononic in nature in this system.

To summarize these findings, MD simulations show that the average frictional force dramatically decreases in system with third-body molecules, especially at high loads. Simulations also provide information about the tribochemistry that can occur between lubricants and sliding surfaces. Furthermore, the presence of small molecules on thin metal films can influence film properties, such as resistivity.

13.5 Lubrication

As discussed at the beginning of the last section, friction at sliding solid interfaces results from the conversion of the work of sliding into other, less ordered forms. For adhering systems, the work of sliding may be converted into damage within the bulk. For very weakly adhesive forces, friction can still occur through the conversion of

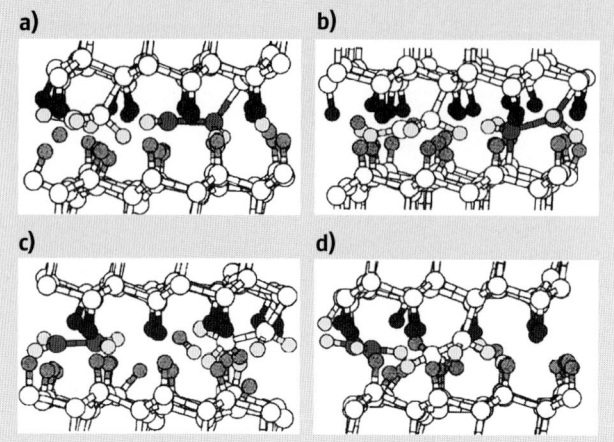

Fig. 13.29. Snapshots from a MD simulation of the sliding of two hydrogen-terminated diamond (111) surfaces against one another in the [11$\bar{2}$] direction. One of the surface has two ethyl fragments chemisorbed to it. The simulation shows how the sliding friction can induce chemistry at the interface (After [173])

work to heat at the interface with no permanent damage to the surface. The latter regime, which is achieved through the presence of lubricating materials, is the topic of this section.

The simplest type of lubricating thin film consists of small molecules that are analogous to wear debris that can "roll" between the sliding surfaces or that represent very short-chain bonded lubricants. These conditions were addressed in Sect. 13.4.2. The rest of this section will consider the effects of larger particles, including liquids, nanoparticles, self-assembled thin films, and solid thin films.

13.5.1 Liquid Films

The behavior of liquids consisting of spherical molecules during sliding has been well-characterized experimentally using the SFA and MD simulations by *Berman* et al. [178]. The SFA experiments consider liquid layers that vary in thickness from one to three layers. The stick–slip motion at the interface increases in a quantized fashion as the number of lubricant layers decrease. When the sliding surfaces are not moving, there are no external forces in the system and the solid–lubricant interactions are strong enough to force the liquid molecules to form a close-packed structure that is solid-like in nature. As a result, the two surfaces bond to each other through the lubricant. When the surfaces start to slide, lateral shear forces are introduced that steadily increase. This causes the molecules in the liquid to undergo small lateral displacements that changes the film thickness. If the shear forces increase past a critical value, the film will disorder, or melt, which allows the surfaces to slide easily past each other, still in a quantized manner. When the shear force falls below the critical value, the liquid again becomes solid-like and the slipping stops. At this point,

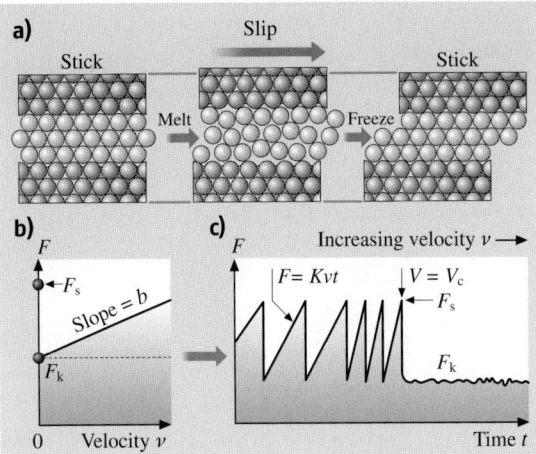

Fig. 13.30. The stick/slip transition that occurs for thin films of liquid between two sliding solid surfaces. F is the intrinsic friction and F_s is the friction where the liquid is in the rigid state, F_k is the friction where the liquid is in the liquidlike state (After [178])

the liquid can no longer be characterized by simple viscous forces. This sequence of events is nicely illustrated in Fig. 13.30.

Molecular dynamics simulations by *Persson* et al. [179] show the mechanism by which this sharp transition occurs. Specifically, in the case of sliding on insulating crystal surfaces, the solid-state lubricant may be in a "superlubric" state where the friction becomes negligible. It is clear from the simulations, however, that even a low concentration of surface defects will convert the lubricant from the solid state to the fluid state. When sliding occurs on metallic surfaces above cryogenic temperatures, the electronic contributions to friction are no longer zero and no superlubric state is possible.

If the applied normal pressure is large enough, the fluid molecules can be squeezed out from between the two confining surfaces [180]. The fact that liquid molecules close to a stiff surface are strongly layered in the direction perpendicular to the surface explains the experimental observation of a $(n \rightarrow n-1)$ layer transition, where n is number of monolayers, that is observed as the normal load increases [181]. Nucleation theory can be used to calculate the critical pressure and determine the spreading dynamics of the $(n-1)$ island.

If the liquid molecules are non-spherical, it is more difficult for them to align themselves and solidify. MD simulations show that spherical molecules have higher critical velocities than branched molecules. For example, MD simulations by *Thompson* and *Robbins* [182] show that when the molecules are branched, it is not the friction that changes with sliding speed but rather the amount of time various parts of the system spend in the sticking and sliding modes. The critical velocity can also depend on the number of liquid layers in the film, the structure and relative orientation of the two sliding surfaces, the applied load, and stiffness of the surfaces.

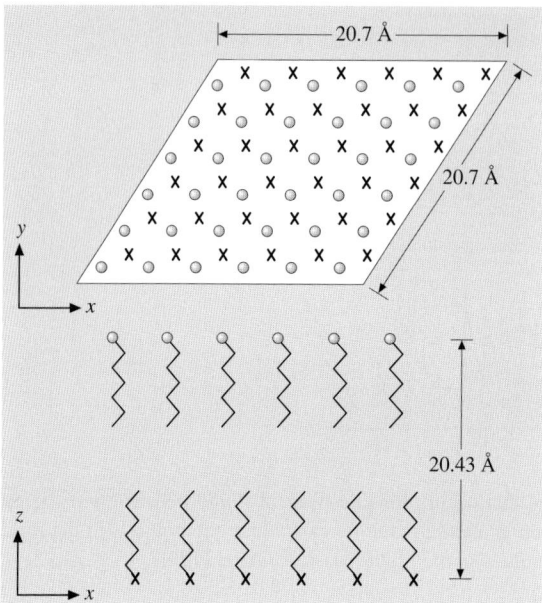

Fig. 13.31. Top and side views of the alkane chains attached to surfaces that are sliding against each other (After [139])

Boundary layer friction can also be affected by the reactivity of the liquid molecules, as shown by *Persson* [183]. Inert molecules interact only weakly with the sliding surfaces. As a result, as the rate of sliding increases, the molecules convert from the solid state to the liquid state in an abrupt manner. On the other hand, when the molecules are reactive, they have strong interactions with the surfaces and consequently undergo a gradual transition from the solid to the liquid state. If the molecules are attached to one of the surfaces, the transitions can be abrupt, especially if there are large separations between the chains.

Most lubricating liquids consist of long-chain hydrocarbons. MD simulations by *Glosli* et al. [139] were used to model sliding between two ordered monolayers of alkane chains attached to two rigid substrates, shown schematically in Fig. 13.31. Energy dissipation occurs by a discontinuous plucking mechanism (sudden release of shear strain) or a viscous mechanism (continuous collisions of atoms of opposite films), depending on the interfacial interaction strength. The "pluck" occurs when mechanical energy stored as strain is converted into thermal energy. Therefore, at low temperatures, the friction force is low. In contrast, at higher temperatures, some of the energy of sliding is dissipated through phonon excitations, increasing the frictional force. However, this trend reverses again at the highest temperatures where the molecules move so much that they slide easily over the surfaces and the frictional force decreases. These results are summarized in Figs. 13.32 and 13.33.

Landman et al. [184] also used MD simulations to study the sliding of two Au(111) surfaces with pyramidal asperities with n-$C_{16}H_{34}$ lubricant molecules

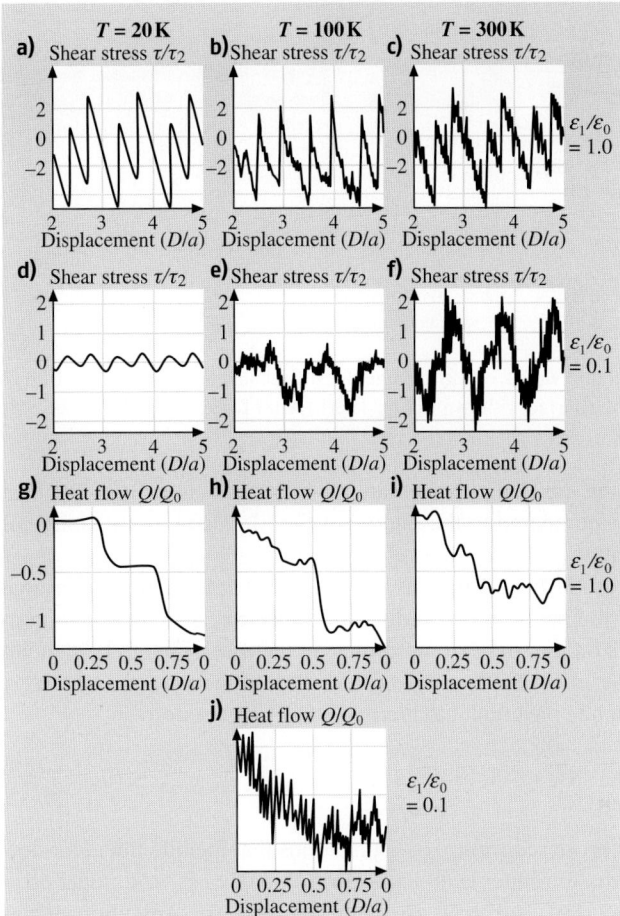

Fig. 13.32. Data from MD simulations of the sliding of the surfaces shown in Fig. 13.31. (a)–(f) The shear stress and (g)–(j) the heat flow as a function of sliding for normal and reduced interfacial strengths. The plots show how the calculated values change with system temperature (After [139])

trapped between them, as shown in Fig. 13.34. The sliding rate in the simulations is about 10 m/s, which is the same order of magnitude as the scanning speed in a computer disk. As the asperities approach each other, the hydrocarbon molecules began to form layers, which is reflected in the oscillations in the frictional force shown in Fig. 13.35. When the asperities overlap in height and approach each other laterally, the pressure of the lubricant molecules increases to about 4 GPa and causes the Au asperities to deform.

Similar simulations by *Manias* et al. examined the shear of entangled oligomer chains on sliding surfaces, as shown in Fig. 13.36 [185]. Slip takes place within

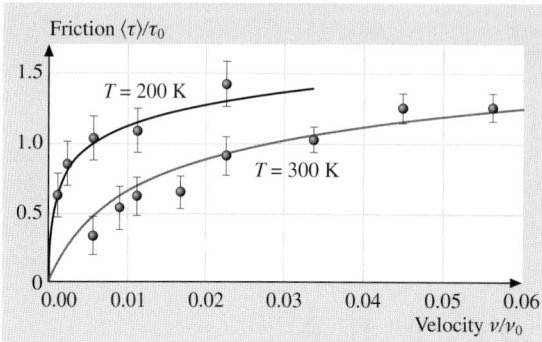

Fig. 13.33. A plot of calculated values of the average interfacial shear stress as a function of the sliding velocity of the two surfaces shown in Fig. 13.31 (After [139])

the film through changes in the chain conformations. Increased viscosity is found at the film-surface interface compared to the middle of the film, leading to a range of viscosities across the film.

In short, experiments and MD simulations show similar stick/slip transitions that occur for thin films of liquid between two sliding solid surfaces. The frictional properties can vary depending on the shape of molecules. In the case of long-chain molecules, the temperature can affect the frictional force because the mechanical energy stored in long-chains can be converted into thermal energy by friction.

13.5.2 Nanoparticles

Nanoparticles have been proposed for use in applications that include fibers for new composite materials, wires for nanometer-scale electronic devices, and novel catalysts or catalytic supports [186]. They have also been proposed as novel lubricant materials as they may function as "nano-ballbearings" with exceptionally low friction coefficients. This has spurred the study of the frictional properties of C_{60} [187–199], carbon nanotubes [200–207], and MoO_3 nanoparticles on MoS_2 surfaces [208, 209], among others [210].

There is a spread in the experimentally measured friction coefficients for C_{60} as shown in Table 13.1. These variations may be caused by differences in the experimental methods used, thickness of nanoparticle layer or island, atmosphere (argon versus air, levels of humidity), or transfer of C_{60} to FFM tips in some cases. Thus, there is much that remains to be clarified about the tribological behavior of C_{60} films.

Some experimental studies show evidence of C_{60} molecules rolling against the substrate, each other, or the sliding surfaces [187, 192, 194, 197, 199]. However, others hypothesize that the low friction of C_{60} films is due, in part, to blunting of the tip by transfer of fullerene molecules to the tip apex. For instance, when the nanomechanical properties of thin C_{60} films grown epitaxially on GeS (001) are measured experimentally [211], the local friction coefficient for the fullerene islands is much smaller than on the bare substrate, which is a layered material. However, in another

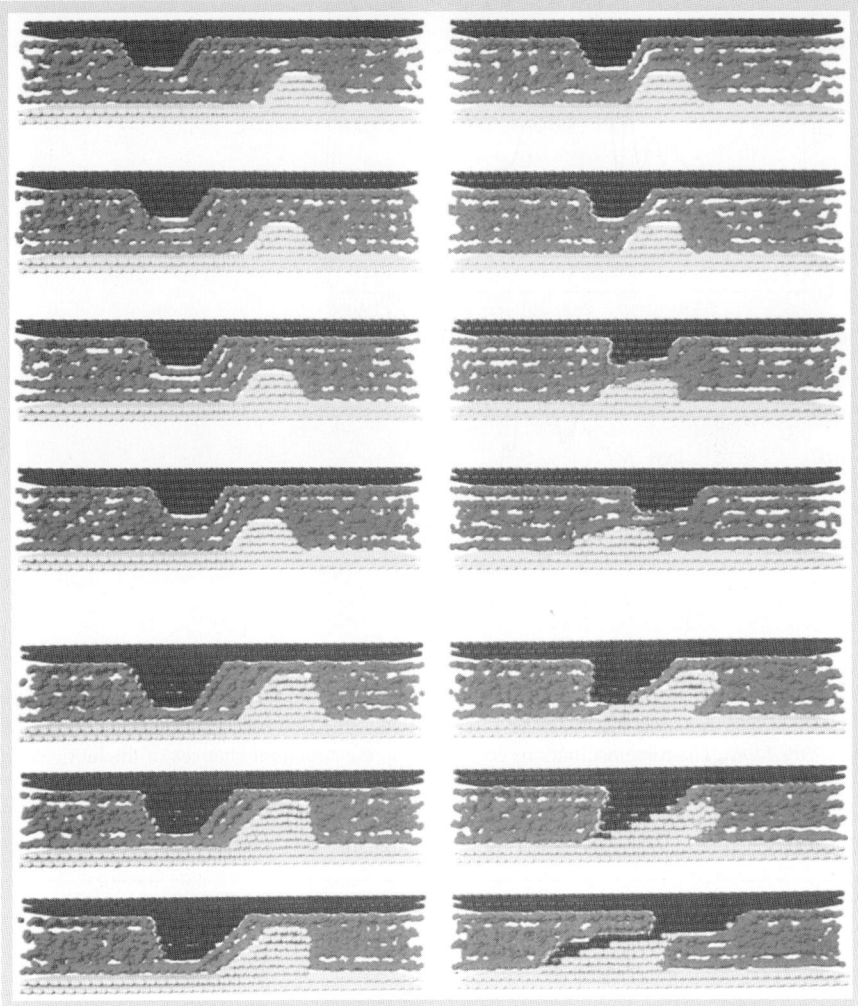

Fig. 13.34. Stills from a MD simulation where Au(111) surfaces with surface roughness slide over one another while separated by hexadecane molecules. The scanning velocity is 10 m/s. Layering of the lubricant and asperity deformation occurs as the sliding continues. The *top four rows* show the results when the asperity heights are separated by 4.6 Å. The *bottom three rows* show the results when the asperity heights are separated by −6.7 Å (After [184])

study, using and SFM, of C_{60} island films on NaCl surfaces [212], friction is higher on the C_{60} islands than on the bare NaCl(100) surface by a ratio of 1:3 for NaCl:C_{60}.

Fullerene films are found experimentally to have dissipation energies and shear strengths that are a full order of magnitude lower than the values that are typical for boundary lubricants [213]. Experimental testing of the frictional properties of fullerenes has revealed low mechanical stability when they are only physisorbed

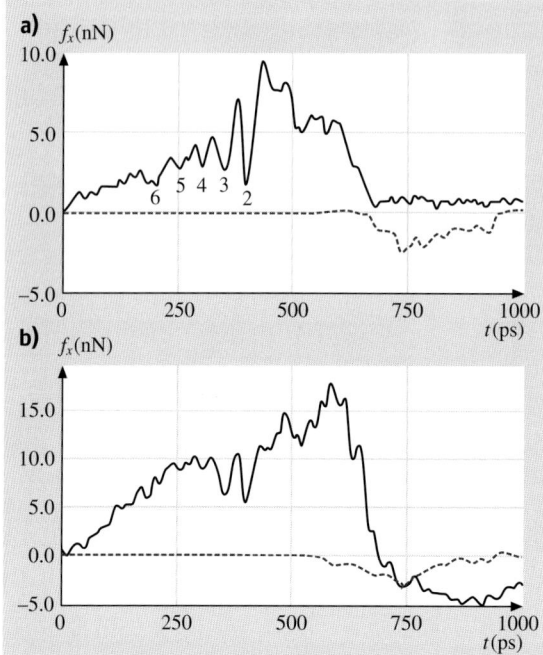

Fig. 13.35. The lateral force (f_x) and normal force (f_z) from the MD simulations shown in Fig. 13.34 as a function of time. The forces between the two metal surfaces are shown by the *dashed line*. The force oscillations correspond to the structural changes of the lubricant in Fig. 13.34 (After [184])

Table 13.1. Experimentally measured friction coefficients for C_{60}

System	Experimental method	Friction coefficient	Reference
Sublimed C_{60} films on Au-coated mica	AFM	0.06	[195]
Sublimed C_{60} films on Au-coated mica	Steel ball	0.08	[187]
C_{60} islands on NaCl	AFM	0.15 ± 0.05	[191]
Epitaxial monolayers of C_{60} on GeS	AFM	0.665 ± 0.012	[194]
Sublimed C_{60} on GeS	AFM	0.67 ± 0.22, 0.82 ± 0.33	[196]
C_{60} on Si and mica	AFM	0.9	[188]

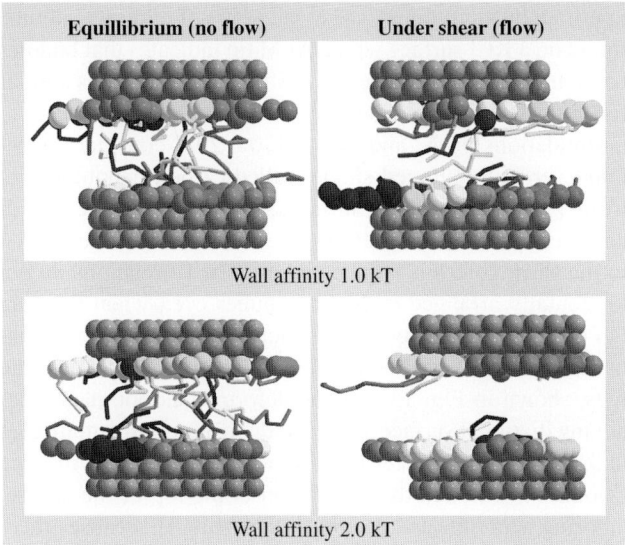

Fig. 13.36. Changes in the conformation of adsorbed hydrocarbon chains on weakly (*top*) and strongly (*bottom*) physisorbing surfaces at equilibrium and under shear (After [185])

on a solid surface [214], where progressive wear and transfer of fullerene materials occurrs readily. Measurements with a FFM show that under certain conditions, adsorbed fullerene films deteriorate at pressures as low as about 0.1 GPa [215]. Therefore, the challenge is to obtain mechanically stable, ordered molecular films of fullerenes that are firmly attached to a solid substrate.

Legoas et al. [216] investigated the experimentally observed low-friction system of C_{60} molecules positioned on highly oriented pylolitic graphite using MD simulations. The results show that decreasing the van der Waals interaction between a C_{60} monolayer and graphite sheets, and the characteristic movements of graphite flakes over C_{60} monolayers, can explain the ultra-low friction of C_{60} molecules and graphite sheets.

Related molecular dynamics simulations by *Buldum* and *Lu* and *Schall* and *Brenner* [200, 203] indicate that carbon nanotubes roll when their honeycomb lattice is "in registry" with the honeycomb lattice of the graphite. When this registry is not present, the carbon nanotubes slide in response to being pushed by a AFM tip. These MD simulation findings were subsequently confirmed in experimental studies by *Flavo* et al. [202]. Experimental studies of multi-walled carbon nanotubes on graphite [206] show similar evidence of the rolling of the outer tube when pushed. The carbon nanotubes are observed to roll more easily under Ar than in air.

It is likely that in any actual lubrication application would utilize bundles of carbon nanotubes rather than individual nanotubes. This is not only because individual carbon nanotubes are too small to be useful outside of perhaps nanometer-scale machines, but also because carbon nanotubes naturally form bundles that are difficult to disperse [186]. An experimental study by *Miura* et al. [207] of carbon nanotube bundles being pushed around on a KCl surface with a AFM tip indicates that bundles of single-walled carbon nanotubes can be induced to roll in a manner that is similar to the rolling observed for multi-walled nanotubes.

Molecular dynamics simulations by *Ni* and *Sinnot* [204, 205] consider the responses of horizontally and vertically aligned single-walled carbon nanotubes between two, hydrogen-terminated diamond (111) surfaces, where the top one is sliding relative to the bottom one. For both orientations, the movement of the carbon nanotubes in response to the shear forces is simple sliding. The simulations do not predict rolling of the horizontally arranged carbon nanotubes even when they are aligned with each other in two-layer and three-layer structures. Rather, at low compressive forces, shown in Fig. 13.37a, the nanotube bundle slides as a single unit and at high compressive forces, shown in Fig. 13.37b, the deformed carbon nanotubes closest to the topmost moving diamond surface start to slide in a motion reminiscent of the movement of a tank or bulldozer wheel belt. However, when these moving carbon nanotube atoms would have turned the first corner at the top of the ellipse, they encounter the neighboring nanotube and cannot slide past it. This causes them to deform even further, form cross-links with one another, and, in some cases, move in the reverse direction to the sliding motion of the diamond surface. This causes the large oscillations in the normal and lateral forces plotted in Fig. 13.37c.

At high compression between the two diamond surfaces, the responses of the horizontally arranged carbon nanotubes are substantially different from the responses of the vertically arranged nanotubes, as can be seen by comparing Figs. 13.37b and 13.38a. The flexibility of the vertical, capped carbon nanotubes allows the tubes to bend and buckle. As the buckle is forming, the normal force decreases than stabilizes in the buckled structure shown in Fig. 13.38a. As the topmost diamond surface slides, the buckled nanotubes swing around the buckle "neck", which helps dissipate the applied stresses. For this reason, the magnitudes of the lateral forces are not significantly different for the vertical nanotubes at low and high compression, as indicated in Fig. 13.38b.

When the ratio of the frictional (lateral) force to the normal force is taken to calculate friction coefficients for these systems, high, non-intuitive values were obtained. As outlined in [204], this is because the actual contact area of the nanotubes are not proportional to the sliding force. In the case of the horizontal nanotube bundles, the tubes are able to deform and significantly change their contact area with the sliding surface with minimal change in the normal force, as shown in Fig. 13.37c. In the case of the vertical nanotubes, the contact area remains approximately the same regardless of the initial loading force because of the flexibility of the nanotubes. This causes the lateral forces to change only slightly with significant changes in the normal force, as shown in Fig. 13.38b. This analysis indicates that care must be taken in calculating friction coefficients for nanotubes systems. Recent experiments by *Dickrell* et al. [217] show good agreement with these predictions.

Experiments using AFM and STM by *Sheehan* et al. and *Wang* et al. [208, 209] show that molybdenum oxide (MoO_3) nanocrystals on single-crystal molybdenum disulfide (MoS_2) surfaces can slide easily along the specific direction due to the

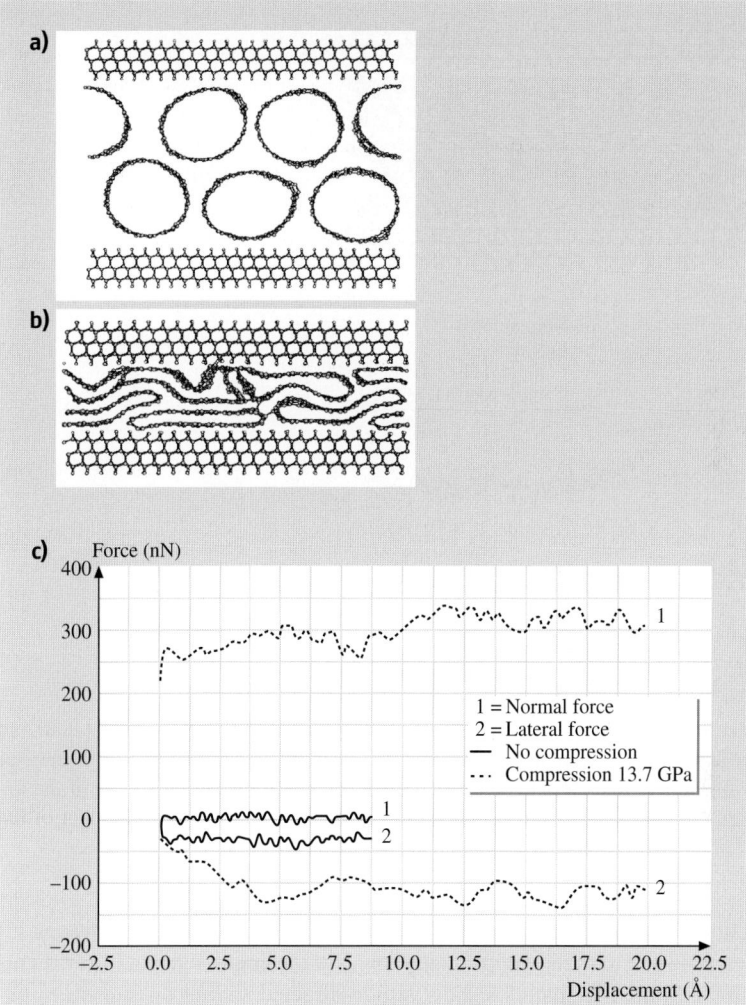

Fig. 13.37. (**a**),(**b**) Snapshots from simulations that examine the sliding of the potential of the topmost diamond surface on horizontally arranged nanotubes at different compressions; (**a**) is at a pressure of ≈ 0 GPa; (**b**) is with a pressure of 13.7 GPa. (**c**) Plots of the normal and lateral components of force during sliding of the top diamonds surface on horizontally arranged nanotubes as a function of the displacement of the top diamond surface with respect to the diamond surface on the bottom

atomic structure of the nanocrystal-substrate interface constraining the motion of the nanocrystals to the lattice rows of the substrate, a phenomenon termed lattice-directed sliding. The energy per unit area required to move the MoO_3 nanocrystals can be decreased by an order of magnitude compared to the energy required to slide macroscopic MoS_2-bearing contact depending on the preferred sliding direction.

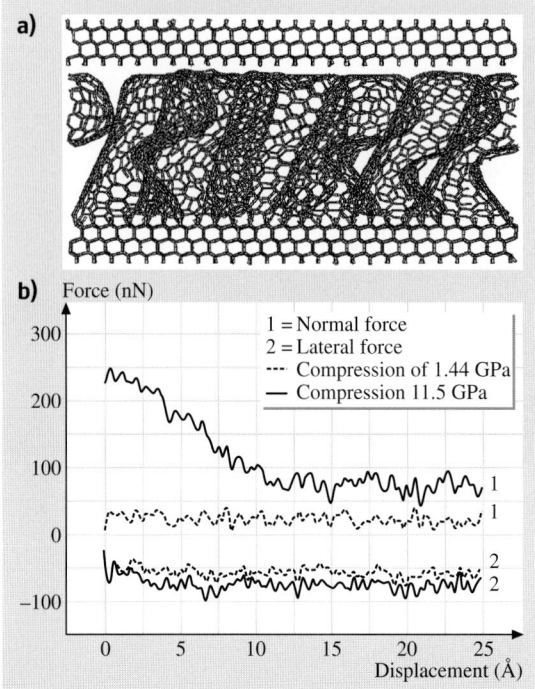

Fig. 13.38. (a) Snapshots from simulations that examine the sliding of the topmost diamond surface on vertically arranged nanotubes with one set of capped ends compressed at a pressure of 11.5 GPa. (b) Plots of the normal and lateral components of force during sliding of the top diamonds surface on vertically arranged nanotubes as a function of the displacement of the top diamond surface with respect to the diamond surface on the bottom

This section shows that nanoparticles and nanotubes show some promise as lubricating materials due to their exceptionally low friction coefficients in experiments and simulations. Some nanopaticles show lattice-directed sliding on substrates because of their unique atomic structures. Nanotubes may be used in bundles as a result of their small size and the associated difficulties in dispersing them.

13.5.3 Self-Assembled Thin Films

Some solid polymers can act like liquid polymer films if there is room for the backbone to move in response to the sliding frictional forces. Films of this type are called "self-assembled". There have been numerous experimental studies of friction on self-assembled LB films that are deposited on solid surfaces with AFM/FFM. These experiments show the relationships among elastic compliance, topography and friction on thin LB films [218]. For example, AFM and related methods are sensitive enough to detect differences in the adhesive interactions between the microscope tips and the CH_3 to CF_3 end groups [82]. The experiments show that the fluorocarbon domains generally exhibit higher friction than the hydrocarbon films, which is attributed to the lower elasticity modulus of the fluorocarbon films that results in a larger contact area between the tip and the sample [218–220].

The frictional properties of self-assembled films can be affected by molecular disorder of the alkyl chains at the surface if the layers are not packed too closely together [221]. When the film is indented, disorder occurs in the chains, which then compress as the tip continues to press against them. If the tip presses hard enough, the film can harden due to the repulsive forces between the compressed chains. In contrast, if the films have been tilted, they bend or deform when the tip pushes on them in a mostly elastic fashion. This results in long lubrication lifetimes before failure of the lubricant occurs. In general, at low contact loads (about 10^{-8} N), wear can occur at defect sites, such as steps, due to shear forces. However, wear can also occur due to particularly strong adhesive forces between the film and the surface [222].

Molecular dynamics simulations of self-assembled monolayers sliding by *Mikulski* and *Harrison* [223, 224] show that periodicities observed in a number of system quantities are a result of the tight packing of the monolayer and the commensurate structure of the diamond counterface. The packing and commensurability of the system force synchronized motion of the chains during sliding contact. Both highly and loosely packed systems give a similar average friction at low loads. Under high loads, the tightly packed monolayer shows significantly lower friction than the loosely packed monolayer. While the movement of chains is greatly constricted in both systems, the tightly packed monolayer under high loads is clearly more uniform in geometry and more constrained with respect to the movement of individual chains than the loosely packed monolayer. This means that efficient packing of the chains is responsible for the lower friction of tight packed monolayers under high load. This is supported by the fact that sliding initiates larger bond-length fluctuations in the loosely packed system, which ultimately leads to more energy dissipation via vibration.

Recent simulations by *Harrison* and coworkers [225] examined the link between friction and disorder in monolayers composed of *n*-alkane chains. The tribological behavior of tightly packed, pure monolayers composed of chains containing 14 carbon atoms was compared to mixed monolayers that randomly combine equal amounts of 12 and 16 carbon-atom chains. When sliding in the direction of chain cant under repulsive (positive) loads, pure monolayers consistently show lower friction than mixed monolayers. This trend was also observed in AFM experiments of mixed-length alkanethiols [226] and spiroalkanedithiols on Au [227]. The distribution of contact forces between individual monolayer chain groups and the tip showed pure and mixed monolayers resist tip motion in the same way. (Contact force was defined as the force between the tip and a $-CH_3$- or a $-CH_2$-group in the alkane chains.) In contrast, the contact forces "pushing" the tip along differ in the two monolayers. The pure monolayers exhibit a high level of symmetry between resisting and pushing forces, which results in a lower net friction than the mixed monolayers. In other words, the ordered, densely packed nature of the pure monolayers allows the energy stored when the monolayer is resisting tip motion (positive forces) to be regained efficiently when the monolayer "pushes" on the tip (negative forces). The shape of the distribution of negative contact forces in the mixed monolayers is different from the shape of the positive distribution. Thus, mechanical energy is not efficiently channeled back into the mixed monolayer as the tip passes over the chains. The increased range of motion of the protruding tails of the chains allows for the enhanced dissipation of energy.

Both systems exhibit a marked friction anisotropy. The contact force distribution changes dramatically as a result of the change in sliding direction, resulting in an increase in friction. Upon continued sliding in the direction perpendicular to chain cant, both types of monolayers are capable of transitioning to a state where the chains are primarily oriented with the cant along the sliding direction. A large change in the distribution of contact forces and a reduction in friction accompany this transition.

Recently, MD simulations were used to examine the response of monolayers composed of alkyne chains, which contain diacetylene moieties, to compression and shear [228]. There are the only simulations to date where both compression and shear result in cross-linking, or polymerization, between chains. Irregular polymerization patterns appear among the carbon backbones. The vertical positioning of the diacetylene moieties within the alkyne chains (spacer length) and the sliding direction have an influence on the pattern of cross-linking and friction. In addition, chemical reactions between the chains of the monolayer and the amorphous carbon tip occur when diacetylene moieties are located at the ends of the chains closest to the tip. These adhesive interactions increase friction.

Zhang et al. [229] studied the friction of alkanethiol self-assembled monolayers on Au(111) using hybrid molecular and dynamic element model simulations at the AFM/FFM experimental time scale. They investigated various parameters that could influence frictional properties, such as chain length, terminal group, scan direction, and scan velocity. The simulations show that the frictional force decreases as the chain lengths increase and is smallest when scanned along the tilt direction. The simulations also show that a maximum friction coefficient occurs for hydrophobic $-CH_3$-terminated monolayers that decreases for hydrophilic $-OH$-terminated monolayers as scan velocity increases. The friction coefficient saturated to a constant value at high scan velocities for both surfaces.

In short, both experiments and simulations show the relationship among elastic properties, degree of molecular disorder, and friction of self-assembled thin films.

13.5.4 Solid Thin Films

Liquid lubricants allow surfaces to slide over each other at high loads with a minimum of resistance from friction. Solid materials can also fulfill these functions and, when they do, are considered to be solid lubricants. Generally, these are solids with friction coefficients of 0.3 or less and low wear rates.

Bowden and *Tabor* showed how thin solid films can reduce friction using the following expression [230]

$$F_f = AF_s + F_p , \qquad (13.15)$$

where F_f is the friction force, F_p is the plowing term, A is the area of contact and F_S is the shear strength of the interface. If the surfaces under the solid film are very stiff, A and F_p will decrease, thereby decreasing friction. However, if the surfaces are soft, F_S will be reduced while the other parameters will increase. The properties specific to the film will have an effect on friction as well. For example, if the films are less than 1 μm thick, the surface asperities will be able to break through the film to eventually cause wear between the surfaces under normal circumstances. However, if the lubricant film is too thick, there will be increased plowing of the film that causes the frictional forces to increase. Lastly, it has been pointed out that the lubricant must not delaminate in response to the frictional forces, so strong bonds between the lubricant and the surface are required for the lubricant to be effective.

The most common materials used as solid lubricants have layered structures such as graphite or MoS_2, that, as discussed above, experience low friction. Nevertheless, not all layered structures are lubricants. For instance, mica gives a relatively high coefficient of friction (> 1). It is also important to note that it is not necessary for the lubricant film to have a layered structure of have low friction. For example, DLC has some of the lowest coefficients of friction measured and yet does not have a layered structure.

Molecular dynamics simulations by *Gao* et al. [231] were conducted to investigate the atomic-scale friction and wear when hydrogen-terminated diamond (111) counterfaces are in sliding contact with diamond (111) surfaces coated with amorphous, hydrogen-free carbon films, namely DLC films. Two films, with approximately the same ratio of sp^3-to-sp^2 carbon but different thicknesses, were examined. Both systems give a similar average friction in the load range examined. Above a critical load, a series of tribochemical reactions occur resulting in a significant restructuring of the film. This restructuring is analogous to the "run-in" observed in macroscopic friction experiments and reduces the friction. The contribution of adhesion between the probe (counterface) and the sample to friction was examined by varying the saturation of the counterface. Decreasing the degree of counterface saturation, by reducing the hydrogen termination, increases the friction. Finally, the contribution of long-range interactions to friction was examined by using two potential energy functions that differ only in their long-range forces to examine friction in the same system.

More recently, *Gao* et al. [232] used MD simulations to examine the effects of sp^2-to-sp^3 carbon ratio and surface hydrogen on the mechanical and tribological properties of amorphous carbon films. This work showed that, in addition to the sp^2-to-sp^3 ratio of carbon, the three-dimensional structure of the films is important when determining their mechanical properties. For example, it is possible to have high sp^2 carbon content, which is normally associated with softer films, and large elastic constants. This occurs when sp^2 ringlike structures are oriented perpendicular to the compression direction. The hydrogen-free films in this work were layered, which led to novel mechanical behavior that influences the shape of the friction versus load data.

There have also been numerous detailed experiments with the QCM of the sliding of liquid and solid noble-gas materials over metal surfaces [20, 233]. Many of

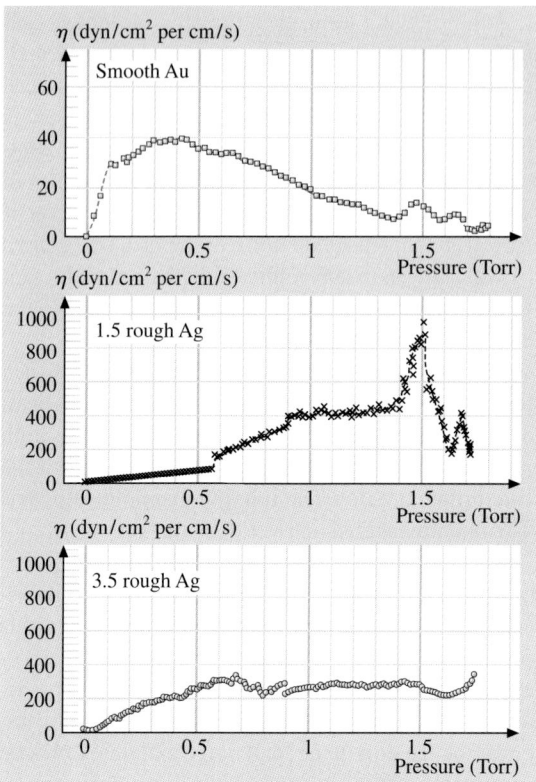

Fig. 13.39. Measured interfacial viscosity versus pressure for the sliding of Kr films on smooth Au surfaces and rough Ag surfaces. The experiments were performed with a QCM (After [20])

the results are non-intuitive and have stimulated much work in the area of atomic-scale friction. For instance, solid layers are observed to be much more sensitive to surface morphology than liquids. This was conclusively shown when solid films of Kr were observed to slide more easily on smooth Au than liquid Kr films, while the reverse was true for the Ag substrate with a respective surface area 1.5 times that of a geometrically flat surface (Fig. 13.39). In addition, the force required to slide one and two atom thick solid films of Xe on the Ag(111) surface was measured and it was determined that the friction of the bilayer is about 25% greater than the friction of the single layer.

Molecular dynamics simulations by *Cieplak* et al. [234] have also been performed to study the movement of Kr films on defect free surfaces of Au(111). They do good job of explaining the results obtained with the QCM, although electronic mechanisms for energy dissipation are not included. In particular, they reveal that the more mobile liquid Kr atoms move around until they adsorbed in the sites between the surface atoms. The solid Kr atoms were more constrained and could not adsorb as easily.

Electronic friction is usually not expected to be much larger for two layers of Xe than for a monolayer. However, computational results have shown that phonon-dissipation mechanisms might be much greater for two layers compared to a monolayer. When the QCM was used to slide molecularly thin Xe films along Ag(111) [233, 235], it was discovered that the friction due to sliding depends greatly on the Xe film's lattice spacing, since Xe can form an uncompressed close-packed triangular solid monolayer that is highly incommensurate with the Ag surface. The film accommodates more atoms by compressing until it reaches the spacing of solid bulk Xe and forms a second layer gas phase. As stated above, a comparison of the force required to slide one- or two-atom-thick solid Xe films across the surface showed that the friction in the latter case is about 25% greater than in the case of a monolayer. This confirms the results of recent MD simulations by *Persson* [236] that suggest that phonon friction can be much greater for bilayers than for single monolayers due to the increased number of vibrational modes available to dissipate mechanical energy. The experiments also show that the magnitude of the frictional force is determined mostly by the first layer, which is a strong indicator of the localized nature of interface frictional energy dissipation mechanisms.

Finally, much work has been done to study adhesion and friction of solid, thin polymer films [237]. For example, molecular scale stick–slip was measured on two layers with an FFM [238]. Frictional domains are observed and the frictional contrast is measured as the sliding direction is varied. The measurements with FFMs [239] on surfaces covered with thin organic films demonstrate that a single monolayer is sufficient to completely change the frictional behavior of surfaces [81, 219].

It has been suggested that there is a relationship between frictional forces and the adhesion strength of polymers [240] that has been verified by recent experiments [241, 242] and simulations [243]. The results show that the relationship deviates from the expected behavior when shear and rupture take place at different locations. When these two processes occur internally, there is a relationship between the frictional force during uniform shear and the excess work during rupture, which has also been seen in the experiments. The simulations also showed that solid polymer films shear and rupture through phase transformations, and while these transitions are taking place, individual atoms can move at velocities near the speed of sound. This contributes to the energy dissipation in a manner that is independent of the sliding speed. In the experiments, shear produces rapid internal arrangements that strengthened the film followed by atomic-scale stick–slip at the interface. This interfacial sliding reduces the interfacial shear strength and adhesive failure occurred through cavitation, plastic flow and bridge rupture. The results indicate that bridge rupture increases as the polymer chain length gets longer and the chains can tangle with each other.

Lateral force microscopy studies of oriented solid polymers can be performed with molecular resolution [244]. A high degree of orientation is obtained in the polymer layer by pressing and sliding a clean, heated mica surface across the film surface.

Stick–slip is observed in scanning directions opposite to the chain directions while smooth films are observed along the chain directions.

13.6 Conclusions

In this chapter, relatively complete discussion of the contribution that molecular dynamics and related simulations are making in the area of nanotribology is presented. The examples discussed above make it clear that these approaches are providing exciting insights into friction, wear, and related processes at the atomic scale that could not have been obtained in any other way. Furthermore, the synergy between these simulations and new experimental techniques such as the surface force apparatus and proximal probe microscopes is producing a revolution in our understanding of the origin of friction at its most fundamental atomic level.

Acknowledgement

S.-J. H. and S. B. S. acknowledge support from the Air Force through grant FA9550-04-1-0367 and from the National Science Foundation supported Network for Computational Nanotechnology (EEC-0228390). J. A. H. acknowledges support from the Air Force through grants NMIPR045203577 and NMIPR045203924 and from the Office of Naval Research via grant N0001404WX20212. D. W. B. acknowledges support from the Office of Naval Research through grant N00014-04-2006 and from the National Science Foundation through grant DMR-0304299.

References

1. D. Dowson. *History of Tribology*. Longman, 1979.
2. K. L. Johnson, K. Kendell, and A. D. Roberts. Surface energy, the contact of elastic solids. *Proc. R. Soc. Lond. A*, 324:301–313, 1971.
3. J. Krim. Friction at the atomic scale. *Sci. Am.*, 275:74–80, 1996.
4. J. Krim. Atomic-scale origins of friction. *Langmuir*, 12:4564–4566, 1996.
5. J. Krim. Experimental probes of atomic scale friction. *Comments Cond. Mater. Phys.*, 17:263–280, 1995.
6. A. P. Sutton. Deformation mechanisms, electronic conductance, friction of metallic nanocontacts. *Curr. Opin. Solid. State M*, 1:827–833, 1996.
7. C. M. Mate. Force microscopy studies of the molecular origins of friction, lubrication. *IBM J. Res. Dev.*, 39:617–627, 1995.
8. A. M. Stoneham, M. M. D. Ramos, and A. P. Sutton. How do they stick together – the statics, dynamics of interfaces. *Philos. Mag. A*, 67:797–811, 1993.
9. I. L. Singer. Friction and energy dissipation at the atomic scale: A review. *J. Vac. Sci. Technol. A*, 12:2605–2616, 1994.
10. B. Bhushan, J. N. Israelachvili, and U. Landman. Nanotribology – friction, wear, lubrication at the atomic-scale. *Nature*, 374:607–616, 1995.
11. B. Bhushan. *Handbook of Micro/Nanotechnology*. CRC, 1995.

12. J. B. Sokoloff. Theory of atomic level sliding friction between ideal crystal interfaces. *J. Appl. Phys.*, 72:1262–1270, 1992.
13. W. Zhong, G. Overney, and D. Tomanek. Theory of atomic force microscopy on elastic surfaces, the structure of surfaces 111. In S. Y. Tong, M. S. Van Hove, X. Xide, and K. Takayanagi, editors, *Proceedings of the third International Conf. on the Structure of Surfaces*, page 243, Berlin, 1991. Springer-Verlag.
14. I. L. Singer and H. M. Pollock. *Macroscopic, Microscopic Processes*. Kluwer, 1992.
15. J. N. Israelachvili. *Intermolecular and Surface Forces: With Applications to Colloidal and Biological Systems*. Academic Press, 1992.
16. G. Binnig, C. F. Quate, and C. Gerber. Atomic force microscope. *Phys. Rev. Lett.*, 56:930–933, 1986.
17. C. M. Mate, G. M. Mcclelland, R. Erlandsson, and S. Chiang. Atomic-scale friction of a tungsten tip on a graphite surface. *Phys. Rev. Lett.*, 59:1942–1945, 1987.
18. G. J. Germann, S. R. Cohen, G. Neubauer, G. M. Mcclelland, H. Seki, and D. Coulman. Atomic scale friction of a diamond tip on diamond (100)-surface, (111)-surface. *J. Appl. Phys.*, 73:163–167, 1993.
19. R. W. Carpick and M. Salmeron. Scratching the surface: Fundamental investigations of tribology with atomic force microscopy. *Chem. Rev.*, 97:1163–1194, 1997.
20. J. Krim, D. H. Solina, and R. Chiarello. Nanotribology of a Kr monolayer – a quartz-crystal microbalance study of atomic-scale friction. *Phys. Rev. Lett.*, 66:181–184, 1991.
21. W. Zhong and D. Tomanek. First-principles theory of atomic-scale friction. *Phys. Rev. Lett.*, 64:3054–3057, 1990.
22. G. A. Tomlinson. A molecular theory of friction. *Philos. Mag. Ser. 7*, 7:905–939, 1929.
23. F. C. Frenkel and T. Kontorova. On the theory of plastic demortation and twinning. *Zh. Eksp. Teor. Fiz.*, 8:1340, 1938.
24. J. B. Sokoloff. Theory of dynamical friction between idealized sliding surfaces. *Surf. Sci.*, 144:267–272, 1984.
25. J. B. Sokoloff. Theory of energy-dissipation in sliding crystal-surfaces. *Phys. Rev. B*, 42:760–765, 1990.
26. J. B. Sokoloff. Possible nearly frictionless sliding for mesoscopic solids. *Phys. Rev. Lett.*, 71:3450–3453, 1993.
27. B. N. J. Persson and E. Tosatti. *Physics of Sliding Friction*. Kluwer, 1996.
28. J. S. Helman, W. Baltensperger, and J. A. Holyst. Simple-model for dry friction. *Phys. Rev. B*, 49:3831–3838, 1994.
29. B. N. J. Persson, D. Schumacher, and A. Otto. Surface resistivity, vibrational damping in adsorbed layers. *Chem. Phys. Lett.*, 178:204–212, 1991.
30. C. W. Gear. *Numerical Initial Value Problems in Ordinary Differential Equations*. Prentice-Hall, 1971.
31. W. G. Hoover. *Molecular Dynamics*. Springer, 1986.
32. D. W. Heermann. *Computer Simulation Methods in Theoretical Physics*. Springer, 1986.
33. M. P. Allen and D. J. Tildesley. *Computer Simulation of Liquids*. Oxford, 1987.
34. J. M. Haile. *Molecular Dynamics Simulation: Elementary Methods*. Wiley, 1992.
35. D. W. Brenner and B. J. Garrison. Gas surface-reactions – molecular-dynamics simulations of real systems. *Adv. Chem. Phys.*, 76:281–334, 1989.
36. D. W. Brenner. The art, science of an analytic potential. *Phys. Status Solidi B*, 217:23–40, 2000.
37. R. Car and M. Parrinello. Unified approach for molecular-dynamics, density-functional theory. *Phys. Rev. Lett.*, 55:2471–2474, 1985.

38. M. Menon and R. E. Allen. New technique for molecular-dynamics computer-simulations – Hellmann–Feynman theorem, subspace Hamiltonian approach. *Phys. Rev. B*, 33:7099–7101, 1986.
39. O. F. Sankey and R. E. Allen. Atomic forces from electronic energies via the hellmann–feynman theorem, with application to semiconductor (110) surface relaxation. *Phys. Rev. B*, 33:7164–7171, 1986.
40. P. Ewald. Die berechnung optischer und elektrostatischer gitterpotentiale. *Ann. Phys.*, 64:253–287, 1921.
41. D. M. Heyes. Electrostatic potentials, fields in infinite point-charge lattices. *J. Chem. Phys.*, 74:1924–1929, 1981.
42. F. H. Stillinger and T. A. Weber. Computer-simulation of local order in condensed phases of silicon. *Phys. Rev. B*, 31:5262–5271, 1985.
43. J. N. Murrell, S. Carter, S. C. Farantos, P. Huxley, and A. J. C. Varandas. *Molecular Potential Energy Functions*. Wiley, 1984.
44. M. W. Finnis and J. E. Sinclair. A simple empirical n-body potential for transition-metals. *Philos. Mag. A*, 50:45–55, 1984.
45. S. M. Foiles, M. I. Baskes, and M. S. Daw. Embedded-atom-method functions for the fcc metals Cu, Ag, Au, Ni, Pd, Pt, and their alloys. *Phys. Rev. B*, 33:7983–7991, 1986.
46. F. Ercolessi, M. Parrinello, and E. Tosatti. Au(100) reconstruction in the glue model. *Surf. Sci.*, 177:314–328, 1986.
47. F. Ercolessi, E. Tosatti, and M. Parrinello. Au (100) surface reconstruction. *Phys. Rev. Lett.*, 57:719–722, 1986.
48. A. P. Sutton. *Electronic Structure of Materials*. Clarendon, 1993.
49. G. C. Abell. Empirical chemical pseudopotential theory of molecular and metallic bonding. *Phys. Rev. B*, 31:6184–6196, 1985.
50. J. Tersoff. New empirical-model for the structural-properties of silicon. *Phys. Rev. Lett.*, 56:632–635, 1986.
51. J. Tersoff. Modeling solid-state chemistry – interatomic potentials for multicomponent systems. *Phys. Rev. B*, 39:5566–5568, 1989.
52. D. W. Brenner. Tersoff-type potentials for carbon, hydrogen, oxygen. *Mater. Res. Soc. Symp. Proc.*, 141:59–65, 1989.
53. D. W. Brenner. Empirical potential for hydrocarbons for use in simulating the chemical vapor-deposition of diamond films. *Phys. Rev. B*, 42:9458–9471, 1990.
54. K. E. Khor and S. Dassarma. Proposed universal interatomic potential for elemental tetrahedrally bonded semiconductors. *Phys. Rev. B*, 38:3318–3322, 1988.
55. D. W. Brenner. Relationship between the embedded-atom method, Tersoff potentials. *Phys. Rev. Lett.*, 63:1022–1022, 1989.
56. B. G. Dick and A. W. Overhauser. Theory of the dielectric constant of alkali halide crystals. *Phys. Rev.*, 112:90–103, 1958.
57. F. H. Streitz and J. W. Mintmire. Electrostatic potentials for metal-oxide surfaces and interfaces. *Phys. Rev. B*, 50:11996–12003, 1994.
58. S. Ogata, H. Iyetomi, K. Tsuruta, F. Shimojo, R. K. Kalia, A. Nakano, and P. Vashishta. Variable-charge interatomic potentials for molecular-dynamics simulations of TiO_2. *J. Appl. Phys.*, 86:3036–3041, 1991.
59. M. I. Baskes. Application of the embedded atom method to covalent materials: A semi-empirical potential for silicon. *Phys. Rev. Lett.*, 59:2666–2669, 1987.
60. M. I. Baskes, J. S. Nelson, and A. F. Wright. Semiempirical modified embedded atom potentials for silicon and germanium. *Phys. Rev. B*, 40:6085–6100, 1989.
61. M. I. Baskes. Modified embedded-atom potentials for cubic materials and impurities. *Phys. Rev. B*, 46:2727–2742, 1992.

62. T. Ohira, Y. Inoue, K. Murata, and J. Murayama. Magnetite scale cluster adhesion on metal oxide surfaces: Atomistic simulation study. *Appl. Surf. Sci.*, 171:175–188, 2001.
63. J. H. R. Clarke, W. Smith, and L. V. Woodcok. Short range effective potentials for ionic fluids. *J. Chem. Phys.*, 84:2290–2294, 1986.
64. D. Wolf, P. Keblinski, S. R. Phillpot, and J. Eggebrecht. Exact method for the simulation of coulombic systems by spherically truncated, pairwise 1/r summation. *J. Chem. Phys.*, 110:8254–8282, 1999.
65. A. Yasukawa. Using an extended tersoff interatomic potential to analyze the static-fatigue strength of SiO_2 under athmospheric influence. *JSME Int. J. A*, 39:313–320, 1996.
66. T. Iwasaki and H. Miura. Molecular dynamics analysis of adhesion strength of interfaces between thin films. *J. Mater. Res.*, 16:1789–1794, 2001.
67. L. V. Woodcock. Isothermal molecular dynamics calculations for liquid salts. *Chem. Phys. Lett.*, 10:257, 1971.
68. T. Schneider and E. Stoll. Molecular-dynamics study of a three-dimensional one-component model for distortive phase-transitions. *Phys. Rev. B*, 17:1302–1322, 1978.
69. K. Kremer and G. S. Grest. Dynamics of entangled linear polymer melts – a molecular-dynamics simulation. *J. Chem. Phys.*, 92:5057–5086, 1990.
70. S. A. Adelman and J. D. Doll. Generalized Langevin equation approach for atom-solid-surface scattering – general formulation for classical scattering off harmonic solids. *J. Chem. Phys.*, 64:2375–2388, 1976.
71. S. A. Adelman. Generalized Langevin equations, many-body problems in chemical dynamics. *Adv. Chem. Phys.*, 44:143–253, 1980.
72. J. C. Tully. Dynamics of gas-surface interactions – 3d generalized Langevin model applied to fcc, bcc surfaces. *J. Chem. Phys.*, 73:1975–1985, 1980.
73. H. J. C. Berendsen, J. P. M. Postman, W. F. Van Gunsteren, A. DiNola, and J. R. Haak. Molecular dynamics with coupling to an external bath. *J. Chem. Phys.*, 81:3684–3690, 1984.
74. S. Nose. A unified formulation of the constant temperature molecular-dynamics methods. *J. Chem. Phys.*, 81:511–519, 1984.
75. S. Nose. A molecular-dynamics method for simulations in the canonical ensemble. *Mol. Phys.*, 52:255–268, 1984.
76. G. J. Martyna, M. L. Klein, and M. Tuckerman. Nose–Hoover chains – the canonical ensemble via continuous dynamics. *J. Chem. Phys.*, 97:2635–2643, 1992.
77. M. Schoen, C. L. Rhykerd, D. J. Diestler, and J. H. Cushman. Shear forces in molecularly thin-films. *Science*, 245:1223–1225, 1989.
78. J. E. Curry, F. S. Zhang, J. H. Cushman, M. Schoen, and D. J. Diestler. Transient coexisting nanophases in ultrathin films confined between corrugated walls. *J. Chem. Phys.*, 101:10824–10832, 1994.
79. D. J. Adams. Grand canonical ensemble Monte-Carlo for a Lennard-Jones fluid. *Mol. Phys.*, 29:307–311, 1975.
80. D. A. Bonnell. *Scanning Tunneling Microscopy and Spectroscopy: Theory, Techniques, and Applications*. VCH, 1993.
81. E. Meyer, R. Overney, D. Brodbeck, L. Howald, R. Luthi, J. Frommer, and H. J. Guntherodt. Friction, and wear of Langmuir–Blodgett-films observed by friction force microscopy. *Phys. Rev. Lett.*, 69:1777–1780, 1992.
82. N. A. Burnham, D. D. Dominguez, R. L. Mowery, and R. J. Colton. Probing the surface forces of monolayer films with an atomic-force microscope. *Phys. Rev. Lett.*, 64:1931–1934, 1990.

83. N. A. Burnham and R. J. Colton. Measuring the nanomechanical properties and surface forces of materials using an atomic force microscope. *J. Vac. Sci. Technol. A*, 7:2906–2913, 1989.
84. U. Landman, W. D. Luedtke, N. A. Burnham, and R. J. Colton. Atomistic mechanisms, dynamics of adhesion, nanoindentation, and fracture. *Science*, 248:454–461, 1990.
85. D. G. and A. H. Cottrell. *Electron Theory in Alloy Design*. Institute of Materials, 1992.
86. H. Raffi-Tabar and A. P. Sutton. Long-range Finnis-Sinclair potentials for fcc metallic alloys. *Philos. Mag. Lett.*, 63:217–224, 1991.
87. U. Landman, W. D. Luedtke, and E. M. Ringer. Atomistic mechanisms of adhesive contact formation and interfacial processes. *Wear*, 153:3–30, 1992.
88. O. Tomagnini, F. Ercolessi, and E. Tosatti. Microscopic interaction between a gold tip and a Pb(100) surface. *Surf. Sci.*, 287/288:1041–1045, 1991.
89. N. Ohmae. Field ion microscopy of microdeformation induced by metallic contacts. *Philos. Mag. A*, 74:1319–1327, 1996.
90. N. A. Burnham and R. J. Colton. Measuring the nanomechanical properties and surface forces of materials using an atomic force microscope. *J. Vac. Sci. Technol. A*, 7:2906–2913, 1996.
91. N. A. Burnham, R. J. Colton, and H. M. Pollock. Interpretation of force curves in force microscopy. *Nanotechnol.*, 4:64–80, 1993.
92. N. Agrait, G. Rubio, and S. Vieira. Plastic deformation in nanometer scale contacts. *Langmuir*, 12:4505–4509, 1996.
93. R. W. Carpick, N. Agrait, D. F. Ogletree, and M. Salmeron. Variation of the interfacial shear strength, adhesion of ananometer-sized contact. *Langmuir*, 12:3334–3340, 1996.
94. O. Tomagnini, F. Ercolessi, and E. Tosatti. Microscopic interaction between a gold tip, and a pb(110) surface. *Surf. Sci.*, 287:1041–1045, 1993.
95. J. W. M. Frenken, H. M. Vanpinxteren, and L. Kuipers. New views on surface melting obtained with stm and ion-scattering. *Surf. Sci.*, 283:283–289, 1993.
96. J. Belak and I. F. Stowers. A molecular dynamics model of the orthogonal cutting process. *Proc. Am. Soc. Precis. Eng.*, pages 76–79, 1990.
97. T. Yokohata and K. Kato. Mechanism of nanoscale indentation. *Wear*, 168:109–114, 1993.
98. M. Fournel, E. Lacaze, and M. Schott. Tip–surface interactions in stm experiments on au(111): Atomic-scale metal friction. *Europhys. Lett.*, 34:489–494, 1996.
99. C. L. Kelchner, S. J. Plimpton, and J. C. Hamilton. Dislocation nucleation and defect structure during surface indentation. *Phys. Rev. B*, 58:11085–11088, 1998.
100. J. L. Costakramer, N. Garcia, P. Garciamochales, and P. A. Serena. Nanowire formation in macroscopic metallic contacts – quantum-mechanical conductance tapping a table top. *Surf. Sci.*, 342:1144, 1995.
101. A. I. Yanson, I. K. Yanson, and J. M. van Ruitenbeek. Crossover from electronic to atomic shell structure in alkali metal nanowires. *Phys. Rev. Lett.*, 8721, 2001.
102. A. I. Yanson, J. M. van Ruitenbeek, and I. K. Yanson. Shell effects in alkali metal nanowires. *Low Temp. Phys.*, 27:807–820, 2001.
103. H. Raffi-Tabar and Y. Kawazoe. Dynamics of atomically thin layers-surface interactions in tip-substrate geometry. *Jpn. J. Appl. Phys. 1*, 32:1394–1400, 1993.
104. H. Raffi-Tabar, J. B. Pethica, and A. P. Sutton. Influence of adsorbate monolayer on the nano-mechanics of tip-substrate interactions. *Mater. Res. Soc. Symp. Proc.*, 239:313–318, 1992.
105. K. Komvopoulos and W. Yan. Molecular dynamics simulation of single and repeated indentation. *J. Appl. Phys.*, 82:4823–4830, 1997.

106. J. A. Zimmerman, C. L. Kelchner, P. A. Klein, J. C. Hamilton, and S. M. Foiles. Surface step effects on nanoindentation. *Phys. Rev. Lett.*, 8716, 2001.
107. S. Kokubo. On the change in hardness of a plate caused by bending. *Science Reports of the Tohoku Imperial University*, 21:256–267, 1932.
108. G. Sines and R. Calson. Hardness measurements for determination of residual stresses. *ASTM Bull.*, 180:35–37, 1952.
109. G. U. Oppel. Biaxial elasto-plastic analysis of load and residual stresses. *Exp. Mech.*, 21:135–140, 1964.
110. T. R. Simes, S. G. Mellor, and D. A. Hills. A note on the influence of residual-stress on measured hardness. *J. Strain Analysis Eng. Design*, 19:135–137, 1984.
111. T. Y. Tsui, W. C. Oliver, and G. M. Pharr. Influences of stress on the measurement of mechanical properties using nanoindentation. 1. experimental studies in an aluminum alloy. *J. Mater. Res.*, 11:752–759, 1996.
112. A. Bolshakov, W. C. Oliver, and G. M. Pharr. Influences of stress on the measurement of mechanical properties using nanoindentation. 2. finite element simulations. *J. Mater. Res.*, 11:760–768, 1996.
113. J. D. Schall and D. W. Brenner. Atomistic simulation of the influence of pre-existing stress on the interpretation of nanoindentation data. *J. Mater. Sci. B*, 19:3172–3180, 2004.
114. U. Landman, W. D. Luedtke, and M. W. Ribarsky. Structural and dynamical consequences of interactions in interfacial systems. *J. Vac. Sci. Technol. A*, 7:2829–2839, 1989.
115. J. S. Kallman, W. G. Hoover, C. G. Hoover, A. J. Degroot, S. M. Lee, and F. Wooten. Molecular-dynamics of silicon indentation. *Phys. Rev. B*, 47:7705–7709, 1993.
116. K. Minowa and K. Sumino. Stress-induced amorphization of a silicon crystal by mechanical scratching. *Phys. Rev. Lett.*, 69:320–322, 1992.
117. J. A. Harrison, C. T. White, R. J. Colton, and D. W. Brenner. Nanoscale investigation of indentation, adhesion, and fracture of diamond (111) surfaces. *Surf. Sci.*, 271:57–67, 1992.
118. K. Enke, H. Dimigen, and H. Hubsch. Frictional-properties of diamond-like carbon layers. *Appl. Phys. Lett.*, 36:291–292, 1980.
119. K. Enke. Some new results on the fabrication of and the mechanical, electrical, and optical-properties of i-carbon layers. *Thin Solid Films*, 80:227–234, 1981.
120. S. Miyake, S. Takahashi, I. Watanabe, and H. Yoshihara. Friction and wear behavior of hard carbon-films. *Asle Trans.*, 30:121–127, 1987.
121. S. B. Sinnott, R. J. Colton, C. T. White, O. A. Shenderova, D. W. Brenner, and J. A. Harrison. Atomistic simulations of the nanometer-scale indentation of amorphous-carbon thin films. *J. Vac. Sci. Technol. A*, 15:936–940, 1997.
122. J. N. Glosli, M. R. Philpott, and G. M. McClelland. Molecular dynamics simulation of mechanical deformation of ultra-thin amorphous carbon films. *Mater. Res. Soc. Symp. Proc.*, 383:431–435, 1995.
123. T. Y. Tsui, G. M. Pharr, W. C. Oliver, C. S. Bhatia, C. T. White, S. Anders, A. Anders, and I. G. Brown. Nanoindentation and nanoscratching of hard carbon coatings for magnetic disks. *Mater. Res. Soc. Symp. Proc.*, 383:447, 1995.
124. A. Garg, J. Han, and S. B. Sinnott. Interactions of carbon-nanotubule proximal probe tips with diamond and graphene. *Phys. Rev. Lett.*, 81:2260–2263, 1998.
125. A. L. Shluger, R. T. Williams, and A. L. Rohl. Lateral and friction forces originating during force microscope scanning of ionic surfaces. *Surf. Sci.*, 343:273–287, 1995.
126. K. Cho and J. D. Joannopoulos. Mechanical hysteresis on an atomic-scale. *Surf. Sci.*, 328:320–324, 1995.

127. V. Licht, E. Ernst, and N. Huber. Simulation of the hertzian contact damage in ceramics. *Model. Simul. Mater. Sci. Eng.*, 11:477–486, 2003.
128. K. J. Tupper and D. W. Brenner. Compression-induced structural transition in a self-assembled monolayer. *Langmuir*, 10:2335–2338, 1994.
129. K. J. Tupper, R. J. Colton, and D. W. Brenner. Simulations of self-assembled monolayers under compression – effect of surface asperities. *Langmuir*, 10:2041–2043, 1994.
130. S. A. Joyce, R. C. Thomas, J. E. Houston, T. A. Michalske, and R. M. Crooks. Mechanical relaxation of organic monolayer films measured by force microscopy. *Phys. Rev. Lett.*, 68:2790–2793, 1992.
131. C. M. Mate. Atomic-force-microscope study of polymer lubricants on silicon surfaces. *Phys. Rev. Lett.*, 68:3323–3326, 1992.
132. V. N. Koinkar and B. Bhushan. Micro/nanoscale studies of boundary layers of liquid lubricants for magnetic disks. *J. Appl. Phys.*, 79:8071–8075, 1996.
133. A. B. Tutein, S. J. Stuart, and J. A. Harrison. Indentation analysis of linear-chain hydrocarbon monolayers anchored to diamond. *J. Phys. Chem. B*, 103:11357–11365, 1999.
134. I. L. Singer. A thermochemical model for analyzing low wear-rate materials. *Surf. Coat. Tech.*, 49:474–481, 1991.
135. I. L. Singer, S. Fayeulle, and P. D. Ehni. Friction and wear behavior of tin in air – the chemistry of transfer films and debris formation. *Wear*, 149:375–394, 1991.
136. J. A. Harrison, C. T. White, R. J. Colton, and D. W. Brenner. Molecular-dynamics simulations of atomic-scale friction of diamond surfaces. *Phys. Rev. B*, 46:9700–9708, 1992.
137. J. A. Harrison, R. J. Colton, C. T. White, and D. W. Brenner. Effect of atomic-scale surface-roughness on friction – a molecular-dynamics study of diamond surfaces. *Wear*, 168:127–133, 1993.
138. J. A. Harrison, C. T. White, R. J. Colton, and D. W. Brenner. Atomistic simulations of friction at sliding diamond interfaces. *Mater. Res. Soc. Bull.*, 18:50–53, 1993.
139. J. N. Glosli and G. M. Mcclelland. Molecular-dynamics study of sliding friction of ordered organic monolayers. *Phys. Rev. Lett.*, 70:1960–1963, 1993.
140. A. Koike and M. Yoneya. Molecular dynamics simulations of sliding friction of Langmuir–Blodgett monolayers. *J. Chem. Phys.*, 105:6060–6067, 1996.
141. J. E. Hammerberg, B. L. Holian, and S. J. Zhuo. Studies of sliding friction in compressed copper. *Conf. Am. Phys. Soc. Topical Group Shock Compression of Condensed Matter*, Part 1:370, 1995.
142. M. R. Sorensen, K. W. Jacobsen, and P. Stoltze. Simulations of atomic-scale sliding friction. *Phys. Rev. B*, 53:2101–2113, 1996.
143. U. Landman, W. D. Luedtke, and A. Nitzan. Dynamics of tip substrate interactions in atomic force microscopy. *Surf. Sci.*, 210:177, 1989.
144. M. D. Perry and J. A. Harrison. Friction between diamond surfaces in the presence of small third-body molecules. *J. Phys. Chem. B*, 101:1364–1373, 1997.
145. A. Buldum and S. Ciraci. Atomic-scale study of dry sliding friction. *Phys. Rev. B*, 55:2606–2611, 1997.
146. A. P. Sutton and J. B. Pithica. Inelastic flow processes in nanometer volumes of solids. *J. Phys. Condens. Matter*, 2:5317–5326, 1990.
147. S. Akamine, R. C. Barrett, and C. F. Quate. Improved atomic force microscope images using microcantilevers with sharp tips. *Appl. Phys. Lett.*, 57:316–318, 1990.
148. J. A. Nieminen, A. P. Sutton, and J. B. Pethica. Static junction growth during frictional sliding of metals. *Acta Metall. Mater.*, 40:2503–2509, 1992.
149. J. A. Niemienen, A. P. Sutton, J. B. Pethica, and K. Kaski. Mechanism of lubrication by a thin solid film on a metal surface. *Mod. Sim. Mater. Sci. Eng.*, 1:83–90, 1992.

150. V. V. Pokropivny, V. V. Skorokhod, and A. V. Pokropivny. Atomistic mechanism of adhesive wear during friction of atomic sharp tungsten asperity over (114) bcc-iron surface. *Mater. Lett.*, 31:49–54, 1997.
151. A. Dayo, W. Alnasrallah, and J. Krim. Superconductivity-dependent sliding friction. *Phys. Rev. Lett.*, 80:1690–1693, 1998.
152. R. Erlandsson, G. Hadziioannou, C. M. Mate, G. M. Mcclelland, and S. Chiang. Atomic scale friction between the muscovite mica cleavage plane and a tungsten tip. *J. Chem. Phys.*, 89:5190–5193, 1988.
153. K. L. Johnson. *Contact Mechanics*. Cambridge Univ. Press, 1985.
154. J. B. Pethica. Interatomic forces in scanning tunneling microscopy – giant corrugations of the graphite surface – comment. *Phys. Rev. Lett.*, 57:3235–3235, 1986.
155. H. Tang, C. Joachim, and J. Devillers. Interpretation of afm images – the graphite surface with a diamond tip. *Surf. Sci.*, 291:439–450, 1993.
156. S. Fujisawa, Y. Sugawara, and S. Morita. Localized fluctuation of a two-dimensional atomic-scale friction. *Jpn. J. Appl. Phys. 1*, 35:5909–5913, 1996.
157. S. Fujisawa, Y. Sugawara, S. Ito, S. Mishima, T. Okada, and S. Morita. The two-dimensional stick-slip phenomenon with atomic resolution. *Nanotechnol.*, 3:138–142, 1993.
158. S. Fujisawa, Y. Sugawara, S. Morita, S. Ito, S. Mishima, and T. Okada. Study on the stick-slip phenomenon on a cleaved surface of the muscovite mica using an atomic-force lateral force microscope. *J. Vac. Sci. Technol. B*, 12:1635–1637, 1994.
159. S. Morita, S. Fujisawa, and Y. Sugawara. Spatially quantized friction with a lattice periodicity. *Surf. Sci. Rep.*, 23:1–41, 1996.
160. J. A. Ruan and B. Bhushan. Atomic-scale and microscale friction studies of graphite, diamond using friction force microscopy. *J. Appl. Phys.*, 76:5022–5035, 1994.
161. R. W. Carpick, N. Agrait, D. F. Ogletree, and M. Salmeron. Measurement of interfacial shear (friction) with an ultrahigh vacuum atomic force microscope. *J. Vac. Sci. Technol. B*, 14:2772–2772, 1996.
162. B. Samuels and J. Wilks. The friction of diamond sliding on diamond. *J. Mater. Sci.*, 23:2846–2864, 1988.
163. S. Miyake, T. Miyamoto, and R. Kaneko. Increase of nanometer-scale wear of polished chemical-vapor-deposited diamond films due to nitrogen ion implantation. *Nucl. Instrum. Methods B*, 108:70–74, 1996.
164. R. J. A. van den Oetelaar and C. F. J. Flipse. Atomic-scale friction on diamond(111) studied by ultra-high vacuum atomic force microscopy. *Surf. Sci.*, 384:828, 1997.
165. D. Mulliah, S. D. Kenny, and R. Smith. Modeling of stick–slip phenomena using molecular dynamics. *Phys. Rev. B*, 69:205407, 2004.
166. A. G. Khurshudov, K. Kato, and H. Koide. Nano-wear of the diamond AFM probing tip under scratching of silicon, studies by AFM. *Tribol. Lett.*, 2:345–354, 1996.
167. A. Khurshudov and K. Kato. Volume increase phenomena in reciprocal scratching of polycarbonate studied by atomic-force microscopy. *J. Vac. Sci. Technol. B*, 13:1938–1944, 1995.
168. M. D. Perry and J. A. Harrison. Molecular dynamics studies of the frictional properties of hydrocarbon materials. *Langmuir*, 12:4552–4556, 1996.
169. M. D. Perry and J. A. Harrison. Molecular dynamics investigations of the effects of debris molecules on the friction and wear of diamond. *Thin Solid Films*, 291:211–215, 1996.
170. J. A. Harrison, C. T. White, R. J. Colton, and D. W. Brenner. Investigation of the atomic-scale friction and energy-dissipation in diamond using molecular-dynamics. *Thin Solid Films*, 260:205–211, 1995.

171. J. A. Harrison, C. T. White, R. J. Colton, and D. W. Brenner. Effects of chemically-bound, flexible hydrocarbon species on the frictional-properties of diamond surfaces. *J. Phys. Chem.*, 97:6573–6576, 1993.
172. J. A. Harrison, R. J. Colton, C. T. White, and D. W. Brenner. Atomistic simulation of the nanoindentation of diamond and graphite surfaces. *Mater. Res. Soc. Sym. Proc.*, 239:573–578, 1992.
173. J. A. Harrison and D. W. Brenner. Simulated tribochemistry – an atomic-scale view of the wear of diamond. *J. Am. Chem. Soc.*, 116:10399–10402, 1994.
174. Z. Feng and J. E. Field. Friction of diamond on diamond and chemical vapor-deposition diamond coatings. *Surf. Coat. Technol.*, 47:631–645, 1991.
175. B. N. J. Persson. Applications of surface resistivity to atomic scale friction, to the migration of hot adatoms, and to electrochemistry. *J. Chem. Phys.*, 98:1659–1672, 1993.
176. B. N. J. Persson and A. I. Volokitin. Electronic friction of physisorbed molecules. *J. Chem. Phys.*, 103:8679–8683, 1995.
177. H. Grabhorn, A. Otto, D. Schumacher, and B. N. J. Persson. Variation of the dc-resistance of smooth and atomically rough silver films during exposure to C_2H_6, C_2H_4. *Surf. Sci.*, 264:327–340, 1992.
178. A. D. Berman, W. A. Ducker, and J. N. Israelachvili. Origin and characterization of different stick–slip friction mechanisms. *Langmuir*, 12:4559–4563, 1996.
179. B. N. J. Persson. Theory of friction – dynamical phase-transitions in adsorbed layers. *J. Chem. Phys.*, 103:3849–3860, 1995.
180. B. N. J. Persson and E. Tosatti. Layering transition in confined molecular thin-films – nucleation and growth. *Phys. Rev. B*, 50:5590–5599, 1994.
181. H. Yoshizawa and J. Israelachvili. Fundamental mechanisms of interfacial friction. 2. stick-slip friction of spherical and chain molecules. *J. Phys. Chem.*, 97:11300–11313, 1993.
182. P. A. Thompson and M. O. Robbins. Origin of stick–slip motion in boundary lubrication. *Science*, 250:792–794, 1990.
183. B. N. J. Persson. Theory of friction: Friction dynamics for boundary lubricated surfaces. *Phys. Rev. B*, 55:8004–8012, 1997.
184. U. Landman, W. D. Luedtke, and J. P. Gao. Atomic-scale issues in tribology: Interfacial junctions and nano-elastohydrodynamics. *Langmuir*, 12:4514–4528, 1996.
185. E. Manias, G. Hadziioannou, and G. ten Brinke. Inhomogeneities in sheared ultrathin lubricating films. *Langmuir*, 12:4587–4593, 1996.
186. S. B. Sinnott and R. Andrews. Carbon nanotubes: Synthesis, properties, applications. *Crit. Rev. Solid State Mater. Sci.*, 26:145–249, 2001.
187. B. Bhushan, B. K. Gupta, G. W. Van Cleef, C. Capp, and J. V. Coe. Sublimed C_{60} films for tribology. *Appl. Phys. Lett.*, 62:3253–3255, 1993.
188. T. Thundat, R. J. Warmack, D. Ding, and R. N. Compton. Atomic force microscope investigation of C_{60} adsorbed on silicon and mica. *Appl. Phys. Lett.*, 63:891–893, 1993.
189. C. M. Mate. Nanotribology studies of carbon surfaces by force microscopy. *Wear*, 168:17–20, 1993.
190. R. Lüthi, E. Meyer, and H. Haefke. Sled-type motion on the nanometer scale: Determination of dissipation and cohesive energies of C_{60}. *Science*, 266:1979–1981, 1993.
191. R. Lüthi, H. Haefke, E. Meyer, L. Howald, H.-P. Lang, G. Gerth, and H. J. Güntherodt Frictional and atomic-scale study of C_{60} thin films by scanning force microscopy. *Z. Phys. B*, 95:1–3, 1994.
192. Q.-J. Xue, X.-S. Zhang, and F.-Y. Yan. Study of the structural transformations of C_{60}/C_{70} crystals during friction. *Chin. Sci. Bull.*, 39:819–822, 1994.

193. W. Allers, U. D. Schwarz, G. Gensterblum, and R. Wiesendanger. Low-load friction behavior of epitaxial C_{60} monolayers. *Z. Phys. B*, 99:1–2, 1995.
194. U. D. Schwarz, W. Allers, G. Gensterblum, and R. Wiesendanger. Low-load friction behavior of epitaxial C_{60} monolayers under Hertzian contact. *Phys. Rev. B*, 52:14976–14984, 1995.
195. J. Ruan and B. Bhushan. Nanoindentation studies of sublimed fullerene films using atomic force microscopy. *J. Mater. Res.*, 8:3019–3022, 1996.
196. U. D. Schwarz, O. Zworner, P. Koster, and R. Wiesendanger. Quantiative analysis of the frictional properties of solid materials at low loads. i. carbon compounds. *Phys. Rev. B*, 56:6987–6996, 1997.
197. S. Okita, M. Ishikawa, and K. Miura. Nanotribological behavior of C_{60} films at an extremely low load. *Surf. Sci.*, 442:959, 1999.
198. S. Okita and K. Miura. Molecular arrangement in C_{60} and C_{70} films on graphite and their nanotribological behavior. *Nano Lett.*, 1:101–103, 2001.
199. K. Miura, S. Kamiya, and N. Sasaki. C_{60} molecular bearings. *Phys. Rev. Lett.*, 90:055509, 2003.
200. A. Buldum and J. P. Lu. Atomic scale sliding and rolling of carbon nanotubes. *Phys. Rev. Lett.*, 83:5050–5053, 1999.
201. M. R. Falvo, R. M. Taylor, A. Helser, V. Chi, F. P. Brooks, S. Washburn, and R. Superfine. Nanometer-scale rolling and sliding of carbon nanotubes. *Nature*, 397:236–238, 1999.
202. M. R. Falvo, J. Steele, R. M. T. II, and R. Superfine. Gearlike rolling motion mediated by commensurate contact: Carbon nanotubes on hopg. *Phys. Rev. B*, 62:R10664–R10667, 2000.
203. J. D. Schall and D. W. Brenner. Molecular dynamics simulations of carbon nanotube rolling and sliding on graphite. *Mol. Simulat.*, 25:73–80, 2000.
204. B. Ni and S. B. Sinnott. Tribological properties of carbon nanotube bundles. *Surf. Sci.*, 487:87–96, 2001.
205. B. Ni and S. B. Sinnott. Mechanical and tribological properties of carbon nanotubes investigated with atomistic simulations. In *Nanotubes and related materials*, pages A17.3.1–A17.3.5. MRS Symposia Proceedings, Materials Research Society, 2001.
206. K. Miura, T. Takagi, S. Kamiya, T. Sahashi, and M. Yamauchi. Natural rolling of zigzag multiwalled carbon nanotubes on graphite. *Nano Lett.*, 1:161–163, 2001.
207. K. Miura, M. Ishikawa, R. Kitanishi, M. Yoshimura, K. Ueda, Y. Tatsumi, and N. Minami. Bundle structure and sliding of single-walled carbon nanotubes observed by friction-force microscopy. *Appl. Phys. Lett.*, 78:832–834, 2001.
208. P. E. Sheehan and C. M. Lieber. Nanotribology and nanofabrication of MoO_3 structures by atomic force microscopy. *Science*, 272:1158–1161, 1996.
209. J. Wang, K. C. Rose, and C. M. Lieber. Load-independent friction: MoO_3 nanocrystal lubricants. *J. Phys. Chem. B*, 103:8405–8408, 1999.
210. Q. Ouyang and K. Okada. Nano-ball bearing effect of ultra-fine particles of cluster diamond. *Appl. Surf. Sci.*, 78:309–313, 1994.
211. W. Allers, C. Hahn, M. Lohndorf, S. Lukas, S. Pan, U. D. Schwarz, and R. Wiesendanger. Nanomechanical investigations and modifications of thin films based on scanning force methods. *Nanotechnol.*, 7:346–350, 1996.
212. R. Luthi, H. Haefke, E. Meyer, L. Howald, H. P. Lang, G. Gerth, and H. J. Guntherodt. Frictional and atomic-scale study of C_{60} thin-films by scanning force microscopy. *Z. Phys. B*, 95:1–3, 1994.
213. R. Luthi, E. Meyer, H. Haefke, L. Howald, W. Gutmannsbauer, and H. J. Guntherodt. Sled-type motion on the nanometer-scale – determination of dissipation and cohesive energies of C_{60}. *Science*, 266:1979–1981, 1994.

214. B. Bhushan, B. K. Gupta, G. W. Vancleef, C. Capp, and J. V. Coe. Fullerene (C_{60}) films for solid lubrication. *Tribology*, 36:573–580, 1993.
215. U. D. Schwarz, W. Allers, G. Gensterblum, and R. Wiesendanger. Low-load friction behavior of epitaxial C_{60} monolayers under hertzian contact. *Phys. Rev. B*, 52:14976–14984, 1995.
216. S. B. Legoas, R. Giro, and D. S. Galvao. Molecular dynamics simulations of C_{60} nanobearings. *Chem. Phys. Lett.*, 386:425–429, 2004.
217. P. L. Dickrell, W. G. Sawyer, D. W. Hahn, S. B. Sinnott, B. Yurdumakan, A. Dhinojwala, N. R. Raravikar, L. S. Schadler, and P. M. Ajayan. Tribology of oriented carbon nanotube layers: Large frictional anisotropy and super adhesive behavior. *Tribol. Lett.*, 18:59–62, 2005.
218. R. M. Overney, T. Bonner, E. Meyer, M. Reutschi, R. Luthi, L. Howald, J. Frommer, H. J. Guntherodt, M. Fujihara, and H. Takano. Elasticity, wear, and friction properties of thin organic films observed with atomic-force microscopy. *J. Vac. Sci. Technol. B*, 12:1973–1976, 1994.
219. R. M. Overney, E. Meyer, J. Frommer, D. Brodbeck, R. Luthi, L. Howald, H. J. Guntherodt, M. Fujihara, H. Takano, and Y. Gotoh. Friction measurements on phase-separated thin-films with a modified atomic force microscope. *Nature*, 359:133–135, 1992.
220. R. M. Overney, E. Meyer, J. Frommer, H. J. Guntherodt, M. Fujihira, H. Takano, and Y. Gotoh. Force microscopy study of friction and elastic compliance of phase-separated organic thin-films. *Langmuir*, 10:1281–1286, 1994.
221. M. GarciaParajo, C. Longo, J. Servat, P. Gorostiza, and F. Sanz. Nanotribological properties of octadecyltrichlorosilane self-assembled ultrathin films studied by atomic force microscopy: Contact and tapping modes. *Langmuir*, 13:2333–2339, 1997.
222. R. M. Overney, H. Takano, M. Fujihira, E. Meyer, and H. J. Guntherodt. Wear, friction and sliding speed correlations on Langmuir–Blodgett-films observed by atomic-force microscopy. *Thin Solid Films*, 240:105–109, 1994.
223. P. T. Mikulski and J. A. Harrison. Periodicities in the properties associated with the friction of model self-assembled monolayers. *Tribol. Lett.*, 10:29–35, 2001.
224. P. T. Mikulski and J. A. Harrison. Packing-density effects on the friction of n-alkane monolayers. *J. Am. Chem. Soc.*, 123:6873–6881, 2001.
225. P. T. Mikulski, G.-T. Gao, G. M. Chateauneuf, and J. A. Harrison. Contact forces at the sliding interface: Mixed vs. pure model alkane monolayers. *J. Chem. Phys.*, 2004.
226. E. Barrena, C. Ocal, and M. Salmeron. Acomparative afm study of the structural and frictional properties of mixed and single component films of alkanethiols on Au(111). *Surf. Sci.*, 482:1216–1221, 2001.
227. S. Lee, Y. S. Shon, R. Colorado, R. L. Guenard, T. R. Lee, and S. S. Perry. The influence of packing densities, surface order on the frictional properties of alkanethiol self-assembled monolayers (SAMs) on gold: A comparison of SAMS derived from normal and spiroalkanedithiols. *Langmuir*, 16:2220–2224, 2000.
228. G. M. Chateauneuf, P. T. Mikulski, G. T. Gao, and J. A. Harrison. Compression- and shear-induced polymerization in model diacetylene-containing monolayers. *J. Phys. Chem. B*, 108:16626–16635, 2004.
229. L. Z. Zhang, Y. S. Leng, and S. Y. Jiang. Tip-based hybrid simulation study of frictional properties of self-assembled monolayers: Effects of chain length, terminal group, scan direction, and scan velocity. *Langmuir*, 19:9742–9747, 2003.
230. F. P. Bowden and D. Tabor. *The Friction and Lubrication of Solids*. Part 2. Clarendon, 1964.

231. G. T. Gao, P. T. Mikulski, and J. A. Harrison. Molecular-scale tribology of amorphous carbon coatings: Effects of film thickness, adhesion, and long-range interactions. *J. Am. Chem. Soc.*, 124:7202–7209, 2002.
232. G. T. Gao, P. T. Mikulski, G. M. Chateauneuf, and J. A. Harrison. The effects of film structure and surface hydrogen on the properties of amorphous carbon films. *J. Phys. Chem. B*, 107:11082–11090, 2003.
233. C. Daly and J. Krim. Sliding friction of solid xenon monolayers and bilayers on ag(111). *Phys. Rev. Lett.*, 76:803–806, 1996.
234. M. Cieplak, E. D. Smith, and M. O. Robbins. Molecular-origins of friction – the force on adsorbed layers. *Science*, 265:1209–1212, 1994.
235. C. Daly and J. Krim. Friction and damping of Xe/Ag(111). *Surf. Sci.*, 368:49–54, 1996.
236. B. N. J. Persson. *Physics of Sliding Friction*. Kluwer, 1996.
237. A. R. C. Baljon and M. O. Robbins. Adhesion and friction of thin films. *Mater. Res. Soc. Bull.*, 22:22–26, 1997.
238. R. M. Overney, H. Takano, M. Fujihira, W. Paulus, and H. Ringsdorf. Anisotropy in friction and molecular stick–slip motion. *Phys. Rev. Lett.*, 72:3546–3549, 1994.
239. E. Meyer, R. Luthi, L. Howald, W. Gutmannsbauer, H. Haefke, and H.-J. Guntherodt. Friction force microscopy on well defined surfaces. *Nanotechnol.*, 7:340–344, 1996.
240. A. N. Gent and J. Schultz. Effect of wetting liquids on strength of adhesion of viscoelastic materials. *J. Adhesion*, 3:281–283, 1972.
241. A. N. Gent and S. M. Lai. Interfacial bonding, energy-dissipation, and adhesion. *J. Polym. Sci. Pol. Phys.*, 32:1543–1555, 1994.
242. A. Mayer, T. Pith, G. H. Hu, and M. Lambla. Effect of the structure of latex-particles on adhesion. 3. analogy between peel adhesion and rheological properties of acrylic copolymers. *J. Polym. Sci. Pol. Phys.*, 33:1793–1801, 1995.
243. A. R. C. Baljon and M. O. Robbins. Energy dissipation during rupture of adhesive bonds. *Science*, 271:482–484, 1996.
244. G. J. Vancso, S. Forster, H. Leist, G. Liu, and D. Trifonova. Anisotropic stick-slip friction of highly oriented thin films of poly(tetrafluoroethylene) at the molecular level. *Tribol. Lett.*, 2:231–246, 1996.

14

Mechanics of Biological Nanotechnology

Rob Phillips, Prashant K. Purohit, and Jané Kondev

Summary. One of the most compelling areas to be touched by nanotechnology is biological science. Indeed, we will argue that there is a fascinating interplay between these two subjects, with biology as a key beneficiary of advances in nanotechnology as a result of a new generation of single molecule experiments that complement traditional assays involving statistical assemblages of molecules. This interplay runs in both directions, with nanotechnology continually receiving inspiration from biology itself. The goal of this chapter is to highlight some representative examples of the exchange between biology and nanotechnology and to illustrate the role of nanomechanics in this field and how mechanical models have arisen in response to the emergence of this new field. Primary attention will be given to the particular example of the processes that attend the life cycle of bacterial viruses. Viruses feature many of the key lessons of biological nanotechnology, including self assembly, as evidenced in the spontaneous formation of the protein shell (capsid) within which the viral genome is packaged, and a motor-mediated biological process, namely, the packaging of DNA in this capsid by a molecular motor that pushes the DNA into the capsid. We argue that these processes in viruses are a compelling real-world example of nature's nanotechnology and reveal the nanomechanical challenges that will continue to be confronted at the nanotechnology-biology interface.

14.1 Introduction

Nanotechnology is the seat of a broad variety of interdisciplinary activity with applications as diverse as optoelectronics, microfluidics, and medicine. However, one of the richest interdisciplinary areas only now beginning to be harvested is the interface between the biological sciences and nanotechnology. The aim of the present chapter is threefold. First, we wish to illustrate the synergy that exists between biological nanotechnology (e.g., molecular motors, transmembrane pumps, etc.) and biologically inspired nanotechnology (e.g., synthetic proteins, artificial viruses aimed at delivering drugs, biofunctionalized cantilevers, etc.). Secondly, through a series of order-of-magnitude estimates and an associated examination of the units for describing spatial dimensions, temporal processes, forces at the nanoscale, and the energy budget in nanoscale systems, we aim to build an intuitive sense of the workings of nanotechnology in the biological setting. Finally, through several specific case studies, we hope to illustrate some of the challenges faced in modeling the mechanical processes of biological nanotechnology and show how such challenges have been met thus far and how they might be met in the future.

14.2 Science at the Biology–Nanotechnology Interface

14.2.1 Biological Nanotechnology

Though the innovations leading to the adoption of the expression "nanotechnology" are indeed impressive, the perspective of this chapter is that one need only look inward to the way that our muscles move, to how we digest and synthesize molecules of dazzling complexity, to the way in which we think the thoughts that permit us to fill the shelves of libraries with scientific journals to realize that the greatest nanotechnology of all is that which is revealed in the living world. That is, one of the central thrusts is the idea that the microscopic workings of life offer an inspiring vision of nanotechnology. Clearly, the diversity of the examples of "nanotechnology" seen in the living world can (and do) fill the pages of learned texts. A guided tour of the machinery of the cell can be found in *Alberts* et al. [1], while a vision of the cell as an assemblage of "protein machines" has been argued for by *Alberts* [2]. The ambition of the present section is to provide several cursory examples of the nanotechnological marvels that power the living realm.

14.2.2 Self-Assembly as Biological Nanotechnology

One of the most intriguing nanotechnological tricks of the living world is the central role played by self-assembly in such systems. Whether we consider the spontaneous assembly of viruses, either in test tubes or in the interior of an infected cell, or the fusion of one membrane-bound region to another through vesicle fusion, spontaneous formation of functional "materials" is a key part of the biological repertoire. To be more concrete, we note that self-assembly in biological systems takes place in a number of different guises. First, the self-assembly of linear assemblies is a key part of the cytoskeletal assembly process with G-actin associating to form actin filaments and, similarly, tubulin monomers joining to form microtubules. This is also the process that is used by certain bacteria such as *Listeria* for locomotion. Both of these examples are described more fully in [3]. This same type of process is taken to the next level of sophistication in simple viruses such as tobacco mosaic virus that involve not only the assembly of protein monomers, but also the genetic message in the form of long RNA molecules. A second broad class of self-assembly processes is associated with the formation of containers such as liposomes or viral capsids. In the case of liposomes, lipid molecules such as phosphatidylcholine spontaneously organize in a way that sequesters the hydrophobic tails from the surrounding water. Similarly, protein subunits spontaneously assemble to form viral capsids [4]. These structures play a variety of roles in the biological realm, from serving as containers for different macromolecules to providing for concentration gradients of small ions that are maintained and utilized by molecular motors such as ATPsynthase [1].

Beyond the simplistic description of self-assembly advanced above, a second key feature of biological self-assembly must also be considered. In particular, a variety of self-assembly processes in the biological realm are templated (or coded). As is well-known, proteins are a hugely versatile class of molecules all based upon the same fundamental building blocks. Interestingly, the enormous diversity of protein action is founded upon 20 distinct amino acid building blocks, and the template for assembling a given protein is carried in the form of messenger RNA, which is then read and used as the basis for protein synthesis by the ribosome. One of the most exciting developments in modern nanotechnology is the attempt to exploit templated self-assembly processes for the purposes of creating new materials. An example in the context of protein-based materials can be found in the article of *van Hest* and *Tirrell* [5], where the machinery of the cell has been tricked into incorporating artificial amino acids into synthetic proteins.

14.2.3 Molecular Motors as Biological Nanotechnology

One of the hallmarks of life is change and motion. At the cellular level, such motion is effected through a dizzying variety of mechanisms, most of which when viewed at the molecular level are seen to be the result of the action of molecular motors [3]. For example, muscle contraction reflects the coherent action of huge numbers of myosin motor molecules as they march along actin filaments. Similarly, the motion of certain bacteria can be traced to the rotation of a rotary motor embedded in the cell wall, which is attached to filaments known as flagella [6]. In a similar vein, just as our modern society is replete with examples of systems aimed at allowing for communication and transport between widely separated geographic locations, so, too, has the living world had to answer these same challenges. As will be shown in the section of this chapter aimed at making estimates of various scales and processes in nanotechnology, one of the key mechanisms for communication and transport is diffusion. That is, a chemical concentration at one place can make its presence known at a distance x with a characteristic time, $t_{\text{diffusive}} \approx x^2/2D$, where D is the diffusion constant. However, as will be shown later in the chapter, there are many cellular processes that cannot wait as long as $t_{\text{diffusion}}$. As a result, a host of molecular motors and an associated transport system (the elements of the cytoskeleton) permit active transport. For example, kinesin is a motor molecule that transports material along long, relatively rigid polymeric assemblies known as microtubules [1].

Even more incredible are rotary machines like those associated with ATPsynthase and bacterial flagella. ATPsynthase is a membrane-embedded machine that rotates and, in so doing, synthesizes new ATP molecules (adenosine triphosphate, energy currency of the cell) [1, 7–9]. Similarly, the bacterial flagellar motor rotates with the result that the attached bacterial flagellum rotates and thereby induces motion of the cell [3, 6]. The exquisite details of the construction of these devices are themselves breathtaking. The rotary motors in bacteria are constructed from several components that are much like a rotor and stator and perform periodic motions by deriving energy from the flow of protons across a membrane [6]. These rotary motors are very powerful, as evidenced by F_1-ATPsynthase, which can generate a torque large enough to rotate a molecule of actin 100 times its own length [10]. In addition, bacterial motors have another layer of sophistication in that stimuli from the external environment, which are sensed by the bacteria through the pores on its membrane, can change the direction of rotation of the motor in a process known as chemotaxis [3]. Feedback and signal transmission are implemented in engineering devices by means of complex circuitry, whereas in the living world these functions are often accomplished by means of chemistry and conformational changes in large molecules.

All of these molecular machines involve a rich interplay between chemistry, thermodynamics, and mechanics. From a structural perspective, most molecular motors are proteins with different subunits performing different functions [11]. Often there is some region within the motor where chemical energy is derived from the hydrolysis of adenosine triphosphate (ATP). This chemical energy is then converted to mechanical energy through a conformational change in the protein. These examples serve to call the reader's attention to the importance and variety of motor molecules found throughout the living world and which almost any sensible definition of nanotechnology would have to include as particulary sophisticated examples.

14.2.4 Molecular Channels and Pumps as Biological Nanotechnology

From a structural perspective, one of the most intriguing features of cellular systems is their division into a number of separate membrane-bound compartments. We have already touched upon this compartmentalization in the context of self-assembly, and note that the role of membranes and the proteins bound within them is described clearly in *Alberts* et al. [1]. The presence of such compartments, often marked by large concentration gradients with respect to the surrounding medium, hints at another nanotechnological wonder of the living world, namely, the presence of a wide range of transmembrane channels that mediate the exchange of material between these different compartments. Certain passive versions of these channels are gated by various mechanisms such as the arrival of signaling molecules or tension in the membrane within which they are found [12–14]. Once the channel is in the open state, ions pass through passively, by diffusion. Active versions of ion channels that transfer ions such as Na^+ and Ca^{2+} are similarly critical for the functioning of a cell, both in the case of unicellular and multicellular organisms. For example, in the case of Na^+ ions, typical concentrations within the cell can be as much as a factor of 10–20 less than those in the extracellular milieu. Such concentration gradients imply the need for sophisticated active "devices" which can do work against such gradients. One of the most remarkable of machines of this type is the Na^+–K^+ pump [1]. This machine is powered by hydrolysis of ATP (i.e., the consumption of ATP fuel), and it can pump ions up a potential gradient. In particular, this pump hydrolyzes ATP and pumps Na^+ ions *out* of the cell against a very steep concentration gradient while pumping K^+ ions *in* again against a steep gradient.

The point of this brief discussion has been to illustrate the first of several perspectives that we will bring to bear on the question of the biology-nanotechnology interface. Thus far, we have noted that nature is replete with examples of macromolecules and macromolecular assemblies that perform nanotechnological tasks and in this capacity serve as examples of biological nanotechnology. Next, we wish to examine the ways in which biological phenomena can inspire nanotechnology itself.

14.2.5 Biologically Inspired Nanotechnology

As noted in the beginning of this chapter, biological systems are nanotechnologically relevant for several reasons. First, as shown above, the living world is full of examples of nanotechnological devices. However, a second key point is that nanotechnology has been driven and inspired by the example of biological systems and the need (for example, in medicine) to influence biological systems at the scale of a single cell. In addition, preliminary steps have been taken to harness the nanotechnology of biological systems and use it to perform useful functions.

One compelling example of a proof-of-principle, biologically inspired device emerged from work aimed at exploring the function of ATPsynthase, the rotary device already described above. Fluorescently labelled filamentous proteins from the cytoskeleton, known as actin, were attached to the putative rotary component of the F_1 subunit. The point of this exercise is that such filaments are observable using light microscopy. It was then observed that the long actin filament rotated like a propeller when the ATPsynthase performed ATP hydrolysis [15].

A second example, also involving ATPsynthase, is suggested by a set of beautiful experiments done by *Racker* and others [16, 17]. The idea is illustrated schematically in Fig. 14.1. Two different protein machines are "reconstituted" in an artificial membrane-bound region known as a liposome. One of these devices is known as bacteriorhodopsin and has the capacity to pump hydrogen ions when it is exposed to light. Since both the bacteriorhodopsin and ATPsynthase are embedded in the same membrane, the ATPsynthase can then exploit the light-induced proton gradient to perform ATP synthesis. Again, these experiments were undertaken not for their role as possible devices, but rather to probe the nature of various molecular machines. Nevertheless, we view them as a provocative demonstration of both the manipulation and use of such machines in artificial environments. As such, they provide an inspiring vision of the possibilities for biologically inspired nanotechnology.

Another fascinating example of biologically inspired nanotechnology is that of biofunctionalized cantilevers (see Fig. 14.2). A typical example is provided by the experiments of *Wu* et al. [18], who have demonstrated the use of a biofunctionalized cantilever as a scheme for detecting small concentrations of biologically interesting molecules such as prostrate-specific antigen and single-stranded DNA. One surface of the cantilever is coated with an antibody, and then it is placed in environments containing different concentrations of the antigen. The key ideas from a mechanical perspective are: i) the difference in surface energy between the top and bottom surfaces of the cantilever induces spontaneous bending and ii) the binding of molecules of interest to target molecules initially present on the surface leads to surface energy differences and bending that can be detected by optical means. In this way the concentration of the molecules of interest can be measured. The specificity and sensitivity of the method makes it viable for use in the laboratory, as well as for

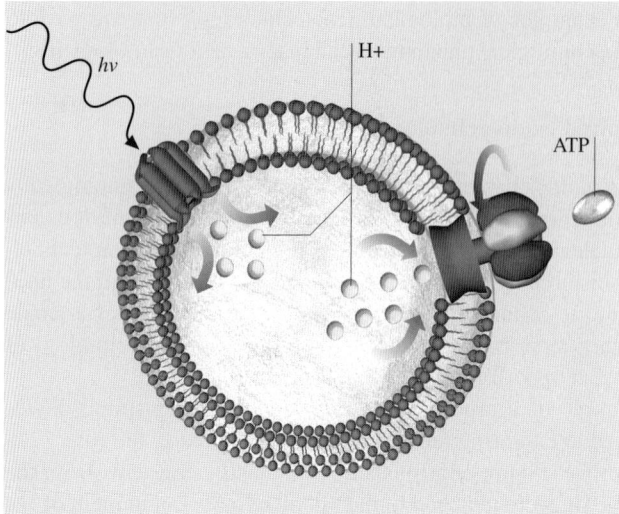

Fig. 14.1. Schematic of the experiment in which bacteriorhodopsin (*top of figure*) and ATP-synthase (*right of figure*) are artificially reconstituted in a liposome, and they act in unison to produce ATP molecules

commercial purposes. We return to this example in our discussion of the modeling challenges posed by problems at the interface between biology and nanotechnology.

14.2.6 Nanotechnology and Single Molecule Assays in Biology

One of the key refrains of the nanotechnological era is Feynman's quip that "there is plenty of room at the bottom" [19]. The benefits of miniaturization are evident at every turn in applications ranging from our cars to our computers. Associated with the development of the inspiring new techniques made possible by nanotechnology has been the emergence of a host of scientific opportunities. One of the arenas to benefit from these new techniques is biology. As scientists and engineers have taken the plunge to the nanotechnological "bottom" foreseen by Feynman, opportunities have constantly arisen to manipulate biological systems in ways that were heretofore unimagined, culminating in a new era of single molecule biology. Single molecule experiments have, in fact, presented us with a view of Feynman's "room at the bottom" as being filled with very complicated machines whose functioning makes life possible.

Single molecule assays complement statistical/collective studies involving a large number of molecular actors by revealing the prominent role of fluctuations at the sub-cellular scale. For example, a photospectrometer measures the optical response of a huge collection of molecules, whereas optical tweezers pull on a single molecule of DNA and enable us to follow the changes in conformation or the breakage of bonds. In fact, experimental methods are so advanced that it is now possible to manipulate a single molecule even as we watch it on a screen as it jiggles around in

Fig. 14.2. Schematic of the use of biofunctionalized cantilevers as a tool for detecting molecules of biological interest (figure courtesy of Arun Majumdar, Berkeley)

different conformations. In what follows, we give several examples of how nanotechnology has reached out to help create single molecule biology and in the process has led to the advent of new quantitative opportunities to investigate biological systems.

Atomic-Force Microscopy

One of the tools that has revolutionized nanotechnology, in general, and single molecule biology, in particular, is the atomic-force microscope. The AFM has helped create the field of single molecule force spectroscopy [20]. We note that mechanics has a long tradition of using force-extension data (much like the electrical engineer uses current-voltage data) to probe the inner workings of various materials. It is now possible to apply forces of known magnitude on a macromolecule and study how it deforms under the force. This furnishes structural information and provides insights into the energy landscape the molecule needs to navigate as it undergoes force-induced conformational changes. The energy of deformation associated with such molecules is primarily determined by weak forces such as hydrogen bonds, van der Waals contacts, and hydration effects. On a more philosophical note, these experiments force us to think in terms of forces and not energy, complementing the traditional views held in molecular biology, and they can lead to many new insights about the relation between structure and function in proteins, polynucleotides, and other macromolecular entities.

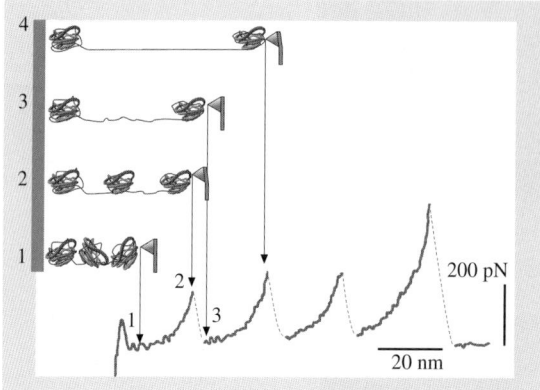

Fig. 14.3. Schematic illustrating the way that the AFM has been harnessed as a nanotechnological analogue of the Instron machine for the measurement of the force-extension properties of single molecules. Four snapshots in the life history of a globular protein subject to loading are shown, as well as the measured force-extension curve. Each sawtooth corresponds to the unfolding of a single protein domain. (Figure adapted from [22])

The AFM has been used in a wide variety of single molecule experiments on many of the key classes of molecules found in the living world, including nucleic acids, proteins, and carbohydrates. One fascinating example is the use of atomic force microscopy to examine the mechanical properties of the muscle protein titin [21]. The experiment is illustrated schematically in Fig. 14.3, which shows that there is a series of force-extension signatures (increasing force and extension followed by a precipitous load drop) that correspond to the unbinding of the individual domains that make up this protein.

Optical Tweezers

Another instrument that has been used with great success in the realm of biophysics for the purposes of performing nanomechanical measurements on macromolecules and their assemblies is the optical tweezer [23, 24]. While the AFM is relatively stiff and applies large forces (on the order of 100–1000 pN), optical tweezers are compliant and can measure smaller forces (on the order of 0.1–10 pN). Some of the most interesting experiments performed with optical tweezers concern the functioning of molecular motors, which by themselves are marvels of nanotechnology. For example, *Svoboda* et al. [25] attached kinesin to an optically trapped bead and observed its movement along a microtubule. An example of the type of data to emerge from such experiments is shown in Fig. 14.4. One of the conclusions to emerge from such experiments is that kinesin can exert forces on the order of 5–7 pN before it stalls. Such experiments also permit an examination of the effect of changing the concentration of ATP on the functioning of kinesin [26]. Similar experiments have been performed on RNA polymerase as it advances along DNA to deduce not only the stall force, but also its velocity as a function of the constraining force [27]. Such measurements pro-

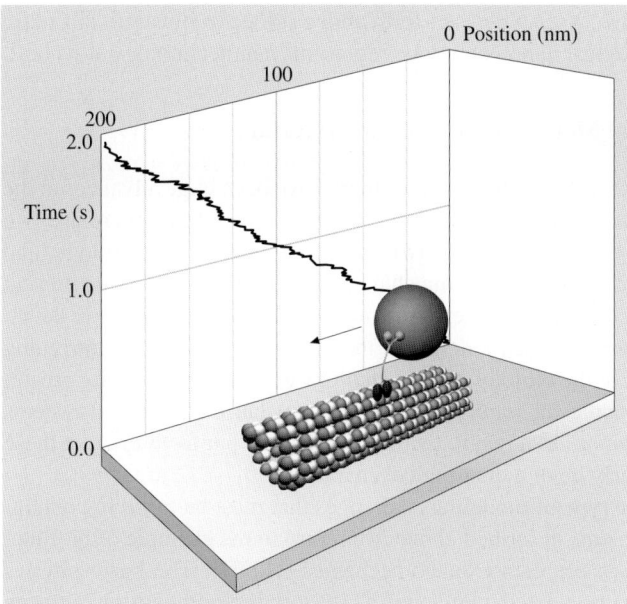

Fig. 14.4. Schematic illustrating the use of optical tweezers to measure the speed of the molecular motor kinesin in its journey along a microtubule

vide a mechano-chemical basis of biological function and go a long way in revealing connections between chemical kinetics and mechanical processes at the molecular level.

As noted in the abstract, one of the most fascinating examples of the biology-nanotechnology interface is that of bacterial viruses, known as bacteriophages. The life cycle of a large class of bacteriophages is characterized by self-assembly processes that lead to the formation of the protein shell of the virus followed by active packaging of the viral DNA within this shell by a motor. The structure of this so-called "viral portal motor" has recently been solved using X-ray crystallography [28]. In a recent experiment, *Smith* et al. used optical tweezers to study the characteristics of the DNA packaging process of the ϕ-29 bacteriophage [29]. One of the conclusions of this experiment is that the motor has to act against an increasing resistive force as more and more of the DNA is packed inside the capsid. From a quantitative perspective, this experiment yields the force and the rate of packing as a function of the fraction of the genome packed. It is also important to note that bacteriophages are not only an obscure subject of quiet enquiry, but are also the basis of a huge range of cloning products (see, for example, the lambda ZAP vectors of Stratagene) used for doing experiments with recombinant DNA, and more generally, viruses are being explored as the basis of gene therapy.

Our discussion thus far has been aimed at providing a rough overview of the vast landscape that sits at the interface between biology and nanotechnology. It is hoped that the few representative case studies set forth above suffice to illustrate our basic thesis, namely, that biological nanotechnology represents nanotechnology at its best.

14.2.7 The Challenge of Modeling the Bio-Nano Interface

As highlighted in the previous two subsections, there have been huge advances at the interface between nanotechnology and biology. We have argued that there are two distinct representations of the interface between biology and nanotechnology, and each has its own associated set of modeling challenges. The argument of the present discussion is that another key part of the infrastructure that must attend these developments is that associated with the modeling of these systems. One of the intriguing ways in which modeling at the biology-nanotechnology interface is assuming greater importance is that with increasing regularity, experimental data on biological systems is of a quantitative character. As a result, the models that are put forth to greet these experiments must similarly be of a quantitative character [30].

As an example of the type of modeling challenges that must be faced in contemplating the types of problems described above, we return to the example of biofunctionalized cantilevers as a problem in nanomechanics [18, 31]. The basic physics behind the use of biofunctionalized cantilevers as sensors is a competition between elastic bending energy and the surface free energy difference between the upper and lower faces of the cantilever. The face with the lower free energy per unit area tends to increase its area by bending the cantilever. The amount of bending, on the other hand, is limited by the elastic energy cost. The utility of this device derives from the fact that the difference in the surface free energy is affected by specific binding of target molecules to probe molecules that are initially deposited on one side of the cantilever.

To provide a quantitative model of the biofunctionalized cantilever, we construct an energy functional that takes into account the elastic energy of beam bending and the surface energy. Both contributions to the total energy can be written as functionals of $u(x)$, the deflection of the cantilever. Note that in the case in which the two surfaces have the same free energy per unit area, $\gamma_{up} = \gamma_{down}$, the equilibrium configuration corresponds to $u(x) = 0$. The case of interest here is that in which the two surfaces have different energies.

We recall that the energy associated with beam bending is of the form [32]

$$E_{bend}[u(x)] = \frac{EI}{2} \int_0^L [u''(x)]^2 \, dx, \tag{14.1}$$

where E is Young's modulus, L is the length of the beam, and $I = wt^3/12$ is the areal moment of inertia; t and w are the beam thickness and width. The main approximation we have made in writing (14.1) is that the cross section of the beam remains unchanged by the bending process. The surface contribution to the total energy is associated with the changes of the areas of the upper and lower surfaces by virtue of the beam deforming. In particular, we have

$$E_{\text{surf}}[u(x)] = \gamma_{\text{up}} w \left[L - \frac{t}{2} \int_0^L u''(x) dx \right]$$
$$+ \gamma_{\text{down}} w \left[L + \frac{t}{2} \int_0^L u''(x) dx \right], \quad (14.2)$$

where the terms in parentheses are the arc lengths along the top and bottom surface of the beam, respectively. The physical content of this functional is the idea that if $u'' > 0$ (concave upward) then the upper surface area will shrink while the lower surface area will increase. Neglecting uninteresting constant terms, the total energy functional is

$$E_{\text{tot}}[u(x)] = \frac{EI}{2} \int_0^L [u''(x)]^2 \, dx$$
$$+ \frac{tw}{2} (\gamma_{\text{down}} - \gamma_{\text{up}}) \int_0^L u''(x) dx . \quad (14.3)$$

Our goal is to find the displacement profile $u(x)$ that yields the minimum value for the total energy. The general mathematical framework for effecting this minimization is the calculus of variations. For the functional in (14.3), a more direct route to the result can be obtained by completing the square. Namely, we note that the E_{tot} involves a term quadratic in $u''(x)$ and a second term that is linear in the same function. As a result, the total energy can be rewritten in the form

$$E_{\text{tot}}[u(x)] = \frac{EI}{2} \int_0^L dx \left[u''(x) + \frac{tw\Delta\gamma}{2EI} \right]^2 , \quad (14.4)$$

where we have introduced $\Delta\gamma = \gamma_{\text{down}} - \gamma_{\text{up}}$, and we neglect an uninteresting constant term. Clearly the $u(x)$ that minimizes the energy is one for which the integrand vanishes everywhere, or

$$u''(x) + \frac{tw\Delta\gamma}{2EI} = 0 . \quad (14.5)$$

At this point, we are left with a standard differential equation whose solution, given the boundary conditions $u(0) = 0$ and $u'(0) = 0$, is

$$u(x) = -\frac{tw}{4EI} \Delta\gamma x^2 . \quad (14.6)$$

The physical meaning of this solution is hinted at by observing that if $\gamma_{\text{down}} > \gamma_{\text{up}}$ then the beam curves downward with the result that the area of the lower surface is reduced.

Measurements of Wu et al. [31] indicate an upward cantilever deflection when single stranded DNA (ssDNA), up to 20 nucleotides in length, hybridizes with complementary strands of ssDNA, which were deposited initially so as to functionalize the beam. This effect might be attributed to the difference in elastic properties of ssDNA and its double stranded counterpart. Namely, under physiological conditions, ssDNA has a persistence length equal to two nucleotides, while dsDNA is much stiffer (due to hydrogen bonding and base stacking interactions between the two strands) and has a persistence length of 150 nucleotides. Therefore, deposition of the flexible ssDNA molecules initially leads to bending of the cantilever downwarddue to entropic repulsion between the ssDNA chains. After hybridization, rigid dsDNA strands are formed, there is no longer any entropy to be gained by increasing the area of the top surface, and the beam bends back upwards. Remarkably, Wu et al. demonstrate that their biofunctionalized cantilever is sensitive to ssDNA differing in length by a single nucleotide!

To gain further insight into the physics of the biofunctionalized cantilever, it is instructive to examine some quantitative aspects of the experiment using (14.6). Namely, Wu et al. find a deflection of $u(L) = 12$ nm when a 20-nucleotide strand of ssDNA hybridizes with a 20-nucleotide-long complementary target. Using the quoted numbers for the cantilever, $E = 180$ GPa, $L = 200$ μm, $t = 0.5$ μm, $w = 20$ μm, leads to $\Delta\gamma = 4.5$ fJ/μm^2. From this result we can estimate the change in surface free energy due to beam bending,

$$\Delta E_{\text{surf}} \approx wL\Delta\gamma . \tag{14.7}$$

We find $\Delta E_{\text{surf}} = -18$ pJ, which, given the quoted areal chain density of 6×10^{12} cm^{-2}, leads to a decrease in free energy of 75 pN nm, or $18\,k_B T$, per chain. This is comparable to the entropy of a 20 nt ssDNA, which is on the order of $10\,k_B$, lending support to the idea that entropic repulsion between ssDNA strands is implicated in cantilever bending.

While this example gives a feel for the way in which quantitative models have been put forth to respond to biologically inspired nanotechnologies, the remainder of the chapter will emphasize attempts to construct nanomechanical models of biological nanotechnology itself. We turn first to a discussion of the various scales that arise in thinking about the biology-nanotechnology interface and then conclude with several modeling examples.

14.3 Scales at the Bio-Nano Interface

Every scientific discipline has a preferred set of units that lends itself to building intuition concerning the system at hand. For example, an astronomer thinks of distances between stars in light years, not kilometers. Though most of us have an intuitive sense of the meaning of a kilometer, by the time we add more than six zeros, all intuition is lost. At terrestrial scales, we talk of distances between cities in terms of the flying time or the driving time between them and usually not in terms of hundreds of miles. Similar choices must be faced in the biological setting. For example, a biologist might characterize the complexity of an organism by the size (in kilo base-pairs) of the genome and not by the organism's physical size. The aim of the present section is to highlight some of the key scales and units that reveal themselves at the interface between biology and nanotechnology. Indeed, we go further and assert that as yet we are still in the process of searching for the most suitable units to characterize the biology-nanotechnology interface. Although our attempt to determine such units and scales might involve seemingly complicated interconversions, such as measuring distances in terms of time (via diffusion), or measuring concentrations in terms of distances, we hold that the approximate *numerical* characterization of the scales of interest is of crucial importance to the endeavor of considering nanomechanics at the biology-nanotechnology interface. In particular, the right choice of units can assist us in building intuition about these systems. Our goal in this section is to emphasize the scales in length, time, force, energy, and power that are relevant in contemplating the nanomechanics of biological systems.

14.3.1 Spatial Scales and Structures

We begin with a discussion of the length scales that arise in contemplating nanomechanics at the biology-nanotechnology interface. In this case, the prefix *nano* leads justifiably to a consideration of the nanometer as one possible choice as the fundamental unit of length. However, to prepare ourselves for the question of how best to describe the dimensions of the spatial structures of interest here, it is important to consider the *hierarchy* of length scales that arise in the nano-bio arena. After examining this hierarchy of structures, we reformulate these length scales in terms of the volumes of these structures measured in units of the volume of a typical bacterial cell, and conclude the present section with a discussion of the way that chemical concentrations can also be interpreted as determining a length scale.

As noted above, a first step in developing intuition concerning the spatial scales found at the biology-nanotechnology interface is through reference to the hierarchy of scales and structures that arise in this arena. The shortest distance in this hierarchy of scales that will interest us is that associated with the size of individual atoms. We recall that the size of a hydrogen atom is roughly 0.1 nm. The scale characterizing the linkage between atoms is that of typical bond lengths that range from roughly 0.1–0.3 nm. A step further in this hierarchy brings us to the basic building blocks of the biological world such as amino acids, nucleotides, individual sugars, lipid molecules, etc. A typical length scale that characterizes these building blocks is the nanometer itself. For example, we have already made reference to the importance of lipid-bilayer membranes in bounding different regions of the cell. Phosphatidylcholine is one of the molecular building blocks of many such membranes. It has a polar head group and a hydrocarbon tail with an overall dimension of 2–3 nm. Similarly, the dimensions of single amino acids and nucleotides are on the order of 1 nm as well.

As is well-known, individual nucleotides are assembled to form nucleic acids such as DNA, amino acids are assembled to form proteins, sugars combine to form polysaccharides, and lipids self-assemble to form membranes. One way of estimating the size of the resulting molecules is by taking the scale of the individual units and scaling up with the number of such units. The various molecular actors of relevance to the present discussion can also be characterized by a length, scale known as the persistence length, which gives a rough description of the length over which the molecule behaves as a stiff rod. For example, the persistence length of DNA is $\approx$

50 nm, while that of the cytoskeletal filaments is in the range of 15 μm for actin and 6 mm for microtubules [11]. We note that in our later discussions of viruses as a profound example of biological nanotechnology, the ratio of the persistence length of DNA to the size of the viral capsid (the container within which the DNA is packaged) will serve as a measure of the energetic cost of packing the DNA within the capsid and will signal the need for molecular motors to take active part in the packaging process.

The next scale above that of the various macromolecular building blocks are assemblies of such molecules in the form of various molecular machines which are some of the most compelling examples of nature's nanotechnology. The ability to begin formulating mechanistic models of such machines is founded upon key advances in both X-ray crystallography and cryo-electron microscopy [33, 34]. Examples of such machines and their associated dimensions include: the machine responsible for making the message carried by DNA readable by the protein synthesis machinery, namely, RNA polymerase ($\approx$ 15 nm) [35], the machine that produces ATP, the energy currency of the cell, namely, ATPsynthase ($\approx$ 10 nm) [9], and the machine that carries out protein synthesis, namely, the ribosome ($\approx$ 25 nm) [36]. A second class of assemblies of particular relevance to the present article are viruses, representative examples of which include lambda phage ($\approx$ 27 nm), tobacco mosaic virus ($\approx$ 250 nm in length), and the HIV virus ($\approx$ 110–125 nm) [37].

From the standpoint of cellular function, the next level of structural organization is associated with the various organelles within the cell, structures such as the cell nucleus (3–10 μm), the mitochondria that serve as the power plant of the cell ($\approx$ 1000 nm), the Golgi apparatus wherein modifications are made to newly synthesized macromolecular components ($\approx$ 1000 nm), and so on [1]. These organelles should be thought of as factories in which many molecular motors of the type described earlier do their job simultaneously. At larger scales yet, and constituting a higher level of overall organization, life as nanotechnology is revealed as cells themselves, the fundamental unit of life that is self-replicating and self-sustaining. We will make special reference to one particular bacterial cell, namely, *E. coli* with typical dimensions of 1 μm. This should be contrasted with a typical eucaryotic cell such as yeast, which has linear dimensions roughly a factor of 10 larger than the *E. coli* cell.

We note that the various biological structures described above should be seen with reference to the sorts of man-made structures to which they are interfaced. For example, earlier in the chapter we mentioned the importance of optical tweezers as a means of communicating forces to macromolecules and their assemblies. Typical dimensions for the optical beads used in optical tweezers are on the scale of 500–1000 nm. We note that though such dimensions are characteristic of organelles, they are much larger than the individual molecules they are used to study. A second way in which individual macromolecules are communicated with is through small tips such as those found on an AFM. In this case, the size of the tip can be understood through reference to its radius of curvature, which is typically of the order of 50 nm.

Table 14.1. Sizes of entities in comparison to the size of *E. coli*

Entity	Size (nm)	# fitting in an *E. coli*
Amino acid	1	5.0×10^9
Nucleotide	1	5.0×10^9
Monosaccharide	1	5.0×10^9
Phosphatidyl choline	2–3	2.0×10^8
Proteins	5–6	3.0×10^7
Ribosome	20	1.0×10^6
Mitochondria	500–1000	3–4
T4 phage	100	5000

Our discussion thus far has centered on the use of a single characteristic length to describe the spatial extent of biological structures. In our quest to develop intuition about the typical spatial scales found at the biology-nanotechnology interface we note that a second way to evaluate such scales is through reference to the typical *volumes* of the various structures of interest. As noted in the introduction to this section, we claim that a key way to develop intuition is by making sure to use appropriate and revealing units. For the present purposes, we argue that one useful unit of volume is that of an *E. coli* bacterium, which if idealized as a cylinder with diameter 1 µm and height 2 µm, has a volume of $\approx 1.5 \times 10^9$ nm^3. In particular, we will measure the sizes of the various entities described above in terms of how many of them can fit into a single *E. coli* bacterium. Our estimates of the comparative sizes of various structures of interest are given in Table 14.1. Note that we do not claim that this is literally how many of each of these entities are found in an *E. coli* bacterium, but rather we seek to give an impression of the relative volumes of some of the different structures that arise when contemplating biological nanotechnology.

The idea of counting the number of molecules that fit into a given volume is actually quite standard. This is exactly what chemists do when they invoke the notion of concentration: measuring the strength of an acid or base using pH amounts to expressing the concentration of H$^+$ ions in solution. Similarly, when we refer to the molarity of a given solution, it is a statement of how many copies of a given molecule will be found per liter of solution. These ideas are pertinent for our examination of the scales that arise in contemplating biological nanotechnology. As an example, we consider the action of molecular pumps such as that which maintains the concentration gradient in Ca^{2+} ions between the cellular interior and the extracellular medium. Note that this is described in detail in chapter 12 of *Alberts* et al. [1]. The Ca^{2+} pump is responsible for insuring that the intercellular concentration of Ca^{2+} ions is 10^{-7} smaller than that in the cellular exterior. One explanation for this low a concentration of Ca^{2+} in the cell is the role played by these ions in signaling other activities. The low concentration might serve as a scheme for increasing the signal-to-noise ratio. The reported intracellular concentration of such ions is roughly 10^{-7} mM. It is interesting to ask how many Ca^{2+} ions this corresponds to in a eucaryotic cell. The volume of a eucaryotic cell measuring 20 µm in diameter is about

4×10^{-12} liters, which translates into ≈ 250 Ca^{2+} ions in the cytoplasm. For the case of *E. coli*, which has a volume approximately 1000 smaller, this concentration would correspond to one ion for every four cells. This number gives us far greater insight than the standard molar (M = mol/liter) method of expressing concentrations. In a similar vein, we argue that distances between individual ions/molecules in solution are perhaps a better way of thinking about pH and molarity when using them in the context of biological nanosystems. To this end, Fig. 14.5 shows a plot of pH and molarity versus average distance d in nm between individual ions/molecules. Note that, in addition, we have also translated these concentrations into number of copies per *E. coli* cell and the time for an ion to diffuse over the average separation distance. For the time estimate we make use of the typical diffusion constant for small ions in water at $25\,°C$, $D = 2000\,\mu m^2/s$.

We have seen that the world of biological nanotechnology utilizes length scales that span a rather wide range, from fractions of a nanometer to tens of microns. This provides a challenge to modeling, whereby methods with atomic-scale resolution need to be combined in a consistent and seamless way with coarse-grained continuum descriptions to provide a complete picture. An example of such modeling will be provided in later sections when we take up the mechanical aspect of the viral life cycle.

14.3.2 Temporal Scales and Processes

We have seen that a study of spatial scales is a first step in our quest to understand the units that are suited to characterizing the biology-nanotechnology interface. We note that our understanding of spatial structures both in the living world, as well as those used in intriguing man-made technologies such as microelectronics have been built around important advances such as electron microscopy, X-ray crystallography, nuclear magnetic resonance, etc. that have revolutionized structural biology and materials science alike. On the other hand, one of the key challenges that remain in really appreciating the structure-properties paradigm, which is as central to biology as it is to materials science, is the need to acknowledge the dynamical evolution of these structures.

We note that just as there is a hierarchy of spatial scales that are important to consider in describing biological nanotechnology, so too is there a hierarchy of *temporal processes* that also demand a careful consideration of relevant units. An impressive representation of the temporal hierarchy that must be faced in contemplating biological systems is given by *Chan* and *Dill* [38] in their Fig. 5. They organize their temporal hierarchy according to factors of 1000, starting at the femtosecond time scale and ending with time scales representative of the cell cycle itself. One of the elementary dynamical processes undergone by all of the molecular actors of the living world is thermal vibration. Vibrations of atoms occur at the time scales of 10^{-15} to 10^{-12} s. As will be noted later, from a modeling perspective, this depressing fact manifests itself in the necessity of using time steps of the same order when integrating the equations of motion describing molecular dynamics. The reason that this is unfortunate is that almost everything of dynamical interest occurs at time scales

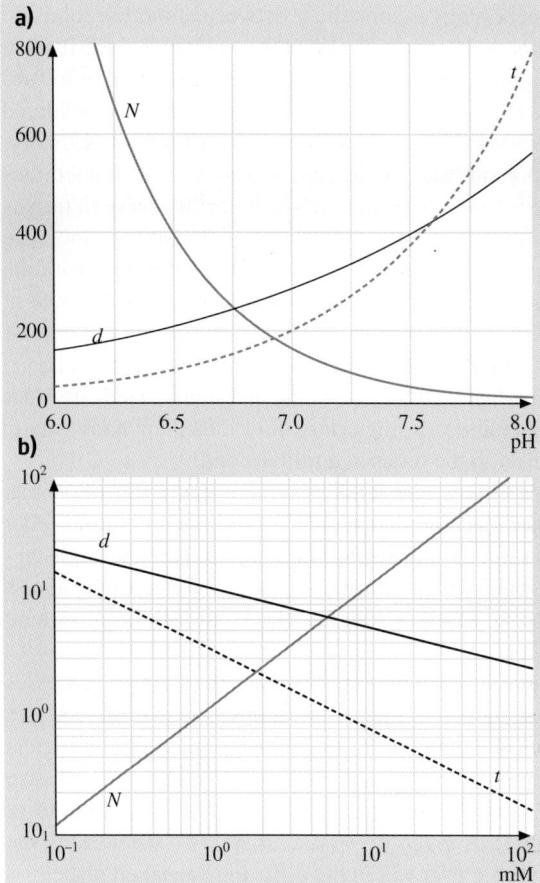

Fig. 14.5. (a) Representations of pH in terms of average distance between ions (d in nm), the number of ions per one *E. coli* cell volume (N), and the typical time for an ion to diffuse over a distance equal to their average separation (t in 100 ns). (b) Concentration unit millimolar (mM) represented in terms of d (nm), N (millions of molecules), and t (10 ns)

much longer than the femtosecond time scale characteristic of molecular jiggling, thus making molecular dynamics investigations computationally very expensive.

We follow *Chan* and *Dill* [38] in examining how successive thousandfold increases in our temporal resolution smears out features such as vibrational motion and brings into focus more interesting processes characteristic of macromolecular function. Indeed, the side chains of amino acids rotate with a characteristic period on the order of 10^{-9} s. Another thousandfold increase in time scale begins to bring key biological processes such as polymerization into view. Just as a length scale on the order of 10–100 nm is perhaps most characteristic of the length scales of biological nanotechnology, the time scale that is most relevant for considering the processes associated with biological nanotechnology is something between a microsecond and a millisecond. To drive home this point, we reconsider the machines described in the previous section from the length scale perspective, but now with an eye toward how fast they perform their function. Several of the molecular machines considered in the previous section mediate polymerization reactions. For example, RNA polymerase produces messenger RNA molecules by the repeated addition of nucleotides to a chain of ever increasing length. Such messenger RNA molecules serve in turn as the template for synthesis of new proteins by the ribosome, which reads the message contained in the RNA and adds individual amino acids onto the protein. From the temporal perspective of the present section, it is of interest to examine the rate at which new monomers are added to these polymers – roughly 10 per s in the case of RNA polymerase [39], and 2 per s in the case of the ribosome in eucaryotic cells [1] (bacterial ribosomes are 10 times faster). Once proteins are synthesized, in order to assume their full cellular responsibilities, they must fold into their native state, a process that occurs on time scales on the order of a millisecond.

In addition to translational machines like RNA polymerase and the ribosome, there are a host of intriguing rotary machines such as ATP synthase and the flagellar rotary motor. The flagellar motor, which is the means of locomotion for several bacteria, rotates at about 100 rpm [6]. ATP synthase makes use of a proton gradient across a lipid bilayer to provide rotation at a rate of roughly 6000 rpm [40], approaching that of turbojet engines.

Thus far, our discussion has emphasized the rate of a variety of active processes of importance to biological nanotechnology. It is of interest to contrast these scales with those pertaining to diffusion. We note that in many instances in the biological setting the time scale of interest is that determined by the time it takes for diffusive communication of two spatially separated regions. In particular, as noted earlier in the chapter, the operative time scale is given by $t_{\text{diffusive}} = x^2/2D$, where D is the diffusion constant and x is the distance over which the diffusion has taken place.

To get a sense for all of the time scales described in this section, Table 14.2 presents a variety of results considered above in tabular form, but now with all times measured in units of the time it takes for ATPsynthase to make one rotation (≈ 10 ms). Our reason for adopting ATPsynthase rotation as defining a unit of time is that it exhibits motions associated with one of the most important of life's processes, namely, the synthesis of new ATP molecules. In addition, our later discussion of units associated with both energy and power will once again appeal to the central role played by ATP in biological nanotechnology.

14.3.3 Force and Energy Scales: The Interplay of Deterministic and Thermal Forces

Another set of scales of great relevance and importance to the nano-bio interface concerns the nature of the forces that act in this setting. Thus far, we have examined the spatial extent and the cycle time of a variety of examples of biological nanotechnology. As a next step in our examination of scales, we consider the forces and energies associated with these structures. Two outstanding sources for developing a feel for

Table 14.2. Time scales of various events in biological nanotechnology measured in units of the average time taken for ATPsynthase to make a single rotation

Process	Time scale (units of ATPase rotation)	Ref.
1/(frequency of amino acid addition) – eukaryotic ribosome	50	[1]
1/(frequency of monomer addition) – RNA polymerase	10	[39]
Time between motion reset of *E. coli*	10	[6]
Kinesin step	1	[26]
Diffusion time for protein (ion) to cross an *E. coli*	0.5 (0.02)	[11]
Vibrational period of nanocantilever	$10^{-3} - 10^{-5}$	[41]

the relevant numbers are the books of *Wrigglesworth* [42] and *Smil* [43]. Perhaps the most compelling feature when thinking about forces at these scales is the interplay between deterministic and thermal forces. To substantiate this claim, we note that at room temperature the fundamental energetic quantity is $k_B T \approx 4.1 \, \text{pN nm}$. The reason this number is of interest to the current endeavor is the realization that many of the molecular motors that have thus far been investigated act with forces on the piconewton scale over distances of nanometers.

The fact that the piconewton is the relevant unit of force can be gleaned from a simple estimate. Namely, consider a typical skeletal muscle in the human arm. The cross-sectional size of the muscle is of order 3 cm. The muscle consists of cylindrical rods of protein called myofibrils, which are roughly 2 μm in diameter, while the myofibrils themselves are made of strands of actin and myosin filaments, which total some 60 nm in diameter [1]. This gives 10^{12} myosin filaments per cross-sectional area of the muscle. As each myosin filament over the length of a single sarcomere (the contractile unit of myofibrils, some 2.5 μm long) contains some 300 myosin heads, lifting a 30 kg load corresponds to a force of 1 pN per myosin head. This is certainly an underestimate, since not all myosin heads are attached to the actin filament at the same time. Since our estimate leaves out many details, we might wonder how it compares to the measured forces exerted by molecular motors. Sophisticated experiments with optical tweezers have revealed that actin-myosin motors stall at a force of around 5 pN [44]; RNA polymerase stalls at a force of about 20 pN [39]; the portal motor of the ϕ-29 bacteriophage exerts forces up to 50 pN [29].

As noted earlier, RNA polymerase is a molecular motor that moves along DNA while transcribing genes into messenger RNA. The DNA itself is an elastic object that is deformed by forces exerted on it by various proteins. At forces less than 0.1 pN its response is that of an entropic spring with a stretch modulus of 0.1 pN, while at forces exceeding 10 pN its stretch modulus is determined by hydrogen bonding of the base pairs and is roughly 1000 pN [45]. All the above mentioned data reinforces the argument that piconewton is the relevant unit of force in the nanomechanical world of the cell.

The concept of "stress" is closely related to that of force. Stress is a continuum mechanical concept of force per unit area, and it has been used with great success in solid and fluid mechanics. We extend the idea of stress to the nanomechanical level to see what numbers we arrive at. From the data above we can deduce that a single myosin fiber sustains a stress of about 10^{-2} pN/nm^2, which is the same as 10^{-2} MPa. Migratory animal cells such as fibroblasts, which are responsible for scavenging and destroying undesirable products in tissues, can generate a maximum stress on their substrate on the order of 3.0×10^{-2} MPa [46]. DNA can sustain stresses in the excess of 20 pN/nm^2, or 20 MPa, at which point an interesting structural transition accompanied by an overall increase of its contour length is observed [45]. Engineering materials such as steel and aluminium, on the other hand, can sustain stresses of about 100 MPa. This goes to show that nanotechnology in the context of biological systems is built from rather soft materials.

It is of interest to translate our intuition concerning piconewton forces into corresponding energetic terms. The kinesin motor advances 8 nm in each step against forces as high as 5 pN [25]. This translates to a work done on the order of 40 pN nm. A myosin motor suffers displacements of about 15 nm with forces in the piconewton range. ATP hydrolysis (to ADP) releases energy (at pH 7 and room temperature) of about 50 pN nm [1]. When a titin molecule is pulled, it unfolds under forces of 30–300 pN, causing discrete expansions of 10–30 nm [22], implying energies in the range of 300–9000 pN nm. Experimental data provides a compelling argument in favor of thinking of the pN nm as a unit of energy for nanomechanics. However, the observation that $k_B T = 4.1$ pN nm at room temperature gives important insight, since it reveals that thermal forces and entropic effects play a competing role in biological nanomechanics. This provides nature with unique design challenges, whereby molecular motors that can perform useful work must do so in the presence of strong thermal fluctuations for the normal functioning of the cell. Operation of motors in such a noisy environment is governed by laws that are probabilistic in nature. This is manifest in single molecule experiments that observe motors stalling and sometimes reversing direction.

Energy conversion is crucial to any developmental or evolutionary process. The steam engine powered the industrial revolution. However, long before thinking beings with man-made machines founded the industrial revolution, power generation had already become a central part of life's nanotechnology. Indeed, the development of ATP-synthase is one of the cornerstones for the evolution of higher life-forms. An inevitable concomitant of evolution is the necessity for faster and more efficient operations. The industrial revolution led to the emergence of bigger and faster modes of transport; biological evolution led to the emergence of complex and intelligent organisms. Invariably, a machine or an organism is limited in its abilities by the speed at which it can convert one form of energy (usually chemical) to other forms of energy (usually mechanical). This is why a study of their "power plants" becomes important.

Power plants (or engines) are usually characterized by their force-velocity curves and compared using their power-to-weight (P/W) ratio. For example, the myosin motor has a P/W ratio of 2×10^4 W/kg, the bacterial flagellar motor stands at 100 W/kg, an internal combustion engine is at about 300 W/kg, and a turbojet engine stands at 3000 W/kg [47]. These figures tell us that linear motors like myosin and kinesin are extremely powerful machines.

14.4 Modeling at the Nano-Bio Interface

We have already provided a number of different views of the biology-nanotechnology interface, all of which reveal the insights that can emerge from model building. Indeed, one of the key thrusts of this entire chapter is the view that as the type of data that emerges concerning biological systems becomes increasingly quantitative, it must be responded to with models that are also quantitative [30]. The plan of this section is to show how atomistic and continuum analyses each offer insights into problems of nanotechnological significance, but under some circumstances, both are found wanting and it is only through a synthesis of both types of models that certain problems will surrender. The plan of this section is to examine the advantages and difficulties associated with adopting both atomistic and continuum perspectives and then to hint at the benefits of seeking mixed representations.

14.4.1 Tension Between Universality and Specificity

One of the key insights concerning model building in nanomechanics, whether we are talking about the nanoscale tribological questions pertinent to magnetic recording or the operations of molecular motors, is that such questions live in the no-man's-land between traditional continuum analysis at one extreme and all-atom approaches such as molecular dynamics on the other. Indeed, there is much discussion about the breakdown of continuum mechanics in modeling the mechanics of systems at the nanoscale. This dichotomy between continuum theories, which treat matter as continuously distributed, and atomic-level models, which explicitly acknowledge the graininess of matter, can be restated in a different (and perhaps more enlightening) way. In particular, it is possible to see atomistic and continuum theories as offering complementary views of the same underlying physics. Continuum models are suitable for characterizing those features of a system that can be thought of as averages over the underlying microscopic fluctuations. By way of contrast, atomistic models reveal the details that a continuum model will never capture, and in particular, they shed light on the specificity of the problem at hand.

The perspective adopted here is that continuum models and atomistic models each reveal important features of a given problem. For example, in contemplating the competition between fracture and plasticity at crack tips, a continuum analysis provides critical insights into the nature of the elastic fields surrounding a defect such as a crack. These fields adopt a fundamental and universal form at large scales with all detailed material features buried in simple material parameters. By way of contrast, the precise details of the dissipative processes occurring at a crack tip (in particular, the competition between bond breaking with the creation of new free surface and dislocation nucleation) require detailed atomic-level descriptions of the energetics of bond stretching and breaking. In the biological setting, similar remarks can be made. In certain instances, the description of biological polymers as random coils suffices and yields insights into features such as the mean size of the polymer chain as a function of its length. On the other hand, if our objective is mechanistic understanding of processes such as how phosphorylation of a particular protein induces conformational change, this is an intrinsically atomic-level question. The language we invoke to describe this dichotomy is the use of the terms *universality* and *specificity*, where, as described above, insights of a universal character refer to those features of systems that are generic, while specificity refers to the features of systems that depend upon precise details such as whether or not a particular molecule is bound at a particular site. This fundamental tension between atomistic and continuum perspectives is elaborated in *Phillips* et al. [48].

14.4.2 Atomic-Level Analysis of Biological Systems

As already described in the introduction, one of the intriguing roles of nanotechnology in the biological setting is that it has brought the Instron technology of traditional solid mechanics to the nanoscale and has permitted the investigation of the force-extension characteristics of nanoscale systems (macromolecules and their assemblies, in particular). As noted earlier, mechanical force spectroscopy [20] is emerging as a profound tool for exploring the connection between structure, force, and chemistry, in much the same way that conventional stress-strain tests provided insights into the connection between structure and properties of conventional materials. Figure 14.3 gives one such example in the case of the muscle protein titin. Similar insights have been obtained through systematic examination of the force-extension properties of DNA [49].

The objective of this section is to call the reader's attention to the types of modeling that can be done from an all-atom perspective. What exactly is meant by a model in this setting? We begin by noting that for the purposes of the general discussion given here, the same basic ideas are present whether one is modeling tribological processes such as the sliding of adjacent surfaces, or attempting to examine the operation of a protein machine such as ATPsynthase. The set of degrees of freedom considered by the atomic-scale modeler is the full set of atomic positions $(\mathbf{R}_1, \mathbf{R}_2, \cdots \mathbf{R}_N)$, which we also refer to as $\{\mathbf{R}_i\}$. For most purposes, one proceeds through reference to a *classical* potential energy function, $E_{\text{tot}}(\{\mathbf{R}_i\})$, which is a rule that assigns an energy for every configuration $\{\mathbf{R}_i\}$. While it would be most appealing to be able to perform a full quantum mechanical analysis, such calculations are computationally prohibitive. Given the potential energy, the forces on each and every atom can be computed where, for example, the force in the αth Cartesian direction on the ith atom is given by

$$F_{i,\alpha} = -\frac{\partial E_{\text{tot}}}{\partial R_{i,\alpha}}. \tag{14.8}$$

For those interested in finding the energy minimizers, such forces can be used in conjunction with methods such as the conjugate gradient method or the Newton–

Raphson method. Alternatively, many questions are of a dynamic character, and in these cases, Newton's equations of motion are integrated, thus permitting an investigation of the temporal evolution of the system of interest. We note that we have neglected to discuss subtleties of how one maintains the system at constant temperature, and we leave such subtleties to the curious reader, who can learn more about them in *Frenkel* and *Smit* [50].

To give a flavor of where such calculations can lead we note that the same force-extension characteristics already shown in Fig. 14.3 have been computed in a molecular dynamics simulation [51]. One of the insights to emerge from these calculations was the particular dynamical pathway, namely, the breaking of a particular collection of hydrogen bonds during the rupture process of each of the immunoglobulin domains. We further note that in the case of titin there has been an especially pleasing synergy between the atomic-scale calculations and the corresponding force-extension measurements. In particular, in response to the suggestion that it was a particular set of hydrogen bonds that were impugned in the rupture process, mutated versions of the titin protein were created in which the number of such hydrogen bonds was changed with the result that the rupture force was changed according to expectation [52]. This can be seen as a primitive example of the ultimate goal of tailoring new materials (both biological and otherwise) through appropriate computer modeling.

14.4.3 Continuum Analysis of Biological Systems

We have already noted that there are many appealing features to strictly continuum analyses of material systems. In particular, models based upon continuum mechanics result in a mathematical formulation that permits us to uncork the traditional tools associated with partial differential equations and functional analysis. In keeping with our argument that it is the role of models to serve our intuition, the ability to write down continuum models raises the possibility of obtaining analytic solutions to problems of interest, an eventuality that is nearly impossible once the all-atom framework has been adopted.

Macromolecules as Elastic Rods

Whether we contemplate the information carrying nucleic acids, the workhorse proteins, or energy storing sugars, ultimately, the molecular business of the living world is dominated by long chain molecules. For the model builder, such molecules suggest two complementary perspectives, each of which contains a part of the whole truth. On the one hand, polymer physicists have gained huge insights [53] by thinking of long chain molecules as random walks. Stated simply, the key virtue of the random-walk description of long chain molecules is that it reflects the overwhelming importance of entropy in governing the geometric conformations that can be adopted by macromolecules. The other side of the same coin considers long chain molecules as elastic rods with a stiffness that governs their propensity for bending. We will take up this perspective in great detail in the final section of this chapter when we consider the energetics of DNA packaging in viruses.

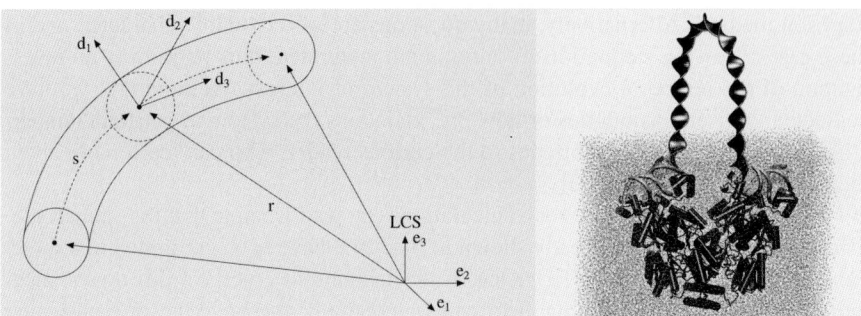

Fig. 14.6. Illustration of the way that the properties of a particular DNA-protein binding interaction were modeled by virtue of an elastic treatment of the looped region of the DNA [48]

For our present purposes, we examine one case study in treating macromolecules as elastic rods. The example of interest here is chosen in part because it reflects another fascinating aspect of biological nanotechnology, namely, regulation and control. It is well-known that genes are switched on and off as they are needed. These topics are described beautifully in the work of *Ptashne* [54]. The basic idea can be elucidated through the example of a particular set of genes in *E. coli*: the *lac* operon. There are a set of enzymes exploited by this bacterial cell when it needs to digest the sugar lactose. The gene that codes for these enzymes is only turned on when lactose is present and certain other sugars are absent. The nanotechnological solution to the problem of regulating this gene has been solved by nature through the presence of a molecule known as the *lac* repressor, which binds onto the DNA in the region of this *lac* gene and prevents the gene from being expressed. More specifically, the *lac* repressor binds onto two different regions of the DNA molecule simultaneously, forming a loop in the intervening region, and thus rendering that region inoperative for transcription.

We have belabored the difficulties that attend the all-atom simulation of most problems of nanotechnological interest – system sizes are too small and simulated times are too short. On the other hand, clearly it is of great interest to probe the atomic-level dynamics of the way that molecules such as *lac* repressor interact with DNA and thereby serve as gene regulators. A compromise position adopted by *Balaeff* et al. [55] was to use elasticity theory to model the structure and energetics of DNA, which then served effectively as a boundary condition for their all-atom calculation of the properties of the *lac* repressor/DNA complex. The simulation cell is shown in Fig. 14.6, which also shows the looped region of DNA that serves as a boundary condition for the atomic-level model of the complex.

Membranes as Elastic Media

One way to classify biological structures is along the lines of their dimensionality. In the previous discussion, we examined the sense in which many of the key macromolecules of the living world can be thought of as one-dimensional rods. The next level in this dimensional hierarchy is to consider the various membranes that compartmentalize the cell and which can be examined from the perspective of two-dimensional elasticity. Just as there are a huge number of modeling questions to be posed concerning the structure and function of long chain molecules such as nucleic acids and proteins, there is a similar list of questions that attend the presence of a host of different membranes throughout the cell. To name a few, we first remind the reader that such membranes are full of various proteins that serve in a variety of different capacities, some of which depend upon the mechanical state of the membrane itself. In a different vein, there has been great interest in examining the factors that give rise to the equilibrium shapes of cells [56, 57]. In both instances, the logic associated with building models of these phenomena centers on the construction of an elastic Hamiltonian that captures the energetics of deformations expressed in terms of surface area, curvature, and variations in thickness. Also, just as there has been great progress in measuring the properties of single proteins and nucleic acids, there has been considerable progress in examining the force response of membranes as well [58].

The prospect of bringing the tools of traditional continuum theory to bear on problems of biological significance is indeed a daunting one. As noted earlier, to our way of thinking one of the biggest challenges posed by biological problems is the dependence of these systems on detailed molecular structures and dynamics – what we have earlier characterized as "specificity", one of the hallmarks of biological action.

14.5 Nature's Nanotechnology Revealed: Viruses as a Case Study

Over the course of this chapter, we have presented different facets of the relationship between biology and nanotechnology and the modeling challenges they encompass. For example, we have noted that two key aspects of nature's nanotechnologies are the exploitation of self-assembly processes for the construction of molecular machines and the role of active processes mediated by molecular motors. We have also argued for the synergistic role of nanotechnology in producing new methods for the experimental analysis of biological systems with examples ranging from new molecular dyes to the use of optical tweezers. Finally, we have described some of the modeling challenges posed by contemplating the biology-nanotechnology interface and the sorts of coarse-grained models that have arisen to meet these challenges. In this final section, we present a discussion of viruses as a case study at the confluence of these different themes.

Though the importance of viruses from a health perspective are well-known even to the casual observer, they are similarly important both from the technological perspective (as will be explained below) and as a compelling and profound example of nature's nanotechnology par excellence. To appreciate the sense in which viruses serve as a compelling example of biological nanotechnology, we begin by reviewing the nature of the viral life cycle with special reference to the case of bacterial viruses (i.e., viruses that infect bacteria), known as bacteriophages. For concreteness, we consider the life cycle of a bacteriophage such as the famed lambda phage that in-

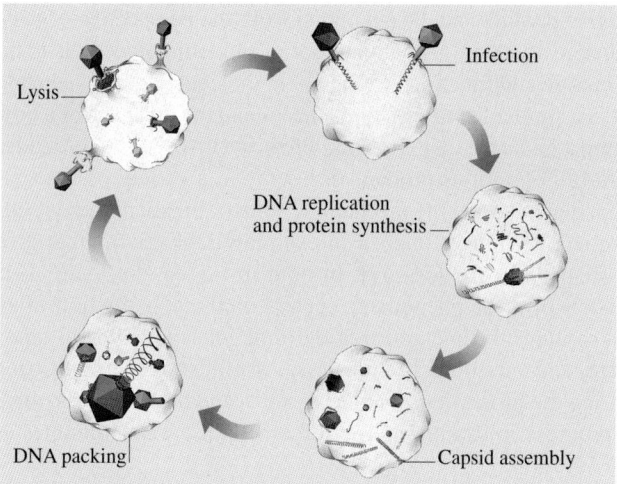

Fig. 14.7. Schematic representation of the viral life cycle illustrating the various nanotechnological actions in this cycle. During the infection stage, the viral genome is delivered to the victimized bacterial cell. During the process labeled DNA replication and protein synthesis, the machinery of the host cell is enlisted to produce renegade DNA and proteins to be used in the assembly of new viral particles. The capsid assembly process involves the attachment of the various protein building blocks that make up the capsid. DNA packing refers to the active packaging of DNA into the newly formed capsids by a molecular motor. Finally, once the assembly process is complete, the infectious phage particles are released from the victimized cell to repeat their infectious act elsewhere

fects *E. coli*. The life cycle of such viruses is shown schematically in Fig. 14.7. Upon an encounter with the *E. coli* host, the virus attaches to a receptor (protein) embedded in the bacterial membrane and ejects its DNA into the host cell, leaving an empty capsid as refuse from the process. As an aside, it is worth noting that experiments such as the famed Hershey–Chase experiment used tagged DNA and tagged proteins on viruses to settle the question of whether proteins or nucleic acids are the carriers of genetic information. The outcome of these experiments was the conclusion that DNA is the genetic material. For the purposes of the present discussion, the other interesting outcome of the Hershey–Chase experiment is that it provides insights into the mechanistic process associated with delivery of the viral genome.

Once the viral genome has been delivered to the host cell (we now oversimplify), the replication machinery of *E. coli* is hijacked to do the virus' bidding. In particular, the genes coded for in the viral DNA are expressed, and the proteins needed to make copies of the phage particle (i.e., the ingredients for an eventual self-assembly process) are created. Interestingly, part of the gene products associated with this process are the components of the molecular motor, which is responsible for packaging the replicated viral DNA into the new protein capsids that will eventually become the next generation of viruses. Indeed, once assembled, this motor takes the replicated viral DNA and packs it into the viral capsid. Recall our insistence that another of the important themes presented by biological examples of nanotechnology is the huge role played by active processes that reflect mechanochemical coupling. Once the packaging process is completed, and the remaining parts of the self-assembly process have been effected (such as the attachment of the viral sheath and legs whereby the phage attaches to host cells), enzymes are released that breakdown the cell wall of the infected cell with the ultimate result that what started as a single bacteriophage has in less than an hour become on the order of 100 new phage particles ready to infect new cells.

As noted above, a second sense in which viruses are deserving of case study status in the current chapter is the role viruses such as lambda phage play in biotechnology itself. For example, from the standpoint of both cloning and the construction of genomic libraries, the use of viruses is commonplace. From a more speculative perspective, viruses are also drawing increasing attention from the standpoint of gene therapy (as a way of delivering DNA to specific locations) and, more generally, as small-scale containers. To substantiate these assertions, we briefly consider the construction of genomic libraries and the role played by viruses in these manipulations. We pose the following question: Given that the length of the human genome is on the order of 10^9 base pairs, how can one organize and store all of this genetic information for the purposes of experiments such as the sequencing of the human genome? One answer to this challenge is the use of bacteriophage to deliver phage DNA to *E. coli* cells, but with the subtlety that the delivered DNA fragments have ligated within them fragments of the human genome whose lengths are on the order of 10 kb (kb = kilo base-pairs). The particulars of this procedure involve first cutting the DNA into fragments of roughly 10 kb in length using a class of enzymes known as restriction enzymes. The result of this operation is that the original genome is now separated into a random collection of fragments. These fragments are then mixed with the original lambda phage DNA that has also been cut at a single site such that the genomic fragments and the lambda fragments are complementary. Using a second enzyme known as ligase, the genomic DNA is joined to the original lambda phage fragments so that the resulting DNA resembles the original lambda DNA, but now with a 10 kb fragment inserted in the middle. These cloned lambda DNA molecules are then packaged into the lambda phage using a packaging reaction (see the website www.stratagene.com for an example of these products). The resulting lambda phage, now fully packed with cloned DNA, are used to infect *E. coli* cells, and the cloned DNA, once in the bacterial cell, circularizes into a DNA fragment known as a plasmid. The plasmid is then passed from one generation of *E. coli* to the next. This is the so-called prophage pathway in which, unlike the lytic pathway shown in Fig. 14.7, the bacteriophage is latent and does not destroy the cell [1]. Note that different *E. coli* cells are infected with viruses containing different cloned fragments. As a result, the collection of all such *E. coli* cells constitute a library of the original genomic DNA.

We round out our introductory discussion on viruses with a discussion of the compelling recent experiments that have been performed to investigate the problem of viral packing and, similarly, how ideas like those described in this chapter can be used to model these processes. *Smith* et al. [29] used optical tweezers to measure the force applied by the packaging motor of the ϕ-29 bacteriophage during the DNA packaging process. In particular, they measured the force and rate of packaging as a function of the amount of DNA packed into the viral head with the result that as more DNA is packed, the resistive force due to the packed DNA increases, and the packing rate is reduced.

As noted above, the viral problem is interesting not only because it exemplifies many of the features of biological nanotechnology introduced throughout this chapter, but also illustrates the way in which model building has arisen in response to experimental insights. The various competing energies that are implicated in the DNA packing process have been described by *Riemer* and *Bloomfield* [59]. The energetics of viral packing is characterized by a number of different factors, including: i) the entropic-spring effect that causes the DNA in solution to adopt a more spread out configuration than that in the viral capsid, ii) the energetics of elastic bending, which results from inducing curvature in the DNA on a scale that is considerably smaller than the persistence length of $\xi_p \approx 50$ nm, and iii) those factors related to the presence of charge both on the DNA itself and in the surrounding solution. As shown by *Riemer* and *Bloomfield* [59], the entropic contribution is smaller by a factor of 10 or more relative to the bending energies and those mediated by the charges on the DNA and the surrounding solution, and hence we make no further reference to it. As a result, just like in earlier work [60, 61], we examine the interplay of elastic and interaction forces, though we neglect surface terms originating from DNA-capsid and DNA-solvent interactions.

We note that the viral packing process involves DNA segments with lengths on the order of $10\,\mu$m and takes place on the time scale of minutes. As a result, from a modeling perspective, it is clear that such problems are clearly out of reach of conventional molecular dynamics. As a result, we exploit a continuum description of the DNA packing process with the proviso that such models will ultimately need to be refined to account for the sequence dependence of the elasticity of DNA.

A mathematical description of the energetics of viral packing must account for two competing factors, namely, the energy cost to bend the DNA and place it in the capsid and the repulsive interaction between adjacent DNA segments that are too close together. The structural picture of the packaged DNA is inspired by experiments that indicate that the DNA is packed in concentric rings from large radii inwards [62–64]. The bending energy cost of accumulating hoops within the capsid is given by

$$E_{\text{el}} = \pi \xi_p k_\text{B} T \sum_i \frac{N(R_i)}{R_i}, \qquad (14.9)$$

where $N(R_i)$ is the number of hoops that are packed at the radius R_i [59]. The basic idea of this expression is that we are adding up the elastic energy on a hoop by hoop basis, with each hoop penalized by the usual energy cost to bend an elastic beam into a circular arc. The presence of $N(R_i)$ reflects the fact that because of the shape of the capsid, as the radius gets smaller the DNA can pack higher up into the capsid, thus increasing the number of allowed hoops. To make analytic progress with the expression for the stored elastic energy given above, we convert it into an integral of the form

$$E_{\text{el}} = \frac{\pi \xi_p k_B T}{\sqrt{3} d_s/2} \int_R^{R_{\text{out}}} \frac{N(R')}{R'} dR' . \tag{14.10}$$

The summation $\sum_i$ has been replaced by an integral $\int_R^{R_{\text{out}}} R'/(\sqrt{3} d_s/2)$, where the integration bounds are the inner and outer radii of the inner spool and $\sqrt{3} d_s/2$ is the horizontal spacing between adjacent strands of the DNA. The geometrical factor $\sqrt{3}/2$ owes its presence to the hexagonal packing of the DNA strands. R is the radius of the innermost stack of hoops, R_{out} is the radius of the capsid, $N(R')$ is the number of hoops at radius R', d_s is the spacing between adjacent hoops, and ξ_p is the persistence length of DNA. We note that this expression is a manifestation of the first term in the energy functional in (14.3), used earlier in the context of the biofunctionalized cantilever and now specialized to the geometry of a partially filled capsid. Just as the energy functional in the cantilever context reflected a competition between different contributions to the total energy (in that case, elastic bending energy and surface terms), the DNA packing problem reflects a similar competition, this time between the elastic bending and the interactions between nearby DNA segments by virtue of the charge along the DNA. In particular, the interaction energy between the DNA hoops packed in the viral capsid scales with the length and is given by

$$E_{\text{int}} = F_0 L (c^2 + c d_s) \exp(-d_s/c) , \tag{14.11}$$

where c and F_0 are constants for a given solvent and L is the length of the packed DNA. This form of the interaction energy was shown to be very robust for a variety of solvents containing monovalent and divalent cations in a series of experiments by *Rau* et al. [65, 66] and *Parsegian* et al. [67]. The forces arising from this kind of an energy are purely repulsive in character. Interestingly, in this setting, rather than using temperature as a control parameter, it is much more common to change the circumstances in such experiments by tuning concentrations of various chemical constituents. In particular, when trivalent or tetravalent cations are present in solution there is an effective attractive interaction between DNA segments. In that case, the forces are repulsive only when d_s is smaller than a certain critical separation d_0 (and attractive otherwise) and the form of the interaction energy changes to

$$E_{\text{int}} = F_0 L (c^2 + c d_s) \exp\left(\frac{d_0 - d_s}{c}\right) . \tag{14.12}$$

Once both the elastic and interaction contributions to the energy have been reckoned, we are in a position to compute the total energy. If we recognize that the length L of the packed DNA is given by

$$L = \frac{2}{\sqrt{3} d_s} \int_R^{R_{\text{out}}} 2\pi R' N(R') dR' , \tag{14.13}$$

then the total energy for the packed DNA, in terms of L and d_s, is given by

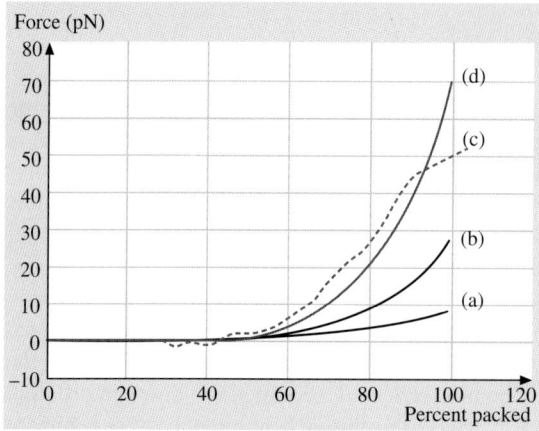

Fig. 14.8. Force as a function of the percent packed for a cylindrical capsid under purely repulsive solvent conditions. The dimensions of the capsid and the length of genome packed were chosen to represent the $\phi 29$ phage: $R_{\text{out}} = 210\,\text{Å}$, $z = 540\,\text{Å}$, and $L = 6.6\,\mu\text{m}$. Curve (c) shows the experimental results of *Smith* et al. [29], while theoretical curves (a), (b), and (d) are given by (14.15), with $F_0 = 10 \times 55\,000$, $40 \times 55\,000$, and $100 \times 55\,000\,\text{pN/nm}^2$, respectively

$$E(L, d_s) = \frac{\pi \xi_p k_B T}{\sqrt{3} d_s/2} \int_R^{R_{\text{out}}} \frac{N(R')}{R'} dR' \qquad (14.14)$$

$$+ F_0(c^2 + cd_s)\exp(-d_s/c)\frac{2}{\sqrt{3}d_s}\int_R^{R_{\text{out}}} 2\pi R' N(R') dR' \ .$$

The spacing at a given length L is determined by requiring that the packed DNA be in a minimum energy configuration, which is equivalent to asking that $\partial E/\partial d_s = 0$. Hence, for a given geometry and length packed, one obtains d_s and thereby calculates the energy. Thus the energy is known as a function of the length packed, and to compute the force as a function of length of DNA packed one need only differentiate this energy with respect to the length packed. The result of this procedure is an expression for the force of the form

$$F(L) = F_0(c^2 + cd_s)\exp\left(-\frac{d_s}{c}\right) + \frac{\xi_p k_B T}{2R^2} \ . \qquad (14.15)$$

This is a generic expression valid for a capsid of any shape. The effect of the geometry is captured in the variation of d_s as a function of the length packed, as well as through the inner packing radius R. The resulting force-packing curves are shown in Fig. 14.8. In particular, this figure shows the packing force as a function of a fraction of the genome packed for a number of different solutions, with the different curves revealing the large role played by positive counterions in dictating the overall energetics of these processes.

On the basis of calculations like those given above, *Kindt* et al. [61] estimate that the pressure inside the capsid of the phage is on the order of 35 atmospheres. A crude estimate by *Smith* et al. [29] gives a figure close to 60 atmospheres. These numbers are intriguing in their own right, but more importantly they demonstrate the promise that proteins and other biological materials hold as candidates for engineering materials. To further explore the structural integrity of viruses in their role as pressurized protein shells, we now turn to an examination of the rupture stress of viral capsids.

Throughout this chapter, one of our main arguments has been the idea that whether it is our ambition to model conventional materials or their biological counterparts, modeling at the nanoscale implies challenges that can not be met either by purely continuum ideas or by traditional all-atom thinking. Indeed, one set of powerful methods is built around the attempt to borrow those features of continuum and atomistic thinking that are most robust, while rejecting those that are inapplicable at the atomic scale. As an example of this type of thinking, we examine another question drawn from the viruses-as-nanotechnology setting. In particular, since viruses have been considered a means of transporting material other than the genetic material of the virus itself, it is of interest to consider the maximum internal pressure viral capsids can sustain without rupture, as this will determine how much material can be safely packaged. To that end, we use continuum mechanics to estimate the stresses within the capsid walls. These stresses are then mapped onto atomic-level forces by appealing to the details of the protein structure of the monomers making up the capsid and a knowledge of the forces that link them. By relating the continuum and atomistic calculations, we then determine the maximum sustainable internal pressure.

We imagine the capsid to be a hollow sphere loaded by a pressure p_i from inside and a pressure p_o from outside. The inner and outer radii are R_i and R_o, respectively. As a representative example, bacteriophage GA is characterized geometrically by $R_i = 123$ Å and $R_o = 145$ Å. Evaluation of a number of different capsid types suggests that treating capsids as though they have a mean thickness of roughly 15 Å suffices for the level of modeling being considered here. For the purposes of computing the internal stresses within the capsid, we begin with a statement of equilibrium from continuum mechanics that requires that at every point in the capsid

$$\nabla \cdot \sigma = \mathbf{0}, \tag{14.16}$$

where σ is the stress tensor comprising three normal stresses and three shear stresses. For a problem with spherical symmetry, like that considered here, the stresses reduce to a radial stress σ_R and a circumferential stress σ_T. Solution of the equilibrium equations results in stresses of the form [32]

$$\sigma_R = \frac{C}{r^3} + D, \quad \sigma_T = -\frac{C}{2r^3} + D. \tag{14.17}$$

By using the boundary conditions $\sigma_R|_{r=R_i} = -p_i$ and $\sigma_R|_{r=R_o} = -p_o$, the constants C and D can be determined with the result

$$\sigma_R = \frac{p_o R_o^3 \left(r^3 - R_i^3\right)}{r^3 \left(R_i^3 - R_o^3\right)} + \frac{p_i R_i^3 \left(R_o^3 - r^3\right)}{r^3 \left(R_i^3 - R_o^3\right)}, \tag{14.18}$$

$$\sigma_T = \frac{p_o R_o^3 \left(2r^3 + R_i^3\right)}{2r^3 \left(R_i^3 - R_o^3\right)} - \frac{p_i R_i^3 \left(2r^3 + R_o^3\right)}{2r^3 \left(R_i^3 - R_o^3\right)}. \tag{14.19}$$

The stress σ_T is our primary concern, since it acts so as to tear the sphere apart. By looking at the expressions above, we can see that this stress is maximum at $r = R_i$ and the maximum value is given by

$$\sigma_T^{max} = \frac{3 p_o R_o^3 - p_i \left(2R_i^3 + R_o^3\right)}{2 \left(R_i^3 - R_o^3\right)}. \tag{14.20}$$

We note that elasticity theory in and of itself is unable to comment on σ_T^{max}, since this is effectively a material parameter that characterizes the contacts between the various protein monomers that make up the capsid. As a result, we first examine how the rupture strength depends upon capsid dimensions in abstract terms and then turn to a concrete estimate of σ_T^{max} itself from several complementary perspectives. If (14.20) is rewritten using $p_i = p_o + \Delta p$ and $R_o = R_i + \Delta R$, and, further, it is realized that for typical capsid dimensions $\Delta R/R_i \ll 1$, it can be shown by rearranging (14.20) and by considering the case in which σ_T^{max} is much larger than p_o, the maximum sustainable internal pressure is given by

$$p_i^{max} = \frac{2 \sigma_T^{max} \Delta R}{R_i}, \tag{14.21}$$

where p_i^{max} is really the quantity of interest, namely, the maximum sustainable internal pressure.

To make concrete progress to the point where we can actually estimate the rupture stress in atm, we need to consider capsids for which the structure is known and for which, at least approximately, the bonds between the monomers making up the capsid are understood. We note that in the language of fracture mechanics, what we seek is a cohesive surface model that provides a measure of the energy of interaction between two surfaces as a function of their separation [68]. There are a number of different ways to go about estimating the effective interaction between the monomers making up the capsid, one of which is by appealing to atomic-level calculations like those made by *Reddy* et al. [69]. The Viper web-site has systematized such information for a number of capsids, and one of the avenues we take to estimate σ_T^{max} is to appeal directly to their calculations. To that end, we assume that the energy of interaction per unit area as a function of separation x between adjacent monomers making up the capsid can be written in the form

$$E(x) = V_0 \left[\frac{1}{4} \left(\frac{x^*}{x}\right)^{12} - \frac{1}{2} \left(\frac{x^*}{x}\right)^6 \right]. \tag{14.22}$$

The motivation for this functional form is the idea that the energy of interaction between adjacent monomeric units making up the capsid is the result of van der Waals contacts, and hence our cohesive surface law has inherited the properties of the underlying atomic force fields. To proceed to an estimate of $\sigma_T^{\max}$ itself, we must determine the parameters in the cohesive model described above. To that end, we note that *Reddy* et al. [69] have computed the association energies of various inequivalent contacts throughout a number of different icosahedral capsids. Their calculations result in a roughly constant value of $\approx -45\,\text{cal/mole Å}^2$ as the association energy, which in the language of our cohesive potential results in $V_0 \approx 12.5\,\text{pN/Å}$. This may be seen by noting that the association energy is given by $E(x^*) = -V_0/4$.

Once we have fixed x^*, which amounts to choosing a particular value for the equilibrium separation between two monomers, the material parameter $\sigma_T^{\max}$ is obtained by evaluating $\partial E(x)/\partial x$ at a value of x corresponding to the point of inflection ($\partial^2 E(x)/\partial x^2 = 0$) in the cohesive surface function. For the cohesive surface function used above, this results in a maximum stress of the form

$$\sigma_T^{\max} = \frac{7^{7/6} \times 18}{13^{13/6}} \frac{V_0}{x^*}. \tag{14.23}$$

With $\sigma_T^{\max}$ in hand, the maximum sustainable pressure is obtained from (14.21), and the results are shown in Fig. 14.9. These estimates suggest that the pressures within capsids as a result of packed DNA while large, are still smaller than our estimated rupture stresses. To more completely examine this question we have undertaken finite element elasticity calculations to examine the stresses in capsids exhibiting irregularities in both shape and thickness. In addition, further atomistic analysis of $\sigma_T^{\max}$ is needed with special reference to its dependence on distance between protein units. It would also be of interest to examine mutant versions of the monomeric units making up the capsid to see the implications of such mutations for $\sigma_T^{\max}$.

14.6 Concluding Remarks

One of the most compelling areas to be touched by nanotechnology is biological science. Indeed, we have argued that there is a fascinating interplay between these two subjects, with biology as a key beneficiary of advances in nanotechnology as a result of a new generation of single molecule experiments that complement traditional assays involving statistical assemblages of molecules. This interplay runs in both directions with nanotechnology continually receiving inspiration from biology itself. The goal of this chapter has been to highlight some representative examples of the interplay between biology and nanotechnology and to illustrate the role of nanomechanics in this field and how mechanical models have arisen in response to the emergence of this new field. Primary attention has been given to the particular example of the processes that attend the life cycle of bacterial viruses. Viruses feature many of the key lessons of biological nanotechnology, including self-assembly, as evidenced in the spontaneous formation of the protein shell (capsid) within which the viral genome is packaged and a motor-mediated biological process, namely, the

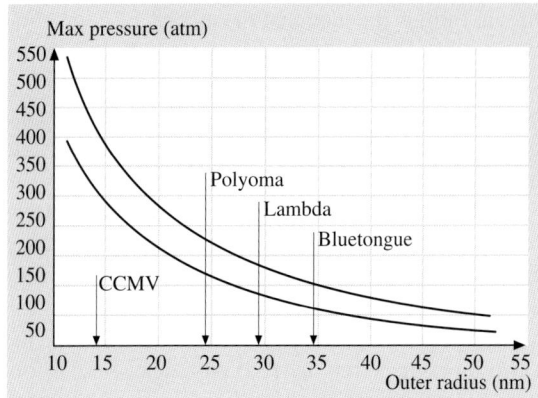

Fig. 14.9. Rupture pressure as a function of capsid radius for $x^* = 3\,\text{Å}$ (*upper curve*) and $x^* = 4\,\text{Å}$ (*lower curve*). The width of the capsid walls was set to $\Delta R = 1.5$ nm, while $V_0 = 180\,\text{cal/mol\AA}^2$

packaging of DNA in this capsid by a molecular motor that pushes the DNA into the capsid. We argue that these processes in viruses are a compelling real-world example of nature's nanotechnology and reveal the nanomechanical challenges that will continue to be confronted at the nanotechnology-biology interface.

Though this chapter represents something of a departure from the rest of the chapters in this volume, it advances the view that biological nanotechnology serves as an inspiring vision of what nanotechnology can offer and reflects the authors' views that biologically inspired nanotechnology will play an ever increasing role as Feynman's view that "there is plenty of room at the bottom" continues to play out. Moreover, developing simple models of nature's nanotechnology can provide important insights into viable strategies for making machines at the nanoscale.

We have argued that the advance of single molecule technology presents both scientific and technological possibilities. Both classes of questions imply significant modeling demands, just as earlier advances in other settings, such as in materials for microelectronics applications, did. We claim that these modeling challenges are perhaps more acute as a result of the vicious chemical and structural nonhomogeneity of biological systems. As a result, we see an ever increasing role for modeling methods that aim to keep atomic-level specificity where needed, while rejecting such resolution elsewhere.

Acknowledgement

We happily acknowledge useful discussions with Kai Zinn, Jon Widom, Bill Gelbart, Andy Spakowitz, Zhen-Gang Wang, Ken Dill, Carlos Bustamante, Tom Powers, Larry Friedman, Jack Johnson, Pamela Bjorkman, Paul Wiggins, Steve Williams, Wayne Falk, Adrian Parsegian, Alasdair Steven and Steve Quake. RP and PP acknowledge support of the NSF through grant number 9971922, the NSF supported

CIMMS center, and the support of the Keck Foundation. JK is supported by the NSF under grant number DMR-9984471 and is a Cottrell Scholar of Research Corporation.

References

1. B. Alberts, D. Bray, A. Johnson, J. Lewis, M. Raff, K. Roberts, and P. Walter. *Essential Cell Biology*. Garland, 1997.
2. B. Alberts. The cell as a collection of protein machines: Preparing the next generation of molecular biologists. *Cell*, 92:291, 1998.
3. D. Bray. *Cell Movements From Molecules to Motility*. Garland, 2001.
4. S. J. Flint, L. W. Enquist, R. M. Krug, V. R. Racaniello, and A. M. Skalka. *Principles of Virology*. ASM, 2000.
5. J. C. M. van Hest and D. A. Tirrell. Protein-based materials, toward a new level of structural control. *Chem. Commun.*, 19:1807–1904, 2001.
6. H. Berg. Motile behavior of bacteria. *Phys. Today*, 53:24, 2000.
7. J. E. Walker. Atp synthesis by rotary catalysis. *Angew. Chem. Int. Edn.*, 37:2308, 1998.
8. P. D. Boyer. The atp synthase – a splendid molecular machine. *Annu. Rev. Biochem.*, 66:717, 1997.
9. M. Yoshida, E. Muneyuki, and T. Hisabori. Atp synthase – a marvelous rotary engine of the cell. *Nature Rev. Mol. Cell Bio.*, 2:669, 2001.
10. R. Yasuda, H. Noji, K. Kinosita, and M. Yoshida. F_1-atpase is a highly efficient molecular motor that rotates with discrete 120° steps. *Cell*, 93:1117, 1998.
11. J. Howard. *Mechanics of Motor Proteins and the Cytoskeleton*. Sinauer, 2001.
12. B. Hille. *Ion Channels of Excitable Membranes*. Sinauer, 2001.
13. S. Sukharev, S. R. Durell, and H. R. Guy. Structural models of the mscl gating mechanism. *Biophys. J.*, 81:917, 2001.
14. S. Sukharev, M. Betanzos, C.-S. Chiang, and H. R. Guy. The gating mechanism of the large mechanosensitive channel mscl. *Nature*, 409:720, 2001.
15. K. Kinosita, R. Yasuda, and H. Noji. F_1-atpase: A highly efficient rotary atp machine. *Essays Biochem.*, 35:3, 2000.
16. E. Racker and W. Stoeckenius. Reconstitution of purple membrane vesicles catalyzing light-driven proton uptake and adenosine triphosphate formation. *J. Biol. Chem.*, 249:662, 1974.
17. G. Groth and J. E. Walker. Atp synthase from bovine heart mitochondria: reconstitution into unilamellar phospholipid vesicles of the pure enzyme in a functional state. *Biochem. J.*, 318:351, 1996.
18. G. Wu, R. H. Datar, K. M. Hansen, T. Thundat, R. J. Cote, and A. Majumdar. Bioassay of prostrate-specific antigen (psa) using microcantilevers. *Nature Biotech.*, 19:856, 2001.
19. R. P. Feynmann. There's plenty of room at the bottom. *Eng. Sci.*, 23:22–36, 1960.
20. E. Evans. Probing the relation between force-lifetime and chemistry in single molecular bonds. *Annu. Rev. Biophys. Biomol. Struct.*, 30:105, 2001.
21. T. E. Fisher, P. E. Marszalek, and J. M. Fernandez. Stretching single molecules into novel conformations using the atomic force microscope. *Nature Struct. Bio.*, 7:719, 2000.
22. M. Rief, M. Gautel, F. Oesterhelt, J. M. Fernandez, and H. Gaub. Reversible unfolding of individual titin immunoglobulin domains by afm. *Science*, 276:1109, 1997.
23. K. Svoboda and S. M. Block. Biological applications of optical forces. *Annu. Rev. Biophys. Biomol. Struct.*, 23:247, 1994.

24. C. Bustamante, J. C. Macosko, and G. J. L. Wuite. Grabbing the cat by the tail: Manipulating molecules one by one. *Nature Rev. Mol. Cell Bio.*, 1:130, 2000.
25. K. Svoboda and S. M. Block. Force and velocity measured for single kinesin molecules. *Cell*, 77:773, 1994.
26. M. J. Schnitzer and S. M. Block. Kinesin hydrolyses one atp per 8-nm step. *Nature*, 388:386, 1997.
27. B. Maier, T. R. Strick, V. Croquette, and D. Bensimon. Study of dna motors by single molecule micromanipulation. *Single Mol.*, 1:145, 2000.
28. A. A. Simpson, Y. Tao, P. G. Leiman, M. O. Badasso, Y. He, P. J. Jardine, N. H. Olson, M. C. Morais, S. Grimes, D. L. Anderson, T. S. Baker, and M. G. Rossmann. Structure of the bacteriophage ϕ-29 dna packaging motor. *Nature*, 408:745, 2000.
29. D. E. Smith, S. J. Tans, S. B. Smith, S. Grimes, D. L. Anderson, and C. Bustamante. The bacteriophage ϕ-29 portal motor can package dna against a large internal force. *Nature*, 413:748, 2001.
30. D. Bray. Reasoning for results. *Nature*, 412:863, 2001.
31. G. Wu, H. Ji, K. Hansen, T. Thundat, R. Datar, R. Cote, M. F. Hagan, A. K. Charkraborty, and A. Majumdar. Origin of nanomechanical cantilever motion generated from biomolecular interactions. *Proc. Nat. Acad. Sci.*, 98:1560, 2001.
32. L. D. Landau and E. M. Lifshitz. *Theory of Elasticity*. Pergamon, 1986.
33. D. H. Bamford, R. J. C. Gilbert, J. M. Grimes, and D. I. Stuart. Macromolecular assemblies: greater than their parts. *Curr. Op. Struc. Biology*, 11:107, 2001.
34. J. Frank. Single-particle imaging of macromolecules by cryo-electron microscopy. *Annu. Rev. Biophys. Biomol. Struct.*, 31:303, 2002.
35. S. A. Darst. Bacterial rna polymerase. *Curr. Op. Struc. Biology*, 11:155, 2001.
36. N. Ban, P. Nissen, J. Hansen, P. B. Moore, and T. A. Steitz. The complete atomic structure of the large ribosomal subunit at 2.4 Å resolution. *Science*, 289:905, 2000.
37. T. S. Baker, N. H. Olson, and S. D. Fuller. Adding the third dimension to virus life cycles: Three-dimensional reconstruction of icosahedral viruses from cryo-electron micrographs. *Microbiol. Mol. Bio. Rev.*, 63:862, 1999.
38. H.-S. Chan and K. A. Dill. The protein folding problem. *Phys. Today*, 46:24, February 1993.
39. M. D. Wang, M. J. Schnitzer, H. Yin, R. Landick, J. Gelles, and S. M. Block. Force and velocity measured for single molecules of rna polymerase. *Science*, 282:902, 1998.
40. K. Adachi, R. Yasuda, H. Noji, H. Itoh, Y. Harada, M. Yoshida, and K. Kinosita. Stepping rotation of f_1-atpase visualized through angle-resolved single-fluorophore imaging. *Proc. Nat. Acad. Sci.*, 97:7243, 2000.
41. M. L. Roukes. Nanoelectromechanical systems, (2000.
42. J. Wrigglesworth. *Energy and Life*. Taylor and Francis, 1997.
43. V. Smil. *Energies*. MIT, 1999.
44. J. T. Finer, R. M. Simmons, and J. A. Spudich. Single myosin molecule mechanics: Piconewton forces and nanometre steps. *Nature*, 368:113, 1994.
45. S. B. Smith, Y. Cui, and C. Bustamante. Overstretching b-dna: The elastic response of individual double-stranded and single-sranded dna molecules. *Science*, 271:795, 1996.
46. S. Munevar, Y. Wang, and M. Dembo. Traction force microscopy of migrating normal and h-ras transformed 3t3 fibroblasts. *Biophys. J.*, 80:1744, 2001.
47. L. Mahadevan and P. Matsudaira. Motility powered by supramolecular springs and ratchets. *Science*, 288:95, 2000.
48. R. Phillips, M. Dittrich, and K. Schulten. Quasicontinuum representations of atomic-scale mechanics: From proteins to dislocations. *Ann. Rev. Mat. Sci.*, 32:219, 2002.

49. T. Strick, J.-F. Allemand, V. Croquette, and D. Bensimon. Twisting and stretching single dna molecules. *Prog. Biophys. Mol. Bio.*, 74:115, 2000.
50. D. Frenkel and B. Smit. *Understanding Molecular Simulation*. Academic, 1996.
51. H. Lu, B. Isralewitz, A. Krammer, V. Vogel, and K. Schulten. Unfolding of titin immunoglobulin domains by steered molecular dynamics simulation. *Biophys. J.*, 75:662, 1998.
52. M. Carrion-Vazquez, A. F. Oberhauser, T. E. Fisher, P. E. Marszalek, H. Li, and J. M. Fernandez. Mechanical design of proteins studies by single-molecule force spectroscopy and protein engineering. *Prog. Biophys. Mol. Bio.*, 74:63, 2000.
53. A. Y. Grosberg and A. R. Khokhlov. *Giant Molecules*. Academic, 1997.
54. M. Ptashne and A. Gann. *Genes and Signals*. Cold Spring Harbor Laboratory Press, 2002.
55. A. Balaeff, L. Mahadevan, and K. Schulten. Elastic rod model of a dna loop in the lac operon. *Phys. Rev. Lett.*, 83:4900, 1999.
56. U. Seifert. Configurations of fluid membranes and vesicles. *Adv. Phys.*, 46:13, 1997.
57. D. Boal. *Mechanics of the Cell*. Cambridge Univ. Press, 2002.
58. D. Leckband and J. Israelachvili. Intermolecular forces in biology. *Q. Rev. Biophys.*, 34:105, 2001.
59. S. C. Riemer and V. A. Bloomfield. Packaging of dna in bacteriophage heads: Some considerations on energetics. *Biopolymers*, 17:785, 1978.
60. T. Odijk. Hexagonally packed dna within bacteriophage t7 stabilized by curvature stress. *Biophys. J.*, 75:1223, 1998.
61. J. Kindt, S. Tzlil, A. Ben-Shaul, and W. Gelbart. Dna packaging and ejection forces in bacteriophage. *Proc. Nat. Acad. Sci.*, 98:13671, 2001.
62. K. E. Richards, R. C. Williams, and R. Calendar. Mode of dna packing within bacteriophage heads. *J. Mol. Bio.*, 78:255, 1973.
63. M. E. Cerritelli, N. Cheng, A. H. Rosenberg, C. E. McPherson, F. P. Booy, and A. C. Steven. Encapsidated conformation of bacteriophage t7 dna. *Cell*, 91:271, 1997.
64. N. H. Olson, M. Gingery, F. A. Eiserling, and T. S. Baker. The structure of isomeric capsids of bacteriophage t4. *Virology*, 279:385, 2001.
65. D. C. Rau, B. Lee, and V. A. Parsegian. Measurement of the repulsive force between polyelectrolyte molecules in ionic solution: Hydration forces between parallel dna double helices. *Proc. Natl. Acad. Sci.*, 81:2621, 1984.
66. D. C. Rau and V. A. Parsegian. Direct measurement of the intermolecular forces between counterion-condensed dna double helices. *Biophys. J.*, 61:246, 1992.
67. V. A. Parsegian, R. P. Rand, N. L. Fuller, and D. C. Rau. Osmotic stress for the direct measurement of intermolecular forces. *Meth. Enzymol.*, 127:400, 1986.
68. R. Phillips. *Crystals, Defects and Microstructures*. Cambridge Univ. Press, 2001.
69. V. S. Reddy, H. A. Giesing, R. T. Morton, A. Kumar, C. B. Post, C. L. Brooks, and J. E. Johnson. Energetics of quasiequivalence: Computational analysis of protein-protein interactions in icosahedral viruses, 1998.

15

Mechanical Properties of Nanostructures

Bharat Bhushan

Summary. Knowledge of the mechanical properties of nanostructures is necessary for designing realistic MEMS/NEMS devices. Microelectromechanical systems (MEMS) refer to microscopic devices that have a characteristic length of less than 1 mm but more than 1 μm and combine electrical and mechanical components. Nanoelectromechanical systems (NEMS) refer to nanoscopic devices that have a characteristic length of less than 1 μm and combine electrical and mechanical components.

Elastic and inelastic properties are needed to predict deformation from an applied load in the elastic and inelastic regimes, respectively. The strength property is needed to predict the allowable operating limit. Some of the properties of interest are hardness, elastic modulus, bending strength, fracture toughness, and fatigue strength.

Structural integrity is of paramount importance in all devices. Load applied during the use of devices can result in component failure. Cracks can develop and propagate under tensile stresses, leading to failure. Atomic force microscopy and nanoindenters can be used satisfactorily to evaluate the mechanical properties of micro/nanoscale structures for use in MEMS/NEMS.

The most commonly used materials are single-crystal silicon and silicon-based materials, e.g., SiO_2 and polysilicon films deposited by low-pressure chemical vapor deposition. An early study showed silicon to be a mechanically resilient material in addition to its favorable electronic properties. Single-crystal SiC deposited on large-area silicon substrates is used for high-temperature micro/nanosensors and actuators. Amorphous alloys can be formed on both metal and silicon substrates by sputtering and plating techniques, providing more flexibility in surface-integration. Electroless deposited Ni-P amorphous thin films have been used to construct microdevices, especially using the so-called LIGA techniques. Micro/nanodevices need conductors to provide power, as well as electrical/magnetic signals to make them functional. Electroplated gold films have found wide applications in electronic devices because of their ability to make thin films and process simply. Use of SiC, Ni-P, and Au films, together with silicon and silicon-based materials, opens up new design opportunities for MEMS/NEMS devices.

This chapter presents a review of mechanical property measurements on the nanoscale of various materials of interest and stress and deformation analyses of nanostructures.

15.1 Introduction

Microelectromechanical systems (MEMS) microelectromechanical systemsrefer to microscopic devices that have a characteristic length of less than 1 mm but more than 1 μm and combine electrical and mechanical components. Nanoelectromechanical systems (NEMS) refer to nanoscopic devices that have a characteristic length of less than 1 μm and combine electrical and mechanical components. To put the dimensions in perspective, individual atoms are typically fraction of a nanometer in diameter, DNA molecules are about 2.5 nm wide, biological cells are in the range of thousands of nm in diameter, and human hair is about 75 μm in diameter. The mass of a micromachined silicon structure can be as low as 1 nN, and NEMS can be built with a mass as low as 10^{-20} N with cross sections of about 10 nm. In comparison, the mass of a drop of water is about 10 μN and the mass of an eyelash is about 100 nN. The acronym MEMS originated in the United States. The term commonly used in Europe and Japan is micro/nanodevices, which is used in a much broader sense. MEMS/NEMS terms are also now used in a broad sense. A micro/nanosystem, a term commonly used in Europe, is referred to as an intelligent miniaturized system comprising sensing, processing, and/or actuating functions.

A wide variety of MEMS, including Si-based devices, chemical and biological sensors and actuators, and miniature non-silicon structures (e.g., devices made from plastics or ceramics) have been fabricated with dimensions in the range of a couple to a few thousand microns (see e.g., [1–9]). A variety of NEMS have also been produced [10–15]. Two of the largest "killer" industrial applications of MEMS are accelerometers (about 85 million units in 2002) and digital micromirror devices (about $ 400 million in revenues in 2002). BIOMEMS and BIONEMS are increasingly used in commercial applications. The largest applications of BIOMEMS include silicon-based disposable blood pressure sensor chips for blood pressure monitoring (about 20 million units in 2002) and a variety of biosensors.

Structural integrity is of paramount importance in all devices. Load applied during the use of devices can result in component failure. Cracks can develop and propagate under tensile stresses, leading to failure [16, 17]. Friction/stiction and wear limit the lifetimes and compromise the performance and reliability of the devices involving relative motion [4, 18]. Most MEMS/NEMS applications demand extreme reliability. Stress and deformation analyses are carried out for an optimal design. MEMS/NEMS designers require mechanical properties on the nanoscale. Mechanical properties include elastic, inelastic (plastic, fracture, or viscoelastic), and strength. Elastic and inelastic properties are needed to predict deformation from an applied load in the elastic and inelastic regimes, respectively. The strength property is needed to predict the allowable operating limit. Some of the properties of interest are hardness, elastic modulus, bending strength (fracture stress), fracture toughness, and fatigue strength. Micro/nanostructures have some surface topography and local scratches dependent upon the manufacturing process. Surface roughness and local scratches may compromise the reliability of the devices and their effect needs to be studied.

Most mechanical properties are size-dependent [11, 16, 17]. Several researchers have measured mechanical properties of silicon and silicon-based milli- to microscale structures using tensile tests and bending tests [19–28], resonant structure tests for measurement of elastic properties [29], fracture toughness tests [20, 22, 30–34], and fatigue tests [32, 35, 36]. Most recently, a few researchers have measured mechanical properties of nanoscale structures using atomic force microscopy (AFM) [37, 38] and nanoindentation [39, 40]. For stress and deformation analyses of simple geometries and boundary condition, analytical models can be used. For analysis of complex geometries, numerical models are needed. The conventional finite element method (FEM) can be used down to a few tens of nanometers, although its applicability is questionable at the nanoscale. FEM has been used for simulation and prediction of residual stresses and strains induced in MEMS devices during fabrication [41], to perform fault analysis in order to study MEMS faulty behavior [42], to compute mechanical strain resulting from the doping of silicon [43], and to analyze micromechanical experimental data [22, 44, 45] and nanomechanical experimental data [38]. FEM analysis of nanostructures has been carried out to analyze the effect of types of surface roughness and scratches on stresses in nanostructures [46, 47].

The most commonly used materials for MEMS/NEMS are single-crystal silicon and silicon-based materials (e.g., SiO_2 and polysilicon films deposited by low pressure chemical vapor deposition (LPCVD) process) [11]. An early study showed silicon to be a mechanically resilient material in addition to its favorable electronic properties [48]. Single-crystal 3C-SiC (cubic or β-SiC) films deposited by atmospheric pressure chemical vapor deposition (APCVD) process on large-area silicon substrates are produced for high-temperature micro/nanosensor and actuator applications [49–51]. Amorphous alloys can be formed on both metal and silicon substrates by sputtering and plating techniques, providing more flexibility in surface-integration. Electroless deposited Ni-P amorphous thin films have been used to construct microdevices, especially using the so-called LIGA techniques [11, 40]. Micro/nanodevices need conductors to provide power, as well as electrical/magnetic signals to make them functional. Electroplated gold films have found wide applications in electronic devices, because of their ability to make thin films and process simplicity [40]. Use of SiC, Ni-P, and Au films together with silicon and silicon-based materials open up new design opportunities for MEMS/NEMS devices.

This chapter presents a review of mechanical property measurements on the nanoscale of various materials of interest, and stress and deformation analyses of nanostructures.

15.2 Experimental Techniques for Measurement of Mechanical Properties of Nanostructures

15.2.1 Indentation and Scratch Tests Using Micro/Nanoindenters

A nanoindenter is commonly used to measure hardness, elastic modulus, and fracture toughness and to perform micro/nanoscratch studies to get a measure of scratch/wear resistance of materials [11, 52].

Hardness and Elastic Modulus

The nanoindenter monitors and records the dynamic load and displacement of the three-sided pyramidal diamond (Berkovich) indenter during indentation with a force resolution of about 75 nN and displacement resolution of about 0.1 nm. Hardness and elastic modulus are calculated from the load-displacement data [11]. The peak indentation load depends on the mechanical properties of the specimen; a harder material requires higher load for a reasonable indentation depth.

Fracture Toughness

The indentation technique for fracture toughness measurement on the microscale is based on the measurement of the lengths of median-radial cracks produced by indentation. A Vickers indenter (a four-sided diamond pyramid) is used in a microhardness test. A load on the order of 0.5 N is typically used in making the Vickers indentations. The indentation impressions are examined using an optical microscope with Nomarski interference contrast to measure the length of median-radial cracks, c. The fracture toughness ($K_{\mathrm{ICintegrated\ circuit}}$) is calculated by the following relation [53]:

$$K_{\mathrm{IC}} = \alpha \left(\frac{E}{H}\right)^{1/2} \left(\frac{P}{c^{3/2}}\right), \tag{15.1}$$

where α is an empirical constant depending on the geometry of the indenter, H and E are hardness and elastic moduli, and P is the peak indentation load. For Vickers indenters, α has been empirically found based on experimental data and is equal to 0.016 [11]. Both E and H values are obtained from the nanoindentation data. The crack length is measured from the center of the indent to the end of the crack using an optical microscope. For one indent, all crack lengths are measured. The crack length c is obtained from the average values of several indents.

Scratch Resistance

In micro/nanoscratch studies, a conical diamond indenter with a tip radius of about 1 μm and an included angle of 60° is drawn over the sample surface, and the load is ramped up until substantial damage occurs [11]. The coefficient of friction is monitored during scratching. In order to obtain scratch depths during scratching, the surface profile of the sample surface is first obtained by translating the sample at a low load of about 0.2 mN, which is insufficient to damage a hard sample surface. The 500-μm-long scratches are made by translating the sample while ramping the loads on the conical tip over different loads depending on the material hardness. The actual depth during scratching is obtained by subtracting the initial profile from the scratch depth measured during scratching. In order to measure the scratch depth after the scratch, the scratched surface is profiled at a low load of 0.2 mN and is subtracted from the actual surface profile before scratching.

15.2.2 Bending Tests of Nanostructures Using an AFM

Quasi-static bending tests of fixed nanobeam arrays are carried out using an AFM [37, 38, 54]. A three-sided pyramidal diamond tip (with a radius of about 200 nm) mounted on a rectangular stainless steel cantilever is used for the bending tests. The beam stiffness is selected based on the desired load range. The stiffness of the cantilever beams for application of normal load up to 100 µN is about 150–200 N/m.

The wafer with nanobeam array is fixed onto a flat sample chuck using double-stick tape [38]. For the bending test, the tip is brought over the nanobeam array with the help of the sample stage of the AFM and a built-in high magnification optical microscope (Fig. 15.1). For the fine positioning of the tip over a chosen beam, the array is scanned in contact mode at a contact load of about 2–4 µN, which results in negligible damage to the sample. After scanning, the tip is located at one end of a chosen beam. To position the tip at the center of the beam span, the tip is moved to the other end of the beam by giving the X-piezo an offset voltage. The value of this offset is determined after several such attempts have been made in order to minimize effects of piezo drift. Half of this offset is then applied to the X-piezo after the tip is positioned at one end of the beam, which usually results in the tip being moved to the center of the span. Once the tip is positioned over the center of the beam span, the tip is held stationary without scanning, and the Z-piezo is extended by a known distance, typically about 2.5 µm, at a rate of 10 nm/s, as shown in Fig. 15.1. During this time, the vertical deflection signal (V_{AFM}), which is proportional to the deflection of the cantilever (D_{tip}), is monitored. The displacement of the piezo is equal to the sum of the displacements of the cantilever and nanobeam. Hence, the displacement of the nanobeam (D_{beam}) under the point of load can be determined as

$$D_{beam} = D_{piezo} - D_{tip}. \tag{15.2}$$

The load (F_{beam}) on the nanobeam is the same as the load on the tip/cantilever (F_{tip}) and is given by

$$F_{beam} = F_{tip} = D_{tip} \times k, \tag{15.3}$$

where k is the stiffness of the tip/cantilever. In this manner, a load displacement curve for each nanobeam can be obtained.

The photodetector sensitivity of the cantilever needs to be calibrated to obtain D_{tip} in nm. For this calibration, the tip is pushed against a smooth diamond sample by moving the Z-piezo over a known distance. For the hard diamond material, the actual deflection of the tip can be assumed to be the same as the Z-piezo travel (D_{piezo}), and the photodetector sensitivity (S) for the cantilever setup is determined as

$$S = D_{piezo}/V_{AFM} \text{ nm/V}. \tag{15.4}$$

In the measurements, D_{tip} is given as $V_{AFM} \times S$.

Since a sharp tip would result in an undesirable large local indentation, *Sundararajan* and *Bhushan* [38] used a worn (blunt) diamond tip. Indentation experiments using this tip on a silicon substrate yielded a residual depth of less than 8 nm

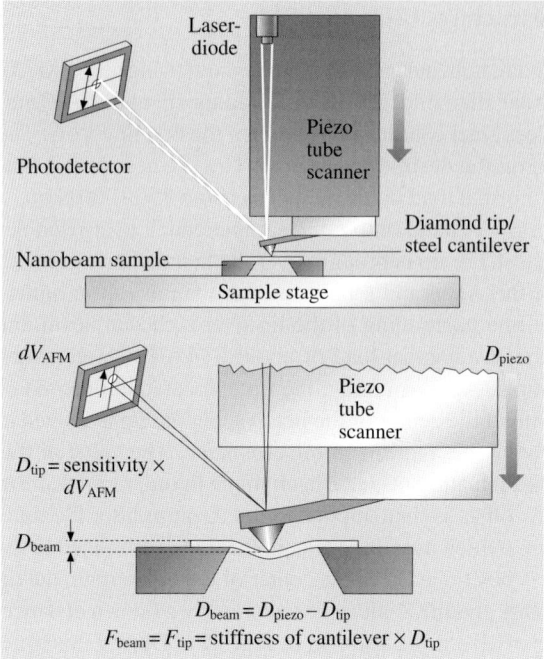

Fig. 15.1. Schematic showing the details of a nanoscale bending test. The AFM tip is brought to the center of the nanobeam and the piezo is extended over a known distance. By measuring the tip displacement, a load-displacement curve of the nanobeam can be obtained [38]

at a maximum load of 120 µN, which is negligible compared to displacements of the beams (several hundred nm). Hence, we can assume that negligible local indentation or damage is created during the bending process of the beams, and that the displacement calculated from (15.2) is entirely that of the beam structure.

Elastic Modulus and Bending Strength

Elastic modulus and bending strength (fracture stress) of the beams can be estimated by equations based on the assumption that the beams follow linear elastic theory of an isotropic material. This is probably valid since the beams have high length-to-width (ℓ/w) and length-to-thickness (ℓ/t) ratios and also since the length direction is along the principal stress direction during the test. For a fixed elastic beam loaded at the center of the span, the elastic modulus is expressed as

$$E = \frac{\ell^3}{192I} m, \qquad (15.5)$$

where ℓ is the beam length, I is the area moment of inertia for the beam cross section, and m is the slope of the load-displacement curve during bending [55]. The area moment of inertia for a beam with a trapezoidal cross section is calculated from the following equation:

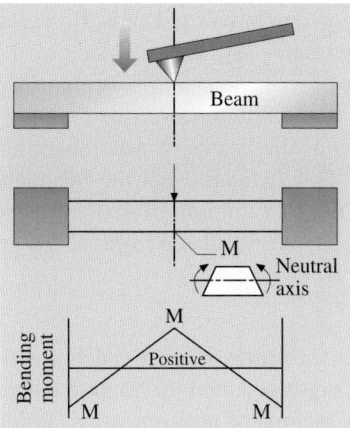

Fig. 15.2. A schematic of the bending moments generated in the beam during a quasi-static bending experiment with the load at the center of the span. The maximum moments occur under the load and at the fixed ends. Due to the trapezoidal cross section, the maximum tensile bending stresses occur at the top surfaces at the fixed ends

$$I = \frac{w_1^2 + 4w_1w_2 + w_2^2}{36(w_1 + w_2)} t^3, \tag{15.6}$$

where w_1 and w_2 are the upper and lower widths, respectively, and t is the thickness of the beam. According to linear elastic theory, for a centrally loaded beam, the moment diagram is shown in Fig. 15.2. The maximum moments are generated at the ends (negative moment) and under the loading point (positive moment), as shown in Fig. 15.2. The bending stresses generated in the beam are proportional to the moments and are compressive or tensile about the neutral axis (line of zero stress). The maximum tensile stress (σ_b, which is the bending strength or fracture stress) is produced on the top surface at both the ends and is given by [55]

$$\sigma_b = \frac{F_{max}\ell e_1}{8I}, \tag{15.7}$$

where F_{max} is the applied load at failure and e_1 is the distance of the top surface from the neutral plane of the beam cross section and is given by [55]

$$e_1 = \frac{t(w_1 + 2w_2)}{3(w_1 + w_2)}. \tag{15.8}$$

Although the moment value at the center of the beam is the same as at the ends, the tensile stresses at the center (generated on the bottom surface) are less than those generated at the ends, as per (15.7), because the distance from the neutral axis to the bottom surface is less than e_1. This is because of the trapezoidal cross section of the beam, which results in the neutral axis being closer to the bottom surface than the top (Fig. 15.2).

In the preceding analysis, the beams were assumed to have fixed ends. However, in the nanobeams used by *Sundararajan* and *Bhushan* [38], the underside of the beams was pinned over some distance on either side of the span. Hence, a finite element model of the beams was created to see if the difference in the boundary conditions affected the stresses and displacements of the beams. It was found that the difference in the stresses was less than 1%. This indicates that the boundary conditions near the ends of the actual beams are not that different from that of the fixed ends. Therefore, the bending strength values can be calculated from (15.7).

Fracture Toughness

Fracture toughness is another important parameter for brittle materials such as silicon. In the case of the nanobeam arrays, these are not best suited for fracture toughness measurements, because they do not possess regions of uniform stress during bending. *Sundararajan* and *Bhushan* [38] developed a methodology, and its steps are outlined schematically in Fig. 15.3a. First, a crack of known geometry is introduced in the region of maximum tensile bending stress, i.e., on the top surface near the ends of the beam. This is achieved by generating a scratch at high normal load across the width (w_1) of the beam using a sharp diamond tip (radius < 100 nm). A typical scratch thus generated is shown in Fig. 15.3b. By bending the beam as shown, a stress concentration will be formed under the scratch. This will lead to failure of the beam under the scratch once a critical load (fracture load) is attained. The fracture load and relevant dimensions of the scratch are input into the FEM model, which is used to generate the fracture stress plots. Figure 15.3c shows an FEM simulation of one such experiment, which reveals that the maximum stress does occur under the scratch.

If we assume that the scratch tip acts as a crack tip, a bending stress will tend to open the crack in Mode I. In this case, the stress field around the crack tip can be described by the stress intensity parameter K_I (for Mode I) for linear elastic materials [56]. In particular, the stresses corresponding to the bending stresses are described by

$$\sigma = \frac{K_I}{\sqrt{2\pi r}} \cos\left(\frac{\theta}{2}\right)\left[1 + \sin\left(\frac{\theta}{2}\right)\sin\left(\frac{3\theta}{2}\right)\right] \tag{15.9}$$

for every point $p(r, \theta)$ around the crack tip, as shown in Fig. 15.4. If we substitute the fracture stress (σ_f) to the left hand side of (15.9), then the K_I value can be substituted with its critical value, which is the fracture toughness $K_{IC\,\text{integrated circuit}}$. Now, the fracture stress can be determined for the point ($r = 0, \theta = 0$), i.e., right under the crack tip as explained above. However, we cannot substitute $r = 0$ in (15.9). The alternative is to substitute a value for r, which is as close to zero as possible. For silicon, a reasonable number is the distance between neighboring atoms in the (111) plane, the plane along which silicon exhibits the lowest fracture energy. This value was calculated from silicon unit cell dimensions of 0.5431 nm [57] to be 0.4 nm (half of the face diagonal). This assumes that Si displays no plastic zone around the crack tip, which is reasonable since in tension, silicon is not known to display much plastic deformation

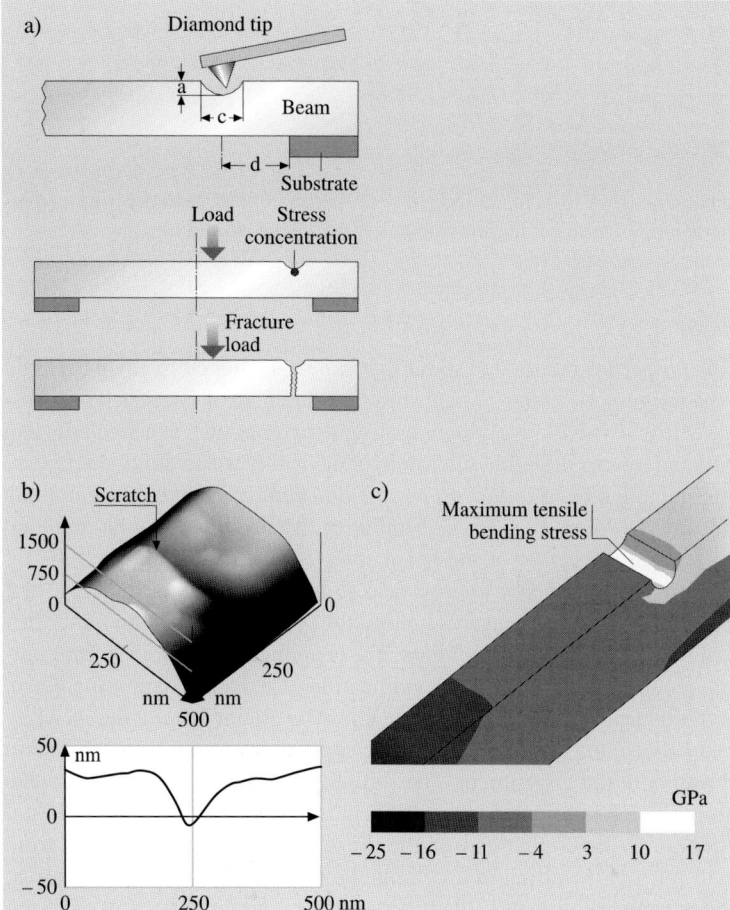

Fig. 15.3. (a) Schematic of technique to generate a defect (crack) of known dimensions in order to estimate fracture toughness. A diamond tip is used to generate a scratch across the width of the beam. When the beam is loaded as shown, a stress concentration is formed at the bottom of the scratch. The fracture load is then used to evaluate the stresses using FEM. (b) AFM 3-D image and 2-D profile of a typical scratch. (c) Finite element model results verifying that the maximum bending stress occurs at the bottom of the scratch [38]

at room temperature. *Sundararajan* and *Bhushan* [38] used values $r = 0.4$–1.6 nm (i.e., distances up to four times the distance between the nearest neighboring atoms) to estimate the fracture toughness for both Si and SiO$_2$ according to the following equation:

$$K_{\text{ICintegrated circuit}} = \sigma_f \sqrt{2\pi r} \qquad r = 0.4\text{–}1.6\,\text{nm}\,. \tag{15.10}$$

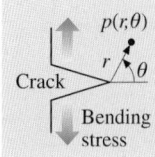

Fig. 15.4. Schematic of crack tip and coordinate systems used in (15.9) to describe a stress field around the crack tip in terms of the stress intensity parameter, K_I [38]

Fatigue Strength

In addition to the properties mentioned so far that can be evaluated from quasi-static bending tests, the fatigue properties of nanostructures are also of interest. This is especially true for MEMS/NEMS involving vibrating structures such as oscillators and comb drives [58] and hinges in digital micromirror devices [59]. To study the fatigue properties of the nanobeams, *Sundararajan* and *Bhushan* [38] applied monotonic cyclic stresses using an AFM, Fig. 15.5a. Similar to the bending test, the diamond tip is first positioned at the center of the beam span. In order to ensure that the tip is always in contact with the beam (as opposed to impacting it), the piezo is first extended by a distance D_1, which ensures a minimum stress on the beam. After this extension, a cyclic displacement of amplitude, D_2, is applied continuously until failure of the beam occurs. This results in the application of a cyclic load to the beam. The maximum frequency of the cyclic load that could be attained using the AFM by *Sundararajan* and *Bhushan* [38] was 4.2 Hz. The vertical deflection signal of the tip is monitored throughout the experiment. The signal follows the pattern of the piezo input up to failure, which is indicated by a sudden drop in the signal. During initial runs, piezo drift was observed that caused the piezo to gradually move away from the beam (i.e., to retract), resulting in a continuous decrease in the applied normal load. In order to compensate for this, the piezo is given a finite extension of 75 nm every 300 s, as shown in Fig. 15.5a. This results in keeping the applied loads fairly constant. The normal load variation (calculated from the vertical deflection signal) from a fatigue test is shown in Fig. 15.5b. The cyclic stress amplitudes (corresponding to D_2) and fatigue lives are recorded for every sample tested. Values for D_1 are set such that minimum stress levels are about 20% of the bending strengths.

15.2.3 Bending Tests Using a Nanoindenter

Quasi-static bending tests of micro/nanostructures are also carried out using a nanoindenter [39, 40]. The advantage of the nanoindenter is that loads up to about 500 mN, higher than that in AFM (up to about 100 µN), can be used for structures requiring high loads for experiments. To avoid the indenter tip pushing into the specimen, a blunt tip is used in the bending and fatigue tests. *Li* et al. [40] used a diamond conical indenter with a radius of 1 µm and an included angle of 60°. The load position used was at the center of the span for the bridge beams and 10 µm off from the free end of the cantilever beams. An optical microscope with a magnification of 1500×

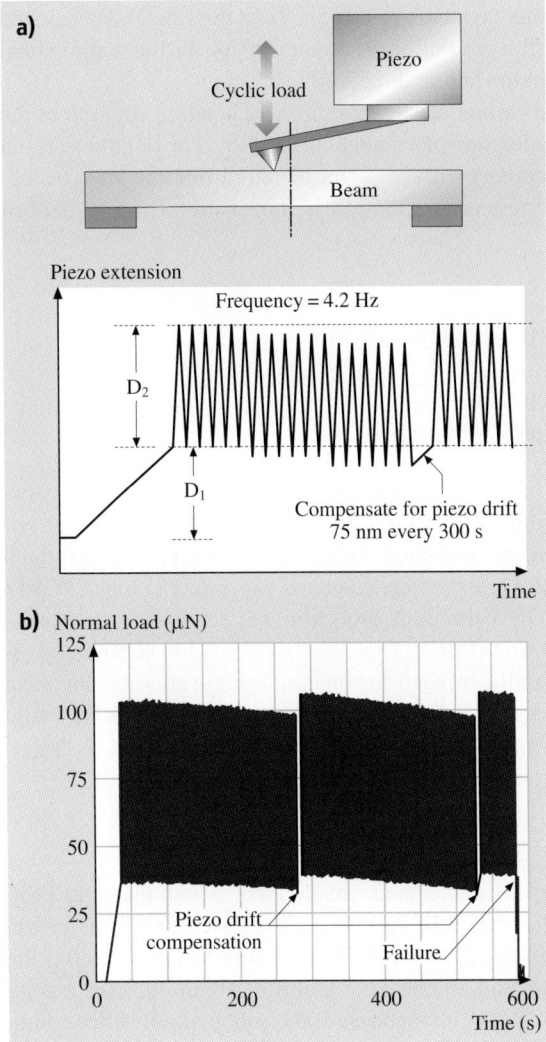

Fig. 15.5. (a) Schematic showing the details of the technique to study fatigue behavior of the nanobeams. The diamond tip is located at the middle of the span, and a cyclic load at 4.2 Hz is applied to the beam by forcing the piezo to move in the pattern shown. An extension is made every 300 s to compensate for the piezo drift to ensure that the load on the beam is kept fairly constant. (b) Data from a fatigue experiment on a nanobeam until failure. The normal load is computed from the raw vertical deflection signal. The compensations for piezo drift keep the load fairly constant [38]

or an in situ AFM is used to locate the loading position. Then the specimen is moved under the indenter location with a resolution of about 200 nm in the longitudinal direction and less than 100 nm in the lateral direction.

Using the analysis presented earlier, elastic modulus and bending strength of the beams can be obtained from the load-displacement curves [40]. For fatigue tests, an oscillating load is applied and contact stiffness is measured during the tests. A significant drop in the contact stiffness during the test is a measure of the number of cycles to failure [39].

15.3 Experimental Results and Discussion

15.3.1 Indentation and Scratch Tests of Various Materials Using Micro/Nanoindenters

Studies have been conducted on five different materials: undoped single-crystal Si(100), undoped polysilicon film, SiO_2 film, SiC film, electroless deposited Ni-11.5 wt% P amorphous film, and electroplated Au film [40, 50, 51]. A 3-μm thick polysilicon film was deposited by a low pressure chemical vapor deposition (LPCVD) process on a Si(100) substrate. The 1-μm thick SiO_2 film was deposited by a plasma enhanced chemical vapor deposition (PECVD) process on a Si(111) substrate. A 3-μm thick 3C-SiC film was epitaxially grown using an atmospheric pressure chemical vapor deposition (APCVD) process on a Si(100) substrate. A 12-μm thick Ni-P film was electroless plated on a 0.8-mm-thick Al-4.5 wt% Mg alloy substrate. A 3-μm-thick Au film was electroplated on a Si(100) substrate.

Hardness and Elastic Modulus

Hardness and elastic modulus measurements are made using a nanoindenter [40]. The hardness and elastic modulus values of various materials at a peak indentation depth of 50 nm are summarized in Fig. 15.6 and Table 15.1. The SiC film exhibits the highest hardness of about 25 GPa and an elastic modulus of about 395 GPa among the samples examined, followed by the undoped Si(100), undoped polysilicon film, SiO_2 film, Ni-P film, and Au film. The hardness and elastic modulus data of the undoped Si(100) and undoped polysilicon film are comparable. For the metal alloy films, the Ni-P film exhibits higher hardness and elastic modulus than the Au film.

Fracture Toughness

The optical images of Vickers indentations made using a microindenter at a normal load of 0.5 N held for 15 s on the undoped Si(100), undoped polysilicon film, and SiC film are shown in Fig. 15.7 [51]. The SiC film exhibits the smallest indentation mark, followed by the undoped polysilicon film and undoped Si(100). These Vickers indentation depths are smaller than one-third of the film thickness. Thus, the influence of the substrate on the fracture toughness of the films can be ignored. In addition to the indentation marks, radial cracks are observed emanating from the indentation

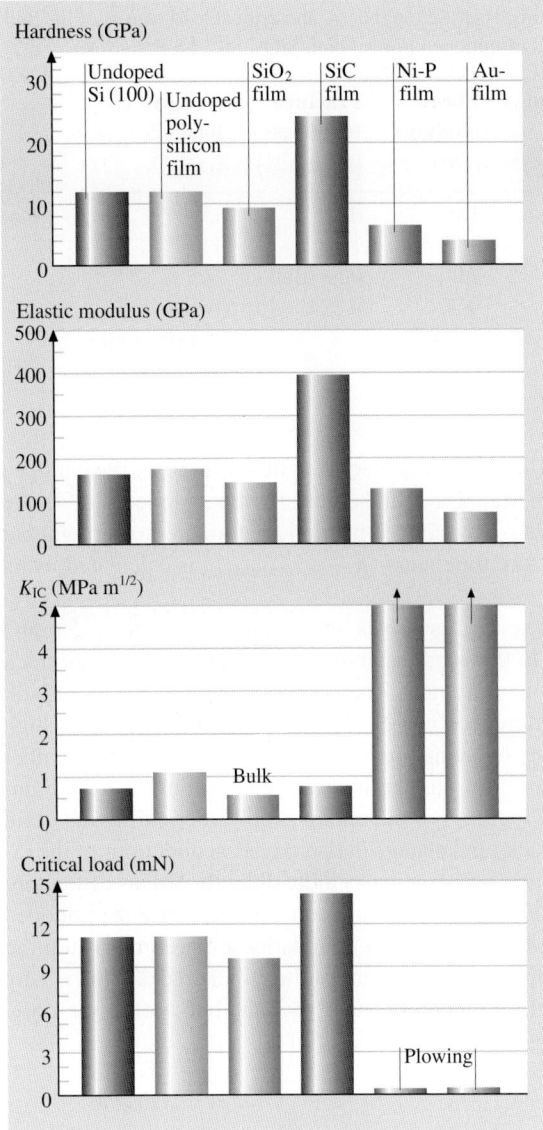

Fig. 15.6. Bar chart summarizing the hardness, elastic modulus, fracture toughness, and critical load (from scratch tests) results of the bulk undoped single-crystal Si(100) and thin films of undoped polysilicon, SiO_2, SiC, Ni-P, and Au [40]

Table 15.1. Hardness, elastic modulus, fracture toughness, and critical load results of the bulk single-crystal Si(100) and thin films of undoped polysilicon, SiO_2, SiC, Ni-P, and Au

Samples	Hardness (GPa)	Elastic modulus (GPa)	Fracture toughness ($MPa\,m^{1/2}$)	Critical load (mN)
Undoped Si(100)	12	165	0.75	11
Undoped polysilicon film	12	167	1.11	11
SiO_2 film	9.5	144	0.58 (Bulk)	9.5
SiC film	24.5	395	0.78	14
Ni-P film	6.5	130		0.4 (Plowing)
Au film	4	72		0.4 (Plowing)

corners. The SiC film shows the longest radial crack length, followed by the undoped Si(100) and undoped polysilicon film. The radial cracks for the undoped Si(100) are straight, whereas those for the SiC and undoped polysilicon film are not straight but go in a zigzag manner. The fracture toughness ($K_{\mathrm{IC\,integrated\,circuit}}$) is calculated using (15.1).

The fracture toughness values of all samples are summarized in Fig. 15.6 and Table 15.1. The SiO_2 film used in this study is about 1 μm thick, which is not thick enough for a fracture toughness measurement. The fracture toughness values of bulk silica are listed instead for a reference. The Ni-P and Au films exhibit very high fracture toughness values that cannot be measured by indentation methods. For other samples, the undoped polysilicon film has the highest value, followed by the undoped Si(100), SiC film, and SiO_2 film. For the undoped polysilicon film, the grain boundaries can stop the radial cracks and change the propagation directions of the radial cracks, making the propagation of these cracks more difficult. Values of fracture toughness for the undoped Si(100) and SiC film are comparable. Since the undoped Si(100) and SiC film are single crystal, no grain boundaries are present to stop the radial cracks and change the propagation directions of the radial cracks. This is why the SiC film shows a lower fracture toughness value than the bulk polycrystal SiC materials of $3.6\,MPa\,m^{1/2}$ [60].

Scratch Resistance

The scratch resistance of various materials has been studied using a nanoindenter by *Li* et al. [40]. Figure 15.8 compares the coefficient of friction and scratch depth profiles as a function of increasing normal load and optical images of three regions over scratches: at the beginning of the scratch (indicated by A on the friction profile), at the point of initiation of damage at which the coefficient of friction increases to a high value or increases abruptly (indicated by B on the friction profile), and towards the end of the scratch (indicated by C on the friction profile) for all samples. Note that the ramp loads for the Ni-P and Au range are from 0.2 to 5 mN, whereas the ramp loads for other samples are from 0.2 to 20 mN. All samples exhibit a continuous in-

Fig. 15.7. Optical images of Vickers indentations made at a normal load of 0.5 N held for 15 s on the undoped Si(100), undoped polysilicon film, and SiC film [51]

crease in the coefficient of friction with increasing normal load from the beginning of the scratch. The continuous increase in the coefficient of friction during scratching is attributed to the increasing plowing of the sample by the tip with increasing normal load, as shown in the SEM images in Fig. 15.8. The abrupt increase in the coefficient of friction is associated with catastrophic failure, as well as significant plowing of the tip into the sample. Before the critical load, the coefficient of friction of the undoped polysilicon, SiC, and SiO_2 films increased at a slower rate and was smoother than that of the other samples. The undoped Si(100) exhibits some bursts in the friction profiles before the critical load. At the critical load, the SiC and undoped polysilicon films exhibit a small increase in the coefficient of friction, whereas the undoped Si(100) and undoped polysilicon film exhibit a sudden increase in the coefficient of friction. The Ni-P and Au films show a continuous increase in the coefficient of friction, indicating the behavior of a ductile metal. The bursts in the friction profile might result from the plastic deformation and material pile-up in front of the

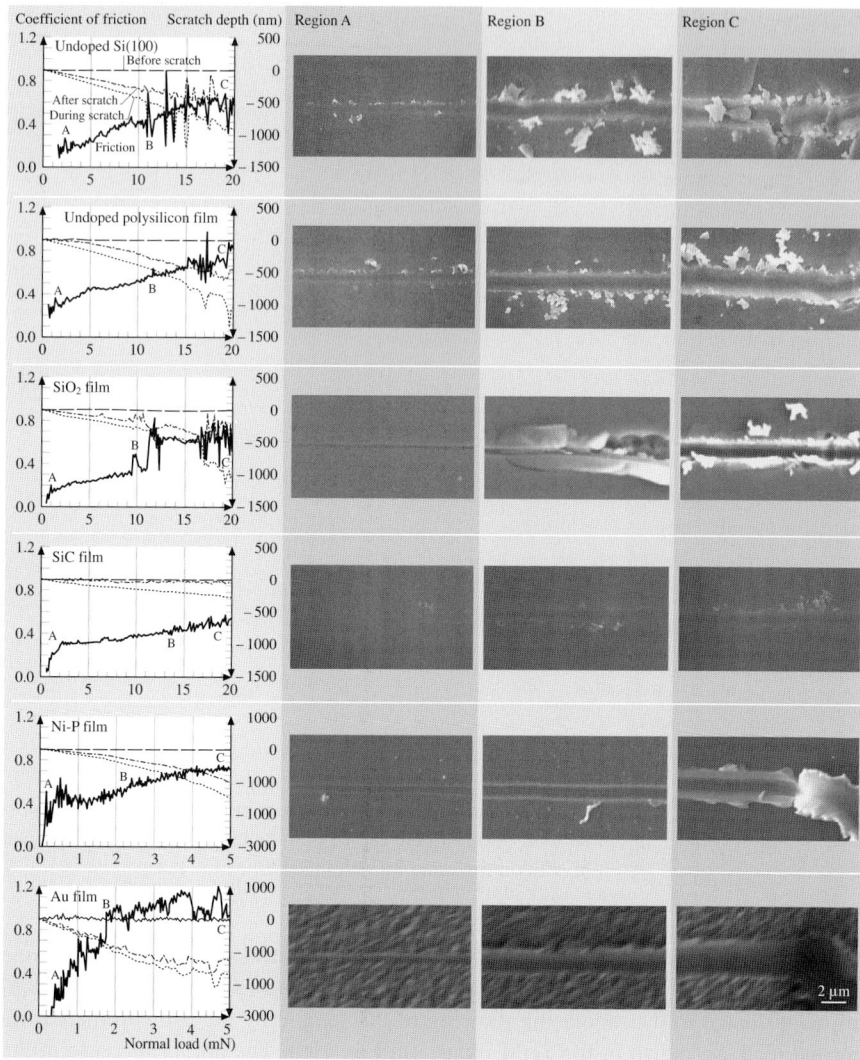

Fig. 15.8. Coefficient of friction and scratch depth profiles as a function of increasing normal load and optical images of three regions over scratches: at the beginning of the scratch (indicated by A on the friction profile), at the point of initiation of damage at which the coefficient of friction increases to a high value or increases abruptly (indicated by B on the friction profile), and towards the end of the scratch (indicated by C on the friction profile) for all samples [40]

scratch tip. The Au film exhibits a higher coefficient of friction than the Ni-P film, because it has lower hardness and elastic modulus values than the Ni-P film.

The SEM images show that below the critical loads the undoped Si(100) and undoped polysilicon film were damaged by plowing, associated with the plastic flow of the material and formation of debris on the sides of the scratch. For the SiC and SiO$_2$ films, in region A, a plowing scratch track was found without any debris on the side of the scratch, which is probably responsible for the smoother curve and slower increase in the coefficient of friction before the critical load. After the critical load, for the SiO$_2$ film, delamination of the film from the substrate occurred, followed by cracking along the scratch track. For the SiC film, only several small debris particles were found without any cracks on the side of the scratch, which is responsible for the small increase in the coefficient of friction at the critical load. For the undoped Si(100), cracks were found on the side of the scratch right from the critical load and up, which is probably responsible for the big bursts in the friction profile. For the undoped polysilicon film, no cracks were found on the side of the scratch at the critical load. This might result from grain boundaries, which can stop the propagation of cracks. At the end of the scratch, some of the surface material was torn away and cracks were found on the side of the scratch in the undoped Si(100). A couple of small cracks were found in the undoped polysilicon and SiO$_2$ films. No crack was found in the SiC film. Even at the end of the scratch, less debris was found in the SiC film. A curly chip was found at the end of the scratch in both Ni-P and Au films. This is a typical characteristic of ductile metal alloys. The Ni-P and Au films were damaged by plowing right from the beginning of the scratch with material pile-up at the side of the scratch.

The scratch depth profiles obtained during and after the scratch on all samples with respect to initial profile, after the cylindrical curvature is removed, are plotted in Fig. 15.8. Reduction in scratch depth is observed after scratching as opposed to during scratching. This reduction in scratch depth is attributed to an elastic recovery after removal of the normal load. The scratch depth after scratching indicates the final depth, which reflects the extent of permanent damage and plowing of the tip into the sample surface and is probably more relevant for visualizing the damage that can occur in real applications. For the undoped Si(100), undoped polysilicon film, and SiO$_2$ film, there is a large scatter in the scratch depth data after the critical loads, which is associated with the generation of cracks, material removal, and debris. The scratch depth profile is smooth for the SiC film. It is noted that the SiC film exhibits the lowest scratch depth among the samples examined. The scratch depths of the undoped Si(100), undoped polysilicon film, and SiO$_2$ film are comparable. The Ni-P and Au films exhibit much larger scratch depths than other samples. The scratch depth of the Ni-P film is smaller than that of the Au film.

The critical loads estimated from friction profiles for all samples are compared in Fig. 15.6 and Table 15.1. The SiC film exhibits the highest critical load of about 14 mN, as compared to other samples. The undoped Si(100) and undoped polysilicon film show comparable critical load of about 11 mN, whereas the SiO$_2$ film shows a low critical load of about 9.5 mN. The Ni-P and Au films were damaged by plowing right from the beginning of the scratch.

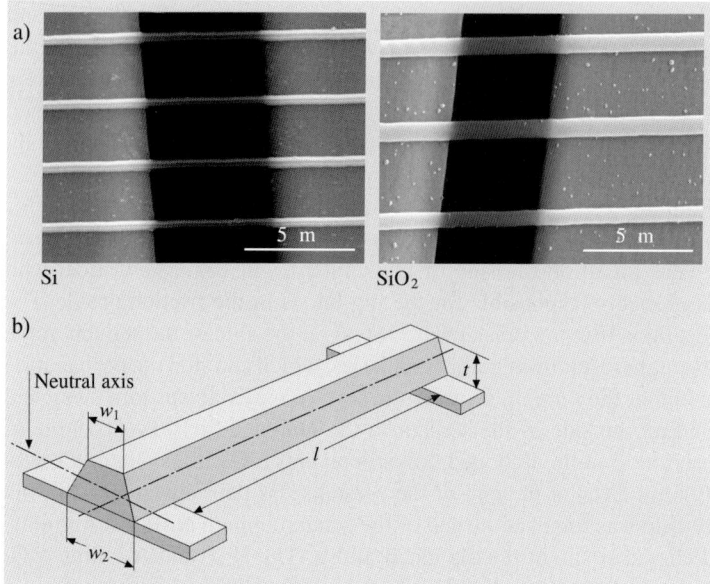

Fig. 15.9. (a) SEM micrographs of nanobeam arrays, and (b) a schematic of the shape of a typical nanobeam. The trapezoidal cross section is due to the anisotropic wet etching during the fabrication [54]

15.3.2 Bending Tests of Nanobeams Using an AFM

Bending tests have been performed on Si and SiO_2 nanobeam arrays [38, 54]. The single-crystal silicon bridge nanobeams were fabricated by bulk micromachining incorporating enhanced-field anodization using an AFM [37]. The Si nanobeams are oriented along the [110] direction in the (001) plane. Subsequent thermal oxidation of the beams results in the formation of SiO_2 beams. The cross section of the nanobeams is trapezoidal owing to the anisotropic wet etching process. SEM micrographs of Si and SiO_2 nanobeam arrays and a schematic of the shape of a typical nanobeam are shown in Fig. 15.9. The actual widths and thicknesses of nanobeams were measured using an AFM in tapping mode prior to tests using a standard Si tapping mode tip (tip radius < 10 nm). Surface roughness measurements of the nanobeam surfaces in tapping mode yielded a σ of 0.7 ± 0.2 nm and peak-to-valley (P–V) distance of 4 ± 1.2 nm for Si and a σ of 0.8 ± 0.3 nm and a P–V distance of 3.1 ± 0.8 nm for SiO_2. Prior to testing, the Si nanobeams were cleaned by immersing them in a "piranha etch" solution (3 : 1 solution by volume of 98% sulphuric acid and 30% hydrogen peroxide) for 600 s to remove any organic contaminants.

Bending Strength

Figure 15.10 shows typical load-displacement curves for Si and SiO_2 beams that were bent to failure [38, 54]. The upper width (w_1) of the beams is indicated in the

Table 15.2. Summary of measured parameters from quasi-static bending tests

Sample	Elastic modulus E (GPa)		Bending strength σ_b (GPa)		Fracture toughness K_{IC} (MPa $\sqrt{m}$)		
	Measured	Bulk value	Measured	Reported (microscale)	Estimated	Reported (microscale)	Bulk value
Si	182±11	169[1]	18±3	< 10[3]	1.67 ± 0.4	0.6–1.65[5]	0.9[6]
SiO$_2$	85±13	73[2]	7.6±2	< 2[4]	0.60 ± 0.2	0.5–0.9[4]	–

[1] Si(110), [61]   [2] [62]   [3] [19, 20, 22–25, 45, 63, 64]
[4] [34]   [5] [30–33]   [6] [57]

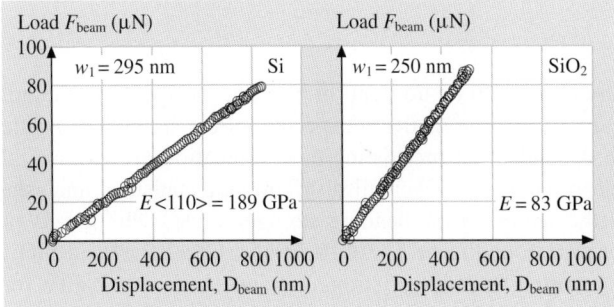

Fig. 15.10. Typical load-displacement curves of silicon and SiO$_2$ nanobeams. The curves are linear until sudden failure, indicative of brittle fracture of the beams. The elastic modulus (E) values calculated from the curves are shown. The dimensions of the Si beam were w_1 = 295 nm, w_2 = 484 nm, and t = 255 nm, while those of the SiO$_2$ beam were w_1 = 250 nm, w_2 = 560 nm, and t = 425 nm [54]

figure. Also indicated are the elastic modulus values obtained from the slope of the load-displacement curve (15.5). All the beams tested showed linear elastic behavior followed by abrupt failure, which is suggestive of brittle fracture. Figure 15.11 shows the scatter in the values of elastic modulus obtained for both Si and SiO$_2$ along with the average values (± standard deviation). The scatter in the values may be due to differences in the orientation of the beams with respect to the trench and the loading point being a little off center with respect to the beam span. The average values are a little higher than the bulk values (169 GPa for Si[110] and 73 GPa for SiO$_2$ in Table 15.2). However, the values of E obtained from (15.5) have an error of about 20% due to the uncertainties in beam dimensions and spring constant of the tip/cantilever (which affects the measured load). Hence the elastic modulus values on the nanoscale can be considered comparable to bulk values.

Most of the beams when loaded quasi-statically at the center of the span broke at the ends, as shown in Fig. 15.12a, which is consistent with the fact that maximum tensile stresses occur on the top surfaces near the ends. (See FEM stress distribution results in Fig. 15.12b.) Figure 15.13 shows the values of bending strength obtained

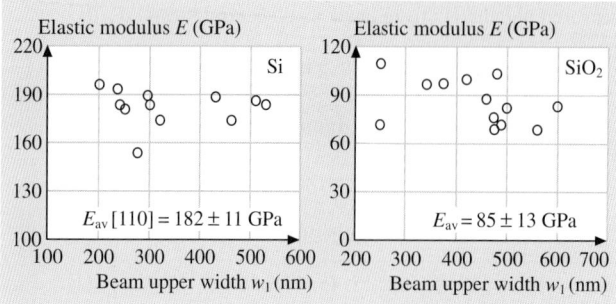

Fig. 15.11. Elastic modulus values measured for Si and SiO$_2$. The average values are shown. These are comparable to bulk values, which demonstrates that elastic modulus shows no specimen size dependence [38]

for different beams. There appears to be no trend in bending strength with the upper width (w_1) of the beams. The large scatter is expected for the strength of brittle materials, since they are dependent on preexisting flaw population in the material and, hence, are statistical in nature. The Weibull distribution, a statistical analysis, can be used to describe the scatter in the bending strength values. The means of the Weibull distributions were found to be 17.9 GPa and 7.6 GPa for Si and SiO$_2$, respectively. Previously reported numbers of strengths range from 1 to 6 GPa for silicon [19, 20, 22–25, 27, 45, 63, 64] and about 1 GPa for SiO$_2$ [34] microscale specimens. This clearly indicates that bending strength shows a specimen size dependence. Strength of brittle materials is dependent on preexisting flaws in the material. Since for nanoscale specimens the volume is smaller than for micro- and macroscale specimens, the flaw population will be smaller as well, resulting in higher values of strength.

Fracture Toughness

Estimates of fracture toughness calculated using (15.10) for Si and SiO$_2$ are shown in Fig. 15.14 [38]. The results show that the K_{IC} estimate for Si is about 1–2 MPa $\sqrt{m}$, whereas for SiO$_2$ the estimate is about 0.5–0.9 MPa $\sqrt{m}$. These values are comparable to values reported by others on larger specimens for Si [30–33] and SiO$_2$ [34]. The high values obtained for Si could be due to the fact that the scratches, despite being quite sharp, still have a finite radius of about 100 nm. The bulk value for silicon is about 0.9 MPa $\sqrt{m}$ (Table 15.2). Fracture toughness is considered a material property and is believed to be independent of specimen size. The values obtained in this study, given its limitations, appear to show that fracture toughness is comparable, if not a little higher on the nanoscale.

Fatigue Strength

Fatigue strength measurements of Si nanobeams have been carried out by *Sundararajan* and *Bhushan* [38] using an AFM and *Li* and *Bhushan* [39] using a nanoindenter.

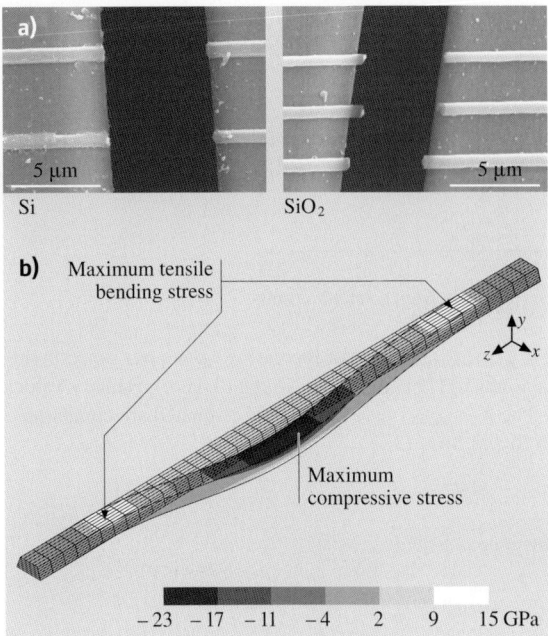

Fig. 15.12. (a) SEM micrographs of nanobeams that failed during quasi-static bending experiments. The beams failed at or near the ends, which is the location of maximum tensile bending stress [54], and (b) bending stress distribution for silicon nanobeam, indicating that the maximum tensile stresses occur on the top surfaces near the fixed ends

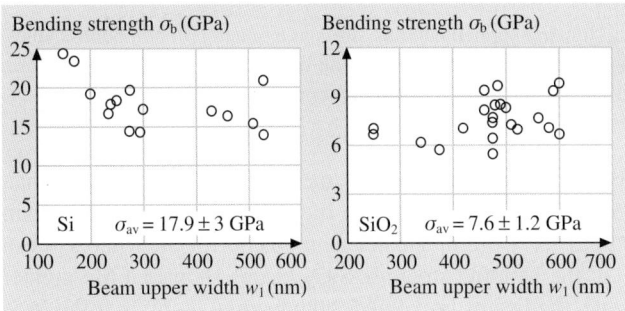

Fig. 15.13. Bending strength values obtained from bending experiments. Average values are indicated. These values are much higher than values reported for microscale specimens, indicating that bending strength shows a specimen size effect [38]

Various stress levels were applied to nanobeams by *Sundararajan* and *Bhushan* [38]. The minimum stress was 3.5 GPa for Si beams and 2.2 GPa for SiO_2 beams. The frequency of applied load was 4.2 Hz. In general, the fatigue life decreased with increasing mean stress as well as increasing stress amplitude. When the stress am-

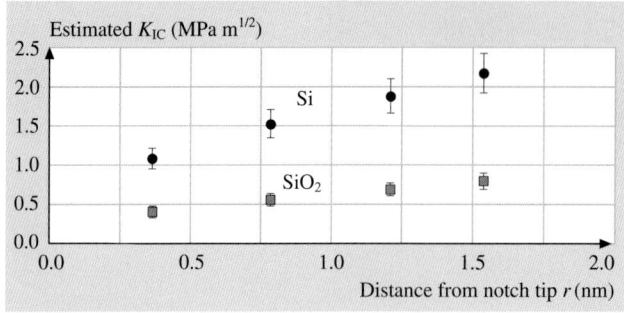

Fig. 15.14. Fracture toughness ($K_{\text{ICintegrated circuit}}$) values for increasing values of r corresponding to distance between neighboring atoms in {111} planes of silicon (0.4 nm). Hence r values between 0.4 and 1.6 nm are chosen. The $K_{\text{ICintegrated circuit}}$ values thus estimated are comparable to values reported by others for both Si and SiO$_2$ [38]

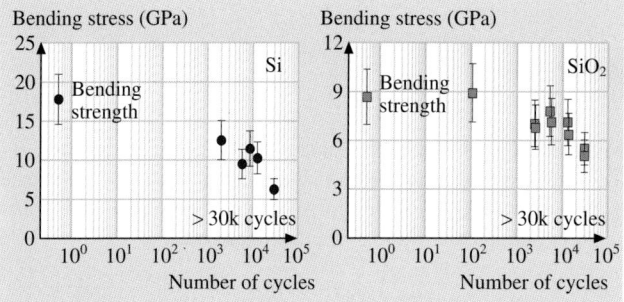

Fig. 15.15. Fatigue test data showing applied bending stress as a function of number of cycles. A single load-unload sequence is considered 1 cycle. The bending strength data points are, therefore, associated with $\frac{1}{2}$ cycle, since failure occurs upon loading [38]

plitude was less than 15% of the bending strength, the fatigue life was greater than 30 000 cycles for both Si and SiO$_2$. However, the mean stress had to be less than 30% of the bending strength for a life of greater than 30 000 for Si, whereas even at a mean stress of 43% of the bending strength, SiO$_2$ beams showed a life greater than 30 000. During fatigue, the beams broke under the loading point or at the ends when loaded at the center of the span. This was different from the quasi-static bending tests, where the beams broke at the ends almost every time. This could be due to the fact that the stress levels under the load and at the ends are not that different and fatigue crack propagation could occur at either location. Figure 15.15 shows a nanoscale S-N curve with bending stress (S) as a function of fatigue in cycles (N) with an apparent endurance life at lower stress. This study clearly demonstrates that fatigue properties of nanoscale specimens can be studied.

SEM Observations of Fracture Surfaces

Figure 15.16 shows SEM images of the fracture surfaces of nanobeams broken during quasi-static bending, as well as fatigue [38]. In the quasi-static cases, the maximum tensile stresses occur on the top surface, so it is reasonable to assume that fracture initiated at or near the top surface and propagated downward. The fracture surfaces of the beams suggest a cleavage type of fracture. Silicon beam surfaces show various ledges or facets, which is typical for crystalline brittle materials. Silicon usually fractures along the (111) plane, due to this plane having the lowest surface energy to overcome by a propagating crack. However, failure has also been known to occur along the (110) planes in microscale specimens, despite the higher energy required compared to the (111) planes [45]. The plane normal to the beam direction in these samples is the (110) plane, while (111) planes will be oriented at 35° from the (110) plane. The presence of facets and irregularities on the silicon surface in Fig. 15.16a suggest that it is a combination of these two types of fractures that has occurred. Since the stress levels are very high for these specimens, it is reasonable to assume that crack propagating forces will be high enough to result in (110) type failures.

In contrast, the silicon fracture surfaces under fatigue, shown in Fig. 15.16b, appear very smooth without facets or irregularities. This is suggestive of low energy fracture, i.e., of (111) type fracture. We do not see evidence of fatigue crack propagation in the form of steps or striations on the fracture surface. We believe that for the stress levels applied in these fatigue experiments, failure in silicon occurred via cleavage associated with "static fatigue" type of failures.

SiO_2 shows very smooth fracture surfaces for both quasi-static bending and fatigue. This is in contrast to the hackled surface one might expect for the brittle failure of an amorphous material on the macroscale. However, in larger scale fracture surfaces for such materials, the region near the crack initiation usually appears smooth or mirror-like. Since the fracture surface here is so small and very near the crack initiation site, it is not unreasonable to see such a smooth surface for SiO_2 on this scale. There appears to be no difference between the fracture surfaces obtained by quasi-static bending and fatigue for SiO_2.

Summary of Mechanical Properties Measured Using Quasi-Static Bending Tests

Table 15.2 summarizes the various properties measured via quasi-static bending in this study [38]. Also shown are bulk values of the parameters, along with values reported on larger scale specimens by other researchers. Elastic modulus and fracture toughness values appear to be comparable to bulk values and show no dependence on specimen size. However, bending strength shows a clear specimen size dependence with nanoscale numbers being twice as large as numbers reported for larger-scale specimens.

15.3.3 Bending Tests of Microbeams Using a Nanoindenter

Bending tests have been performed on Ni-P and Au microbeams [40]. The Ni-P cantilever microbeams were fabricated by focused ion beam machining technique.

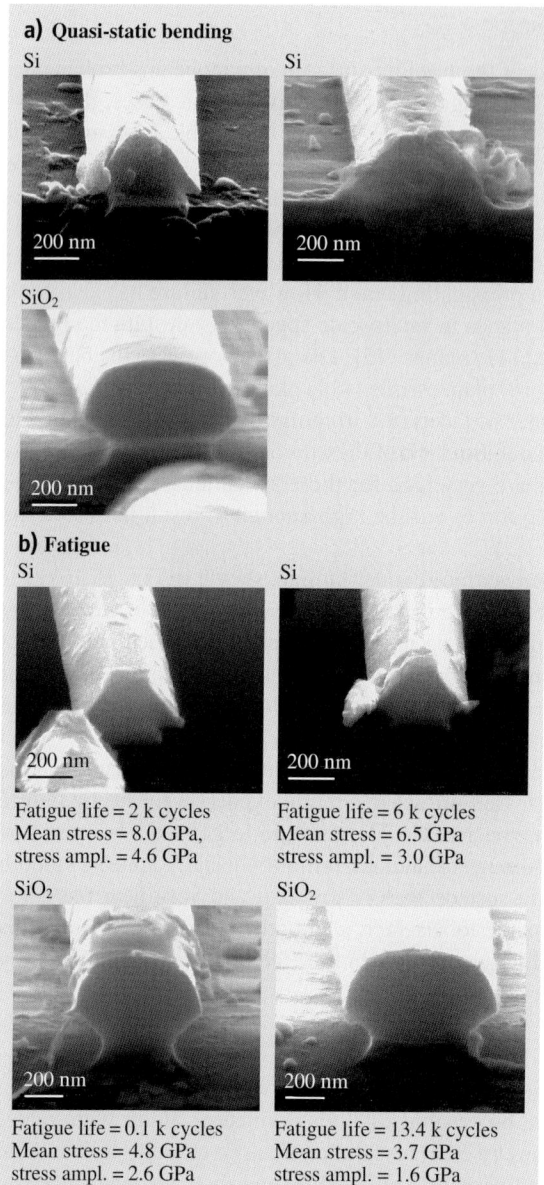

Fig. 15.16. SEM micrographs of fracture surfaces of silicon and SiO_2 beams subjected to (**a**) quasi-static bending and (**b**) fatigue [38]

The dimensions were $10 \times 12 \times 50\,\mu m^3$. Notches with a depth of 3 μm and a tip radius of 0.25 μm were introduced in the microbeams to facilitate failure at a lower load in the bending tests. The Au bridge microbeams were fabricated by electroplating technique.

Figure 15.17 shows the SEM images, load displacement curve, and FEM stress contour for the notched Ni-P cantilever microbeam that was bent to failure [40]. The distance between the loading position and the fixed end is 40 μm. The 3-μm-deep notch is 10 μm from the fixed end. The notched beam showed linear behavior followed by abrupt failure. The FEM stress contour shows that there is higher stress concentration at the notch tip. The maximum tensile stress σ_m at the notch tip can be analyzed using Griffith fracture theory as follows [53]:

$$\sigma_m \approx 2\sigma_o \left(\frac{c}{\rho}\right)^{1/2}, \qquad (15.11)$$

where σ_o is the average applied tensile stress on the beam, c is the crack length, and ρ is the crack tip radius. Therefore, elastic-plastic deformation will first occur locally at the end of the notch tip, followed by abrupt fracture failure after the σ_m reaches the ultimate tensile strength of Ni-P, even though the rest of the beam is still in the elastic regime. The SEM image of the fracture surface shows that the fracture started right from the notch tip with plastic deformation characteristic. This indicates that although local plastic deformation occurred at the notch tip area, the whole beam failed catastrophically. The present study shows that FEM simulation can predict well the stress concentration and helps in understanding the failure mechanism of the notched beams.

Figure 15.18 shows the SEM images, load-displacement curve and FEM stress contour for the Au bridge microbeam that was deformed by the indenter [40]. The recession gap between the beam and substrate is about 7 μm, which is not large enough to break the beam at the load applied. From the load-displacement curve, we note that the beam experienced elastic-plastic deformation. The FEM stress contour shows that the maximum tensile stress is located at the fixed ends, whereas the minimum compressive stress is located around the center of the beam. The SEM image shows that the beam has been permanently deformed. No crack was found on the beam surface. The present study shows a possibility for mechanically forming the Au film into the shape as needed. This may help in designing/fabricating functionally complex smart micro/nanodevices that need conductors for power supply and input/output signals.

15.4 Finite Element Analysis of Nanostructures with Roughness and Scratches

Micro/nanostructures have some surface topography and local scratches dependent upon the manufacturing process. Surface roughness and local scratches may compromise the reliability of the devices and their effect needs to be studied. Finite element modeling is used to perform parametric analysis to study the effect of surface roughness and scratches in different well-defined forms on tensile stresses that are responsible for crack propagation [46, 47]. The analysis has been carried out on trapezoidal

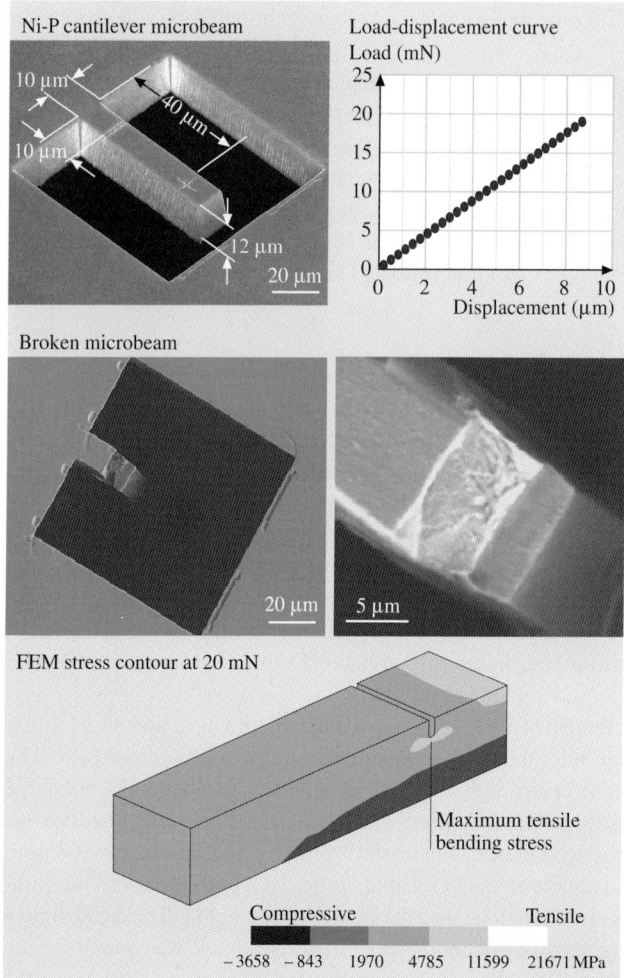

Fig. 15.17. SEM micrographs of the new and broken beams, load-displacement curve, and FEM stress contour for the notched Ni-P cantilever microbeam [40]

beams supported at the bottom whose data (on Si and SiO$_2$ nanobeams) have been presented earlier.

The finite element analysis has been carried out using the static analysis of ANSYS 5.7, which calculates the deflections and stresses produced by the applied loading. The type of element selected for the present study was the SOLID95, which allows the use of different shapes without much loss of accuracy. This element is 3-D with 20 nodes, each node having three degrees of freedom that imply translation in the x, y, and z directions. Each nanobeam cross section is divided into six elements along the width and the thickness and 40 elements along the length. SOLID95

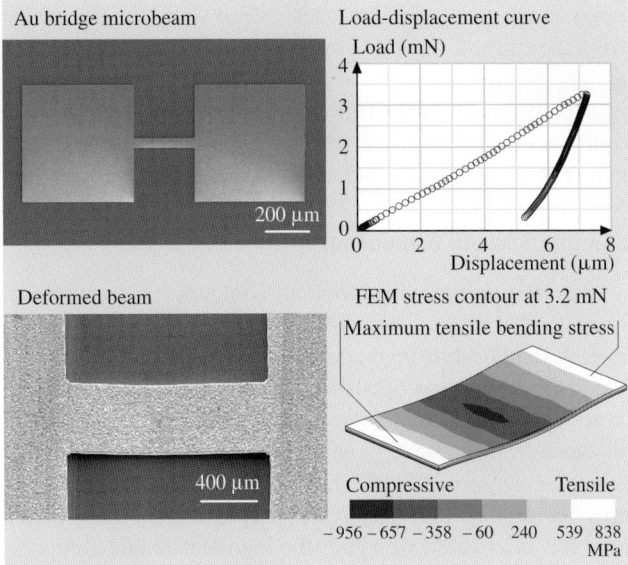

Fig. 15.18. SEM micrographs of the new and deformed beams, load-displacement curve, and FEM stress contour for the Au bridge microbeam [40]

has plasticity, creep, stress stiffening, large deflection, and large strain capabilities. The large displacement analysis is used for large loads. The mesh is kept finer near the asperities and the scratches in order to take into account variation in the bending stresses. The beam materials studied are made of single-crystal silicon (110) and SiO_2 films whose data have been presented earlier. Based on bending experiments presented earlier, the beam materials can be assumed to be linearly elastic isotropic materials. Young's modulus of elasticity (E) and Poisson's ratio (v) for Si and SiO_2 are 169 GPa [61] and 0.28 [57], and 73 GPa [62] and 0.17 [62], respectively. A sample nanobeam of silicon was chosen for performing most of the analysis, as silicon is the most widely used MEMS/NEMS material. The cross section of the fabricated beams used in the experiment is trapezoidal and supported at the bottom, Fig. 15.9. The following dimensions are used: w_1 = 200 nm, w_2 = 370 nm, t = 255 nm, and ℓ = 6 µm. In the boundary conditions, the displacements are constrained in all directions on the bottom surface for 1 µm from each end. A point load applied at the center of the beam is simulated with the load being applied at three closely located central nodes on the beam used. It has been observed from the experimental results that the Si nanobeam breaks at around 80 µN. Therefore, in this analysis, a nominal load of 70 µN is selected. At this load, deformations are large and a large displacement option is used.

To study the effect of surface roughness and scratches on the maximum bending stresses the following cases were studied. First, the semicircular and grooved asperities in the longitudinal direction with defined geometrical parameters are analyzed,

Fig. 15.19a. Next, semicircular asperities and scratches placed along the transverse direction at a distance c from the end and separated by pitch p from each other are analyzed, Fig. 15.19b. Lastly, the beam material is assumed to be either purely elastic, elastic-plastic, or elastic-perfectly plastic. In the following section, we begin with the stress distribution in smooth nanobeams followed by the effect of surface roughness in the longitudinal and transverse directions and scratches in the transverse direction.

15.4.1 Stress Distribution in a Smooth Nanobeam

Figure 15.20 shows the stress and vertical displacement contours for a nanobeam supported at the bottom and loaded at the center [46, 47]. As expected, the maximum tensile stress occurs at the ends, while the maximum compressive stress occurs under the load at the center. Stress contours obtained at a section of the beam from the front and side are also shown. In the beam cross section, the stresses remain constant at a given vertical distance from one side to another and change with a change in vertical location. This can be explained by the fact that the bending moment is constant at a particular cross section, so the stress is only dependent on the distance from the neutral axis. However, in cross section A-A the high tensile and compressive stresses are localized near the end of the beam at top and bottom, respectively, whereas the lower values are spread out away from the ends. High value of tensile stresses occurs near the ends because of high bending moment.

15.4.2 Effect of Roughness in the Longitudinal Direction

The roughness in the form of semicircular and grooved asperities in the longitudinal direction on the maximum bending stresses are analyzed [47]. The radius R and depth L are kept fixed at 25 nm, while the number of asperities is varied and their effect is observed on the maximum bending stresses. Figure 15.21 shows the variation of maximum bending stresses as a function of asperity shape and the number of asperities. The maximum bending stresses increase as the asperity number increases for both semicircular and grooved asperities. This can be attributed to the fact that as the asperity number increases, the moment of inertia decreases for that cross section. Also, the distance from the neutral axis increases because the neutral axis shifts downwards. Both these factors lead to the increase in the maximum bending stresses, and this effect is more pronounced in the case of semicircular asperity as it exhibits a higher value of maximum bending stress than that in grooved asperity. Figure 15.21 shows the stress contours obtained at a section of the beam from the front side for both cases when we have a single semicircular asperity and when four adjacent semicircular asperities are present. Trends are similar to those observed earlier for a smooth nanobeam (Fig. 15.20).

15.4.3 Effect of Roughness in the Transverse Direction and Scratches

We analyze semicircular asperities when placed along the transverse direction followed by the effect of scratches on the maximum bending stresses in varying numbers and pitch [47]. In the analysis of semicircular transverse asperities, three cases were considered, which included a single asperity and asperities throughout the nanobeam surface separated by pitch equal to 50 nm and 100 nm. In all of these cases, c value was kept equal to 0 nm. Figure 15.22 shows that the value of maximum tensile stress is 42 GPa, which is much larger than the maximum tensile stress value with no asperity of 16 GPa, or when the semicircular asperity is present in the longitudinal direction. It is also observed that the maximum tensile stress does not vary with the number of asperities or the pitch, while the maximum compressive stress does increase dramatically for the asperities present throughout the beam surface from its value when a single asperity is present. Maximum tensile stress occurs at the ends, and an increase in p does not add any asperities at the ends, whereas asperities are added in the central region where compressive stresses are maximum. The semicircular asperities present at the center cause the local perturbation in the stress distribution at the center of the asperity where load is being applied, leading to a high value of maximum compressive stress [65]. Figure 15.22 also shows the stress contours obtained at a section of the beam from the front and side for both cases when there is a single semicircular asperity and when asperities are present throughout the beam surface at a pitch equal to 50 nm. Trends are similar to those observed earlier for a smooth nanobeam (Fig. 15.20).

In the study pertaining to scratches, the number of scratches are varied along with the variation in the pitch. Furthermore, the load is applied at the center of the beam and at the center of the scratch near the end, as for all the cases. In all of these, c value was kept equal to 50 nm and L value was equal to 100 nm with h value being 20 nm. Figure 15.23 shows that the value of maximum tensile stress remains almost the same with the number of scratches for both types of loading – that is, when load is applied at the center of the beam and at the center of the scratch near the end. This is because the maximum tensile stress occurs at the beam ends no matter where the load gets applied. But the presence of scratch does increase the maximum tensile stress compared to its value for a smooth nanobeam, although the number of scratches no longer matter as the maximum tensile stress occurring at the nanobeam end is unaffected by the presence of more scratches beyond the first scratch in the direction toward the center. The value of the tensile stress is much lower when the load is applied at the center of scratch, and it can be explained as follows. The negative bending moment at the end near the applied load decreases with load offset after two-thirds of the length of the beam [66]. Since this negative bending moment is responsible for tensile stresses, their behavior with the load offset is the same as the negative bending moment. Also, the value of maximum compressive stress when load is applied at the center of the nanobeam remains almost the same as the center geometry is unchanged due to the number of scratches and, hence, the maximum compressive stress occurring below the load at the center is same. On the other hand, when the load is applied at the center of the scratch we observe that the maximum compressive stress increases dramatically because the local perturbation in the stress distribution at the center of the scratch where load is being applied leads to a high value of maximum compressive stress [65]. It increases further with the number of scratches and then levels off. This can be attributed to the fact that when there is another scratch present close to the scratch near the end, the stress concentration is

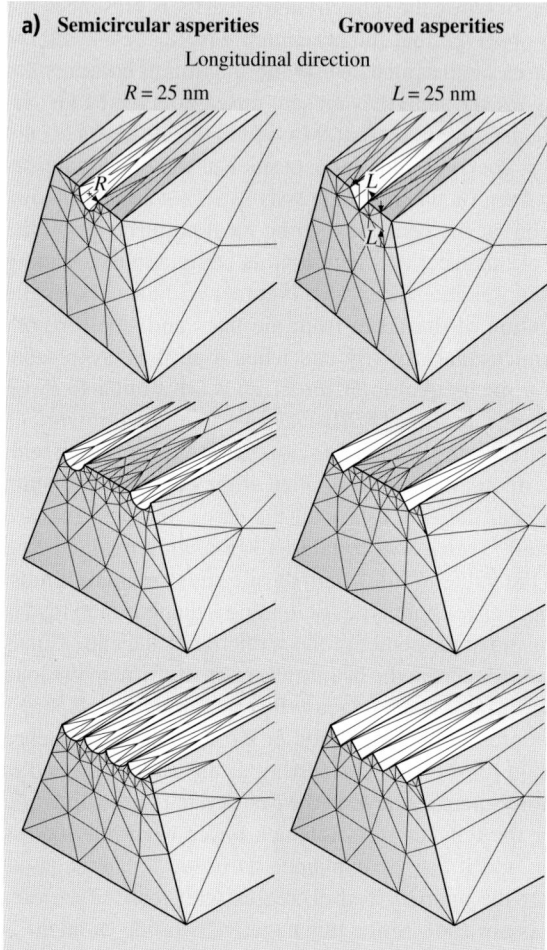

Fig. 15.19. (a) Plots showing the geometries of modeled roughness – semicircular and grooved asperities along the nanobeam length with defined geometrical parameters. (b) (*See on next page.*) Schematic showing semicircular asperities and scratches in the transverse direction, followed by the illustration of the mesh created on the beam with fine mesh near the asperities and scratches. Also shown are the semicircular asperities and scratches at different pitch values

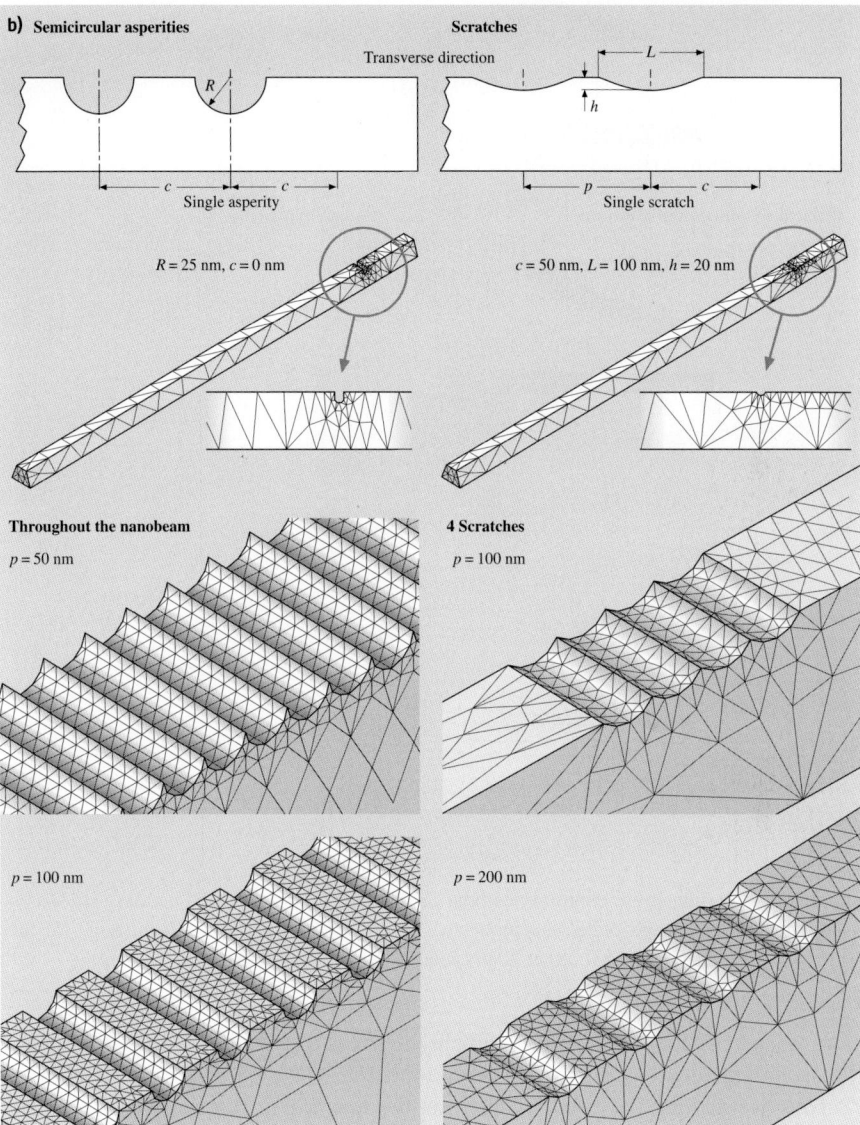

Fig. 15.19. continue

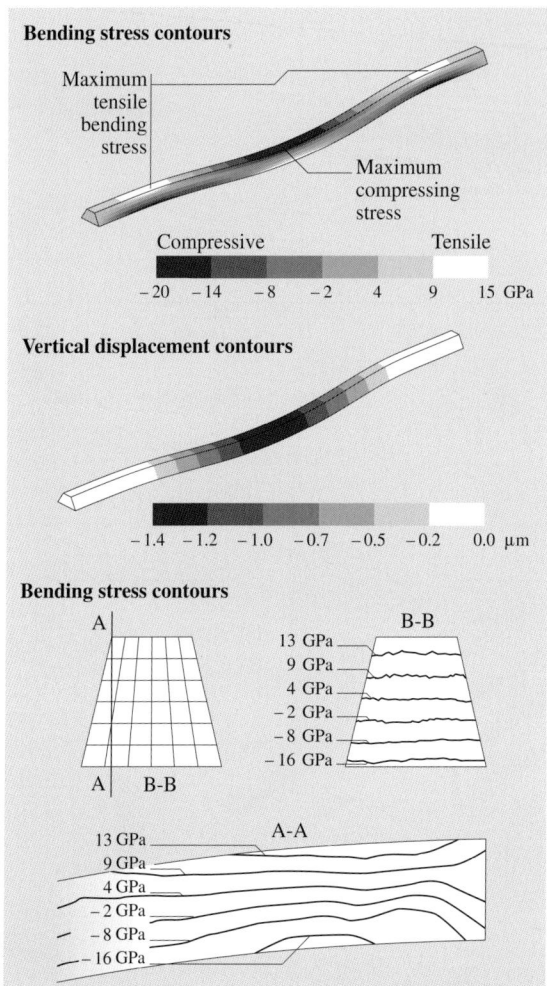

Fig. 15.20. Bending stress contours, vertical displacement contours, and bending stress contours after loading trapezoidal Si nanobeam (w_1 = 200 nm, w_2 = 370 nm, t = 255 nm, ℓ = 6 μm, E = 169 GPa, v = 0.28) at 70 μN load [47]

more, as the effect of local perturbation in the stress distribution is more significant. However, this effect is insignificant when more than two scratches are present.

Now we address the effect of pitch on the maximum compressive stress when the load is applied at the center of the scratch near the end. When the pitch is up to a value of 200 nm the maximum compressive stress increases with the number of scratches, as discussed earlier. On the other hand, when the pitch value goes beyond 225 nm this effect is reversed. This is because the presence of another scratch no longer affects the local perturbation in the stress distribution at the scratch near the end. Instead, more scratches at a fair distance distribute the maximum compressive

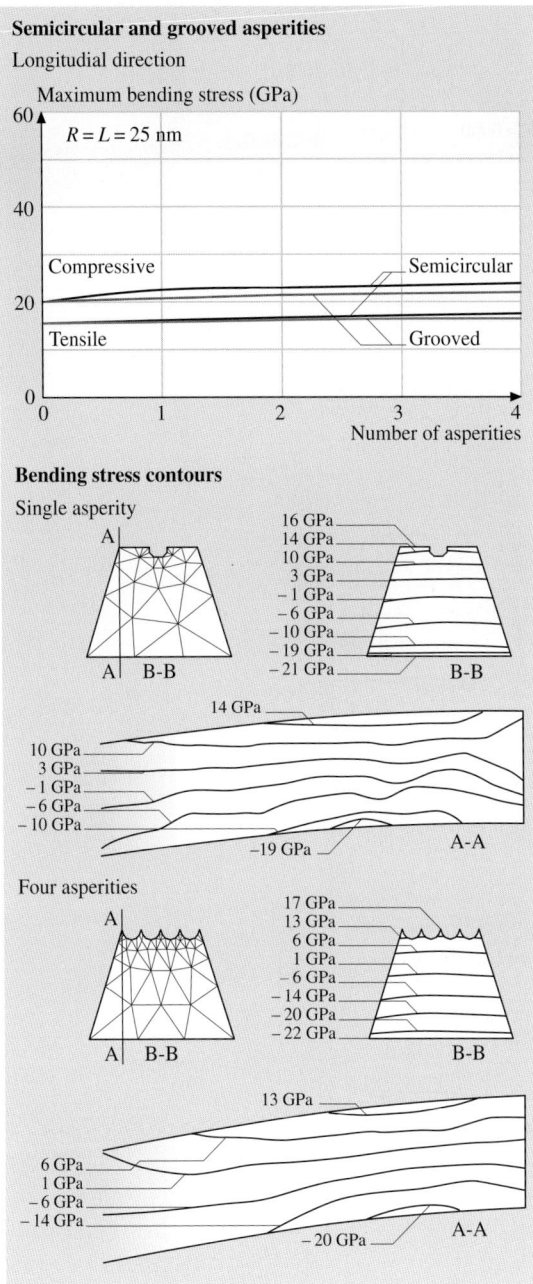

Fig. 15.21. Effect of longitudinal semicircular and grooved asperities in different numbers on maximum bending stresses after loading trapezoidal Si nanobeams (w_1 = 200 nm, w_2 = 370 nm, t = 255 nm, ℓ = 6 μm, E = 169 GPa, v = 0.28, load = 70 μN). Bending stress contours obtained in the beam with semicircular single asperity and four adjacent asperities of R = 25 nm [47]

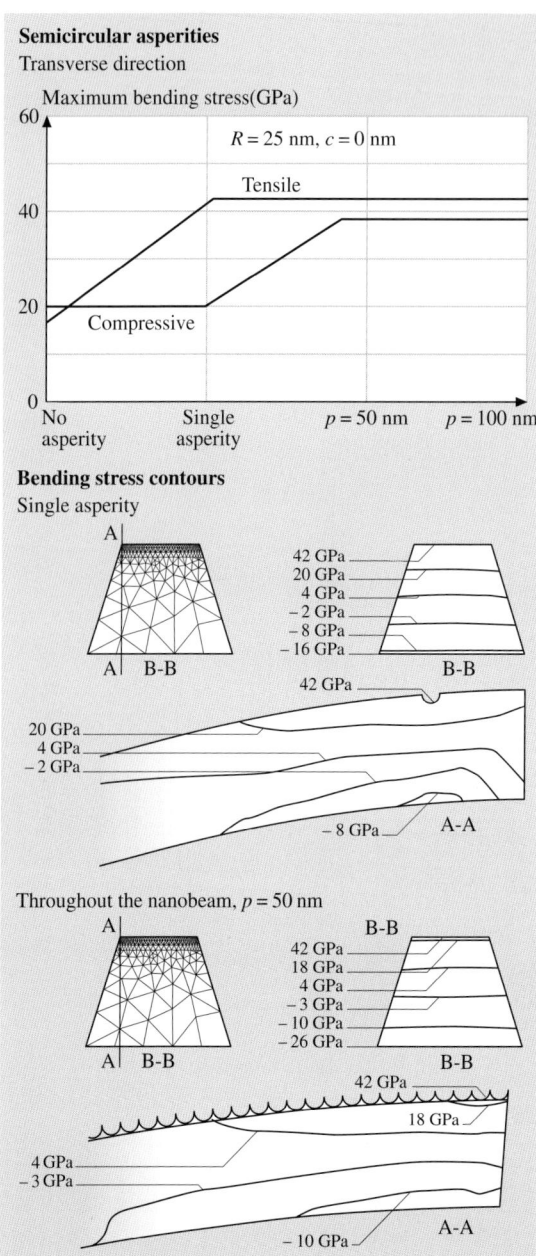

Fig. 15.22. Effect of transverse semicircular asperities located at different pitch values on the maximum bending stresses after loading trapezoidal Si nanobeams ($w_1 = 200$ nm, $w_2 = 370$ nm, $t = 255$ nm, $\ell = 6\,\mu$m, $E = 169$ GPa, $\nu = 0.28$, load = 70 μN). Bending stress contours obtained in the beam with semicircular single asperity and semicircular asperities throughout the nanobeam surface at $p = 50$ nm [47]

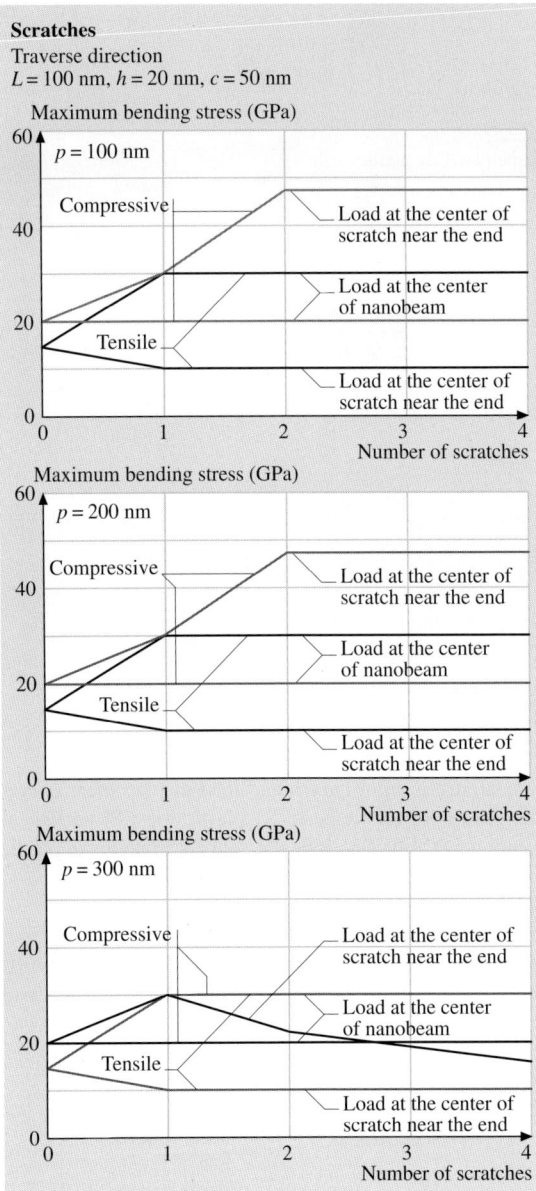

Fig. 15.23. Effect of number of scratches along with the variation in the pitch on the maximum bending stresses after loading trapezoidal Si nanobeams ($w_1 = 200$ nm, $w_2 = 370$ nm, $t = 255$ nm, $\ell = 6\,\mu$m, $E = 169$ GPa, $\nu = 0.28$, load $= 70\,\mu$N). Also shown is the effect of load when applied at the center of the beam and at the center of the scratch near the end [47]

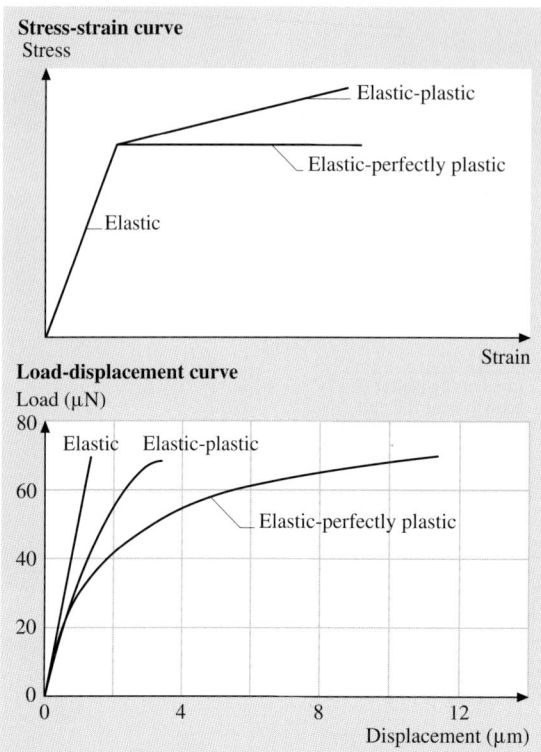

Fig. 15.24. Schematic representation of stress-strain curves and load-displacement curves for material when it is elastic, elastic-plastic, or elastic-perfectly plastic for a Si nanobeam ($w_1 = 200$ nm, $w_2 = 370$ nm, $t = 255$ nm, $\ell = 6$ μm, $E = 169$ GPa, tangent modulus in plastic range $= 0.5$ E, $\nu = 0.28$) [47]

stress at the scratch near the end and the stress starts going down. Such observations of maximum bending stresses can help in identifying the number of asperities and scratches allowed separated by an optimum distance from each other.

15.4.4 Effect on Stresses and Displacements for Materials That Are Elastic, Elastic-Plastic, or Elastic-Perfectly Plastic

This section deals with the beam modeled as elastic, elastic-plastic, and elastic-perfectly plastic to observe the variation in the stresses and displacements from an elastic model used so far [47]. Figure 15.24 shows the typical stress-strain curves for the three types of deformation regimes and their corresponding load-displacement curves obtained from the model of an Si nanobeam that are found to exhibit the same trends.

Table 15.3 shows the comparison of maximum von Mises stress and maximum displacements for both smooth nanobeam and nanobeam with a defined roughness,

Table 15.3. Stresses and displacements for materials that are elastic, elastic-plastic, or elastic-perfectly plastic (load = 70 μN, w_1 = 200 nm, w_2 = 370 nm, t = 255 nm, ℓ = 6 μm, R = 25 nm, E = 169 GPa, tangent modulus in plastic range = $0.5E$, v = 0.28)

	Elastic		Elastic-plastic		Elastic-perfectly plastic	
	Smooth nanobeam	Single semicircular longitudinal asperity	Smooth nano-beam	Single semicircular longitudinal asperity	Smooth nano-beam	Single semicircular longitudinal asperity
Max. von Mises stress (GPa)	18.2	19.3	13.5	15.2	7.8	9.1
Max. displacement (μm)	1.34	1.40	3.35	3.65	11.5	12.3

which is single semicircular longitudinal asperity of R value equal to 25 nm for the three different models. It is observed that the maximum value of stress is obtained at a given load for elastic material, whereas the displacement is maximum for elastic-perfectly plastic material. Also, the pattern that the maximum bending stress value increases for a rough nanobeam still holds true in the other models as well.

15.5 Closure

Mechanical properties of nanostructures are necessary in designing realistic MEMS/NEMS devices. Most mechanical properties are scale-dependent. Micro/nanomechanical properties, hardness, elastic modulus, and scratch resistance of bulk materials of undoped single-crystal silicon (Si) and thin films of undoped polysilicon, SiO_2, SiC, Ni-P, and Au are presented. It is found that the SiC film exhibits higher hardness, elastic modulus, and scratch resistance as compared to other materials.

Bending tests have been performed on the Si and SiO_2 nanobeams and Ni-P and Au microbeams using an AFM and a depth-sensing nanoindenter, respectively. The bending tests were used to evaluate elastic modulus, bending strength (fracture stress), fracture toughness (K_{IC}), and fatigue strength of the beam materials. The Si and SiO_2 nanobeams exhibited elastic linear response with sudden brittle fracture. The notched Ni-P beam showed linear deformation behavior followed by abrupt failure. The Au beam showed elastic-plastic deformation behavior. Elastic modulus values of 182 ± 11 GPa for Si(110) and 85 ± 3 GPa for SiO_2 were obtained, which are comparable to bulk values. Bending strength values of 18 ± 3 GPa for Si and 7.6 ± 2 GPa for SiO_2 were obtained, which are twice as large as values reported on larger scale specimens. This indicates that bending strength shows a specimen size dependence. Fracture toughness value estimates obtained were 1.67 ± 0.4 MPa $\sqrt{m}$ for Si and 0.60 ± 0.2 MPa $\sqrt{m}$ for SiO_2, which are also comparable to values obtained on larger specimens. At stress amplitudes less than 15% of their bending strength and at mean stresses of less than 30% of the bending strength, Si and SiO_2

displayed an apparent endurance life of greater than 30 000 cycles. SEM observations of the fracture surfaces revealed a cleavage type of fracture for both materials when subjected to bending, as well as fatigue. The AFM and nanoindenters used in this study can be satisfactorily used to evaluate the mechanical properties of micro/nanoscale structures for use in MEMS/NEMS.

FEM simulations are used to predict the stress and deformation in nanostructures. The FEM has been used to analyze the effect of the type of surface roughness and scratches on stresses and deformation of nanostructures. We find that roughness affects the maximum bending stresses. The maximum bending stresses increase as the asperity number increases for both semicircular and grooved asperities in the longitudinal direction. When the semicircular asperity is present in the transverse direction the maximum tensile stress is much larger than the maximum tensile stress value with no asperity or when the semicircular asperity is present in the longitudinal direction. This observation suggests that the asperity in the transverse direction is more detrimental. The presence of scratches increases the maximum tensile stress. The maximum tensile stress remains almost the same with the number of scratches for two types of loading, that is, when load is applied at the center of the beam or at the center of the scratch near the end, although the value of the tensile stress is much lower when the load is applied at the center of the scratch. This means that the load applied at the ends is less damaging. This analysis shows that FEM simulations can be useful to designers to develop the most suitable geometry for nanostructures.

References

1. R. S. Muller, R. T. Howe, S. D. Senturia, R. L. Smith, and R. M. White. *Microsensors.* IEEE, 1990.
2. I. Fujimasa. *Micromachines: A New Era in Mechanical Engineering.* Oxford Univ. Press, 1996.
3. W. S. Trimmer, editor. *Micromachines and MEMS, Classic and Seminal Papers to 1990.* IEEE, 1997.
4. B. Bhushan. *Tribology Issues and Opportunities in MEMS.* Kluwer, 1998.
5. G. T. A. Kovacs. *Micromachined Transducers Sourcebook.* WCB McGraw-Hill, 1998.
6. S. D. Senturia. *Microsystem Design.* Kluwer, 2001.
7. M. Gad-el-Hak. *The MEMS Handbook.* CRC, 2002.
8. T. R. Hsu. *MEMS and Microsystems.* McGraw-Hill, 2002.
9. M. Madou. *Fundamentals of Microfabrication: The Science of Miniaturization.* CRC, 2nd edition, 2002.
10. K. E. Drexler. *Nanosystems: Molecular Machinery, Manufacturing and Computation.* Wiley, 1992.
11. B. Bhushan. *Handbook of Micro/Nanotribology.* CRC, 2nd edition, 1999.
12. G. Timp, editor. *Nanotechnology.* Springer, 1999.
13. E. A. Rietman. *Molecular Engineering of Nanosystems.* Springer, 2001.
14. H. S. Nalwa, editor. *Nanostructures Materials and Nanotechnology.* Academic, 2002.
15. W. A. Goddard, D. W. Brenner, S. E. Lyshevski, and G. J. Iafrate. *Handbook of Nanoscience, Engineering, and Technology.* CRC, 2003.
16. B. Bhushan. *Principles and Applications of Tribology.* Wiley, 1999.

17. B. Bhushan. *Introduction to Tribology*. Wiley, 2002.
18. B. Bhushan. *Macro- and microtribology of MEMS materials*, pages 1515–1548. CRC, 2001.
19. S. Johansson, J. A. Schweitz, L. Tenerz, and J. Tiren. Fracture testing of silicon microelements in-situ in a scanning electron microscope. *J. Appl. Phys.*, 63:4799–4803, 1988.
20. F. Ericson and J. A. Schweitz. Micromechanical fracture strength of silicon. *J. Appl. Phys.*, 68:5840–5844, 1990.
21. E. Obermeier. Mechanical and thermophysical properties of thin film materials for mems: Techniques and devices. *Micromechan. Struct. Mater. Res. Symp. Proc.*, 444:39–57, 1996.
22. C. J. Wilson, A. Ormeggi, and M. Narbutovskih. Fracture testing of silicon microcantilever beams. *J. Appl. Phys.*, 79:2386–2393, 1995.
23. W. N. Sharpe, Jr., B. Yuan, and R. L. Edwards. A new technique for measuring the mechanical properties of thin films. *J. Microelectromech. Syst.*, 6:193–199, 1997.
24. K. Sato, T. Yoshioka, T. Anso, M. Shikida, and T. Kawabata. Tensile testing of silicon film having different crystallographic orientations carried out on a silicon chip. *Sens. Actuators A*, 70:148–152, 1998.
25. S. Greek, F. Ericson, S. Johansson, M. Furtsch, and A. Rump. Mechanical characterization of thick polysilicon films: Young's modulus and fracture strength evaluated with microstructures. *J. Micromech. Microeng.*, 9:245–251, 1999.
26. D. A. LaVan and T. E. Buchheit. Strength of polysilicon for mems devices. *Proc. SPIE*, 3880:40–44, 1999.
27. E. Mazza and J. Dual. Mechanical behavior of a μm-sized single crystal silicon structure with sharp notches. *J. Mech. Phys. Solids*, 47:1795–1821, 1999.
28. T. Yi and C. J. Kim. Measurement of mechanical properties for mems materials. *Meas. Sci. Technol.*, 10:706–716, 1999.
29. H. Kahn, M. A. Huff, and A. H. Heuer. Heating effects on the young's modulus of films sputtered onto micromachined resonators. *Microelectromech. Struct. Mater. Res. Symp. Proc.*, 518:33–38, 1998.
30. S. Johansson, F. Ericson, and J. A. Schweitz. Influence of surface-coatings on elasticity, residual-stresses, and fracture properties of silicon microelements. *J. Appl. Phys.*, 65:122–128, 1989.
31. R. Ballarini, R. L. Mullen, Y. Yin, H. Kahn, S. Stemmer, and A. H. Heuer. The fracture toughness of polysilicon microdevices: A first report. *J. Mater. Res.*, 12:915–922, 1997.
32. H. Kahn, R. Ballarini, R. L. Mullen, and A. H. Heuer. Electrostatically actuated failure of microfabricated polysilicon fracture mechanics specimens. *Proc. R. Soc. London A*, 455:3807–3823, 1999.
33. A. M. Fitzgerald, R. H. Dauskardt, and T. W. Kenny. Fracture toughness and crack growth phenomena of plasma-etched single crystal silicon. *Sens. Actuators A*, 83:194–199, 2000.
34. T. Tsuchiya, A. Inoue, and J. Sakata. Tensile testing of insulating thin films: Humidity effect on tensile strength of SiO_2 films. *Sens. Actuators A*, 82:286–290, 2000.
35. J. A. Connally and S. B. Brown. Micromechanical fatigue testing. *Exp. Mech.*, 33:81–90, 1993.
36. K. Komai, K. Minoshima, and S. Inoue. Fracture and fatigue behavior of single-crystal silicon microelements and nanoscopic afm damage evaluation. *Microsyst. Technol.*, 5:30–37, 1998.
37. T. Namazu, Y. Isono, and T. Tanaka. Evaluation of size effect on mechanical properties of single-crystal silicon by nanoscale bending test using afm. *J. Microelectromech. Syst.*, 9:450–459, 2000.
38. S. Sundararajan and B. Bhushan. Development of afm-based techniques to measure mechanical properties of nanoscale structures. *Sens. Actuators A*, 101:338–351, 2002.

39. X. Li and B. Bhushan. Fatigue studies of nanoscale structures for mems/nems applications using nanoindentation techniques. *Surf. Coat. Technol.*, 163-164:521–526, 2003.
40. X. Li, B. Bhushan, K. Takashima, C. W. Baek, and Y. K. Kim. Mechanical characterization of micro/nanoscale structures for mems/nems applications using nanoindentation techniques. *Ultramicroscopy*, 97:481–494, 2003.
41. T. Hsu and N. Sun. Residual stresses/strains analysis of mems, 1998.
42. A. Kolpekwar, C. Kellen, and R. D. (Shawn) Blanton. Fault model generation for mems, 1998.
43. H. A. Rueda and M. E. Law. Modeling of strain in boron-doped silicon cantilevers, 1998.
44. M. Heinzelmann and M. Petzold. Fem analysis of microbeam bending experiments using ultra-micro indentation. *Comput. Mater. Sci.*, 3:169–176, 1994.
45. C. J. Wilson and P. A. Beck. Fracture testing of bulk silicon microcantilever beams subjected to a side load. *J. Microelectromech. Syst.*, 5:142–150, 1996.
46. B. Bhushan and G. B. Agrawal. Stress analysis of nanostructures using a finite element method. *Nanotechnology*, 13:515–523, 2002.
47. B. Bhushan and G. B. Agrawal. Finite element analysis of nanostructures with roughness and scratches. *Ultramicroscopy*, 97:495–501, 2003.
48. K. E. Petersen. Silicon as a mechanical material. *Proc. IEEE*, 70:420–457, 1982.
49. B. Bhushan, S. Sundararajan, X. Li, C. A. Zorman, and M. Mehregany. *Micro/nanotribological studies of single-crystal silicon and polysilicon and SiC films for use in MEMS devices*, pages 407–430. Kluwer, 1998.
50. S. Sundararajan and B. Bhushan. Micro/nanotribological studies of polysilicon and SiC films for mems applications. *Wear*, 217:251–261, 1998.
51. X. Li and B. Bhushan. Micro/nanomechanical characterization of ceramic films for microdevices. *Thin Solid Films*, 340:210–217, 1999.
52. B. Bhushan and X. Li. Nanomechanical characterization of solid surfaces and thin films. *Int. Mater. Rev.*, 48:125–164, 2003.
53. B. R. Lawn, A. G. Evans, and D. B. Marshall. Elastic/plastic indentation damage in ceramics: The median/radial system. *J. Am. Ceram. Soc.*, 63:574, 1980.
54. S. Sundararajan, B. Bhushan, T. Namazu, and Y. Isono. Mechanical property measurements of nanoscale structures using an atomic force microscope. *Ultramicroscopy*, 91:111–118, 2002.
55. R. J. Roark. *Formulas for Stress and Strain*. McGraw-Hill, 6th edition, 1989.
56. R. W. Hertzberg. *Deformation and Fracture Mechanics of Engineering Materials*. Wiley, 3rd edition, 1989.
57. Anonymous. Properties of silicon. *EMIS Datareviews Series No. 4*, 1988.
58. C. T. C. Nguyen and R. T. Howe. An integrated cmos micromechanical resonator high-q oscillator. *IEEE J. Solid State Circ.*, 34:440–455, 1999.
59. L. J. Hornbeck. A digital light processing^{TM} update – status and future applications. *Proc. SPIE*, 3634:158–170, 1999.
60. M. Tanaka. Fracture toughness and crack morphology in indentation fracture of brittle materials. *J. Mater. Sci.*, 31:749, 1996.
61. B. Bhushan and S. Venkatesan. Mechanical and tribological properties of silicon for micromechanical applications: A review. *Adv. Info. Storage Syst.*, 5:211–239, 1993.
62. B. Bhushan and B. K. Gupta. *Handbook of Tribology: Materials, Coatings, and Surface Treatments*. McGraw-Hill, 1991.
63. T. Tsuchiya, O. Tabata, J. Sakata, and Y. Taga. Specimen size effect on tensile strength of surface-micromachined polycrystalline silicon thin films. *J. Microelectromech. Syst.*, 7:106–113, 1998.

64. T. Yi, L. Li, and C. J. Kim. Microscale material testing of single crystalline silicon: Process effects on surface morphology and tensile strength. *Sens. Actuators A*, 83:172–178, 2000.
65. S. P. Timoshenko and J. N. Goodier. *Theory of Elasticity*. McGraw-Hill, 3rd edition, 1970.
66. J. E. Shigley and L. D. Mitchell. *Mechanical Engineering Design*. McGraw-Hill, 4th edition, 1993.

16

Scale Effect in Mechanical Properties and Tribology

Bharat Bhushan and Michael Nosonovsky

Summary. A model, which explains scale effects in mechanical properties and tribology is presented. Mechanical properties are scale dependent based on the strain gradient plasticity and the effect of dislocation-assisted sliding. Both single asperity and multiple asperity contacts are considered. The relevant scaling length is the nominal contact length – contact diameter for a single-asperity contact, and scan length for multiple-asperity contacts. For multiple asperity contacts, based on an empirical power-rule for scale dependence of roughness, contact parameters are calculated. The effect of load on the contact parameters and the coefficient of friction is also considered. During sliding, adhesion and two- and three-body deformation, as well as ratchet mechanism, contribute to the dry friction force. These components of the friction force depend on the relevant real areas of contact (dependent on roughness and mechanical properties), average asperity slope, number of trapped particles, and shear strength during sliding. Scale dependence of the components of the coefficient of friction is studied. A scale dependent transition index, which is responsible for transition from predominantly elastic adhesion to plastic deformation has been proposed. Scale dependence of the wet friction, wear, and interface temperature has been also analyzed. The proposed model is used to explain the trends in the experimental data for various materials at nanoscale and microscale, which indicate that nanoscale values of coefficient of friction are lower than the microscale values due to an increase of the three-body deformation and transition from elastic adhesive contact to plastic deformation.

16.1 Nomenclature

$a, \bar{a}, \bar{a}_0, a_{\max}, \bar{a}_{\max}$: Contact radius, mean contact radius, macroscale value of mean contact radius, maximum contact radius, mean value of maximum contact radius.
$A_a, A_r, A_{ra}, A_{re}, A_{re0}, A_{rp}, A_{rp0}, A_{ds}, A_{dp}$: Apparent area of contact, real area of contact, real area of contact during adhesion, real area of elastic contact, macroscale value of real area of elastic contact, real area of plastic contact, macroscale value of real area of plastic contact, real area of contact during asperity summit deformation, area of contact with particles

b: Burgers vector

c: Constant, specified by crystal structure
C_0: Constant required for normalization of $p(d)$
$d, d_e, d_n, d_{ln}, \bar{d}, \bar{d}_0$: Particle diameter, minimum for exponential distribution, mean for normal distribution, exponential of mean of $\ln(d)$ for log-normal distribution, mean trapped particles diameter, macroscale value of mean trapped particles diameter
D: Interface zone thickness
E_1, E_2, E^*: Elastic moduli of contacting bodies, effective elastic modulus
$F, F_a, F_d, F_{ae}, F_{ap}, F_a, F_{ds}, F_{dp}, F_m, F_{m0}$: Friction force, friction force due to adhesion, friction force due to deformation, friction force during elastic adhesional contact, plastic adhesional contact, summit deformation, particles deformation respectively, meniscus force for wet contact, macroscale value of meniscus force

G: Elastic shear modulus
h: Indentation depth
h_f: Liquid film thickness
H, H_0: Hardness, hardness in absence of strain gradient
k, k_0: Wear coefficient, macroscale value of wear coefficient
l_s, l_d: Material-specific characteristic length parameters
$L, L_{lwl}, L_{lc}, L_s, L_d$: Length of the nominal contact zone, long wavelength limit for roughness parameters, long wavelength limit for contact parameters, length parameters related to l_s and l_d
L_p: Peclet number
m, n: Indices of exponents for scale-dependence of σ and β^*
n_{tr}: Number of trapped particles divided by the total number of particles
p_a, p_{ac}: Apparent pressure, critical apparent pressure
$p(d), p_{tr}(d)$: Probability density function for particle size distribution, probability density function for trapped particle size distribution
$P(d)$: Cumulative probability distribution for particle size
$R, R_p, \bar{R}_p, \bar{R}_{p0}$: Effective radius of summit tips, radius of summit tip, mean radius of summit tips, macroscale value of the mean radius of summit tips
$R(\tau)$: Autocorrelation function
s: Spacing between slip steps on the indentation surface
s_d: Separation distance between reference planes of two surfaces in contact
N, N_0: Total number of contacts, macroscale value of total number of contacts
T, T_0: Maximum flash temperature rise, macroscale value of temperature rise
x: Sliding distance
v: Volume of worn material
V: Sliding velocity
W: Normal load
z, z_{min}, z_{max}: Random variable, minimum and maximum value of z
α: Probability for a particle in the border zone to leave the contact region
β^*, β_0^*: Correlation length, macroscale value of correlation length
γ: Surface tension
Γ: Gamma function

ε: Strain

η: Density of particles per apparent area of contact

η_{int}, η_{cr}: Density of dislocation lines per interface area, critical density of dislocation lines per interface area

κ: Curvature

κ_t: Thermal diffusivity

θ: Contact angle between the liquid and surface

θ_i: Indentation angle

θ_r: Roughness angle

μ, μ_a, μ_{ae}, μ_{ae0}, μ_{ap}, μ_{ap0}, μ_d, μ_{ds}, μ_{ds0}, μ_{dp}, μ_{dp0}, μ_r, μ_{r0}, μ_{re}, μ_{re0}, μ_{rp}, μ_{rp0}, μ_{wet}: Coefficient of friction, coefficient of adhesional friction, coefficient of adhesional elastic friction, macroscale value of coefficient of adhesional elastic friction, coefficient of adhesional plastic friction, macroscale value of coefficient of adhesional plastic friction, coefficient of deformation friction, coefficient of summits deformation friction, macroscale value of coefficient of summits deformation friction, coefficient of particles deformation friction, macroscale value of coefficient of particles deformation friction, ratchet component of the coefficient of friction, macroscale value of ratchet component of the coefficient of friction, ratchet component of the coefficient of elastic friction, macroscale value of ratchet component of the coefficient of elastic friction, ratchet component of the coefficient of plastic friction, macroscale value of ratchet component of the coefficient of plastic friction, and coefficient of wet friction

v_1, v_2: Poisson's ratios of contacting bodies

ρc_p: Volumetric specific heat

σ, σ_0, σ_e, σ_n, σ_{ln}: Standard deviation of rough surface profile height, macroscale value of standard deviation of rough surface profile height, standard deviation for the exponential distribution, standard deviation for the normal distributions, standard deviation for ln(d) of the log normal distribution

ρ, ρ_G, ρ_S: Total density of dislocation lines per volume, density of GND per volume, density of SSD per volume

ϕ, ϕ_0: Transition index, macroscale value of transition index

τ, τ_0: Spatial parameter, value at which the autocorrelation function decays

τ_a, τ_{a0}, τ_Y, τ_{Y0}, τ_{ds}, τ_{ds0}, τ_{dp}, τ_{dp0}, τ_p: Adhesional shear strength during sliding, macroscale value of adhesional shear strength, shear yield strength, shear yield strength in absence of strain gradient, shear strength during summits deformation, macroscale value of shear strength during summits deformation, shear strength during particles deformation, macroscale value of shear strength during particles deformation, Peierls stress

16.2 Introduction

Microscale and nanoscale measurements of tribological properties, which became possible due to the development of the Surface Force Apparatus (SFA), Atomic Force

Microscope (AFM), and Friction Force Microscope (FFM) demonstrate scale dependence of adhesion, friction, and wear as well as mechanical properties including hardness [1–4]. Advances of micro/nanoelectromechanical systems (MEMS/NEMS) technology in the past decade make understanding of scale effects in adhesion, friction, and wear especially important, since surface to volume ratio grows with miniaturization and surface phenomena dominate. Dimensions of MEMS/NEMS devices range from about 1 mm to few nm.

Experimental studies of scale dependence of tribological phenomena have been conducted recently. AFM experiments provide data on nanoscale [5–10] whereas microtriboapparatus [11, 12] and SFA [13] provide data on microscale. Experimental data indicate that wear mechanisms and wear rates are different at macro- and micro/nanoscales [14, 15]. During sliding, the effect of operating conditions such as load and velocity on friction and wear are frequently manifestations of the effect of temperature rise on the variable under study. The overall interface temperature rise is a cumulative result of numerous flash temperature rises at individual asperity contacts. The temperature rise at each contact is expected to be scale dependent, since it depends on contact size, which is scale dependent.

Friction is a complex phenomenon, which involves asperity interactions involving adhesion and deformation (plowing), Fig. 16.1. Adhesion and plastic deformation imply energy dissipation, which is responsible for friction. A contact between two bodies takes place on high asperities, and the real area of contact (A_r) is a small fraction of the apparent area of contact [16]. During contact of two asperities, a lateral force may be required for asperities of a given slope to climb against each other. This mechanism is known as ratchet mechanism, and it also contributes to the friction. Wear and contaminant particles present at the interface, referred as the "third body", also contribute to friction, Fig. 16.2a. In addition, during contact, even at low humidity, a meniscus is formed. Generally, any liquid that wets or has a small contact angle on surfaces will condense from vapor into cracks and pores on surfaces as bulk liquid and in the form of annular-shaped capillary condensate in the contact zone. Figure 16.2b shows a random rough surface in contact with a smooth surface with a continuous liquid film on the smooth surface. The presence of the liquid film of the condensate or pre-existing film of the liquid can significantly increase the adhesion between the solid bodies [16]. The effect of meniscus is scale-dependent.

A quantitative theory of scale effects in friction should consider scale effect on physical properties relevant to these contributions. However, conventional theories of contact and friction lack characteristic length parameters, which would be responsible for scale effects. The linear elasticity and conventional plasticity theories are scale-invariant and do not include any material length scales. A strain gradient plasticity theory has been developed, for microscale deformations, by *Fleck* et al. [17], *Nix* and *Gao* [18] and *Hutchinson* [19]. Their theory predicts a dependence of mechanical properties on the strain gradient, which is scale dependent: the smaller is the size of the deformed region, the greater is the gradient of plastic strain, and, the greater is the yield strength and hardness.

A comprehensive model of scale effect in friction including adhesion, two- and three-body deformations and the ratchet mechanism, has recently been proposed by

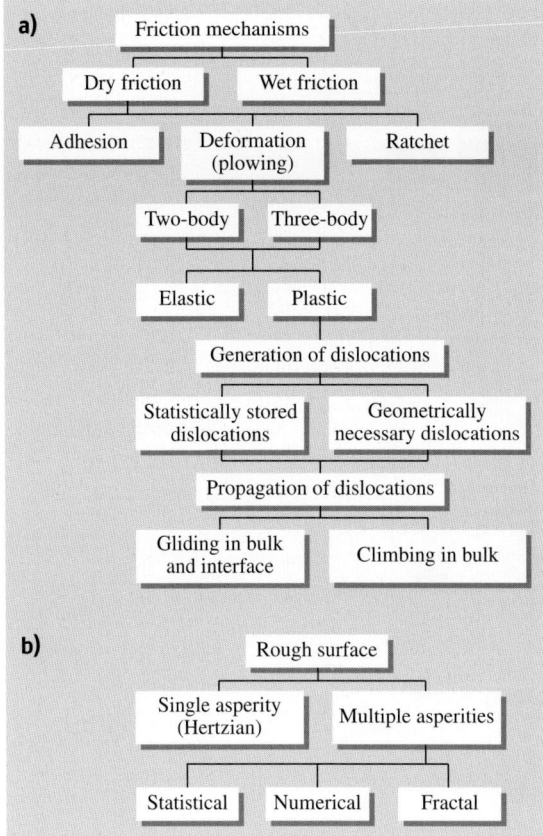

Fig. 16.1. (a) A block diagram showing friction mechanisms and generation and propagation of dislocations during sliding, (b) a block diagram of rough contact models

Bhushan and *Nosonovsky* [20–22] and *Nosonovsky* and *Bhushan* [23]. The model for adhesional friction during single and multiple asperity contact was developed by *Bhushan* and *Nosonovsky* [20] and is based on the strain gradient plasticity and dislocation assisted sliding (gliding dislocations at the interface or microslip). The model for the two-body and three-body deformation was proposed by *Bhushan* and *Nosonovsky* [21] and for the ratchet mechanism by *Nosonovsky* and *Bhushan* [23]. The model has been extended for wet contacts, wear and interface temperature by *Bhushan* and *Nosonovsky* [22]. The detailed model is presented in this chapter.

The chapter is organized as follows. In the next section of this chapter, the scale effect in mechanical properties is considered, including yield strength and hardness based on the strain gradient plasticity and shear strength at the interface based on the dislocation assisted sliding (microslip). In the fourth section, scale effect in surface roughness and contact parameters is considered, including the real area of contact, number of contacts, and mean size of contact. Load dependence of contact para-

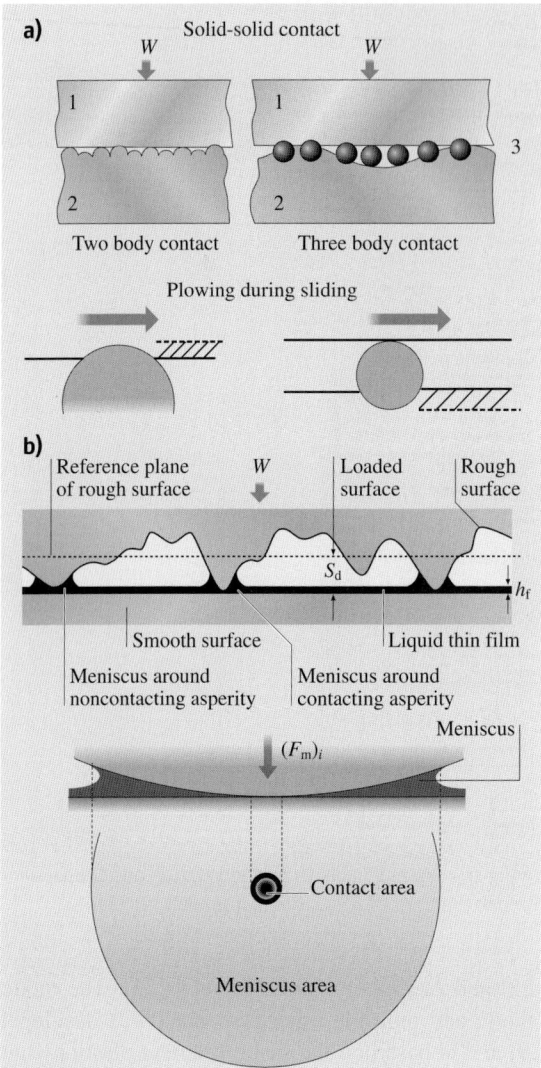

Fig. 16.2. Schematics of (**a**) two-bodies and three-bodies during dry contact of rough surfaces, (**b**) formation of menisci during wet contact

meters is also studied in this section. In the fifth section, scale effect in friction is considered, including adhesion, two- and three-body deformation, ratchet mechanism, meniscus analysis, total value of the coefficient of friction and comparison with the experimental data. In the sixth and seventh sections, scale effects in wear and interface temperature are analyzed, respectively.

16.3 Scale Effect in Mechanical Properties

In this section, scale dependence of hardness and shear strength at the interface is considered. A strain gradient plasticity theory has been developed, for microscale deformations, by *Fleck* et al. [17], *Nix* and *Gao* [18], *Hutchinson* [19], and others, which is based on statistically stored and geometrically necessary dislocations (to be described later). Their theory predicts a dependence of mechanical properties on the strain gradient, which is scale dependent: the smaller is the size of the deformed region, the greater is the gradient of plastic strain, and, the greater is the yield strength and hardness. *Gao* et al. [24] and *Huang* et al. [25] proposed a mechanism-based strain gradient (MSG) plasticity theory, which is based on a multiscale framework, linking the microscale (10–100 nm) notion of statistically stored and geometrically necessary dislocations to the mesoscale (1–10 µm) notion of plastic strain and strain gradient. *Bazant* [26] analyzed scale effect based on the MSG plasticity theory in the limit of small scale, and found that corresponding nominal stresses in geometrically similar structures of different sizes depend on the size according to a power exponent law.

It was recently suggested also, that relative motion of two contacting bodies during sliding takes place due to dislocation-assisted sliding (microslip), which results in scale-dependent shear strength at the interface [20]. Scale effects in mechanical properties (yield strength, hardness, and shear strength at the interface) based on the strain gradient plasticity and dislocation-assisted sliding models are considered in this section.

16.3.1 Yield Strength and Hardness

Plastic deformation occurs during asperity contacts because a small real area of contact results in high contact stresses, which are often beyond the limits of the elasticity. As stated earlier, during loading, generation and propagation of dislocations is responsible for plastic deformation. Because dislocation motion is irreversible, plastic deformation provides a mechanism for energy dissipation during friction. The strain gradient plasticity theories [17–19] consider two types of dislocations: randomly created Statistically Stored Dislocations (SSD) and Geometrically Necessary Dislocations (GND). The GND are required for strain compatibility reasons. Randomly created SSD during shear and GND during bending are presented in Fig. 16.3a. The density of the GND (total length of dislocation lines per volume) during bending is proportional to the curvature κ and to the strain gradient

$$\rho_G = \frac{\kappa}{b} = \frac{1}{b}\frac{\partial \varepsilon}{\partial z} \propto \nabla \varepsilon, \qquad (16.1)$$

where ε is strain, b is the Burgers vector, and $\nabla \varepsilon$ is the strain gradient.

The GND during indentation, Fig. 16.3b, are located in a certain sub-surface volume. The large strain gradients in small indentations require GND to account for the large slope at the indented surface. SSD, not shown here, also would be created and would contribute to deformation resistance, and are function of strain rather

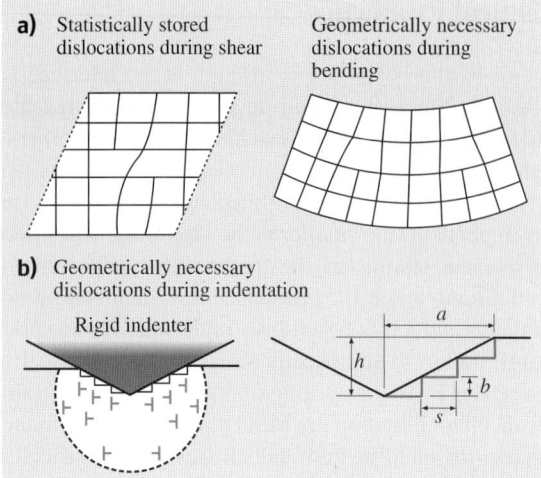

Fig. 16.3. (**a**) Illustration of statistically stored dislocations during shear and geometrically necessary dislocations during bending, (**b**) geometrically necessary dislocations during indentation

than strain gradient. According to *Nix* and *Gao* [18], we assume that indentation is accommodated by circular loops of GND with Burgers vector normal to the plane of the surface. If we think of the individual dislocation loops being spaced equally along the surface of the indentation, then the surface slope

$$\tan\theta_i = \frac{h}{a} = \frac{b}{s},\qquad(16.2)$$

where θ_i is the angle between the surface of the conical indenter and the plane of the surface, a is the contact radius, h is the indentation depth, b is the Burgers vector, and s is the spacing between individual slip steps on the indentation surface (Fig. 16.3b). They reported that for geometrical (strain compatibility) considerations, the density of the GND is

$$\rho_G = \frac{3}{2bh}\tan^2\theta_i = \frac{3}{2b}\left(\frac{\tan\theta_i}{a}\right) = \frac{3}{2b}\nabla\varepsilon.\qquad(16.3)$$

Thus ρ_G is proportional to strain gradient (scale dependent) whereas the density of SSD, ρ_S is dependent upon the average strain in the indentation, which is related to the slope of the indenter ($\tan\theta_i$). Based on experimental observations, ρ_S is approximately proportional to strain [17].

According to the Taylor model of plasticity [27], dislocations are emitted from Frank–Read sources. Due to interaction with each other, the dislocations may become stuck in what is called the Taylor network, but when externally applied stress reaches the order of Peierls stress for the dislocations, they start to move and the plastic yield is initiated. The magnitude of the Peierls stress τ_p is proportional to the dislocation's Burgers vector b divided by a distance between dislocation lines s [27, 28]

$$\tau_p = Gb/(2\pi s), \tag{16.4}$$

where G is the elastic shear modulus. An approximate relation of the shear yield strength τ_Y to the dislocations density at a moment when yield is initiated is given by [27]

$$\tau_{Y0} = cGb/s = cGb\sqrt{\rho}, \tag{16.5}$$

where c is a constant on the order of unity, specified by the crystal structure and ρ is the total length of dislocation lines per volume, which is a complicated function of strain ε and strain gradient ($\nabla\varepsilon$)

$$\rho = \rho_S(\varepsilon) + \rho_G(\nabla\varepsilon). \tag{16.6}$$

The shear yield strength τ_Y can be written now as a function of SSD and GND densities [27]

$$\tau_Y = cGb\sqrt{\rho_S + \rho_G} = \tau_{Y0}\sqrt{1 + (\rho_G/\rho_S)}, \tag{16.7}$$

where

$$\tau_{Y0} = cGb\sqrt{\rho_S} \tag{16.8}$$

is the shear yield strength value in the limit of small ρ_G/ρ_S ratio (large scale) that would arise from the SSD, in the absence of GND. Note that the ratio of the two densities is defined by the problem geometry and is scale dependent. Based on the relationships for ρ_G (16.3) and ρ_S, the ratio ρ_G/ρ_S is inversely proportional to a and (16.7) reduces to

$$\tau_Y = \tau_{Y0}\sqrt{1 + (l_d/a)}, \tag{16.9}$$

where l_d is a plastic deformation length that characterizes depth dependence on shear yield strength. According to *Hutchinson* [19], this length is physically related to an average distance a dislocation travels, which was experimentally determined to be between 0.2 μm and 5 μm for copper and nickel. Note that l_d is a function of the material and the asperity geometry and is dependent on SSD.

Using von Mises yield criterion, hardness $H = 3\sqrt{3}\tau_Y$. From (16.9) the hardness is also scale-dependent [18]

$$H = H_0\sqrt{1 + (l_d/a)}, \tag{16.10}$$

where H_0 is hardness in absence of strain gradient. Equation (16.9) provides dependence of the resistance force to deformation upon the scale in a general case of plastic deformation [20].

Scale dependence of yield strength and hardness has been well established experimentally. *Bhushan* and *Koinkar* [29] and *Bhushan* et al. [30] measured hardness of single-crystal silicon (100) up to a peak load of 500 μN. *Kulkarni* and *Bhushan* [31] measured hardness of single crystal aluminium (100) up to 2000 μN and *Nix* and

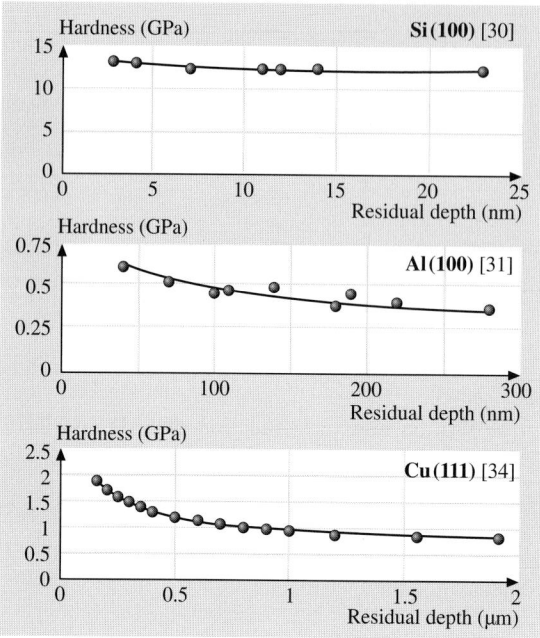

Fig. 16.4. Indentation hardness as a function of residual indentation depth for Si(100) [30], Al(100) [31], Cu(111) [34]

Gao [18] presented data for single crystal copper; using a three-sided pyramidal (Berkovich) diamond tip. The hardness on nanoscale is found to be higher than on microscale (Fig. 16.4). Similar results have been reported in other tests, including indentation tests for other materials [32–34], torsion and tension experiments on copper wires [17, 19], and bending experiments on silicon and silica beams [35].

16.3.2 Shear Strength at the Interface

Mechanism of slip involves motion of large number of dislocations, which is responsible for plastic deformation during sliding. Dislocations are generated and stored in the body and propagate under load. There are two modes of possible line (or edge) dislocation motion: gliding, when dislocation moves in the direction of its Burgers vector b by a unit step of its magnitude, and climbing, when dislocation moves in a direction, perpendicular to its Burgers vector (Fig. 16.5a). Motion of dislocations can take place in the bulk of the body or at the interface. Due to periodicity of the lattice, a gliding dislocation experiences a periodic force, known as the Peierls force [28]. The Peierls force is responsible for keeping the dislocation at a central position between symmetric lattice lines and it opposes dislocation's gliding (Fig. 16.5b). Therefore, an external force should be applied to overcome Peierls force resistance against dislocation's motion. *Weertman* [36] showed that a dislocation or a group of dislocations can glide uniformly along an interface between two

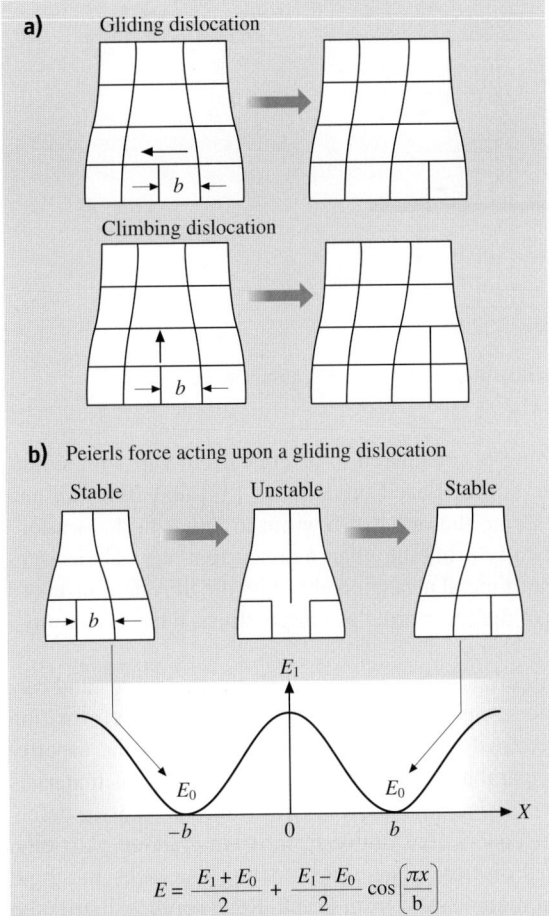

Fig. 16.5. (a) Schematics of gliding and climbing dislocations motion by a unit step of Burgers vector b. (b) Origin of the periodic force acting upon a gliding dislocation (Peierls force). Gliding dislocation passes locations of high and low potential energy

bodies of different elastic properties. In continuum elasticity formulation, this motion is equivalent to a propagating interface slip pulse, however the physical nature of this deformation is plastic, because dislocation motion is irreversible. The local plastic deformation can occur at the interface due to concentration of dislocations even in the predominantly elastic contacts. Gliding of a dislocation along the interface results in a relative displacement of the bodies for a distance equal to the Burgers vector of the dislocation, whereas a propagating set of dislocations effectively results in dislocation-assisted sliding, or microslip (Fig. 16.6).

Several types of microslip are known in the tribology literature [16], the dislocation-assisted sliding is one type of microslip, which propagates along the interface.

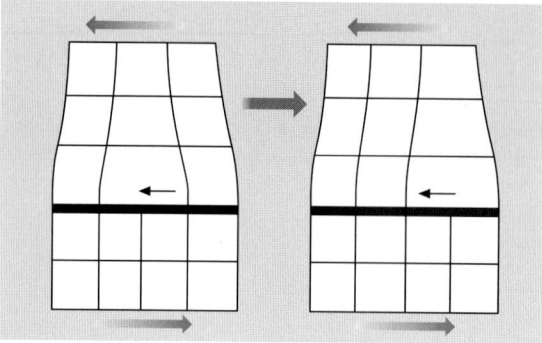

Fig. 16.6. Schematic showing microslip due to gliding dislocations at the interface

Conventional mechanism of sliding is considered to be concurrent slip with simultaneous breaking of all adhesive bonds. Based on *Johnson* [37] and *Bhushan* and *Nosonovsky* [20], for contact sizes on the order of few nm to few μm, dislocation-assisted sliding is more energetically profitable than a concurrent slip. Their argument is based on the fact that experimental measurements with the SFA demonstrated that, for mica, frictional stress is of the same order as Peierls stress, which is required for gliding of dislocations.

Polonsky and *Keer* [38] considered the pre-existing dislocation sources and carried out a numerical microcontact simulation based on contact plastic deformation representation in terms of discrete dislocations. They found that when the asperity size decreases and becomes comparable with the characteristic length of materials microstructure (distance between dislocation sources), resistance to plastic deformation increases, which supports conclusions drawn from strain gradient plasticity. *Deshpande* et al. [39] conducted discrete plasticity modeling of cracks in single crystals and considered dislocation nucleation from Frank–Read sources distributed randomly in the material. Pre-existing sources of dislocations, considered by all of these authors, are believed to be a more realistic reason for increasing number of dislocations during loading, rather than newly nucleated dislocations [27]. In general, dislocations are emitted under loads from pre-existing sources and propagate along slip lines (Fig. 16.7). As shown in the figure, in regions of higher loads, number of emitted dislocations is higher. Their approach was limited to numerical analysis of special cases.

Bhushan and *Nosonovsky* [20] considered a sliding contact between two bodies. Slip along the contact interface is an important special case of plastic deformation. The local dislocation-assisted microslip can exist even if the contact is predominantly elastic due to concentration of dislocations at the interface. Due to these dislocations, the stress at which yield occurs at the interface is lower than shear yield strength in the bulk. This means that average shear strength at the interface is lower than in the bulk.

An assumption that all dislocations produced by externally applied forces are distributed randomly throughout the volume would result in vanishing small proba-

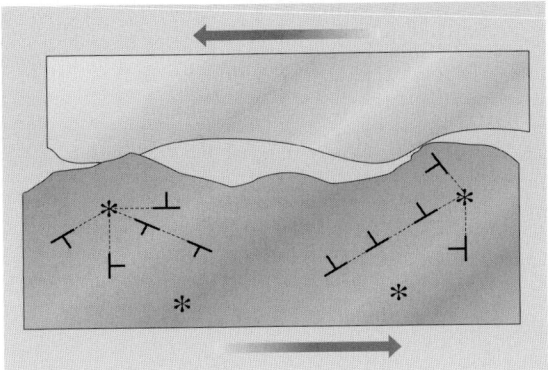

Fig. 16.7. Generation of dislocations from sources (∗) during plowing due to plastic deformation

bility for a dislocation to be exactly at the interface. However, many traveling (gliding and climbing) dislocations will be stuck at the interface as soon as they reach it. As a result of this, a certain number of dislocations will be located at the interface. In order to account for a finite dislocation density at the interface, *Bhushan* and *Nosonovsky* [20] assumed, that the interface zone has a finite thickness D. Dislocations within the interface zone may reach the contact surface due to climbing and contribute into the microslip. In the case of a small contact radius a, compared to interface zone thickness D, which is scale dependent, and is approximately equal to a. However, in the case of a large contact radius, the interface zone thickness is approximately equal to the average distance dislocations can climb l_s. An illustration of this is provided in Fig. 16.8. The depth of the subsurface volume, from which dislocations have a high chance to reach the interface is limited by l_s and by a, respectively, for the two cases considered here. Based on these geometrical considerations, an approximate relation can be written as

$$D = \frac{al_s}{l_s + a} . \tag{16.11}$$

The interface density of dislocations (total length of dislocation lines per interface area) is related to the volume density as

$$\eta_{\text{int}} = \rho D = \rho \left(\frac{al_s}{l_s + a} \right) . \tag{16.12}$$

During sliding, dislocations must be generated at the interface with a certain critical density $\eta_{\text{int}} = \eta_{\text{cr}}$. The corresponding shear strength during sliding can be written following (16.9) as

$$\tau_a = \tau_{a0} \sqrt{1 + (l_s/a)} , \tag{16.13}$$

where

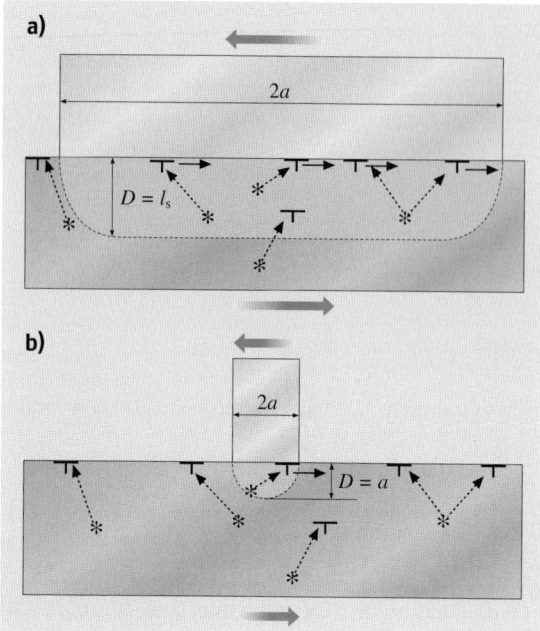

Fig. 16.8. Gliding dislocations at the interface generated from sources (∗). Only dislocations generated within the interface zone can reach the interface. (**a**) For a large contact radius a, thickness of this zone D is approximately equal to an average distance dislocations climb l_s. (**b**) For small contact radius a, the thickness of the interface zone is approximately equal to a

$$\tau_{a0} = cGb\sqrt{\frac{\eta_{cr}}{l_s}} \tag{16.14}$$

is the shear strength during sliding in the limit of $a \gg l_s$.

Equation (16.13) gives scale-dependence of the shear strength at the interface and is based on the following assumptions. First, it is assumed that only dislocations in the interface zone of thickness D, given by (16.11), contribute into sliding. Second, it is assumed, that a critical density of dislocations at the interface η_{cr} is required for sliding. Third, the shear strength is equal to the Peierls stress, which is related to the volume density of the dislocations $\rho = \eta/D$ according to (16.4), with the typical distance between dislocations $s = 1/\sqrt{\rho}$. It is noted, that proposed scaling rule for the dislocation assisted sliding mechanism (16.13) has a similar form to that for the yield strength (16.9), since both results are consequences of scale dependent generation and propagation of dislocations under load [20].

16.4 Scale Effect in Surface Roughness and Contact Parameters

During multiple-asperity contact, scale dependence of surface roughness is a factor which contributes to scale dependence of the real area of contact. Roughness parameters are known to be scale dependent [16], which results, during the contact of two bodies, in scale dependence of the real area of contact, number of contacts and mean contact size. The contact parameters also depend on the normal load, and the load dependence is similar to the scale dependence [23]. Both effects are analyzed in this section.

16.4.1 Scale Dependence of Roughness and Contact Parameters

A random rough surface with Gaussian height distribution is characterized by the standard deviation of surface height σ and the correlation length β^* [16]. The correlation length is a measure of how quickly a random event decays and it is equal to the length, over which the autocorrelation function drops to a small fraction of the value at the origin. The correlation length can be considered as a distance, at which two points on a surface have just reached the condition where they can be regarded as being statistically independent. Thus, σ is a measure of height distribution and β^* is a measure of spatial distribution.

A surface is composed of a large number of length scales of roughness that are superimposed on each other. According to AFM measurements on glass-ceramic disk surface, both σ and β^* initially increase with the scan size and then approach a constant value, at certain scan size (Fig. 16.9). This result suggests that disk roughness has a long wavelength limit, L_{lwl}, which is equal to the scan size at which the roughness values approach a constant value [16]. It can be assumed that σ and β^* depend on the scan size according to an empirical power rule

$$\sigma = \sigma_0 \left(\frac{L}{L_{lwl}}\right)^n, \qquad L < L_{lwl},$$

$$\beta^* = \beta_0^* \left(\frac{L}{L_{lwl}}\right)^m, \qquad L < L_{lwl}, \qquad (16.15)$$

where n and m are indices of corresponding exponents and σ_0 and β_0^* are macroscale values [20]. Based on the data, presented in Fig. 16.9, it is noted that for glass-ceramic disk, long-wavelength limit for σ and β^* is about 17 µm and 23 µm, respectively. The difference is expected to be due to measurement errors. An average value $L_{lwl} = 20$ µm is taken here for calculations. The values of the indices are found as $m = 0.5$, $n = 0.2$, and the macroscale values are $\sigma_0 = 5.3$ nm, $\beta_0^* = 0.37$ µm [23].

For two random surfaces in contact, the length of the nominal contact size L defines the characteristic length scale of the problem. The contact problem can be simplified by considering a rough surface with composite roughness parameters in contact with a flat surface. The composite roughness parameters σ and β^* can be obtained based on individual values for the two surfaces [16]. For Gaussian surfaces, the contact parameters of interest, to be discussed later, are the real area of contact A_r, number of contacts N, and mean contact radius $\bar{a}$. The long wavelength limit for scale dependence of the contact parameters L_{lc}, which is not necessarily equal to that of the roughness, L_{lwl}, will be used for normalization of length parameters. The scale dependence of the contact parameters exists if $L < L_{lc}$ [23].

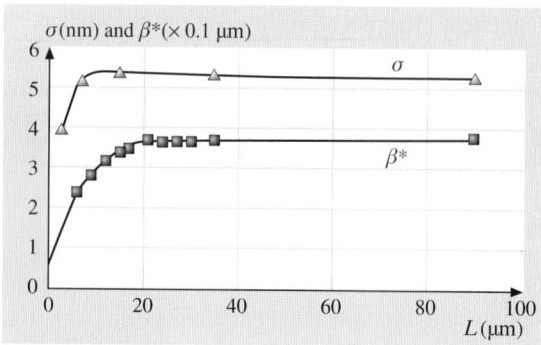

Fig. 16.9. Roughness parameters as a function of scan size for a glass-ceramic disk measured using AFM [16]

The mean of surface height distribution corresponds to so-called reference plane of the surface. Separation s_d is a distance between reference planes of two surfaces in contact, normalized by σ. For a given s_d and statistical distribution of surface heights, the total real area of contact (A_r), number of contacts (N), and elastic normal load W_e can be found, using statistical analysis of contacts. The real area of contact, number of contacts and elastic normal load are related to the separation distance s_d [40]

$$A_r \propto F_A(s_d),$$
$$N \propto \frac{1}{(\beta^*)^2} F_N(s_d),$$
$$W_e \propto \frac{E^*\sigma}{\beta^*} F_W(s_d), \qquad (16.16)$$

where $F_A(s_d)$, $F_N(s_d)$, and $F_W(s_d)$, are integral functions defined by *Onions* and *Archard* [40]. It should be noted, that A_r and N as a function of s_d are prescribed by the contact geometry (σ, β^*) and do not depend on whether the contact is elastic or plastic. Based on Onions and Archard data, it is observed that the ratio F_W/F_A is almost constant for moderate $s_d < 1.4$ and increases slightly for $s_d > 1.4$. The ratio F_A/F_N decreases rapidly with s_d and becomes almost constant for $s_d > 2.0$. For moderate loads, the contact is expected to occur on the upper parts of the asperities ($s_d > 2.0$), and a linear proportionality of $F_A(s_d)$, $F_N(s_d)$, and $F_W(s_d)$ can be assumed [20].

Based on (16.16) and the observation that F_W/F_A is almost constant, for moderate loads, A_{re} (the real area of elastic contact), N, and $\bar{a}$ are related to the roughness, based on the parameter L_{lc}, as

$$A_{\text{re}} \propto \frac{\beta^*}{\sigma E^*} W = A_{\text{re}0}\left(\frac{L}{L_{\text{lc}}}\right)^{m-n}, \qquad L < L_{\text{lc}}, \qquad (16.17)$$

$$N \propto \frac{W}{\beta^* \sigma E^*} = N_0 \left(\frac{L}{L_{\text{lc}}}\right)^{-m-n}, \qquad L < L_{\text{lc}}, \qquad (16.18)$$

$$\bar{a} \propto \beta^* = \sqrt{\frac{A_{\text{r}}}{N}} = \bar{a}_0 \left(\frac{L}{L_{\text{lc}}}\right)^m, \qquad L < L_{\text{lc}}. \qquad (16.19)$$

The mean radius of summit tips $\overline{R}_{\text{p}}$ is given, according to *Whitehouse* and *Archard* [41]

$$\overline{R}_{\text{p}} \propto \frac{(\beta^*)^2}{\sigma} = \overline{R}_{\text{p}0}\left(\frac{L}{L_{\text{lwl}}}\right)^{2m-n}, \qquad L < L_{\text{lwl}}, \qquad (16.20)$$

where $\bar{a}_0$, N_0 and $\overline{R}_{\text{p}0}$ are macroscale values, E^* is the effective elastic modulus of contacting bodies [22], which is related to the elastic moduli E_1, E_2 and Poisson's ratios ν_1, ν_2 of the two bodies as $1/E^* = (1 - \nu_1^2)/E_1 + (1 - \nu_2^2)/E_2$ and which is known to be scale independent, and variables with the subscript "0" are corresponding macroscale values (for $L \geq L_{\text{lc}}$).

Dependence of the real area of plastic contact A_{rp} on the load is given by

$$A_{\text{rp}} = \frac{W}{H}, \qquad (16.21)$$

where H is hardness. According to the strain gradient plasticity model [17, 18], the yield strength τ_Y is given by (16.9) and hardness H is given by (16.10). In the case of plastic contact, the mean contact radius can be determined from (16.19), which is based on the contact geometry and independent of load [20]. Assuming the contact radius as its mean value from (16.19) based on elastic analysis, and combining (16.10), (16.19) and (16.21), the real area of plastic contact is given as

$$A_{\text{rp}} = \frac{W}{H_0 \sqrt{1 + (l_{\text{d}}/\bar{a})}} = \frac{W}{H_0 \sqrt{1 + (L_{\text{d}}/L)^m}}, \qquad L < L_{\text{lc}}, \qquad (16.22)$$

where L_{d} is a characteristic length parameter related to l_{d}, $\bar{a}$, and L_{lc} [20]

$$L_{\text{d}} = L_{\text{lc}}\left(\frac{l_{\text{d}}}{\bar{a}_0}\right)^{1/m}. \qquad (16.23)$$

The scale dependence of A_{re}, N, and $\bar{a}$ is presented in Fig. 16.10.

16.4.2 Dependence of Contact Parameters on Load

The effect of short and long wavelength details of rough surfaces on contact parameters also depends on the normal load. For low loads, the ratio of real to apparent areas of contact $A_{\text{r}}/A_{\text{a}}$, is small, contact spots are small, and long wavelength details are irrelevant. For higher $A_{\text{r}}/A_{\text{a}}$, long wavelength details become important, whereas

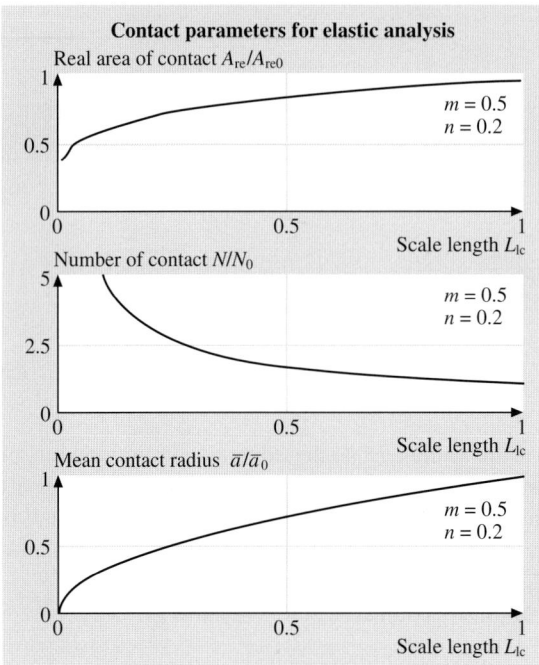

Fig. 16.10. Scale length dependence of normalized contact parameters ($m = 0.5$, $n = 0.2$) (**a**) real area of contact, (**b**) number of contacts, and (**c**) mean contact radius

small wavelength details of the surface geometry become irrelevant. The effect of increased load is similar to the effect of increased scale length [23].

In the preceding subsections, it was assumed that the roughness parameters are scale-dependent for $L < L_{lwl}$, whereas the contact parameters are scale-dependent for $L < L_{lc}$. The upper limit of scale dependence for the contact parameters, L_{lc}, depends on the normal load, and it is reasonable to assume that L_{lc} is a function of A_r/A_a, and the contact parameters are scale-dependent when A_r/A_a is below a certain critical value. It is convenient to consider the apparent pressure p_a, which is equal to the normal load divided by the apparent area of contact [23].

For elastic contact, based on (16.15) and (16.17), this condition can be written as

$$\frac{A_{re}}{A_a} \propto \frac{\beta^* p_a}{\sigma} = p_a \frac{\beta_0^*}{\sigma_0} \left(\frac{L}{L_{lwl}}\right)^{m-n} < p_{ac}, \tag{16.24}$$

where p_{ac} is a critical apparent pressure, below which the scale dependence occurs [23]. From (16.24) one can find

$$L < L_{lwl} \left(\frac{\beta_0^*}{\sigma_0} \frac{p_a}{p_{ac}}\right)^{1/(n-m)}. \tag{16.25}$$

The right-hand expression in (16.24) is defined as L_{lc}

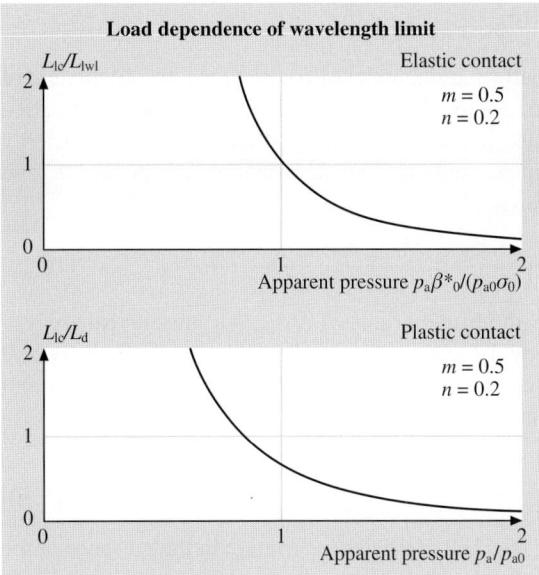

Fig. 16.11. Dependence of the normalized long wavelength limit for contact parameters on load (normalized apparent pressure) for elastic and plastic contacts, ($m = 0.5, n = 0.2$)

$$L_{lc} = L_{lwl} \left(\frac{\beta_0^* \, p_a}{\sigma_0 \, p_{ac}} \right)^{1/(n-m)} . \quad (16.26)$$

For plastic contact, based on (16.22)

$$\frac{A_{rp}}{A_a} \propto \frac{p_a}{\sqrt{1 + (L_d/L)^m}} < p_{ac} . \quad (16.27)$$

In a similar manner to the elastic case, (16.27) yields [23]

$$L_{lc} = L_d \left[\left(\frac{p_a}{p_{ac}} \right)^2 - 1 \right]^{-1/m} . \quad (16.28)$$

Load dependence of the long wavelength limit for contact parameters, L_{lc} is presented in Fig. 16.11 for an elastic contact based on (16.28), and for a plastic contact based on (16.28), for $m = 0.5, n = 0.2$ [23]. The load (apparent pressure) is normalized by $\beta_0^*/(p_{ac}\sigma_0)$ for the elastic contact and by p_{ac} for the plastic contact. In the case of elastic contact, it is observed, that the long wavelength limit decreases with increasing load. For a problem, characterized by a given scale length L, increase of load will result in decrease of L_{lc} and, eventually, the condition $L < L_{lc}$ will be violated; thus the contact parameters, including the coefficient of friction, will reach the macroscale values. Decrease of L_{lc} with increasing load is also observed in the case of plastic contact, the data presented for $p_a/p_{ac} > 1$.

16.5 Scale Effect in Friction

According to the adhesion and deformation model of friction [16], the coefficient of dry friction μ can be presented as a sum of adhesion component μ_a and deformation (plowing) component μ_d. The later, in the presence of particles, is a sum of asperity summits deformation component μ_{ds} and particles deformation component μ_{dp}, so that the total coefficient of friction is [21]

$$\mu = \mu_a + \mu_{ds} + \mu_{dp} = \frac{F_a + F_{ds} + F_{dp}}{W} = \frac{A_{ra}\tau_a + A_{ds}\tau_{ds} + A_{dp}\tau_{dp}}{W}, \quad (16.29)$$

where W is the normal load, F is the friction force, A_{ra}, A_{ds}, A_{dp} are the real areas of contact during adhesion, two body deformation and with particles, respectively, and τ is the shear strength. The subscripts a, ds, and dp correspond to adhesion, summit deformation and particle deformation.

In the presence of meniscus, the friction force is given by

$$F = \mu(W + F_m), \quad (16.30)$$

where F_m is the meniscus force [16]. The coefficient of friction in the presence of the meniscus force, μ_{wet}, is calculated using only the applied normal load, as normally measured in the experiments [22]

$$\mu_{wet} = \mu\left(1 + \frac{F_m}{W}\right) = \frac{A_{ra}\tau_a + A_{ds}\tau_{ds} + A_{dp}\tau_{dp}}{W}\left(1 + \frac{F_m}{W}\right). \quad (16.31)$$

The (16.31) shows that μ_{wet} is greater than μ, because F_m is not taken into account for calculation of the normal load in the wet contact.

It was shown by *Greenwood* and *Williamson* [42] and by subsequent modifications of their model, that for contacting surfaces with common statistical distributions of asperity heights, the real area of contact is almost linearly proportional to the normal load. This linear dependence, along with (16.29), result in linear dependence of the friction force on the normal load, or coefficient of friction being independent of the normal load. For a review of the numerical analysis of rough surface contacts, see *Bhushan* [43, 44] and *Bhushan* and *Peng* [45]. The statistical and numerical theories of contact involve roughness parameters – e.g. the standard deviation of asperity heights and the correlation length [16]. The roughness parameters are scale dependent. In contrast to this, the theory of self-similar (fractal) surfaces solid contact developed by *Majumdar* and *Bhushan* [46] does not include length parameters and are scale-invariant in principle. The shear strength of the contacts in (16.29) is also scale dependent. In addition to the adhesional contribution to friction, elastic and plastic deformation on nano- to macroscale contributes to friction [16]. The deformations are also scale dependent.

16.5.1 Adhesional Friction

The adhesional component of friction depends on the real area of contact and adhesion shear strength. Here we derive expressions for scale dependence of adhesional friction during single-asperity and multiple-asperity contacts.

Single-Asperity Contact

The scale length during single-asperity contact is the nominal contact length, which is equal to the contact diameter $2a$. In the case of predominantly elastic contacts, the real area of contact A_{re} depends on the load according to the Hertz analysis [47]

$$A_{re} = \pi a^2, \tag{16.32}$$

and

$$a = \left(\frac{3WR}{4E^*}\right)^{1/3}, \tag{16.33}$$

where R is effective radius of curvature of summit tips, and E^* is the effective elastic modulus of the two bodies. In the case of predominantly plastic contact, the real area of contact A_{rp} is given by (16.21), whereas the hardness is given by (16.10)

Combining (16.10), (16.13), (16.29), and (16.32), the adhesional component of the coefficient of friction can be determined for the predominantly elastic contact as

$$\mu_{ae} = \mu_{ae0}\sqrt{1+(l_s/a)} \tag{16.34}$$

and for the predominantly plastic contact as

$$\mu_{ap} = \mu_{ap0}\sqrt{\frac{1+(l_s/a)}{1+(l_d/a)}}, \tag{16.35}$$

where μ_{ae0} and μ_{ap0} are corresponding values at the macroscale [20].

The scale dependence of adhesional friction in single-asperity contact is presented in Fig. 16.12a. In the case of single asperity elastic contact, the coefficient of friction increases with decreasing scale (contact diameter), because of an increase in the adhesion strength, according to (16.34). In the case of single asperity plastic contact, the coefficient of friction can increase or decrease with decreasing scale, because of an increased hardness or increase in adhesional strength. The competition of these two factors is governed by l_d/l_s, according to (16.35). There is no direct way to measure l_d and l_s. We will see later, from experimental data, that the coefficient of friction tends to decrease with decreasing scale, therefore, it must be assumed that $l_d/l_s > 1$ for the data reported in the paper [20].

Multiple-Asperity Contact

The adhesional component of friction depends on the real area of contact and adhesion shear strength. Scale dependence of the real area of contact was considered in the preceding section. Here we derive expressions for scale-dependence of the shear strength at the interface during adhesional friction. It is suggested by *Bhushan* and *Nosonovsky* [20] that, for many materials, dislocation-assisted sliding (microslip) is the main mechanism, which is responsible for the shear strength. They considered

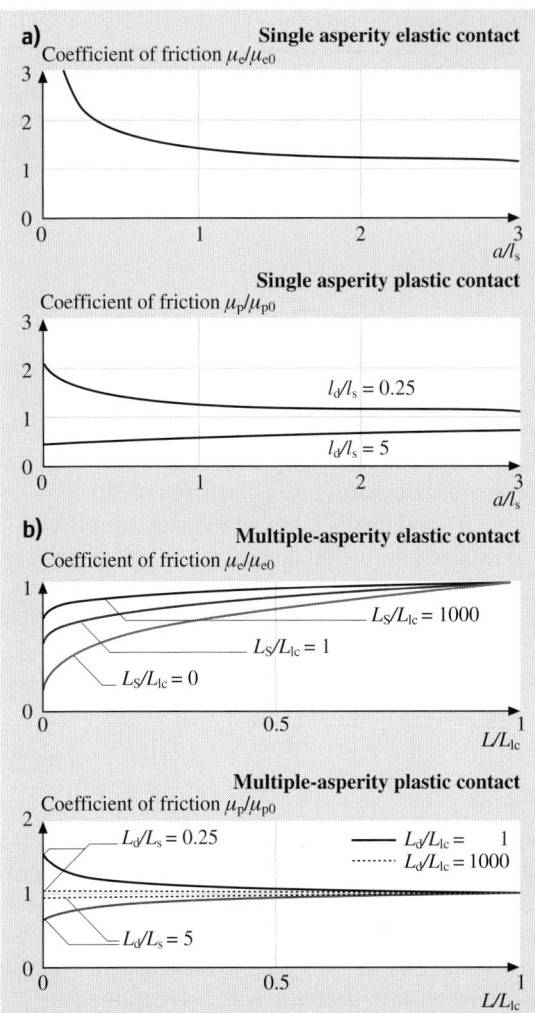

Fig. 16.12. Normalized results for the adhesional component of the coefficient of friction, as a function of scale (a/l_s for single asperity contact and L/L_{lc} for multi-asperity contact). In the case of single asperity plastic contact, data are presented for two values of l_d/l_s. In the case of multi-asperity contact, data are presented for $m = 0.5$, $n = 0.2$. For multi-asperity elastic contact, data are presented for three values of L_S/L_{lc}. For multi-asperity plastic contact, data are presented for two values of L_d/L_s

dislocation assisted sliding based on the assumption, that contributing dislocations are located in a subsurface volume. The thickness of this volume is limited by the distance which dislocations can climb l_s (material parameter) and by the radius of contact a. They showed that τ_a is scale dependent according to (16.13). Assuming the contact radii equal to the mean value given by (16.19)

$$\tau_a = \tau_{a0}\sqrt{1+(L_s/L)^m}, \quad L < L_{lc}, \qquad (16.36)$$

where

$$L_s = L_{lc}\left(\frac{l_s}{\bar{a}_0}\right)^{1/m}. \qquad (16.37)$$

In the case of absence of the microslip (e.g., for an amorphous material), it should be assumed in (16.34–16.36), $L_s = l_s = 0$.

Based on (16.9, 16.17, 16.24, 16.29, 16.36, 16.37), the adhesional component of the coefficient of friction in the case of elastic contact μ_{ae} and in the case of plastic contact μ_{ap}, is given as [20]

$$\begin{aligned}\mu_{ae} &= \frac{\tau_a A_{re}}{W} = \frac{\tau_{a0} A_{re0}}{W}\left(\frac{L}{L_{lc}}\right)^{m-n}\sqrt{1+(L_s/L)^m} \\ &= \frac{\mu_{ae0}}{\sqrt{1+(l_s/\bar{a}_0)}}\left(\frac{L}{L_{lc}}\right)^{m-n}\sqrt{1+(L_s/L)^m} \quad L < L_{lc};\end{aligned} \qquad (16.38)$$

$$\begin{aligned}\mu_{ap} &= \frac{\tau_{a0}}{H_0}\sqrt{\frac{1+(L_s/L)^m}{1+(L_d/L)^m}} \\ &= \mu_{ap0}\sqrt{\frac{1+(l_d/\bar{a}_0)}{1+(l_s/\bar{a}_0)}}\sqrt{\frac{1+(L_s/L)^m}{1+(L_d/L)^m}} \quad L < L_{lc},\end{aligned} \qquad (16.39)$$

where μ_{ae0} and μ_{ap0} are values of the coefficient of friction at macroscale ($L \geq L_{lc}$).

The scale dependence of adhesional friction in multiple-asperity elastic contact is presented in Fig. 16.12b, which is based on (16.38), for various values of L_s/L_{lc}. The change of scale length L affects the coefficient of friction in two different ways: through the change of A_{re} (16.17) and τ_a (16.36) below L_{lc}. Further, τ_a is controlled by the ratio L_s/L. Based on (16.36), for small ratio of L_s/L_{lc}, scale effects on τ_a is insignificant for $L/L_{lc} > 0$. As it is seen from Fig. 16.12b by comparison of the curve with $L_s/L_{lc} = 0$ (insignificant scale effect on τ_a), $L_s/L_{lc} = 1$, and $L_s/L_{lc} = 1000$ (significant scale effect on τ_a), the results for the normalized coefficient of friction are close, thus, the main contribution to the scaling effect is due to change of A_{re}. In the case of multiple-asperity plastic contact, the results, based on (16.39), are presented in Fig. 16.12b for $L_d/L_s = 0.25$, $L_d/L_s = 5$ and $L_d/L_{lc} = 1$ and $L_d/L_{lc} = 1000$. The change of scale affects the coefficient of friction through the change of A_{rp} (16.34), which is controlled by L_d, and τ_a (16.36), which is controlled by L_s. It can be observed from Fig. 16.12b, that for $L_d > L_s$, the change of A_{rp} prevails over the change of τ_a, with decreasing scale, and the coefficient of friction decreases. For $L_d < L_s$, the change of τ_a prevails, with decreasing scale, and the coefficient of friction increases [20]. Expressions for the coefficient of adhesional friction are presented in Table 16.1.

16.5.2 Two-Body Deformation

Based on the assumption that multiple asperities of two rough surfaces in contact have conical shape, the two-body deformation component of friction can be deter-

Table 16.1. Scaling factors for the coefficient of adhesional friction [20]

Single asperity elastic contact	Single asperity plastic contact	Multiple-asperity elastic contact	Multiple-asperity plastic contact
$\mu_e =$ $\mu_{e0}\sqrt{1+(l_s/a)}$	$\mu_e =$ $\mu_{e0}\sqrt{\frac{1+(l_s/a)}{1+(l_d/a)}}$	$\mu_e =$ $\mu_{e0}C_E L^{m-n} \times \sqrt{1+(L_s/L)^m}$	$\mu_p =$ $\mu_{p0}C_P\sqrt{\frac{1+(L_s/L)^m}{1+(L_d/L)^m}}$

mined as

$$\mu_{ds} = \frac{2\tan\theta_r}{\pi}, \qquad (16.40)$$

where θ_r is the roughness angle (or attack angle) of a conical asperity [16, 48]. Mechanical properties affect the real area of contact and shear strength and these cancel out in (16.29).

The roughness angle is scale-dependent and is related to the roughness parameters [41]. Based on statistical analysis of a random Gaussian surface,

$$\tan\theta_r \propto \frac{\sigma}{\beta^*}. \qquad (16.41)$$

From (16.40) it can be interpreted that stretching the rough surface in the vertical direction (increasing vertical scale parameter σ) increases $\tan\theta_r$, and stretching in the horizontal direction (increasing vertical scale parameter β^*) decreases $\tan\theta_r$.

Using (16.40) and (16.41), the scale dependence of the two-body deformation component of the coefficient of friction is given as [21]

$$\mu_{ds} = \frac{2\sigma_0}{\pi\beta^*}\left(\frac{L}{L_{lc}}\right)^{n-m} = \mu_{ds0}\left(\frac{L}{L_{lc}}\right)^{n-m}, \quad L < L_{lc}, \qquad (16.42)$$

where μ_{ds0} is the value of the coefficient of summits deformation component of the coefficient of friction at macroscale ($L \geq L_{lc}$),

The scale dependence for the two-body deformation component of the coefficient of friction is presented in Fig. 16.13 for $m = 0.5$, $n = 0.2$. The coefficient of friction increases with decreasing scale, according to (16.42). This effect is a consequence of increasing average slope or roughness angle.

16.5.3 Three-Body Deformation Friction

In this sections of the paper, size distribution of particles will be idealized according to the exponential, normal, and log normal density functions, since these distributions are the most common in nature and industrial applications (see Appendix A.1). The probability for a particle of a given size to be trapped at the interface depends on the size of the contact region. Particles at the edge of the region of contact are likely to leave the contact area, whereas those in the middle are likely to be trapped. The ratio

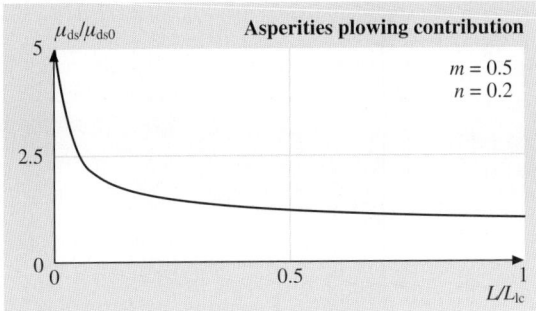

Fig. 16.13. Normalized results for the two-body deformation component of the coefficient of friction

of the edge region area to the total apparent area of contact increases with decreasing scale size. Therefore, the probability for a particle to be trapped decreases, as well as the three-body component of the coefficient of friction [21].

Let us consider a square region of contact of two rough surfaces with a length L (relevant scale length), with the density of debris of η particles per unit area (Fig. 16.14). We assume that the particles have the spherical form and that $p(d)$ is the probability density function of particles size. It is also assumed that, for a given diameter, particles at the border region of the contact zone of the width $d/2$ are likely to leave the contact zone, with a certain probability α, whereas particles at the center of the contact region are likely to be trapped. It should be noted, that particles in the corners of the contact region can leave in two different directions, therefore, for them the probability to leave is 2α. The total nominal contact area is equal to L^2, the area of the border region, without the corners, is equal to $4(L-d)d/2$, and the area of the corners is equal to d^2.

The probability density of size distribution for the trapped particles $p_{tr}(d)$ can be calculated by multiplying $p(d)$ by one minus the probability of a particle with diameter d to leave; the later is equal to the ratio of the border region area, multiplied by a corresponding probability of the particle to leave, divided by the total contact area [21]

$$p_{tr}(d) = p(d)\left(1 - \frac{2\alpha(L-d)d + 2\alpha d^2}{L^2}\right) = p(d)\left(1 - \frac{2\alpha d}{L}\right), \quad d < \frac{L}{2\alpha}. \quad (16.43)$$

The ratio of the number of trapped particles to the total number of particles, average radius of a trapped particle $\overline{d}$, and average square of trapped particles $\overline{d^2}$, as functions of L, can be calculated as

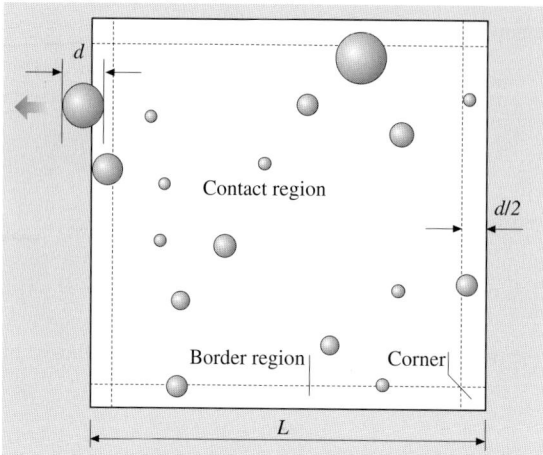

Fig. 16.14. Schematics of debris at the contact zone and at its border region. A particle of diameter d in the border region of $d/2$ is likely to leave the contact zone

$$n_{tr} = \frac{\int_0^{L/2} p_{tr}(d)dd}{\int_0^{\infty} p(d)dd} = \frac{\int_0^{L/2} p(d)\left(1 - \frac{2\alpha d}{L}\right)dd}{\int_0^{\infty} p(d)dd},$$

$$\bar{d} = \frac{\int_0^{L/2} dp_{tr}(d)dd}{\int_0^{L/2} p_{tr}(d)dd},$$

$$\bar{d^2} = \frac{\int_0^{L/2} d^2 p_{tr}(d)dd}{\int_0^{L/2} p_{tr}(d)dd}. \qquad (16.44)$$

Let us assume an exponential distribution of particles' size (16.83) with $d_e = 0$. Substituting (16.83) into (16.44) and integrating yields for the ratio of trapped particles [21]

$$n_{tr} = \frac{\int_0^{L/(\alpha 2)} \frac{1}{\sigma_e}\exp\left(-\frac{d}{\sigma_e}\right)\left(1 - \frac{2\alpha d}{L}\right)dd}{\int_0^{\infty} \frac{1}{\sigma_e}\exp\left(-\frac{d}{\sigma_e}\right)dd} = \exp\left(-\frac{d}{\sigma_e}\right)\left.\frac{\sigma_e - L/(2\alpha) + d}{L/(2\alpha)}\right|_0^{L/(2\alpha)}$$

$$= \frac{2\alpha\sigma_e}{L}\left[\exp\left(-\frac{L}{2\alpha\sigma_e}\right) - 1\right] + 1 \qquad (16.45)$$

whereas the mean diameter of the trapped particles is

$$\bar{d} = \frac{\int_0^{L/(2\alpha)} d\exp\left(-\frac{d}{\sigma_e}\right)\left(1 - \frac{2\alpha d}{L}\right)dd}{\int_0^{L/(2\alpha)} \exp\left(-\frac{d}{\sigma_e}\right)\left(1 - \frac{2\alpha d}{L}\right)dd}$$

$$= \sigma_e \frac{\exp\left(-\frac{L}{2\alpha\sigma_e}\right)\left(1 + \frac{4\alpha\sigma_e}{L}\right) + 1 - \frac{4\alpha\sigma_e}{L}}{\frac{2\alpha\sigma_e}{L}\left[\exp\left(-\frac{L}{2\alpha\sigma_e}\right) - 1\right] + 1} \qquad (16.46)$$

and the mean square radius of the trapped particles is

$$\overline{d^2} = \frac{\int_0^{L/(2\alpha)} d^2 \exp\left(-\frac{d}{\sigma_e}\right)\left(1 - \frac{2\alpha d}{L}\right) dd}{\int_0^{L/(2\alpha)} \exp\left(-\frac{d}{\sigma_e}\right)\left(1 - \frac{2\alpha d}{L}\right) dd}$$

$$= \sigma_e^2 \frac{\exp\left(-\frac{L}{2\alpha\sigma_e}\right)\left(\frac{L}{2\alpha\sigma_e} + 4 + \frac{12\alpha\sigma_e}{L}\right) + 2 - \frac{12\alpha\sigma_e}{L}}{\frac{2\alpha\sigma_e}{L}\left(\exp\left(-\frac{L}{2\alpha\sigma_e}\right) - 1\right) + 1}. \qquad (16.47)$$

For the normal and log normal distributions, similar calculations can be conducted numerically.

The area, supported by particles can be found as the number of trapped particles $\eta L^2 n_{tr}$ multiplied by average particle contact area

$$A_{dp} = \eta L^2 n_{tr} \frac{\pi \overline{d^2}}{4}, \qquad (16.48)$$

where $\overline{d^2}$ is mean square of particle diameter, η is particle density per apparent area of contact (L^2) and n_{tr} is a number of trapped particles divided by the total number of particles [21].

The plowing deformation is plastic and, assuming that particles are harder than the bodies, the shear strength τ_{dp} is equal to the shear yield strength of the softer body τ_Y which is given by the (16.9) with $a = \overline{d}/2$. Combining (16.29) with (16.9) and (16.48)

$$\mu_{dp} = \frac{A_{dp}\tau_{dp}}{W} = \eta \frac{L^2}{W} \frac{\pi \overline{d^2}}{4} n_{tr} \tau_{Y0} \sqrt{1 + 2l_d/\overline{d}} = \mu_{dp0} n_{tr} \frac{\overline{d^2}}{\overline{d_0^2}} \frac{\sqrt{1 + 2l_d/\overline{d}}}{\sqrt{1 + 2l_d/\overline{d_0}}}, \qquad (16.49)$$

where $\overline{d}$ is mean particle diameter, $\overline{d_0}$ is the macroscale value of mean particle diameter, and μ_{dp0} is macroscale ($L \to \infty$, $n_{tr} \to 1$) value of the third-body deformation component of the coefficient of friction given as

$$\mu_{dp0} = \eta \frac{L^2}{W} \frac{\pi \overline{d_0^2}}{4} \tau_{Y0} \sqrt{1 + 2l_d/\overline{d_0}}. \qquad (16.50)$$

Scale dependence of the three-body deformation component of the coefficient of friction is presented in Fig. 16.15, based on (16.49). The number of trapped particles divided by the total number of particles, as well as the three-body deformation component of the coefficient of friction, are presented as a function of scale size divided by α for the exponential, normal, and log normal distributions. The dependence of μ_d/μ_{d0} is shown as a function of $L/(\alpha\sigma_e)$ for the exponential distribution and normal distribution, for $d_n = d_e = 2\sigma_e$ and $l_d/\sigma_e = 1$, whereas for the log normal distribution the results are presented as a function of L/α, for $\langle \ln d_{ln}\rangle = 2$, $\sigma_{ln} = 1$, and $l_d/\sigma_{ln} = 1$. This component of the coefficient of friction decreases for all of the three distributions. The results are shown for $l_d/\sigma_{ln} = 1$, however, variation of l_d/σ_{ln} in the

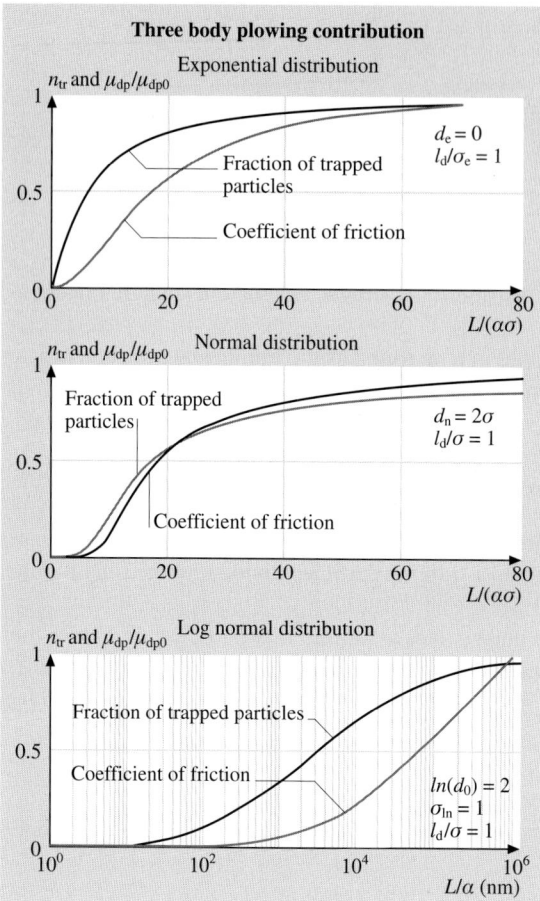

Fig. 16.15. The number of trapped particles divided by the total number of particles and three-body deformation component of the coefficient of friction, normalized by the macroscale value, for three different distributions of debris size: (a) exponential (b) normal (c) and lognormal distributions

range between 0.1 and 10 does not change significantly the shape of the curve. The decrease of the three-body deformation friction force with decreasing scale results with this component being small at the nanoscale.

16.5.4 Ratchet Mechanism

Surface roughness can have an appreciable influence on friction during adhesion. If one of the contacting surfaces has asperities of much smaller lateral size, such that a small tip slides over an asperity, having the average angle θ_r (so called ratchet mechanism), the corresponding component of the coefficient of friction is given by

$$\mu_r = \mu_a \tan^2 \theta_r , \qquad (16.51)$$

where μ_r is the ratchet mechanism component of friction [16]. Combining (16.15, 16.41, 16.38, 16.39) yields for the scale dependence of the ratchet component of the coefficient of friction in the case of elastic, μ_{re}, and plastic contact, μ_{rp}

$$\mu_{re} = \mu_{ae} \left[\frac{2\sigma_0}{\pi \beta_0^*}\left(\frac{L}{L_{lc}}\right)^{n-m}\right]^2$$

$$= \frac{\mu_{re0}}{\sqrt{1+(l_s/\bar{a}_0)}}\left(\frac{L}{L_{lc}}\right)^{n-m}\sqrt{1+(L_s/L)^m}, \quad L < L_{lc}, \tag{16.52}$$

$$\mu_{rp} = \mu_{ap}\left[\frac{2\sigma_0}{\pi\beta_0^*}\left(\frac{L}{L_{lc}}\right)^{n-m}\right]^2$$

$$= \mu_{rp0}\left(\frac{L}{L_{lc}}\right)^{2(n-m)}\sqrt{\frac{1+(l_d/\bar{a}_0)}{1+(l_s/\bar{a}_0)}}\sqrt{\frac{1+(L_s/L)^m}{1+(L_d/L)^m}}, \quad L < L_{lc}, \tag{16.53}$$

where μ_{re0} and μ_{rp0} are the macroscale values of the ratchet component of the coefficient of friction for elastic and plastic contact correspondingly [23].

Scale dependence of the ratchet component of the coefficient of friction, normalized by the macroscale value, is presented in Fig. 16.16, for scale independent adhesional shear strength, τ_a = const, (L_s = 0) and for scale dependent τ_a (L_s = $10L_d$), based on (16.51) and (16.52). The ratchet component during adhesional elastic friction, μ_{re}, is presented in Fig. 16.16a. It is observed, that, with decreasing scale, μ_{re} increases. The ratchet component during adhesional plastic friction, μ_{rp}, is presented in Fig. 16.16b. It is observed, that, for L_s = 0, with decreasing scale, μ_{rp} increases [23].

16.5.5 Meniscus Analysis

During contact, if a liquid is introduced at the point of asperity contact, the surface tension results in a pressure difference across a meniscus surface, referred to as capillary pressure or Laplace pressure. The attractive force for a sphere in contact with a plane surface is proportional to the sphere radius R_p, for a sphere close to a surface with separation s or for a sphere close to a surface with continuous liquid film [16]

$$F_m \propto R_p. \tag{16.54}$$

The case of multiple-asperity contact is shown in Fig. 16.1b. Note, that both contacting and near-contacting asperities wetted by the liquid film contribute to the total meniscus force. A statistical approach can be used to model the contact. In general, given the interplanar separation s_d, the mean peak radius $\overline{R_p}$, the thickness of liquid film h_f, the surface tension γ, liquid contact angle between the liquid and surface θ, and the total number of summits in the nominal contact area N,

$$F_m = 2\pi \overline{R_p} \gamma (1 + \cos\theta) N. \tag{16.55}$$

In (16.54), γ and θ are material properties, which are not expected to depend on scale, whereas $\overline{R_p}$ and N depend on surface topography, and are scale-dependent, according to (16.18) and (16.20).

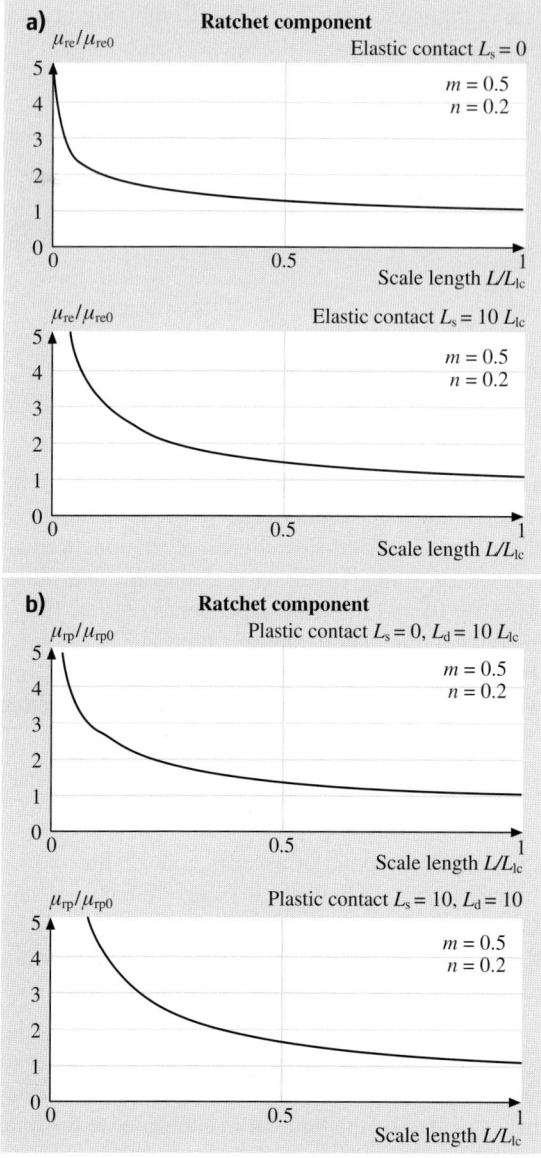

Fig. 16.16. Normalized results for the ratchet component of the coefficient of friction, as a function of scale, for scale independent ($L_s = 0$) and scale dependent ($L_s = 10L_{lc}$) shear strength ($m = 0.5$, $n = 0.2$). (**a**) Elastic contact, (**b**) Plastic contact ($L_d = 10L_{lc}$)

$$F_m \propto \overline{R_p} N = F_{m0} \left(\frac{L}{L_{lwl}} \right)^{m-2n}, \quad L < L_{lwl}, \tag{16.56}$$

where F_{m0} is the macroscale value of the meniscus force ($L \geq L_{lwl}$).

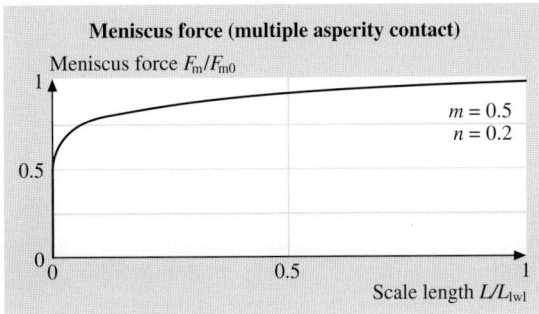

Fig. 16.17. Meniscus force for $m = 0.5$, $n = 0.2$

Scale dependence of the meniscus force is presented in Fig. 16.17, based on (16.56) for $m = 0.5$, $n = 0.2$. It may be observed that, depending on the value of D, the meniscus force may increase or decrease with decreasing scale size.

16.5.6 Total Value of Coefficient of Friction and Transition from Elastic to Plastic Regime

During transition from elastic to plastic regime, contribution of each of the three components of the coefficient of friction in (16.29) changes. In the elastic regime, the dominant contribution is expected to be adhesion involving elastic deformation, and in the plastic regime the dominant contribution is expected to be deformation. Therefore, in order to study transition from elastic to plastic regime, the ratios of deformation to adhesion component should be considered. The expression for the total value of the coefficient of friction, which includes meniscus force contribution, based on (16.29) and (16.31) can be rewritten as [21]

$$\mu_{\text{wet}} = \mu_a \left(1 + \frac{\mu_{ds}}{\mu_a} + \frac{\mu_{dp}}{\mu_a}\right)\left(1 + \frac{F_m}{W}\right). \tag{16.57}$$

The ratchet mechanism component is ignored here since it is present only in special cases. Results in the preceding subsection provide us with data about the adhesion and two-body and three-body deformation components of the coefficient of friction, normalized by their values at the macroscale. However, that analysis does not provide any information about their relation to each other or about transition from the elastic to plastic regime. In order to analyze the transition from pure adhesion involving elastic deformation to plastic deformation, a transition index ϕ can be considered [21]. The transition index is equal to the ratio of average pressure in the elastic regime (normal load per real area of elastic contact) to hardness or simply the ratio of the real area of plastic contact divided by the real area of elastic contact

$$\phi = \frac{W}{A_{re} H} = \frac{A_{rp}}{A_{re}}. \tag{16.58}$$

Using (16.17) and (16.22), the scale-dependence of ϕ is

$$\phi = \frac{W}{A_{re0}(L/L_{lc})^{m-n} H_0 \sqrt{1+(L_s/L)^m}}$$
$$= \phi_0 \frac{\sqrt{1+(l_s/\bar{a})}(L/L_{lc})^{n-m}}{\sqrt{1+(L_s/L)^m}}, \quad L < L_{lc}, \tag{16.59}$$

where ϕ_0 is the macroscale value of the transition index [21].

With a low value of ϕ close to zero, the contacts are mostly elastic and only adhesion contributes to the coefficient of friction involving elastic deformation. Whereas with increasing ϕ approaching unity, the contacts become predominantly plastic and deformation becomes a dominant contributor. It can be argued that A_{ds}/A_{re} and A_{dp}/A_{re} will also be a direct function of ϕ, and in the paper these will be assumed to have linear relationship.

Next, the ratio of adhesion and deformation components of the coefficient of friction in terms of ϕ is obtained. In this relationship, τ_{ds} and τ_{dp} are equal to the shear yield strength, which is proportional to hardness and can be obtained from (16.9), using (16.19) and (16.36)

$$\frac{\mu_{ds}}{\mu_{ae}} = \frac{A_{ds} \tau_{ds}}{A_{re} \tau_a} \propto \phi \frac{\tau_{ds}}{\tau_a} = \phi \frac{\tau_{ds0} \sqrt{1+(L_d/L)^m}}{\tau_{a0} \sqrt{1+(L_s/L)^m}}, \quad L < L_{lc}, \tag{16.60}$$

$$\frac{\mu_{dp}}{\mu_{ae}} = \frac{A_{dp} \tau_{dp}}{A_{re} \tau_a} \propto \phi \frac{\tau_{dp}}{\tau_a} = \phi \frac{\tau_{Y0} \sqrt{1+2l_d/\bar{d}}}{\tau_{a0} \sqrt{1+(L_s/L)^m}}, \quad L < L_{lc}. \tag{16.61}$$

The sum of adhesion and deformation components [21]

$$\mu_{wet} = \mu_{ae} \left[1 + \phi \left(\frac{\tau_{ds0} \sqrt{1+(L_d/L)^m}}{\tau_{a0} \sqrt{1+(L_s/L)^m}} + \frac{\tau_{Y0} \sqrt{1+2l_d/\bar{d}}}{\tau_{a0} \sqrt{1+(L_s/L)^m}} \right) \right]$$
$$\left[1 + \frac{F_{m0}}{W} \left(\frac{L}{L_{lwl}} \right)^{m-2n} \right], \quad L < L_{lc}. \tag{16.62}$$

Note that ϕ itself is a complicated function of L, according to (16.59).

Scale dependence of the transition index, normalized by the macroscale value, is presented in Fig. 16.18, based on (16.59). It is observed that, for $L_s = 0$, the transition index decreases with increasing scale. For $L_s = 10 L_{lc}$, the same trend is observed for $m > 2n$, but, in the case $m < 2n$, ϕ decreases. An increase of the transition index means that the ratio of plastic to elastic real areas of contact increases. With decreasing scale, the mean radius of contact decreases, causing hardness enhancement and decrease of the plastic area of contact. Based on this, the model may predict an increase or decrease of the transition index, depending on whether elastic or plastic area decreases faster.

The dependence of the coefficient of friction on ϕ is illustrated in Fig. 16.19, based on (16.62). It is assumed in the figure that the slope for the dependence of μ_{dp} on ϕ is greater than the slope for the dependence of μ_{ds} on ϕ. For ϕ close to zero, the contact is predominantly elastic, whereas for ϕ approaching unity the contact is predominantly plastic.

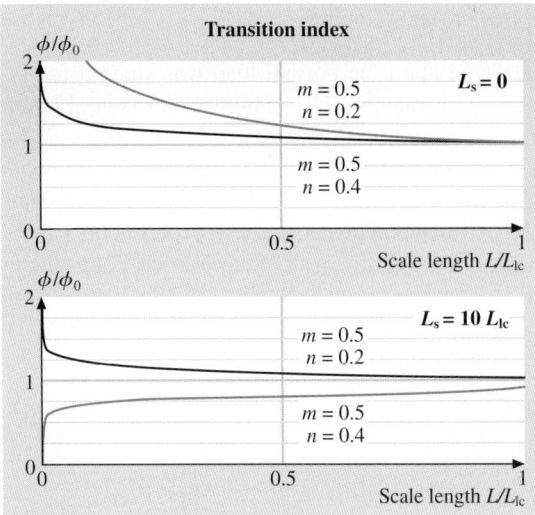

Fig. 16.18. The transition index as a function of scale. Presented for $m = 0.5$, $n = 0.2$ and $m = 0.5$, $n = 0.4$

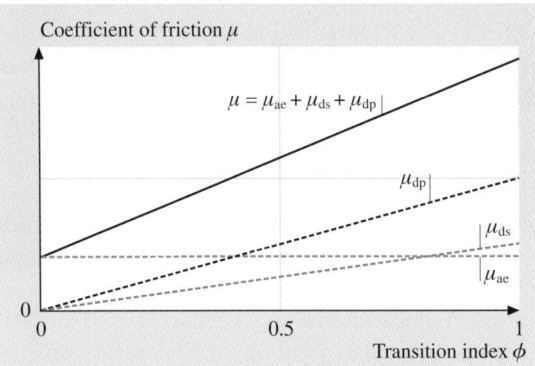

Fig. 16.19. The coefficient of friction (dry contact) as a function of the transition index for given scale length L. With increasing ϕ and onset of plastic deformation, both μ_{ds} and μ_{dp} grow, as a result of this, the total coefficient of friction μ grows as well

16.5.7 Comparison with the Experimental Data

Experimental data on friction at micro- and nanoscale are presented in this subsection and compared with the model. First, a single-asperity predominantly elastic contact is considered [20], then transition to plastic deformation involving multiple asperity contacts is analyzed [23].

Single-Asperity Predominantly-Elastic Contact

Nanoscale dependence of friction force upon the normal load was studied for Pt-coated AFM tip versus mica in ultra-high vacuum (UHV) by *Carpick* et al. [7], for Si tip versus diamond and amorphous carbon by *Schwarz* et al. [8] and for Si_3N_4 tip on Si, SiO_2, and diamond by *Bhushan* and *Kulkarni* [6] (Fig. 16.20a). *Homola* et al. [13] conducted SFA experiments with mica rolls with a single contact zone (before onset of wear), Fig. 16.21b. Contacts relevant in these experiments can be considered as single-asperity, predominantly elastic in all of these cases. For a single-asperity elastic contact of radius a, expression for μ is given by (16.17). For the limit of a small contact radius $a \ll l_s$, the (16.13) combined with the Hertzian dependence of the contact area upon the normal load (16.33) yields

$$F_e \approx \pi a^2 \tau_0 \sqrt{l_s/a} \propto a^{3/2} \propto W^{1/2} . \tag{16.63}$$

If an adhesive pull-off force W_0 is large, (16.63) can be modified as

$$F_e = C_0 \sqrt{W + W_0} , \tag{16.64}$$

where C_0 is a constant. Friction force increases with square root of the normal load, opposed to the two third exponent in scale independent analysis.

The results in Fig. 16.20 demonstrate a reasonable agreement of the experimental data with the model. The platinum-coated tip versus mica [7] has a relatively high pull-off force and the data fit with $C_0 = 23.7\,(nN)^{1/2}$ and $W_0 = 170\,nN$. For the silicon tip vs. amorphous carbon and natural diamond, the fit is given by $C_0 = 8.0$, $19.3\,(nN)^{1/2}$ and small W_0. For the virgin Si(111), SiO_2, and natural diamond sliding versus Si_3N_4 tip [8], the fit is given by $C_0 = 0.40, 0.76, 0.86\,(nN)^{1/2}$ for Si(111), SiO_2, and diamond, respectively and small W_0. For two mica rolls [13], the fit is given by $C_0 = 10\,N^{1/2}$ and $W_0 = 0.5\,N$ [20].

AFM experiments provide data on nanoscale, whereas SFA experiments provide data on microscale. Next we study scale dependence on the shear strength based on these data. In the AFM measurements by *Carpick* et al. [7], the average shear strength during sliding for Pt–mica interface was reported as 0.86 GPa, whereas the pull-off contact radius was reported as 7 nm. In the SFA measurements by *Homola* et al. [13], the average shear strength during sliding for mica–mica interface was reported as 25 MPa, whereas the contact area during high loads was on the order of $10^{-8}\,m^2$, which corresponds to a contact radius on the order $100\,\mu m$. To normalize shear strength, we need shear modulus. The shear modulus for mica is $G_{mica} = 25.7\,GPa$ [49] and for Pt is $G_{Pt} = 63.7\,GPa$ [50]. For mica–Pt interface, the effective shear modulus is

$$G = 2G_{mica}G_{Pt}/(G_{mica} + G_{Pt}) = 36.6\,GPa . \tag{16.65}$$

This yields the value of the shear stress normalized by the shear modulus $\tau_a/G = 2.35 \times 10^{-2}$ for *Carpick* et al. [7] AFM data and 9.73×10^{-4} for the SFA data. These values are presented in the Fig. 16.21 together with the values predicted by the model for assumed values of $l_s = 1\,\mu m$ and $10\,\mu m$. It can be seen that the model (16.13) provides an explanation of adhesional shear strength increase with a scale decrease [20].

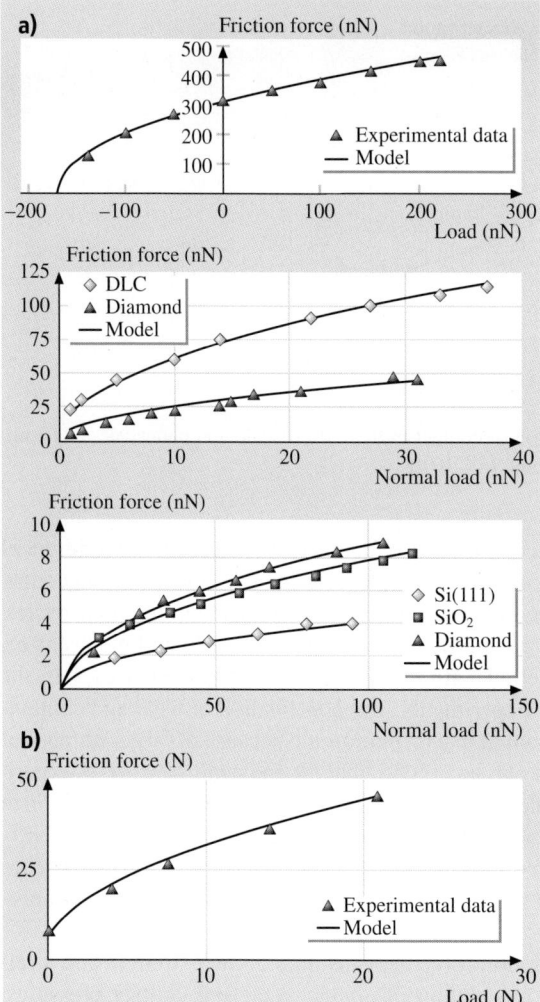

Fig. 16.20. Summary of (**a**) AFM data (*upper*: Pt-coated tip on mica in UHV [7], *middle*: Si tip on DLC and diamond in UHV [8], *lower*: Si_3N_4 tip on various materials [6]) and (**b**) SFA data (on mica vs. mica in dry air [13]) for friction force as a function of normal load

Transition to Predominantly Plastic Deformation Involving Multiple Asperity Contacts

Next, we analyze the effect of transition from predominantly elastic adhesion to predominantly plastic deformation involving multiple asperity contacts [23]. The data on nano- and microscale friction for various materials, are presented in Table 16.2, based on *Ruan* and *Bhushan* [5], *Liu* and *Bhushan* [11], and *Bhushan* et al. [12], for Si(100), graphite (HOPG), highly oriented pyrolytic graphitenatural diamond, and

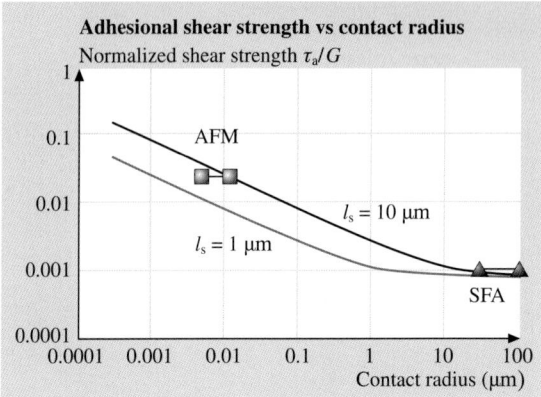

Fig. 16.21. Shear stress as a function of contact radius. Microscale and nanoscale data compared with the model for $l_s = 1\,\mu\text{m}$ and $l_s = 10\,\mu\text{m}$

diamond-like carbon (DLC). There are several factors responsible for the differences in the coefficients of friction at micro- and nanoscale. Among them are the contributions from ratchet mechanism, meniscus effect, wear and contamination particles, and transition from elasticity to plasticity. The ratchet mechanism and meniscus effect result in an increase of friction with decreasing scale and cannot explain the decrease of friction found in the experiments. The contribution of wear and contamination particles is more significant at macro/microscale because of larger number of trapped particles (Fig. 16.15). It can be argued, that for the nanoscale AFM experiments the contacts are predominantly elastic and adhesion is the main contribution to the friction, whereas for the microscale experiments the contacts are predominantly plastic and deformation is an important factor. Therefore, transition from elastic contacts in nanoscale contacts to plastic deformation in microscale contacts is an important effect [23].

According to (16.29), the friction force depends on the shear strength and a relevant real area of contact. For calculation of contact radii and contact pressures, the elastic modulus, Poisson's ratio, and hardness for various samples, are required and presented in Table 16.2 [50–55]. In the nanoscale AFM experiments a sharp tip was slid against a flat sample. The apparent contact size and mean contact pressures are calculated based on the assumption, that the contacts are single asperity, elastic contacts (contact pressures are small compared to hardness). Based on the Hertz equation [47], for spherical asperity of radius R in contact with a flat surface, with an effective elastic modulus E^*, under normal load W, the contact radius a and mean apparent contact pressure p_a are given by

$$a = \left(\frac{3WR}{4E^*}\right)^{1/3}, \tag{16.66}$$

$$p_a = \frac{W}{\pi a^2}. \tag{16.67}$$

Table 16.2. Micro- and nanoscale values of the coefficient of friction, typical physical properties of specimens, and calculated apparent contact radii and apparent contact pressures at loads used in micro- and nanoscale measurements. For calculation purposes it is assumed that contacts on micro- and nanoscale are single-asperity elastic contacts [23]

Specimen	Coefficient of friction		Elastic modulus	Poisson's ratio	Hardness	Apparent contact radius at test load for		Mean apparent pressure at test load for	
	Micro-scale	Nano-scale	(GPa)		(GPa)	Micro-scale (μm)	Nano-scale (nm)	Micro-scale (GPa)	Nano-scale (GPa)
Si(100) wafer	0.47[a]	0.06[c]	130[e,f]	0.28[f]	9–10[e,f]	0.8–2.2	1.6–3.4	0.05–0.13[a]	1.3–2.8[c]
Graphite (HOPG)	0.1[b]	0.006[c]	9–15[g] (9)	– (0.25)	0.01[j]	62	3.4–7.4	0.082[b]	0.27–0.58[c]
Natural diamond	0.2[b]	0.05[c]	1140[h]	0.07[h]	80–104[g,h]	21	1.1–2.5	0.74[b]	2.5–5.3[c]
DLC film	0.19[a]	0.03[d]	280[i]	0.25[i]	20–30[i]	0.7–2.0	1.3–2.9	0.06–0.16[a]	1.8–3.8[d]

[a] 500 μm radius Si(100) ball at 100–2000 μN and 720 μm/s in dry air [12]
[b] 3 mm radius Si$_3$N$_4$ ball (Elastic modulus 310 GPa, Poisson's ratio 0.22 [50]), at 1 N and 800 μm/s [5]
[c] 50 nm radius Si$_3$N$_4$ tip at load range from 10–100 nN and 0.5 nm/s, in dry air [5]
[d] 50 nm radius Si$_3$N$_4$ tip at load range from 10–100 nN in dry air [12]
[e] [51], [f] [52], [g] [50], [h] [53], [i] [54], [j] [55]

The surface energy effect [16] was neglected in (16.66) and (16.67), because the experimental value of the normal adhesion force was small, compared to W [5]. The calculated values of a and p_a for the relevant normal load are presented in Table 16.2 [23].

In the microscale experiments, a ball was slid against a nominally flat surface. The contact in this case is multiple-asperity contact due to the roughness, and the contact pressure of the asperity contacts is higher than the apparent pressure. For calculation of a characteristic scale length for the multiple asperity contacts, which is equal to the apparent length of contact, (16.66) was also used. Apparent radius and mean apparent contact pressure for microscale contacts at relevant load ranges are also presented in Table 16.2 [23].

A quantitative estimate of the effect of the shear strength and the real area of contact on friction is presented in Table 16.3. The friction force at mean load (average of maximum and minimum loads) is shown, based on the experimental data presented in Table 16.2. For microscale data, the real area of contact was estimated based on the assumption that the contacts are plastic and based on (16.33) for mean loads given in Table 16.2. For nanoscale data, the apparent area of contact was on the order of several square nanometers, and it was assumed that the real area of contact is comparable with the mean apparent area of contact, which can be calculated for the mean apparent contact radius, given in Table 16.2. The estimate provides with the upper limit of the real area of contact. The lower limit of the shear strength is calculated as friction force, divided by the upper limit of the real area of contact, and presented in Table 16.3 [23]. Based on the data in Table 16.3, for Si(100), natural diamond and DLC film, the microscale value of shear strength is about two orders of magnitude

Table 16.3. Mean friction force, the real area of contact and lower limit of shear strength [23]

Specimen	Friction force at mean load [a]		Upper limit of real area of contact at mean load		Lower limit of mean shear strength (GPa)	
	Microscale (mN)	Nanoscale (nN)	Microscale[b] (μm^2)	Nanoscale[c] (nm^2)	Microscale[d]	Nanoscale[d]
Si(100) wafer	0.49	3.3	0.11	19	4.5	0.17
Graphite (HOPG)	100	0.33	10^5	92	0.001	0.004
Natural diamond	200	2.7	10.9	10	18.4	0.27
DLC film	0.2	1.7	0.042	14	4.8	0.12

[a] Based on the data from Table 16.2. Mean load at microscale is 1050 μN for Si(100) and DLC film and 1 N for HOPG and natural diamond, and 55 nN for all samples at nanoscale
[b] For plastic contact, based on hardness values from Table 16.2. Scale-dependent hardness value will be higher at relevant scale, presented values of the real area of contact are an upper estimate
[c] Upper limit for the real area is given by the apparent area of contact calculated based on the radius of contact data from Table 16.2
[d] Lower limit for the mean shear strength is obtained by dividing the friction force by the upper limit of the real area of contact

higher, than the nanoscale value, which indicates, that transition from adhesion to deformation mechanism of friction and the third-body effect are responsible for an increase of friction at microscale. For graphite, this effect is less pronounced due to molecularly smooth structure of the graphite surface [23].

Based on data available in the literature [6], load dependence of friction at nano/microscale as a function of normal load is presented in Fig. 16.22. Coefficient of friction was measured for Si_3N_4 tip versus Si, SiO_2, and natural diamond using an AFM. They reported that for low loads the coefficient of friction is independent of load and increases with increasing load after a certain load. It is noted that the critical value of loads for Si and SiO_2 corresponds to stresses equal to their hardness values, which suggests that transition to plasticity plays a role in this effect. The friction values at higher loads for Si and SiO_2 approach to that of macroscale values. This result is consistent with predictions of the model for plastic contact (Fig. 16.11), which states that, with increasing normal load, the long wavelength limit for the contact parameters decreases. This decrease results in violation of the condition $L < L_{lc}$, and the contact parameters and the coefficient of friction reach the macroscale values, as was discussed earlier. It must be noted, that the values of $m = 0.5$ and $n = 0.2$ are taken based on available data for the glass-ceramic disk (Fig. 16.9), these parameters depend on material and on surface preparation and may be different for Si, SiO_2, and

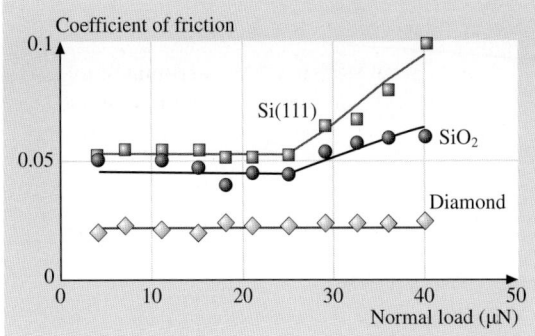

Fig. 16.22. Coefficient of friction as a function of normal load [6]

natural diamond, however, no experimental data on scale dependence of roughness parameters for the materials of interest are available.

16.6 Scale Effect in Wear

The amount of wear during adhesive or abrasive wear involving plastic deformation is proportional to the load and sliding distance x, divided by hardness [16]

$$v = k_0 \frac{Wx}{H}, \tag{16.68}$$

where v is volume of worn material and k_0 is a nondimensional wear coefficient. Using (16.10) and (16.19), the relationships can be obtained for scale dependence of the coefficient of wear in the case of the fractal surface and power-law dependence of roughness parameters

$$v = k \frac{Wx}{H_0} \tag{16.69}$$

$$\text{and} \quad k = \frac{k_0}{\sqrt{1 + (l_\mathrm{d}/\overline{a})}} = \frac{k_0}{\sqrt{1 + (L_\mathrm{d}/L)^m}}, \, L < L_{\mathrm{lwl}}, \tag{16.70}$$

where k is scale-dependent wear coefficient, and k_0 corresponds to the macroscale limit of the value of k [22].

Scale dependence of the wear coefficient is presented in Fig. 16.23 for $m = 0.5$ and $n = 0.2$, based on (16.70). It is observed, that the wear coefficient decreases with decreasing scale; this is due to the fact that the hardness increases with decreasing mean contact size.

16.7 Scale Effect in Interface Temperature

Frictional sliding is a dissipative process, and frictional energy is dissipated as heat over asperity contacts. Therefore, a high amount of heat per unit area is generated

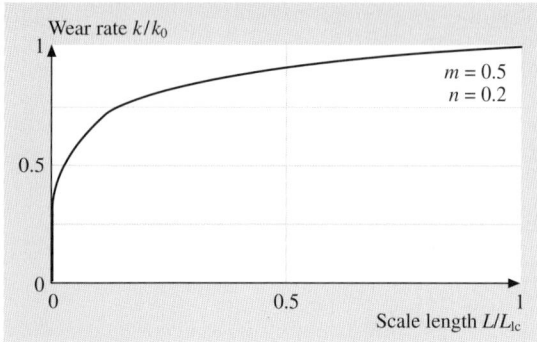

Fig. 16.23. The wear coefficient as a function of scale, presented for $m = 0.5$, $n = 0.2$

during sliding. A contact is formed and destroyed as one asperity passes the other at a given velocity. When an asperity comes into contact with another asperity, the real area of contact starts to grow, when the asperities are directly above each other, the area is at maximum, as they move away from each other, the area starts to get smaller. There are number of contacts at a given time during sliding. For each individual asperity contact, a flash temperature rise can be calculated. High temperature rise affects mechanical and physical properties of contacting bodies.

For thermal analysis, a dimensionless Peclet number is used

$$L_p = \frac{6 V a_{max}}{16 \kappa_t}, \qquad (16.71)$$

where V is sliding velocity, a_{max} is maximum radius of contact for a given contact spot, and κ_t is thermal diffusivity. This parameter indicates whether the sliding is high-speed or low-speed. If $L_p > 10$, the contact falls into the category of high speed; if $L_p < 0.5$, it falls into the category of low speed; if $0.5 \leq L_p \leq 10$, a transition regime should be considered [16]. For high L_p, there is not enough time for the heat to flow to the sides during the lifetime of the contact and the heat flows only in the direction, perpendicular to the sliding surface. Based on the numerical calculations for flash temperature rise of as asperity contact for adhesional contact [16], the following relation holds for the maximum temperature rise T, normalized by the rate at which heat is generated q, divided by the volumetric specific heat ρc_p

$$\frac{T \rho c_p V}{q} = 0.95 \left(\frac{2 V a_{max}}{\kappa_t} \right)^{1/2}, \qquad L_p > 10$$

$$= 0.33 \left(\frac{2 V a_{max}}{\kappa_t} \right), \qquad L_p < 0.5 . \qquad (16.72)$$

The rate at which heat generated per time per unit area depends on the coefficient of friction μ, sliding velocity V, apparent normal pressure p_a, and ratio of the apparent to real areas of contact (A_a/A_r)

$$q = \mu p_a V \frac{A_a}{A_r}. \tag{16.73}$$

Based on (16.72) and (16.73),

$$\frac{T \rho c_p}{p_a} = 0.95 \frac{A_r}{A_a} \mu \left(\frac{2 V a_{max}}{K_t} \right)^{1/2}, \qquad L_p > 10$$

$$= 0.33 \frac{A_r}{A_a} \mu \left(\frac{2 V a_{max}}{K_t} \right), \qquad L_p < 0.5. \tag{16.74}$$

For a multiple asperity contact, mean temperature in terms of average of maximum contact size can be written as

$$\frac{\overline{T} \rho c_p}{p_a} = 0.95 \frac{A_r}{A_a} \mu \left(\frac{2 V \overline{a}_{max}}{K_t} \right)^{1/2}, \qquad L_p > 10$$

$$= 0.33 \frac{A_r}{A_a} \mu \left(\frac{2 V \overline{a}_{max}}{K_t} \right), \qquad L_p < 0.5. \tag{16.75}$$

In (16.75) $\overline{a}_{max}$, μ and A_a/A_r are scale dependent parameters. During adhesional contact, the maximum radius $\overline{a}_{max}$ is proportional to the contact radius $\overline{a}$, and the scale dependence for $\overline{a}_{max}$ is given by (16.19), for μ by (16.38–16.39), and for A_{re} and A_{rp} by (16.17) and (16.21). The scale dependence of q, involving μ and A_r, and $\overline{a}_{max}$ in (16.72) can be considered separately and then combined. For the sake of simplicity, we only consider the scale dependence of $\overline{a}_{max}$. For the empirical rule dependence of surface roughness parameters and the fractal model, in the case of high and low velocity, (16.75) yields [22]

$$\frac{\overline{T} \rho c_p V}{q} = 0.95 \left(\frac{2 V C_A L^m}{K} \right)^{1/2}, \qquad L < L_{lwl}, L_p > 10$$

$$= 0.33 \left(\frac{2 V C_A L^m}{K} \right), \qquad L < L_{lwl}, L_p < 0.5. \tag{16.76}$$

Scale dependence for the ratio of the flash temperature rise to the amount of heat generated per unit time per unit area, for a given sliding velocity, as a function of scale, is presented in Fig. 16.24, based on (16.76), for the high-speed and low-speed cases. For the empirical rule dependence of roughness parameters, the results are shown for $m = 0.5$, $n = 0.2$.

16.8 Closure

A model, which explains scale effects in mechanical properties (yield strength, hardness, and shear strength at the interface) and tribology (surface roughness, contact parameters, friction, wear, and interface temperature), has been presented in this chapter.

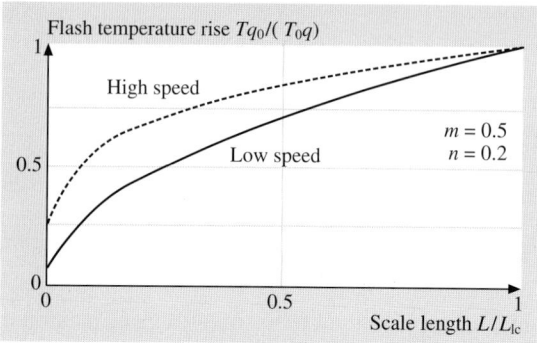

Fig. 16.24. Ratio of the flash temperature rise to the amount of heat generated per unit time per unit area, for a given sliding velocity, as a function of scale. Presented for $m = 0.5, n = 0.2$

Both mechanical properties and roughness parameters are scale-dependent. According to the strain gradient plasticity, the scale dependence of the so-called geometrically necessary dislocations causes enhanced yield strength and hardness with decreasing scale. The shear strength at the interface is scale dependent due to the effect of dislocation-assisted sliding. An empirical rule for scale dependence of the roughness parameters has been proposed, namely, it was assumed, that the standard deviation of surface height and autocorrelation length depend on scale according to a power law when scale is less than the long wavelength limit value.

Both single asperity and multiple asperity contacts were considered. For multiple asperity contacts, based on the empirical power-rule for scale dependence of roughness, contact parameters were calculated. The effect of load on the contact parameters was also studied. The effect of increasing load is similar to that of increasing scale because it results in increased relevance of longer wavelength details of roughness of surfaces in contact.

During sliding, adhesion and two- and three-body deformation, as well as ratchet mechanism, contribute to the friction force. These components of the friction force depend on the relevant real areas of contact (dependent on roughness, mechanical properties, and load), average asperity slope, number of trapped particles, and relevant shear strength during sliding. The relevant scaling length is the nominal contact length – contact diameter ($2a$) for a single-asperity contact, only considered in adhesion, and scan length (L) for multiple-asperity contacts, considered in adhesion and deformation.

For the adhesional component of the coefficient of friction, the shear yield strength and hardness increase with decreasing scale. In the case of elastic contact, the real area of contact is scale independent for single-asperity contact, and may increase or decrease depending on roughness parameters, for multiple-asperity contact. In the case of plastic contact, enhanced hardness results in a decrease in the real area of contact. The adhesional shear strength at the interface may remain constant or increase with decreasing scale, due to dislocation-assisted sliding (or microslip). The model predicts that the adhesional component of the coefficient of friction may increase or decrease with scale, depending on the material parameters and roughness. The coefficient of friction during two-body deformation and the ratchet component depend on the average slope of the rough surface. The average slope increases with scale due to scale dependence of the roughness parameters. As a result, the two-body deformation component of the coefficient of friction increases with decreasing scale. The three-body component of the coefficient of friction depends on the concentrations of particles, trapped at the interface, which decreases with decreasing scale.

The transition index, which is responsible for transition from predominantly elastic adhesional friction to plastic deformation was proposed and was found to change with scale, due to scale dependence of roughness parameters. For the transition index close to zero, the contact is predominantly elastic and the dominant contribution to friction is adhesion involving elastic deformation. The increase of the transition index leads to an increase in plastic deformation with increasing contribution of the deformation component of friction, which results in larger value of the total coefficient of friction.

In presence of the meniscus force, the measured value of the coefficient of friction is greater than the value of the coefficient of dry friction. The difference is especially important for small loads, when the normal load is comparable with the meniscus force. The meniscus force depends on peak radii and may either increase or decrease with scale, depending on the surface parameters.

The wear coefficient and the ratio of the maximum flash temperature rise to the amount of heat generated per unit time per unit area, for a given sliding velocity, as a function of scale, decrease with decreasing scale due to decrease in the mean contact size.

The proposed model is used to explain the trends in the experimental data for various materials at nanoscale and microscale, which indicate that nanoscale values of coefficient of friction are lower than the microscale values (Tables 16.2 and 16.3). The two factors responsible for this trend are the increase of the three-body deformation and transition from elastic adhesive contact to plastic deformation. Experimental data show that the coefficient of friction increases with increasing load after a certain load and reaches the macroscale value. This is due to the onset of plastic deformation with increasing load and the effect of load on contact parameters, which affect the coefficient of friction.

References

1. B. Bhushan. *Handbook of Micro/Nanotribology*. CRC, 2nd edition, 1999.
2. B. Bhushan. Nanoscale tribophysics and tribomechanics. *Wear*, 225–229:465–492, 1999.
3. B. Bhushan. *Springer Handbook of Nanotechnology*. Springer, 2004.
4. B. Bhushan, J. N. Israelachvili, and U. Landman. Nanotribology: Friction, wear and lubrication at the atomic scale. *Nature*, 374:607–616, 1995.
5. J. Ruan and B. Bhushan. Atomic-scale friction measurements using friction force microscopy: Part i – general principles and new measurement technique. *ASME J. Tribol.*, 116:378–388, 1994.

6. B. Bhushan and A. V. Kulkarni. Effect of normal load on microscale friction measurements. *Thin Solid Films*, 278:49–56, 1996.
7. R. W. Carpick, N. Agrait, D. F. Ogletree, and M. Salmeron. Measurement of interfacial shear (friction) with an ultrahigh vacuum atomic force microscope. *J. Vac. Sci. Technol. B*, 14:1289–1295, 1996.
8. U. D. Schwarz, O. Zwörner, P. Köster, and R. Wiesendanger. Quantitative analysis of the frictional properties of solid materials at low loads. 1. carbon compounds. *Phys. Rev. B*, 56:6987–6996, 1997.
9. B. Bhushan and S. Sundararajan. Micro/nanoscale friction and wear mechanisms of thin films using atomic force and friction force microscopy. *Acta Mater.*, 46:3793–3804, 1998.
10. B. Bhushan and C. Dandavate. Thin-film friction and adhesion studies using atomic force microscopy. *J. Appl. Phys.*, 87:1201–1210, 2000.
11. H. Liu and B. Bhushan. Adhesion and friction studies of microelectromechanical systems/nanoelectromechanical systems materials using a novel microtriboapparatus. *J. Vac. Sci. Technol. A*, 21:1538–1538, 2003.
12. B. Bhushan, H. Liu, and S. M. Hsu. Adhesion and friction studies of silicon and hydrophobic and low friction films and investigation of scale effects. *ASME J. Tribol.*, 126:583–590, 2004.
13. A. W. Homola, J. N. Israelachvili, P. M. McGuiggan, and M. L. Gee. Fundamental experimental studies in tribology: The transition from interfacial friction of undamaged molecularly smooth surfaces to normal friction with wear. *Wear*, 136:65–83, 1990.
14. V. N. Koinkar and B. Bhushan. Scanning and transmission electron microscopies of single-crystal silicon microworn/machined using atomic force microscopy. *J. Mater. Res.*, 12:3219–3224, 1997.
15. X. Zhao and B. Bhushan. Material removal mechanisms of single-crystal silicon on nanoscale and at ultralow loads. *Wear*, 223:66–78, 1998.
16. B. Bhushan. *Introduction to Tribology*. Wiley, 2002.
17. N. A. Fleck, G. M. Muller, M. F. Ashby, and J. W. Hutchinson. Strain gradient plasticity: Theory and experiment. *Acta Metall. Mater.*, 42:475–487, 1994.
18. W. D. Nix and H. Gao. Indentation size effects in crystalline materials: A law for strain gradient plasticity. *J. Mech. Phys. Solids*, 46:411–425, 1998.
19. J. W. Hutchinson. Plasticity at the micron scale. *Int. J. Solids Struct.*, 37:225–238, 2000.
20. B. Bhushan and M. Nosonovsky. Scale effects in friction using strain gradient plasticity and dislocation-assisted sliding (microslip). *Acta Mater.*, 51:4331–4345, 2003.
21. B. Bhushan and M. Nosonovsky. Comprehensive model for scale effects in friction due to adhesion and two- and three-body deformation (plowing). *Acta Mater.*, 52:2461–2474, 2004.
22. B. Bhushan and M. Nosonovsky. Scale effects in dry and wet friction, wear, and interface temperature. *Nanotechnol.*, 15:749–761, 2004.
23. M. Nosonovsky and B. Bhushan. Scale effect in dry friction during multiple asperity contact. *ASME J. Tribol.*, 127:37–46, 2005.
24. H. Gao, Y. Huang, W. D. Nix, and J. W. Hutchinson. Mechanism-based strain-gradient plasticity – i. theory. *J. Mech. Phys. Solids*, 47:1239–1263, 1999.
25. Y. Huang, H. Gao, W. D. Nix, and J. W. Hutchinson. Mechanism-based strain-gradient plasticity – ii. analysis. *J. Mech. Phys. Solids*, 48:99–128, 2000.
26. Z. P. Bazant. Scaling of dislocation-based strain-gradient plasticity. *J. Mech. Phys. Solids*, 50:435–448, 2002.
27. J. Friedel. *Dislocations*. Pergamon, 1964.
28. J. Weertman and J. R. Weertman. *Elementary Dislocations Theory*. MacMillan, 1966.

29. B. Bhushan and A. V. Koinkar. Nanoindentation hardness measurements using atomic force microscopy. *Appl. Phys. Lett.*, 64:1653–1655, 1994.
30. B. Bhushan, A. V. Kulkarni, W. Bonin, and J. T. Wyrobek. Nano/picoindentation measurement using a capacitive transducer system in atomic force microscopy. *Philos. Mag.*, 74:1117–1128, 1996.
31. A. V. Kulkarni and B. Bhushan. Nanoscale mechanical property measurements using modified atomic force microscopy. *Thin Solid Films*, 290-291:206–210, 1996.
32. N. Gane and J. M. Cox. tsnote hier *Philos. Mag.*, 22:881, 1970.
33. M. A. Stelmashenko, M. G. Walls, L. M. Brown, and Y. V. Miman. Microindentation on W and Mo oriented single crystal an sem study. *Acta Met. Mater.*, 41:2855–2865, 1993.
34. K. W. McElhaney, J. J. Vlassak, and W. D. Nix. Determination of indenter tip geometry and indentation contact area of depth-sensing indentation experiments. *J. Mater. Res.*, 13:1300–1306, 1998.
35. S. Sundararajan and B. Bhushan. Development of afm-based techniques to measure mechanical properties of nanoscale structures. *Sensors Actuator A*, 101:338–351, 2002.
36. J. J. Weertman. Dislocations moving uniformly on the interface between isotropic media of different elastic properties. *J. Mech. Phys. Solids*, 11:197–204, 1963.
37. K. L. Johnson. Adhesion and friction between a smooth elastic spherical asperity and a plane surface. *Proc. R. Soc. London A*, 453:163–179, 1997.
38. I. A. Polonsky and L. M. Keer. Scale effects of elastic-plastic behavior of microscopic asperity contact. *ASME J. Tribol.*, 118:335–340, 1996.
39. V. S. Deshpande, A. Needleman, and E. Van der Giessen. Discrete dislocation plasticity modeling of short cracks in single crystals. *Acta Mater.*, 51:1–15, 2003.
40. R. A. Onions and J. F. Archard. The contact of surfaces having a random structure. *J. Phys. D*, 6:289–304, 1973.
41. D. J. Whitehouse and J. F. Archard. The properties of random surfaces of significance in their contact. *Proc. R. Soc. London A*, 316:97–121, 1970.
42. J. A. Greenwood and J. B. P. Williamson. Contact of nominally flat surfaces. *Proc. R. Soc. London A*, 295:300–319, 1966.
43. B. Bhushan. Contact mechanics of rough surfaces in tribology: Single asperity contact. *Appl. Mech. Rev.*, 49:275–298, 1996.
44. B. Bhushan. Contact mechanics of rough surfaces in tribology: Multiple asperities contact. *Tribol. Lett.*, 4:1–35, 1998.
45. B. Bhushan and W. Peng. Contact modeling of multilayered rough surfaces. *Appl. Mech. Rev.*, 55:435–480, 2002.
46. A. Majumdar and B. Bhushan. Fractal model of elastic-plastic contact between rough surfaces. *ASME J. Tribol.*, 113:1–11, 1991.
47. K. L. Johnson. *Contact Mechanics*. Clarendon, 1985.
48. E. Rabinowicz. *Friction and Wear of Materials*. Wiley, 2nd edition, 1995.
49. H. R. Clauser (Ed.). *The Encyclopedia of Engineering Materials and Processes*. Reinhold, 1963.
50. B. Bhushan and B. K. Gupta. *Handbook of Tribology: Materials, Coatings, and Surface Treatments*. McGraw-Hill, New York 1991; Krieger, Malabar, 1997.
51. B.. Bhushan and S. Venkatesan. Mechanical and tribological properties of silicon for micromechanical applications: A review. *Adv. Info. Storage Syst.*, 5:211–239, 1993.
52. INSPEC. *Properties of Silicon*. EMIS Data Rev. Ser. No. 4. INSPEC, Institution of Electrical Engineers, 2002.
53. J. E.. Field (Ed.). *The Properties of Natural and Synthetic Diamond*. Academic, 1992.

54. B. Bhushan. Chemical, mechanical and tribological characterization of ultra-thin and hard amorphous carbon coatings as thin as 3.5 nm: Recent developments. *Diam. Relat. Mater.*, 8:1985–2015, 1999.
55. National Carbon Comp. *The Industrial Graphite Engineering Handbook.* National Carbon Company, 1959.
56. C. Bernhardt. *Particle Size Analysis.* Chapman Hall, 1994.
57. J. L. Devoro. *Probability and Statistics for Engineering and the Sciences.* Duxbury, 1995.
58. B. S. Everitt. *The Cambridge Dictionary of Statistics.* Cambridge Univ. Press, 1998.
59. D. Zwillinger and S. Kokoska. *CRC Standard Probability and Statistics Tables and Formulas.* CRC, 2000.
60. S. Wolfram. *The Mathematica Book.* Wolfram Media, 5th edition, 2003.
61. J. S. Bendet and A. G. Piersol. *Engineering Applications of Correlation and Spectral Analysis.* Wiley, 2nd edition, 1986.
62. R. D. Cadle. *Particle Size – Theory and Industrial Applications.* Reinhold, 1965.
63. G. Herdan. *Small Particle Statistics.* Butterworth, 1960.
64. Y. Xie and B. Bhushan. Effect of particle size, polishing pad and contact pressure in free abrasive polishing. *Wear*, 200:281–295, 1996.
65. J. L. Xuan, H. S. Cheng, and R. J. Miller. Generation of submicrometer particles in dry sliding. *ASME J. Tribol.*, 112:664–691, 1990.
66. M. Mizumoto and K. Kato. *Size distribution and number of wear particles generated by the abrasive sliding of a model asperity in the SEM-tribosystem*, pages 523–530. Elsevier, 1992.
67. A. S. Shanbhag, H. O. Bailey, D. S. Hwang, C. W. Cha, N. G. Eror, and H. E. Rubash. Quantitative analysis of ultrahigh molecular weight polyethylene (uhmwpe) wear debris associated with total knee replacements. *J. Biomed. Mater. Res.*, 53:100–110, 2000.
68. T. M. Hunt. *Handbook of Wear Debris Analysis and Particle Detection in Liquids.* Elsevier Applied Science, 1993.
69. W. W. Seifert and V. C. Westcott. A method for the study of wear particles in lubricating oil. *Wear*, 21:27–42, 1972.
70. D. Scott and V. C. Westcott. Predictive maintenance by ferrography. *Wear*, 44:173–182, 1977.
71. D. P. Anderson. *Wear Particle Atlas.* Spectro Inc. Industrial Tribology Systems, 2nd edition, 1991.

A Statistics of Particle Size Distribution

A.1 Statistical Models of Particle Size Distribution

Particle size analysis is an important field for different areas of engineering, environmental, and biomedical studies. In general, size distribution of particles depends on how the particles were formed and sorted. Several statistical distributions, which govern distribution of random variables including particle size, have been suggested (Fig. 16.25), [56–60]. Statistical distributions commonly used are either the probability density (or frequency) function (PDF), $p(z)$, or cumulative distribution function (CDF), $P(h)$. $P(h)$ associated with random variable $z(x)$, which can take any value between $-\infty$ and $+\infty$ or $z_{\min}$ and $z_{\max}$, is defined as the probability of the event $z(x) \leq z'$ and is written as [61]

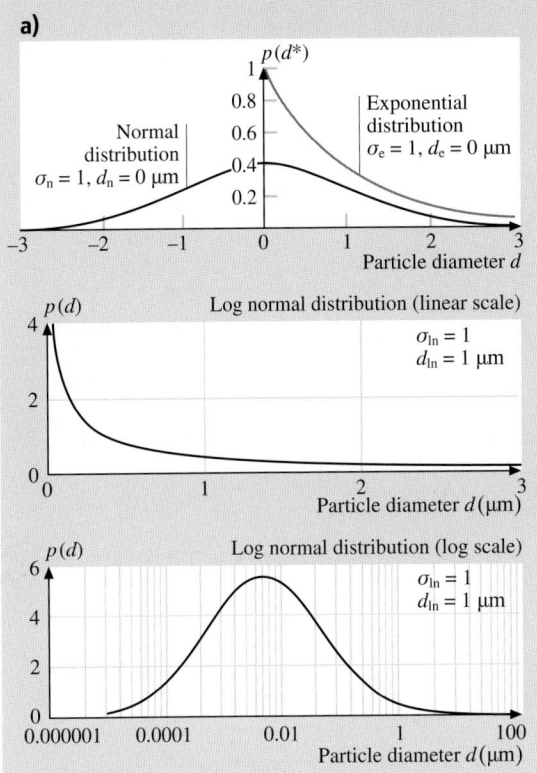

Fig. 16.25. Common statistical distributions of particle size. (**a**) Propbability density distributions. (**b**) Cumulative distributions

$$P(z) = Prob(z \leq z') \tag{16.77}$$

with $P(-\infty) = 0$ and $P(\infty) = 1$.

The pdf is the slope of the CDF given by its derivative

$$P(z) = \frac{dP(z)}{dz} \tag{16.78}$$

or

$$P(z \leq z') = P(z') = \int_{-\infty}^{z'} p(z)dz . \tag{16.79}$$

Furthermore, the total area under the probability density function must be unity; that is, it is certain that the value of z at any x must fall somewhere between plus and minus infinity or z_{min} and z_{max}. The definition of $p(z)$ is phrased as that the random variable $z(x)$ is distributed as $p(z)$.

The probability density (or frequency) function, $p(d)$, in the exponential form is the simplest distribution mathematically

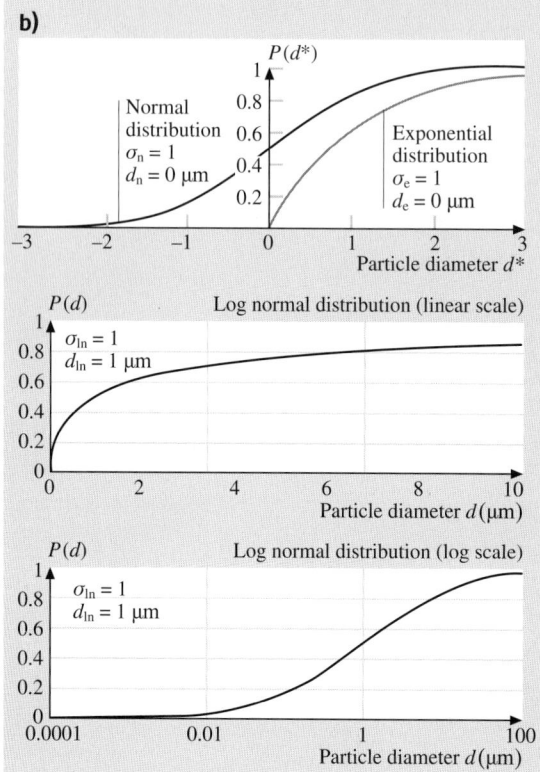

Fig. 16.25. continue

$$p(d) = \frac{1}{\sigma_e} \exp\left(-\frac{d - d_e}{\sigma_e}\right), \quad d \geq d_0, \tag{16.80}$$

where d is particle diameter, σ_e is standard deviation, and d_e is minimum value (for this distribution). For convenience, the density function can be normalized by σ_e in terms of a normalized variable, d^*, equal to $(d - d_e)/\sigma_e$

$$p(d^*) = \exp(-d^*), \quad d^* \geq 0 \tag{16.81}$$

which has zero minimum and unity standard deviation. The cumulative distribution function $P(d')$ is given as

$$P(d') = P(d^* \leq d') = 1 - \exp(-d'). \tag{16.82}$$

The Gaussian or normal distribution is used to represent data for a wide collection of random physical phenomena in practice such as surface roughness. The probability density and cumulative distribution functions are given as

$$p(d) = \frac{1}{\sqrt{2\pi}\sigma_n} \exp\left(-\frac{(d - d_n)^2}{2\sigma_n^2}\right), \quad -\infty < d < \infty, -\infty < d_n < \infty, \sigma_e > 0,$$

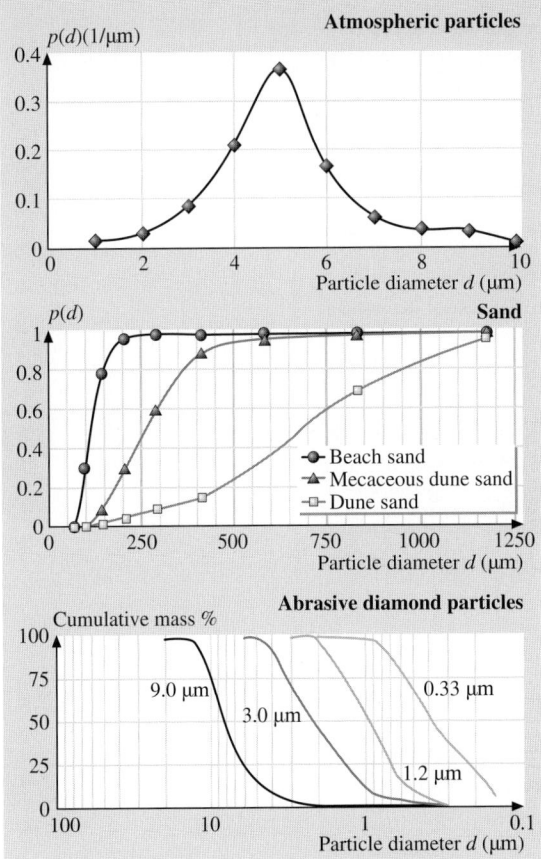

Fig. 16.26. (a) Experimental data for atmospheric [62], sand [63], and abrasive diamond [64] particle size distribution. (b) Experimental data for wear particle size distribution (steel–steel [65], steel–diamond [66], steel–polyethylene [67]). (c) Change with time of wear debris production rate during lubricated sliding as a function of particle size [68]

$$\tag{16.83}$$

where d_n is the mean value. The integral of $p(d)$ in the interval $-\infty < d < \infty$ is equal to 1. In terms of the normalized variables, (16.82) reduces to

$$p(d^*) = \frac{1}{\sqrt{2\pi}} \exp\left(-\frac{d^{*2}}{2}\right) \tag{16.84}$$

and

$$P(d') = P(d^* \leq d') = \frac{1}{\sqrt{2\pi}} \int_{-\infty}^{d'} \exp\left[-(d^*)^2/2\right] dd^* = \text{erf}(d'), \tag{16.85}$$

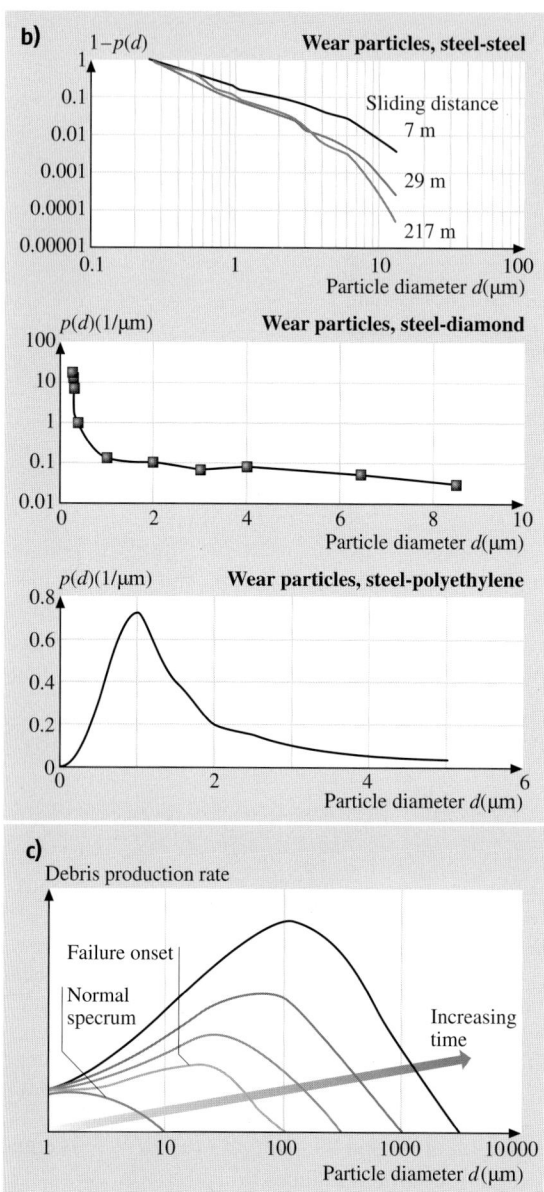

Fig. 16.26. continue

where erf(d') is called the "error function" and its values are listed in most statistical handbooks. The pdf is bell-shaped and the CDF is S-shaped.

For particle size distribution, of interest here, the diameter cannot be less than zero. For this condition, (16.83) must be modified by using a constant on the right side

$$p(d) = \frac{C_0}{\sqrt{2\pi}\sigma_e} \exp\left(-\frac{(d-d_n)^2}{2\sigma_e^2}\right), \quad 0 \leq d < \infty, \tag{16.86}$$

where

$$C_0 = \left[\frac{1}{\sqrt{2\pi}} \int_{-d_0/\sigma}^{\infty} \exp\left(-\frac{t^2}{2}\right) dt\right]^{-1}.$$

The constant is calculated by integrating $p(d)$ in the interval $0 \leq d \leq \infty$ and equating to one

$$\int_0^\infty p(d)\mathrm{d}d = 1. \tag{16.87}$$

The log normal distribution is commonly used to describe particle size distribution. A variable d is log normally distributed if $\ln d$ is normally distributed. Log normal probability density function for variable d, for which $\ln(d)$ has a Gaussian distribution with a mean $\ln(d)_{\ln}$) and standard deviation $\sigma_{\ln}$, is given as

$$p(d) = \frac{1}{\sqrt{2\pi}\sigma_{\ln}}\left(\frac{1}{d}\right)\exp\left(-\frac{[\ln(d/d_{\ln})]^2}{2\sigma_{\ln}^2}\right), \quad 0 < d < \infty. \tag{16.88}$$

The mean of the log normal distribution is $\exp\left(\ln d_{\ln} + \sigma_{\ln}^2/2\right)$, the standard deviation is $\exp\left(2\ln d_{\ln} + \sigma_{\ln}^2\right)\left[\exp(\sigma_{\ln}^2) - 1\right]$, the skewness is $\left[\exp(\sigma_{\ln})^2 + 2\right]\left[\exp(\sigma_{\ln}^2) - 1\right]^{1/2}$, and kurtosis is $\exp\left[4(\sigma_{\ln})^2\right] + 2\exp\left[3(\sigma_{\ln})^2\right] + 3\exp\left[2(\sigma_{\ln})^2\right] - 3$ [58]. The case where $d_{\ln} = 0$ is called the standard log normal distribution. The density function can be normalized by $\sigma_{\ln}$ in terns of a normalized variable, $d^* = (\ln d - d_{\ln})/\sigma_{\ln}$

$$p(d^*) = \frac{1}{\sqrt{2\pi}}\left(\frac{1}{d^*}\right)\exp\left(-\frac{(d^*)^2}{2}\right) \tag{16.89}$$

and

$$P(d') = P(d^* \leq d') = \frac{1}{2}\left[1 + \mathrm{erf}\left(\frac{d'}{\sqrt{2}}\right)\right]. \tag{16.90}$$

The log normal distribution of particle size occurs when the dispersion is attained by comminution (milling, grinding, crushing). The size distribution of pulverized silica, granite, calcite, limestone, quartz, soda, ash, alumina, clay, as well as of wear particles is often governed by the log-normal distribution [63]. A size distribution is usually presented either as probability density or frequency $p(d)$, or as cumulative percent (percent of particles greater than given size) $P(d)$, or as cumulative mass vs. particle size. All these presentations are interrelated [63].

A.2 Typical Particle Size Distribution Data

Typical experimental data for size distributions of atmospheric (dust), sand, and abrasive diamond particles are presented in Fig. 16.26a. It can be seen, that the atmospheric particles [62] follow the normal distribution function. The dune sand is low in heavy mineral content, so the curve is concaved downward. Micaceous dune sand is sorted by gravity slide on a sharp mountain slope and appears to follow log normal distribution, as many distributions of sediments, which are sorted by gravity. Whereas beach sand distribution curve is concaved upward, due to richness in smaller size component [63]. The abrasive diamond particles follow the log normal distribution [64].

Size distribution of wear particles has been studied actively since 1970s, when the ferrography was introduced [69, 70]. The data for wear particles is presented in Fig. 16.26b. *Xuan* et al. [65] studied the size distribution of submicrometer particles during sliding of steel-steel using a Falex 3, pin-on-disk machine, using a laser particle counter for various sliding distances. They found a distribution, which is close to the log normal function. *Mizumoto* and *Kato* [66] studied size distribution of particles generated during pin-on-disk test, for diamond, sapphire, silicon carbide, and tungsten carbide pins vs. steel disk, using a laser particle counter. They found that the probability density function is exponential for particles greater than 1 μm diameter, however for smaller particles a linear law was assumed. *Shanbhag* et al. [67] studied wear particles for ultrahigh molecular weight polyethylene (UHMWPE) versus titanium in biomedical applications (total knee replacement) using a scanning electron microscope. They found that the distribution is close to that of the normal distribution. Numerous data for wear particles are presented by *Anderson* [71]. *Hunt* [68] discusses various techniques of debris measurement and analysis in lubricants. A typical change in wear debris generation rate, which occurs with time, is presented in Fig. 16.26c. Change in the size distribution of wear particles in lubricant indicates an onset of mechanical failure.

Part III

Moleculary-Think Films for Lubrication

17

Nanotribology of Ultrathin and Hard Amorphous Carbon Films

Bharat Bhushan

Summary. Diamond material and its smooth coatings are used for very low wear and relatively low friction. Major limitations of the true diamond coatings are that they need to be deposited at high temperatures, can only be deposited on selected substrates, and require surface finishing. Hard amorphous carbon, commonly known as diamond-like carbon or DLC coatings, exhibit mechanical, thermal, and optical properties close to that of diamond. These can be deposited with a large range of thicknesses by using a variety of deposition processes on a variety of substrates at or near room temperature. The coatings reproduce substrate topography, avoiding the need of post-finishing. Friction and wear properties of some DLC coatings can be very attractive for tribological applications. The largest industrial application of these coatings is in magnetic storage devices.

The prevailing atomic arrangement in the DLC coatings is amorphous or quasi-amorphous with small diamond, graphite, and other unidentifiable micro- or nanocrystallites. Most DLC coatings, except those produced by filtered cathodic arc, contain from a few to about 50 at% hydrogen. Sometimes hydrogen is deliberately incorporated in the sputtered and ion plated coatings to tailor their properties.

EELS and Raman spectroscopies can be successfully used for chemical characterization of amorphous carbon coatings. The prevailing atomic arrangement in the DLC coatings is amorphous or quasi-amorphous with small diamond (sp^3), graphite (sp^2) and other unidentifiable micro- or nanocrystallites. Most DLC coatings except those produced by filtered cathodic arc contain from a few to about 50 at% hydrogen. Sometimes hydrogen is deliberately incorporated in the sputtered and ion plated coatings to tailor their properties.

Amorphous carbon coatings deposited by various techniques exhibit different mechanical and tribological properties. The nanoindenter can be successfully used for measurement of hardness, elastic modulus, fracture toughness, and fatigue life. Microscratch and microwear experiments can be performed using either a nanoindenter or an AFM. Thin coatings deposited by filtered cathodic arc, ion beam, and ECR-CVD hold a promise for tribological applications. Coatings as thin as 5 nm or even thinner provide wear protection. Microscratch, microwear, and accelerated wear testing, if simulated properly, can be successfully used to screen coating candidates for industrial applications. In the examples shown in this chapter, trends observed in the microscratch, microwear, and accelerated macrofriction wear tests are similar to that found in functional tests.

In this chapter, the state-of-the-art of recent developments in the chemical, mechanical, and tribological characterization of ultrathin amorphous carbon coatings is presented.

17.1 Introduction

Carbon exists in both crystalline and amorphous forms and exhibits both metallic and nonmetallic characteristics [1–3]. Crystalline carbon includes graphite, diamond, and a family of fullerenes, Fig. 17.1. The graphite and diamond are infinite periodic network solids with a planar structure, whereas the fullerenes are a molecular form of pure carbon with a finite network with a nonplanar structure. Graphite has a hexagonal, layered structure with weak interlayer bonding forces and exhibits excellent lubrication properties. The graphite crystal may be visualized as infinite parallel layers of hexagons stacked 0.34 nm apart with a 0.1415-nm interatomic distance between the carbon atoms in the basal plane. The atoms lying in the basal planes are trigonally coordinated and closely packed with strong σ (covalent) bonds to its three carbon neighbors using the hybrid sp^2 orbitals. The fourth electron lies in a p_z orbital lying normal to the σ bonding plane and forms a weak π bond by overlapping side to side with a p_z orbital of an adjacent atom to which carbon is attached by a σ bond. The layers (basal planes) themselves are relatively far apart and the forces that bond them are weak van der Waals forces. These layers can align themselves parallel to the direction of the relative motion and slide over one another with relative ease, thus providing low friction. Strong interatomic bonding and packing in each layer is thought to help reduce wear. The operating environment has a significant influence on lubrication, i.e., low friction and low wear, properties of graphite. It lubricates better in a humid environment than a dry one, resulting from adsorption of water vapor and other gases from the environment, which further weakens the interlayer bonding forces and results in easy shear and transfer of the crystallite platelets to the mating surface. Thus, transfer plays an important role in controlling friction and wear. Graphite oxidizes at high operating temperatures and can be used up to about 430 °C.

One of the fullerene molecules is C_{60}, commonly known as Buckyball. Since the C_{60} molecules are very stable and do not require additional atoms to satisfy chemical bonding requirements, they are expected to have low adhesion to the mating surface and low surface energy. Since C_{60} molecules with a perfect spherical symmetry are weakly bonded to other molecules, C_{60} clusters get detached readily, similar to other layered lattice structures, and either get transferred to the mating surface by mechanical compaction, or are present as loose wear particles that may roll like tiny ball bearings in a sliding contact, resulting in low friction and wear. The wear particles are expected to be harder than as-deposited C_{60} molecules, because of their phase transformation at high asperity contact pressures present in a sliding interface. The low surface energy, spherical shape of C_{60} molecules, weak intermolecular bonding, and high load bearing capacity offer potential for various mechanical and tribological applications. The sublimed C_{60} coatings and fullerene particles as an additive to mineral oils and greases have been reported to be good solid lubricants comparable to graphite and MoS_2 [4–6].

Diamond crystallizes in the modified face centered cubic (*fcc*) structure with an interactomic distance of 0.154 nm. The diamond cubic lattice consists of two interpenetrating fcc lattices displaced by one-quarter of the cube diagonal. Each carbon

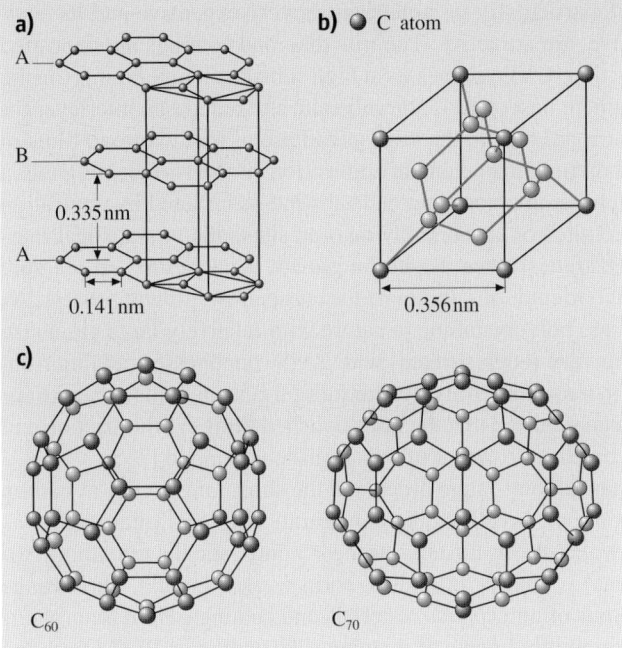

Fig. 17.1. The structure of three known forms of crystalline carbon (**a**) hexagonal structure of graphite, (**b**) modified face-centered cubic (fcc) structure, two interpenetrating fcc lattices displaced by one-quarter of the cube diagonal, of diamond (each atom is bonded to four others that form the corners of the pyramidal structure called tetrahedron), and (**c**) the structure of the two most common forms of fullerenes-soccer ball C_{60} and rugby ball C_{70} molecules

atom is tetrahedrally coordinated, making strong σ (covalent) bonds to its four carbon neighbors using the hybrid sp^3 atomic orbitals, which accounts for its highest hardness (80–104 GPa) and thermal conductivity (900–2100 W/mK, on the order of five times that of copper) of any known solid, and a high electrical resistivity, optical transmission, and a large optical band gap. It is relatively chemically inert, and it exhibits poor adhesion with other solids with consequent low friction and wear. Its high thermal conductivity allows dissipation of frictional heat during sliding and protects the interface, and the dangling carbon bonds on the surface react with the environment to form hydrocarbons that act as good lubrication films. These are some of the reasons for low friction and wear of the diamond. Diamond and its coatings find many industrial applications: tribological applications (low friction and wear), optical applications (exceptional optical transmission, high abrasion resistance), and thermal management or heat sink applications (high thermal conductivity). The diamond can be used to high temperatures, and it starts to graphitize at about 1000 °C in ambient air and at about 1400 °C in vacuum. Diamond is an attractive material for cutting tools, as an abrasive for grinding wheels and lapping compounds, and other extreme wear applications.

The natural diamond, particularly in large sizes, is very expensive and its coatings, a low cost alternative, are attractive. The true diamond coatings are deposited by chemical vapor deposition (CVD) processes at high substrate temperatures (on the order of 800 °C). They adhere best on a silicon substrate and require an interlayer for other substrates. A major roadblock to the widespread use of true diamond films in tribological, optical, and thermal management applications is the surface roughness. Growth of the diamond phase on a non-diamond substrate is initiated by nucleation either at randomly seeded sites or at thermally favored sites, due to statistical thermal fluctuation at the substrate surface. Based on growth temperature and pressure conditions, favored crystal orientations dominate the competitive growth process. As a result, the grown films are polycrystalline in nature with relatively large grain size (> 1 µm) and terminate in very rough surfaces with RMS roughnesses ranging from a few tenths of a micron to tens of microns. Techniques for polishing these films have been developed. It has been reported that the laser polished films exhibit friction and wear properties almost comparable to that of bulk polished diamond [7, 8].

Amorphous carbon has no long-range order, and the short-range order of carbon atoms can have one or more of three bonding configurations: – sp^3 (diamond), sp^2 (graphite), or sp^1 (with two electrons forming strong σ bonds, and the remaining two electrons left in orthogonal p_y and p_z orbitals to form weak π bonds). Short-range order controls the properties of amorphous materials and coatings. Hard amorphous carbon (a-C) coatings, commonly known as diamond-like carbon or DLC (implying high hardness) coatings, are a class of coatings that are mostly metastable amorphous materials, but include a micro- or nanocrystalline phase. The coatings are random networks of covalently bonded carbon in hybridized tetragonal (sp^3) and trigonal (sp^2) local coordination with some of the bonds terminated by hydrogen. These coatings have been successfully deposited by a variety of vacuum deposition techniques on a variety of substrates at or near room temperature. These coatings generally reproduce substrate topography and do not require any post-finishing. However, these coatings mostly adhere best on silicon substrates. The best adhesion is obtained on substrates that form carbides, e.g., Si, Fe, and Ti. Based on depth profile analyses using Auger and XPS of DLC coatings deposited on silicon substrates, it has been reported that a substantial amount of silicon carbide (on the order of 5–10 nm in thickness) is present at the carbon-silicon interface for the coatings with good adhesion and high hardness (e.g., [9]). For good adhesion of DLC coatings to other substrates, in most cases, an interlayer of silicon is required except for cathodic arc deposited coatings.

There is significant interest in DLC coatings because of their unique combination of desirable properties. These properties include high hardness and wear resistance, chemical inertness to both acids and alkalis, lack of magnetic response, and an optical band gap ranging from zero to a few eV, depending upon the deposition conditions. These are used in a wide range of applications, including tribological, optical, electronic, and biomedical applications [1, 10, 11]. The high hardness, good friction and wear properties, versatility in deposition and substrates, and no requirements of post-finishing make them very attractive for tribological applications. Two primary examples include overcoats for magnetic media (thin-film disks and ME

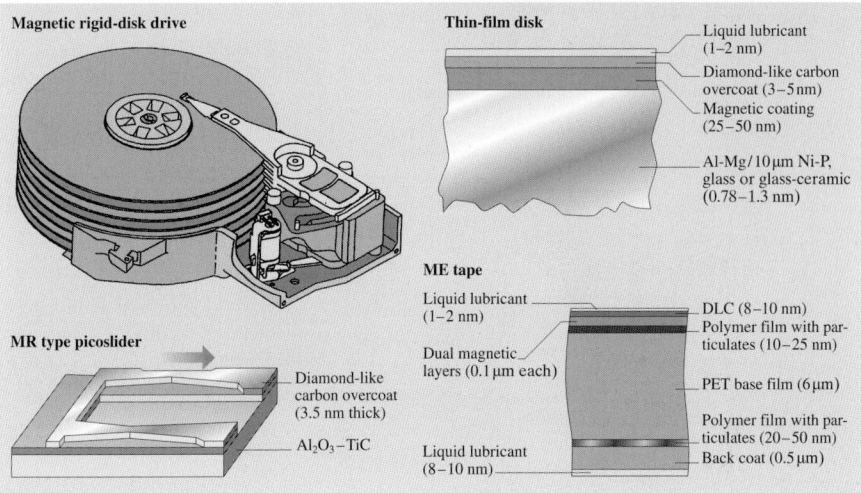

Fig. 17.2. Schematic of a magnetic rigid-disk drive and MR type picoslider and cross-sectional schematics of a magnetic thin-film rigid disk and a metal evaporated (ME) tape

tapes) and MR type magnetic heads for magnetic storage devices, Fig. 17.2 [12–20] and the emerging field of microelectromechanical systems, Fig. 17.3 [21–24]. The largest industrial application of the family of amorphous carbon coatings, typically deposited by DC/RF magnetron sputtering, plasma-enhanced chemical vapor deposition, or ion beam deposition techniques, is in magnetic storage devices. These are employed to protect against wear and corrosion, magnetic coatings on thin-film rigid disks and metal evaporated tapes, and the thin-film head structure of a read/write disk head (Fig. 17.2). To maintain low physical spacing between the magnetic element of a read/write head and the magnetic layer of a media, thicknesses ranging from 3 to 10 nm are employed. Mechanical properties affect friction wear and therefore need to be optimized. In 1998, Gillette introduced Mach 3 razor blades with ultrathin DLC coatings, which has the potential of becoming a very large industrial application. DLC coatings are also used in other commercial applications such as glass windows of supermarket laser barcode scanners and sunglasses. These coatings are actively pursued in microelectromechanical systems (MEMS) MEMScomponents [23].

In this chapter, a state-of-the-art review of recent developments in the chemical, mechanical, and tribological characterization of ultrathin amorphous carbon coatings is presented. An overview of the most commonly used deposition techniques is presented, and followed by typical chemical and mechanical characterization data and typical tribological data both from coupon level testing and functional testing.

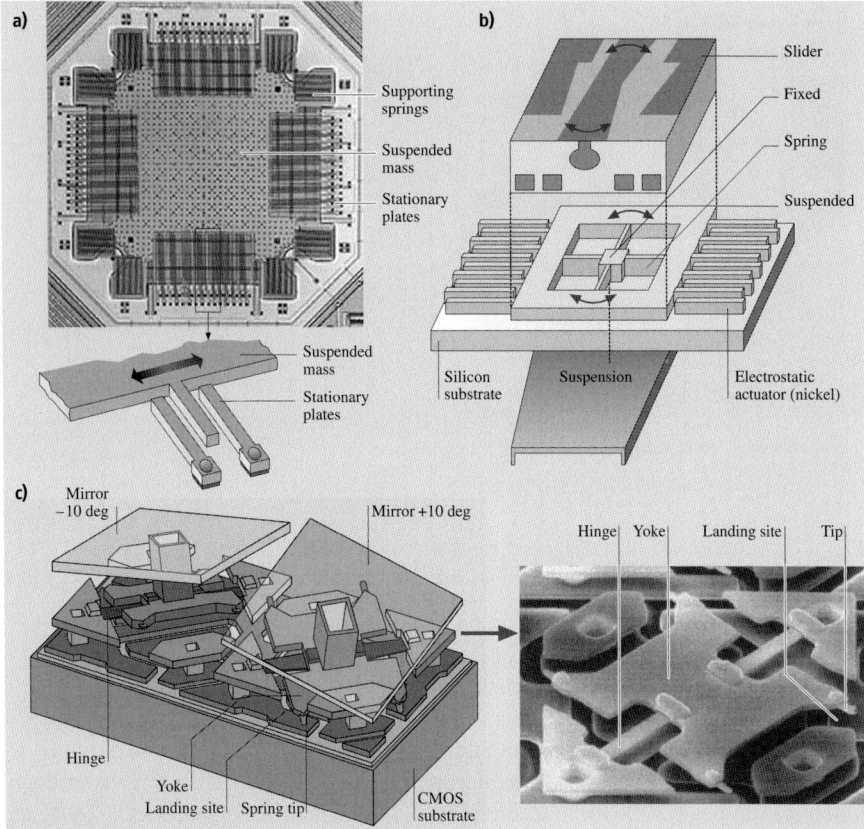

Fig. 17.3. Schematics of (**a**) a capacitive-type silicon accelerometer for automotive sensory applications, (**b**) digital micrometer devices for high-projection displays, and (**c**) polysilicon rotary microactuator for magnetic disk drives

17.2 Description of Commonly Used Deposition Techniques

The first hard amorphous carbon coatings were deposited by a beam of carbon ions produced in an argon plasma on room-temperature substrates, as reported by *Aisenberg* and *Chabot* [25]. Subsequent confirmation by *Spencer* et al. [26] led to the explosive growth of this field. Following the first work, several alternative techniques have been developed. The amorphous carbon coatings have been prepared by a variety of deposition techniques and precursors, including evaporation, DC, RF or ion beam sputtering, RF or DC plasma-enhanced chemical vapor deposition (PECVD), electron cyclotron resonance chemical vapor deposition (ECR-CVD), direct ion beam deposition, pulsed laser vaporization and vacuum arc, from a variety of carbon-bearing solids or gaseous source materials [1, 27]. Coatings with both graphitic and diamond-like properties have been produced. Evaporation and ion plating techniques have been used to produce coatings with graphitic properties (low hardness, high electrical conductivity, very low friction, etc.), and all techniques have been used to produce coatings with diamond-like properties.

The structure and properties of a coating are dependent upon the deposition technique and parameters. High-energy surface bombardment has been used to produce harder and denser coatings. It is reported that sp^3/sp^2 fractions are in the decreasing order for cathodic arc deposition, pulsed laser vaporization, direct ion beam deposition, plasma-enhanced chemical vapor deposition, ion beam sputtering, and DC/RF sputtering [12, 28, 29]. A common feature to these techniques is that the deposition is energetic, i.e., carbon species arrive with an energy significantly greater than that represented by the substrate temperature. The resultant coatings are amorphous in structure, with hydrogen content up to 50%, and display a high degree of sp^3 character. From the results of previous investigations, it has been proposed that deposition of sp^3-bonded carbon requires that the depositing species have kinetic energies on the order of 100 eV or higher, well above those obtained in thermal processes like evaporation (0–0.1 eV). The species must then be quenched into the metastable configuration via rapid energy removal. Excess energy, such as that provided by substrate heating, is detrimental to the achievement of a high sp^3 fraction. In general, a high fraction of the sp^3-bonded carbon atoms in an amorphous network results in a higher hardness [29–36]. The mechanical and tribological properties of a carbon coating depend on the sp^3/sp^2-bonded carbon ratio, the amount of hydrogen in the coating, and adhesion of the coating to the substrate, which are influenced by the precursor material, kinetic energy of the carbon species prior to deposition, deposition rate, substrate temperature, substrate biasing, and the substrate itself [29, 33, 35, 37–46]. The kinetic energies and deposition rates involved in selected deposition processes used in the deposition of DLC coatings are compared in Table 17.1 [1, 28].

In the studies by *Gupta* and *Bhushan* [12, 47], *Li* and *Bhushan* [48, 49], and *Sundararajan* and *Bhushan* [50], DLC coatings ranging typically in thickness from 3.5 nm to 20 nm were deposited on single-crystal silicon, magnetic Ni-Zn ferrite, and Al_2O_3-TiC substrates (surface roughness ≈ 1–3 nm RMS) by filtered cathodic arc (FCA) deposition, (direct) ion beam deposition (IBD), electron cyclotron resonance chemical vapor deposition (ECR-CVD), plasma-enhanced chemical vapor deposition (PECVD), and DC/RF planar, magnetron sputtering (SP) deposition techniques [51]. In this chapter, we will limit the presentation of data of coatings deposited by FCA, IBD, ECR-CVD, and SP deposition techniques.

17.2.1 Filtered Cathodic Arc Deposition Technique

In the filtered cathodic arc deposition of carbon coating [29, 52–59], a vacuum arc plasma source is used to form carbon film. In the FCA technique used by *Bhushan* et al. (e.g., [12]), energetic carbon ions are produced by a vacuum arc discharge between a planar graphite cathode and grounded anode, Fig. 17.4a. The cathode is a 6-mm-diameter high-density graphite disk mounted on a water-cooled copper block. The arc is driven at an arc current of 200 A, arc duration of 5 ms, and arc repetition rate of 1 Hz. The plasma beam is guided by a magnetic field that transports current between the electrodes to form tiny, rapidly moving spots on the cathode surface.

Table 17.1. Summary of the most commonly used deposition techniques and the kinetic energy of depositing species and deposition rates

Deposition technique	Process	Kinetic energy (eV)	Deposition rate (nm/s)
Filtered cathodic arc (FCA)	Energetic carbon ions produced by a vacuum arc discharge between a graphite cathode and grounded anode	100–2500	0.1–1
Direct ion beam (IB)	Carbon ions produced from methane gas in an ion source and accelerated toward a substrate	50–500	0.1–1
Plasma-enhanced chemical vapor deposition (PECVD)	Hydrocarbon species, produced by plasma decomposition of hydrocarbon gases (e.g., acetylene), are accelerated toward a DC-biased substrate	1–30	1–10
Electron cyclotron resonance plasma chemical vapor deposition (ECR-CVD)	Hydrocarbon ions, produced by plasma decomposition of ethylene gas in the presence of a plasma in electron cyclotron resonance condition, are accelerated toward a RF-biased substrate	1–50	1–10
DC/RF sputtering	Sputtering of graphite target by argon ion plasma	1–10	1–10

The source is coupled to a 90° bent magnetic filter to remove the macroparticles produced concurrently with the plasma in the cathode spots. The ion current density at the substrate is in the range of 10–50 mA/cm^2. The base pressure is less than 10^{-4} Pa. Compared with electron beam evaporation with auxiliary discharge, much higher plasma density is achieved with the aid of powerful arc discharge. In this process, the cathodic material suffers a complicated transition from the solid phase to an expanding, nonequilibrium plasma via liquid and dense, equilibrium non-ideal plasma phases [58]. The carbon ions in the vacuum arc plasma have a direct kinetic energy of 20–30 eV. The high voltage pulses are applied to the substrate mounted on a water-cooled sample holder, and ions are accelerated through the sheath and arrive at the substrate with an additional energy given by the potential difference between the plasma and the substrate. The substrate holder is pulse biased to a negative voltage up to −2 kV with a pulse duration of 1 μs. The negative biasing of −2 kV corresponds to 2 keV kinetic energy of the carbon ions. The use of a pulse bias instead of a DC bias has advantages of applying a much higher voltage and building a surface potential on a nonconducting film. The energy of the ions is varied during the deposition. For the first 10% of the deposition the substrates are pulsed biased to −2 keV with a pulse duty cycle of 25%, i.e., for 25% of the time the energy is 2 keV, for the remaining 75% it is 20 eV, which is the "natural" energy of carbon ions in a vacuum discharge. For the last 90% of the deposition the pulsed bias voltage is reduced to −200 eV with a pulse bias duty cycle of 25%, i.e., the energy is 200 eV for 25% and 20 eV for 75% of the deposition. The high energy at the beginning leads to a good intermixing and adhesion of the films, whereas the lower energy at the later stage leads to hard films. Under the conditions described, the deposition rate at the substrate is about 0.1 nm/s, which is slow. Compared with most gaseous plasma, the cathodic arc plasma is nearly fully ionized, and the ionized carbon atoms have high kinetic energy of carbon ions which help achieving a high fraction of sp^3-bonded carbon ions, which in turn result in a high hardness and higher interfacial adhesion. *Cuomo et al.* [42] have reported that based on electron energy loss spectroscopy (EELS) analysis, the sp^3-bonded carbon fraction of cathodic arc coating is 83% compared to 38% for the ion beam sputtered carbon. These coatings are reported to be *nonhydrogenated*.

This technique does not require an adhesion underlayer for non-silicon substrates. However, adhesion of the DLC coatings on the electrically insulating substrate is poor, as negative pulse biasing forms an electrical sheath that accelerates depositing ions to the substrate and enhances the adhesion of the coating to the substrate with associated ion implantation. It is difficult to build potential on an insulating substrate, and lack of biasing results in poor adhesion.

17.2.2 Ion Beam Deposition Technique

In the direct ion beam deposition of carbon coating [60–64], as used by *Bhushan* et al. (e.g., [12]), the carbon coating is deposited from an accelerated carbon ion beam. The sample is pre-cleaned by ion etching. For the case of non-silicon substrates, a 2–3-nm-thick amorphous silicon adhesion layer is deposited by ion beam sputtering using an ion beam of a mixture of methane and argon at 200 V. For the carbon deposition, the chamber is pumped to about 10^{-4} Pa, and methane gas is fed through the cylindrical ion source and is ionized by energetic electrons produced by a hot-wire filament, Fig. 17.4b. Ionized species then pass through a grid with a bias voltage of about 50 eV, where they gain a high acceleration energy and reach a hot-wire filament, emitting thermionic electrons that neutralize the incoming ions. The discharging of ions is important when insulating ceramics are used as substrates. The species are then deposited on a water-cooled substrate. Operating conditions are adjusted to give an ion beam with an acceleration energy of about 200 eV and a current density of about 1 mA/cm². At this operating condition, the deposition rate is about 0.1 nm/s, which is slow. Incidentally, tough and soft coatings are deposited at a high acceleration energy of about 400 eV and at a deposition rate of about 1 nm/s. The ion beam deposited carbon coatings are reported to be hydrogenated (30–40 at% hydrogen).

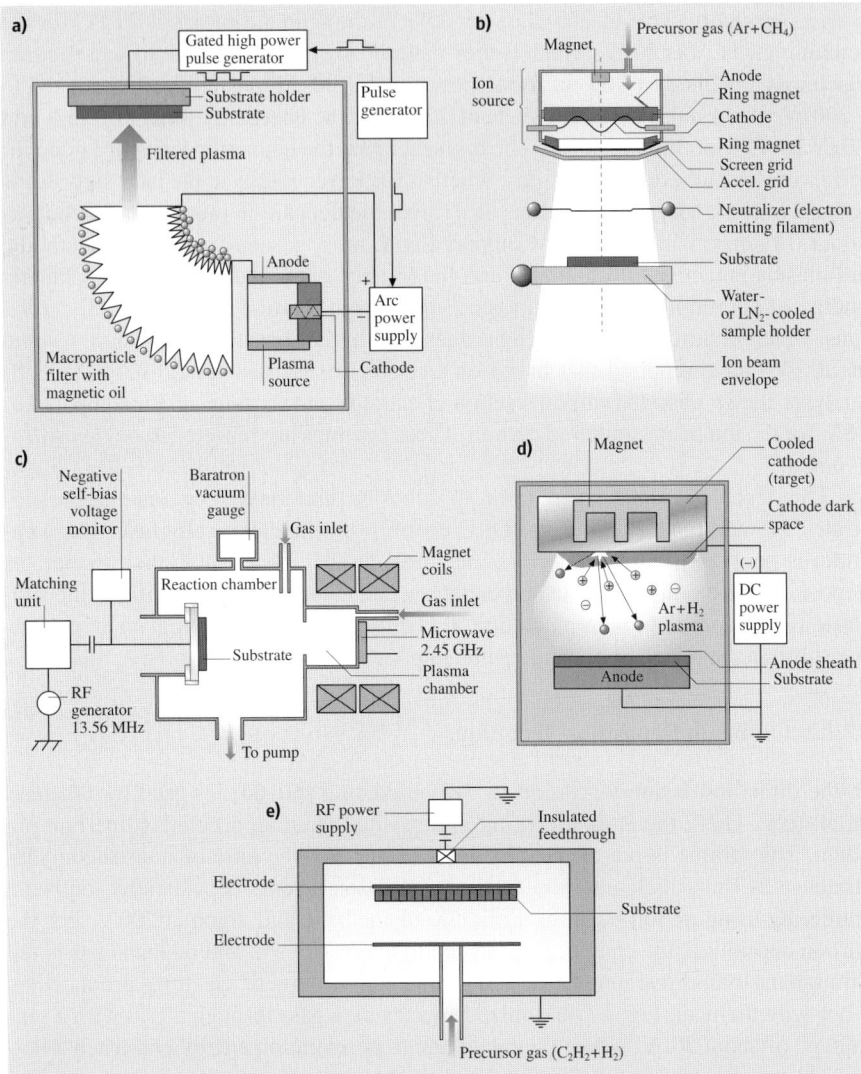

Fig. 17.4. Schematic diagrams of deposition by (**a**) filtered cathodic arc deposition, (**b**) ion beam deposition, (**c**) electron cyclotron resonance chemical vapor deposition (ECR-CVD), (**d**) DC planar magnetron sputtering, and (**e**) plasma-enhanced chemical vapor deposition (PECVD)

17.2.3 Electron Cyclotron Resonance Chemical Vapor Deposition Technique

ECR plasma's lack of electrodes and ability to create high densities of charged and excited species at low pressures ($\leq 10^{-4}$ Torr) make it an attractive processing discharge in the coating depositions [65]. In the ECR-CVD deposition process of carbon coating described by *Suzuki* and *Okada* [66] and used by *Li* and *Bhushan* [48, 49] and *Sundararajan* and *Bhushan* [50], microwave power is generated by a magnetron operating in continuous mode at a frequency of 2.45 GHz, Fig. 17.4c. The plasma chamber functions as a microwave cavity resonator. The magnetic coils arranged around the plasma chamber generate a magnetic field of 875 G, necessary for electron cyclotron resonance condition. The substrate is placed on a stage that is connected capacitively to a 13.56 MHz RF generator. The process gas is introduced into the plasma chamber and the hydrocarbon ions generated are accelerated by a negative self-bias voltage, which is generated by applying RF power to the substrate. Both the substrate stage and plasma chamber are water-cooled. The process gas used is 100% ethylene and its flow rate is held constant at 100 sccm. The microwave power is 100–900 W. The RF power is 30–120 W. The pressure during deposition is kept close to the optimum value of 5.5×10^{-3} Torr. Before the deposition, the substrates are cleaned using Ar ions generated in the ECR plasma chamber.

17.2.4 Sputtering Deposition Technique

In DC planar magnetron sputtered carbon coating [13, 33, 37, 40, 67–71], the carbon coating is deposited by the sputtering of graphite target with Ar ion plasma. In the glow discharge, positive ions from the plasma strike the target with sufficient energy to dislodge the atoms by momentum transfer, which are intercepted by the substrate. First, an about 5-nm-thick amorphous silicon adhesion layer is deposited by sputtering if the deposition is to be carried out on a non-silicon surface. In the process used by *Bhushan* et al. (e.g., [12]), the coating is deposited by the sputtering of a graphite target with Ar ion plasma at 300 W power for a 200-mm-diameter target and pressure of about 0.5 Pa (6 mTorr), Fig. 17.4d. Plasma is generated by applying a DC potential between the substrate and a target. *Bhushan* et al. [35] reported that sputtered carbon coating contains about 35 at% hydrogen. Hydrogen comes from the hydrocarbon contaminants present in the deposition chamber. To produce hydrogenated carbon coating with a larger concentration of hydrogen, deposition is carried out in Ar and hydrogen plasma.

17.2.5 Plasma-Enhanced Chemical Vapor Deposition Technique

In the RF-PECVD deposition of carbon coating, as used by *Bhushan* et al. (e.g., [12]), carbon coating is deposited by adsorption of most free radicals of hydrocarbon to the substrate and chemical bonding to other atoms on the surface. The hydrocarbon species are produced by the RF plasma decomposition of hydrocarbon precursors such as acetylene (C_2H_2), Fig. 17.4e [27, 69, 72–75]. Instead of requiring thermal energy in thermal CVD, the energetic electrons in the plasma (at pressures ranging from 1 to 5×10^2 Pa, typically less than 10 Pa) can activate almost any reaction among the gases in the glow discharge at relatively low substrate temperatures ranging from 100 to 600 °C (typically less than 300 °C). To deposit the coating on nonsilicon substrates, an about 4-nm-thick amorphous silicon adhesion layer, used to improve adhesion, is first deposited under similar conditions from a gas mixture of 1%

silane in argon [76]. In the process used by Bhushan and coworkers [12], the plasma is sustained in a parallel-plate geometry by a capacitive discharge at 13.56 MHz, at a surface power density on the order of 100 mW/cm^2. The deposition is performed at a flow rate on the order of 6 sccm and a pressure on the order of 4 Pa (30 mTorr) on a cathode-mounted substrate maintained at a substrate temperature of 180 °C. The cathode bias is held fixed at about −120 V with an external DC power supply attached to the substrate (powered electrode). The carbon coatings deposited by PECVD usually contain hydrogen up to 50% [35, 77].

17.3 Chemical Characterization and Effect of Deposition Conditions on Chemical Characteristics and Physical Properties

The chemical structure and properties of amorphous carbon coatings are a function of deposition conditions. It is important to understand the relationship of the chemical structure of amorphous carbon coatings to the properties in order to define useful deposition parameters. Amorphous carbon films are metastable phases formed when carbon particles are condensed on a substrate. The prevailing atomic arrangement in the DLC coatings is amorphous or quasi-amorphous with small diamond (sp^3), graphite (sp^2), and other unidentifiable micro- or nanocrystallites. The coating is dependent upon the deposition process and its conditions contain varying amounts of sp^3/sp^2 ratio and hydrogen. The sp^3/sp^2 ratio of DLC coatings ranges typically from 50% to close to 100% with an increase in hardness with the sp^3/sp^2 ratio. Most DLC coatings, except those produced by filtered cathodic arc, contain from a few to about 50 at% hydrogen. Sometimes hydrogen and nitrogen are deliberately added to produce hydrogenated (a-C:H) and nitrogenated amorphous carbon (a-C:N) coatings, respectively. Hydrogen helps to stabilize sp^3 sites (most of the carbon atoms attached to hydrogen have a tetrahedral structure); therefore, the sp^3/sp^2 ratio for hydrogenated carbon is higher [30]. Optimum sp^3/sp^2 in a random covalent network composed of sp^3 and sp^2 carbon sites (N_{sp^2} and N_{sp^3}) and hydrogen is [30]:

$$\frac{N_{sp^3}}{N_{sp^2}} = \frac{6X_H - 1}{8 - 13X_H}, \tag{17.1}$$

where X_H is the atom fraction of hydrogen. The hydrogenated carbon has a larger optical band gap, higher electrical resistivity (semiconductor), and a lower optical absorption or high optical transmission. The hydrogenated coatings have a lower density, probably because of the reduction of cross-linking due to hydrogen incorporation. However, hardness decreases with an increase of the hydrogen, even though the proportion of sp^3 sites increases (that is, as the local bonding environment becomes more diamond-like) [78, 79]. It is speculated that the high hydrogen content introduces frequent terminations in the otherwise strong 3-D network, and hydrogen increases the soft polymeric component of the structure more than it enhances the cross-linking sp^3 fraction.

A number of investigations have been performed to identify the microstructure of amorphous carbon films using a variety of techniques such as Raman spectroscopy, EELS, nuclear magnetic resonance, optical measurements, transmission electron microscopy, and X-ray photoelectron spectroscopy [33]. The structure of diamond-like amorphous carbon is amorphous or quasi-amorphous with small graphitic (sp^2) and tetrahedrally coordinated (sp^3) and other unidentifiable nanocrystallites (typically on the order of a couple nm, randomly oriented) [33, 80, 81]. These studies indicate that the chemical structure and physical properties of the coatings are quite variable, depending on the deposition techniques and film growth conditions. It is clear that both sp^2 and sp^3-bonded atomic sites are incorporated in diamond-like amorphous carbon coatings and that the physical and chemical properties of the coatings depend strongly on their chemical bonding and microstructure. Systematic studies have been conducted to carry out chemical characterization and investigate how the physical and chemical properties of amorphous carbon coatings vary as a function of deposition parameters (e.g., [33, 35, 40]). EELS and Raman spectroscopy are commonly used to characterize the chemical bonding and microstructure. Hydrogen concentration of the coatings is obtained by means of forward recoil spectrometry (FRS). A variety of physical properties of the coatings relevant to tribological performance are measured.

To present the typical data obtained for characterization of typical amorphous carbon coatings and their relationships to physical properties, we present data on several sputtered coatings, RF-PECVD amorphous carbon and microwave-PECVD (MPECVD) diamond coatings [33, 35, 40]. The sputtered coatings were DC magnetron sputtered at a chamber pressure of 10 mTorr under sputtering power densities of 0.1 and 2.1 W/cm^2 in a pure Ar plasma, labeled as W1 and W2, respectively. They were prepared at a power density of 2.1 W/cm^2 with various hydrogen fractions of 0.5, 1, 3, 5, 7 and 10% of Ar/H, the gas mixtures labeled as H1, H2, H3, H4, H5, and H6, respectively.

17.3.1 EELS and Raman Spectroscopy

EELS and Raman spectra of four sputtered (W1, W2, H1, and H3) and one PECVD carbon samples were obtained. Figure 17.5 shows the EELS spectra of these carbon coatings. EELS spectra for bulk diamond and polycrystalline graphite in an energy range up to 50 eV are also shown in Fig. 17.5. One prominent peak is seen at 35 eV in diamond, while two peaks are seen at 27 eV and 6.5 eV in graphite, which are called ($\pi + \sigma$) and (π) peaks, respectively. These peaks are produced by the energy loss of transmitted electrons to the plasmon oscillation of the valence electrons. The $\pi + \sigma$ peak in each coating is positioned at a lower energy region than that of graphite. The π peaks in the W series and PECVD samples are also seen at a lower energy region than that of the graphite. However, the π peaks in the H-series are comparable to or higher than that of graphite (see Table 17.2). The plasmon oscillation frequency is proportional to the square root of the corresponding electron density to a first approximation. Therefore, the samples in the H-series most likely have a higher density of π electrons than the other samples.

Table 17.2. Experimental results from EELS and Raman spectroscopy [35]

Sample	EELS peak position		Raman peak position		Raman FWHM[a]		I_D/I_G^d
	π (eV)	$\pi + \sigma$ (eV)	G-band[b] (cm^{-1})	D-band[c] (cm^{-1})	G-band (cm^{-1})	D-band (cm^{-1})	
Sputtered a-C coating (W1)	5.0	24.6	1541	1368	105	254	2.0
Sputtered a-C coating (W2)	6.1	24.7	1560	1379	147	394	5.3
Sputtered a-C:H coating (H1)	6.3	23.3	1542	1334	95	187	1.6
Sputtered a-C:H coating (H3)	6.7	22.4	e	e	e	e	e
PECVD a-C:H coating	5.8	24.0	1533	1341	157	427	1.5
Diamond coating	...	...	1525[f]	1333[g]	...	8[g]	...
Graphite (for reference)	6.4	27.0	1580	1358	37	47	0.7
Diamond (for reference)	...	37.0	...	1332[g]	...	2[g]	...

[a] Full width at half maximum width
[b] Peak associated with sp^2 "graphite" carbon
[c] Peak associated with sp^2 "disordered" carbon (not sp^3-bonded carbon)
[d] Intensity ratio of the D-band to the G-band
[e] Fluorescence
[f] Includes D and G band, signal too weak to analyze
[g] Peak position and width for diamond phonon

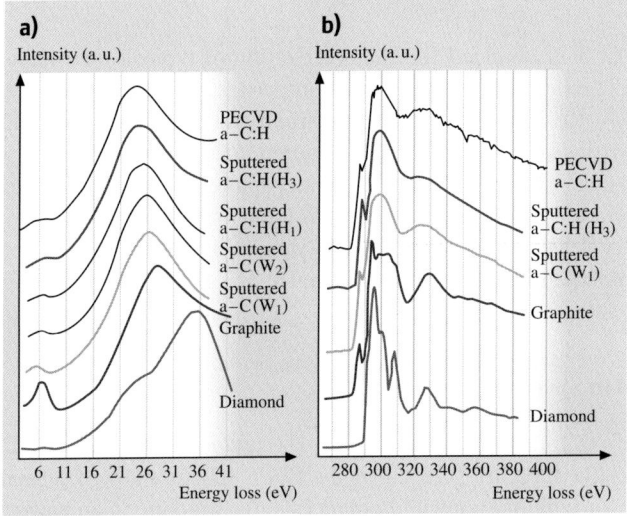

Fig. 17.5. (a) Low energy, and (b) high energy EELS of DLC coatings produced by DC magnetron sputtering and RF-PECVD techniques. Data for bulk diamond and polycrystalline graphite are included for comparison [35]

Amorphous carbon coatings contain (mainly) a mixture of sp^2- and sp^3-bonds, even though there is some evidence for the presence of sp-bonds as well [82]. The PECVD coatings and the H-series coatings in this study have nearly the same mass density, as seen in Table 17.4, to be presented later, but the former have a lower concentration of hydrogen (18.1%) than the H-series (35–39%), as seen in Table 17.3, to be presented later. The relatively low energy position of π peaks of PECVD coat-

Table 17.3. Experimental results of FRS analysis [35]

Sample	Ar/H ratio	C (at% ± 0.5)	H (at% ± 0.5)	Ar (at% ± 0.5)	O (at% ± 0.5)
Sputtered a-C coating (W2)	100/0	90.5	9.3	0.2	...
Sputtered a-C:H coating (H2)	99/1	63.9	35.5	0.6	...
Sputtered a-C:H coating (H3)	97/3	56.1	36.5	...	7.4
Sputtered a-C:H coating (H4)	95/5	53.4	39.4	...	7.2
Sputtered a-C:H coating (H5)	93/7	58.2	35.4	0.2	6.2
Sputtered a-C:H coating (H6)	90/10	57.3	35.5	...	7.2
PECVD a-C:H coating	99.5% CH_4	81.9	18.1	...	...
Diamond coating	H_2-1 mole % CH_4	94.0	6.0	...	...

Table 17.4. Experimental results of physical properties [35]

Sample	Mass density (g/cm^3)	Nano-hardness (GPa)	Elastic modulus (GPa)	Electrical resistivity (Ohm-cm)	Compressive residual stress (GPa)
Sputtered a-C coating (W1)	2.1	15	141	1300	0.55
Sputtered a-C:H coating (W2)	1.8	14	136	0.61	0.57
Sputtered a-C:H coating (H1)	...	14	96	...	> 2
Sputtered a-C:H coating (H3)	1.7	7	35	> 10^6	0.3
PECVD a-C:H coating	1.6–1.8	33–35	~ 200	> 10^6	1.5–3.0
Diamond coating	...	40–75	370–430	...	...
Graphite (for reference)	2.267	Soft	9–15	$5 \times 10^{-5 a}$, $4 \times 10^{-3 b}$	0
Diamond (for reference)	3.515	70–102	900–1050	$10^7 – 10^{20}$	0

^{a} Parallel to layer planes
^{b} Perpendicular to layer planes

ings, compared to those of the H-series, indicates that the PECVD coatings contain a higher fraction of sp^3-bonds than the sputtered hydrogenated carbon coatings (H-series).

Figure 17.5b shows the EELS spectra associated with the inner-shell (K-shell) ionization. Again, the spectra for diamond and polycrystalline graphite are included for comparison. Sharp peaks are observed at 285.5 eV and 292.5 eV in graphite, while no peak is seen at 285.5 eV in diamond. The general features of the K-shell EELS spectra for the sputtered and PECVD carbon samples resemble those of graphite, but with the higher energy features smeared. The observation of the peak at 285.5 eV in the sputtered and PECVD coatings also indicates the presence of sp^2-bonded atomic sites in the coatings. All these spectra peak at 292.5 eV, similar to the spectra of graphite, but the peak in graphite is sharper.

Raman spectra from samples W1, W2, H1, and PECVD are shown in Fig. 17.6. Raman spectra could not be observed in specimens H2 and H3 due to high flourescence signals. The Raman spectra of single crystal diamond and polycrstalline graphite are also shown for comparison in Fig. 17.6. The results of the spectral fits are summarized in Table 17.2. We will focus on the G-band position, which has been shown to be related to the fraction of sp^3-bonded sites. Increasing the power density in the amorphous carbon coatings (W1 and W2) results in a higher G-band frequency,

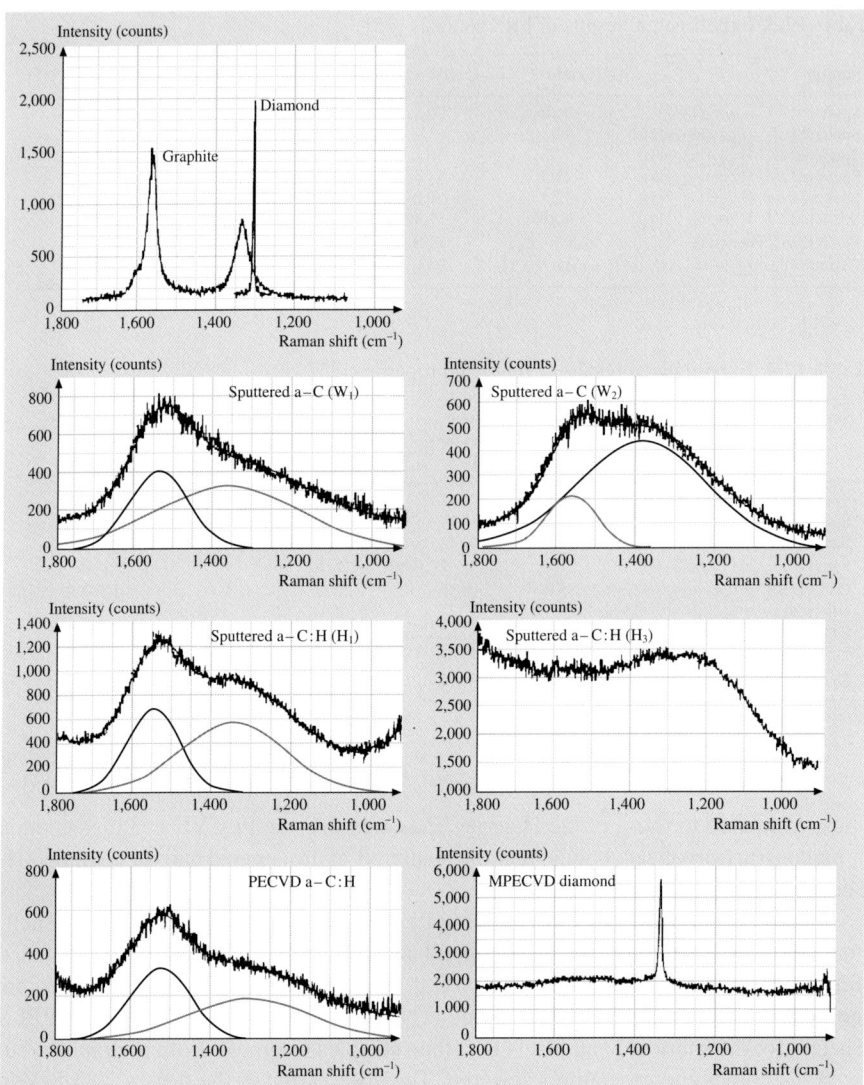

Fig. 17.6. Raman spectra of DLC coatings produced by DC magnetron sputtering and RF-PECVD techniques and a diamond film produced by MPECVD technique. Data for bulk diamond and microcrystalline graphite are included for comparison [35]

implying a smaller fraction of sp^3-bonding in W2 than in W1. This is consistent with higher density of W1. H1 and PEVCD have still lower G-band positions than W1, implying an even higher fraction of sp^3-bonding, which is presumably caused by the incorporation of H atoms into the lattice. The high hardness of H3 might be attributed to efficient sp^3 cross-linking of small, sp^2-ordered domains.

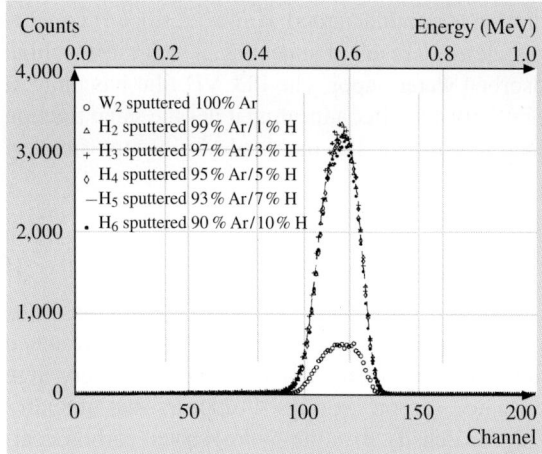

Fig. 17.7. FRS spectra of six DLC coatings produced by DC magnetron sputtering [35]

The Raman spectrum of a MPECVD diamond coating is shown in Fig. 17.6. The diamond Raman peak is at $1333\,\text{cm}^{-1}$ with a line width of $7.9\,\text{cm}^{-1}$. There is a small broad peak around $1525\,\text{cm}^{-1}$, which is attributed to a small amount of a-C:H. This impurity peak is not intense enough to fit to separate G- and D-bands. The diamond peak frequency is very close to that of natural diamond ($1332.5\,\text{cm}^{-1}$, e.g. Fig. 17.6), indicating that the coating is not under stress [83]. The large line width compared to thatof natural diamond ($2\,\text{cm}^{-1}$) indicates that the microcrystallites likely have a high concentration of defects [84].

17.3.2 Hydrogen Concentrations

FRS analysis of six sputtered (W2, H2, H3, H4, H5, and H6) coatings, one PECVD coating, and one diamond coating was performed. Figure 17.7 shows an overlay of the spectra from the six sputtered samples. Similar spectra were obtained from the PECVD and the diamond films. Table 17.3 shows the H and C fractions, as well as the amount of impurities (Ar and O) in the films in atomic %. Most apparent is the large fraction of H in the sputtered films. Regardless of how much H_2 is in the Ar sputtering gas, the H content of the coatings is about the same, ~ 35 at%. Interestingly, there is still $\sim 10\%$ H present in the coating sputtered in pure Ar (W2). It is interesting to note that Ar is present only in coatings grown under low ($< 1\%$) H content in the Ar carrier gas. The presence of O in the coatings, combined with the fact that the coatings were prepared approximately nine months before the FRS analysis, caused suspicion that they had absorbed water vapor, and that this may be the cause for the H peak in specimen W2.

All samples were annealed for 24 h at 250 °C in a flowing He furnace and then reanalyzed. Surprisingly, the H content of all coatings measured increased slightly, even though the O content decreased, and W2 still had a substantial amount of H_2.

This slight increase in H concentration is not understood. However, since H concentration did not decrease with the oxygen as a result of annealing, it suggests that high H concentration is not due to adsorbed water vapor. The PECVD film has more H ($\sim 18\%$) than the sputtered films initially, but after annealing it has the same fraction as specimen W2, the film sputtered in pure Ar. The diamond film has the smallest amounts of hydrogen, as seen in Table 17.3.

17.3.3 Physical Properties

Physical properties of the four sputtered (W1, W2, H1, and H3) coatings, one PECVD coating, one diamond coating, and bulk diamond and graphite are presented in Table 17.4. The hydrogenated carbon and the diamond coatings have very high resistivity compared to unhydrogenated carbon coatings. It appears that unhydrogenated carbon coatings have a higher density than the hydrogenated carbon coatings, although both groups are less dense than graphite. The density depends on the deposition technique and the deposition parameters. It appears that unhydrogenated sputtered coatings deposited at low power exhibit the highest density. Nanohardness of hydrogenated carbon is somewhat lower than that of the unhydrogenated carbon coatings. PECVD coatings are significantly harder than sputtered coatings. The nanohardness and modulus of elasticity of the diamond coatings are very high compared to that of DLC coatings, even though the hydrogen content is similar. The compressive residual stresses of the PECVD coatings are substantially higher than those of sputtered coatings, which is consistent with the hardness results.

Figure 17.8a shows the effect of hydrogen in the plasma on the residual stresses and the nanohardness for sputtered coatings W2 and H1 to H6. The coatings made with H_2 flow between 0.5 and 1.0% delaminate very quickly, even when only a few tens of nm thick. In pure Ar and at H_2 flows greater than 1%, the coatings appear to be more adhesive. The tendency of some coatings to delaminate can be caused by intrinsic stress in the coating, which is measured by substrate bending. All of the coatings in the figure are in compressive stress. The maximum stress occurs between 0 and 1% H_2 flow, but the stress cannot be quantified in this range because the coatings instantly delaminate upon exposure to air. At higher hydrogen concentrations the stress gradually diminishes. A generally decreasing trend is observed in the hardness of the coatings as hydrogen content increased. The hardness decreases slightly, going from 0% H_2 to 0.5% H_2, and then decreases sharply. These results are probably lower than the true values because of local delamination around the indentation point. This is especially likely for the 0.5% and 1.0% coatings where delamination is visually apparent, but may also be true to a lesser extent for the other coatings. Such an adjustment would bring the hardness profile into closer correlation with the stress profile. *Weissmantel* et al. [68] and *Scharff* et al. [85] observed a downturn in hardness for high bias and low pressure of hydrocarbon gas in ion plated carbon coating, and, therefore, presumably low hydrogen content in support of the above contention.

Figure 17.8b shows the effect of sputtering power (with no hydrogen added in the plasma) on the residual stresses and nanohardness for various sputtered coatings. As

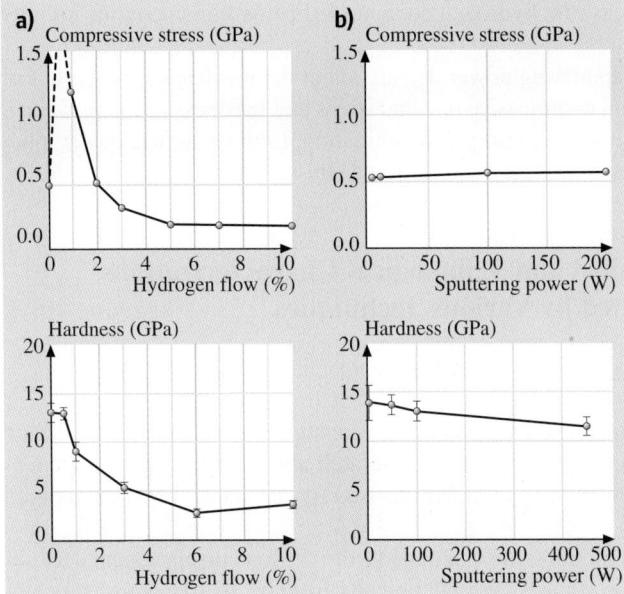

Fig. 17.8. Residual compressive stresses and nanohardness (**a**) as a function of hydrogen flow rate; sputtering power is 100 W and target diameter is 75 mm (power density = 2.1 W/cm²), and (**b**) as a function of sputtering power over a 75-mm-diameter target with no hydrogen added in the plasma [40]

the power decreases, compressive stress does not seem to change while nanohardness slowly increases. The rate of change becomes more rapid at very low power levels.

The addition of H_2 during sputtering of carbon coatings increases H concentration in the coating. Hydrogen causes the character of the C-C bonds to shift from sp^2 to sp^3, and the rising number of C-H bonds, which ultimately relieves stress and produces a softer "polymer-like" material. Low power deposition, like the presence of hydrogen, appears to stabilize the formation of sp^3 C-C bonds, increasing hardness. These coatings have relieved stress and led to better adhesion. An increase in temperature during deposition at high power density results in graphitization of the coating material, responsible for a decrease in hardness with an increase in power density. Unfortunately, low power also means impractically low deposition rates.

17.3.4 Summary

Based on the EELS and Raman data, all DLC coatings have both sp^2 and sp^3 bondings. The sp^2/sp^3 bonding ratio depends on the deposition techniques and parameters. The DLC coatings deposited by sputtering and PECVD contain significant concentrations of hydrogen, while the diamond coating contains only small amounts of hydrogen impurity. Sputtered coatings with no deliberate addition of hydrogen in the plasma contain a significant amount of hydrogen. Regardless of how much hydrogen is in the Ar sputtering gas, the hydrogen content of the coatings increases initially with no further increase.

Hydrogen flow and sputtering power density affect the mechanical properties of these coatings. Maximum compressive residual stress and hardness occur between 0 and 1% hydrogen flow, resulting in rapid delamination. Low sputtering power moderately increases hardness while relieving residual stress.

17.4 Micromechanical and Tribological Characterizations of Coatings Deposited by Various Techniques

17.4.1 Micromechanical Characterization

Common mechanical characterizations include measurement of hardness and elastic modulus, fracture toughness, fatigue life, and scratch and wear testing. Nanoindentation and atomic force microscopy (AFM) are used for mechanical characterization of ultrathin films.

Hardness and elastic modulus are calculated from the load displacement data obtained by nanoindentation at loads ranging typically from 0.2 to 10 mN using a commercially available nanoindenter [23, 86]. This instrument monitors and records the dynamic load and displacement of the three-sided pyramidal diamond (Berkovich) indenter during indentation. For the fracture toughness measurement of ultrathin films ranging from 100 nm to a few μm, a nanoindentation-based technique is used in which through-thickness cracking in the coating is detected from a discontinuity observed in the load-displacement curve and energy released during the cracking is obtained from the curve [87–89]. Based on the energy released, fracture mechanics analysis is then used to calculate fracture toughness. An indenter with a cube-corner tip geometry is preferred because the through-thickness cracking of hard films can be accomplished at lower loads. In fatigue measurement, a conical diamond indenter having a tip radius of about one micron is used and load cycles of a sinusoidal shape are applied [90, 91]. The fatigue behavior of coatings is studied by monitoring the change in contact stiffness, which is sensitive to damage formation.

Hardness and Elastic Modulus

For materials that undergo plastic deformation, high hardness and elastic modulus are generally needed for low friction and wear, whereas for brittle materials, high fracture toughness is needed [2, 3, 21]. DLC coatings used for many applications are hard and brittle, and values of hardness and fracture toughness need to be optimized.

Representative load-displacement plots of indentations made at 0.2 mN peak indentation load on 100-nm-thick DLC coatings deposited by the four deposition techniques on single-crystal silicon substrate are compared in Fig. 17.9. The indentation depths at the peak load range from about 18 to 26 nm, smaller than that of the coating thickness. Many of the coatings exhibit a discontinuity or pop-in marks in the loading curve, which indicate a sudden penetration of the tip into the sample. A nonuniform

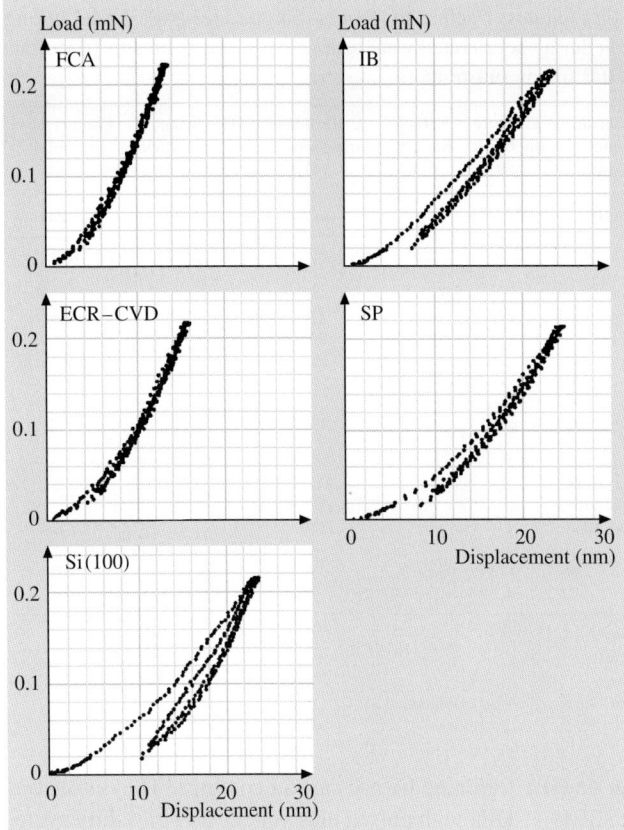

Fig. 17.9. Load versus displacement plots for various 100-nm-thick amorphous carbon coatings on single-crystal silicon substrate and bare substrate

penetration of the tip into a thin coating possibly results from formation of cracks in the coating, formation of cracks at the coating-substrate interface, or debonding or delamination of the coating from the substrate.

The hardness and elastic modulus values at a peak load of 0.2 mN on the various coatings and single-crystal silicon substrate are summarized in Table 17.5 and Fig. 17.10 [47, 49, 89, 90]. Typical values for the peak and residual indentation depths ranged from 18 to 26 nm and 6 to 12 nm, respectively. The FCA coatings exhibit the highest hardness of 24 GPa and elastic modulus of 280 GPa of various coatings, followed by the ECR-CVD, IB, and SP coatings. Hardness and elastic modulus have been known to vary over a wide range with sp^3-to-sp^2 bonding ratio, which depends on the kinetic energy of the carbon species and amount of hydrogen [6, 30, 47, 92, 93]. The high hardness and elastic modulus of the FCA coatings are attributed to the high kinetic energy of carbon species involved in the FCA deposition [12, 47]. *Anders* et al. [57] also reported a high hardness, measured by

Table 17.5. Hardness, elastic modulus, fracture toughness, fatigue life, critical load during scratch, coefficient of friction during accelerated wear testing and residual stresses of various DLC coatings on single-crystal silicon substrate

Coating	Hardness[a] [48] (GPa)	Elastic modulus[a] [48] (GPa)	Fracture toughness[a] [89] (MPa m$^{1/2}$)	Fatigue life[b], N_f^d [90] $\times 10^4$	Critical load during scratch[b] [48] (mN)	Coefficient of friction during accelerated wear testing[b] [48]	Compressive residual stress[c] [47] (GPa)
Cathodic arc carbon coating (a-C)	24	280	11.8	2.0	3.8	0.18	12.5
Ion beam carbon coating (a-C:H)	19	140	4.3	0.8	2.3	0.18	1.5
ECR-CVD carbon coating (a-C:H)	22	180	6.4	1.2	5.3	0.22	0.6
DC sputtered carbon coating (a-C:H)	15	140	2.8	0.2	1.1	0.32	2.0
Bulk graphite (for comparison)	Very soft		9–15	–	–	–	–
Diamond (for comparison)	80–104	900–1050	–	–	–	–	–
Si(100) substrate	11	220	0.75	–	0.6	0.55	0.02

[a] Measured on 100-nm-thick coatings
[b] Measured on 20-nm-thick coatings
[c] Measured on 400-nm-thick coatings
[d] N_f was obtained at a mean load of 10 μN and a load amplitude of 8 μN

nanoindentation, of about 45 GPa for cathodic arc carbon coatings. They observed a change in hardness from 25 to 45 GPa with pulsed bias voltage and bias duty cycle. The high hardness of cathodic arc carbon was attributed to the high percentage (more than 50%) of sp^3 bonding. *Savvides* and *Bell* [94] reported an increase in hardness from 12 to 30 GPa and an elastic modulus from 62 to 213 GPa with an increase of sp^3-to-sp^2 bonding ratio, from 3 to 6, for a C:H coating deposited by low-energy ion-assisted unbalanced magnetron sputtering of a graphite target in an Ar-H$_2$ mixture.

Bhushan et al. [35] reported hardnesses of about 15 and 35 GPa and elastic moduli of about 140 and 200 GPa, measured by nanoindentation, for a-C:H coatings deposited by DC magnetron sputtering and RF-plasma-enhanced chemical vapor deposition techniques, respectively. The high hardness of RF-PECVD a-C:H coatings is attributed to a higher concentration of sp^3 bonding than in a sputtered hydrogenated a-C:H coating. Hydrogen is believed to play a crucial role in the bonding configuration of carbon atoms by helping to stabilize tetrahedra-coordination (sp^3 bonding) of carbon species. *Jansen* et al. [78] suggested that the incorporation of hydrogen efficiently passivates the dangling bonds and saturates the graphitic bonding to some extent. However, a large concentration of hydrogen in the plasma in sputter deposition is undesirable. *Cho* et al. [33] and *Rubin* et al. [40] observed that hardness decreased from 15 to 3 GPa with increased hydrogen content. *Bhushan* and *Doerner* [95] re-

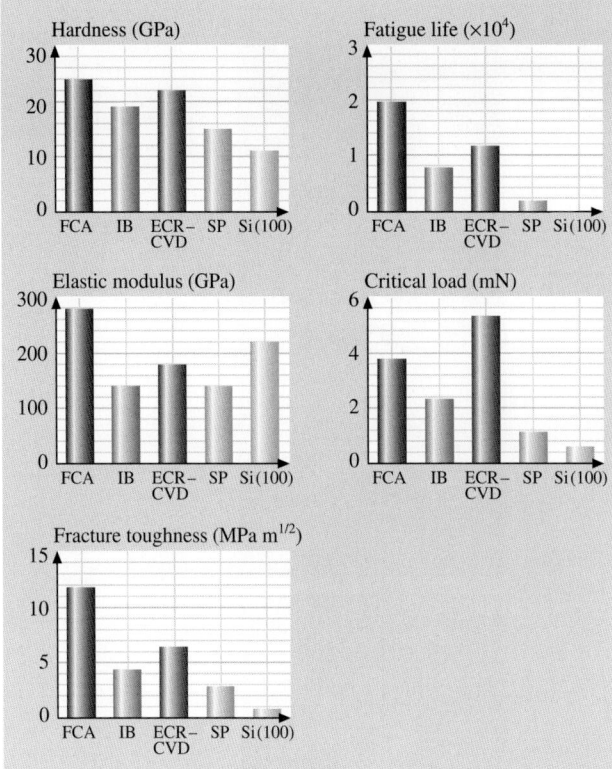

Fig. 17.10. Bar charts summarizing the data of various coatings and single-crystal silicon substrate. Hardness, elastic modulus, and fracture toughness were measured on 100-nm-thick coatings, and fatigue life and critical load during scratch were measured on 20-nm-thick coatings

ported a hardness of about 10–20 GPa and an elastic modulus of about 170 GPa, measured by nanoindentation, for 100-nm-thick DC-magnetron sputtered a-C:H on the silicon substrate.

Residual stresses measured using a well-known curvature measurement technique are also presented in Table 17.5. The DLC coatings are under significant compressive internal stresses. Very high compressive stresses in FCA coatings are believed to be partly responsible for their high hardness. However, high stresses result in coating delamination and buckling. For this reason, the coatings that are thicker than about 1 μm have a tendency to delaminate from the substrates.

Fracture Toughness

Representative load-displacement curves of indentations on the 400-nm-thick cathodic arc carbon coating on silicon at various peak loads are shown in Fig. 17.11. Steps are found in all curves, as shown by arrows in Fig. 17.11a. In the 30 mN SEM

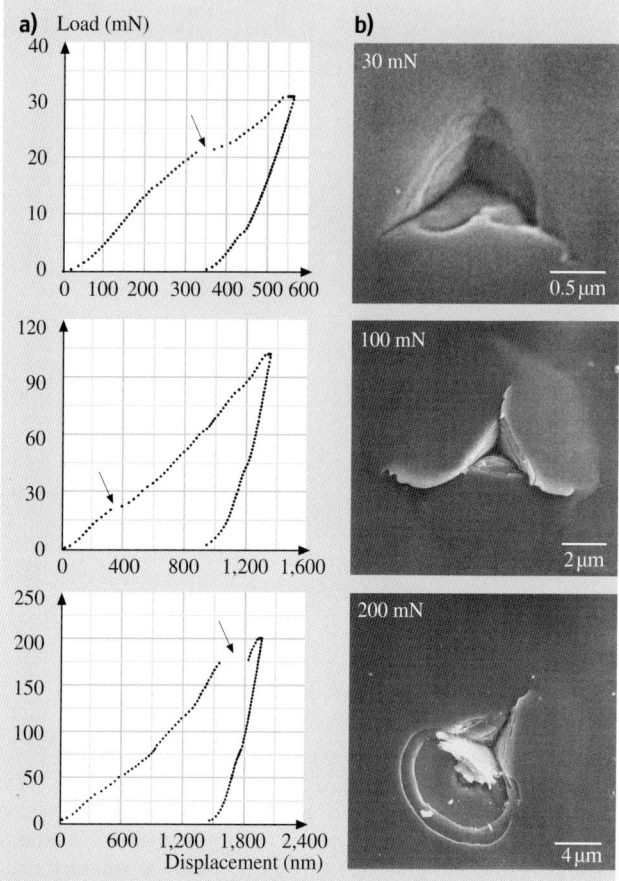

Fig. 17.11. (a) Load-displacement curves of indentations made at 30, 100, and 200 mN peak indentation loads using the cube corner indenter, and (b) the SEM micrographs of indentations on the 400-nm-thick cathodic arc carbon coating on silicon. *Arrows* indicate steps during loading portion of the load-displacement curve [87]

micrograph, in addition to several radial cracks, ring-like through-thickness cracking is observed with small lips of material overhanging the edge of indentation. The step at about 23 mN in the loading curves of indentations made at 30 and 100 mN peak indentation loads results from the ring-like through-thickness cracking. The step at 175 mN in the loading curve of the indentation made at 200 mN peak indentation load is caused by spalling and second ring-like through-thickness cracking.

Based on *Li* et al. [87], the fracture process progresses in three stages: (1) ring-like through-thickness cracks form around the indenter by high stresses in the contact area, (2) delamination and buckling occur around contact area at the coating/substrate interface by high lateral pressure, (3) second ring-like through-thickness cracks and spalling are generated by high bending stesses at the edges of the buckled coating, see Fig. 17.12a. In the first stage, if the coating under the indenter is separated from the bulk coating via the first ring-like through-thickness cracking, a corresponding step will be present in the loading curve. If discontinuous cracks form and the coating under the indenter is not separated from the remaining coating, no step appears in the loading curve, because the coating still supports the indenter and the indenter cannot suddenly advance into the material. In the second stage, for the coating used in the present study, the advances of the indenter during the radial cracking, delamination, and buckling are not big enough to form steps in the loading curve, because the coating around the indenter still supports the indenter, but generate discontinuities that change the slope of the loading curve with increasing indentation loads. In the third stage, the stress concentration at the end of the interfacial crack cannot be relaxed by the propagation of the interfacial crack. With an increase in indentation depth, the height of the bulged coating increases. When the height reaches a critical value, the bending stresses caused by the bulged coating around the indenter will result in the second ring-like through-thickness crack formation and spalling at the edge of the buckled coating, as shown in Fig. 17.12a, which leads to a step in the loading curve. This is a single event and results in the separation of the part of the coating around the indenter from the bulk coating via cracking through coatings. The step in the loading curve results totally from the coating cracking and not from the interfacial cracking or the substrate cracking.

The area under the load-displacement curve is the work performed by the indenter during elastic-plastic deformation of the coating/substrate system. The strain energy release in the first/second ring-like cracking and spalling can be calculated from the corresponding steps in the loading curve. Fig. 17.12b shows a modeled load-displacement curve. OACD is the loading curve and DE is the unloading curve. The first ring-like through-thickness crack should be considered. It should be emphasized that the edge of the buckled coating is far from the indenter and, therefore, it does not matter if the indentation depth exceeds the coating thickness, or if deformation of the substrate occurs around the indenter when we measure fracture toughness of the coating from the released energy during the second ring-like through-thickness cracking (spalling). Suppose that the second ring-like through-thickness cracking occurs at AC. Now, let us consider the loading curve OAC. If the second ring-like through-thickness crack does not occur, it can be understood that OA will be extended to OB to reach the same displacement as OC. This means that the crack formation changes the loading curve OAB into OAC. For point B, the elastic-plastic energy stored in the coating/substrate system should be OBF. For point C, the elastic-plastic energy stored in the coating/substrate system should be OACF. Therefore, the energy difference before and after the crack generation is the area of ABC, i.e., this energy stored in ABC will be released as strain energy to create the ring-like through-thickness crack. According to the theoretical analysis by *Li* et al. [87], the fracture toughness of thin films can be written as

$$K_{\mathrm{Ic}} = \left[\left(\frac{E}{(1-v^2)\,2\pi C_{\mathrm{R}}}\right)\left(\frac{U}{t}\right)\right]^{1/2}, \qquad (17.2)$$

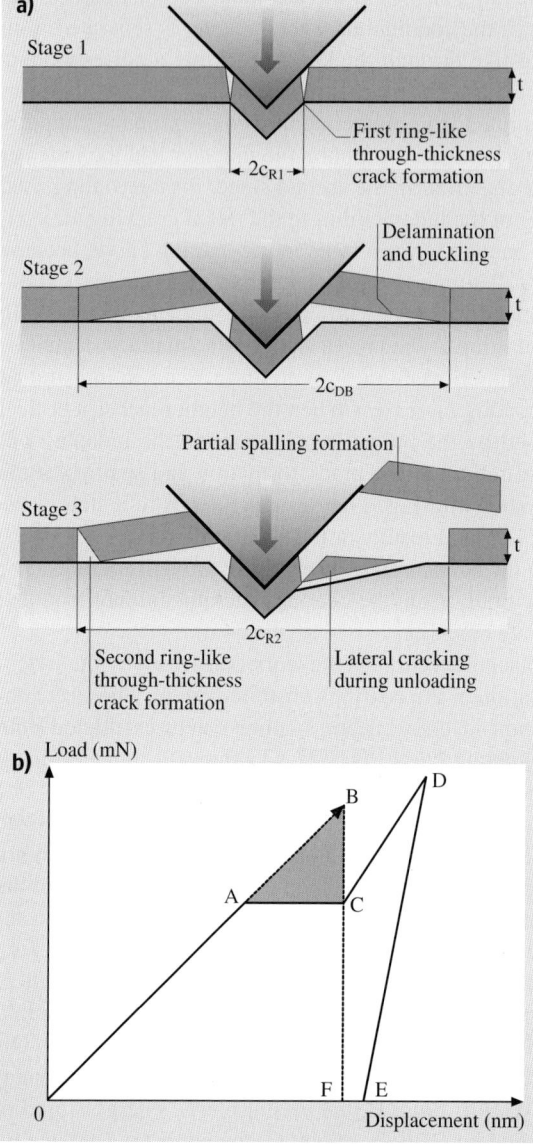

Fig. 17.12. (a) Schematic of various stages in nanoindentation fracture for the coatings/substrate system, and (b) schematic of a load-displacement curve showing a step during the loading cycle and associated energy release

where E is the elastic modulus, v is the Poisson's ratio, $2\pi C_R$ is the crack length in the coating plane, t is the coating thickness, and U is the strain energy difference before and after cracking.

Using (17.2), the fracture toughness of the coatings is calculated. The loading curve is extrapolated along the tangential direction of the loading curve from the starting point of the step up to reach the same displacement as the step. The area between the extrapolated line and the step is the estimated strain energy difference before and after cracking. C_R is measured from SEM micrographs or AFM images of indentations. The second ring-like crack is where the spalling occurs. For example, for the 400-nm-thick cathodic arc carbon coating data presented in Fig. 17.11, U of 7.1 nNm is assessed from the steps in Fig. 17.11a at the peak indentation loads of 200 mN. For C_R of 7.0 μm from Fig. 17.11b, E = 300 GPa measured using a nanoindenter and an assumed value of 0.25 for v, fracture toughness values are calculated as 10.9 MPa $\sqrt{m}$ [87, 88]. The fracture toughness and related data for various 100-nm-thick DLC coatings are presented in Fig. 17.10 and Table 17.5.

Nanofatigue

Delayed fracture resulting from extended service is called fatigue [96]. Fatigue fracturing progresses through a material via changes within the material at the tip of a crack, where there is a high stress intensity. There are several situations: cyclic fatigue, stress corrosion, and static fatigue. Cyclic fatigue results from cyclic loading of machine components. In a low-flying slider in a magnetic head-disk interface, isolated asperity contacts occur during use and the fatigue failure occurs in the multilayered thin-film structure of the magnetic disk [13]. In many MEMS components, impact occurs and the failure mode is cyclic fatigue. Asperity contacts can be simulated using a sharp diamond tip in an oscillating contact with the component.

Figure 17.13 shows the schematic of a fatigue test on a coating/substrate system using a continuous stiffness measurement (CSM) technique. Load cycles are applied to the coating, resulting in a cyclic stress. P is the cyclic load, P_{mean} is the mean load, P_o is the oscillation load amplitude, and ω is the oscillation frequency. The following results can be obtained: (1) endurance limit, i.e., the maximum load below which there is no coating failure for a preset number of cycles; (2) number of cycles at which the coating failure occurs; and (3) changes in contact stiffness measured by using the unloading slope of each cycle, which can be used to monitor the propagation of the interfacial cracks during cyclic fatigue process.

Figure 17.14a shows the contact stiffness as a function of the number of cycles for 20-nm-thick FCA coatings cyclically deformed by various oscillation load amplitudes with a mean load of 10 μN at a frequency of 45 Hz. At 4 μN load amplitude, no change in contact stiffness was found for all coatings. This indicates that 4 μN load amplitude is not high enough to damage the coatings. At 6 μN load amplitude, an abrupt decrease in contact stiffness was found at a certain number of cycles for each coating, indicating that fatigue damage had occurred. With increasing load amplitude, the number of cycles to failure, N_f, decreases for all coatings. Load amplitude versus N_f, a so-called S-N curve, is plotted in Fig. 17.14b. The critical load amplitude, below which no fatigue damage occurs (an endurance limit), was identified for

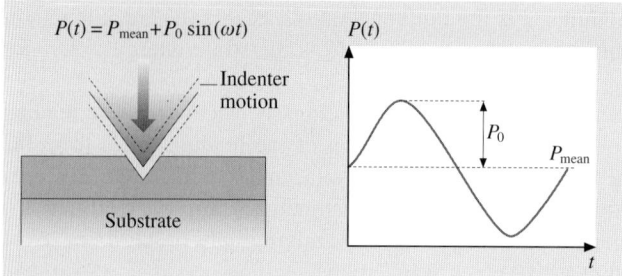

Fig. 17.13. Schematic of a fatigue test on a coating/substrate system using the continuous stiffness measurement technique

each coating. This critical load amplitude, together with mean load, are of critical importance to the design of head-disk interfaces or MEMS/NEMS device interfaces.

To compare the fatigue lives of different coatings studied, the contact stiffness as a function of the number of cycles for 20-nm-thick FCA, IB, ECR-CVD and SP coatings cyclically deformed by an oscillation load amplitude of 8 µN with a mean load of 10 µN at a frequency of 45 Hz is shown in Fig. 17.14c. FCA coating has the longest N_f, followed by ECR-CVD, IB, and SP coatings. In addition, after the N_f, the contact stiffness of the FCA coating shows a slower decrease than the other coatings. This indicates that after the N_f, the FCA coating had less damage than the others. The fatigue behavior of FCA and ECR-CVD coatings of different thicknesses is compared in Fig. 17.14d. For both coatings, N_f decreases with decreasing coating thickness. At 10 nm, FCA and ECR-CVD have almost the same fatigue life. At 5 nm, ECR-CVD coating shows a slightly longer fatigue life than FCA coating. This indicates that even for nanometer-thick DLC coatings their microstructure and residual stresses are not uniform across the thickness direction. Thinner coatings are more influenced by interfacial stresses than thicker coating.

Figure 17.15a shows the high magnification SEM images of 20-nm-thick FCA coatings before, at, and after N_f. In the SEM images, the net-like structure is the gold film coated on the DLC coating, which should be ignored in analyzing the indentation fatigue damage. Before the N_f, no delamination or buckling was found except the residual indentation mark at magnifications up to 1 200 000× using SEM. This suggests that only plastic deformation occurred before the N_f. At the N_f, the coating around the indenter bulged upwards, indicating delamination and buckling. Therefore, it is believed that the decrease in contact stiffness at the N_f results from the delamination and buckling of the coating from the substrate. After the N_f, the buckled coating was broken down around the edge of the buckled area, forming a ring-like crack. The remaining coating overhung at the edge of the buckled area. It is noted that the indentation size increases with the increasing number of cycles. This indicates that deformation, delamination and buckling, and ring-like crack formation occurred over a period.

The schematic in Fig. 17.15b shows various stages in the indentation fatigue damage for a coating/substrate system. Based on this study, three stages in the in-

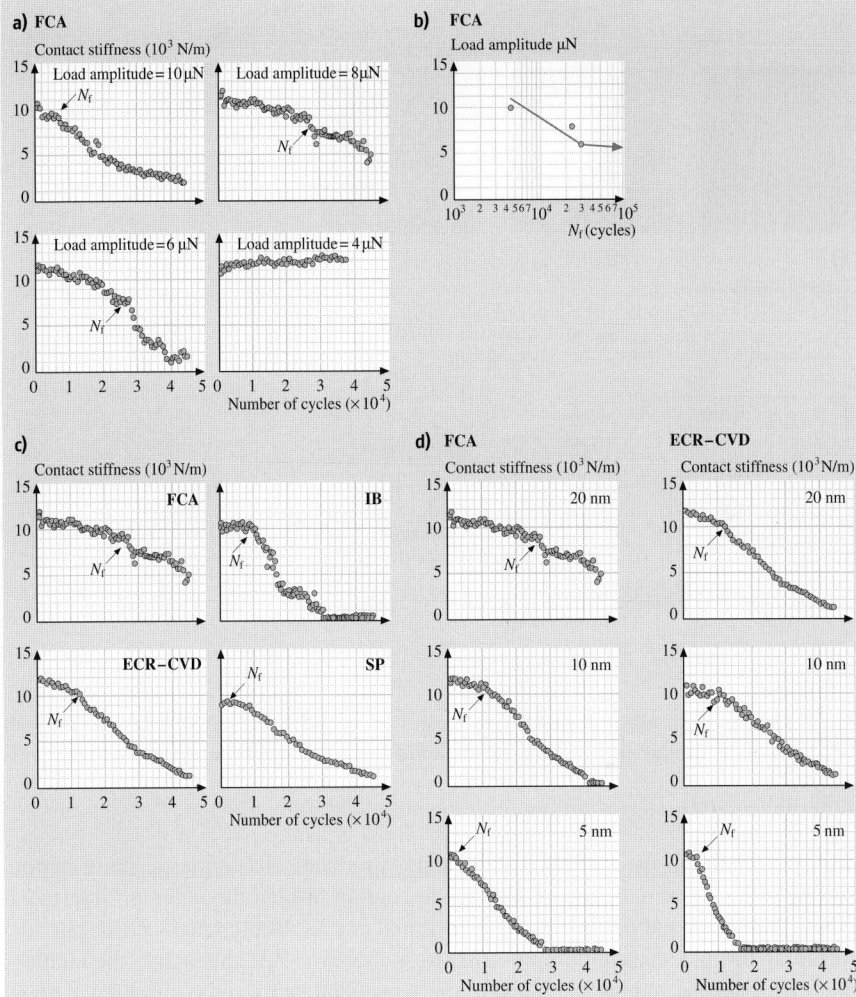

Fig. 17.14. (a) Contact stiffness as a function of the number of cycles for 20-nm-thick FCA coatings cyclically deformed by various oscillation load amplitudes with a mean load of 10 μN at a frequency of 45 Hz; (b) plot of load amplitude versus N_f; (c) contact stiffness as a function of the number of cycles for four different 20-nm-thick coatings with a mean load of 10 μN and load amplitude of 8 μN at a frequency of 45 Hz; and (d) contact stiffness as a function of the number of cycles for two coatings of different thicknesses at a mean load of 10 μN and load amplitude of 8 μN at a frequency of 45 Hz

dentation fatigue damage appear to exist: (1) indentation-induced compression; (2) delamination and buckling; (3) ring-like crack formation at the edge of the buckled coating. The deposition process often induces residual stresses in coatings. The model shown in Fig. 17.15b considers a coating with the uniform biaxial residual

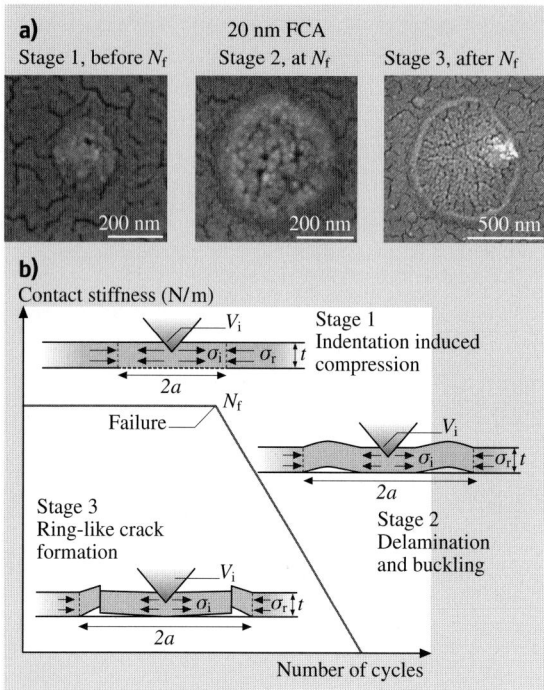

Fig. 17.15. (a) High magnification SEM images of a coating before, at, and after N_f, and (b) schematic of various stages in the indentation fatigue damage for a coating/substrate system [90]

compression σ_r. In the first stage, indentation induces elastic/plastic deformation, exerting an outward acting pressure on the coating around the indenter. Interfacial defects like voids and impurities act as original cracks. These cracks propagate and link up as the indentation compressive stress increases. At this stage, the coating, which is under the indenter and above the interfacial crack (with a crack length of $2a$), still maintains a solid contact with the substrate; the substrate still fully supports the coating. Therefore, this interfacial crack does not lead to an abrupt decrease in contact stiffness, but gives rise to a slight decrease in contact stiffness, as shown in Fig. 17.14. The coating above the interfacial crack is treated as a rigidly clamped disk. We assume that the crack radius, a, is large compared with the coating thickness t. Since the coating thickness ranges from 5 to 20 nm, this assumption is easily satisfied in this study (radius of the delaminated and buckled area, shown in Fig. 17.15a, is on the order of 100 nm). The compressive stress caused by indentation is given as [97]:

$$\sigma_i = \frac{E}{(1-v)}\varepsilon_i = \frac{EV_i}{2\pi t a^2 (1-v)}, \qquad (17.3)$$

where v and E are the Poisson's ratio and elastic modulus of the coating, V_i is the indentation volume, t is the coating thickness, and a is the crack radius. With an increasing number of cycles, the indentation volume V_i increases. Therefore, the indentation compressive stress σ_i increases accordingly. In the second stage, buckling occurs during the unloading segment of the fatigue testing cycle when the sum of indentation compressive stress σ_i and the residual stress σ_r exceed the critical buckling stress σ_b for the delaminated circular section as given by [98]

$$\sigma_b = \frac{\mu^2 E}{12(1-v^2)} \left(\frac{t}{a}\right)^2, \tag{17.4}$$

where the constant μ equals 42.67 for a circular clamped plate with a constrained center point and 14.68 when the center is unconstrained. The buckled coating acts as a cantilever. In this case, the indenter indents a cantilever rather than a coating/substrate system. This ultrathin coating cantilever has much less contact stiffness than the coating/substrate system. Therefore, the contact stiffness shows an abrupt decrease at the N_f. In the third stage, with an increasing number of cycles, the delaminated and buckled size increases, resulting in a further decrease in contact stiffness since the cantilever beam length increases. On the other hand, a high bending stress acts at the edge of the buckled coating. The larger the buckled size, the higher the bending stress. The cyclically bending stress causes fatigue damage at the end of the buckled coating, forming a ring-like crack. The coating under the indenter is separated from the bulk coating (caused by the ring-like crack at the edge of the buckled coating) and the substrate (caused by the delamination and buckling in the second stage). Therefore, the coating under the indenter is not constrained, but is free to move with the indenter during fatigue testing. At this point, the sharp nature of the indenter is lost, because the coating under the indenter gets stuck on the indenter. The indentation fatigue experiment results in the contact of a relative huge blunt tip with the substrate. This results in a low contact stiffness value.

Compressive residual stresses result in delamination and buckling. A coating with a higher adhesion strength and a less compressive residual stress is required for a higher fatigue life. Interfacial defects should be avoided in the coating deposition process. We know that the ring-like crack formation occurs in the coating. Formation of fatigue cracks in the coating depends upon the hardness and fracture toughness. Cracks are more difficult to form and propagate in the coating with higher strength and fracture toughness.

It is now accepted that long fatigue life in a coating/substrate almost always involves "living with crack", that the threshold or limit condition is associated with the non-propagation of exiting cracks or defects, even though these cracks may be undetectable [96]. For all coatings studied, at 4 µN, contact stiffness does not change much. This indicates that delamination and buckling did not occur within the number of cycles tested in this study. This is probably because the indentation-induced compressive stress was not high enough to allow the cracks to propagate and link up under the indenter, or the sum of indentation compressive stress σ_i and the residual stress σ_r did not exceed the critical buckling stress σ_b.

Figure 17.10 and Table 17.5 summarize the hardness, elastic modulus, fracture toughness, and fatigue life of all coatings studied. A good correlation exists between fatigue life and other mechanical properties. Higher mechanical properties result in a longer fatigue life. The mechanical properties of DLC coatings are controlled by the sp^3-to-sp^2 ratio. The sp^3-bonded carbon exhibits the outstanding properties of diamond [51]. A higher deposition kinetic energy will result in a larger fraction of sp^3-bonded carbon in an amorphous network. Thus, the higher kinetic energy for the FCA could be responsible for its better carbon structure and higher mechanical properties [48–50, 99]. Higher adhesion strength between the FCA coating and substrate makes the FCA coating more difficult to delaminate from the substrate.

17.4.2 Microscratch and Microwear Studies

For microscratch studies, a conical diamond indenter (e.g., having a tip radius of about one micron and included angle of 60°) is drawn over the sample surface, and the load is ramped up (typically from 2 mN to 25 mN) until substantial damage occurs. The coefficient of friction is monitored during scratching. Scratch-induced damage of coating, specifically fracture or delamination, can be monitored by in situ friction force measurements and by optical and SEM imaging of the scratches after tests. A gradual increase in friction is associated with plowing, and an abrupt increase in friction is associated with fracture or catastrophic failure [100]. The load, corresponding to the abrupt increase in friction or an increase in friction above a certain value (typically 2× the initial value), provides a measure of scratch resistance or adhesive strength of a coating and is called "critical load". The depth of scratches with increasing scratch length or normal load is measured using an AFM, typically with an area of $10 \times 10\,\mu m$ [48, 49, 101].

The microscratch and microwear studies are also conducted using an AFM [23, 50, 99, 102, 103]. A square pyramidal diamond tip (tip radius ~ 100 nm) or a three-sided pyramidal diamond (Berkovich) tip with an apex angle of 60° and a tip radius of about 100 nm mounted on a platinum-coated, rectangular stainless steel cantilever of stiffness of about 40 N/m is scanned orthogonal to the long axis of the cantilever to generate scratch and wear marks. During the scratch test, the normal load is either kept constant or increased (typically from 0 to 100 µN) until damage occurs. Topography images of the scratch are obtained in situ with the AFM at a low load. By scanning the sample during scratching, wear experiments can be conducted. Wear at a constant load is monitored as a function of the number of cycles. Normal loads ranging from 10–80 µN are typically used.

Microscratch

Scratch tests conducted with a sharp diamond tip simulate a sharp asperity contact. In a scratch test, the cracking or delamination of a hard coating is signaled by a sudden increase in the coefficient of friction [23]. The load associated with this event is called the "critical load".

Wu [104], *Bhushan* et al. [70], *Gupta* and *Bhushan* [12, 47], and *Li* and *Bhushan* [48, 49, 101] have used a nanoindenter to perform microscratch studies

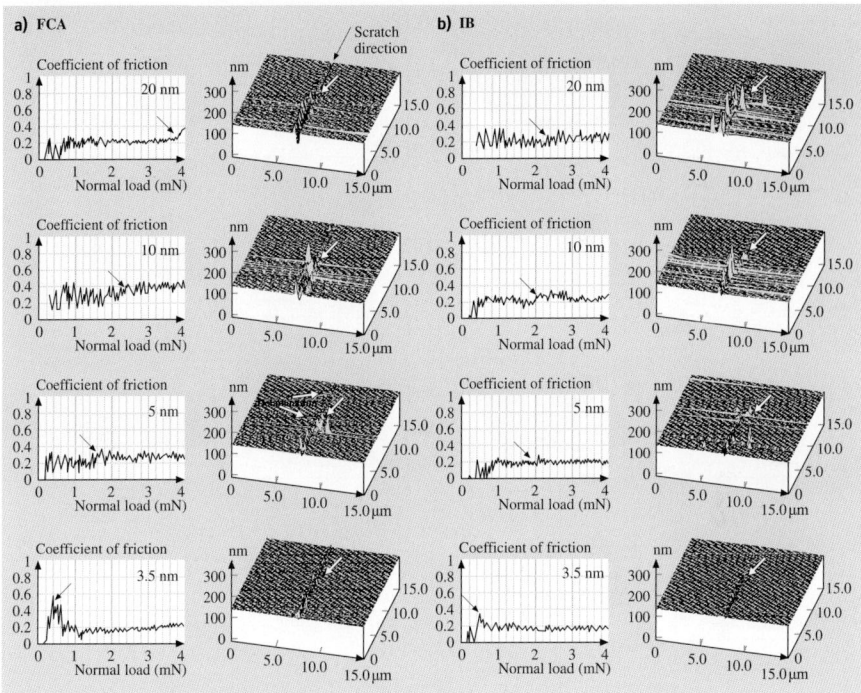

Fig. 17.16. Coefficient of friction profiles as a function of normal load and corresponding AFM surface height maps of regions over scratches at the respective critical loads (indicated by the arrows in the friction profiles and AFM images) for coatings of different thicknesses deposited by various deposition techniques: (**a**) FCA, (**b**) IB, (**c**) ECR-CVD, (**d**) SP

(mechanical durability) of various carbon coatings. The coefficient of friction profiles as a function of increasing normal load and AFM surface height maps of regions over scratches at the respective critical loads (indicated by the arrows in the friction profiles and AFM images) made on the various coatings of different thicknesses and single-crystal silicon substrate using a conical tip are compared in Figs. 17.16 and 17.17. *Bhushan* and *Koinkar* [102], *Koinkar* and *Bhushan* [103], *Bhushan* [23], and *Sundararajan* and *Bhushan* [50, 99] have used an AFM to perform microscratch studies. Data for various coatings of different thicknesses and silicon substrate using a Berkovich tip are compared in Figs. 17.18 and 17.19. Critical loads for various coatings tested using a nanoindenter and AFM are summarized in Fig. 17.20. The selected data for 20-nm-thick coatings obtained using nanoindenter are also presented in Fig. 17.10 and Table 17.5.

It can be seen that a well-defined critical load exists for each coating. The AFM images clearly show that below the critical loads the coatings were plowed by the scratch tip, associated with the plastic flow of materials. At and after the critical loads, debris (chips) or buckling was observed on the sides of scratches. Delamination or buckling can be observed around or after the critical loads, which suggests

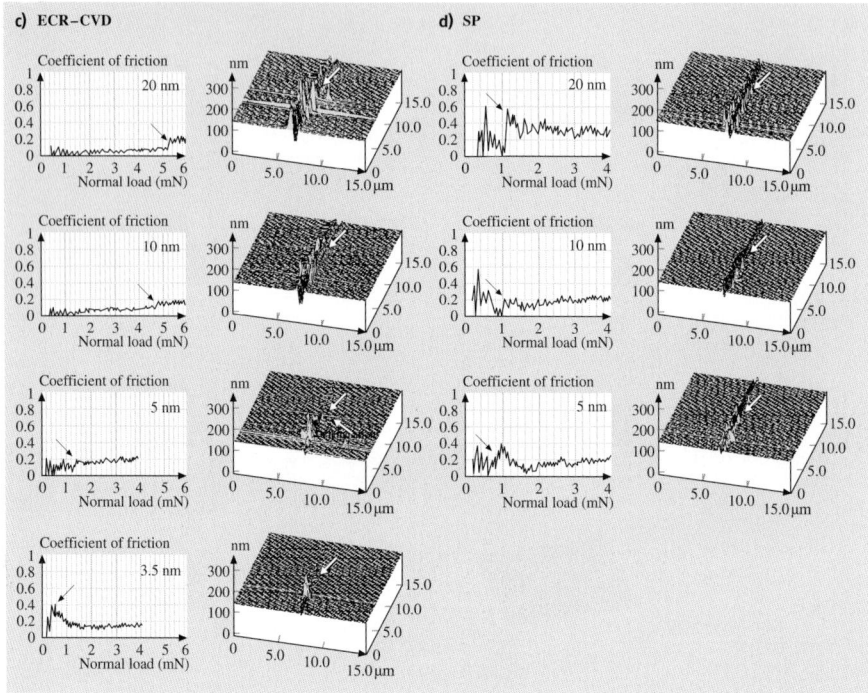

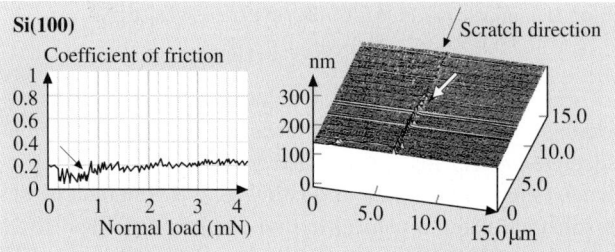

Fig. 17.17. Coefficient of friction profiles as a function of normal load and corresponding AFM surface height maps of regions over scratches at the respective critical loads (indicated by the arrows in the friction profiles and AFM images) for Si(100)

that the damage starts from delamination and buckling. For the 3.5- and 5-nm-thick FCA coatings, before the critical loads small debris is observed on the sides of scratches. This suggests that the thinner FCA coatings are not so durable. It is obvious that for a given deposition method, the critical loads increase with increasing coating thickness. This indicates that the critical load is determined not only by the adhesion strength to the substrate, but also by the coating thickness. We note that the

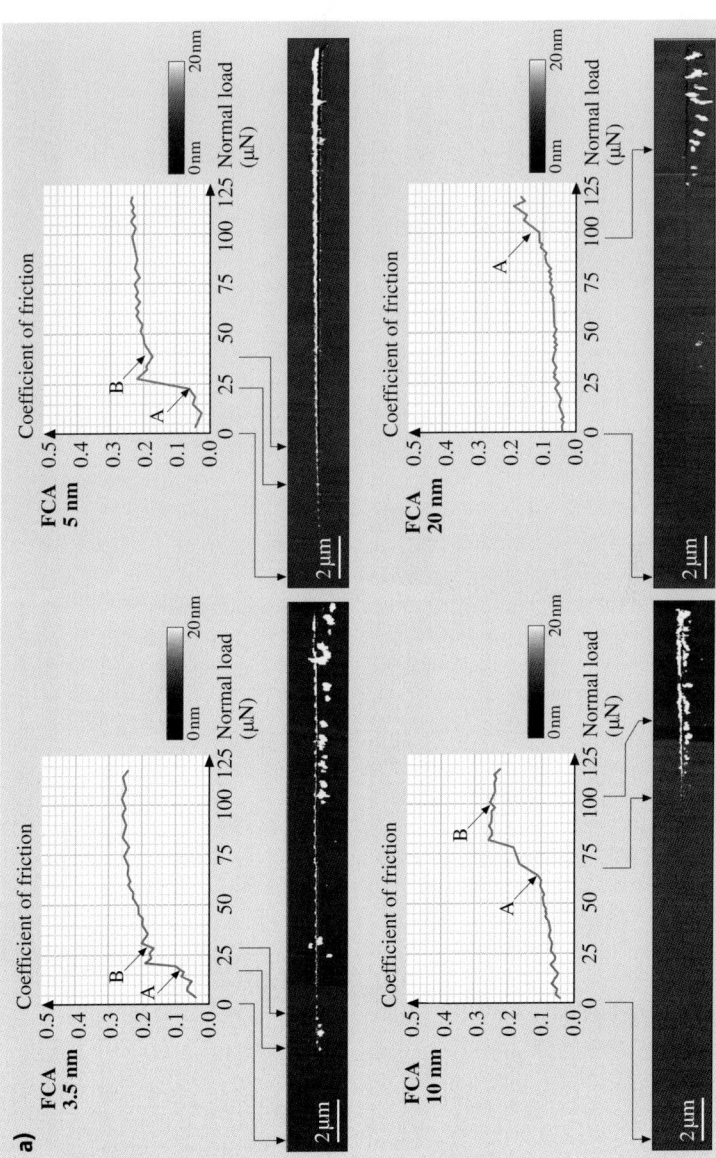

Fig. 17.18. Coefficient of friction profiles during scratch as a function of normal load and corresponding AFM surface height maps for (**a**) FCA, (**b**) ECR-CVD, and (**c**) SP coatings [99]

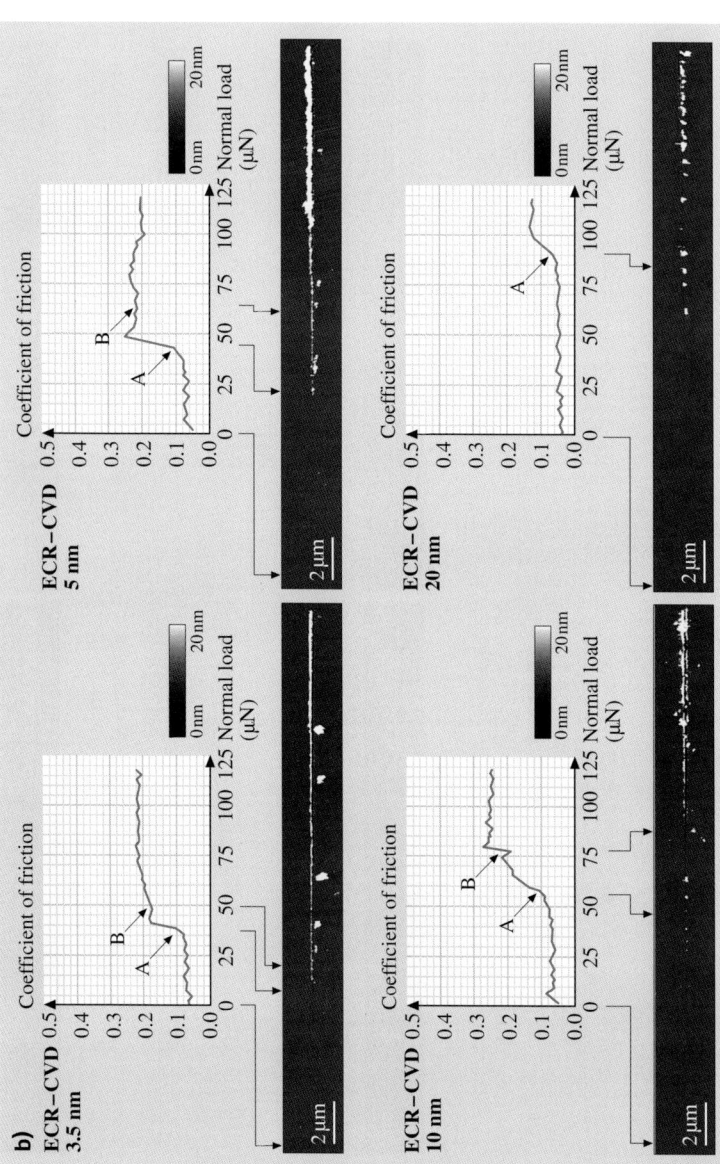

Fig. 17.18. continue

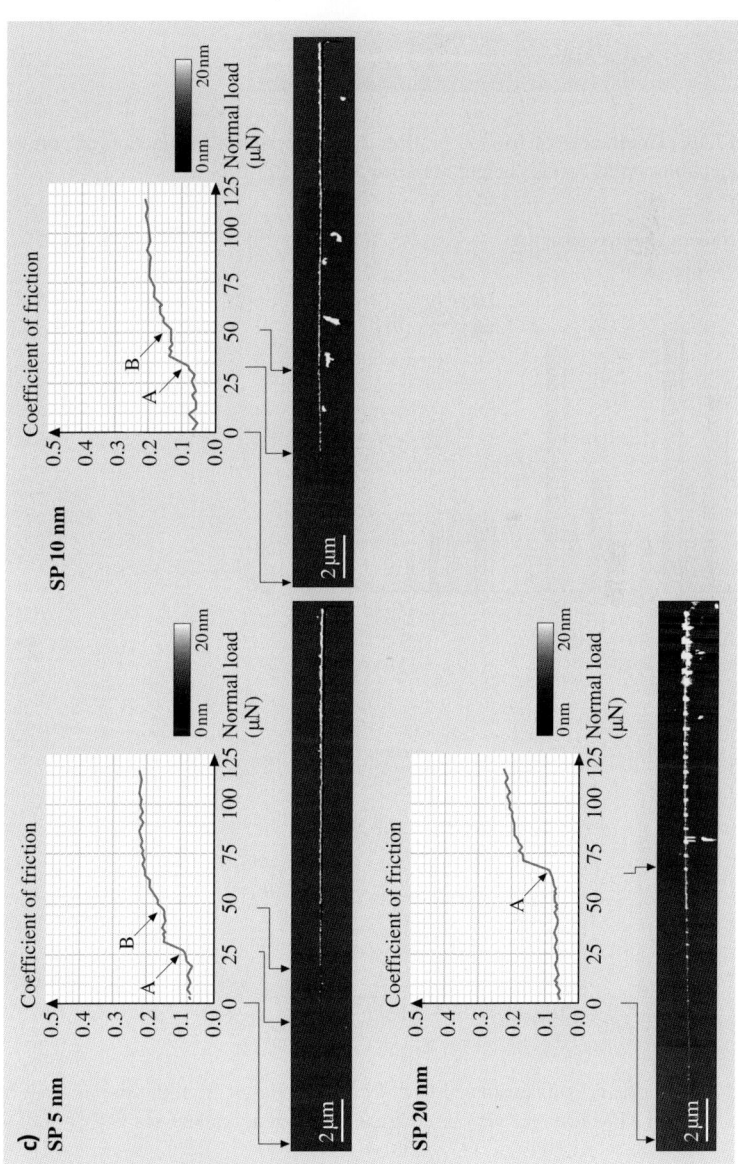

Fig. 17.18. continue

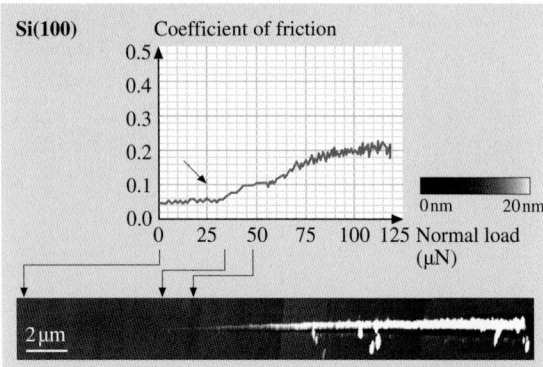

Fig. 17.19. Coefficient of friction profiles during scratch as a function of normal load and corresponding AFM surface height maps for Si(100) [99]

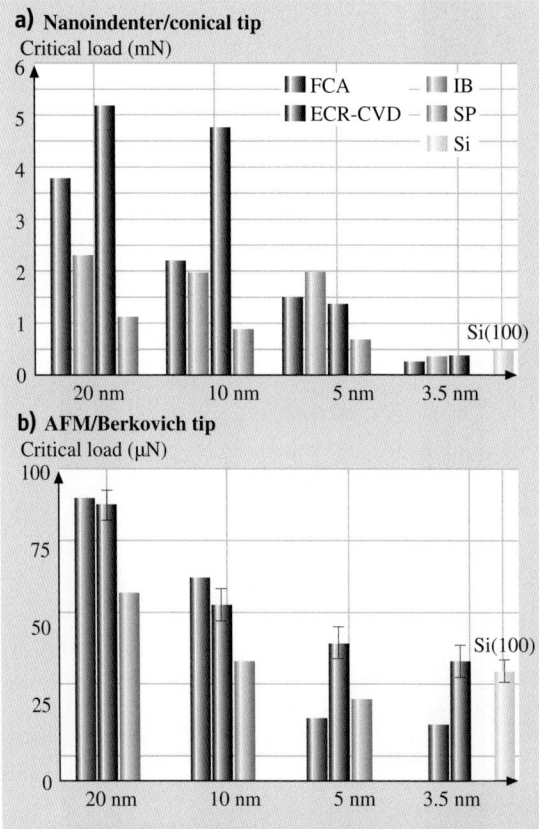

Fig. 17.20. Critical loads estimated from the coefficient of friction profiles from (**a**) nanoindenter and (**b**) AFM tests for various coatings of different thicknesses and Si(100) substrate

debris generated on the thicker coatings is larger than that generated on the thinner coatings. For a thicker coating, it is more difficult to be broken; the broken coating chips (debris) for a thicker coating are larger than those for the thinner coatings. The difference in the residual stresses of the coatings of different thicknesses could also affect the size of debris. The AFM image shows that the silicon substrate was damaged by plowing, associated with the plastic flow of materials. At and after the critical load, small and uniform debris is observed and the amount of debris increases with increasing normal load.

Since at the critical load, the damage mechanism appears to be the onset of plowing, higher hardness and fracture toughness of a coating will therefore result in a higher load required for deformation and hence a higher critical load. Figure 17.21 shows critical loads of the various coatings, obtained with AFM tests, as a function of the coating hardness and fracture toughness (from Table 17.5). It can be seen that, in general, higher coating hardness and fracture toughness result in a higher critical load. The only exceptions are the FCA coatings at 5- and 3.5-nm coating thickness, which show the lowest critical loads despite their high hardness and fracture toughness. The brittleness of the thinner FCA coatings may be one reason for their low critical loads. The mechanical properties of coatings that are less than 10 nm thick are unknown. The FCA process may result in coatings with low hardness at such low thickness due to differences in coating stoichiometry and structure compared to the coatings of higher thickness. Also, at these thicknesses stresses at the coating-substrate interface may affect adhesion and load-carrying capacity of the coatings.

Based on the experimental results, a schematic of scratch damage mechanisms of the DLC coatings used in this study is shown in Fig. 17.22. Below the critical load, if a coating has a good combination of strength and fracture toughness, plowing associated with the plastic flow of materials is responsible for the damage of coating (Fig. 17.22a). Whereas if a coating has a lower fracture toughness, cracking could occur during plowing, associated with the formation of small debris (Fig. 17.22b). When normal load increases up to the critical load, delamination or buckling will occur at the coating/substrate interface (Fig. 17.22c). A further increase in normal load will result in the breakdown of coating via through coating thickness cracking, as shown in Fig. 17.22d. Therefore, adhesion strength plays a crucial role in the determination of critical load. If a coating has stronger adhesive strength with the substrate, the coating is more difficult to delaminate, which will result in a higher critical load. The interfacial and residual stresses of a coating could greatly affect the delamination and buckling [1]. The coating with higher interfacial and residual stresses is more easily delaminated and buckled, which will result in a low critical load. It has been reported earlier that the FCA coatings have higher residual stresses than the other coatings [47]. Interfacial stresses play a more important role when a coating gets thinner. A large mismatch in elastic modulus between the FCA coatings and the silicon substrate may cause large interfacial stresses. This may be why the thinner FCA coatings show relatively low critical loads compared with the thicker FCA coatings, even though the FCA coatings have a higher hardness and elastic modulus. The brittleness of the thinner FCA coatings may be another reason for the lower critical loads. The strength and fracture toughness of a coating also affect the critical

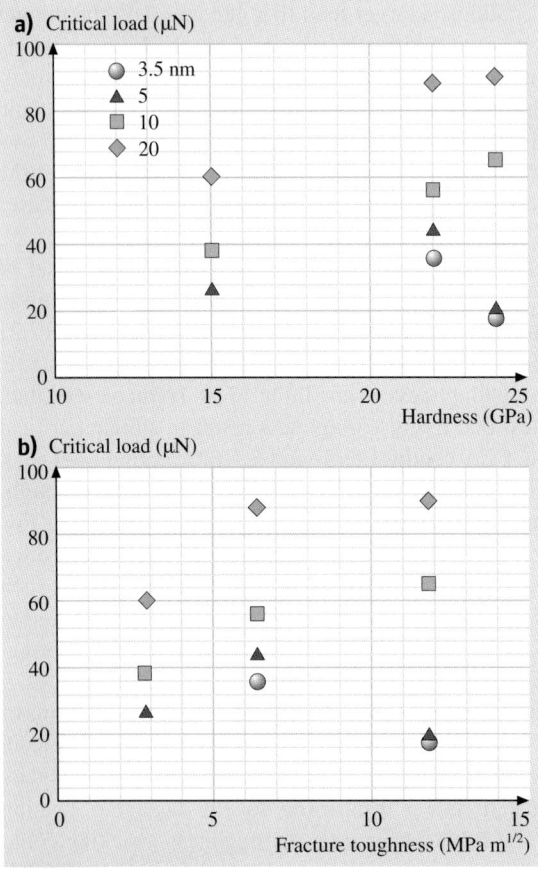

Fig. 17.21. Measured critical loads estimated from the coefficient of friction profiles from AFM tests as a function of (**a**) coating hardness and (**b**) fracture toughness. Coating hardness and fracture toughness values were obtained using a nanoindenter on 100-nm-thick coatings (Table 17.5)

load. Higher strength and fracture toughness will make the coating more difficult to be broken after delamination and buckling. The high scratch resistance/adhesion of FCA coatings is attributed to an atomic intermixing at the coating-substrate interface because of high kinetic energy (2 keV) plasma formed during the cathodic arc deposition process [57]. The atomic intermixing at the interface provides a graded compositional transition between the coating and the substrate materials. In all other coatings used in this study, the kinetic energy of the plasma was insufficient for atomic intermixing.

Gupta and *Bhushan* [12, 47] and *Li* and *Bhushan* [48, 49] measured scratch resistance of DLC coatings deposited on Al_2O_3-TiC, Ni-Zn ferrite, and single-crystal silicon substrates. For good adhesion of DLC coating to other substrates, in most

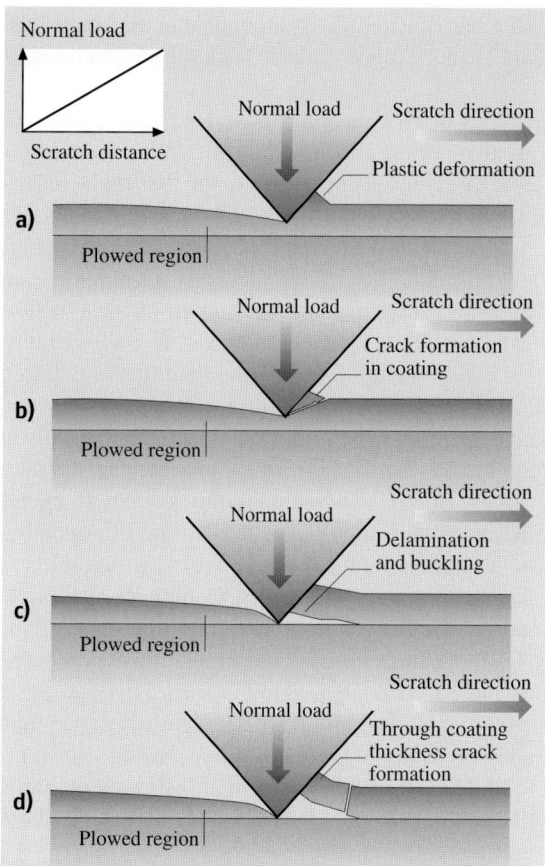

Fig. 17.22. Schematic of scratch damage mechanisms of the DLC coatings: (**a**) plowing associated with the plastic flow of materials, (**b**) plowing associated with the formation of small debris, (**c**) delamination and buckling at the critical load, and (**d**) breakdown via through coating thickness cracking at and after the critical load [48]

cases, an interlayer of silicon is required except for cathodic arc deposited coatings. The best adhesion with cathodic arc carbon coating is obtained on electrically conducting substrates such as Al_2O_3-TiC and silicon, as compared to Ni-Zn ferrite.

Microwear

Microwear studies can be conducted using an AFM [23]. For microwear studies, a three-sided pyramidal single-crystal natural diamond tip with an apex angle of about 80° and a tip radius of about 100 nm is used at relatively high loads of 1–150 µN. The diamond tip is mounted on a stainless steel cantilever beam with a normal stiffness of about 30 N/m. The sample is generally scanned in a direction orthogonal to the long axis of the cantilever beam (typically at a rate of 0.5 Hz). The tip is mounted on the beam such that one of its edges is orthogonal to the beam axis. In wear studies, typically an area of 2 µm × 2 µm is scanned for a selected number of cycles.

Microwear studies of various types of DLC coatings have been conducted [50, 102, 103]. Fig. 17.23a shows a wear mark on uncoated Si(100). Wear occurs uniformly and material is removed layer by layer via plowing from the first cycle, resulting in the constant friction force seen during the wear (Fig. 17.24a). Figure 17.23b shows AFM images of the wear marks on all 10 nm coatings. It is seen that coatings wear nonuniformly. Coating failure is sudden and accompanied by a sudden rise in the friction force (Fig. 17.24b). Figure 17.24 shows the wear depth of Si(100) substrate and various DLC coatings at two different loads. FCA and ECR-CVD, 20-nm-thick coatings show excellent wear resistance up to 80 µN, the load that is required for the IB 20 nm coating to fail. In these tests, "failure" of a coating results when the wear depth exceeds the quoted coating thickness. The SP 20 nm coating fails at the much lower load of 35 µN. At 60 µN, the coating hardly provides any protection. Moving on to the 10 nm coatings, ECR-CVD coating requires about 45 cycles at 60 µN to fail as compared to IB and FCA, which fail at 45 µN. The FCA coating exhibits slight roughening in the wear track after the first few cycles, which leads to an increase in the friction force. The SP coating continues to exhibit poor resistance, failing at 20 µN. For the 5 nm coatings, the load required to fail the coatings continues to decrease. But IB and ECR-CVD still provide adequate protection as compared to bare Si(100) in that order, failing at 35 µN compared to FCA at 25 µN and SP at 20 µN. Almost all the 20, 10, and 5 nm coatings provide better wear resistance than bare silicon. At 3.5 nm, FCA coating provides no wear resistance, failing almost instantly at 20 µN. The IB and ECR-CVD coating show good wear resistance at 20 µN compared to bare Si(100). But IB lasts only about 10 cycles and ECR-CVD about 3 cycles at 25 µN.

The wear tests highlight the differences in the coatings more vividly than the scratch tests. At higher thicknesses (10 and 20 nm), the ECR-CVD and FCA coatings appear to show the best wear resistance. This is probably due to higher hardness of the coatings (see Table 17.5). At 5 nm, IB coating appears to be the best. FCA coatings show poorer wear resistance with decreasing coating thickness. This suggests that the trends in hardness seen in Table 17.5 no longer hold at low thicknesses. SP coatings showed consistently poor wear resistance at all thicknesses. The IB 3.5 nm coating does provide reasonable wear protection at low loads.

17.4.3 Macroscale Tribological Characterization

So far, data on mechanical characterization and microscratch and microwear studies using a nanoindenter and an AFM have been presented. Mechanical properties affect tribological performance of the coatings, and microwear studies simulate a single asperity contact, which helps in developing a fundamental understanding of the wear process. These studies are useful in screening various candidates, as well as understanding the relationships between deposition conditions and properties of various

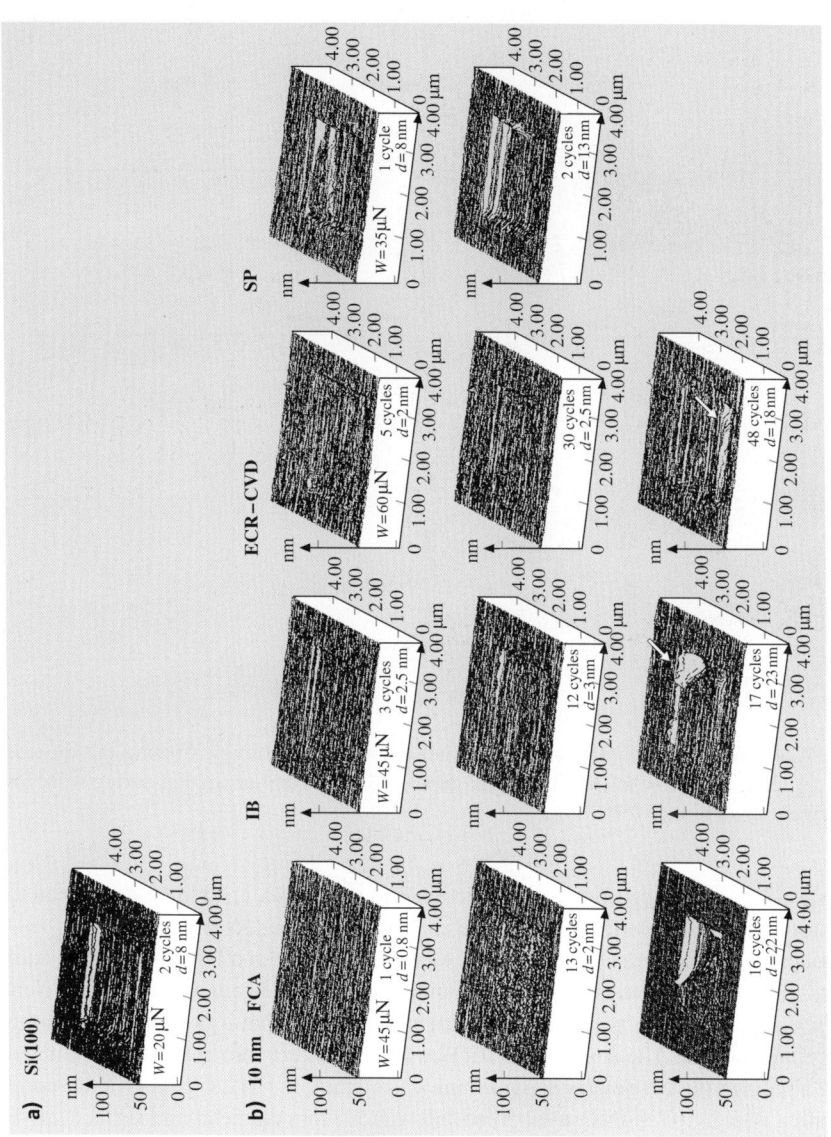

Fig. 17.23. AFM images of wear marks on (**a**) bare Si(100), and (**b**) all 10-nm-thick DLC coatings [50]

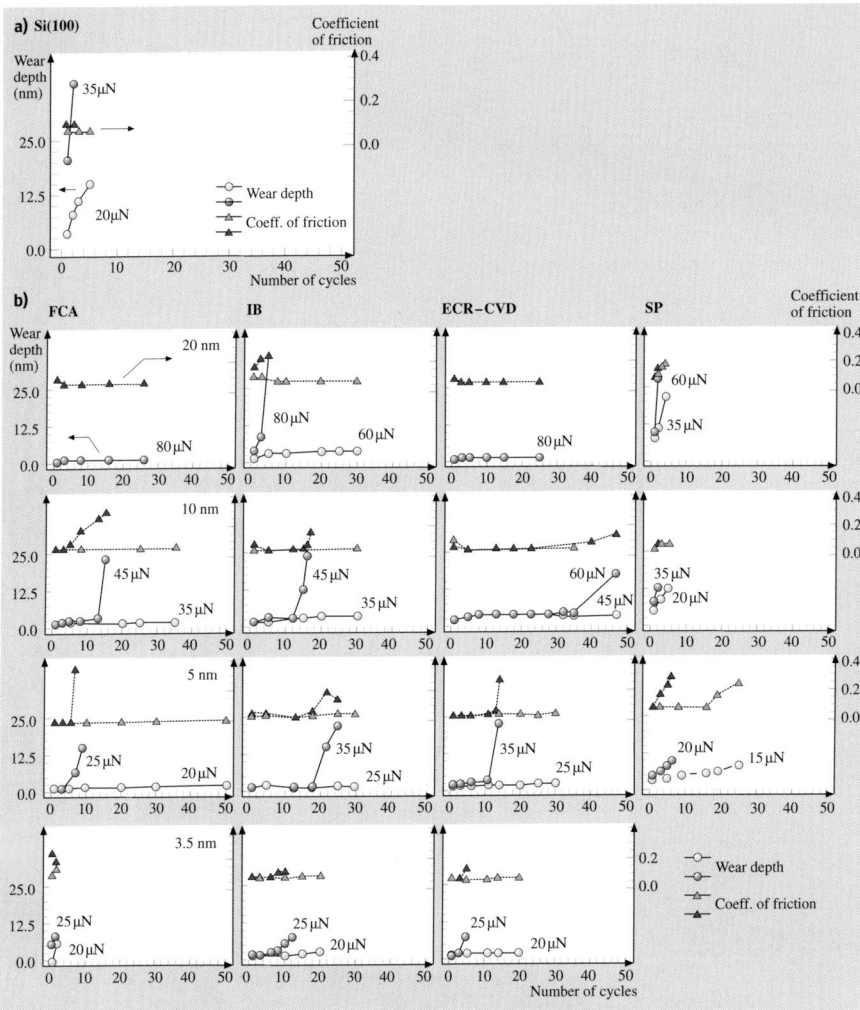

Fig. 17.24. Wear data of (**a**) bare Si(100) and (**b**) all DLC coatings. Coating thickness is constant along each row in (**b**). Both wear depth and coefficient of friction during wear for a given cycle are plotted [50]

samples. As a next step, macroscale friction and wear tests need to be conducted to measure tribological performance of the coatings.

Macroscale accelerated friction and wear tests to screen a large number of candidates and functional tests on selected candidates have been conducted. An accelerated test is designed to accelerate the wear process such that it does not change the failure mechanism. The accelerated friction and wear tests are generally conducted using a ball-on-flat tribometer under reciprocating motion [70]. Typically, a diamond tip with a 20 μm tip radius or a sapphire ball with a 3 mm diameter and surface finish of about 2 nm RMS is slid against the coated substrates at selected loads. Coefficient of friction is monitored during the tests.

Functional tests are conducted using an actual machine under close-to-actual operating conditions for which coatings are developed. Generally, the tests are accelerated somewhat to fail the interface in a short time.

Accelerated Friction and Wear Tests

Li and *Bhushan* [48] conducted accelerated friction and wear tests on DLC coatings deposited by various deposition techniques using a ball-on-flat tribometer. Average values of coefficient of friction are presented in Table 17.5. The optical micrographs of wear tracks and debris formed on all samples when slid against a sapphire ball after a sliding distance of 5 m are presented in Fig. 17.25. The normal load used for the 20- and 10-nm-thick coatings is 200 mN, and the normal load used for the 5- and 3.5-nm-thick coatings and silicon substrate is 150 mN.

Among the 20-nm-thick coatings, the SP coating exhibits a higher coefficient of friction of about 0.3 than the other coatings, which have comparable values of coefficient of friction of about 0.2. The optical micrographs show that the SP coating has a larger wear track and more debris than the IB coatings. No wear track and debris were found on the 20-nm-thick FCA and ECR-CVD coatings. The optical micrographs of 10-nm-thick coatings show that the SP coating was severely damaged, showing a large wear track with scratches and lots of debris. The FCA and ECR-CVD coatings show smaller wear tracks and less debris than the IB coatings.

For the 5-nm-thick coatings, the wear tracks and debris of the IB and ECR-CVD coatings are comparable. The bad wear resistance of the 5-nm-thick FCA coating is in good agreement with the low scratch critical load, which may be due to the higher interfacial and residual stresses as well as brittleness of the coating.

At 3.5 nm, all coatings exhibit wear. The FCA coating provides no wear resistance, failing instantly like the silicon substrate. Large block-like debris is observed on the sides of the wear track of the FCA coating. This indicates that large region delamination and buckling occurred during sliding, resulting in large block-like debris. This block-like debris, in turn, scratched the coating, making the coating damage even more severe. The IB and ECR-CVD coatings are able to provide some protection against wear at 3.5 nm.

In order to better evaluate the wear resistance of various coatings, based on the optical examination of the wear tracks and debris after tests, the wear damage index bar chart of the various coatings of different thicknesses and an uncoated silicon substrate is presented in Fig. 17.26. Among the 20- and 10-nm-thick coatings, the SP coatings show the worst damage, followed by FCA/ECR-CVD. At 5 nm, the FCA and SP coatings show the worst damage, followed by IB and ECR-CVD coatings. All 3.5-nm-thick coatings show the same heavy damage as the uncoated silicon substrate.

The wear damage mechanisms of thick and thin DLC coatings studied are believed to be as illustrated in Fig. 17.27. At the early stages of sliding, deformation zone and Hertzian and wear fatigue cracks formed beneath the surface extend within the coating on subsequent sliding [1]. Formation of fatigue cracks depend on the

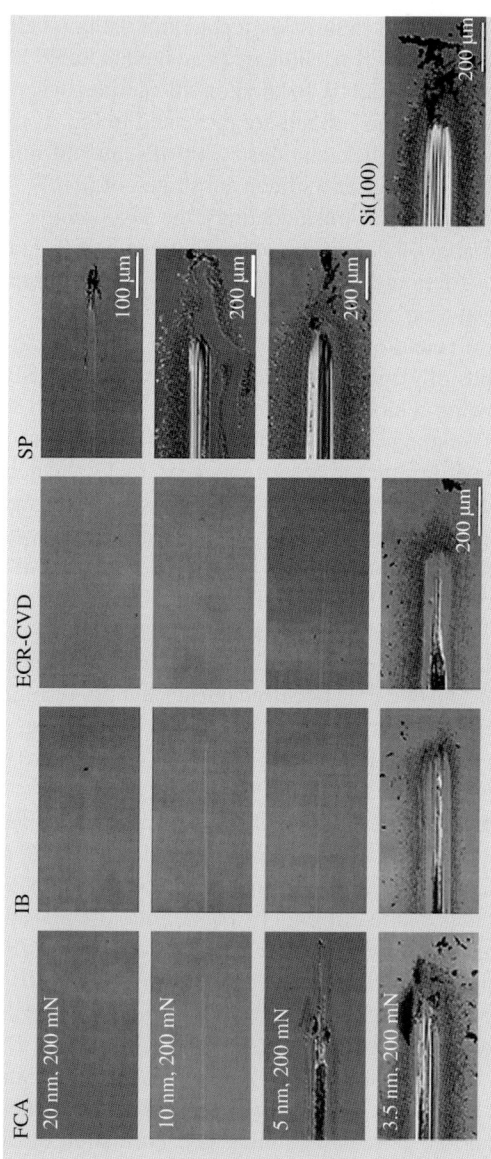

Fig. 17.25. Optical micrographs of wear tracks and debris formed on the various coatings of different thicknesses and silicon substrate when slid against a sapphire ball after sliding distance of 5 m. The end of the wear track is on the right-hand side of the image

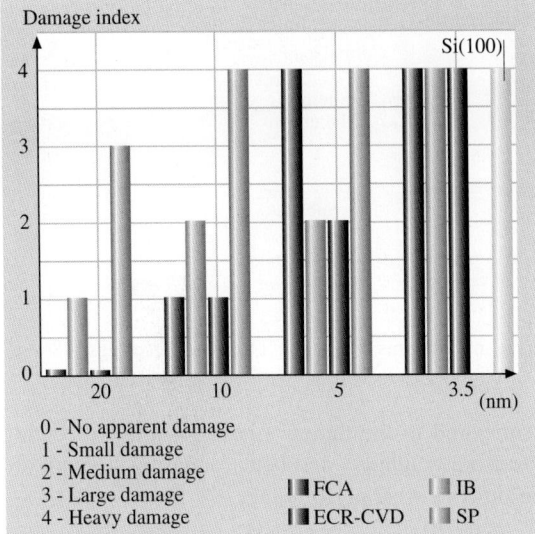

Fig. 17.26. Wear damage index bar chart of the various coatings of different thicknesses and Si(100) substrate based on optical examination of the wear tracks and debris

hardness and subsequent cycles. These are controlled by the sp^3-to-sp^2 ratio. For thicker coating, the cracks generally do not penetrate the coating. For a thinner coating, the cracks easily propagate down to the interface with the aid of the interfacial stresses and get diverted along the interface just enough to cause local delamination of the coating. When this happens, the coating experiences excessive plowing. At this point, the coating fails catastrophically, resulting in a sudden rise in the coefficient of friction. All 3.5-nm-thick coatings failed much quicker than the thicker coatings. It appears that these thin coatings have very low load-carrying capacity and, therefore, the substrate undergoes deformation almost immediately. This generates stresses at the interface that weaken the coating adhesion and lead to delamination of the coating. Another reason may be that the thickness is insufficient to produce a coating comprised of the DLC structure. Instead, the bulk may be made up of a matrix characteristic of the interface region where atomic mixing occurs with the substrate and/or any interlayer used. This would also result in poor wear resistance and silicon-like behavior of the coating, especially in the case of FCA coatings, which show the worst performance at 3.5 nm. In contrast to the other coatings, all SP coatings show the worst wear performance at any thickness (Fig. 17.25). This may be due to their poor mechanical properties, such as lower hardness and scratch resistance, compared to the other coatings.

Comparison of Figs. 17.20 and 17.26 shows a very good correlation between the wear damage and scratch critical loads. Less wear damage corresponds to higher scratch critical load. Based on the data, thicker coatings do show better scratch and wear resistance than thinner coatings. This is probably due to better load-carrying

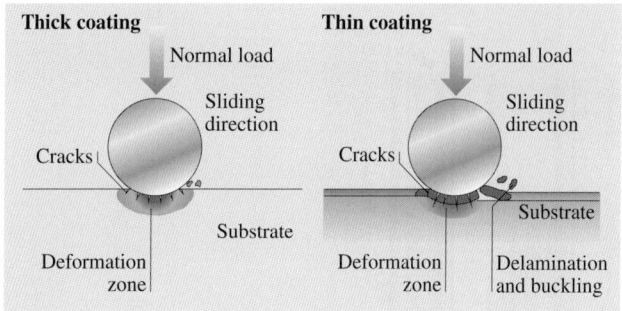

Fig. 17.27. Schematic of wear damage mechanisms of thick and thin DLC coatings [48]

capacity of the thick coatings compared to the thinner ones. For a given coating thickness, higher hardness and fracture toughness and better adhesion strength are believed to be responsible for the superior wear performance.

Effect of Environment

Friction and wear performance of amorphous carbon coatings are found to be strongly dependent on the water vapor content and partial gas pressure in the test environment. The friction data for an amorphous carbon film on the silicon substrate sliding against steel are presented as a function of the partial pressure of water vapor in Fig. 17.28 [1, 13, 69, 105, 106]. Friction increases dramatically above a relative humidity of about 40%. At high relative humidity, condensed water vapor forms meniscus bridges at the contacting asperities, and the meniscii result in an intrinsic attractive force that is responsible for an increase in friction. For completeness, the data for the coefficient of friction of bulk graphitic carbon are also presented in Fig. 17.28. Note that friction decreases with an increase in the relative humidity [107]. Graphitic carbon has a layered lattice crystal structure. Graphite absorbs polar gases (H_2O, O_2, CO_2, NH_3, etc.) at the edges of the crystallites, which weakens the interlayer bonding forces facilitating interlayer slip and results in lower friction [1].

To better study the effect of environment for carbon coated magnetic disks, a number of tests have been conducted in controlled environments. *Marchon* et al. [108] conducted tests in alternating environments of oxygen and nitrogen gases, Fig. 17.30. The coefficient of friction increases as soon as oxygen is added to the test environment, whereas in a nitrogen environment the coefficient of friction reduces slightly. Tribochemical oxidation of the DLC coating in the oxidizing environment is responsible for an increase in the coefficient of friction, implying wear. *Dugger* et al. [109], *Strom* et al. [110], *Bhushan* and *Ruan* [111], and *Bhushan* et al. [71] conducted tests with DLC coated magnetic disks (with about 2-nm-thick perfluoropolyether lubricant film) in contact with Al_2O_3-TiC sliders, in different gaseous environments including a high vacuum of 2×10^{-7} Torr, Fig. 17.29. The wear lives are shortest in high vacuum and longest in most atmospheres of nitrogen and argon

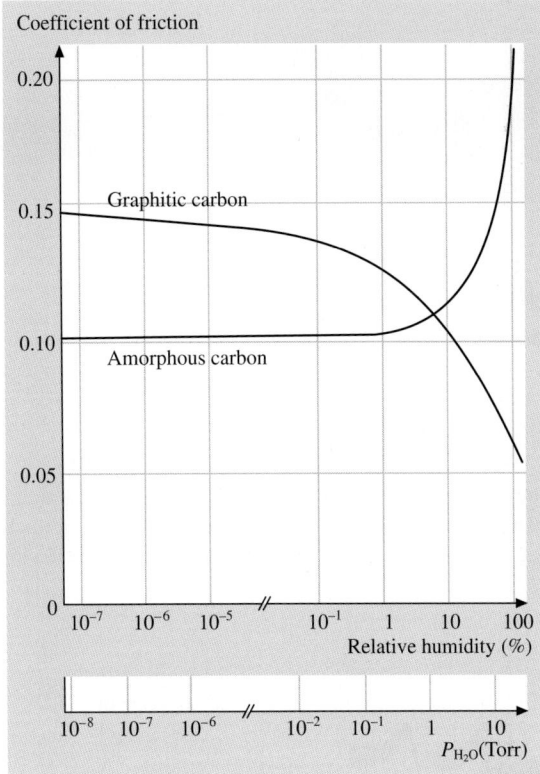

Fig. 17.28. Coefficient of friction as a function of relative humidity and water vapor partial pressure of RF-plasma deposited amorphous carbon coating and bulk graphitic carbon coating sliding against a steel ball

with the following order (from best to worst): argon or nitrogen, Ar + H_2O, ambient, Ar + O_2, and Ar + H_2O, vacuum. From this sequence of wear performance, we can see that having oxygen and water in an operating environment worsens wear performance of the coatings, but having nothing in it (vacuum) is the worst of all. Indeed, failure mechanisms differ in various environments. In high vacuum, intimate contact between disk and slider surfaces results in significant wear. In ambient air, Ar + O_2, and Ar + H_2O, tribochemical oxidation of the carbon overcoat is responsible for interface failure. For experiments performed in pure argon and nitrogen, mechanical shearing of the asperities cause the formation of debris, which is responsible for the formation of scratch marks on the carbon surface as could be observed with an optical microscope [71].

Functional Tests

Magnetic thin-film heads made with Al_2O_3-TiC substrate are used in magnetic storage applications [13]. A multilayered thin-film pole-tip structure present on the head

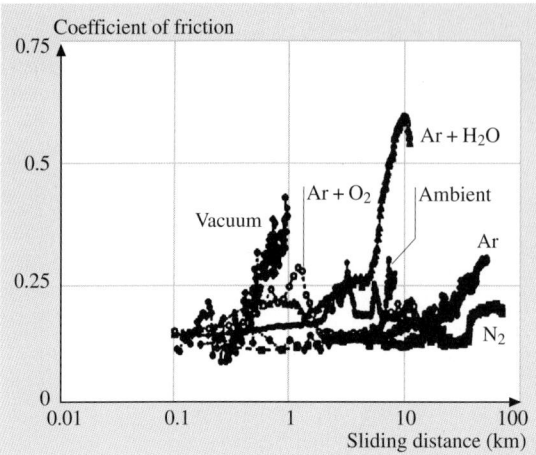

Fig. 17.29. Durability measured by sliding a Al_2O_3-TiC magnetic slider against a magnetic disk coated with 20-nm-thick DC sputtered amorphous carbon coating and 2-nm-thick perfluoropolyether film, measured at a speed of 0.75 m/s and 10 g load. Vacuum refers to 2×10^{-7} Torr [71]

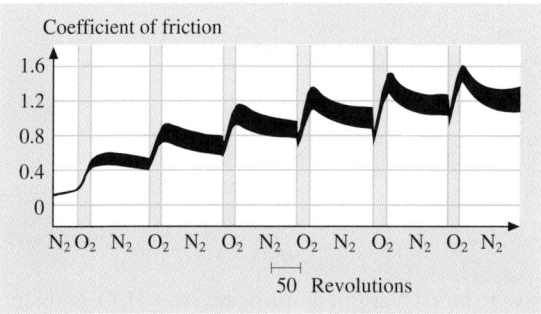

Fig. 17.30. Coefficient of friction as a function of sliding distance for a ceramic slider against a magnetic disk coated with 20-nm-thick DC magnetron sputtered DLC coating, measured at a speed of 0.06 m/s and 10 g load. The environment is alternated between oxygen and nitrogengases [108]

surface wears more rapidly than the much harder Al_2O_3-TiC substrate. Pole-tip recession (PTR) is a serious concern in magnetic storage [15–19, 112]. Two of the diamond-like carbon coatings superior in mechanical properties – ion beam and cathodic arc carbon – were deposited on the air bearing surfaces of Al_2O_3-TiC head sliders [15]. The functional tests were conducted by running a metal-particle (MP) tape in a computer tape drive. Average PTR as a function of sliding distance data are presented in Fig. 17.31. We note that PTR increases for the uncoated head, whereas

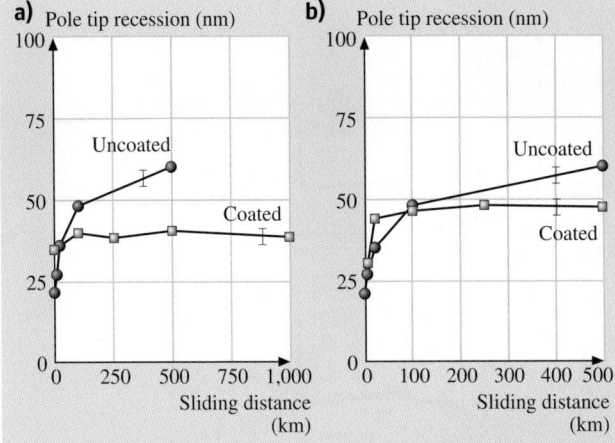

Fig. 17.31. Pole tip recession as a function of sliding distance measured with an AFM for (**a**) uncoated and 20-nm-thick ion beam carbon coated, and (**b**) uncoated and 20-nm-thick cathodic arc carbon coated Al_2O_3-TiC heads run against MP tapes [15]

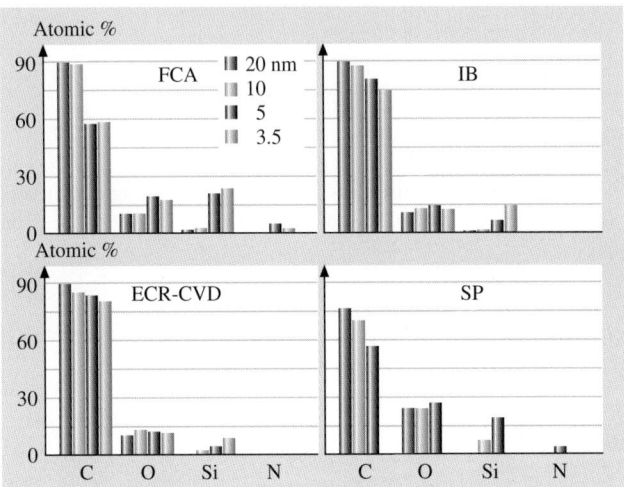

Fig. 17.32. Quantified XPS data for various DLC coatings on Si(100) substrate [50]. Atomic concentrations are shown

for the coated heads there is a slight increase in PTR in early sliding followed by little change. Thus, coatings provide protection.

Micromechanical and accelerated and functional tribological data presented here clearly suggest that there is a good correlation between the scratch resistance and

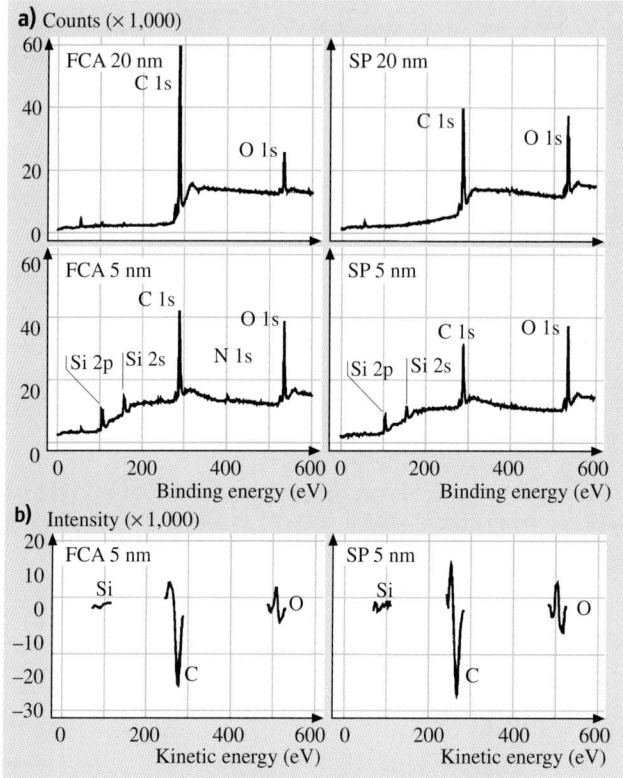

Fig. 17.33. (a) XPS spectra for FCA and SP coatings of 5 nm and 20 nm coating thicknesses on Si(100) substrate, and (b) AES spectra for FCA and SP coatings at 5 nm thickness on Si(100) substrate [50]

wear resistance measured using accelerated tests and functional tests. Thus, scratch tests can be successfully used to screen coatings for wear applications.

17.4.4 Coating Continuity Analysis

Ultrathin coatings less than 10 nm may not uniformly coat the sample surface. In other words, coating may be discontinuous and deposited in the form of islands on the microscale. A possible reason for poor wear protection and the nonuniform failure of the coatings may be due to poor coverage of the thin coatings on the substrate. Coating continuity can be studied by using surface analytical techniques such as Auger and/or XPS analyses. Any discontinuity with coating thickness less than the sampling depth of the instrument will detect locally the substrate species [49, 50, 102].

The results of XPS analysis on various coatings deposited on Si(100) substrates, over a 1.3 mm^2 region (single point measurement with spot diameter of 1300 μm) are shown in Fig. 17.32. The sampling depth is about 2–3 nm. The poor SP coatings show much less carbon content (< 75% atomic concentration) as does the poor 5 nm and 3.5 nm FCA coatings (< 60%) compared to the IB and ECR-CVD coatings. Silicon is detected in all 5 nm coatings. From the data it is hard to say if the Si is from the substrate or from exposed regions due to coating discontinuity. Based on the sampling depth any Si detected in 3.5 nm coatings would likely be from the substrate. The other interesting observation is that all poor coatings (all SP and FCA 5 and 3.5 nm) show almost twice the oxygen content than that of the other coatings. Any oxygen present may be due to leaks in the deposition chamber and is present in the form of silicon oxides.

AES measurements averaged over a scan area of $900\,\mu m^2$ were conducted on FCA and SP 5 nm coatings at six different regions on each sample. Very little silicon was detected on this scale, and the detected peaks were characteristic of oxides. The oxygen levels were comparable to that seen for the good coatings by XPS. These contrast the XPS measurements at a larger scale, suggesting that the coating possesses discontinuities only at isolated areas and that the 5 nm coatings are generally continuous on the microscale. Figure 17.33 shows representative XPS and AES spectra of selected samples.

References

1. B. Bhushan and B. K. Gupta. *Handbook of Tribology: Materials, Coatings, and Surface Treatments*. Krieger, reprint edition, 1997.
2. B. Bhushan. *Principles and Applications of Tribology*. Wiley, 1999.
3. B. Bhushan. *Introduction to Tribology*. Wiley, 2002.
4. B. Bhushan, B. K. Gupta, G. W. VanCleef, C. Capp, and J. V. Coe. Fullerene (c_{60}) films for solid lubrication. *Tribol. Trans.*, 36:573–580, 1993.
5. B. K. Gupta, B. Bhushan, C. Capp, and J. V. Coe. Material characterization and effect of purity and ion implantation on the friction and wear of sublimed fullerene films. *J. Mater. Res.*, 9:2823–2838, 1994.
6. B. K. Gupta and B. Bhushan. Fullerene particles as an additive to liquid lubricants and greases for low friction and wear. *Lubr. Eng.*, 50:524–528, 1994.
7. B. Bhushan, V. V. Subramaniam, A. Malshe, B. K. Gupta, and J. Ruan. Tribological properties of polished diamond films. *J. Appl. Phys.*, 74:4174–4180, 1993.
8. B. Bhushan, B. K. Gupta, and V. V. Subramaniam. Polishing of diamond films. *Diam. Films Technol.*, 4:71–97, 1994.
9. P. Sander, U. Kaiser, M. Altebockwinkel, L. Wiedmann, A. Benninghoven, R. E. Sah, and P. Koidl. Depth profile analysis of hydrogenated carbon layers on silicon by x-ray photoelectron spectroscopy, auger electron spectroscopy, electron energy-loss spectroscopy, and secondary ion mass spectrometry. *J. Vac. Sci. Technol. A*, 5:1470–1473, 1987.
10. A. Matthews and S. S. Eskildsen. Engineering applications for diamond-like carbon. *Diam. Relat. Mater.*, 3:902–911, 1994.
11. A. H. Lettington. Applications of diamond-like carbon thin films. *Carbon*, 36:555–560, 1998.
12. B. K. Gupta and B. Bhushan. Mechanical and tribological properties of hard carbon coatings for magnetic recording heads. *Wear*, 190:110–122, 1995.

13. B. Bhushan. *Tribology and Mechanics of Magnetic Storage Devices*. Springer, 2nd edition, 1996.
14. B. Bhushan. *Mechanics and Reliability of Flexible Magnetic Media*. Springer, 2nd edition, 2000.
15. B. Bhushan, S. T. Patton, R. Sundaram, and S. Dey. Pole tip recession studies of hard carbon-coated thin-film tape heads. *J. Appl. Phys.*, 79:5916–5918, 1996.
16. J. Xu and B. Bhushan. Pole tip recession studies of thin-film rigid disk head sliders ii: Effects of air bearing surface and pole tip region designs and carbon coating. *Wear*, 219:30–41, 1998.
17. W. W. Scott and B. Bhushan. Corrosion and wear studies of uncoated and ultra-thin dlc coated magnetic tape-write heads and magnetic tapes. *Wear*, 243:31–42, 2000.
18. W. W. Scott and B. Bhushan. Loose debris and head stain generation and pole tip recession in modern tape drives. *J. Info. Storage Proc. Syst.*, 2:221–254, 2000.
19. W. W. Scott, B. Bhushan, and A. V. Lakshmikumaran. Ultrathin diamond-like carbon coatings used for reduction of pole tip recession in magnetic tape heads. *J. Appl Phys*, 87:6182–6184, 2000.
20. B. Bhushan. *Macro- and microtribology of magnetic storage devices*, pages 1413–1513. CRC, 2001.
21. B. Bhushan. Nanotribology and nanomechanics of mems devices.
22. B. Bhushan, editor. *Tribology Issues and Opportunities in MEMS*. Kluwer, 1998.
23. B. Bhushan. *Handbook of Micro/Nanotribology*. CRC, 2nd edition, 1999.
24. B. Bhushan. *Macro- and microtribology of MEMS materials*, pages 1515–1548. CRC, 2001.
25. S. Aisenberg and R. Chabot. Ion beam deposition of thin films of diamond like carbon. *J. Appl. Phys.*, 49:2953–2958, 1971.
26. E. G. Spencer, P. H. Schmidt, D. C. Joy, and F. J. Sansalone. Ion beam deposited polycrystalline diamond-like films. *Appl. Phys. Lett.*, 29:118–120, 1976.
27. A. Grill and B. S. Meyerson. *Development and status of diamondlike carbon*, pages 91–141. Wiley, 1994.
28. Y. Catherine. *Preparation techniques for diamond-like carbon*, pages 193–227. Plenum, 1991.
29. J. J. Cuomo, D. L. Pappas, J. Bruley, J. P. Doyle, and K. L. Seagner. Vapor deposition processes for amorphous carbon films with sp^3 fractions approaching diamond. *J. Appl. Phys.*, 70:1706–1711, 1991.
30. J. C. Angus and C. C. Hayman. Low pressure metastable growth of diamond and diamondlike phase. *Science*, 241:913–921, 1988.
31. J. C. Angus and F. Jensen. Dense diamondlike hydrocarbons as random covalent networks. *J. Vac. Sci. Technol. A*, 6:1778–1782, 1988.
32. D. C. Green, D. R. McKenzie, and P. B. Lukins. The microstructure of carbon thin films. *Mater. Sci. Forum*, 52-53:103–124, 1989.
33. N. H. Cho, K. M. Krishnan, D. K. Veirs, M. D. Rubin, C. B. Hopper, B. Bhushan, and D. B. Bogy. Chemical structure and physical properties of diamond-like amorphous carbon films prepared by magnetron sputtering. *J. Mater. Res.*, 5:2543–2554, 1990.
34. J. C. Angus. Diamond and diamondlike films. *Thin Solid Films*, 216:126–133, 1992.
35. B. Bhushan, A. J. Kellock, N. H. Cho, and J. W. Ager III. Characterization of chemical bonding and physical characteristic of diamond-like amorphous carbon and diamond films. *J. Mater. Res.*, 7:404–410, 1992.
36. J. Robertson. Properties of diamond-like carbon. *Surf. Coat. Technol.*, 50:185–203, 1992.

37. N. Savvides and B. Window. Diamondlike amorphous carbon films prepared by magnetron sputtering of graphite. *J. Vac. Sci. Technol. A*, 3:2386–2390, 1985.
38. J. C. Angus, P. Koidl, and S. Domitz. *Carbon thin films*, pages 89–127. CRC, 1986.
39. J. Robertson. Amorphous carbon. *Adv. Phys.*, 35:317–374, 1986.
40. M. Rubin, C. B. Hooper, N. H. Cho, and B. Bhushan. Optical and mechanical properties of dc sputtered carbon films. *J. Mater. Res.*, 5:2538–2542, 1990.
41. G. J. Vandentop, M. Kawasaki, R. M. Nix, I. G. Brown, M. Salmeron, and G. A. Somorjai. Formation of hydrogenated amorphous carbon films of controlled hardness from a methane plasma. *Phys. Rev. B*, 41:3200–3210, 1990.
42. J. J. Cuomo, D. L. Pappas, R. Lossy, J. P. Doyle, J. Bruley, G. W. Di Bello, and W. Krakow. Energetic carbon deposition at oblique angles. *J. Vac. Sci. Technol. A*, 10:3414–3418, 1992.
43. D. L. Pappas, K. L. Saegner, J. Bruley, W. Krakow, and J. J. Cuomo. Pulsed laser deposition of diamondlike carbon films. *J. Appl. Phys.*, 71:5675–5684, 1992.
44. H. J. Scheibe and B. Schultrich. Dlc film deposition by laser-arc and study of properties. *Thin Solid Films*, 246:92–102, 1994.
45. C. Donnet and A. Grill. Friction control of diamond-like carbon coatings. *Surf. Coat. Technol.*, 94-95:456, 1997.
46. A. Grill. Tribological properties of diamondlike carbon and related materials. *Surf. Coat. Technol.*, 94-95:507, 1997.
47. B. K. Gupta and B. Bhushan. Micromechanical properties of amorphous carbon coatings deposited by different deposition techniques. *Thin Solid Films*, 270:391–398, 1995.
48. X. Li and B. Bhushan. Micro/nanomechanical and tribological characterization of ultrathin amorphous carbon coatings. *J. Mater. Res.*, 14:2328–2337, 1999.
49. X. Li and B. Bhushan. Mechanical and tribological studies of ultra-thin hard carbon overcoats for magnetic recording heads. *Z. Metallkd.*, 90:820–830, 1999.
50. S. Sundararajan and B. Bhushan. Micro/nanotribology of ultra-thin hard amorphous carbon coatings using atomic force/friction force microscopy. *Wear*, 225-229:678–689, 1999.
51. B. Bhushan. Chemical, mechanical, and tribological characterization of ultra-thin and hard amorphous carbon coatings as thin as 3.5 nm: Recent developments. *Diam. Relat. Mater.*, 8:1985–2015, 1999.
52. I. I. Aksenov and V. E. Strel'Nitskii. Wear resistance of diamond-like carbon coatings. *Surf. Coat. Technol.*, 47:252–256, 1991.
53. D. R. McKenzie, D. Muller, B. A. Pailthorpe, Z. H. Wang, E. Kravtchinskaia, D. Segal, P. B. Lukins, P. J. Martin, G. Amaratunga, P. H. Gaskell, and A. Saeed. Properties of tetrahedral amorphous carbon prepared by vacuum arc deposition. *Diam. Relat. Mater.*, 1:51–59, 1991.
54. R. Lossy, D. L. Pappas, R. A. Roy, and J. J. Cuomo. Filtered arc deposition of amorphous diamond. *Appl. Phys. Lett.*, 61:171–173, 1992.
55. I. G. Brown, A. Anders, S. Anders, M. R. Dickinson, I. C. Ivanov, R. A. MacGill, X. Y. Yao, and K. M. Yu. Plasma synthesis of metallic and composite thin films with atomically mixed substrate bonding. *Nucl. Instrum. Meth. Phys. Res. B*, 80-81:1281–1287, 1993.
56. P. J. Fallon, V. S. Veerasamy, C. A. Davis, J. Robertson, G. A. J. Amaratunga, W. I. Milne, and J. Koskinen. Properties of filtered-ion-beam-deposited diamond-like carbon as a function of ion energy. *Phys. Rev. B*, 48:4777–4782, 1993.
57. S. Anders, A. Anders, I. G. Brown, B. Wei, K. Komvopoulos, J. W. Ager III, and K. M. Yu. Effect of vacuum arc deposition parameters on the properties of amorphous carbon thin films. *Surf. Coat. Technol.*, 68-69:388–393, 1994.

58. S. Anders, A. Anders, I. G. Brown, M. R. Dickinson, and R. A. MacGill. Metal plasma immersion ion implantation and deposition using arc plasma sources. *J. Vac. Sci. Technol. B*, 12:815–820, 1994.
59. S. Anders, A. Anders, and I. G. Brown. Transport of vacuum arc plasma through magnetic macroparticle filters. *Plasma Sources Sci.*, 4:1–12, 1995.
60. D. M. Swec, M. J. Mirtich, and B. A. Banks. Ion beam and plasma methods of producing diamondlike carbon films.
61. A. Erdemir, M. Switala, R. Wei, and P. Wilbur. A tribological investigation of the graphite-to-diamond-like behavior of amorphous carbon films ion beam deposited on ceramic substrates. *Surf. Coat. Technol.*, 50:17–23, 1991.
62. A. Erdemir, F. A. Nicols, X. Z. Pan, R. Wei, and P. J. Wilbur. Friction and wear performance of ion-beam deposited diamond-like carbon films on steel substrates. *Diam. Relat. Mater.*, 3:119–125, 1993.
63. R. Wei, P. J. Wilbur, and M. J. Liston. Effects of diamond-like hydrocarbon films on rolling contact fatigue of bearing steels. *Diam. Relat. Mater.*, 2:898–903, 1993.
64. A. Erdemir and C. Donnet. *Tribology of diamond, diamond-like carbon, and related films*, pages 871–908. CRC, 2001.
65. J. Asmussen. Electron cyclotron resonance microwave discharges for etching and thin-film deposition. *J. Vac. Sci. Technol. A*, 7:883–893, 1989.
66. J. Suzuki and S. Okada. Deposition of diamondlike carbon films using electron cyclotron resonance plasma chemical vapor deposition from ethylene gas. *Jpn. J. Appl. Phys.*, 34:L1218–L1220, 1995.
67. B. A. Banks and S. K. Rutledge. Ion beam sputter deposited diamond like films. *J. Vac. Sci. Technol.*, 21:807–814, 1982.
68. C. Weissmantel, K. Bewilogua, K. Breuer, D. Dietrich, U. Ebersbach, H. J. Erler, B. Rau, and G. Reisse. Preparation and properties of hard i-c and i-bn coatings. *Thin Solid Films*, 96:31–44, 1982.
69. H. Dimigen and H. Hubsch. Applying low-friction wear-resistant thin solid films by physical vapor deposition. *Philips Tech. Rev.*, 41:186–197, 1983/84.
70. B. Bhushan, B. K. Gupta, and M. H. Azarian. Nanoindentation, microscratch, friction and wear studies for contact recording applications. *Wear*, 181-183:743–758, 1995.
71. B. Bhushan, L. Yang, C. Gao, S. Suri, R. A. Miller, and B. Marchon. Friction and wear studies of magnetic thin-film rigid disks with glass-ceramic, glass and aluminum-magnesium substrates. *Wear*, 190:44–59, 1995.
72. L. Holland and S. M. Ojha. Deposition of hard and insulating carbonaceous films of an rf target in butane plasma. *Thin Solid Films*, 38:L17–L19, 1976.
73. L. P. Andersson. A review of recent work on hard i-c films. *Thin Solid Films*, 86:193–200, 1981.
74. A. Bubenzer, B. Dischler, B. Brandt, and P. Koidl. R.f. plasma deposited amorphous hydrogenated hard carbon thin films, preparation, properties and applications. *J. Appl. Phys.*, 54:4590–4594, 1983.
75. A. Grill, B. S. Meyerson, and V. V. Patel. Diamond-like carbon films by rf plasma-assisted chemical vapor deposition from acetylene. *IBM J. Res. Develop.*, 34:849–857, 1990.
76. A. Grill, B. S. Meyerson, and V. V. Patel. Interface modification for improving the adhesion of a-c:h to metals. *J. Mater. Res.*, 3:214, 1988.
77. A. Grill, V. V. Patel, and B. S. Meyerson. Optical and tribological properties of heat-treated diamond-like carbon. *J. Mater. Res.*, 5:2531–2537, 1990.

78. F. Jansen, M. Machonkin, S. Kaplan, and S. Hark. The effect of hydrogenation on the properties of ion beam sputter deposited amorphous carbon. *J. Vac. Sci. Technol. A*, 3:605–609, 1985.
79. S. Kaplan, F. Jansen, and M. Machonkin. Characterization of amorphous carbon-hydrogen films by solid-state nuclear magnetic resonance. *Appl. Phys. Lett.*, 47:750–753, 1985.
80. H. C. Tsai, D. B. Bogy, M. K. Kundmann, D. K. Veirs, M. R. Hilton, and S. T. Mayer. Structure and properties of sputtered carbon overcoats on rigid magnetic media disks. *J. Vac. Sci. Technol. A*, 6:2307–2315, 1988.
81. B. Marchon, M. Salmeron, and W. Siekhaus. Observation of graphitic and amorphous structures on the surface of hard carbon films by scanning tunneling microscopy. *Phys. Rev. B*, 39:12907–12910, 1989.
82. B. Dischler, A. Bubenzer, and P. Koidl. Hard carbon coatings with low optical-absorption. *Appl. Phys. Lett.*, 42:636–638, 1983.
83. D. S. Knight and W. B. White. Characterization of diamond films by raman spectroscopy. *J. Mater. Res.*, 4:385–393, 1989.
84. J. W. Ager, D. K. Veirs, and C. M. Rosenblatt. Spatially resolved raman studies of diamond films grown by chemical vapor deposition. *Phys. Rev. B*, 43:6491–6499, 1991.
85. W. Scharff, K. Hammer, O. Stenzel, J. Ullman, M. Vogel, T. Frauenheim, B. Eibisch, S. Roth, S. Schulze, and I. Muhling. Preparation of amorphous i-C films by ion-assisted methods. *Thin Solid Films*, 171:157–169, 1989.
86. B. Bhushan and X. Li. Nanomechanical characterization of solid surfaces and thin films. *Int. Mater. Rev.*, 48:125–164, 2003.
87. X. Li, D. Diao, and B. Bhushan. Fracture mechanisms of thin amorphous carbon films in nanoindentation. *Acta Mater.*, 45:4453–4461, 1997.
88. X. Li and B. Bhushan. Measurement of fracture toughness of ultra-thin amorphous carbon films. *Thin Solid Films*, 315:214–221, 1998.
89. X. Li and B. Bhushan. Evaluation of fracture toughness of ultra-thin amorphous carbon coatings deposited by different deposition techniques. *Thin Solid Films*, 355-356:330–336, 1999.
90. X. Li and B. Bhushan. Development of a nanoscale fatigue measurement technique and its application to ultrathin amorphous carbon coatings. *Scripta Mater.*, 47:473–479, 2002.
91. X. Li and B. Bhushan. Nanofatigue studies of ultrathin hard carbon overcoats used in magnetic storage devices. *J. Appl. Phys.*, 91:8334–8336, 2002.
92. J. Robertson. Deposition of diamond-like carbon. *Philos. Trans. R. Soc. London A*, 342:277–286, 1993.
93. S. J. Bull. Tribology of carbon coatings: Dlc, diamond and beyond. *Diam. Relat. Mater.*, 4:827–836, 1995.
94. N. Savvides and T. J. Bell. Microhardness and young's modulus of diamond and diamondlike carbon films. *J. Appl. Phys.*, 72:2791–2796, 1992.
95. B. Bhushan and M. F. Doerner. Role of mechanical properties and surface texture in the real area of contact of magnetic rigid disks. *ASME J. Tribol.*, 111:452–458, 1989.
96. S. Suresh. *Fatigue of Materials*. Cambridge Univ. Press, 1991.
97. D. B. Marshall and A. G. Evans. Measurement of adherence of residual stresses in thin films by indentation. i. mechanics of interface delamination. *J. Appl. Phys.*, 15:2632–2638, 1984.
98. A. G. Evans and J. W. Hutchinson. On the mechanics of delamination and spalling in compressed films. *Int. J. Solids Struct.*, 20:455–466, 1984.

99. S. Sundararajan and B. Bhushan. Development of a continuous microscratch technique in an atomic force microscope and its application to study scratch resistance of ultrathin hard amorphous carbon coatings. *J. Mater. Res.*, 16:437–445, 2001.
100. B. Bhushan and B. K. Gupta. Micromechanical characterization of ni-p coated aluminum-magnesium, glass and glass-ceramic substrates and finished magnetic thin-film rigid disks. *Adv. Info. Storage Syst.*, 6:193–208, 1995.
101. X. Li and B. Bhushan. Micromechanical and tribological characterization of hard amorphous carbon coatings as thin as 5 nm for magnetic recording heads. *Wear*, 220:51–58, 1998.
102. B. Bhushan and V. N. Koinkar. Microscale mechanical and tribological characterization of hard amorphous coatings as thin as 5 nm for magnetic disks. *Surf. Coat. Technol.*, 76-77:655–669, 1995.
103. V. N. Koinkar and B. Bhushan. Microtribological properties of hard amorphous carbon protective coatings for thin-film magnetic disks and heads. *Proc. Inst. Mech. Eng., Part J*, 211:365–372, 1997.
104. T. W. Wu. Microscratch and load relaxation tests for ultra-thin films. *J. Mater. Res.*, 6:407–426, 1991.
105. R. Memming, H. J. Tolle, and P. E. Wierenga. Properties of polymeric layers of hydrogenated amorphous carbon produced by plasma-activated chemical vapor deposition: tribological and mechanical properties. *Thin Solid Films*, 143:31–41, 1986.
106. C. Donnet, T. Le Mogne, L. Ponsonnet, M. Belin, A. Grill, and V. Patel. The respective role of oxygen and water vapor on the tribology of hydrogenated diamond-like carbon coatings. *Tribol. Lett.*, 4:259, 1998.
107. F. P. Bowden and J. E. Young. Friction of diamond, graphite and carbon and the influence of surface films. *Proc. R. Soc. Lond.*, 208:444–455, 1951.
108. B. Marchon, N. Heiman, and M. R. Khan. Evidence for tribochemical wear on amorphous carbon thin films. *IEEE Trans. Magn.*, 26:168–170, 1990.
109. M. T. Dugger, Y. W. Chung, B. Bhushan, and W. Rothschild. Friction, wear, and interfacial chemistry in thin film magnetic rigid disk files. *ASME J. Tribol.*, 112:238–245, 1990.
110. B. D. Strom, D. B. Bogy, C. S. Bhatia, and B. Bhushan. Tribochemical effects of various gases and water vapor on thin film magnetic disks with carbon overcoats. *ASME J. Tribol.*, 113:689–693, 1991.
111. B. Bhushan and J. Ruan. Tribological performance of thin film amorphous carbon overcoats for magnetic recording rigid disks in various environments. *Surf. Coat. Technol.*, 68/69:644–650, 1994.
112. B. Bhushan, G. S. A. M. Theunissen, and X. Li. Tribological studies of chromium oxide films for magnetic recording applications. *Thin Solid Films*, 311:67–80, 1997.

18

Self-Assembled Monolayers for Controlling Adhesion, Friction and Wear

Bharat Bhushan and Huiwen Liu

Summary. Reliability of micro- and nanodevices, as well as magnetic storage devices require the use of lubricant films for the protection of sliding surfaces. To minimize high adhesion, friction, and because of small clearances in the devices, these films should be molecularly thick. Liquid films of low surface tension or certain hydrophobic solid films can be used. Ordered molecular assemblies with high hydrophobicity can be engineered using chemical grafting of various polymer molecules with suitable functional head groups, spacer chains and nonpolar surface terminal groups.

The classical approach to lubrication uses multi-molecular layers of liquid lubricants. Boundary lubricant films are formed by either physisorption, chemisorption, or chemical reaction. The physiosorbed films can be either monomolecularly or polymolecularly thick. The chemisorbed films are monomolecular, but stoichiometric films formed by chemical reaction can be multilayered. A good boundary lubricant should have a high degree of interaction between its molecules and the sliding surface. As a general rule, liquids are good lubricants when they are polar and thus able to grip on solid surfaces (or be adsorbed).

In this chapter, we focus on self-assembled monolayers (SAMs) for high hydrophobicity and/or low adhesion, friction, and wear. SAMs are produced by various organic precursors. We first present a primer to organic chemistry followed by an overview on suitable substrates, head groups, spacer chains, and end groups in the molecular chains and an overview of tribological properties of SAMs. The adhesion, friction, and wear properties of SAMs, having alkyl and biphenyl spacer chains with different surface terminal and head groups, are surveyed. The friction data are explained using a molecular spring model in which the local stiffness and intermolecular force govern its frictional performance. Based on the nanotribological studies of SAM films by AFM, they exhibit attractive hydrophobic and tribological properties.

18.1 Introduction

Reliability of micro/nanodevices, also commonly referred to as micro/nanoelectromechanical systems (MEMS/NEMS), nanoelectromechanical systems microelectromechanical systems as well as magnetic storage devices (which include magnetic rigid disk drives, flexible disk drives, and tape drives) require the use of molecularly thick films for protection of sliding surfaces [1–8]. A solid or liquid film is generally necessary for acceptable tribological properties of sliding interfaces. However, a small quantity of high surface tension liquid present between smooth surfaces can substantially increase the adhesion, friction, and wear as a result of formation of menisci or adhesive bridges [9, 10]. It becomes a major concern in micro/nanoscale devices operating at ultralow loads, as the liquid mediated adhesive force may be on the same order as the external load.

The source of the liquid film can be either preexisting film of liquid and/or capillary condensates of water vapor from the environment. If the liquid wets the surface ($0 \leq \theta < 90°$, where θ is the contact angle between the liquid-vapor interface and the liquid-solid interface for a liquid droplet sitting on a solid surface, Fig. 18.1a), the liquid surface is thereby constrained to lie parallel to the surface [11], and the complete liquid surface must therefore be concave in shape, Fig. 18.1b. Surface tension results in a pressure difference across any meniscus surface, referred to as capillary pressure or Laplace pressure, and is negative for a concave meniscus [9, 10]. The negative Laplace pressure results in an intrinsic attractive (adhesive) force, which depends on the interface roughness (local geometry of interacting asperities and number of asperities), the surface tension, and the contact angle. During sliding, frictional effects need to be overcome, not only because of external load, but also because of intrinsic adhesive force. Measured value of high static friction force contributed largely by liquid mediated adhesion (meniscus contribution) is generally referred to as "stiction". Basically, there are three ways to minimize the effect of liquid-mediated adhesion: increase in surface roughness, use of hydrophobic (water-fearing) rather than hydrophilic (water-loving) surfaces, and the use of a liquid with low surface tension [9, 10, 12, 13].

As an example, bulk silicon and polysilicon films used in the construction of micro/nanodevices can be dipped in hydrofluoric (HF) to make them hydrophobic (e.g., *Maboudian* [14]; *Scherge* and *Schaefer* [15]). In HF etching of silicon, hydrogen passivates the silicon surface by saturating the dangling bonds, and it results in a hydrogen-terminated silicon surface that is responsible for less adsorption of water. However, after exposure of the treated surface to the environment, it reoxidizes, which can adsorb water, and thus the surface again becomes hydrophilic.

The surfaces can also be treated or coated with a liquid with relatively low surface tension or a certain solid film to make them hydrophobic and/or to control adhesion, friction, and wear. The liquid lubricant film should be thin (about half of the composite roughness of the interface) to minimize liquid-mediated adhesion contribution [9, 10, 12, 13]. Thus, for ultra-smooth surfaces with RMS roughness on the order of a few nm, molecularly thick liquid films are required for liquid lubrication. The classical approach to lubrication uses freely supported multi-molecular layers of liquid lubricants [2, 4, 9, 10, 16–19]. Boundary lubricant films are formed by either physisorption, chemisorption, or chemical reaction. The physiosorbed films can be either monomolecularly or polymolecularly thick. The chemisorbed films are monomolecular, but stoichiometric films formed by chemical reaction can be multilayered. In general, stability and durability of surface films decrease in the following order: chemically reacted films, chemisorbed films, and physisorbed films. A good boundary lubricant should have a high degree of interaction between its molecules and the sliding surface. As a general rule, liquids are good lubricants when they are

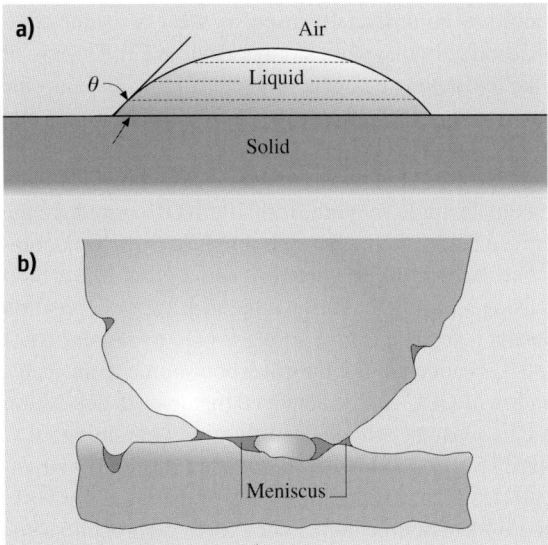

Fig. 18.1. (a) Schematic of a sessile-drop on a solid surface and the definition of contact angle, and (b) formation of meniscus bridges as a result of liquid present at an interface. The direct measurement of contact angle is most widely made from sessile-drops. The angle is generally measured by aligning a tangent with the drop profile at the point of contact with the solid surface using a telescope equipped with a goniometer eyepiece

polar and thus able to grip on solid surfaces (or be adsorbed). Polar lubricants contain reactive functional end groups. Boundary lubrication properties are also dependent on the molecular conformation and lubricant spreading. It should be noted that the liquid films with a thickness on the order of a few nm, may be discontinuous and may deposit in an island form of nonuniform thickness with a lateral resolution on the nanometerscale.

Solid films are also commonly used for controlling hydrophobicity and/or adhesion, friction, and wear. Hydrophobic films have nonpolar surface terminal groups (to be described later) that repel water. These films have low surface energy (15–30 dyn/cm) and high contact angle ($\theta \geq 90°$), which minimize wetting (e.g., *Zisman* [20]; *Schrader* and *Loeb* [21]; *Neumann* and *Spelt* [22]). It should be noted that these films do not totally eliminate wetting. Multi-molecularly thick (few tenths of a nm) films of conventional solid-lubricants have been studied. *Hansma* et al. [23] reported the deposition of multi-molecularly thick, highly-oriented PTFE films from the melt or vapor phase or from solution by a mechanical deposition technique by dragging the polymer at controlled temperature, pressure, and speed against a smooth glass substrate. *Scandella* et al. [24] reported that the coefficient of nanoscale friction of MoS_2 platelets on mica, obtained by exfoliation of lithium intercalated MoS_2 in water, was a factor of 1.4 less than that of mica itself. However, MoS_2 is reactive to water and its friction and wear properties degrade with an increase in humidity [9, 10]. Amorphous diamond-like carbon (DLC) coatings can be produced with extremely high hardness and are commercially used as wear-resistant coatings [25, 26]. Their largest application is in magnetic storage devices [2]. Doping of the DLC matrix with elements like hydrogen, nitrogen, oxygen, silicon, and fluorine influences their hydrophobicity and tribological properties [25, 27, 28]. Nitrogen and oxygen reduce the contact angle (or increase the surface energy), due to the strong polarity that is formed when these elements are bonded to carbon. On the other hand, silicon and fluorine increase the contact angle ranging from 70–100° (or reduce the surface energy to 20–40 dyn/cm), making them hydrophobic [29, 30]. Nanocomposite coatings with a diamond-like carbon (a-C:H) network and a glass-like a-Si:O network are generally deposited using a plasma-enhanced chemical vapor deposition (PECVD) technique in which plasma is formed from a siloxane precursor using a hot filament. For a fluorinated DLC, CF_4 is added as the fluorocarbon source to an acetylene plasma. In addition, fluorination of DLC can be achieved by the post-deposition treatment of DLC coatings in a CF_4 plasma. Silicon- and fluorine-containing DLC coatings mainly reduce their polarity due to the loss of sp^2-bonded carbon (polarization potential of the involved π electrons) and dangling bonds of the DLC network. As silicon and fluorine are unable to form double bonds, they force carbon into a sp^3 bonding state [30]. Friction and wear properties of both silicon-containing and fluorinated DLC coatings have been reported to be superior to that of conventional DLC coatings [31, 32]. However, DLC coatings require line of sight deposition process, which prevents deposition on complex geometries. Furthermore, it has been reported that some self-assembled monolayers (SAMs) are superior to DLC coatings in hydrophobicity and tribological performance [33].

Organized and dense molecular-scale layers of, preferably, long-chain organic molecules are known to be superior lubricants on both macro- and micro/nanoscales as compared with freely supported multi-molecular layers [4, 34–38]. Common techniques to produce molecular-scale organized layers are Langmuir–Blodgett (LB) deposition and chemical grafting of organic molecules to realize SAMs [39, 40]. In the LB technique, organic molecules from suitable amphiphilic molecules are first organized at the air-water interface and then physisorbed on a solid surface to form mono- or multi-molecular layers [41]. In the case of SAMs, the functional groups of molecules chemisorb on a solid surface, which results in the spontaneous formation of robust, highly ordered, oriented, and dense monolayers [40]. In both cases, the organic molecules used have well distinguished amphiphilic properties (hydrophilic functional head and a hydrophobic aliphatic tail) so that an adsorption of such molecules on an active inorganic substrate leads to their firm attachment to the surface. Direct organization of SAMs on the solid surfaces allows coating in the tight areas such as the bearing and journal surfaces in an assembled bearing. The weak adhesion of classical LB films to the substrate surface restricts their lifetimes during sliding, whereas certain SAMs can be very durable [34]. As a result, SAMs are of great interest in tribological applications.

Much of the research in the application of SAMs has been carried out for the so-called soft lithographic technique [42, 43]. This is a non-photolithographic technique. Photolithography is based on a projection-printing system used for projection of an image from a mask to a thin-film photoresist and its resolution is limited by optical diffraction limits. In the soft lithography, an elastomeric stamp or mold is used to generate micropatterns of SAMs either by contact printing (known as microcontact printing or μCP [44]), by embossing (imprinting) [45], or by replica molding [46], which circumvents the diffraction limits of photolithography. The stamps are generally cast from photolithographically generated patterned masters, and the stamp material is generally polydimethylsiloxane (PDMS). In μCP, the ink is a SAM precursor to produce nm-thick resists with lines thinner than 100 nm. Although soft lithography requires little capital investment, it is doubtful that it would ever replace well established photolithography. However, μCP and embossing techniques may be used to produce microdevices that are a substantially cheaper and more flexible choice of material for construction than conventional photolithography (e.g., SAMs and non-SAM entities for μCP and elastomers for embossing).

Other applications for SAMs are in the areas of bio/chemical and optical sensors, for use as drug-delivery vehicles and in the construction of electronic components [39, 47]. Bio/chemical sensors require the use of highly sensitive organic layers with tailored biological properties that can be incorporated into electronic, optical, or electrochemical devices. Self-assembled microscopic vesicles are being developed to ferry potentially lifesaving drugs to cancer patients. By getting organic, metal, and phosphonate molecules (complexes of phosphorous and oxygen atoms) to assemble themselves into conductive materials, these can be produced as self-made sandwiches for use as electronic components. The application of interest here is an application requiring hydrophobicity and/or low friction and wear.

An overview of molecularly thick layers of liquid lubricants and conventional solid lubricant can be found in various references [2, 4, 9, 10, 18, 19, 25]. In this chapter, we focus on SAMs for high hydrophobicity and/or, low adhesion, friction, and wear. SAMs are produced by various organic precursors. We first present a primer to organic chemistry followed by an overview on suitable substrates, spacer chains, and end groups in the molecular chains, an overview of tribological properties of SAMs, and some concluding remarks.

18.2 A Primer to Organic Chemistry

All organic compounds contain the carbon (C) atom. Carbon, in combination with hydrogen, oxygen, nitrogen, and sulfur results in a large number of organic compounds. The atomic number of carbon is six, and its electron structure is $1s^2\, 2s^2\, 2p^2$. Two stable isotopes of carbon, ^{12}C and ^{13}C, exist. With four electrons in its outer shell, carbon forms four covalent bonds with each bond resulting from two atoms sharing a pair of electrons. The number of electron pairs that two atoms share determines whether or not the bond is single or multiple. In a single bond, only one pair of electrons is shared by the atoms. Carbon can also form multiple bonds by sharing two or three pairs of electrons between the atoms. For example, the double bond formed by sharing two electron pairs is stronger than a single bond, and it is shorter than a single bond. An organic compound is classified as saturated if it contains only

Table 18.1. Relative electronegativity of selected elements

Element	Relative electronegativity
F	4.0
O	3.5
N	3.0
Cl	3.0
C	2.5
S	2.5
P	2.1
H	2.1

the single bond and as unsaturated if the molecules possess one or more multiple carbon-carbon bonds.

18.2.1 Electronegativity/Polarity

When two different kinds of atoms share a pair of electrons a bond is formed in which electrons are shared unequally, one atom assumes a partial positive charge and the other a negative charge with respect to each other. This difference in charge occurs because the two atoms exert unequal attraction for the pair of shared electrons. The attractive force that an atom of an element has for shared electrons in a molecule or polyatomic ion is known as its electronegativity. Elements differ in their electronegativities. A scale of relative electronegatives, in which the most electronegative element, fluorine, is assigned a value of 4.0, was developed by Linus Pauling. The relative electronegativities of the elements in the periodic table can be found in most undergraduate chemistry textbooks (e.g., *Hein* et al. [48]). Relative electronegativity of the nonmetals is high compared to that of metals. Relative electronegativity of selected elements of interest with high values is presented in Table 18.1.

The polarity of a bond is determined by the difference in electronegativity values of the atoms forming the bond. If the electronegativities are the same, the bond is nonpolar and the electrons are shared equally. In this type of bond, there is no separation of positive and negative charge between atoms. If the atoms have greatly different electronegativities, the bond is very polar. A dipole is a molecule that is electrically asymmetrical, causing it to be oppositely charged at two points. As an example, both hydrogen and chlorine need one electron to form stable electron configurations. They share a pair of electrons in hydrogen chloride, HCl. Chlorine is more electronegative and, therefore, has a greater attraction for the shared electrons than hydrogen. As a result, the pair of electrons is displaced toward the chlorine atom, giving it a partial negative charge and leaving the hydrogen atom with a partial positive charge, Fig. 18.2. However, the entire molecule, HCl, is electrically neutral. The hydrogen atom with a partial positive charge (exposed proton on one end) can be easily attracted to the negative charge of other molecules, and this is responsible

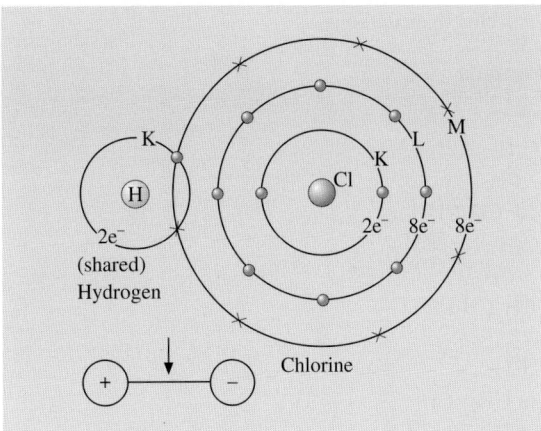

Fig. 18.2. Schematic representation of the formation of a polar HCl molecule

for the polarity of the molecule. A partial charge is usually indicated by δ, and the electronic structure of HCl is given as:

$$\overset{\delta+}{H}:\overset{\delta-}{\ddot{\underset{..}{Cl}}}: \ .$$

Similar to the HCl molecule, HF is polar, and both behave as a small dipole. On the other hand, methane (CH_4), carbon tetrachloride (CCl_4), and carbon dioxide (CO_2) are nonpolar. In CH_4 and CCl_4, the four C–H and C–Cl polar bonds are identical, and because these bonds emanate from the center to the corners of a tetrahedron in the molecule, the effect of their polarities cancels out one another. CO_2 (O=C=O) is nonpolar, because carbon-oxygen dipoles cancel each other by acting in opposite directions. Symmetric molecules are generally nonpolar. Water (H–O–H) is a polar molecule. If the atoms in water were linear as in CO_2, two O–H dipoles would cancel each other, and the molecule would be nonpolar. However, water has a bent structure with an angle of 105° between the two bonds, which is responsible for the water being a polar molecule.

18.2.2 Classification and Structure of Organic Compounds

Tables 18.2–18.3 present selected organic compounds grouped into classes.

Hydrocarbons

Hydrocarbons are compounds that are composed entirely of carbon and hydrogen atoms bonded to each other by covalent bonds. Saturated hydrocarbons (alkanes) contain single bonds. Unsaturated hydrocarbons that contain a carbon-carbon double bond are called alkenes, and ones with a triple bond are called alkynes. Unsaturated hydrocarbons that contain an aromatic ring, e.g., benzene ring, are called aromatic hydrocarbons.

Table 18.2. Names and formulas of selected hydrocarbons

Name	Formula
Saturated hydrocarbons	
Straight-chain alkanes	C_nH_{2n+2}
e.g., methane,	CH_4
ethane	C_2H_6 or CH_3CH_3
Alkyl groups	C_nH_{2n+1}
e.g., methyl,	$-CH_3$
ethyl	$-CH_2CH_3$
Unsaturated hydrocarbons	
Alkenes	$(CH_2)_n$
e.g., ethene,	C_2H_4 or $CH_2=CH_2$
propene	C_3H_6 or $CH_3CH=CH_2$
Alkynes	
e.g., acetylene	$HC\equiv CH$
Aromatic hydrocarbons	
e.g., benzene	(benzene ring structure)

Table 18.3. Names and formulas of selected organic nitrogen compounds (amides and amines)

Name	Formula
Amides	$RCONH_2$ or $R-\overset{\overset{O}{\|\|}}{C}-NH_2$
e.g., methanamide (formamide),	$HCONH_2$
ethanamide (acetamide)	CH_3CONH_2
Amines	RNH_2 or $R-N\overset{H}{\underset{H}{\diagdown}}$
	R_2NH
	R_3N
e.g., methylamine,	CH_3NH_2
ethylamine	$CH_3CH_2NH_2$

The letter R represents an alkyl group or aromatic group.

Saturated Hydrocarbons: Alkanes The alkanes, also known as paraffins, are saturated hydrocarbons, straight- or branched-chain hydrocarbons with only single covalent bonds between the carbon atoms. The general molecular formula for alkanes is C_nH_{2n+2}, where n is the number of carbon atoms in the molecule. Each carbon atom is connected to four other atoms by four single covalent bonds. These bonds are separated by angles of 109.5° (the angle by lines drawn from the center of a regular tetrahedron to its corners). Alkane molecules contain only carbon-carbon and carbon-hydrogen bonds, which are symmetrically directed toward the corners of a tetrahedron. Therefore, alkane molecules are essentially nonpolar.

Common alkyl groups have the general formula C_nH_{2n+1} (one hydrogen atom less than the corresponding alkane). The missing H atom may be detached from any carbon in the alkane. The name of the group is formed from the name of the corresponding alkane by replacing -ane with -yl ending. Some examples are shown in Table 18.2.

Unsaturated Hydrocarbons Unsaturated hydrocarbons consist of three families of compounds that contain fewer hydrogen atoms than the alkane with the corresponding number of carbon atoms and contain multiple bonds between carbon atoms. These include alkenes (with carbon-carbon double bonds), alkynes (with carbon-carbon triple bonds), and aromatic compounds (with benzene rings that are arranged in a six-member ring with one hydrogen atom bonded to each carbon atom and three carbon-carbon double bonds). Some examples are shown in Table 18.2.

Alcohols, Ethers, Phenols, and Thiols

Organic molecules with certain functional groups are synthesized for desirable properties. Alcohols, ethers, and phenols are derived from the structure of water by replacing the hydrogen atoms of water with alkyl groups (R) or aromatic (Ar) rings. For example, phenol is a class of compounds that have a hydroxy group attached to an aromatic ring (benzene ring). Organic compounds that contain the −SH group are analogs of alcohols and are known as thiols. Some examples are shown in Table 18.4.

Aldehydes and Ketones

Both aldehydes and ketones contain the carbonyl group, $>C=O$, a carbon-oxygen double bond. Aldehydes have at least one hydrogen atom bonded to the carbonyl group, whereas ketones have only an alkyl or aromatic group bonded to the carbonyl group. The general formula for the saturated homologous series of aldehydes and ketones is $C_nH_{2n}O$. Some examples are shown in Table 18.5.

Carboxyl Acids and Esters

The functional group of the carboxylic acids is known as a carboxyl group, represented as −COOH. Carboxylic acids can be either aliphatic (RCOOH) or aromatic (ArCOOH). The carboxylic acids with even numbers of carbon atoms, n, ranging from 4 to about 20 are called fatty acids (e.g., $n = 10, 12, 14, 16$ and 18 are called capric acid, lauric acid, myristic acid, palmitic acid, and stearic acid, respectively).

Table 18.4. Names and formulas of selected alcohols, ethers, phenols, and thiols

Name	Formula
Alcohols	R–OH
e.g., methanol,	CH$_3$OH
ethanol	CH$_3$CH$_2$OH
Ethers	R–O–R′
e.g., dimethyl ether,	CH$_3$–O–CH$_3$
diethyl ether	CH$_3$CH$_2$–O–CH$_2$CH$_3$
Phenols	C$_6$H$_5$OH or C$_6$H$_5$–OH (phenyl ring with OH)
Thiols	–SH
e.g., methanethiol	CH$_3$SH

The letters R- and R′- represent an alkyl group. The R- groups in ethers can be the same or different and can be alkyl or aromatic (Ar) groups.

Table 18.5. Names and formulas of selected aldehydes and ketones

Name	Formula
Aldehydes	RCHO or R–C(=O)–H
	ArCHO or Ar–C(=O)–H
e.g., methanal or formaldehyde ethanal or acetaldehyde	HCHO CH$_3$CHO
Ketones	RCOR′ or R–C(=O)–R′
	RCOAr or R–C(=O)–Ar
	ArCOAr or Ar–C(=O)–Ar
e.g., butanone or methyl-ethyl ketone	CH$_3$COCH$_2$CH$_3$

The letters R and R′ represent alkyl groups, and Ar represents an aromatic group.

Table 18.6. Names and formulas of selected carboxylic acids and esters

Name	Formula
Carboxylic acid*	RCOOH or R—C(=O)—OH
	ArCOOH or Ar—C(=O)—OH
e.g., methanoic acid (formic acid),	HCOOH
ethanoic acid (acetic acid),	CH_3COOH
octadecanoic acid (stearic acid)	$CH_3(CH_2)_{16}COOH$
Esters**	RCOOR' or R—C(=O)—O—R' (acid, alcohol)
e.g., methyl propanoate	$CH_3CH_2COOCH_3$

* The letter R represents an alkyl group and Ar represents an aromatic group.
** The letter R represents hydrogen, an alkyl group, or an aromatic group and R' represents an alkyl group or aromatic group

Esters are alcohol derivates of carboxylic acids. Their general formula is RCOOR', where R may be a hydrogen, alkyl group, or aromatic group, and R' may be an alkyl group or aromatic group but not a hydrogen. Esters are found in fats and oils. Some examples are shown in Table 18.6.

Amides and Amines

Amides and amines are organic compounds containing nitrogen. Amides are nitrogen derivates of carboxylic acids. The carbon atom of a carbonyl group is bonded directly to a nitrogen atom of a $-NH_2$, $-NHR$, or $-NR_2$ group. The characteristic structure of amide is $RCONH_2$.

An amine is a substituted ammonia molecule that has a general structure of RNH_2, R_2NH, or R_3N, where R is an alkyl or aromatic group. Some examples are shown in Table 18.3.

18.2.3 Polar and Nonpolar Groups

Table 18.7 summarizes polar and nonpolar groups commonly used in the construction of hydrophobic and hydrophilic molecules. Table 18.8 lists the relative polarity of selected polar groups [49]. Thiol, silane, carboxylic acid, and alcohol (hydroxyl) groups are the most commonly used polar anchor groups for their attachment to surfaces. Methyl and trifluoromethyl are commonly used end groups for hydrophobic film surfaces.

Table 18.7. Some examples of polar (hydrophilic) and nonpolar (hydrophobic) groups

Name	Formula
Polar	
Alcohol (hydroxyl)	$-OH$
Carboxyl acid	$-COOH$
Aldehyde	$-COH$
Ketone	$R-\overset{\overset{O}{\|\|}}{C}-R$
Ester	$-COO-$
Carbonyl	$>C=O$
Ether	$R-O-R$
Amine	$-NH_2$
Amide	$-\overset{\overset{O}{\|\|}}{C}-NH_2$
Phenol	C$_6$H$_5$–OH
Thiol	$-SH$
Trichlorosilane	$SiCl_3$
Nonpolar	
Methyl	$-CH_3$
Trifluoromethyl	$-CF_3$
Aryl (benzene ring)	C$_6$H$_5$–

The letter R represents an alkyl group.

18.3 Self-Assembled Monolayers: Substrates, Head Groups, Spacer Chains, and End Groups

SAMs are formed as a result of spontaneous, self-organization of functionalized organic molecules onto the surfaces of appropriate substrates into stable, well-defined structures, Fig. 18.3. The final structure is close to or at thermodynamic equilibrium, and as a result, it tends to form spontaneously and rejects defects. SAMs consist of three building groups: a head group that binds strongly to a substrate, a surface terminal (tail) group that constitutes the outer surface of the film, and a spacer chain (backbone chain) that connects the head and surface terminal groups. The SAMs are named based on the surface terminal group followed by spacer chain and the head group (or type of compound formed at the surface). For a SAM to control hydropho-

Table 18.8. Organic groups listed in the increasing order of polarity

Alkanes
Alkenes
Aromatic hydrocarbons
Ethers
Trichlorosilanes
Aldehydes, ketones, esters, carbonyls
Thiols
Amines
Alcohols, phenols
Amides
Carboxylic acids

bicity, adhesion, friction, and wear, it should be strongly adherent to the substrate, and the surface terminal group of the organic molecular chain should be nonpolar. For strong attachment of the organic molecules to the substrate, the head group of the molecular chain should contain a polar end group resulting in the exothermic process (energies on the order of tens of kcal/mol), i.e., it results in an apparent pinning of the head group to a specific site on the surface through a chemical bond. Furthermore, molecular structure and any crosslinking would have a significant effect on their friction and wear performance. The substrate surface should have a high surface energy (hydropolic), so that there will be a strong tendency for molecules to adsorb on the surface. The surface should be highly functional with polar groups and dangling bonds (generally unpaired electrons), so that they can react with organic molecules and provide a strong bond. Because of the exothermic head group–substrate interactions, molecules try to occupy every available binding site on the surface, and during this process they generally push together molecules that have already adsorbed. The process results in the formation of crystalline molecular assemblies. The interactions between molecular chains are van der Waals or electrostatic-type with energies on the order of a few (< 10) kcal/mol, exothermic. The molecular chains in SAMs are not perpendicular to the surface; the tilt angle depends on the anchor group, as well as on the substrate and the spacer group. For example, the tilt angle for alkanethiolate on Au is typically about 30–35° angle with respect to substrate normal.

SAMs are usually produced by immersing a substrate in a solution containing precursor (ligand) that is reactive to the substrate surface, or by exposing the substrate to a vapor of the reactive chemical species [39]. Table 18.9 lists selected systems that have been used for formation of SAMs [43]. The spacer chain of SAM is mostly an alkyl chain ($-C_nH_{2n+1}$), or made of a derivatized alkyl group. By attaching different terminal groups at the surface, the film surface can be made to attract or repel water. The commonly used surface terminal group of a hydrophobic film with low surface energy, in the case of a single alkyl chain, is a nonpolar methyl ($-CH_3$) or trifluoromethyl ($-CF_3$) group. For a hydrophilic film, the commonly used surface terminal groups are alcohol ($-OH$) or carboxylic acid ($-COOH$) groups. Surface active head groups most commonly used are thiol ($-SH$), silane (e.g., trichlorosilane

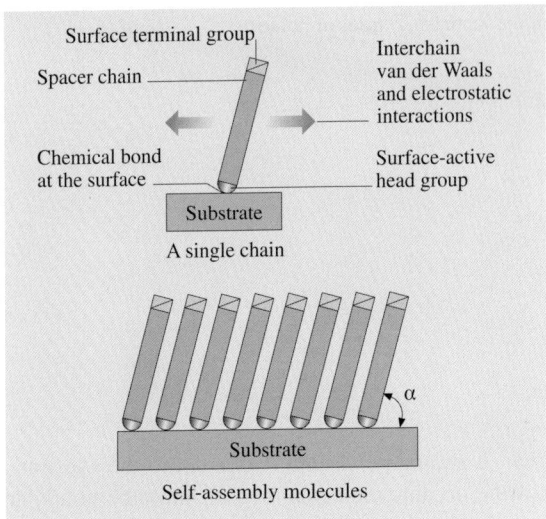

Fig. 18.3. Schematic of a self-assembled monolayer on a surface and the associated forces

or $-SiCl_3$), and carboxyl ($-COOH$) groups. The substrates most commonly used are gold, silver, platinum, copper, hydroxylated (activated) surfaces of SiO_2 on Si, Al_2O_3 on Al, glass, and hydrogen-terminated single-crystal silicon (H−Si). Hydroxylation of oxide surfaces is important to make them hydrophilic. Thermally grown silica can be activated through a sulfochromic treatment. The sample is dipped in a solution consisting of 100 mL of concentrated H_2SO_4, 5 mL of water, and 2 g of potassium bichromate for 3–15 min [34]. The sample is then rinsed with flowing pure water. This results in a surface with silanol groups ($-Si-OH$) that is hydrophilic, Fig. 18.4. Bulk silicon, polysilicon film, or SiO_2 film surfaces can also be treated to produce activated hydrophilic silica surface by immersion in about three parts H_2SO_4 and one part H_2O_2 at temperatures ranging from ambient to 80 °C. For organic molecules to pack together and provide a better ordering, a substrate for given molecules should be selected such that the cross-sectional diameter of the spacer chains of the molecule is equal to or smaller than the distance between the anchor groups attached to the substrate. As an example, epitaxial Au film on glass, mica, or single-crystal silicon, produced by e-beam evaporation, is commonly used because it can be deposited on smooth surfaces as a film that is atomically flat and defect free. For the case of alkanethiolate film, the advantage of Au substrate over SiO_2 substrate is that it results in better ordering, because the cross-sectional diameter of the alkane molecule is slightly smaller than the distance between sulfur atoms attached to the Au substrate (∼ 0.53 nm). The thickness of the film can be controlled by varying the length of the hydrocarbon chain, and the surface properties of the film can be modified by the terminal group.

Some of the SAMs have been widely reported. SAMs of long-chain fatty acids, $C_nH_{2n+1}COOH$ or $(CH_3)(CH_2)_nCOOH$ ($n = 10, 12, 14$ or 16), on glass or alumina

Table 18.9. Selected substrates and precursors that have commonly been used for formation of SAMs

Substrate	Precursor	Binding with substrate
Au	RSH (thiol)	RS–Au
Au	ArSH (thiol)	ArS–Au
Au	RSSR′ (disulfide)	RS–Au
Au	RSR′ (sulfide)	
Si/SiO$_2$, glass	RSiCl$_3$ (trichlorosilane)	Si–O–Si (Siloxane)
Si/Si–H	RCOOH (carboxyl)	R–Si
Metal oxides (e.g., Al$_2$O$_3$, SnO$_2$, TiO$_2$)	RCOOH (carboxyl)	RCOO–...MO$_n$

R represents alkane (C$_n$H$_{2n+2}$) and Ar represents aromatic hydrocarbon. It consists of various surface active head groups and mostly with methyl terminal group

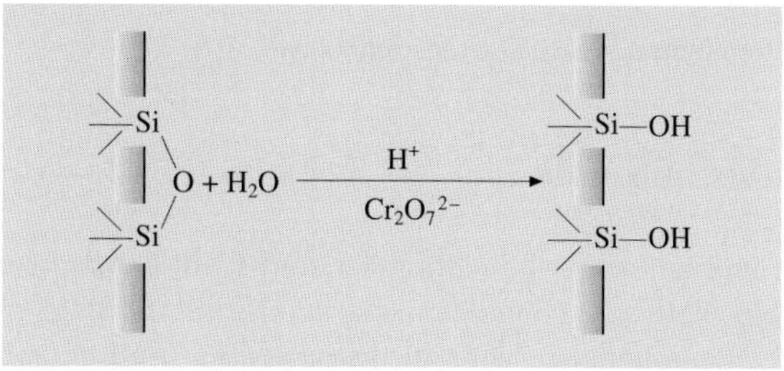

Fig. 18.4. Schematic showing the hydroxylation process occurring on a silica surface through a sulfochromic treatment

have been used in films studies since the 1950s [16, 20]. Probably the most studied SAMs to date are n-alkanethiolate monolayers CH$_3$(CH$_2$)$_n$S–, prepared from adsorption of alkanethiol –(CH$_2$)$_n$SH solution onto an Au film (n-alkyl and n-alkane are used interchangeably) [35–37, 43] and n-alkylsiloxane monolayersproduced by adsorption of n-alkyltrichlorosilane –(CH$_2$)$_n$SiCl$_3$ solution onto hydroxylated Si/SiO$_2$ substrate [50] with siloxane (Si–O–Si) binding, Fig. 18.5. *Jung* et al. [51] have produced organosulfur monolayers – decanethiol (CH$_3$)(CH$_2$)$_9$SH, and didecyl CH$_3$(CH$_2$)$_9$–S–(CH$_2$)$_9$CH$_3$ on Au films. *Geyer* et al. [52], *Bhushan* and *Liu* [35], *Liu* et al. [36], and *Liu* and *Bhushan* [37] have produced monolayers of 1,1′-biphenyl-4-thiol (BPT) on Au surface, where the spacer chain of the film consists of two phenyl rings with a hydrogen end group. *Bhushan* and *Liu* [35] and *Liu* and *Bhushan* [37] have also reported monolayers of 4,4′-dihydroxybiphenyl on Si surface.

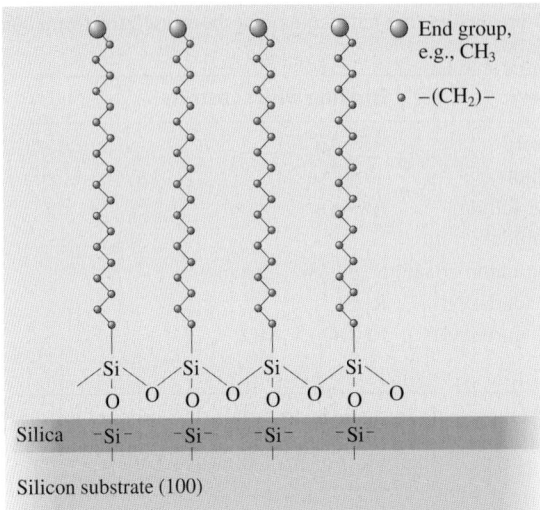

Fig. 18.5. Schematics of a methyl-terminated, n-alkylsiloxane monolayer on Si/SiO$_2$

18.4 Tribological Properties of SAMs

The basis for molecular design and tailoring of SAMs must start from a complete knowledge of the interrelationships between the molecular structure and tribological properties of SAMs, as well as a deep understanding of the adhesion, friction, and wear mechanisms of SAMs at the molecular level. Friction and wear properties of SAMs have been studied on macro- and nanoscales. Macroscale tests are conducted using a so-called pin-on-disk tribotester apparatus in which a ball specimen slides against a lubricated flat specimen [9, 10]. Nanoscale tests are conducted using an atomic force/friction force microscope (AFM/FFM) [4, 9, 10]. In the AFM/FFM experiments, a sharp tip of radius ranging from about 5–50 nm slides against a SAM specimen. A Si$_3$N$_4$ tip is commonly used for friction studies and a natural diamond tip is commonly used for scratch, wear, and indentation studies.

In early studies, the effect of chain length of the carbon atoms of fatty acid monolayers on the coefficient of friction and wear on the macroscale was studied by *Bowden* and *Tabor* [16] and *Zisman* [20]. *Zisman* reported that for the monolayers deposited on a glass surface sliding against a stainless steel surface, there is a steady decrease in friction with increasing chain length. At a significantly long chain length, the coefficient of friction reaches a lower limit, Fig. 18.6. He further reported that monolayers having a chain length below 12 carbon atoms behave as liquids (poor durability), those with chain length of 12–15 carbon atoms behave like a plastic solid (medium durability), whereas those with chain lengths above 15 carbon atoms behave like a crystalline solid (high durability). Investigations by *Ruhe* et al. [53] indicated that the lifetime of the alkylsilane monolayer coating on a silicon surface increases greatly with an increase in the chain length

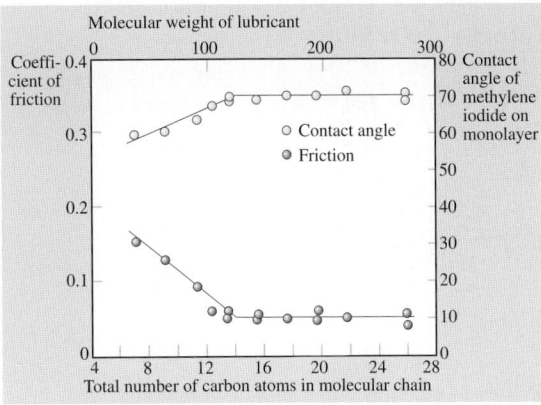

Fig. 18.6. Effect of chain length (or molecular weight) on coefficient of macroscale friction of stainless steel sliding on glass lubricated with a monolayer of fatty acid and contact angle of methyl iodide on condensed monolayers of fatty acid on glass [20]

of the alkyl substituent. *DePalma* and *Tillman* [54] showed that a monolayer of n-octadecyltrichlorosilane (n-$C_{18}H_{37}SiCl_3$, OTS) is an effective lubricant on silicon, followed by n-undecyltrichlorosilane (n-$C_{11}H_{23}SiCl_3$, UTS) and (tridecafluoro-1,1,2,2-tetrahydrooct-1-yl) trichlorosilane (n-$C_6F_{13}CH_2CH_2SiCl_3$, FTS). *Ando* et al. [55] suggested that γ-(N,N-dioctadecylsuccinylamino) propyltriethoxysilane monolayer is a candidate lubricant. The film exhibited a low coefficient of kinetic friction of 0.1, without stick-slip. However, tests in all these investigations were carried out using a pin-on-disk tribotester under relatively large normal loads up to 0.15 N. Thus, the relevance of these tests is questionable for micro/nanodevice application.

With the development of AFM techniques, researchers have successfully characterized the nanotribological properties of self-assembled monolayers [1, 4, 34–38]. Studies by *Bhushan* et al. [34] showed that C_{18} alkylsiloxane films exhibit the lowest coefficient of friction and can withstand a much higher normal load during sliding compared to LB films, soft Au films, and hard SiO_2 coatings.

McDermott et al. [56] studied the effect of length of alkyl chains on the frictional properties of methyl-terminated n-alkylthiolate $CH_3(CH_2)_nS-$ films chemisorbed on Au(111) using an AFM. They reported that the longer chain monolayers exhibit a markedly lower friction and a reduced propensity to wear than shorter chain monolayers, Fig. 18.7. These results are in good agreement with the macroscale results by *Zisman* [20]. They also conducted infrared reflection spectroscopy to measure the bandwidth of the methylene stretching mode (v_a (CH_2)), which exhibits a qualitative correlation with the packing density of the chains. They found that the chain structures of monolayers prepared with longer chain lengths are more ordered and more densely packed in comparison to those of monolayers prepared with shorter chain lengths. They further reported that the ability of the longer chain monolayers to retain molecular-scale order during shear leads to a lower observed friction. Mono-

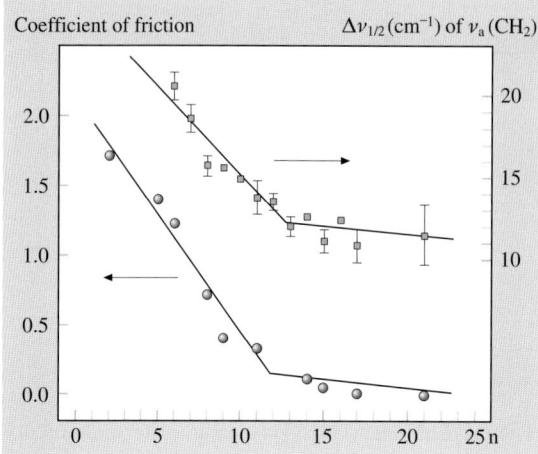

Fig. 18.7. Effect of chain length of methyl-terminated, n-alkanethiolate over Au film AuS(CH$_2$)$_n$CH$_3$ on the coefficient of microscale friction and peak bandwidth at half maximum ($\Delta \nu_{1/2}$) for the bandwidth of the methylene stretching mode [ν_a (CH$_2$)] [56]

layers with a chain length of more than 12 carbon atoms, preferably 18 or more, are desirable for tribological applications. (Incidentally, the monolayer with 18 carbon atoms, octadecanethiol films, are commonly studied.)

Xiao et al. [57] and *Lio* et al. [58] also studied the effect of the length of the alkyl chains on the frictional properties of n-alkanethiolate films on gold and n-alkylsilane films on mica. Friction was found to be particularly high with short chains of less than eight carbon atoms. Thiols and silanes exhibit a similar friction force for the same n when $n > 11$; while for $n < 11$, silanes exhibit higher friction, larger than that for thiols by a factor of about 3 for $n = 6$. The increase in friction was attributed to the large number of dissipative modes in the less ordered chains that occurs when going from a thiol to a silane anchor or when decreasing n. Longer chains ($n > 11$), stabilized by van der Waals attraction, form more compact and rigid layers and act as better lubricants. *Schonherr* and *Vancso* [59] also correlated the magnitude of friction with the order of the alkane chains. The disorder of the short-chain hydrocarbon disulfides SAMs was found to result in a significant increase in the magnitude of friction.

Fluorinated carbon (fluorocarbon) molecules are commonly used for lubrication [2, 9, 10]. *Kim* et al. [60, 61] studied the influence of methyl-, isopropyl-, and trifluoromethyl-terminated alkanethiols on the friction properties of SAMs. They reported a factor of three increases in the friction in going from the methyl-terminated SAM to the trifluoromethyl-terminated SAM. They suggested that fluorinated monolayers exhibit higher frictional properties due to tighter packing at the interface, which arises from the larger van der Waals radii of the fluorine atoms. Subsequent steric and rotational factors between adjacent terminal groups give rise to long-range multi-molecular interactions in the plane of the −CF$_3$ groups. When these energetic barriers are overcome in the film structure, more energy is imparted to the film during sliding and results in higher friction for the fluorinated films.

Tsukruk et al. [62] and *Lee* et al. [63] studied the nanotribological properties of C_{60}-terminated alkyltrichlorosilanes and alkanedisulfides self-assembled monolayers. Both of their studies observed that the frictional forces of the C_{60}-terminated films are higher than those on normal methyl-terminated SAMs. However, *Tsukruk* et al. [62] reported that the C_{60}-terminated alkyltrichlorosilanes SAMs exhibit high wear resistance. *Tsukruk* and *Bliznyuk* [64] also studied the adhesion and friction between a Si sample and Si_3N_4 tip, in which both surfaces were modified by $-CH_3$, $-NH_2$, and $-SO_3H$ terminated silane-based SAMs. Various polymer molecules for the backbone were used. They reported that a very broad maximum adhesive force in the pH range of from 4 to 8 with a minimum adhesion at pH > 9 and pH < 3 were observed for all of the studied mating surfaces. This observation can be understood by considering a balance of electrostatic and van der Waals interaction between composite surfaces with multiple isoelectric points. The friction coefficient of NH_2/NH_2 and SO_3H/SO_3H mating SAMs is very high in aqueous solutions. Cappings of NH_2 modified surfaces (3-aminopropyltriethoxysilane) with rigid and soft polymer layers resulted in a significant reduction in adhesion to a level lower than that of untreated surfaces [65]. *Fujihira* et al. [66] studied the influence of surface terminal groups of SAMs and functional tip on adhesive force. It was found that the adhesive forces measured in air increase in the order of CH_3/CH_3, $CH_3/COOH$, and $COOH/COOH$.

Bhushan and *Liu* [35], *Liu* et al. [36], and *Liu* and *Bhushan* [37, 38] have studied the influence of spacer chains, surface terminal groups, and head groups on adhesion, friction, and wear properties of SAMs. They have explained the friction mechanisms using a molecular spring model in which local stiffness and the intermolecular forces govern the friction properties. They studied the influence of relative humidity, temperature, and velocity on adhesion and friction. They also investigated the wear mechanisms of SAMs by a continuous microscratch AFM technique.

To date, the nanotribological properties of alkyltrichlorosilanes, alkanethiols, and biphenyl thiol SAMs have been widely studied. In this chapter, we review, in some detail, the nanotribological properties of five kinds of SAMs, having alkyl and biphenyl spacer chains with different surface terminal groups ($-CH_3$, $-COOH$, and $-OH$), and head groups ($-SH$ and $-OH$), which have been investigated by AFM at various operating conditions, Fig. 18.8 [35–38]. Hexadecane thiol (HDT), 1,1′-biphenyl-4-thiol (BPT) and crosslinked BPT (BPTC), and 16-mercaptohexadecanoic acid thiol (MHA) were deposited on Au(111) substrates. A 4,4′-dihydroxybiphenyl (DHBp) film was deposited on a hydrogenated Si(111) substrate. Figure 18.8 shows that HDT and MHA have the same head groups and spacers, but different surface terminal groups; their surface terminals are $-CH_3$ and $-COOH$, respectively. BPT and DHBp have the same spacers, but the head and surface terminal groups of DHBp are $-OH$ instead of $-SH$ and $-CH$ in BPT. Crosslinked BPTC was produced by irradiation of BPT monolayers with low energy electrons.

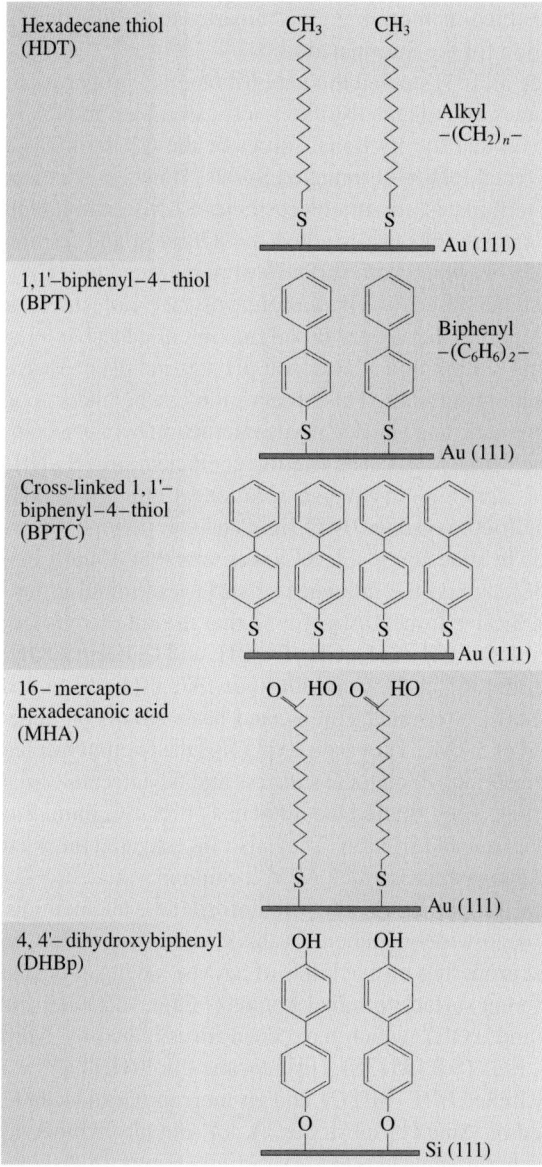

Fig. 18.8. Schematics of the structures of HDT, BPT, MHA, DHBp, and BPTC SAMs [35]

Table 18.10. The R_a roughness, thickness, tilt angles, and spacer chain lengths of SAMs

Samples	R_a roughness[1] (nm)	Thickness[2] (nm)	Tilt angle[2] (deg.)	Spacer length[3] (nm)
Si(111)	0.07			
Au(111)	0.37			
HDT	0.92	1.89	30	1.91
BPT	0.36	1.25	15	0.89
BPTC	0.14	1.14	25	0.89
MHA	0.37	2.01	30	1.91
DHBp	0.25	1.13	–	0.89

[1] Measured by an AFM with 1 μm × 1 μm scan size using a Si_3N_4 tip under 3.3 nN normal load.
[2] The thickness and tilt angles of BPT, BPTC, and DHBp are reported by *Geyer* et al. [52]. The thickness and tilt angles of HDT and MHA are reported by *Ulman* [40].
[3] The spacer chain lengths of alkylthiols were calculated by the method reported by *Miura* et al. [68]. The spacer chain lengths of biphenyl thiols were calculated by the data reported by *Ratajczak-Sitarz* et al. [69].

18.4.1 Surface Roughness and Friction Images of SAMs Films

Surface height and friction force images of SAMs were recorded simultaneously on an area of 1 μm × 1 μm by an AFM, Fig. 18.9 [35]. The topography of Au(111) film and SAMs deposited on Au(111) substrates appear to be granular. A good correlation between the surface height and the corresponding friction force images was observed. It was noticed that the friction force changed with surface height in the same direction (upwards and downwards) in both trace and retrace friction profiles of the friction loop. Thus, the change in friction force corresponds to transitions in surface slope [4, 35, 36, 67]. Figure 18.9 also indicates that the topography and friction images of Si(111) and DHBp/Si(111) are very similar; they exhibit featureless surfaces and their friction force images do not show any abrupt changes.

For further analysis presented later in this chapter, the roughness, thickness, tilt angles, and spacer chain lengths of Si(111), Au(111), and various SAMs are listed in Table 18.10 [35]. The roughness values of BPT and MHA are very close to that of Au(111). But the roughness of BPTC is lower than that of Au(111) and BPT; this is caused by electron irradiation. Table 18.10 indicates that the roughness values of HDT and DHBp are much higher than their substrate roughness of Au(111) and Si(111), respectively. This is caused by local aggregation of organic compounds on the substrates during SAMs deposition. Table 18.10 also indicates that the thickness of biphenyl thiol SAMs is generally thinner than the alkylthiol, which is responsible for shorter spacer chain in biphenyl thiol.

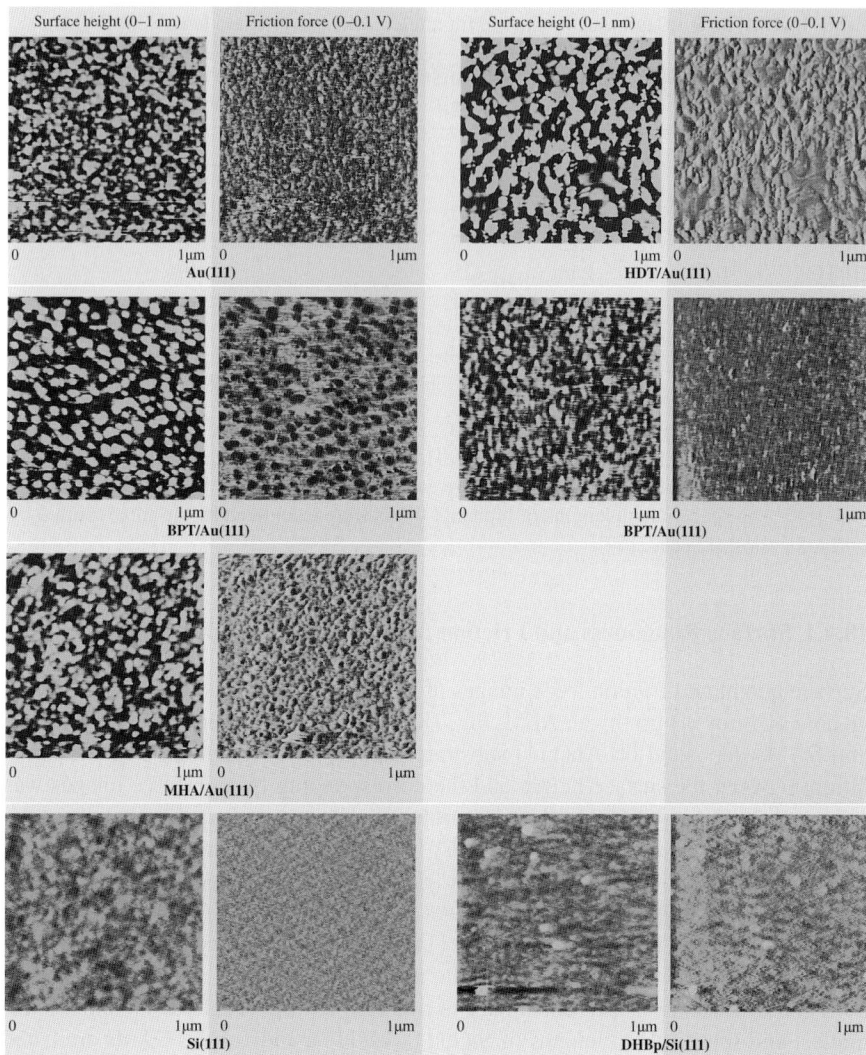

Fig. 18.9. AFM gray-scale surface height and friction force images of Au(111), Si(111), and various SAMs [35]

18.4.2 Adhesion, Friction, and Work of Adhesion

The average values and standard deviation of the adhesive force and coefficients of friction measured by contact mode AFM are presented in Fig. 18.10 [35]. Based on the data, the adhesive force and coefficient of friction of SAMs are less than their corresponding substrates. Among various films, HDT exhibits the lowest values. The ranking of adhesive forces F_a is in the following order: $F_{a\text{-Au}} > F_{a\text{-Si}} > F_{a\text{-DHBp}} \approx F_{a\text{-MHA}} > F_{a\text{-BPT}} > F_{a\text{-BPTC}} > F_{a\text{-HDT}}$. And the ranking of the coeffi-

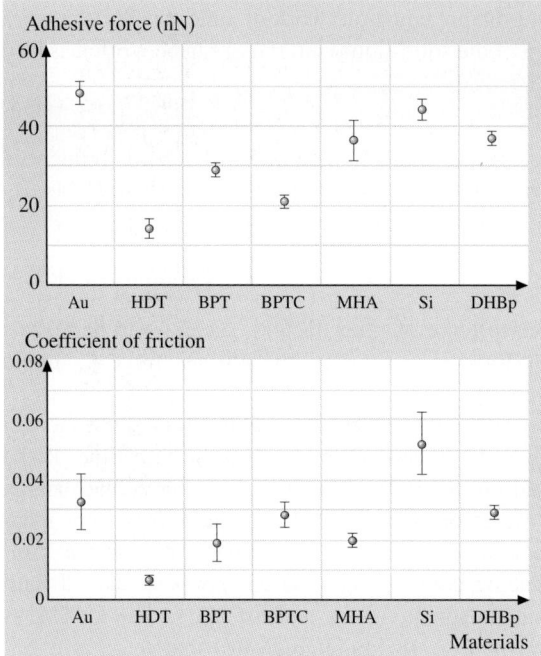

Fig. 18.10. Adhesive forces and coefficients of friction of Au(111), Si(111), and various SAMs [35]

cients of friction μ is in the following order: $\mu_{Si} > \mu_{Au} > \mu_{DHBp} > \mu_{BPTC} > \mu_{BPT} \approx \mu_{MHA\,16-mercaptohexadecanoicacidthiol} > \mu_{HDT}$. The ranking of various SAMs for adhesive force and coefficient of friction is similar. It suggests that alkylthiol and biphenyl SAMs can be used as effective molecular lubricants for micro/nanodevices fabricated from silicon.

In micro/nano scale contact, liquid capillary condensation is one of the factors that influence adhesion and friction. In the case of a sphere in contact with a flat surface, the attractive Laplace force caused by water capillary is:

$$F_L = 2\pi R \gamma_{la}(\cos\theta_1 + \cos\theta_2), \quad (18.1)$$

where R is the radius of the sphere, γ_{la} is the surface tension of the liquid against air, θ_1 and θ_2 are the contact angles between liquid and flat and spherical surfaces, respectively [9, 10]. In an AFM adhesive study, the tip-flat sample contact is just like a sphere in contact with a flat surface and the liquid is water. Since a single tip was used in the adhesion measurements, $\cos\theta_2$ can be treated as a constant. Therefore,

$$\begin{aligned} F_L &= 2\pi R \gamma_{la}(1 + \cos\theta_1) - 2\pi R \gamma_{la}(1 - \cos\theta_2) \\ &= 2\pi R \gamma_{la}(1 + \cos\theta_1) - C, \end{aligned} \quad (18.2)$$

where C is a constant.

Based on the following Young-Dupre equation, work of adhesion W_a (the work required to pull apart the unit area of the solid-liquid interface) can be written as [70]

$$W_a = \gamma_{la}(1 + \cos\theta_1) .\tag{18.3}$$

It indicates that W_a is determined by the contact angle of SAMs, i.e., is influenced by the surface chemistry properties (polarization and hydrophobicity) of SAMs. By substituting (18.3) into (18.2), F_L can be expressed as

$$F_L = 2\pi R W_a - C .\tag{18.4}$$

When the influence on the adhesive force of other factors, such as van der Waals force, is very small, then adhesive force $F_a \approx F_L$. Thus the adhesive force F_a should be proportional to work of adhesion W_a.

Contact angle is a measure of the wettability of a solid by a liquid and determines the W_a value. The value of the contact angle changes with time. If the droplet expands, the contact angle is referred to as the advancing contact angle. The contact angles of distilled water on Si(111), Au(111),and SAMs were measured using a contact angle goniometer [35]. After dropping water on the sample surfaces, the dynamic advancing contact angle (DACA) was measured with a frequency of 60s for spontaneous spreading for at least 420s. The DACA as a function of time data is given in Fig. 18.11a. It indicates that there is a linear relationship between DACA and rest time. As stated by many authors, the static advancing contact angle (SACA) (at $t = 0$) should be used for surface characterization [71–73]. This SACA can be obtained by simple extrapolation of the data in Fig. 18.11a to $t = 0$. The SACA of Si(111), Au(111), and SAMs is summarized in Fig. 18.11b. For water, $\gamma_{la} = 72.6\,\text{mJ/m}^2$ at $22\,°C$, and by using (18.3), the W_a data are obtained and presented in Fig. 18.11c. The W_a can be ranked in the following order: $W_{a\text{-Si}}(117.0) > W_{a\text{-DHBp}}(98.8) \approx W_{a\text{-MHA}}(101.8) \approx W_{a\text{-Au}}(97.1) > W_{a\text{-BPT}}(86.8) > W_{a\text{-BPTC}}(82.1) > W_{a\text{-HDT}}(61.4)$. Excluding $W_{a\text{-Au}}$, this order exactly matches the order of adhesion force in Fig. 18.10. The relationship between F_a and W_a is summarized in Fig. 18.12 [35]. It indicates that the adhesive force F_a (nN) increases with work of adhesion W_a (mJ/m^2) by the following linear relationship:

$$F_a = 0.57 W_a - 22 .\tag{18.5}$$

These experimental results agree well with the modeling prediction presented earlier in (18.4). It proves that on the nanoscale at ambient condition the adhesive force of SAMs is mainly influenced by the water capillary force. Comparing the results of Figure 18.12 with Figure 18.8, it is found that MHA and DHBp, which have polar surface terminal groups (−COOH and −OH), lead to larger W_a and eventually larger adhesive forces. Though both HDT and BPT do not have polar surface groups, the surface terminal of HDT has a symmetrical structure, which causes a smaller electrostatic attractive force and yields smaller adhesive force than BPT. It is believed that the easy attachment of Au on the tip should be one of the reasons that cause the large adhesive force, which do not fit the linear relationship described by (18.5).

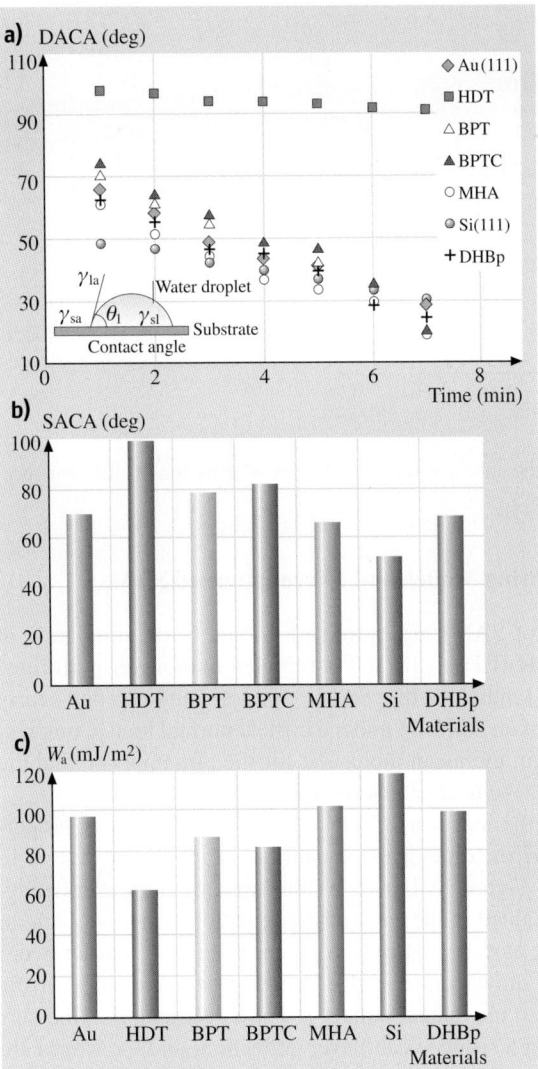

Fig. 18.11. (**a**) Variation of dynamic advancing contact angle (DACA) with time, (**b**) the static advancing contact angle (SACA), and (**c**) work of adhesion of Au(111), Si(111), and various SAMs. All of the points in this figure represent the mean value of six measurements. The uncertainty associated with the average contact angle is within ±2°. The insert schematic in (**a**) shows the contact angle of the water droplet on the sample surface [35]

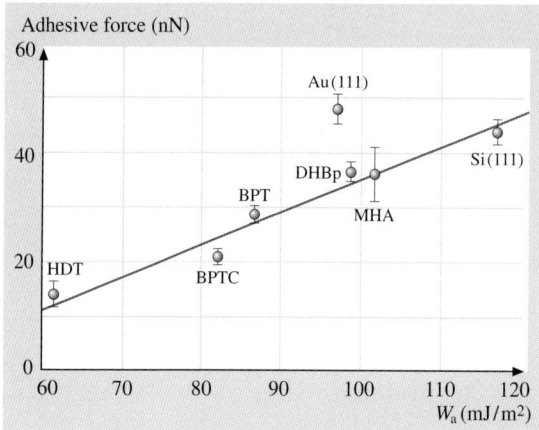

Fig. 18.12. Relationship between the adhesive forces and work of adhesion of different specimen [35]

18.4.3 Stiffness, Molecular Spring Model, and Micropatterned SAMs

Next, the friction mechanisms of SAMs are examined. Monte Carlo simulation of the mechanical relaxation of $CH_3(CH_2)_{15}SH$ self-assembled monolayer performed by *Siepman* and *McDonald* [74] indicated that SAMs compress and respond nearly elastically to microindentation of an AFM tip under a critical normal load. Compression can lead to major changes in the mean molecular tilt (i.e., orientation), but the original structure is recovered as the normal load is removed.

To study the difference in the stiffness of various films, the stiffness properties were measured by an AFM in force-distance calibration mode, as well as in force modulation mode [4, 37]. Figure 18.13a shows the slope of cantilever deflection versus piezo movement obtained in the force-distance calibration mode. For an infinite hard material, since the surface can not be compressed, the cantilever deflection should equal the piezo movement distance, which means the slope of the cantilever deflection versus piezo movement equals 1. For a soft material, the surface can be compressed, causing a reduced cantilever deflection, and the slopes are smaller than 1. In this study the cantilever has been initially calibrated to slope = 1 against a clean, rigid Si(111) surface. Figure 18.13a indicates that the slopes of all of the SAMs are less than 1, which suggests that all of the SAMs can be compressed by a Si_3N_4 tip during the loading process. The slope value of SAMs can be ranked in the following order: $S_{DHBp} > S_{BPT} > S_{MHA16-mercaptohexadecanoicacidthiol} \approx S_{HDT}$. This order reflects the influence of molecular structure on the compression properties of SAMs. Since BPT and DHBp have rigid benzene structure, they are more difficult to compress than HDT and MHA. Figure 18.13b shows the variation of the displacement with normal load during indentation mode. It also clearly indicates that SAMs can be compressed. At a given normal load, long carbon chain structure SAMs such as HDT and MHA are easy to compress compared to rigid benzene-ring structure SAMs such as BPT and DHBp. *Garcia-Parajo* et al. [75] have also reported the

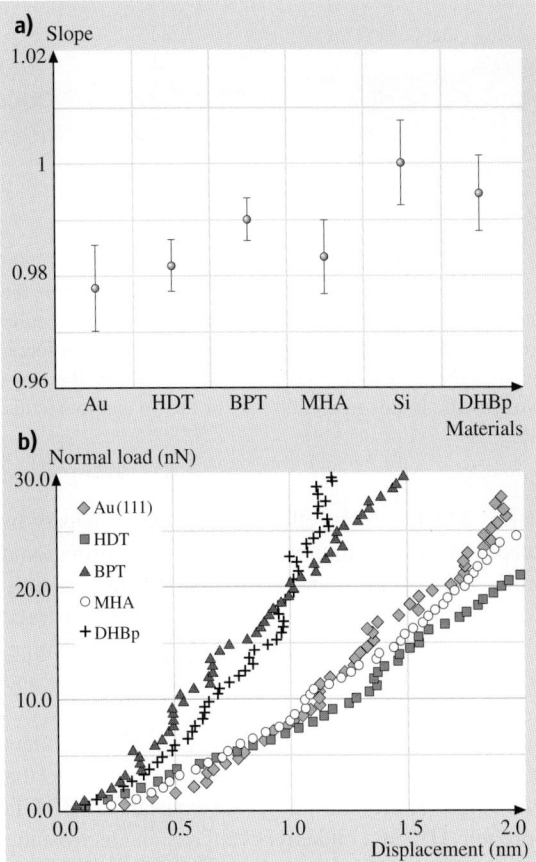

Fig. 18.13. (a) Slopes of cantilever deflection versus piezo displacement measured by force distance calibration mode AFM, and (b) normal load versus displacement curves of Au(111), Si(111), and various SAMs

compression and relaxation of octadecyltrichlorosilane (OTS) film in their loading and unloading tests.

In order to explain the frictional difference of SAMs, based on the friction and stiffness measurements by AFM and the Monte Carlo simulation, a molecular spring model is presented in Fig. 18.14. It is believed that the self-assembled molecules on a substrate are just like assembled molecular springs anchored to the substrates [35]. A Si_3N_4 tip sliding on the surface of SAMs is like a tip sliding on the top of "molecular springs or brush". The molecular spring assembly has compliant features and can experience compression and orientation under normal load. The orientation of the "molecular springs or brush" reduces the shearing force at the interface, which, in turn, reduces the friction force. The possibility of orientation is determined by the spring constant of a single molecule (local stiffness), as well as the interaction

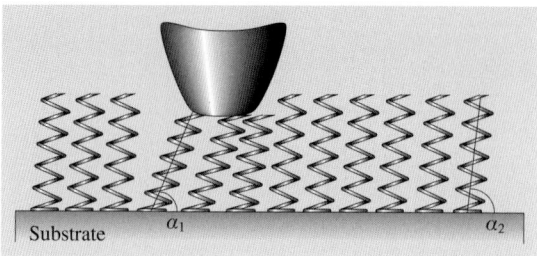

Fig. 18.14. Molecular spring model of SAMs. In this figure, $\alpha_1 < \alpha_2$, which is caused by the orientation under the normal load applied by an AFM tip. The orientation of the molecular springs reduces the shearing force at the interface, which, in turn, reduces the friction force. The molecular spring constant, as well as the inter-molecular forces can determine the magnitude of the coefficients of friction of SAMs. In this figure, the size of the tip and molecular springs are not drawn exactly to scale [35]

between the neighboring molecules, which can be reflected by packing density or packing energy. It should be noted that the orientation can lead to conformational defects along the molecular chains, which lead to energy dissipation. In the study of BPT by AFM, it was found that after the first several scans, the friction force is significantly reduced, but the surface height does not show any apparent change. This suggests that the molecular orientation can be facilitated by initial sliding and is reversible [38].

Based on the stiffness measurement results presented in Fig. 18.13 and the view of molecular structures in Fig. 18.14, biphenyl is a more rigid structure due to the contribution of two rigid benzene rings. Therefore, the spring constant of BPT is larger than that of HDT. The hydrogen (H^+) in a biphenyl chain has an electrostatic attractive force with the π electrons in the neighboring benzene ring. Thus, the intermolecular force between biphenyl chains is stronger than that for alkyl chains. The larger spring constant of BPT and stronger intermolecular force require a larger external force to allow it to orient, thus causing higher coefficient of friction. For MHA and DHBp, their basic chain structures are very close to HDT and BPT, respectively. But their surface terminals are different. The polar −COOH and −OH external functional groups in MHA and DHBp increase the adhesive force, thus leading to higher friction force than HDT and BPT, respectively. The crosslinking of BPT leads to a larger packing energy for BPTC. Therefore, it requires a larger external force to allow BPTC orientation, i.e., the coefficient of BPTC is higher than BPT.

An elegant way to demonstrate the influence of molecular stiffness on friction is to investigate SAMs with different structures on the same wafer. For this purpose, a micropatterned SAM was prepared. First, the biphenyldimethylchlorosilane (BDCS) was deposited on silicon by the typical self-assembly method [37]. Then the film was partially crosslinked using mask technique by low energy electron irradiation. Finally, the micropatterned BDCS films were realized that had the as-deposited and crosslinked coatings on the same wafer. The local stiffness properties of these micropatterned samples were investigated by force modulation AFM technique [76].

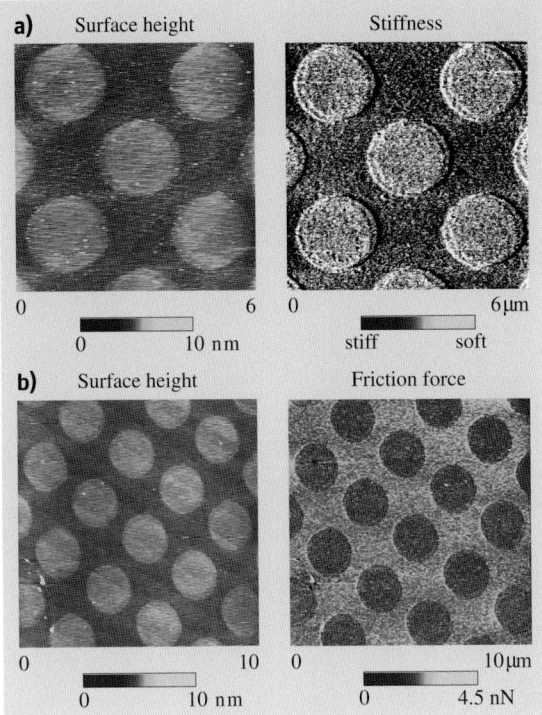

Fig. 18.15. (a) AFM gray-scale surface height and stiffness images, and (b) AFM gray-scale surface height and friction force images of micropatterned BDCS [37]

The variation in the deflection amplitude provides a measure of the relative local stiffness of the surface. Surface height, stiffness, and friction images of the micropatterned biphenyldimethylchlorosilane (BDCS) specimen are obtained and presented in Fig. 18.15 [37]. The circular areas correspond to the as-deposited film, and the remaining area to the crosslinked film. Figure 18.15a indicates that crosslinking caused by the low energy electron irradiation leads to about a 0.5 nm decrease of the surface height of BDCS films. The corresponding stiffness images indicate that the crosslinked area has a higher stiffness than the as-deposited area. Figure 18.15b indicates that the as-deposited area (higher surface height area) has lower friction force. Obviously, the data of the micropatterned sample prove that the local stiffness of SAMs has an influence on their friction performance. Higher stiffness leads to larger friction force. These results provide a strong proof of the suggested molecular spring model.

In summary, it has been found that SAMs exhibit compliance and can experience compression and orientation under normal load. The orientation of SAMs reduces the shear stress at the interface; therefore SAMs can serve as good lubricants. The molecular spring constant (local stiffness), as well as the intermolecular forces can influence the magnitude of the coefficients of friction of SAMs.

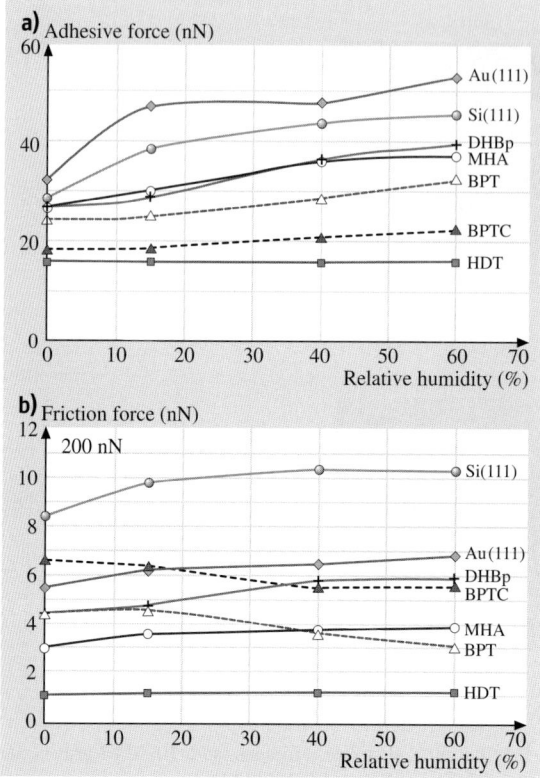

Fig. 18.16. The influence of relative humidity on the (**a**) adhesive force and (**b**) friction forces at 200 nN of Si(111), Au(111), and various SAMs [35]

18.4.4 Influence of Humidity, Temperature, and Velocity on Adhesion and Friction

The influence of relative humidity on adhesion and friction was studied in an environmentally controlled chamber at 22 °C [35]. The results are presented in Fig. 18.16. It shows that for Si(111), Au(111), DHBp and MHA, the adhesive and friction forces increase with relative humidity. For BPT and BPTC, the adhesive force only slightly increases with relative humidity when the relative humidity is higher than 40%, but it is very interesting that their friction decreases in the same range. For HDT, in the testing range, the adhesive and friction force are not sensitive to changes in relative humidity.

It has already been shown that the adhesive force is linearly related to the W_a value of the surfaces in Fig. 18.12. DHBp and MHA have polar surface terminals and larger W_a value. This must therefore lead to the larger adhesive forces in higher relative humidity, which, in turn, increase the friction force. In comparison, HDT has a nonpolar $-CH_3$ surface terminal and a very small W_a value. Thus, the adhesive

Table 18.11. Melting point of typical organic compounds similar to HDT and BPT SAMs

Compound type	Molecular formula	Melting points (°C) [77]
Linear carbon chain	$CH_3(CH_2)_{14}CH_2OH$	50
Benzene ring	3-$HO-C_6H_4C_6H_5$	78
	2-$HO-C_6H_4C_6H_5$	58–60

force and friction force of HDT are not sensitive to the change of humidity. The W_a values of BPT and BPTC are between Si(111) and HDT (see Fig. 18.11c); therefore their adhesive forces show a slight increase when RH > 40%. The fact that BPT and BPTC show lower friction force when RH > 40% suggests that a thin adsorbed water layer can act as a lubricant [35, 37]. The reason why the adsorbed water layer acts as a lubricant may be related to the thickness of the absorbed layer.

Results of the influence of temperature on adhesion and friction of Si(111) and two selected typical SAMs – HDT and BPT – are presented in Fig. 18.17 [37]. The friction data are presented at various normal loads. In all cases, the normal load does not change the general trends. The data show that for Si(111), once the temperature is higher than 75 °C, the adhesive force decreases with temperature. The friction force of silicon decreases slightly with temperature from room temperature to 100 °C; once the temperature is higher than 100 °C, the friction force shows an apparent decrease. It is believed that desorption of the adsorbed water layer on Si(111) and the reduction of surface tension of water at high temperature are responsible for the decrease in adhesion and friction [77]. For HDT, in the testing range (22 °C–125 °C) its adhesion and friction show very slight change with temperature. For BPT, its adhesive force initially increases with temperature from 22 °C–75 °C, but when the temperature is higher than 75 °C, the adhesive force remains steady within the experimental error. The variations in the friction force with temperature for BPT exhibit the maximum value, and the corresponding critical temperature is related to the normal load. The reason why BPT shows maximum friction force is believed to be caused by the melting of film at high temperature. When the temperature is below the melting point, an increase of temperature leads to the softening of SAM, which increases the real contact area and, consequently, the friction force. Once the temperature is higher than the melting point, the lubrication regime is changed from boundary lubrication in a solid SAM to liquid lubrication in the melted SAM, and therefore the friction force is decreased. To date, the melting points of HDT and BPT SAMs are unknown. But from the literature [77], the melting points of typical linear carbon chain and benzene ring compounds are presented in Table 18.11. It is found that the reported melting of the benzene ring compounds is somewhat close to the frictional transition temperature of BPT measured by AFM. For HDT, their very compliant carbon chain can serve as an excellent lubricant even in solid state (i.e., at or near room temperature), so clear temperature effect is not observed as in the case of HDT.

Figure 18.18 shows the influence of velocity on the friction force of SAMs at different normal loads [37]. In all cases, normal load does not change the general

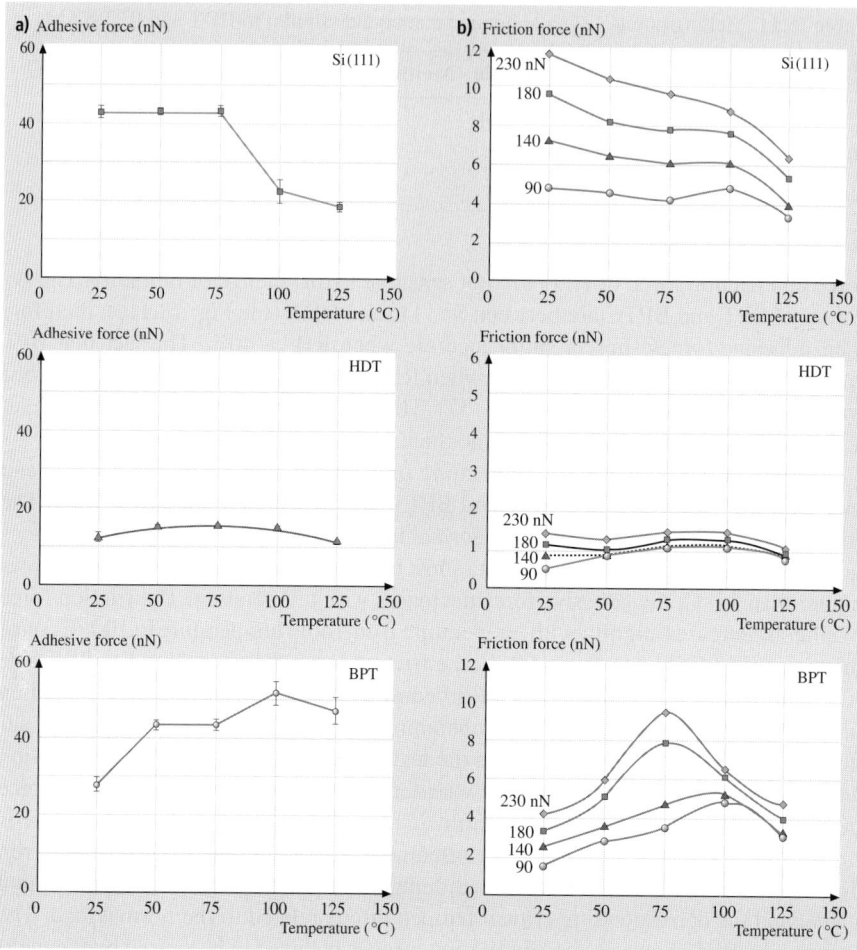

Fig. 18.17. The influence of temperature on (**a**) adhesive forces and (**b**) friction forces of Si(111), HDT, and BPT [37]

trends. It is shown that for a silicon wafer, cleaned by Piranha solution just before the friction test, the friction force decreases linearly with increasing velocity. But for Si(111) without cleaning just before the AFM test, the friction force decreases only at high velocity. It is believed that this difference is caused by the adsorbed water and/or contamination on the surface during storage. Figure 18.18 also indicates that the velocity effects of SAMs depend on the molecular structures of SAMs. For SAMs that have compliant long carbon spacer chains, such as HDT and MHA, the friction force increases at high velocity, while for SAMs that have rigid biphenyl chains, such as BPT, BPTC, and DHBp, the friction force changes in the opposite way. The mechanisms responsible for the variation of the friction forces of SAMs with velocity are believed to be related to the viscoelastic properties of SAMs.

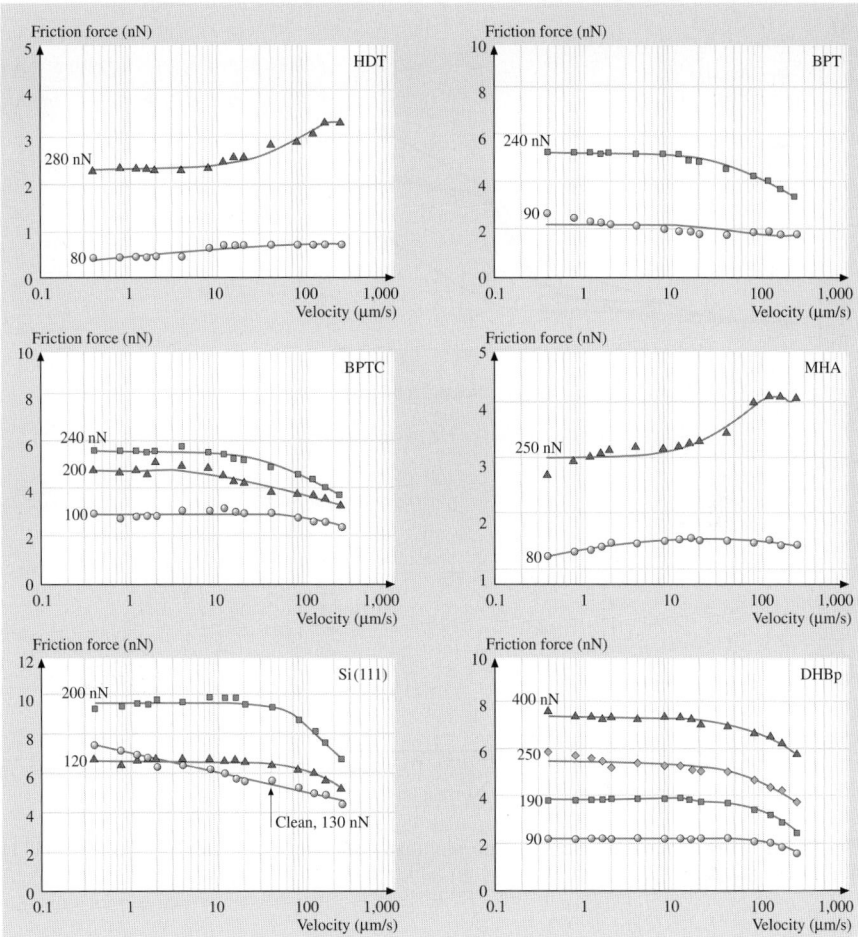

Fig. 18.18. The influence of velocity on friction forces of Si(111) and SAMs [37]

18.4.5 Wear and Scratch Resistance of SAMs

Wear resistance was studied on an area of 1 μm × 1 μm using a diamond tip in an AFM. The variation of wear depth with normal loads is presented in Fig. 18.19 [35]. It clearly shows that in the whole testing range, DHBp on Si(111) exhibits much better wear resistance compared to Si(111), Au(111), and other SAMs that were deposited on the Au(111) substrate. For the SAMs deposited on Au(111), HDT exhibits the best wear resistance. For all of the tested SAMs, in the wear depth as a function of normal load curves, there appears a critical normal load, which is marked by arrows in Fig. 18.19. When the normal load is smaller than the critical normal load, the monolayer only shows slight height change in the scan areas. When the normal load is higher than the critical value, the height change of SAMs increase dramati-

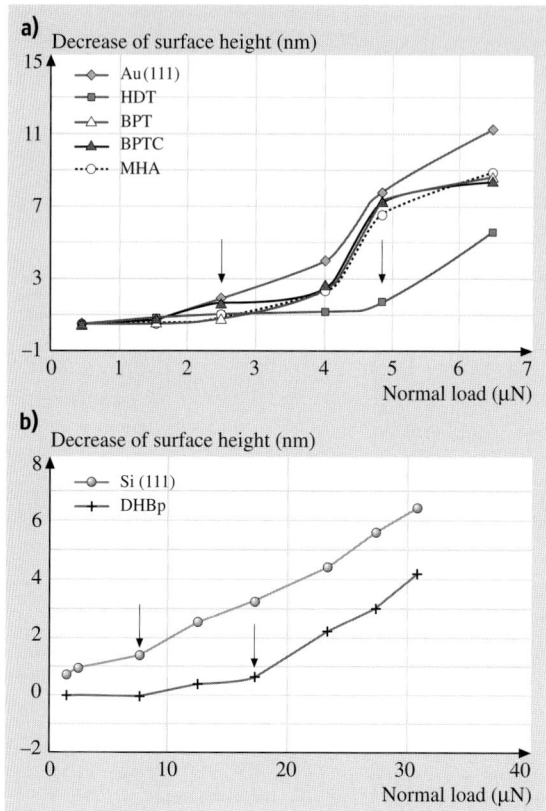

Fig. 18.19. Wear depth as a function of normal load after one scan cycle [35]. (**a**) on Au(111) and SAMs deposited on Au(111)/Si(111) (**b**) on Si(111) and DHPp deposited on Si(111)

cally. Relocation and accumulation of BPT molecules have been observed during the initial several scans, which lead to the formation of a larger terrace. Wear studies of a single BPT terrace indicate that the wear life of BPT increases exponentially with terrace size [36, 37].

Scratch resistances of Si(111), Au(111), and SAMs were studied by a continuous AFM microscratch technique. Figure 18.20a shows coefficient of friction profiles as a function of increasing normal load and corresponding tapping mode AFM surface height images of the scratches captured on Si(111), Au(111), and SAMs [37]. Figure 18.20a indicates that there is an abrupt increase in the coefficients of friction for all of the tested samples. The normal load associated with this event is termed the critical load (indicated by the arrows labeled "A"). At the initial stages of the scratch, all the samples exhibit a low coefficient of friction, indicating that the friction force is dominated by the shear component. This is in agreement with analysis of the AFM images, which shows negligible damage on the surfaces prior to the critical load. At the critical load, a clear groove is formed, accompanied by the formation of material pileup at the sides of the scratch. This suggests that the initial damage that occurs at the critical load is due to ploughing associated with plastic deformation, which causes a sharp rise in the coefficient of friction. Beyond the critical load, debris can be seen in addition to material pile-up at the sides of the scratch. Figure 18.20b summarizes the critical loads for the various samples obtained in this study. It clearly indicates that all SAMs can increase the critical load of the corresponding substrate. DHBp, which is deposited on Si(111), shows superior scratch resistance in all of the tested samples.

Mechanisms responsible for a sudden drop in the decrease in surface height with an increase in load during wear and scratch tests need to be understood. *Barrena* et al. [78] observed that the height of self-assembled alkylsilanes decrease in discrete amounts with normal load. This step-like behavior is due to the discrete molecular tilts, which are dictated by the geometrical requirements of the close packing of molecules. Only certain angles are allowed due to the zigzag arrangement of the carbon atoms. The relative height of the monolayer under pressure can be calculated by the following equation:

$$\left(\frac{h}{L}\right) = \left[1 + \left(\frac{na}{d}\right)^2\right]^{-\frac{1}{2}}, \tag{18.6}$$

where L is the total length of the molecule, h is the height of the SAMs in the tilt configuration (monolayer thickness), a is the distance between alternate carbon atoms in the molecule, d is the separation of the molecules, and n is the step number. For HDT and MHA, only the head groups are different in their work and this study. The spacer carbon chains in alkanethiol and alkylsilane are exactly the same. So the same a (0.25 nm) and d (0.47 nm) values are used in the calculation for HDT and MHA. The calculated and measured relative heights of HDT and MHA are listed in Table 18.12. When the normal loads are smaller than the critical values in Fig. 18.19, the measured relative height values of HDT and MHA are very close to the calculated values. This means that HDT and MHA underwent step tilting below critical normal loads.

The residual SAMs thickness after wear under critical normal load was measured by profiling the worn film using AFM. The results are listed in Table 18.13. For an alkanethiol monolayer, the relationship between the monolayer thickness h and intercept length L_0 can be expressed as (see Fig. 18.21):

$$h = b\cos(\alpha)n + L_0, \tag{18.7}$$

where b is the length of the projection of the C−C bond onto the main chain axis (b = 0.127 nm for alkanethiol), n is the chain length defined by $CH_3(CH_2)_nSH$, and α is the tilt angle [68]. For BPT and BPTC, based on the same principle and using the bond lengths reported in reference [69], the L_0 values are also calculated, Table 18.13. It indicates that the measured residual thickness values of SAMs under critical load are very close to the calculated intercept length L_0 values. It means that under the critical normal load, the Si_3N_4 tip approaches the interface and SAMs wear severely away from the substrate. This is due to the interface chemical adsorption

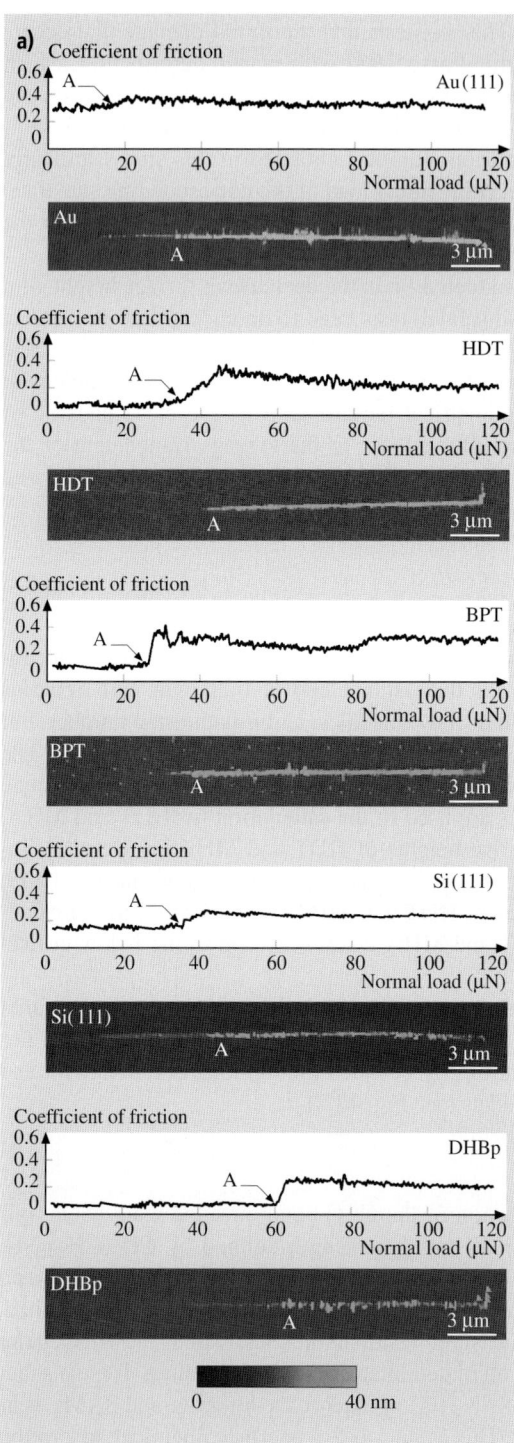

Fig. 18.20. (a) Coefficient of friction profiles during scratch as a function of normal load and corresponding AFM surface height images, and (b) critical loads estimated from the coefficient of friction profile and AFM images for Au(111), HDT/Au(111), BPT/Au(111), Si(111), and DHBp/Si(111) [37]

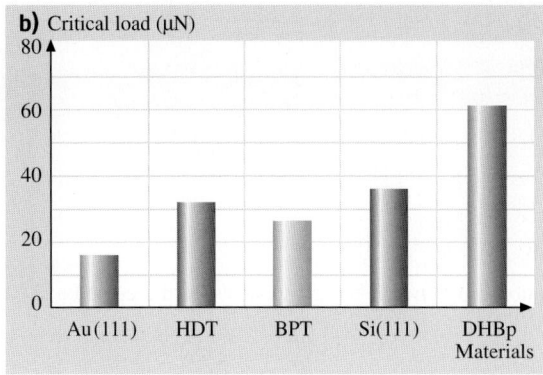

Fig. 18.20. continue

Table 18.12. Calculated $\left[1+\left(\frac{na}{d}\right)^2\right]^{-1/2}$ and measured $\left(\frac{h}{L}\right)$ relative heights of HDT and MHA self-assembled monolayers [35]

Steps (n)	Calculated[1] $\left[1+\left(\frac{na}{d}\right)^2\right]^{-1/2}$	Measured[2] $\left(\frac{h}{L}\right)$ HDT	MHA
1	0.883		
2	0.685	0.674[3]	0.643[3]
3	0.531	0.532[3]	0.552[3]
4	0.425	0.416[3]	
5	0.352	0.354[3]	
6	0.299		

[1] Calculations are based on the assumption that the molecules tilt in discrete steps (n) upon compression with a diamond AFM tip [78].
[2] Measured values are the mean values of three tests.
[3] These measured values correspond to the normal loads of 0.50 µN, 1.57 µN, 2.53 µN, and 4.03 µN, respectively.

bond strength (HS−Au and Si−O) being generally smaller than the other chemical bond strengths in SAMs spacer chains (see Table 18.14). Based on this discussion, it is believed that the reason why DHBp has the best wear resistance is due to the rigid biphenyl ring structure (compared to linear carbon chain in alkylthiol), the hard Si(111) substrate (compared to Au(111) substrate), and the strong interface Si−O bond strength (compared to the weak S−Au bond strength in the other SAMs, see Table 18.14).

According to the wear and scratch results reported here and the above discussion, the transition of the wear mechanisms of SAMs with increasing normal load is illustrated in Fig. 18.22. Below the critical normal load, SAMs undergo step orientation, at the critical load SAMs wear away from the substrate due to the weak interface

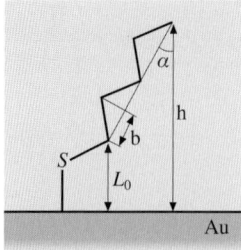

Fig. 18.21. Illustration of the relationship between the components of the equation $h = b\cos(\alpha)n + L_0$ [35]

Table 18.13. Calculated L_0 and measured residual film thickness for SAMs under critical load

	L_0^1 (nm)	**Residual thickness**[2] (nm)
HDT	0.24	0.25
BPT	0.39	0.42
BPTC	0.33	0.38
MHA	0.36	0.29
DHBp	--	0.52

[1] Calculated by the equation of $h = b\cos(\alpha)n + L_0$ [68]
[2] Measured by AFM using diamond tip under critical normal load. All of the data are the mean values of three tests

bond strengths, while above the critical normal load severe wear takes place on the substrate. In order to improve wear resistance, the interface bond must be enhanced; a rigid spacer chain and a hard substrate are also preferred.

18.5 Closure

Exposure of devices to humid environments results in condensates of water vapor from the environment. Condensed water, or a preexisting film of liquid, forms concave meniscus bridges between the mating surfaces. The negative Laplace pressure present in the meniscus results in an adhesive force that depends on the interface roughness, surface tension, and contact angle. The adhesive force can be significant in an interface with ultra-smooth surfaces, and it can be on the same order as the external load if the latter is small, such as in micro/nanodevices. Surfaces with high hydrophobicity can be produced by surface treatment. In many applications, hydrophobic films are expected to provide low adhesion, friction, and wear. To minimize high adhesion, friction, and/or because of small clearances in micro/nanodevices, these films should be molecularly thick. Liquid films of low surface tension or certain

Table 18.14. Bond strengths of the chemical bonds in SAMs [35]

SAMs	Bond	Bond strength[1] (kJ/mol)
Thiol on Au	S–Au	184 [58]
Hydroxyl on Si	O–Si	242.7 [79]
HDT	H–CH_2	464.8
	H–CH	421.7
	CH_3-t-C_4H_9	425.9±8
	t-C_4H_9–SH	286.2 ± 6.3
MHA	O=CO	532.2 ± 0.4
	H–$OCOC_2H_5$	445.2±8
	HO–$OCH_2C(CH_3)$	193.7 ± 7.9
	$C_6H_5CH_2$–COOH	280
BPT	CH_3–CH_3	376.0 ± 2.1
	$H_2C=CH_2$	733±8
	C_6H_5–SH	361.9±8
DHBp	C_6H_5–OH	361.9±8

[1] Most of the data are cited from [77], except where indicated. For MHA and DHBp, the bonds that are common as in HDT and DHBp are not repeated

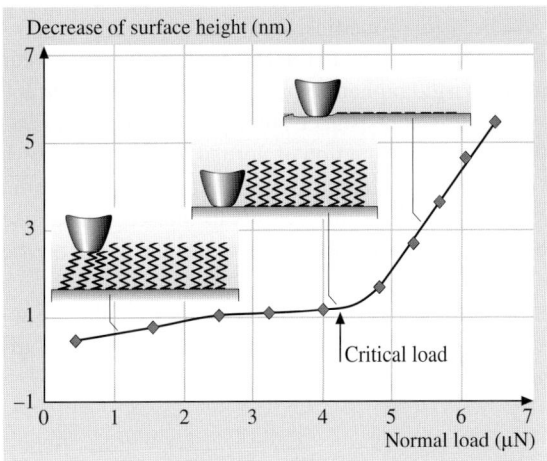

Fig. 18.22. Illustration of the wear mechanisms of SAMs with increasing normal load [37]

hydrophobic solid films can be used. Ordered molecular assemblies with high hydrophobicity can be engineered using chemical grafting of various polymer molecules with suitable functional head groups and nonpolar surface terminal groups.

The adhesion, friction, and wear properties of SAMs, having alkyl and biphenyl spacer chains with different surface terminal groups (−CH$_3$, −COOH, and −OH) and head groups (−SH and −OH), studied using an AFM, are reported in this chapter. It is found that the adhesive force varies linearly with the W_a value of SAMs, which indicates that capillary condensation of water plays an important role to the adhesion of SAMs on nanoscale at ambient conditions. It has been found that HDT exhibits the smallest adhesive force and friction force, because of low W_a of −CH$_3$ surface terminal groups, and high-compliance long carbon spacer chain. The friction data are explained using a molecular spring model, in which the local stiffness and intermolecular force governs its frictional performance. The results of the stiffness and friction characterization of the micropatterned sample with different structures support this model. The influence of relative humidity on adhesion and friction of SAMs is dominated by the thickness of the adsorbed water layer. At higher humidity, water can either increase friction through increased adhesion by meniscus effect in the contact zone or reduce friction through an enhanced water-lubricating effect. In the case of Si(111), the desorption of the adsorbed water layer and reduction of surface tension of water with increasing temperature reduce the adhesive force and friction force. The increase of temperature does not show an apparent influence on HDT, but strongly influences the adhesion and friction properties of BPT, which are believed to be related to its melting. The effect of velocity on friction properties of SAMs depends on their molecular structures. For alkylthiol SAMs, their friction is increased by increasing velocity, while for biphenyl SAMs it changes in the opposite way. The mechanism responsible for the variation of the friction forces of SAMs with velocity is believed to be related to their viscoelastic properties. Wear and continuous microscratch tests show that among the SAMs on Au(111), HDT exhibits the best wear resistance. DHBp on Si(111), due to its rigid biphenyl spacer chains, strong interface bonds, and a hard substrate, has the best wear resistance among all of the tested samples. For all of the SAMs, the wear depth as a function of normal load curves shows critical normal loads. Below the critical normal load SAMs undergo step orientation, at the critical load SAMs wear away from the substrate due to the weak interface bond strengths, while above the critical normal load severe wear take place on the substrate.

Based on the nanotribological studies of SAM films by AFM, they exhibit attractive hydrophobic and tribological properties. SAM films should find many tribological applications, including in micro/nanodevices.

References

1. B. Bhushan, J. N. Israelachvili, and U. Landman. Nanotribology: Friction, wear and lubrication at the atomic scale. *Nature*, 374:607–616, 1995.
2. B. Bhushan. *Tribology and Mechanics of Magnetic Storage Devices*. Springer, 2nd edition, 1996.
3. B. Bhushan. *Tribology Issues and Opportunities in MEMS*. Kluwer, 1998.
4. B. Bhushan. *Handbook of Micro/Nanotribology*. CRC, 2nd edition, 1999.

5. K. F. Man, B. H. Stark, and R. Ramesham. *A Resource Handbook for MEMS Reliability.* California Institute of Technology, 1998.
6. S. Kayali, R. Lawton, and B. H. Stark. Mems reliability assurance activities at jpl. *EEE Links*, 5:10–13, 1999.
7. D. M. Tanner, N. F. Smith, L. W. Irwin, et al. *MEMS Reliability: Infrastructure, Test Structure, Experiments, and Failure Modes.* Sandia National Laboratories, 2000.
8. S. Arney. Designing for mems reliability. *MRS Bull.*, 26:296–299, 2001.
9. B. Bhushan. *Principles and Applications of Tribology.* Wiley, 1999.
10. B. Bhushan. *Introduction to Tribology.* Wiley, 2002.
11. A. Ulman, editor. *Characterization of Organic Thin Films.* Butterworth-Heineman, 1995.
12. B. Bhushan. Contact mechanics of rough surfaces in tribology: Multiple asperity contact. *Tribol. Lett.*, 4:1–35, 1998.
13. B. Bhushan and W. Peng. Contact mechanics of multilayered rough surfaces. *Appl. Mech. Rev.*, 55:435–480, 2002.
14. R. Maboudian. Surface processes in mems technology. *Surf. Sci. Rep.*, 30:209–269, 1998.
15. M. Scherge and J. A. Schaefer. *Surface modification and mechanical properties of bulk silicon*, pages 529–537. Kluwer, 1998.
16. F. P. Bowden and D. Tabor. *The Friction and Lubrication of Solids, Part I.* Clarendon, 1950.
17. V. N. Koinkar and B. Bhushan. Microtribological studies of unlubricated and lubricated surfaces using atomic force/friction force microscopy. *J. Vac. Sci. Technol. A*, 14:2378–2391, 1996.
18. B. Bhushan and Z. Zhao. Macro- and microscale tribological studies of molecularly-thick boundary layers of perfluoropolyether lubricants for magnetic thin-film rigid disks. *J. Info. Storage Proc. Syst.*, 1:1–21, 1999.
19. H. Liu and B. Bhushan. Nanotribological characterization of molecularly-thick lubricant films for applications to mems/nems by afm. *Ultramicroscopy*, 97:321–340, 2003.
20. W. A. Zisman. *Friction, durability and wettability properties of monomolecular films on solids*, pages 110–148. Elsevier, 1959.
21. M. E. Schrader and G. I. Loeb, editors. *Modern Approaches to Wettability.* Plenum, 1992.
22. A. W. Neumann and J. K. Spelt, editors. *Applied Surface Thermodynamics.* Dekker, 1996.
23. H. Hansma, F. Motamedi, P. Smith, P. Hansma, and J. C. Wittman. Molecular resolution of thin, highly oriented poly(tetrafluoroethylene) films with the atomic force microscope. *Polymer Commun.*, 33:647–649, 1992.
24. L. Scandella, A. Schumacher, N. Kruse, R. Prins, E. Meyer, R. Luethi, L. Howald, and H. J. Guentherodt. Tribology of ultra-thin MoS_2 platelets on mica: Studies by scanning force microscopy. *Thin Solid Films*, 240:101–104, 1994.
25. B. Bhushan. Chemical, mechanical and tribological characterization of ultra-thin and hard amorphous carbon coatings as thin as 3.5 nm: Recent developments. *Diam. Relat. Mater.*, 8:1985–2015, 1999.
26. A. Erdemir and C. Donnet. *Tribology of diamond, diamond-like carbon, and related films*, pages 871–908. CRC, 2001.
27. V. F. Dorfman. Diamond-like nanocomposites (dln). *Thin Solid Films*, 212:267–273, 1992.
28. M. Grischke, K. Bewilogua, K. Trojan, and H. Dimigan. Application-oriented modification of deposition process for diamond-like carbon based coatings. *Surf. Coat. Technol.*, 74-75:739–745, 1995.
29. R. S. Butter, D. R. Waterman, A. H. Lettington, R. T. Ramos, and E. J. Fordham. Production and wetting properties of fluorinated diamond-like carbon coatings. *Thin Solid Films*, 311:107–113, 1997.

30. M. Grischke, A. Hieke, F. Morgenweck, and H. Dimigan. Variation of the wettability of dlc coatings by network modification using silicon and oxygen. *Diam. Relat. Mater.*, 7:454–458, 1998.
31. C. Donnet, J. Fontaine, A. Grill, V. Patel, C. Jahnes, and M. Belin. Wear-resistant fluorinated diamondlike carbon films. *Surf. Coat. Technol.*, 94-95:531–536, 1997.
32. D. J. Kester, C. L. Brodbeck, I. L. Singer, and A. Kyriakopoulos. Sliding wear behavior of diamond-like nanocomposite coatings. *Surf. Coat. Technol.*, 113:268–273, 1999.
33. H. Liu and B. Bhushan. Adhesion and friction studies of microelectromechanical systems/nanoelectromechanical systems materials using a novel microtriboapparatus. *J. Vac. Sci. Technol. A*, 21:1528–1538, 2003.
34. B. Bhushan, A. V. Kulkarni, V. N. Koinkar, M. Boehm, L. Odoni, C. Martelet, and M. Belin. Microtribological characterization of self-assembled and langmuir–blodgett monolayers by atomic and friction force microscopy. *Langmuir*, 11:3189–3198, 1995.
35. B. Bhushan and H. Liu. Nanotribological properties and mechanisms of alkylthiol and biphenyl thiol self-assembled monolayers studied by atomic force microscopy. *Phys. Rev. B*, 63:245412-1–245412-11, 2001.
36. H. Liu, B. Bhushan, W. Eck, and V. Stadler. Investigation of the adhesion, friction, and wear properties of biphenyl thiol self-assembled monolayers by atomic force microscopy. *J. Vac. Sci. Technol. A*, 19:1234–1240, 2001.
37. H. Liu and B. Bhushan. Investigation of nanotribological properties of alkylthiol and biphenyl thiol self-assembled monolayers. *Ultramicroscopy*, 91:185–202, 2002.
38. H. Liu and B. Bhushan. Orientation and relocation of biphenyl thiol self-assembled monolayers. *Ultramicroscopy*, 91:177–183, 2002.
39. A. Ulman. *An Introduction to Ultrathin Organic Films: From Langmuir–Blodgett to Self-Assembly*. Academic, 1991.
40. A. Ulman. Formation and structure of self-assembled monolayers. *Chem. Rev.*, 96:1533–1554, 1996.
41. J. A. Zasadzinski, R. Viswanathan, L. Madsen, J. Garnaes, and D. K. Schwartz. Langmuir–blodgett films. *Science*, 263:1726–1733, 1994.
42. J. Tian, Y. Xia, and G. M. Whitesides. *Microcontact printing of SAMs*, volume 24, pages 227–254. Academic, 1998.
43. Y. Xia and G. M. Whitesides. Soft lithography. *Angew. Chem. Int. Edn.*, 37:550–575, 1998.
44. A. Kumar and G. M. Whitesides. Features of gold having micrometer to centimeter dimensions can be formed through a combination of stamping with an elastomeric stamp and an alkanethiol ink followed by chemical etching. *Appl. Phys. Lett.*, 63:2002–2004, 1993.
45. S. Y. Chou, P. R. Krauss, and P. J. Renstrom. Imprint lithography with 25-nanometer resolution. *Science*, 272:85–87, 1996.
46. Y. Xia, E. Kim, X. M. Zhao, J. A. Rogers, M. Prentiss, and G. M. Whitesides. Complex optical surfaces formed by replica molding against elastomeric masters. *Science*, 273:347–349, 1996.
47. R. F. Service. Self-assembly comes together. *Science*, 265:316–318, 1994.
48. M. Hein, L. R. Best, S. Pattison, and S. Arena. *Introduction to General, Organic, and Biochemistry*. Brooks/Cole, 6th edition, 1997.
49. J. R. Mohrig, C. N. Hammond, T. C. Morrill, and D. C. Neckers. *Experimental Organic Chemistry*. Freeman, 1998.
50. S. R. Wasserman, Y. T. Tao, and G. M. Whitesides. Structure and reactivity of alkylsiloxane monolayers formed by reaction of alkylchlorosilanes on silicon substrates. *Langmuir*, 5:1074–1089, 1989.

51. C. Jung, O. Dannenberger, Y. Xu, M. Buck, and M. Grunze. Self-assembled monolayers from organosulfur compounds: A comparison between sulfides, disulfides, and thiols. *Langmuir*, 14:1103–1107, 1998.
52. W. Geyer, V. Stadler, W. Eck, M. Zharnikov, A. Golzhauser, and M. Grunze. Electron-induced crosslinking of aromatic self-assembled monolayers: Negative resists for nanolithography. *Appl. Phys. Lett.*, 75:2401–2403, 1999.
53. J. Ruhe, V. J. Novotny, K. K. Kanazawa, T. Clarke, and G. B. Street. Structure and tribological properties of ultrathin alkylsilane films chemisorbed to solid surfaces. *Langmuir*, 9:2383–2388, 1993.
54. V. DePalma and N. Tillman. Friction and wear of self-assembled tricholosilane monolayer films on silicon. *Langmuir*, 5:868–872, 1989.
55. E. Ando, Y. Goto, K. Morimoto, K. Ariga, and Y. Okahata. Frictional properties of monolayers of silane compounds. *Thin Solid Films*, 180:287–291, 1989.
56. M. T. McDermott, J. B. D. Green, and M. D. Porter. Scanning force microscopic exploration of the lubrication capabilities of n-alkanethiolate monolayers chemisorbed at gold: Structural basis of microscopic friction and wear. *Langmuir*, 13:2504–2510, 1997.
57. X. Xiao, J. Hu, D. H. Charych, and M. Salmeron. Chain length dependence of the frictional properties of alkylsilane molecules self-assembled on mica studied by atomic force microscopy. *Langmuir*, 12:235–237, 1996.
58. A. Lio, D. H. Charych, and M. Salmeron. Comparative atomic force microscopy study of the chain length dependence of frictional properties of alkanethiol on gold and alkylsilanes on mica. *J. Phys. Chem. B*, 101:3800–3805, 1997.
59. H. Schonherr and G. J. Vancso. Tribological properties of self-assembled monolayers of fluorocarbon and hydrocarbon thiols and disulfides on Au(111) studied by scanning force microscopy. *Mater. Sci. Eng. C*, 8-9:243–249, 1999.
60. H. I. Kim, T. Koini, T. R. Lee, and S. S. Perry. Systematic studies of the frictional properties of fluorinated monolayers with atomic force microscopy: Comparison of CF_3- and CH_3-terminated groups. *Langmuir*, 13:7192–7196, 1997.
61. H. I. Kim, M. Graupe, O. Oloba, T. Koini, S. Imaduddin, T. R. Lee, and S. S. Perry. Molecularly specific studies of the frictional properties of monolayer films: A systematic comparison of CF_3-, $(ch_3)_2$ch-, and CH_3-terminated films. *Langmuir*, 15:3179–3185, 1999.
62. V. V. Tsukruk, M. P. Everson, L. M. Lander, and W. J. Brittain. Nanotribological properties of composite molecular films: C_{60} anchored to a self-assembled monolayer. *Langmuir*, 12:3905–3911, 1996.
63. S. Lee, Y. S. Shon, T. R. Lee, and S. S. Perry. Structural characterization and frictional properties of C_{60}-terminated self-assembled monolayers on Au(111). *Thin Solid Films*, 358:152–158, 2000.
64. V. V. Tsukruk and V. N. Bliznyuk. Adhesive and friction forces between chemically modified silicon and silicon nitride surfaces. *Langmuir*, 14:446–455, 1998.
65. V. V. Tsukruk, T. Nguyen, M. Lemieux, J. Hazel, W. H. Weber, V. V. Shevchenko, N. Klimenko, and E. Sheludko. *Tribological properties of modified MEMS surfaces*, pages 607–614. Kluwer, 1998.
66. M. Fujihira, Y. Tani, M. Furugori, U. Akiba, and Y. Okabe. Chemical force microscopy of self-assembled monolayers on sputtered gold films patterned by phase separation. *Ultramicroscopy*, 86:63–73, 2001.
67. S. Sundararajan and B. Bhushan. Topography-induced contribution to friction forces measured using an atomic force/friction force microscopy. *J. Appl. Phys.*, 88:4825–4831, 2000.

68. Y. F. Miura, M. Takenga, T. Koini, M. Graupe, N. Garg, R. L. Graham, and T. R. Lee. Wettability of self-assembled monolayers generated from CF_3-terminated alkanethiols on gold. *Langmuir*, 14:5821–5825, 1998.
69. M. Ratajczak-Sitarz, A. Katrusiak, Z. Kaluski, and J. Garbarczyk. 4,4′-biphenyldithiol. *Acta Crystallogr. C*, 43:2389–2391, 1987.
70. J. N. Israelachvili. *Intermolecular and Surface Forces*. Academic, 2nd edition, 1992.
71. R. J. Good and C. J. V. Oss. *Modern Approaches to Wettability – Theory and Applications*. Plenum, 1992.
72. Y. Unno, H. Kawamura, H. Kita, and S. Sekiyama. Friction characteristics of sintered Si_3N_4 in oil-lubricated medium, 1995.
73. M. H. V. C. Adao, B. J. V. Saramago, and A. C. Fernandes. Estimation of the surface properties of styrene–acrylonitrile random copolymers from contact angle measurements. *J. Coll. Interface Sci.*, 217:94–106, 1999.
74. J. I. Siepman and I. R. McDonald. Monte carlo simulation of the mechanical relaxation of a self-assembled monolayer. *Phys. Rev. Lett.*, 70:453–456, 1993.
75. M. Garcia-Parajo, C. Longo, J. Servat, P. Gorostiza, and F. Sanz. Nanotribological properties of octadecyltrichlorosilane self-assembled ultrathin films studied by atomic force microscopy: Contact and tapping modes. *Langmuir*, 13:2333–2339, 1997.
76. D. DeVecchio and B. Bhushan. Localized surface elasticity measurements using an atomic force microscope. *Rev. Sci. Instrum.*, 68:4498–4505, 1997.
77. D. R. Lide. *CRC Handbook of Chemistry and Physics*. CRC, 75th edition, 1994.
78. E. Barrena, S. Kopta, D. F. Ogletree, D. H. Charych, and M. Salmeron. Relationship between friction and molecular structure: Alkysilane lubricant films under pressure. *Phys. Rev. Lett.*, 82:2880–2883, 1999.
79. T. Hoshino, C. Yayoi, and K. Inage. Adsorption of atomic and molecular oxygen and desorption of silicon monoxide on Si(111) surface. *Phys. Rev. B*, 59:2332–2340, 1999.

19

Nanoscale Boundary Lubrication Studies

Bharat Bhushan and Huiwen Liu

Summary. Boundary films are formed by physisorption, chemisorption, and chemical reaction. With physisorption, no exchange of electrons takes place between the molecules of the adsorbate and those of the adsorbant. The physisorption process typically involves van der Waals forces, which are relatively weak. In chemisorption, there is an actual sharing of electrons or electron interchange between the chemisorbed species and the solid surface. The solid surfaces bond very strongly to the adsorption species through covalent bonds. Chemically reacted films are formed by the chemical reaction of a solid surface with the environment. The physisorbed film can be either monomolecularly or polymolecularly thick. The chemisorbed films are monomolecular, but stoichiometric films formed by chemical reaction can have a large film thickness. In general, the stability and durability of surface films decrease in the following order: chemically reacted films, chemisorbed films, and physisorbed films. A good boundary lubricant should have a high degree of interaction between its molecules and the sliding surface. As a general rule, liquids are good lubricants when they are polar and, thus, able togrip solid surfaces (or be adsorbed). In this chapter, we focus on PF-PEs. We first introduce details of the commonly used PFPE lubricants; then present a summary of nanodeformation, molecular conformation, and lubricant spreading studies; followed by an overview of nanotribological properties of polar and nonpolar PFPEs studied by atomic force microscopy (AFM) and some concluding remarks.

19.1 Introduction

Boundary films are formed by physisorption, chemisorption, and chemical reaction. With physisorption, no exchange of electrons takes place between the molecules of the adsorbate and those of the adsorbant. The physisorption process typically involves van der Waals forces, which are relatively weak. In chemisorption, there is an actual sharing of electrons or electron interchange between the chemisorbed species and the solid surface. The solid surfaces bond very strongly to the adsorption species through covalent bonds. Chemically reacted films are formed by the chemical reaction of a solid surface with the environment. The physisorbed film can be either monomolecularly or polymolecularly thick. The chemisorbed films are monomolecular, but stoichiometric films formed by chemical reaction can have a large film thickness. In general, the stability and durability of surface films decrease in the following order: chemically reacted films, chemisorbed films, and physisorbed films. A good boundary lubricant should have a high degree of interaction between its molecules and the sliding surface. As a general rule, liquids are good lubricants when they are polar and, thus, able to grip solid surfaces (or be adsorbed). Polar lubricants contain reactive functional groups with low ionization potential, or groups having high polarizability [1–3]. Boundary lubrication properties of lubricants are also dependent upon the molecular conformation and lubricant spreading [4–7].

Self-assembled monolayers (SAMs), Langmuir–Blodgett (LB) films, and perfluoropolyether (PFPE) films can be used as boundary lubricants [2, 3, 8–10]. PFPE films are commonly used for lubrication of magnetic rigid disks and metal evaporated magnetic tapes to reduce friction and wear of a head-medium interface [10]. PFPEs are well suited for this application because of the following properties: low surface tension and low contact angle, which allow easy spreading on surfaces and provide a hydrophobic property; chemical and thermal stability, which minimizes degradation under use; low vapor pressure, which provides low out-gassing; high adhesion to substrate via organofunctional bonds; and good lubricity, which reduces the interfacial friction and wear [10–12]. While the structure of the lubricants employed at the head-medium interface has not changed substantially over the past decade, the thickness of the PFPE film used to lubricate the disk has steadily decreased from multilayer thicknesses to the sub-monolayer thickness regime [11, 13]. Molecularly thick PFPE films are also being considered for lubrication purposes of the evolving microelectromechanical systems (MEMS) MEMSindustry [14]. It is well-known that the properties of molecularly thick liquid films confined to solid surfaces can be dramatically different from those of the corresponding bulk liquid. In order to efficiently develop lubrication systems that meet the requirements of the advanced rigid disk drive and MEMS industries, the impact of thinning the PFPE lubricants on the resulting of nanotribology should be fully understood [15, 16]. It is also important to understand lubricant-substrate interfacial interactions and the influence of the operating environment on the nanotribological performance of molecularly thick PFPEs.

An overview of nanotribological properties of SAMs and LB films can be found in many references such as [17]. In this chapter, we focus on PFPEs. We first introduce details of the commonly used PFPE lubricants; then present a summary of nanodeformation, molecular conformation, and lubricant spreading studies; followed by an overview of nanotribological properties of polar and nonpolar PFPEs studied by atomic force microscopy (AFM) and some concluding remarks.

19.2 Lubricants Details

Properties of two commonly used PFPE lubricants (Z-15 and Z-DOL) are reviewed here. Their molecular structures are shown schematically in Fig. 19.1. Z-15 has nonpolar $-CF_3$ end groups, whereas Z-DOL is a polar lubricant with hydroxyl ($-OH$) end groups. Their typical properties are summarized in Table 19.1. It shows that Z-15 and Z-DOL almost have the same density and surface tension. But Z-15 has larger molecular weight and higher viscosity. Both have low surface tension, low vapor pressure, low evaporation weight loss, and good oxidative stability [10, 12]. Generally, single-crystal Si(100) wafer with a native oxide layer was used as a substrate for deposition of molecularly thick lubricant films for nanotribological characterization. Z-15 and Z-DOL films can be deposited directly on the Si(100) wafer by dip coating technique. The clean silicon wafer is vertically submerged into a dilute solution of lubricant in hydrocarbon solvent (HT-70) for a certain time. The silicon wafers are vertically pulled up from the solution with a motorized stage at a constant speed for deposition of desired thicknesses of Z-15 and Z-DOL lubricants. The lubricant film thickness obtained in dip coating is a function of the concentration and pulling-up speed, among other factors. The Z-DOL film is bonded to the silicon substrate by heating the as-deposited Z-DOL samples in an oven at 150 °C for about 30 s.

Table 19.1. Typical properties of Z-15 and Z-DOL (data from Montefluous S.p.A., Milan, Italy)

	Z-15	Z-DOL (2000)
Formula	$CF_3-O-(CF_2-CF_2-O)_m-(CF_2-O)_n-CF_3^*$	$HO-CH_2-CF_2-O-(CF_2-CF_2-O)_m-(CF_2-O)_n-CF_2-CH_2-OH^*$
Molecular weight (Daltons)	9100	2000
Density *(ASTM D891)* 20 °C (g/cm^3)	1.84	1.81
Kinematic viscosity *(ASTM D445)* (cSt)		
20 °C	148	85
38 °C	90	34
99 °C	25	–
Viscosity index *(ASTM D2270)*	320	–
Surface tension *(ASTM D1331)* (dyn/cm) 20 °C	24	24
Vapor pressure (torr)		
20 °C	1.6×10^{-6}	2×10^{-5}
100 °C	1.7×10^{-5}	6×10^{-4}
Pour point *(ASTM D972)* °C	−80	–
Evaporation weight loss *(ASTM D972)* 149 °C, 22 h (%)	0.7	–
Oxidative stability (°C)	–	320
Specific heat (cal/g °C) 38 °C	0.21	–

* $m/n \sim 2/3$

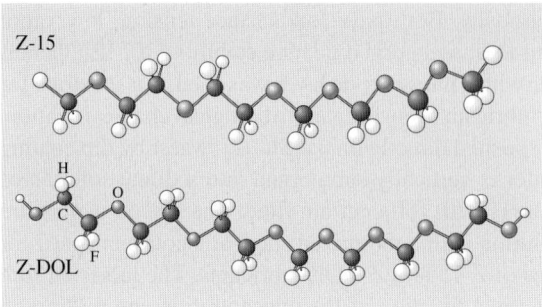

Fig. 19.1. Schematics of the molecular structures of Z-15 and Z-DOL. In this figure the m/n value, shown in Table 19.1, equals 2/3

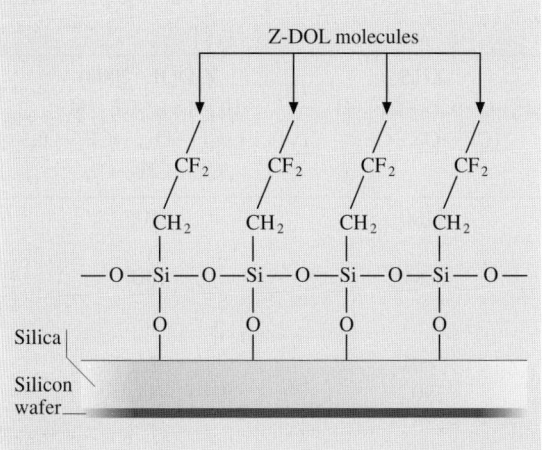

Fig. 19.2. Schematic of Z-DOL molecules that are chemically bonded on Si(100) substrate surface (which has native oxide) after thermal treatment at 150 °C for 30 min

The native oxide layer of Si(100) wafer reacts with the −OH groups of the lubricants during thermal treatment [18–21]. Subsequently, fluorocarbon solvent (FC-72) washing of the thermal treated specimen removes loosely absorbed species, leaving chemically bonded phase on the substrate. The chemical bonding between Z-DOL molecules and silicon substrate is illustrated in Fig. 19.2. The bonded and washed Z-DOL film is referred to as Z-DOL(BW) in this chapter. The as-deposited Z-15 and Z-DOL films are mobile phase lubricants (i.e., liquid-like lubricant), whereas the Z-DOL(BW) films are fully bonded soft solid phase (i.e., solid-like) lubricants. This will be further discussed in the next section.

19.3 Nanodeformation, Molecular Conformation, and Lubricant Spreading

Nanodeformation behavior of Z-DOL lubricants was studied using an AFM by *Blackman* et al. [22, 23]. Before bringing a tungsten tip into contact with a molecular overlayer, it was brought into contact with a bare clean-silicon surface, Fig. 19.3. As the sample approaches the tip, the force initially is zero, but at point A the force suddenly becomes attractive (top curve), which increases until at point B, where the sample and tip come into intimate contact and the force becomes repulsive. As the sample is retracted, a pull-off force of 5×10^{-8} N (point D) is required to overcome adhesion between the tungsten tip and the silicon surface. When an AFM tip is brought into contact with an unbonded Z-DOL film, a sudden jump into adhesive contact is also observed. A much larger pull-off force is required to overcome the adhesion. The adhesion is initiated by the formation of a lubricant meniscus surrounding the tip. This suggests that the unbonded Z-DOL lubricant shows liquid-like behavior. However, when the tip was brought into contact with a lubricant film, which was firmly bonded to the surface, the liquid-like behavior disappears. The initial attractive force (point A) is no longer sudden, as with the liquid film, but, rather, gradually increases as the tip penetrates the film.

According to *Blackman* et al. [22, 23], if the substrate and tip were infinitely hard with no compliance and/or deformation in the tip and sample supports, the line for B to C would be vertical with an infinite slope. The tangent to the force-distance curve at a given point is referred to as the stiffness at that point and was determined by fitting a least-squares line through the nearby data points. For silicon, the deformation is reversible (elastic), since the retracting (outgoing) portion of the curve (C to D) follows the extending (ingoing) portion (B to C). For bonded lubricant film, at the point where slope of the force changes gradually from attractive to repulsive, the stiffness changes gradually, indicating compression of the molecular film. As the load is increased, the slope of the repulsive force eventually approaches that of the bare surface. The bonded film was found to respond elastically up to the highest loads of 5 µN that could be applied. Thus, bonded lubricant behaves as a soft polymer solid.

Figure 19.4 illustrates two extremes for the conformation on a surface of a linear liquid polymer without any reactive end groups and at submonolayer coverages [4, 6]. At one extreme, the molecules lie flat on the surface, reaching no more than their chain diameter δ above the surface. This would be the case if a strong attractive interaction exists between the molecules and the solid. On the other extreme, when a weak attraction exists between polymer segments and the solid, the molecules adopt conformation close to that of the molecules in the bulk, with the microscopic thickness equal to about the radius of gyration R_g. *Mate* and *Novotny* [6] used AFM to study conformation of 0.5–1.3 nm-thick Z-15 molecules on clean Si(100) surfaces. They found that the thickness measured by AFM is thicker than that measured by ellipsometry with the offset ranging from 3–5 nm. They found that the offset was the same for very thin submonolayer coverages. If the coverage is submonolayer and inadequate to make a liquid film, the relevant thickness is then the height (h_e) of the molecules extended above the solid surface. The offset should be

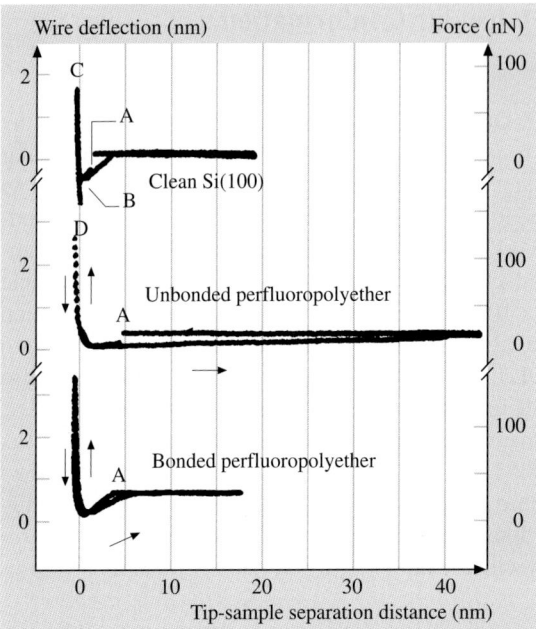

Fig. 19.3. Wire deflection (normal load) as a function of tip-sample separation distance curves comparing the behavior of clean Si(100) surface to a surface lubricated with free and unbonded PFPE lubricant, and a surface where the PFPE lubricant film was thermally bonded to the surface [22]

equal to $2h_e$, assuming that the molecules extend the same height above both the tip and silicon surfaces. They therefore concluded that the molecules do not extend more than 1.5–2.5 nm above a solid or liquid surface, much smaller than the radius of gyration of the lubricants ranging between 3.2 and 7.3 nm, and to the approximate cross-sectional diameter of 0.6–0.7 nm for the linear polymer chain. Consequently, the height that the molecules extend above the surface is considerably less than the diameter of gyration of the molecules and only a few molecular diameters in height, implying that the physisorbed molecules on a solid surface have an extended, flat conformation. They also determined the disjoining pressure of these liquid films from AFM measurements of the distance needed to break the liquid meniscus that forms between the solid surface and the AFM tip. (Also see [7].) For a monolayer thickness of about 0.7 nm, the disjoining pressure is about 5 MPa, indicating strong attractive interaction between the liquid molecules and the solid surface. The disjoining pressure decreases with increasing film thickness in a manner consistent with a strong attractive van der Waals interaction between the liquid molecules and the solid surface.

Rheological characterization shows that the flow activation energy of PFPE lubricants is weakly dependent on chain length and is strongly dependent on the functional end groups [24]. PFPE lubricant films that contain polar end groups have lower

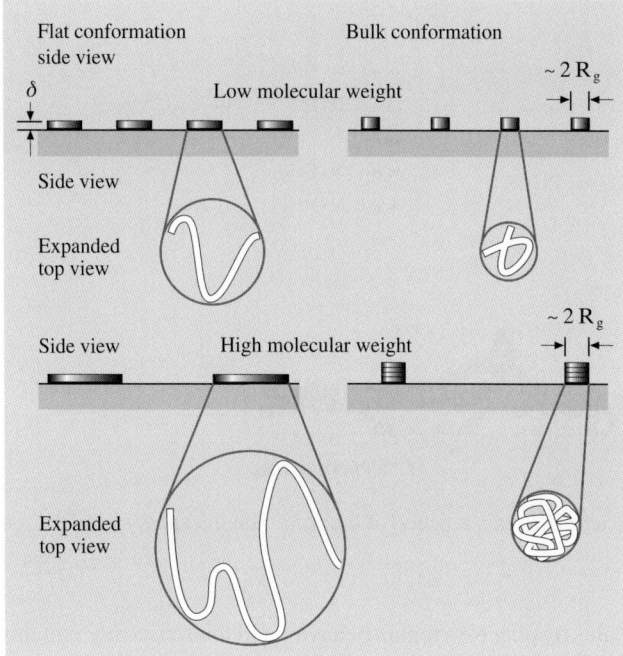

Fig. 19.4. Schematic representation of two extreme liquid conformations at the surface of the solid for low and high molecular weights at low surface coverage. δ is the cross-sectional diameter of the liquid chain, and R_g is the radius of gyration of the molecules in the bulk [6]

mobility than those with nonpolar end groups of similar chain length [25]. The mobility of PFPE also depends on the surface chemical properties of the substrate. The spreading of Z-DOL on amorphous carbon surface has been studied as a function of hydrogen or nitrogen content in the carbon film, using scanning microellipsometry [26]. The diffusion coefficient data presented in Fig. 19.5 is thickness-dependent. It shows that the surface mobility of Z-DOL increased as the hydrogen content increased, but decreased as nitrogen content increased. The enhancement of Z-DOL surface mobility by hydrogenation may be understood from the fact that the interactions between Z-DOL molecules and the carbon surface can be significantly weakened, due to a reduction of the number of high-energy binding sites on the carbon surface. The stronger interactions between the Z-DOL molecules and carbon surface, as nitrogen content in the carbon coating increases, leads to the lowering of Z-DOL surface mobility.

Molecularly thick films may be sheared at very high shear rates, on the order of 10^8–10^9 s^{-1} during sliding, such as during magnetic disk drive operation. During such shear, lubricant stability is critical to the protection of the interface. For proper lubricant selection, viscosity at high shear rates and associated shear thinning need to be understood. Viscosity measurements of eight different types of PFPE films

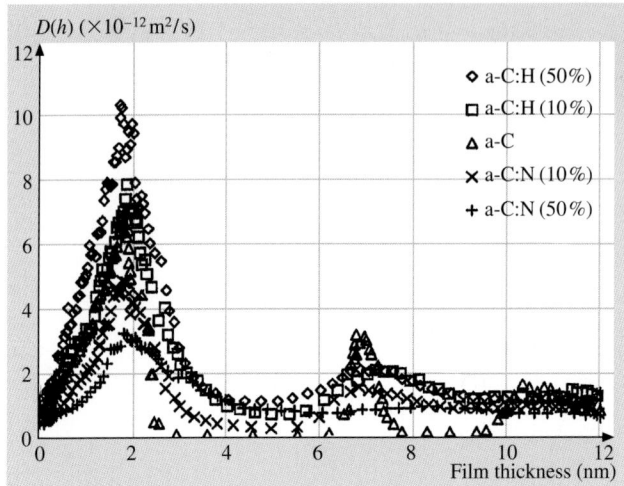

Fig. 19.5. Diffusion coefficient $D(h)$ as a function of lubricant film thickness for Z-DOL on different carbon films [26]

show that all eight lubricants display Newtonian behavior and their viscosity remains constant at a shear rate up to 10^7 s^{-1} [27, 28].

19.4 Boundary Lubrication Studies

With the development of AFM techniques, studies have been carried out to investigate the nanotribological performance of PFPEs. *Mate* [29, 30], *O'Shea* et al. [31, 32], *Bhushan* et al. [15, 33], *Koinkar* and *Bhushan* [20, 34], *Bhushan* and *Sundararajan* [35], *Bhushan* and *Dandavate* [36], and *Liu* and *Bhushan* [21] used an AFM to provide insight into how PFPE lubricants function at the molecular level. *Mate* [29, 30] conducted friction experiments on bonded and unbonded Z-DOL and found that the coefficient of friction of the unbonded Z-DOL is about two times larger than the bonded Z-DOL (also see [31, 32]). *Koinkar* and *Bhushan* [20, 34] and *Liu* and *Bhushan* [21] studied the friction and wear performance of a Si(100) sample lubricated with Z-15, Z-DOL, and Z-DOL(BW) lubricants. They found that using Z-DOL(BW) could significantly improve the adhesion, friction, and wear performance of Si(100). They also discussed the lubrication mechanisms on the molecular level. *Bhushan* and *Sundararajan* [35] and *Bhushan* and *Dandavate* [36] studied the effect of tip radius and relative humidity on the adhesion and friction properties of Si(100) coated with Z-DOL(BW).

In this section, we review, in some detail, the adhesion, friction, and wear properties of two kinds of typical PFPE lubricants of Z-15 and Z-DOL at various operating conditions (rest time, velocity, relative humidity, temperature, and tip radius). The experiments were carried out using a commercial AFM system with pyramidal Si$_3$N$_4$

and diamond tips. An environmentally controlled chamber and a thermal stage were used to perform relative humidity and temperature effect studies.

19.4.1 Friction and Adhesion

To investigate the friction properties of Z-15 and Z-DOL(BW) films on Si(100), the friction force versus normal load curves were measured by making friction measurements at increasing normal loads [21]. The representative results of Si(100), Z-15, and Z-DOL(BW) are shown in Fig. 19.6. An approximately linear response of all three samples is observed in the load range of 5–130 nN. The friction force of solid-like Z-DOL(BW) is consistently smaller than that for Si(100), but the friction force of liquid-like Z-15 lubricant is higher than that of Si(100). *Sundararajan* and *Bhushan* [37] have studied the static friction force of silicon micromotors lubricated with Z-DOL by AFM. They also found that liquid-like lubricants of Z-DOL significantly increase the static friction force, whereas solid-like Z-DOL(BW) coating can dramatically reduce the static friction force. This is in good agreement with the results of *Liu* and *Bhushan* [21]. In Fig. 19.6, the nonzero value of the friction force signal at zero external load is due to the adhesive forces. It is well-known that the following relationship exists between the friction force F and external normal load W [2, 3]

$$F = \mu(W + W_a) , \qquad (19.1)$$

where μ is the coefficient of friction and W_a is the adhesive force. Based on this equation and the data in Fig. 19.6, we can calculate the μ and W_a values. The coefficients of friction of Si(100), Z-15, and Z-DOL are 0.07, 0.09, and 0.04, respectively. Based on (19.1), the adhesive force values are obtained from the horizontal intercepts of the friction force versus normal load curves at a zero value of friction force. Adhesive force values of Si(100), Z-15, and Z-DOL are 52 nN, 91 nN, and 34 nN, respectively.

The adhesive forces of these samples were also measured using a force calibration plot (FCP) technique. In this technique, the tip is brought into contact with the sample and the maximum force, needed to pull the tip and sample apart, is measured as the adhesive force. Figure 19.7 shows the typical FCP curves of Si(100), Z-15, and Z-DOL(BW) [21]. As the tip approaches the sample within a few nanometers (A), an attractive force exists between the tip and the sample surfaces. The tip is pulled toward the sample, and contact occurs at point B on the graph. The adsorption of water molecules and/or presence of liquid lubricant molecules on the sample surface can also accelerate this so-called "snap-in", due to the formation of meniscus of the water and/or liquid lubricant around the tip. From this point on, the tip is in contact with surface, and as the piezo extends further, the cantilever is further deflected. This is represented by the slope portion of the curve. As the piezo retracts, at point C the tip goes beyond the zero deflection (flat) line, because of the attractive forces, into the adhesive force regime. At point D, the tip snaps free of the adhesive forces and is again in free air. The adhesive force (pull-off force) is determined by multiplying the cantilever spring constant (0.58 N/m) by the horizontal distance between

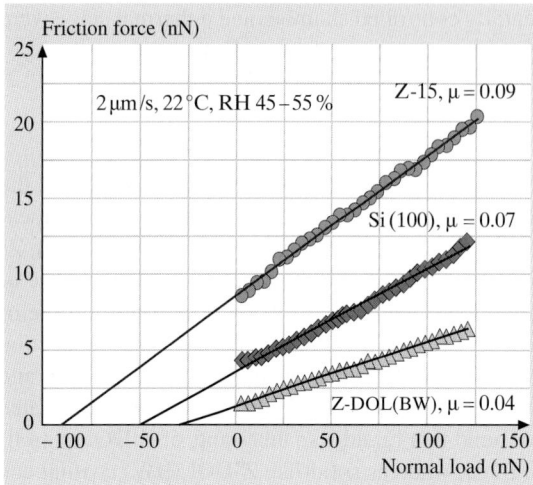

Fig. 19.6. Friction force versus normal load curves for Si(100), 2.8-nm-thick Z-15 film, and 2.3-nm-thick Z-DOL(BW) film at 2 µm/s, and in ambient air sliding against a Si_3N_4 tip. Based on these curves, coefficient of friction μ and adhesion force of W_a can be calculated [21]

points C and D, which corresponds to the maximum cantilever deflection toward the samples before the tip is disengaged. Incidentally, the horizontal shift between the loading and unloading curves results from the hysteresis of the PZT tube.

The adhesive forces of Si(100), Z-15, and Z-DOL(BW) measured by FCP and friction force versus normal load plot are summarized in Fig. 19.8 [21]. The results measured by these two methods are in good agreement. Figure 19.8 shows that the presence of mobile Z-15 lubricant film increases the adhesive force as compared to that of Si(100). In contrast, the presence of solid phase Z-DOL(BW) film reduces the adhesive force as compared to that of Si(100). This result is in good agreement with the results of *Blackman* et al. [22] and *Bhushan* and *Ruan* [38]. Sources of adhesive forces between the tip and the sample surfaces are van der Waals attraction and long-range meniscus force [2, 3, 16]. Relative magnitudes of the forces from the two sources are dependent on various factors, including the distance between the tip and the sample surface, their surface roughness, their hydrophobicity, and relative humidity [39]. For most surfaces with some roughness, meniscus contribution dominates at moderate to high humidities.

The schematic (bottom) in Fig. 19.8 shows relative size and sources of meniscus. The native oxide layer (SiO_2) on the top of Si(100) wafer exhibits hydrophilic properties, and some water molecules can be adsorbed on this surface. The condensed water will form meniscus as the tip approaches the sample surface. In the case of a sphere (such as a single-asperity AFM tip) in contact with a flat surface, the attractive Laplace force (F_L) caused by capillary is:

$$F_L = 2\pi R \gamma_{la}(\cos\theta_1 + \cos\theta_2) \,, \tag{19.2}$$

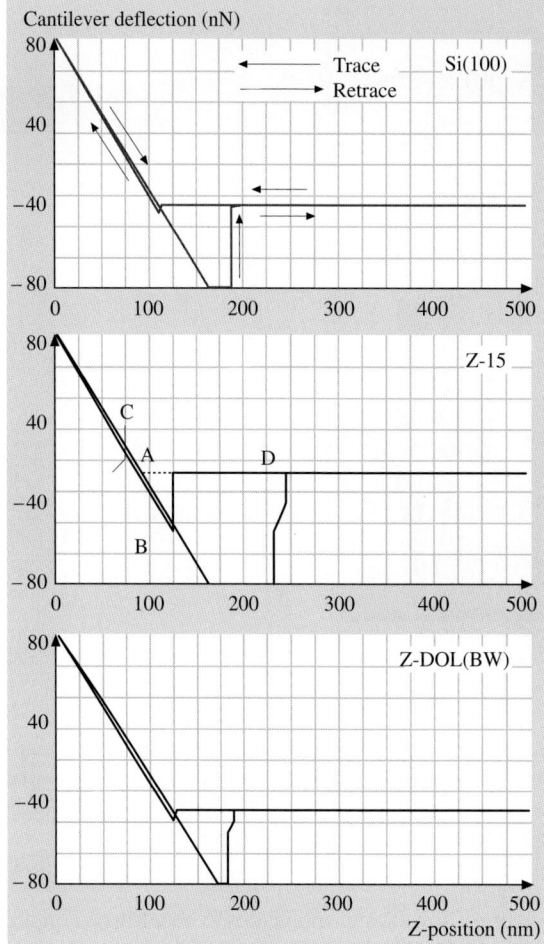

Fig. 19.7. Typical force calibration plots of Si(100), 2.8-nm-thick Z-15 film, and 2.3-nm-thick Z-DOL(BW) film in ambient air. The adhesive forces can be calculated from the horizontal distance between points C and D, and the cantilever spring constant of 0.58 N/m [21]

where R is the radius of the sphere, γ_{la} is the surface tension of the liquid against air, θ_1 and θ_2 are the contact angles between liquid and flat and spherical surfaces, respectively [2, 3, 40]. As the surface tension value of Z-15 (24 dyn/cm) is smaller than that of water (72 dyn/cm), the larger adhesive force in Z-15 cannot only be caused by the Z-15 meniscus. The nonpolarized Z-15 liquid does not have complete coverage and strong bonding with Si(100). In the ambient environment, the condensed water molecules will permeate through the liquid Z-15 lubricant film and compete with the lubricant molecules present on the substrate. The interaction of the liquid lubricant with the substrate is weakened, and a boundary layer of the liquid lubricant

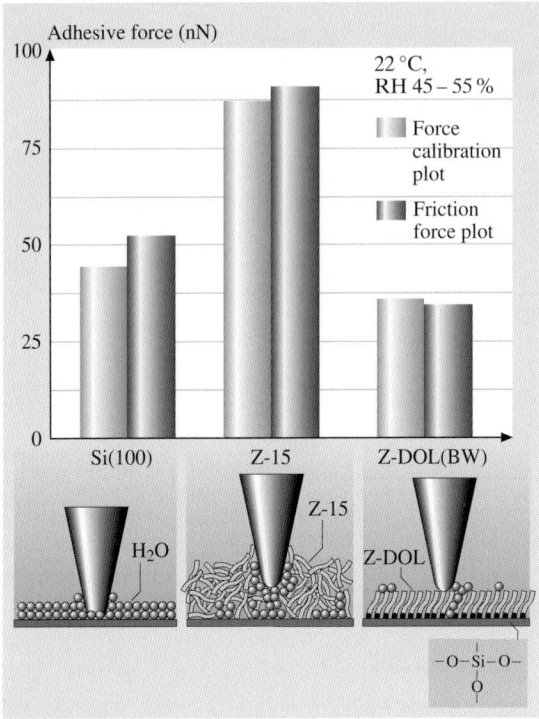

Fig. 19.8. Summary of the adhesive forces of Si(100), 2.8-nm-thick Z-15 film, and 2.3-nm-thick Z-DOL(BW) film measured by force calibration plots and friction force versus normal load plots in ambient air. The schematic (*bottom*) shows the effect of meniscus formed between the AFM tip and the sample surface on the adhesive and friction forces [21]

forms puddles [20, 34]. This dewetting allows water molecules to be adsorbed on the Si(100) surface as aggregates along with Z-15 molecules. And both of them can form meniscus while the tip approaches the surface. In addition, as the Z-15 film is pretty soft compared to the solid Si(100) surface, penetration of the tip in the film occurs while pushing the tip down. This leads to a large area of the tip involved to form the meniscus at the tip-liquid (water aggregates along with Z-15) interface. These two factors of the liquid-like Z-15 film result in higher adhesive force. It should also be noted that Z-15 has a higher viscosity compared to that of water. Therefore, Z-15 film provides higher resistance to sliding motion and results in a larger coefficient of friction. In the case of Z-DOL(BW) film, both of the active groups of Z-DOL molecules are strongly bonded on Si(100) substrate through the thermal and washing treatment. Thus, the Z-DOL(BW) film has relatively low free surface energy and cannot be displaced readily by water molecules or readily adsorb water molecules. Thus, the use of Z-DOL(BW) can reduce the adhesive force. We further note that the bonded Z-DOL molecules can be orientated under stress (behaves as a soft polymer solid), which facilitates sliding and reduces coefficient of friction.

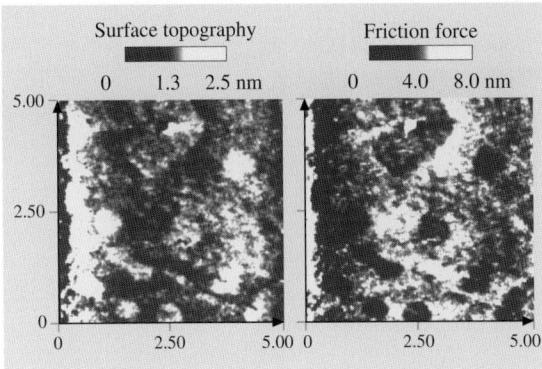

Fig. 19.9. Gray-scale plots of the surface topography and friction force obtained simultaneously for unbonded 2.3-nm-thick Demnum-type PFPE lubricant film on silicon [20]

These studies suggest that if the lubricant films exist as liquid-like, such as Z-15 films, they easily form meniscus (by themselves and the adsorbed water molecules), and thus have higher adhesive force and higher friction force. Whereas if the lubricant film exists in solid-like phase, such as Z-DOL(BW) films, they are hydrophobic with low adhesion and friction.

In order to study the uniformity of lubricant film and its influence on friction and adhesion, friction force mapping and adhesive force mapping of PFPE have been carried out by *Koinkar* and *Bhushan* [34] and *Bhushan* and *Dandavate* [36], respectively. Figure 19.9 shows gray-scale plots of surface topography and friction force images obtained simultaneously for unbonded Demnum-type PFPE lubricant film on silicon [34]. The friction force plot shows well distinguished low and high friction regions corresponding roughly to high and low surface height regions in the topography image (thick and thin lubricant regions). A uniformly lubricated sample does not show such a variation in friction. Figure 19.10 shows the gray-scale plots of the adhesive force distribution for silicon samples coated uniformly and nonuniformly with Z-DOL lubricant. It can be clearly seen that there exists a region that has an adhesive force distinctly different from the other region for the nonuniformly coated sample. This implies that the liquid film thickness is nonuniform, giving rise to a difference in the meniscus forces.

19.4.2 Rest Time Effect

It is well-known that in the computer rigid disk drive, the stiction force increases rapidly with an increase in rest time between the head and magnetic medium disk [10, 11]. Considering that the stiction and friction are two of the major issues that lead to the failure of computer rigid disk drives and MEMS, it is very important to find out if the rest time effect also exists on the nanoscale. First, the rest time effect on the friction force, adhesive force, and coefficient of Si(100) sliding against Si_3N_4 tip was studied, Fig. 19.11a [21]. It was found that the friction and adhesive forces

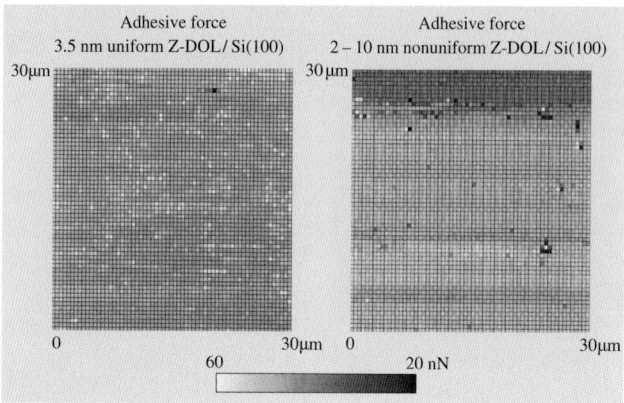

Fig. 19.10. Gray-scale plots of the adhesive force distribution of a uniformly coated, 3.5-nm-thick unbonded Z-DOL film on silicon and 3- to 10-nm-thick unbonded Z-DOL film on silicon that was deliberately coated nonuniformly by vibrating the sample during the coating process [36]

logarithmically increase up to a certain equilibrium time after which they remain constant. Figure 19.11a also shows that the rest time does not affect the coefficient of friction. These results suggest that the rest time can result in growth of the meniscus, which causes a higher adhesive force, and in turn, a higher friction force. But in the whole testing range the friction mechanisms do not change with the rest time. Similar studies were also performed on Z-15 and Z-DOL(BW) films. The results are summarized in Fig. 19.11b [21]. It is seen that similar time effect has been observed on Z-15 film, but not on Z-DOL(BW) film.

AFM tip in contact with a flat sample surface is generally believed to represent a single asperity contact. Therefore, a Si_3N_4 tip in contact with $Si(100)$ or Z-15/Si(100) can be modeled as a sphere in contact with a flat surface covered by a layer of liquid (adsorbed water and/or liquid lubricant), Fig. 19.12a. Meniscus forms around the contacting asperity and grows with time until equilibrium occurs [41]. The meniscus force, which is the product of meniscus pressure and meniscus area, depends on the flow of liquid phase toward the contact zone. The flow of the liquid toward the contact zone is governed by the capillary pressure P_c, which draws liquid into the meniscus, and the disjoining pressure Π, which tends to draw the liquid away from the meniscus. Based on the Young and Laplace equation, the capillary pressure, P_c, is:

$$P_c = 2K\gamma, \tag{19.3}$$

where $2K$ is the mean meniscus curvature ($= K_1 + K_2$, where K_1 and K_2 are the curvatures of the meniscus in the contact plane and perpendicular to the contact plane) and γ is the surface tension of the liquid. *Mate* and *Novotny* [6] have shown that the disjoining pressure decreases rapidly with increasing liquid film thickness in a manner consistent with a strong van der Waals attraction. The disjoining pressure, Π, for

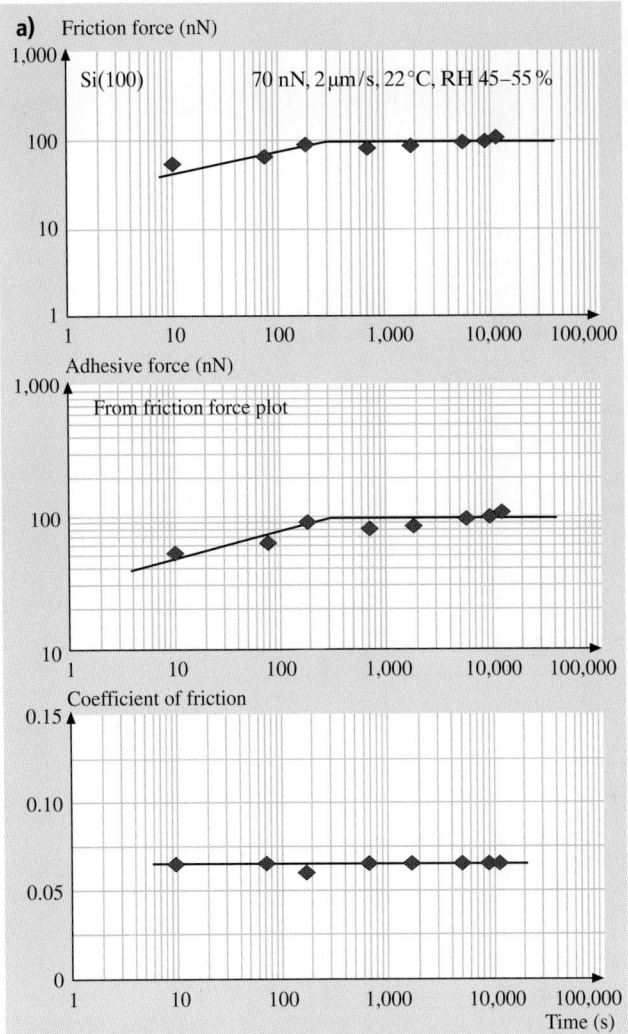

Fig. 19.11. (**a**) Rest time effect on friction force, adhesive force, and coefficient of friction of Si(100). (**b**) (*see next page*) Summary of the rest time effect on friction force, adhesive force, and coefficient of friction of Si(100), 2.8-nm-thick Z-15 film, and 2.3-nm-thick Z-DOL(BW) film. All of the measurements were carried out at 70 nN, 2 μm/s, and in ambient air [21]

these liquid films can be expressed as:

$$\Pi = \frac{A}{6\pi h^3}, \tag{19.4}$$

where A is the Hamaker constant and h is the liquid film thickness. The driving forces that cause the lubricant flow that results in an increase in the meniscus force

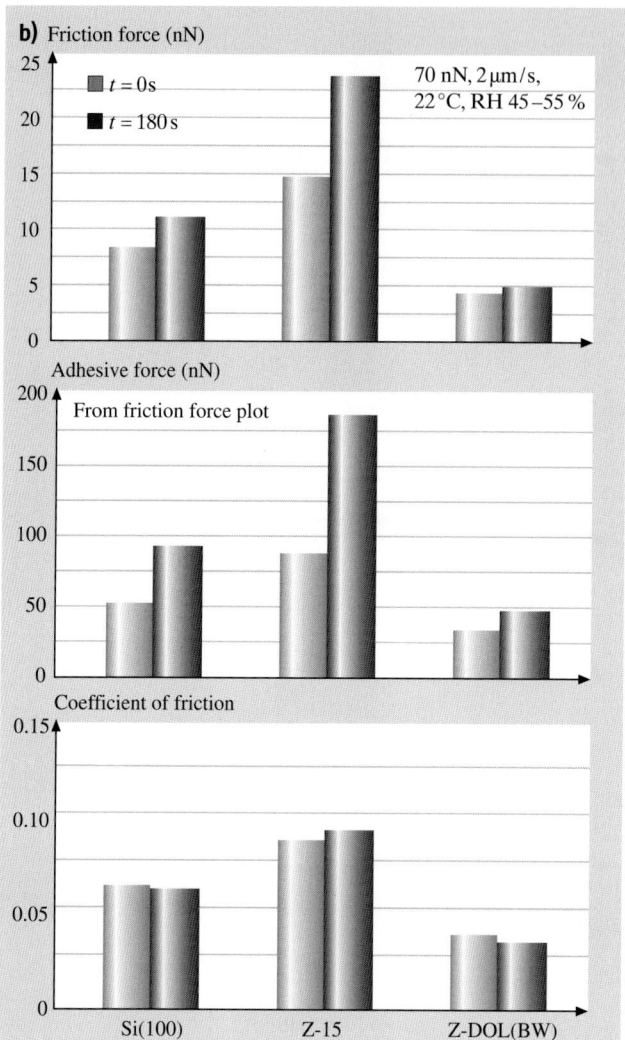

Fig. 19.11. continue

are the disjoining pressure gradient, due to a gradient in film thickness, and capillary pressure gradient, due to curved liquid-air interface. The driving pressure, P, can then be written as:

$$P = -2K\gamma - \Pi \ . \tag{19.5}$$

Based on these three basic relationships, the following differential equation has been derived by *Chilamakuri* and *Bhushan* [41], which can describe the meniscus at time t:

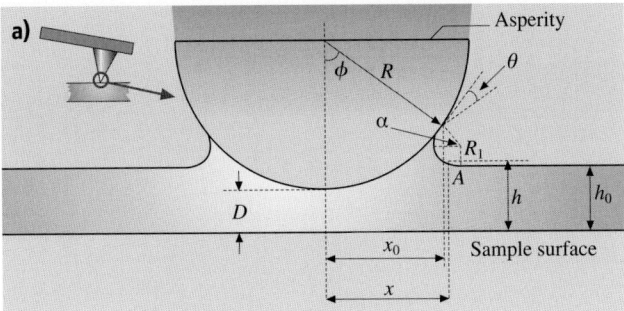

Fig. 19.12. (a) Schematic of a single asperity in contact with a smooth flat surface in the presence of a continuous liquid film when ϕ is large. (b) Results of the single asperity model. Effect of viscosity of the liquid, radius of the asperity, and film thickness is studied with respect to the time-dependent meniscus force [41]

$$2\pi x_0 \left(D + \frac{x_0^2}{2R} - h_0\right)\frac{\mathrm{d}x_0}{\mathrm{d}t}$$
$$= \frac{2\pi h_0^3 \gamma}{3\eta}\frac{(1+\cos\theta)}{D+a-h_0} - \frac{Ax_0}{3\eta h}\cot\alpha \,, \tag{19.6}$$

where η is the viscosity of the liquid and a is given as

$$a = R(1-\cos\phi) \sim \frac{R\phi^2}{2} \sim \frac{x_0^2}{2R}\,. \tag{19.7}$$

The differential equation (19.4) was solved numerically using Newton's iteration method. The meniscus force at any time t less than the equilibrium time is proportional to the meniscus area and meniscus pressure $(2K\gamma)$, and it is given by

$$f_m(t) = 2\pi R\gamma(1+\cos\theta)\left(\frac{x_0}{(x_0)_{\mathrm{eq}}}\right)^2\left(\frac{K}{K_{\mathrm{eq}}}\right), \tag{19.8}$$

where $(x_0)_{\mathrm{eq}}$ is the value of x_0 at the equilibrium time

$$\left[(x_0)_{\mathrm{eq}}\right]^2 = 2R\left[\frac{-6\pi h_0^3\gamma(1+\cos\theta)}{A}+(h_0-D)\right]. \tag{19.9}$$

This modeling work (at the microscale) showed that the meniscus force initially increases logarithmically with the rest time up to a certain equilibrium time after which it remains constant. Equilibrium time decreases with an increase in liquid film thickness, a decrease in viscosity, and a decrease in the tip radius, Fig. 19.12b. This early numerical modeling work and the data at the nanoscale in Fig. 19.11a are in good agreement.

19.4.3 Velocity Effect

To investigate the velocity effect on friction and adhesion, the friction force versus normal load relationships of Si(100), Z-15, and Z-DOL(BW) at different veloc-

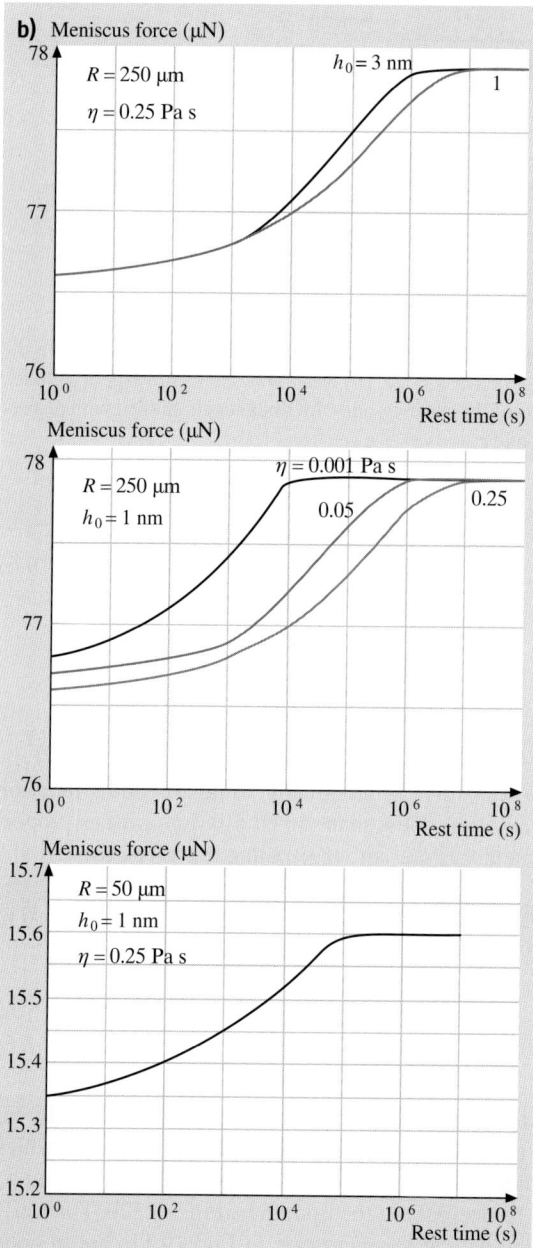

Fig. 19.12. countinue

ities were measured, Fig. 19.13 [21]. Based on these data, the adhesive force and coefficient of friction values can be calculated by (19.1). The variation of friction force, adhesive force, and coefficient of friction of Si(100), Z-15, and Z-DOL(BW) as a function of velocity are summarized in Fig. 19.14. It indicates that for silicon wafer, the friction force decreases logarithmically with increasing velocity. For Z-15, the friction force decreases with increasing velocity up to $10\,\mu m/s$, after which it remains almost constant. The velocity has a much smaller effect on the friction force of Z-DOL(BW); it reduced slightly only at very high velocity. Figure 19.14 also indicates that the adhesive force of Si(100) is increased when the velocity is higher than $10\,\mu m/s$. The adhesive force of Z-15 is reduced dramatically with a velocity increase up to $20\,\mu m/s$, after which it is reduced slightly. And the adhesive force of Z-DOL(BW) also decreases at high velocity. In the testing range of velocity, only the coefficient of friction of Si(100) decreases with velocity, but the coefficients of friction of Z-15 and Z-DOL(BW) almost remain constant. This implies that the friction mechanisms of Z-15 and Z-DOL(BW) do not change with the variation of velocity.

The mechanisms of the effect of velocity on the adhesion and friction are explained based on the schematics shown in Fig. 19.14 (right). For Si(100), tribochemical reaction plays a major role. Although at high velocity the meniscus is broken and does not have enough time to rebuild, the contact stresses and high velocity lead to tribochemical reactions of Si(100) wafer and Si_3N_4 tip, which have native oxide (SiO_2) layers with water molecules. The following reactions occur:

$$SiO_2 + 2H_2O = Si(OH)_4 \qquad (19.10)$$

$$Si_3N_4 + 16H_2O = 3Si(OH)_4 + 4NH_4OH \,. \qquad (19.11)$$

The $Si(OH)_4$ is removed and continuously replenished during sliding. The $Si(OH)_4$ layer between the tip and Si(100) surface is known to be of low shear strength and causes a decrease in friction force and coefficient of friction in the lateral direction [42–46]. The chemical bonds of Si−OH between the tip and Si(100) surface induce large adhesive force in the normal direction. For Z-15 film, at high velocity the meniscus formed by condensed water and Z-15 molecules is broken and does not have enough time to rebuild. Therefore, the adhesive force and, consequently, friction force is reduced. For Z-DOL(BW) film, the surface can adsorb few water molecules in ambient condition, and at high velocity these molecules are displaced, which is responsible for a slight decrease in friction force and adhesive force. Even at high velocity range, the friction mechanisms for Z-15 and Z-DOL(BW) films still areshearing of the viscous liquid and molecular orientation, respectively. Thus the coefficients of friction of Z-15 and Z-DOL(BW) do not change with velocity.

Koinkar and *Bhushan* [20, 34] have suggested that in the case of samples with mobile films, such as condensed water and Z-15 films, alignment of liquid molecules (shear thinning) is responsible for the drop in friction force with an increase in scanning velocity. This could be another reason for the decrease in friction force with velocity for Si(100) and Z-15 film in this study.

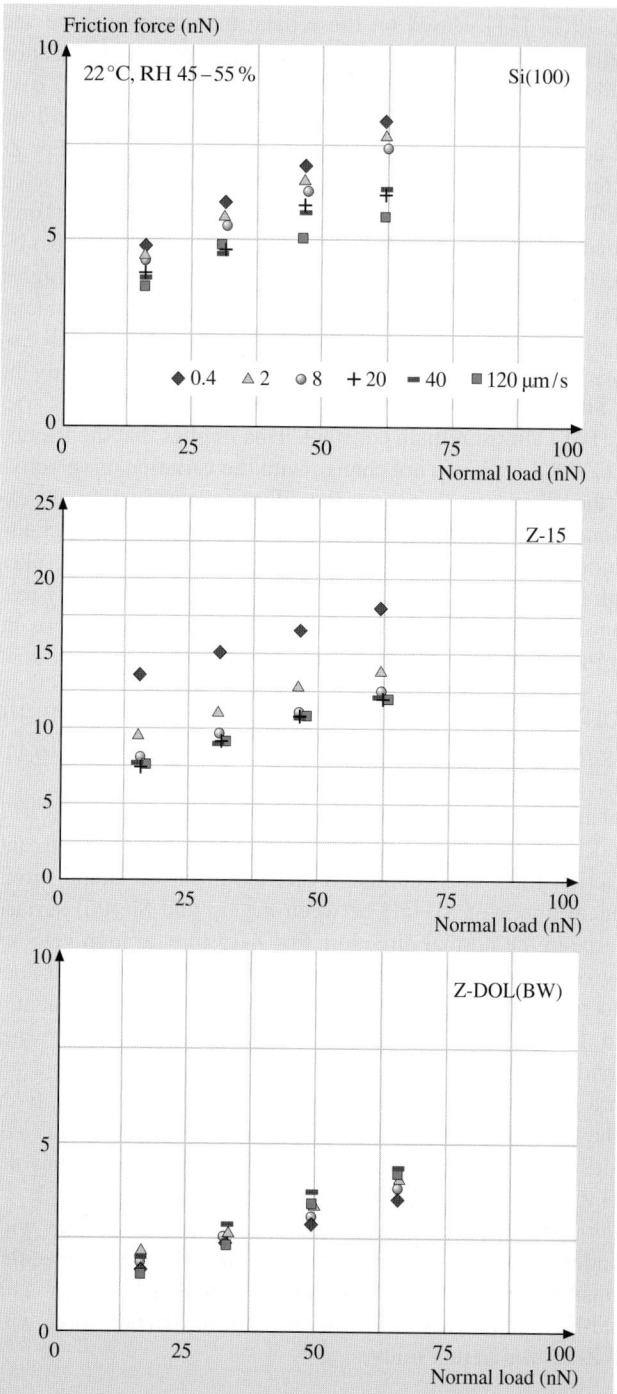

Fig. 19.13. Friction forces versus normal load data of Si(100), 2.8-nm-thick Z-15 film, and 2.3-nm-thick Z-DOL(BW) film at various velocities in ambient air [21]

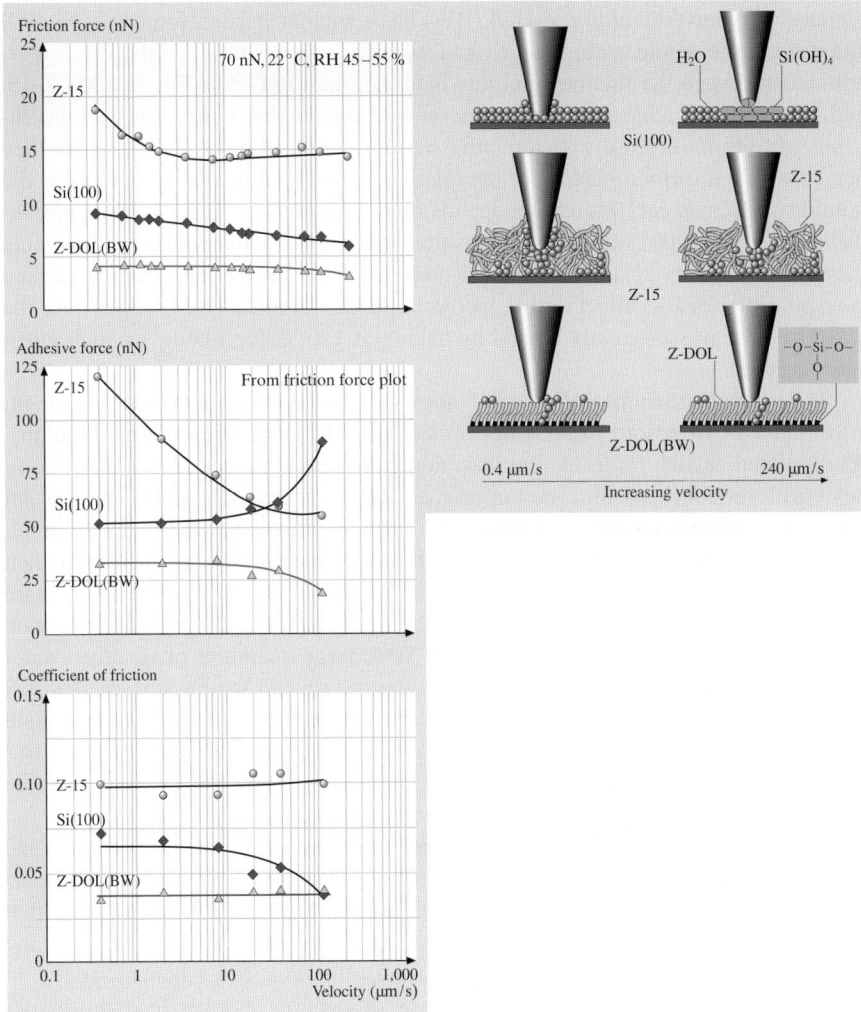

Fig. 19.14. The influence of velocity on the friction force, adhesive force, and coefficient of friction of Si(100), 2.8-nm-thick Z-15 film, and 2.3-nm-thick Z-DOL(BW) film at 70 nN, in ambient air. The schematic (*right*) shows the change of surface composition (by tribochemical reaction) and change of meniscus with increasing velocity [21]

19.4.4 Relative Humidity and Temperature Effect

The influence of relative humidity on friction and adhesion was studied in an environmentally controlled chamber. The friction force was measured by making measurements at increasing relative humidity, the results are presented in Fig. 19.15 [21]. It shows that for Si(100) and Z-15 film, the friction force increases with a relative humidity increase up to RH 45%, and then it shows a slight decrease with a further increase in relative humidity. Z-DOL(BW) has a smaller friction force than Si(100) and Z-15 in the whole testing range. And its friction force shows a relatively apparent increase when the relative humidity is higher than RH 45%. For Si(100), Z-15, and Z-DOL(BW), adhesive forces increase with relative humidity. And their coefficients of friction increase with a relative humidity up to RH 45%, after which they decrease with a further increase of the relative humidity. It is also observed that the humidity effect on Si(100) really depends on the history of the Si(100) sample. As the surface of Si(100) wafer readily adsorbs water in air, without any pre-treatment the Si(100) used in our study almost reaches its saturate stage of adsorbing water and is responsible for less effect during increasing relative humidity. However, once the Si(100) wafer was thermally treated by baking at 150 °C for 1 hour, a bigger effect was observed.

The schematic (right) in Fig. 19.15 shows that because its high free surface energy Si(100) can adsorb more water molecules with increasing relative humidity. As discussed earlier, for Z-15 film in a humid environment, condensed water competes with the lubricant film present on the sample surface. Obviously, more water molecules can also be adsorbed on a Z-15 surface with increasing relative humidity. Themore adsorbed water molecules in the case of Si(100), along with lubricant molecules in Z-15 film, form a bigger water meniscus, which leads to an increase of friction force, adhesive force, and coefficient of friction of Si(100) and Z-15 with humidity. But at very high humidity of RH 70%, large quantities of adsorbed water can form a continuous water layer that separates the tip and sample surface, and acts as a kind of lubricant, which causes a decrease in the friction force and coefficient of friction. For Z-DOL(BW) film, because of its hydrophobic surface properties, water molecules can only be adsorbed at high humidity ($\geq$ RH 45%), which causes an increase in the adhesive force and friction force.

The effect of temperature on friction and adhesion was studied using a thermal stage attached to the AFM. The friction force was measured at increasing temperature from 22–125 °C. The results are presented in Fig. 19.16 [21]. It shows that the increasing temperature causes a decrease of friction force, adhesive force, and coefficient of friction of Si(100), Z-15, and Z-DOL(BW). The schematic (right) in Fig. 19.16 indicates that at high temperature, desorption of water leads to the decrease of friction force, adhesive force, and coefficient of friction for all of the samples. Besides that, the reduction of surface tension of water also contributes to the decrease of friction and adhesion. For Z-15 film, the reduction of viscosity at high temperature has an additional contribution to the decrease of friction. In the case of Z-DOL(BW) film, molecules are more easily oriented at high temperature, which may also be responsible for the low friction.

Using a surface force apparatus, *Yoshizawa* and
Israelachvili [47, 48] have shown that a change in the velocity or temperature induces phase transformation (from crystalline solid-like to amorphous, then to liquid-like) in surfactant monolayers, which are responsible for the observed changes in the friction force. Stick-slip is observed in a low velocity regime of a few μm/s, and adhesion and friction first increase followed by a decrease in the temperature range of 0–50 °C. Stick-slip at low velocity and adhesion and friction curves peaking at

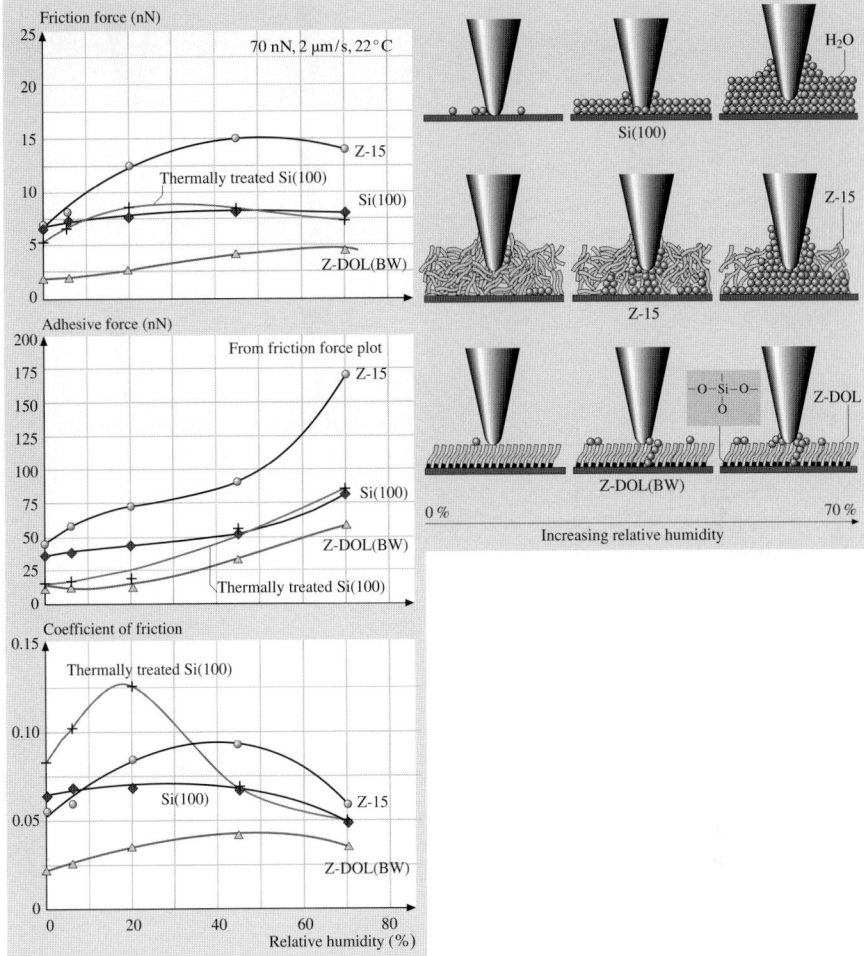

Fig. 19.15. The influence of relative humidity (RH) on the friction force, adhesive force, and coefficient of friction of Si(100), 2.8-nm-thick Z-15 film, and 2.3-nm-thick Z-DOL(BW) film at 70 nN, 2 μm/s, and in 22 °C air. Schematic (*right*) shows the change of meniscus with increasing relative humidity. In this figure, the thermally treated Si(100) represents the Si(100) wafer that was baked at 150 °C for 1 hour in an oven (in order to remove the adsorbed water) just before it was placed in the 0% RH chamber [21]

some particular temperature (observed in their study), have not been observed in the AFM study. It suggests that the phase transformation may not happen in this study, because PFPEs generally have very good thermal stability [10, 12].

As a brief summary, the influence of velocity, relative humidity, and temperature on the friction force of Z-15 film is presented in Fig. 19.17. The changing trends are also addressed in this figure.

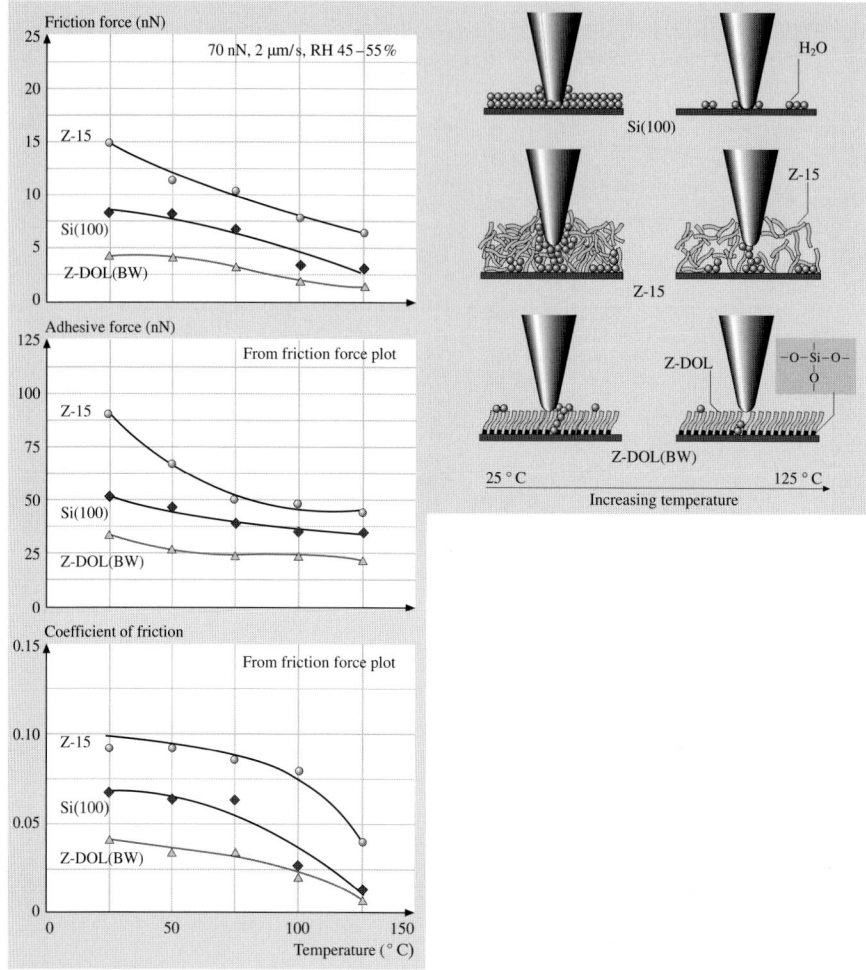

Fig. 19.16. The influence of temperature on the friction force, adhesive force, and coefficient of friction of Si(100), 2.8-nm-thick Z-15 film, and 2.3-nm-thick Z-DOL(BW) film at 70 nN, at 2 μm/s, and in RH 40–50 % air. The schematic (*right*) shows that at high temperature, desorption of water decreases the adhesive forces. And the reduced viscosity of Z-15 leads to the decrease of coefficient of friction. High temperature facilitates orientation of molecules in Z-DOL(BW) film, which results in lower coefficient of friction [21]

19.4.5 Tip Radius Effect

The tip radius and relative humidity affect adhesion and friction for dry and lubricated surfaces [35, 36]. Figure 19.18a shows the variation of single point adhesive force measurements as a function of tip radius on a Si(100) sample for several humidities. The adhesive force data are also plotted as a function of relative humidity for various tip radii. Figure 19.18a indicates that the tip radius has little effect on the

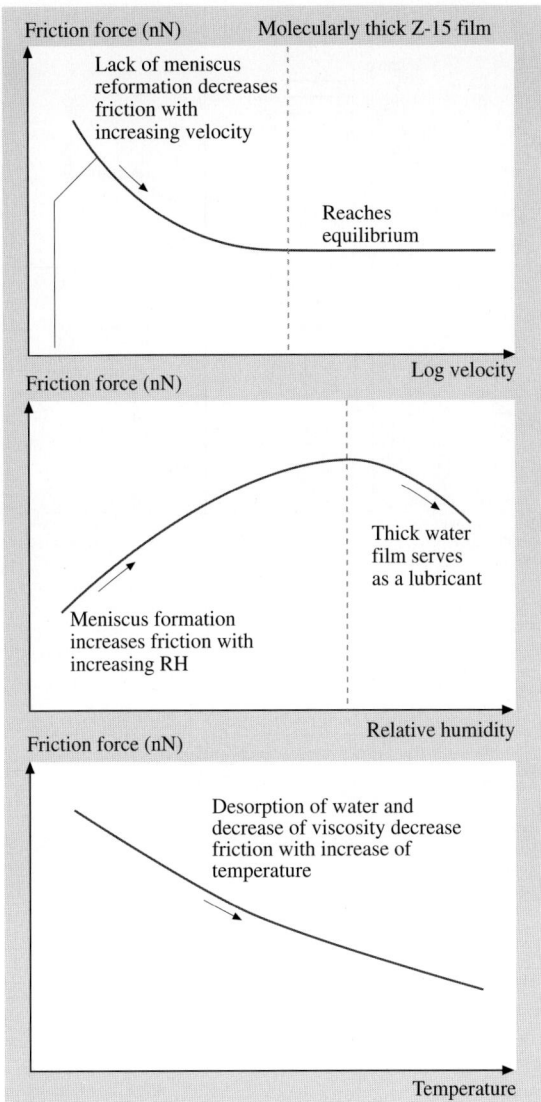

Fig. 19.17. Schematic shows the change of friction force of molecularly thick Z-15 films with log velocity, relative humidity, and temperature [21]

adhesive forces at low humidities, but the adhesive force increases with tip radius at high humidity. Adhesive force also increases with an increase in humidity for all tips. The trend in adhesive forces as a function of tip radii and relative humidity, in Fig. 19.18a, can be explained by the presence of meniscus forces, which arise from capillary condensation of water vapor from the environment. If enough liquid is present to form a meniscus bridge, the meniscus force should increase with an

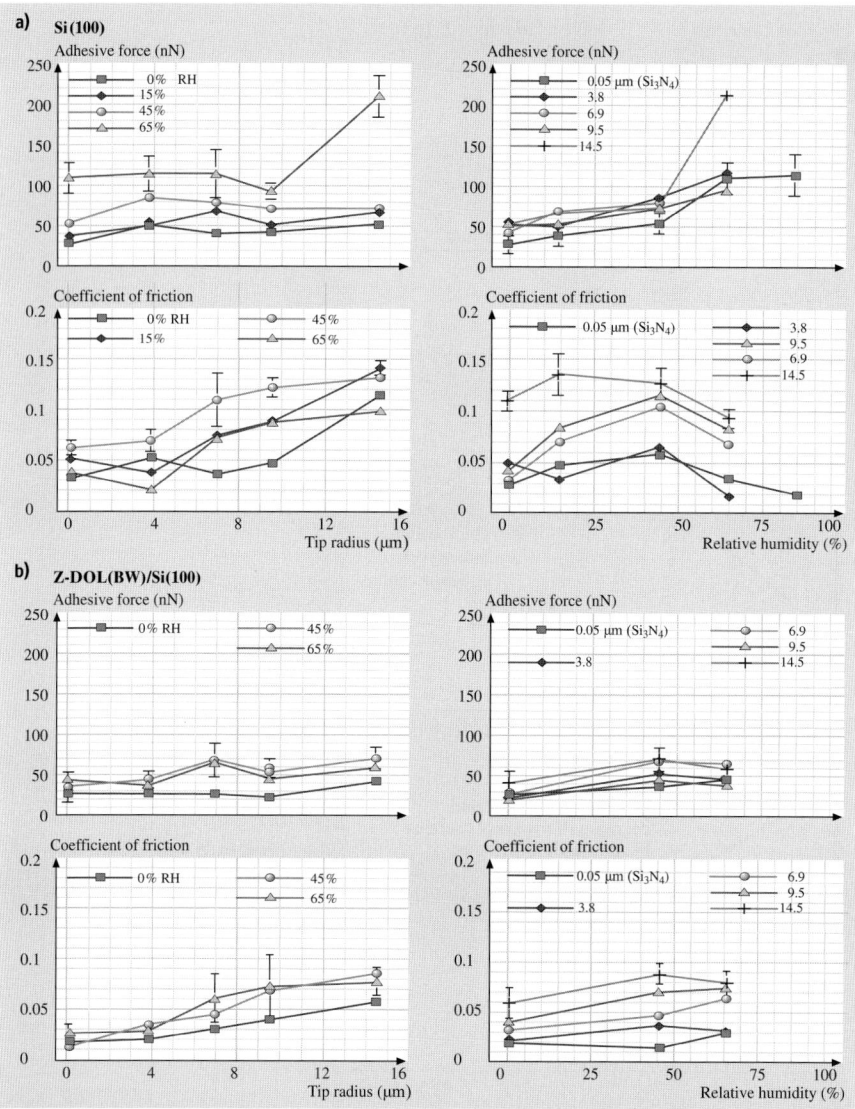

Fig. 19.18. Adhesive force and coefficient of friction as a function of tip radius at several humidities and as a function of relative humidity at several tip radii on (**a**) Si(100) and (**b**) 0.5-nm Z-DOL(BW) films [35]

increase in tip radius based on (19.2). This observation suggests that thickness of the liquid film at low humidities is insufficient to form continuous meniscus bridges and to affect adhesive forces in the case of all tips.

Figure 19.18a also shows the variation in coefficient of friction as a function of tip radius at a given humidity and as a function of relative humidity for a given tip radius

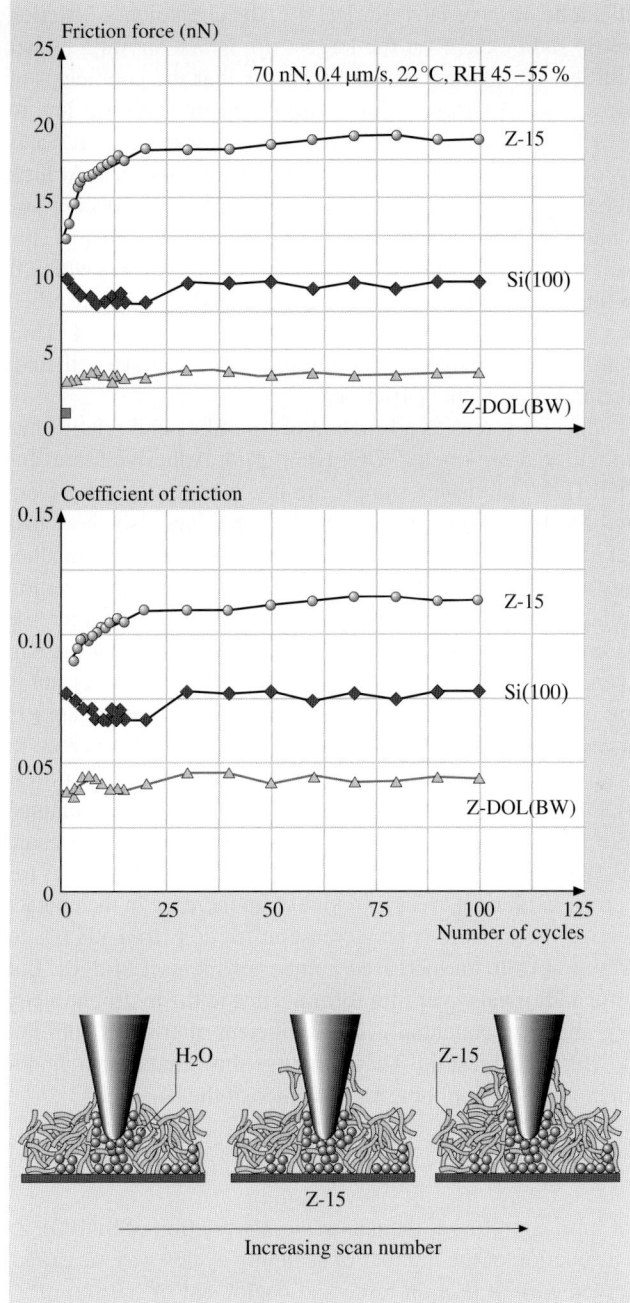

Fig. 19.19. Friction force and coefficient of friction versus number of sliding cycles for Si(100), 2.8-nm-thick Z-15 film, and 2.3-nm-thick Z-DOL(BW) film at 70 nN, 0.8 μm/s, and in ambient air. Schematic (*bottom*) shows that some liquid Z-15 molecules can be attached on the tip. The molecular interaction between the attached molecules on the tip with the Z-15 molecules in the film results in an increase of the friction force with multiplescanning [21]

on the Si(100) sample. It can be observed that for RH 0%, the coefficient of friction is about the same for the tip radii except for the largest tip, which shows a higher value. At all other humidities, the trend consistently shows that the coefficient of friction increases with tip radius. An increase in friction with tip radius at low to moderate humidities arises from increased contact area (i.e., higher van der Waals forces) and higher values of shear forces required for a larger contact area. At high humidities, similar to adhesive force data, an increase with tip radius occurs due to of both contact area and meniscus effects. It can be seen that for all tips, the coefficient of friction increases with humidity to about RH 45%, beyond which it starts to decrease. This is attributed to the fact that at higher humidities, the adsorbed water film on the surface acts as a lubricant between the two surfaces [21]. Thus the interface is changed at higher humidities, resulting in lower shear strength and, hence, lower friction force and coefficient of friction.

Figure 19.18b shows adhesive forces as a function of tip radius and relative humidity on Si(100) coated with 0.5-nm-thick Z-DOL(BW) film. Adhesive forces for all the tips with the Z-DOL(BW) lubricated sample are much lower than those experienced on unlubricated Si(100), shown in Fig. 19.18a. The data also show that even at a monolayer thickness of the lubricant there is very little variation in adhesive forces with tip radius at a given humidity. For a given tip radius, the variation in adhesive forces with relative humidity indicates that these forces slightly increase from RH 0% to RH 45%, but remain more or less the same with a further increase in humidity. This is seen even with the largest tip, which indicates that the lubricant is indeed hydrophobic; there is some meniscus formation at humidities higher than RH 0%, but the formation is very minimal and does not increase appreciably even up to RH 65%. Figure 19.18b also shows coefficient of friction for various tips at different humidities for the Z-DOL(BW) lubricated sample. Again, all the values obtained with the lubricated sample are much lower than the values obtained on unlubricated Si(100), shown in Fig. 19.18a. The coefficient of friction increases with tip radius for all humidities, as was seen on unlubricated Si(100), due to an increase in the contact area. Similar to the adhesive forces, there is an increase in friction from RH 0% to RH 45%, due to a contribution from an increased number of menisci bridges. But thereafter there is very little additional water film forming, due to the hydrophobicity of the Z-DOL(BW) layer, and, consequentially, the coefficient of friction does not change appreciably, even with the largest tip. These findings show that even a monolayer of Z-DOL(BW) offers good hydrophobic performance of the surface.

19.4.6 Wear Study

To study the durability of lubricant films at the nanoscale, the friction of Si(100), Z-15, and Z-DOL(BW) as a function of the number of scanning cycles was measured, Fig. 19.19 [21]. As observed earlier, friction force and coefficient of friction of Z-15 is higher than that of Si(100), and Z-DOL(BW) has the lowest values. During cycling, friction force and coefficient of friction of Si(100) show a slight variation during the initial few cycles then remain constant. This is related to the removal of the top adsorbed layer. In the case of Z-15 film, the friction force and coefficient of friction show an increase during the initial few cycles and then approach higher and stable values. This is believed to be caused by the attachment of the Z-15 molecules onto the tip. The molecular interaction between these attached molecules to the tip and molecules on the film surface is responsible for an increase in the friction. But after several scans, this molecular interaction reaches the equilibrium, and after that, friction force and coefficient of friction remain constant. In the case of Z-DOL(BW) film, the friction force and coefficient of friction start out low and remain low during the entire test for 100 cycles. It suggests that Z-DOL(BW) molecules do not get attached or displaced as readily as Z-15.

Koinkar and *Bhushan* [20, 34] conducted wear studies using a diamond tip at high loads. Figure 19.20 shows the plots of wear depth as a function of normal force, and Fig. 19.21 shows the wear profiles of the worn samples at 40 µN normal load. The 2.3-nm-thick Z-DOL(BW) lubricated sample exhibits better wear resistance than unlubricated and 2.9-nm-thick Z-15 lubricated silicon samples. Wear resistance of a Z-15 lubricated sample is little better than that of unlubricated sample. The Z-15 lubricated sample shows debris inside the wear track. Since the Z-15 is a liquid lubricant, debris generated is held by the lubricant and they become sticky. The debris moves inside the wear track and does damage, Fig. 19.20. These results suggest that using Z-DOL(BW) can improve the wear resistance of substrate.

To study the effect of the degree of chemical bonding, the durability tests were conducted on both fully bonded and partially bonded Z-DOL films. Durability results for Z-DOL(BW) and Z-DOL bonded and unwashed (Z-DOL(BUW), a partially bonded film that contains both bonded and mobile phase lubricants) with different film thicknesses are shown in Fig. 19.22 [34]. Thicker films, such as Z-DOL(BUW), with a thickness of 4.0 nm (bonded/mobile = 2.3 nm/1.7 nm) exhibit behavior similar to 2.3-nm-thick Z-DOL(BW) film. Figure 19.22 also indicates that Z-DOL(BW) and Z-DOL(BUW) films with a thinner film thickness exhibit a higher friction value. Comparing 1.0-nm-thick Z-DOL(BW) with 3.0-nm-

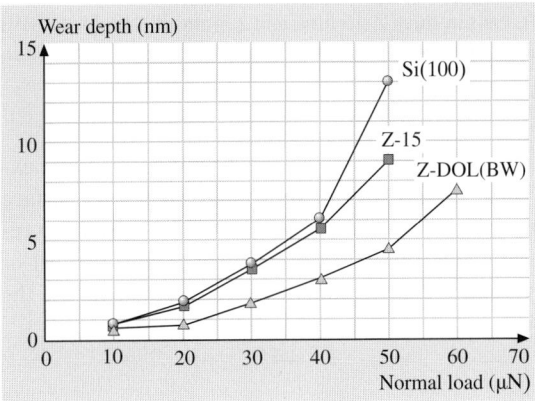

Fig. 19.20. Wear depth as a function of normal load using a diamond tip for Si(100), 2.9-nm-thick Z-15 film, and 2.3-nm-thick Z-DOL(BW) after one cycle [20]

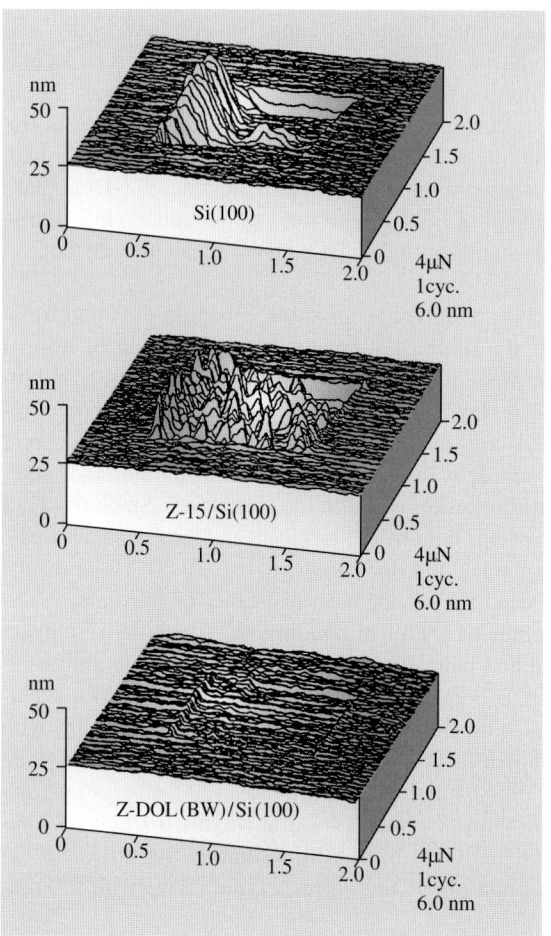

Fig. 19.21. Wear profiles for Si(100), 2.9-nm-thick Z-15 film, and 2.3-nm-thick Z-DOL(BW) film after wear studies using a diamond tip. Normal force used and wear depths are listed in the figure [20]

thick Z-DOL(BUW) (bonded/mobile = 1.0 nm/2.0 nm), the Z-DOL(BUW) film exhibits a lower and stable friction value. This is because the mobile phase on a surface acts as a source of lubricant replenishment. A similar conclusion has also been reported by *Ruhe* et al. [19], *Bhushan* and *Zhao* [13], and *Eapen* et al. [49]. All of them indicate that using partially bonded Z-DOL films can dramatically reduce the friction and improve the wear life.

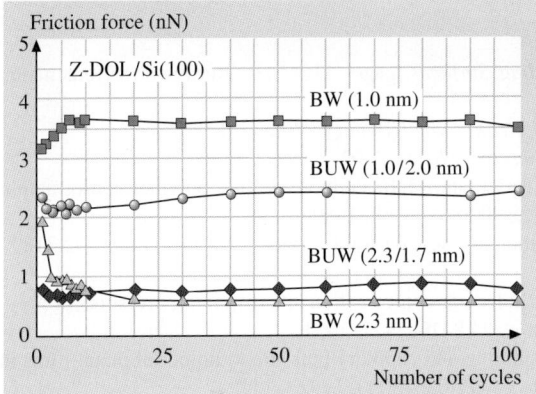

Fig. 19.22. Friction force as a function of number of cycles using a Si_3N_4 tip at a normal load of 300 nN for Z-DOL(BW) and Z-DOL(BUW) films with different film thicknesses [34]

19.5 Closure

Nanodeformation study has shown that fully bonded Z-DOL lubricants behave as soft polymer solids, while the unbonded lubricants behave liquid-like. AFM studies have shown that the physisorbed nonpolar molecules on a solid surface have an extended, flat conformation. The spreading property of PFPE is strongly dependent on the molecular end groups and substrate chemistry.

Using solid-like Z-DOL(BW) film can reduce the friction and adhesion of Si(100), while using liquid-like lubricant of Z-15 shows a negative effect. Si(100) and Z-15 film show apparent time effect. The friction and adhesion forces increase as a result of growth of meniscus up to an equilibrium time, after which they remain constant. Using Z-DOL(BW) film can prevent time effects. High velocity leads to the rupture of meniscus and prevents its reformation, which leads to a decrease of friction and adhesive forces of Z-15 and Z-DOL(BW). The influence of relative humidity on the friction and adhesion is dominated by the amount of the adsorbed water molecules. Increasing humidity can either increase friction through increased adhesion by water meniscus, or reduce friction through an enhanced water-lubricating effect. Increasing temperature leads to desorption of water layer, decrease of water surface tension, decrease of viscosity, and easier orientation of the Z-DOL(BW) molecules. These changes cause a decrease of friction force and adhesion at high temperature. During cycling tests, the molecular interaction between the attached Z-15 molecules to the tip and the Z-15 molecules on the film surface causes the initial rise of friction. Wear tests show that Z-DOL(BW) can improve the wear resistance of silicon. Partially bonded PFPE film appears to be more durable than fully bonded films.

These results suggest that partially/fully bonded films are good lubricants for devices operating in different environments and under varying conditions.

References

1. B. Bhushan. *Magnetic Recording Surfaces*, pages 116–133. Butterworth-Heinemann, 1993.
2. B. Bhushan. *Principles and Applications of Tribology*. Wiley, 1999.
3. B. Bhushan. *Introduction to Tribology*. Wiley, 2002.
4. V. J. Novotny, I. Hussla, J. M. Turlet, and M. R. Philpott. Liquid polymer conformation on solid surfaces. *J. Chem. Phys.*, 90:5861–5868, 1989.
5. V. J. Novotny. Migration of liquid polymers on solid surfaces. *J. Chem. Phys.*, 92:3189–3196, 1990.
6. C. M. Mate and V. J. Novotny. Molecular conformation and disjoining pressures of polymeric liquid films. *J. Chem. Phys.*, 94:8420–8427, 1991.
7. C. M. Mate. Application of disjoining and capillary pressure to liquid lubricant films in magnetic recording. *J. Appl. Phys.*, 72:3084–3090, 1992.
8. G. G. Roberts. *Langmuir–Blodgett Films*. Plenum, 1990.
9. A. Ulman. *An Introduction to Ultrathin Organic Films*. Academic, 1991.
10. B. Bhushan. *Tribology and Mechanics of Magnetic Storage Devices*. Springer, 2nd edition, 1996.
11. B. Bhushan. *Macro- and microtribology of magnetic storage devices*, pages 1413–1513. CRC, 2001.
12. Anonymous. Fomblin z perfluoropolyethers, 2002.
13. B. Bhushan and Z. Zhao. Macroscale and microscale tribological studies of molecularly thick boundary layers of perfluoropolyether lubricants for magnetic thin-film rigid disks. *J. Info. Storage Proc. Syst.*, 1:1–21, 1999.
14. B. Bhushan. *Tribology Issues and Opportunities in MEMS*. Kluwer, 1998.
15. B. Bhushan, J. N. Israelachvili, and U. Landman. Nanotribology: Friction, wear and lubrication at the atomic scale. *Nature*, 374:607–616, 1995.
16. B. Bhushan. *Handbook of Micro/Nanotribology*. CRC, 2nd edition, 1999.
17. B. Bhushan. *Self-assembled monolayers for controlling hydrophobicity and/or friction and wear*, pages 909–929. CRC, 2001.
18. J. Ruhe, G. Blackman, V. J. Novotny, T. Clarke, G. B. Street, and S. Kuan. Thermal attachment of perfluorinated polymers to solid surfaces. *J. Appl. Polym. Sci.*, 53:825–836, 1994.
19. J. Ruhe, V. Novotny, T. Clarke, and G. B. Street. Ultrathin perfluoropolyether films – influence of anchoring and mobility of polymers on the tribological properties. *ASME J. Tribol.*, 118:663–668, 1996.
20. V. N. Koinkar and B. Bhushan. Microtribological studies of unlubricated and lubricated surfaces using atomic force/friction force microscopy. *J. Vac. Sci. Technol. A*, 14:2378–2391, 1996.
21. H. Liu and B. Bhushan. Nanotribological characterization of molecularly-thick lubricant films for applications to mems/nems by afm. *Ultramicroscopy*, 97:321–340, 2003.
22. G. S. Blackman, C. M. Mate, and M. R. Philpott. Interaction forces of a sharp tungsten tip with molecular films on silicon surface. *Phys. Rev. Lett.*, 65:2270–2273, 1990.
23. G. S. Blackman, C. M. Mate, and M. R. Philpott. Atomic force microscope studies of lubricant films on solid surfaces. *Vacuum*, 41:1283–1286, 1990.
24. C. A. Kim, H. J. Choi, R. N. Kono, and M. S. Jhon. Rheological characterization of perfluoropolyether lubricant. *Polym. Prepr.*, 40:647–649, 1999.
25. M. Ruths and S. Granick. Rate-dependent adhesion between opposed perfluoropoly(alkylether) layers: Dependence on chain-end functionality and chain length. *J. Phys. Chem. B*, 102:6056–6063, 1998.

26. X. Ma, J. Gui, K. J. Grannen, L. A. Smoliar, B. Marchon, M. S. Jhon, and C. L. Bauer. Spreading of pfpe lubricants on carbon surfaces: Effect of hydrogen and nitrogen content. *Tribol. Lett.*, 6:9–14, 1999.
27. U. Jonsson and B. Bhushan. Measurement of rheological properties of ultrathin lubricant films at very high shear rates and near-ambient pressure. *J. Appl. Phys.*, 78:3107–3109, 1995.
28. C. Hahm and B. Bhushan. High shear rate viscosity measurement of perfluoropolyether lubricants for magnetic thin-film rigid disks. *J. Appl. Phys.*, 81:5384–5386, 1997.
29. C. M. Mate. Atomic-force-microscope study of polymer lubricants on silicon surface. *Phys. Rev. Lett.*, 68:3323–3326, 1992.
30. C. M. Mate. Nanotribology of lubricated and unlubricated carbon overcoats on magnetic disks studied by friction force microscopy. *Surf. Coat. Technol.*, 62:373–379, 1993.
31. S. J. O'Shea, M. E. Welland, and T. Rayment. Atomic force microscope study of boundary layer lubrication. *Appl. Phys. Lett.*, 61:2240–2242, 1992.
32. S. J. O'Shea, M. E. Welland, and J. B. Pethica. Atomic force microscopy of local compliance at solid–liquid interface. *Chem. Phys. Lett.*, 223:336–340, 1994.
33. B. Bhushan, T. Miyamoto, and V. N. Koinkar. Microscopic friction between a sharp diamond tip and thin-film magnetic rigid disks by friction force microscopy. *Adv. Info. Storage Syst.*, 6:151–161, 1995.
34. V. N. Koinkar and B. Bhushan. Micro/nanoscale studies of boundary layers of liquid lubricants for magnetic disks. *J. Appl. Phys.*, 79:8071–8075, 1996.
35. B. Bhushan and S. Sundararajan. Micro/nanoscale friction and wear mechanisms of thin films using atomic force and friction force microscopy. *Acta Mater.*, 46:3793–3804, 1998.
36. B. Bhushan and C. Dandavate. Thin-film friction and adhesion studies using atomic force microscopy. *J. Appl. Phys.*, 87:1201–1210, 2000.
37. S. Sundararajan and B. Bhushan. Static friction and surface roughness studies of surface micromachined electrostatic micromotors using an atomic force/friction force microscope. *J. Vac. Sci. Technol. A*, 19:1777–1785, 2001.
38. B. Bhushan and J. Ruan. Atomic-scale friction measurements using friction force microscopy: Part ii – application to magnetic media. *ASME J. Tribol.*, 116:389–396, 1994.
39. T. Stifter, O. Marti, and B. Bhushan. Theoretical investigation of the distance dependence of capillary and van der waals forces in scanning probe microscopy. *Phys. Rev. B*, 62:13667–13673, 2000.
40. J. N. Israelachvili. *Intermolecular and Surface Forces*. Academic, 2nd edition, 1992.
41. S. K. Chilamakuri and B. Bhushan. A comprehensive kinetic meniscus model for prediction of long-term static friction. *J. Appl. Phys.*, 15:4649–4656, 1999.
42. H. Ishigaki, I. Kawaguchi, M. Iwasa, and Y. Toibana. Friction and wear of hot pressed silicon nitride and other ceramics. *ASME J. Tribol.*, 108:514–521, 1986.
43. T. E. Fischer. Tribochemistry. *Annu. Rev. Mater. Sci.*, 18:303–323, 1988.
44. K. Mizuhara and S. M. Hsu. *Tribochemical reaction of oxygen and water on silicon surfaces*, pages 323–328. Elsevier, 1992.
45. S. Danyluk, M. McNallan, and D. S. Park. *Friction and wear of silicon nitride exposed to moisture at high temperatures*, pages 61–79. Dekker, 1994.
46. V. A. Muratov and T. E. Fischer. Tribochemical polishing. *Annu. Rev. Mater. Sci.*, 30:27–51, 2000.
47. H. Yoshizawa, Y. L. Chen, and J. N. Israelachvili. Fundamental mechanisms of interfacial friction i: Relationship between adhesion and friction. *J. Phys. Chem.*, 97:4128–4140, 1993.

48. H. Yoshizawa and J. N. Israelachvili. Fundamental mechanisms of interfacial friction ii: Stick slip friction of spherical and chain molecules. *J. Phys. Chem.*, 97:11300–11313, 1993.
49. K. C. Eapen, S. T. Patton, and J. S. Zabinski. Lubrication of microelectromechanical systems (mems) using bound and mobile phase of fomblin z-dol. *Tibol. Lett.*, 12:35–41, 2002.

Part IV

Applications

20

Micro/Nanotribology and Micro/Nanomechanics of Magnetic Storage Devices

Bharat Bhushan

Summary. A magnetic recording process involves relative motion between a magnetic medium (tape or disk) against a stationary or rotating read/write magnetic head. For ever-increasing, high areal recording density, the linear flux density (number of flux reversals per unit distance) and the track density (number of tracks per unit distance) should be as high as possible. The size of a single bit dimension for current devices is typically less than $1000\,\text{nm}^2$. This dimension places stringent restrictions on the defect size present on the head and medium surfaces.

Reproduced (read-back) magnetic signal amplitude decreases with a decrease in the recording wavelength and/or the track width. The signal loss results from the magnetic coating thickness, read gap length, and head-to-medium spacing (clearance or flying height). It is known that the signal loss as a result of spacing can be reduced exponentially by reducing the separation between the head and the medium. The need for increasingly higher recording densities requires that surfaces be as smooth as possible and the flying height (physical separation or clearance between a head and a medium) be as low as possible. The ultimate objective is to run two surfaces in contact (with practically zero physical separation) if the tribological issues can be resolved. Smooth surfaces in near contact lead to an increase in adhesion, friction, and interface temperatures, and closer flying heights lead to occasional rubbing of high asperities and increased wear. Friction and wear issues are resolved by appropriate selection of interface materials and lubricants, by controlling the dynamics of the head and medium, and the environment. A fundamental understanding of the tribology (friction, wear, and lubrication) of the magnetic head/medium interface, both on macro- and micro/nanoscales, becomes crucial for the continued growth of this more than $60 billion a year magnetic storage industry.

In this chapter, initially, the general operation of drives and the construction and materials used in magnetic head and medium components are described. Then the micro/nanotribological and micro/nanomechanics studies including surface roughness, friction, adhesion, scratching, wear, indentation, and lubrication relevant to magnetic storage devices are presented.

20.1 Introduction

20.1.1 Magnetic Storage Devices

Magnetic storage devices used for storage and retrieval are tape, flexible (floppy) disk and rigid disk drives. These devices are used for audio, video and data-storage applications. Magnetic storage industry is some $60 billion a year industry with $20 billion for audio and video recording (almost all tape drives/media) and $40 billion for data storage. In the data-storage industry, magnetic rigid disk drives/media, tape drives/media, flexible disk drives/media, and optical disk drive/media account for about $25 B, $6 B, $3 B, and $6 B, respectively. Magnetic recording and playback involves the relative motion between a magnetic medium (tape or disk) against a read-write magnetic head. Heads are designed so that they develop a (load-carrying) hydrodynamic air film under steady operating conditions to minimize head-medium contact. However, physical contact between the medium and head occurs during starts and stops, referred to as contact-start-stops (CSS) technology [1–4]. In the modern magnetic storage devices, the flying heights (head-to-medium separation) are on the order of 5 to 20 nm and roughnesses of head and medium surfaces are on the order of 1–2 nm RMS. The need for ever-increasing recording densities requires that surfaces be as smooth as possible and the flying heights be as low as possible. Smooth surfaces lead to an increase in adhesion, friction, and interface temperatures, and closer flying heights lead to occasional rubbing of high asperities and increased wear. High stiction (static friction) and wear are the limiting technology to future of this industry. Head load/unload (L/UL) technology has recently been used as an alternative to CSS technology in rigid disk drives that eliminates stiction and wear failure mode associated with CSS. In an L/UL drive, a lift tab extending from the suspension load beam engages a ramp or cam structure as the actuator moves beyond the outer radius of the disk. The ramp lifts (or unloads) the head stack from the disk surfaces as the actuator moves to the parking position. Starting and stopping the disk only occur with the head in the unloaded state. Several contact or near contact recording devices are at various stages of development. High stiction and wear are the major impediments to the commercialization of the contact recording [1–7].

Magnetic media fall into two categories: particulate media, where magnetic particles (γ-Fe_2O_3, Co–γFe_2O_3, CrO_2, Fe or metal (MP), or barium ferrite) are dispersed in a polymeric matrix and coated onto a polymeric substrate for flexible media (tape and flexible disks); thin-film media, where continuous films of magnetic materials are deposited by vacuum deposition techniques onto a polymer substrate for flexible media or onto a rigid substrate (typically aluminium and more recently glass or glass ceramic) for rigid disks. The most commonly used thin magnetic films for tapes are evaporated Co–Ni (82–18 at.%) or Co–O dual layer. Typical magnetic films for rigid disks are metal films of cobalt-based alloys (such as sputtered Co–Pt–Ni, Co–Ni, Co–Pt–Cr, Co–Cr and Co–NiCr). For high recording densities, trends have been to use thin-film media. Magnetic heads used to date are either conventional thin-film inductive, magnetoresistive (MR) and giant MR (GMR) heads. The air-bearing surfaces (ABS) of tape heads are generally cylindrical in shape. For dual-sided flexible-disk heads, two heads are either spherically contoured and slightly offset (to reduce normal pressure) or are flat and loaded against each other. The rigid-disk heads are supported by a leaf spring (flexure) suspension. The ABS of heads are almost made of Mn–Zn ferrite, Ni–Zn ferrite, Al_2O_3–TiC and calcium titanate. The ABS of some conventional heads are made of plasma sprayed coatings of hard materials such as Al_2O_3–TiO_2 and ZrO_2 [2–4].

20 Micro/Nanotribology and Micro/Nanomechanics of Magnetic Storage Devices

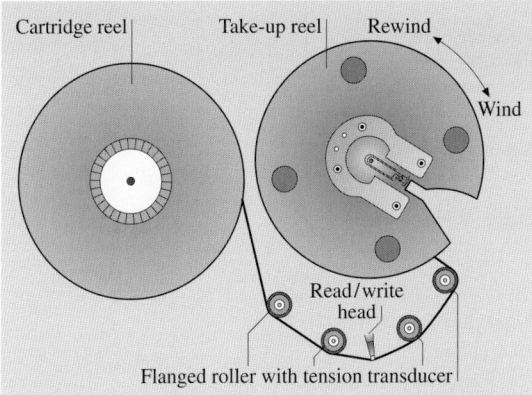

Fig. 20.1. Schematic of tape path in an IBM Linear Tape Open (LTO) tape drive

Figure 20.1 shows the schematic illustrating the tape path with details of tape guides in a data-processing linear tape drive (IBM LTO Gen1) which uses a rectangular tape cartridge. Figure 20.2a shows the sectional views of particulate and thin-film magnetic tapes. Almost exclusively, the base film is made of semicrystalline biaxially-oriented poly (ethylene terephthalate) (or PET) or poly (ethylene 2,6 naphthalate) (or PEN) or Aramid. The particulate coating formulation consists of binder (typically polyester polyurethane), submicron accicular shaped magnetic particles (about 50 nm long with an aspect ratio of about 5), submicron head cleaning agents (typically alumina) and lubricant (typically fatty acid ester). For protection against wear and corrosion and low friction/stiction, the thin-film tape is first coated with a diamondlike carbon (DLC) overcoat deposited by plasma enhanced chemical vapor deposition, topically lubricated with primarily a perfluoropolyether lubricant. Figure 20.2b shows the schematic of an 8-track (along with 2 servo tracks) thin-film read-write head with MR read and inductive write. The head steps up and down to provide 384 total data tracks across the width of the tape. The ABS is made of Al_2O_3–TiC. A tape tension of about 1 N over a 12.7-mm wide tape (normal pressure ≈ 14 kPa) is used during use. The RMS roughnesses of ABS of the heads and tape surfaces typically are 1–1.5 nm and 5–8 nm, respectively.

Figure 20.3 shows the schematic of a data processing rigid disk drive with 21.6-, 27.4-, 48-, 63.5-, 75-, and 95-mm form factor. Nonremovable stack of multiple disks mounted on a ball bearing or hydrodynamic spindle, are rotated by an electric motor at constant angular speed ranging from about 5000 to in excess of 15 000 RPM, dependent upon the disk size. Head slider-suspension assembly (allowing one slider for each disk surface) is actuated by a stepper motor or a voice coil motor using a rotary actuator. Figure 20.4a shows the sectional views of a thin-film rigid disk. The substrate for rigid disks is generally a non heat-treatable aluminium-magnesium alloy 5086, glass or glass ceramic. The protective overcoat commonly used for thin-film disks is sputtered DLC, topically lubricated with perfluoropolyether type of lubricants. Lubricants with polar-end groups are generally used for thin-film disks in order

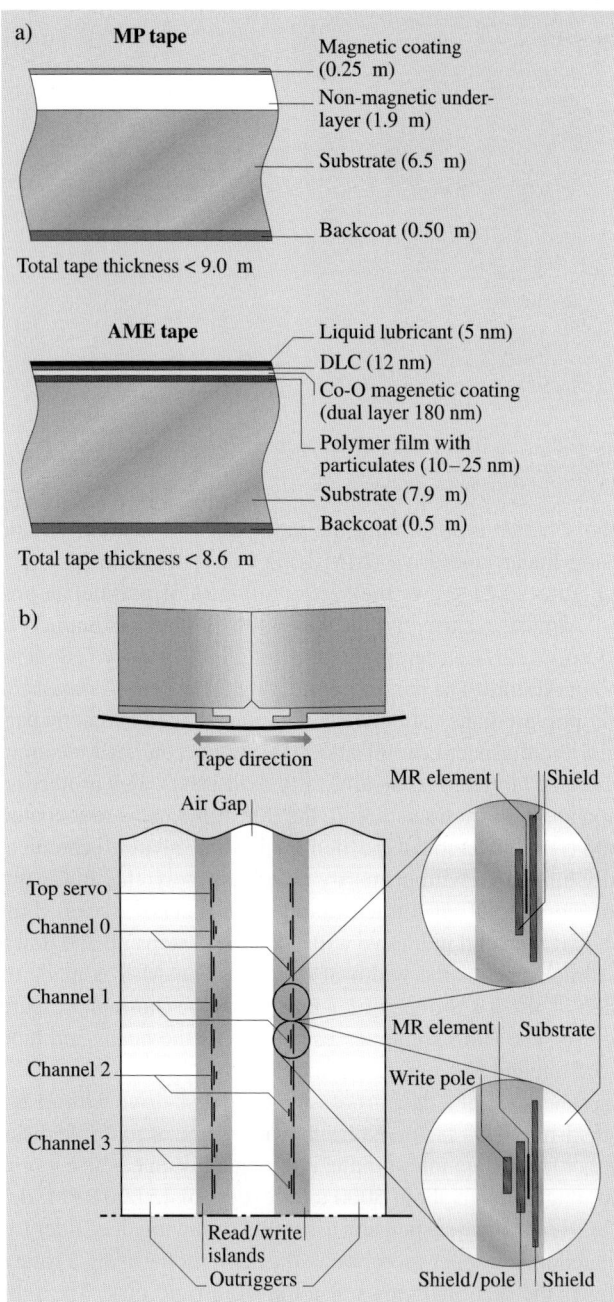

Fig. 20.2. (a) Sectional views of particulate and thin-film magnetic tapes, and (b) schematic of a magnetic thin-film read/write head for an IBM LTO Gen 1 tape drive

Fig. 20.3. Schematic of a data-processing magnetic rigid disk drive

to provide partial chemical bonding to the overcoat surface. The disks used for CSS technology are laser textured in the landing zone. Figure 20.4b shows the schematic of two thin-film head picosliders with a step at the leading edge, and GMR read and inductive write. "Pico" refers to the small sizes of 1.25 mm × 1 mm. These sliders use Al_2O_3-TiC ($[-w\%t]7030$) as the substrate material with multilayered thin-film head structure coated and with about 3.5-nm thick DLC coating to prevent the thin film structure from electrostatic discharge. The seven pads on the padded slider are made of DLC and are about 40 μm in diameter and 50 nm in height. A normal load of about 3 g is applied during use.

20.1.2 Micro/Nanotribology and Micro/Nanomechanics and Their Applications

The micro/nanotribological studies are needed to develop fundamental understanding of interfacial phenomena on a small scale and to study interfacial phenomena in [8–12]. Magnetic storage devices operate under low load and encounter isolated asperity interactions. These use multilayered thin film structure and are generally lubricated with molecularly-thin films. Micro/nanotribological and micro/nanomechanical techniques are ideal to study the friction and wear processes of micro/nanoscale and molecularly thick films. These studies are also valuable in fundamental understanding of interfacial phenomena in macrostructures to provide a bridge between science and engineering. At interfaces of technological applications, contact occurs at multiple asperity contacts. A sharp tip of tip-based microscopes (atomic force/friction force microscopes or AFM/FFM) sliding on a surface simulates a single asperity contact, thus allowing high-resolution measurements of surface interactions at a single asperity contacts. AFMs/FFMs are now commonly used for tribological studies.

In this chapter, we present state-of-the-art of micro/nanotribology and micro/nanomechanics of magnetic storage devices including surface roughness, friction, adhesion, scratching, wear, indentation, and lubrication.

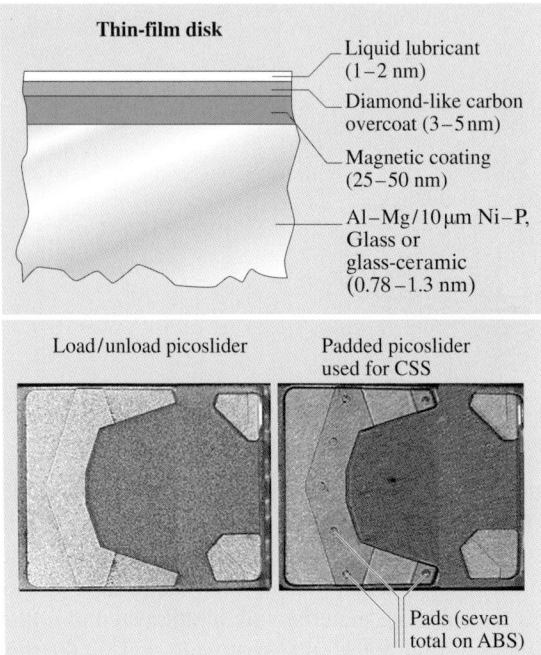

Fig. 20.4. (a) Sectional views of a thin-film magnetic rigid disk, and (b) schematic of two picosliders - load/unload picoslider and padded picoslider used for CSS

20.2 Experimental

20.2.1 Description of AFM/FFM

AFM/FFMs used in the tribological studies have been described in several papers [10, 12, 13]. Briefly, in one of the commercial designs, the sample is mounted on a PZT tube scanner to scan the sample in the x-y-plane and to move the sample in the vertical (z) direction (Fig. 20.5). A sharp tip at the end of a flexible cantilever is brought in contact with the sample and the sample is scanned in a raster pattern (Fig. 20.6). Normal and frictional forces being applied at the tip–sample interface are simultaneously measured using a laser beam deflection technique. Surface roughness is measured either in the contact mode or the so-called tapping mode (intermittent contact mode). For surface roughness and friction measurements, a microfabricated square pyramidal Si_3N_4 tip with a tip radius of about 30 nm attached to a cantilever beam (with a normal beam stiffness of about 0.5 N/m) for contact mode or a square-pyramidal etched single-crystal silicon tip with a rectangular silicon cantilever beam (Fig. 20.7) is generally used at normal loads ranging from 10 to 150 nN. A preferred method of measuring friction and calibration procedures for conversion of voltages corresponding to normal and friction forces to force units, are described by *Bhushan* [10, 12, 13]. The samples are typically scanned over scan areas ranging

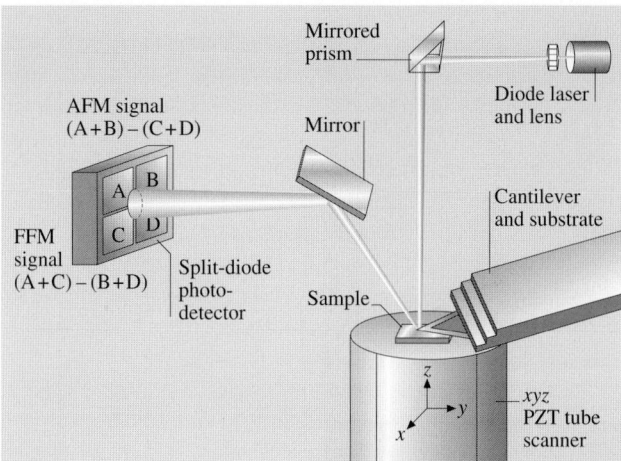

Fig. 20.5. Principles of operation of a commercial small sample AFM/FFM

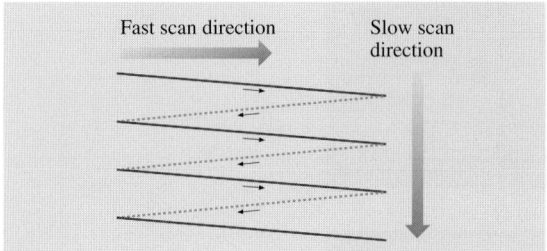

Fig. 20.6. Schematic of triangular pattern trajectory of the AFM tip as the sample is scanned in two dimensions. During imaging, data are recorded only during scans along the solid scan lines

from 50 nm × 50 nm to 10 µm × 10 µm, in a direction orthogonal to the long axis of the cantilever beam [14]. The scan rate is on the order of 1 Hz. For example, for this rate, the sample scanning speed would be 1 µm/s for a 500 nm × 500 nm scan area. Adhesive force measurements are performed in the so-called friction calibration mode. In this technique, the tip is brought in contact with the sample and then pulled away. The force required to pull the tip off the sample is a measure of adhesive force.

In nanoscale wear studies, the sample is initially scanned twice, typically at 10 nN to obtain the surface profile, then scanned twice at a higher load of typically 100 nN to wear and to image the surface simultaneously, and then rescanned twice at 10 nN to obtain the profile of the worn surface. For magnetic media studied by *Bhushan* and *Ruan* [15], no noticeable change in the roughness profiles was observed between the initial two scans at 10 nN and between profiles scanned at 100 nN and the final scans at 10 nN. Therefore any changes in the topography between the initial

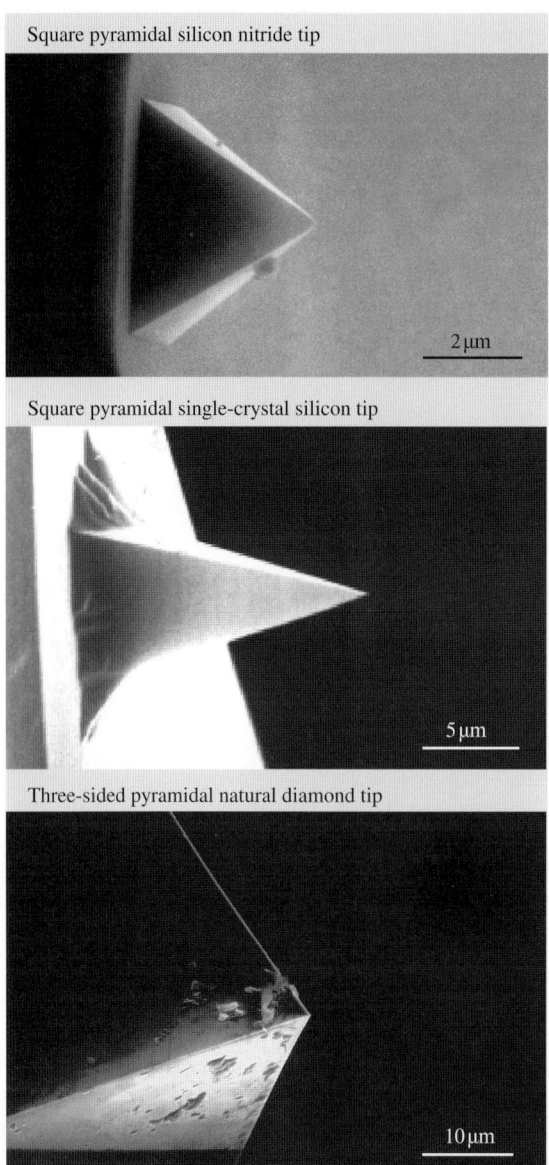

Fig. 20.7. SEM micrographs of a square-pyramidal PECVD Si_3N_4 tip with a triangular cantilever beam (*top*), a square-pyramidal etched single-crystal silicon tip with a rectangular silicon cantilever beam (*middle*), and a three-sided pyramidal natural diamond tip with a square stainless steel cantilever beam (*bottom*)

scans at 10 nN and the scans at 100 nN (or the final scans at 10 nN) are believed to occur as a result of local deformation of the sample surface. In picoindentation studies, the sample is loaded in contact with the tip. During loading, tip deflection (normal force) is measured as a function of vertical position of the sample. For a rigid sample, the tip deflection and the sample traveling distance (when the tip and sample come into contact) are equal. Any decrease in the tip deflection as compared to vertical position of the sample represents indentation. To ensure that the curvature in the tip deflection-sample traveling distance curve does not arise from PZT hysteresis, measurements on several rigid samples, including single-crystal natural diamond (IIa), were made by *Bhushan* and *Ruan* [15]. No curvature was noticed for the case of rigid samples. This suggests that any curvature for other samples should arise from the indentation of the sample.

For microscale scratching, microscale wear and nanoscale indentation hardness measurements, a three-sided pyramidal single-crystal natural diamond tip with an apex angle of 80° and a tip radius of about 100 nm (determined by scanning electron microscopy imaging) (Fig. 20.7) is used at relatively high loads (1 µN–150 µN). The diamond tip is mounted on a stainless steel cantilever beam with normal stiffness of about 25 N/m [16–19]. For scratching and wear studies, the sample is generally scanned in a direction orthogonal to the long axis of the cantilever beam (typically at a rate of 0.5 Hz). For wear studies, typically an area of 2 µm × 2 µm is scanned at various normal loads (ranging from 1 to 100 µN) for selected number of cycles. Scratching can also be performed at ramped loads [20]. For nanoindentation hardness measurements, the scan size is set to zero and then normal load is applied to make the indents. During this procedure, the diamond tip is continuously pressed against the sample surface for about two seconds at various indentation loads. Sample surface is scanned before and after the scratching, wear, or nanoindentation, to obtain the initial and the final surface topography, at a low normal load of about 0.3 µN using the same diamond tip. An area larger than the scratched, worn or indentation region is scanned to observe the scratch, wear scars, or indentation marks. Nanohardness is calculated by dividing the indentation load by the projected residual area of the indents [17]. Nanohardness and Young's modulus of elasticity (stiffness) at shallow depths as low as 5 nm are measured using a depth-sensing capacitance transducer system in an AFM [19].

Indentation experiments provide a single-point measurement of the Young's modulus of elasticity (stiffness), localized surface elasticity as well as phase contrast maps (to obtain viscoelastic properties map) can be obtained using dynamic force microscopy in which an oscillating tip is scanned over the sample surface in contact under steady and oscillating load [21–24]. Recently, a torsional resonance (TR) mode has been introduced [25, 26] which provides higher resolution. Stiffness and phase contrast maps can provide magnetic particle/polymer distributions in magnetic tapes as well as lubricant film thickness distribution.

Boundary lubrication studies are conducted using either Si_3N_4 or diamond tips [27–30]. The coefficient of friction is monitored as a function of sliding cycles.

All measurements are carried out in the ambient atmosphere (22±1 °C, 45±5% RH, and Class 10000).

20.2.2 Test Specimens

Data on various head slider materials and magnetic media are presented in the chapter. Al_2O_3–TiC (70–30 w/o) and polycrystalline and single-crystal (110) Mn–Zn ferrite are commonly used for construction of disk and tape heads. Al_2O_3, a single-phase material, is also selected for comparisons with the performance of Al_2O_3–TiC, a two-phase material. An α-type SiC is also selected which is a candidate slider material because of its high thermal conductivity and attractive machining and friction and wear properties [31]. Single crystal silicon has also been used in some head sliders but its use is discontinued [32].

Two thin-film rigid disks with polished and textured substrates, with and without a bonded perfluoropolyether are selected. These disks are 95 mm in diameter made of Al–Mg alloy substrate (1.3 mm thick) with a 10 µm thick electroless plated Ni–P coating, 75 nm thick ($Co_{79}Pt_{14}Ni_7$) magnetic coating, 20-nm thick amorphous carbon or diamondlike carbon (DLC) coating (microhardness $\approx$ 1 500 kg/mm^2 as measured using a Berkovich indenter), and with or without a top layer of perfluoropolyether lubricant with polar end groups (Z-DOL) coating. The thickness of the lubricant film is about 2 nm. The metal particle (MP) tape is a 12.7 mm wide and 13.2 µm thick (PET base thickness of 9.8 µm, magnetic coating of 2.9 µm with metal magnetic particles and nonmagnetic particles of Al_2O_3 and Cr_2O_3, and back coating of 0.5 µm). The barium ferrite (BaFe) tape is a 12.7 mm wide and 11 µm thick (PET base thickness of 7.3 µm, magnetic coating of 2.5 µm with barium ferrite magnetic particles and nonmagnetic particles of Al_2O_3, and back coating of 1.2 µm). Metal evaporated (ME) tape is a 12.7 mm wide tape with 10 µm thick base, 0.2 µm thick evaporated Co–Ni magnetic film and about 10-nm thick perfluoropolyether lubricant and a backcoat. PET film is a biaxially-oriented, semicrystalline polymer with particulates. Two sizes of nearly spherical particulates are generally used in the construction of PET: submicron ($\approx$ 0.5 µm) particles of typically carbon and larger particles (2–3 µm) of silica.

20.3 Surface Roughness

Solid surfaces, irrespective of the method of formation, contain surface irregularities or deviations from the prescribed geometrical form. When two nominally flat surfaces are placed in contact, surface roughness causes contact to occur at discrete contact points. Deformation occurs at these points, and may be either elastic or plastic, depending on the nominal stress, surface roughness and material properties. The sum of the areas of all the contact points constitutes the real area that would be in contact, and for most materials at normal loads, this will be only a small fraction of the area of contact if the surfaces were perfectly smooth. In general, real area of contact must be minimized to minimize adhesion, friction and wear [12].

Characterizing surface roughness is therefore important for predicting and understanding the tribological properties of solids in contact. Various measurement techniques are used to measure surface roughness. The AFM is used to measure surface roughness on length scales from nanometers to micrometers. A second technique is noncontact optical profiler (NOP) which is a noncontact technique and does not damage the surface. The third technique is stylus profiler (SP) in which a sharp tip is dragged over the sample surface. These techniques differ in lateral resolution. Roughness plots of a glass-ceramic disk measured using an AFM (lateral resolution $\approx$ 15 nm), NOP (lateral resolution $\approx$ 1 μm) and SP (lateral resolution of $\approx$ 0.2 μm) are shown in Fig. 20.8a. Figure 20.8b compares the profiles of the disk obtained with different instruments at a common scale. The figures show that roughness is found at scales ranging from millimeter to nanometer scales. Measured roughness profile is dependent on the lateral and normal resolutions of the measuring instrument [33–37]. Instruments with different lateral resolutions measure features with different scale lengths. It can be concluded that a surface is composed of a large number of length of scales of roughness that are superimposed on each other.

Surface roughness is most commonly characterized by the standard deviation of surface heights which is the square roots of the arithmetic average of squares of the vertical deviation of a surface profile from its mean plane. Due to the multiscale nature of surfaces, it is found that the variances of surface height and its derivatives and other roughness parameters depend strongly on the resolution of the roughness measuring instrument or any other form of filter, hence not unique for a surface [35–38] (Fig. 20.9). Therefore, a rough surface should be characterized in a way such that the structural information of roughness at all scales is retained. It is necessary to quantify the multiscale nature of surface roughness.

An unique property of rough surfaces is that if a surface is repeatedly magnified, increasing details of roughness are observed right down to nanoscale. In addition, the roughness at all magnifications appear quite similar in structure, as qualitatively shown in Fig. 20.10. The statistical self-affinity is due to similarity in appearance of a profile under different magnifications. Such a behavior can be characterized by fractal analysis [35, 39]. The main conclusion from these studies are that a fractal characterization of surface roughness is *scale independent* and provides information of the roughness structure at all length scales that exhibit the fractal behavior.

Structure function and power spectrum of a self-affine fractal surface follow a power law and can be written as (Ganti and Bhushan model)

$$S(\tau) = C\eta^{(2D-3)}\tau^{(4-2D)}, \tag{20.1}$$

$$P(\omega) = \frac{c_1 \eta^{(2D-3)}}{\omega^{(5-2D)}}, \tag{20.2a}$$

and

$$c_1 = \frac{\Gamma(5-2D)\sin[\pi(2-D)]}{2\pi}C. \tag{20.2b}$$

The fractal analysis allows the characterization of surface roughness by two parameters D and C which are instrument-independent and unique for each surface.

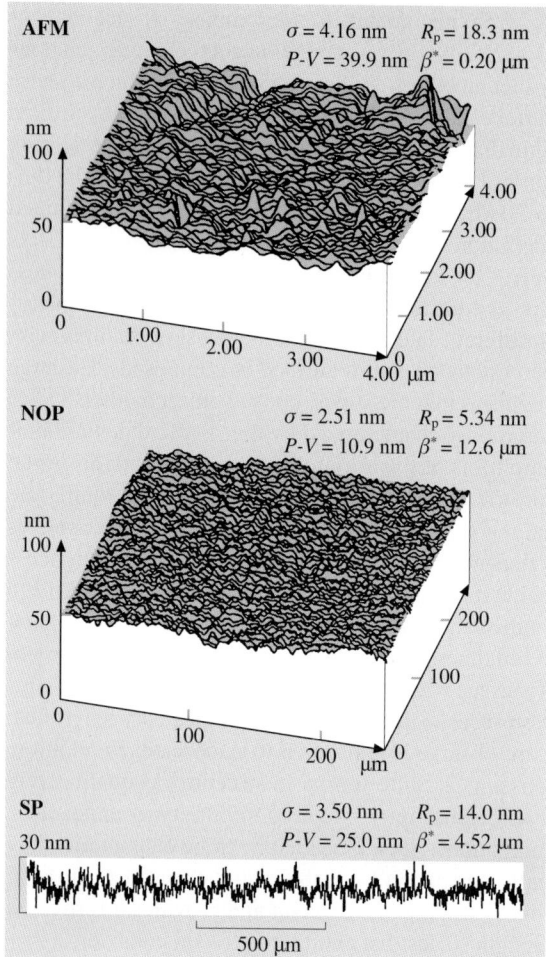

Fig. 20.8. Surface roughness plots of a glass-ceramic disk (**a**) measured using an atomic force microscope (lateral resolution ≈ 15 nm), noncontact optical profiler (NOP) (lateral resolution ≈ 1 μm) and stylus profiler (SP) with a stylus tip of 0.2-μm radius (lateral resolution ≈ 0.2 μm), and (**b**) measured using an AFM (≈ 150 nm), SP (≈ 0.2 μm), and NOP (≈ 1 μm) and plotted on a common scale [36]

D (ranging from 1 to 2 for surface profile) primarily relates to relative power of the frequency contents and C to the amplitude of all frequencies. η is the lateral resolution of the measuring instrument, τ is the size of the increment (distance), and ω is the frequency of the roughness. Note that if $S(\tau)$ or $P(\omega)$ are plotted as a function of τ or ω, respectively, on a log-log plot, then the power law behavior would result into a straight line. The slope of line is related to D and the location of the spectrum along the power axis is related to C.

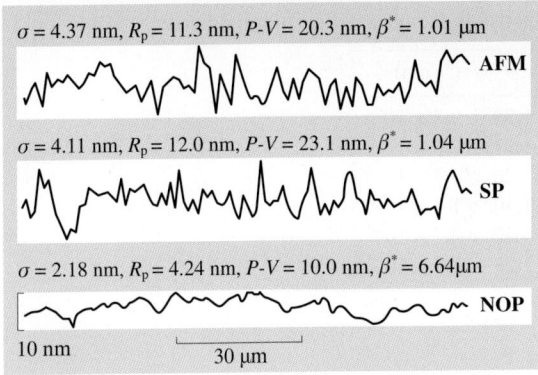

Fig. 20.8. continue

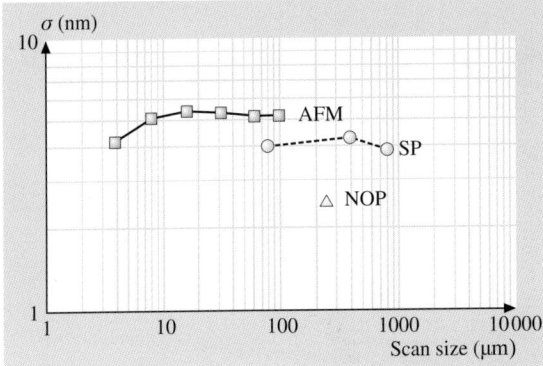

Fig. 20.9. Scale dependence of standard deviation of surface heights for a glass-ceramic disk, measured using atomic force microscope (AFM), stylus profiler (SP), and noncontact optical profiler (NOP)

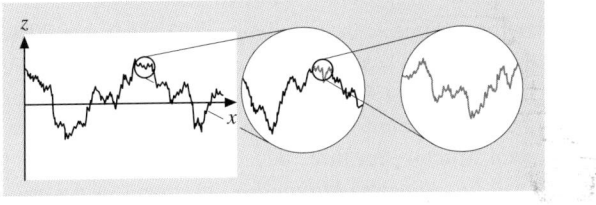

Fig. 20.10. Qualitative description of statistical self-affinity for a surface profile

Figure 20.11 present the structure function of a thin-film rigid disk measured using AFM, non-contact optical profiler (NOP), and stylus profiler (SP). Horizontal shift in the structure functions from one scan to another, arises from the change in the

Table 20.1. Surface roughness parameters for a polished thin-film rigid disk

Scan size (μm×μm)	σ (nm)	D	C (nm)
1 (AFM)	0.7	1.33	9.8×10^{-4}
10 (AFM)	2.1	1.31	7.6×10^{-3}
50 (AFM)	4.8	1.26	1.7×10^{-2}
100 (AFM)	5.6	1.30	1.4×10^{-2}
250 (NOP)	2.4	1.32	2.7×10^{-4}
4000 (NOP)	3.7	1.29	7.9×10^{-5}

AFM – Atomic force microscope,
NOP – Noncontact optical profiler

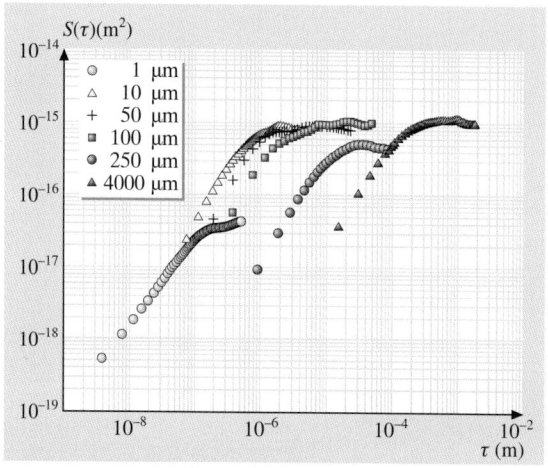

Fig. 20.11. Structure functions for the roughness data measured at various scan sizes using AFM (scan sizes: 1 μm × 1 μm, 10 μm × 10 μm, 50 μm × 50 μm, and 100 μm), NOP (scan size: 250 μm × 250 μm), and SP (scan length: 4000 μm), for a magnetic thin-film rigid disk [35]

lateral resolution. D and C values for various scan lengths are listed in Table 20.1. We note that fractal dimension of the various scans is fairly constant (1.26 to 1.33); however, C increases/decreases monotonically with σ for the AFM data. The error in estimation of η is believed to be responsible for variation in C. These data show that the disk surface follows a fractal structure for three decades of length scales.

Majumdar and *Bhushan* [40] and *Bhushan* and *Majumdar* [41] developed a fractal theory of contact between two rough surfaces. This model has been used to predict whether contacts experience elastic or plastic deformation and to predict the statistical distribution of contact points. For a review of contact models, see *Bhushan* [42, 43] and *Bhushan* and *Peng* [44].

Based on the fractal model of elastic–plastic contact, whether contacts go through elastic or plastic deformation is determined by a critical area which is a function

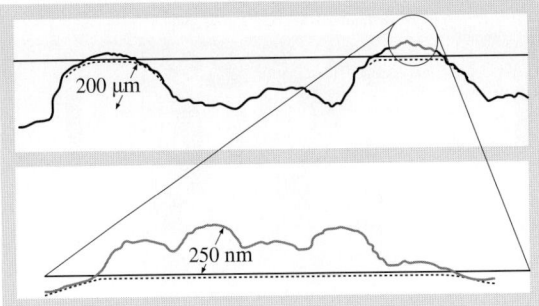

Fig. 20.12. Schematic of local asperity deformation during contact of a rough surface, upper profile measured by an optical profiler and lower profile measured by AFM, typical dimensions are shown for a polished thin-film rigid disk against a flat head slider surface [33]

of D, C, hardness and modulus of elasticity of the mating surfaces. If contact spot is smaller than the critical area, it goes through the plastic deformations and large spots go through elastic deformations. The critical contact area for inception of plastic deformation for a thin-film disk was reported by *Majumdar* and *Bhushan* [40] to be about 10^{-27} m^2, so small that all contact spots can be assumed to be elastic at moderate loads.

The question remains as to how large spots become elastic when they must have initially been plastic spots. The possible explanation is shown in Fig. 20.12. As two surfaces touch, the nanoasperities (detected by AFM type of instruments) first coming into contact have smaller radii of curvature and are therefore plastically deformed instantly, and the contact area increases. When load is increased, nanoasperities in the contact merge, and the load is supported by elastic deformation of the large scale asperities or microasperities (detected by optical profiler type of instruments) [33].

Majumdar and *Bhushan* [40] and *Bhushan* and *Majumdar* [41] have reported relationships for cumulative size distribution of the contact spots, portions of the real area of contact in elastic and plastic deformation modes, and the load-area relationships.

20.4 Friction and Adhesion

Friction and adhesion of magnetic head sliders and magnetic media have been measured by *Bhushan* and *Koinkar* [16, 45–48], *Bhushan* and *Ruan* [15], *Ruan* and *Bhushan* [14], *Bhushan* et al. [27], *Bhushan* [2, 4, 10], *Koinkar* and *Bhushan* [28, 29, 38, 49], *Kulkarni* and *Bhushan* [50], and *Li* and *Bhushan* [51, 52], and *Sundararajan* and *Bhushan* [53].

Koinkar and *Bhushan* [28, 29] and *Poon* and *Bhushan* [36, 37] reported that RMS roughness and friction force increase with an increase in scan size at a given scanning velocity and normal force. Therefore, it is important that while reporting friction force values, scan sizes and scanning velocity should be mentioned. *Bhushan* and

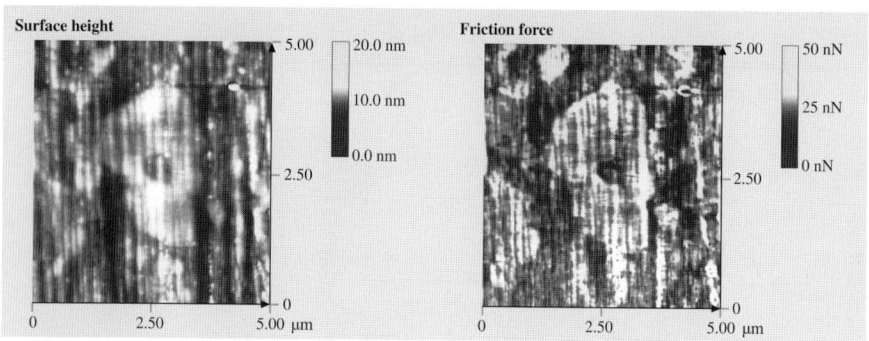

Fig. 20.13. Gray scale surface roughness ($\sigma = 0.97$ nm) and friction force map (mean = 7.0 nN, $\sigma = 0.60$ nN) for Al_2O_3–TiC (70 to 30 wt%) at a normal load of 166 nN

Sundararajan [54] reported that friction and adhesion forces are a function of tip radius and relative humidity (also see [29]). Therefore, relative humidity should be controlled during the experiments. Care also should be taken to ensure that tip radius does not change during the experiments.

20.4.1 Magnetic Head Materials

Al_2O_3–TiC is a commonly used slider material. In order to study the friction characteristics of this two phase material, friction of Al_2O_3–TiC (70–30 wt%) surface was measured. Figure 20.13 shows the surface roughness and friction force profiles [28]. TiC grains have a Knoop hardness of about $2800\,kg/mm^2$ which is higher than that of Al_2O_3 grains of about $2100\,kg/mm^2$. Therefore, TiC grains do not polish as much and result in a slightly higher elevation (about 2–3 nm higher than that of Al_2O_3 grains). Based on friction force measurements, TiC grains exhibit higher friction force than Al_2O_3 grains. The coefficients of friction of TiC and Al_2O_3 grains are 0.034 and 0.026, respectively and the coefficient of friction of Al_2O_3–TiC composite is 0.03. Local variation in friction force also arises from the scratches present on the Al_2O_3–TiC surface. Thus, local friction values of a two phase materials can be measured. *Ruan* and *Bhushan* [55] reported that local variation in the coefficient of friction of cleaved HOP graphite was significant which arises from structural changes occurring during the cleaving process. The cleaved HOPG surface is largely atomically smooth but exhibits line shaped regions in which the coefficient of friction is more than an order of magnitude larger. These measurements suggest that friction measurements can be used for structural mapping of the surfaces.

Surface roughness and coefficient of friction of various head slider materials were measured by *Koinkar* and *Bhushan* [29]. For typical values, see Table 20.2. Macroscale friction values for all samples are higher than microscale friction values as there is less plowing contribution in microscale measurements [10, 12, 13].

Table 20.2. Surface roughness(σ and P–V distance), micro- and macro-scale friction, micro-scratching/wear, and nano- and microhardness data for various samples

Sample	Surface roughness (1 μm × 1 μm)		Coefficient of friction			Scratch depth at 60 μN (nm)	Wear depth at 60 μN (nm)	Hardness	
				Macroscale[b]				Nano at 2 mN (GPa)	Micro at 2 mN (GPa)
	σ (nm)	$P-V^a$ (nm)	Micro-scale	Initial	Final				
Al_2O_3	0.97	9.9	0.03	0.18	0.2–0.6	3.2	3.7	24.8	15.0
Al_2O_3–TiC	0.80	9.1	0.05	0.24	0.2–0.6	2.8	22.0	23.6	20.2
Polycrystalline Mn–Zn ferrite	2.4	20.0	0.04	0.27	0.24–0.4	9.6	83.6	9.6	5.6
Single--crystal (110) Mn–Zn ferrite	1.9	13.7	0.02	0.16	0.18–0.24	9.0	56.0	9.8	5.6
SiC (α-type)	0.91	7.2	0.02	0.29	0.18–0.24	0.4	7.7	26.7	21.8

[a] Peak-to-valley distance
[b] Obtained using silicon nitride ball with 3 mm diameter in a reciprocating mode at a normal load of 10 mN, reciprocating amplitude of 7 mm and average sliding speed of 1 mm/s. Initial coefficient of friction values were obtained at first cycle (0.007 m sliding distance) and final values at a sliding distance of 5 m

20.4.2 Magnetic Media

Bhushan and coworkers measured friction properties of magnetic media including polished and textured thin-film rigid disks, metal particle (MP), barium ferrite (BaFe) and metal evaporated (ME) tapes, and poly(ethylene terephthalate)(PET) tape substrate [2, 4, 10]. For typical values of coefficients of friction of thin-film rigid disks and MP, BaFe and ME tapes, PET tape substrate (Table 20.3). In the case of magnetic disks, similar coefficients of friction are observed for both lubricated and unlubricated disks, indicating that most of the lubricant (though partially thermally bonded) is squeezed out from between the rubbing surfaces at high interface pressures, consistent with liquids being poor boundary lubricant [13]. Coefficient of friction values on a microscale are much lower than those on the macroscale. When measured for the small contact areas and very low loads used in microscale studies, indentation hardness and modulus of elasticity are higher than at the macroscale (data to be presented later). This reduces the real area of contact and the degree of wear. In addition, the small apparent areas of contact reduces the number of particles trapped at the interface, and thus minimize the "plowing" contribution to the friction force [8, 14, 18].

Local variations in the microscale friction of rough surfaces can be significant. Figure 20.14 shows the surface height map, the slopes of surface map taken along the sliding direction, and friction force map for a thin-film disk [8, 15, 18, 28, 38, 45–47, 56]. We note that there is no resemblance between the coefficient of friction profiles and the corresponding roughness maps, e.g., high or low points on the friction profile do not correspond to high or low points on the roughness profiles. By comparing

Table 20.3. Surface roughness (σ), microscale and macro-scale friction, and nanohardness data of thin-film magnetic rigid disk, magnetic tape and magnetic tape substrate (PET) samples

Sample	σ (nm)			Coefficient of microscale friction		Coefficient of macro-scale friction		Nano-hardness (GPa)/ Normal load (μN)
	NOP 250 μm × 250 μm[a]	AFM 1 μm × 1 μm[a]	AFM 10 μm × 10 μm[a]	1 μm × 1 μm[a]	10 μm × 10 μm[a]	Mn–Zn ferrite	Al$_2$O$_3$–TiC	
Polished, unlubricated disk	2.2	3.3	4.5	0.05	0.06	–	0.26	21/100
Polished, lubricated disk	2.3	2.3	4.1	0.04	0.05	–	0.19	–
Textured, lubricated disk	4.6	5.4	8.7	0.04	0.05	–	0.16	–
Metal-particle tape	6.0	5.1	12.5	0.08	0.06	0.19	–	0.30/50
Barium-ferrite tape	12.3	7.0	7.9	0.07	0.03	0.18	–	0.25/25
Metal-evaporated tape	9.3	4.7	5.1	0.05	0.03	0.18	–	0.7 to 4.3/75
PET tape substrate	33	5.8	7.0	0.05	0.04	0.55	–	0.3/20 and 1.4/20[b]

[a] Scan area; NOP – Noncontact optical profiler; AFM – Atomic force microscope
[b] Numbers are for polymer and particulate regions, respectively

the slope and friction profiles, we observe a strong correlation between the two. (For a clearer correlation, see gray-scale plots of slope and friction profiles for FFM tip sliding in either directions, in Fig. 20.15 to be presented in the next paragraph). We have shown that this correlation holds for various magnetic tapes, silicon, diamond, and other materials. This correlation can be explained by a "ratchet" mechanism; based on this mechanism, the local friction is a function of the local slope of the sample surface [10, 12, 13]. The friction is high at the leading edge of asperities and low at the trailing edge. In addition to the slope effect, the collision of tip encountering an asperity with a positive slope produces additional torsion of the cantilever beam leading to higher measured friction force. When encountering an asperity with a negative slope, however, there is no collision effect and hence no effect on friction. The ratchet mechanism and the collision effects thus explain the correlation between the slopes of the roughness maps and friction maps observed in Fig. 20.14.

To study the directionality effect on the friction, gray scale plots of coefficients of local friction of the thin-film disk as the FFM tip is scanned in either direction are shown in Fig. 20.15. Corresponding gray scale plots of slope of roughness maps are also shown in Fig. 20.15. The left hand set of the figures corresponds to the tip sliding from the left towards right (trace direction), and the middle set corresponds to the tip sliding from the right towards left (retrace direction). It is important to take

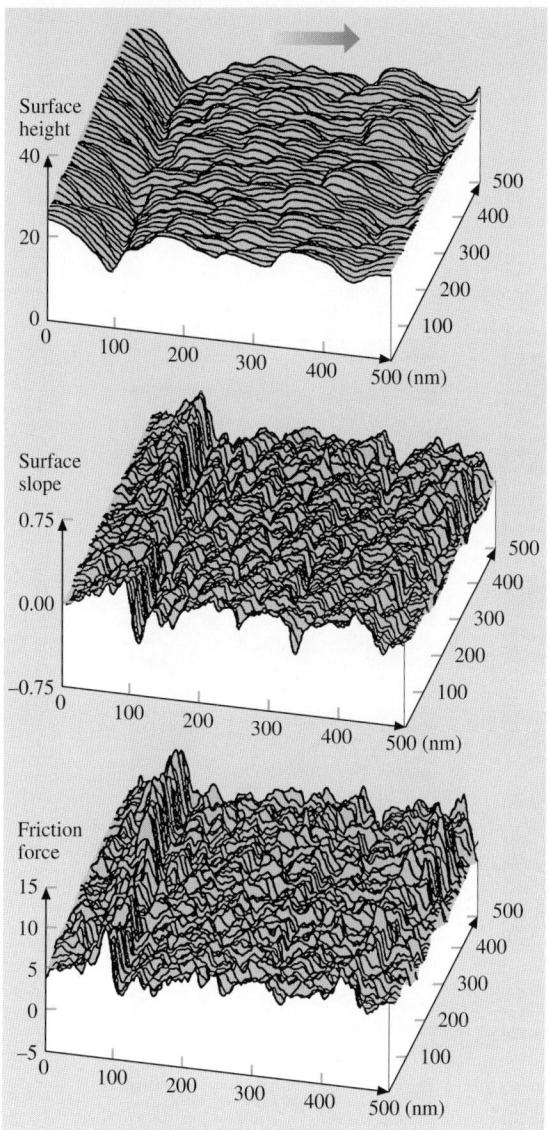

Fig. 20.14. (a) Surface height map ($\sigma = 4.4$ nm), (b) slope of the roughness profiles taken in the sample sliding direction (the *horizontal axis*) (mean = 0.023, $\sigma = 0.18$), and (c) friction force map (mean = 6.2 nN, $\sigma = 2.1$ nN) for a textured thin-film rigid disk at a normal load of 160 nN [18]

into account the sign change of surface slope and friction force which occur in the trace and retrace directions. In order to facilitate comparison of directionality effect on friction, the last set of the figures in the right hand column show the data with sign

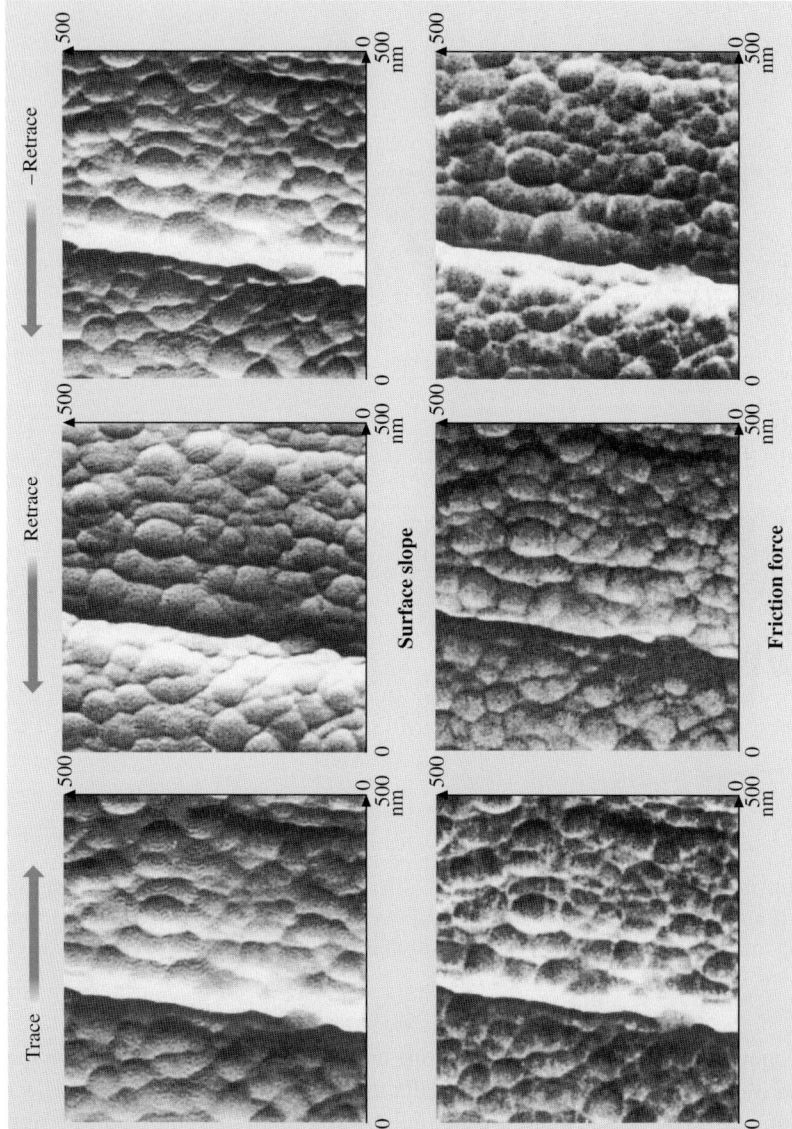

Fig. 20.15. Gray-scale plots of the slope of the surface roughness and the friction force maps for a textured thin-film rigid disk with FFM tip sliding in different directions. *Higher points* are shown by *lighter color*

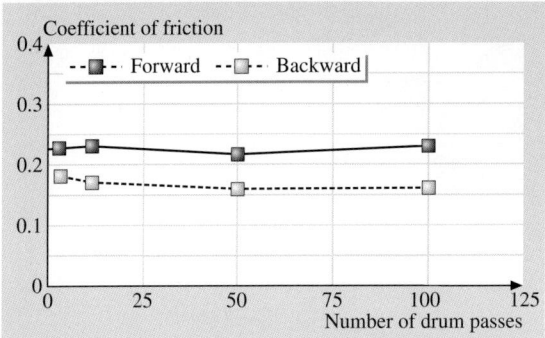

Fig. 20.16. Coefficient of macro-scale of friction as a function of sliding cycles for a metal-particle (MP) tape sliding over an aluminium drum in a reciprocating mode in both directions. Normal load = 0.5 N over 12.7 mm wide tape, sliding speed = 60 mm/s [7]

of surface slope and friction data in the retrace direction reversed. Now we compare trace and -retrace data. It is clear that the friction experienced by the tip is dependent upon the scanning direction because of surface topography.

The directionality effect in friction on a macroscale is generally averaged out over a large number of contacts. It has been observed in some magnetic tapes. In a macro-scale test, a 12.7 mm wide MP tape was wrapped over an aluminium drum and slid in a reciprocating motion with a normal load of 0.5 N and a sliding speed of about 60 mm/s. Coefficient of friction as a function of sliding distance in either direction is shown in Fig. 20.16. We note that coefficient of friction on a macro-scale for this tape is different in different directions.

20.5 Scratching and Wear

20.5.1 Nanoscale Wear

Bhushan and *Ruan* [15] conducted nanoscale wear tests on MP tapes at a normal load of 100 nN. Figure 20.17 shows the topography of the MP tape obtained at two different loads. For a given normal load, measurements were made twice. There was no discernible difference between consecutive measurements for a given normal load. However, as the load increased from 10 to 100 nN, topographical changes were observed; material (indicated by an arrow) was pushed toward the right side in the sliding direction of the AFM tip relative to the sample. The material movement is believed to occur as a result of plastic deformation of the tape surface. Similar behavior was observed on all tapes studied. Magnetic tape coating is made of magnetic particles and polymeric binder. Any movement of the coating material can eventually lead to loose debris. Debris formation is an undesirable situation as it may contaminate the head which may increase friction and/or wear between the head and tape, in addition to the deterioration of the tape itself. With disks, they did not notice any deformation under a 100 nN normal load.

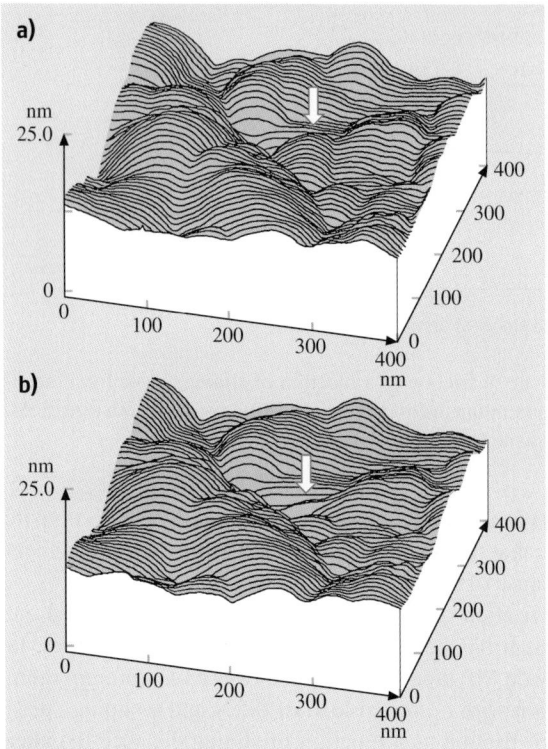

Fig. 20.17. Surface roughness maps of a metal-particle (MP) tape at applied normal load of (**a**) 10 nN and (**b**) 100 nN. Location of the change in surface topography as a result of nanowear is indicated by *arrows* [15]

20.5.2 Microscale Scratching

Microscratches have been made on various potential head slider materials (Al_2O_3, Al_2O_3–TiC, Mn–Zn ferrite and SiC), and various magnetic media (unlubricated polished thin-film disk, MP, BaFe, ME tapes, and PET substrates) and virgin, treated and coated Si(111) wafers at various loads [16, 18, 28, 45–49, 53, 57, 58]. As mentioned earlier, the scratches are made using a diamond tip.

Magnetic Head Materials

Scratch depths as a function of load and representative scratch profiles with corresponding 2-D gray scale plots at various loads after a single pass (unidirectional scratching) for Al_2O_3, Al_2O_3–TiC, polycrystalline and single-crystal Mn–Zn ferrite and SiC are shown in Figs. 20.18 and 20.19, respectively. Variation in the scratch depth along the scratch is about ±15%. The Al_2O_3 surface could be scratched at a normal load of 40 µN. The surface topography of polycrystalline Al_2O_3 shows the presence of porous holes on the surface. The 2-D gray scale plot of scratched Al_2O_3

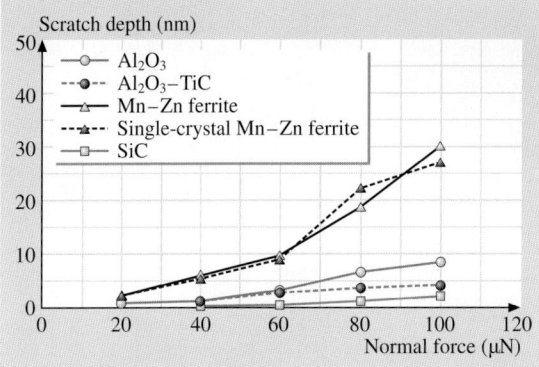

Fig. 20.18. Scratch depth as a function of normal load after one unidirectional cycle for Al_2O_3, Al_2O_3–TiC, polycrystalline Mn–Zn ferrite, single-crystal Mn–Zn ferrite and SiC [28]

surface shows one porous hole between scratches made at normal loads of 40 μN and 60 μN. Regions with defects or porous holes present, exhibit lower scratch resistance (see region marked by the arrow on 2-D gray scale plot of Al_2O_3). The Al_2O_3–TiC surface could be scratched at a normal load of 20 μN. The scratch resistance for TiC grains is higher than that of Al_2O_3 grains. The scratches generated at normal loads of 80 μN and 100 μN show that scratch depth of Al_2O_3 grains is higher than that of TiC grains (see corresponding gray scale plot for Al_2O_3–TiC). Polycrystalline and single-crystal Mn–Zn ferrite could be scratched at a normal load of 20 μN. The scratch width is much larger for the ferrite specimens as compared with other specimens. For SiC, there is no measurable scratch observed at a normal load of 20 μN. At higher normal loads, very shallow scratches are produced. Table 20.2 presents average scratch depth at 60 μN normal load for all specimens. SiC has the highest scratch resistance followed by Al_2O_3–TiC, Al_2O_3 and polycrystalline and single-crystal Mn–Zn ferrite. Polycrystalline and single-crystal Mn–Zn ferrite specimens exhibit comparable scratch resistance.

Magnetic Media

Scratch depths as a function of load and scratch profiles at various loads after ten scratch cycles for unlubricated, polished disk and MP tape are shown in Figs. 20.20 and 20.21. We note that scratch depth increases with an increase in the normal load. Tape could be scratched at about 100 nN. With disk, gentle scratch marks under 10 μN load were barely visible. It is possible that material removal did occur at lower load on an atomic scale which was not observable with a scan size of 5 μm square. For disk, scratch depth at 40 μN is less than 10 nm deep. The scratch depth increased slightly at the load of 50 μN. Once the load is increased in excess of 60 μN, the scratch depth increased rapidly. These data suggest that the carbon coating on the disk surface is much harder to scratch than the underlying thin-film magnetic film. This is expected since the carbon coating is harder than the magnetic material used in the construction of the disks.

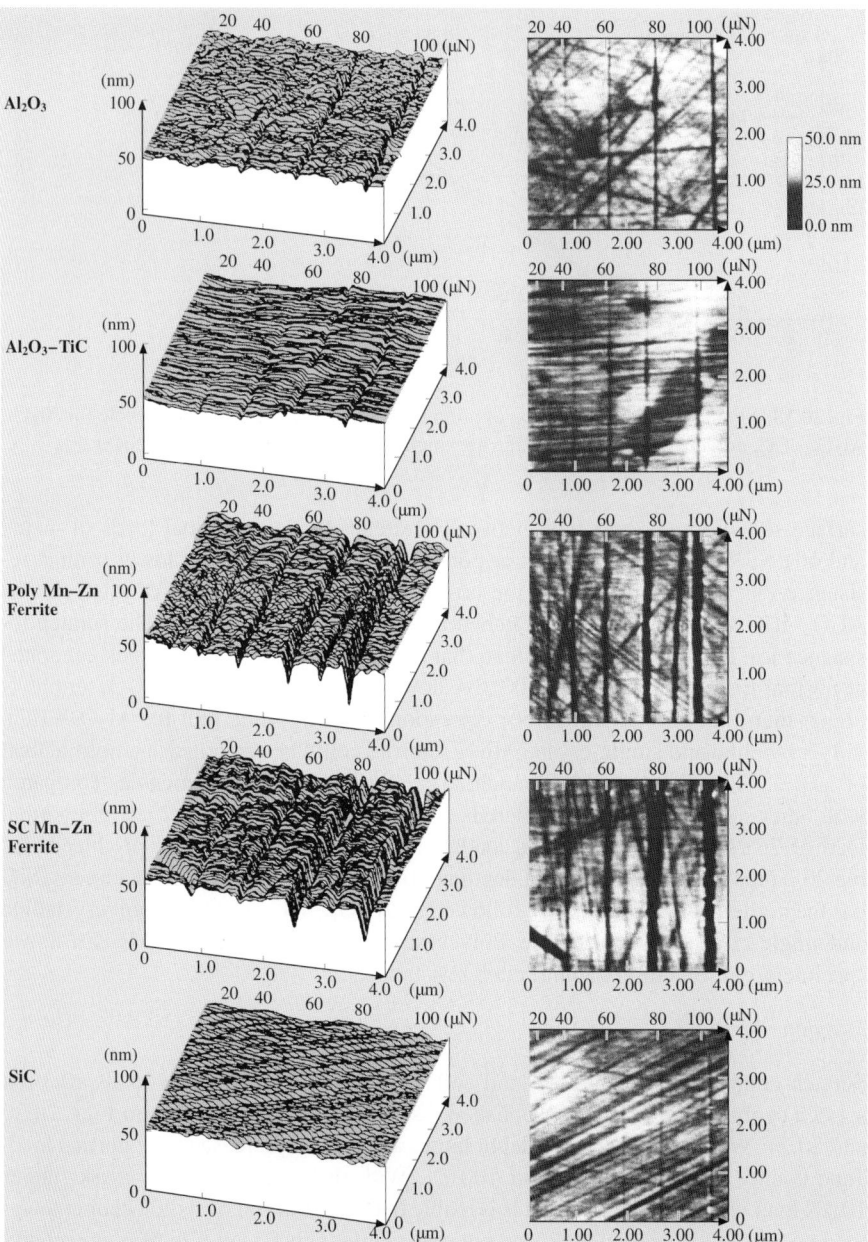

Fig. 20.19. Surface profiles (*left column*) and 2-D gray scale plots (*right column*) of scratched Al_2O_3, Al_2O_3–TiC, polycrystalline Mn–Zn ferrite, single-crystal Mn–Zn ferrite, and SiC surfaces. Normal loads used for scratching for one unidirectional cycle are listed in the figure [28]

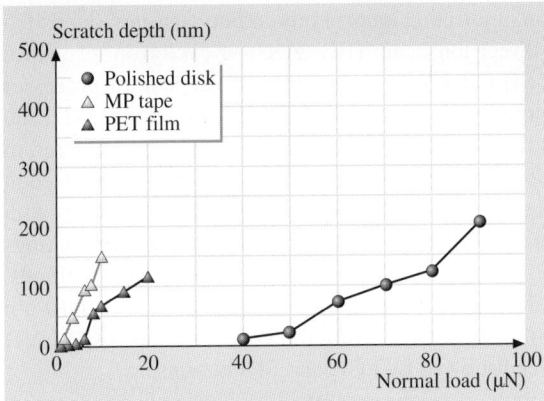

Fig. 20.20. Scratch depth as a function of normal load after ten scratch cycles for unlubricated polished thin-film rigid disk, MP tape and PET film [18, 45, 47]

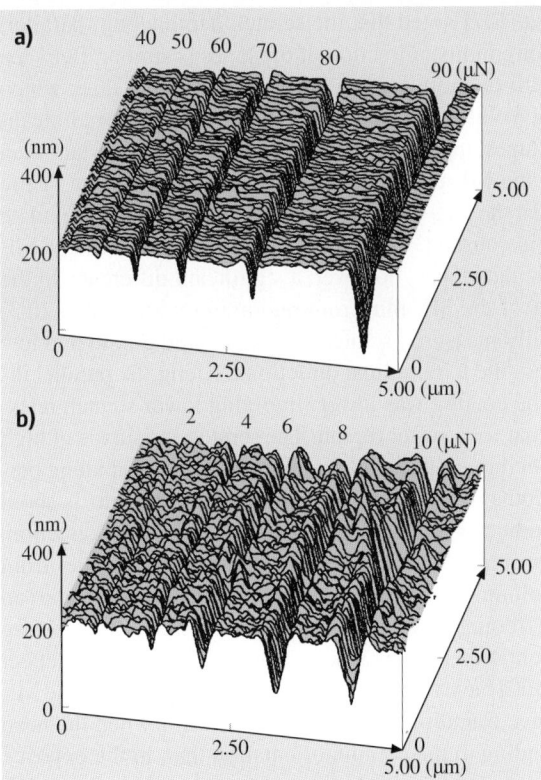

Fig. 20.21. Surface profiles for scratched (**a**) unlubricated polished thin-film rigid disk and (**b**) MP tape. Normal loads used for scratching for ten cycles are listed in the figure [18]

Microscratch characterization of ultrathin amorphous carbon coatings, deposited by filtered cathodic arc (FCA) direct ion beam (IB), electron cyclotron resonance plasma chemical vapor deposition (ECR-CVP), and sputter (SP) deposition processes have been conducted using a nanoindenter and an AFM [20, 48, 49, 51–53, 59]. Data on various coatings of different thicknesses using a Berkovich tip are compared in Fig. 20.22. Critical loads for various coatings and silicon substrate are summarized in Fig. 20.23. It is clear that, for all deposition methods, the critical load increases with increasing coating thickness due to better load-carrying capacity of thicker coatings as compared to the thinner ones. In comparison of the different deposition methods, ECR-CVD and FCA coatings show superior scratch resistance at 20- and 10-nm thicknesses compared to SP coatings. As the coating thickness reduces, ECR-CVD exhibits the best scratch resistance followed by FCA and SP coatings.

Since tapes scratch readily, for comparisons in scratch resistance of various tapes, *Bhushan* and *Koinkar* [47] made scratches on selected three tapes only with one cycle. Figure 20.24 presents the scratch depths as a function of normal load after one cycle for three tapes – MP, BaFe and ME tapes. For the MP and BaFe particulate tapes, *Bhushan* and *Koinkar* [47] noted that the scratch depth along (parallel) and across (perpendicular) the longitudinal direction of the tapes is similar. Between the two tapes, MP tape appears to be more scratch resistant than BaFe tape, which depends on the binder, pigment volume concentration (PVC) and the head cleaning agent (HCA) contents. ME tapes appear to be much more scratch resistant than the particulate tapes. However, the ME tape breaks up catastrophically in a brittle mode at a normal load higher than the 50 µN (Fig. 20.25), as compared to particulate tapes in which the scratch rate is constant. They reported that the hardness of ME tapes is higher than that of particulate tapes, however, a significant difference in the nanoindentation hardness values of the ME film from region to region (Table 20.3) was observed. They systematically measured scratch resistance in the high and low hardness regions along and across the longitudinal directions. Along the parallel direction, load required to crack the coating was lower (implying lower scratch resistance) for a harder region, than that for a softer region. The scratch resistance of high hardness region along the parallel direction is slightly poorer than that for along perpendicular direction. Scratch widths in both low and high hardness regions is about half ($\approx 2\,\mu\mathrm{m}$) than that in perpendicular direction ($\approx 1\,\mu\mathrm{m}$). In the parallel direction, the material is removed in the form of chips and lateral cracking also emanates from the wear zone. ME films have columnar structure with the columns lined up with an oblique angle of on the order of about 35° with respect to the normal to the coating surface [3, 60]. The column orientation may be responsible for directionality effect on the scratch resistance. *Hibst* [60] have reported the directionality effect in the ME tape-head wear studies. They have found that the wear rate is lower when the head moves in the direction corresponding to the column orientation than in the opposite direction.

PET films could be scratched at loads of as low as about $2\,\mu\mathrm{N}$ (Fig. 20.26). Figure 20.26a shows scratch marks made at various loads. Scratch depth along the scratch does not appear to be uniform. This may occur because of variations in the mechanical properties of the film. *Bhushan* and *Koinkar* [45] also conducted scratch

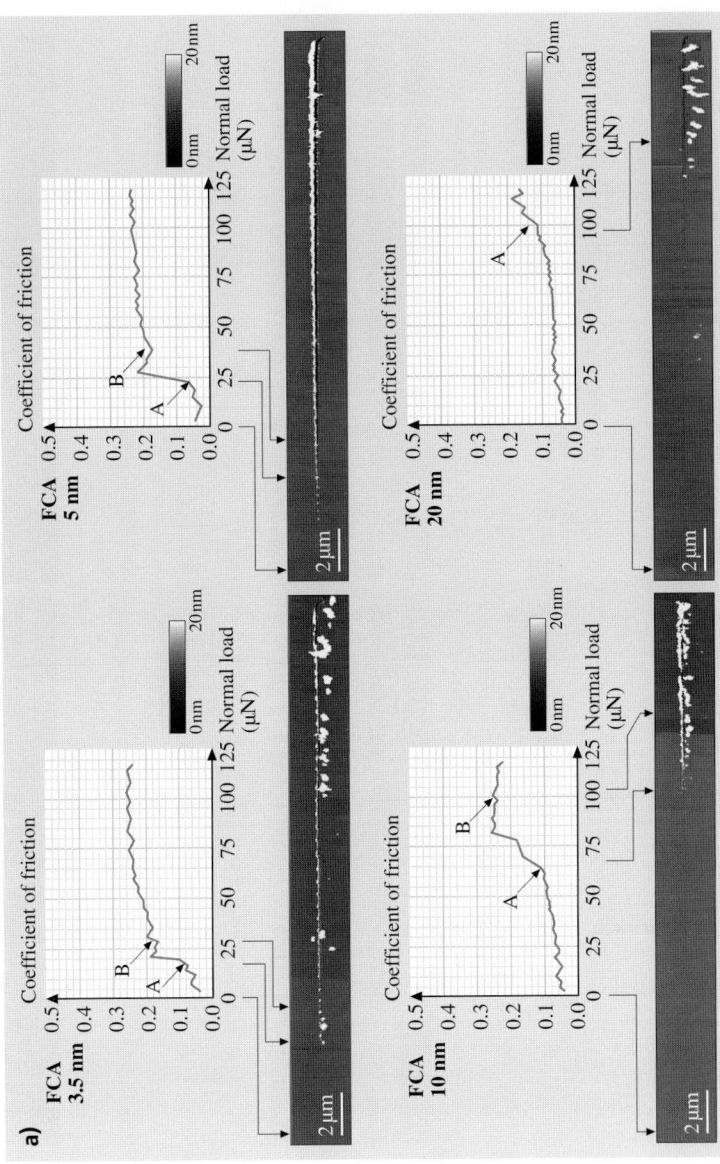

Fig. 20.22. Coefficient of Friction profiles during scratch as a function of normal load and corresponding AFM surface height images for (**a**) FCA, (**b**) ECR-CVD, and (**c**) SP-coatings [20]

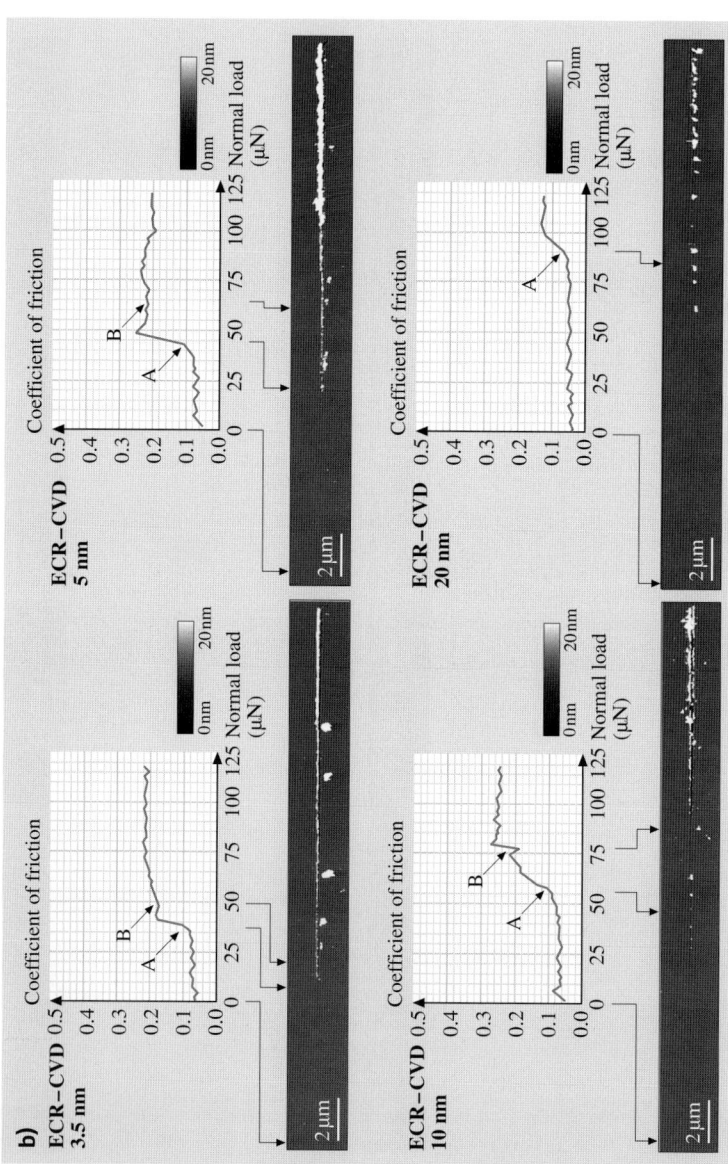

Fig. 20.22. continue

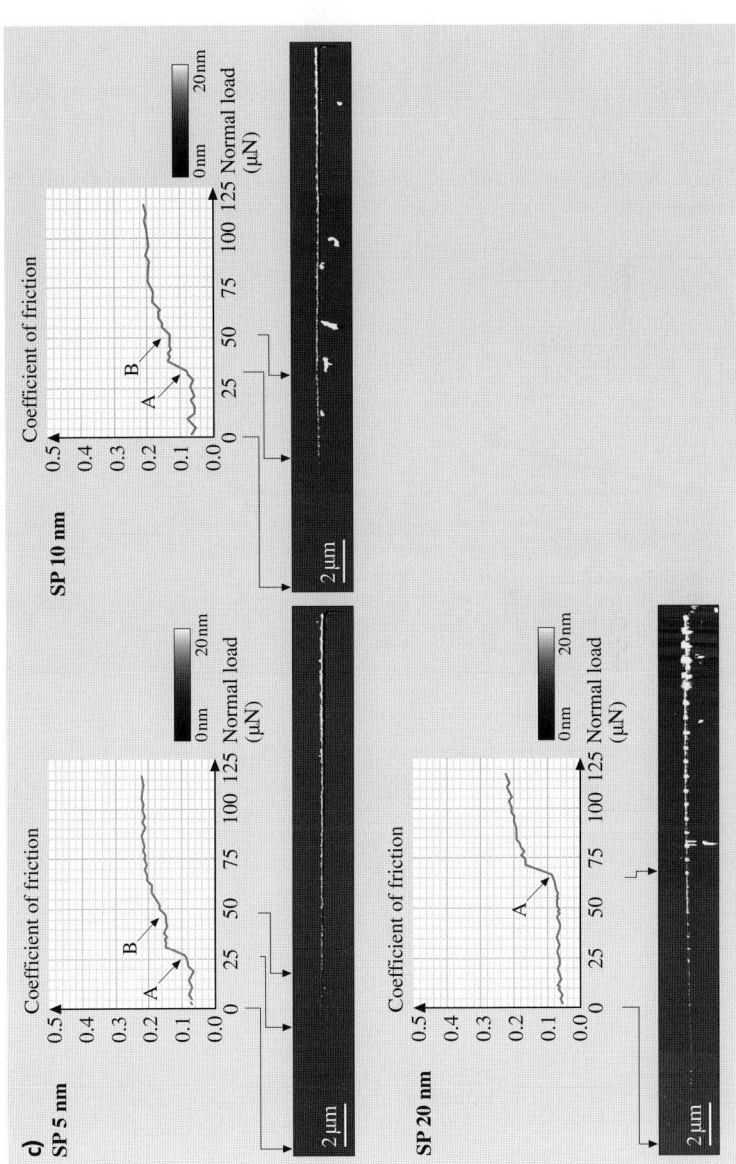

Fig. 20.22. continue

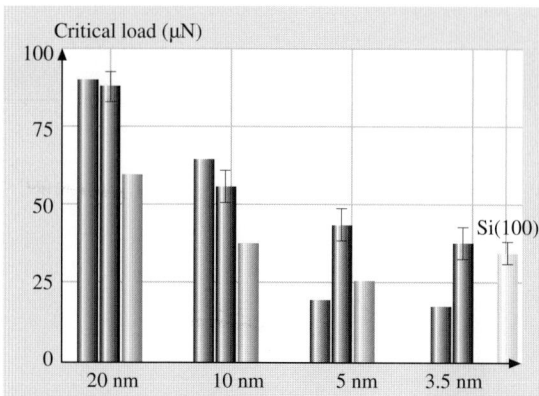

Fig. 20.23. Summary of critical loads estimated from the coefficient of friction profiles and AFM images [20]

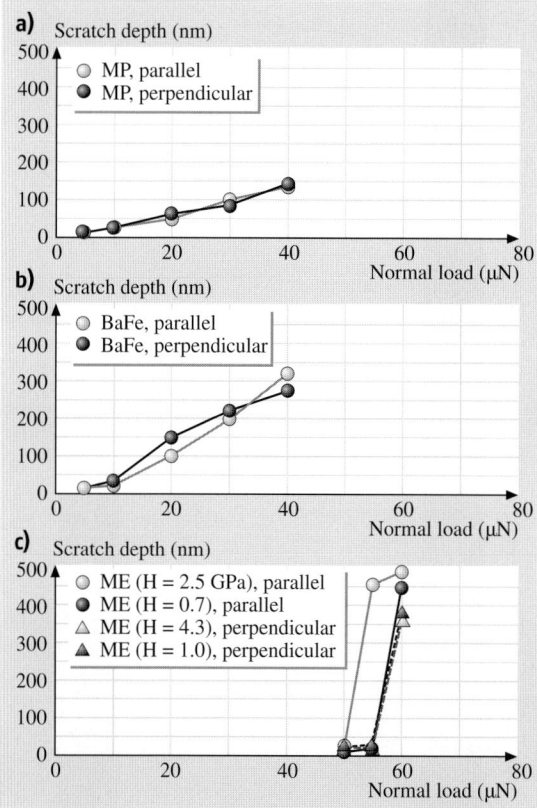

Fig. 20.24. Scratch depth as a function of normal load after one scratch cycle for (**a**) MP, (**b**) BaFe, and (**c**) ME tapes along parallel and perpendicular directions with respect to the longitudinal axis of the tape [47]

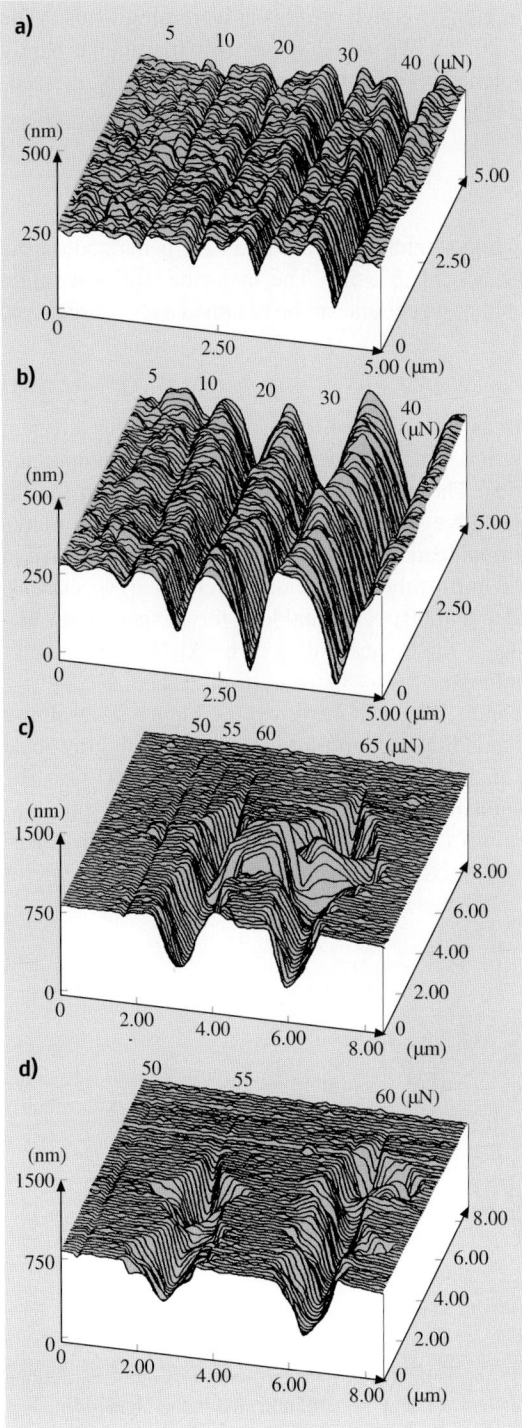

Fig. 20.25. Surface maps for scratched (**a**) MP, (**b**) BaFe, (**c**) ME (H = 0.7 GPa), and (**d**) ME (H= 2.5 GPa) tapes along parallel direction. Normal loads used for scratching for one cycle are listed in the figure [47]

studies in the selected particulate regions. Scratch profiles at increasing loads in the particulate region are shown in Fig. 20.26b. We note that the bump (particle) is barely scratched at 5 µN and it can be scratched readily at higher loads. At 20 µN, it essentially disappears.

20.5.3 Microscale Wear

By scanning the sample (in 2D) while scratching, wear scars are generated on the sample surface [16, 18, 28, 45–49, 53, 54, 57, 58]. The major benefit of a single cycle wear test over a scratch test is that wear data can be obtained over a large area.

Magnetic Head Materials

Figure 20.27 shows the wear depth as a function of load for one cycle for different slider materials. Variation in the wear depth in the wear mark is dependent upon the material. It is generally within ±5%. The mean wear depth increases with the increase in normal load. The representative surface profiles showing the wear marks (central 2 µm×2 µm region) at a normal load of 60 µN for all specimens are shown in Fig. 20.28. The material is removed uniformly in the wear region for all specimens. Table 20.2 presents average wear depth at 60 µN normal load for all specimens. Microwear resistance of SiC and Al_2O_3 is the highest followed by Al_2O_3–TiC, single-crystal and polycrystalline Mn–Zn ferrite.

Next, wear experiments were conducted for multiple cycles. Figure 20.29 shows the 2-D gray scale plots and corresponding section plot (on tope of each gray scale plot), taken at a location shown by an arrow for Al_2O_3 (left column) and Al_2O_3–TiC (right column) specimen obtained at a normal load of 20 µN and at a different number of scan cycles. The central regions (2 µm × 2 µm) show the wear mark generated after a different number of cycles. Note the difference in the vertical scale of

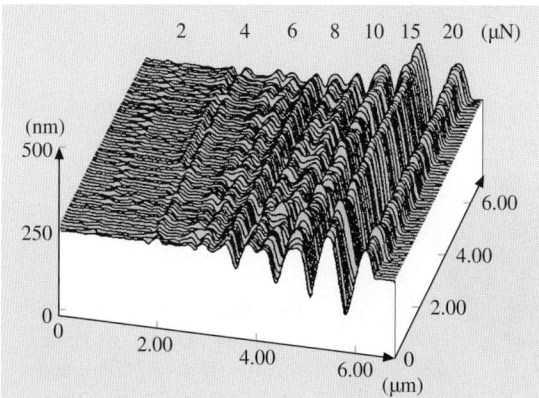

Fig. 20.26. Surface profiles for scratched PET film (**a**) polymer region, (**b**) ceramic particulate region. The loads used for various scratches at ten cycles are indicated in the plots [45]

20 Micro/Nanotribology and Micro/Nanomechanics of Magnetic Storage Devices 997

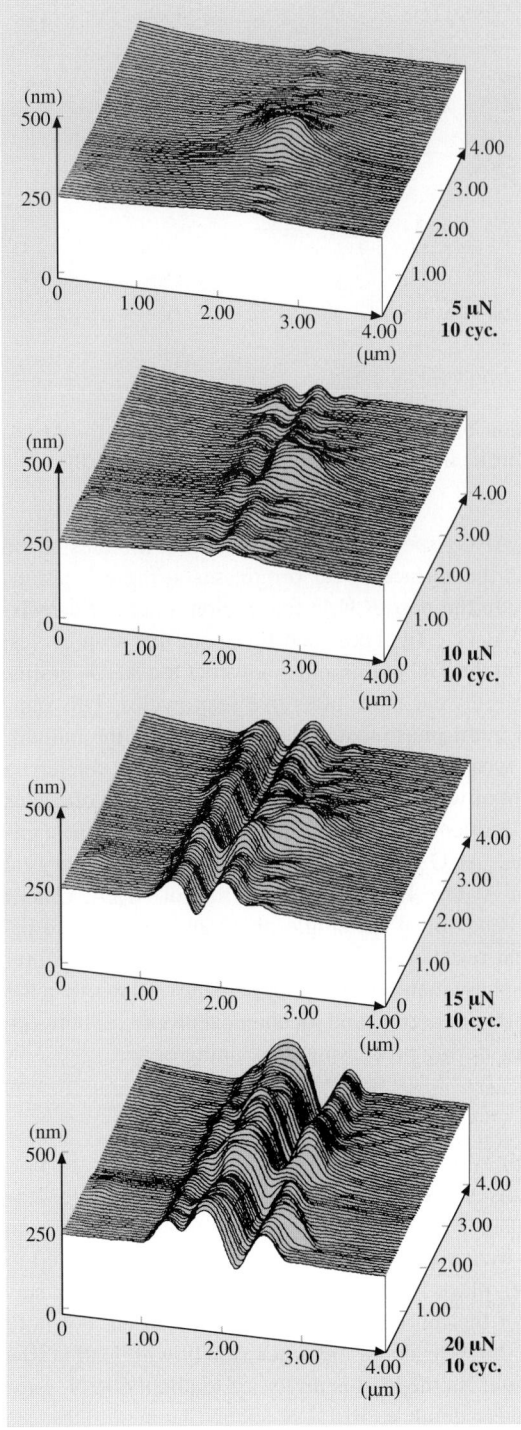

Fig. 20.26. continue

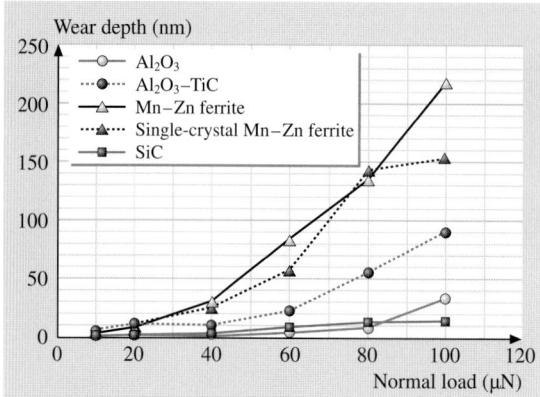

Fig. 20.27. Wear depth as a function of normal load after one scan cycle for Al_2O_3, Al_2O_3–TiC, polycrystalline Mn–Zn ferrite, single-crystal Mn–Zn ferrite and SiC [28]

gray scale and section plots. The Al_2O_3 specimen shows that wear initiates at the porous holes or defects present on the surface. Wear progresses at these locations as a function of number of cycles. In the porous hole free region, microwear resistance is higher. In the case of the Al_2O_3–TiC specimen for about five scan cycles, the microwear resistance is higher at the TiC grains and is lower at the Al_2O_3 grains. The TiC grains are removed from the wear mark after five scan cycles. This indicates that microwear resistance of multi-phase materials depends upon the individual grain properties. Evolution of wear is uniform within the wear mark for ferrite specimens. Figure 20.30 shows plot of wear depth as a function of number of cycles at a normal load of 20 μN for all specimens. The Al_2O_3 specimen reveals highest microwear resistance followed by SiC, Al_2O_3–TiC, polycrystalline and single-crystal Mn–Zn ferrite. Wear resistance of Al_2O_3–TiC is inferior to that of Al_2O_3. Chu et al. [61] studied friction and wear behavior of the single-phase and multi-phase ceramic materials and found that wear resistance of multi-phase materials was poorer than single-phase materials. Multi-phase materials have more material flaws than the single-phase material. The differences in thermal and mechanical properties between the two phases may lead to cracking during processing, machining or use.

Magnetic Media

Figure 20.31 shows the wear depth as a function of load for one cycle for the polished, unlubricated and lubricated disks [18]. Figure 20.32 shows profiles of the wear scars generated on unlubricated disk. The normal force for the imaging was about 0.5 μN and the loads used for the wear were 20, 50, 80 and 100 μN as indicated in the figure. We note that wear takes place relatively uniformly across the disk surface and essentially independent of the lubrication for the disks studied. For both lubricated and unlubricated disks, the wear depth increases slowly with load at low loads with almost the same wear rate. As the load is increased to about 60 μN, wear increases rapidly with load. The wear depth at 50 μN is about 14 nm, slightly less

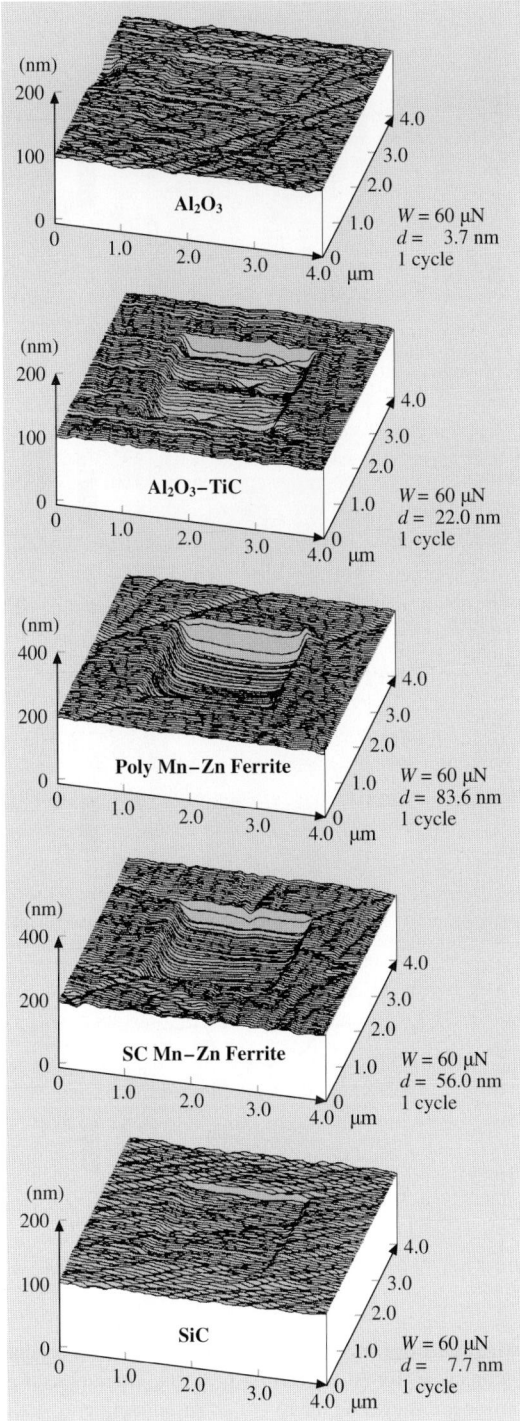

Fig. 20.28. Surface profiles showing the worn region (center $2\,\mu m \times 2\,\mu m$) after one scan cycles at a normal load of $60\,\mu N$ for Al_2O_3, Al_2O_3-TiC, polycrystalline Mn–Zn ferrite, single-crystal Mn–Zn ferrite and SiC [28]

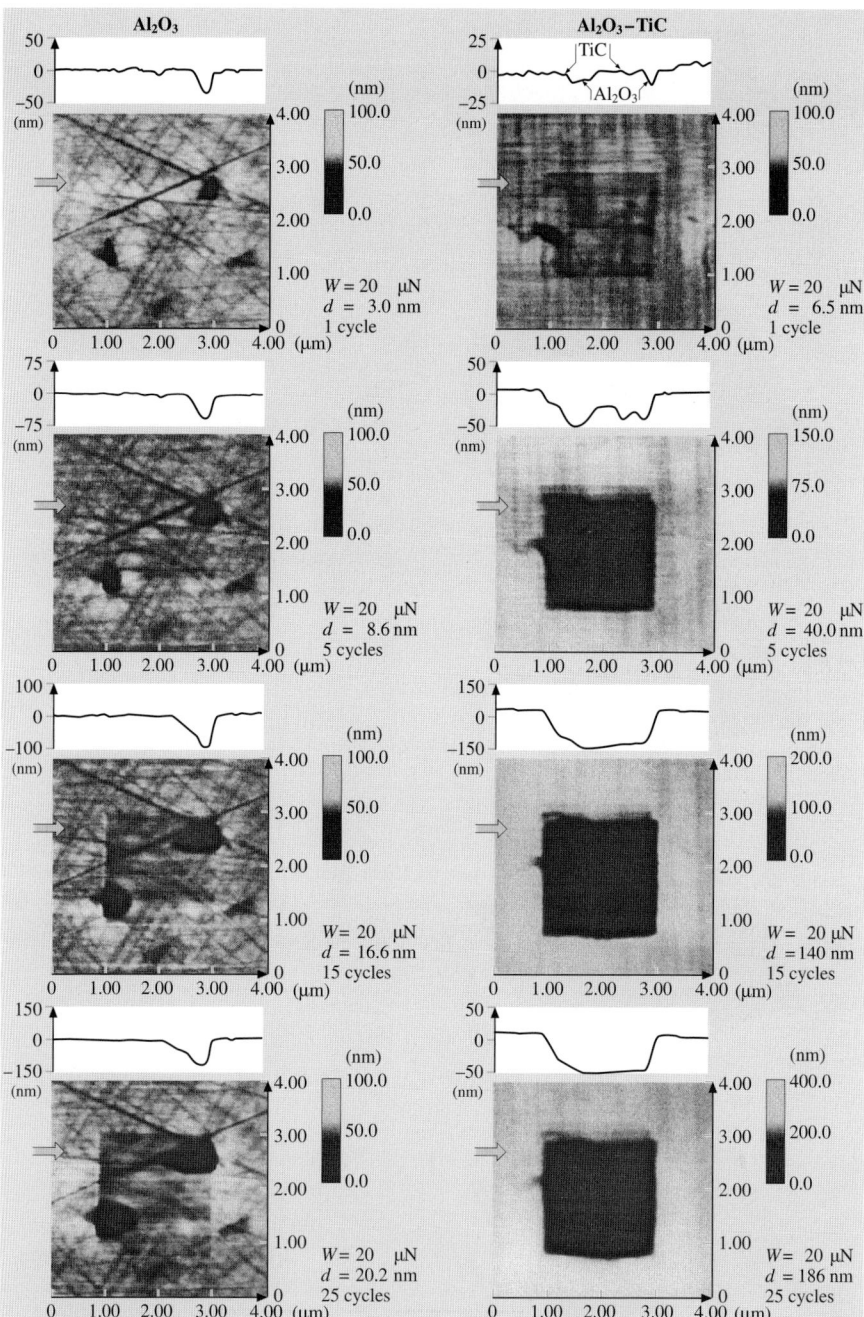

Fig. 20.29. Gray scale 2-D plots showing the worn region (center $2\,\mu\text{m} \times 2\,\mu\text{m}$) at a normal load of $20\,\mu\text{N}$ and different number of scan cycles for Al_2O_3 and Al_2O_3–TiC. The 2-D section plots taken at a location shown by an arrow are shown on the top of corresponding gray scale plot. Note the change in vertical scale for gray scale and 2-D section plots [28]

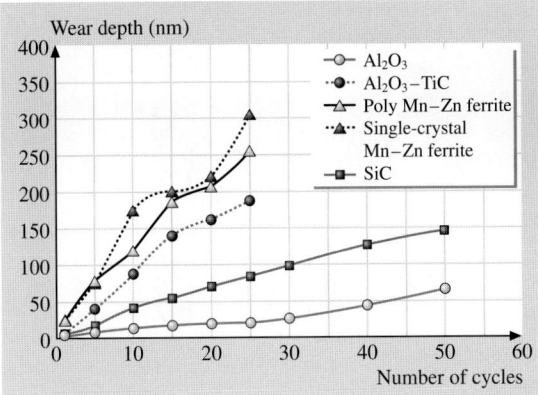

Fig. 20.30. Wear depth as a function of number of cycles at a normal load of 20 μN for Al_2O_3, Al_2O_3–TiC, polycrystalline Mn–Zn ferrite, single-crystal (SC) Mn–Zn ferrite and SiC [28]

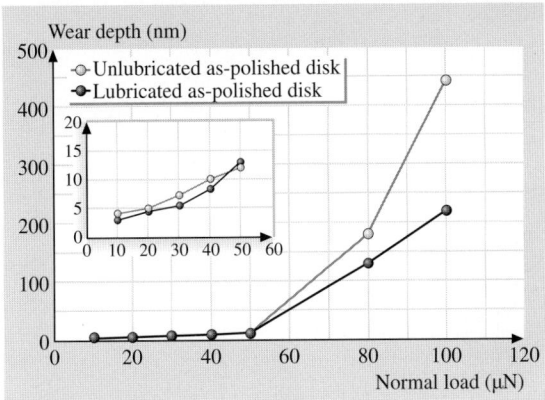

Fig. 20.31. Wear depth as a function of normal load for polished, lubricated and unlubricated thin-film rigid disks after one cycle [18]

than the thickness of the carbon film. The rapid increase of wear with load at loads larger than 60 μN is an indication of the breakdown of the carbon coating on the disk surface.

Figure 20.33 shows the wear depth as a function of number of cycles for the polished disks (lubricated and unlubricated). Again, for both unlubricated and lubricated disks, wear initially takes place slowly with a sudden increase between 40 and 50 cycles at 10 μN. The sudden increase occurred after 10 cycles at 20 μN. This rapid increase is associated with the breakdown of the carbon coating. The wear profiles at various cycles are shown in Fig. 20.34 for a polished, unlubricated disk at a normal load of 20 μN. Wear is not uniform and the wear is largely initiated at the texture

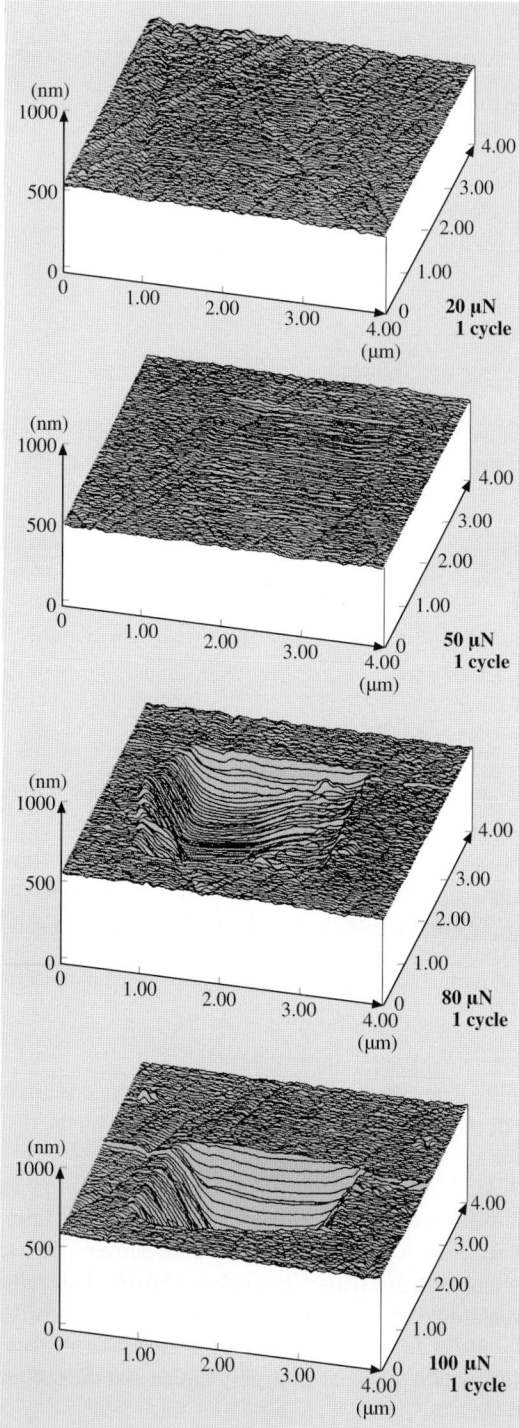

Fig. 20.32. Surface maps of a polished, unlubricated thin-film rigid disk showing the worn region (center $2\,\mu m \times 2\,\mu m$) after one cycle. The normal loads are indicted in the figure [18]

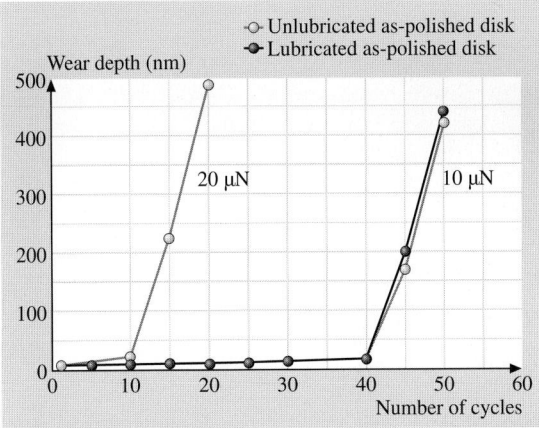

Fig. 20.33. Wear depth as a function of number of cycles for polished, lubricated and unlubricated thin-film rigid disks at 10 μN and for polished, unlubricated disk at 20 μN [18]

grooves present on the disk surface. This indicates that surface defects strongly affect the wear rate.

Hard amorphous carbon coatings are used to provide wear and corrosion resistance to magnetic disks and MR/GMR magnetic heads. A thick coating is desirable for long durability; however, to achieve ever increasing high recording densities, it is necessary to use as thin a coating as possible. Microwear data on various amorphous carbon coatings of different thicknesses have been conducted by *Bhushan* and *Koinkar* [48], *Koinkar* and *Bhushan* [49], and *Sundararajan* and *Bhushan* [53]. Figure 20.35 shows a wear mark on an uncoated Si(100) and various 10 nm thick carbon coatings. It is seen that Si(100) wears uniformly, whereas carbon coatings wear nonuniformly. Carbon coating failure is sudden and accompanied by a sudden rise in friction force. Figure 20.36 shows the wear depth of Si(100) substrate and various coatings at two different loads. FCA and ECR-CVD, 20 nm thick coatings show excellent wear resistance up to 80 μN, the load that is required for the IB 20 nm coating to fail. In these tests, failure of a coating results when the wear depth exceeds the quoted coating thickness. The SP 20 nm coating fails at the much lower load of 35 μN. At 60 μN, the coating hardly provides any protection. Moving on to the 10 nm coatings, ECR-CVD coating requires about 40 cycles at 60 μN to fail as compared to IB and FCA, which fail at 45 μN. the FCA coating exhibits slight roughening in the wear track after the first few cycles, which leads to an increase in the friction force. The SP coating continues to exhibit poor resistance, failing at 20 μN. For the 5 nm coatings, the load required to fail the coatings continues to decrease. But IB and ECR-CVD still provide adequate protection as compared to bare Si(100) in that order, failing at 35 μN compared to FCA at 25 μN and SP at 20 μN. Almost all the 20, 10, and 5 nm coatings provide better wear resistance than bare silicon. At 3.5 nm, FCA coating provides no wear resistance, failing almost instantly at 20 μN.

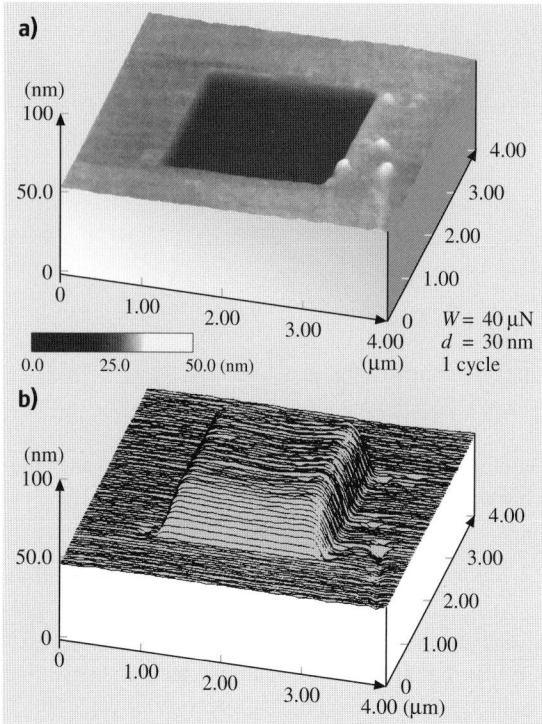

Fig. 20.34. Surface maps of a polished, unlubricated thin-film rigid disk showing the worn region (center $2\,\mu m \times 2\,\mu m$) at $20\,\mu N$. The number of cycles are indicated in the figure [18]

The IB and ECR-CVD coating show good wear resistance at $20\,\mu N$ compared to bare Si(100). But IB lasts only about 10 cycles and ECR-CVD about 3 cycles at $25\,\mu N$.

The wear tests highlight the differences in the coatings more vividly than the scratch tests data presented earlier. At higher thicknesses (10 and 20 nm), the ECR-CVD and FCA coatings appear to show the best wear resistance. This is probably due to higher hardness of the coatings (see data presented later). At 5 nm, IB coating appears to be the best. FCA coatings show poorer wear resisting with decreasing coating thickness. SP coatings showed consistently poor wear resistance at all thicknesses. The IB 2.5 nm coating does provide reasonable wear protection at low loads.

Wear depths as a function of normal load for MP, BaFe and ME tapes along the parallel direction are plotted in Fig. 20.37 [47]. For the ME tape, there is negligible wear until the normal load of about $50\,\mu N$, above this load the magnetic coating fails rapidly. This observation is consistent with the scratch data. Wear depths as a function of number of cycles for MP, BaFe, and ME tapes are shown in Fig. 20.38. For the MP and BaFe particulate tapes, wear rates appear to be independent of the particulate density. Again as observed in the scratch testing, wear rate of BaFe tapes is higher than that for MP tapes. ME tapes are much more wear resistant than the particulate tapes. However, the failure of ME tapes are catastrophic as observed in

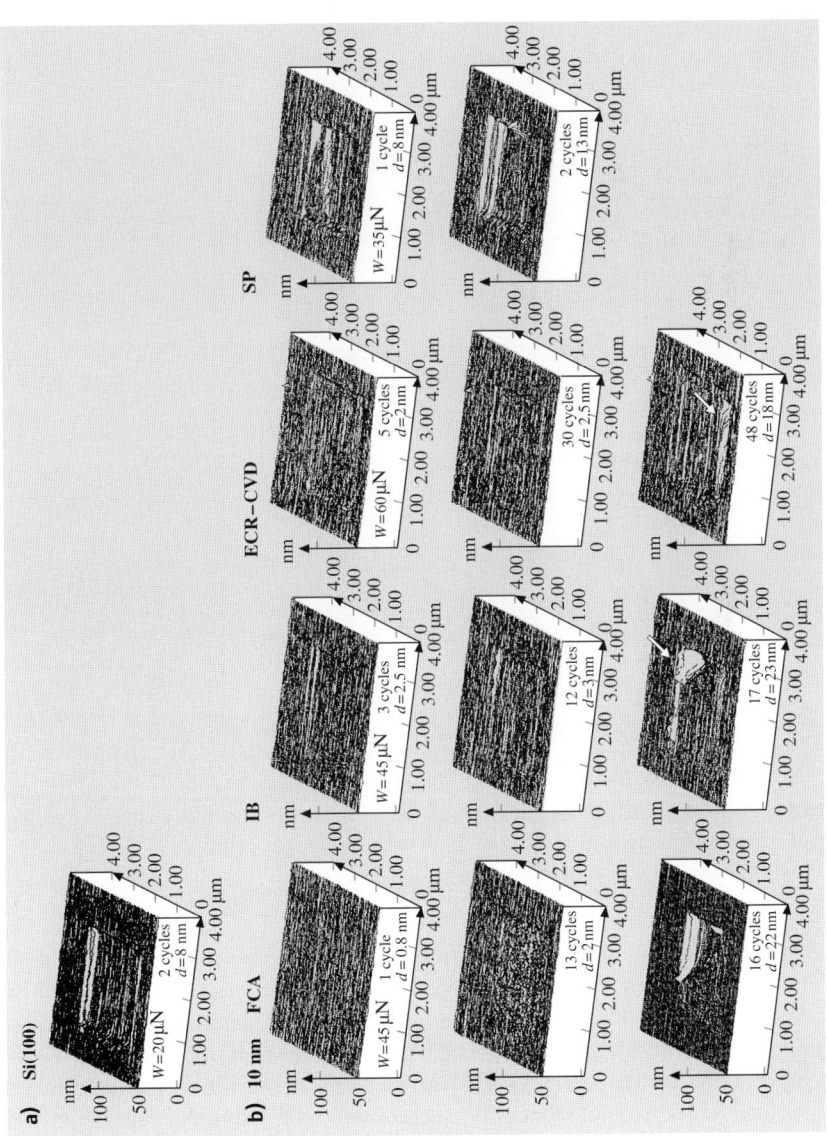

Fig. 20.35. AFM images of wear marks on (a) bare Si(100), and (b) all 10 nm thick amorphous carbon coatings [53]

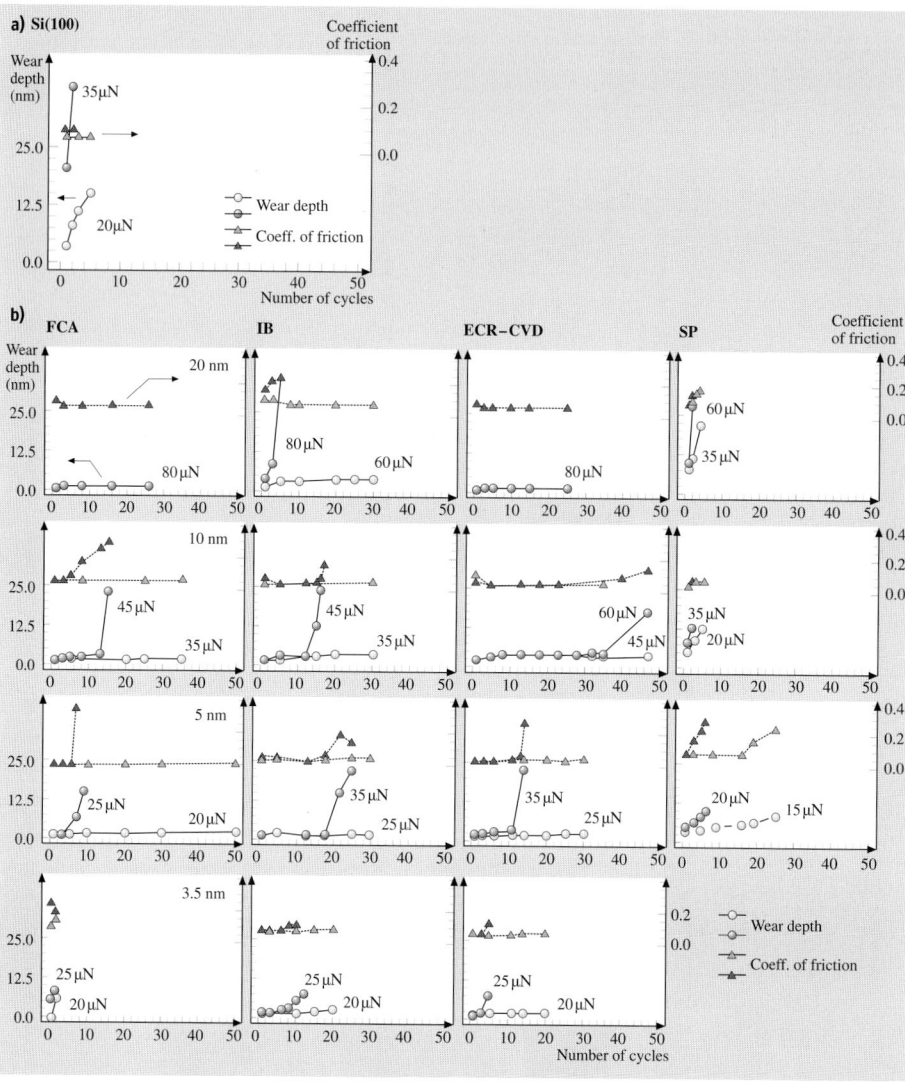

Fig. 20.36. Wear data of (**a**) bare Si(100) and (**b**) all amorphous carbon coatings. Coating thickness is constant along each row in (**b**). Both wear depth and coefficient of friction during wear for a given cycle are plotted [53]

scratch testing. Wear studies were performed along and across the longitudinal tape direction in high and low hardness regions. At the high hardness regions of the ME tapes, failure occurs at lower loads. Directionality effect again may arise from the columnar structure of the ME films [3, 60]. Wear profiles at various cycles at a normal load of 2 µN for MP and at 20 µN for ME tapes are shown in Fig. 20.39. For the particulate tapes, we note that polymer gets removed before the particulates do

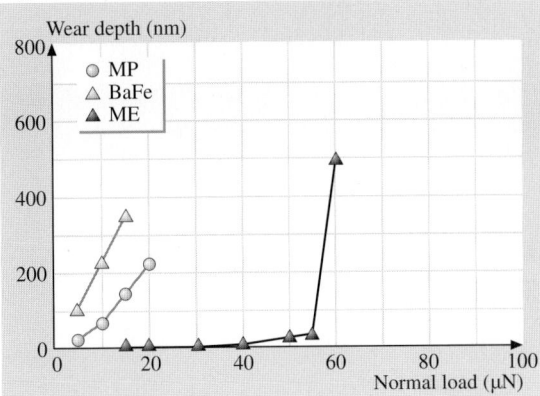

Fig. 20.37. Wear depth as a function of normal load for three tapes in the parallel direction after one cycle [47]

(Fig. 20.39a). Based on the wear profiles of the ME tape shown in Fig. 20.39a, we note that most wear occurs between 50 to 60 cycles which shows the catastrophic removal of the coating. It was also observed that wear debris generated during wear test in all cases is loose and can easily be removed away from the scan area at light loads ($\approx 0.3\,\mu\text{N}$).

The average wear depth as a function of load for a PET film is shown in Fig. 20.40. Again, the wear depth increases linearly with load. Figure 20.41 shows the average wear depth as a function of number of cycles. The observed wear rate is approximately constant. PET tape substrate consists of particles sticking out on its surface to facilitate winding. Figure 20.42 shows the wear profiles as a function of number of cycles at 1 µN load on the PET film in the nonparticulate and particulate regions [45]. We note that polymeric materials tear in microwear tests. The particles do not wear readily at 1 µN. Polymer around the particles is removed but the particles remain intact. Wear in the particulate region is much smaller than that in the polymer region. We will see later that nanohardness of the particulate region is about 1.4 GPa compared to 0.3 GPa in the nonparticulate region (Table 20.3).

20.6 Indentation

20.6.1 Picoscale Indentation

Bhushan and *Ruan* [15] measured indentability of magnetic tapes at increasing loads on a picoscale, Fig. 20.43. In this figure, the vertical axis represents the cantilever tip deflection and the horizontal axis represents the vertical position (Z) of the sample. The "extending" and "retracting" curves correspond to the sample being moved toward or away from the cantilever tip, respectively. In this experiment, as the sample surface approaches the AFM tip fraction of a nm away from the sample (point A),

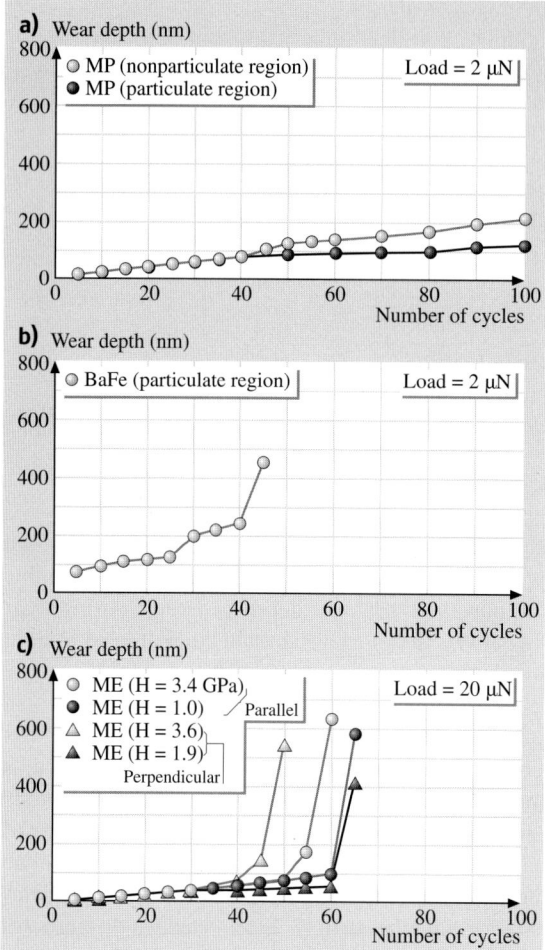

Fig. 20.38. Wear depth as a function of number of cycles for (**a**) MP, (**b**) BaFe, and (**c**) ME tapes in different regions at normal loads indicated in the figure. Note a higher load used for the ME tape in (**c**) [47]

the cantilever bends toward the sample (part B) because of attractive forces between the tip and sample. As we continue the forward position of the sample, it pushes the cantilever back through its original rest position (point of zero applied load) entering the repulsive region (or loading portion) of the force curve. As the sample is retracted, the cantilever deflection decreases. At point D in the retracting curve, the sample is disengaged from the tip. Before the disengagement, the tip is pulled toward the sample after the zero deflection point of the force curve (point C) because of attractive forces (van der Waals forces and longer range meniscus forces). A thin layer of liquid, such as liquid lubricant and condensations of water vapor from ambient, will give rise to capillary forces that act to draw the tip towards sample at small separations. The horizontal shift between the loading and unloading curves results from the hysteresis in the PZT tube.

The left portion of the curve shows the tip deflection as a function of the sample traveling distance during sample–tip contact, which would be equal to each other for a rigid sample. However, if the tip indents into the sample, the tip deflection would be less than the sample traveling distance, or in other words, the slope of the line would be less than 1. In Fig. 20.43, we note that line in the left portion of the figure is curved with a slope of less than 1 shortly after the sample touches the tip, which suggests that the tip has indented the sample. Later, the slope is equal to 1 suggesting that the tip no longer indents the sample. This observation indicates that the tape surface is soft locally (polymer rich) but it is hard (as a result of magnetic particles) underneath. Since the curves in extending and retracting modes are identical, the indentation is elastic up to at a maximum load of about 22 nN used in the measurements.

According to *Bhushan* and *Ruan* [15], during indentation of rigid disks, the slope of the deflection curves remained constant as the disks touch and continue to push the AFM tip. The disks were not indented.

20.6.2 Nanoscale Indentation

Indentation hardness with a penetration depth as low as 5 nm can be measured using AFM. *Bhushan* and *Koinkar* [17] measured hardness of thin-film disks at load of 80, 100, and 140 µN. Hardness values were 20 GPa (10 nm), 21 GPa (15 nm) and 9 GPA (40 nm); the depths of indentation are shown in the parenthesis. The hardness value at 100 µN is much higher than at 140 µN. This is expected since the indentation depth is only about 15 nm at 100 µN which is smaller than the thickness of carbon coating (≈ 30 nm). The hardness value at lower loads is primarily the value of the carbon coating. The hardness value at higher loads is primarily the value of the magnetic film, which is softer than the carbon coating [2]. This result is consistent with the scratch and wear data discussed previously.

For the case of hardness measurements made on magnetic thin film rigid disk at low loads, the indentation depth is on the same order at the variation in the surface roughness. For accurate measurements of indentation size and depth, it is desirable to subtract the original (unindented) profile from the indented profile. *Bhushan* et al. [18] developed an algorithm for this purpose. Because of hysteresis, a translational shift in the sample plane occurs during the scanning period, resulting in a shift between images captured before and after indentation. Therefore, the image for perfect overlap needs to be shifted before subtraction can be performed. To accomplish this objective, a small region on the original image was selected and the corresponding region on the indented image was found by maximizing the correlation between the two regions. (Profiles were plane-fitted before subtraction.) Once two regions were identified, overlapped areas between the two images were determined and the original image was shifted with the required translational shift and this subtracted from the indented image. An example of profiles before and after subtraction is shown in Fig. 20.44. It is easier to measure the indent on the subtracted image. At a normal

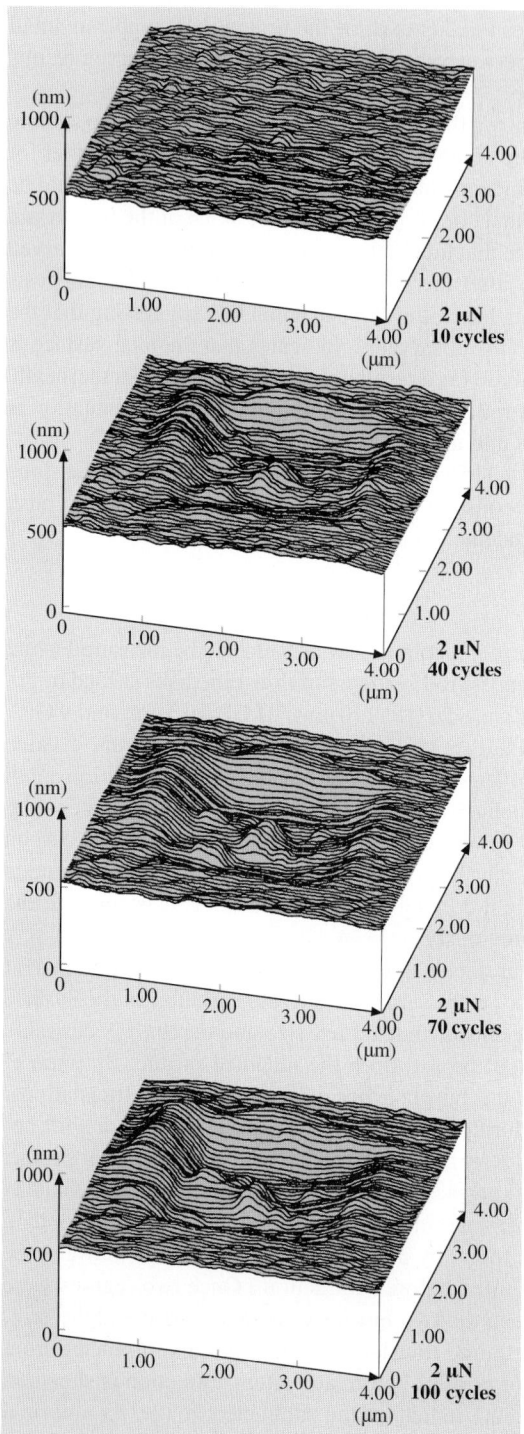

Fig. 20.39. Surface maps showing the worn region (center $2\,\mu m \times 2\,\mu m$) after various cycles of wear at (**a**) $2\,\mu N$ for MP (particulate region) and at (**b**) $20\,\mu N$ for ME (H= 3.4 GPa, parallel direction) tapes. Note a different *vertical scale* for the bottom profile of (**b**) [47]

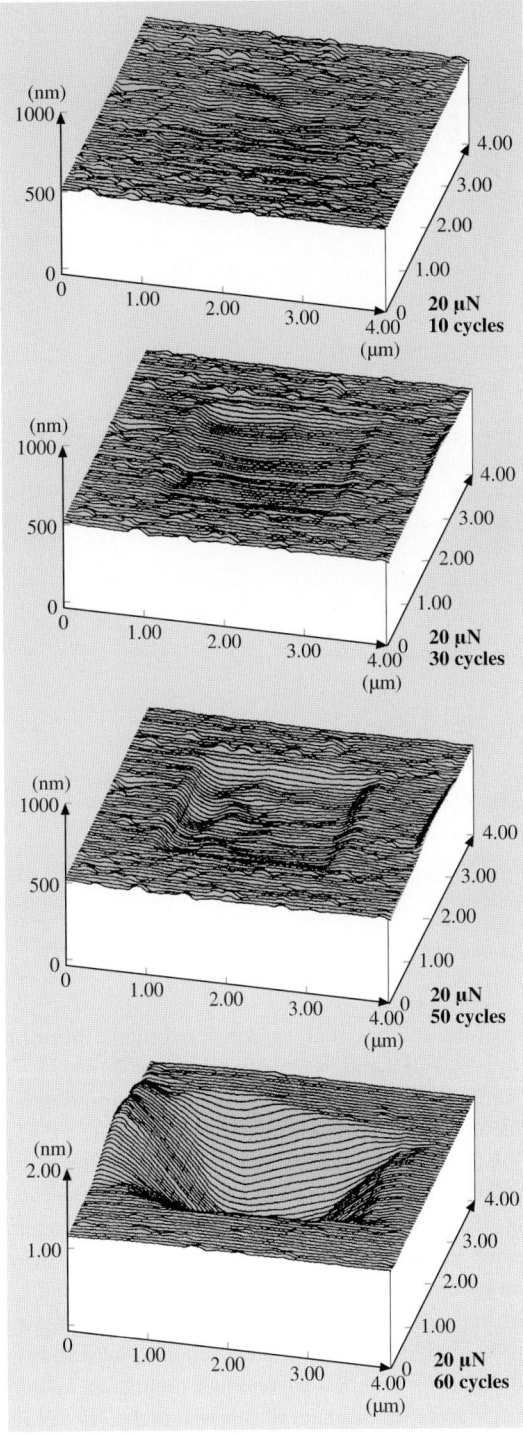

Fig. 20.39. continue

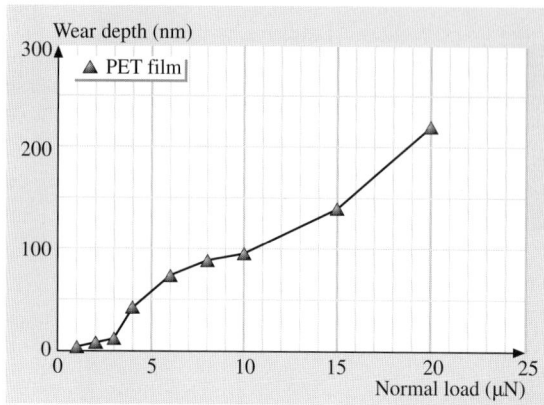

Fig. 20.40. Wear depth as a function of normal load (after one cycle) for a PET film [45]

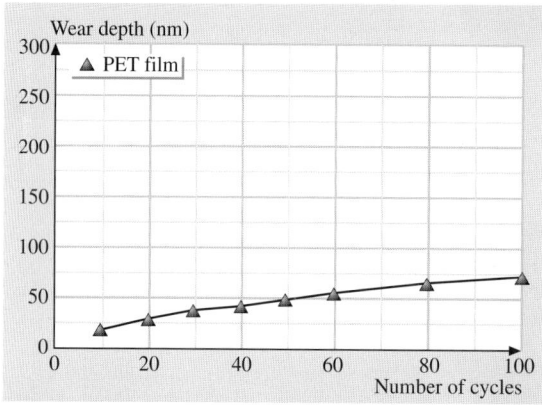

Fig. 20.41. Wear depth as a function of number of cycles at 1 µN for a PET film [45]

load of 140 mN the hardness value of an unlubricated, as-polish magnetic thin film rigid disk (rms roughness = 3.3 nm) is 9.0 GPa and indentation depth is 40 nm.

For accurate measurement of nanohardness at very shallow indentation depths, depth-sensing capacitance transducer system in an AFM is used [19]. Figure 20.45a shows the hardness as a function of residual depth for three types of 100 nm thick amorphous carbon coatings deposited on silicon by sputtering, ion beam and cathodic arc processes [50]. Data on uncoated silicon are also included for comparisons. The cathodic arc carbon coating exhibits highest hardness of about 24.9 GPa, whereas the sputtered and ion beam carbon coatings exhibit hardness values of 17.2 and 15.2 GPa respectively. The hardness of Si(100) is 13.2 GPa. High hardness of cathodic arc carbon coating explains its high wear resistance, reported earlier. Figure 20.45b shows the elastic modulus as a function of residual depth for various samples. The cathodic arc coating exhibits the highest elastic modulus. Its elastic modulus decreases with an increasing residual depth, while the elastic moduli for the other carbon coatings remain almost constant. In general, hardness and elastic modulus of coatings are strongly influenced by their crystalline structure, stoichiometry and growth characteristics which depend on the deposition parameters. Mechanical properties of carbon coatings have been known to change over a wide range with sp^3–sp^2 bonding ratio and amount of hydrogen. Hydrogen is believed to play an important role in the bonding configuration of carbon atoms by helping to stabilize tetrahedral coordination of carbon atoms. Detailed mechanical characterization of amorphous carbon coatings is presented by *Li* and *Bhushan* [51, 52] and *Bhushan* [59].

20.6.3 Localized Surface Elasticity

By using an AFM in a so-called force modulation mode, it is possible to quantitatively measure the elasticity of soft and compliant materials with penetration depths of less than 100 nm [21, 22]. This technique has been successfully used to get localized elasticity maps of particulate magnetic tapes. Elasticity map of a tape can be used to identify relative distribution of hard magnetic/nonmagnetic ceramic particles and the polymeric binder on the tape surface which has an effect on friction and stiction at the head–tape interface. Figure 20.46 shows surface height and elasticity maps on an MP tape. The elasticity image reveals sharp variations in surface elasticity due to the composite nature of the film. As can be clearly seen, regions of high elasticity do not always correspond to high or low topography. Based on a Hertzian elastic-contact analysis, the static indentation depth of these sample during the force modulation scan is estimated to be about 1 nm. The contrast seen is influenced most strongly by material properties in the top few nanometers, independent of the composite structure beneath the surface layer. The trend in number of stiff regions has been correlated to reduced stiction problems in tapes [62].

Figure 20.47 shows the surface topography and phase image of an alumina particle embedded in the MP tape using a so-called TR mode [25, 26]. The cross-section view of the particle obtained from the topographic image is shown at the bottom as a visual aid. The edges of the particle show up darker in the TR phase angle image, which suggests that it is less viscoelastic compared to the background. The magnetic particles on top of the alumina particle are clearly visible in the TR phase image. These have a brighter contrast, which is the same as that of the background. Phase contrast mapping appears to privide better resolution than stiffness mapping for magnetic tapes.

20.7 Lubrication

The boundary films are formed by physical adsorption, chemical adsorption, and chemical reaction. The physisorbed film can be either monomolecular or polymolecular thick. The chemisorbed films are monomolecular, but stoichiometric films formed by chemical reaction can have a large film thickness. In general, the stability and durability of surface films decrease in the following order: chemical reaction films, chemisorbed films and physisorbed films. A good boundary lubricant

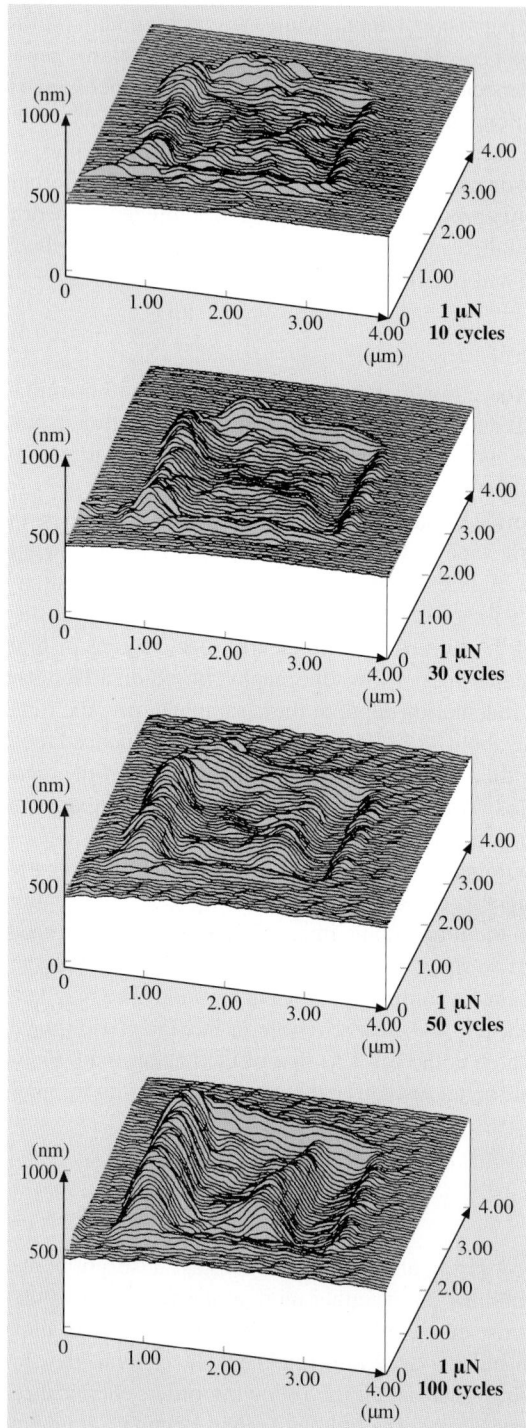

Fig. 20.42. Surface maps showing the worn region (center 2 μm × 2 μm) at 1 μN for a PET film (**a**) in the polymer region, (**b**) in the particulate region. The number of cycles are indicated in the figure [45]

20 Micro/Nanotribology and Micro/Nanomechanics of Magnetic Storage Devices 1015

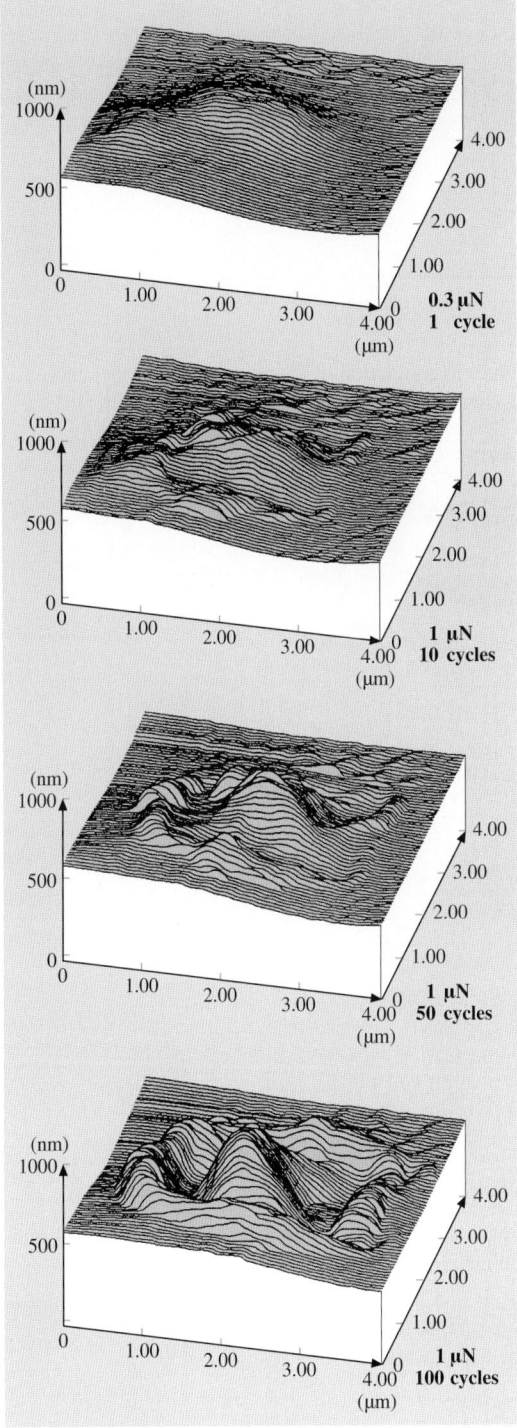

Fig. 20.42. continue

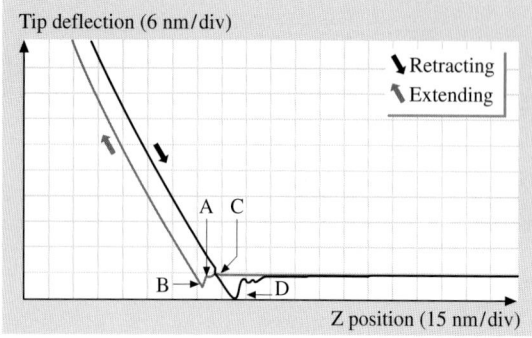

Fig. 20.43. Tip deflection (normal force) as a function of Z (separation distance) curve for a metal-particle (MP) tape. The spring constant of the cantilever used was 0.4 N/m [15]

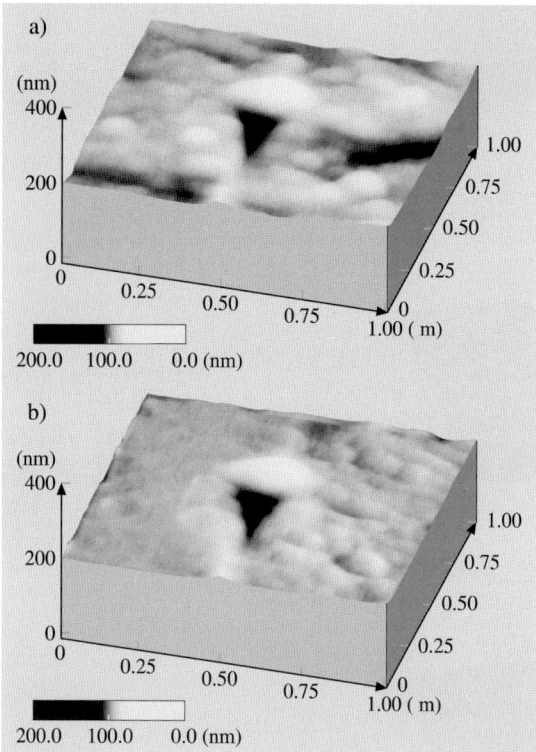

Fig. 20.44. Images with nanoindentation marks generated on a polished, unlubricated thin-film rigid disk at 140 μN (**a**) before subtraction, and (**b**) after subtraction [18]

20 Micro/Nanotribology and Micro/Nanomechanics of Magnetic Storage Devices   1017

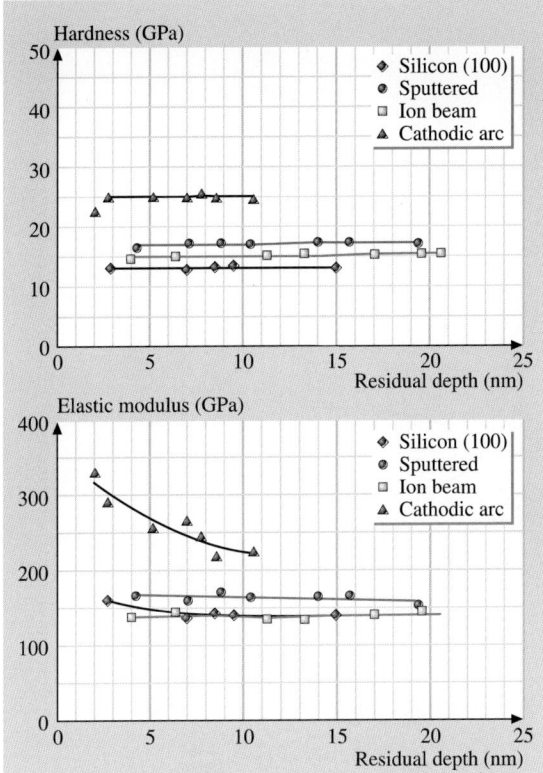

Fig. 20.45. Nanohardness and elastic modulus as a function of residual indentation depth for Si(100) and 100 nm thick coatings deposited by sputtering, ion beam and cathodic arc processes [50]

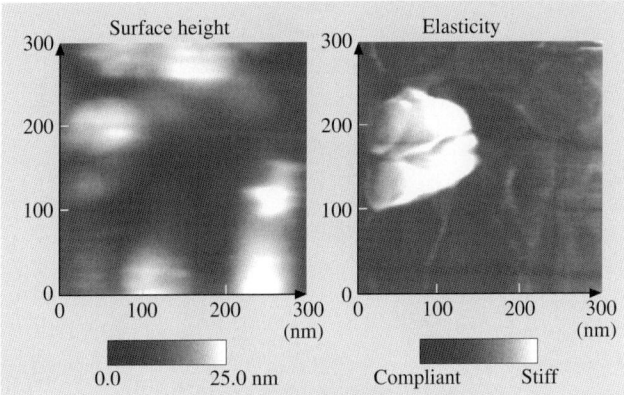

Fig. 20.46. Surface height and elasticity maps for a metal-particle tape A ($\sigma = 6.72$ nm and $P-V = 31.7$ nm). σ and $P-V$ refer to standard deviation of surface heights and peak-to-valley distance, respectively. The *grayscale* on the elasticity map is arbitrary [21]

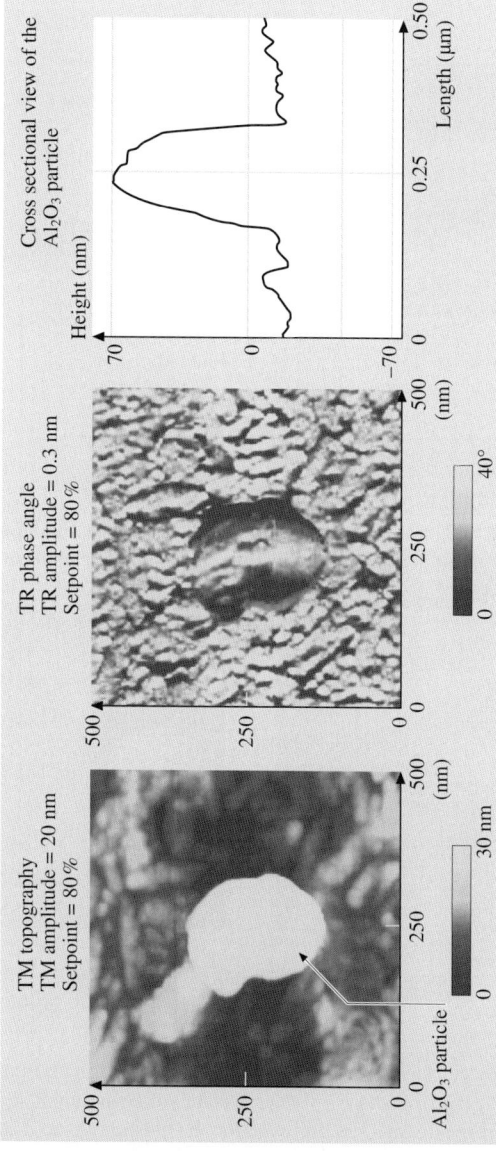

Fig. 20.47. Tapping mode (TM) topography and TR phase angle image of an alumina particle that is used as a head cleaning agent for MP tape. A cross-sectional view of the particle is also shown [25]

should have a high degree of interaction between its molecules and the sliding surface. As a general rule, liquids are good lubricants when they are polar and thus able to grip solid surfaces (or be adsorbed). Polar lubricants contain reactive functional groups with low ionization potential or groups having high polarizability [5]. Boundary lubrication properties of lubricants are also dependent upon the molecular conformation and lubricant spreading [63–66].

Mechanical interactions between the magnetic head and the medium in magnetic storage devices are minimized by the lubrication of the magnetic medium [2, 3]. The primary function of the lubricant is to reduce the wear of the medium and to ensure that friction remains low throughout the operation of the drive. The main challenge, though, in selecting the best candidate for a specific surface is to find a material that provides an acceptable wear protection for the entire life of the product, which can be several years in duration. There are many requirements that a lubricant must satisfy in order to guarantee an acceptable life performance. An optimum lubricant thickness is one of these requirements. If the lubricant film is too thick, excessive stiction and mechanical failure of the head-disk is observed. On the other hand, if the film is too thin, protection of the interface is compromised, and high friction and excessive wear will result in catastrophic failure. An acceptable lubricant must exhibit properties such as chemical inertness, low volatility, high thermal, oxidative and hydrolytic stability, shear stability, and good affinity to the magnetic medium surface.

Fatty acid esters are excellent boundary lubricants, and esters such as tridecyl stearate, butyl stearate, butyl palmitate, buryl myristate, stearic acid, and myrstic acid are commonly used as internal lubricants, roughly 1–7% by weight of the magnetic coating in particulate flexible media (tapes and particulate flexible disks) [2, 3]. The fatty acids involved include those with acid groups with an even number of carbon atoms between C_{12} and C_{22}, with alcohols ranging from C_3 to C_{13}. These acids are all solids with melting points above the normal surface operating temperature of the magnetic media. This suggests that the decomposition products of the ester via lubrication chemistry during a head-flexible medium contact may be the key to lubrication.

Topical lubrication is used to reduce the wear of rigid disks and thin-film tapes [67]. Perfluoropolyethers (PFPEs) are chemically the most stable lubricants with some boundary lubrication capability, and are most commonly used for topical lubrication of rigid disks. PFPEs commonly used include Fomblin Z lubricants, made by Solvay Solexis Inc., Milan, Italy; and Demnum S, made by Diakin, Japan; and their difunctional derivatives containing various reactive end groups, e.g., hydroxyl or alcohol (Fomblin Z-DOL and Z-TETROL), piperonyl (Fomblin AM 2001), isocyanate (Fomblin Z-DISOC), and ester (Demnum SP). Fomblin Y and Krytox 143AD (made by Dupont USA) have been used in the past for particulate rigid disks. The difunctional derivatives are referred to as reactive (polar) PFPE lubricants. The chemical structures, molecular weights, and viscosities of various types of PFPE lubricants are given in Table 20.4. We note that rheological properties of thin-films of lubricants are expected to be different from their bulk properties. Fomblin Z and Demnum S are linear PFPE, and Fomblin Y and Krytox 143 AD are branched PFPE,

Table 20.4. Chemical structure, molecular weight, and viscosity of perfluoropolyether lubricants

Lubricant	Formula	Molecular weight (Daltons)	Kinematic viscosity cSt(mm^2/s)
Fomblin Z-25	$CF_3-O-(CF_2-CF_2-O)_m-(CF_2-O)_n-CF_3$	12 800	250
Fomblin Z-15	$CF_3-O-(CF_2-CF_2-O)_m-(CF_2-O)_n-CF_3$ ($m/n \approx 2/3$)	9100	150
Fomblin Z-03	$CF_3-O-(CF_2-CF_2-O)_m-(CF_2-O)_n-CF_3$	3600	30
Fomblin Z-DOL	$HO-CH_2-CF_2-O-(CF_2-CF_2-O)_m-(CF_2-O)_n-CF_2-CH_2-OH$	2000	80
Fomblin AM2001	Piperonyl$-O-CH_2-CF_2-O-(CF_2-CF_2-O)_m-(CF_2-O)_n-CF_2-O-$piperonyl[a]	2300	80
Fomblin Z-DISOC	$O-CN-C_6H_3-(CH_3)-NH-CO-CF_2-O-(CF_2-CF_2-O)_n-(CF_2-O)_m-CF_2-CO-NH-C_6H_3-(CH_3)-N-CO$	1500	160
Fomblin YR	$CF_3-O-(\underset{\underset{F}{\|}}{C}-CF_2-O)_m(CF_2-O)_n-CF_3$ ($m/n \approx 40/1$)	6800	1600
Demnum S-100	$CF_3-CF_2-CF_2-O-(CF_2-CF_2-CF_2-O)_m-CF_2-CF_3$	5600	250
Krytox 143AD	$CF_3-CF_2-CF_2-O-(\underset{\underset{F}{\|}}{\overset{\overset{CF_3}{\|}}{C}}-CF_2-O_m)-CF_2-CF_3$	2600	–

[a] 3,4-methylenedioxybenzyl

where the regularity of the chain is perturbed by $-CF_3$ side groups. The bulk viscosity of Fomblin Y and Krytox 143 AD is almost an order of magnitude higher than the Z type. Fomblin Z is thermally decomposed more rapidly than Y [5]. The molecular diameter is about 0.8 nm for these lubricant molecules. The monolayer thickness of these molecules depends on the molecular conformations of the polymer chain on the surface [64, 65].

The adsorption of the lubricant molecules on a magnetic disk surface is due to van der Waals forces, which are too weak to offset the spin-off losses, or to arrest displacement of the lubricant by water or other ambient contaminants. Considering that these lubricating films are on the order of a monolayer thick and are required to function satisfactorily for the duration of several years, the task of developing a workable interface is quite formidable. An approach aiming at alleviating these shortcomings

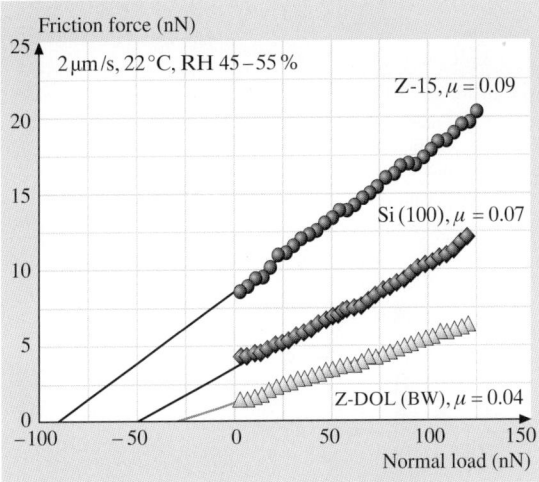

Fig. 20.48. Friction force versus normal load curves for Si(100), 2.8 nm thick Z-15 film, and 2.3 nm thick Z-DOL(BW) film at 2 μm/s, and in ambient air sliding against a Si_3N_4 tip. Based on these curves, coefficient of friction (μ) and adhesive force can be calculated [30]

is to enhance the attachment of the molecules to the overcoat, which, for most cases, is sputtered carbon. There are basically two approaches which have been shown to be successful in bonding the monolayer to the carbon. The first relies on exposure of the disk lubricated with neutral PFPE to various forms of radiation, such as low-energy X-ray [68], nitrogen plasma [69], or far ultraviolet (e.g., 185 nm) [70]. Another approach is to use chemically active PFPE molecules, where the various functional (reactive) end groups offer the opportunity of strong attachments to specific interface. These functional groups can react with surfaces and bond the lubricant to the disk surface, which reduces its loss due to spin off and evaporation. Bonding of lubricant to the disk surface depends upon the surface cleanliness. After lubrication, the disk is generally heated at 150 °C for 30 min to 1 h to improve the bonding. If only a bonded lubrication is desired, the unbonded fraction can be removed by washing it off for 60 s with a non-freon solvent (FC-72). Their main advantage is their ability to enhance durability without the problem of stiction usually associated with weakly bonded lubricants (*Bhushan*, 1996a).

20.7.1 Boundary Lubrication Studies

Koinkar and *Bhushan* [29] and *Liu* and *Bhushan* [30] studied friction, adhesion, and durability of Z-15 and Z-DOL (bonded and washed, BW) lubricants on Si(100) surface. To investigate the friction properties of Si(100), Z-15, and Z-DOL(BW), the friction force versus normal load curves were obtained by making friction measurements at increasing normal loads, Fig. 20.48. An approximately linear response of all three samples is observed in the load range of 5–130 nN. From the horizontal intercept at zero value of friction force, adhesive force can be obtained. The adhesive

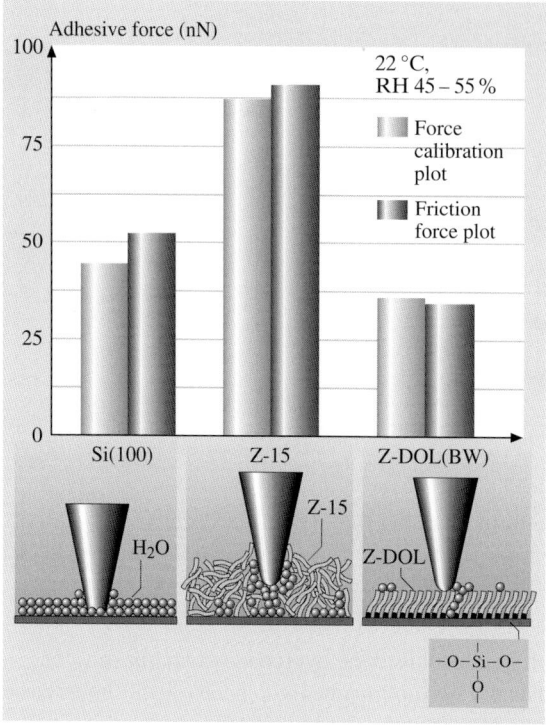

Fig. 20.49. Summary of the adhesive forces of Si(100), 2.8 nm thick Z-15 film, and 2.3 nm thick Z-DOL(BW) film. The schematic (*bottom*) shows the effect of meniscus formation between the AFM tip and the sample surface on the adhesive and friction forces [30]

forces for three samples were also measured using the force calibration plot technique. The adhesive force data obtained by the two techniques are summarized in Fig. 20.49, and the trends in the data obtained by two techniques are similar. The friction force and adhesive force of solid-like Z-DOL(BW) are consistently smaller than that for Si(100), but these values of liquid-like Z-15 lubricant is higher than that of Si(100). The presence of mobile Z-15 lubricant film increases adhesive force as compared to that of the Si(100) by meniscus formation. Whereas, the presence of Z-DOL(BW) film reduces the adhesive force because of absence of mobile liquid. See schematics at the bottom of Fig. 20.49. It is well known that in computer rigid disk drives, the stiction force increases rapidly with an increase in rest time between head and the disk [2]. The effect of rest time of 180 s on the friction force, adhesive force, and coefficient of friction for three samples are summarized in Fig. 20.50. It is seen that time effect is present in Si(100) and Z-15 with mobile liquid present. Whereas, time effect is not present for Z-DOL(BW) because of the absence of mobile liquid.

To study lubricant depletion during microscale measurements, nanowear studies were conducted using Si_3N_4 tips. Measured friction as a function of number of cycles for Si(100) and silicon surface lubricated with Z-15 and Z-DOL (BW) lubricants are

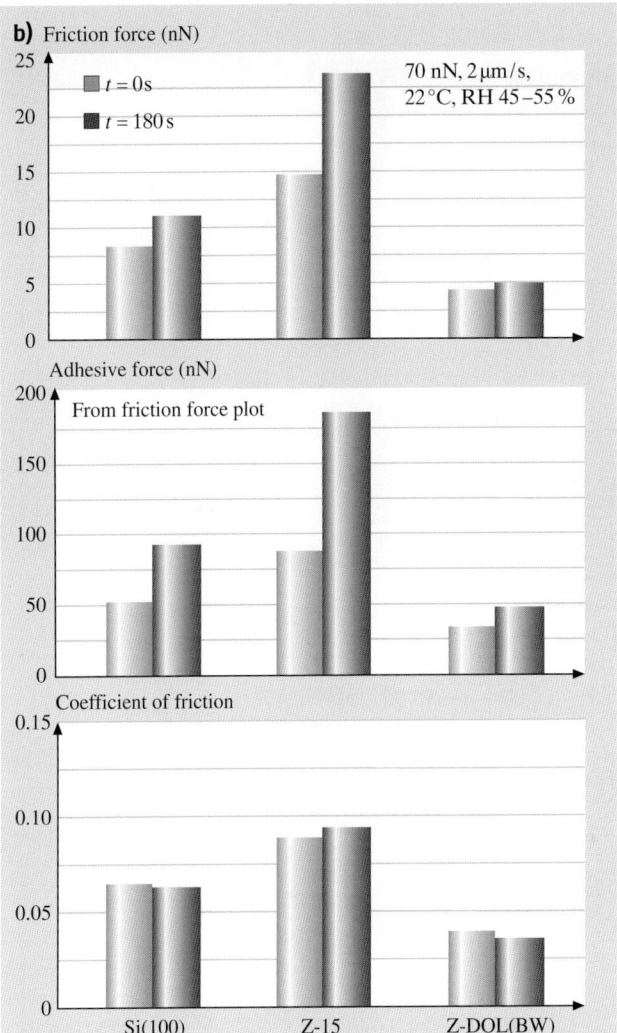

Fig. 20.50. Summary of rest time effect on friction force, adhesive force, and coefficient of friction of Si(100), 2.8 nm thick Z-15 film, and 2.3 nm thick Z-DOL(BW) film [30]

presented in Fig. 20.51. An area of 2 μm × 2 μm was scanned at a normal force of 70 nN. As observed before, friction force and coefficient of friction of Z-15 is higher than that of Si(100) with the lowest values for Z-DOL(BW). During cycling, friction force and coefficient of friction of Si(100) show a slight decrease during initial few cycles, then remain constant. This is related to the removal of the top adsorbed layer. In the case of Z-15 film, the friction force and coefficient of friction show an increase during the initial few cycles and then approach to higher and stable values. This is believed to be caused by the attachment of the Z-15 molecules onto the tip. The

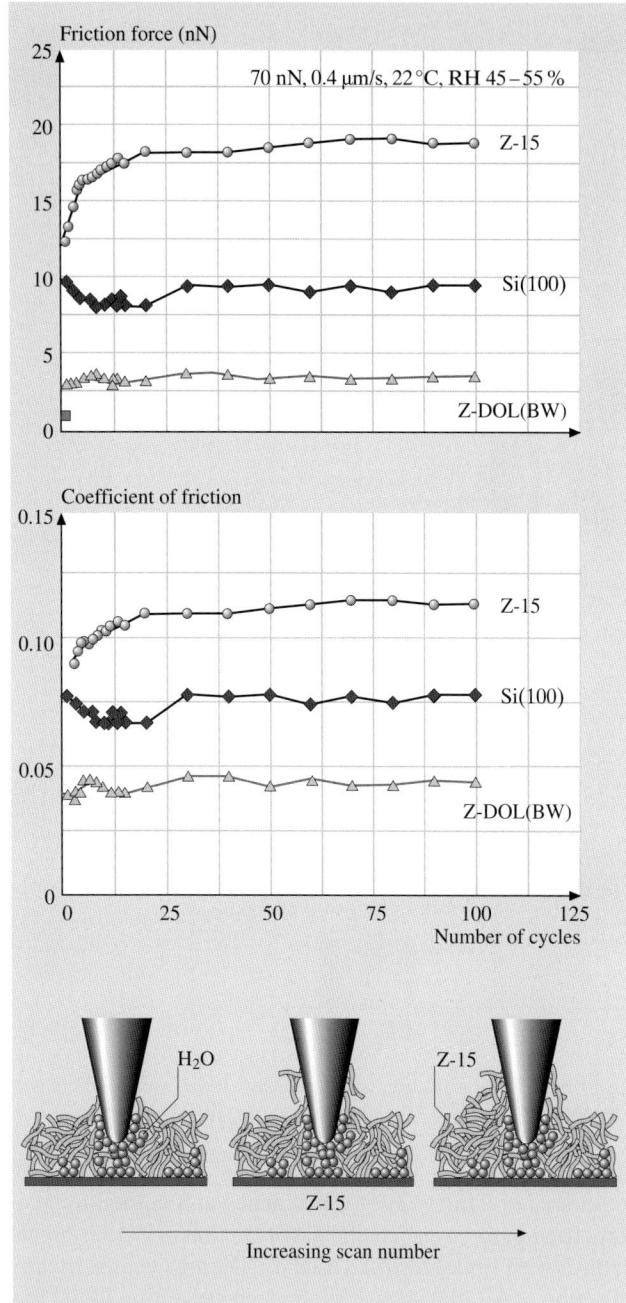

Fig. 20.51. Friction force and coefficient of friction versus number of sliding cycles for Si(100), 2.8 nm thick Z-15, and 2.3 nm thick Z-DOL (BW) film at 70 nN, 0.8 μm/s, and in ambient air. Schematic bottom shows that some liquid Z-15 molecules can be attached onto the tip. The molecular interaction between the attached molecules onto the tip with the Z-15 molecules in the film results in an increase of the friction force with multiscanning [30]

molecular interaction between these attached molecules to the tip and molecules on the film surface is responsible for an increase in the friction. But after several scans, this molecular interaction reaches to the equilibrium and after that friction force and coefficient of friction remain constant. In the case of Z-DOL (BW) film, the friction force and coefficient of friction start out to be low and remain low during the entire test for 100 cycles. It suggests that Z-DOL (BW) molecules do not get attached or displaced as readily as Z-15.

20.8 Closure

Atomic force microscope/friction force microscope (AFM/FFM) have been successfully used for measurements of surface roughness, friction, adhesion, scratching, wear, indentation, and lubrication on a micro to nanoscales. Commonly measured roughness parameters are scale dependent, requiring the need of scale-independent fractal parameters to characterize surface roughness. A generalized fractal analysis is presented which allows the characterization of surface roughness by two scale-independent parameters. Local variation in microscale friction force is found to correspond to the local surface slope suggesting that a ratchet mechanism is responsible for this variation. Directionality in the friction is observed on both micro- and macro-scales because of surface topography. Microscale friction is found to be significantly smaller than the macro-scale friction as there is less plowing contribution in microscale measurements.

Wear rates for particulate magnetic tapes and polyester tape substrates are approximately constant for various loads and test durations. However, for magnetic disks and magnetic tapes with a multilayered thin-film structure, the wear of the diamondlike amorphous carbon overcoat in the case of disks and magnetic layer in the case of tapes, is catastrophic. Breakdown of thin-films can be detected with AFM. Evolution of the wear has also been studied using AFM. We find that the wear is initiated at nanoscratches. Amorphous carbon films as thin as 3.5 nm are deposited as continuous films and exhibit some wear life. Wear life increases with an increase in film thickness. Carbon coatings deposited by cathodic arc and ECR-CVD processes are superior in wear and mechanical properties followed by ion beam and sputtering processes. AFM has been modified for nanoindentation hardness measurements with depth of indentation as low as 5 nm. Scratching and indentation on nanoscales are the powerful ways of evaluation of the mechanical integrity of ultrathin films.

AFM/FFM friction experiments show that lubricants with polar (reactive) end groups dramatically increase the load or contact pressure that a liquid film can support before solid-solid contact and thus exhibit long durability. The lubricants with the absence of mobile liquid exhibit low friction and adhesion and don't exhibit rest time effect.

References

1. B. Bhushan. *Tribology of magnetic storage systems*, volume 3, pages 325–374. CRC, 1994.
2. B. Bhushan. *Tribology and Mechanics of Magnetic Storage Devices*. Springer, 2nd edition, 1996.
3. B. Bhushan. *Mechanics and Reliability of Flexible Magnetic Media*. Springer, 2nd edition, 2000.
4. B. Bhushan. *Macro- and microtribology of magnetic storage devices*, volume 2, Materials, Coatings, and Industrial Applications, pages 1413–1513. CRC, 2001.
5. B. Bhushan. *Magnetic recording surfaces*, pages 116–133. Butterworth-Heinemann, 1993.
6. B. Bhushan. *Nanotribology and its applications to magnetic storage devices and MEMS*, pages 367–395. Kluwer, 1995.
7. B. Bhushan. Micro/nanotribology and its application to magnetic storage devices and mems. *Tribol. Int.*, 28:85–95, 1995.
8. B. Bhushan, J. N. Israelachvili, and U. Landman. Nanotribology: Friction, wear and lubrication at the atomic scale. *Nature*, 374:607–616, 1995.
9. B. Bhushan. *Micro/Nanotribology and its Applications*. NATO ASI Ser. 330. Kluwer, 1997.
10. B. Bhushan. *Handbook of Micro/Nanotribology*. CRC, 2nd edition, 1999.
11. B. Bhushan. *Fundamentals of Tribology and Bridging the Gap Between the Macro- and Micro/Nanoscales*. NATO Sci. Ser. II. Kluwer, 2001.
12. B. Bhushan. *Introduction to Tribology*. Wiley, 2002.
13. B. Bhushan (Ed.). *Springer Handbook of Nanotechnology*. Springer, 2004.
14. J. Ruan and B. Bhushan. Atomic-scale friction measurements using friction force microscopy: Part i – general principles and new measurement techniques. *ASME J. Tribol.*, 116:378–388, 1994.
15. B. Bhushan and J. Ruan. Atomic-scale friction measurements using friction force microscopy: Part ii – application to magnetic media. *ASME J. Tribol.*, 116:389–396, 1994.
16. B. Bhushan and V. N. Koinkar. Tribological studies of silicon for magnetic recording applications. *J. Appl. Phys*, 75:5741–5746, 1994.
17. B. Bhushan and V. N. Koinkar. Nanoindentation hardness measurements using atomic force microscopy. *Appl. Phys. Lett.*, 64:1653–1655, 1994.
18. B. Bhushan, V. N. Koinkar, and J. Ruan. Microtribology of magnetic media. *Proc. Inst. Mech. Eng., J. Eng. Tribol.*, 208:17–29, 1994.
19. B. Bhushan, A. V. Kulkarni, W. Bonin, and J. T. Wyrobek. Nanoindentation and picoindentation measurements using a capacitance transducer system in atomic force microscopy. *Philos. Mag.*, 74:1117–1128, 1996.
20. S. Sundararajan and B. Bhushan. Development of a continuous microscratch technique in an atomic force microscopy and its applications to study scratch resistance of ultra-thin hard amorphous carbon coatings. *J. Mater. Res.*, 16:437–445, 2001.
21. D. DeVecchio and B. Bhushan. Localized surface elasticity measurements using an atomic force microscope. *Rev. Sci. Instrum.*, 68:4498–4505, 1997.
22. V. Scherer, B. Bhushan, U. Rabe, and W. Arnold. Local elasticity and lubrication measurements using atomic force and friction force microscopy at ultrasonic frequencies. *IEEE Trans. Mag.*, 33:4077–4079, 1997.
23. W. W. Scott and B. Bhushan. Use of phase imaging in atomic force microscopy for measurement of viscoelastic contrast in polymer nanocomposites and molecularly-thick lubricant films. *Ultramicrosc.*, 97:151–169, 2003.

24. B. Bhushan and J. Qi. Phase contrast imaging of nanocomposites and molecularly-thick lubricant films in magnetic media. *Nanotechnol.*, 14:886–895, 2003.
25. T. Kasai, B. Bhushan, L. Huang, and C. Su. Topography and phase imaging using the torsional resonance mode. *Nanotechnol.*, 15:731–742, 2004.
26. B. Bhushan and T. Kasai. A surface topography-independent friction measurement technique using torsional resonance mode in an afm. *Nanotechnol.*, 15:923–935, 2004.
27. B. Bhushan, T. Miyamoto, and V. N. Koinkar. Microscopic friction between a sharp diamond tip and thin-film magnetic rigid disks by friction force microscopy. *Adv. Info. Storage Syst.*, 6:151–161, 1995.
28. V. N. Koinkar and B. Bhushan. Microtribological studies of Al_2O_3, Al_2O_3−TiC, polycrystalline and single-crystal Mn−Zn ferrite and SiC head slider materials. *Wear*, 202:110–122, 1996.
29. V. N. Koinkar and B. Bhushan. Microtribological studies of unlubricated and lubricated surfaces using atomic force/friction force microscopy. *J. Vac. Sci. Technol. A*, 14:2378–2391, 1996.
30. H. Liu and B. Bhushan. Nanotribological characterization of molecularly-thick lubricant films for applications to mems/nems by afm. *Ultramicrosc.*, 97:321–340, 2003.
31. B. Bhushan. Magnetic slider/rigid disk substrate materials and disk texturing techniques – status and future outlook. *Adv. Info. Storage Syst.*, 5:175–209, 1993.
32. B. Bhushan, M. Dominiak, and J. P. Lazzari. Contact- start-stop studies with silicon planar head sliders against thin-film disks. *IEEE Trans. Mag.*, 28:2874–2876, 1992.
33. B. Bhushan and G. S. Blackman. Atomic force microscopy of magnetic rigid disks and sliders and its applications to tribology. *ASME J. Tribol.*, 113:452–458, 1991.
34. P. I. Oden, A. Majumdar, B. Bhushan, A. Padmanabhan, and J. J. Graham. Afm imaging, roughness analysis and contact mechanics of magnetic tape and head surfaces. *ASME J. Tribol.*, 114:666–674, 1992.
35. S. Ganti and B. Bhushan. Generalized fractal analysis and its applications to engineering surfaces. *Wear*, 180:17–34, 1995.
36. C. Y. Poon and B. Bhushan. Comparison of surface roughness measurements by stylus profiler, afm and non-contact optical profiler. *Wear*, 190:76–88, 1995.
37. C. Y. Poon and B. Bhushan. Surface roughness analysis of glass-ceramic substrates and finished magnetic disks, and Ni−P coated Al−Mg and glass substrates. *Wear*, 190:89–109, 1995.
38. V. N. Koinkar and B. Bhushan. Effect of scan size and surface roughness on microscale friction measurements. *J. Appl. Phys.*, 81:2472–2479, 1997.
39. A. Majumdar and B. Bhushan. Role of fractal geometry in roughness characterization and contact mechanics of surfaces. *ASME J. Tribol.*, 112:205–216, 1990.
40. A. Majumdar and B. Bhushan. Fractal model of elastic-plastic contact between rough surfaces. *ASME J. Tribol.*, 113:1–11, 1991.
41. B. Bhushan and A. Majumdar. Elastic-plastic contact model for bifractal surfaces. *Wear*, 153:53–64, 1992.
42. B. Bhushan. Contact mechanics of rough surfaces in tribology: Single asperity contact. *Appl. Mech. Rev.*, 49:275–298, 1996.
43. B. Bhushan. Contact mechanics of rough surfaces in tribology: Multiple asperity contact. *Tribol. Lett.*, 4:1–35, 1998.
44. B. Bhushan and W. Peng. Contact mechanics of multilayered rough surfaces. *Appl. Mech. Rev.*, 55:435–480, 2002.
45. B. Bhushan and V. N. Koinkar. Microtribology of pet polymeric films. *Tribol. Trans.*, 38:119–127, 1995.

46. B. Bhushan and V. N. Koinkar. Macro and microtribological studies of CrO_2 video tapes. *Wear*, 180:9–16, 1995.
47. B. Bhushan and V. N. Koinkar. Microtribology of metal particle, barium ferrite and metal evaporated magnetic tapes. *Wear*, 181–183:360–370, 1995.
48. B. Bhushan and V. N. Koinkar. Microscale mechanical and tribological characterization of hard amorphous carbon coatings as thin as 5 nm for magnetic disks. *Surf. Coat. Tech.*, 76-77:655–669, 1995.
49. V. N. Koinkar and B. Bhushan. Microtribological properties of hard amorphous carbon protective coatings for thin-film magnetic disks and heads. *Proc. Inst. Mech. Eng., J. Eng. Tribol.*, 211:365–372, 1997.
50. A. V. Kulkarni and B. Bhushan. Nanoindentation measurements of amorphous carbon coatings. *J. Mater. Res.*, 12:2707–2714, 1997.
51. X. Li and B. Bhushan. Micro/nanomechanical and tribological characterization of ultra-thin amorphous carbon coatings. *J. Mater. Res.*, 14:2328–2337, 1999.
52. X. Li and B. Bhushan. Mechanical and tribological studies of ultra-thin hard carbon overcoats for magnetic recording heads. *Z. Metallk.*, 90:820–830, 1999.
53. S. Sundararajan and B. Bhushan. Micro/nanotribology of ultra-thin hard amorphous carbon coatings using atomic force/friction force microscopy. *Wear*, 225-229:678–689, 1999.
54. B. Bhushan and S. Sundararajan. Micro/nanoscale friction and wear mechanisms of thin films using atomic force and friction force microscopy. *Acta Mater.*, 46:3793–3804, 1998.
55. J. Ruan and B. Bhushan. Frictional behavior of highly oriented pyrolytic graphite. *J. Appl. Phys.*, 76:8117–8120, 1994.
56. S. Sundararajan and B. Bhushan. Topography-induced contributions to friction forces measured using an atomic force/friction force microscope. *J. Appl. Phys.*, 88:4825–4831, 2000.
57. B. Bhushan and V. N. Koinkar. Microtribological studies of doped single-crystal silicon and polysilicon films for mems devices. *Sensor Actuator A*, 57:91–102, 1997.
58. S. Sundararajan and B. Bhushan. Micro/nanotribological studies of polysilicon and SiC films for mems applications. *Wear*, 217:251–261, 1998.
59. B. Bhushan. Chemical, mechanical and tribological characterization of ultra-thin and hard amorphous carbon coatings as thin as 3.5 nm: Recent developments. *Diam. Relat. Mater.*, 8:1985–2015, 1999.
60. H. Hibst. *Metal evaporated tapes and Co−Cr media for high definition video recording*, pages 137–159. Kluwer, 1993.
61. M. Y. Chu, B. Bhushan, and L. DeJonghe. Wear behavior of ceramic sliders in sliding contact with rigid magnetic thin-film disks. *Tribol. Trans.*, 35:603–610, 1992.
62. B. Bhushan, S. Sundararajan, W. W. Scott, and S. Chilamakuri. Stiction analysis of magnetic tapes. *IEEE Trans. Mag.*, 33:3211–3213, 1997.
63. V. J. Novotny, I. Hussla, J. M. Turlet, and M. R. Philpott. Liquid polymer conformation on solid surfaces. *J. Chem. Phys.*, 90:5861–5868, 1989.
64. V. J. Novotny. Migration of liquid polymers on solid surfaces. *J. Chem. Phys.*, 92:3189–3196, 1990.
65. C. M. Mate and V. J. Novotny. Molecular conformation and disjoining pressures of polymeric liquid films. *J. Chem. Phys.*, 94:8420–8427, 1991.
66. C. M. Mate. Application of disjoining and capillary pressure to liquid lubricant films in magnetic recording. *J. Appl. Phys.*, 72:3084–3090, 1992.
67. B. Bhushan and Z. Zhao. Macro- and microscale studies of molecularly-thick boundary layers of perfluoropolyether lubricants for magnetic thin-film rigid disks. *J. Info. Storage Proc. Syst.*, 1:1–21, 1999.

68. R. Heideman and M. Wirth. Transforming the lubricant on a magnetic disk into a solid fluorine compound. *IBM Technol. Disclosure Bull.*, 27:3199–3205, 1984.
69. A. M. Homola, L. J. Lin, and D. D. Saperstein. Process for bonding lubricant to a thin film magnetic recording disk, 1990.
70. D. D. Saperstein and L. J. Lin. Improved surface adhesion and coverage of perfluoropolyether lubricant following far-uv irradiation. *Langmuir*, 6:1522–1524, 1990.

21 Micro/Nanotribology of MEMS/NEMS Materials and Devices

Bharat Bhushan

Summary. The field of microelectromechanical systemsMEMS/NEMS nanoelectromechanical systemshas expanded considerably over the last decade. The length scale and large surface-to-volume ratio of the devices result in very high retarding forces such as adhesion and friction that seriously undermine the performance and reliability of the devices. These tribological phenomena need to be studied and understood at the micro- to nanoscales. In addition, materials for MEMS/NEMS must exhibit good microscale tribological properties. There is a need to develop lubricants and identify lubrication methods that are suitable for MEMS/NEMS. Using AFM-based techniques, researchers have conducted micro/nanotribological studies of materials and lubricants for use in MEMS/NEMS. In addition, component level testing has also been carried out to aid in better understanding the observed tribological phenomena in MEMS/NEMS.

Macroscale and microscale tribological studies of silicon and polysilicon films have been performed. The effects of doping and oxide films and environment on the tribological properties of these popular MEMS/NEMS materials have also been studied. SiC film is found to be a good tribological material for use in high-temperature MEMS/NEMS devices. Hexadecane thiol self-assembled monolayers and bonded perfluoropolyether lubricants appear to be well suited for lubrication of microdevices under a range of environmental conditions. DLC coatings can also be used for low friction and wear. Surface roughness measurements of micromachined polysilicon surfaces have been made using an AFM. The roughness distribution on surfaces is strongly dependent on the fabrication process. Roughness should be optimized for low adhesion, friction, and wear. Adhesion and friction of microstructures can be measured using novel apparatuses. Adhesion and friction measurements on silicon-on-silicon confirm AFM measurements that hexadecane thiol and bonded perfluoropolyether films exhibit superior adhesion and friction properties. Static friction force measurements of micromotors have been performed using an AFM. The forces are found to vary considerably with humidity. A bonded layer of perfluoropolyether lubricant is found to satisfactorily reduce the friction forces in the micromotor.

AFM/FFM-based techniques can be satisfactorily used to study and evaluate micro/nanoscale tribological phenomena related to MEMS/NEMS devices.

This chapter presents a review of macro- and micro/nanoscale tribological studies of materials and lubrication studies for MEMS/NEMS and component-level studies of stiction phenomena in MEMS/NEMS devices.

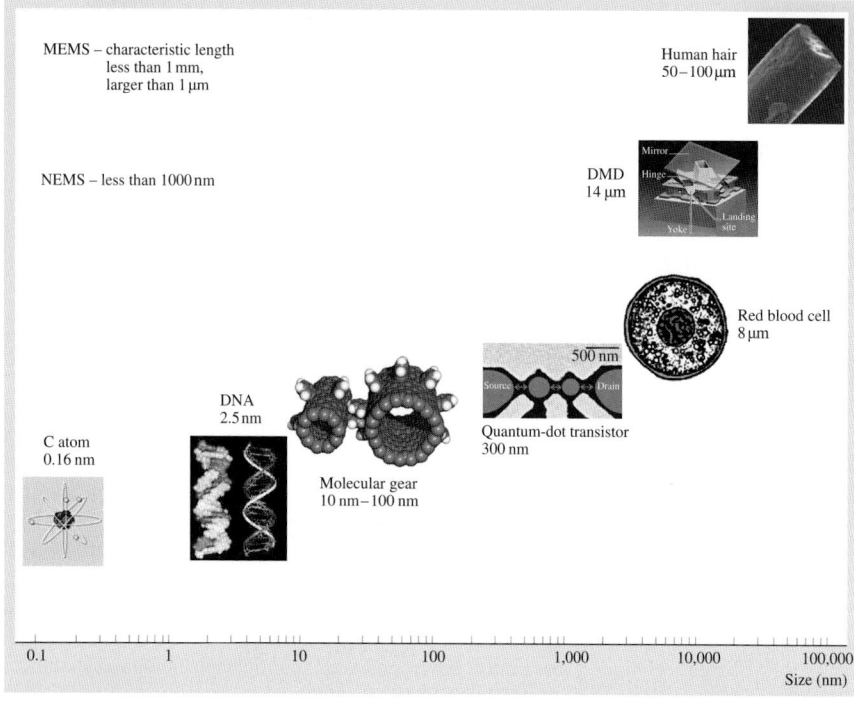

Fig. 21.1. Dimensions of MEMS and NEMS in perspective. An example of molecular dynamic simulations of carbon-nanotube-based gear is obtained from [1], quantum-dot transistor is obtained from [2], and DMD is obtained from [3]

21.1 Introduction

Microelectromechanical systems (MEMS) refer to microscopic devices that have a characteristic length of less than 1 mm but more than 1 μm and combine electrical and mechanical components. Nanoelectromechanical systems (NEMS) refer to nanoscopic devices that have a characteristic length of less than 1 μm and combine electrical and mechanical components. In mesoscale devices, if the functional components are on the micro- or nanoscale, they may be referred to as MEMS or NEMS, respectively. To put the dimensions and masses in perspective, see Fig. 21.1 and Table 21.1. The acronym MEMS originated in the United States. The term commonly used in Europe and Japan is micro/nanodevices, which is used in a much broader sense. MEMS/NEMS terms are also now used in a broad sense. A micro/nanosystem, a term commonly used in Europe, is referred to as an intelligent miniaturized system comprising sensing, processing, and/or actuating functions.

Fabrication techniques include top-down methods, in which one builds down from the large to the small, and the bottom-up methods, in which one builds up from the small to the large. Top-down methods include micro/nanomachining methods and methods based on lithography, as well as nonlithographic miniaturization for MEMS

Table 21.1. Dimensions and masses in perspective

A Dimensions in Perspective

NEMS characteristic length	< 1000 nm
MEMS characteristic length	< 1 mm and > 1 μm
Individual atoms	typically fraction of a nm in diameter
DNA molecules	~ 2.5 nm wide
Molecular gear	~ 50 nm
Biological cells	in the range of thousands of nm in diameter
Human hair	~ 75 000 nm in diameter

B Masses in Perspective

NEMS built with cross sections of about 10 nm	as low as 10^{-20} N
Micromachine silicon structure	as low as 1 nN
Water droplet	~ 10 μN
Eyelash	~ 100 nN

and NEMS fabrication. In the bottom-up methods, also referred to as nanochemistry, the devices and systems are assembled from their elemental constituents for NEMS fabrication, much like the way nature uses proteins and other macromolecules to construct complex biological systems. The bottom-up approach has the potential to go far beyond the limits of top-down technology by producing nanoscale features through synthesis and subsequent assembly. Furthermore, the bottom-up approach offers the potential to produce structures with enhanced and/or completely new functions. It allows the combination of materials with distinct chemical composition, structure, and morphology.

MEMS and emerging NEMS are expected to have a major impact on our lives, comparable to that of semiconductor technology, information technology, or cellular and molecular biology [4, 5]. MEMS/NEMS are used in mechanical, information/communication, chemical, and biological applications. The MEMS industry in 2000 was worth about
$15 billion and, with a projected 10–20% annual growth rate, is expected to be more than $100 billion by the end of this decade [6]. Growth of Si MEMS may slow down and non-silicon MEMS may pick up during this decade. The NEMS industry is expected to expand in this decade, mostly in biomedical applications, as well as in nanoelectronics or molecular electronics. The NEMS industry was worth about $100 million in 2002 and integrated NEMS are expected to be more than $25 billion by the end of this decade. Due to the enabling nature of these systems, and because of the significant impact they can have on both commercial and defense applications, industry, as well as the federal government have taken a special interest in seeing growth nurtured in this field. Micro- and nanosystems are the next logical step in the "silicon revolution."

21.2 Introduction to MEMS

The advances in silicon photolithographic process technology beginning in the 1960s led to the development of MEMS in the early 1980s. More recently, lithographic processes have been developed to process nonsilicon materials. The lithographic processes are being complemented with non-lithographic processes for fabrication of components or devices made from plastics or ceramics. Using these fabrication processes, researchers have fabricated a wide variety of miniaturized devices, including Si-based devices, chemical and biological sensors and actuators, and miniature non-silicon structures (e.g., devices made from plastics or ceramics) with dimensions in the range of a couple to a few thousand microns (see e.g., [7–15]). MEMS for mechanical applications include acceleration, pressure, flow, gas sensors, linear and rotary actuators, and other microstructures or microcomponents such as electric motors, gear trains, gas turbine engines, nozzles, fluid pumps, fluid valves, switches, grippers, and tweezers. MEMS for chemical applications include chemical sensors and various analytical instruments. Microoptoelectromechanical systems, or MOEMS, include micromirror arrays and fiber optic connectors. BIOMEMS include biochips. Radio frequency MEMS or RF-MEMS include inductors, capacitors, and antennas. High-aspect-ratio MEMS (HARMEMS) have also been introduced.

The fabrication techniques for MEMS devices include lithographic and non-lithographic techniques. The lithographic techniques fall into three basic categories: bulk micromachining, surface micromachining, and LIGA (a German acronym for Lithographie Galvanoformung Abformung), a German term for lithography, electroforming, and plastic molding. The first two approaches, bulk and surface micromachining, use planar photolithographic fabrication processes developed for semiconductor devices in producing two-dimensional (2-D) structures [10, 15–17]. The various steps involved in these two fabrication processes are shown schematically in Fig. 21.2. Bulk micromachining employs anisotropic etching to remove sections through the thickness of a single-crystal silicon wafer, typically 250 to 500 µm thick. Bulk micromachining is a proven high-volume production process and is routinely used to fabricate microstructures such as accelerometers, pressure sensors, and flow sensors. In surface micromachining, structural and sacrificial films are alternatively deposited, patterned and etched to produce a freestanding structure. These films are typically made of low-pressure chemical vapor deposition (LPCVD) polysilicon film with 2 to 20 µm thickness. Surface micromachining is used to produce sensors, actuators, and complex microdevices such as micromirror arrays, motors, gears, and grippers.

The LIGA process is based on the combined use of X-ray lithography, electroforming, and molding processes. The steps involved in the LIGA process are shown schematically in Fig. 21.3. LIGA is used to produce high-aspect-ratio MEMS (HARMEMS) devices that are up to 1 mm in height and only a few microns in width or length [18]. The LIGA process yields very sturdy 3-D structures due to their increased thickness. One of the limitations of silicon microfabrication processes originally used for fabrication of MEMS devices is the lack of suitable materials that can be processed. With LIGA, a variety of non-silicon materials such as

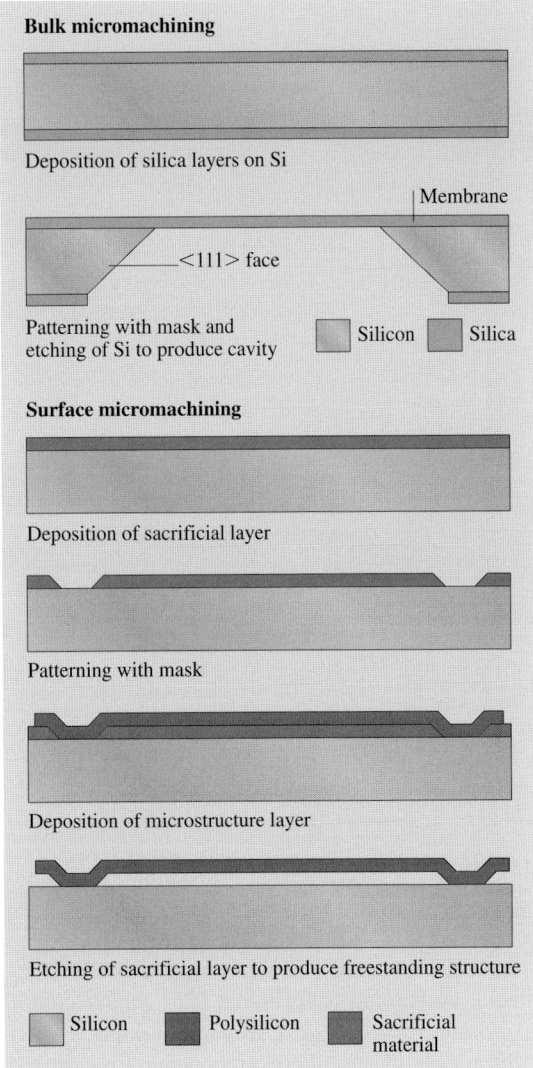

Fig. 21.2. Schematics of process steps involved in bulk micromachining and surface micromachining fabrication of MEMS

metals, ceramics, and polymers can be processed. Nonlithographic micromachining processes, primarily in Europe and Japan, are also being used for fabrication of millimeter-scale devices using direct material microcutting or micromechanical machining (such as microturning, micromilling, and microdrilling), or removal by energy beams (such as microspark erosion, focused ion beam, laser ablation, and laser polymerization) [19, 20]. Hybrid technologies, including LIGA and high-precision micromachining techniques, have been used to produce miniaturized motors, gears,

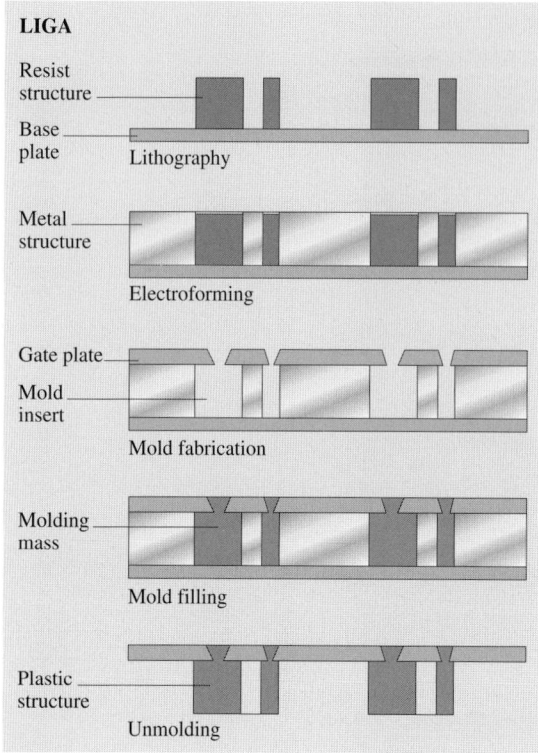

Fig. 21.3. Schematic of process steps involved in LIGA fabrication of MEMS

actuators, and connectors [21–24]. These millimeter-scale devices may find more immediate applications.

A microfabrication technique, so-called "soft lithography," is a nonlithographic technique [25, 26] in which an elastomeric stamp or mold is used to generate micropatterns by replica molding [27], embossing (imprinting) [28], or by contact printing (known as microcontact printing or µCP) [29]. Replica molding is commonly used for mass-produced, disposable plastic microcomponents, for example, microfluidic chips, generally made of poly(dimethylsiloxane) (PDMS) and polymethyl methacrylate (PMMA). The elastomeric stamps are generally cast from photolithographically generated micropatterned masters. This technique is substantially cheaper and more flexible in choice of materials for construction than conventional photolithography.

To assemble microsystems, microrobots are used. Microrobotics include building blocks such as steering links, microgrippers, conveyor system, and locomotive robots [13].

A variety of MEMS devices have been produced and some are commercially used [8, 10–12, 14, 15]. A variety of sensors are used in industrial, consumer, defense, and biomedical applications. Various microstructures or microcomponents are used in micro-instruments and other industrial applications such as micromirror arrays. Two of the largest "killer" industrial applications are accelerometers (about 85 million units in 2002) and digital micromirror devices (about $400 million in sales in 2001). Integrated capacitive-type, silicon accelerometers have been used in airbag deployment in automobiles since 1991 [30, 31]. Accelerometer technology was about a billion-dollar-a-year industry in 2001 dominated by Analog Devices, followed by Motorola and Bosch. Commercial digital light processing (DLP) equipments, using digital micromirror devices (DMD), were launched in 1996 by Texas Instruments for digital projection displays in portable and home theater projectors, as well as table top and projection TVs [32–34]. More than 1.5 million projectors were sold before 2002. Other major industrial applications include pressure sensors, inkjet printer heads and optical switches. Silicon-based piezoresistive pressure sensors for manifold absolute pressure sensing for engines were launched in 1991 by NovaSensor, and their annual sales were about 25 million units in 2002. Capacitive pressure sensors for tire pressure measurements were launched by Motorola. Annual sales of inkjet printer heads with microscale functional components were about 400 million units in 2002. Other applications of MEMS devices include chemical sensors, gas sensors, infrared detectors and focal plane arrays for earth observation, space science, and missile defense applications, pico-satellites for space applications, and many hydraulic, pneumatic, and other consumer products. MEMS devices are also being pursued in magnetic storage systems [35], where they are being developed for super-compact and ultrahigh-recording-density magnetic disk drives. Several integrated head/suspension microdevices have been fabricated for contact recording applications [36, 37]. High-bandwidth servo-controlled microactuators have been fabricated for ultrahigh-track-density applications that serve as the fine-position control element of a two-stage, coarse/fine servo system, coupled with a conventional actuator [38–40].

BIOMEMS are used increasingly in commercial and defense applications (e.g., [8, 41–44]). Applications of BIOMEMS include biofluidic chips (microfluidic chips or bioflips or simply biochips) for chemical and biochemical analyses (biosensors) in medical diagnostics (e.g., DNA, RNA, proteins, cells, blood pressure and assays, and toxin identification) and implantable pharmaceutical drug delivery. The biosensors, also referred to as lab-on-a-chip, integrate sample handling, separation, detection, and data analysis onto one platform. Biosensors are designed to either detect a single or class of (bio)chemicals or system-level analytical capabilities for a broad range of (bio)chemical species known as micro total analysis systems (μTAS). The chips rely on microfluidics and involve manipulation of tiny amounts of fluids in microchannels using microvalves. The chips consist of several basic components, including microfluidic channels and reservoirs, microvalves, micropumps, flow sensors, and biosensors. The test fluid is injected into the chip generally using an external pump or a syringe for analyses. Some chips have been designed with an integrated electrostatically actuated diaphragm-type micropump. Micropumps both with and without valves are used. The sample, which can have volume measured in nanoliters, flows through microfluidic channels via an electric potential and capillary action using microvalves (having various designs, including membrane type) for various analyses. The fluid is preprocessed and then analyzed using a biosensor. For a review on micropumps and microvalves, see [11, 45, 46]. Silicon-based disposable blood-pressure sensor chips were introduced in the early 1990s by NovaSensor for blood pressure monitoring (about 20 million units in 2002). A blood sugar monitor, referred to as GlucoWatch, was introduced in 2002. It automatically checks blood sugar every 10 min by detecting glucose through the skin, without having to draw blood. If glucose is out of the acceptable range, it sounds an alarm so the diabetic can address the problem quickly. A variety of biosensors, many using plastic substrates, are manufactured by various companies, including ACLARA, Agilent Technologies, Calipertech, and I-STAT.

After the tragedy of Sept. 11, 2001, concern over biological and chemical warfare has led to the development of handheld units with bio- and chemical sensors for detection of biological germs, chemical or nerve agents, and mustard agents, and chemical precursors to protect subways, airports, water supply, and population at large [47].

Other BIOMEMS applications include minimal invasive surgery, including endoscopic surgery, laser angioplasty, and microscopic surgery. Implantable artificial organs can also be produced.

Micro-instruments and micromanipulators are used to move, position, probe, pattern, and characterize nanoscale objects and nanoscale features. Miniaturized analytical equipment include gas chromatography and mass spectrometry. Other instruments include micro-STM, where STM stands for scanning tunneling microscope.

In some cases, MEMS devices are used primarily for their miniature size. While in others, as in the case of air bags, they are used because of their low-cost manufacturing techniques, since semiconductor-processing costs have reduced drastically over the last decade, allowing the use of MEMS in many fields.

21.3 Introduction to NEMS

NEMS nanoelectromechanical systemsare produced by nanomachining in a typical top-down approach (from large to small) and bottom-up approach (from small to large) largely relying on nanochemistry [48–53]. The top-down approach relies on fabrication methods, including advanced integrated-circuit (IC) lithographic methods – electron-beam lithography and STM writing by removing material atom by atom. The bottom-up approach includes chemical synthesis, the spontaneous "self-assembly" of molecular clusters (molecular self-assembly) from simple reagents in solution, or biological molecules (e.g., DNA) as building blocks to produce 3-D nanostructures, quantum dots (nanocrystals) of arbitrary diameter (about $10-10^5$ atoms), molecular beam epitaxy (MBE) and organometallic vapor phase epitaxy (OMVPE) to create specialized crystals one atomic or molecular layer at a time, and manipulation of individual atoms by an atomic force microscope or atom optics. The self-assembly must be encoded. That is, one must be able to precisely assemble one object next to another to form a designed pattern. A variety of nonequilibrium plasma chemistry techniques are also used to produce layered nanocomposites, nanotubes, and nanoparticles. The NEMS field, in addition to fabrication of nanosystems, has provided the impetus to develop experimental and computation tools.

Examples of NEMS include nanocomponents, nanodevices, nanosystems, and nanomaterials such as microcantilevers with integrated sharp nanotips for STM and atomic force microscopy (AFM), AFM array (millipede) for data storage, AFM tips for nanolithography, dip-pen nanolithography for printing molecules, biological (DNA) motors, molecular gears, molecularly thick films (e.g., in giant magnetoresistive or GMR heads, and magnetic media), nanoparticles (e.g., nanomagnetic particles in magnetic media), nanowires, carbon nanotubes, quantum wires (QWRs), quantum boxes (QBs), and quantum transistors. BIONEMS include nanobiosensors – a microarray of silicon nanowires, roughly a few nm in size, to selectively bind and detect even a single biological molecule such as DNA or protein by using nanoelectronics to detect the slight electrical charge caused by such binding, or a microarray of carbon nanotubes to electrically detect glucose, implantable drug-delivery devices – e.g., micro/nanoparticles with drug molecules encapsulated in functionized shells for a site-specific targeting applications and a silicon capsule with a nanoporous membrane filled with drugs for long-term delivery – , nanodevices for sequencing single molecules of DNA in the Human Genome Project, cellular growth using carbon nanotubes for spinal cord repair, nanotubes for nanostructured materials for various applications such as spinal fusion devices, organ growth, and growth of artificial tissues using nanofibers.

Nanoelectronics can be used to build computer memory using individual molecules or nanotubes to store bits of information, molecular switches, molecular or nanotube transistors, nanotube flat-panel displays, nanotube integrated circuits, fast logic gates, switches, nanoscopic lasers, and nanotubes as electrodes in fuel cells.

21.4 Tribological Issues in MEMS/NEMS

Tribological issues are important in MEMS/NEMS requiring intended and/or unintended relative motion. In MEMS/NEMS, various forces associated with the device scale down with size. When the length of the machine decreases from 1 mm to 1 μm, the area decreases by a factor of a million and the volume decreases by a factor of a billion. As a result, surface forces such as adhesion, friction, meniscus forces, viscous drag forces, and surface tension that are proportional to area become a thousand times larger than the forces proportional to the volume such as inertial and electromagnetic forces. In addition to the consequence of a large surface-to-volume ratio, since MEMS/NEMS are designed for small tolerances, physical contact becomes more likely, which makes them particularly vulnerable to adhesion between adjacent components. Since the start-up forces and torques involved in MEMS/NEMS operation available to overcome retarding forces are small, the increase in resistive forces such as adhesion and friction become a serious tribological concern that limits the life and reliability of MEMS/NEMS [10]. A large lateral force required to initiate relative motion between two surfaces, large static friction, is referred to as "stiction," which has been studied extensively in tribology of magnetic storage systems [35, 54, 55]. A large normal force required to separate two surfaces is also referred to as stiction. Adhesion, friction/stiction (static friction), wear, and surface contamination affect MEMS/NEMS performance and in some cases, can even prevent devices from working. Some examples of devices that experience tribological problems follow.

21.4.1 MEMS

Figure 21.4a shows examples of several microcomponents that can encounter the aforementioned tribological problems. The polysilicon electrostatic micromotor has 12 stators and a four-pole rotor and is produced by surface micromachining. The rotor diameter is 120 μm and the air gap between the rotor and stator is 2 μm [56]. It is capable of continuous rotation up to speeds of 100 000 RPM. The intermittent contact at the rotor-stator interface and physical contact at the rotor-hub flange interface result in wear issues, and high stiction between the contacting surfaces limits the repeatability of operation, or may even prevent the operation altogether. Next, a bulk micromachined silicon stator/rotor pair is shown with bladed rotor and nozzle guide vanes on the stator with dimensions less than a mm [57]. These are being developed for high-temperature micro-gas turbine generators with an operating speed up to 1 million RPM. Erosion of blades and vanes and wear of bearings used in the turbine generators are some of the concerns. Next, is an SEM micrograph of a surface micromachined polysilicon six-gear chain from Sandia National Lab. (For more examples of previous versions, see [58, 59].) As an example of non-silicon components, a milligear system produced using the LIGA process for a DC brushless permanent magnet millimotor (diameter = 1.9 mm, length = 5.5 mm) with an integrated milligear box [21–23] is also shown. The gears are made of metal (electroplated Ni-Fe), but can also be made from injected polymer materials (e.g., Polyoxy-methylene, or POM) using the LIGA process. Even though the torque transmitted at the gear teeth is small, on the order of a fraction of a nNm, because of small dimensions of gear teeth, the bending stresses are large where the teeth mesh. Tooth breakage and wear at the contact of gear teeth are concerns. Next, in a micromachined flow modulator, several micromachined flow channels are integrated in series with electrostatically actuated microvalves [60]. The flow channels lead to a central gas outlet hole drilled in the glass substrate. Gas enters the device through a bulk micromachined gas inlet hole in the silicon cap. After passing through an open microvalve, the gas flows parallel to the glass substrate through flow channels and exits the device through an outlet. The normally open valve structure consists of a freestanding double-end-clamped beam, which is positioned beneath the gas inlet orifice. When deflected electrostatically upwards, the beam seals against the inlet orifice, and the valve is closed. In these microvalves used for flow control, the mating valve surfaces should be smooth enough to seal while maintaining a minimum roughness to ensure low adhesion [54, 55, 61]. High adhesion (stiction) is a major issue.

Figure 21.4b shows a polysilicon, multiple microgear speed reduction unit and its components after laboratory wear tests conducted in air [62]. These units have been developed for an electrostatically driven microactuator (microengine) developed at

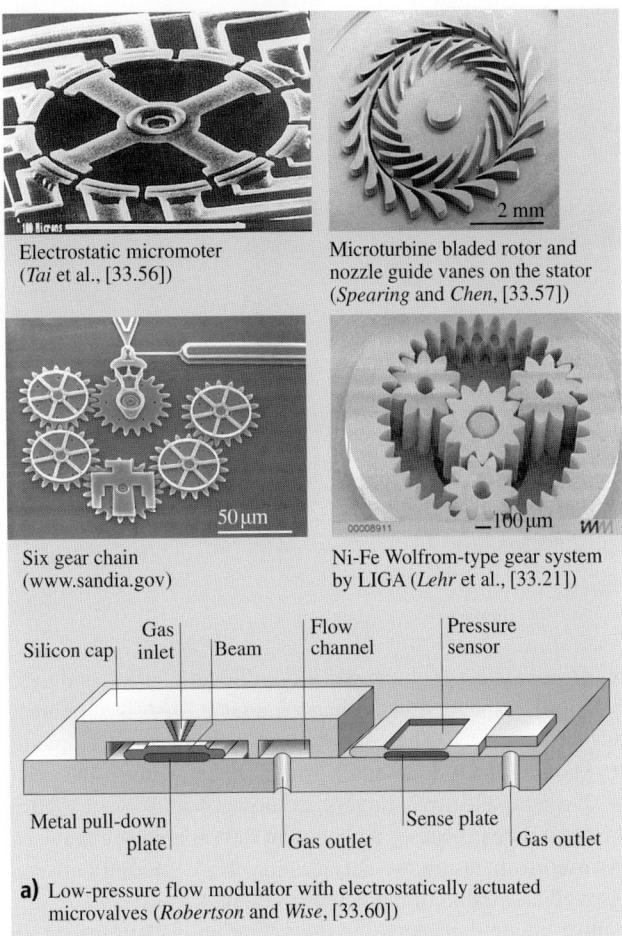

Fig. 21.4. Examples of MEMS devices and components that experience tribological problems: (**a**) several microcomponent and (**b**) a polysilicon, multiple microgear speed reduction unit

Sandia National Lab for operation in the kHz frequency range. Wear of various components is clearly observed in the figure. Humidity was shown to be a strong factor in the wear of rubbing surfaces. In order to improve the wear characteristics of rubbing surfaces, 20-nm-thick tungsten (W) coating using chemical vapor deposition (CVD) technique was used [63]. Tungsten-coated microengines tested for reliability showed improved wear characteristics with longer lifetimes than polysilicon microengines.

Commercially available MEMS devices also exhibit tribological problems. Figure 21.5 shows an integrated capacitive-type silicon accelerometer fabricated using surface micromachining by Analog Devices, a couple of mm in dimension, which is used for the deployment of airbags in automotives and more recently for the consumer electronics market [30, 64, 65]. These accelerometers are now being devel-

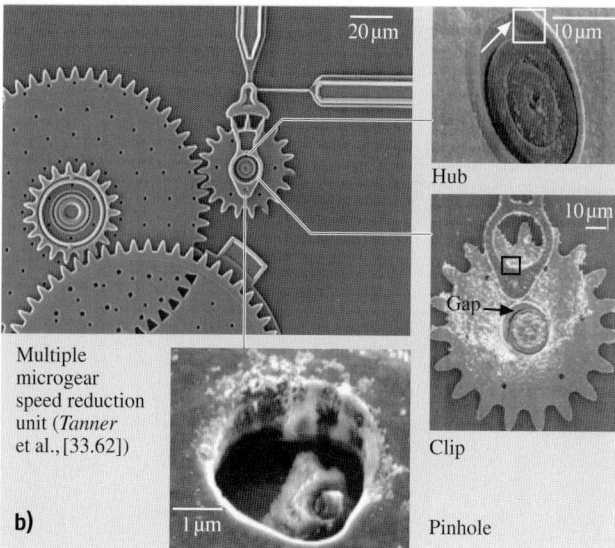

Fig. 21.4. continue

oped for many other applications such as vehicle stability, rollover control, and gyro sensor applications. The central suspended beam mass (about 0.7 μg) is supported on the four corners by spring structures. The central beam has interdigitated, cantilevered electrode fingers (about 125 μm long and 3 μm thick) on all four sides that alternate with those of the stationary electrode fingers, as shown, with about a 1.3 μm gap. Lateral motion of the central beam causes a change in the capacitance between these electrodes, which is used to measure the acceleration. Here stiction between the adjacent electrodes, as well as stiction of the beam structure
with the underlying substrate are detrimental to the operation of the sensor [30, 64]. Wear during unintended contacts of these polysilicon fingers is also a problem. A molecularly thick diphenyl siloxane lubricant film with high resistance to temperature and oxidation, applied by a vapor deposition process is used on the electrodes to reduce stiction and wear.

Figure 21.5 also shows two digital micromirror device (DMD) pixels used in digital light processing (DLP) technology for digital projection displays in portable and home theater projectors, as well as table-top and projection TVs [32–34]. The entire array (chip set) consists of a large number of rotatable aluminium micromirrors (digital light switches) that are fabricated on top of a CMOS static random access memory integrated circuit. The surface micromachined array consists of half of a million to more than two million of these independently controlled, reflective micromirrors (mirror size on the order of 14 μm square and 15 μm pitch) that flip backward and forward at a frequency on the order of 5000 times per s. For the binary operation, a micromirror/yoke structure mounted on torsional hinges is rotated ±10° (with respect to the horizontal plane) as a result of electrostatic attraction between the

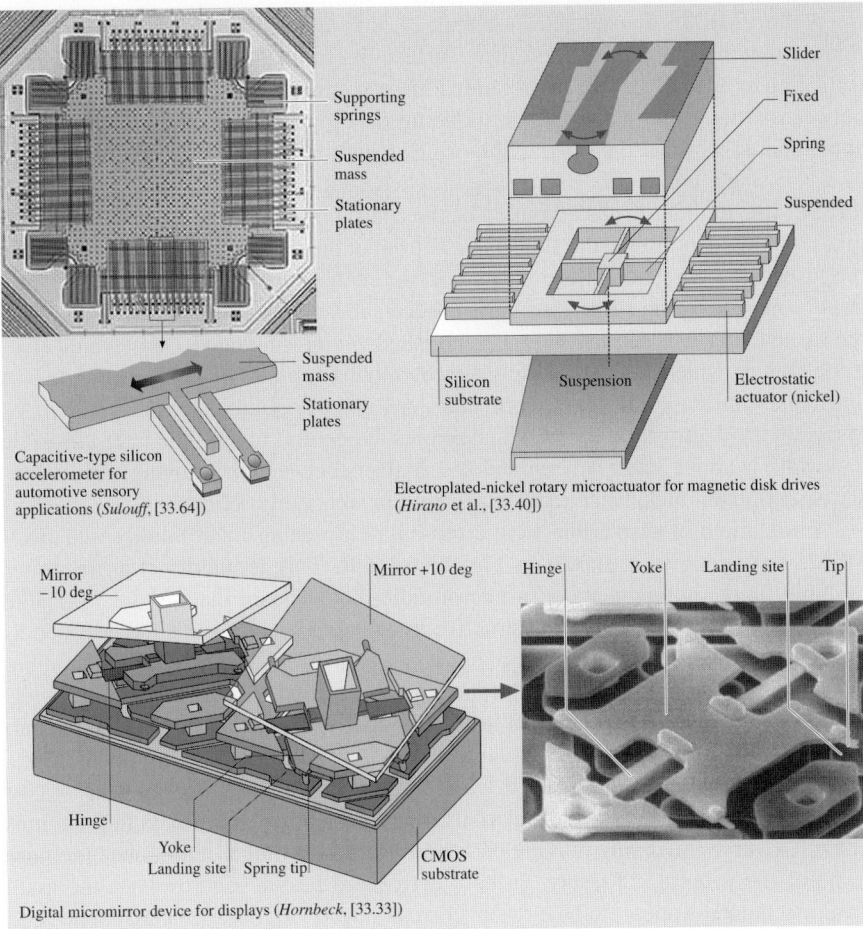

Fig. 21.5. Examples of commercial use MEMS devices that experience tribological problems

micromirror structure and the underlying memory cell and is limited by a mechanical stop. Contact between cantilevered spring tips at the end of the yoke (four present on each yoke) with the underlying stationary landing sites is required for true digital (binary) operation. Stiction and wear during a contact between aluminium alloy spring tips and landing sites, hinge memory (metal creep at high operating temperatures), hinge fatigue, shock and vibration failure, and sensitivity to particles in the chip package and operating environment are some of the important issues affecting the reliable operation of a micromirror device [66–68]. Perfluorodecanoic acid (PFDA) self-assembled monolayers are used on the tip and landing sites to reduce stiction and wear. The spring tip is employed in order to use the spring stored energy to pop up the tip during pull-off. A lifetime estimate of over 100 000 operating hours with no degradation in image quality is the norm.

A third MEMS device, shown in Fig. 21.5, is an electrostatically driven rotary microactuator for a magnetic disk drive surface-micromachined by a multilayer electroplating method [40]. This high-bandwidth servo-controlled microactuator, located between a slider and a suspension, is being developed for ultrahigh-track-density applications, which serves as the fine-position and high-bandwidth control element of a two-stage, coarse/fine servo system when coupled with a conventional actuator [38–40]. A slider is placed on top of the central block of a microactuator, which gives rotational motion to the slider. The bottom of the silicon substrate is attached to the suspension. The radial flexure beams in the central block give the rotational freedom of motion to the suspended mass (slider), and the electrostatic actuator drives the suspended mass. Actuation is accomplished via interdigitated, cantilevered electrode fingers, which are alternatingly attached to the central body of the moving part and to the stationary substrate to form pairs. A voltage applied across these electrodes results in an electrostatic force, which rotates the central block. The inter-electrode gap width is about 2 μm. Any unintended contacts between the moving and stationary electroplated-nickel electrodes may result in wear and stiction.

An example of disposable plastic lab-on-a-chip using microfluidics and BIO-MEMS technologies is shown in Fig. 21.6a [69]. This biofluidic chip integrates multiple fluidic components on a microfluidic motherboard that can be worn like a wristwatch. It enables the unobstructive assessment of a human's medical condition through continuous blood sampling. It can be used for the analysis of a number of different compounds, including lactate, glucose, carbon dioxide, and oxygen levels in the blood and for the detection of infectious diseases. The test fluid is injected into the microchannels using a micropump, and the flow is regulated using microvalves. A magnetic bead approach is used for identification of target biomolecules based on sandwich immunoassay and electrochemical detection. Adhesion in micropumps and microvalves involving moving parts and adhesion of fluid in microchannels are some of the tribological issues. In both active and passive microvalves, the main wear mechanism appears to be erosion, corrosion, and fatigue. As the magnetic bead particulates are pumped through the microfluidic system, they could erode the valve material. Also, several different reagents are used to perform the necessary analyses. The interaction of these reagents with the valve material could cause corrosive wear. Finally, because the valves are continuously being activated, fatigue failure becomes a concern.

Stiction/friction and wear clearly limit the lifetimes and compromise the performance and reliability of microdevices. Figure 21.7 summarizes tribological problems encountered in some of the MEMS devices and components just discussed. In addition, tribological issues are present in the processes used for fabrication of MEMS/NEMS. For example, in surface micromachining, the suspended structures can sometimes collapse and permanently adhere to the underlying substrate due to meniscus effects during the final rinse and dry process, as shown in Fig. 21.7 [70]. Adhesion is caused by water molecules adsorbed on the adhering surfaces and/or because of formation of adhesive bonds by silica residues that remain on the surfaces after the water has evaporated. This so called release stictionis overcome by using dry release methods (e.g., CO_2 critical point drying or sublimation methods [71]).

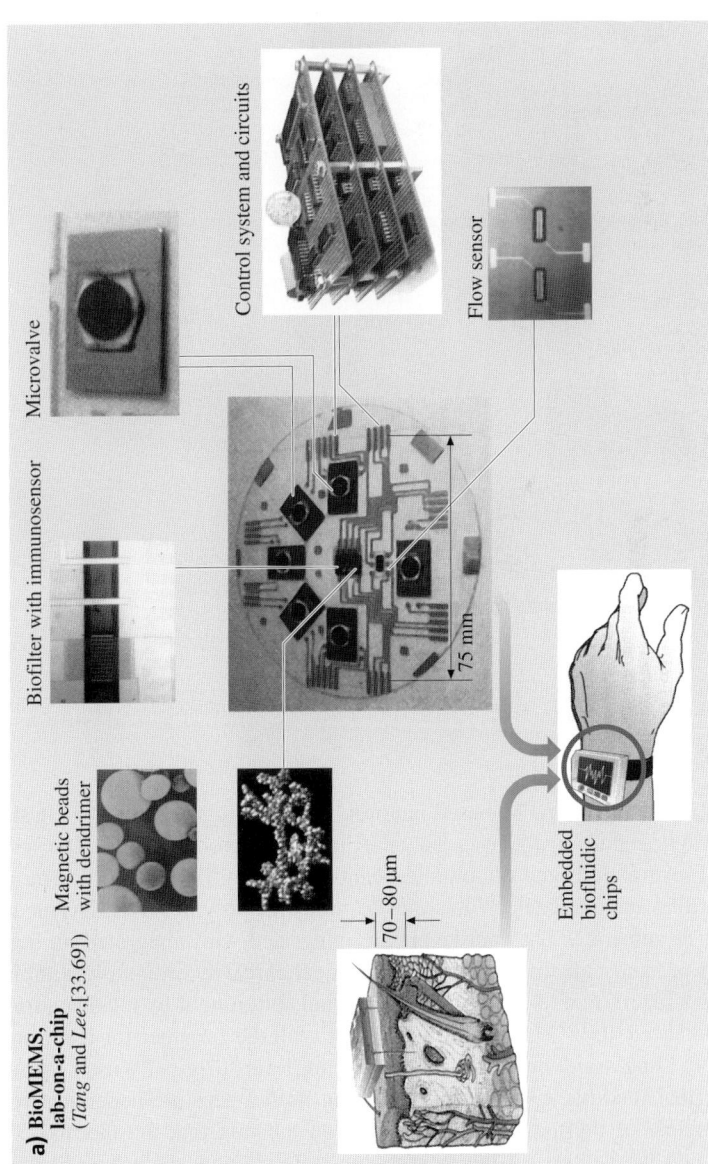

Fig. 21.6. Schematics of (**a**) a MEMS-based biofluidic chip, commonly known as disposable lab-on-a-chip, and (**b**) (*see next page*) a NEMS-based submicroscopic drug-delivery device, and an intravascular micro/nanoparticles for search-and-destroy diseased blood cells

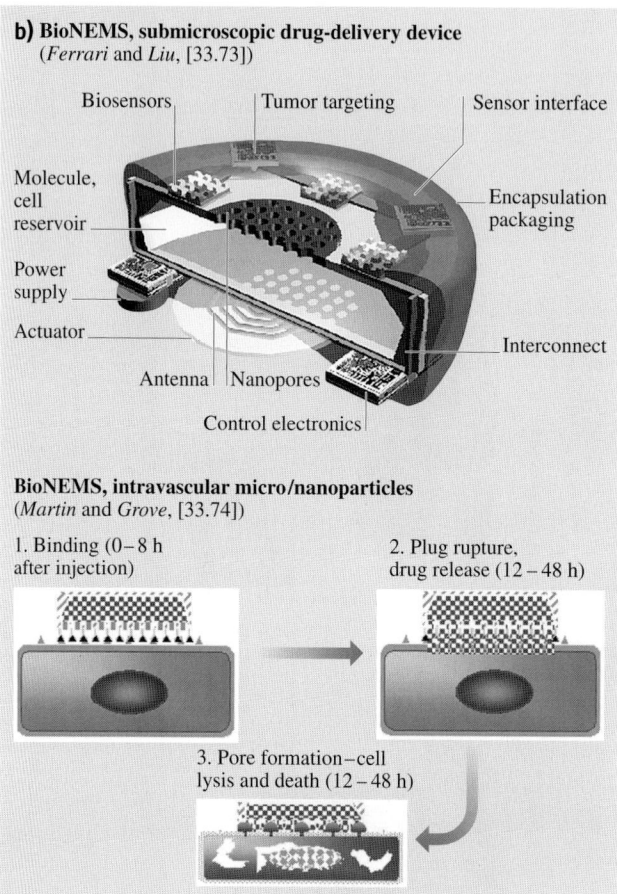

Fig. 21.6. continue

21.4.2 NEMS

Figure 21.8 shows an AFM-based nanoscale data storage system for ultrahigh density magnetic recording that experiences tribological problems [72]. The system uses arrays of several thousand silicon microcantilevers ("millipede") for thermomechanical recording and playback on an about 40-nm-thick polymer (PMMA) medium with a harder Si substrate. The cantilevers are integrated with integrated tip heaters that have tips of nanoscale dimensions. Thermomechanical recording is a combination of applying a local force to the polymer layer, and softening it by local heating. The tip heated to about 400 °C is brought in contact with the polymer for recording. Imaging and reading are done using the heater cantilever, originally used for recording, as a thermal read-back sensor by exploiting its temperature-dependent resistance. The principle of thermal sensing is based on the fact that the thermal conductivity between the heater and the storage substrate changes according to the spacing

21 Micro/Nanotribology of MEMS/NEMS Materials and Devices 1047

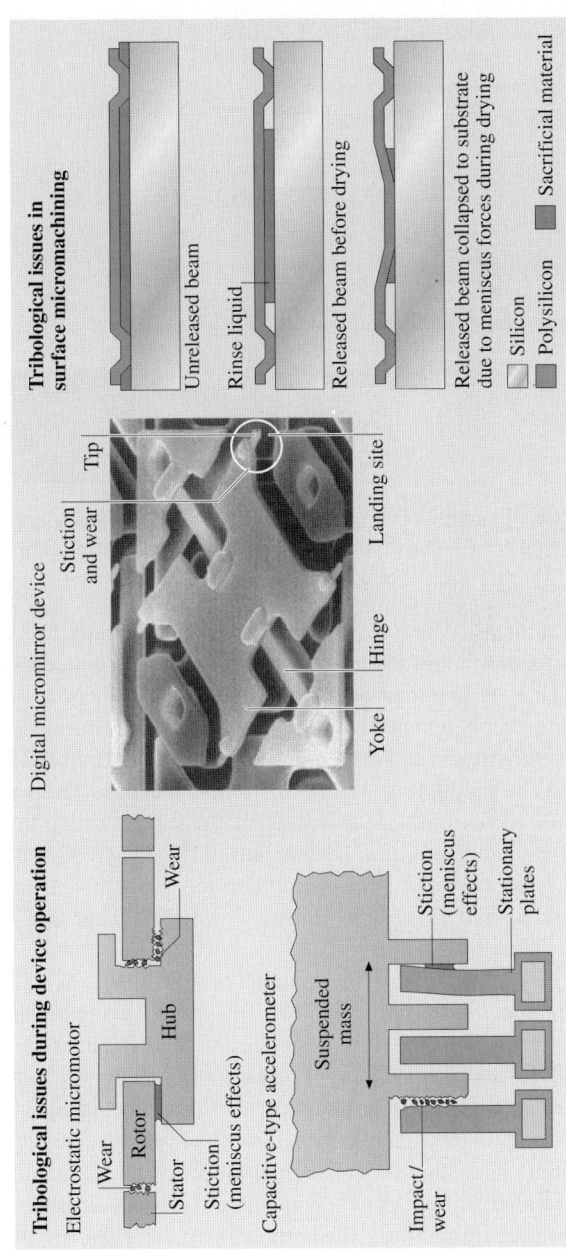

Fig. 21.7. Summary of tribological issues in MEMS device operation and fabrication via surface micromachining

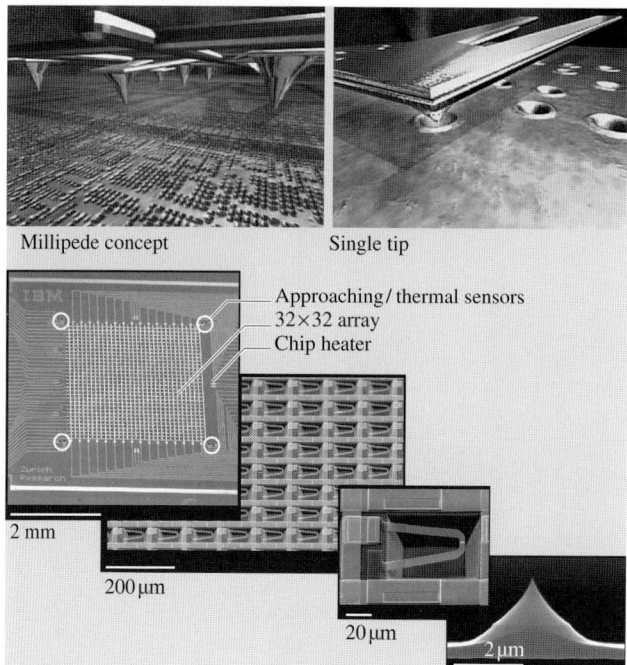

Fig. 21.8. Example of a NEMS device – AFM-based nanoscale data storage system with 32 × 32 tip array – that experiences tribological problems [72]

between them. When the spacing between the heater and sample is reduced as the tip moves into a bit, the heater's temperature and hence its resistance will decrease. Thus, changes in temperature of the continuously heated resistor are monitored while the cantilever is scanned over data bits, providing a means of detecting the bits. Erasing for subsequent rewriting is carried out by thermal reflow of the storage field by heating the medium to 150 °C for a few seconds. The smoothness of the reflown medium allows multiple rewriting of the same storage field. Bit sizes ranging between 10 and 50 nm have been achieved by using a 32 × 32 (1024) array write/read chip (3 mm × 3 mm). It has been reported that tip wear occurs through the contact between tip and Si substrate during writing. Tip wear is considered a major factor in device reliability.

Figure 21.6b shows two examples of BIONEMS [73, 74]. The figure shows a conceptual model of a submicroscopic drug delivery device with the ability to localize in the areas of need. It is a silicon capsule with a nanoporous membrane filled with drugs for long-term delivery. It uses a nanomembrane with pores as small as 6 nm that are used as flux regulators for the long-term release of drugs. The nanomembrane also protects therapeutic substances from attack by the body's immune system. The figure also shows a conceptual model of an intravascular drug delivery device – micro/nanoparticle used for search-and-destroy disease cells. With the lateral dimensions of 1 μm or less, the particle is smaller than any blood cells. These particles can be injected into the blood stream and travel freely through the circulatory system. In order to direct these drug-delivery micro/nanoparticles to cancer sites, their external surfaces are chemically modified to carry molecules that have lock-and-key binding specificity with molecules that support a growing cancer mass. As soon as the particles dock on the cells, a compound is released that forms a pore on the membrane of the cells, which leads to cell death and ultimately to that of the cancer mass that was nourished by the blood vessel. Adhesion between micro/nanodevices and disease cells is required.

21.4.3 Tribological Needs

The MEMS/NEMS need to be designed to perform expected functions typically in the ms to ps range. Expected life of the devices for high speed contacts can vary from a few hundred thousand to many billions of cycles, e.g., over a hundred billion cycles for DMDs, which puts serious requirements on materials [10, 62, 75–78]. Most mechanical properties are known to be scale-dependent [79]. The properties of nanoscale structures need to be measured [80]. Tribology is an important factor affecting the performance and reliability of MEMS/NEMS [10, 49, 81]. There is a need for developing a fundamental understanding of adhesion, friction/stiction, wear, and the roles of surface contamination and environment [10]. MEMS/NEMS materials need to exhibit good mechanical and tribological properties on the micro/nanoscale. There is a need to develop lubricants and identify lubrication methods that are suitable for MEMS/NEMS. Component-level studies are required to provide a better understanding of the tribological phenomena occurring in MEMS/NEMS. The emergence of micro/nanotribology and atomic force microscopy-based techniques has provided researchers with a viable

approach to address these problems [49, 82]. This chapter presents a review of macro- and micro/nanoscale tribological studies of materials and lubrication studies for MEMS/NEMS and component-level studies of stiction phenomena in MEMS/NEMS devices.

21.5 Tribological Studies of Silicon and Related Materials

Materials of most interest for planar fabrication processes using silicon as the structural material are undoped and boron-doped (p^+-type) single-crystal silicon for bulk micromachining and phosphorus (n^+-type) doped and undoped LPCVD polysilicon films for surface micromachining. Silicon-based devices lack high-temperature capabilities with respect to both mechanical and electrical properties. Researchers have been pursuing SiC as a material for high-temperature microsensor and microactuator applications, for some time [83, 84]. SiC is a likely candidate for such applications, since it has long been used in high temperature electronics, high frequency and high power devices. SiC can also be desirable for high-frequency micromechanical resonators, in the GHz range, because of its high modulus of elasticity and, consequently, high resonant frequency. Table 21.2 compares selected bulk properties of

Table 21.2. Selected bulk properties[a] of 3C (β- or cubic) SiC and Si(100)

Sample	Density (kg/m^3)	Hardness (GPa)	Elastic modulus (GPa)	Fracture toughness (MPa m$^{1/2}$)	Thermal conductivity[b] (W/m K)	Coeff. of thermal expansion[b] ($\times 10^{-6}/°C$)	Melting point (°C)	Band gap (eV)
β-SiC	3210	23.5–26.5	440	4.6	85–260	4.5–6	2830	2.3
Si(100)	2330	9–10	130	0.95	155	2–4.5	1410	1.1

[a] Unless stated otherwise, data shown were obtained from [87]
[b] Obtained from [88]

SiC and Si(100). Researchers have found low cost techniques of producing single-crystal 3C-SiC (cubic or β-SiC) films via epitaxial growth on large-area silicon substrates for bulk micromachining [85] and polycrystalline 3C-SiC films on polysilicon and silicon dioxide layers for surface micromachining of SiC [86].

As will be shown, bare silicon exhibits inadequate tribological performance and needs to be coated with a solid and/or liquid overcoat or surface treated (e.g., oxidation and ion implantation, commonly used in semiconductor manufacturing), which exhibit lower friction and wear. SiC films exhibit good tribological performance. Both macroscale and microscale tribological properties of virgin and treated/coated silicon, polysilicon films, and SiC are presented next.

21.5.1 Tribological Properties of Silicon and the Effect of Ion Implantation

Friction and wear of single-crystalline and polycrystalline silicon samples, as well as the effect of ion implantation with various doses of C^+, B^+, N_2^+, and Ar^+ ion species at 200 keV energy to improve their friction and wear properties have been studied [89–91]. The coefficient of macroscale friction and wear factor of virgin single-crystal silicon and C^+-implanted silicon samples as a function of ion dose are presented in Fig. 21.9 [89]. The macroscale friction and wear tests were conducted using a ball-on-flat tribometer. Each data bar represents the average value of four to six measurements. The coefficient of friction and wear factor for bare silicon are very high and decrease drastically with ion dose. Silicon samples bombarded above the ion dose of 10^{17} C^+ cm^{-2} exhibit extremely low values of coefficients of friction (typically 0.03 to 0.06 in air) and wear factor (reduced by as much as four orders of magnitude). *Gupta* et al. [89] reported that a decrease in coefficient of friction and wear factor of silicon as a result of C^+ ion bombardment occurred because of formation of silicon carbide, rather than amorphization of silicon. *Gupta* et al. [90] also reported an improvement in friction and wear with B^+ ion implantation.

Microscale friction measurements were performed using an atomic force/friction force microscope (AFM/FFM) [49]. Table 21.3 shows values of surface roughness and coefficients of macroscale and microscale friction for virgin and doped silicon. There is a decrease in coefficients of microscale and macroscale friction values as a result of ion implantation. When measured for small contact areas and very low

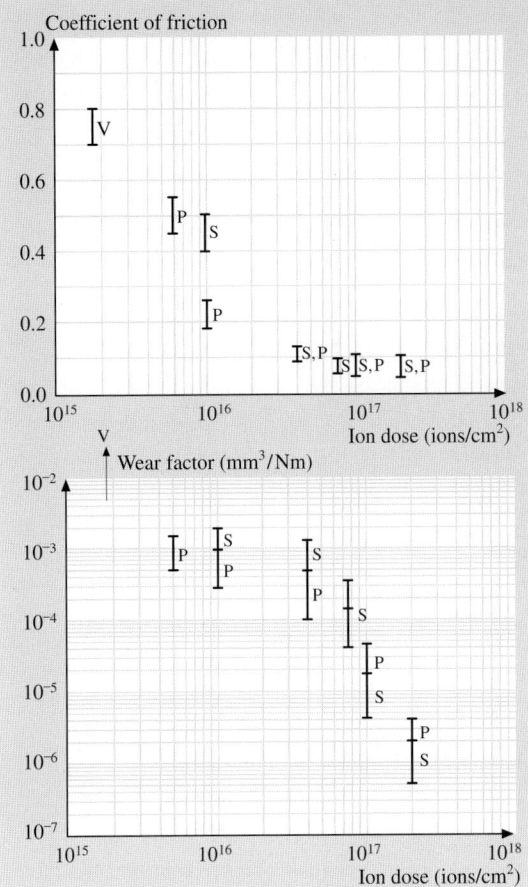

Fig. 21.9. Influence of ion doses on the coefficient of friction and wear factor on C$^+$ ion bombarded single-crystal and polycrystalline silicon slid against an alumina ball. V corresponds to virgin single-crystal silicon, while S and P denote tests that correspond to doped single- and polycrystalline silicon, respectively [89]

loads used in microscale studies, the indentation hardness and elastic modulus are higher than at the macroscale. This, added to the effect of the small apparent area of contact reducing the number of trapped particles on the interface, results in less plowing contribution and lower friction in the case of microscale friction measurements. Results of microscale wear resistance studies of ion-implanted silicon samples studied using a diamond tip in an AFM [92] are shown in Fig. 21.10a, b. For tests conducted at various loads on Si(111) and C$^+$-implanted Si(111), it is noted that wear resistance of an implanted sample is slightly poorer than that of virgin silicon up to about 80 μN. Above 80 μN, the wear resistance of implanted Si improves. As one continues to run tests at 40 μN for a larger number of cycles, the implanted sample

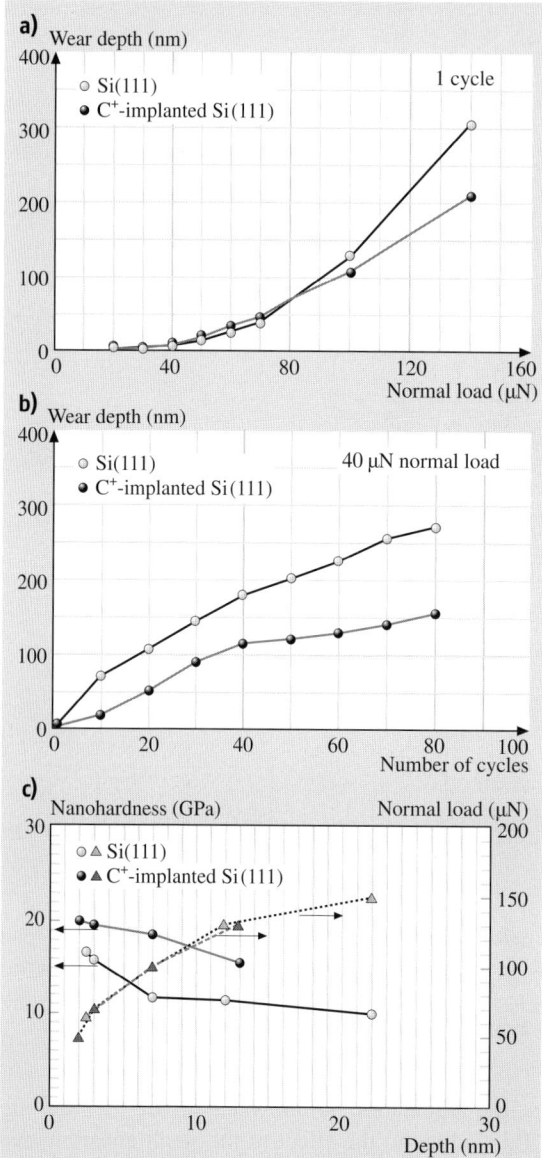

Fig. 21.10. Wear depth as a function of (**a**) load (after one cycle), and (**b**) cycles (normal load = 40 mN) for Si(111) and C$^+$-implanted Si(111). (**c**) Nanohardness and normal load as a function of indentation depth for virgin and C$^+$-implanted Si(111) [92]

exhibits higher wear resistance than the unimplanted sample. Damage from the implantation in the top layer results in poorer wear resistance. However, the implanted zone at the subsurface is more wear resistant than the virgin silicon.

Table 21.3. Surface roughness and micro- and macroscale coefficients of friction of selected samples

Material	RMS roughness (nm)	Coefficient of microscale friction[a]	Coefficient of macroscale friction[b]
Si(111)	0.11	0.03	0.33
C$^+$-implanted Si(111)	0.33	0.02	0.18

[a] Versus Si$_3$N$_4$ tip, tip radius of 50 nm in the load range of 10–150 nN (2.5–6.1 GPa) at a scanning speed of 5 μm/s over a scan area of 1 μm × 1 μm in an AFM
[b] Versus Si$_3$N$_4$ ball, ball radius of 3 mm at a normal load of 0.1 N (0.3 GPa) at an average sliding speed of 0.8 mm/s using a tribo-meter

Hardness values of virgin and C$^+$-implanted Si(111) at various indentation depths (normal loads) are presented in Fig. 21.10c [92]. The hardness at a small indentation depth of 2.5 nm is 16.6 GPa and drops to a value of 11.7 GPa at a depth of 7 nm and a normal load of 100 μN. Higher hardness values obtained in low-load indentation may arise from the observed pressure-induced phase transformation during the nanoindentation [93, 94]. An additional increase in the hardness at an even lower indentation depth of 2.5 nm reported here may arise from the contribution of complex chemical films (not from native oxide films) present on the silicon surface. At small volumes there is a lower probability of encountering material defects (dislocations, etc.). Furthermore, according to the strain gradient plasticity theory advanced by *Fleck* et al. [95], large strain gradients inherent in small indentations lead to the accumulation of geometrically necessary dislocations that cause enhanced hardening. These are some of the plausible explanations for an increase in hardness at smaller volumes. If the silicon material were to be used at very light loads such as in microsystems, the high hardness of surface films would protect the surface until it is worn.

From Fig. 21.10c, hardness values of C$^+$-implanted Si(111) at a normal load of 50 μN is 20.0 GPa with an indentation depth of about 2 nm, which is comparable to the hardness value of 19.5 GPa at 70 μN, whereas measured hardness value for virgin silicon at an indentation depth of about 7 nm (normal load of 100 μN) is only about 11.7 GPa. Thus, ion implantation with C$^+$ results in an increase in hardness in silicon. Note that the surface layer of the implanted zone is much harder compared to the subsurface and may be brittle, leading to higher wear on the surface. The subsurface of the implanted zone is harder than the virgin silicon, resulting in higher wear resistance, which is also observed in the results of the macroscale tests conducted at high loads.

21.5.2 Effect of Oxide Films on Tribological Properties of Silicon

Macroscale friction and wear experiments have been performed using a magnetic disk drive with bare, oxidized, and implanted pins sliding against amorphous-carbon coated magnetic disks lubricated with a thin layer of perfluoropolyether lubricant [96–99]. Representative profiles of the variation of the coefficient of friction with a number of sliding cycles for Al_2O_3-TiC slider and bare and dry-oxidized silicon pins are shown in Fig. 21.11. For bare Si(111), after an initial increase, the coefficient of friction drops to a steady state value of 0.1 following the increase, as seen in Fig. 21.11. The rise in the coefficient of friction for the Si(111) pin is associated with the transfer of amorphous carbon from the disk to the pin and the oxidation-enhanced fracture of pin material followed by tribochemical oxidation of the transfer film, while the drop is associated with the formation of a transfer coating on the pin, as shown in Fig. 21.12. Dry-oxidized Si(111) exhibits excellent characteristics, and no significant increase was observed over 50 000 cycles (Fig. 21.11). This behavior has been attributed to the chemical passivity of the oxide and lack of transfer of DLC from the disk to the pin. The behavior of PECVD-oxide (data are not presented here) was comparable to that of dry oxide, but for the wet oxide there was some variation in the coefficient of friction (0.26 to 0.4). The difference between dry and wet oxide was attributed to increased porosity of the wet oxide [96]. Since tribochemical oxidation was determined to be a significant factor, experiments were conducted in dry nitrogen [97, 98]. The variation of the coefficient of friction for a silicon pin sliding against a thin-film disk in dry nitrogen is shown in Fig. 21.11. It is seen that in a dry nitrogen environment, the coefficient of friction of Si(111) sliding against a disk decreased from an initial value of about 0.2 to 0.05 with continued sliding. Based on SEM and chemical analysis, this behavior has been attributed to the formation of a smooth amorphous-carbon/lubricant transfer patch and suppression of oxidation in a dry nitrogen environment. Based on macroscale tests using disk drives, it was found that the friction and wear performance of bare silicon is not adequate. With dry-oxidized or PECVD SiO_2-coated silicon, no significant friction increase or interfacial degradation was observed in ambient air.

Table 21.4 and Fig. 21.13 show surface roughness, microscale friction and scratch data, and nanoindentation hardness for various silicon samples [92]. Scratch experiments were performed using a diamond tip in an AFM. Results on polysilicon samples are also shown for comparison. Coefficients of microscale friction values for all the samples are about the same. These samples could be scratched at $10\,\mu N$ load. Scratch depth increased with normal load. Crystalline orientation of silicon has little influence

on scratch resistance, because natural oxidation of silicon in ambient masks the expected effect of crystallographic orientation. PECVD-oxide samples showed the best scratch resistance, followed by dry-oxidized, wet-oxidized, and ion-implanted samples. Ion implantation with C^+ does not appear to improve scratch resistance.

Wear data on the silicon samples are also presented in Table 21.1 [92]. PECVD-oxide samples showed superior wear resistance followed by dry-oxidized, wet-oxidized, and ion-implanted samples. This agrees with the trends seen in scratch resistance. In PECVD, ion bombardment during deposition improves the coating properties such as suppression of columnar growth, freedom from pinhole, decrease in crystalline size, and increase in density, hardness, and substrate-coating adhesion. These effects may help in improving the mechanical integrity of the sample surface.

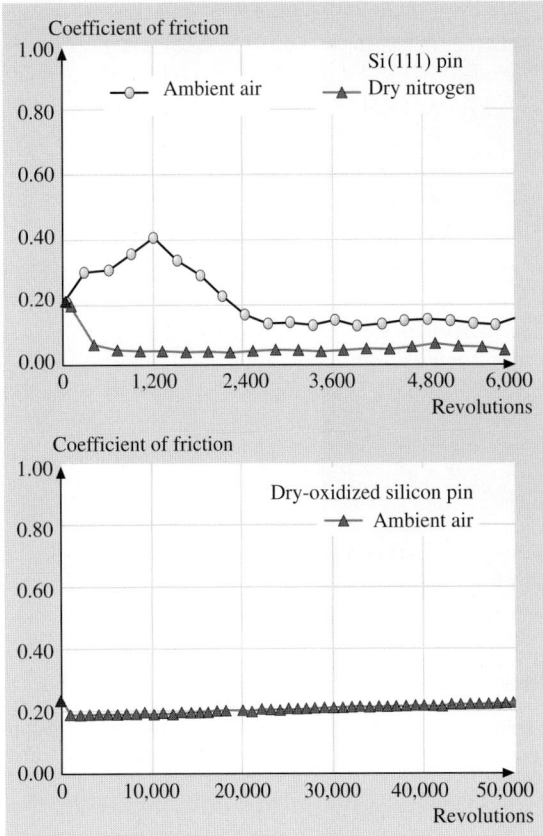

Fig. 21.11. Coefficient of friction as a function of number of sliding revolutions in ambient air for Si(111) pin in ambient air and dry nitrogen and dry-oxidized silicon pin in ambient air [96]

Table 21.4. RMS, microfriction, microscratching/microwear, and nanoindentation hardness data for various virgin, coated, and treated silicon samples

Material	RMS roughness[a] (nm)	Coefficient of microscale friction[b]	Scratch depth[c] at 40 µN (nm)	Wear depth[c] at 40 µN (nm)	Nanohardness[c] at 100 µN (GPa)
Si(111)	0.11	0.03	20	27	11.7
Si(110)	0.09	0.04	20		
Si(100)	0.12	0.03	25		
Polysilicon	1.07	0.04	18		
Polysilicon (lapped)	0.16	0.05	18	25	12.5
PECVD-oxide coated Si(111)	1.50	0.01	8	5	18.0
Dry-oxidized Si(111)	0.11	0.04	16	14	17.0
Wet-oxidized Si(111)	0.25	0.04	17	18	14.4
C^+-implanted Si(111)	0.33	0.02	20	23	18.6

[a] Scan size of 500 nm × 500 nm using AFM.
[b] Versus Si_3N_4 tip in AFM/FFM, radius 50 nm; at 1 µm × 1 µm scan size.
[c] Measured using an AFM with a diamond tip of radius 100 nm.

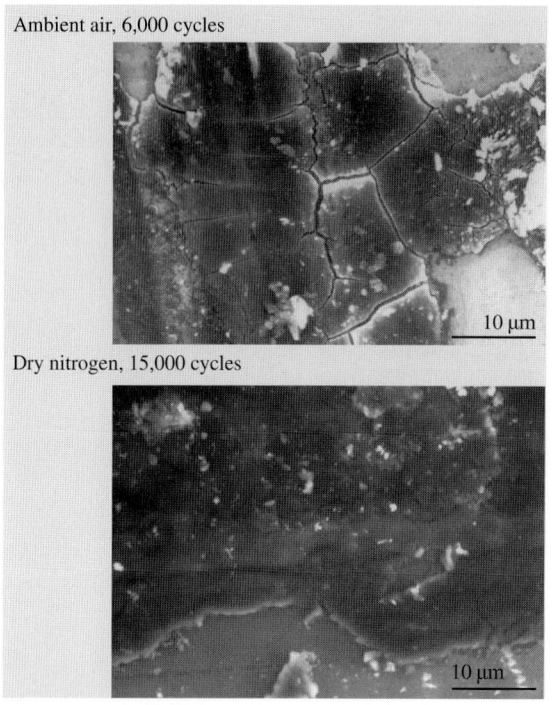

Fig. 21.12. Scanning electron micrographs of Si(111) after sliding against a magnetic disk in a rigid disk drive in ambient air for 6000 cycles and in dry nitrogen after 15 000 cycles

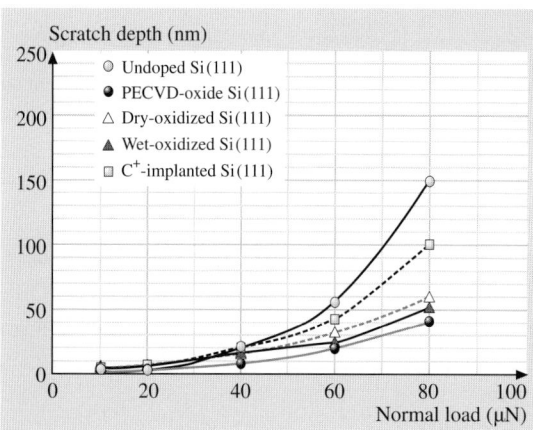

Fig. 21.13. Scratch depth as a function of normal load after 10 cycles for various silicon samples: virgin, treated, and coated [92]

Table 21.5. Summary of micro/nanotribological properties of the sample materials

Sample	RMS roughness[a] (nm)	P–V distance[a] (nm)	Coefficient of friction Micro[b]	Coefficient of friction Macro[c]	Scratch depth[d] (nm)	Wear depth[e] (nm)	Nano-hardness[f] (GPa)	Young's modulus[f] (GPa)	Fracture toughness[g], K_{IC} MPa m$^{1/2}$
Undoped Si(100)	0.09	0.9	0.06	0.33	89	84	12	168	0.75
Undoped polysilicon film (as deposited)	46	340	0.05						
Undoped polysilicon film (polished)	0.86	6	0.04	0.46	99	140	12	175	1.11
n$^+$-Type polysilicon film (as deposited)	12	91	0.07						
n$^+$-Type polysilicon film (polished)	1.0	7	0.02	0.23	61	51	9	95	0.89
SiC film (as deposited)	25	150	0.03						
SiC film (polished)	0.89	6	0.02	0.20	6	16	25	395	0.78

[a] Measured using AFM over a scan size of 10 μm × 10 μm
[b] Measured using AFM/FFM over a scan size of 10 μm × 10 μm
[c] Obtained using a 3-mm-diameter sapphire ball in a reciprocating mode at a normal load of 10 mN and average sliding speed of 1 mm/s after 4 m sliding distance
[d] Measured using AFM at a normal load of 40 μN for 10 cycles, scan length of 5 μm
[e] Measured using AFM at normal load of 40 μN for 1 cycle, wear area of 2 μm × 2 μm
[f] Measured using nanoindenter at a peak indentation depth of 20 nm
[g] Measured using microindenter with Vickers indenter at a normal load of 0.5 N

Coatings and treatments improved nanohardness of silicon. Note that dry-oxidized and PECVD films are harder than wet-oxidized films, as these films may be porous. High hardness of oxidized films may be responsible for measured high scratch/wear resistance.

21.5.3 Tribological Properties of Polysilicon Films and SiC Film

Studies have also been conducted on undoped polysilicon film, heavily doped (n$^+$-type) polysilicon film, heavily doped (p$^+$-type) single-crystal Si(100), and 3C-SiC (cubic or β-SiC)film [10, 100, 101]. The polysilicon films studied here are different from the ones discussed previously.

Table 21.5 presents a summary of the tribological studies conducted on polysilicon and SiC films. Values for single-crystal silicon are shown for comparison. Polishing of the as-deposited polysilicon and SiC films drastically affects the roughness, as the values reduce by two orders of magnitude. Si(100) appears to be the smoothest followed by polished undoped polysilicon and SiC films, which have comparable roughness. The doped polysilicon film shows higher roughness than the undoped sample, which is attributed to the doping process. Polished SiC film shows the lowest friction followed by polished and undoped polysilicon film, which strongly supports the candidacy of SiC films for use in MEMS/NEMS devices. Macroscale friction measurements indicate that SiC film exhibits one of the lowest friction values as compared to the other samples. The doped polysilicon sample shows low friction on the macroscale as compared to the undoped polysilicon sample, possibly due to the doping effect.

Figure 21.14a shows a plot of scratch depth vs. normal load for various samples [10, 100]. Scratch depth increases with increasing normal load. Figure 21.15 shows AFM 3-D maps and averaged 2-D profiles of the scratch marks on the various samples. It is observed that scratch depth increases almost linearly with the normal load. Si(100) and the doped and undoped polysilicon films show similar scratch resistance. It is clear from the data that the SiC film is much more scratch resistant than the other samples. Figure 21.14b shows results from microscale wear tests on the various films. For all the materials, the wear depth increases almost linearly with increasing number of cycles. This suggests that the material is removed layer by layer in all the materials. Here also SiC film exhibits lower wear depths than the other samples. Doped polysilicon film wears less than the undoped film. Higher fracture toughness and hardness of SiC compared to Si(100) are responsible for its lower wear. Also the higher thermal conductivity of SiC (see Table 21.2) compared to the other materials leads to lower interface temperatures, which generally results in less degradation of the surface [35, 54, 55]. Doping of the polysilicon does not affect the scratch/wear resistance and hardness much. The measurements made on the doped sample are affected by the presence of grain boundaries. These studies indicate that SiC film exhibits desirable tribological properties for use in MEMS devices. Recently, researchers have fabricated SiC micromotors and have reported satisfactory operation at high temperatures [102].

21.6 Lubrication Studies for MEMS/NEMS

Several studies of liquid perfluoropolyether (PFPE) lubricant films, self-assembled monolayers (SAMs), and hard diamond-like carbon (DLC) coatings have been carried out for the purpose of minimizing adhesion, friction, and wear [49, 66, 99, 103–108]. Many variations of these films are hydrophobic (low surface tension and high contact angle) and have low shear strength, which provide low adhesion, friction, and wear. Relevant details are presented here.

21.6.1 Perfluoropolyether Lubricants

The classical approach to lubrication uses freely supported multimolecular layers of liquid lubricants [49, 54, 55]. The liquid lubricants are sometimes chemically bonded to improve their wear resistance. Partially chemically bonded, molecularly thick perfluoropolyether (PFPE) lubricants are widely used for lubrication of magnetic storage media [35] and are suitable for MEMS/NEMS devices.

Adhesion, friction, and durability experiments have been performed on virgin Si(100) surfaces and silicon surfaces lubricated with two commonly used PFPE lubricants – Z-15 (with -CF_3 nonpolar end groups) and Z-DOL (with -OH polar end groups) [49, 104, 105, 108]. Z-DOL film was thermally bonded at 150 °C for 30 min and an unbonded fraction was removed by a solvent (BW) [35]. The thicknesses of Z-15 and Z-DOL (BW) films were 2.8 nm and 2.3 nm, respectively. Nanoscale measurements were made using an AFM. The adhesive forces of Si(100), Z-15 and

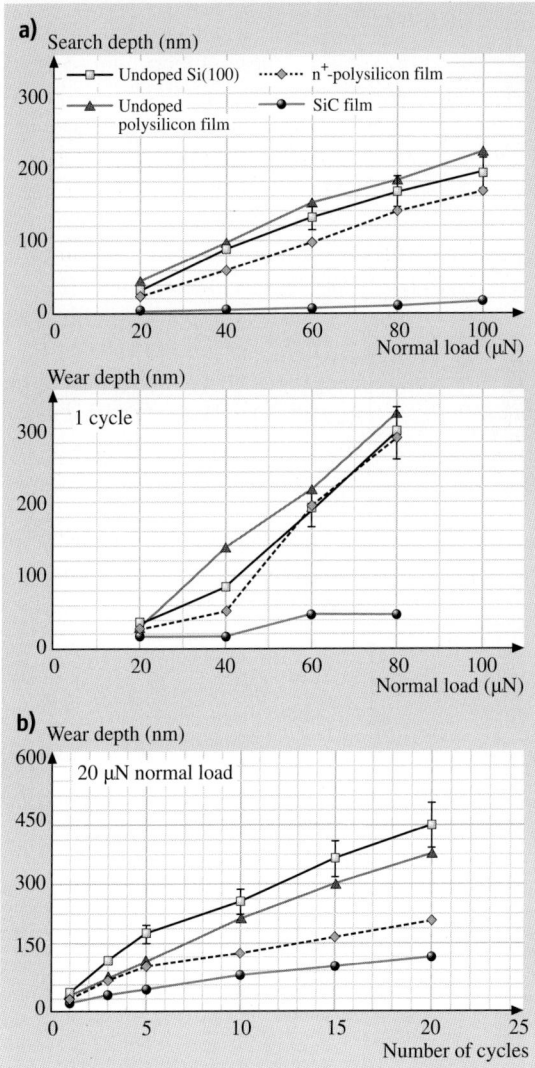

Fig. 21.14. (a) Scratch depths for 10 cycles as a function of normal load and (b) wear depths as a function of normal load and as a function of number of cycles for various samples [10]

Z-DOL (BW) measured by force calibration plot and friction force versus normal load plot are summarized in Fig. 21.16. The results measured by these two methods are in good agreement. Figure 21.16 shows that the presence of mobile Z-15 lubricant film increases the adhesive force compared to that of Si(100) by meniscus formation [54, 55, 109]. The presence of solid phase Z-DOL (BW) film reduces the adhesive force compared to that of Si(100) due to the absence of mobile liquid. The schematic (bottom) in Fig. 21.16 shows the relative size and sources of menis-

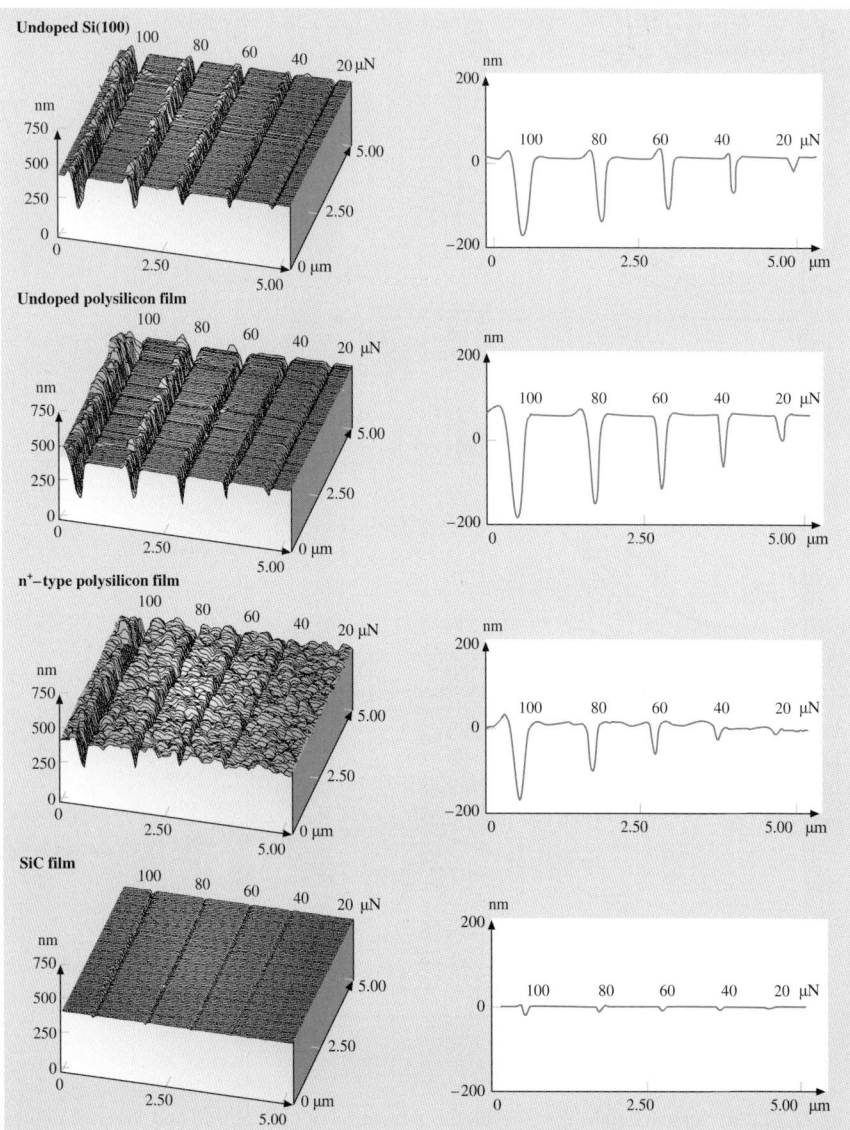

Fig. 21.15. AFM 3-D maps and averaged 2-D profiles of scratch marks on various samples [10]

cus. It is well-known that the native oxide layer (SiO_2) on the top of Si(100) wafer exhibits hydrophilic properties, and some water molecules can be adsorbed on this surface. The condensed water will form meniscus as the tip approaches the sample surface. The larger adhesive force in Z-15 is not only caused by the Z-15 meniscus. The nonpolarized Z-15 liquid does not have good wettability and strong bonding with Si(100). In the ambient environment, the condensed water molecules from the environment permeate through the liquid Z-15 lubricant film and compete with the lubricant molecules presented on the substrate. The interaction of the liquid lubricant with the substrate is weakened, and a boundary layer of the liquid lubricant forms puddles [104, 105]. This dewetting allows water molecules to be adsorbed on the Si(100) surface as aggregates, along with Z-15 molecules. And both of them can form meniscus while the tip approaches the surface. Thus, the dewetting of liquid Z-15 film results in higher adhesive force and poorer lubrication performance. In addition, as the Z-15 film is pretty soft compared to the solid Si(100) surface, penetration of the tip in the film occurs while pushing the tip downwards. This leads to the large area of the tip involved to form the meniscus at the tip-liquid (mixture of Z-15 and water) interface. It should also be noted that Z-15 has a higher viscosity compared to water, and therefore, Z-15 film provides higher resistance to motion and coefficient of friction. In the case of Z-DOL (BW) film, the active groups of Z-DOL molecules are mostly bonded on Si(100) substrate, thus the Z-DOL (BW) film has low free surface energy and cannot be displaced readily by water molecules, or readily adsorb water molecules. Thus, the use of Z-DOL (BW) can reduce the adhesive force.

To study the relative humidity effect on friction and adhesion, the variations of friction force, adhesive force, and coefficient of friction of Si(100), Z-15, and Z-DOL (BW) as a function of relative humidity are shown in Fig. 21.17. It shows that for Si(100) and Z-15 film, the friction force increases with a relative humidity increase of up to 45% and then slightly decreases with a further increase in relative humidity. Z-DOL (BW) has a smaller friction force than Si(100) and Z-15 in the whole testing range. And its friction force shows a relative apparent increase when the relative humidity is higher than 45%. For Si(100), Z-15, and Z-DOL (BW), the adhesive forces increase with relative humidity. And their coefficients of friction increase with a relative humidity up to 45%, after which they decrease with a further increase in relative humidity. It is also observed that the humidity effect on Si(100) really depends on the history of the Si(100) sample. As the surface of Si(100) wafer readily adsorbs water in air, without any pre-treatment the Si(100) used in our study almost reaches its saturated stage of adsorbed water and is responsible for less effect during increasing relative humidity. However, once the Si(100) wafer was thermally treated by baking at 150 °C for 1 h, a bigger effect was observed.

The schematic (right) in Fig. 21.17 shows that Si(100), because its high freesurface energy, can adsorb more water molecules with increasing relative humidity. As discussed earlier, for the Z-15 film in the humid environment, the condensed water from the humid environment competes with the lubricant film present on the sample surface, and interaction of the liquid lubricant film with the silicon substrate is weakened, and a boundary layer of the liquid lubricant forms puddles. This dewetting allows water molecules to be adsorbed on the Si(100) substrate mixed with Z-15 molecules [104, 105]. Obviously, more water molecules can be adsorbed on Z-15 surface while increasing relative humidity. The more adsorbed water molecules in the case of Si(100), along with lubricant molecules in Z-15 film case, form bigger water meniscus, which leads to an increase in friction force, adhesive force, and

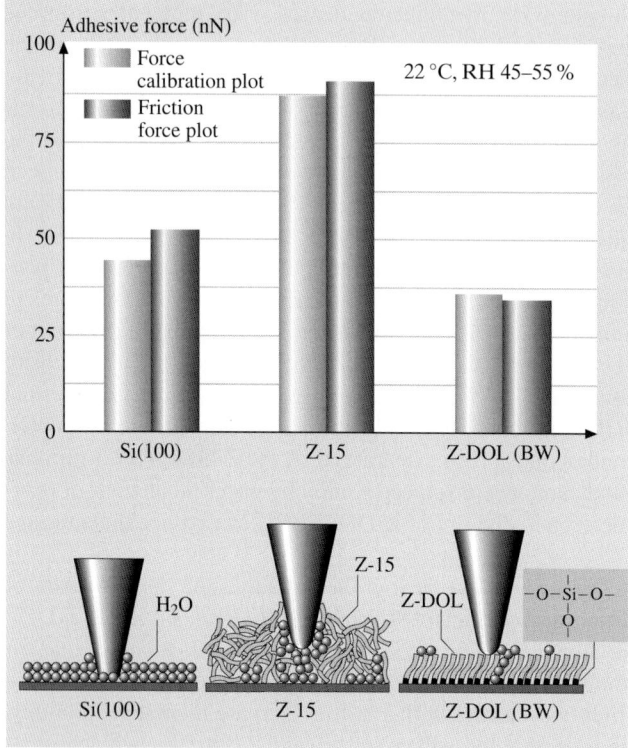

Fig. 21.16. Summary of the adhesive forces of Si(100) and Z-15 and Z-DOL (BW) films measured by force calibration plots and friction force versus normal load plots in ambient air. The schematic (*bottom*) showing the effect of meniscus, formed between the AFM tip and the surface sample, on the adhesive and friction forces [108]

coefficient of friction of Si(100) and Z-15 with humidity. But at a very high humidity of 70%, large quantities of adsorbed water can form a continuous water layer that separates the tip and sample surface and acts as a kind of lubricant, which causes a decrease in the friction force and coefficient of friction. For Z-DOL (BW) film, because of its hydrophobic surface properties, water molecules can be adsorbed at a humidity higher than 45% and cause an increase in the adhesive and friction forces.

To study the durability of lubricant films at the nanoscale, the friction of Si(100), Z-15, and Z-DOL (BW) as a function of the number of scanning cycles are shown in Fig. 21.18. As observed earlier, friction force and coefficient of friction of Z-15 is higher than that of Si(100) with the lowest values for Z-DOL (BW). During cycling, friction force and coefficient of friction of Si(100) show a slight decrease during the initial few cycles, then remain constant. This is related to the removal of the top adsorbed layer. In the case of Z-15 film, the friction force and coefficient of friction show an increase during the initial few cycles and then approach higher and stable values. This is believed to be caused by the attachment of the Z-15 molecules onto

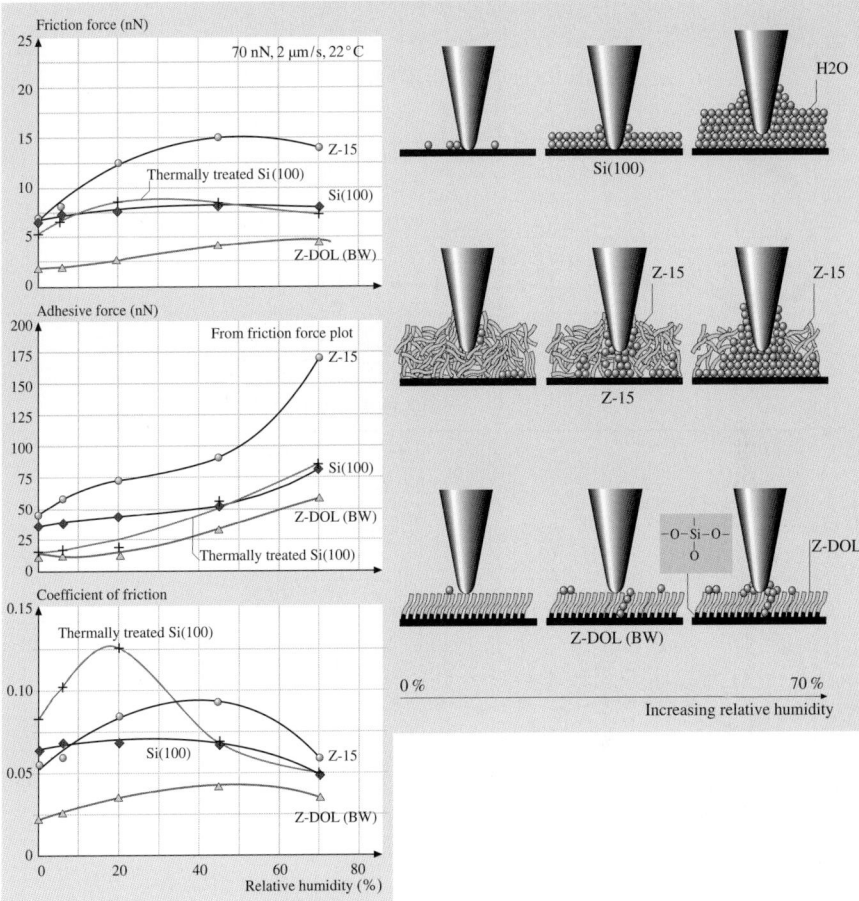

Fig. 21.17. The influence of relative humidity on the friction force, adhesive force, and coefficient of friction of Si(100), Z-15, and Z-DOL (BW) films at 70 nN, 2 µm/s, and in 22 °C air. Schematic (*right*) shows the change of meniscus with increasing relative humidity. In this figure, the thermally treated Si(100) represents the Si(100) wafer that was baked at 150 °C for 1 h in an oven (in order to remove the adsorbed water) just before it was placed in the 0% RH chamber [108]

the tip. The molecular interaction between these attached molecules and molecules on the film surface is responsible for an increase in friction. But after several scans, this molecular interaction reaches the equilibrium, and after that, friction force and coefficient of friction remain constant. In the case of Z-DOL (BW) film, the friction force and coefficient of friction start out low and remain low during the entire test for 100 cycles, suggesting that Z-DOL (BW) molecules do not get attached or displaced as readily as Z-15.

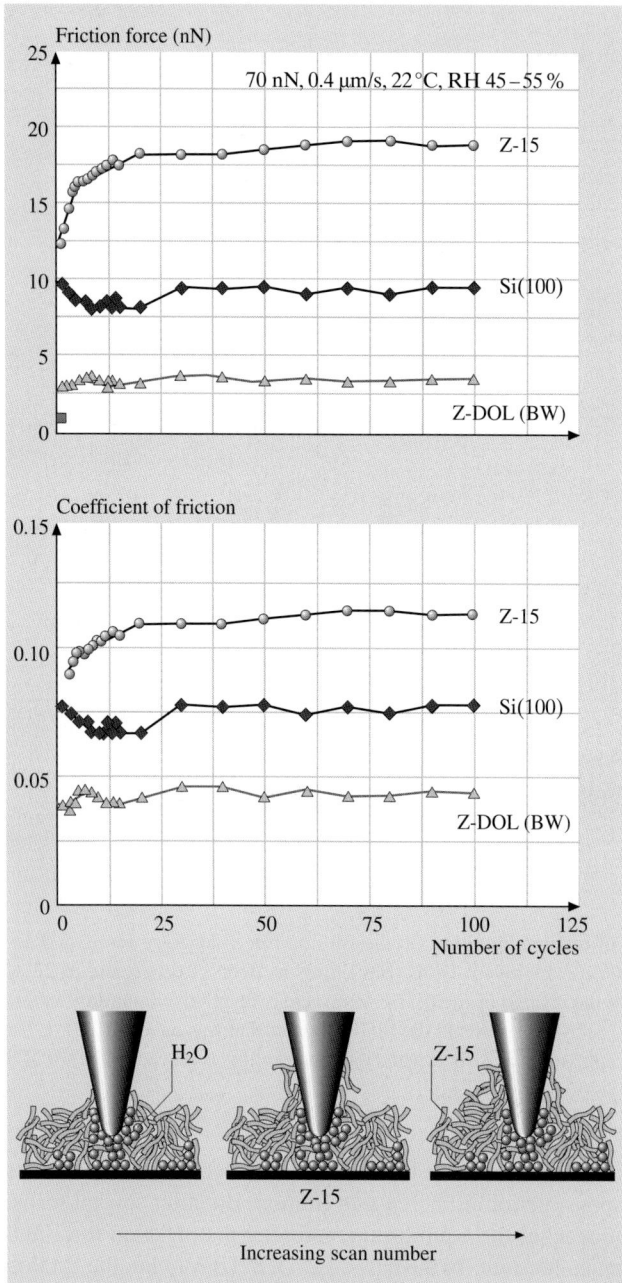

Fig. 21.18. Friction force and coefficient of friction versus number of sliding cycles for Si(100) and Z-15 and Z-DOL (BW) films at 70 nN, 0.4 μm/s, and in ambient air. Schematic (*bottom*) shows that some liquid Z-15 molecules can be attached onto the tip. The molecular interaction between the attached molecules with the Z-15 molecules in the film results in an increase of the friction force with multi-scanning [108]

21.6.2 Self-Assembled Monolayers (SAMs)

A preferred method of lubrication of MEMS/NEMS is by the deposition of organized and dense molecular-scale layers of long-chain molecules, as they have been shown to be superior lubricants [54, 55, 103, 106, 107, 110, 111]. Two common methods of producing monolayers are the Langmuir–Blodgett (L-B) deposition and self-assembled monolayers (SAMs) by chemical grafting of molecules. L-B films are physically bonded to the substrate by weak van der Waals forces, while SAMs are bonded covalently to the substrate and provide high durability.

SAMs can be formed spontaneously by the immersion of an appropriate substrate into a solution of an active surfactant in an organic solvent. SAMs offer the flexibility and advantage of molecular tailoring to obtain a variety of different tribological and mechanical properties. For example, researchers have shown that by changing the head groups, tail groups, chain lengths, or types of bonds within a chain, varying degrees of friction, adhesion and/or compliance can be obtained (e.g., [106, 107]). These studies indicate that the basis for molecular design and tailoring of SAMs must include a complete understanding of the interrelationships between the molecular structure and tribological properties of SAMs, as well as a deep understanding of the friction and wear mechanisms of SAMs at the molecular level.

Bhushan and *Liu* [106] and *Liu* and *Bhushan* [107] studied nanotribological properties of four different kinds of alkylthiol and biphenyl thiol monolayers with different surface terminals, spacer chains, and head groups using AFM/FFM techniques. These monolayers, along with a schematic of their structures and substrates, are listed in Fig. 21.19. Surface roughness, adhesion, and microscale friction studies were carried out using an AFM/FFM. The average values and standard deviation of the adhesive force and coefficient of microscale friction for the various SAMs are summarized in Fig. 21.20. It shows that SAMs can reduce the adhesive and friction forces. Alkylthiol and biphenyl thiol SAMs can be used as effective molecular lubricants for MEMS/NEMS fabricated from silicon. In order to explain the frictional differences in SAMs, a molecular spring model is presented in Fig. 21.21. It is suggested that the chemically adsorbed self-assembled molecules on a substrate are just like assembled molecular springs anchored to the substrate. An AFM tip sliding on the surface of a SAM is like a tip sliding on the top of "molecular springs or brush." The molecular assembly has compliant features and can experience orientation and compression under load. The orientation of the molecular springs under normal load reduces the shearing force at the interface, which, in turn, reduces the friction force. The possibility of orientation is determined by the spring constant of a single molecule (local stiffness), as well as the interaction between the neighboring molecules, which is reflected by the packing density or packing energy. It is noted that HDT exhibits the lowest adhesive and friction forces. Based on local stiffness measurements [107] and the view of molecular structures, biphenyl is a more rigid structure due to the contribution by two benzene rings. Therefore, the spring constant of BPT is larger than that of HDT. A more compliant film (HDT) exhibits lower friction force than a rigid film (BPT/BPTC). Crosslinking of chains in BPTC resulted in higher co-

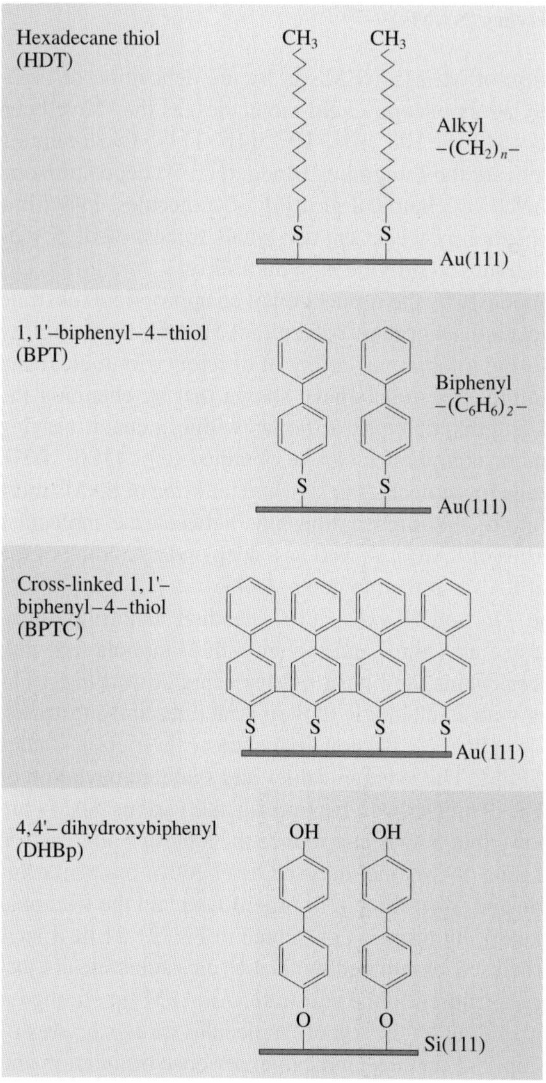

Fig. 21.19. Structure of SAMs studied using AFM/FFM techniques

efficient of friction. It should be noted that the orientation can lead to conformational defects along the molecular chains that lead to energy dissipation.

The influence of relative humidity on adhesion and friction was studied in an environmentally controlled chamber. The results are given in Fig. 21.22, which show that for Si(111), Au(111), and DHBp, the adhesive and frictional forces increase with relative humidity. For BPT and BPTC (cross-linked BPT), the adhesive forces slightly increase with relative humidity when the relative humidity is higher than 40%, but it is very interesting that their coefficients of friction decrease slightly in

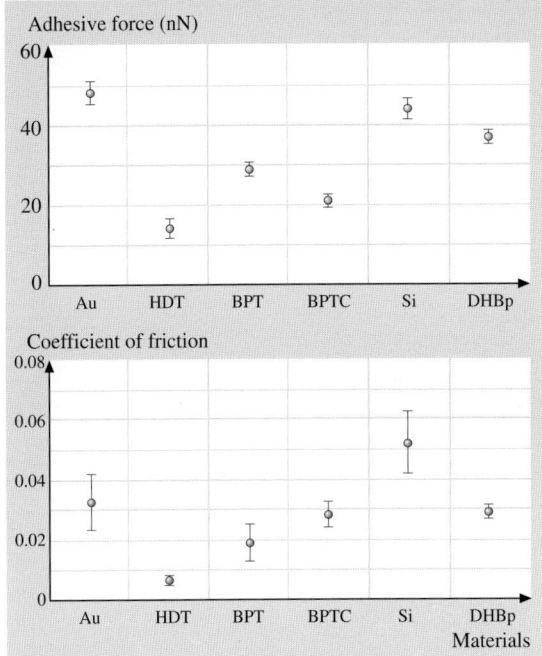

Fig. 21.20. Adhesive force and coefficient of friction values for the various SAMs studied

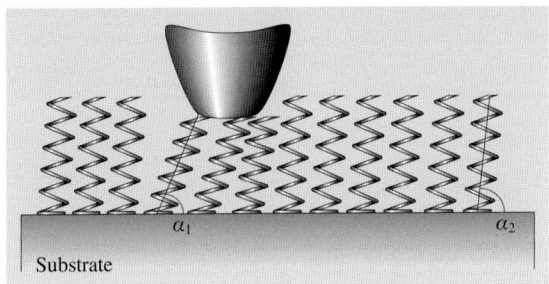

Fig. 21.21. Molecular spring model of SAMs. In this figure, $\alpha_1 < \alpha_2$, which is caused by the further orientation under the normal load applied by an asperity tip [106]

the same range. For HDT, over the testing range, the adhesive and friction force do not seem to be sensitive to the change in relative humidity. The influence of relative humidity on adhesive and frictional forces can be mainly understood by comparing their surface terminal polarization properties and work of adhesion. Meniscus forces are related linearly to the works of adhesion of
a surface, which in the case of SAMs, is dependent on the surface terminals. Polar surface terminals such as DHBp on Si(111) result in higher work of adhesion and

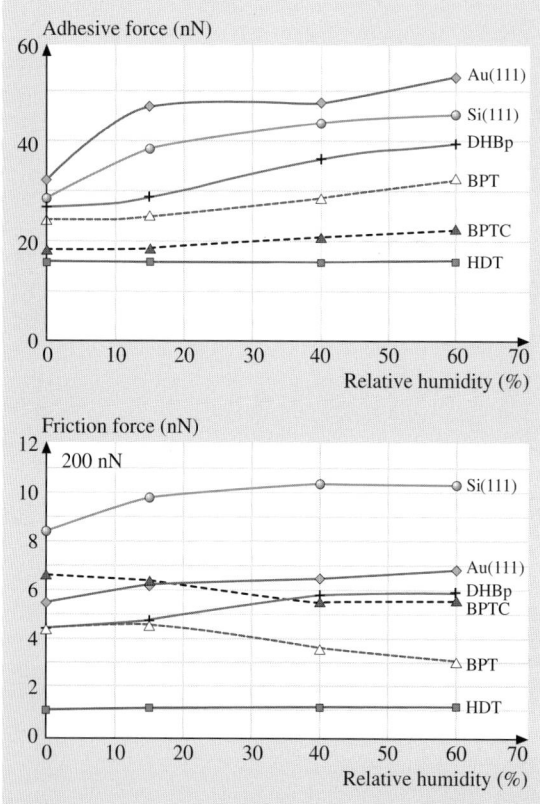

Fig. 21.22. The influence of relative humidity on the adhesive and friction forces at 200 nN of various SAMs studied [106]

hence higher adhesive forces. Larger adhesive forces result in higher friction forces. Nonpolar surface terminals (HDT) have very small work of adhesion and hence low adhesive and friction forces. In higher humidity, water capillary condensation can either increase friction through increased adhesion in the contact zone, or reduce friction through an enhanced water-lubricating effect.

Figure 21.23 shows results of microscale wear tests on SAMs using a diamond tip in an AFM. DHBp on Si(111) shows the best wear resistance. For the SAMs deposited on gold substrate, HDT exhibits the best wear resistance. Scratch tests corroborated the wear results. The wear resistance of SAMs is influenced by the interfacial bond strength, the molecular structure of the spacers, and the substrate hardness. Based on the wear data, the SAMs with high-compliance long carbon chains, in addition to low friction, exhibit the best wear resistance [106, 107]. In wear experiments, the wear depth as a function of normal load curves shows critical normal load (Fig. 21.23). Below the critical normal load, SAMs undergo orientation. At the critical load, SAMs wear away from the substrate due to weak interfacial

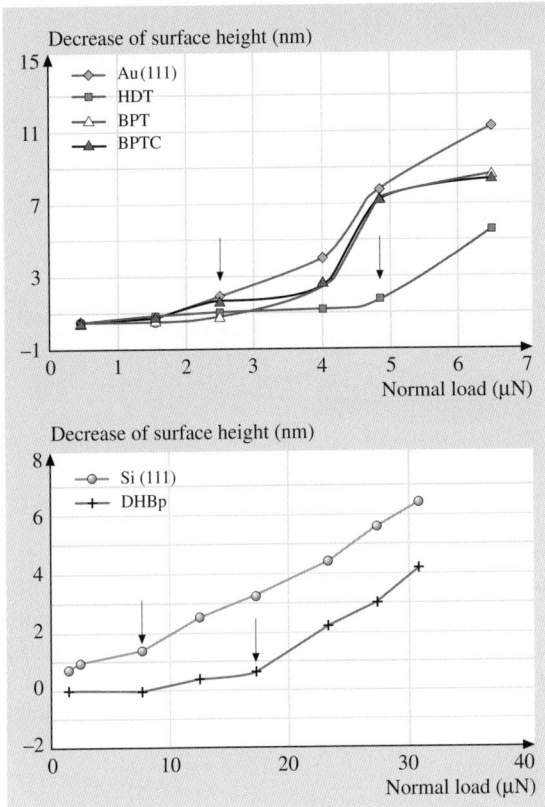

Fig. 21.23. Wear depth as a function of normal load for various SAMs [106]

bond strength, while above the critical normal load, severe wear takes place on the substrate, Fig. 21.24.

According to these results, it is suggested that a dual SAM with a compliant layer on a stiff layer, deposited on hydrogenated silicon, may have optimized tribological performance for MEMS/NEMS applications.

21.6.3 Hard Diamond-like Carbon (DLC) Coatings

Hard amorphous carbon (a-C), commonly known as DLC (implying high hardness), coatings are deposited by a variety of deposition techniquesincluding filtered cathodic arc, ion beam, electron cyclotron resonance chemical vapor deposition (ECR-CVD), plasma-enhanced chemical vapor deposition (PECVD), and sputtering [83, 99]. These coatings are used in a wide range of applications, including tribological, optical, electronic, and biomedical applications. Ultrathin coatings (3.5 to 10 nm thick) are employed to protect against wear and corrosion in magnetic storage applications (thin-film rigid disks, metal evaporated tapes) and thin-film read/write

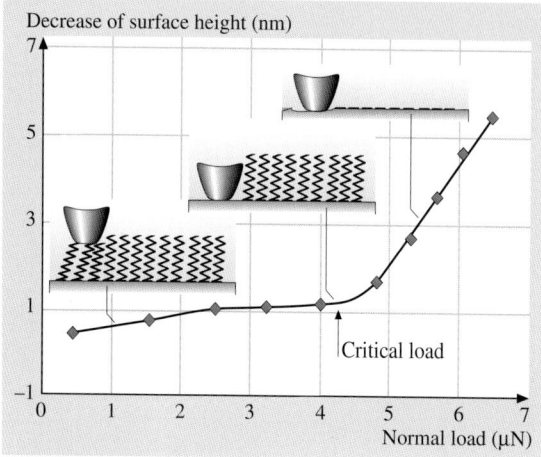

Fig. 21.24. Illustration of the wear mechanism of SAMs with increasing normal load [107]

head (Gillette Mach 3 razor blades, glass windows, and sunglasses). The coatings exhibit low friction, high hardness and wear resistance, chemical inertness to both acids and alkalis, lack of magnetic response, and optical band gap ranging from zero to a few eV, depending on the deposition technique and its conditions. Selected data on DLC coatings relevant for MEMS/NEMS applications are presented in a following section on adhesion measurements.

21.7 Component-Level Studies

21.7.1 Surface Roughness Studies of Micromotor Components

Most of the friction forces resisting motion in the micromotor are concentrated near the rotor-hub interface, where continuous physical contact occurs. Surface roughness of the surfaces usually has a strong influence on the friction characteristics on the micro/nanoscale. A catalog of roughness measurements on various components of a MEMS device does not exist in the literature. Using an AFM, measurements on various component surfaces was made for the first time by *Sundararajan* and *Bhushan* [112].

Table 21.6 shows various surface roughness parameters obtained from $5 \times 5\,\mu\text{m}$ scans of the various component surfaces of several unlubricated micromotors using the AFM in tapping mode. A surface with a Gaussian height distribution should have a skewness of zero and kurtosis of three. Although the rotor and stator top surfaces exhibit comparable roughness parameters, the underside of the rotors exhibits lower RMS roughness and peak-to-valley distance values. More importantly, the rotor underside shows negative skewness and lower kurtosis than the topsides both of which are conducive to high real area of contact and hence high friction [54, 55]. The rotor

Table 21.6. Surface roughness parameters and microscale coefficient of friction for various micromotor component surfaces measured using an AFM. Mean and $\pm 1\sigma$ values are given

	RMS roughness[a] (nm)	Peak-to-valley distance[a] (nm)	Skewness[a], Sk	Kurtosis[a], K	Coefficient of microscale friction[b] μ
Rotor topside	21 ± 0.6	225 ± 23	1.4 ± 0.30	6.1 ± 1.7	0.07 ± 0.02
Rotor underside	14 ± 2.4	80 ± 11	−1.0 ± 0.22	3.5 ± 0.50	0.11 ± 0.03
Stator topside	19 ± 1	246 ± 21	1.4 ± 0.50	6.6 ± 1.5	0.08 ± 0.01

[a] Measured from a tapping mode AFM scan of size $5\,\mu m \times 5\,\mu m$ using a standard Si tip scanning at $5\,\mu m/s$ in a direction orthogonal to the long axis of the cantilever
[b] Measured using an AFM in contact mode at $5\,\mu m \times 5\,\mu m$ scan size using a standard Si_3N_4 tip scanning at $10\,\mu m/s$ in a direction parallel to the long axis of the cantilever

underside also exhibits higher coefficient of microscale friction than the rotor topside and stator, as shown in Table 21.6. Figure 21.25 shows representative surface height maps of the various surfaces of a micromotor measured using the AFM in tapping mode. The rotor underside exhibits varying topography from the outer edge to the middle and inner edge. At the outer edges, the topography shows smaller circular asperities, similar to the topside. The middle and inner regions show deep pits with fine edges that may have been created by the etchants used for etching the sacrificial layer. It is known that etching can affect the surface roughness of surfaces in surface micromachining. The residence time of the etchant near the inner region is high, responsible for larger pits. Figure 21.26 shows roughness of the surface directly beneath the rotors (the base polysilicon layer). There appears to be a difference in the roughness between the portion of this surface that was initially underneath the rotor (region B) during fabrication and the portion that was away from the rotor and hence always exposed (region A). The former region shows lower roughness than the latter region. This suggests that the surfaces at the rotor-hub interface that come into contact at the end of the fabrication process exhibit large real areas of contact that result in high friction.

21.7.2 Adhesion Measurements

Surface force apparatus (SFA) and AFMs are used to measure adhesion between two surfaces on micro- to nanoscales. In the SFA, the adhesion of liquid films sandwiched between two curved and smooth surfaces is measured. In an AFM, as discussed earlier, adhesion between a sharp tip and the surface of interest is measured. To measure adhesion between two beams in the mesoscopic length scale, a cantilever beam array (CBA) technique is used [113–116]. The technique utilizes an array of micromachined polysilicon beams (for Si MEMS applications) anchored to the substrate at one end and with different lengths parallel to the surface. It relies on the peeling and detachment of cantilever beams. Change in free energy, or reversible work done to separate unit areas of two surfaces from contact is called work of adhesion. To mea-

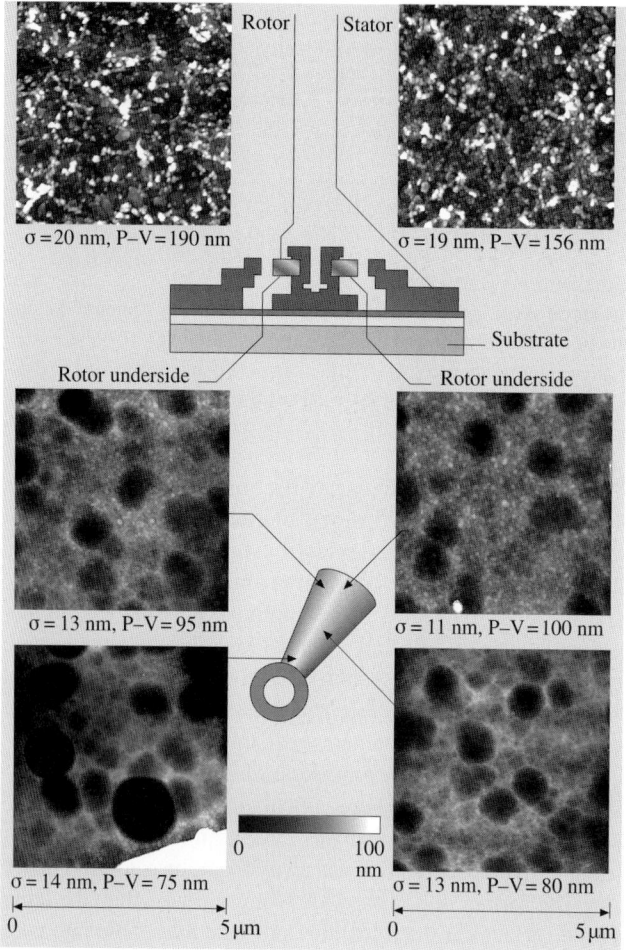

Fig. 21.25. Representative AFM surface height images obtained in tapping mode (5 μm×5 μm) scan size) of various component surfaces of a micromotor. RMS roughness and peak-to-valley (P–V) values of the surfaces are given. The underside of the rotor exhibits drastically different topography from the topside [112]

sure the work of adhesion, electrostatic actuation is used to bring all beams in contact with the substrate, Fig. 21.27 [113, 116]. Once the actuation force is removed, the beams begin to peel themselves off the substrate, which can be observed with an optical interference microscope. For beams shorter than a characteristic length, the so-called detachment length, their stiffness is sufficient to completely free them from the substrate underneath. Beams larger than the detachment length remain adhered. The beams at the transition region start to detach and remain attached to the substrate just at the tips. For this case, by equating the elastic energy stored within the beam and the beam-substrate interfacial energy, the work of adhesion, W_{ad}, can be

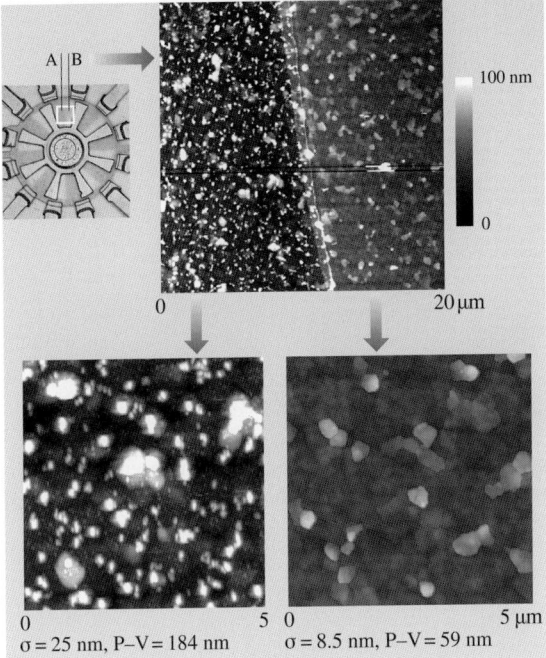

Fig. 21.26. Surface height images of polysilicon regions directly below the rotor. Region A is away from the rotor, while region B was initially covered by the rotor prior to the release etch of the rotor. During this step, slight movement of the rotor caused region B to be exposed [112]

calculated by the following equation [116]:

$$W_{ad} = \frac{3Ed^2t^3}{8l_d^4}, \quad (21.1)$$

where E is the Young's modulus of the beam, d is the spacing between the undeflected beam and the substrate, t is the beam thickness, and l_d is the detachment length. The technique has been used to screen methods for adhesion reduction in polysilicon microstructures.

To measure adhesion, friction, and wear between two microcomponents, a microtriboapparatus has been used. Figure 21.28 shows a schematic of a microtriboapparatus capable of adopting MEMS components [117]. In this apparatus, an upper specimen mounted on a soft cantilever beam comes in contact with a lower specimen mounted on a lower specimen holder. The apparatus consists of two piezos (x- and z-piezos), and four fiber optic sensors (x- and z-displacement sensors, and x- and z-force sensors). For adhesion and friction studies, z- and x-piezos are used to bring the upper and lower specimens in contact and to apply a relative motion in the lateral direction, respectively. The x- and z-displacement sensors are used to measure the lateral position of the lower specimen and vertical position of the upper specimen, respectively. The x- and z-force sensors are used to measure friction force and nor-

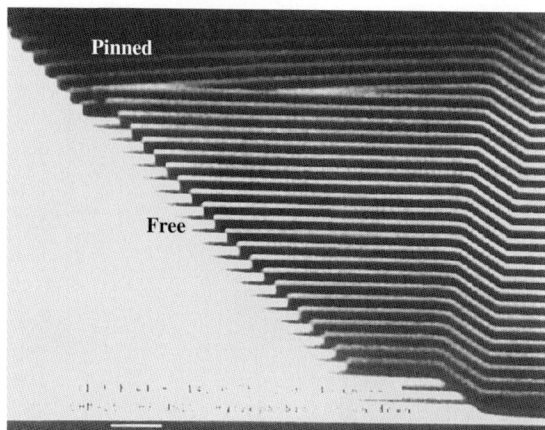

Fig. 21.27. SEM micrograph of a micromachined array of polysilicon cantilever beams of increasing length. The micrograph shows the onset of pinning for beams longer than 34 μm [113]

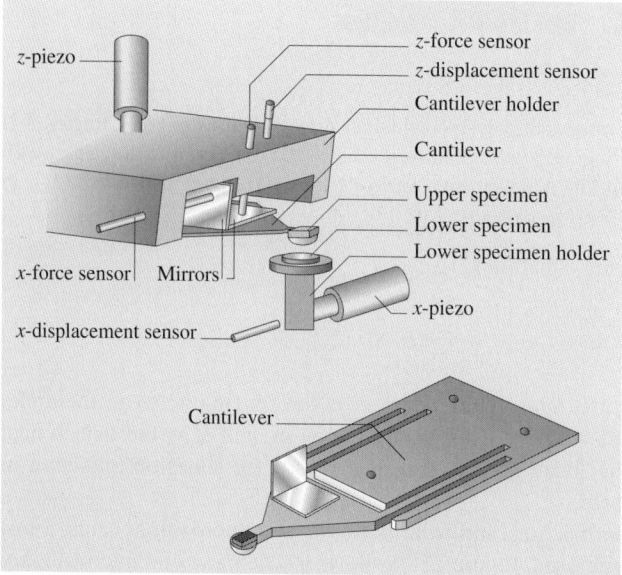

Fig. 21.28. Schematic of the microtriboapparatus, including specially designed cantilever (with two perpendicular mirrors attached on the end), lower specimen holder, two piezos (x- and z-piezos), and four fiber optic sensors (x- and z-displacement sensors and x- and z-force sensors) [117]

mal load/adhesive force between these two specimens, respectively, by monitoring the deflection of the cantilever.

As most of the MEMS/NEMS devices are fabricated from silicon, the study of silicon-on-silicon contacts is important. This contact was simulated by a flat single-crystal Si(100) wafer (phosphorus doped) specimen sliding against a single crystal Si(100) ball (1 mm in diameter, 5×10^{17} atoms/cm^3 boron doped) mounted on a stainless steel cantilever [117, 118]. Both of them have a native oxide layer on their surfaces. The other materials studied were 10-nm-thick DLC deposited by filtered cathodic arc deposition on Si(100), 2.3-nm-thick chemically bonded PFPE (Z-DOL, BW) on Si(100), and hexadecane thiol (HDT) monolayer on evaporated Au(111) film to investigate their anti-adhesion performance.

It is well-known that in the computer rigid disk drives, the adhesive force increases rapidly with an increase in rest time between a magnetic head and a magnetic disk [35]. Considering that adhesion and friction are the major issues that lead to the failure of MEMS/NEMS devices, the rest time effect on microscale on Si(100), DLC, PFPE, and HDT was studied, and the results are summarized in Fig. 21.29a. It is found that the adhesive force of Si(100) increases logarithmically with the rest time to a certain equilibrium time ($t = 1000$ s), after which it remains constant. Figure 21.29a also shows that the adhesive force of DLC, PFPE, and HDT does not change with rest time. Single-asperity contact modeling of the dependence of meniscus force on the rest time has been carried out by *Chilamakuri* and *Bhushan* [119], and the modeling results (Fig. 21.29b) verify experimental observations. Due to the presence of thin film adsorbed water on Si(100), meniscus forms around the contacting asperities and grows with time until equilibrium occurs, which causes the rest time effect on its adhesive force. The adhesive forces of DLC, PFPE, and HDT do not change with rest time, which suggests that either water meniscus is not present on their surfaces, or it does not increase with time.

The measured adhesive forces of Si(100), DLC, PFPE, and HDT at rest time of 1 s are summarized in Fig. 21.30. It shows that the presence of solid films of DLC, PFPE, and HDT greatly reduces the adhesive force of Si(100), whereas HDT film has the lowest adhesive force. It is well-known that the native oxide layer (SiO$_2$) on the top of Si(100) wafer exhibits hydrophilic properties, and water molecules produced by capillary condensation of water vapor from the environment can be adsorbed easily on this surface. The condensed water will form meniscus as the upper specimen approaches the lower specimen surface. The meniscus force is a major contributor to the adhesive force. In the case of DLC, PFPE, and HDT, the films are found to be hydrophobic based on contact angle measurements, and the amount of condensed water vapor is low compared to that on Si(100). It should be noted that the measured adhesive force is generally higher than that measured in AFM, because the larger radius of the Si(100) ball compared to that of an AFM tip induces larger meniscus and van der Waals forces.

To investigate the velocity effect on friction, the friction force as a function of velocity was measured and is summarized in Fig. 21.31a. It indicates that for Si(100) the friction force initially decreases with increasing velocity until equilibrium occurs. Figure 21.31a also indicates that the velocity almost has no effect on the friction properties of DLC, PFPE, and HDT. This implies that the friction mechanisms of DLC, PFPE, and HDT do not change with the variation in velocity. For Si(100),

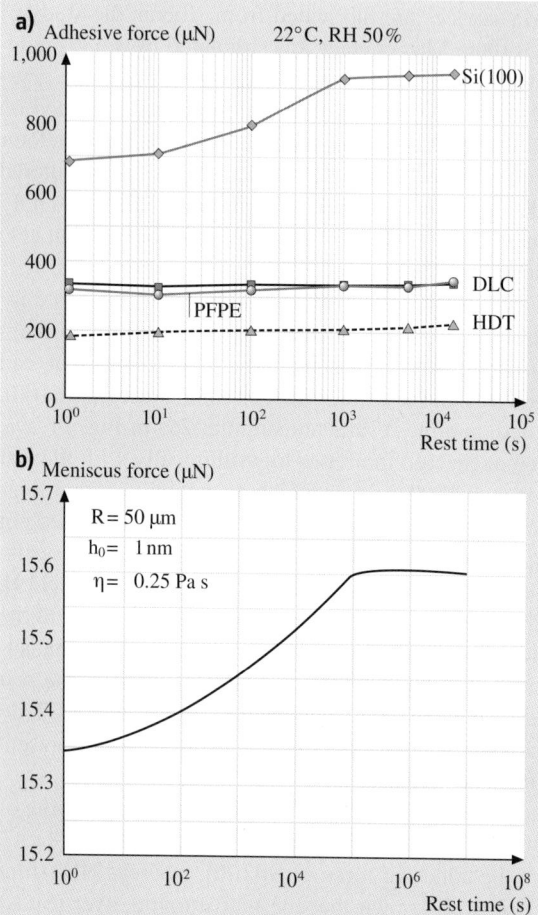

Fig. 21.29. (a) The influence of rest time on the adhesive force of Si(100), DLC, chemically bonded PFPE, and HDT, and (b) single asperity contact modeling results of the rest time effect on the meniscus force for an asperity of R in contact with a flat surface with a water-film thickness of h_0 and absolute viscosity of η_0 [119]

at high velocity, the meniscus is broken and does not have enough time to rebuild. In addition, it is believed that tribochemical reaction plays an important role. The high velocity leads to tribochemical reactions of Si(100), which has native oxide SiO_2, with water molecules to form a $Si(OH)_4$ film. This film is removed and continuously replenished during sliding. The $Si(OH)_4$ layer at the sliding surface is known to be of low shear strength. The breaking of the water meniscus and the formation of a $Si(OH)_4$ layer result in a decrease in friction force of Si(100). For DLC, PFPE, and HDT, surfaces exhibit hydrophobic properties and can only adsorb a few water molecules in ambient conditions. The above mentioned meniscus breaking and tribo-

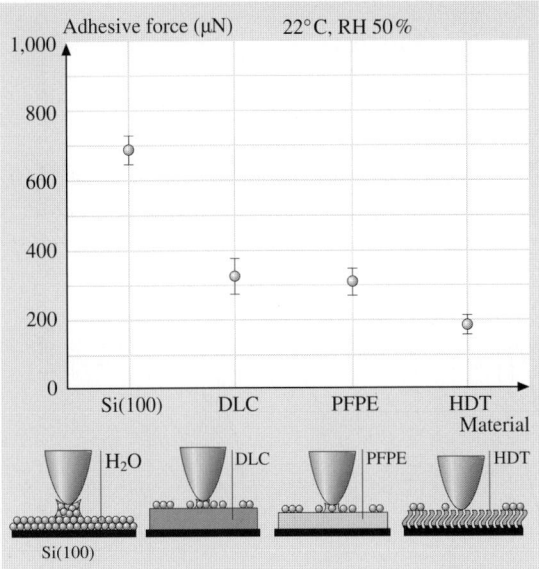

Fig. 21.30. Adhesive forces of Si(100), DLC, chemically bonded PFPE, and HDT at ambient conditions and a schematic showing the relative size of water meniscus on different specimens

chemical reaction mechanisms do not exist for these films. Therefore, their friction force does not change with velocity.

The influence of relative humidity was studied in an environmentally controlled chamber. The adhesive and friction forces were measured at increasing relative humidity, and the results are summarized in Fig. 21.31b. The figure shows that for Si(100), the adhesive force increases with relative humidity, but the adhesive forces of DLC and PFPE only show slight increases when humidity is higher than 45%, while the adhesive force of HDT does not change with humidity. Figure 21.31b also shows that for Si(100), the friction force increases with a relative humidity increase up to 45%, and then it shows a slight decrease with a further increase in the relative humidity. For PFPE, there is an increase in the friction force when humidity is higher than 45%. In the whole testing range, relative humidity does not have any apparent influence on the friction properties of DLC and HDT. In the case of Si(100), the initial increase of relative humidity up to 45% causes more adsorbed water molecules and forms bigger water meniscus, which leads to an increase of friction force. But at a very high humidity of 65%, large quantities of adsorbed water can form a continuous water layer that separates the tip and sample surfaces and acts as a kind of lubricant, which causes a decrease in the friction force. For PFPE, dewetting of lubricant film at humidity larger than 45% results in an increase in the adhesive and friction forces. DLC and HDT surfaces show hydrophobic properties, and increasing relative humidity does not play much of a role in their friction forces.

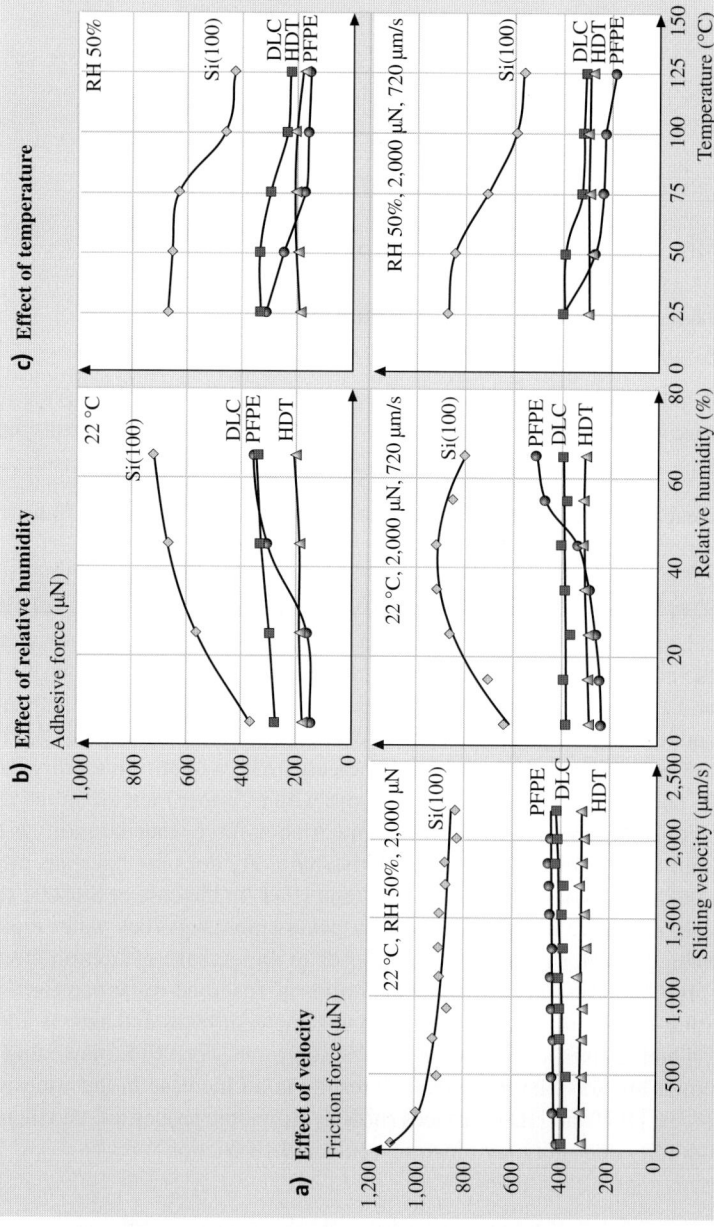

Fig. 21.31. The influence of (**a**) sliding velocity on the friction forces, (**b**) relative humidity on the adhesive and friction forces, and (**c**) temperature on the adhesive and friction forces of Si(100), DLC, chemically bonded PFPE, and HDT

The influence of temperature was studied using a heated stage. The adhesive force and friction force were measured at increasing temperatures from 22 °C to 125 °C. The results are presented in Fig. 21.31c. It shows that once the temperature is higher than 50 °C, increasing temperature causes a significant decrease in adhesive and friction forces of Si(100) and a slight decrease in the cases of DLC and PFPE. But the adhesion and friction forces of HDT do not show any apparent change with test temperature. At high temperature, desorption of water, and reduction of surface tension of water lead to the decrease of adhesive and friction forces of Si(100), DLC, and PFPE. However, in the case of HDT film, as only a few water molecules are adsorbed on the surface, the above mentioned mechanisms do not play a big role. Therefore, the adhesive and friction forces of HDT do not show any apparent change with temperature. Figure 21.31 shows that in the whole velocity, relative humidity, and temperature test range, the adhesive forces and friction forces of DLC, PFPE, and HDT are always smaller than that of Si(100), whereas HDT has the smallest value.

To summarize, several methods can be used to reduce adhesion in microstructures. MEMS surfaces can be coated with hydrophobic coatings such as PFPEs, SAMs, and passivated DLC coatings. It should be noted that other methods to reduce adhesion include the formation of dimples on the contact surfaces to reduce contact area [10, 54, 55, 115, 116]. Furthermore, an increase in hydrophobicity of the solid surfaces (high contact angle approaching 180°) can be achieved by using surfaces with suitable roughness, in addition to lowering their surface energy [120, 121]. The hydrophobicity of surfaces is dependent on a subtle interplay between surface chemistry and mesoscopic topography. The self-cleaning mechanism, or so-called "Lotus-effect", is closely related to the ultra-hydrophobic properties of these biological surfaces, which usually show microsculptures in specific dimensions.

21.7.3 Static Friction Force (Stiction) Measurements in MEMS

In MEMS devices involving parts in relative motion to each other, such as micromotors, large friction forces become the limiting factor to their successful operation and reliability. It is generally known that most micromotors cannot be rotated as manufactured and require some form of lubrication. It is therefore critical to determine the friction forces present in such MEMS devices. To measure in situ the static friction of a rotor-bearing interface in a micromotor, *Tai* and *Muller* [122] measured the starting torque (voltage) and pausing position for different starting positions under a constant-bias voltage. A friction-torque model was used to obtain the coefficient of static friction. To measure the in situ kinetic friction of the turbine and gear structures, *Gabriel* et al. [123] used a laser-based measurement system to monitor the steady-state spins and decelerations. *Lim* et al. [124] designed and fabricated a polysilicon microstructure to in situ measure the static friction of various films. The microstructure consisted of a shuttle suspended above the underlying electrode by a folded beam suspension. A known normal force was applied, and lateral force was measured to obtain the coefficient of static friction. *Beerschwinger* et al. developed a cantilever-deflection rig to measure friction of LIGA-processed micro-

Table 21.7. Published data on coefficient of static friction measurements of MEMS devices and structures

Reference	Test method	Device/structure	Material pairs	Environment	Coefficient of static friction
[122]	Starting voltage	IC-processed micromotor	PolySi/Si$_3$N$_4$	Air	0.20–0.40
[124]	Electrostatic loading	Comb-drive microstructure	PolySi/PolySi PolySi/Si$_3$N$_4$	Air	4.9 ± 1.0 2.5 ± 0.5
[127]	Pull-off force	Silicon microbeams	SiO$_2$/SiO$_2$	Air	2.1 ± 0.8
[126]	Cantilever/fiber deflection rig	LIGA micromotors	Ni/Alumina	Air	0.6–1.2

motors [125, 126]. Table 21.7 presents static friction coefficients of various MEMS devices evaluated by various researchers. Most of these techniques employ indirect methods to determine the friction forces, or involve fabrication of complex structures.

A novel technique to measure the static friction force (stiction) encountered in surface micromachined polysilicon micromotors using an AFM has been developed by *Sundararajan* and *Bhushan* [112]. Continuous physical contact occurs during rotor movement (rotation) in the micromotors between the rotor and lower hub flange. In addition, contact occurs at other locations between the rotor and the hub surfaces and between the rotor and the stator. Friction forces will be present at these contact regions during motor operation. Although the actual distribution of these forces is not known, they can be expected to be concentrated near the hub, where there is continuous contact. If we, therefore, represent the static friction force of the micromotor as a single force F_s acting at point P_1 (as shown in Fig. 21.32a), then the magnitude of the frictional torque about the center of the motor (O) that must be overcome before rotor movement can be initiated is:

$$T_s = F_s l_1 \,, \tag{21.2}$$

where l_1 is the distance OP_1, which is assumed to be the average distance from the center at which the friction force F_s occurs. Now consider an AFM tip moving against a rotor arm in a direction perpendicular to the long axis of the cantilever beam (the rotor arm edge closest to the tip is parallel to the long axis of the cantilever beam), as shown in Fig. 21.32a. When the tip encounters the rotor at point P_2, the tip will twist, generating a lateral force between the tip and the rotor (event A in Fig. 21.32b). This reaction force will generate a torque about the center of the motor. Since the tip is trying to move farther in the direction shown, the tip will continue to twist to a maximum value at which the lateral force between the tip and the rotor becomes high enough such that the resultant torque T_f about the center of the motor equals the static friction torque T_s. At this point, the rotor will begin to rotate, and the twist of the cantilever decreases sharply (event B in Fig. 21.32b). The twist of the cantilever is measured in the AFM as a change in the lateral deflection signal

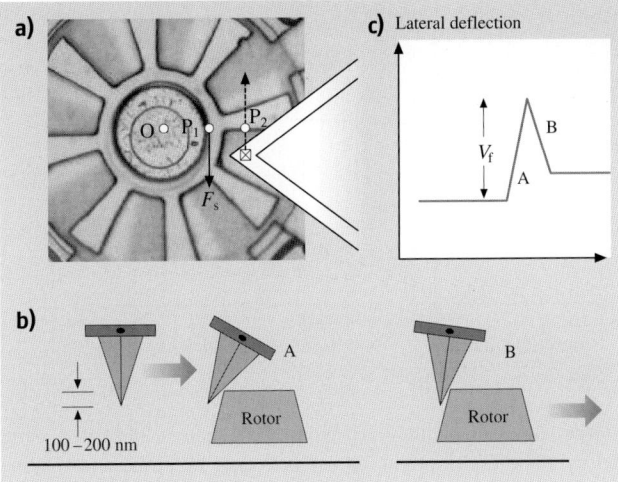

Fig. 21.32. (a) Schematic of the technique used to measure the force, F_s, required to initiate rotor movement using an AFM/FFM. (b) As the tip is pushed against the rotor, the lateral deflection experienced by the rotor due to the twisting of the tip prior to rotor movement is a measure of static friction force, F_s, of the rotors. (c) Schematic of lateral deflection expected from the above mentioned experiment. The peak V_f is related to the state of the rotor [112]

(in volts), which is the underlying concept of friction force microscopy (FFM). The change in the lateral deflection signal corresponding to the above mentioned events as the tip approaches the rotor is shown schematically in Fig. 21.32b. The value of the peak V_f is a measure of the force exerted on the rotor by the tip just before the static friction torque is matched and the rotor begins to rotate.

Using this technique, the viability of PFPE lubricants for micromotors has been investigated and the effect of humidity on the friction forces of unlubricated and lubricated devices was studied as well. Figure 21.33 shows static friction forces normalized over the weight of the rotor of unlubricated and lubricated micromotors as a function of rest time and relative humidity. Rest time here is defined as the time elapsed between the first experiment conducted on a given motor (solid symbol at time zero) and subsequent experiments (open symbols). Each open symbol data point is an average of six measurements. It can be seen that for the unlubricated motor and the motor lubricated with a bonded layer of Z-DOL (BW), the static friction force is highest for the first experiment and then drops to an almost constant level. In the case of the motor with an as-is mobile layer of Z-DOL, the values remain very high up to 10 days after lubrication. In all cases, there is negligible difference in the static friction force at 0% and 45% RH. At 70% RH, the unlubricated motor exhibits a substantial increase in the static friction force, while the motor with bonded Z-DOL shows no increase in static friction force due to the hydrophobicity of the lubricant layer. The motor with an as-is mobile layer of the lubricant shows consistently high values of static friction force that vary little with humidity.

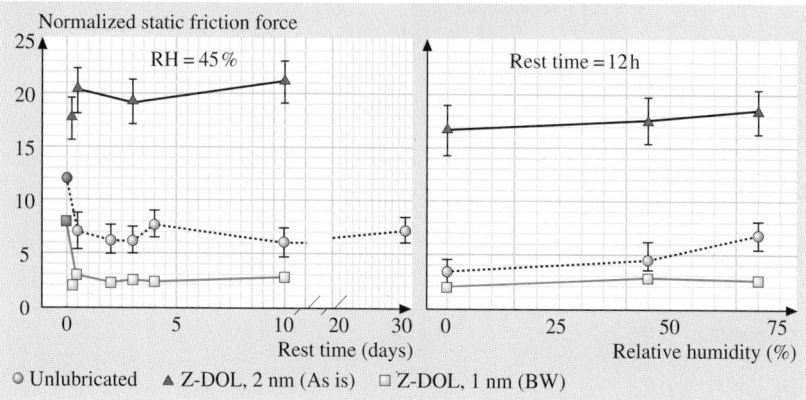

Fig. 21.33. Static friction force values of unlubricated motors and motors lubricated using PFPE lubricants, normalized over the rotor weight, as a function of rest time and relative humidity. Rest time is defined as the time elapsed between a given experiment and the first experiment in which motor movement was recorded (time 0). The motors were allowed to sit at a particular humidity for 12 h prior to measurements [112]

21.7.4 Mechanisms Associated with Observed Stiction Phenomena in Micromotors

Figure 21.34 summarizes static friction force data for two motors, M1 and M2, along with schematics of the meniscus effects for the unlubricated and lubricated surfaces. Capillary condensation of water vapor from the environment results in the formation of meniscus bridges between contacting and near-contacting asperities of two surfaces in close proximity to each other, as shown in Fig. 21.34. For unlubricated surfaces, more menisci are formed at higher humidity, resulting in a higher friction force between the surfaces. The formation of meniscus bridges is supported by the fact that the static friction force for unlubricated motors increases at high humidity (Fig. 21.34). Solid bridging may occur near the rotor-hub interface due to silica residues after the first etching process. In addition, the drying process after the final etch can result in liquid bridging formed by the drying liquid due to meniscus force at these areas [54, 55, 113, 115]. The initial static friction force therefore will be quite high, as evidenced by the solid data points in Fig. 21.34. Once the first movement of the rotor permanently breaks these solid and liquid bridges, the static friction force of the motors will drop (as seen in Fig. 21.34) to a value dictated predominantly by the adhesive energies of rotor and hub surfaces, the real area of contact between these surfaces and meniscus forces due to water vapor in the air, at which point, the effect of lubricant films can be observed. Lubrication with a mobile layer, even a thin one, results in veryhigh static friction forces due to meniscus effects of the lubricant liquid itself at and near the contact regions. It should be noted that a motor submerged in a liquid lubricant would result in a fully flooded lubrication regime. In this case, there is no meniscus contribution and only the viscous contribution to the friction forces would be relevant. However, submerging the device in

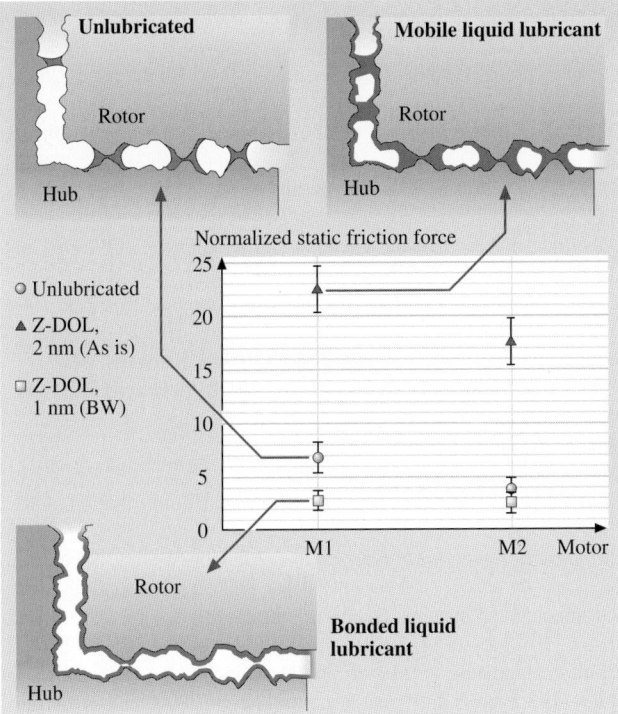

Fig. 21.34. Summary of effect of liquid and solid lubricants on static friction force of micromotors. Despite the hydrophobicity of the lubricant used (Z-DOL), a mobile liquid lubricant (Z-DOL as is) leads to very high static friction force due to increased meniscus forces, whereas a solid-like lubricant (bonded Z-DOL, BW) appears to provide some amount of reduction in static friction force

a lubricant may not be a practical method. A solid-like hydrophobic lubricant layer (such as bonded Z-DOL) results in favorable friction characteristics of the motor. The hydrophobic nature of the lubricant inhibits meniscus formation between the contact surfaces and maintains low friction even at high humidity (Fig. 21.34). This suggests that solid-like hydrophobic lubricants are ideal for lubrication of MEMS, while mobile lubricants result in increased values of static friction force.

References

1. NASA Ames Research Center, Mottett Field, CA, USA, http://www.ipt.arc.nasa.gov/gallery.html.
2. W. G. van der Wiel, S. De Franceschi, J. M. Elzerman, T. Fujisawa, S. Tarucha, and L. P. Kouwenhoven. Electron transport through double quantum dots. *Rev. Mod. Phys.*, 75:1–22, 2003.
3. Texas Instruments DLP Products, Plano, TX, USA, http://www.dlp.com.

4. Anonymous. *Microelectromechanical Systems: Advanced Materials and Fabrication Methods.* National Academy Press, 1997.
5. M. Roukes. Nanoelectromechanical systems face the future. *Phys. World*, pages 25–31, February 2001.
6. M. A. Huff. A distributed mems processing environment. *http://www.memsexchange.org/*, 2002.
7. R. S. Muller, R. T. Howe, S. D. Senturia, R. L. Smith, and R. M. White. *Microsensors.* IEEE, 1990.
8. I. Fujimasa. *Micromachines: A New Era in Mechanical Engineering.* Oxford Univ. Press, 1996.
9. W. S. Trimmer, editor. *Micromachines and MEMS: Classic and Seminal Papers to 1990.* IEEE, 1997.
10. B. Bhushan, editor. *Tribology Issues and Opportunities in MEMS.* Kluwer, 1998.
11. G. T. A. Kovacs. *Micromachined Transducers Sourcebook.* WCB McGraw-Hill, 1998.
12. S. D. Senturia. *Microsystem Design.* Kluwer, 2001.
13. M. Gad-el-Hak. *The MEMS Handbook.* CRC, 2002.
14. T. R. Hsu. *MEMS and Microsystems.* McGraw-Hill, 2002.
15. M. Madou. *Fundamentals of Microfabrication: The Science of Miniaturization.* CRC, 2nd edition, 2002.
16. R. C. Jaeger. *Introduction to Microelectronic Fabrication.* Addison-Wesley, 5th edition, 1988.
17. J. W. Judy. Microelectromechanical systems (mems): Fabrication, design, and applications. *Smart Mater. Struct.*, 10:1115–1134, 2001.
18. E. W. Becker, W. Ehrfeld, P. Hagmann, A. Maner, and D. Munchmeyer. Fabrication of microstructures with high aspect ratios and great structural heights by synchrotron radiation lithography, galvanoforming, and plastic moulding (liga process). *Microelectron. Eng.*, 4:35–56, 1986.
19. C. R. Friedrich and R. O. Warrington. *Surface characterization of non-lithographic micromachining*, pages 73–84. Kluwer, 1998.
20. M. Madou. *Facilitating choices of machining tools and materials for miniaturization science: A review*, pages 31–51. Kluwer, 1998.
21. H. Lehr, S. Abel, J. Doppler, W. Ehrfeld, B. Hagemann, K. P. Kamper, F. Michel, Ch. Schulz, and Ch. Thurigen. Microactuators as driving units for microrobotic systems. *SPIE Proc. Microrobotics: Components and Applications*, 2906:202–210, 1996.
22. H. Lehr, W. Ehrfeld, B. Hagemann, K. P. Kamper, F. Michel, Ch. Schulz, and Ch. Thurigen. Development of micro-millimotors. *Min. Invas. Ther. Allied Technol.*, 6:191–194, 1997.
23. F. Michel and W. Ehrfeld. *Microfabrication technologies for high performance microactuators*, pages 53–72. Kluwer, 1998.
24. M. Tanaka. Development of desktop machining microfactory. *Riken Rev.*, 34:46–49, 2001.
25. Y. Xia and G. M. Whitesides. Soft lithography. *Angew. Chem. Int. Edn.*, 37:550–575, 1998.
26. H. Becker and C. Gaertner. Polymer microfabrication methods for microfluidic analytical applications. *Electrophoresis*, 21:12–26, 2000.
27. Y. Xia, E. Kim, X. M. Zhao, J. A. Rogers, M. Prentiss, and G. M. Whitesides. Complex optical surfaces formed by replica molding against elastomeric masters. *Science*, 273:347–349, 1996.
28. S. Y. Chou, P. R. Krauss, and P. J. Renstrom. Imprint lithography with 25-nanometer resolution. *Science*, 272:85–87, 1996.

29. A. Kumar and G. M. Whitesides. Features of gold having micrometer to centimeter dimensions can be formed through a combination of stamping with an elastomeric stamp and an alkanethiol ink followed by chemical etching. *Appl. Phys. Lett.*, 63:2002–2004, 1993.
30. T. A. Core, W. K. Tsang, and S. J. Sherman. Fabrication technology for an integrated surface-micromachined sensor. *Solid State Technol.*, 36:39–47, 1993.
31. J. Bryzek, K. Peterson, and W. McCulley. Micromachines on the march. *IEEE Spectrum*, pages 20–31, May 1994.
32. L. J. Hornbeck and W. E. Nelson. *Bistable deformable mirror device*, pages 107–110. OSA, 1988.
33. L. J. Hornbeck. A digital light processing(tm) update – status and future applications. *Proc. SPIE*, 3634:158–170, 1999.
34. L. J. Hornbeck. The dmd^{TM} projection display chip: A mems-based technology. *MRS Bull.*, 26:325–328, 2001.
35. B. Bhushan. *Tribology and Mechanics of Magnetic Storage Devices*. Springer, 2nd edition, 1996.
36. H. Hamilton. Contact recording on perpendicular rigid media. *J. Mag. Soc. Jpn.*, 15:(Suppl. S2) 483–481, 1991.
37. T. Ohwe, Y. Mizoshita, and S. Yonoeka. Development of integrated suspension system for a nanoslider with an mr head transducer. *IEEE Trans. Magn.*, 29:3924–3926, 1993.
38. L. S. Fan and S. Woodman. Batch fabrication of mechanical platforms for high-density data storage.
39. D. A. Horsley, M. B. Cohn, A. Singh, R. Horowitz, and A. P. Pisano. Design and fabrication of an angular microactuator for magnetic disk drives. *J. Microelectromech. Syst.*, 7:141–148, 1998.
40. T. Hirano, L. S. Fan, D. Kercher, S. Pattanaik, and T. S. Pan. Hdd tracking microactuator and its integration issues, 2000.
41. P. Gravesen, J. Branebjerg, and O. S. Jensen. Microfluidics – a review. *J. Micromech. Microeng.*, 3:168–182, 1993.
42. C. Lai Poh San and E. P. H. Yap, editors. *Frontiers in Human Genetics*. World Scientific, 2001.
43. C. H. Mastrangelo and H. Becker (Eds.). Microfluidics and biomems. *Proc. SPIE*, 4560, 2001.
44. H. Becker and L. E. Locascio. Polymer microfluidic devices. *Talanta*, 56:267–287, 2002.
45. S. Shoji and M. Esashi. Microflow devices and systems. *J. Micromech. Microeng.*, 4:157–171, 1994.
46. P. Woias. Micropumps – summarizing the first two decades.
47. M. Scott. Mems and moems for national security applications.
48. K. E. Drexler. *Nanosystems: Molecular Machinery, Manufacturing and Computation*. Wiley, 1992.
49. B. Bhushan. *Handbook of Micro/Nanotribology*. CRC, 2nd edition, 1999.
50. G. Timp, editor. *Nanotechnology*. Springer, 1999.
51. E. A. Rietman. *Molecular Engineering of Nanosystems*. Springer, 2001.
52. H. S. Nalwa, editor. *Nanostructured Materials and Nanotechnology*. Academic, 2002.
53. W. A. Goddard, D. W. Brenner, S. E. Lyshevski, and G. J. Iafrate. *Handbook of Nanoscience, Engineering, and Technology*. CRC, 2003.
54. B. Bhushan. *Principles and Applications of Tribology*. Wiley, 1999.
55. B. Bhushan. *Introduction to Tribology*. Wiley, 2002.

56. Y. C. Tai, L. S. Fan, and R. S. Muller. Ic-processed micro-motors: Design, technology and testing.
57. S. M. Spearing and K. S. Chen. Micro-gas turbine engine materials and structures. *Ceramic Eng. Sci. Proc.*, 18:11–18, 2001.
58. M. Mehregany, K. J. Gabriel, and W. S. N. Trimmer. Integrated fabrication of polysilicon mechanisms. *IEEE Trans. Electron. Dev.*, 35:719–723, 1988.
59. E. J. Garcia and J. J. Sniegowski. Surface micromachined microengine. *Sens. Actuators A*, 48:203–214, 1995.
60. J. K. Robertson and K. D. Wise. An electrostatically actuated integrated microflow controller. *Sens. Actuators A*, 71:98–106, 1998.
61. B. Bhushan. Nanotribology and nanomechanics of mems devices, 1996.
62. D. M. Tanner, N. F. Smith, L. W. Irwin, et al. *MEMS Reliability: Infrastructure, Test Structures, Experiments, and Failure Modes.* Sandia National Laboratories, 2000.
63. S. S. Mani, J. G. Fleming, J. A. Walraven, J. J. Sniegowski, et al. Effect of w coating on microengine performance, 2000.
64. R. E. Sulouff. *MEMS opportunities in accelerometers and gyros and the microtribology problems limiting commercialization*, pages 109–120. Kluwer, 1998.
65. Analog Devices Inc., Berkeley, CA, USA, http://www.analog.com.
66. S. A. Henck. Lubrication of digital micromirror devices. *Tribol. Lett.*, 3:239–247, 1997.
67. M. R. Douglass. Lifetime estimates and unique failure mechanisms of the digital micromirror devices (dmd), 1998.
68. M. R. Douglass. Dmd reliability: A mems success story.
69. W. C. Tang and A. P. Lee. Defense applications of mems. *MRS Bull.*, 26:318–319. Also see www.darpa.mil/mto/mems, 2001.
70. H. Guckel and D. W. Burns. Fabrication of micromechanical devices from polysilicon films with smooth surfaces. *Sens. Actuators*, 20:117–122, 1989.
71. G. T. Mulhern, D. S. Soane, and R. T. Howe. Supercritical carbon dioxide drying of microstructures, 1993.
72. P. Vettiger, J. Brugger, M. Despont, U. Drechsler, U. Duerig, and W. Haeberle. Ultrahigh density, high data-rate nems based afm data storage system. *Microelectron. Eng.*, 46:11–27, 1999.
73. M. Ferrari and J. Liu. The engineered course of treatment. *Mech. Eng.*, pages 44–47, December 2001.
74. F. J. Martin and C. Grove. Microfabricated drug delivery systems: Concepts to improve clinical benefits. *Biomed. Microdev.*, 3:97–108, 2001.
75. K. F. Man, B. H. Stark, and R. Ramesham. *A Resource Handbook for MEMS Reliability*, Rev. A. California Institute of Technology, 1998.
76. S. Kayali, R. Lawton, and B. H. Stark. Mems reliability assurance activities at jpl. *EEE Links*, 5:10–13, 1999.
77. S. Arney. Designing for mems reliability. *MRS Bull.*, 26:296–299, 2001.
78. K. F. Man. Mems reliability for space applications by elimination of potential failure modes through testing and analysis (2001).
79. B. Bhushan, A. V. Kulkarni, W. Bonin, and J. T. Wyrobek. Nano/picoindentation measurement using a capacitance transducer system in atomic force microscopy. *Philos. Mag.*, 74:1117–1128, 1996.
80. S. Sundararajan and B. Bhushan. Development of afm-based techniques to measure mechanical properties of nanoscale structures. *Sens. Actuators A*, 101:338–351, 2002.
81. B. Bhushan, editor. *Modern Tribology Handbook.* CRC, 2001.

82. B. Bhushan, J. N. Israelachvili, and U. Landman. Nanotribology: Friction, wear and lubrication at the atomic scale. *Nature*, 374:607–616, 1995.
83. J. S. Shor, D. Goldstein, and A. D. Kurtz. Characterization of n-type β-SiC as a piezoresistor. *IEEE Trans. Electron. Dev.*, 40:1093–1099, 1993.
84. M. Mehregany, C. A. Zorman, N. Rajan, and C. H. Wu. Silicon carbide mems for harsh environments. *Proc. IEEE*, 86:1594–1610, 1998.
85. C. A. Zorman, A. J. Fleischmann, A. S. Dewa, M. Mehregany, C. Jacob, S. Nishino, and P. Pirouz. Epitaxial growth of 3C-SiC films on 4 in. diam Si(100) silicon wafers by atmospheric pressure chemical vapor deposition. *J. Appl. Phys.*, 78:5136–5138, 1995.
86. C. A. Zorman, S. Roy, C. H. Wu, A. J. Fleischman, and M. Mehregany. Characterization of polycrystalline silicon carbide films grown by atmospheric pressure chemical vapor deposition on polycrystalline silicon. *J. Mater. Res.*, 13:406–412, 1998.
87. B. Bhushan and B. K. Gupta. *Handbook of Tribology: Materials, Coatings and Surface Treatments*. Krieger, 1997.
88. J. F. Shackelford, W. Alexander, and J. S. Park, editors. *CRC Material Science and Engineering Handbook*. CRC, 2nd edition, 1994.
89. B. K. Gupta, J. Chevallier, and B. Bhushan. Tribology of ion bombarded silicon for micromechanical applications. *ASME J. Tribol.*, 115:392–399, 1993.
90. B. K. Gupta, B. Bhushan, and J. Chevallier. Modification of tribological properties of silicon by boron ion implantation. *Tribol. Trans.*, 37:601–607, 1994.
91. B. K. Gupta and B. Bhushan. Nanoindentation studies of ion implanted silicon. *Surf. Coat. Technol.*, 68-69:564–570, 1994.
92. B. Bhushan and V. N. Koinkar. Tribological studies of silicon for magnetic recording applications. *J. Appl. Phys.*, 75:5741–5746, 1994.
93. G. M. Pharr. *The anomalous behavior of silicon during nanoindentation*, volume 239, pages 301–312. Materials Research Soc., 1991.
94. D. L. Callahan and J. C. Morris. The extent of phase transformation in silicon hardness indentation. *J. Mater. Res.*, 7:1612–1617, 1992.
95. N. A. Fleck, G. M. Muller, M. F. Ashby, and J. W. Hutchinson. Strain gradient plasticity: Theory and experiment. *Acta Metall. Mater.*, 42:475–487, 1994.
96. B. Bhushan and S. Venkatesan. Friction and wear studies of silicon in sliding contact with thin-film magnetic rigid disks. *J. Mater. Res.*, 8:1611–1628, 1993.
97. S. Venkatesan and B. Bhushan. The role of environment in the friction and wear of single-crystal silicon in sliding contact with thin-film magnetic rigid disks. *Adv. Info Storage Syst.*, 5:241–257, 1993.
98. S. Venkatesan and B. Bhushan. The sliding friction and wear behavior of single-crystal, polycrystalline and oxidized silicon. *Wear*, 171:25–32, 1994.
99. B. Bhushan. Chemical, mechanical, and tribological characterization of ultra-thin and hard amorphous carbon coatings as thin as 3.5 nm: Recent developments. *Diam. Relat. Mater.*, 8:1985–2015, 1999.
100. S. Sundararajan and B. Bhushan. Micro/nanotribological studies of polysilicon and SiC films for mems applications. *Wear*, 217:251–261, 1998.
101. X. Li and B. Bhushan. Micro/nanomechanical characterization of ceramic films for microdevices. *Thin Solid Films*, 340:210–217, 1999.
102. A. A. Yasseen, C. H. Wu, C. A. Zorman, and M. Mehregany. Fabrication and testing of surface micromachined polycrystalline SiC micromotors. *IEEE Electron. Dev. Lett.*, 21:164–166, 2000.
103. B. Bhushan, A. V. Kulkarni, V. N. Koinkar, M. Boehm, L. Odoni, C. Martelet, and M. Belin. Microtribological characterization of self-assembled and langmuir–blodgett

monolayers by atomic force and friction force microscopy. *Langmuir*, 11:3189–3198, 1995.
104. V. N. Koinkar and B. Bhushan. Micro/nanoscale studies of boundary layers of liquid lubricants for magnetic disks. *J. Appl. Phys.*, 79:8071–8075, 1996.
105. V. N. Koinkar and B. Bhushan. Microtribological studies of unlubricated and lubricated surfaces using atomic force/friction force microscopy. *J. Vac. Sci. Technol. A*, 14:2378–2391, 1996.
106. B. Bhushan and H. Liu. Nanotribological properties and mechanisms of alkylthiol and biphenyl thiol self-assembled monolayers studied by afm. *Phys. Rev. B*, 63:245412:1–11, 2001.
107. H. Liu and B. Bhushan. Investigation of nanotribological properties of self-assembled monolayers with alkyl and biphenyl spacer chains. *Ultramicroscopy*, 91:185–202, 2002.
108. H. Liu and B. Bhushan. Nanotribological characterization of molecularly-thick lubricant films for applications to mems/nems by afm. *Ultramicroscopy*, 97:321–340, 2003.
109. T. Stifter, O. Marti, and B. Bhushan. Theoretical investigation of the distance dependence of capillary and van der waals forces in scanning force microscopy. *Phys. Rev. B*, 62:13667–13673, 2000.
110. B. Bhushan. *Self-assembled monolayers for controlling hydrophobicity and/or friction and wear*, pages 909–929. CRC, 2001.
111. H. Liu, B. Bhushan, W. Eck, and V. Stadler. Investigation of the adhesion, friction, and wear properties of biphenyl thiol self-assembled monolayers by atomic force microscopy. *J. Vac. Sci. Technol. A*, 19:1234–1240, 2001.
112. S. Sundararajan and B. Bhushan. Static friction and surface roughness studies of surface micromachined electrostatic micromotors using an atomic force/friction force microscope. *J. Vac. Sci. Technol. A*, 19:1777–1785, 2001.
113. C. H. Mastrangelo and C. H. Hsu. Mechanical stability and adhesion of microstructures under capillary forces – part ii: Experiments. *J. Microelectromech. Syst.*, 2:44–55, 1993.
114. M. P. De Boer and T. A. Michalske. Accurate method for determining adhesion of cantilever beams. *J. Appl. Phys.*, 86:817, 1999.
115. R. Maboudian and R. T. Howe. Critical review: Adhesion in surface micromechanical structures. *J. Vac. Sci. Technol. B*, 15:1–20, 1997.
116. C. H. Mastrangelo. *Surface force induced failures in microelectromechanical systems*, pages 367–395. Kluwer, 1998.
117. H. Liu and B. Bhushan. Adhesion and friction studies of microelectromechanical systems/nanoelectromechanical systems materials using a novel microtriboapparatus. *J. Vac. Sci. Technol. A*, 21:1528–1538, 2003.
118. B. Bhushan, H. Liu, and S. M. Hsu. Adhesion and friction studies of silicon and hydrophobic and low friction films and investigation of scale effects. *ASME J. Tribol.*, page in press.
119. S. K. Chilamakuri and B. Bhushan. A comprehensive kinetic meniscus model for prediction of long-term static friction. *J. Appl. Phys.*, 15:4649–4656, 1999.
120. J. Kijlstra, K. Reihs, and A. Klamt. Roughness and topology of ultra-hydrophobic surfaces. *Colloids Surf. A*, 206:521–529, 2002.
121. D. Quere and P. Aussillous. Non-stick droplets. *Chem. Eng. Technol.*, 25:925–928, 2002.
122. Y. C. Tai and R. S. Muller. Frictional study of ic processed micromotors. *Sens. Actuators A*, 21-23:180–183, 1990.
123. K. J. Gabriel, F. Behi, R. Mahadevan, and M. Mehregany. In situ friction and wear measurement in integrated polysilicon mechanisms. *Sens. Actuators A*, 21-23:184–188, 1990.

124. M. G. Lim, J. C. Chang, D. P. Schultz, R. T. Howe, and R. M. White. Polysilicon microstructures to characterize static friction, 1990.
125. U. Beerschwinger, S. J. Yang, R. L. Reuben, M. R. Taghizadeh, and U. Wallrabe. Friction measurements on liga-processed microstructures. *J. Micromech. Microeng.*, 4:14–24, 1994.
126. D. Matheison, U. Beerschwinger, S. J. Young, R. L. Rueben, M. Taghizadeh, S. Eckert, and U. Wallrabe. Effect of progressive wear on the friction characteristics of nickel liga processed rotors. *Wear*, 192:199–207, 1996.
127. R. Maboudian. Adhesion and friction issues associated with reliable operation of mems. *MRS Bull.*, pages 47–51, June 1998.

22

Mechanical Properties of Micromachined Structures

Harold Kahn

Summary. To be able to accurately design structures and make reliability predictions, in any field it is necessary first to know the mechanical properties of the materials that make up the structural components. In the fields of microelectromechanical systems (MEMS) MEMSand nanoelectromechanical systems (NEMS), nanoelectromechanical systemsthe devices are necessarily very small. The processing techniques and microstructures of the materials in these devices may differ significantly from bulk structures. Also, the surface-area-to-volume ratio in these structures is much higher than in bulk samples, and so the surface properties become much more important. In short, it cannot be assumed that mechanical properties measured using bulk specimens will apply to the same materials when used in MEMS and NEMS. This chapter will review the techniques that have been used to determine the mechanical properties of micromachined structures, especially residual stress, strength, and Young's modulus. The experimental measurements that have beenperformed will then be summarized, in particular the values obtained for polycrystalline silicon (polysilicon).

22.1 Measuring Mechanical Properties of Films on Substrates

In order to determine accurately the mechanical properties of very small structures, it is necessary to test specimens made from the same materials, processed in the same way, and of the same approximate size. Not surprisingly it is often difficult to handle specimens this small. One solution is to test the properties of films remaining on substrates. Micro- and nanomachined structures are typically fabricated from films that are initially deposited onto a substrate, are subsequently patterned and etched into the appropriate shapes, and then finally released from the substrate. If the testing is performed on the continuous film, before patterning and release, the substrate can be used as an effective "handle" for the specimen (in this case, the film). Of course since the films are adhered to the substrate, the types of tests possible are severely limited.

22.1.1 Residual Stress Measurements

One common measurement easily performed on films attached to substrates is residual film stress. The curvature of the substrate is measured before and after film deposition. Curvature can be measured in a number of ways. The most common technique is to scan a laser across the surface (or scan the substrate beneath the laser) and detect the angle of the reflected signal. Alternatively, profilometry, optical interferometry, or even atomic force microscopy can be used. As expected, tools that map a surface or perform multiple linear scans can give more accurate readings than tools that measure only a single scan.

Assuming that the film is thin compared to the substrate, the average residual stress in the film, σ_f, is given by the Stoney equation,

$$\sigma_f = \frac{1}{6} \frac{E_s}{(1-\nu_s)} \frac{t_s^2}{t_f} \left(\frac{1}{R_1} - \frac{1}{R_2} \right), \tag{22.1}$$

where the subscripts f and s refer to the film and substrate, respectively; t is thickness, E is Young's modulus, ν is Poisson's ratio, and R is radius of curvature before (R_1) and after (R_2) film deposition [1]. For the typical (100)-oriented silicon substrate, $E/(1-\nu)$ (also known as the biaxial modulus) is equal to 180.5 GPa, independent of in-plane rotation [2]. This investigation can be performed on the as-deposited film or after any subsequent annealing step, provided no changes occur to the substrate.

This measurement will reveal the average residual stress of the film. Typically, however, the residual stresses of deposited films will vary through the film thickness. One way to detect this, using substrate curvature techniques, is to etch away a fraction of the film and repeat the curvature measurement. This can be iterated any number of times to obtain a residual stress profile for the film [3]. Alternatively, tools have been designed to measure the substrate curvature during the deposition process itself, to obtain information on how the stresses evolve [4].

An additional feature of some of these tools is the ability to heat the substrates while performing the stress measurement. An example of the results obtained in such an experiment is shown in Fig. 22.1 [5], for an aluminium film on a silicon substrate. The slope of the heating curve gives the difference in thermal expansion between the film and the substrate. When the heating curve changes slope and becomes nearly horizontal, the yield strength of the film has been reached.

22.1.2 Mechanical Measurements Using Nanoindentation

Aside from residual stress, it is difficult to measure mechanical properties of films adhered to substrates without the measurement being affected by the presence of the substrate. Recent developments in nanoindentation equipment have allowed this technique to be used in some cases. With specially designed tools, indentation can be performed using very low loads. If the films being investigated are thick and rigid enough, measurements can be made that are not influenced by the presence of the substrate. Of course, this can be verified by depositing the same film onto different substrates. By continuously monitoring the displacement as well as the load during indentation, a variety of properties can be measured, including hardness and Young's modulus [6]. This area is covered in more detail in a separate chapter.

For brittle materials, cracks can be generated by indentation, and strength information can be gathered. But the exact stress fields created during the indentation

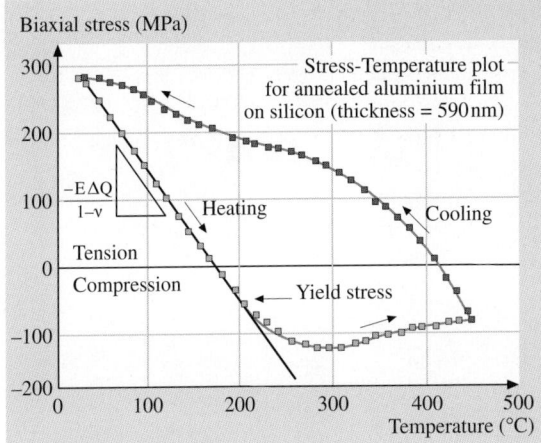

Fig. 22.1. Typical results for residual stress as a function of temperature for an aluminium film on a silicon substrate [5]. The stresses were determined by measuring the curvature of the substrate before and after film deposition, using the reflected signal of a laser scanned across the substrate surface

process are not exactly known, and, therefore, quantitative values for strength are difficult to determine. Anisotropic etching of single crystal silicon has been done to create 30-μm-tall structures that were then indented to examine fracture toughness [7], however this is not possible with most materials.

22.2 Micromachined Structures for Measuring Mechanical Properties

Certainly the most direct way to measure the mechanical properties of small structures is to fabricate structures that would be conducive to such tests. Fabrication techniques are sufficiently advanced that virtually any design can be realized, at least in two dimensions. Two basic types of devices are used for mechanical property testing: "passive" structures and "active" structures.

22.2.1 Passive Structures

As mentioned previously, the main difficulty in testing very small specimens is in handling. One way to circumvent this problem is by using passive structures. These structures are designed to act as soon as they are released from the substrate and to provide whatever information they are designed to supply without further manipulation. For all of these passive structures, the forces acting on them come from the structural materials' own internal residual stresses. For devices on the micron scale or smaller, gravitational forces can be neglected, and therefore internal stresses are the only source of actuation force.

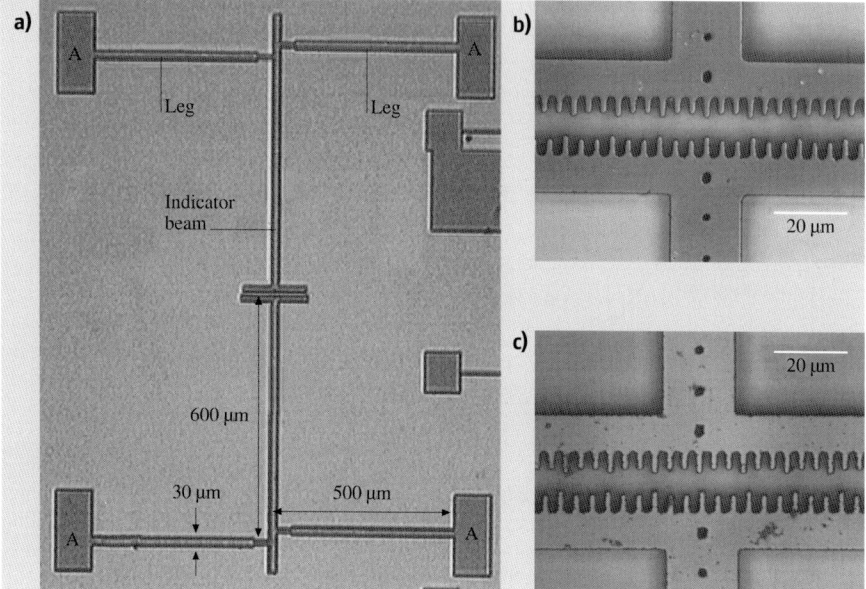

Fig. 22.2. (a) Microstrain gauge fabricated from polysilicon. (b) shows a close-up of the Vernier scale before release, and (c) shows the same area after release

Stress Measurements

Since the internal residual stresses act upon the passive devices when they are released, it is natural to design a device that can be used to measure residual stresses. One such device, a rotating microstrain gauge, is shown in Fig. 22.2. There are many different microstrain gauge designs, but all operate under the same principle. In Fig. 22.2a, the large pads, labeled A, will remain anchored to the substrate when the rest of the device is released. Upon release, the device will expand or contract in order to relieve its internal residual stresses. A structure under tension will contract, and a structure under compression will expand. For the structure in Fig. 22.2, a compressive stress will cause the legs to lengthen. Since the two opposing legs are not attached to the central beam at the same point but are offset, they will cause the central beam to rotate when they expand. The device in Fig. 22.2 contains two independent gauges that point to one another. At the ends of the two central beams are two parts of a Vernier scale. By observing this scale, one can measure the rotation of the beams.

If the connections between the legs and the central beams were simple pin connections, the strain, ε, of the legs (the fraction of expansion or contraction) could be determined simply by the measured rotation and the geometry of the device, namely

$$\varepsilon = \frac{d_{\text{beam}} d_{\text{offset}}}{2 L_{\text{central}} L_{\text{leg}}}, \tag{22.2}$$

where d_{beam} is the lateral deflection of the end of one central beam, d_{offset} is the distance between the connections of the opposing legs, $L_{central}$ is the length of the central beam (measured to the center point between the leg connections), and L_{leg} is the length of the leg. But since the entire device was fabricated from a single polysilicon film, this cannot be the case; there must be some bending occurring at the connections. As a result, to get an accurate determination of the strain relieved upon release, finite element analysis (FEA) of the structure must be performed. This is a common situation for microdevices. FEA is a powerful tool in determining the displacements and stresses of nonideal geometries. One drawback is that Young's modulus of the material must be known in order to do the FEA as well as to convert the measured strain into a stress value. But Young's moduli for many micromachined materials are known or can be measured using other techniques.

Other devices, besides rotating strain gauges, are designed to measure residual stresses. One of the most simple is a doubly clamped beam, a long, narrow beam of constant width and thickness that is anchored to the substrate at both ends. If the beam contains a tensile stress, it will remain straight. But if the beam contains a compressive stress, it will buckle if its length exceeds a critical value, l_{cr}, according to the Euler buckling criterion [8],

$$\varepsilon_r = -\frac{\pi^2}{3}\left(\frac{h}{l_{cr}}\right)^2, \qquad (22.3)$$

where ε_r is the residual strain in the beam, and h is the width or thickness of the beam, whichever is less. To determine the residual strain, a series of doubly clamped beams are fabricated of varying lengths. In this way, the critical length, l_{cr}, for buckling can be deduced after release. One problem with this technique is that during the release process, any turbulence in the solution will lead to enhanced buckling of beams, and a low value for l_{cr} will be obtained.

For films with tensile stresses, a similar analysis can be done using ring-and-beam structures, also called Guckel rings after their inventor, Henry Guckel. A schematic of this design is shown in Fig. 22.3 [8]. A tensile stress in the outer ring will cause it to contract. This will lead to a compressive stress in the central beam, even though the material was originally tensile before release. The amount of compression in the central beam can be determined analytically from the geometry of the device and the residual strain of the material. Again, by changing the length of the central beam, the l_{cr} will be determined, and then the residual strain can be deduced.

Stress Gradient Measurements

For structures fabricated from thin deposited films, the stress gradient can be just as important as the stress itself. Figure 22.4 shows a portion of a silicon microactuator. The device is designed to be completely planar; however, stress gradients in the film cause the structures to bend. This figure illustrates the importance of characterizing and controlling stress gradients, and it also demonstrates that the easiest structure for measuring stress gradients is a simple cantilever beam. By measuring the end

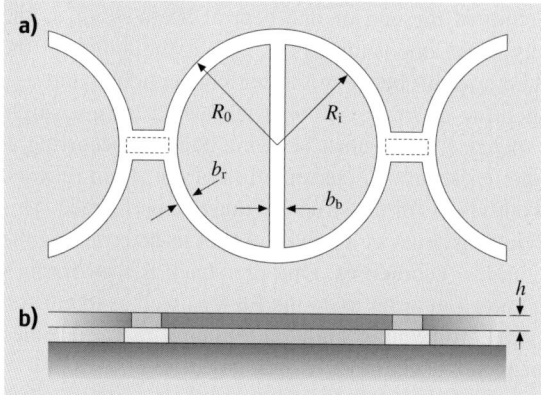

Fig. 22.3. Schematic (**a**) top view and (**b**) side view of Guckel ring structures [8]. The *dashed lines* in (**a**) indicate the anchors

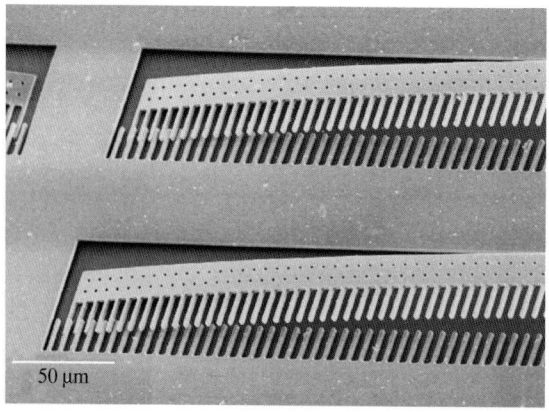

Fig. 22.4. Scanning electron micrograph (SEM) of a portion of a silicon microactuator. Residual stress gradients in the silicon cause the structure to bend

deflection δ, of a cantilever beam of length l, and thickness t, the stress gradient, $d\sigma/dt$ is determined by [9]

$$\frac{d\sigma}{dt} = \frac{2\delta}{l^2} \frac{E}{1-\nu}. \tag{22.4}$$

The magnitude of the end deflection can be measured by microscopy, optical interferometry, or any other technique.

Another useful structure for measuring stress gradients is a spiral. For this structure, the end of the spiral not anchored to the substrate will move out-of-plane. Also the diameter of the spiral will contract, and the free end of the spiral will rotate when released [10].

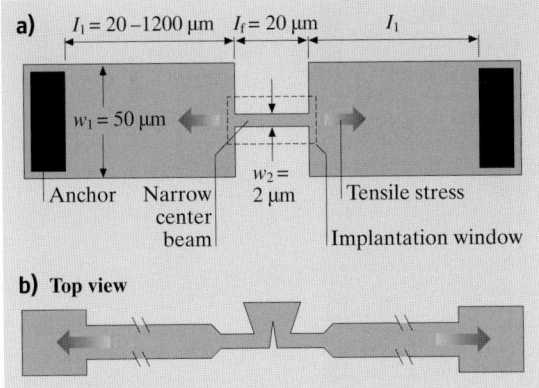

Fig. 22.5. Schematic designs of doubly clamped beams with stress concentrations for measuring strength. (**a**) was fabricated from polysilicon [11], and (**b**) was fabricated from Si_xN_y [12]

Strength and Fracture Toughness Measurements

As mentioned above, if a doubly clamped beam contains a tensile stress, it will remain taut when released because it cannot relieve any of its stress by contracting. This tensile stress can be thought of as a tensile load being applied at the ends of the beam. If this tensile load exceeds the tensile strength of the material, the beam will break. Since the tensile stress can be measured, as discussed in the Sect. 22.2.1, this technique can be used to gather information on the strength of materials. Figure 22.5 shows two different beam designs that have been used to measure strength. The device shown in Fig. 22.5a was fabricated from a tensile polysilicon film [11]. Different beams were designed with varying lengths of the wider regions (marked l_1 in the figure). In this manner, the load applied to the narrow center beam was varied, even though the entire film contained a uniform residual tensile stress. For l_1 greater than a critical value, the narrow center beam fractured, giving a measurement for the tensile strength of polysilicon.

The design shown in Fig. 22.5b was fabricated from a tensile Si_xN_y film [12]. As seen in the figure, a stress concentration was designed into the beam, to ensure the fracture strength would be exceeded. In this case, a notch was etched into one side of the beam. Since the stress concentration is not symmetric with regard to the beam axis, this results in a large bending moment at that position, and the test measures the bend strength of the material. Again, like the beams shown in Fig. 22.5a, the geometry of various beams fabricated from the same film were varied, to vary the maximum stress seen at the notch. By seeing which beams fracture at the stress concentration after release, the strength can be determined.

Through a similar technique but by using an atomically sharp pre-crack instead of a stress concentration, the fracture toughness of a material can be determined. Sharp pre-cracks can be introduced into micromachined structures before release by placing a Vickers indent on the substrate, near the device; the radial crack formed

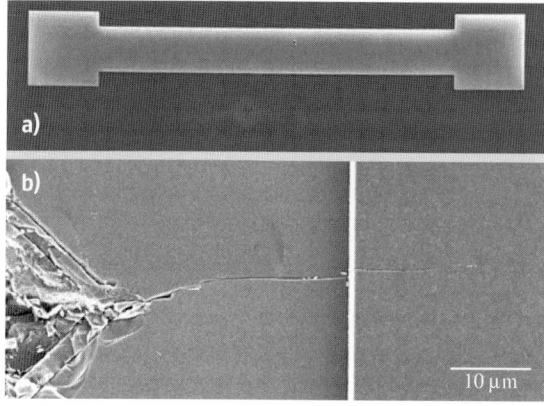

Fig. 22.6. (a) SEM of a 500 μm long polysilicon beam with a Vickers indent placed near its center; (b) higher magnification SEM of the area near the indent showing the pre-crack traveling from the substrate into the beam [14]

by the indent will propagate into the overlying structure [13]. Accordingly, the beam with a sharp pre-crack, shown in Fig. 22.6, was fabricated using polysilicon [14]. Due to the stochastic nature of indentation, the initial pre-crack length will vary from beam to beam. Because of this, even though the geometry of the beams remains identical, the stress intensity, K, at the pre-crack tip will vary. Upon release, only those pre-cracks whose K exceeds the fracture toughness of the material, K_{Ic}, will propagate, and in this way upper and lower bounds for K_{Ic} for the material can be determined.

For all of the beams discussed in this section, finite element analysis is required to determine the stress concentrations and stress intensities. Even though approximate analytical solutions may exist for these designs, the actual fabricated structure will not have idealized geometries. For example, corners will never be perfectly sharp, and cracks will never be perfectly straight. This reinforces the idea that FEA is a powerful tool in determining mechanical properties of very small structures.

22.2.2 Active Structures

As discussed above, it is very convenient to design structures that act upon release to provide information on the mechanical properties of the structural materials. This is not always possible, however. For example, those passive devices just discussed rely on residual stresses to create the changes (rotation or fracture) that occur upon release, but many materials do not contain high residual stresses as deposited, or the processing scheme of the device precludes the generation of residual stresses. Also, some mechanical properties, such as fatigue resistance, require motion in order to be studied. Active devices are therefore used. These are acted upon by a force (the source of this force can be integrated into the device itself or can be external to the

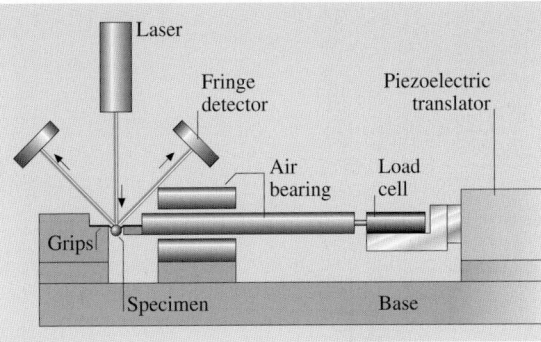

Fig. 22.7. Schematic of a measurement system for tensile loading of micromachined specimens [15]

device) in order to create a change, and by the response to the force, the mechanical properties are studied.

Young's Modulus Measurements

Young's modulus, E, is a material property critical to any structural device design. It describes the elastic response of a material and relates stress, σ, and strain, ε, by

$$\sigma = E\varepsilon . \tag{22.5}$$

In bulk samples, E is often measured by loading a specimen in tension and measuring displacement as a function of stress for a given length. While this is considerably more difficult for small structures, such as those fabricated from thin deposited films, it can be achieved with careful experimental techniques. Figure 22.7 shows the schematic of one such measurement system [15]. The fringe detectors in the figure detect the reflected laser signal from two gold lines deposited onto the polysilicon specimen, which act as gage markers. In this manner, the strain in the specimen during loading can be monitored. Besides gold lines, Vickers indents placed in a nickel specimen can also serve as gage markers [16], or a speckle interferometry technique [17] can be used to determine strain in the specimen. Once the stress-versus-strain behavior is measured, the slope of the curve is equal to E.

In addition to the tensile test, Young's modulus can be determined by other measures of stress-strain behavior. As seen in Fig. 22.8, a cantilever beam can be bent by pushing on the free end with a nanoindenter [18]. The nanoindenter can monitor the force applied and the displacement, and simple beam theory can convert the displacement into strain to obtain E. A similar technique, shown schematically in Fig. 22.9 [19], involves pulling downward on a cantilever beam by means of an electrostatic force. An electrode is fabricated into the substrate beneath the cantilever beam, and a voltage is applied between the beam and the bottom electrode. The force acting on the beam is equal to the electrostatic force corrected to include the effects of fringing fields acting on the sides of the beam, namely

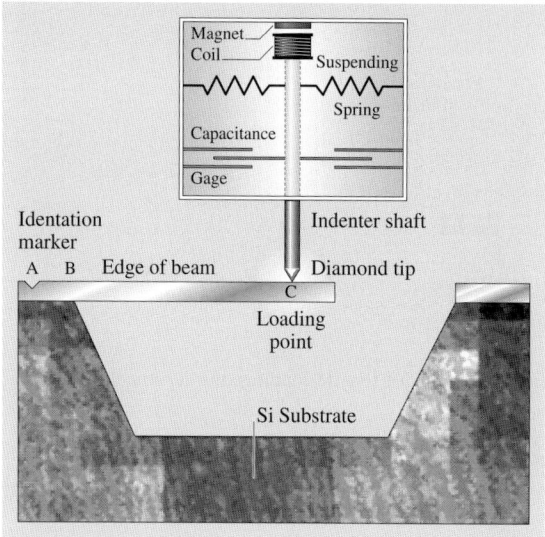

Fig. 22.8. Schematic of a nanoindenter loading mechanism pushing on the end of a cantilever beam [18]

$$F(x) = \frac{\varepsilon_0}{2}\left(\frac{V}{g+z(x)}\right)^2\left(1+\frac{0.65[g+z(x)]}{w}\right), \tag{22.6}$$

where $F(x)$ is the electrostatic force at x, ε_0 is the dielectric constant of air, g is the gap between the beam and the bottom electrode, $z(x)$ is the out-of-plane deflection of the beam, w is the beam width, and V is the applied voltage [19]. In this work, the deflection of the beam as a function of position is measured using optical interferometry. These measurements combine to give stress-strain behavior for the cantilever beam. An extension of this technique uses doubly clamped beams instead of cantilever beams. In this case, the deflection of the beam at a given electrostatic force depends on the residual stress in the material as well as Young's modulus. This method can also be used, therefore, to measure residual stresses in doubly clamped beams.

Another device possible to fabricate from a thin film and also possible to use for investigating stress-strain behavior is a suspended membrane, seen in Fig. 22.10 [20]. As shown in the schematic figure, the membrane is exposed to an elevated pressure on one side, causing it to bulge in the opposite direction. The deflection of the membrane is measured by optical or other techniques and related to the strain in the membrane. These membranes can be fabricated in any shape, typically square or circular. Both analytical solutions and finite element analyses have been done to relate the deflection to the strain. Like the doubly clamped beams, both Young's modulus and residual stress play a role in the deflected shape. Both of these mechanical properties can therefore be determined by the pressure-versus-deflection performance of the membrane.

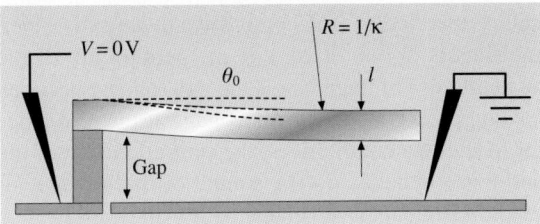

Fig. 22.9. Schematic of a cantilever beam bending test using an electrostatic voltage to pull the beam toward the substrate [19]

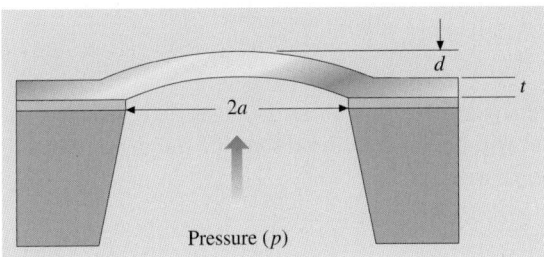

Fig. 22.10. Schematic cross section of a microfabricated membrane [20]

Another measurement besides stress-strain behavior that can reveal Young's modulus of a material is the determination of the natural resonance frequency. For a cantilever, the resonance frequency, f_r, for free undamped vibration is given by

$$f_r = \frac{\lambda_i^2 t}{4\pi l^2}\left(\frac{E}{3\rho}\right)^{1/2}, \quad (22.7)$$

where ρ, l, and t are density, length, and thickness of the cantilever; λ_i is the eigenvalue, where i is an integer that describes the resonance mode number; for the first mode $\lambda_1 = 1.875$ [21]. By measuring f_r and knowing the geometry and density, therefore, E can be determined. The cantilever can be vibrated by a number of techniques, including a laser, loudspeaker, or piezoelectric shaker. The frequency that produces the highest amplitude of vibration is the resonance frequency.

A micromachined device that uses an electrostatic comb drive and an AC signal to generate the vibration of the structure is known as a lateral resonator [22]. One example is shown in Fig. 22.11 [23]. When a voltage is applied across either set of interdigitated comb fingers shown in Fig. 22.11, an electrostatic attraction will be generated due to the increase in capacitance as the overlap between the comb fingers increases. The force, F, generated by the comb drive is given by

$$F = \frac{1}{2}\frac{\partial C}{\partial x}V^2 = n\varepsilon\frac{h}{g}V^2, \quad (22.8)$$

where C is capacitance, x is distance traveled by one comb-drive toward the other, n is the number of pairs of comb fingers in one drive, ε is the permittivity of the fluid between the fingers, h is the height of the fingers, g is the gap spacing between the fingers, and V is the applied voltage [22]. When an AC voltage at the resonance frequency is applied across either of the two comb drives, the central portion of the device will vibrate. In fact, since force depends on the square of the voltage for electrostatic actuation, for a time, t, dependent drive voltage, $v_D(t)$, of

$$v_D(t) = V_P + v_d \sin(\omega t), \tag{22.9}$$

where V_P is the DC bias and v_d is the AC drive amplitude, the time dependent portion of the force will scale with

$$2\omega V_P v_d \cos(\omega t) + \omega v_d^2 \sin(2\omega t) \tag{22.10}$$

[22]. Therefore, if an AC drive signal is used with no DC bias, at resonance, the frequency of the AC drive signal will be one half the resonance frequency. For this device, the resonance frequency, f_r, will be

$$f_r = \frac{1}{2\pi} \left(\frac{k_{sys}}{M} \right)^{1/2}, \tag{22.11}$$

where k_{sys} is the spring constant of the support beams and M is the mass of the portion of the device that vibrates. The spring constant is given by

$$k_{sys} = 24EI/L^3, \tag{22.12}$$

$$I = \frac{hw^3}{12}, \tag{22.13}$$

where I is the moment of inertia of the beams, and L, h, and w are the length, thickness, and width of the beams. By combining these equations and measuring f_r, therefore, E can be determined.

One distinct advantage of the lateral resonator technique and the electrostatically pulled cantilever technique for measuring Young's modulus is that they require no external loading sources. Portions of the devices are electrically contacted, and a voltage is applied. For the pure tension tests, such as shown in Fig. 22.7, the specimen must be attached to a loading system, which for the very small specimens discussed here, can be extremely difficult, and any misalignment or eccentricity in the test could lead to unreliable results. But the advantage of the externally loaded technique is that there are no limitations on the type of materials that can be tested. Conductivity is not a requirement, nor is any compatibility with electrical actuation.

Strength and Fracture Toughness Measurements

As one might expect, any of the techniques discussed in the previous section that strain specimens in order to measure Young's modulus can also be used to measure fracture strength. Simply, the load is increased until the specimen breaks. As long as

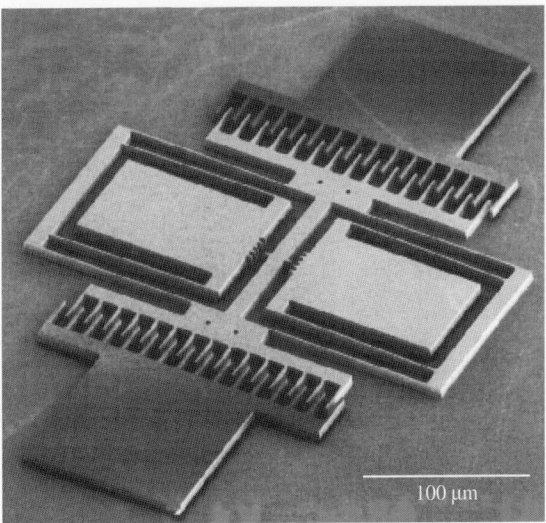

Fig. 22.11. SEM of a polysilicon lateral resonator [23]

either the load or the strain is measured at fracture, and the geometry of the specimen is known, the maximum stress required for fracture, σ_{crit}, can be determined, either through analytical analysis or FEA. Depending on the geometry of the test, σ_{crit} will represent the tensile or bend strength of the material.

If the available force is limited, or if it is desired to have a localized fracture site, stress concentrations can be added to the specimens. These are typically notches micromachined into the edges of specimens. Focused ion beams have also been used to carve stress concentrations into fracture specimens.

All of the external loading schemes, such as those shown in Figs. 22.7 and 22.8, have been used to measure fracture strength. Also, the electrostatically loaded doubly clamped beams can be pulled until they fracture. In this case, there is one complication. The electrostatic force is inversely proportional to the distance between the electrodes, and at a certain voltage, called the "pull-in voltage," the attraction between the beam and the substrate will become so great that the beam will immediately be pulled into contact with the bottom electrode. As long as the fracture takes place before the pull-in voltage is reached, the experiment will give valid results.

Other loading techniques have been used to generate fracture of microspecimens. Figure 22.12 [24] shows one device designed to be pushed by the end of a micromanipulated needle. The long beams that extend from the sides of the central shuttle come into contact with anchored posts, and, at a critical degree of bending, the beams will break off. Since the applied force cannot be measured in this technique, the experiment is optically monitored continuously during the test, and the image of the beams just before fracture is analyzed to determine σ_{crit}.

Another loading scheme that has been demonstrated for micromachined specimens utilizes scratch drive actuators to load the specimens [25]. These types of

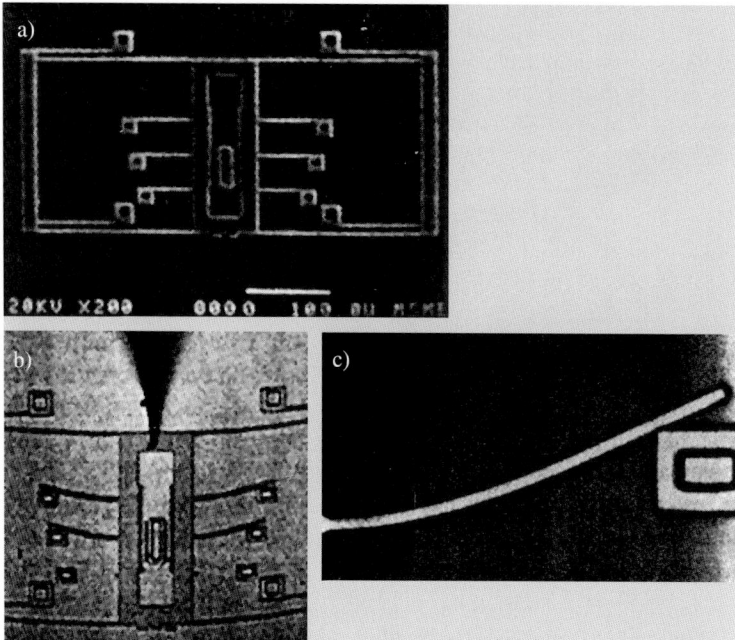

Fig. 22.12. (a) SEM of a device for measuring bend strength of polysilicon beams; (b) image of a test in process; (c) higher magnification view of one beam shortly before breaking [24]

actuators work like inchworms, traveling across a substrate in discrete advances as an electrostatic force is repeatedly applied between the actuator and the substrate. The stepping motion can be made in the nm scale, depending on the frequency of the applied voltage, and so it can be an acceptable approximation to continuous loading. One advantage of this scheme is that very large forces can be generated by relatively small devices. The exact forces generated cannot be measured, so like the technique that used micromanipulated pushing, the test is continuously observed to determine the strain at fracture. Another advantage of this technique is that, like the lateral resonator and the electrically pulled cantilever, the loading takes place on-chip, and, therefore, the difficulties associated with attaching and aligning an external loading source are eliminated.

Another on-chip actuator used to load microspecimens is shown in Fig. 22.13, along with three different microspecimens [13]. Devices have been fabricated with each of the three microspecimens integrated with the same electrostatic comb-drive actuator. In all three cases, when a DC voltage is applied to the actuator, it moves downward, as oriented in Fig. 22.13. This pulls down on the left end of each of the three microspecimens, which are anchored on the right. The actuator contains 1486 pairs of comb fingers. The maximum voltage that can be applied is limited by the breakdown voltage of the medium in which the test takes place. In air, that limits the voltage to less than 200 V. As a result, given a finger height of 4 μm and a gap of 2 μm, and using (22.8), the maximum force generated by this actuator is limited to about 1 mN. Standard optical photolithography has a minimum feature size of about 2 μm. As a result, the electrostatic actuator cannot generate sufficient force to perform a standard tensile test on MEMS structural materials such as polysilicon. The microspecimens shown in Fig. 22.13 are therefore designed such that the stress is amplified.

The specimen shown in Fig. 22.13b is designed to measure bend strength. It contains a micromachined notch with a root radius of 1 μm. When the actuator pulls downward on the left end of this specimen, the notch serve as a stress concentration, and when the stress at the notch root exceeds σ_{crit}, the specimen fractures. The specimen in Fig. 22.13c is designed to test tensile strength. When the left end of this specimen is pulled downward, a tensile stress is generated in the upper thin horizontal beam near the right end of the specimen. As the actuator continues to move downward, the tensile stress in this beam will exceed the tensile strength, causing fracture. Finally, the specimen in Fig. 22.13d is similar to that in Fig. 22.13b, except that the notch is replaced by a sharp pre-crack that was produced by the Vickers indent placed on the substrate near the specimen. When this specimen is loaded, a stress intensity, K, is generated at the crack tip. When the stress intensity exceeds a critical value, K_{Ic}, the crack propagates. K_{Ic} is also referred to as the fracture toughness.

The force generated by the electrostatic actuator can be calculated using (22.8). But (22.8) assumes a perfectly planar, two-dimensional device. In fact, when actuated, the electric fields extend out of the plane of the device, and so (22.8) is just an approximation. Instead, like many of the techniques discussed in this section, the test is continuously monitored, and the actuator displacement at the time of fracture is recorded. Then FEA is used to determine the magnitude of the stress or stress intensity seen by the specimen at the point of fracture.

Fatigue Measurements

A benefit of the electrostatic actuator shown in Fig. 22.13 is that, besides monotonic loading, it can generate cyclic loading. This allows the fatigue resistance of materials to be studied. Simply by using an AC signal instead of a DC voltage, the device can be driven at its resonant frequency. The amplitude of the resonance depends on the magnitude of the AC signal. This amplitude can be increased until the specimen breaks; this will investigate the low-cycle fatigue resistance. Otherwise the amplitude of resonance can be left constant at a level below that is required for fast fracture, and the device will resonate indefinitely until the specimen breaks; this will investigate high-cycle fatigue. It should be noted that the resonance frequency of such a device is about 10 kHz. Therefore, it is possible to stress a specimen over 10^9 cycles in less than a day. In addition to simple cyclic loading, if a DC bias is added to the AC signal, a mean stress can be superimposed on the cyclic load. In this way, nonsymmetric cyclic loading (either with a large tensile stress alternating with a small compressive stress, or vice versa) can be studied.

Another device that can be used to investigate fatigue resistance in MEMS materials is shown in Fig. 22.14 [26]. In this case, a large mass is attached to the end

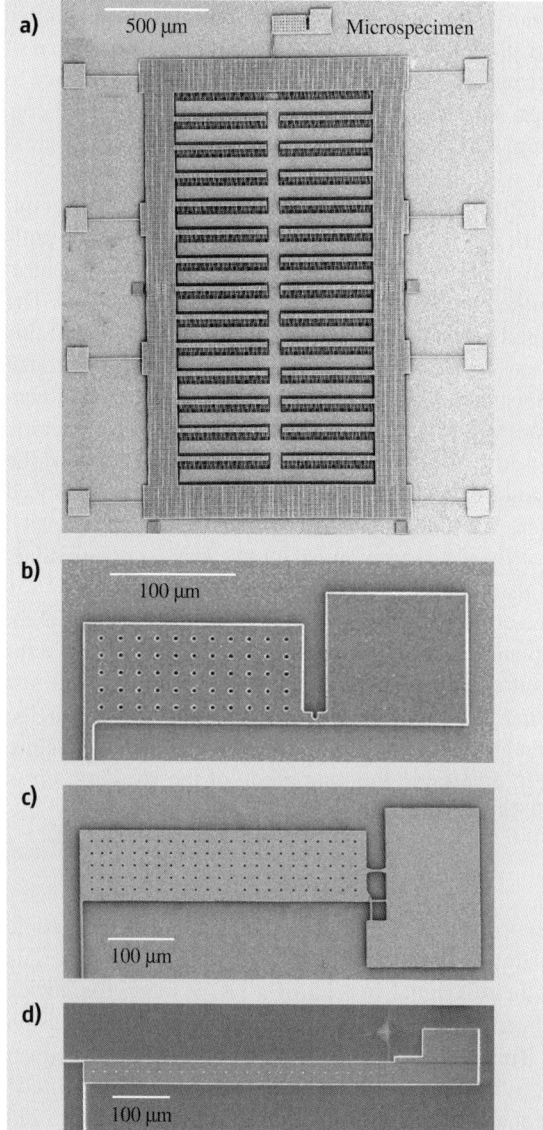

Fig. 22.13. (a) SEM of a micromachined device for conducting strength tests; the device consists of a large comb drive electrostatic actuator integrated with a microspecimen; (b)–(d) SEMs of various microspecimens for testing bend strength, tensile strength, and fracture toughness, respectively [13]

of a notched cantilever beam. The mass contains two comb drives on opposite ends. When an AC signal is applied to one comb drive, the device will resonate, cyclically loading the notch. The comb drive on the opposite side is used as a capacitive

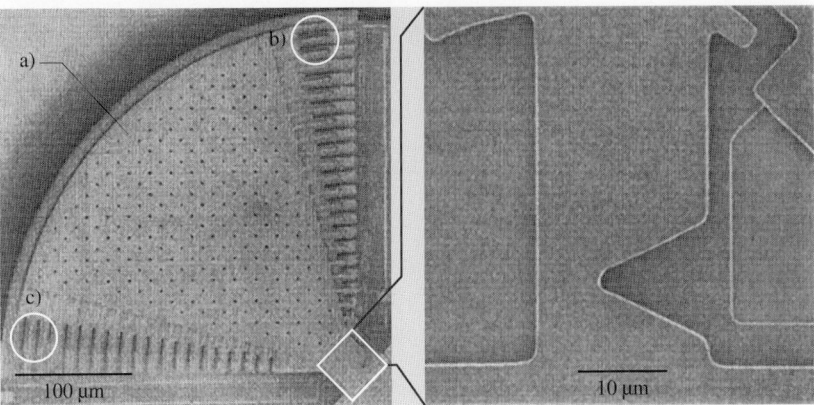

Fig. 22.14. SEM of a device for investigating fatigue; the image on the right is a higher magnification view of the notch near the base of the moving part of the structure [26] (a) mass, b) comb drive actuator, c) capacitive displacement sensor)

displacement sensor. This device contains many fewer comb fingers than the device shown in Fig. 22.13. As a result, it can apply cyclic loads by exploiting the resonance frequency of the device but cannot supply sufficient force to achieve monotonic loading.

Fatigue loading has also been studied using the same external loading techniques shown in Fig. 22.7. In this case, the frequency of the cyclic load is considerably lower, since the resonance frequency of the device is not being utilized. This leads to longer high-cycle testing times. Since the force is essentially unlimited, however, this technique allows a variety of frequencies to be studied to determine their effect on the fatigue behavior.

Friction Measurements

Friction is another property that has been studied in micromachined structures. To study friction, of course, two surfaces must be brought into contact with each other. This is usually avoided at all costs for these devices because of the risk of stiction. (Stiction is the term used when two surfaces that come into contact adhere so strongly that they cannot be separated.) Even so, a few devices have been designed to investigate friction. One of these is shown schematically in Fig. 22.15 [27]. It consists of a movable structure with a comb drive on one end and a cantilever beam on the other. Beneath the cantilever, on the substrate, is planar electrode. The device is moved to one side using the comb drive. Then a voltage is applied between the cantilever beam and the substrate electrode. The voltage on the comb drive is then released. The device would normally return to its original position, to relax the deflection in the truss suspensions, but the friction between the cantilever and the substrate electrode holds it in place. The voltage to the substrate electrode is slowly decreased until the device starts to slide. Knowing the electrostatic force generated by the substrate electrode and the stiffness of the truss suspensions, the static friction can be determined. For

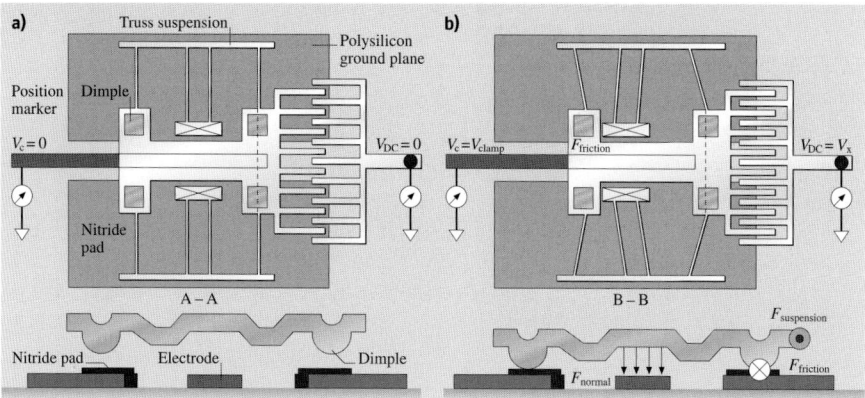

Fig. 22.15. Schematic drawings of a device used to study friction. (**a**) shows top and side views of the device in its original position, and (**b**) shows views of the device after it has been displaced using the comb drive and clamped using the substrate electrode [27]

this device, bumps were fabricated on the bottoms of the cantilever beams. This limited the surface area that came into contact with the substrate and so lowered the risk of stiction.

Another device designed to study friction is shown in Fig. 22.16 [28]. This technique uses a hinged cantilever. The portion near the free end acts as the friction test structure, and the portion near the anchored end acts as the driver. The friction test structure is attracted to the substrate by means of electrostatic actuation, and when a second electrostatic actuator pulls down the driver, the friction test structure slips forward by a length proportional to the forces involved, including the frictional force. This distance, however, has a maximum of 30 nm, so all measurements must be exceedingly accurate in order to investigate a range of forces. This test structure can be used to determine the friction coefficients for surfaces with and without lubricating coatings.

22.3 Measurements of Mechanical Properties

All of the techniques discussed in Sects. 22.1 and 22.2 have been used to measure the mechanical properties of MEMS and NEMS materials. As a general rule, the results from the various techniques have agreed well with each other, and the argument becomes which of the measurement techniques is easiest and most reliable to perform. It is crucial to keep in mind, however, that certain properties, such as strength, are process dependent, and so the results taken at one laboratory will not necessarily match those taken from another. This will be discussed in more detail in Sect. 22.3.1.

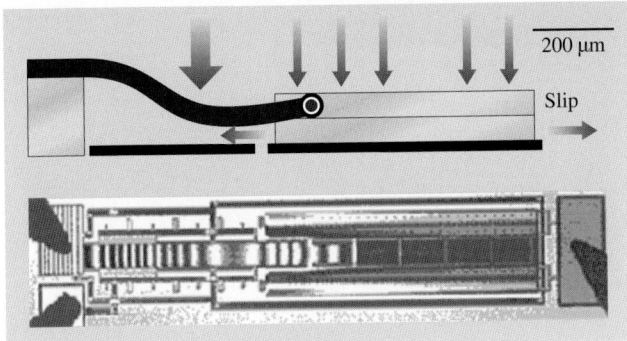

Fig. 22.16. Schematic cross section and top-view optical micrograph of a hinged-cantilever test structure for measuring friction in micromachined devices [28]

22.3.1 Mechanical Properties of Polysilicon

In current MEMS technology, the most widely used structural material is polysilicon deposited by low-pressure chemical vapor deposition (LPCVD). One reason for the prevalence of polysilicon is the large body of processing knowledge for this material that has been developed by the integrated circuit community. Another reason, of course, is that polysilicon possesses a number of qualities beneficial for MEMS devices, particularly its high strength and Young's modulus. As a natural result, most of the mechanical properties investigations on MEMS materials have focused on polysilicon.

Residual Stresses in Polysilicon

The residual stresses of LPCVD polysilicon have been thoroughly characterized using both the wafer curvature technique, discussed in Sect. 22.1.1, and the microstrain gauges, discussed in Sect. 22.2.1. The results from both techniques give consistent values. Figure 22.17 summarizes the residual stress measurements as a function of deposition temperature taken from five different investigations at five different laboratories [29]. All five sets of data show the same trend. The stresses change from compressive at the lowest deposition temperatures to tensile at intermediate temperatures and back to compressive at the highest temperatures. The exact transition temperatures vary somewhat among the different investigations, probably due to differences in the deposition conditions: the silane or dichlorosilane pressure, the gas flow rate, the geometry of the deposition system, and the temperature uniformity. But in each data set the transitions are easily discernible. The origin of these residual stress changes lies with the microstructure of the LPCVD polysilicon films.

As with all deposited films, the microstructure of LPCVD polysilicon films is dependent on the deposition conditions. In general, the films are amorphous at the lowest growth temperatures (lower than ~ 570 °C), display fine (~ 0.1 μm diameter) grains at intermediate temperatures (~ 570 °C to ~ 610 °C), and contain columnar (110)-textured grains with a thin fine-grained nucleation layer at the substrate

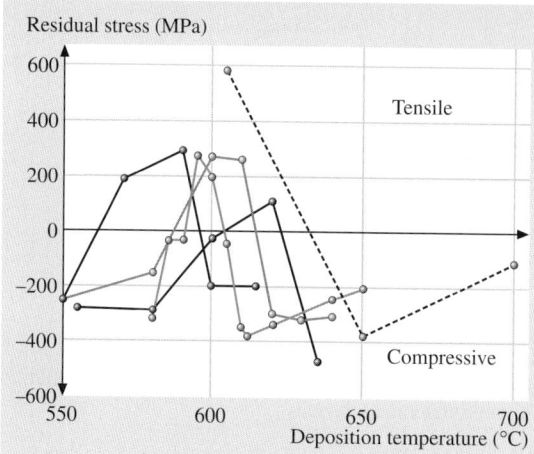

Fig. 22.17. Results for residual stress of LPCVD polysilicon films taken from five different investigations [29]. The data from each investigation are connected by a line

interface at higher temperatures (~ 610 °C to ~ 700 °C) [29]. The fine-grained microstructure results from the homogeneous nucleation and growth of silicon grains within an as-deposited amorphous silicon film. In this regime, the deposition rate is just slightly faster than the crystallization rate. The as-deposited films will be crystalline near the substrate interface and amorphous at the free surface. (The amorphous fraction can be quickly crystallized by annealing above 610 °C.) The columnar microstructure seen at the higher growth temperatures results from the formation of crystalline silicon films as deposited, with growth being fastest in the ⟨110⟩ directions.

The origin of the tensile stress in the fine-grained polysilicon arises from the volume decrease that accompanies the crystallization of the as-deposited amorphous material. The origins of the compressive stresses in the amorphous and columnar films are less well understood. One proposed explanation for compressive stress generation during thin film growth postulates that an increase in the surface chemical potential is caused by the deposition of atoms from the vapor; the increase in surface chemical potential induces atoms to flow into newly formed grain boundaries, creating a compressive stress in the film [30].

Stress gradients are also typical of LPCVD polysilicon films. The partially amorphous films contain large stress gradients since they are essentially bilayers of compressive amorphous silicon on top of tensile fine-grained polysilicon. The fully crystalline films also exhibit stress gradients. The columnar compressive films are most highly stressed at the film-substrate interface, with the compressive stresses decreasing as the film thickness increases; the fine-grained films are less tensile at the film-substrate interface, with the tensile stresses increasing as the film thickness increases [29]. Both stress gradients are associated with microstructural variations. For the columnar films, the initial nucleation layer corresponds to a very high compressive stress, which decreases as the columnar morphology develops. For the fine-grained films, the region near the film-substrate interface has a slightly smaller average grain size, due to heterogeneous nucleation at the interface. This region displays a slightly lower tensile stress than the rest of the film, since the increased grain boundary area reduces the local density.

Young's Modulus of Polysilicon

The Young's modulus of polysilicon films has been measured using all the techniques discussed in Sect. 22.2.2. A good review of the experimental results taken from bulge testing, tensile testing, beam bending, and lateral resonators are contained in [31]. All of the reported results are in reasonable agreement, varying from 130 to 175 GPa, though many values are reported with a relatively high experimental scatter. The main origin of the error in these results is the uncertainties involving the geometries of the small specimens used to make the measurements. For example, from (22.13), the Young's modulus determined by the lateral resonators depends on the cube of the tether beam width, typically about 2 μm. In general, the beam width and other dimensions can be measured with scanning electron microscopy to within about 0.1 μm; however, the width of the beams is not perfectly constant along the entire length or even through the thickness. These uncertainties in geometry lead to uncertainties in modulus.

In addition, the various experimental measurements lie close to the Voigt and Reuss bounds for Young's modulus calculated using the elastic stiffnesses and compliances for single crystal silicon [31]. This strongly implies that Young's modulus of micro- and nanomachined polysilicon structures will be the same as for bulk samples made from polysilicon. This is not unexpected, since Young's modulus is a material property. It is related to the interatomic interactions and should have no dependence on the geometry of the sample. It should be noted that polysilicon can display a preferred crystallographic orientation depending on the deposition conditions, and that this could affect the Young's modulus of the material, since the Young's modulus of silicon is not isotropic. But the anisotropy is fairly small for the cubic silicon.

A more recent investigation that utilized electrostatically actuated cantilevers and interferometric deflection detection yielded a Young's modulus of 164 GPa [19]. They found the grains in their polysilicon films to be randomly oriented and calculated the Voigt and Reuss bounds to be 163.4–164.4 GPa. This appears to be a very reliable value for randomly oriented polysilicon.

Fracture Toughness and Strength of Polysilicon

Using the device shown in Fig. 22.13a and the specimen shown in Fig. 22.13d, the fracture toughness, K_{Ic}, of polysilicon has been shown to be 1.0 ± 0.1 MPa m$^{1/2}$ [32]. Several different polysilicon microstructures were tested, including fine-grained, columnar, and multilayered. Amorphous silicon was also investigated. All of the microstructures displayed the same K_{Ic}. This indicates that, like Young's modulus, fracture toughness is a material property, independent of the material microstructure or the geometry of the sample.

A tensile test, such as shown in Fig. 22.7 but using a sample with indentation-induced sharp pre-cracks, reveals a K_{Ic} of 0.86 MPa m$^{1/2}$ [33]. The passive, residual stress loaded beams with sharp pre-cracks, shown in Fig. 22.6, gave a K_{Ic} of 0.81 MPa m$^{1/2}$ [14].

Given that K_{Ic} is a material property for polysilicon, measured fracture strength, σ_{crit}, is related to K_{Ic} by

$$K_{Ic} = c\sigma_{crit}(\pi a)^{1/2}, \tag{22.14}$$

where a is the crack-initiating flaw size, and c is a constant of order unity. The value for c will depend on the exact size, shape, and orientation of the flaw; for a semicircular flaw, c is equal to 0.71 [34]. Therefore, any differences in the reported fracture strength of polysilicon will be the result of changes in a.

A good review of the experimental results available in the literature for polysilicon strength is contained in [35]. The tensile strength data vary from about 0.5 to 5 GPa. Like many brittle materials, the measured strength of polysilicon is found to obey Weibull statistics. This implies that the polysilicon samples contain a random distribution of flaws of various sizes, and that the failure of any particular specimen will occur at the largest flaw that experiences the highest stress. One consequence of this behavior is that, since larger specimens have a greater probability of containing larger flaws, they will exhibit decreased strengths. More specifically, it was found that the most important geometrical parameter is the surface area of the sidewalls of a polysilicon specimen [35]. The sidewalls, as opposed to the top and bottom surfaces, are those surfaces created by etching the polysilicon film. This is not surprising since LPCVD polysilicon films contain essentially no flaws within the bulk, and the top and bottom surfaces are typically very smooth.

As a result, the etching techniques used to create the structures will have a strong impact on the fracture strength of the material. For single-crystal silicon specimens it was found that the choice of etchant could change the observed tensile strength by a factor of two [36]. In addition, the bend strength of amorphous silicon was measured to be twice that of polysilicon for specimens processed identically [32]. It was found that the reactive ion etching used to fabricate the specimens produced much rougher sidewalls on the polysilicon than on the amorphous silicon.

Fatigue of Polysilicon

Fatigue failure involves fracture after a number of load cycles, when each individual load is not sufficient by itself to generate catastrophic cracking in the material. For ductile materials, such as metals, fatigue occurs due to accumulated damage at the site of maximum stress and involves local plasticity. As a brittle material, polysilicon would not be expected to be susceptible to cyclic fatigue. But fatigue has been observed for polysilicon tensile samples [33], polysilicon bend specimens with notches [26, 37], and polysilicon bend specimens with sharp cracks [38]. The exact origins of the fatigue behavior are still subject to debate. But some aspects of the experimental data are that the fatigue lifetime does not depend on the loading frequency [33], the fatigue behavior is affected by the ambient [14, 38], and the fatigue depends on the ratio of compressive to tensile stresses seen in the load cycle [14].

Friction of Polysilicon

The friction of polysilicon structures has been measured using the techniques described in Sect. 22.2.2. The measured coefficient of friction was found to vary from 4.9 [27] to 7.8 [28].

22.3.2 Mechanical Properties of Other Materials

As discussed above, of all the materials used for MEMS and NEMS, polysilicon has generated the most interest as well as the most research in mechanical properties characterization. But measurements have been taken on other materials, and these are summarized in this section.

As discussed in Sect. 22.2.2, one advantage of the externally loaded tension test, as shown in Fig. 22.7, is that essentially any material can be tested using this technique. As such, tensile strengths have been measured to be 0.6 to 1.9 GPa for SiO_2 [39] and 0.7 to 1.1 GPa for titanium [40]. The yield strength for electrodeposited nickel was found to vary from 370 to 900 MPa, depending on annealing temperature [16]. In addition, the yield strength was strongly affected by the current density during the electrodeposition process. Both the annealing and current density effects were correlated to changes in the microstructure of the material. Young's moduli were determined to be 100 GPa for titanium [40] and 215 GPa for electrodeposited nickel [16].

The technique of bending cantilever beams, shown in Fig. 22.8, can also be performed on a variety of materials. The yield strength and Young's modulus of gold were found to be 260 MPa and 57 GPa, respectively, using this method [18]. Another technique that can be used with a number of materials is the membrane deflection method, shown in Fig. 22.10. A polyimide membrane gave a residual stress of 32 MPa, a Young's modulus of 3.0 GPa, and an ultimate strain of about four percent [20]. Membranes were also fabricated from polycrystalline SiC films with two different grain structures [41]. The film with (110)-texture columnar grains had a residual stress of 434 MPa and a Young's modulus of 349 GPa. The film with equiaxed (110)- and (111)-textured grains had a residual stress of 446 MPa and a Young's modulus of 456 GPa.

Other devices used for measuring mechanical properties require more complicated micromachining, namely patterning, etching, and release, in order to operate. These devices are more difficult to fabricate with materials not commonly used as MEMS structural materials. The following examples, however, demonstrate work in this area. The structure shown in Fig. 22.5b was fabricated from Si_xN_y and revealed a nominal fracture toughness of $1.8\,\mathrm{MPa\,m^{1/2}}$ [12]. Lateral resonators of the type shown in Fig. 22.11 were processed using polycrystalline SiC, and Young's modulus was determined to be 426 GPa [42]. The device shown in Fig. 22.12 was fabricated from polycrystalline germanium and used to measure a bend strength of 1.5 GPa for unannealed Ge and 2.2 GPa for annealed Ge [43].

References

1. G. G. Stoney. The tension of metallic films deposited by electrolysis. *Proc. R. Soc. Lond. A*, 82:172–175, 1909.
2. W. Brantley. Calculated elastic constants for stress problems associated with semiconductor devices. *J. Appl. Phys.*, 44:534–535, 1973.
3. A. Ni, D. Sherman, R. Ballarini, H. Kahn, B. Mi, S. M. Phillips, and A. H. Heuer. Optimal design of multilayered polysilicon films for prescribed curvature. *J. Mater. Sci.*, page in press.
4. J. A. Floro, E. Chason, S. R. Lee, R. D. Twesten, R. Q. Hwang, and L. B. Freund. Real-time stress evolution during $Si_{1-x}Ge_x$ heteroepitaxy: Dislocations, islanding, and segregation. *J. Electron. Mater.*, 26:969–979, 1997.
5. W. Nix. Mechanical properties of thin films. *Metall. Trans. A*, 20:2217–2245, 1989.
6. X. Li and B. Bhusan. Micro/nanomechanical characterization of ceramic films for microdevices. *Thin Solid Films*, 340:210–217, 1999.
7. M. P. de Boer, H. Huang, J. C. Nelson, Z. P. Jiang, and W. W. Gerberich. Fracture toughness of silicon and thin film micro-structures by wedge indentation. *Mater. Res. Soc. Symp. Proc.*, 308:647–652, 1993.
8. H. Guckel, D. Burns, C. Rutigliano, E. Lovell, and B. Choi. Diagnostic microstructures for the measurement of intrinsic strain in thin films. *J. Micromech. Microeng.*, 2:86–95, 1992.
9. F. Ericson, S. Greek, J. Soderkvist, and J.-A. Schweitz. High sensitivity surface micromachined structures for internal stress and stress gradient evaluation. *J. Micromech. Microeng.*, 7:30–36, 1997.
10. L. S. Fan, R. S. Muller, W. Yun, R. T. Howe, and J. Huang. Spiral microstructures for the measurement of average strain gradients in thin films, 1990.
11. M. Biebl and H. von Philipsborn. Fracture strength of doped and undoped polysilicon, 1995.
12. L. S. Fan, R. T. Howe, and R. S. Muller. Fracture toughness characterization of brittle films. *Sens. Actuators A*, 21-23:872–874, 1990.
13. H. Kahn, N. Tayebi, R. Ballarini, R. L. Mullen, and A. H. Heuer. Wafer-level strength and fracture toughness testing of surface-micromachined mems devices. *Mater. Res. Soc. Symp. Proc.*, 605:25–30, 2000.
14. H. Kahn, R. Ballarini, J. J. Bellante, and A. H. Heuer. Fatigue failure in polysilicon is not due to simple stress corrosion cracking. *Science*, 298:1215–1218, 2002.
15. W. N. Sharpe Jr., B. Yuan, and R. L. Edwards. A new technique for measuring the mechanical properties of thin films. *J. Microelectromech. Syst.*, 6:193–199, 1997.
16. H. S. Cho, W. G. Babcock, H. Last, and K. J. Hemker. Annealing effects on the microstructure and mechanical properties of liga nickel for mems. *Mater. Res. Soc. Symp. Proc.*, 657, 2001.
17. W. Suwito, M. L. Dunn, S. J. Cunningham, and D. T. Read. Elastic moduli, strength, and fracture initiation at sharp notches in etched single crystal silicon microstructures. *J. Appl. Phys.*, 85:3519–3534, 1999.
18. T. P. Weihs, S. Hong, J. C. Bravman, and W. D. Nix. Mechanical deflection of cantilever microbeams: A new technique for testing the mechanical properties of thin films. *J. Mater. Res.*, 3:931–942, 1988.
19. B. D. Jensen, M. P. de Boer, N. D. Masters, F. Bitsie, and D. A. La Van. Interferometry of actuated microcantilevers to determine material properties and test structure nonidealities in mems. *J. Microelectromech. Syst.*, 10:336–346, 2001.

20. M. G. Allen, M. Mehregany, R. T. Howe, and S. D. Senturia. Microfabricated structures for the in situ measurement of residual stress, young's modulus, and ultimate strain of thin films. *Appl. Phys. Lett.*, 51:241–243, 1987.
21. L. Kiesewetter, J.-M. Zhang, D. Houdeau, and A. Steckenborn. Determination of young's moduli of micromechanical thin films using the resonance method. *Sens. Actuators A*, 35:153–159, 1992.
22. W. C. Tang, T.-C. H. Nguyen, and R. T. Howe. Laterally driven polysilicon resonant microstructures. *Sens. Actuators A*, 20:25–32, 1989.
23. H. Kahn, S. Stemmer, K. Nandakumar, A. H. Heuer, R. L. Mullen, R. Ballarini, and M. A. Huff. Mechanical properties of thick, surface micromachined polysilicon films, 1996.
24. P. T. Jones, G. C. Johnson, and R. T. Howe. Fracture strength of polycrystalline silicon. *Mater. Res. Soc. Symp. Proc.*, 518:197–202, 1998.
25. P. Minotti, R. Le Moal, E. Joseph, and G. Bourbon. Toward standard method for microelectromechanical systems material measurement through on-chip electrostatic probing of micrometer size polysilicon tensile specimens. *Jpn. J. Appl. Phys.*, 40:L120–L122, 2001.
26. C. L. Muhlstein, E. A. Stach, and R. O. Ritchie. A reaction-layer mechanism for the delayed failure of micron-scale polycrystalline silicon structural films subjected to high-cycle fatigue loading. *Acta Mater.*, 50:3579–3595, 2002.
27. M. G. Lim, J. C. Chang, D. P. Schultz, R. T. Howe, and R. M. White. Polysilicon microstructures to characterize static friction, 1990.
28. B. T. Crozier, M. P. de Boer, J. M. Redmond, D. F. Bahr, and T. A. Michalske. Friction measurement in mems using a new test structure. *Mater. Res. Soc. Symp. Proc.*, 605:129–134, 2000.
29. J. Yang, H. Kahn, A. Q. He, S. M. Phillips, and A. H. Heuer. A new technique for producing large-area as-deposited zero-stress lpcvd polysilicon films: The multipoly process. *J. Microelectromech. Syst.*, 9:485–494, 2000.
30. E. Chason, B. W. Sheldon, L. B. Freund, J. A. Floro, and S. J. Hearne. Origin of compressive residual stress in polycrystalline thin films. *Phys. Rev. Lett.*, 88, 2002.
31. S. Jayaraman, R. L. Edwards, and K. J. Hemker. Relating mechanical testing and microstructural features of polysilicon thin films. *J. Mater. Res.*, 14:688–697, 1999.
32. R. Ballarini, H. Kahn, N. Tayebi, and A. H. Heuer. Effects of microstructure on the strength and fracture toughness of polysilicon: A wafer level testing approach. *ASTM STP*, 1413:37–51, 2001.
33. J. Bagdahn, J. Schischka, M. Petzold, and W. N. Sharpe Jr. Fracture toughness and fatigue investigations of polycrystalline silicon. *Proc. SPIE*, 4558:159–168, 2001.
34. I. S. Raju and J. C. Newman Jr. Stress intensity factors for a wide range of semi-elliptical surface cracks in finite-thickness plates. *Eng. Fract. Mech.*, 11:817–829, 1979.
35. J. Bagdahn, W. N. Sharpe Jr., and O. Jadaan. Fracture strength of polysilicon at stress concentrations. *J. Microelectromech. Syst.*, pages 302–312, 2003.
36. T. Yi, L. Li, and C.-J. Kim. Microscale material testing of single crystalline silicon: Process effects on surface morphology an dtensile strength. *Sens. Actuators A*, 83:172–178, 2000.
37. H. Kahn, R. Ballarini, R. L. Mullen, and A. H. Heuer. Electrostatically actuated failure of microfabricated polysilicon fracture mechanics specimens. *Proc. R. Soc. Lond. A*, 455:3807–3923, 1999.
38. W. W. Van Arsdell and S. B. Brown. Subcritical crack growth in silicon mems. *J. Microelectromech. Syst.*, 8:319–327, 1999.

39. T. Tsuchiya, A. Inoue, and J. Sakata. Tensile testing of insulating thin films; humidity effect on tensile strength of sio_2 films. *Sens. Actuators A*, 82:286–290, 2000.
40. H. Ogawa, K. Suzuki, S. Kaneko, Y. Nakano, Y. Ishikawa, and T. Kitahara. Measurements of mechanical properties of microfabricated thin films, 1997.
41. S. Roy, C. A. Zorman, and M. Mehregany. The mechanical properties of polycrystalline silicon carbide films determined using bulk micromachined diaphragms. *Mater. Res. Soc. Symp. Proc.*, 657, 2001.
42. A. J. Fleischman, X. Wei, C. A. Zorman, and M. Mehregany. Surface micromachining of polycrystalline sic deposited on sio_2 by apcvd. *Mater. Sci. Forum*, 264-268:885–888, 1998.
43. A. E. Franke, E. Bilic, D. T. Chang, P. T. Jones, T.-J. King, R. T. Howe, and G. C. Johnson. Post-cmos integration of germanium microstructures, 1999.

The Editor

Dr. Bharat Bhushan received an M.S. in mechanical engineering from the Massachusetts Institute of Technology in 1971, an M.S. in mechanics and a Ph.D. in mechanical engineering from the University of Colorado at Boulder in 1973 and 1976, respectively, an MBA from Rensselaer Polytechnic Institute at Troy, NY in 1980, a Doctor Technicae from the University of Trondheim at Trondheim, Norway in 1990, a Doctor of Technical Sciences from the Warsaw University of Technology at Warsaw, Poland in 1996, and Doctor Honouris Causa from the National Academy of Sciences at Gomel, Belarus in 2000. He is a registered professional engineer (mechanical). He is presently an Ohio Eminent Scholar and The Howard D. Winbigler Professor in the Department of Mechanical Engineering, Graduate Research Faculty Advisor in the Department of Materials Science and Engineering, and the Director of the Nanotribology Laboratory for Information Storage & MEMS/NEMS (NLIM) at the Ohio State University, Columbus, Ohio. He is an internationally recognized expert of tribology and mechanics on the macro- to nanoscales, and is one of the most prolific authors in these areas. He is considered by some a pioneer of the tribology and mechanics of magnetic storage devices and a leading researcher in the fields of nanotribology and nanomechanics using scanning probe microscopy and applications to micro/nanotechnology. He has authored five technical books, more than 50 handbook chapters, more than 500 technical papers in refereed journals, and more than 60 technical reports. He has edited more than 25 books, and holds 16 US patents. He is co-editor of Springer's NanoScience and Technology and co-editor of Microsystem Technologies-Micro- & Nanosystems and Information Storage & Processing Systems (formerly called the Journal of Information Storage and Processing Systems). He has given more than 250 invited presentations on five continents and more than 60 keynote/plenary addresses at major international conferences.

Dr. Bhushan is an accomplished organizer. He organized the first symposium on Tribology and Mechanics of Magnetic Storage Systems in 1984 and the first international symposium on Advances in Information Storage Systems in 1990, both of which are now held annually. He is the founder of an ASME Information Storage and Processing Systems Division founded in 1993 and served as the founding chair from 1993–1998. His biography has been listed in over two dozen Who's Who books including Who's Who in the World and has received more than two dozen awards for his contributions to science and technology from professional societies, industry, and US government agencies. He is also the recipient of various international fellowships including the Alexander von Humboldt Research Prize for Senior Scientists, Max Planck Foundation Research Award for Outstanding Foreign Scientists, and the Fulbright Senior Scholar Award. He is a foreign member of the International Academy of Engineering (Russia), Byelorussian Academy of Engineering and Technology and the Academy of Triboengineering of Ukraine, an honorary member of the Society of Tribologists of Belarus, a fellow of ASME, IEEE, STLE, and the New York Academy of Sciences, and a member of ASEE, Sigma Xi and Tau Beta Pi.

Dr. Bhushan has previously worked for the R&D Division of Mechanical Technology Inc., Latham, NY; the Technology Services Division of SKF Industries Inc., King of Prussia, PA; the General Products Division Laboratory of IBM Corporation, Tucson, AZ; and the Almaden Research Center of IBM Corporation, San Jose, CA.

Index

1, 1′-biphenyl-4-thiol (BPT), 903
1/f noise, 144
16-mercaptohexadecanoic acid thiol (MHA), 903
2-D FKT model, 495
2-D-histogram technique, 512
2-DEG
   2-DEG
     two-dimensional electron gas, 227
2-mercaptoethylamine HCl, 287
2-pyridyldithiopropionyl (PDP), 286
3-D bulk state, 209
3-D-force measurements, 99
4, 4′-dihydroxybiphenyl (DHBp), 903

abrasive wear, 515, 521, 811
Abrikosov lattice, 231
accelerated friction, 871
acceleration energy, 835
acetylene (C_2H_2), 837
acoustic emission, 580
actin-myosin motor, 711
activation energy barrier, 294
active linearization, 104
active structures, 1098
actuator
   miniaturized, 1036
adatoms, 154
adenosine triphosphate (ATP), 696
adhesion, 41, 316, 329, 341, 792, 803, 837, 906, 1031
   control, 886
   force, 246, 393, 400, 419, 422, 424
   force, quantized, 409, 453

hysteresis, 425, 426, 434
hysteresis, relation to friction, 434–436, 438, 442, 443, 454
influence of humidity on, 914
measurement, 1071
mechanics, 419, 420, 422, 423
performance, 936
primary minimum, 391, 406, 411
rate-dependent, 425, 426
adhesion-controlled friction, 430, 431, 433, 437
adhesional friction
   scale dependence, 793
adhesive
   coating, 288
adhesive force, 58, 511, 536, 937, 1061
   increase, 953
   intrinsic, 886
   mapping, 381, 941
   measurement, 323
adiabatic limit, 294
adsorbate, 173, 589
adsorbed
   insulator substrate, 167
   water, 950
   water film, 956
adsorption, 285
AES (Auger electron spectroscopy)
   measurement, 879
AFAM (atomic force acoustic microscopy), 42
AFM (atomic force microscope), 699, 735
   Binnig design, 54

calculated sensitivity, 92
cantilever, 52, 84
  commercial, 54
  contact mode, 50
  control electronics, 106
  design optimization, 79
  designs, 125
  feedback loop, 105
  for UHV application, 489
  image, 154
  imaging signal, 142
  instrumentation, 78
  manufacturers, 54
  microscratch technique, 918
  probe construction, 64
  probes, 128
  resolution, 122
  set-up, 245, 250
  spectroscopy, 256
  surface height map, 861
  test, 865
  tip, 130, 142, 246, 248, 285, 290, 1075
  tip containing antibodies, 305
  tip radius, 349
  tip sensor design, 285
Ag, 161
  Ag
    on Nb(110), 210
  Ag
    trimer, 158
Ag, 157
air damping, 276
air induced oscillations, 79
air/water interface, 288
Al_2O_3-TiC head, 877
Al_2O_3
  Al_2O_3
    grains, 331
  Al_2O_3
    ultrafiltration membrane, 548
Al_2O_3-TiC composite, 331
Al_2O_3-TiC slider, 1054
Al_2O_3, 264, 485
alcohol, 893
aldehydes, 893
alkali halides, 161
alkane, 891
alkanethiolate
  film, 898

alkanethiols, 173
all-fiber interferometer, 191
alloy, 537
aluminium oxide, 264, 485
AM (amplitude modulation)
  AFM, 145
  mode, 253
amide, 895
amine, 895
Amontons law, 250, 348, 429, 430, 433, 437, 445, 510
amorphous
  surfaces, 391
amorphous carbon, 830
  chemical structure, 838
  coatings, 838, 847
amplitude feedback, 268
amplitude modulation (AM), 145, 254
  mode, 51
  SFM (scanning force microscope), 218
anchor group, 897
angle of twist formula, 92
anion-terminated tip, 164
anisotropy, 604
  of friction, 499
annealing effect, 1113
antibody–antigen, 284
  complex, 297
  recognition, 307
antiferromagnetic spin ordering, 172
APCVD (atmospheric pressure chemical vapor deposition), 742
arc discharge, 833
Arg-Gly-Asp (RGD), 303
artifacts, 188
as-deposited film, 1110
asperity, 773, 781, 792
  contact, 777
assembled nanotube probes, 128
association process, 284
atmospheric pressure chemical vapor deposition (APCVD), 742
atomic
  interaction force, 248
  manipulation, 194
  motion imaging, 195
atomic force acoustic microscopy (AFAM), 42

atomic force microscope (AFM), 41, 50, 118, 141, 172, 284, 316, 392, 394–396, 453, 457, 576, 699, 776, 900, 929, 930
atomic resolution, 48, 136, 144, 146, 148, 206, 284, 290
  imaging, 152, 156–162, 263
atomic-scale
  dissipation, 525
  force measurement, 77
  friction, 329
  hysteresis, 273
  image, 43, 70, 219
atomic-scale friction
  simulation, 625
atomistic computer simulation, 162, 624
ATP hydrolysis, 697
ATP(adenosintriphosphate) synthase, 695
attraction
  long-range, 398, 416
attractive force–distance profile, 292
attractive interaction, 933
Au film, 745
Au microbeam, 757
Au microbeam, 753
chemAu microbeam, 767
Au(111), 911, 914
Au(111), 200
austenite, 543
average
  distance, 708
  lifetime, 293
average asperity, 773
axial strength, 554

backbone chain, 896
bacteriophage, 701, 717
bacteriorhodopsin, 698
BaF_2, 164
batch fabrication
  techniques, 120
batch nanotube tip fabrication, 133
BDCS (biphenyldimethylchlorosilane), 913
beam bending energy, 702
beam deflection, 252
beam failure, 751
beam-deflection FFM, 485
Bell's formula, 295
bending
  moment, 737, 759

stiffness, 82
strength, 738, 749, 751, 767
stress, 738, 758
test, 735, 736, 740
Berendsen thermostat, 632
Berkovich
  indenter, 734
  indenter tip, 561
  pyramid, 590
  tip, 859
$Bi_2Sr_2CaCu_2O_{8+\delta}$ (BSCCO), 212
bias voltage, 46
binary compound, 162
bio force probe (BFP), 290
bio–nano interface, 702
bio-surfaces, 304
biofluidic chip, 1037
biofunctionalized cantilever, 697, 699, 702
biological
  device, 697
  evolution, 712
  nanotechnology, 693, 726
biomaterials, 615
biomechanics, 558
BioMEMS (Biological or Biomedical Microelectromechanical Systems), 732, 1037
BioNEMS (Biological or Biomedical Nanoelectromechanical Systems), 732, 1039
biotin, 285
biotin-avidin
  spectrum, 302
biotin-directed IgG, 285
biotinylated AFM tip, 304
biphenyl, 195
birefringent crystal, 88
bistability, 259
blister test, 582
Bloch states, 206
Bloch wave, 206
block-like debris, 871
Boltzmann ansatz, 294
bond
  breakage, 295
  lifetime, 294
  rupture, 302
  strength, 398
bonded lubricant film, 933

bonded PFPE, 1076
bonding
  energy, 143
bone
  lamellae, 558
  lamellation, 564
  tissue, 558
boron
  ion implantation, 50
boundary
  film formation, 929
  lubricant, 929, 930
  lubricant film, 886
  lubrication, 371, 427–430, 445, 450, 451, 453, 915, 936
  lubrication measurement, 329
bovine serum albumin (BSA), 285
BPT (1,1′-biphenyl-4-thiol), 920, 1066
BPTC (cross-linked BPT), 1066
bridging of polymer chains, 416, 418
broken beams, 756
broken coating chip, 865
brush, *see* polymer brush
Buckingham potential, 630
buckling, 851, 855, 1095
  force, 128
  stress, 857
bulge test, 582
bulk
  atoms, 144
  conduction band, 201
  diamond, 830, 840, 842
  graphitic carbon, 874
  state, 209
  xenon, 223
Burgers vector, 782

C_{60}, 829
  C_{60}
    film, 48
    island, 502
  C_{60}
    island, 500
    multilayered film, 173
C_{60}-terminated film, 903
C_{70}, 829
C_{60}, 173, 511, 828
Ca ion, 164
cadherin-mediated adhesion, 303

CaF_2(111), 165
CaF_2, 164
CaF_2 tip, 520
CaF_2(111), 164
CaF_2(111) surface, 164, 520
calculus of variations, 703
calibration, 594, 595
calorimetric experiments, 293
cantilever, 143, 150, 255, 292
  Q-factor, 219
  axis, 191
  base, 275
  beam, 1080
  beam array (CBA), 1071
  biofunctionalized, 697, 699, 702
  deflection, 53, 119, 144, 147, 191, 269, 291, 484, 485, 702, 910
  deflection calculation, 73
  driven, 252
  effective mass, 487
  eigenfrequency, 187
  elasticity, 491
  flexible, 52
  foil, 120
  material, 65, 143
  motion, 50
  mount, 58
  oscillation, 255, 274
  resonance, 245
  resonance behavior, 83
  resonance frequency, 292, 1101
  spring, 250
  spring constant, 298
  stainless steel, 858
  stiffness, 61, 327
  thickness, 485, 486
  tip, 250
  triangular, 65, 82, 92
  untwisted, 57
cantilever-based probes, 218
capacitance detection, 95
capacitive
  detection, 103
  detector, 51
  displacement sensor, 1107
  forces, 268
capillary
  force, 391, 411, 419, 421, 536
capsid wall, 726

carbon, 828
  crystalline, 828, 829
  film, 936
  magnetron sputtered, 837
  spacer chain, 916
carbon coating
  unhydrogenated, 844
carbon nanotube (CNT), 206, 544, 545
  mechanical properties, 126
  tip, 70, 122, 125, 321
carbon–carbon distance, 223
carboxyl acid, 893
carboxylates (RCOO$^-$), 173
carrier
  gas, 843
Casimir force, 390
catalysis, 161
catalyst, 545
catalytically grown MWNT, 555
cathodic arc carbon, 848
cell adhesion, 304
$CeO_2(111)$, 166
ceramic
  slider, 876
CFM (chemical force microscopy), 304
change of meniscus, 1063
characteristic distance, 456
charge density wave (CDW), 210
charge exchange interactions, 390, 399, 400
charge fluctuation forces, *see* ion correlation forces
charge transfer interactions, *see* charge exchange interactions
chemical
  bond, 143
  bonding, 145, 932
  bonding force, 52
  characterization, 838
  force, 142
  force microscopy (CFM), 304
  heterogeneity, 425
  interaction force, 247
  vapor deposition (CVD), 830
chemisorption, 929
clustered acetate, 177
Co clusters, 206
CO on Cu(110), 194
Co-O film on PET, 360
co-transporter, 304

coating
  continuity, 878
  damage, 871
  failure, 868
  hardness, 866
  hydrophobic, 1079
  mass density, 840
  microstructure, 839
  thickness, 853, 856
coating–substrate interface, 866
coefficient
  effective, 511
coefficient of friction, 321, 333, 347, 375,
      see friction coefficient, 443, 488, 744,
      746, 773, 778, 791, 794, 803, 814,
      848, 858, 871, 875, 907, 940, 947,
      1050, 1063
  Si(100), 372
  scale dependence, 796, 801
  Z-15, 372
  Z-DOL (BW), 372
coefficient of friction relationship, 72
cognitive ligands, 284
coherence length, 201
cohesive
  potential, 725
  surface model, 724
cold welding, 402, 403, 444
collagen fibers, 565
colloidal
  probe, 396
colossal magneto resistive effect, 229
comb-drive, 1101
combined AFM nanoindenter device, 559
commercial MEMS devices, 1043
communicating
  force, 706
compact bone
  lamellae, 563
  mean hardness, 566
complex
  bonds, 299
component
  failure, 732
  surface, 1070
composite, 537
compressive stress, 845, 856, 1110
computer simulation
  forces, 403, 408, 409, 414, 417, 420, 453

friction, 403, 434, 437, 443, 444, 448, 453, 455, 456, 461, 462, 464, 467
concentration
  critical, 518
confinement, 390, 391, 428, 449, 465, 466
conformation, 933
conformational defect, 912
connector
  miniaturized, 1036
constant
  amplitude (CA), 267
  amplitude FM mode, 273
  current mode, 46
  excitation mode (CE), 267
  force mode, 105
  height mode, 46, 164
constant-energy simulation, 636
contact
  AFM (atomic force microscope), 252
  AFM dynamic method, 537
  angle, 888
  angle goniometer, 908
  angle of SAM, 908
  conductance, 515
  elastic, 602
  mechanics, *see* adhesion mechanics
  mode, 119
  multi-asperity, 794
  multiple-asperity, 773, 792, 807
  pressure, 808, 809
  printing, 889
  radius, 562
  resistance, 580
  single-asperity, 792, 794, 806, 809, 814
  stiffness, 561, 580, 594, 599, 853
  stress, 779, 947
  value theorem, 440
contact area, 396, 517, 520, 561, 596, 612
  apparent macroscopic, 429, 441, 445
  true molecular, 395, 397, 422, 423, 430, 433, 434, 441, 467
contact-start-stops (CSS), 966
contamination, 53, 808
continuous stiffness measurement (CSM), 853
continuum
  analysis, 715
  model, 495
  theory, 404, 407, 410, 411, 427, 445, 446

contrast, 220
  formation, 162
control
  system, 99, 105
controlled desorption, 195
controlled geometry (CG), 49
CoO, 168
Cooper pairs, 210
copper tip, 521
Couette flow, 429, 464
Coulomb
  force, 160
  law of friction, 505
Cr coating, 215
Cr(001), 206
crack spacing, 359
cracks, 850
creep, 188, 581
  effect, 368
critical
  concentration, 518
  degree of bending, 1103
  load, 848
  normal load, 917
  position, 491
  shear stress, 428, 430, 442, 453, 455
  temperature, 200, 219
  velocity, 454, 458, 462, 463
crosslinked BPT (BPTC), 903
crosstalk, 106
cryostat, 188
crystal
  structure, 53
  surfaces, 123
crystalline
  carbon, 828, 829
  surfaces, 391
CSM (continuous stiffness measurement), 853
Cu(001)), 162
Cu(100), 499
Cu(100), 497
Cu(111), 499
Cu(111), 200, 497, 525
Cu(111) surface states, 206
Cu(111) tip, 521
cube corner, 590, 605
cuprates, 215
current density effect, 1113

cut-off distance, 400, 435
CVD (chemical vapor deposition), 830
cyclic
  fatigue, 853, 1112
cytoplasmic surface, 288
cytoskeletal filament, 706

D_2O, 203
damage, 403, 427–456
damage mechanism, 865
damped harmonic oscillator, 259
damping, 539
  effective, 276
  effects, 245
  pneumatic, 191
data
  storage system, 1046
Deborah number, 451
debris, 871
Debye length, 405, 408
defect motion, 188
defect nucleation, 601
defect production, 518
deflection, 255
  measurement, 144
  noise, 219
deformation, 395, 397
  during sliding, 782
  elastic, 244, 422, 446, 447, 511
  length, 781
  of microtubule, 556
  plastic, 403, 445
deformed beam, 757
degrees of freedom, 714
delamination, 605, 844, 851, 855
Demnum-type PFPE, 381
  lubricant film, 941
dental enamel, 615
depletion
  attraction, 390, 416
  interaction, 416
  stabilization, 390, 416
deposition
  rate, 835
  techniques, 832, 834
Derjaguin approximation, 392, 408, 411
Derjaguin–Landau–Verwey–
      Overbeek (DLVO) theory, 406
Derjaguin–Muller–Toporov (DMT)

  model, 511
  theory, 422
design rule, 80
detection
  systems, 53, 84
device
  scaling, 1039
dewetting, 940, 1061, 1077
DFM (dynamic force microscopy), 218, 278, 305, 320
DFS (dynamic force spectroscopy), 297
DHBp (4, 4′-dihydroxybiphenyl), 903, 920, 1066
diamond, 120, 131, 555, 594, 828
  coating, 844, 887
  film, 830
  like amorphous carbon coating, 839
  like carbon (DLC), 359, 1058
  tip, 326, 519, 536, 559, 858, 867, 900, 937, 1054
diamond-like carbon (DLC), 808
dielectric
  breakdown, 581
differential scanning calorimetry (DSC), 540
diffusion
  coefficient, 936
  parameters, 195
  thermally activated, 188
diffusive
  communication, 710
  relaxation, 294
digital
  feedback, 105
  light processing (DLP), 1037
  micromirror device (DMD), 1037, 1042
  signal processor (DSP), 46
dilation, 434, 452, 455, 456, 466
dimension, 143
dimensions of spatial structures, 705
dimer structure, 156
dimer-adatom-stacking (DAS), 153
dip coating technique, 931
dipole molecule, 890
directly growing nanotubes, 130
disjoining pressure, 942
dislocation, 783, 785
  geometrically necessary, 779
  gliding, 783
  line tension, 604

microslip, 784
motion, 428
nucleation, 591
sliding, 783
dislocation-assisted sliding, 773, 779, 783, 814
dispersion
  force, 177
displacement
  resolution, 567
  vertical, 758
disposable plastic lab-on-a-chip, 1044
dissipation, 272
  force, 261
  measurement, 483
dissipative tip-sample interaction, 273
dissociation, 284
distortion, 102
dithio-bis(succinimidylundecanoate), 285
dithio-phospholipids, 288
DLC
  coating, 360, 830, 833, 848, 867, 877, 1069
  coating microstructure, 854
DLVO interactions, 397, 406, 408, 410
DMT (Derjaguin–Muller–Toporov) model, 589
DNA, 173, 267, 272, 285, 705
  packing, 720
domain
  pattern, 229
doped
  polysilicon film, 1057
  silicon, 1051
DOS structure, 200
double-layer
  interaction, 390, 405, 407, 408
doubly clamped beam, 1100
dried tissue properties, 558
drive amplitude, 276
driven cantilever, 252
driving frequency, 84
dry
  nitrogen environment, 1054
dry surfaces, 430, 459
  forces, 398, 402
  friction, 429, 430, 441, 456, 460
Dupré equation, 427
dynamic

interactions, 390–392, 428, 446, 449, 453
mode, 51
operation mode, 145
oscillation force, 253
dynamic AFM, 145, 249, 250, 252, 263, 272, 275
dynamic force microscopy (DFM), 255, 278, 305, 320
dynamic force spectroscopy (DFS), 297
dynamic friction, *see* kinetic friction

E. coli
  bacterium, 707
  cells, 719
EBD (electron beam deposited) tips, 125
ECR (electron cyclotron resonance) CVD, 832, 859, 1069
  coating, 868
EELS (electron resonance loss spectroscopy), 839
EFC (electrostatic force constant), 563
effect of surface roughness, 757, 758
effective
  coefficient of friction, 511
  damping constant, 276
  force gradient, 264
  ligand concentration, 297
  mass, 219
  shear stress, 513
  spring constant, 492, 507
  tether length, 297
  viscosity, 450, 464
elastic
  adhesion, 773
  contact, 602, 803
  deformation, 803
  force, 720
  Hamiltonian, 717
  modulus, 539, 594, 736, 742, 750, 846, 848
  properties, 555
  properties of ssDNA, 703
elastic deformation, 244, 511
elastic-plastic deformation, 767
elasticity, 219, 537
  limits of, 779
  of DNA, 720
  two-dimensional, 717

elastohydrodynamic lubrication, 428, 429, 446, 447, 450, 451
electric force gradient, 62
electrochemical
    AFM, 63
    deposition, 557
    etch, 137
    etching, 120
    STM, 48
electrochemically etched tips, 137
electromagnetic
    forces, 398
electromigration, 195
electron
    beam deposition (EBD), 125
    cyclotron resonance chemical vapor deposition (ECR-CVD), 832, 836
    energy loss spectroscopy (EELS), 835
    interactions, 209
    tunneling, 43
electron-beam lithography, 556
electron-electron interaction, 200
electron-phonon interaction, 210
electronegativity, 890
electronic
    noise, 54
electrostatic
    binding, 288
    interaction, 155, 169, 267
    potential, 163, 227
    short-range interaction, 163
electrostatic force, 142, 224, 390–410
    constant (EFC), 559
    interaction, 155
    microscopy, 226
embedded-atom method, 628
embossing, 889
end deflection, 1096
end group, 933
endothelial cell surfaces, 303
energy
    barrier, 294
    conservation, 274
    dissipation, 273, 425–427, 431, 432, 434, 435, 437, 440, 441, 443, 456, 525, 912
    resolution, 198
    scales, 710
engineering materials, 723
entangled states, 428

entropic force, 417
enzymes, 294
epitope mapping, 307
equilibrium
    interactions, *see* static interactions
    true (full) or restricted, 416
ester, 893
ethanolamine, 286
ether, 893
Euler
    buckling criterion, 1095
    equation, 80
examples of NEMS, 1039
exchange
    force, 170
    force interaction, 169
    interaction energy, 170
exchangeable carrier plate, 490
excitation
    external, 253
excitation amplitude, 262
external
    excitation, 253
    excitation frequency, 270
    noise, 188
    normal load, 937
    vibrations, 47
Eyring
    model, 506

Fab molecule, 285
face-centered cubic (fcc), 829
failure
    mechanism, 875
    of MEMS/NEMS device, 1075
Fano resonance, 206
fatigue, 853
    behavior, 741
    crack, 752, 871
    damage, 853
    failure, 1112
    life, 848
    measurement, 1105
    properties, 740, 752
    resistance, 1098
    strength, 750
    test, 752, 857
fatty acid monolayer, 900
FCA (filtered cathodic arc), 859

coating, 847, 855
Fe(NO$_3$)$_3$, 133
Fe-N/Ti-N multilayer, 557
Fe-coated tip, 171
feedback
  circuit, 54, 267, 268
  loop, 107, 119, 244, 254, 262, 268, 298, 488
  network, 44
  signal, 145
FeO, 168
Fermi
  level, 198
  points, 210
ferromagnetic
  probe, 227
  tip, 232
FFM (friction force microscopy), 485
  dynamic mode, 520
  on atomic scale, 507
  tip, 331, 492, 497
FIB (focused ion beam) tips, 125
FIB-milled probe, 70
fiber
  optical interferometer, 88
field
  emission tip, 128
  ion microscope (FIM), 248
film–substrate interface, 1110
films on substrate, 1091
filtered cathodic arc (FCA) deposition, 833–855
FIM (field ion microscope), 248
  technique, 267
finite
  element analysis, 756
  element method (FEM), 733
  element modeling (FEM), 595
first principle calculation, 160
first principle simulation, 220
FKT (Frenkel–Kontorova–Tomlinson) model, 494
flagellar motor rotation rate, 710
flat punch, 592
flexible cantilever, 52
flocculation, 406
Flory temperature, 415
flow
  activation energy, 934

fluoride, 161, 164
fluoride (111) surface, 165
fluorinated DLC, 888
FM (frequency modulation)
  AFM, 267
  AFM images, 220
  mode, 263
focused ion beam (FIB), 49, 68, 125, 585, 612
foil cantilever, 120
force
  between macroscopic bodies, 401, 407
  between surfaces in liquids, 403
  calibration, 485
  calibration mode, 364
  calibration plot (FCP), 937, 1062
  cantilever-based, 218
  curve, 134
  detection, 80
  effective gradient, 264
  elastic, 720
  extension curve, 700
  induced unbinding, 301
  long-range, 147, 162, 163, 224
  mapping, 305
  measurement, 144
  measuring techniques, 392, 394, 397
  modulation mode (FMM), 537
  modulation technique, 327
  packing curves, 722
  repulsive, 292
  resolution, 218, 292
  scales, 710
  sensing tip, 99
  sensitivity, 292
  sensor, 52, 148, 489
  spectroscopy (FS), 105, 219, 224, 284
  undulation, 390, 417
  velocity curve, 712
force field spectroscopy
  three-dimensional, 225
force-displacement
  curve, 559, 568
force-distance
  calibration, 910
  curve, 59, 246, 247, 254, 536, 547
  cycle, 291
  diagram, 274
formate

(HCOO⁻), 173
  covered surface, 173
formation of SAM, 897
four-quadrants photodetectors, 484
fracture, 604
  strength, 1112
  stress, 738
  surfaces, 753
  toughness, 734, 738, 744, 750, 752, 846, 848, 849, 851, 865, 866
  toughness measurement, 1097, 1102
Frank-Read source, 602
free surface energy, 950
freezing-melting transitions, 457, 460, 462
Frenkel–Kontorova–Tomlinson (FKT), 494
frequency
  measurement precision, 98
  modulation (FM), 145
  modulation (FM) mode, 51
  modulation AFM (FM-AFM), 147
  modulation SFM (FM-SFM), 187
  shift calculation, 149
  shift curve, 265, 267
friction, 41, 316, 329, 341, 483, 623, 776, 829, 906, 1031
  anisotropy, 499
  characteristics, 1070
  coefficient, 428–431, 437, 441–443, 445, 449, 451, 462, 465
    load dependence, 349
  control, 886
  deformation model, 792
  directionality effect, 340
  effect of humidity on, 341
  effect of tip on, 341
  experiments on atomic scale, 495
  global internal, 544
  image, 499
  kinetic, 430, 432, 434, 451, 454, 455, 459, 460, 462, 464
  lateral, 249
  loop, 492, 495, 515
  macroscale, 901, 1053
  map, 448, 450, 451, 498
  measurement, 1107
  measurement methods, 70
  measuring techniques, 395, 396, 441, 448, 453
  mechanism, 333, 777, 903, 910, 942, 1075

  molecular dynamics simulation, 519
  performance, 936
  scale dependent, 348
  scale effect, 776
  torque, 1080
  torque model, 1079
  total coefficient, 792
friction force, 52, 57, 320, 346, 501, 858, 902, 915, 938, 1061, 1063
  calibration, 58, 76
  curve, 106
  decrease of , 950
  image, 906
  magnitude, 74
  map, 338, 513
  mapping, 941
  microscope (FFM), 41, 316, 318, 900
  microscopy (FFM), 381, 483
  Z-15, 372
  Z-DOL (BW), 372
friction force microscope (FFM), 776
friction models
  cobblestone model, 431, 432, 435
  Coulomb model, 431
  creep model, 460
  distance-dependent model, 460
  interlocking asperity model, 431
  phase transitions model, 458, 461, 462
  rate-and-state dependent, 462
  rough surfaces model, 460
  surface topology model, 460
  velocity-dependent model, 458, 460
fringe detector, 1099
fullerene, 828
fundamental resonant frequency, 83
fused silica, 595, 597, 604
  hardness, 563

g-factor, 203
GaAs/AlGaAs
  GaAs/AlGaAs heterostructures, 227
gage marker, 1099
gain, 269
gain control circuit, 148
γ-modified geometry, 598
gap stability, 45
GaP(110), 221
Gd on Nb(110), 210

gear
    miniaturized, 1035
genomic libraries, 719
geometrically necessary dislocation (GND), 779, 814
geometry effects in nanocontacts, 508
germanium, 605
giant MR (GMR), 966
glass
    transition temperature, 428, 443, 540
glassiness, 465
gliding, 784
    dislocation, 782, 783
global internal friction, 544
gold, 49
    coated tip, 285
Grahame equation, 405
grain
    boundary, 402, 403, 1111
    size, 830
graphene, 545
graphite, 187, 191, 331, 495, 519, 828, 829
    cathode, 833
    surface, 266
graphite(0001), 206
Griffith fracture theory, 755
grinding, 136
growth
    of nanotubes, 546

H_2O, 203
Hall resistance, 227
Hall–Petch behavior, 613
Hamaker constant, 143, 265, 399–442, 943
Hamilton–Jacobi method, 150
hard amorphous carbon coatings, 832
hard-core or steric repulsion, 390
hardness, 326, 742, 838, 842, 846, 848, 1092
    scale dependence, 779, 781
harmonic oscillator, 250
harpooning interaction, 390, 402
HDT (hexadecanethiol), 903, 920, 1065, 1076
head
    group, 896
heavy ion bombardment, 231
Hershey–Chase experiment, 718
Hertz model, 586
Hertz-plus-offset relation, 511

Hertzian contact model, 589
heterodyne interferometer, 86
hexadecanethiol (HDT), 903, 920, 1065, 1076
hierarchy
    of structures, 705
    of temporal processes, 708
high speed scanning, 270
high temperature superconductivity (HTCS), 194, 212, 231
high-aspect-ratio
    MEMS (HARMEMS), 1034
    tips, 125, 127
high-resolution
    FM-AFM, 222
    imaging, 177
    spectroscopy, 198
    tips, 130
high-temperature operation STM, 189
higher orbital tip states, 200
highest resolution images, 136
highly oriented pyrolytic graphite (HOPG), 59, 329
Hill coefficient, 303
hinged cantilever, 1108
homodyne interferometer, 84
honeycomb
    chained trimer (HCT), 157
    pattern, 165
Hooke's law, 218, 244, 250, 275, 291
HOPG, 807
horizontal coupling, 47
HtBDC (hexa-tert-butyl-decacyclene) on Cu(110), 195
HTCS (high temperature superconductivity), 194, 212, 231
Huber-Mises, 588
human bone tissue, 558
humidity, 507
hybrid
    nanotube tip fabrication, 133
hybridization, 160
hydration forces, 390–439
hydration regulation, 413
hydrocarbon, 891
    precursors, 837
    unsaturated, 893
hydrodynamic
    forces, 390, 397

radius, 415
hydrofluoric (HFhydrofluoric acid), 886
hydrogen, 156
   bonding, 414, 415
   concentrations, 843
   content, 844
   end group, 899
   flow rate, 845
   termination, 156
hydrogenated carbon, 838
hydrogenated coating, 838
hydrophilic
   surfaces, 391, 413
hydrophobic
   coating, 1079
   force, 411, 413
   surfaces, 391, 413
hydroxylation, 898
Hysitron, 559
hysteresis, 102, 135, 147, 188, 229, 259, 274, 568
   loop, 54, 107

IB (ion beam)
   coating, 868
IBD (ion beam deposition), 833, 835, 859
IC (integrated circuit), 1038
IgG
   repetitive, 297
image
   effects, 613, 614
   processing software, 108
   topography, 51, 174
imaging
   bandwidth, 64
   electronic wave functions, 206
   signal noise, 147, 151
in situ
   sharpening of the tips, 45
in vivo properties, 559
InAs, 191
InAs(110), 201, 220
indentation, 316, 364, 634
   ceramics, 641
   creep process, 368
   depth, 365, 562, 846, 851, 1053
   fatigue damage, 856
   hardness, 42, 365
   induced compression, 855

   modulus, 564
   size, 365
   size effect (ISE), 603
   technique, 734
indenting metal surfaces, 636
inelastic tunneling, 194, 195
infected cell, 719
influence of humidity
   on adhesion, 914
initial contact, 602
InP(110), 152
instability position, 261
instability, jump-to-position, 146
insulator, 162
integrated
   circuit (IC), 1038
   tip, 67, 120
interaction
   energy, 721
   force, 220, 720
interatomic
   attractive force, 72
   bonding, 155
   force, 52
   force constants, 144
   interaction, 1111
   spring constant, 52
intercalated MoS_2, 887
intercellular adhesion molecule-1 (ICAM-1), 304
intercellular concentration, 707
intercept length, 919
interchain interactions, 296
interdiffusion, 425
interdigitation, 425
interface
   scale-dependence of the shear, 793
interface temperature, 776, 777, 811
   scale dependence, 773
interfacial
   defects, 857
   energy (tension), see surface energy (tension)
   friction, see also boundary lubrication, 430, 431, 435, 440, 445, 453, 930
   stress, 865
interferometeric detection sensitivity, 89
intermediate or mixed lubrication, 427–429, 448–450

intermittent contact mode, 120, 260
intermolecular force, 292, 297
internal stress, 1093
intraband transitions, 200
intramolecular
  forces, 296
intrinsic
  adhesive force, 886
  damping, 275
  stress, 844
iodine, 195
iodobenzene, 195
ion
  correlation forces, 405, 408
  displacement, 163
  implantation, 367, 1050
  implanted silicon, 1051
  plating techniques, 832
  source, 835
ion beam
  deposition (IBD), 833, 835
  sputtered carbon, 835
ion correlation forces, 390
ionic bond, 390
isolated nanotubes, 133
itinerant nanotube levels, 206

JKR (Johnson–Kendall–Roberts)
  model, 588
  relation, 511
  theory, 422
Joule
  dissipation, 526
  heating, 229
jump-to-contact, 145, 224, 247

K-shell EELS spectra, 841
KBr, 162
KBr(100), 498
KBr(100), 516
$KCl_{0.6}Br_{0.4}(001)$ surface, 163
Kelvin
  equation, 419
Kelvin probe force microscopy (KPFM), 178, 324
ketones, 893
key scale at bio-nano interface, 704
kinesin, 695, 701
kinetic

friction, 430, 432, 434, 451, 454, 455, 459, 460, 462, 464
  friction ultralow, 463
  off-rate, 299, 303
  processes, 602
  rate constant, 293, 299
knife-edge blocking, 103
KOH, 49
Kondo
  effect, 187, 203
  temperature, 203

lab-on-a-chip, 1037
  disposable plastic, 1044
lac repressor, 716
Lamb waves, 584
Landau
  levels, 188, 227
  quantization, 203
Langevin thermostat, 631
Langmuir–Blodgett (LB), 930
  deposition, 378, 888
  film (ethyl-2,3-dihydroxyoctadecanoate), 272
Laplace
  force, 907, 938
  pressure, 391, 419, 886
laser
  beam deflection method, 250
  deflection sensing, 249
  deflection technique, 55
lateral
  contact stiffness, 483
  cracks, 585
  deflection, 550
  deflection signal, 1080
  displacement, 547
  force, 484, 504, 547, 1080
  force calculation, 92
  force microscope (LFM), 41, 318, 483
  force microscopy (LFM), 172, 396
  friction, 249
  resolution, 45, 51, 286, 320, 538
  resonator, 1113
  spring constant, 64, 491
  stiffness, 106, 126, 513
lattice
  imaging, 52
length scale, 705

Lennard-Jones potential, 222, 255
leukocyte function-associated antigen-1 (LFA-1), 304
LiF, 162
LiF(100) surface, 520
lifetime broadening, 194, 200
lifetime-force relation, 298
Lifshitz theory, 399, 400, 404
lift mode, 63, 324
LIGA, 1034
   fabrication of MEMS, 1036
   technique, 733
ligand concentration
   effective, 297
ligand-receptor interaction, 285
light beam deflection galvanometer, 89
linear
   variable differential transformers (LVDT), 103
lipid, 705
lipid film, 507
liposome, 698
liquid
   capillary condensation, 907
   film thickness, 378
   helium, 188
   helium operation STM, 189
   lubricant, 371, 886
   lubricated surfaces, 430, 441, 443, 459
   mediated adhesion, 886
   nitrogen (LN), 189
   perfluoropolyether (PFPE), 1058
   solid interface, 886
   vapor interface, 886
load
   critical, 848
load contribution to friction, 431, 433, 435
load dependence of friction, 511
load-carrying capacity, 873
load-controlled friction, 430, 445
load-displacement, 368
load-displacement curve, 749, 755, 850, 851
loading curve, 598
loading rate, 295
local deformation, 325, 349
local deformation of material, 359
local density of states (LDOS), 198
local mechanical spectroscopy, 540
local stiffness, 106, 1065

London
   dispersion interaction, 390, 398, 399
   penetration depth, 212, 231
long-range
   attraction, 398, 416
   force, 147, 162, 163, 224
   tip-molecule force, 176
long-term
   measurements, 187
   stability, 186
longitudinal
   piezo-resistive effect, 95
loss modulus, 601
low pressure chemical vapor deposition (LPCVD), 742
low temperature
   AFM/STM, 53
   NC-AFM, 170
   SFM (LTSFM), 191
low-cycle fatigue resistance, 1105
low-noise measurement, 490
low-temperature microscope operation, 187
low-temperature scanning tunneling spectroscopy (LTSTM), 194
low-temperature SPM (LTSPM), 186
LPCVD (low pressure chemical vapor deposition), 742, 1109
LTSFM, 191
LTSTM, 194
lubricant
   meniscus, 933
   spreading, 933
lubrication
   elastohydrodynamic, 429, 446, 447, 450, 451
   intermediate or mixed, 427, 429, 448–450
   method, 1049
LVDT
   linear variable differential transformers, 103

macromolecular
   building block, 706
   interaction, 283
macromolecules as elastic rods, 715
macroscale friction, 901, 1053
magnetic
   Ni, 212
   dipole interaction, 170

disk, 876
disk drive, 832
quantum flux, 231
storage device, 831, 885
tape, 271, 340, 359, 930
thin-film head, 875
tip, 270
magnetic force, 142
gradient, 62
microscope (MFM), 62
microscopy (MFM), 42, 227
magnetic resonance
force microscopy (MRFM), 233
magnetically coated tip, 305
magnetoresistive (MR), 966
magnetostatic interaction, 229
magnetron sputtered carbon, 837
manganites, 215
manipulation of individual atoms, 42
manually assembled MWNT tips, 129
martensite, 541
mass
effective, 219
mass of cantilever, 50
material
hardness, 734
structural, 1098
MBI (multiple beam interferometry), 394
MD simulations
nanoindentation, 635
mean-field theory, 407
measurement of hardness, 327
mechanical
coupling, 457
dissipation in nanoscopic device, 483
instability, 247
relaxation, 255, 910
resonance, 463
spectroscopy, 538
surface relaxations, 273
mechanical properties, 733, 1091
characterization, 1113
of bone, 558
of carbon nanotubes, 546, 554
of DLC coating, 858
mechanically cut tips, 136
mechanics of cantilevers, 78
mechanism-based strain gradient (MSG), 779

melting point of SAM, 915
membrane
deflection method, 1113
proteins, 122
membrane-embedded machine, 695
membranes as elastic media, 716
memory
distance, 456
MEMS, 125, 732, 885, 1031
applications, 1034
components, 1073
device, 1036, 1041
device operation, 1047
technology, 1109
tribological problems, 1040
MEMS/NEMS, 371
meniscus
bridge, 886, 954
force, 346, 942
of liquid, 419, 421
meniscus effect, 808
messenger RNA, 710
metal
catalyst, 130
evaporated (ME) tape, 831
oxide, 161, 166
porphyrin (Cu-TBPP), 173
metal-catalyzed
chemical vapor deposition (CVD), 130
metal-deposited Si surface, 157
metal-particle (MP) tape, 876
methyl-terminated SAM, 902
methylene stretching mode, 901
Meyer's law, 611
MFM sensitivity, 62
Mg-terminated tip, 163
MgO tip, 163
MgO(001) surface, 171
mica, 59, 498
muscovite, 516
surface, 501
micro-Raman spectroscopy, 586
micro/nanoelectromechanical systems
(MEMS/NEMS), 776
microactuator, 1040
microcantilever, 1046
microcomponent, 1040
microcontact printing (μCP), 889
microcrystalline graphite, 842

microdevice, 1095
microelectronics, 161, 708
microfabricated
   cantilever, 118, 249
   silicon cantilevers, 143
microfabrication
   techniques, 118
microfriction, 339
microindentation, 577
micromachined
   notch, 1105
micromanipulator, 1038
micromirror, 1042
micromotor, 1082
micropatterned SAM, 910
micropipette aspiration, 397
microscale
   friction, 1065, 1071
   material removal, 354
   scratching, 323
   wear, 323, 354
   wear test, 1068
microscope eigenfrequency, 79
microscratch, 858
   test, 557
microscratching measurement, 351
microstrain gauge, 1094, 1109
microtriboapparatus, 1074
microtubule, 694
   bending, 555
   buckling, 555
microwear, 858, 867
milligear system, 1040
millipede, 1046
miniaturized
   actuator, 1036
   connector, 1036
   gear, 1035
   motor, 1035
minimal detectable depression, 121
misfit
   angle, 495
mismatch of crystalline surfaces, 423, 466
mitochondria, 706
mixed lubrication, *see* intermediate or mixed lubrication
Mn on Nb(110), 210
Mn on W(110), 215
modeling
   at the nanoscale, 713
   mechanics of systems, 713
   tribological process, 714
modulus
   elastic, 539, 594, 736, 742, 750, 846, 848
modulus of elasticity, 42, 347
molarity, 708
molecular
   beam epitaxy (MBE), 1038
   building blocks, 705
   chain, 896
   conformation, 933
   dynamics (MD), 284, 708
   dynamics (MD) calculation, 497
   dynamics simulation (MDS), 302, 483, 519, 715
   interaction, 957
   motor, 555, 694
   pump, 696, 707
   recognition force microscopy (MRFM), 284
   recognition force spectroscopy principles, 293
   resolution, 198
   scale layer, 888
   shape, 410, 411, 449, 456, 463–465
   spring, 378
   spring model, 903, 910
   stiffness, 912
molecular dynamics
   interatomic potential, 626
molecular dynamics simulation
   many-body expansion, 628
   surface structur, 640
moment of inertia, 1102
monolayer
   thickness, 919
Monte Carlo simulation, 910
Morse potential, 265
MoS_2 friction, 500
MoS_2 friction, 498, 547
motion
   unperturbed, 149
motor
   actin-myosin, 711
   flagellar, 710
   micro, 1082
   miniaturized, 1035
   molecular, 713

1136   Index

myosin, 695, 701
MRFM
   magnetic resonance force microscopy, 233
μCP (microcontact printing), 889
multilayer, 613
   Fe-N/Ti-N, 557
   thin-film, 875
multimode AFM, 62
multiple beam interferometry (MBI), 394
multiple-asperity contact, 348
multiplication of dislocations, 602
multiwall carbon nanotubes (MWCNT), 70, 126, 545
muscle contraction, 695
muscovite mica, 516
MWCNT
   multiwall carbon nanotubes, 70, 126, 545
MWNT (multiwall nanotube)
   catalytically grown, 555
   purified, 128
   tips
      manually assembled, 129
      Young's modulus, 547
myosin motor, 712

N-hydroxysuccinimidyl (NHS), 286
n-octadecyltrichlorosilane (n-$C_{18}H_{37}SiCl_3$, OTS), 901
N-succinnimidyl-3-(S-acethylthio)propionate (SATP), 286
n-undecyltrichlorosilane (n-$C_{11}H_{23}SiCl_3$, UTS), 901
Na on Cu(111), 198
NaCl
   NaCl
      island on Cu(111), 525
      island on Cu(111), 526
   NaCl
      islands, 274
   NaCl
      thin film on Cu(111), 163
NaCl, 162
NaCl(001), 162
NaCl(100), 502, 503, 513
NaCl(100), 501
NaF, 162, 498
nano-positioning, 99

nanoasperity, 341, 589
nanobeam, 547
   array, 735, 748
nanochemistry, 1033
nanocluster, 166
nanocomposite
   coating, 888
nanocrystallites, 838
nanodeformation, 933, 959
nanoelectromechanical systems (NEMS), 732, 1032
nanofabrication, 42, 362
nanofatigue, 853
nanohardness, 588, 844, 845, 1057
nanoindentation, 557, 576, 592, 593, 846, 1092
   curve, 560
   measurement, 326
nanoindenter, 733, 740, 866, 1099
nanomachining, 42
nanomagnetism, 215
nanomaterial, 537
nanoparticle, 122
nanorheology, 449
nanoscale
   mechanical properties, 535
   supramolecular assembly, 178
   wear, 349
Nanoscope I, 46
nanoscratch, 357
   studies, 734
nanostructures
   mechanical properties, 733, 1091
nanotechnological
   actions, 718
nanotechnology
   in the living world, 694
nanotribology, 930
nanotube
   AFM tips, 130
   buckling, 127
   length, 134
   strain, 554
   surface energy, 132
   tip fabrication, 131
   tips, surface growth, 132
native oxide layer, 938
Nb superconductor, 200
$NbSe_2$, 212, 515

NC-AFM (noncontact atomic force microscopy), 162, 218
near-field
  scanning optical microscopy (NSOM), 118
  technique, 220
near-surface mechanical properties, 541
neighbouring indents, 563
NEMS, 885, 1031, 1038, 1091
  nanoelectromechanical systems, 732, 1032
nested-arc structure, 566
Newton's first law, 250
Newton–Raphson method, 715
Newtonian flow, 428, 446, 448, 450, 451
Ni(001) tip on Cu(100), 522
Ni(111) tip on Cu(110), 522
Ni-P
  Ni-P
    beam, 767
  Ni-P
    cantilever, 755
  Ni-P
    film, 742, 745
  Ni-P
    microbeam, 753, 767
Ni-P
  microbeam, 756
Ni^{2+}-chelating, 298
NIH3T3 fibroblast cell, 304
NiO, 168, 170, 191
NiO(001), 162, 166, 170
NiO(001) surface, 171
NiTi
  NiTi
    transformation behavior, 543
NiTi, 541
nitrogenated carbon, 838
noble metal surfaces, 203
noise, 143, 152, 249
  1/f, 144
  electronic, 54
  external, 188
  performance, 151
  source, 93
  vertical, 151
Nomarski-interferometer, 88
non-Newtonian flow, 427, 428, 446, 448, 450

nonconducting film, 834
nonconductive
  materials, 172
  sample, 47
  surface, 264
noncontact
  dynamic force microscopy, 524
  friction, 226, 526
  imaging, 51
  mode, 120, 145
noncontact atomic force microscopy (NC-AFM), 122, 141, 152, 154, 162, 172, 524
nondestructive contact-mode measurement, 521
nonequilibrium interactions, *see* dynamic interactions
nonlinear
  force, 259
  spring, 307
nonliquidlike behavior, 449
nonmagnetic Zn, 212
nonpolar group, 895
nonspherical tip, 512
nonwetting, 413
normal friction, 427, 437, 444, 445
normal load, 938
  external, 937
normalized frequency shift, 265
NTA (nitrilotriacetate) - His_6, 287
nucleotides, 705

$(OCN)^-$, 163
OH-terminated tip, 163
on-chip
  actuator, 1104
  operation, 144
optical
  deflection systems, 118
  detector, 51
  head, 58
  head mount, 59
  trap, 290
  tweezers (O§T), 397, 700
  tweezers (OT), 290
optical lever, 89
  angular sensitivity, 93
  deflection method, 106
  optimal sensitivity, 93

optimal beam waist, 93
order
  in-plane, 423, 453, 465
  long-range, 428
  out-of-plane, 409–411, 423, 428, 453, 464
  parameter, 428, 456
organic
  compounds, 891
organometallic vapor phase epitaxy (OMVPE), 1038
Orowan strengthening, 613, 614
oscillating tip, 57, 275
oscillation
  amplitude, 147, 249, 251, 254, 262
  loop, 264
oscillatory force, 390, 391, 409–412, 414, 423, 424
osmotic
  interactions, 390, 405, 416, 417
  pressure, 406, 416, 440
  stress technique, 397
OTS (octadecyltrichlorsilane), 911
oxide layer, 603
oxide-sharpened tips, 124
oxygen
  content, 879

P-selectin glycoprotein ligand-1 (PSGL-1), 302
packing density
  chains, 901
paraboloid load displacement, 592
paraffins, 893
particle deformation, 792
passive
  linearization, 103
  structure, 1093
Pb on Ge(111), 210
PDP (2-pyridyldithiopropionyl), 286
peak indentation load, 853
PECVD (plasma enhanced chemical vapor deposition), 888, 1069
  carbon sample, 839
PEG (polyethylene glycol)
  crosslinkers, 289
Peierls force, 782
Peierls instability, 210
perfluorodecanoic acid (PFDA), 1043
perfluoropolyether (PFPE), 371, 930

periodic
  potential, 493
permanent dipole moment, 177
perpendicular scan, 74
persistence length, 302
perturbation approach, 150
perylene, 173
PFPE (perfluoropolyether), 934, 1075
  lubricant, 930, 941, 1058, 1081
phase
  angle image, 370
  curve, 258
  images, 277
  lag spectrum, 543
  measurement, 277
  signal change, 257
  transformation, 580, 605
phenol, 893
phonon
  excitation, 273
photo-crosslinker, 288
photodetector sensitivity, 735
photoemission spectroscopy (PES), 200
photothermal effect, 268
physisorbed protein layer, 285
physisorption, 929
pick-up
  SWNT tips, 135
  tip method, 134
piezo
  ceramic material, 102
  effect, 100
  element, 262
  excitation, 250
  hysteresis, 188
  relaxation, 188
  stacks, 104
  tube, 100, 101
piezoelectric
  drive, 44
  leg, 490
  positioning elements, 186
  scanner, 119, 188, 559
piezoresistive
  cantilever, 94
  coefficients, 94
  detection, 94, 144
piezoscanner, 107, 136
piezotranslator, 106

piezotube calibration, 47
pile-up, nanoindent, 585, 595, 596, 612, 615
piling-up behavior, 557
pin-on-disk tribotester, 900
pinning, 231
plasma enhanced chemical vapor deposition (PECVD), 65, 742, 832, 837
plasmon mode, 198
plastic
  contact regime, 347
  deformation, 356, 854, 919
plastic deformation, 773, 779
platinum-iridium, 49
PMMA (poly(methyl methacrylate)), 551, 1046
pneumatic
  damping, 191
point probes, 576
Poisson's ratio, 594
polar
  group, 895
  lubricant, 887, 930
polishing, 830, 1057
poly(ethylene glycol) (PEG), 286
poly(methyl methacrylate) (PMMA), 551, 1046
polycrystalline
  graphite, 840
polydimethylsiloxane (PDMS), 889
polyethylene terephthalate (PET), 359
polymer, 540
  blend, 278, 537
  brush, 417, 418, 439, 447
  mushroom, 417
polymeric
  liquids (melts), 415, 447, 448
  magnetic tape, 365
polypropylene, 173, 278
polysilicon, 1073
  fatigue, 1112
  film, 1049, 1097
  fracture strength, 1112
  fracture toughness, 1111
  friction, 1113
  mechanical properties, 1109
  microstructure, 1111
  residual stress, 1109
  Young's modulus, 1111
polytetrafluoroethylene (PTFE), 497

polyurethane, 278
pop-ins, 580
pop-outs, 580
position
  accuracy, 107
power dissipation, 277, 525
power to weight ratio, 712
pre-crack length, 1098
pressure, 942
primary minimum adhesion, 391, 406, 411
probe
  FIB-milled, 70
probe tip, 118, 119
  performance, 121
probe-sample distance, 256, 263
probe-surface distance, 252
process
  gas, 837
propagation of cracks, 359
properties of a coating, 833
protection of sliding surface, 885
protein, 286
protofilament, 555
protrusion force, 390, 417
Pt alloy tip, 50
Pt-Ir
  tip, 49
PTCDA, 173
PTFE (polytetrafluoroethylene)
  coated Si-tip, 499
  film, 887
pull-hold-release cycle, 297
pull-off force, *see* adhesion force, 488, 933
pulse-etching, 129
purified MWNT, 128
pyramidal
  etch, 120
PZT (led zirconate titanate)
  scanner, 320
  tube scanner, 46, 55

Q-Control, 268, 270
Q-factor, 146, 224, 251
quad photodetector, 55
quadrant detector, 91
quality factor Q, 64, 80, 148, 250
quantum
  corrals, 206
  Hall regime, 227

well, 198
quasi-optical experiment, 198
quasi-static
 bending test, 749, 753
 mode, 146
 strain rate, 560

radial cracking, 851
radius of gyration, 415
Raleigh's method, 79
Raman
 spectra, 841
 spectroscopy, 839
randomly oriented polysilicon, 1111
ratchet mechanism, 773, 776, 778, 800, 803, 808
Rayleigh wave, 583
RbBr, 162
RC-oscillators, 98
reaction kinetics, 294
readout
 electronics, 91
receptor-ligand
 bond, 293
 interaction, 296, 390
 unbinding, 294
 unbinding force, 285
recognition force microscopy (RFM), 284
recognition force spectroscopy (RFS), 296
recognition imaging, 304
rectangular cantilever, 65, 526
reduced modulus, 580
reflection interference contrast microscopy (RICM), 398
relative humidity (RH)
 effects of, 375, 949, 1061
relaxation
 mechanical, 255, 910
relaxation time, 428–437, 451, 456, 465, 466
release
 long-term of drugs, 1048
reliability, 885
 of MEMS, 1049
remote detection system, 91
repetitive IgG, 297
replica molding, 889
repulsive force, 292
residual
 film stress, 1091

 stress, 583, 595, 612, 844, 848, 857
 thickness, 919
resolution, 121
 vertical, 51
resonance
 curve detection, 64
 frequency, 250
 mechanical, 463
resonant
 frequency, 119
response time, 270
rest time, 941
retardation effect, 399, 400
RF magnetron sputtering, 831
rheological
 characterization, 934
 model, 275, 539
rhombhedral R-8, 605
rigid
 biphenyl chain, 916
 disk drive, 930, 941
RMS roughness, 1072
RNA polymerase, 706
rolling friction, 402, 425
root mean square, 218
rotary microactuator, 1044
rotating strain gauge, 1095
roughness, 410, 411, 423–425, 467, 773, 786
 contact parameter, 777
 parameter, 788, 811
  scale dependence, 813
 scale dependence, 787
rule of mixtures, 611, 613
rupture forces, 302

s-like tip states, 200
SAM (self-assembled monolayer)
 melting point of, 915
sample holder, 189
saturated hydrocarbons, 893
scale dependence
 area of contact, 793
scale effect
 tribology, 773
scaling length, 773, 814
scan
 area, 58
 direction, 75
 frequency, 58

head, 189
range, 101, 103
rate, 47
size, 59
speed, 103
scanner, piezo, 224
scanning
    acoustic microscopy (SAM), 537, 583
    capacitance microscopy (SCM), 42
    chemical potential microscopy (SCPM), 42
    electrochemical microscopy (SEcM), 42
    electron microscope (SEM), 120, 354
    electrostatic force microscopy (SEFM), 42
    head, 46
    ion conductance microscopy (SICM), 42
    Kelvin probe microscopy (SKPM), 42
    local-acceleration microscopy (SLAM), 537
    magnetic microscopy (SMM), 42
    nanoindentation (SN), 557
    nanoindenter, 536
    near field optical microscopy (SNOM), 42
    speed, 58
    system, 99
    thermal microscopy (SThM), 42
    tunneling microscope (STM), 141, 160, 244, 316
    tunneling microscopy (STM), 41, 118, 136, 162
    velocity, 351
scanning force
    acoustic microscopy (SFAM), 42
    microscopy (SFM), 42, 218, 537
    spectroscopy (SFS), 218
scanning probe
    microscope (SPM), 42
    microscopy (SPM), 118, 141, 537
schematic diagram of T-SLAM, 539
scratch
    critical load, 873
    damage mechanism, 865, 867
    depth, 744, 747, 1056
    depth profile, 746
    drive actuator, 1103
    profile, 351
    resistance, 734, 744, 917, 1058
    test, 351, 858

scratch-induced damage, 858
scratching measurement, 42
self-assembled
    growth, 195
    microscopic vesicles, 889
    monolayers (SAMs), 285, 378, 603, 888, 896, 930, 1058, 1065
self-assembly
    in biological systems, 694
self-enhancing instability, 256
self-excitation
    modes, 262
    scheme, 267
self-lubrication, 520, 522
semiconductor, 152
    quantum dots, 206
    surface, 152
sensitivity, 94, 97, 150, 270, 290, 485
sessile-drop, 887
SFA (surface forces apparatus), 1071
SH-group, 285
shallow indentation measurement, 564
sharpened metal wire tip, 136
shear
    force, see kinetic friction, static friction
    melting, 428, 455
    modulus, 485
    rate, 935
    strength, 508
    stress, 913
    thinning, 446, 449–451, 461
shear flow
    detachment (SFD), 289
shear strength
    scale dependence, 779
shear strength at the interface
    scale dependence, 786
shear stress
    critical, 430, 442, 453, 455
    effective, 513
shearing force, 911
shell model, 630
short-cut carbon nanotubes, 209
short-range
    chemical force, 224, 265
    chemical interaction, 169
    contribution, 142
    electrostatic attraction, 164
    electrostatic interaction, 162

1142 Index

energy dissipation, 273
interatomic force, 252
magnetic interaction, 162, 169, 170, 172
shot noise, 93
Si
  Si
    adatom, 155
  cantilever, 269
  Si
    cantilever, 120, 258
  Si
    MEMS, 1033
  nanobeam, 762
  Si
    nanobeam, 748, 767
  Si
    tip, 121, 152
  Si
    trimer, 158
  Si
    wafer, 120
Si, 121, 124
Si_3N_4
  tip, 938, 959
Si_3N_4
  Si_3N_4
    cantilever, 120
  Si_3N_4
    layer, 120
  Si_3N_4
    tip, 67, 121, 373, 900, 910, 937, 942
Si(001)2 × 1, 156
Si(001)2 × 1-H surface, 157
Si(100), 864, 932, 939, 940, 943, 952, 954, 955, 1076
  Si(100)
    wafer, 931
Si(100), 345, 352, 868, 1050
Si(100)2 × 1:H monohydride, 155
Si(111), 911, 914, 1056
Si(111), 350, 365, 915
Si(111) surface, 173
Si(111) $\sqrt{3} \times \sqrt{3}$-Ag, 159
Si(111)(7 × 7), 499
Si(111)(7 × 7), 220, 274
Si(111)(7 × 7) surface, 148, 162, 263, 497
Si(111) $\sqrt{3} \times \sqrt{3}$-Ag, 157
$Si(OH)_4$ film, 1076
Si-SiO_2 stress formation, 124

Si-Ag covalent bond, 161
Si-based devices, 732
Si-wafer, 260
Si
  wafer, 258
Si_3N_4, 124, 321
SiC
  SiC
    bulk properties, 1050
  SiC
    film, 742
SiC, 1049
silica
  thermally grown, 898
silicon, 52, 94, 145, 585, 586, 605
  AFM tip, 288
  cantilever, 144, 485
  chemistry, 288
  fractures, 753
  grain, 1110
  micromotor, 937
  oxide, 286
  surface, 286
  tip, 154, 220, 266, 489
  wafer, 250
silicon nitride (Si_3N_4), 52, 285
  cantilever, 119
  tip, 122, 489
silicon-on-silicon contact, 1075
silicon-terminated tip, 220
siloxane, 888
simultaneous imaging, 162
  of a $TiO_2(110)$ surface, 168
single asperity, 508
  AFM tip, 938
  contact, 348, 1075
single crystal
  Si(100), 742, 931
  aluminium (100), 357
  silicon, 324, 1049, 1093
  silicon cantilever, 67
single crystalline iron, 557
single molecular motor, 290
single molecule
  assays, 698
  biology, 699
  detection, 286
  experiment, 725
  studies, 297

single receptor-ligand pair, 291
single stranded DNA (ssDNA), 703
single-particle wave function, 198
single-wall carbon nanotube (SWCNT), 70, 125, 545
sink-in, 585, 595, 596, 612, 615
  behavior, 557
sintering, 399, 402
SiO_2
  SiO_2
    film, 742
    nanobeam, 749
  SiO_2
    nanobeam, 748, 767
SiO_2, 120, 124, 372
SiOH groups, 288
SLAM, 538
sliding, 773, 814
  contact, 519
  direction, 453
  distance, 459
  of tip, 484
  velocity, 437, 453, 455, 506
slipping and sliding regimes, 455
small specimens handling, 1093
smooth sliding, 453
SnO_2(110), 166
soft
  cantilever, 249
  coatings, 610
  lithography, 889
solid boundary lubricated surfaces, 429, 430, 441–443, 459
solid xenon, 223
solid-like
  behavior, 465
  Z-DOL(BW), 937
SOLID95, 756
solidification, 443, 448
solvation
  forces, 390, 408, 409, 414, 423, 424
sonolubrication, 341
SP, 859
sp^3 bonding, 848
sp^3-bonded carbon, 858
spacer chain, 896, 897, 899
  length, 905
specificity, 714
spectroscopic resolution in STS, 194

spin
  density waves, 212
  quantization, 201
SPM (scanning probe microscopy), 118
  methods, 554
  tip, 248
spring
  sheet cantilever, 68
  system, 45
spring constant, 126, 143, 148, 155, 218, 245, 246, 275, 291, 294, 487, 550, 1102
  calculation, 77
  changes, 187
  effective, 492, 507
  lateral, 64, 491
  measurement, 77
  vertical, 64
sputtered coatings
  physical properties, 844
sputtering
  deposition, 837
  power, 844
SrF_2, 164
$SrTiO_3$(100), 169
$SrTiO_3$(100), 166, 168
$SrTiO_3$, 168
stainless steel cantilever, 858
static
  advancing contact angle (SACA), 908
  AFM, 144, 274
  deflection AFM, 245
  friction, 430, 436, 443, 451, 453, 455, 459, 460, 462, 466, 1107
  friction force, 937, 1079
  interactions, 391, 449
  mode, 144, 218, 249
  mode AFM, 91
statistically stored dislocation (SSD), 779
steady-state sliding, *see* kinetic friction
step tilting, 919
stepping motor, 490
stick slip, 449
stick–slip, 430, 436, 444, 450, 452, 453, 458, 462, 950
  mechanism, 491
sticking regime, 455
stiction, 443, 886, 1042
  phenomena in micromotors, 1082

stiffness, 150, 537, 867
    torsional, 82
stiffness measurement
    continuous (CSM), 853
STM (scanning tunneling microscope), 41, 162
    cantilever, 49
    cantilever material, 65
    principle, 43
    probe construction, 49
    tip, 42, 189
Stoney equation, 1092
storage
    modulus, 601
strain
    energy difference, 853
strain gradient plasticity, 773, 777, 784, 814
stray capacitance, 97
strength, 1091
streptavidin, 288
    mutants, 296
stress, 712, 1091
    distribution, 758
    field, 1092
    gradient, 1095
    maximum, 1097
    measurement, 1094
    tensor, 723
stress–strain
    behavior, 1100
    curve, 554, 766
stresses within the capsid walls, 723
stretch modulus, 711
Stribeck curve, 429, 449
strip domain, 229
structural
    forces, 390, 408, 411, 423
    integrity, 732
    material, 1098
sub-Angstrom deflections, 119
sub-nanonewton precision, 248
submonolayer
    coverage, 933
substrate, 896, 898
substrate curvature technique, 1092
subsurface water irrigation, 559
summit deformation, 792
super modulus, 582
superconducting
    gap, 212
    magnetic levitation, 45
    matrix, 231
superconductivity, 194
superconductors, 210
    type-I, 212
    type-II, 212
superlubricity, 499
surface
    amorphous, 391
    asperity effect, 340
    atom layer, 146
    band, 201
    charge, 408
    charge density, 405–407
    crystalline, 391
    elasticity, 368
    energy, 828
    energy (tension), 394, 400, 422, 425, 431, 432, 435, 438
    forces, 389
    forces apparatus (SFA), 289, 316, 392, 394, 395, 400, 441, 453, 457
    free energy, 702, 704
    hydrophobic, 391, 413
    micromechanical properties, 315
    microtribological properties, 315
    mobility, 935
    nanomechanical properties, 315
    nanotribological properties, 315
    potential, 195, 405, 407, 408, 834
    potential measurement, 324
    slope, 905
    state lifetime, 200
    stiffness, 327
    structure, 166, 391, 410, 411, 423, 428, 456, 463
    tension, 347, 886, 907, 930
    terminal group, 896
    topography, 484, 755
surface force apparatus (SFA), 775
surface roughness, 136, 557, 563, 748, 755, 905, 1050, 1070
    map, 331
    measurement, 318
    scale dependence, 786
surface structure, 640
surfactant monolayers, 394, 417, 426, 427, 437, 442, 445, 452, 454, 459

suspended membrane, 1100
sustainable internal pressure
  maximum, 724
SWNT (single wall nanotube)
  rope, 545, 549

T-SLAM, 538
tailoring of SAMs, 900, 1065
tapping amplitude, 370
tapping-mode, 52, 120, 253, 254, 320
  AFM, 254, 258
  etched silicon probe (TESP), 67
  tip assembly, 329
technical PVC, 540
Teflon layer, 521
temperature
  critical, 200, 219
  dependence of friction, 508
  domain, 540
  effect, 949
temporal scales, 708
tensile
  load, 1097
  maximum stress, 759
  stress, 737, 1095, 1110
  test, 1112
test environment, 874
tethering
  flexible, 307
thermal
  drift, 187, 189
  effect, 501
  energy, 292
  expansion coefficient, 145
  fluctuation, 487
  fluctuation forces, 417
  force, 711
  frequency noise, 187
  noise, 292
  vibration amplitude, 127
thermomechanical
  noise, 218
theta condition, 415
thick film lubrication, *see* elastohydrodynamic lubrication
thickness of nanotubes, 547
thin film, 598, 609
thin-film
  lubrication, 316

thiol, 267, 893
three-body deformation, 773, 800, 814
three-dimensional force field spectroscopy, 225
through-thickness
  cracking, 850
Ti atom, 173
TiAl alloy, 557
TiC grains, 331
tight-binding approximation, 170
tilt angle of SAMs, 919
TiO_2
  substrate, 175
TiO_2, 168, 203
$TiO_2(100)$, 166
$TiO_2(110)$, 176
TiO_2-terminated layer, 168
tip
  anion-terminated, 164
  apex, 220, 286, 525
  artifact, 154
  atom(s), 188
  atomic structure, 137
  deflection, 323
  Fe-coated, 171
  ferromagnetic, 232
  geometry, 127
  material, 137
  mount, 47
  oscillation, 119
  oscillation amplitude, 524
  performance, 136
  preparation in UHV, 489
  preparation method, 49
  properties, 136
  radius effect, 64, 341, 858, 952, 956
  surface, 286
  vibration, 260
  vibration amplitude, 538
tip–cantilever assembly, 125
tip–liquid interface, 372, 940
tip–molecule
  distance, 195
  gap, 177
  long-range force, 176
tip–sample, 245
  contact, 271
  dissipative interaction, 273

distance, 146, 159, 164, 168, 221, 245, 259, 329
electric field, 526
energy dissipation, 148
force, 119, 146, 218, 224, 246, 249, 252, 260, 262
force gradient, 151
interaction, 151, 218, 232, 252, 253, 264, 524
interaction potential, 267
interface, 119
junction, 273
potential, 150
separation, 271
system, 539
tip–surface
distance, 177, 273
interaction, 305, 520
potential, 491
separation, 291
tip-bound
antigens, 297
tip-broadened image, 121
tip-induced atomic relaxation, 220
tip-induced quantum dot, 203
tip-terminated atom, 166
titin, 298
Tomlinson model, 483, 491
finite temperature, 501
one-dimensional, 491
two-dimensional, 492
tooth enamel, 558
topographic
images, 51
topographical asymmetry, 172
topography measurement, 57
torsional resonance (TR), 973
torus model calculation, 101
total internal reflection microscopy (TIRM), 398
trabecular bone, 564
transition index
scale dependence, 804
transition metal oxides, 168
transitions between smooth and stick–slip sliding, 455
transmembrane channel, 696
transmission electron microscopy (TEM), 122, 354, 585

transversal piezo-resistive effect, 95
traveling direction of the sample, 76
Tresca criterion, 588
triangular cantilever, 65, 82, 92
tribochemical
oxidation, 1054
reaction, 373, 947
tribological
characterization of coating, 846
tribological issues
in MEMS, 1039
in NEMS, 1039
tribological performance, 1050
of coatings, 868
tribological properties
of SAM, 900
of silicon, 1050
tribology
scale effect, 773
tribometer, 1050
(tridecafluoro-1,1,2,2-tetrahydrooct-1-yl) trichlorosilane (n-$C_6F_{13}CH_2CH_2SiCl_3$, FTS), 901
triethoxysilane, 506
trifluoromethyl-terminated SAM, 902
trimer tip apex, 248
true atomic resolution, 148, 152, 220
tubulin monomer, 694
tungsten, 49
sphere, 486
tip, 49, 273
tunneling
current, 43, 146, 148, 203
detector, 51
tip, 215
two-body deformation, 795
two-dimensional
elasticity, 717
electron gas (2-DEG), 227
electron system (2-DES), 209

UHV environment, 495
UHV-AVM, 490
ultrahigh vacuum (UHV), 152, 191, 262, 489
ultrasonic
lubrication, 341
transducer, 538
ultrathin DLC coatings, 831
unbinding

force, 284, 292
force distribution, 298
pathway, 303
unconstrained binding, 286
universality, 714
unlubricated sample, 957
unlubricated surfaces, *see* dry surfaces
upward cantilever deflection, 703

V-shaped cantilever, 67, 487
van der Waals
  attraction, 341
  contact, 725
  force, 119, 142, 162, 173, 177, 245, 390, 391, 398–401, 403, 404, 408, 410
  interaction, 169, 255, 903, 934
  surfaces, 221
vanadium carbide, 581
variable force mode, 105
variable temperature STM
  setup, 189
variable-temperature SLAM (T-SLAM), 538
velocity
  critical, 454, 458, 462, 463
velocity dependence of friction, 506
velocity effect, 945
Verlet algorithm, 255
vertical coupling, 47
vertical rms-noise, 219
Verwey transition temperature, 229
vibration, 708
  external, 47
Vickers
  hardness, 556, 562, 588
  indentations, 745
  indenter, 734
vinylidene fluoride, 173
viral life cycle, 717, 718
virgin single-crystal silicon, 1051
viruses, 706
viscoelastic
  effect, 273
  properties, 568
viscoelasticity, 537, 600, 615
  mapping, 327, 368
viscous
  damping, 275
  force, 429, 445, 446, 453
voltage bias, 137

vortex in superconductor, 229

W tip, 50
wafer
  curvature technique, 1109
water
  capillary force, 908
  meniscus, 261
water vapor, 581, 589
  content, 874
Watson-Crick base pairing, 296
wear, 316, *see* damage, 623, 829, 848
  coefficient, 811
  contribution to friction, 516
  control, 886
  damage, 873
  damage mechanism, 874
  debris, 356
  depths, 1059
  measurement, 42
  mechanism, 357
  performance, 936
  process, 868
  region, 359
  resistance, 868, 871, 917, 921, 957, 1051, 1068
  scale dependence, 773, 811
  study, 957
  test, 871
Weibull
  distribution, 750
  statistics, 1112
weight function, 150
wet friction
  scale dependence, 773
wettability, 908
wetting, 425
wire cantilever, 64
work
  hardening, 603, 612
  of adhesion, 906, 908, 1071
  of indentation, 599

xenon, 191

yield point, 428, 429, 453
  load, 602
yield strength
  scale dependence, 781

yield stress, 603
Young's modulus, 77, 106, 126, 187, 510, 547, 562, 757, 1091, 1092
  measurement, 1099
Young–Dupre equation, 908

Z-15, 372, 930, 932, 951–953, 955, 1064
  film, 372, 958
  lubricant, 1059
  lubricant film, 938
  properties, 931
Z-DOL, 372, 930, 932, 942
  partially bonded film, 957
  properties, 931
Z-DOL(BW), 940, 951, 954, 955, 958, 1064
  film, 372, 938, 1059
zinc, 604